Fundamentals and Practice in Statistical Thermodynamics

Fundamentals and Practice in Statistical Thermodynamics

Jianzhong Wu
University of California, Riverside

John M. Prausnitz
University of California, Berkeley

WILEY

Published by John Wiley & Sons, Inc., Hoboken, New Jersey.
Published simultaneously in Canada.

Library of Congress Cataloging-in-Publication Data

Names: Wu, Jianzhong, author. | Prausnitz, John M., author.
Title: Fundamentals and practice in statistical thermodynamics / Jianzhong Wu, John M. Prausnitz.
Description: Hoboken, New Jersey : Wiley, [2024] | Includes index.
Identifiers: LCCN 2023046574 (print) | LCCN 2023046575 (ebook) | ISBN 9781394161423 (cloth) | ISBN 9781394161430 (adobe pdf) | ISBN 9781394161447 (epub)
Subjects: LCSH: Statistical thermodynamics.
Classification: LCC QC311.5 .W85 2024 (print) | LCC QC311.5 (ebook) | DDC 530.13/2—dc23/eng/20231026
LC record available at https://lccn.loc.gov/2023046574
LC ebook record available at https://lccn.loc.gov/2023046575

Cover Design: Wiley
Cover Image: © Jorg Greuel/Getty Images

Set in 9.5/12.5pt STIXTwoText by Straive, Chennai, India

To Jane, Hamlin and Sean

Contents

Preface

About 25 years ago, I was assigned to prepare a textbook introducing statistical mechanics to senior undergraduates and first-year graduate students in chemical engineering. The first draft was finished when I left Berkeley at the end of 2000. Although the content was hardly the same as what would be presented in the current form, the goal has never been changed, i.e., a systematic exposition of the fundamental principles with an emphasis on the practical needs, particularly in the emerging areas of engineering.

In terms of practice, chemical engineering is a broad field, encompassing nearly every aspect of industrial processes concerning the transformation of natural resources into energy, materials, and chemical products (and beyond). Historically, such transformation starts with fossil fuels through macroscopic operations whereby thermodynamic principles can help improve the efficiency of energy conversion, chemical separation, and reactions. As fossil fuels are phasing out from traditional power generation and other industrial processes, recent few decades have seen extensive developments in alternative routes to chemical production and materials synthesis using renewable energy and sustainable feedstock. To make the carbon-neutral or carbon-negative processes that are not just environmentally friendly but economically practical, statistical mechanics offers useful theoretical tools to understand, correlate, and predict the thermophysical properties of matter essential for the materials design and industrial optimization from a molecular perspective.

This text introduces the fundamental principles of statistical mechanics that bridge the microscopic details associated with the individual elements of thermodynamic systems with their macroscopic behavior including cooperative phenomena and phase transition. The self-contained text offers a comprehensive description of the microscopic origins of thermodynamic variables, the physical significance of microstates according to the first principles of quantum mechanics as well as phenomenological models, and the dynamic equations dictating the time evolution of microstates in different statistical ensembles (Chapters 1 and 2). Applications of the fundamental principles to non-interacting quantum and classical systems are discussed in the context of both idealized and realistic models with a balance of pedagogy and practical use (Chapters 3 and 4). For non-ideal molecular systems, the fundamental principles of statistical mechanics can be implemented with simulation methods systematically, i.e., through molecular dynamics (Chapter 2) or Monte Calo simulation (Chapter 6). To understand cooperative phenomena and phase transition in thermodynamic systems, we use the Ising model as a pedagogical platform to elucidate important concepts such as correlation length, order parameters, mean-field approximation, thermal fluctuations, and universality (Chapter 5). For quantitative predictions of the structure and thermodynamic properties of chemical systems such as colloids, polymers, and electrolyte solutions, special attention is given to the development of liquid-state methods, field-theoretical tools, and solution theories (Chapters 7–9). The book need not be read in order; each chapter or section can be studied independently.

Dozens of exercise problems (with detailed solutions), many extracted from the recent literature, are included after each chapter to elucidate the diverse applications of statistical thermodynamics to chemical and biological systems as well as materials engineering. For self-supporting, the text also includes many appendices to provide the mathematical details, and online Supplementary Materials introducing the basics of quantum mechanics, variational methods for classical mechanics, electronic density functional theory, and intermolecular interactions.

The text may be adopted for advanced thermodynamics courses from various engineering disciplines. It may also be useful for students from physics, materials science, and chemistry majors to understand the fundamentals of statistical mechanics, cooperative phenomena, and phase transformations in the context of practical applications. Besides, the book should be of value to researchers who are interested in the computational design of chemicals and materials.

I am indebted to the following individuals for critiquing portions of the manuscript: Dusan Bratko, Marcus Müller, Lloyd Lee, Joachim Dzubiella, John O'Connell, Keith Gubbins, Abbas Firoozabadi, An-Chang Shi, Mikhail Anisimov, Rui Qiao, Jeff Chen, Roland Roth, Danel Borgis, Marc-Olivier Coppens, Jian Qin, Thomas Truskett, Chi Wu, Dadong Yan, and Jianwen Jiang. I am also grateful to my coworkers, Musen Zhou, Alejandro Gallegos and Runtong Pan, for their technical assistance and valuable feedback, and to the editors from the publisher, Michael Leventhal, Judy Howarth, Elizabeth Amaladoss, and Govind Nagaraj and his production team for their help and commitment to quality textbook publishing. Finally, I thank my wife, Hong, for her company, encouragement, and understanding.

August 2023 *Jianzhong Wu*
University of California, Riverside

About the Companion Website

This book is accompanied by a companion website:

www.wiley.com/go/jianzhongwu/fundamentalsandpractice

The website includes Supplementary Materials, MatLab Codes and PPT files.

1

Microscopic Origin of Thermodynamics

The real-life applications of thermodynamics hinge on experimental measurements and/or theoretical predictions of thermodynamic properties for macroscopic systems of practical interest. The experimental approach is exemplified by the traditional applications of engineering and chemical thermodynamics as evidenced in the extensive use of thermodynamic tables, diagrams, and semi-empirical correlations. Conversely, the theoretical approach is based on statistical mechanics, which is of central concern in this book.

This introductory chapter presents the key hypotheses of statistical mechanics to describe the thermodynamic properties of an equilibrium system from a microscopic perspective. After a brief overview recapitulating the essential ingredients of classical thermodynamics and fundamental relations linking different thermodynamic variables, we introduce *internal energy* and *entropy* – two most fundamental quantities of thermodynamics – in terms of the properties of individual particles, i.e., the *microscopic constituents* of a thermodynamic system. The *statistical nature* of thermodynamic variables will be elucidated in the context of ensemble, ergodicity, and microstates. In addition, we will discuss *flexibility* in the statistical descriptions of individual particles and microstates of a thermodynamic system, as well as *additivity* and *relativity* pertaining to both internal energy and entropy when assessed from microscopic perspectives.

We assume that the readers are already familiar with the fundamentals of classical thermodynamics including their applications to chemical and phase equilibria. However, the previous exposure of statistical mechanics is not expected. The supplementary material gives a brief overview of classical and quantum mechanics for those who are unfamiliar with this subject. While an advanced understanding of quantum mechanics is not a prerequisite, basic concepts such as Hamiltonian, quantum states, and the Schrödinger equation will be used to describe particle energy and the microscopic constituents of quantum systems (e.g., photons, electrons, and phonons).

1.1 Microscopic Constituents of Thermodynamic Systems

In this section, we discuss the essential ideas of classical thermodynamics from a microscopic perspective and recapitulate the fundamental relationships among different thermodynamic variables such as temperature, pressure, entropy, and energy. In addition, we will elucidate how statistical mechanics helps to understand the macroscopic properties of thermodynamic systems based on the dynamic behavior of their constituent particles.

Fundamentals and Practice in Statistical Thermodynamics, First Edition. Jianzhong Wu and John M. Prausnitz.

Companion website: www.wiley.com/go/jianzhongwu/fundamentalsandpractice

1.1.1 Classical Thermodynamics

Classical thermodynamics is centered around two fundamental laws of nature that are universally applicable to the *collective* behavior of macroscopic systems. The first law follows the conservation of the total energy, and the second law asserts that spontaneous events in nature proceed in a particular direction. These thermodynamic laws were established in the second half of the nineteenth century from repeated observations of natural phenomena underlying transformations among different forms of energy and their connections with the physical properties of matter. Extensive experience over many years renders us confidence that the thermodynamics laws are permanent, unlikely to be refuted by future scientific developments. As famously stated by Albert Einstein,[1] "*A theory is the more impressive the greater the simplicity of its premises is, the more different kinds of things it relates, and the more extended is its area of applicability. Therefore the deep impression which classical thermodynamics made upon me. It is the only physical theory of universal content concerning which I am convinced that, within the framework of the applicability of its basic concepts, it will never be overthrown (for the special attention of those who are skeptics on principle).*"

Closely affiliated with the fundamental laws of thermodynamics are two indispensable quantities, internal energy U and entropy S. As discussed in more detail later in this chapter, both internal energy and entropy are defined by the *microscopic constituents* of a thermodynamic system. More precisely, internal energy refers to the total energy arising from the perpetual motions of *individual particles* and inter-particle interactions. The former is commonly known as *kinetic energy*, and the latter is *potential energy*. Conversely, entropy provides a measure of the uncertainty of a macroscopic system in terms of the dynamic behavior of the individual particles or, without concerning the time, in terms of *possible ways* that individual particles may exist (e.g., particle positions and momenta or wave functions) under a particular thermodynamic condition. Establishing the connection between thermodynamic quantities and the dynamic behavior of individual particles is an essential task of statistical thermodynamics.

Classical thermodynamics is concerned with variations in the *equilibrium* properties of macroscopic systems, i.e., systems consisting of many particles, typically on the order of 10^{23}. For a macroscopic system at equilibrium, all thermodynamic quantities are fixed.[2] In other words, all macroscopic properties of interest are independent of time. Here, time independence means that the duration of observation is sufficiently long compared with the time scale that characterizes the dynamics of individual particles. Thermodynamic laws cannot be applied if one is interested only in a single particle in the vacuum or even a few particles. If a system contains only a few particles, the dynamic behavior can be described with conventional equations from classical or quantum mechanics. It is the enormous number of particles in a thermodynamic system that prevents the direct use of the mechanical equations to describe particle motions at a level of certainty the same as that for a system with only a small number of particles. *Uncertainty at the microscopic level* is intrinsic for all thermodynamic systems.

The individual particles of a thermodynamic system embody not only kinetic and potential energies but also information concerning the precise meaning of microscopic constituents. Thermodynamics makes no assumption on the physical nature of the individual particles, i.e., the thermodynamic laws hold regardless of how the microscopic constituents are characterized or interpreted. Indeed, individual particles in a thermodynamic system are rather diverse; they may

---

1 Einstein A., "Autobiographical Notes", p. 33, in *Albert Einstein: Philosopher-Scientist*, Schilpp P. A., Ed., The Library of Living Philosophers Volume VII. MJF Book, New York, 1970.

2 The macroscopic properties of a steady-state system are also independent of time, but that system is not necessarily at equilibrium. In a steady-state system, there is a net flux of energy or mass, independent of time. However, for equilibrium, it is necessary that there is no net flux of mass or energy is zero.

refer to atoms or molecules as commonly present in different states of matter (i.e., gases, liquids, and solids), or elementary particles such as photons and electrons, or certain aspects of elementary particles (e.g., magnetic spins), or individual units of a molecule (segments), or aggregates of molecules (e.g., colloidal particles). The *flexibility* in interpreting the microscopic constituents of a thermodynamic systems suggests that thermodynamic variables such as internal energy and entropy are intrinsically *relative*, i.e., their absolute values are dependent on the *definition* of individual particles. For example, internal energy and entropy may take different values when an atomic system is described in terms of electrons and atomic nuclei with quantum mechanics or as individual atoms like classical particles. In either case, the changes in thermodynamic properties predicted from a microscopic perspective should be in accordance with experimental observations.

1.1.2 The Fundamental Equation of Thermodynamics

In addition to internal energy and entropy, we use *auxiliary* quantities such as enthalpy H, Helmholtz energy F and Gibbs energy G to describe the macroscopic properties of a thermodynamic system. These auxiliary functions are introduced for the convenience of practical applications when thermodynamic systems are prepared under different circumstances (e.g., fixed total energy or fixed temperature, constant pressure or constant volume, etc.). All auxiliary quantities can be *formally* derived from internal energy and entropy along with variables that specify a thermodynamic system (e.g., temperature, pressure, and total volume). Using the auxiliary functions, we can describe heat effects for constant pressure processes simply in terms of the changes in enthalpy, and the thermodynamic limits of various isothermal processes as well as the conditions of equilibrium with different thermodynamic potentials or free energies.

For a closed system, i.e., a system free of mass transfer with its surroundings, classical thermodynamics asserts that the internal energy U and the entropy S are related through the fundamental equation

$$dU = TdS - PdV. \tag{1.1}$$

Eq. (1.1) can be obtained by applying the first and second laws to a closed system undergoing a reversible process that involves volumetric work $-PdV$ and heat transfer TdS. Additional variables must be introduced to account for other forms of reversible work or the mass transfer of any chemical species between the system and its surroundings. According to multivariable calculus, the differential form in Eq. (1.1) suggests that the internal energy U is an analytical function of entropy S and volume V, i.e., $U = U(S, V)$. In addition, Eq. (1.1) implies that temperature T and pressure P can be expressed as partial derivatives

$$T = \left(\frac{\partial U}{\partial S}\right)_V, \tag{1.2}$$

$$P = -\left(\frac{\partial U}{\partial V}\right)_S. \tag{1.3}$$

In practical applications, a closed system is often specified in terms of temperature and volume or temperature and pressure. Accordingly, it is convenient to describe the thermodynamic properties with such variables. Mathematically, the change of independent variables can be achieved through the Legendre transformation[3]:

$$F \equiv U - \left(\frac{\partial U}{\partial S}\right)_V S = U - TS, \tag{1.4}$$

3 The Legendre transformation of an analytical function results in a new function in which the slope of the original function becomes the independent variable. For example, $f(x) \to g(y) \equiv f(x) - yx$, where $y = f'(x)$, represents the Legendre transformation of function $f(x)$ to $g(y)$. The transformation implies $dg = f'dx - d(yx) = -xdy$, i.e., $-x = g'(y)$, where g is a function with y as the independent variable.

$$H \equiv U - \left(\frac{\partial U}{\partial V}\right)_S V = U + PV, \tag{1.5}$$

$$G \equiv H - \left(\frac{\partial H}{\partial S}\right)_P S = H - TS \tag{1.6}$$

where $F = F(T, V)$ is called Helmholtz energy, $H = H(S, P)$ is enthalpy, and $G = G(T, P)$ is Gibbs energy. Substituting Eqs. (1.4)–(1.6) to (1.1) leads to alternative forms of the fundamental equation

$$dF = -SdT - PdV, \tag{1.7}$$

$$dH = TdS + VdP, \tag{1.8}$$

$$dG = -SdT + VdP. \tag{1.9}$$

The procedure can be generalized to other forms of thermodynamic potentials (e.g., grand potential to be discussed in Chapter 2), and similar fundamental equations can be derived by Legendre transformation.

While the formal relations discussed above are most relevant to gases or liquids in a bulk phase, similar equations can be readily established for two- and one-dimensional systems such as molecules adsorbed at a surface or in narrow micropores. For a two-dimensional system, the thermodynamic properties depend on surface area A instead of volume V. Accordingly, the fundamental equation is given by

$$dU = TdS - \varsigma dA \tag{1.10}$$

where ς denotes the surface pressure, i.e., the variation of energy with the surface area. For a one-dimensional system, the system length is often used as the spatial variable. For example, the fundamental equation for a rubber band can be expressed as

$$dU = TdS + fdL \tag{1.11}$$

where f and L represent the tension and length of the rubber band, respectively. Similar to those for bulk fluids, other forms of the fundamental equation can be derived for both two- and one-dimensional systems.

1.1.3 Statistical Thermodynamics

Statistical mechanics aims to predict the properties of a macroscopic system based on the dynamic behavior of individual particles. To establish connections between micro- and macroscopic properties of a thermodynamic system, statistical mechanics offers rigorous mathematical procedures to evaluate the *statistical* distributions of the microstates of individual particles. Such procedures are universally applicable to all many-body systems, i.e., from systems containing a few particles up to the thermodynamic limit.[4]

In statistical mechanics, a *microstate* refers to a particular way that the individual particles of a thermodynamic system exist. In other words, at any microstate, all variables concerning the dynamic behavior of individual particles are fully specified. If the particles are classical and have a spherical shape, these variables correspond to the positions and momenta of individual spheres. If the particle motions are described by quantum mechanics in terms of the wave function of the entire system, each microstate may be understood as a quantum state. As mentioned above, individual particles in a thermodynamic system are not necessarily correspondent to the

4 In the thermodynamic limit, the system size, including the number of particles, approaches infinity.

fundamental constituents of matter as defined in particle physics. In introducing microstates, we assume that the dimensionality of individual particles is exceedingly small in comparison to that of a macroscopic system. At any moment, a thermodynamic system may reside in one of many microstates.

According to statistical mechanics, the mechanical behavior of individual particles is responsible for all macroscopic properties of a thermodynamic system, including not only internal energy and entropy but also all auxiliary properties such as chemical potential, enthalpy, and the equilibrium constants of chemical reactions. As uncertainty is inevitable in describing the dynamic behavior of individual particles, statistical mechanics predict the macroscopic properties in terms of the *statistical averages* of quantities related to the mechanical behavior of individual particles. By accounting for the probability of a macroscopic system in different microstates, statistical mechanics can predict, in principle, both static and dynamic properties including those corresponding to systems at nonequilibrium conditions. While the procedure is applicable to macroscopic systems at or away from equilibrium in principle, our concern in this book is focused on the microscopic structure and thermodynamic properties of equilibrium systems.

1.1.4 Summary

In this section, we recapitulate some of the most important concepts from classical thermodynamics, the microscopic meaning of key thermodynamic quantities such as internal energy and entropy, and the connection between thermodynamics and statistical mechanics. While thermodynamics is, as Einstein indicated, a theory of principle that is universally applicable to any macroscopic systems, statistical mechanics intends to establish connections between thermodynamic variables with the microscopic constituents of thermodynamics systems. Statistical thermodynamics is a branch of statistical mechanics that is concerned with the properties of equilibrium systems. In this book, we do not distinguish these two terms and use them interchangeably because conventional statistical-mechanical methods are applicable only at or near equilibrium conditions.

1.2 Thermodynamic Relations

In this section, we analyze the mathematical implications of thermodynamic functions and discuss how they may lead to formal thermodynamic relations such as the Gibbs–Duhem and the Maxwell relations. These relations are useful because they facilitate easier application of thermodynamic principles and enable quantitative predictions of thermodynamic properties that may not have obvious microscopic counterpart or be directly accessible through experiments. Exact relations among different thermodynamic quantities are also useful for testing the self-consistency of experimental and/or theoretical results.

1.2.1 Intensive and Extensive Variables

Some thermodynamic variables, such as temperature, pressure, volume, and composition, can be measured directly; while others, such as entropy, internal energy, and chemical potential, are determined indirectly from measurements of other, directly measurable, variables. Both types of thermodynamic variables can be intensive or extensive. An intensive variable is independent of the system size, and an extensive variable is size-dependent.

Examples of intensive variables are temperature T, pressure P, and the chemical potentials of individual chemical species μ_i. These variables are associated with the driving forces for the exchange of energy and mass between the system and its surroundings: T dictates the direction of heat transfer, P is related to the volumetric work, and μ_i reflects the tendency of mass transfer through diffusion or chemical conversion. Extensive variables include internal energy U, entropy S, enthalpy H, Helmholtz energy F, Gibbs energy G, constant-pressure heat capacity C_P or constant-volume heat capacity C_V, the number of moles n_i or mass m_i for species i, and the system volume V. Other extensive variables are possible when we consider nonvolumetric work such as that due to the change in interfacial area or particle motions in an external field. Extensive variables are linearly scaled with the system size, i.e., if a system is replicated a certain number of times, all extensive properties are multiplied by the same factor accordingly.

The linear dependence of extensive variables on the system size implies that they can be converted into intensive variables after being normalized with the total mass, volume, or the number of moles of all chemical species. For example, we can obtain specific internal energy and specific entropy by dividing the total internal energy and the total entropy by the system mass, respectively. Similarly, molar internal energy and molar entropy correspond to the total internal energy and the total entropy divided by the total number of moles, and the internal energy density and the entropy density correspond to the total internal energy and the total entropy divided by the system volume.

1.2.2 The Gibbs Phase Rule

One remarkable feature of classical thermodynamics is that an equilibrium system can be defined by a few thermodynamic variables regardless of its complexity. For example, except for a scaling factor related to the system size, the properties of a uniform system such as liquid water in the bulk phase[5] are completely specified by temperature and pressure. Similarly, a multi-component system may be specified by temperature, pressure, and composition if it exists in a single phase. In practice, the composition may be defined in terms of mole fractions, volume fractions, weight fractions, or the concentrations of different chemical species. These variables are both necessary and sufficient to characterize the properties of a macroscopic system at equilibrium. Once a system is defined by these few thermodynamic variables, all macroscopic properties of the system can be determined, in principle, by either experiment or theoretical calculation.

For a multicomponent system free of chemical reactions, the Gibbs phase rule predicts that a thermodynamic state can be fully specified by N_f *independent* intensive variables

$$N_f = C - \mathcal{P} + 2 \tag{1.12}$$

where C is the number of chemical species or components in the mixture, and $\mathcal{P}$ is the number of coexisting phases. The number of independent variables to define a thermodynamic system is called *the degrees of freedom*. Eq. (1.12) was derived first by J. W. Gibbs by applying the first and second laws of thermodynamics to macroscopic systems at phase and chemical equilibria.[6]

---

5 In thermodynamics, a phase refers to a state of matter such as gas, liquid or solid that can be distinguished from other states in terms of both the microscopic structure and macroscopic properties.

6 Josiah Willard Gibbs published the paper entitled "On the Equilibrium of Heterogeneous Substances" in two parts in the *Transactions of the Connecticut Academy of Arts and (3, 108–248, Oct 1875–May 1876 and 343–524, May 1875–1878). Little known for many years after its publication, the 300-page paper is sometimes referred to as "the principia of thermodynamics" (Wikipedia). Morowitz considered it as "the second most significant document produced in the United States" (Morowitz, H. J., "Let free energy ring", Hospital Practice, 11:189–190, 1976).*

It is worth noting that, in a thermodynamic system, intensive variables may be implicitly related to each other. For example, the molecular fractions of a multicomponent mixture, $x_{i=1,2,...C}$, are intensive variables but not fully independent of each other. The mole fractions are normalized, i.e.,

$$\sum_{i=1}^{C} x_i = 1. \tag{1.13}$$

Therefore, the number of independent variables to describe the composition of an C-component system is $C - 1$.

The Gibbs phase rule does not apply to variables with intrinsic cause and effect relations, such as pressure P and volume V, temperature T and entropy S, chemical potential μ_i and the number of molecules N_i for chemical species i. For example, because liquid water exhibits a maximum density near 4 °C and $P = 1$ atm, specifying pressure and molar volume may not be sufficient to define the thermodynamic state. As shown schematically in Figure 1.1, two thermodynamic states I and II may have the same density and pressure. In thermodynamics, two variables with a cause-and-effect relation are known as a conjugate pair. Typically, a conjugate pair consists of an intensive variable and an extensive variable per unit volume, mass, or mole. The former serves as the thermodynamic driving force for the change of an extensive variable (e.g., P, T and μ_i are the driving forces for the changes in V, S, and N_i, respectively).

While the Gibbs phase rule can be similarly formulated for all thermodynamic systems, Eq. (1.12) is valid only for systems containing multiple bulk phases, i.e., for multi-phase systems free of chemical reactions. In that case, the intensive variables for each phase can be completely specified by temperature, pressure, and chemical composition. In general, a thermodynamic system may also be subject to external potential, i.e., energy related to the positions of individual particles. For example, any thermodynamic system on Earth experiences a gravitational potential due to the mass of its particles. When molecules are confined within porous materials, they are subject to interactions with the solid structure. In such circumstances, the properties of a thermodynamic system are defined not only by the characteristics of the chemical species under bulk conditions but also by the parameters that describe the external potential.

1.2.3 The Gibbs–Duhem Equation

The multiplicative behavior of extensive variables implies that thermodynamic quantities must be interrelated. If the size of a thermodynamic system is scaled by a factor λ, a positive number, all extensive properties of the system must be scaled by the same factor. The linear scaling behavior leads to the Gibbs–Duhem equation, which relates the variation of different thermodynamic quantities.

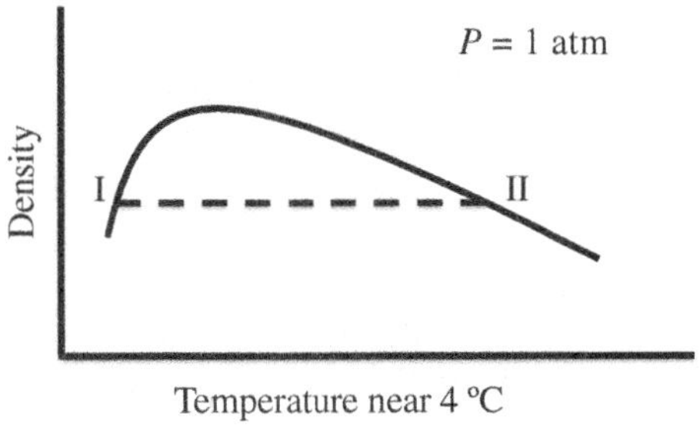

Figure 1.1 Gibbs phase rule predicts the degree of freedom for a one-component system in a single phase is $N_f = 2$. But why are pressure and density insufficient to define liquid water near 4 °C?

To elucidate the mathematical implications of extensive variables, consider a uniform one-component fluid as an example. Similar equations can be readily established for other systems. For a one-component fluid such as a gas or a liquid, the fundamental equation suggests that the internal energy U may be expressed as an analytical function of entropy S, volume V, and the number of molecules N, i.e., $U = U(S,V,N)$. Because these variables are all extensive, their linear scaling by a factor λ leads to

$$\lambda U = U(\lambda S, \lambda V, \lambda N). \tag{1.14}$$

In mathematics, the linear relation among multiplicative variables as shown in Eq. (1.14) is known as a linearly homogeneous function. Accordingly, Euler's theorem predicts

$$U = S\left(\frac{\partial U}{\partial S}\right)_{V,N} + V\left(\frac{\partial U}{\partial V}\right)_{S,N} + N\left(\frac{\partial U}{\partial \mu}\right)_{V,S} = TS - PV + \mu N. \tag{1.15}$$

Eq. (1.15) can be obtained by differentiating both sides of Eq. (1.14) with respect to λ and thermodynamic relations

$$T = \left(\frac{\partial U}{\partial S}\right)_{V,N}, \quad P = -\left(\frac{\partial U}{\partial V}\right)_{S,N}, \quad \text{and} \quad \mu = \left(\frac{\partial U}{\partial N}\right)_{V,S}. \tag{1.16}$$

Recall that the fundamental equation predicts

$$dU = TdS - PdV + \mu dN. \tag{1.17}$$

A combination of Eqs. (1.15) and (1.17) leads to the Gibbs–Duhem equation for a one-component system

$$0 = SdT - VdP + Nd\mu. \tag{1.18}$$

As the Gibbs phase rule predicts that there are only two independent variables for the one-component system, intensive variables T, P, and μ must be related to each other. Specifically, Eq. (1.18) predicts that, at constant temperature, the variation in pressure with respect to chemical potential gives

$$\left(\frac{\partial P}{\partial \mu}\right)_T = \rho \tag{1.19}$$

where $\rho = N/V$ is the number density of molecules. Accordingly, the isothermal compressibility can be written as

$$\kappa_T \equiv \frac{1}{\rho}\left(\frac{\partial \rho}{\partial P}\right)_T = \frac{1}{\rho^2}\left(\frac{\partial \rho}{\partial \mu}\right)_T. \tag{1.20}$$

Eqs. (1.19) and (1.20) allow us to calculate the variation of the chemical potential with respect to pressure or density. Besides, it can also be used to establish, as discussed in later chapters (Sections 2.9 and 7.3), connections between compressibility and fluctuations of thermodynamic quantities.

Other forms of the Gibbs–Duhem equation can be readily derived from Euler's theorem. For example, for a uniform multicomponent system at fixed temperature and pressure, any extensive property can be expressed as a function of the number of moles for each chemical species, $X = X(T, P, n_i)$. The Gibbs–Duhem equation predicts that, at constant T and P,

$$0 = \sum_i n_i d\overline{X}_i \tag{1.21}$$

where $\overline{X}_i = (\partial X/\partial n_i)_{T,P,n_{j\neq i}}$ is called partial molar property. Note that chemical potential for each species is a partial molar property, $\mu_i = \overline{G}_i = (\partial G/\partial n_i)_{T,P,n_{j\neq i}}$. Thus, a special form of the Gibbs–Duhem equation is given by

$$0 = \sum_i n_i d\mu_i. \tag{1.22}$$

Similar equations can be deduced for two-dimensional and one-dimensional systems.

1.2.4 The Maxwell Relations

We can obtain the Maxwell relations by observing that the second-order partial derivatives of a multivariable function are independent of the sequence of differentiation. Using again one-component systems as an example, the second-order partial derivative of $U = U(S,V,N)$ with respect to S and V leads to

$$\left(\frac{\partial T}{\partial V}\right)_{S,N} = -\left(\frac{\partial P}{\partial S}\right)_{V,N}. \tag{1.23}$$

A more useful Maxwell relation can be derived from the fundamental equation

$$dF = -SdT - PdV + \mu dN \tag{1.24}$$

by applying the second-order partial derivative of $F = F(T,V,N)$ with respect to T and V

$$\left(\frac{\partial S}{\partial V}\right)_{T,N} = \left(\frac{\partial P}{\partial T}\right)_{V,N}. \tag{1.25}$$

Other Maxwell relations can be readily derived from Eq. (1.24)

$$\left(\frac{\partial S}{\partial N}\right)_{T,V} = -\left(\frac{\partial \mu}{\partial T}\right)_{V,N}, \tag{1.26}$$

$$\left(\frac{\partial P}{\partial N}\right)_{T,V} = -\left(\frac{\partial \mu}{\partial V}\right)_{T,N}. \tag{1.27}$$

Given an equation of state relating pressure P with temperature T, volume V, and the number of particles N, Eq. (1.25) allows us to predict how the entropy of a closed system, an abstract quantity not directly measurable in experiments, changes with volume at constant temperature.

Many more Maxwell relations can be derived from other forms of the fundamental equation including those for multi-component systems. In practice, the most useful Maxwell relations are those that relate abstract quantities such as entropy and chemical potential to thermodynamic variables that can be measured in the laboratory.

1.2.5 Summary

The Gibbs phase rule predicts the number of independent variables to specify the equilibrium state of a macroscopic system. Because all other variables are dependent on a few variables that define the equilibrium system, thermodynamic quantities are interrelated through a network of equations. While the discussion above is focused on bulk fluids, similar relations can be obtained for other thermodynamic systems including those with different dimensionality. Such relations provide exact connections among different thermodynamic variables. Experimental or theoretical results that violate any exact thermodynamic relation are known as *thermodynamic inconsistency*, which should be avoided in practical applications.

1.3 Microscopic Uncertainty, Ensemble Average, and Ergodicity

In classical thermodynamics, each equilibrium state is defined by a few macroscopic quantities as predicted by the Gibbs phase rule. At any instant, a thermodynamic system may reside in one of its many microstates. Unless there are experimental means to keep track of the dynamic behavior of individual particles, uncertainty is inevitable in describing the microscopic details of a thermodynamic system.

The microscopic uncertainty of thermodynamic systems necessitates that the connection between macroscopic properties and the variables underlying the mechanical properties of individual particles can be established only through *statistical* means. To account for microscopic uncertainty, we introduce in this section ensemble average and ergodicity, two most fundamental concepts in statistical mechanics.

1.3.1 Microscopic Uncertainty

Equilibrium systems are considered thermodynamically equivalent if they exhibit the same macroscopic properties. From a microscopic point of view, however, systems with identical thermodynamic properties can be quite different in terms of the dynamic behavior of individual particles.

To illustrate, Figure 1.2 shows two thermodynamic systems, A and B, containing argon gas with the same temperature, pressure, and total volume. If the containers are identical, we assert that all macroscopic properties of system A are the same as those of B, i.e., any measurement for A would be equivalent to that for B. But at any instant, do they have the same energy? Do individual argon molecules have the same relative positions? Do they have the same momenta? Classical thermodynamics is not concerned with these, and other questions related to the microscopic behavior of individual particles. However, such information is essential in statistical thermodynamics because it aims to predict macroscopic properties based on microscopic details.

One mole of argon gas contains about 6.02×10^{23} molecules. As shown schematically in Figure 1.2, argon molecules are in perpetual motion and interact with each other and with the surface of the containers. At any instant, it is extremely unlikely that all argon molecules in

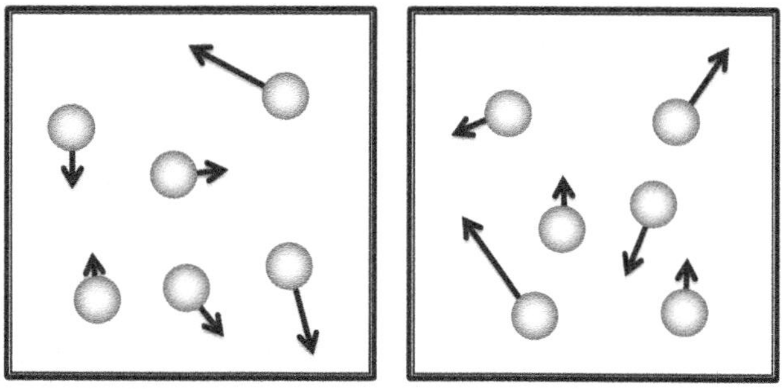

(A) (B)

Figure 1.2 Two thermodynamic systems containing argon gas are considered identical if they have the same temperature, pressure, and total volume. But they may be differentiated from a microscopic perspective because, at any instant, the relative positions, and momenta of argon molecules in system (A) are different from those in system (B). The thermodynamic systems of argon gas are equivalent only on a macroscopic scale, not necessarily on the microscopic scale of individual molecules. Schematically, here each sphere represents an argon molecule, and the arrow denotes the direction and magnitude of momentum.

one system have the positions and momenta the same as those in the other even when they are thermodynamically equivalent. In other words, while the thermodynamic (or macroscopic) state of the argon gas in system A is the same as that in system B, at any instant, the microstate in A is most likely different from that in B.

The above example illustrates that, in application of statistical mechanics to any system of practical interest, we need to define the microstates using a physical model suitable for characterizing the microscopic details. In addition, we must determine the probability distribution for the microstates of the system under consideration. With a precise knowledge of the microstates and their probability distribution, we can then predict macroscopic properties by taking statistical averages.

1.3.2 Ensembles

Ensemble plays a vital role in the statistical description of microstates and in establishing their connections with thermodynamic quantities. Formally, a statistical ensemble, or *ensemble* in short, is referred to as an arbitrarily large number of thermodynamic systems that are macroscopically equivalent.

Figure 1.3 shows a schematic picture of an ensemble of thermodynamic systems consisting of identical particles with the same temperature T, volume V, and particle number N. At any instant, each system has a particular microstate, which may or may not be the same as that of other systems in the ensemble. As the number of macroscopically equivalent systems can be arbitrarily large, the ensemble includes not only all microstates of a real system under study, but also encompasses microstates corresponding to any "mental copies" of the real system, i.e., imaginary systems with thermodynamic properties identical to those of the experimental system under investigation.

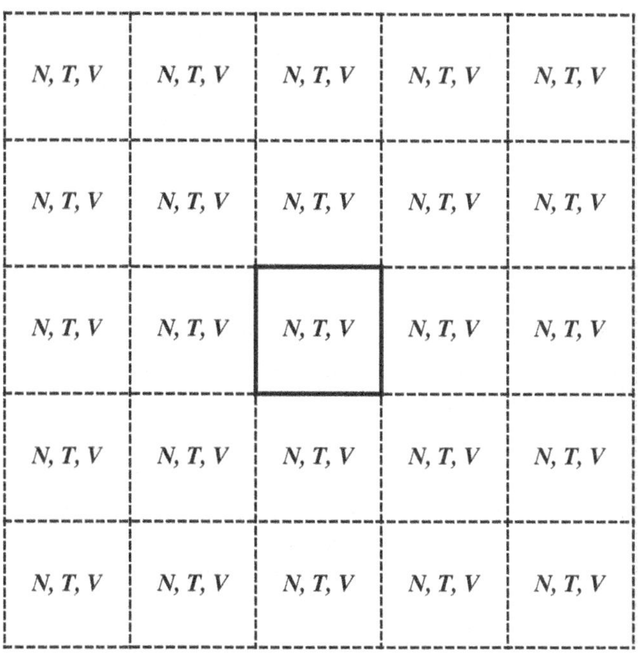

Figure 1.3 An ensemble is defined by an arbitrarily large number of thermodynamic systems with identical macroscopic properties, e.g., temperature T, the number of particles N, and volume V. Here, the box in solid lines represents a thermodynamic system under study; other boxes are mental copies of the real system. All systems have the same T, N, and V. However, at any moment, individual systems may (but need not) exist in microstates different from each other.

By accounting for an arbitrarily large number of systems that are thermodynamically equivalent but not microscopically identical, statistical mechanics provides mathematical means to determine the probability of all microstates in the ensemble and evaluate the average values of both microscopic and macroscopic properties of the real system under study. We will discuss different definitions of microstates in Sections 1.5 and 1.6. Chapter 2 gives formal mathematical procedures to determine the microstate probability distributions and their connections with various thermodynamic quantities based on the dynamic behavior of individual particles. The probability distributions may be alternatively evaluated by counting microstates in many systems consistent with the thermodynamic conditions of interest (e.g., by using a supercomputer), much like polling the approval rate of a politician. This procedure can be implemented with Monte Carlo simulation, which will be discussed in Chapter 6.

1.3.3 Ensemble Averages

An ensemble average refers to the mean value of any quantity affiliated with the microstates of a thermodynamic system. Suppose that M_ν is a property that can be determined for each microstate ν (e.g., the total energy), the ensemble average of M_ν is defined as

$$\langle M \rangle = \sum_\nu p_\nu M_\nu \tag{1.28}$$

where p_ν is the probability of microstate ν in the ensemble, and brackets $\langle \cdots \rangle$ denote an ensemble average.[7] Like any probability distribution, the microstate probability is nonnegative and normalized, i.e.,

$$\sum_\nu p_\nu = 1. \tag{1.29}$$

The ensemble average is applicable to either microscopic or macroscopic quantities, i.e., to the properties of individual particles or of the entire system. For example, internal energy is defined as the ensemble average of the total kinetic and potential energies of all particles over all microstates

$$U = \sum_\nu p_\nu E_\nu \tag{1.30}$$

where E_ν is the total energy of the system at microstate ν. It should be noted that E_ν is associated with the energies of the microscopic constituents (viz., particles) of a thermodynamic system; it does not include the kinetic energies due to the overall motion or potential energies related to the center of mass for the entire system.[8] For a molecular system, E_ν is often represented by the overall kinetic energy due to the translational, vibrational, and rotational motions of individual molecules, the potential energy due to intra- and inter-molecular interactions, and the external energy arising from the molecular interactions with an external field (e.g., an electrical, gravitational, or magnetic potentials).

An ensemble average can also be defined in terms of microscopic quantities. Box 1.1 illustrates the ensemble average of an important microscopic quantity, the probability density of finding a particle at a given position. For a uniform system, this average is trivially known, $\rho = N/V$, where N is the number of particles and V is the system volume. When the system is inhomogeneous

7 In statistics, $\langle M \rangle$ is called expectation, i.e., the expected value of M_ν.

8 If the entire system is in motion relative to some frame of reference, the energy associated with that motion is not included in internal energy. Neither does the internal energy include the gravitational potential related to the elevation of the entire system relative to some frame of reference. While not considered here, it is possible to include macroscopic motion and gravity in statistical thermodynamics.

(e.g., particle distribution under the influence of an external field), the average probability density is known as the density function, which is a fundamental quantity in classical density functional theory (cDFT),[9] an advanced statistical-mechanical method to predict the equilibrium properties of thermodynamic systems.

Box 1.1 Density Function

Consider a thermodynamic system containing N identical particles. Assume these particles are classical such that the microstates of the system can be represented by the particle positions and momenta. At each microstate, the probability density of finding a particle is specified by the Dirac-delta function (see Figure 1.4 for a one-dimensional illustration). The instantaneous local density at position **r** is given by the summation of the probabilities of all particles

$$\hat{\rho}(\mathbf{r}) = \sum_{i=1}^{N} \delta(\mathbf{r} - \mathbf{r}_i).$$

The density function is defined as the ensemble average of the instantaneous local particle density

$$\rho(\mathbf{r}) = \langle \hat{\rho}(\mathbf{r}) \rangle.$$

For a uniform system, the density function is invariant with position, i.e., the particle density is a constant, N/V.

1.3.4 Ergodicity

How do we know that the properties obtained from ensemble averages correspond to those measured in a particular thermodynamic system under study? What are the connections between different systems in the ensemble? Are the instantaneous microstates of the virtual copies of the thermodynamic system under study relevant to that of the real system? To answer these questions, we need the ergodic hypothesis, one of the most fundamental postulates of statistical mechanics.[10]

To understand ergodicity, imagine how the microscopic properties of a particular thermodynamic system vary with microstates. As shown schematically in Figure 1.5, the instantaneous value for any measurable thermodynamic quantity M changes with time because the microstate evolves with the motions of individual particles. If M is measured by some instrument, the experimental

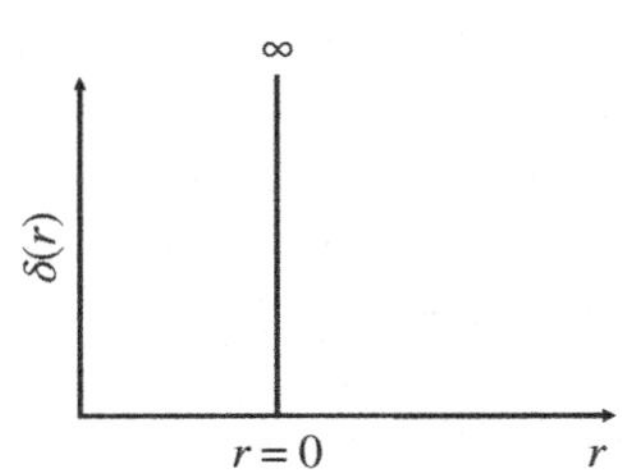

Figure 1.4 The Dirac delta function $\delta(r)$ is everywhere zero except ∞ at $r = 0$. Because $\int \delta(r)dr = 1$, $\delta(r)$ may be understood as a special form of the probability density with the distribution of a continuous variable r localized to a single point.

9 In mathematics, a functional maps a function (or functions) into a number, i.e., a functional is a function with function(s) as the input.
10 There have been many attempts to prove the ergodic hypothesis from the dynamic perspective but a generic proof is yet to be established. For recent developments, see Moore C. C., *PNAS* 112 (7), 1907–1911 (2015).

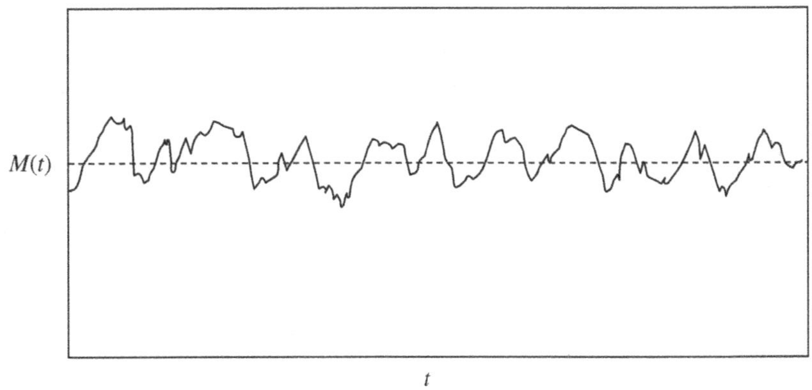

Figure 1.5 Schematic of time variance of dynamic property M of an equilibrium system (e.g., $M(t)$ could be the instantaneous total energy of a tank of argon gas at fixed N, V, and T). While M changes with time, at equilibrium, time average $\overline{M}$ (dashed line) is constant.

result corresponds to the time average of its instantaneous value over duration τ

$$\overline{M} = \frac{1}{\tau}\int_0^{\tau} M(t)\,dt. \tag{1.31}$$

Because thermodynamic quantities are time-invariant, the duration of measurement should be much larger than the time scale relevant to the dynamics of the individual particles. Besides, the experimental result should be independent of the time of measurement.

For a typical molecular system, the time scale relevant to the dynamics of individual atoms is exceedingly small, about 10^{-15} seconds or femtoseconds. Over the duration of a typical measurement for the equilibrium properties of a thermodynamic system (of the order of minutes or hours), the system is allowed to visit virtually all possible microstates. As a result, Eq. (1.31) can be equivalently written as

$$\overline{M} = \lim_{\tau\to\infty}\frac{1}{\tau}\int_0^{\tau} M(t)\,dt. \tag{1.32}$$

As the measurement samples all possible microstates, it is reasonable to conjecture that the microstates distribution is the same as that in the ensemble. In other words, the ensemble average and the time average would yield identical results

$$\langle M\rangle = \lim_{\tau\to\infty}\frac{1}{\tau}\int_0^{\tau} M(t)\,dt. \tag{1.33}$$

The equivalence of time average and ensemble average as given by Eq. (1.33) is called *ergodicity*.[11] Whereas Eq. (1.33) may not be valid for glassy systems that have multiple regions of microstates not mutually accessible to each other, we conjecture that the ergodic hypothesis holds true for all thermodynamic systems at equilibrium. Systems that do not satisfy ergodicity are referred to as non-ergodic.

We may illustrate ergodicity and the conceptual difference between time average and ensemble average by using an analogy. Suppose that we have two dice, each with six facets labeled with numbers 1–6. If we roll the two dice once, we want to know: What is the probability that the face values of the two dice give a seven? We can answer this question in two ways. In the first way, we have one

11 The term "ergodic" was introduced originally by Ludwig Boltzmann based on the Greek words ε'ργον (ergon: "work") and οδός (odos: "path" or "way").

person roll the dice repeatedly, say n times. For each roll, we record the sum of the face values of the dice. For two different dice, there are 36 different outcomes (viz., "microstates"). We can calculate the desired probability by noting how often the two dice give seven and dividing that by n, the total number of rolls. This would give a time-average result because it comes from many rolls, each with the same dice at a different time. Alternatively, we can assemble many people and give each one two dice. The dice given to any one person are identical to those given to any other person. There are n persons in the assembly. At a fixed instant, everyone in the assembly rolls his or her dice. We can then calculate the desired probability by looking at all the rolls and counting the number of rolls that result in seven. We take that number and divide it by n, the number of persons who, at the same time, rolled his or her two dice. In this case, the calculated probability is not the average of many rolls taken by one person over an extended period. Instead, it is the average of many rolls, each taken by one of many persons (an ensemble) at a fixed time. We expect that the two methods will give the same result. Of the 36 possible outcomes, six outcomes give seven; they are: (1,6), (6,1), (2,5), (5,2), (3,4), (4,3). The desired probability is $6/36 = 1/6$.

1.3.5 Summary

In this section, we have discussed microscopic uncertainty as an intrinsic feature of all thermodynamic systems. Because of the microscopic uncertainty, we use statistical ensembles to define the probability distribution for microstates. A fundamental hypothesis in the ensemble approach is ergodicity, which asserts that properties evaluated from ensemble averages are equivalent to corresponding variables in time average as measured in a specific thermodynamic system.

1.4 Entropy and Information

As discussed in Section 1.1, thermodynamics is centered around two fundamental quantities, internal energy and entropy, from which other thermodynamic properties can be derived. While internal energy is linked with the kinetic and potential energies of individual particles, entropy provides a measure of the system *information*, i.e., diverse ways how individual particles may exist. In classical thermodynamics, entropy is often perceived as mysterious because, unlike internal energy, information does not have any obvious meaning without considering the uncertainties in the microscopic constituents of a thermodynamics system. In this section, we introduce two common definitions of entropy in statistical mechanics and explore different interpretations from microscopic perspectives.

1.4.1 The Boltzmann Entropy

In searching for a mechanical theory of heat, Ludwig Boltzmann concluded in 1872 that entropy is an intrinsic property of an equilibrium system reflecting the number of ways that individual particles may exist. According to Boltzmann, entropy is defined as

$$S = k_B \ln W \tag{1.34}$$

where $k_B = 1.381 \times 10^{-23}$ J/K is the Boltzmann constant,[12] and W stands for the number of *accessible* microstates, i.e., microstates that the system may take with nonzero probability. Boltzmann's definition of entropy was carved on his gravestone.

12 Although Boltzmann first linked entropy and probability, the specific constant was introduced by Max Planck in the quantum theory of black-body radiation (see Section 4.3).

Eq. (1.34) provides a simple interpretation of entropy from a microscopic perspective: entropy is related to the number of choices available to the individual particles in a thermodynamic system. Because the particles may exist in different microstates, entropy provides a measure of the microscopic degrees of freedom.

We may illustrate the meaning of the total number of microstates by considering a simple example. Suppose that we have a system containing four *distinguishable* boxes, for example, by their fixed positions. We have four balls that can be randomly placed into these boxes, and we ask: In how many ways (microstates) can we place these four balls into the four boxes such that there is one ball in each box?

First, consider that all balls are different. When we place the first ball, we have four possibilities because all four boxes are initially empty. When we place the second ball, we have three possibilities because one box is already occupied. When we place the third ball, we have two possibilities and when we place the fourth ball, we have only one possibility. Therefore, the number of ways to place the balls is

$$W = 4 \cdot 3 \cdot 2 \cdot 1 = 24. \tag{1.35}$$

If the four balls are identical, i.e., indistinguishable, however, the final arrangements of balls in the four boxes are equivalent, independent of the sequence of filling. In that case, it makes no difference which ball we place first, second, third, or fourth. Therefore, we must divide the apparent W by 4!, giving $W = 1$. In other words, there is only one way to place four identical balls into four different boxes such that one ball is in each box. For identical balls, the number of microstates is unity. In these two scenarios, we say that the entropy for the case with different balls is higher than that with identical balls because the former provides more ways to arrange the balls (viz., more microstates).

But now suppose we have four balls where two are black and two are white; otherwise, the balls are identical. Again, we ask, in how many ways (microstates) can we place the four balls into the four boxes? For this case, we must divide the apparent W, as given by Eq. (1.35), by $2! \times 2!$, because two black balls and two white balls are indistinguishable. For this case, $W = 6$.

Figure 1.6 shows the six ways of filling four different boxes with two identical black balls and two identical white balls such that one ball is in each box. Because entropy is related to the number of microstates, the entropy of two identical black balls and two identical white balls in four different boxes is larger than that of four identical white balls (or four identical black balls) in four boxes.

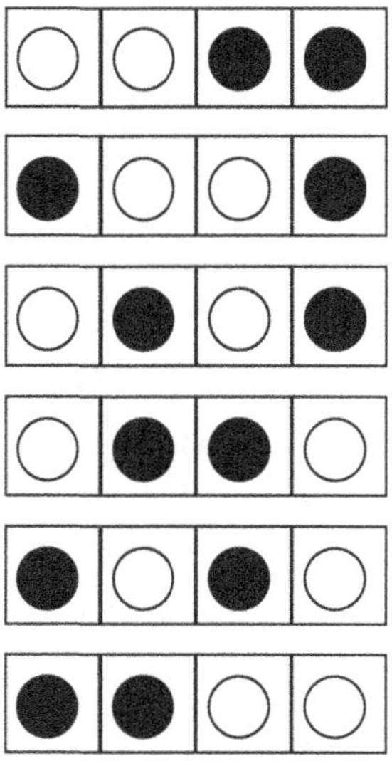

Figure 1.6 Six ways to place two identical white balls and two identical black balls into four different boxes such that one ball is in each box.

Why is it larger? When filling the four boxes with four identical balls, the information is complete; there is only one arrangement. But when filling four boxes with two white balls and two black balls, the information is incomplete. As shown in Figure 1.6, that case has six arrangements. We know that there are two black balls and two white balls in four boxes, but we do not know the detailed arrangement; information concerning the state of the system is incomplete.

The above example illustrates that, according to Boltzmann, entropy is a measure of the number of ways that a set of particles can be arranged. Here, different arrangements are affiliated with the particle identity. When the particles can be rearranged or shuffled among different configurations over time, the entropy of a system with the same balls is smaller than that of a system with different balls because identical balls reduce the number of ways that the balls can be arranged in the boxes. When otherwise identical balls are labeled, the entropy increases because the particles can be arranged in more different ways. In Box 1.2, we provide a more sophisticated example illustrating how entropy is intrinsically connected with such microscopic information.

Box 1.2 Maxwell's Demon

An imaginary being (a.k.a., a demon) was discussed in a letter written by James Clerk Maxwell to his life-long friend Peter Guthrie Tait in 1867, suggesting that the second law of thermodynamics would be violated. As illustrated in Figure 1.7, Maxwell conceived two gas chambers A and B with a shutter in between. An intelligent demon, who could keep track of the speeds of gas molecules, would be able to control the shutter such that high-speed molecules pass in one direction and low-speed molecules in the other. The biased passing of gas molecules through the shuttle would lead to temperature disparity between A and B. If the collisions of gas molecules with the shutter were elastic, and moving the shutter was frictionless, no work would be done by the demon. As a result, heat transfer takes place from elevated temperature to low temperature without any work.

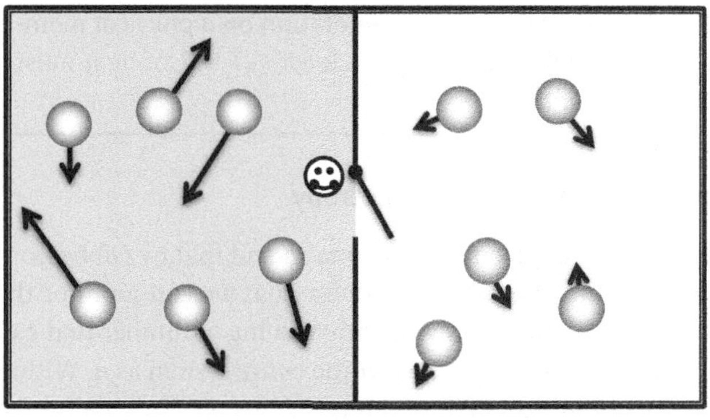

Figure 1.7 Maxwell's demon operated a trap door separating two volumes of gas A and B at initially equal temperature.

The demon's action reduces the total entropy of an isolated system (viz., gas molecules in A and B plus demon) thereby violating the second law of thermodynamics. The second law is saved, even if Maxwell's demon would be invented, only if one recognizes that the entropy decrease is achieved by the acquisition of microscopic information (here, the speed of individual molecules) by the intelligent demon. As illustrated in Box 1.3, Landauer's principle provides an explicit connection between information and thermodynamic quantities.

1.4.2 The Gibbs Entropy

The connection between entropy and microscopic information (more precisely, missing information) was first identified by J.W. Gibbs who provided an alternative definition of entropy[13]

$$S = -k_B \sum_{\nu} p_\nu \ln p_\nu. \tag{1.36}$$

As discussed in Section 1.3, p_ν stands for the probability of a thermodynamic system in microstate ν, and the summation extends to all accessible microstates.

In the context of ensemble average, Eq. (1.36) can be written as

$$S = -k_B \langle \ln p_\nu \rangle. \tag{1.37}$$

Eq. (1.37) suggests that, up to a universal constant k_B (which is needed for unit conversions between energy and temperature), the microscopic counterpart of entropy, $-\ln p_\nu$, can be understood as a measure of the degree of "uncertainty" or "incomplete information". When $p_\nu = 1$, the thermodynamic system has only one microstate ν. In this case, $-k_B \ln p_\nu = 0$ and $S = 0$ means no "uncertainty" at the microscopic level. For $p_\nu < 1$, the microstate of the system is not certain; the degree of "uncertainty" may be quantified in terms of $-\ln p_\nu$; the smaller p_ν is, the larger the system has "uncertainty".[14]

Box 1.3 Landauer's Principle

Rolf Landauer, an IBM scientist who made important contributions to information processing, analyzed the thermodynamics of data processing in 1961.[15] He discovered that information must be embodied in physical states and thus subject to thermodynamic regulations. For any computer operation that manipulates information, e.g., erasing a bit of memory, the entropy will increase due to loss of information about the physical state of individual logical units. The loss of information is manifested as the dissipation of heat accompanying the computer operation.

Landauer's principle states that for any irreversible single-bit operation on a physical memory element in contact with a heat bath at a given temperature, at least $k_B T \ln 2$ of heat must be released from the memory device into environment.

1.4.3 Connection Between Boltzmann's Entropy and Gibbs' Entropy

To establish a connection between the definition of entropy by Boltzmann and that by Gibbs, consider an equilibrium system that may adopt many microstates. Suppose that we can monitor the microstates of the system in $\mathbb{N}$ consecutive steps as time evolves, with $\mathbb{N}$ being a number that can be arbitrarily large. We designate the total number of microstates for the *entire* system as n. Within the duration of observation, the multiplicity of the system in different microstates is equal to the number of different outcomes to toss $\mathbb{N}$ times "a super die" with n facets

$$W_{\mathbb{N}} = \frac{\mathbb{N}!}{\mathbb{N}_1!\mathbb{N}_2!\ldots\mathbb{N}_n!} \tag{1.38}$$

13 Gibbs' definition of entropy was later adopted by Claude Shannon in his theory of signal processing and data analysis, which constitutes an important cornerstone of information theory.

14 Conceptually it is preferable to use a quantity that rises (rather than falls) as information increases. For that purpose, negentropy is defined as the reciprocal of entropy.

15 Landauer R., "Irreversibility and heat generation in the computing process", *IBM J. Res. Develop.* 5 (3), 183–191 (1961).

where $\mathbb{N}_\nu$ is the number of times that the system is in microstate ν, satisfying the normalization condition

$$\sum_{\nu=1}^{n} \mathbb{N}_\nu = \mathbb{N}. \tag{1.39}$$

Because the multiplicity of the outcomes involves $\mathbb{N}$ steps, the average number of microstates at each step is

$$W = W_{\mathbb{N}}^{1/\mathbb{N}}. \tag{1.40}$$

In a special case where all microstates are equally accessible (e.g., a fair super coin), $W_{\mathbb{N}} = n^{\mathbb{N}}$, i.e., the number of accessible microstates at each step is n. In general, evaluation of $W_{\mathbb{N}}^{1/\mathbb{N}}$ requires some specific knowledge of the statistical distribution of the microstates.

According to Boltzmann, the entropy of the system is given by

$$S = k_B \ln W = \frac{k_B}{\mathbb{N}} \ln W_{\mathbb{N}}. \tag{1.41}$$

Using Eq. (1.38) for $W_{\mathbb{N}}$ and Stirling's approximation, $\ln \mathbb{N}! \approx \mathbb{N} \ln \mathbb{N} - \mathbb{N}$, we find that Boltzmann's definition of entropy is the same as that by Gibbs

$$\frac{S}{k_B} = \frac{\left(\ln \mathbb{N}! - \sum_{\nu=1}^{n} \ln \mathbb{N}_\nu!\right)}{\mathbb{N}} = -\sum_{\nu=1}^{n} (\mathbb{N}_\nu/\mathbb{N}) \ln(\mathbb{N}_\nu/\mathbb{N}) = -\sum_{\nu=1}^{n} p_\nu \ln p_\nu \tag{1.42}$$

where $p_\nu = \mathbb{N}_\nu/\mathbb{N}$ is the probability that the system is in microstate ν. Although in general the two definitions of entropy are not always identical (see Chapter 2), this simple example illustrates that both Gibbs' entropy and Boltzmann's entropy are related to the statistical distribution of microstates.

1.4.4 Entropy and Disorder

When thermodynamics is applied to diffusion or mixing processes, the entropy increase may be intuitively understood in terms of the rise of disorder or more randomness. However, a system in a state of higher entropy is not necessarily always equivalent to a state of greater disorder.[16] For certain systems, increasing entropy may serve as a driving force for microscopic ordering rather than disordering.

In the 1940s, Lars Onsager discovered that, at sufficiently high density, a system of rod-like particles exhibits a phase transition from an isotropic phase to a lamellar phase (much like a bunch of chopsticks, the rod-like particles are aligned in the axial direction).[17] The disorder-to-order transition can be explained by a simple thermodynamic model that accounts for the second-virial coefficient for interaction between rod-like particles. The phase transition occurs when the entropy of the ordered lamellar phase is higher than that of the disordered isotropic state. Entropy-driven ordering was also found by Alder and Wainwright in the 1950s using a computer model.[18] At sufficiently high density, spherical particles may crystallize without any attractive force (Figure 1.8). In other words, at sufficiently high density, the entropy of spherical particles in an ordered state may exceed that in a random state (see Section 7.9 for details).

16 As commonly perceived, we define "order" in terms of spatial organization of individual elements. For example, a crystal is referred to as an ordered phase because of the organized arrangement of individual atoms.
17 Onsager L., "The effects of shape on the interaction of colloidal particles", *Ann. NY Acad. Sci.* 51, 627–659 (1949).
18 Alder B. and Wainwright T., "Phase transition for a hard sphere system", *J. Chem. Phys.* 5, 1208–1209 (1957).

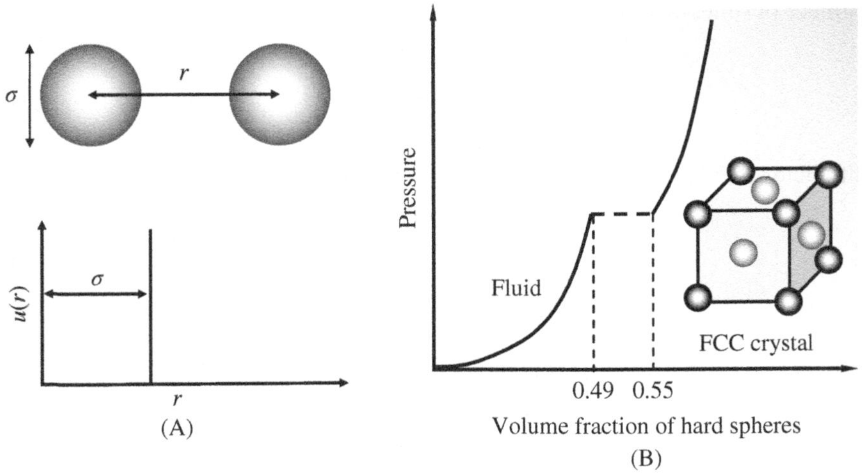

Figure 1.8 At sufficiently high volume fraction, a system of hard spheres maximizes its entropy by formating a face-centered cubic (FCC) structure. A. Schematic of hard-sphere potential between two identical particles; B. The pressure of a uniform hard-sphere system versus the volume fraction. The dashed lines mark the transition between disordered (random) and ordered (FCC) structures.

1.4.5 Entropy and Chaos

In classical thermodynamics, entropy is introduced through heat effects during reversible processes, i.e., idealized procedures for changing the thermodynamic conditions of a macroscopic system. Because the microscopic origin of heat is not intuitive, classical thermodynamics provides little insights on the physical significance of entropy. The lack of a physically comprehensible definition often results in misinterpretations. For example, many thermodynamics texts introduce entropy as a measure of disorder, randomness, or chaos; the higher the entropy, the greater the disorder. Such an interpretation may trace back to Rudolf Clausius who coined the term entropy from Greek, "en tropie," meaning transformation or change. Hermann von Helmholtz, another founding father of thermodynamics, also used German word "Unordnung" (disorder) to describe entropy. The interpretation is misleading because entropy may serve as a driving force of ordering. As discussed above and in more detail in Section 7.9, a thermodynamic state with a higher entropy can be more ordered than that with a lower entropy.

Rudolf Clausius proclaimed in 1865, "The entropy of the universe tends to a maximum." If entropy means disorder, the relentless increase of entropy would degrade our universe to a state of complete randomness or chaos. Historically, this bleak prediction of classical thermodynamics raised concerns among philosophers and laypersons alike. For example, Bertrand Russell wrote pessimistically that "all the labors of the ages, all the devotion, all the inspiration, all the noonday brightness of human genius, are destined to extinction."[19] Yet life emerges on earth, beginning from simple inorganic forms to biomacromolecules and their self-assembly to cells and eventual evolution to human beings with incremental ordered structures. Is entropy necessarily the same as disorder? Is "negative entropy" responsible for the organization in biological systems and the evolution of life?[20] Clarifying the concept of entropy with statistical mechanics will help answer these questions and avoid misunderstanding.

19 Russell B., *The free man's worship*. Routledge, 1903.

20 Erwin Schrödinger stated in his influential book, *What is Life*, that the emergence of life in the universe us due to the input of negative entropy.

1.4.6 Summary

According to statistical mechanics, entropy arises from uncertainties in microscopic details when a macroscopic system is defined by a few thermodynamic variables. As first recognized by Boltzmann, entropy provides a measure of the enormous number of microstates that a thermodynamic system may exist at any instant. Uncertainties in microscopic details are not necessarily equivalent to randomness or chaos – entropy can be a driving force of ordering in macroscopic systems. Mathematically, entropy is naturally affiliated with the information about individual particles in a thermodynamic system.

1.5 *Ab initio* Thermodynamics

When statistical mechanics was originally developed in the late nineteenth century, the subject was primarily an exercise of abstract mathematics. Many concepts were introduced without specifying the physical nature of the particles in thermodynamic systems or how they may interact with each other. Despite its importance, the precise meaning of particles or microstates was obscure. The dilemma was clearly stated in the classical text by J.W. Gibbs,[21] "*we avoid the gravest difficulties when, giving up the attempt to frame hypotheses concerning the constitution of material bodies, we pursue statistical inquiries as a branch of rational mechanics. In the present state of science, it seems hardly possible to frame a dynamic theory of molecular action which shall embrace the phenomena of thermodynamics, of radiation, and of the electrical manifestations which accompany the union of atoms. Yet any theory is obviously inadequate, which does not take account of all these phenomena. Certainly, one is building on an insecure foundation, who rests his work on hypotheses concerning the constitution of matter.*"

Thanks to tremendous developments in theoretical physics, in particular, molecular sciences over the past century, Gibbs' view of statistical mechanics has been long outdated. In this and the next section, we will specify the microscopic details of thermodynamic systems from different perspectives: first principles, atomistic, coarse-grained, and lattice models. This section provides a very brief overview of the so-called first principles or *ab initio* thermodynamics, whereby microstates are defined in terms of the quantum states of electrons and atomic nuclei. All thermodynamic properties can be predicted from the statistical distribution of quantum states.

A detailed description of first-principles methods is beyond the scope of this book. The online Supplementary Material provides a brief introduction of the basic concepts that will be sufficient for understanding the materials discussed in this section. Because quantum states can be predicted without adjustable parameters, a combination of first-principles calculations with statistical mechanics empowers a quantitative description of both the micro- and macro-properties of molecular systems.

1.5.1 Quantum States

Quantum mechanics asserts that elementary particles of matter, such as electrons, protons, and photons, exit in enumerable microstates (viz., quantum states). Each quantum state can be represented by a wave function, and the properties of a many-body system are manifested as the expectation values of observables.

---

21 Gibbs J. W., "*Elementary principles in statistical mechanics*", Dover Publications, reprint edition (2014).

While all microscopic properties can be described in terms of wave functions, exact quantum-mechanical methods are presently too complicated to be numerically soluble for thermodynamic systems of practical interest. Two approximations are commonly adopted in applications of quantum-mechanical methods to materials and chemical systems. First, the Born-Oppenheimer approximation assumes that the electronic degrees of freedom can be decoupled from those corresponding to atomic nuclei such that the atomic nuclei behave like classical particles with negligible size. Here, the degrees of freedom are referred to a complete set of variables to describe the dynamics of individual particles. By classical, we mean that quantum effects such as uncertainty and tunneling are irrelevant in describing particle motions and energy. The second assumption is that, in an atomic system, the electronic properties can be represented by those corresponding to electrons at 0 K in an external field due to interactions with the atomic nuclei. With these two assumptions, the microstates of any atomic system (viz., chemicals and materials) are defined by the quantum states of electrons while the nuclei are treated as classical particles with exact positions and momenta at any moment.

For most atomic systems of practical interest, the Born-Oppenheimer assumption is justified because of the small volume of atomic nuclei and the negligible electron rest mass m_e. On the one hand, the size of nuclear particles is tiny in comparison with the length scale characterizing electrons: the former is about few femtometers (10^{-15} m) but the latter on the order of Angstrom (10^{-10} m). As first discovered by Ernest Rutherford in 1911, "an atom is mostly empty space in which electrons obit". On the other hand, the rest mass of a proton, the smallest nuclear particle, is more than a thousand times the electron rest mass ($m_p/m_e \approx 1836$). The enormous difference in rest mass implies that the electron motion is much faster than that of atomic nuclei, typically by several orders of magnitude. As a result, electrons may be considered in the ground state at any atomic configuration defined by the positions of atomic nuclei.

1.5.2 Density Functional Theory (DFT)

In many practical applications, quantum-mechanical calculations are based on the electronic density functional theory (DFT), a mathematical procedure originally developed by Walter Kohn and others in the 1960s.[22] Rather than finding many-body wave functions directly, DFT seeks to solve quantum-mechanical problems by mapping many-body systems to noninteracting references. In comparison with alternative quantum-mechanical methods, DFT is drastically more efficient from a computational perspective yet applicable to a wide range of materials and chemical systems with reasonable accuracy.

To define the quantum states of electrons in an atomic system, DFT calculations start with the Kohn-Sham (KS) equation[23]

$$\left[-\frac{\hbar^2}{2m_e}\nabla^2 + v^{KS}(\mathbf{r})\right]\psi_\nu(\mathbf{r}) = \varepsilon_\nu \psi_\nu(\mathbf{r}) \tag{1.43}$$

where $\psi_\nu(\mathbf{r})$ stands for the wave function of noninteracting electrons, $\hbar = h/2\pi$, $h = 6.626 \times 10^{-34}$ J s is the Planck constant, $m_e = 9.109 \times 10^{-31}$ kg is the electron rest mass, $v^{KS}(\mathbf{r})$ represents the KS potential for an electron at position $\mathbf{r}$, and ε_ν stands for the electronic energy at a single-particle

22 Hohenberg P. and Kohn W., "Inhomogeneous electron gas", *Phys. Rev.* 136 (3B), B864–B871 (1964); Kohn, W. and Sham, L. J., "Self-consistent equations including exchange and correlation effects", *Phys. Rev.* 140 (4A), A1133–A1138 (1965).

23 An alternative to KS-DFT is provided by orbital-free DFT, which is numerically simpler but theoretically more challenging in formulation of an accurate functional.

quantum state v. Mathematically, Eq. (1.43) resembles the one-particle Schrödinger equation, the simplest possible quantum-mechanical description of electronic systems.

In principle, DFT provides an exact mathematical framework to describe electronic properties at the ground state. Regrettably, the KS potential is only partially known. It includes contributions due to electron-nuclei interactions $v_N(\mathbf{r})$, the electrostatic repulsion among the electrons (viz., the Hartree energy) $v_H(\mathbf{r})$, and an exchange-correlation (xc) potential $v_{xc}(\mathbf{r})$

$$\upsilon^{KS}(\mathbf{r}) = \upsilon_N(\mathbf{r}) + \upsilon_H(\mathbf{r}) + \upsilon_{xc}(\mathbf{r}). \tag{1.44}$$

The first two contributions on the right side of Eq. (1.44) are simply described by Coulomb's law, and the third term, $v_{xc}(\mathbf{r})$, is associated with the multi-body coupling of the Pauli exclusion principle (viz., exchange) and electrostatic interactions (viz., correlations). Because exact formulation of the exchange-correlation potential amounts to solving the many-body Schrödinger equation, approximations are inevitable in practical applications of KS-DFT. Over the years, numerical functionals have been developed to approximate $v_{xc}(\mathbf{r})$ and many software packages are available.[24] In general, the best choice of an xc-potential is dependent on the specific system and/or a given task, more like art than science.

1.5.3 Quantum Molecular Dynamics Simulation (QMD)

One major advantage of the first-principles approach is that it captures chemical reactions as well as intermolecular forces on an equal footing. More important, atomic motions, subsequently the evolution of microstates, can be naturally described by the fundamental laws of physics such that thermodynamic properties can be predicted by averaging over many microstates. The numerical procedure to integrate quantum and statistical mechanical calculations together is known as quantum molecular dynamics simulation (QMD) or *ab initio* molecular dynamics (AIMD). As the computer power ever increases, QMD/AIMD becomes increasingly more feasible and accurate in practical applications.

According to the Born-Oppenheimer assumption, each nuclear particle n experiences a Coulomb force due to interactions with other nucleic particles and with electrons in the surrounding

$$\mathbf{F}_n(\mathbf{r}) = \mathbf{F}_{nm}(\mathbf{r}) - \int d\mathbf{r}'\rho_e(\mathbf{r}')\frac{Z_n e^2(\mathbf{r}-\mathbf{r}')}{4\pi\varepsilon_0|\mathbf{r}-\mathbf{r}'|^3} \tag{1.45}$$

where $\rho_e(\mathbf{r})$ represents the local electronic density, Z_n stands for the nuclear valence, $e = 1.602 \times 10^{-19}$ C is the elementary charge, and $\varepsilon_0 = 8.854 \times 10^{-12}$ J^{-1} C^2 m^{-1} is the dielectric permittivity in a vacuum. In Eq. (1.45), $\mathbf{F}_{nm}(\mathbf{r})$ represents the electrostatic force for the interaction of the nuclear particle with other nuclear charges

$$\mathbf{F}_{nm}(\mathbf{r}) = \sum_m \frac{Z_n Z_m e^2(\mathbf{r}-\mathbf{r}_m)}{4\pi\varepsilon_0|\mathbf{r}-\mathbf{r}_m|^3} \tag{1.46}$$

where the summation applies to all other nuclear particles in the system. The electronic density can be calculated from

$$\rho(\mathbf{r}) = \sum_{i=1}^{N_e}|\psi_i(\mathbf{r})|^2 \tag{1.47}$$

24 Mardirossian N. and Head-Gordon M., "Thirty years of density functional theory in computational chemistry: an overview and extensive assessment of 200 density functionals", *Mol. Phys.* 115, (19), 2315–2372 (2017).

where N_e stands for the total number electrons in the system, and $\psi_i(\mathbf{r})$ is the single-electron wave function obtained from KS-DFT. With the force on each nuclear particle calculated from Eq. (1.45), the atomic motion can be described with the Newton equation

$$\mathbf{F}_n = m_n \mathbf{a}_n \tag{1.48}$$

where m_n stands for the nucleic mass, and $\mathbf{a}_n$ represents the particle acceleration. Once the atomic configuration is changed by numerically integrating the Newton equation, we update the electronic wave functions and, subsequently, the electronic density profile and the momenta of nuclear particles. The QMD/AIMD iteration leads to many microstates from which thermodynamic properties can be calculated from ensemble averages. In Chapter 2, we will discuss the mathematical details to integrate Newton's equation of motion for many particles.

1.5.4 Summary

In practical applications of statistical mechanics, we need an explicit definition of microstates. Otherwise, statistical mechanics would exist merely as an abstract formality, and the theoretical predictions could hardly be comparable with experimental measurements. In this section, we have discussed how "particles", i.e., the individual elements of thermodynamic systems, can be specified from first principles. Based on the Born-Oppenheimer assumption, we can define microstates in terms of the electronic quantum states. Accordingly, thermodynamic properties of an atomic system can be predicted from KS-DFT in combination with molecular dynamics simulation. The so-called *ab initio* thermodynamic procedure will be discussed in detail in Chapter 3 for ideal-gas systems.

1.6 Statistical-Thermodynamic Models

While tremendous progress has been made over the past few decades to predict the properties of materials and chemical systems from first principles, their applications to practical systems remain limited not only because of the computational burden but also because of insufficient numerical accuracy compromised by approximations introduced in quantum-mechanical calculations. As a result, semi-empirical models are commonly used in statistical thermodynamics to represent molecules or individual particles of a thermodynamic system from a classical perspective. In typical semi-empirical models, the microstates are often defined in terms of the positions and momenta of individual particles. The model parameters for describing particle-particle interactions can be calibrated with microscopic and/or macroscopic properties of thermodynamic systems. Based on limited experimental data and/or results from first-principles calculations, the semi-empirical models can predict thermodynamic properties of realistic chemical systems in good agreement with experimental measurements.

In this section, we discuss several types of statistical-thermodynamic models that are commonly employed for representing macroscopic systems, i.e., atomic, solvent-free, and coarse-grained models for materials and molecular systems, and lattice models that are often used to describe cooperative phenomena and universality of phase transition (Chapter 5).

1.6.1 Atomic Models

The materials world is made of a little over 100 types of atomic elements. However, natural phenomena are diverse and fascinating because atoms interact with each other through chemical bonds and longer-ranged forces. In statistical mechanics, atomic models intend to capture bond connectivity and intermolecular interactions, often by treating each atom as a classical particle. The microstates of an atomic system are thus represented by the positions and momenta of classical particles within the framework of Newtonian mechanics.

In a typical atomic model, the total energy of a thermodynamic system includes a kinetic energy (K) associated with the particle motions, and a potential energy (Φ) due to atomic interactions

$$E(\mathbf{r}^N, \mathbf{p}^N) = K(\mathbf{p}^N) + \Phi(\mathbf{r}^N) \tag{1.49}$$

where N represents the total number of atoms, $\mathbf{r}^N \equiv (\mathbf{r}_1, \mathbf{r}_2, \ldots, \mathbf{r}_N)$ is the atomic configuration as defined by the positions of all atoms, and $\mathbf{p}^N \equiv (\mathbf{p}_1, \mathbf{p}_2, \ldots, \mathbf{p}_N)$ represents the momenta of all the atoms. The multidimensional space spanned by the composite vector, $\mathbf{x} \equiv (\mathbf{r}^N, \mathbf{p}^N)$, is called the *phase space*, which defines the degrees of freedom of all particles in the system.

When the atoms are depicted by a classical model, the total kinetic energy is simply determined by the atomic mass (m_i) and momenta ($\mathbf{p}_i$) for each particle

$$K(\mathbf{p}^N) = \sum_{i=1}^{N} |\mathbf{p}_i|^2/2m_i. \tag{1.50}$$

Meanwhile, the potential energy consists of contributions resulting from chemical bonding (Φ^B) and nonbonded (Φ^{NB}) interactions[25]

$$\Phi(\mathbf{r}^N) = \Phi^B(\mathbf{r}^N) + \Phi^{NB}(\mathbf{r}^N). \tag{1.51}$$

Over the years, many semi-empirical functions have been developed to describe the potential energy of atoms in materials and molecular systems. These functions, along with parameters designed to reproduce molecular structure and selected properties of thermodynamic systems, are commonly referred to as *molecular force fields*.

With an analytical expression for the potential energy, we can evaluate the net force on each atom based on its relative position with all other atoms in the system

$$\mathbf{F}_i(\mathbf{r}^N) = -\nabla_{\mathbf{r}_i}\Phi(\mathbf{r}^N). \tag{1.52}$$

Based on an initial condition specifying the positions and momenta of all particles, Eq. (1.52) can be used to predict the evolution of atomic positions (viz., a *trajectory of atomic motions*) by integrating Newton's equation. The numerical procedure is known as classical molecular dynamics simulation, or MD, which will be discussed in Chapter 2.

In comparison with QMD or AIMD, classical MD simulation is computationally less demanding due to the absence of quantum-mechanical calculations. Nevertheless, a supercomputer is often needed for MD simulation of realistic systems because of the system size and time scales. The time scale pertinent to atomic motions is in the order femtoseconds while the diameter of an atom is only a few Ångstrom (1 Å = 10^{-10} m). The computational cost can be drastically reduced by using coarse-grained models where each particle is ascribed to a group of atoms (e.g., functional groups or monomeric segments of a polymer) or molecules.

1.6.2 The Lennard-Jones Model

To gain an understanding of how the microscopic details of individual particles determine the macroscopic properties of a thermodynamic system, it is useful to consider simple models. One

25 Supplementary Materials IV.

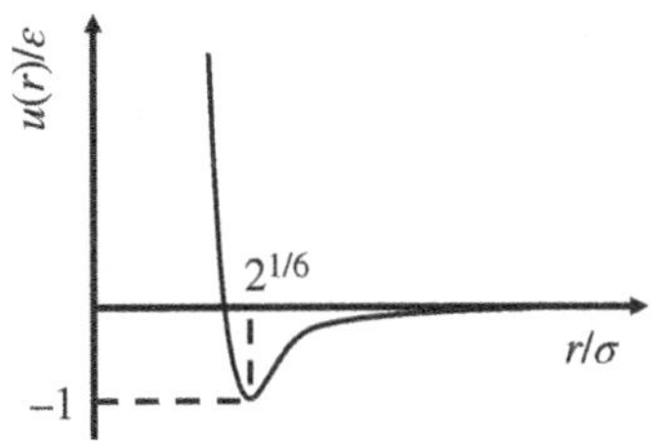

Figure 1.9 In dimensionless units defined by the energy and size parameters of the Lennard-Jones model, the intermolecular potential is a universal function of the distance.

of the simplest atomic models for molecular systems assumes that each molecule is a spherical particle, much like a billiard ball with attraction. The model provides a good description of the thermodynamic properties and phase behavior of simple fluids such as noble gases and systems consisting of small nonpolar molecules (e.g., methane and nitrogen).

In a simple fluid, the intermolecular interaction is often represented by the Lennard-Jones (LJ) potential

$$u(r) = 4\varepsilon\left[\left(\frac{\sigma}{r}\right)^{12} - \left(\frac{\sigma}{r}\right)^{6}\right] \tag{1.53}$$

where r is the center-to-center distance, parameters σ and ε are related to the particle size and the strength of maximum attraction. The second term on the right side of Eq. (1.53) represents the van der Waals attraction, and the first term approximates the short-range repulsion. A similar equation can be written for the pair potential between different types of small nonpolar molecules.

As shown in Figure 1.9, the LJ potential suggests a universal form of intermolecular interaction when the distance and potential energy are normalized by parameters σ and ε, respectively. The universal form offers a statistical-mechanical basis for understanding the theorem of corresponding states first proposed by van der Waals, i.e., in dimensionless variables of k_BT/ε and $\rho\sigma^3$, the thermodynamic properties of a simple fluid can be represented by universal functions of two independent variables as predicted by the Gibbs phase rule.[26]

Tremendous research effort has been devoted to LJ systems because once such systems are understood, it is then possible to apply corrections (viz., perturbations) toward representing chemical systems with more complicated intra- and inter-molecular interactions. This procedure is often used in classical thermodynamics. For example, we use the well-understood properties of an ideal gas as a basis for understanding real gases; we apply corrections through a compressibility factor that can be obtained from a corresponding state correlation or from a realistic equation of state. Similarly, for vapor–liquid equilibria of nonelectrolyte liquid mixtures, we start with an ideal mixture (Raoult's law) and then apply corrections through excess Gibbs energy models for activity coefficients. We will discuss in Chapters 7–9 that simple fluids play a pivotal role in the development of liquid-state theories and equations of state for bulk chemical systems based on statistical mechanics.

1.6.3 Implicit-Solvent Models

In solution thermodynamics, the solute behavior is often of primary interest for both practical and fundamental reasons. For convenience and physical clarity, it is useful to formulate microstates in terms of solute molecules without an explicit consideration of the microscopic details of the solvent. The basic idea is akin to neglecting the electronic degrees of freedom in modeling thermodynamic systems with many atoms. As the electronic effects can be described in terms of molecular structure

26 See, e.g., Tester, J. W. and Modell, M., *Thermodynamics and its applications* Prentice Hall, 1997, p. 244.

and semi-empirical force fields, the solvent effects can be incorporated into particle solvation and the solvent-mediated interactions. Because of their simplicity, the implicit solvent models have been widely adopted in theoretical descriptions of electrochemical systems, aqueous solutions, and colloidal dispersions.

By treating the solvent as a continuous medium, we can define the microstates of a solution or colloidal dispersion in terms of the positions and momenta of individual solute particles, similar to those in a solvent-free system. The particle motions in a solvent are subject to the friction generated by the solvent molecules as well as random forces due to particle collision with the solvent molecules. For historical reasons, the dynamics of particle motion in a solvent is called Brownian motion.[27] Because of the particle-solvent interactions, the total energy of the particles in Brownian motion is not conserved. Subsequently, the evolution of microstates is different from that for solvent-free systems. We will discuss Brownian motion in Section 2.5 and the applications of the solvent-explicit models to solution thermodynamics in Chapter 9.

1.6.4 Lattice Models

To capture the *universal* behavior of natural phenomena emerging from interactions among many particles, we desire an idealized model that retains the essential physics without attention to chemical details. Toward that end, lattice models are often adopted in statistical mechanics. In addition to their broad relevance to practical applications as discussed in Chapter 5, many fundamental concepts of statistical mechanics, such as mean-field approximations, correlations, and universality, are best understood within the framework lattice models.

Unlike electrons and classical particles discussed above, "particles" in a lattice model may not have a precise physical significance. In fact, here particles are simply represented by fundamental constituents of a macroscopic system, which can be constructed in different contexts. For example, in the lattice model, the lattice sites may be affiliated with gas molecules or polymer segments, "spin" orientations, particles with different identities, charge or energy states, etc. If each lattice site has only two states, the lattice model is commonly known as the Ising model, proposed originally by Ernst Ising to describe the magnetization of ferromagnetic materials but applicable to describe many cooperative phenomena in nature. In Chapter 5, we will discuss several applications of the Ising model by mapping the energy states of each lattice site with different physical analogies.

To elucidate the basic ideas, Figure 1.10 shows, for example, a thermodynamic system represented by a square lattice. Here, each site may be affiliated with one of two-energy states. A microstate of the system is specified by the energy state of each lattice site. Because the physical nature of the lattice sites is arbitrary, the model is useful to represent any two-dimensional system consisting of individual elements with two-energy states. For example, these lattice sites may be considered as the discretized positions or orientations of atoms, molecules, and the monomeric segments of polymer chains; they may also be assigned to the conformations of the building blocks of biomacromolecules or to the orientations of spins carried by elementary particles.

Intuitively, the two-dimensional lattice model may also be used to describe the monolayer absorption of gas molecules at a flat surface: the lattice may be identified as a discretized representation of the positions of the adsorbed molecules, and the states of each lattice site correspond to the position being occupied by a single gas molecule or empty. To define the microstates, we may assign

27 Named after Robert Brown, an English Botanist who in 1827 first reported the chaotic motion of pollen grains suspended in liquid water. The observation had a pivotal role in the early development of modern atomic theory.

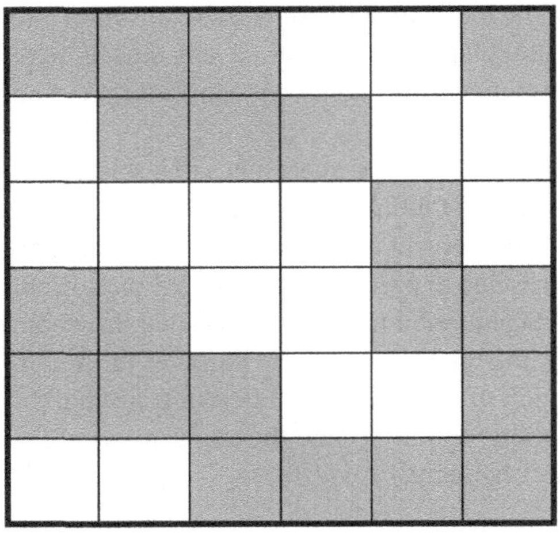

Figure 1.10 In a two-dimensional lattice model, each lattice site takes one of two-energy states as shown in different colors. The dichotomous states may represent, for example, the occupation status of a discretized surface by gas molecules or spin orientations.

any lattice site i with two integers, e.g.,

$$n_i = \begin{cases} 0 & \text{empty} \\ 1 & \text{occupied} \end{cases}. \tag{1.54}$$

Here, we use numbers (0,1) for simplicity; other choices, e.g., (−1,1), are also possible. For each microstate, we have a set of the numbers for $\{n_i\}$ to define the status of the lattice occupation. The system energy can be quantified by assigning each occupied lattice site with energy h, which accounts for the interaction between the surface and a gas molecule, and each unoccupied site with zero energy. An additional parameter, ϵ, may be used to describe nearest-neighbor interactions between gas molecules on the surface, i.e., an energy of ϵ is assigned for each pair of occupied sites directly in contact with each other.

At a given microstate, we have a set of numbers to describe the occupation status of each lattice site, i.e., $n_i = 0$ or 1, for $i = 1, 2, \ldots, N_s$, and N_s is the total number of lattice sites. The total energy at microstate ν can then be written as

$$E_\nu = -\sum_{i=1}^{N_s} n_i h - \frac{1}{2}\sum_{i=1}^{N_s}\sum_{j=1}^{N_n} n_i n_j \varepsilon \tag{1.55}$$

where h stands for a one-body energy, ε denotes a pair interaction energy, N_n represents the number of nearest neighbors for each site, and ½ accounts for the double counting of the nearest-neighbor pairs. In certain cases, we can find the probability distribution of the system in all microstates, p_ν. With this probability, we achieve a *statistical* description of the thermodynamic properties of the system.

We may evaluate ensemble averages using the two-dimensional lattice model for gas adsorption on a planar surface. At microstate ν, the occupation number for each lattice site takes the value of $n_{i,\nu} = 0$ or 1, depending on whether the site is covered by a gas molecule. The ensemble average of the occupation number represents the probability of this lattice site covered by a gas molecule

$$\langle n_i \rangle = \sum_\nu p_\nu n_{i,\nu}. \tag{1.56}$$

If all lattice sites are equivalent, at any moment, the total number of molecules on the surface is

$$N_\nu = \sum_{i=1}^{N_s} n_{i,\nu} \tag{1.57}$$

where N_s is the number of lattice sites. The surface coverage, which can be expressed as an ensemble average of N_ν

$$\theta = \frac{\langle N_\nu \rangle}{N_s} \tag{1.58}$$

is a thermodynamic quantity that can be measured by experiment.

1.6.5 Summary

Both physical and lattice models are commonly used in statistical mechanics. Physical models allow us to evaluate thermodynamic quantities associated with the various forms of kinetic and potential energies. If we are not concerned with dynamic or time-dependent behavior, physical models can be further simplified by ignoring particle motions while capturing only particle-particle interactions as represented by lattice models. The latter are useful to understand cooperative phenomena emerging from many-body interactions in thermodynamic systems. Lattice models are often used to describe the universal behavior of macroscopic systems such as those underlying phase transitions.

1.7 Additivity and Relativity in Thermodynamics

In this section, we discuss two important characteristics of internal energy and entropy, following their definitions in statistical mechanics. These characteristics help us better understand under what conditions extensive thermodynamic quantities are additive, why we can describe the properties of the same thermodynamic system using different statistical-mechanical models, and how we may compare theoretical predictions with experimental data.

1.7.1 Additivity of Extensive Variables

Consider a thermodynamic system consisting of two *independent* subsystems A and B that are macroscopically identical, i.e., the subsystems have the same composition and thermodynamic properties and are not interacting with each other.

As the subsystems are independent, the microstate of the composite system of A and B is defined by the microstates of both subsystems, i.e.,

$$\nu = (\nu_A, \ \nu_B). \tag{1.59}$$

At any instant, the microstate of subsystem A may not be the same as that of subsystem B. Because the composite system combines the degrees of freedom for both systems, the probability of microstate ν is

$$p_\nu = p_{\nu_A} \cdot p_{\nu_B} \tag{1.60}$$

where p_{ν_A} and p_{ν_B} are the microstate probabilities in subsystems A and B, respectively.

With the normalization conditions of microstate probabilities for both sub-systems

$$\sum_{\nu_A} p_{\nu_A} = 1, \tag{1.61}$$

$$\sum_{\nu_B} p_{\nu_B} = 1, \tag{1.62}$$

we can express the total entropy of the composite system as

$$\frac{S}{k_B} = -\sum_{\nu} p_\nu \ln p_\nu = -\sum_{\nu_A}\sum_{\nu_B}(p_{\nu_A}\cdot p_{\nu_B})\ln(p_{\nu_A}\cdot p_{\nu_B})$$
$$= -\sum_{\nu_B} p_{\nu_B}\sum_{\nu_A} p_{\nu_A}\ln p_{\nu_A} - \sum_{\nu_A} p_{\nu_A}\sum_{\nu_B} p_{\nu_B}\ln p_{\nu_B} = \frac{S_A}{k_B}+\frac{S_B}{k_B} \tag{1.63}$$

Eq. (1.63) indicates that, as expected, entropy is additive. Similarly, the total internal energy is given by

$$U = \sum_{\nu} p_\nu E_\nu = \sum_{\nu_A}\sum_{\nu_B}(p_{\nu_A}\cdot p_{\nu_B})[E_{\nu_A}+E_{\nu_B}]$$
$$= \sum_{\nu_B} p_{\nu_B}\sum_{\nu_A} p_{\nu_A}E_{\nu_A} + \sum_{\nu_A} p_{\nu_A}\sum_{\nu_B} p_{\nu_B}E_{\nu_B} = U_A + U_B \tag{1.64}$$

which is also additive.

Additivity holds true for all extensive thermodynamic variables. While for simplicity Eqs. (1.63) and (1.64) are obtained for identical subsystems, they are equally valid when subsystems A and B are not thermodynamically identical (e.g., A and B have different sizes or even different species). Additivity breaks down if the microstates of subsystems A and B are not independent. This would be the case, for example, if particles from subsystem A are interacting with those in subsystem B.

1.7.2 Relativity of Energy and Entropy

Physical models are indispensable in the statistical-mechanical description of thermodynamic properties. However, the selection of such models in practice appears subjective. We may take *ab initio* thermodynamics, atomic models, coarse-grained models, or lattice models for the same thermodynamic system. Are the thermodynamic quantities of a specific system not dependent on the definition of microstates? How do we compare results from different statistical-mechanical models (e.g., atomic against lattice models) for the same thermodynamic system? If the models are subjective, how are the theoretical results compared with experimental data? These questions can be answered by recognizing the relative nature of entropy and internal energy.

With different definitions of microstates, we have different numerical results for both the total number of microstates and the probability distributions. To elucidate their connections, suppose we first define the microstates of a thermodynamic system in terms of quantum states following the most fundamental principles of physics. In such an extreme approach, a microstate would contain information not directly relevant to thermodynamic properties of practical interest. For example, for a typical chemical system, the quantum states of sub-atomic particles within each atom are independent of those from other atoms. Therefore, for each atomic state as defined by, for example, the atomic configuration and kinetic energies, the sub-atomic quantum states may be considered as a subset of microstates. The summation over the microstates of both atomic and sub-atomic degrees of freedom is equivalent to the summation over the atomic states and that over the sub-atomic quantum states, i.e.,

$$\sum_{\nu} = \sum_{a}\sum_{n\in a} \tag{1.65}$$

where subscripts a and n denote atomic and sub-atomic states, respectively. The probability of the system in a particular atomic state is given by the summation over the probabilities of microstates with the sub-atomic states affiliated with the corresponding atomic state

$$p_a = \sum_{n \in a} p_{\{a,n\}} \tag{1.66}$$

where $p_{\{a,\, n\}} = p_\nu$ stands for the probability of the system in a particular microstate with sub-atomic details.

From the probabilities of microstates, we can write the entropy as

$$S = -k_B \sum_{\nu} p_\nu \ln p_\nu = -k_B \sum_{a} \sum_{n \in a} p_{\{a,n\}} \ln p_{\{a,n\}}, \tag{1.67}$$

and the internal energy is

$$U = \sum_{a} \sum_{n \in a} p_{\{a,n\}} E_{\{a,n\}}. \tag{1.68}$$

Using Eq. (1.66) and conditional probability

$$p_{n/a} = p_{\{a,n\}} / p_a, \tag{1.69}$$

we can rewrite the second summation on the right side of Eq. (1.67) as

$$\begin{aligned} \sum_{n \in a} p_{\{a,n\}} \ln p_{\{a,n\}} &= \sum_{n \in a} p_{\{a,n\}} \ln[p_a (p_{\{a,n\}} / p_a)] \\ &= p_a \ln p_a + p_a \sum_{n \in a} p_{n/a} \ln p_{n/a} \end{aligned} \tag{1.70}$$

Thus, the entropy can be decomposed into two terms

$$S = S_g + \sum_{a} p_a S_a \tag{1.71}$$

where subscript "g" and "a" stand for the grouped entropy and atomic entropy with

$$S_g = -k_B \sum_{a} p_a \ln p_a, \tag{1.72}$$

$$S_a = -k_B \sum_{n \in a} p_{n/a} \ln p_{n/a}. \tag{1.73}$$

Eq. (1.71) indicates that the entropy defined according to the microstates of sub-atomic particles is equal to that defined by the atomic states (viz., grouped microstates) plus an additional contribution associated with the sub-atomic degrees of freedom. The latter defines a subsystem entropy for each atom. If the subsystem entropy is a constant independent of the atomic state, the overall entropy becomes

$$S = S_g + S_a. \tag{1.74}$$

Eq. (1.74) suggests that entropy is relative, i.e., its absolute value depends on the relative degrees of freedom accounted for by the microstates.

Similarly, we can show that the total internal energy also satisfies the grouping property

$$U = U_g + U_a \tag{1.75}$$

where the two terms on the right represent the average atomic energy and the average energy of sub-atomic particles at each atomic configuration

$$U_g = \sum_{a} p_a E_a, \tag{1.76}$$

$$U_a = \sum_{n \in a} p_{n/a} E_{n/a}. \tag{1.77}$$

While for illustration the above discussion is concerned with the atomic and sub-atomic degrees of freedom, a similar procedure is applicable to other ways of grouping the microstates (e.g., in terms of clusters of atoms or lattice sites). Different coarse-grained models correspond to separate ways of grouping the microstates into coarse-grained particles. Except a model-dependent constant that cancels out in evaluating the changes of thermodynamic quantities, different statistical-mechanical models are equivalent because they should yield the same relative entropy, relative internal energy, and other relative thermodynamic quantities. As practical applications are concerned only with the changes of thermodynamic properties, different statistical-mechanical models are equivalent in principle; they differ only in grouping the microstates.

1.7.3 Summary

The internal energy and entropy of a thermodynamic system can be described as the sum of those corresponding to sub-systems. This additivity principle, which is also known as the extensiveness of thermodynamic quantities, is valid only when the sub-systems are independent. The relativity of thermodynamic quantities indicates that the variations of internal energy and entropy are model-independent. It should be noted that, in physics literature, the relativity of thermodynamic quantities may also refer to the fact that thermodynamic properties can be described in a frame-independent manner, regardless of the relative motion of the observer. The principle of relativity ensures that the thermodynamic laws are universally applicable, regardless of the reference frame used to describe the system.

1.8 Chapter Summary

This introductory chapter discusses some of the most basic concepts in describing a thermodynamic system from a microscopic perspective such as individual particles, microstates, ensemble, and ergodicity. These concepts allow us to better understand the physical meanings of internal energy and entropy – two most fundamental thermodynamic variables. From internal energy and entropy, we can derive auxiliary thermodynamic functions using the fundamental equations and Legendre transformations. Because both internal energy and entropy satisfy additivity and relativity principles, we may use complementary statistical-mechanical models to define the microstates of a thermodynamic system.

Statistical mechanics is applicable to materials and chemical systems where the dynamics of individual particles are described by fundamental laws of physics, as well as to, in general, many-body systems where particles may not have precise physical significance. In the former case, the microstates are often defined in terms of the degrees of freedom for realistic particles such as atoms or various "coarse-grained" representations of molecules or other basic constituents of a macroscopic system. While information concerning physical details is paramount for understanding the microscopic structure and thermodynamic properties of materials and chemical systems, abstract models are also important in statistical mechanics for their broad applicability to a wide variety of physical phenomena, ranging from cooperativity in many-body systems to critical behavior and the universality of phase transitions. Simple models such as a lattice representation of particle positions are important for pedagogical purposes, for example, to elucidate the basic concepts in statistical mechanics such as cooperativity, fluctuation phenomena, order parameters, phase transition, and broken symmetry. These fundamental concepts are universally applicable to systems containing many particles. The flexibility in choosing the microscopic ingredients of thermodynamic systems, either physical or abstract, makes statistical mechanics a versatile tool for understanding diverse phenomena in nature as well as chemical and material systems of practical interest.

Further Readings

Chandler D., *Introduction to modern statistical mechanics*. Oxford University Press Chapters 1–2, 1987.
Frenkel D., "Order through entropy", *Nat. Mater.* 14, 9–12 (2015).
Gubbins K. E. and Moore J. D., "Molecular modeling of matter: impact and prospects in engineering", *Ind. Eng. Chem. Res.* 49, 3026–3046 (2010).
Jaynes E. T., "Gibbs vs Boltzmann entropies", *Am. J. Phys.* 33, 391–398 (1965).
Moore G. C., "Ergodic theorem, ergodic theory, and statistical mechanics", *PNAS* 112 (7) 1907–1911 (2015).
Tsallis C., "Entropy", *Encyclopedia* 2 (1), 264–300 (2022).

Problems

1.1 Using the phase rule, justify whether the following thermodynamic systems are adequately defined:
(i) a pure liquid of known mass at given temperature and density;
(ii) a pure liquid of known mass at given density and pressure;
(iii) a two-phase system at given temperature, pressure, and the mole fractions of all chemical species in one phase;
(iv) an ionic liquid containing two types of cations and one type of anions at given temperature, pressure, and the ionic composition.

1.2 Based on the fundamental thermodynamic relations, show

$$\left(\frac{\partial C_P}{\partial P}\right)_{T,N} = -T\left(\frac{\partial^2 V}{\partial T^2}\right)_{P,N}.$$

1.3 Acoustic velocity, or the speed of sound, in a fluid can be derived from the first law of thermodynamics for steady-state flow processes

$$d(\hat{H} + v_s^2/2) = 0$$

in combination with the mass-balance equation

$$d(\hat{\rho} v_s) = 0.$$

Here, $\hat{H}$ is the specific enthalpy of the fluid, $\hat{\rho}$ is the mass density, and v_s is the velocity of the elastic wave, i.e., the sound speed.
(i) Derive the following expression from the 1st law and mass balance

$$v_s = \sqrt{\left(\frac{\partial P}{\partial \hat{\rho}}\right)_S}.$$

(ii) Show that the partial derivative at constant entropy S can be expressed in terms of measurable quantities

$$v_s = \sqrt{\frac{\gamma}{\hat{\rho}\kappa_T}},$$

where $\gamma \equiv C_P/C_V$ is the ratio of the constant-pressure and constant-volume heat capacities of the fluid, κ_T is the isothermal compressibility

$$\kappa_T \equiv -\frac{1}{V}\left(\frac{\partial V}{\partial P}\right)_T.$$

(iii) Predict the speed of sound in the atmosphere at 25 °C. Assuming atmosphere is an ideal gas with molar heat capacity $C_P = 29.3$ J/(mol K) and molecular weight $M = 29$ g/mol.

1.4 Consider a one-dimensional model for a thermodynamic system containing N identical particles. Assume that the internal energy is completely determined by temperature T, system total length L, and particle number N. Show that the fundamental equation of thermodynamics can be written as

$$dU = TdS + fdL + \mu dN,$$

where temperature, line tension (i.e., negative of the one-dimensional pressure), and chemical pressure are defined as, respectively,

$$T = \left(\frac{\partial U}{\partial S}\right)_{L,N},$$
$$f = \left(\frac{\partial U}{\partial L}\right)_{S,N},$$
$$\mu = \left(\frac{\partial U}{\partial N}\right)_{L,S}.$$

What are the analogs of the Maxwell relations for this one-dimensional system?

1.5 Two-dimensional models are often used to describe the thermodynamics of molecular adsorption from a gas or a liquid solution on planar surfaces. For a one-component two-dimensional system, the fundamental equation is given by

$$dU = TdS - \varsigma dA + \mu dN,$$

where ς stands for surface pressure. Derive the Gibbs adsorption isotherm:

$$\left(\frac{\partial \varsigma}{\partial \mu}\right)_T = \frac{N}{A}.$$

Can you extend the Gibbs adsorption isotherm to multicomponent systems? What is the surface pressure if there is no adsorption?

1.6 Consider gas adsorption on a planar surface at temperature T. Assume that the gas phase is ideal and that the adsorption can be described by Henry's law

$$N/A = k_H P,$$

where N is the number of gas molecules at the surface, A is the surface area, and k_H is Henry's constant. Show that the surface pressure is given by

$$\varsigma A = N k_B T,$$

where k_B is Boltzmann's constant.

1.7 Consider molecular adsorption from a liquid solution on a planar surface of area A and temperature T. Assume that the liquid phase is an ideal solution and that the adsorption of the solute molecules can be described by Henry's law

$$N/A = k_H C,$$

where N is the *moles* of solute molecules at the surface, k_H is Henry's constant, and C stands for the *molar* concentration of the solute molecules in the bulk solution. Show that the surface pressure is given by

$$\varsigma A = NRT,$$

where R is the gas constant.

1.8 Consider ideal-gas adsorption on a planar surface such that gas molecules at the surface follows a two-dimensional equation of state

$$\varsigma(a - a_0) = k_B T,$$

where a is the surface area per molecule, and a_0 is the surface area occupied by each molecule. Show that the adsorption isotherm follows the Volmer isotherm

$$Pk_H = \frac{\theta}{1-\theta} \exp\left(\frac{\theta}{1-\theta}\right),$$

where $\theta = a_0/a$ stands for the surface coverage, and k_H is Henry's constant defined by $k_H = \theta/P$ as $P \to 0$. How would you extend the Volmer equation to adsorption on a planar surface from an ideal solution?

1.9 Consider ideal-gas adsorption on a planar surface such that gas molecules at the surface follows a two-dimensional analog of the van der Waals equation of state

$$\varsigma = k_B T/(a - a_0) - c/a^2,$$

where a is the surface area per molecule, a_0 is the surface area occupied by each molecule, and c is an energy parameter that accounts for the intermolecular attraction at the surface. Show that the adsorption isotherm follows the Hill–Deboer equation

$$\ln\left[\frac{P(1-\theta)}{\theta}\right] - \frac{\theta}{1-\theta} = -\ln K_1 - \frac{K_2\theta}{k_B T},$$

where $\theta = a_0/a$ stands for the surface coverage, K_1 is defined by $K_1 = \theta/P$ as $P \to 0$ (viz., Henry's constant), and $K_2 = 2c/a_0$.

1.10 A container includes a large number of gas molecules that do not interact with each other. Assume that the container wall is elastic and that the gas molecules can be represented by non-interacting classical particles. Does the system satisfy the ergodic hypothesis? Why?

1.11 The Gibbs equation for entropy can be derived from the hypothesis that entropy is an ensemble average of some microscopic property depending only on the microstate probability, i.e.,

$$S = \sum_\nu p_\nu f(p_\nu),$$

where $f(p_\nu)$ may be understood as the microscopic counterpart of entropy at microstate ν. Show that this hypothesis leads to $f(p_\nu) = -k_B \ln p_\nu$. [Hint: Consider entropy additivity for independent thermodynamic systems.]

1.12 Suppose that a thermodynamic system has an initial distribution of microstates specified by $\{p_\nu^0\}$, where ν denotes microstates, and reaches a new microstate distribution $\{p_\nu\}$ at equilibrium. Show that

$$\sum_\nu p_\nu \ln p_\nu \geq \sum_\nu p_\nu \ln p_\nu^0,$$

where the equality holds if and only if $p_\nu = p_\nu^0$ for all ν. [Hint: $\ln x \leq x - 1$ for all $x > 0$ with equality if and only if $x = 1$.]

1.13 Consider a lattice model for a one-component gas such that each site can accommodate no more than one gas molecule. Assume that each microstate of the system is defined by a particular occupation of the lattice sites by the gas molecules.

(i) What is the number of microstates for the lattice system with n-sites containing N gas molecules?

(ii) Show that the lattice model predicts an entropy

$$S = -nk_B\left[x\ln x + (1-x)\ln(1-x)\right],$$

where $x = N/n$. [Hint: $\ln n! \approx n\ln n - n$.]

(iii) Assume that the system volume is proportional to the number of lattice sites, i.e., $V = nv_0$, where v_0 is the volume for each site. Show that, when $x \to 0$ (i.e., low gas density), the entropy change in response to the volume change from V_1 to V_2 is

$$\Delta S = Nk_B \ln(V_2/V_1).$$

(iv) Show that, when $x \to 0$, the lattice model satisfies the ideal-gas law $PV = Nk_BT$.

(v) How does the entropy change with temperature according to this model?

1.14 Consider a lattice model for a one-component gas such that each lattice site can accommodate no more than one gas molecule. Each gas molecule on the lattice is able to take m orientations with equal energy. Assume that the system volume is proportional to the number of lattice sites, $V = nv_0$, where n is the number of lattice sites, and v_0 is the volume per lattice site.

(i) What is the entropy of the system in terms of the number of gas molecules N and the total number of lattice sites n.

(ii) Show that, when $x = N/n \to 0$ (i.e., low gas density), the entropy change in response to the volume change from V_1 to V_2 is

$$\Delta S = Nk_B \ln(V_2/V_1).$$

(iii) Show that, when $x \to 0$, the lattice model satisfies the ideal-gas law $PV = Nk_BT$.

1.15 Repeat Problem 1.14 by assuming that the gas molecules do not interact with each other and that the m-orientations for each gas molecule on the lattice have different energies $\epsilon_i, i = 1, 2, \ldots, m$. How does the molecular orientation influence the relative entropy of the system?

1.16 Consider a lattice model such that each site can be in one of two energy states, 0 or $\epsilon > 0$. Let n be the total number of lattice sites, and N the number of sites with energy ϵ.

(i) How would you define the microstates of this system?

(ii) Show that the lattice sites with different energies follow the Boltzmann distribution

$$\frac{N}{n-N} = \exp\left(-\frac{\epsilon}{k_B T}\right).$$

(iii) Show that the constant-volume heat capacity of the lattice system is given by

$$C_V = \frac{N(n-N)}{n}\frac{\epsilon^2}{k_B T^2}.$$

(iv) Can the absolute temperature be negative? Why?

1.17 Imagine that Maxwell's demon could control the direction of molecules passing through a hole between identical chambers A and B containing a one-component ideal gas. The demon allows all molecules, fast or slow, to pass from B to A but prevent them from passing from A to B. Eventually all molecules will be concentrated in A and a vacuum will be created in B so that a pressure difference is generated without doing any work. Explain whether the process violates the first or the second law of thermodynamics and discuss how this might be implemented with modern technology.

1.18 In a computer, the Memory Address Register (MAR) is responsible for storing either the memory address from which data will be fetched to the Central Processing Unit (CPU), or the address to which data will be sent and stored. Consider a computer operation that reformats a memory register of n bits. Before the operation, the register as a whole could have existed in any of 2^n states. However, after the operation, the register is left in only one state. To maintain the temperature of the computer at this point, how much heat must be released?

2

Statistical Ensembles and MD Simulation

In Chapter 1, we have learned that an equilibrium system, typically specified by a few macroscopic variables such as temperature and pressure, has uncertainties from a microscopic perspective. Such uncertainties are associated with the dynamic behavior of individual particles and can be quantified in terms of the probability distribution of microstates in an ensemble of macroscopic systems. Thermodynamic systems in a statistical ensemble are macroscopically equivalent, but at any moment, their microstates may be drastically different. The probability of microstates in an ensemble depends not only on the characteristics of individual particles but also on the way a thermodynamic system is prepared, i.e., how certain thermodynamic variables are controlled to maintain a particular equilibrium condition (e.g., at constant volume or constant pressure). In statistical mechanics, those fixed thermodynamic variables are referred to as "constraints," which set restrictions on the microstate probability. As we will see in this chapter, different constraints lead to different ensembles and different microstate distributions.

In this chapter, we derive the probability distributions of microstates in different ensembles, i.e., microcanonical ensemble, canonical ensemble, isothermal–isobaric ensemble, and grand canonical ensemble, using the maximum entropy principle. We will discuss formal mathematical relations among microstate probability, partition function, thermodynamic variables, and thermal fluctuations. In addition, we will elucidate the dynamic behavior of individual particles in different ensembles within the context of molecular dynamics (MD) simulations. Finally, we will discuss the transformation of thermodynamic variables in terms of ensembles and generalize the ensemble method to represent microstates at multiple thermodynamic conditions.

2.1 Microcanonical Ensemble

Conceptually, one of the simplest ways to specify a thermodynamic system is through the identity of individual particles and the boundary of their existence without any other external influence. When the system size is sufficiently large (i.e., in the thermodynamic limit), the boundary effects are insignificant. In that case, we may define an isolated system, i.e., a system free of energy and mass transfer with the surroundings.

An isolated system has a fixed total energy E, volume V, and the number of particles, $N_1, N_2, \ldots$, of different kinds. A microcanonical ensemble consists of an infinitely large number of isolated systems that are thermodynamically equivalent, i.e., systems have the same total energy, volume, particle types, and numbers.

Fundamentals and Practice in Statistical Thermodynamics, First Edition. Jianzhong Wu and John M. Prausnitz.

Companion website: www.wiley.com/go/jianzhongwu/fundamentalsandpractice

2.1.1 The Hypothesis of Equal A Priori Probabilities

One fundamental postulate of statistical mechanics is that, in a microcanonical ensemble, all microstates have the same probability.[1] While rigorous proof of this postulate has yet to be established, the underlying reasoning is not too difficult to perceive. The equal probability of microstates in the microcanonical ensemble is not much different from the equal probability of playing cards, dice, or other games of chance. In a fair game, a reasonable person would agree that rolling a fair cubic die ends up with one of its six facets with equal probability. In other words, the probability distribution is uniform among the six facets. Similarly, the chance of picking a particular card from a well-shuffled pack depends only on the total number of cards, and the probability that a thermodynamic system in the microcanonical ensemble exists in a particular microstate is determined by the total number of microstates.

For a thermodynamic system at equilibrium, we can formally derive the principle of equal *a priori* probabilities based on the second law of thermodynamics. As discussed in Section 1.4, the entropy of an equilibrium system is defined according to the Gibbs equation

$$S = -k_B \sum_{\nu} p_{\nu} \ln p_{\nu} \tag{2.1}$$

where k_B is the Boltzmann constant, and p_{ν} is the probability of the system being in microstate ν. For an isolated system, the conservations of total mass and energy are automatically satisfied during the evolution of microstates. The second law predicts that, at equilibrium, the entropy is maximized with respect to p_{ν} subject to the normalization condition[2]

$$\sum_{\nu} p_{\nu} = 1. \tag{2.2}$$

Using the Lagrange multiplier method, we can find the distribution of microstates corresponding to the maximum entropy

$$\frac{\partial}{\partial p_{\nu}} \left\{ -k_B \sum_{\nu} p_{\nu} \ln p_{\nu} - \alpha \left[\sum_{\nu} p_{\nu} - 1 \right] \right\} = 0. \tag{2.3}$$

where α is called the Lagrange multiplier. After some algebra, Eq. (2.3) leads to

$$-k_B(\ln p_{\nu} + 1) - \alpha = 0 \tag{2.4}$$

which gives

$$p_{\nu} = \exp(-\alpha/k_B - 1). \tag{2.5}$$

Because both k_B and α are constants, Eq. (2.5) indicates that p_{ν} must be the same for all microstates. In other words, the microstates are uniformly distributed in the microcanonical ensemble. A combination of Eqs. (2.5) and (2.2) predicts that p_{ν} is equal to the reciprocal of the total number of microstates W,

$$p_{\nu} = 1/W. \tag{2.6}$$

1 The fundamental postulate can be alternatively stated as follows: the probability distribution of microstates is determined solely by the total energy. This implies that all dynamical properties of the system, including the positions and momenta of the particles, are equally likely to occur as long as the system energy is the same. It is important to note that this postulate applies specifically to the microcanonical ensemble. In other ensembles, such as the canonical ensemble or the grand canonical ensemble, which will be discussed later, the probability distribution is influenced by both the total energy and other relevant variables.

2 Additional constraints on the microstate probability are not applicable to a microcanonical ensemble. This is because, in an isolated system, the dynamics of individual particles adhere to the conservation of total energy and particle mass.

Thus, the Lagrange multiplier is given by

$$\alpha/k_B = \ln W - 1. \tag{2.7}$$

With the hypothesis of equal *a priori* probabilities, we may alternatively obtain the probability of microstates directly from the normalization condition, Eq. (2.2). Substituting Eq. (2.6) into (2.1) reproduces the Boltzmann equation for entropy

$$S = k_B \ln W. \tag{2.8}$$

Whereas the Gibbs and Boltzmann equations yield the same entropy for microcanonical ensembles, Boltzmann's definition of entropy is problematic for other ensembles.

2.1.2 Thermodynamic Quantities

For simplicity, we assume in the following that the system of our interest contains only one type of particle and undergoes no chemical reaction. Similar equations can be extended to multicomponent systems, including those with chemical reactions.

What is the temperature of an isolated system? How do we predict the thermodynamic properties of an isolated system based on the total number of microstates? If a system is truly isolated, temperature loses its physical significance because an isolated system is not experimentally accessible. Nevertheless, all thermodynamic quantities, temperature included, remain well defined, at least mathematically, in the context of the fundamental equations and thermodynamic relations, as discussed in Section 1.2.

As discussed above, the entropy of an isolated system depends on total energy E, the number of particles N, and system volume V. Because microstates in a microcanonical ensemble have the same energy, the internal energy is the same as the total energy

$$U = \langle E \rangle = E. \tag{2.9}$$

Formally, we may write entropy as a function of internal energy U, the number of particles N, and system volume V, i.e., $S = S(U, V, N)$. In differential form, we have

$$dS = \left(\frac{\partial S}{\partial U}\right)_{V,N} dU + \left(\frac{\partial S}{\partial V}\right)_{U,N} dV + \left(\frac{\partial S}{\partial N}\right)_{U,V} dN. \tag{2.10}$$

In comparison with the fundamental equation of thermodynamics

$$dU = TdS - PdV + \mu dN, \tag{2.11}$$

we find

$$\frac{1}{T} = \left(\frac{\partial S}{\partial U}\right)_{V,N}, \tag{2.12}$$

$$\frac{P}{T} = \left(\frac{\partial S}{\partial V}\right)_{U,N}, \tag{2.13}$$

$$-\frac{\mu}{T} = \left(\frac{\partial S}{\partial N}\right)_{U,V}. \tag{2.14}$$

If we have an analytical expression for the number of microstates, $W(E, N, V)$, Eqs. (2.12)–(2.14) allow us to derive the system temperature T, pressure P, and chemical potential μ. Other thermodynamic quantities can also be derived from the thermodynamic network, as discussed in Section 1.2.

Box 2.1 Negative Absolute Temperature

Consider the microcanonical ensemble for a system with N particles placed on N lattice sites. We assume that these particles are indistinguishable and do not interact with each other. However, each particle may take one of two energy states: $-\varepsilon$ or ε, with ε being positive. We ask how the system temperature varies with the total energy. This question can be answered by using Eq. (2.12). While the mathematical procedure is straightforward, the answer might be surprising and has profound implications. It predicts minus absolute temperature!

To elucidate, note first that, for this system, the microstates are defined by the energy states of individual particles. The total energy E, which is the same as the internal energy U of the system in the microcanonical ensemble, is determined by the energy states of individual particles

$$U = E = -n_0\varepsilon + (N - n_0)\varepsilon = (N - 2n_0)\varepsilon \tag{2.15}$$

where n_0 is the number of particles with energy $-\varepsilon$, and $(N - n_0)$ is the number of particles with energy ε.

When the total energy is fixed, $n_0 = (N - E/\varepsilon)/2$ must be constant. Given that the number of particles with energy ε is $N - n_0$ and that with energy $-\varepsilon$ is n_0, the number of microstates can be calculated from the combinatorial number

$$W = \frac{N!}{n_0!(N - n_0)!} \tag{2.16}$$

where the factorials in the denominator account for the fact that the particles are identical. According to the Boltzmann equation, Eq. (2.8), the entropy is given by

$$S = k_B \ln W = k_B \ln \frac{N!}{n_0!(N - n_0)!}. \tag{2.17}$$

Using Stirling's approximation for factorials,

$$\ln N! \approx N \ln N - N, \tag{2.18}$$

We can rewrite Eq. (2.17) as

$$S = -k_B \left[n_0 \ln \frac{n_0}{N} + (N - n_0) \ln \frac{(N - n_0)}{N} \right]. \tag{2.19}$$

Accordingly, the system temperature can be derived from Eq. (2.12)

$$T = \left(\frac{\partial U}{\partial S}\right)_{N,V} = \left(\frac{\partial U}{\partial n_0}\right)_N \left(\frac{\partial S}{\partial n_0}\right)_N^{-1}. \tag{2.20}$$

After differentiating U in Eq. (2.15) and S in Eq. (2.19) with respect to n_0, we find an analytical expression for the absolute temperature

$$T = \frac{2\varepsilon}{k_B} \left[\ln\left(\frac{n_0}{N - n_0}\right) \right]^{-1}. \tag{2.21}$$

Eq. (2.21) is nothing unusual when $n_0 > N - n_0$, i.e., when the number of particles in the low-energy state is larger than that in the high-energy state. In this case, increasing temperature coincides with the increase in the system's entropy and energy. As shown in Figure 2.1A, the energy falls monotonically as more particles reside in the low-energy state. On the other hand, entropy shows a parabolic shape with a maximum when an equal number of particles are located in the low and high-energy states. When $n_0 < N - n_0$, however, Eq. (2.21) predicts $T < 0$, meaning that the absolute temperature is below 0 K.

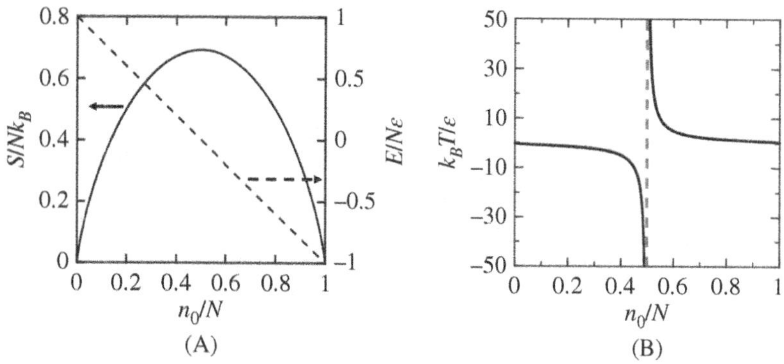

Figure 2.1 A. Entropy (solid curve) and energy (dashed line) of an isolated system of non-interacting particles. B. Temperature versus the fraction of particles in the low-energy state.

The possibility of negative absolute temperatures was first suggested by Lars Onsager in 1949. While the prediction appears absurd because the absolute temperature is commonly interpreted as a measure of the average kinetic energy, which is always non-negative. Nevertheless, it has been demonstrated in the laboratory that an atomic gas may indeed exist at a temperature below zero Kelvin.[3] As shown in Figure 2.1B, when $n_0 < N - n_0$, the more energy the system has, the less entropy there is due to the decrease in the combinatorial number. In general, a negative absolute temperature takes place due to the inversion of the population of the energy states, i.e., when particles in a thermodynamic system have a larger population in the high-energy states than in the low-energy states.

2.1.3 Liouville's Theorem

Named after the nineteenth-century mathematician Joseph Liouville, Liouville's theorem is a fundamental equation in statistical mechanics. The original theorem was derived for first-order differential equations and later extended to thermodynamic systems by using Hamiltonian mechanics.[4] The mathematical result has profound implications for applying the second law of thermodynamics to non-equilibrium systems and, from a practical perspective, is important for describing the dynamic behavior of thermodynamic systems.

Liouville's theorem states that, in a microcanonical ensemble, the probability density of microstates is time-invariant. To gain an understanding of this statement, consider an isolated system of N identical particles in a three-dimensional space. According to classical physics, the dynamics of each particle are fully determined by six continuous variables or coordinates, i.e., three real numbers for the particle position $\mathbf{r} = (x, y, z)$, and three real numbers for the velocity $\mathbf{v} = (v_x, v_y, v_z)$. As the particle mass m is constant, the latter is equivalent to particle momenta $\mathbf{p} \equiv m\mathbf{v} = (p_x, p_y, p_z)$. Therefore, the microstates of the entire system can be specified by $6N$ variables, $(\mathbf{r}_i, \mathbf{p}_i)$,

3 Braun S. et al., "Negative absolute temperature for motional degrees of freedom", *Science* 339 (6115), 52 (2013).
4 Nolte D. D., "The tangled tale of phase space", *Phys. Today* 63 (4), 33 (2010).

$i = 1, 2, \ldots, N$. The 6N-dimensional hyperspace, $\mathbf{x} \equiv \{\mathbf{r}_i, \mathbf{p}_i\}_{i=1,2,\ldots N}$, is conventionally known as *the phase space.*

Given the initial values for the positions and momenta of all particles, we can predict the particle motions using a physical model describing inter-particle interactions. At any moment, the probability density of microstates satisfies the normalization condition

$$\int p(\mathbf{x}, t)\, d\mathbf{x} = 1. \tag{2.22}$$

In Eq. (2.22), the probability density of microstates, $p(\mathbf{x}, t)$, is defined by the number of thermodynamic systems in the ensemble with the microstates in differential volume $d\mathbf{x}$ at time t. The integration amounts to a summation over microstates because they are represented by a continuous variable $\mathbf{x}$ in the phase space.

For any *fixed* volume $\mho$ in the phase space, we may introduce an instantaneous probability,

$$\wp(t) = \int_{\mho} p(\mathbf{x}, t) d\mathbf{x} \tag{2.23}$$

which is related to the number of macroscopically identical systems in the microcanonical ensemble with their microstates confined in phase space volume $\mho$ at time t. Because the total number of thermodynamic systems in the ensemble does not change with time, $\wp(t)$ must be conserved, i.e., the variation in $\wp(t)$ must be attributed to the flux of microstate probability through the surface of phase space volume $\mho$

$$\frac{d\wp(t)}{dt} = -\oint_{\Im_0} p(\mathbf{x}, t)\dot{\mathbf{x}} \cdot d\Im \tag{2.24}$$

where $\Im_0$ stands for the boundary of $\mho$, and $\dot{\mathbf{x}} = d\mathbf{x}/dt$ represents the velocity of a fixed point in the phase space.

Applying the divergence theorem to the right side of Eq. (2.24) gives

$$\frac{d\wp(t)}{dt} = -\int_{\mho} \nabla \cdot [p(\mathbf{x}, t)\dot{\mathbf{x}}]\, d\mathbf{x} \tag{2.25}$$

where ∇ stands for the gradient operator with respect to the phase space variable $\mathbf{x} = \{x_1, x_2, \ldots x_{6N}\}$

$$\nabla = (\partial/\partial x_1, \partial/\partial x_2, \ldots \partial/\partial x_{6N}). \tag{2.26}$$

Because $\mho$ is fixed, we can take the time derivative on the left side of Eq. (2.25) inside the integral

$$\frac{d\wp(t)}{dt} = \frac{d}{dt}\int_{\mho} p(\mathbf{x}, t) d\mathbf{x} = \int_{\mho} \frac{\partial p(\mathbf{x}, t)}{\partial t} d\mathbf{x}. \tag{2.27}$$

Substituting Eq. (2.27) into (2.25) leads to

$$\int_{\mho} \left\{ \frac{\partial p(\mathbf{x}, t)}{\partial t} + \nabla \cdot [p(\mathbf{x}, t)\dot{\mathbf{x}}] \right\} d\mathbf{x} = 0. \tag{2.28}$$

To satisfy Eq. (2.28) for an arbitrary phase space volume $\mho$, the integrand must vanish, i.e.,

$$\frac{\partial p(\mathbf{x}, t)}{\partial t} + \nabla \cdot [p(\mathbf{x}, t)\dot{\mathbf{x}}] = 0. \tag{2.29}$$

Eq. (2.29) indicates that the probability density evolves in the phase space following a conservation equation as described by the equation of continuity. For classical particles, the time derivatives of the particle positions and momenta satisfy the Hamiltonian equations[5]

$$\dot{\mathbf{r}}_i = \frac{\partial \mathcal{H}}{\partial \mathbf{p}_i} \quad \text{and} \quad \dot{\mathbf{p}}_i = -\frac{\partial \mathcal{H}}{\partial \mathbf{r}_i} \text{ for } i = 1,2,\ldots N \tag{2.30}$$

---

5 See Supplementary Materials (I.C and I.D) for the mathematical details of the Hamiltonian equations and Liouville's theorem.

where $\mathcal{H}$ stands for the Hamiltonian function. Accordingly, the divergence of the phase space variables $\nabla \cdot \dot{\mathbf{x}} = 0$ and the second term on the left side of Eq. (2.29) can be simplified as

$$\nabla \cdot [p(\mathbf{x}, t)\dot{\mathbf{x}}] = p(\mathbf{x}, t)\nabla \cdot \dot{\mathbf{x}} + \dot{\mathbf{x}} \cdot \nabla p(\mathbf{x}, t) = \dot{\mathbf{x}} \cdot \nabla p(\mathbf{x}, t). \tag{2.31}$$

Substituting Eq. (2.31) into (2.29) gives

$$\frac{\partial p}{\partial t} + \dot{\mathbf{x}} \cdot \nabla p = 0. \tag{2.32}$$

Eq. (2.32) is known as Liouville's equation. It suggests that the probability density in phase space is time-invariant, i.e., the total differentiation of the microstate probability density with respect to time is exactly equal zero

$$\frac{dp(\mathbf{x}, t)}{dt} = \frac{\partial p(\mathbf{x}, t)}{\partial t} + \dot{\mathbf{x}} \cdot \nabla p(\mathbf{x}, t) = 0. \tag{2.33}$$

Eq. (2.33) is self-evident for isolated systems at equilibrium; it shows that all accessible microstates have the same probability as required by the principle of equal *a priori* probabilities.

Liouville's theorem predicts that, for an isolated system, the accessible volume in phase space does not change with time. In other words, particle motions do not affect the microstate probability in the microcanonical ensemble. Because the mathematical derivation does not explicitly invoke equilibrium,[6] the time invariance of microstate probability appears to suggest that the system entropy would be constant even under non-equilibrium conditions. The puzzling conflict between Liouville's theorem and the second law of thermodynamics has stimulated a long history of debates.[7] From a practical perspective, the probability density of microstates described by Liouville's equation provides a convenient starting point for describing the ensemble averages of dynamic variables

$$\langle M(t) \rangle = \int p(\mathbf{x}, t) M(\mathbf{x}, t) d\mathbf{x} \tag{2.34}$$

where $M(\mathbf{x}, t)$ represents an arbitrary dynamic quantity of a particular system in the microcanonical ensemble (e.g., the positions, velocities, and energies of the particles). As Eq. (2.34) involves time explicitly, the ensemble average of the dynamic variable is meaningful only for equilibrium systems.

2.1.4 Summary

A microcanonical ensemble represents a set of isolated systems that are thermodynamically equivalent. In a microcanonical ensemble, all microstates have equal probability, which is known as the principle of equal *a priori* probabilities in statistical mechanics. For isolated systems at equilibrium, the uniform distribution of microstates is consistent with the prediction of the second law of thermodynamics (viz., the maximum entropy principle) and Liouville's theorem.

2.2 Basics of MD Simulation

In comparison with alternative ways to specify thermodynamic systems, one major advantage of considering isolated systems in the context of microcanonical ensembles is that the dynamic behavior of individual particles can be directly predicted from classical or quantum mechanics.

---

6 Because the microstate probability is defined in terms of an ensemble, application of Liouville's theorem to a specific system implies the ergodic assumption.

7 Albert D. Z., "*Time and Chance*", Harvard University Press (2003).

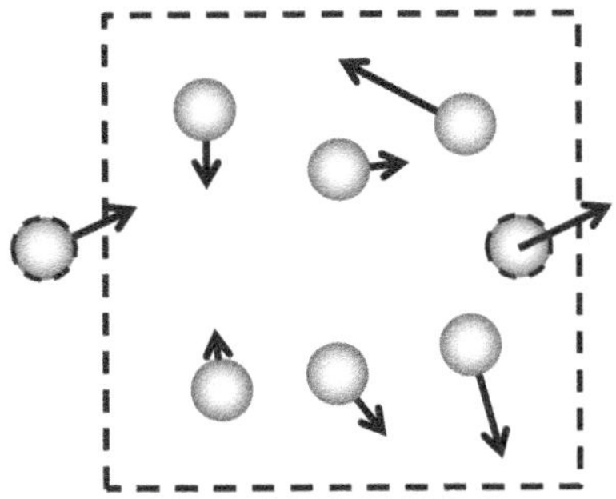

Figure 2.2 Schematic of a simulation box containing spherical particles with different velocities (shown as arrows). The simulation box does not exert any force on the particles except by imposing periodic boundary conditions, i.e., when a particle moves beyond a boundary, it reenters from the opposite side, as illustrated by the spheres marked with dashed circles.

Because the total energy and mass are conserved in an isolated system, we can keep track of the particle positions and momenta using the standard equations of motion. The numerical procedures are called molecular dynamics (MD) simulation, as first demonstrated by Alder and Wainwright with a modern computer.[8] In principle, MD simulation is applicable to all thermodynamic systems, provided there is a suitable model describing the equation of motion for individual particles. In this section, we discuss some basic ingredients of MD simulation for isolated systems. In Sections 2.3–2.4 and 2.7–2.11, we will introduce more sophisticated procedures to impose various macroscopic constraints.

2.2.1 Simulation Box and Numerical Integration

MD simulation has two basic ingredients: one is a simulation box (a.k.a. a simulation cell) containing a certain number of particles, and the other is a numerical algorithm for integrating the equations of motion. A simulation cell mimics a small sample of a macroscopic system such that it reproduces intensive properties identical to those corresponding to the entire system. For simulating a bulk system, the simulation cell is typically a cubic box of side length L, albeit other geometries may also be used to better reflect the internal structure of the simulated system (e.g., a rectangular box for a lipid membrane). Numerical integration is based on a physical model describing the particle motion and inter-particle interactions.

Figure 2.2 shows a schematic representation of an isolated system consisting of spherical particles (e.g., a dilute argon gas). Here, the simulation box contains only seven particles. In practice, the box size and the number of particles simulated are determined by the particle density and the *correlation length*, which will be discussed later in this section. To confine the particles within the simulation cell, we apply the periodic boundary condition (PBC) in all directions. If a particle moves out of the simulation box, for example, $x_i > L$ with x_i being the x-coordinate of particle i confined between $[0, L]$, we assign a new coordinate according to PBC in the x-direction

$$x_i = x_i - L \tag{2.35}$$

such that the particle would be placed back in the simulation box. With the box size L much larger than the correlation length of the system, we assert that all intensive properties calculated from the MD simulation will be the same as those of the macroscopic system. The hypothesis is reasonable because the particle motions are equivalent to those in a macroscopic system consisting of infinite copies of the simulation cell.

Given a physical model describing particle motion and particle–particle interactions, we can keep track of the positions and momenta of all simulated particles using various numerical schemes. For example, Box 2.2 elucidates how the position and momentum of each particle are calculated by

8 Alder B. J. and Wainwright T. E., "Studies in molecular dynamics. I. General method", *J. Chem. Phys.* 31 (2), 459–466 (1959).

the numerical integration of Newton's equation. For a low-density simple fluid, the initial particle positions may be generated by random insertions of particles into the simulation box. At high density, the initial configuration may be generated by placing the particles on a lattice. To avoid particle overlap, the initial configuration should be relaxed by minimizing the total potential energy. To initiate the dynamics, we may assign the particle momenta according to the total kinetic energy. Based on the trajectory of particle positions and momenta generated from MD simulation, we can then calculate thermodynamic properties of the system using ensemble averages.

Box 2.2 The Verlet Algorithm

The velocity Verlet algorithm,[9] named after physicist Loup Verlet, represents a common choice to calculate the trajectories of particles in MD simulation. In this method, the position and momentum of each particle are updated by numerical integration of Newton's equation of motion

$$\mathbf{r}_{n+1} \approx \mathbf{r}_n + \mathbf{v}_n \delta t + \frac{\mathbf{F}_n}{2m}\delta t^2 \tag{2.36}$$

$$\mathbf{v}_{n+1} \approx \mathbf{v}_n + \frac{\mathbf{F}_{n+1} + \mathbf{F}_n}{2m}\delta t \tag{2.37}$$

where n denotes the number of MD steps; δt is a small increment of time or time step size; and **r**, **v**, **F** and m stand for, respectively, the position, velocity, force, and mass of a particular particle. To exclude the translational and rotational motions of the entire system, the particle velocities are defined relative to the frame of the system boundaries (viz., the simulation box).

Eqs. (2.36) and (2.37) can be obtained by the second-order Taylor expansion of the particle position with respect to time. Despite its simplicity, the Verlet algorithm has a few desirable features, such as numerical stability, time reversibility, and the conservation of volume in the phase space. We may estimate the time step δt from the characteristic time of particle motions, which is typically about femtoseconds for atomic systems. The duration of simulation depends on the system *relaxation time* (to be discussed later). With the particle positions and momenta recorded on a computer after each MD step, we obtain a trajectory of microstates from which thermodynamic properties can be evaluated.

2.2.2 Thermodynamic Properties from MD

For an isolated system consisting of N classical particles, the particle positions and momenta are exactly known at any point along the MD trajectory. Because the total energy is conserved during particle motions, the sum of the kinetic and potential energies (thus the total energy) is constant and the same as the internal energy U:

$$E(\mathbf{r}^N, \mathbf{p}^N) = \sum_{i=1}^{N} |\mathbf{p}_i|^2/2m + \Phi(\mathbf{r}^N) = U \tag{2.38}$$

where $\mathbf{p}^N \equiv (\mathbf{p}_1, \mathbf{p}_2, \dots, \mathbf{p}_N)$ denotes the momenta of all particles, $\mathbf{r}^N \equiv (\mathbf{r}_1, \mathbf{r}_2, \dots, \mathbf{r}_N)$ stands for the particle positions (viz., a configuration of the system), m is the particle mass, and $\Phi(\mathbf{r}^N)$ is the total potential energy.

As shown in Appendix 2.A, the system pressure can be calculated from the virial equation

$$P = \frac{1}{3V}\sum_{i=1}^{N} [\langle \mathbf{p}_i^2/m\rangle + \langle \mathbf{r}_i \cdot \mathbf{F}_i\rangle] \tag{2.39}$$

9 Verlet L., "Computer "Experiments" on Classical Fluids. I. Thermodynamical Properties of Lennard-Jones Molecules". *Phys. Rev.*, 159 (1): 98–103 (1967).

where $\mathbf{F}_i = -\nabla_i\Phi(\mathbf{r}^N)$ is the net force on particle i, and $\langle\cdots\rangle$ stands for the ensemble average. In a microcanonical ensemble, all microstates have equal probability. Therefore, the ensemble average of a dynamic quantity is equal to the mean of its instantaneous values over the duration of the MD simulation.

For classical particles, the positions and momenta are independent variables. In the ideal gas limit ($N/V \to 0$), the interaction between particles, and thus the potential energy, becomes negligible. In that case, we can calculate the pressure from the ideal gas law

$$P = \frac{Nk_BT}{V}. \tag{2.40}$$

A comparison of Eqs. (2.39) and (2.40) indicates that the system temperature is proportional to the ensemble average of the total kinetic energy

$$T = \frac{1}{3Nk_B}\sum_{i=1}^{N}\langle \mathbf{p}_i^2/m\rangle. \tag{2.41}$$

How do we calculate the system entropy or free energy from MD? Unfortunately, the number of microstates, thus entropy and free energy, are not directly attainable from MD simulation. Because both particle positions and momenta are continuous variables, the number of microstates is extremely large, even for a system with a small number of particles. With the pressure calculated from the virial theorem for a set of particle densities, we may obtain the Helmholtz energy relative to that of an ideal gas reference state by carrying out the following integration numerically:

$$F = F^{IG}(T,N,V) + \int_V^\infty \left[P - \frac{Nk_BT}{V}\right] dV. \tag{2.42}$$

As to be discussed in Chapter 3, the Helmholtz energy for an ideal gas of spherical particles is given by

$$F^{IG}(T,N,V) = Nk_BT\left[\ln(\rho\Lambda^3) - 1\right] \tag{2.43}$$

where $\rho = N/V$ is the particle density, $\Lambda = h/\sqrt{2\pi mk_BT}$ is the thermal de Broglie wavelength, and h is the Planck constant. From the pressure, Helmholtz energy, and internal energy, we can then calculate entropy and other thermodynamic quantities by numerical means.

2.2.3 Correlation Length and Relaxation Time

In applications of MD simulation, one may wonder: (i) how many particles should be simulated; (ii) how many steps the simulation must run before one starts to gather statistics; and (iii) how many steps one should run to gather enough samples so that reliable results may be obtained. To answer such questions, we need to assess the *correlation length* and *relaxation time* of the thermodynamic system under study. While a generic procedure may not exist for evaluating such quantities, the basic concepts may be explained in terms of the pair distribution functions, i.e., the relative distribution of particles in a thermodynamic system.

For a uniform fluid of spherical particles, the Van Hove function (a.k.a., the time-dependent pair distribution function) is defined as

$$g(r,t) = \frac{V}{4\pi r^2 N(N-1)}\sum_{i\neq j}^{N}\left\langle \delta\left(r - |\mathbf{r}_i(t') - \mathbf{r}_j(t'+t)|\right)\right\rangle \tag{2.44}$$

where the Dirac delta function represents the probability density of finding particle i at position $\mathbf{r}_i(t')$ at time t' while particle j is at position $\mathbf{r}_j(t'+t)$ at time $t'+t$. As mentioned above, the

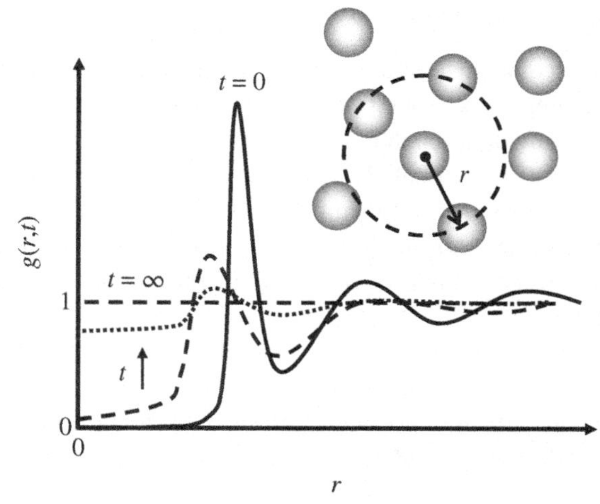

Figure 2.3 Schematic of the time-dependent pair distribution function for a dense fluid as time varies from 0 to ∞. The separation between the peaks is about the same as the particle diameter (not drawn to scale).

ensemble average can be calculated from MD simulation over all possible values of t' and all particle configurations.

In an ideal system, the particles are randomly distributed in space. In that case, the ensemble average in Eq. (2.44) is $4\pi r^2/V$ for any pair of particles, thus $g(r,t) = 1$. The deviation of $g(r,t)$ from unity, i.e., the time-dependent total correlation function $h(r,t) \equiv g(r,t) - 1$, provides a measure of the correlated distribution of particles at distance r due to the particle–particle interactions.

As shown schematically in Figure 2.3, the Van Hove function displays strong peaks at $t = 0$ due to inter-particle interactions. The distance between the neighboring peaks is about the same as the particle diameter, and the consecutive peaks reflect the particle packing effect. At large distances or over a lengthy period, the Van Hove function approaches unity because neighboring particles drift apart. Accordingly, the total correlation function $h(r,t)$ vanishes. The correlation length ξ and relaxation time τ are related to the distance and elapsed time, beyond which $h(r,t)$ becomes insignificant.[10] For a simple fluid remote from the critical point of vapor–liquid transition, the correlation length is no more than a few times the particle diameter σ. Therefore, the relaxation time may be estimated from $\tau \sim \sigma^2/D$ where D stands for the particle diffusivity.

It is worth noting that the definitions of correlation length and relaxation time can vary depending on the specific dynamic quantity of interest. In a given system, multiple correlation lengths may exist, and each of them plays a crucial role in understanding the corresponding macroscopic observable of interest. To gain a comprehensive understanding, it is thus essential to investigate and probe these correlation lengths and relaxation times by conducting several trial simulations with systems of different sizes and durations, often complementarily guided by theoretical analysis. As a rule of thumb, the simulation box should be several times larger than the correlation length ξ, and the simulation time should be much longer than the relaxation time (e.g., by at least one order of magnitude). With those considerations in mind, we can estimate the number of particles to simulate based on the particle density and the number of steps in the simulation by dividing the simulation time by the time step δt.

10 The correlation length and relaxation time can be defined more precisely by the asymptotic behavior of $h(r,t)$. Specifically, the relaxation time τ is obtained by fitting $h(r,t)$ with its asymptotic form $\exp[-(t/\tau)^\gamma]$ at large values of t, and the correlation length ξ is obtained from fitting $rh(r,0)$ with $\exp[-r/\xi]\sin(Kr)$ at large r. (see, e.g., Egami T., *Front. Phys.* 8 (50), doi: 10.3389/fphy.2020.00050(2020).)

2.2.4 Summary

We see in this section that thermodynamic properties can be retrieved from MD simulation by keeping track of the positions and momenta of individual particles. The simulation results hinge not only on a reliable model to describe the dynamic behavior of individual particles but also on efficient numerical algorithms to integrate the equation of motion. While our discussion here is concerned only with simple fluids, similar procedures can be established to simulate more complicated systems, including those based on first-principles calculations for the inter-particle interactions (viz., *ab initio* MD) and chemicals/materials represented by various semi-empirical force fields.

One main limitation of MD simulation is the computational cost, especially when it is applied to large systems of practical interest and for the computation of entropy or free energy, which often involves the numerical integration of simulation results over a range of thermodynamic conditions. Besides, MD simulation is intrinsically limited by the reliability of the physical models that are used to describe particle–particle interactions. With the ever-increasing power of computers, the universal applicability of MD simulation makes it an attractive choice in practical applications of statistical mechanics.

2.3 Canonical Ensemble

The microcanonical ensemble discussed in the earlier sections provides a useful platform for illustrating the fundamental postulates of statistical mechanics, such as the principle of equal *a priori* probabilities. It is also convenient for implementing molecular dynamics simulations, as there are no macroscopic constraints on the dynamic behavior of individual particles. However, the microcanonical ensemble is not the most practical option for many applications of statistical mechanics. Since energy transfer with the environment is nearly inevitable, thermodynamic systems of practical interest are usually specified with a fixed temperature instead of a constant total energy.

A canonical ensemble is defined by a large number of thermodynamically equivalent macroscopic systems with the same temperature, total volume, and particle types and numbers. As implied by the word *canonical*,[11] canonical ensemble represents a standard procedure in practical applications of statistical mechanics and will be frequently used throughout this book. In this section, we introduce a formal mathematical procedure to determine the probability distribution of microstates in canonical ensembles. We will elucidate connections between the microstate probability distribution and the canonical partition function with thermodynamic quantities and fluctuation effects. The implication of constant temperature on the dynamics of individual particles and numerical strategies to fix the constraint in MD simulation will be discussed in the following sections.

2.3.1 The Boltzmann Distribution

To establish the distribution function for microstates in a canonical ensemble, consider a closed system with fixed volume V in contact with a heat bath at a constant temperature T. The heat bath serves as a source or sink of thermal energy, providing or absorbing heat to maintain the desired

11 The word "canonical" means according to the canon, that is, the conventional, well-known and generally accepted code of procedure.

temperature within the system. At equilibrium, there is no *net* flux of energy between the system and the heat bath.

By utilizing a heat bath, it becomes possible to study the behavior of the system at a specific temperature. From a microscopic perspective, the thermal equilibrium implies that the internal energy, i.e., the average energy for the microstates of the system, is constant

$$U = \sum_{\nu} p_{\nu} E_{\nu}. \tag{2.45}$$

In Eq. (2.45), E_ν stands for the total energy at microstate ν, and p_ν denotes the probability of ν in the canonical ensemble. Without changing the internal energy or doing work, the first law asserts no heat transfer between the system and the surroundings. In other words, the system and the thermal bath are at thermal equilibrium and have the same temperature.

To derive p_ν, we use the Gibbs entropy and the Lagrange multiplier method. The procedure is similar to that used for determining the microstate probability in a microcanonical ensemble discussed in Section 2.1. Again, the Gibbs entropy is defined as

$$S = -k_B \sum_{\nu} p_{\nu} \ln p_{\nu}. \tag{2.46}$$

The second law of thermodynamics asserts that, at equilibrium, the entropy is maximized subject to the macroscopic constraints. Here, the constraint is set by thermal equilibrium with a heat bath. In terms of the microstate probabilities, the equilibrium distribution leads to an internal energy commensurate with temperature T while satisfying the normalization condition

$$\sum_{\nu} p_{\nu} = 1. \tag{2.47}$$

Within the constraints for p_ν set by Eqs. (2.45) and (2.47), maximizing the Gibbs entropy, Eq. (2.46), with respect to p_ν gives

$$p_{\nu} = \exp(-1 - \alpha - \beta E_{\nu}) \tag{2.48}$$

where α and β are the Lagrange multipliers.

Substituting Eq. (2.48) into (2.46) gives

$$S/k_B = -\sum_{\nu} p_{\nu} \ln p_{\nu} = \beta\langle E_{\nu}\rangle + \alpha + 1 = \beta U + \alpha + 1. \tag{2.49}$$

From the thermodynamic identity, $(\partial U/\partial S)_V = T$, we have

$$\beta = \left(\frac{\partial S/k_B}{\partial U}\right)_V = \frac{1}{k_B T}. \tag{2.50}$$

Inserting Eqs. (2.48) and (2.50) into the normalization condition, Eq. (2.47), leads to

$$\alpha = -1 + \ln \sum_{\nu} \exp[-E_{\nu}/k_B T]. \tag{2.51}$$

With the help of Eqs. (2.50) and (2.51), we find the probability distribution for microstates in the canonical ensemble

$$p_{\nu} = \frac{\exp(-\beta E_{\nu})}{Q} \tag{2.52}$$

where Q is called the *canonical partition function*, defined as

$$Q \equiv \sum_{\nu} e^{-\beta E_{\nu}}. \tag{2.53}$$

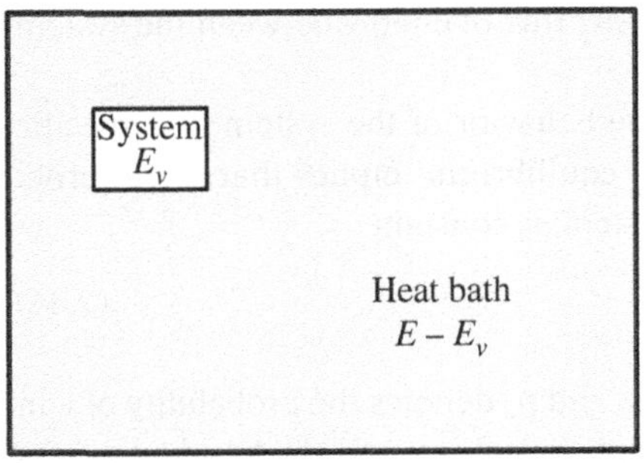

Figure 2.4 A thermodynamic system and a heat bath at thermal equilibrium can be regarded as an isolated macroscopic system. For any microstate, the total energy of the system (E_ν) combined with the energy of the heat bath (E_B) remains constant, i.e., $E = E_\nu + E_B$ is fixed.

The canonical distribution of microstates, Eq. (2.52), is known as the Boltzmann distribution law; and the numerator on the right side of Eq. (2.52) is referred to as the Boltzmann factor. The Boltzmann distribution law indicates that, for a closed system at constant temperature and volume, the equilibrium distribution is biased toward microstates with low energy, i.e., the lower the energy of the microstates, the higher the probability. As the temperature approaches absolute zero, $T \to 0$ K, a closed system would be dominated by the microstate with the lowest energy. If the minimum energy state at 0 K is nondegenerate (viz., a single microstate), we would have $S = 0$ as predicted by the third law of thermodynamics.[12]

Alternatively, the Boltzmann distribution law can be derived from the principle of equal *a priori* probabilities.[13] Imagine that, as shown schematically in Figure 2.4, a thermodynamic system is in equilibrium with a large thermal bath at constant temperature T. Together, the system and the heat bath constitute an isolated system. At any moment, the total energy $E = E_\nu + E_B$ is constant, where E_ν is the system energy at microstate ν and E_B is the energy of the heat bath. Each microstate of the entire system is defined by those of the system and the heat bath.

According to the principle of equal *a priori* probabilities, the probability of the system at a microstate with energy E_ν is proportional to $W_B(E - E_\nu)$, which represents the total number of microstates for the heat bath with energy $E - E_\nu$. The latter can be expressed in terms of a Taylor series

$$\ln W_B(E - E_\nu) = \ln W_B(E) - \frac{d \ln W_B(E)}{dE} E_\nu + \cdots. \tag{2.54}$$

At fixed energy, the entropy is related to the number of microstates according to the Boltzmann equation $S = k_B \ln W$. Eq. (2.54) is thus equivalent to the Taylor expansion of the entropy of the heat bath with respect to energy.

As the thermodynamic properties of the system should be independent of the size of the heat bath, we may consider Eq. (2.54) in the limit $E \gg E_\nu$ such that the Taylor expansion can be truncated after the first-order term. Noting that Eq. (2.50) predicts $\beta = d \ln W_B / dE$, we have from Eq. (2.54)

$$p_\nu \sim W(E_B - E_\nu) \sim e^{-\beta E_\nu}. \tag{2.55}$$

The Boltzmann distribution law, Eq. (2.52), is recovered by applying the normalization condition for p_ν. By focusing on the heat bath, the Taylor expansion approach provides insights into the temperature effects on microstate distributions.

12 The third law of thermodynamics is conventionally stated as follows: the entropy of a perfect crystal is zero at absolute zero temperature. According to the Boltzmann equation for entropy, the statement implies that a perfect crystal has only one microstate at $T = 0$ K.

13 In some older textbooks (e.g., Tolman R., *Principles of statistical mechanics*; Denbigh K.G., *Principles of chemical equilibrium*), the Boltzmann distribution is derived from the method of steepest descents and the sole function method.

2.3.2 Canonical Partition Function

The canonical partition function plays a significant role in linking microscopic properties with thermodynamic variables. On the one hand, Q dictates the distribution of microstates in terms of the Boltzmann distribution law, Eq. (2.52). On the other hand, Q can be directly used to calculate thermodynamic properties.

For example, the internal energy is given by

$$U = \sum_\nu p_\nu E_\nu = \sum_\nu E_\nu \frac{\exp(-\beta E_\nu)}{Q} = -\left(\frac{\partial \ln Q}{\partial \beta}\right)_V. \tag{2.56}$$

A comparison of Eq. (2.56) with the Gibbs–Helmholtz equation

$$U = \left(\frac{\partial \beta F}{\partial \beta}\right)_V \tag{2.57}$$

yields a connection between Q and the Helmholtz energy

$$F = -k_B T \ln Q. \tag{2.58}$$

The entropy is given by

$$S/k_B = \beta(U - F) = -\beta[\partial \ln Q/\partial \beta]_V + \ln Q. \tag{2.59}$$

Other thermodynamic quantities can also be connected with the partition function.

The above equations are formally exact and applicable to all canonical ensembles, i.e., to all closed systems at constant temperature and volume. Regrettably, the formal procedure is not directly useful for most systems of practical interest because the direct evaluation of the canonical partition function is exceedingly difficult. Mathematically, the summation over all possible microstates can be replaced by an integration with respect to the total energy

$$Q = \int dE w(E) e^{-\beta E} \tag{2.60}$$

where $w(E)dE$ corresponds to the number of microstates for the system with energy between E to $E + dE$, and $w(E)$ is called the density of states. If the ground-state energy is set to zero, we have $E \geq 0$ for all microstates. Accordingly, Eq. (2.60) implies that the canonical partition function corresponds to the Laplace transformation of the partition function from the energy domain, $w(E)$, to the temperature domain, $Q(T)$. Because microstates with the same energy have equal probability, Eq. (2.60) is often more useful than the direct application of the Boltzmann distribution law for developing advanced molecular simulation methods to sample the microstates of quasi-ergodic systems.[14]

2.3.3 Thermal Fluctuation

Unlike in a microcanonical ensemble, the total energy is not fixed in a canonical ensemble but fluctuates around its mean value. The energy fluctuation can be attributed to the instantaneous exchange of energy between the system and the heat bath.

We can derive the mean-square deviation of the total energy from the canonical partition function

$$\langle(\delta E)^2\rangle \equiv \langle(E - \langle E\rangle)^2\rangle = \frac{1}{Q}\sum_\nu E_\nu{}^2 e^{-\beta E_\nu} - \left(\frac{1}{Q}\sum_\nu E_\nu e^{-\beta E_\nu}\right)^2 = \frac{\partial^2 \ln Q}{\partial \beta^2}. \tag{2.61}$$

---

14 Quasi-ergodic systems refer to those that have several low-energy configuration regions separated by large energy barriers. Quasi-ergodic systems are often encountered in systems containing multiple phases at equilibrium or that exhibit structure transitions (e.g., polymers and biomacromolecules).

Upon substituting Eq. (2.56) and the definition of the constant-volume heat capacity $C_V \equiv (\partial U/\partial T)_V$ into Eq. (2.61), we obtain a simple relation between the energy fluctuation and heat capacity

$$\langle(\delta E)^2\rangle = k_B T^2 C_V. \tag{2.62}$$

Eq. (2.62) indicates that the constant-volume heat capacity is linearly proportional to the mean-square deviation of energy fluctuation in a canonical ensemble. Because the left side is a square average, Eq. (2.62) suggests that the constant-volume heat capacity can never be negative. The non-negative heat capacity is also required for thermodynamic stability. We note in passing that Eq. (2.62) can be utilized in MD simulation to compute specific heat, which is a measure of the response of the system's energy to changes in temperature. By analyzing the thermal fluctuations of the system, valuable insights can be obtained regarding the correlation effects and thermodynamic properties.

What is the scale of energy fluctuation for a system at fixed temperature, volume, and number of particles? From the Boltzmann distribution law, Eq. (2.52), we can calculate the relative root-mean-square deviation of the total energy

$$\sqrt{\langle(\delta E)^2\rangle}/\langle E\rangle = \sqrt{k_B T^2 C_V}/U. \tag{2.63}$$

Because both heat capacity and internal energy are extensive variables proportional to the number of particles in the system N, Eq. (2.63) predicts that the relative root-mean-square deviation scales as

$$\sqrt{\langle(\delta E)^2\rangle/\langle E\rangle^2} \sim 1/\sqrt{N}. \tag{2.64}$$

In a typical macroscopic system, the number of particles is on the order of $N \sim 10^{23}$. Eq. (2.64) suggests that, in most conditions,[15] the probability that a closed isothermal system has energy appreciably different from its mean energy is extremely small.

2.3.4 Summary

We see from this section that the probability distribution of microstates depends on the thermodynamic variables used to specify a macroscopic system. When the system temperature is fixed, the number of possible microstates is expanded due to energy fluctuations. In the thermodynamic limit, the fluctuation effect has only negligible influence on the expected values of dynamic quantities. Formally, all thermodynamic properties can be derived from the canonical partition function.

2.4 Thermostat Methods

We see from the previous section that when a system is in thermal equilibrium with a heat bath (viz., thermal reservoir), temperature affects the statistical distributions of microstates. However, the thermodynamic variable is not directly shown in conventional mechanical equations for describing the dynamic behavior of individual particles. From a microscopic viewpoint, the physical significance of temperature is manifested either in terms of thermal equilibrium or a constant value corresponding to the ensemble average of the total kinetic and potential energy. Temperature is not a microscopic quantity.

15 We will discuss in Chapter 5 that thermal fluctuation becomes appreciable for a system near a second-order phase transition.

In an isolated system with N classical particles, we learn from Section 2.2 that temperature provides a measure of the ensemble average of the total kinetic energy

$$T = \frac{2}{3k_B N}\sum_{i=1}^{N}\frac{1}{2}m_i\left\langle \mathbf{v}_i^2\right\rangle \tag{2.65}$$

where m_i and $\mathbf{v}_i$ are the mass and the instantaneous velocity of particle i, respectively, and $\langle\cdots\rangle$ denotes an average in the canonical ensemble. Eq. (2.65) suggests that, for a system of classical particles, temperature is a statistical variable that can be regulated through the total kinetic energy, or more specifically, the kinetic energy related to the translational motions of individual particles.

The statistical nature of temperature makes the application of MD simulation to macroscopic systems at constant temperature an intriguing task. On the one hand, both Newtonian and quantum mechanics predict that the total energy is conserved; a direct integration of the equation of motion would lead to microstates corresponding to those in a microcanonical ensemble. On the other hand, it is in general impractical to fix the temperature of a system by explicitly considering the microscopic details of a thermal reservoir, which is often considered immaterial and can be infinitely large. To implement thermal equilibrium, we must thus modify the equation of motion such that the target temperature can be maintained from a statistical perspective. In MD simulation, mathematical procedures particle motions so that temperature can be fixed are commonly known as *thermostats*.

Although *thermostats* may not capture the true dynamics of particle motions in a macroscopic system in equilibrium with a thermal bath, they are useful to generate microstates that are commensurate with the canonical distribution. Based on a trajectory of particle motions generated in accordance with a given temperature, we can evaluate thermodynamic quantities by applying ensemble averages. In this section, we discuss three mathematical strategies to construct thermostats. The dynamic behavior of particles in a solvent, a special type of thermal reservoir, will be discussed in the next two sections.

2.4.1 Velocity Rescaling

Velocity rescaling, also known as the Berendsen thermostat,[16] is a non-physical strategy commonly utilized in MD simulation to control the temperature of a system. In this method, the desired temperature is achieved simply by rescaling the velocities of individual particles.

For a macroscopic system consisting of classical particles, we can set a constant temperature by constraining the ensemble average of the total kinetic energy associated with the translational motions of individual particles. If we are not concerned with the dynamic behavior, we may fix the temperature simply by rescaling the particle velocities so that they yield an average kinetic energy consistent with the preset temperature. More specifically, for a system with N classical particles, we may introduce an "instantaneous" temperature based on the velocities of individual particles

$$\hat{T} = \frac{1}{3k_B N}\sum_{i=1}^{N} m_i \mathbf{v}_i^2. \tag{2.66}$$

The total kinetic energy will be compatible with a preset temperature T by rescaling the particle velocities

$$\mathbf{v}_i^* = \mathbf{v}_i \times \sqrt{T/\hat{T}} \tag{2.67}$$

16 Berendsen H. J. C. et al., "Molecular-dynamics with coupling to an external bath". *J. Chem. Phys.* 81 (8), 3684–3690 (1984).

where the asterisk denotes the adjustment of particle velocities. Substituting Eq. (2.67) into (2.65) indicates that, after velocity rescaling, the macroscopic constraint of a constant temperature is satisfied even when the dynamics of individual particles is described by the conventional equations of motion.

The brutal force velocity rescaling method represents a popular choice in MD simulations due to its simplicity. However, not only is it non-physical to adjust particle velocities by rescaling, but it also prohibits the fluctuation of the total kinetic energy that naturally takes place in an isothermal system. For these reasons, the velocity rescaling method is problematic for calculating equilibrium properties sensitive to thermal fluctuation.[17] Besides, velocity rescaling may generate microstates that are not truly consistent with the canonical ensemble distribution. For example, velocity rescaling may result in an artifact known as the "flying ice cube effect" for molecular systems. Because temperature rescaling is applied only to the particle velocities corresponding to the translational motions, the internal motions of molecules can be sacrificed to compensate for a higher translational kinetic energy. In the extreme case, velocity rescaling may lead to a single molecular conformation, reminiscent of an ice cube flying through space. This artifact affects both the structural and thermodynamic properties calculated by MD simulation. Although the fluctuation of the total kinetic energy may be introduced *ad hoc* by less frequent rescaling, i.e., the velocity adjustment takes place once every few MD steps, the prolonged use of a conventional equation of motion would promote the equal distribution of microstates, thus contradicting the Boltzmann distribution in a canonical ensemble.

2.4.2 Stochastic-Coupling Methods

Originally proposed by Hans Andersen,[18] the stochastic-coupling method fixes temperature in MD simulation by introducing a thermal reservoir that consists of fictitious particles. In this method, the dynamics of simulated particles and their interactions with the fictitious particles conform to the conventional mechanical equations. However, the motion of fictitious particles is not explicitly considered. Instead, they affect only the dynamics of real particles through so-called stochastic coupling, i.e., random collisions with real particles leading to the reassignment of particle velocities. It is assumed that collisions between real and fictitious particles result in a succession of microstates with a constant total energy. The fluctuation in the kinetic energy of the real particles is thus emulated by the variation in the energies of fictitious particles.

Mathematically, the stochastic coupling between real and fictitious particles is achieved by a Poisson process.[19] At each MD step, a real particle is selected from the simulation cell at random to collide with a fictitious particle with probability density

$$p(\tau) = \nu e^{-\nu\tau} \tag{2.68}$$

where ν represents the frequency of collision, and τ represents the interval between successive collisions. After each collision, the real particle is assigned a velocity according to the Maxwell–Boltzmann distribution for the probability density in the three-dimensional velocity space:

$$p(\mathbf{v}) = \left(\frac{m}{2\pi k_B T}\right)^{3/2} \exp\left[-\frac{m|\mathbf{v}|^2}{2k_B T}\right]. \tag{2.69}$$

17 Thermal fluctuations can be introduced by using a stochastic procedure to calculate the rescaling factor such that the kinetic energy follows the canonical distribution (see Bussi G., Donadio D. and Parrinello M., *J. Chem. Phys.* 126, 014101 (2007)).

18 Andersen H. C., "Molecular dynamics simulations at constant pressure and/or temperature", *J. Chem. Phys.*, 72, 2384 (1980).

19 A Poisson process is a mathematical model for a sequence of events where the average frequency of the events is known but the timing of an event to occur is random.

Andersen's thermostat can maintain the system temperature because the particle velocities are assigned according to the desired probability distribution. While any non-zero collision frequency leads to the canonical distribution of microstates, the reassignment of particle velocities at random leads to chaotic motions.

Like that in the velocity rescaling method, the frequency of velocity reassignment should not be too small or too large. If the collision frequency is too small, thermal equilibrium would never be established in a reasonable simulation time. Conversely, too frequent collisions would increase the computational burden and destroy the fluctuation of the kinetic energy. According to Andersen, an optimal collision frequency may be selected according to

$$\nu = \nu_c / N^{2/3} \tag{2.70}$$

where ν_c is the collision frequency among real particles.

2.4.3 Extended-System Methods

To impose a constraint on the long-term average of the total kinetic energy, Shuichi Nosé proposed an extended-system method in which an extra degree of freedom is introduced to represent the effects of a thermal reservoir on particle motions.[20] Intuitively, the additional degrees of freedom may be considered to be those associated with an imaginary particle (a.k.a., *demon*). They are introduced to manipulate the kinetic energy of the real particles being simulated so that they can maintain a desired temperature. While velocity rescaling or Anderson's thermostat leads to an MD trajectory with disruptive particle motions, the extended-system method proposed by Nosé is both deterministic and time-reversible, thereby yielding a smooth trajectory.

To see how Nosé's method works, consider again a classical system with N identical spherical particles. In the presence of an imaginary particle, the fictitious Hamiltonian[21] is given by

$$\mathcal{H}(\mathbf{r}^N, \mathbf{p}^N, s, p_s) = \sum_{i=1}^{N} \frac{|\mathbf{p}_i|^2}{2ms^2} + \Phi(\mathbf{r}^N) + \frac{p_s^2}{2\varpi} + (3N+1)k_B T \ln s \tag{2.71}$$

where $\mathbf{p}^N$ and $\mathbf{r}^N$ denote the particle momenta and positions, respectively, which are the same as those in a regular N-particle system, m is the particle mass, and $\Phi(\mathbf{r}^N)$ is the total potential energy due to particle–particle interactions. The system is "extended" in the sense that the Hamiltonian entails an extra degree of freedom associated with the dimensionless parameter s, and a fictitious momentum $p_s = \varpi ds/dt$.

In Eq. (2.71), parameter s might be understood as the coordinate of an imaginary particle, with ϖ and p_s being its mass and momentum. Parameter ϖ dictates the time scale on which the imaginary particle interacts with the real particles, and the last term on the right side of Eq. (2.71) is chosen such that the canonical distribution of microstates will be created when the particle positions and momenta evolve over time under the influence of the imaginary particle. While the interpretation of the additional terms in the Hamiltonian is intuitively appealing, one should keep in mind that parameters ϖ and p_s do not have real physical significance because the imaginary particle is non-physical. In fact, these parameters do not have the usual units of mass and momentum.

---

20 Nosé S., "A unified formulation of the constant temperature molecular dynamics methods", *J. Chem. Phys.* 81, 511–519 (1984).

21 As discussed in Supplementary Materials, Hamiltonian is a mathematical operator associated with the energy of particles moving along a certain trajectory. In the framework of classical mechanics, Hamiltonian equations describe the variation of particle positions and momenta that are equivalent to those given by Newton's laws.

According to the Hamiltonian mechanics, Eq. (2.71) predicts the following equations of motion for the particles as well as the time evolution of parameters ϖ and p_s:

$$\dot{\mathbf{r}}_i = \frac{\partial \mathcal{H}}{\partial \mathbf{p}_i} = \frac{\mathbf{p}_i}{m_i s^2} \tag{2.72}$$

$$\dot{\mathbf{p}}_i = -\frac{\partial \mathcal{H}}{\partial \mathbf{r}_i} = \mathbf{F}_i \tag{2.73}$$

$$\dot{s} = \frac{\partial \mathcal{H}}{\partial p_s} = \frac{p_s}{\varpi} \tag{2.74}$$

$$\dot{p}_s = -\frac{\partial \mathcal{H}}{\partial s} = \frac{1}{s}\left[\sum_{i=1}^{N} \frac{|\mathbf{p}_i|^2}{m_i s^2} - (3N+1)k_B T\right] \tag{2.75}$$

where the dot sign represents time derivative (e.g., $\dot{s} = ds/dt$).

Like Newton's equations, the Hamiltonian mechanics predicts particle positions and momenta evolving with time at a fixed total energy. However, parameters s and p_s introduce extra degrees of freedom that ensure the MD simulation generates a trajectory consistent with the canonical distribution of microstates. As shown in Appendix 2.B, Nosé's method predicts that the probability density for the positions and momenta of particles in the phase space is consistent with the Boltzmann distribution

$$p(\mathbf{r}^N, \mathbf{p}^N) = \frac{1}{Q} \exp[-\beta E(\mathbf{r}^N, \mathbf{p}^N)] \tag{2.76}$$

where $E(\mathbf{r}^N, \mathbf{p}^N)$ is the total energy of real particles

$$E(\mathbf{r}^N, \mathbf{p}^N) = \sum_{i=1}^{N} |\mathbf{p}_i|^2/2m + \Phi(\mathbf{r}^N) \tag{2.77}$$

and Q is the canonical partition function

$$Q = \frac{1}{N!h^{3N}} \int d\mathbf{r}^N \int d\mathbf{p}^N \exp[-\beta E(\mathbf{r}^N, \mathbf{p}^N)]. \tag{2.78}$$

In Nosé's method, parameter s rescales the velocities of individual particles to maintain the system temperature. While it is consistent with the canonical distribution of microstates, the adoption of virtual variables opens ambiguities in interpreting the dynamic behavior. To avoid such issues, William G. Hoover reformulated the equations of motion by rescaling both the particle momentum and time[22]

$$\mathbf{p}'_i \equiv \mathbf{p}_i/s, \tag{2.79}$$

$$t' \equiv t/s. \tag{2.80}$$

In terms of these rescaled variables, the equations of motion, Eqs. (2.72) and (2.73), are transformed to

$$\frac{d\mathbf{r}_i}{dt'} = \frac{\mathbf{p}'_i}{m_i}, \tag{2.81}$$

$$\frac{d\mathbf{p}'_i}{dt'} = \mathbf{F}_i - \xi \mathbf{p}'_i \tag{2.82}$$

22 Hoover W. G., "Canonical dynamics: equilibrium phase-space distributions", *Phys. Rev. A* 31 (3), 1695–1697 (1985).

where $\xi \equiv p_s/\varpi$ may be understood as a friction coefficient arising from the thermal reservoir. From Eqs. (2.74) and (2.75), we have

$$\frac{ds}{dt'} = s\xi, \tag{2.83}$$

$$\dot{\xi} = \frac{1}{\varpi}\left[\sum_{i=1}^{N}\frac{\left|\mathbf{p}_i'\right|^2}{m_i} - (3N+1)k_BT\right]. \tag{2.84}$$

Eq. (2.83) is not needed in computing the particle trajectories but may be used as a diagnostic tool to check the conservation of energy during MD simulations.

Eqs. (2.81)–(2.84) are known as the Nosé–Hoover equations of motion or the Nosé–Hoover thermostat. Different from Andersen's thermostat, the Nosé–Hoover equations allow the temperature to be controlled without using random numbers. In principle, the Nosé–Hoover thermostat can generate microstates commensurate with the canonical ensemble with arbitrarily small friction coefficients. However, the extended equations of motion often result in strong oscillatory temperature fluctuations. From a practical perspective, too small values of the friction coefficients may cause poor temperature control. Indeed, a microcanonical ensemble is recovered in the limiting case where all friction coefficients are set to zero. If the friction coefficients are chosen too small, the canonical distribution will be obtained only after exceedingly long simulation times. In this case, it leads to systematic energy drifts due to accumulation of numerical errors. Conversely, too large values of the friction coefficients (tight coupling) may cause the large stochastic and frictional forces to perturb the dynamics of the system.

2.4.4 Summary

Various thermostat methods are available to modify the equations of motion such that the temperature can be maintained as a constant value in MD simulation. Because these algorithms are designed to replicate the statistical distribution of the average kinetic energy rather than the true particle motions, they typically generate microstates with probabilities falling in between a canonical and a microcanonical distributions. Moreover, it is known that thermostats may cause an uneven distribution of translational and other modes of kinetic energies. Therefore, the thermostats should be treated with caution in MD simulation, in particular when transport properties are calculated from the non-physical changes of the particle mechanics.[23]

2.5 The Langevin Dynamics

Langevin dynamics describes particle motion in a viscous medium. The mathematical procedure can also be used in MD simulation as a thermostat to control the temperature of a thermodynamic system consisting of particles free of a solvent. In the latter case, a constant temperature is achieved by the proper balance of fictitious friction and stochastic forces applying to individual particles. In this section, we illustrate how the Langevin dynamics can be utilized to account

23 For an in-depth discussion of this topic, see, for example, Maginn E. J. et al., "Best practices for computing transport properties 1. Self-diffusivity and viscosity from equilibrium molecular dynamics", *J. Comp. Mol. Sci.* 1 (1), 6324 (2019).

for solvent effects on particle motions in systems like colloids or liquid solutions. As will be discussed in the next section, Langevin dynamics provides an informative example to elucidate the fluctuation–dissipation theorem, a cornerstone of statistical mechanics for predicting the transport properties of equilibrium systems.

2.5.1 The Langevin Equation

The Langevin equation may be understood as a semi-empirical modification of Newton's equation of motion to account for the friction and random forces acting on a particle due to the presence of a solvent. In its simplest form, the particle position **r** and momentum **p** satisfy the following differential equations:

$$d\mathbf{r}/dt = \mathbf{p}/m, \tag{2.85}$$

$$d\mathbf{p}/dt = -\nabla\varphi(\mathbf{r}) - \gamma_0\mathbf{p} + \xi(t) \tag{2.86}$$

where m stands for the particle mass, $\varphi(\mathbf{r})$ represents a one-body potential, γ_0 is the friction coefficient,[24] and $\xi(t)$ represents a random force or thermal noise. Eq. (2.85) is nothing but the definition of particle momentum, and Eq. (2.86) is a modified Newton's equation to account for the friction and thermal effects. The stochastic nature of the random force is a fundamental aspect of the Langevin equation.

Figure 2.5 provides a schematic representation of particle movement in a solvent, where three force components (frictional, random, and conservative) are depicted in accordance with the Langevin equation. The frictional force arises from the particle motion relative to a continuous medium. The linear relation between the frictional force and particle momentum is obtained from the particle moving in a laminar flow at a steady-state velocity. In other words, the Langevin equation ignores the solvent inertial and hydrodynamic effects. The random force accounts for the particle collision with the solvent molecules in the background. As discussed in the following, the random force is related to the thermal fluctuations of the solvent molecules; thus, it is also called thermal noise. By taking the random force, we assume that the time scale associated with the microscopic degrees of freedom of the solvent molecules is much smaller in comparison with that for the particle motion. Under such conditions, the particle dynamics may be fully decoupled from those corresponding to the solvent molecules. Finally, the conservative force refers to the gradient of the one-body potential $\varphi(\mathbf{r})$, which can be attributed to inter-particle interactions and/or to any external potential. The conservative force dictates the thermodynamic properties of the system.

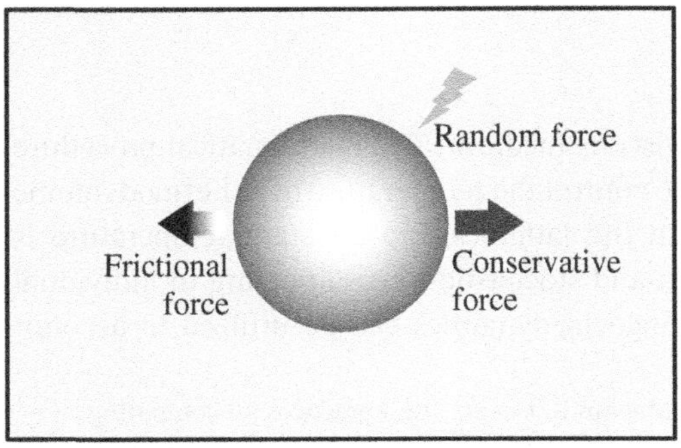

Figure 2.5 When a particle travels within a solvent, it undergoes three types of forces: a conservative force resulting from interactions with external potential or other particles, a frictional force connected to the viscosity of the solvent, and a random force caused by collisions with the solvent molecules.

24 The inverse of the friction coefficient is called mobility.

While the frictional force and the random force make no contributions to the thermodynamic properties of the system, they do play a crucial role in influencing particle dynamics. As a result, these forces can be utilized to control and regulate the system temperature in MD simulation.

2.5.2 Random Force

Langevin dynamics assumes that the particle motion is sufficiently slow such that, at any moment, the solvent is in equilibrium state. Accordingly, the particle dynamics are completely decoupled from the kinetic motions of the solvent molecules. With this assumption, the random force follows a Gaussian distribution with zero mean

$$\langle \xi(t) \rangle = 0, \tag{2.87}$$

and a variance of[25]

$$\langle \xi(t)\xi(t') \rangle = 6\gamma_0 m k_B T \delta(t - t') \tag{2.88}$$

where the angle brackets denote an average over all configurations of the solvent molecules, $\delta(t - t')$ is the Dirac delta function, k_B represents the Boltzmann constant, and T is the absolute temperature. In Eq. (2.88), the Dirac delta function ensures that the random forces are uncorrelated at the time scale of particle motions, i.e., it imposes no memory effect.

As shown in the following, the variance of the random force is related to the average kinetic energy of each particle, which is proportional to $k_B T$ according to the Maxwell–Boltzmann distribution for the particle velocity. Therefore, the Langevin equation provides an intrinsic connection between molecular motions (a microscopic behavior) and system temperature (a macroscopic quantity). When the Langevin equation is used as a thermostat in MD simulation, the coupling between the dissipative energy (viz., friction) and the random force makes the microstates of particles commensurate with a predefined temperature.

In the limit $\gamma_0 \to 0$, both the friction and random forces vanish. In that case, the Langevin equation is reduced to Newton's equation of motion. Conversely, in the limit of large γ_0, the inertial force (i.e., $d\mathbf{p}/dt$) is negligible in comparison with the frictional force, and the particle motion is often referred to as *diffusive* (or *over-damped*). Whereas γ_0 has a physical significance for particle motion in a solvent, this parameter may be optimized in MD simulation to ensure the canonical distribution for microstates of particles.

2.5.3 Particle Velocity Distribution

We may elucidate the physical significance of the frictional and random forces by applying the Langevin equation to a single particle in a solvent (viz., a particle in Brownian motion, named after the botanist Robert Brown who first described the phenomenon in 1827)

$$d\mathbf{p}/dt = -\gamma_0 \mathbf{p} + \xi(t). \tag{2.89}$$

In this case, the one-body potential $\varphi(\mathbf{r})$ vanishes because there are no other particles or external potential. At any instant, the force on the Brownian particle reflects a net effect of the frictional and random forces. While the random force drives the particle in perpetual motion, the frictional force dissipates the particle energy into the solvent and prevents the unlimited increase of the particle velocity.

---

25 Here, ξ is a three-dimensional vector. The coefficient 6 would be replaced by 2 if we consider a one-dimensional random force.

If the random force were constant (e.g., replaced by a constant external force), Eq. (2.89) would predict that it must be balanced by the frictional force $\gamma_0 \mathbf{p}_\infty$ when the particle reaches a terminal velocity with momentum $\mathbf{p}_\infty$. By comparing $\gamma_0 \mathbf{p}_\infty$ with the frictional force predicted by the Stokes law for a spherical particle of radius a moving through a viscous fluid

$$\mathbf{f} = 6\pi a \eta v_\infty, \tag{2.90}$$

we may find a connection between friction coefficient γ_0 and solvent viscosity η

$$\gamma_0 = 6\pi a \eta / m. \tag{2.91}$$

In Eq. (2.90), $\mathbf{v}_\infty = \mathbf{p}_\infty / m$ is the steady-state velocity, and a represents the hydrodynamic radius of the particle. Because the Stokes law is valid only when a steady-state laminar flow is established around the spherical particle, the assumption of a linear response of the frictional force to the particle momentum is inherently adopted in the Langevin equation.

We can derive an analytical solution to Eq. (2.89) following a two-step procedure. Without the random force, Eq. (2.89) would have a simple solution

$$\mathbf{p}(t) = \mathbf{p}(0) \exp(-\gamma_0 t) \tag{2.92}$$

where $\mathbf{p}(0)$ represents the initial momentum at $t = 0$. Eq. (2.92) indicates that the initial momentum dies away over the duration of $1/\gamma_0$. Thus, parameter γ_0 is also known as the damping constant, and $1/\gamma_0$ characterizes a time scale of the particle relaxation.

In the presence of random force $\xi(t)$, we can find a general solution of Eq. (2.89) as[26]

$$\mathbf{p}(t) = \mathbf{p}(0) e^{-\gamma_0 t} + \int_0^t d\tau \xi(\tau) e^{-\gamma_0 (t-\tau)}. \tag{2.93}$$

In comparison with the solution without the random force, Eq. (2.93) includes a convolution term that is independent of the initial condition, i.e., the initial particle momentum $\mathbf{p}(0)$. Because $\mathbf{p}(0)e^{-\gamma_0 t}$ disappears as t increases, the long-time dynamics of the free particle must be fully determined by the convolution term, i.e., by thermal noise $\xi(\tau)$ and damping constant γ_0. The former moves the particle in random directions, while the latter drags the particle motion. The dynamic process resembles a random walk with the duration of $1/\gamma_0$ per step (viz., the time scale of the particle relaxation) and a step length about the order of $\overline{v}/\gamma_0$, where $\overline{v}$ stands for the mean velocity of the particle.[27] The stochastic nature of particle dynamics is a characteristic feature of Brownian motion.

According to Eq. (2.93), the stochastic component of the particle velocity is given by

$$\mathbf{v}^*(t) \equiv \mathbf{v}(t) - \mathbf{v}(0) e^{-\gamma_0 t} = \frac{1}{m} \int_0^t d\tau \xi(\tau) e^{-\gamma_0 (t-\tau)}. \tag{2.94}$$

Because $\xi(\tau)$ is a random variable, the central-limit theorem asserts that $\mathbf{v}^*(t)$ follows the normal distribution.[28] Its mean value and the variance can be readily derived from Eq. (2.88)

$$\langle \mathbf{v}^*(t) \rangle = 0, \tag{2.95}$$

$$\langle \mathbf{v}^*(t) \cdot \mathbf{v}^*(t) \rangle = \frac{6\gamma_0 k_B T}{m} \int_0^t d\tau \int_0^t d\tau' \delta(\tau - \tau') e^{-\gamma_0 (t-\tau)} e^{-\gamma_0 (t-\tau')} = \frac{3k_B T}{m} \left(1 - e^{-2\gamma_0 t}\right). \tag{2.96}$$

26 The Laplace transform of Eq. (2.89) gives $s\mathbf{p}(s) - \mathbf{p}(0) = \gamma_0 \mathbf{p}(s) + \xi(s)$, and the inverse transform of $\mathbf{p}(s) = \mathbf{p}(0)/(s + \gamma_0) + \xi(s)/(s + \gamma_0)$ leads to Eq. (2.93).

27 See Section 3.11 for a more detailed discussion of random walk.

28 The central-limit theorem states that the summation of independent random variables approaches a Gaussian distribution. Here, the summation is replaced by the integration over τ.

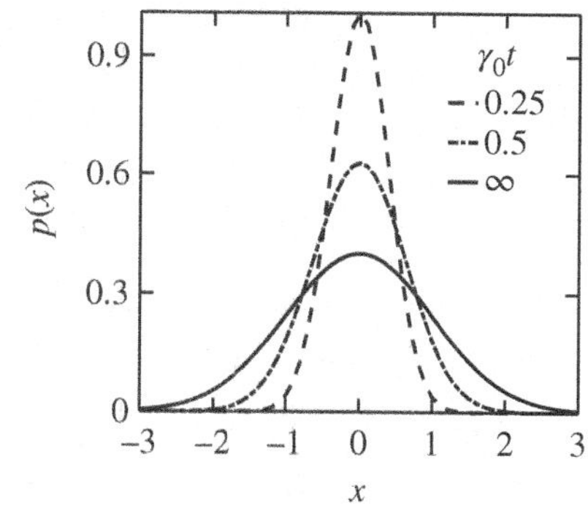

Figure 2.6 Probability density of the stochastic component of particle velocity is predicted by Eq. (2.97). Here $x = v_i/\sqrt{k_BT/m}$, v_i is any component of the velocity vector.

Therefore, the probability density for the particle velocity is given by the Gaussian distribution

$$p(\mathbf{v}^*) = \left[\frac{m}{2\pi k_B T\left(1 - e^{-2\gamma_0 t}\right)}\right]^{3/2} \exp\left[-\frac{m\mathbf{v}^{*2}(t)}{2k_B T\left(1 - e^{-2\gamma_0 t}\right)}\right]. \tag{2.97}$$

Eq. (2.97) indicates that, at a time scale much longer than $1/\gamma_0$, the probability distribution for the velocity of a Brownian particle is the same as the Maxwell–Boltzmann distribution for particles in free space (Figure 2.6). In other words, by a proper choice of the covariance for the random force as given by Eq. (2.88), the Langevin equation results in an average kinetic energy per particle equal to $3k_BT/2$, which is the same as that of the non-interacting particles in vacuum at the same temperature. Therefore, the Langevin dynamics provide an effective means to control the temperature in MD simulation.

2.5.4 The Generalized Langevin Equation

Before closing this section, it is instructive to discuss the generalized Langevin equation so that we may gain some further understanding of the assumptions in Langevin dynamics. A Brownian particle undergoes erratic movements with rapid changes in velocity. However, the friction term in Eq. (2.86) follows from the steady Stokes flow, which is valid only when the surrounding fluid becomes quasi-steady immediately after particle movement. Before complete relaxation, the particle motion is coupled with the hydrodynamics of the solvent flow. To account for the so-called "memory" effects, the Langevin equation can be generalized as

$$d\mathbf{p}/dt = -\nabla\varphi(\mathbf{r}) - \int_0^t d\tau \zeta(t-\tau)\mathbf{p}(\tau) + \xi(t) \tag{2.98}$$

where $\zeta(t)$ represents the memory kernel due to the solvent friction. $\zeta(t)$ determines the memory decay time and the relaxation behavior of the system. Different choices of the memory kernel can lead to different models of dissipation and memory effects.

Similar to the original Langevin model, the random force has a zero average. However, its variance is now correlated with the memory kernel

$$\langle \xi(t)\xi(t')\rangle = 3\zeta(t-t')mk_BT. \tag{2.99}$$

Eq. (2.99) follows the long-term behavior of the average kinetic energy of a single Brownian particle, $3k_BT/2$. In the limit where the solvent can respond to the particle motion instantaneously, the memory kernel reduces to the Dirac delta function

$$\zeta(t) = 2\gamma_0\delta(t). \tag{2.100}$$

Substituting Eq. (2.100) into Eq. (2.98) reproduces the original Langevin equation.

For Brownian motion in an incompressible solvent, the resistance force on a spherical particle of radius a can be derived from the linearized Navier–Stokes equation[29]

$$\mathbf{f}(t) = -6\pi a\eta \mathbf{v} - m\dot{\mathbf{v}}/2 - 6a^2\sqrt{\pi\eta\rho_m}\int_{-\infty}^{t} d\tau \frac{\dot{\mathbf{v}}(\tau)}{\sqrt{t-\tau}} \tag{2.101}$$

where η and ρ_m are the viscosity and mass density of the solvent, respectively. Eq. (2.101) is known as the Basset–Boussinesq–Oseen (BBO) equation. Taking the Laplace transform of Eq. (2.101)

$$\mathbf{f}(s) \equiv \int_0^{\infty} dte^{-st}\mathbf{f}(t) = -\zeta(s)\mathbf{p}(s) = -\left[6\pi a\eta + ms/2 + 6\pi a^2\sqrt{\eta\rho_m s}\right]\mathbf{v}(s), \tag{2.102}$$

we can obtain the memory kernel of friction

$$\zeta(s) = 6a\pi\eta/m + s/2 + 6\pi a^2\sqrt{\eta\rho_m s}/m. \tag{2.103}$$

A comparison of the first and third terms on the right side of Eq. (2.103) suggests that the hydrodynamic effect becomes significant when $s > \eta/(\rho_m a^2)$ or, equivalently, for $t < \rho_m a^2/\eta$. As discussed above, the relaxation time for the Brownian particle is about $1/\gamma_0 = m/(6\pi\eta a)$. Therefore, the solvent inertia or hydrodynamics would have little influence on the Brownian motion when $9\rho_m/2\rho_B \ll 1$, where $\rho_B \equiv 3m/(4\pi a^3)$ is the mass density for the Brownian particle. Typically, solids and liquids have similar densities, so the condition is hardly satisfied for colloidal particles. The hydrodynamic effects are less significant for molecular species due to the smaller particle size.

The generalized Langevin equation is widely used in the study of non-equilibrium statistical mechanics, MD simulation, and the description of systems coupled to complex environments. It provides a more realistic description of their dynamics compared to the traditional Langevin equation.

2.5.5 Summary

The Langevin equation can be used in MD simulation to generate a smooth, non-deterministic, and irreversible trajectory. It is applicable to particle motions in a solvent as well as to atomic systems as a thermostat. When the Langevin equation is applied to particle motions in a solvent, one should keep in mind that the equation of motion is not exact because it ignores the memory effects in the solvent–particle interactions, implying that the time scale corresponding to the microscopic degrees of freedom of the solvent molecules is much smaller than typical time scale associated with the random force on particles. In its application to MD simulation as a thermostat, the Langevin dynamics incorporates stochastic terms to approximate the effects of a thermal reservoir. In comparison with alternative methods, one major advantage of the Langevin model is that it generates a phase-space distribution of microstates independent of the microscopic details of the environment. As the particle momentum eventually follows the Maxwell–Boltzmann distribution, the background serves essentially as a thermal bath to control the system temperature.

2.6 Fluctuation–Dissipation Theorem

The fluctuation–dissipation theorem is a general result of statistical mechanics. It links thermal fluctuations in an equilibrium system with the response of dynamic variables to external stimuli.

29 Parmar M., Haselbacher A. and Balachandar S., "Equation of motion for a sphere in non-uniform compressible flows", *J. Fluid Mech.* 699, 352–375 (2012).

Importantly, the fluctuation–dissipation theorem provides a theoretical basis to predict transport coefficients from MD simulation. The general form of fluctuation–dissipation theorem was proven by Herbert Callen and Theodore Welton in 1951.[30] Here, we illustrate only the essential ideas in the context of the Brownian motion for non-interacting particles.

For a free particle in Brownian motion, we see from the previous section that the kinetic energy stems from its random collision with the solvent molecules. The dynamic behavior reflects a balance between the work done by the solvent molecules and heat dissipation due to the frictional force. Interestingly, the frictional force and the random force have the same origin, i.e., both arise from interactions between the particle and the surrounding solvent molecules.

Albert Einstein was the first to recognize that the random force driving the particle motion (viz., diffusion) and the solvent friction were intrinsically connected.[31] The former is manifested through the fluctuation of the random force due to the particle collision with the solvent molecules, and the latter is manifested as the frictional force, one of the most familiar examples of dissipation behavior. Einstein's theoretical investigation of Brownian motion provides not only considerable evidence for the atomicity of matter but also profound insights into the connection between thermodynamic properties and transport phenomena, later known as the fluctuation–dissipation theorem.

2.6.1 Brownian Motion from the Perspective of Energy Dissipation

To see the connection between the random force moving the particle around (viz., due to thermal fluctuation) and the solvent friction responsible for the generation of heat (viz., energy dissipation or the process by which energy is transferred or transformed into other forms), consider the mean-square displacement of a particle in Brownian motion as a function of time

$$R^2(t) = \langle |\mathbf{r}(t) - \mathbf{r}(0)|^2 \rangle, \tag{2.104}$$

where $\mathbf{r}(0)$ and $\mathbf{r}(t)$ stand for the initial particle position and its location at time t, respectively, and $\langle \cdots \rangle$ represents the "thermal" average, i.e., the particle is in equilibrium with its surroundings. From the perspective of energy dissipation, we may express the particle displacement in terms of the velocity

$$\mathbf{r}(t) - \mathbf{r}(0) = \int_0^t d\tau \mathbf{v}(\tau). \tag{2.105}$$

Substituting Eq. (2.105) into (2.104) leads to

$$R^2(t) = \int_0^t d\tau' \int_0^t d\tau'' \langle \mathbf{v}(\tau') \cdot \mathbf{v}(\tau'') \rangle. \tag{2.106}$$

To find an analytical expression for this, take a time derivative of Eq. (2.106)

$$\frac{dR^2(t)}{dt} = 2\int_0^t d\tau \langle \mathbf{v}(0) \cdot \mathbf{v}(\tau) \rangle = 2\int_0^t d\tau c(\tau) \tag{2.107}$$

where $c(\tau)$ is the velocity autocorrelation function

$$c(t) \equiv \langle \mathbf{v}(0) \cdot \mathbf{v}(t) \rangle. \tag{2.108}$$

30 Callen H.B. and Welton T.A., "Irreversibility and generalized noise", *Phys. Rev.* 83 (1) 34–40 (1951).

31 Einstein A. "On the movement of small particles suspended in stationary liquids required by the molecular-kinetic theory of heat", *Annalen der Physik (in German)*. 322 (8): 549–560 (1905).

Based on $\mathbf{v}(t) = \mathbf{p}(t)/m$ and the momentum of an isolated Brownian particle derived in the previous section

$$\mathbf{p}(t) = \mathbf{p}(0)e^{-\gamma_0 t} + \int_0^t d\tau \xi(\tau)e^{-\gamma_0(t-\tau)}, \tag{2.109}$$

We can find an analytical expression for the velocity autocorrelation

$$\langle \mathbf{v}(0) \cdot \mathbf{v}(t) \rangle = \langle \mathbf{v}(0)^2 \rangle e^{-\gamma_0 t} + \frac{1}{m}\int_0^t d\tau \langle \mathbf{v}(0) \cdot \xi(\tau) \rangle e^{-\gamma_0(t-\tau)}. \tag{2.110}$$

Because the initial velocity and the random force are uncorrelated,[32] and $\langle \xi(\tau) \rangle = 0$, the second term on the right side of Eq. (2.110) disappears.

As discussed in the previous section, the mean-square value of the initial velocity in Eq. (2.110) can be evaluated from the Maxwell–Boltzmann distribution for the particle velocity

$$\langle \mathbf{v}(0)^2 \rangle = \frac{3k_B T}{m}. \tag{2.111}$$

Thus, the velocity autocorrelation function is given by

$$c(t) = \frac{3k_B T}{m} e^{-\gamma_0 t}. \tag{2.112}$$

Substituting Eq. (2.112) into (2.107) leads to

$$\frac{dR^2(t)}{dt} = \frac{6k_B T}{m} \int_0^t d\tau e^{-\gamma_0 t} = \frac{6k_B T}{m\gamma_0}\left(1 - e^{-\gamma_0 t}\right). \tag{2.113}$$

With the initial condition $R^2(0) = 0$, we obtain an analytical expression for the mean-square displacement

$$R^2(t) = \frac{6k_B T}{m\gamma_0}\left(t - 1/\gamma_0 + e^{-\gamma_0 t}/\gamma_0\right). \tag{2.114}$$

Figure 2.7 illustrates the variation of $R^2(t)$ over the duration comparable to $1/\gamma_0$. Initially, the particle motion is dominated by "free flight," resulting in a mean-square displacement that scales as $\sim t^2$. During these very short times, the particle does not experience significant collisions, and the velocity remains constant. For large t, the particle motion follows a random walk,[33] and the

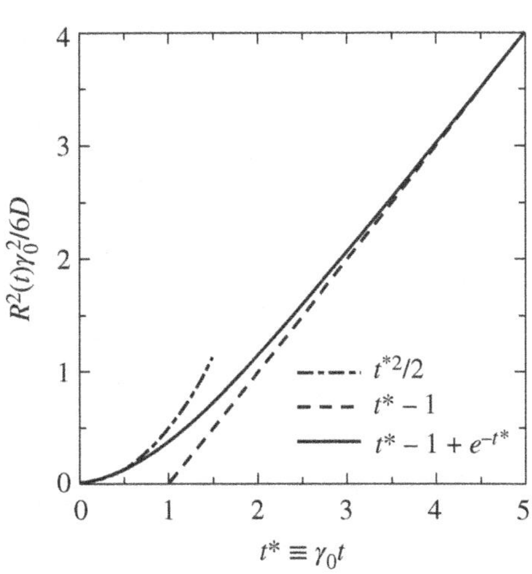

Figure 2.7 Evolution of the mean-square displacement for a particle under Brownian motion. The particle displacement is dictated by damped motion in a short time and diffusive motion in a long time, leading to different scaling behaviors ($\sim t^2$ versus $\sim t$ at short and long time scales, respectively). Approximately, $1/\gamma_0$ represents the time it takes for the transition from the damped motion to diffusion.

32 The consideration of correlation at $t = 0$ would lead to a slightly different expression but all numerical results are virtually indistinguishable.

33 See Section 3.11 for more details.

mean-square displacement scales as $\sim t$. As discussed in the following, we can extract the diffusion coefficient from the slope if the particle motion has been followed over a duration much larger than the relaxation time.

2.6.2 Brownian Motion from Fluctuation Perspective

We now consider the Brownian motion from the perspective of thermal fluctuations. As mentioned above, in the long-term limit, the particle motion resembles a random walk driven by the fluctuating thermal noise. Accordingly, the long-term behavior of particle displacement can be obtained from the diffusion equation

$$\frac{\partial p(\mathbf{r},t)}{\partial t} = D\nabla^2 p(\mathbf{r},t) \tag{2.115}$$

where $p(\mathbf{r}, t)$ stands for the probability density of a particle that is placed at the origin at $t=0$ and reaches position $\mathbf{r}$ at time t, and D is the particle diffusion coefficient. Eq. (2.115) was first recognized by Marian Smoluchowski.

The particle mean-square displacement is then given by

$$R^2(t) = \int d\mathbf{r} r^2 p(\mathbf{r},t). \tag{2.116}$$

Making a derivative of both sides of the above equation with respect to t gives

$$\frac{dR^2(t)}{dt} = \int d\mathbf{r} r^2 \frac{\partial p(\mathbf{r},t)}{\partial t} = D\int d\mathbf{r} r^2 \nabla^2 p(\mathbf{r},t) = 6D. \tag{2.117}$$

where the last equality is obtained from $\nabla^2 p(r,t) = \frac{1}{r^2}\frac{\partial}{\partial r}\left(r^2\frac{\partial p}{\partial r}\right)$ followed by integration by parts. With the initial condition $R^2(0)=0$, we obtain an alternative expression for the mean-square displacement

$$R^2(t) = 6Dt. \tag{2.118}$$

A comparison of Eq. (2.118) and (2.114) at the long-term limit yields the Stokes–Einstein equation

$$D = \frac{k_B T}{m\gamma_0} = \frac{k_B T}{6a\pi\eta}. \tag{2.119}$$

Eq. (2.119) provides a direct link between diffusivity and friction coefficient as predicted by the fluctuation–dissipation theorem. The former arises from the random forces or thermal fluctuations of the solvent molecules, and the latter is related to the energy dissipation due to the same solvent. In other words, particle collisions with the solvent molecules are responsible for both the particle motion and the friction. Because both diffusivity and friction coefficient are experimentally measurable, Eq. (2.119) also allows for determining Avogadro's number, $N_A = R/k_B$, where $R = 8.314\,\mathrm{J/(mol\,K)}$ is the gas constant. Following Einstein's proposal, Jean Perrin first quantified Avogadro's number, $N_A = 6.022\times 10^{23}$ in 1909[34], which was later recognized by the Nobel Prize in Physics for settling the dispute about the atomic nature of matter.

2.6.3 The Green–Kubo Relations

The fluctuation–dissipation theorem can be formulated in many ways. For example, we may obtain another useful equation for diffusion coefficient by comparing Eqs. (2.117) and (2.113)

$$D = \frac{1}{3}\int_0^\infty d\tau c(\tau) = \frac{1}{3}\int_0^\infty d\tau \langle \mathbf{v}(0)\cdot\mathbf{v}(\tau)\rangle. \tag{2.120}$$

34 Perrin J. "Brownian movement and molecular reality", *Annales de chimie et de physique*, 18: 5–114 (1909).

Eq. (2.120) represents one of the Green–Kubo relations, which are important for the theoretical prediction of transport coefficients such as diffusivity, viscosity, and thermal conductivity. Here, the Green–Kubo relation connects diffusivity, a transport coefficient that describes the response of the particle concentration or density to the change in the chemical potential, to the velocity auto-correlation function.

Eq. (2.120) can be generalized to other transport coefficients such as viscosity and thermal conductivity

$$\gamma = K_\gamma \int_0^\infty d\tau \langle \dot{\xi}(0) \cdot \dot{\xi}(\tau) \rangle \tag{2.121}$$

where K_γ is a multiplicative constant, ξ is the mechanical variable associated with the transport coefficient under consideration, and $\dot{\xi}$ represents the time derivative. It is easy to check that, for self-diffusivity given by Eq. (2.120), $K_\gamma = 1/3$ and ξ corresponds to the particle position. Like Eq. (2.120), the generalized Green–Kubo relations can be established by considering the linkage between various driving forces of mass and energy transport and the fluctuation of corresponding quantities in equilibrium systems.[35]

2.6.4 Summary

We conclude this section by noting that statistical mechanics is useful not only for describing thermodynamic quantities, but it is also relevant to understanding non-equilibrium linear response processes. Toward that end, the fluctuation–dissipation theorem provides a general description of the interdependence between fluctuations and the responses of a thermodynamic system when it is weakly displaced from equilibrium. The fluctuation–dissipation theorem provides a theoretical basis for predicting kinetic and transport coefficients from a microscopic perspective. These coefficients are extensively used to describe the kinetics of chemical reactions and transport phenomena.

2.7 Isobaric–Isothermal Ensemble

Many chemical and physical phenomena of practical interest take place at constant temperature and pressure instead of constant volume. It is therefore desirable to construct an ensemble for an isobaric–isothermal system where the total number of particles is fixed but where both the total energy and volume may fluctuate around their mean values. An isobaric–isothermal ensemble refers to the assembly of closed systems with identical temperature, pressure, and chemical composition. It is also known as the NPT ensemble because the number of particles N, pressure P, and temperature T are used as independent variables.

2.7.1 The Partition Function

To derive the probability distribution of microstates in an isobaric–isothermal ensemble, we follow a mathematical procedure like that used for the canonical ensemble. To illustrate, consider a system of classical particles at fixed temperature and pressure, as shown in Figure 2.8. The second law of thermodynamics asserts that, at equilibrium, the entropy attains a maximum, subject to

---

35 Zwanzig R., "Time-correlation functions and transport coefficients in statistical mechanics", *Ann. Rev. Phys. Chem.* 16, 67–102 (1965).

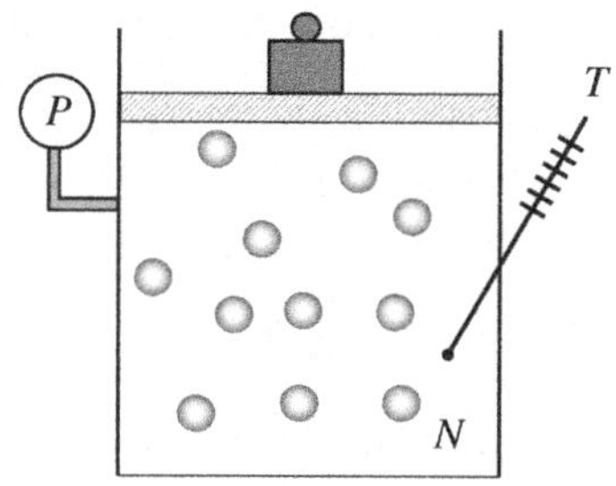

Figure 2.8 Schematic of a system containing N particles at constant temperature T and constant pressure P.

three constraints: constant average total energy, constant average volume, and the normalization of microstate probability[36,37]:

$$U = \sum_i p_\nu E_\nu, \tag{2.122}$$

$$V = \sum_i p_\nu V_\nu, \tag{2.123}$$

$$\sum_\nu p_\nu = 1, \tag{2.124}$$

where p_ν is the probability that the system is in microstate ν, E_ν, and V_ν are the corresponding energy and volume, respectively.

The probability distribution is determined by maximizing the Gibbs entropy

$$S = -k_B \sum_\upsilon p_\upsilon \ln p_\upsilon \tag{2.125}$$

subject to the constraints given by Eqs. (2.122)–(2.124). The conditional maximum can be found by applying the method of Lagrange multipliers

$$-k_B(\ln p_\nu + 1) - \alpha_1 E_\nu - \alpha_2 V_\nu - \alpha_3 = 0 \tag{2.126}$$

where $\alpha_1, \alpha_2, \alpha_3$ can be determined from the normalization of probability and the thermodynamic relations

$$T^{-1} = (\partial S/\partial U)_V \tag{2.127}$$

and

$$P/T = (\partial S/\partial V)_U. \tag{2.128}$$

Following some algebraic rearrangements, we can find the probability of a microstate ν with energy E_ν and volume V_ν

$$p_\nu = \exp(-\beta E_\nu - \beta P V_\nu)/Y \tag{2.129}$$

where Y is called the isobaric–isothermal partition function

$$Y \equiv \sum_\nu \exp(-\beta E_\nu - \beta P V_\nu). \tag{2.130}$$

---

36 At equilibrium, there is no net transfer of energy, by neither heat nor work, between the system and its surroundings. The constraint of constant pressure is equivalent to that of an equilibrium system with a constant average volume.

37 Eq. (2.123) is problematic when the volume of a system is considered as a continuous variable. The issues were discussed by Koper G. J. M. and Reiss H., "Length scale for the constant pressure ensemble: application to small systems and relation to Einstein fluctuation theory", *J. Phys. Chem.* 100, 422 (1996).

2.7.2 Thermodynamic Functions at Constant Temperature and Pressure

The isobaric–isothermal partition function provides a convenient starting point to derive the thermodynamic properties of a closed system at constant temperature and pressure. From the probability distribution, Eq. (2.129), we obtain the system entropy

$$S = -k_B \sum_{\nu} p_\nu \ln p_\nu = \langle E_\nu \rangle / T + P\langle V_\nu \rangle / T + k_B \ln Y. \tag{2.131}$$

Following the definition of the Gibbs energy, $G \equiv U - TS + PV$, we can write entropy as

$$S = U/T + PV/T - G/T. \tag{2.132}$$

Because $U = \langle E \rangle$ and $V = \langle V_\nu \rangle$, a comparison of Eqs. (2.131) and (2.132) yields

$$G = -k_B T \ln Y. \tag{2.133}$$

Like the Helmholtz energy in a canonical ensemble, the Gibbs energy is naturally connected with the isobaric–isothermal partition function.

From the probability distribution of microstates in an isobaric–isothermal ensemble, we can also obtain the average energy and the average volume using Eqs. (2.122) and (2.123), respectively. These average quantities are related to the Gibbs energy by

$$U = -(\partial \ln Y / \partial \beta)_{N,\beta P} = (\partial \beta G / \partial \beta)_{N,\beta P}, \tag{2.134}$$

$$V = -(\partial \ln Y / \partial (\beta P))_{N,\beta} = (\partial G / \partial P)_{N,\beta}. \tag{2.135}$$

Because the partial derivative on the right side of Eq. (2.134) can be written as

$$(\partial \beta G / \partial \beta)_{N,\beta P} = (\partial \beta G / \partial \beta)_{N,P} - P(\partial G / \partial P)_{N,\beta}, \tag{2.136}$$

the enthalpy of the system is then given by

$$H \equiv U + PV = (\partial \beta G / \partial \beta)_{N,P} = -(\partial \ln Y / \partial \beta)_{N,P}. \tag{2.137}$$

Eq. (2.137) is consistent with the familiar Gibbs–Helmholtz equation from classical thermodynamics.

2.7.3 Enthalpy and Volume Fluctuations

In an isothermal–isobaric ensemble, both energy and volume fluctuate around their corresponding equilibrium values. Like that for the instantaneous total energy in a canonical ensemble, the fluctuation of the *instantaneous* enthalpy, $H_\nu \equiv E_\nu - PV_\nu$, is related to the second-order derivative of the partition function:

$$\langle (\delta H)^2 \rangle \equiv \left\langle (H_\nu - \langle H \rangle)^2 \right\rangle = \frac{1}{Y} \sum_{\nu} H_\nu^2 e^{-\beta H_\nu} - \left(\frac{1}{Y} \sum_{\nu} H_\nu e^{-\beta H_\nu} \right)^2 = \left(\frac{\partial^2 \ln Y}{\partial \beta^2} \right)_{P,N}. \tag{2.138}$$

Using Eq. (2.137) and the definition of the constant-pressure heat capacity $C_P \equiv (\partial H / \partial T)_P$, we have

$$\langle (\delta H)^2 \rangle = k_B T^2 C_P. \tag{2.139}$$

Eq. (2.139) indicates that the constant-pressure heat capacity reflects the enthalpy fluctuation in an isothermal–isobaric ensemble. Because the left side of Eq. (2.139) is a square average, it predicts that the constant-pressure heat capacity can never be negative. The non-negative value of heat capacity is required for thermodynamic stability.

In an isothermal–isobaric ensemble, the volume fluctuation is also related to the second-order derivative of the partition function:

$$\langle(\delta V)^2\rangle \equiv \langle(V - \langle V\rangle)^2\rangle = \frac{1}{Y}\sum_{\nu} V_{\nu}{}^2 e^{-\beta H_{\nu}} - \left(\frac{1}{Y}\sum_{\nu} V_{\nu} e^{-\beta H_{\nu}}\right)^2 = \left(\frac{\partial^2 \ln Y}{\partial(\beta P)^2}\right)_{\beta,N}. \tag{2.140}$$

Using Eq. (2.135) and the definition of the isothermal compressibility

$$\kappa \equiv -\frac{1}{V}\left(\frac{\partial V}{\partial P}\right)_{T,N} \tag{2.141}$$

we obtain

$$\langle(\delta V)^2\rangle = k_B T \kappa V. \tag{2.142}$$

Eq. (2.142) shows that the isothermal compressibility is related to the volume fluctuation in the isothermal–isobaric ensemble and, as required by thermodynamic stability, it is always non-negative.

Like the relative energy fluctuation in a canonical ensemble, the relative enthalpy and volume fluctuations in an isothermal–isobaric ensemble vanish when the system size approaches thermodynamic limit:

$$\langle(\delta H)^2\rangle^{1/2}/H = \sqrt{k_B T^2 C_P}/H \sim 1/\sqrt{N}, \tag{2.143}$$

$$\langle(\delta V)^2\rangle^{1/2}/V = \sqrt{k_B T\kappa/V} \sim 1/\sqrt{N}. \tag{2.144}$$

Eqs. (2.143) and (2.144) predict that the fluctuation effects are proportional to $1/\sqrt{N}$, which become less significant as the system size increases.

2.7.4 An Illustrative Example

Before closing this session, it is instructive to discuss a simple application of the isothermal–isobaric ensemble. Point defects refer to the localized anomalies of atomic dimensions that occur in an otherwise perfect crystalline lattice. As shown schematically in Figure 2.9, these anomalies include vacancies, misplaced atoms, or impurity atoms that are deliberately introduced (dopants) to control the electronic properties of a semiconductor or inadvertently incorporated (contaminants) during material growth and processing. Because point defects can have a strong influence on the physicochemical properties of a crystal, they are of great interest in practical applications. For example, the ionic conductivity of a solid electrolyte is dependent on vacant sites to facilitate ion motions.

Here, we apply the isobaric–isothermal ensemble to a particular kind of defect that arises when an atom is removed from a bulk lattice site to the surface of the crystal because of thermal fluctuations. The resulting vacancy is known as the Schottky defect,[38] which is common in crystals of alkali halides. We want to find the equilibrium vacancy concentration of a crystal at a given temperature and pressure.

Consider a crystalline lattice containing N atoms. Let v_0 be the volume of each unit cell and ε_0 be the energy related to the formation of a Schottky vacancy, i.e., the energy required to move an atom from its bulk lattice site to a site on the crystal surface. We assume that the change in the system

38 Named after Walter H. Schottky who played a major role in the early development of the theory of electron and ion emission phenomena.

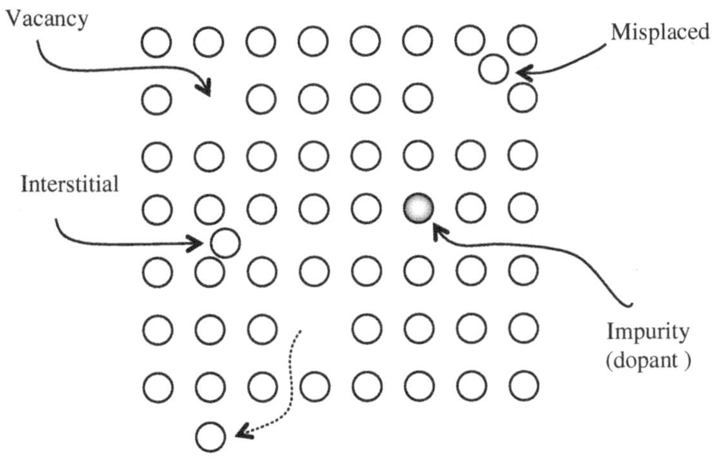

Figure 2.9 Various point defects in a solid crystal. As shown by the particle at the bottom, a Schottky defect (vacancy) is created when an atom is transferred from a bulk lattice site to the surface due to thermal fluctuations. An interstitial defect, also known as the Frenkel defect, is induced by displacement of an atom from its lattice position to an interstitial site.

volume is linearly proportional to the number of vacancies and that the Schottky vacancies do not interact with each other. With these assumptions, we may use a statistical thermodynamic model such that a microstate of the system is specified by the number and arrangements of vacancies in the lattice. At a given temperature and pressure, the distribution of microstates is determined by the isobaric–isothermal partition function

$$Y = \sum_{\nu} \exp(-\beta E_{\nu} - \beta P V_{\nu}) = \sum_{n_{0,\nu}=0}^{\infty} \frac{(N + n_{0,\nu})!}{N! n_{0,\nu}!} \exp[-\beta n_{0,\nu} \varepsilon_0 - \beta P v_0 (N + n_{0,\nu})]. \tag{2.145}$$

In Eq. (2.145), $n_{0,\nu}$ denotes the number of Schottky vacancies in microstate ν, $V_{\nu} = v_0(N + n_{0,\nu})$ is the system volume when it contains $n_{0,\nu}$ Schottky vacancies. The proportionality factor preceding the exponential accounts for the number of ways (degeneracy) we can arrange $n_{0,\nu}$ vacancies at $N + n_{0,\nu}$ lattice sites.

Using the mathematical identity

$$\sum_{k=0}^{\infty} \frac{(N + k)!}{N!k!} x^k = \frac{1}{(1 - x)^{N+1}} \tag{2.146}$$

where $x = \exp(-\beta\varepsilon_0 - \beta P v_0)$, we can evaluate the partition function analytically

$$Y = \frac{\exp(-\beta N P v_0)}{[1 - \exp(-\beta\varepsilon_0 - \beta P v_0)]^{N+1}}. \tag{2.147}$$

Therefore, the average number of vacancies in the crystal is

$$\langle n_0 \rangle = \sum_{\nu} p_{\nu} n_{0\nu} = \left(\frac{\partial \ln Y}{\partial \beta\varepsilon_0}\right)_{N,\beta P} = \frac{(N + 1)\exp(-\beta\varepsilon_0 - \beta P v_0)}{[1 - \exp(-\beta\varepsilon_0 - \beta P v_0)]^{N+2}} \tag{2.148}$$

where p_{ν} is the probability of the microstate in the isobaric–isothermal ensemble.

For example, a silicon crystal has a zincblende structure with lattice constant of 0.543 nm at room temperature. The volume of each vacancy $v_0 = 0.01\ \text{nm}^3$. For this crystal, the energy of formation of a Schottky defect $\varepsilon_0 = 2.6$ eV. At atmospheric pressure, Pv_0 is much smaller than ε_0. Because $N \gg 1$, the denominator in Eq. (2.148) is essentially unity. It follows that, at low or moderate pressure, the fraction of Schottky vacancies in a silicon crystal can be approximated by

$$\langle n_0 \rangle / N \approx \exp(-\beta \varepsilon_0). \tag{2.149}$$

At room temperature, 1 eV $\approx$ 40 $k_B T$. Eq. (2.149) predicts that the fraction of Schottky vacancies in the crystal lattice is about 10^{-44}, a tiny number. However, at higher temperature, the fraction of Schottky vacancies is about 10^{-13} for the same ε_0 at 1000 K. At high-pressure conditions, it is also necessary to consider the effect of Pv_0 in Eq. (2.148) because this term may become comparable to vacancy formation energy ε_0. For example, at 1000 K and 1000 bar, the fraction of Schottky vacancies in silicon crystal is 10^{-5}, which could have significant effect on the electronic properties of the crystal.

Interestingly, the result predicted from Eq. (2.149) is also expected from the Boltzmann distribution law. Why, then, do we need a statistical mechanical derivation for such a simple equation? We need it, first, to provide theoretical support for what we might have intuitively assumed and second, to tell us how the fraction of Schottky defects is influenced by both temperature and pressure.

2.7.5 Summary

An isobaric–isothermal ensemble is a statistical mechanical model that describes a system of particles with constant temperature and pressure. It is also known as the NPT ensemble, where N is the number of particles, P is the pressure, and T is the absolute temperature. This ensemble is useful for studying chemical reactions and phase transitions that occur under constant temperature and pressure. The partition function of the isobaric–isothermal ensemble can be derived from the maximum entropy principle with the constraints of constant average total energy and constant average volume. The fluctuations in total energy and system volume are related to thermodynamic stability.

2.8 Isobaric Molecular Dynamics

Like temperature, pressure is a macroscopic variable that reflects a statistical average of mechanical properties of individual particles (e.g., "instantaneous pressure" as defined by the virial equation). To apply MD simulation to a thermodynamic system at a fixed pressure, we need to modify the equations of motion such that they can generate microstates consistent with a preset value after the ensemble average. The isobaric condition can be satisfied by adopting mathematical strategies (viz., barostats) similar to those used for maintaining a constant temperature. In this section, we discuss two simple procedures to control pressure during MD simulation. In the Berendsen method, the system volume is adjusted during simulation to eliminate any difference between the instantaneous and preset pressures. In Andersen's method, we treat the system volume as a dynamic variable using an imaginary inertia and a pseudo-Newton's equation. A few variants of these methods can also be used in MD simulation to maintain a constant pressure. Interested readers are referred to specialized texts for a more in-depth discussion of this topic.[39]

---

39 For example, Tuckerman M. E., *Statistical mechanics: theory and simulation* Oxford University Press. Chapter 5, 2010.

2.8.1 The Berendsen Algorithm

For a uniform system of spherical particles, we may introduce an instantaneous pressure according to the virial equation (Appendix 2.A):

$$P_v = \frac{1}{3V_v}\sum_i (\mathbf{p}_i^2/m_i + \mathbf{r}_i \cdot \mathbf{F}_i) \tag{2.150}$$

where V_v stands for the instantaneous system volume, $\mathbf{r}_i$ and $\mathbf{p}_i$ denote particle position and momentum, and $\mathbf{F}_i$ represents the instantaneous force on particle i. The pressure of the system corresponds to the ensemble average of the instantaneous pressure, i.e., $P = \langle P_v \rangle$.

To maintain a constant pressure during MD simulation, we allow the system volume to fluctuate by rescaling the atomic positions. In principle, any difference between the instantaneous pressure and the target pressure can be eliminated by changing the system volume according to the compressibility equation:

$$\Delta V = -\kappa V \Delta P \tag{2.151}$$

where $\Delta V = V_v - V$, $\Delta P = P_v - P$, and κ is the isothermal compressibility. Numerically, we can implement the volume change by uniformly scaling the coordinates of all particles by a factor of λ:

$$\mathbf{r}_i' = \lambda \mathbf{r}_i. \tag{2.152}$$

Because the coordinate scaling corresponds to a volume change of $\Delta V = (\lambda^3 - 1)V$, the scaling factor can be solved from the discrepancy in pressure according to Eq. (2.151)

$$\lambda = (1 - \kappa \Delta P)^{1/3}. \tag{2.153}$$

In practice, the direct use of Eq. (2.153) is problematic because the isothermal compressibility is typically unknown before the simulation. Besides, like velocity scaling for temperature control, the coordinate scaling eliminates the fluctuation of the instantaneous pressure, resulting in a microstate distribution inconsistent with that from the isobaric ensemble. Such problems can be partially eliminated by using the so-called Berendsen's algorithm.[40] In this method, the instantaneous pressure is allowed to evolve according to a first-order relaxation equation

$$dP_v(t)/dt = [P - P_v(t)]/\tau_P \tag{2.154}$$

where τ_P is a coupling constant that dictates the relaxation rate. Over a duration of Δt, the difference between the instantaneous pressure and the target pressure can be calculated by integrating Eq. (2.154):

$$\Delta P = P_v(t) - P = [P_v(0) - P]e^{-\Delta t/\tau_P}. \tag{2.155}$$

Accordingly, the pressure difference can be adjusted by rescaling the particle coordinates as shown in Eq. (2.152). While the isothermal compressibility is still involved in determining the scaling factor, its inaccuracy has little consequence for the dynamics.

Although the Berendsen algorithm is not fully consistent with the isobaric constraint, it allows the system to reach equilibrium pressure efficiently. In practical applications, its computational efficiency depends on the judicious selection of coupling parameter τ_P and the frequency of coordinate rescaling as determined by Δt. Because Eq. (2.154) ensures that the instantaneous pressure eventually approaches the target value, the Berendsen method does not require a precise value for the isothermal compressibility of the system.

---

40 Berendsen H. J. C., et al. "Molecular dynamics with coupling to an external bath", *J. Chem. Phys.* 81 (8), 3684–3690 (1984).

2.8.2 Extended System Algorithms

In analogy to the mathematical procedure used in thermostat methods (e.g., the Nosé method), the extended system algorithm introduces an additional degree of freedom to describe the volume changes such that the instantaneous pressure approaches the target value in a statistical manner. This idea was first introduced by Andersen in the context of an extended Lagrangian formalism.[41] To incorporate volume as a dynamic variable in the equations of motion, Andersen proposed to express the particle coordinates in dimensionless form

$$\mathbf{x}_i \equiv V^{-1/3}\mathbf{r}_i. \tag{2.156}$$

Eq. (2.156) suggests that, in an isothermal–isobaric ensemble, the evolution of the particle position includes two contributions, one due to the particle motion as in a conventional equation of motion, and the other due to volume changes, i.e.,

$$\frac{d\mathbf{r}_i}{dt} = \frac{d(\mathbf{x}_i V^{1/3})}{dt} = V^{1/3}\frac{d\mathbf{x}_i}{dt} + \mathbf{x}_i\frac{V^{-2/3}}{3}\frac{dV}{dt}. \tag{2.157}$$

Without the change in volume, the first term on the right side of Eq. (2.157) can be identified as the particle velocity, and the coefficient in the second term can be written as $\mathbf{x}_i V^{1/3}/V = \mathbf{r}_i/V$. As a result, Eq. (2.157) may be rewritten as

$$\dot{\mathbf{r}}_i = \frac{\mathbf{p}_i}{m_i} + \frac{1}{3}\frac{\dot{V}}{V}\mathbf{r}_i. \tag{2.158}$$

Because the phase space volume is conserved, the change in particle coordinates leads to the following expression for the particle momentum

$$\boldsymbol{\pi}_i = V^{1/3}\mathbf{p}_i \tag{2.159}$$

where $\boldsymbol{\pi}_i = m d\mathbf{x}_i/dt$ is the momentum defined by the dimensionless coordinate $\mathbf{x}_i$, such that

$$\mathbf{x}_i \cdot \boldsymbol{\pi}_i = \mathbf{r}_i \cdot \mathbf{p}_i. \tag{2.160}$$

Accordingly, the differential equation for the particle momentum can be written as

$$\dot{\mathbf{p}}_i = \frac{d(\boldsymbol{\pi}_i V^{-1/3})}{dt} = V^{-1/3}\frac{d\boldsymbol{\pi}_i}{dt} - \frac{V^{-4/3}\boldsymbol{\pi}_i}{3}\frac{dV}{dt} = \mathbf{F}_i - \frac{1}{3}\frac{\dot{V}}{V}\mathbf{p}_i \tag{2.161}$$

where the first term on the right side corresponds to the regular force on particle i, and the second term is affiliated with the volume change. As expected, the equations of motion, Eqs. (2.158) and (2.161), reduce to those from Newton's equation when the system volume is fixed.

The essential idea of an extended-system method is to introduce additional degrees of freedom such that macroscopic constraints are enforced by an ensemble average. For a uniform system with constant pressure, the extra degrees of freedom are associated with the volume changes. Intuitively, the rate and acceleration may be described by an analogue of Newton's equation:

$$\ddot{V} = \frac{1}{W}\left[P_v(t) - P\right], \tag{2.162}$$

where W represents an "effective mass" for the volume change (viz., inertia), and the discrepancy between the instantaneous and target pressures serves as a driving force. Andersen demonstrated that the solution to Eqs. (2.161) and (2.162) produces trajectories consistent with the isobaric constraint.

41 Andersen H. C., "Molecular dynamics simulations at constant pressure and/or temperature", *J. Chem. Phys.* 72, 2384 (1980).

It is worth noting that the pressure fluctuation is, in general, much stronger than temperature fluctuation. As a result, the stability of simulation is more sensitive to the coupling strength (here, effective mass). The choice of coupling strength depends on various factors, including the nature of the system, the desired time scales, and the specific research objectives. It often requires careful optimization and testing to determine an appropriate coupling strength for a given system.

2.8.3 Summary

The Berendsen and Andersen barostat methods may serve complementary roles in MD simulation. The former is efficient in equilibrating the system to reach a preset pressure. However, it yields poor statistics for the probability distribution of the microstates. By contrast, Andersen's method ensures that the distribution of microstates is commensurate with an isobaric ensemble, but the instantaneous pressure exhibits slow oscillatory convergence. Thus, the combination of Berendsen and Andersen methods is often considered a good practice. We note in passing that Andersen's method can be extended to inhomogeneous systems wherein pressure is no longer a scalar.[42]

2.9 Grand Canonical Ensemble

The previous sections were concerned with closed systems where the number of particles is fixed but where the system's energy can fluctuate around its mean value. Many thermodynamic systems of practical concern, however, are in contact with an external environment that allows for the exchange of both mass and energy. Common examples of open systems include gas adsorption in porous materials, electrolytes near an electrode surface, and molecules near a substrate as in heterogeneous separation and reaction processes. In statistical mechanics, the most convenient way to define an open system is by specifying temperature (T), volume (V), and chemical potential μ_i for each chemical species i. A grand canonical ensemble consists of an arbitrarily large number of open systems with identical temperature, volume, and chemical potentials of chemical species.

A grand canonical ensemble encompasses all microstates of an open system. To find the microstate probability distribution and the corresponding thermodynamic properties at equilibrium, consider an M-component system at given T, V, and μ_i for each species. Chemical potential is an intensive thermodynamic variable that is associated with the reversible work to transfer molecules to the system from a reference state (e.g., an ideal gas). To help define the chemical potential of individual species, we may imagine that the system coexists with a large reservoir of a bulk system of the same chemical species, shown schematically in Figure 2.10. At equilibrium, the chemical potential of each species is uniform everywhere. For each species in the bulk phase, the chemical potential can be uniquely defined by temperature, pressure, and the chemical composition.

2.9.1 Grand Partition Function

Following a procedure similar to that for a closed system, we can determine the probability distribution for microstates in an open system using the second law of thermodynamics. Like that in a

42 Parrinello M. and Rahman A., "Crystal structure and pair potentials: a molecular-dynamics study", *Phys. Rev. Lett.* 45, 1196–1199 (1980); Nosé S. and Klein M. L., "Constant pressure molecular dynamics for molecular systems", *Mol. Phys.* 50, 1055–1076 (1983).

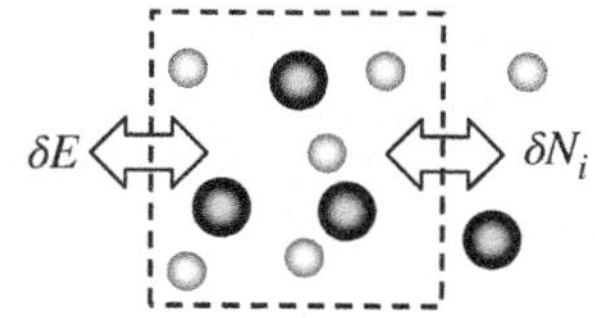

Figure 2.10 A schematic representation of mass and energy fluctuation in an open system with two types of particles. The box with vertical dashed lines indicates that particles are allowed to transfer through the boundary.

canonical or microcanonical ensemble, the entropy is defined by the microstate probabilities

$$S = -k_B \sum_{\nu} p_{\nu} \ln p_{\nu}. \tag{2.163}$$

For an open system at equilibrium, the entropy is maximized subject to the constraints of constant internal energy U, constant average number of particles for each chemical species N_i, and the normalization condition for the probability distribution

$$U = \sum_{\nu} p_{\nu} E_{\nu}, \tag{2.164}$$

$$N_i = \sum_{\nu} p_{\nu} N_{i,\nu}, \quad i = 1, 2, \ldots, M \tag{2.165}$$

$$\sum_{\nu} p_{\nu} = 1 \tag{2.166}$$

where p_{ν} is the probability for the system at microstate ν.

As discussed in the previous sections, the conditional maximum of entropy can be found by using the Lagrange multiplier method. Differentiation of the entropy with respect to the microstate probability subject to the three constraints yields

$$-(\ln p_{\nu} + 1) - \alpha - \beta E_{\nu} - \sum_{i} \gamma_i N_{i,\nu} = 0 \tag{2.167}$$

where α, β, and γ_i are the Lagrange multipliers arising, respectively, from the constraint for the internal energy, Eq. (2.164), the constraint for the average number of molecules for each chemical species in the system, Eq. (2.165), and the normalization condition, Eq. (2.166). Rearrangement of Eq. (2.167) gives

$$p_{\nu} = \exp\left[-\alpha - 1 - \beta E_{\nu} - \sum_{i} \gamma_i N_{i,\nu}\right]. \tag{2.168}$$

Substituting Eq. (2.168) into (2.163) yields

$$S/k_B = -\sum_{\nu} p_{\nu} \ln p_{\nu} = \alpha + 1 + \beta\langle E\rangle + \sum_{i} \gamma_i \langle N_i\rangle. \tag{2.169}$$

Using the thermodynamic relations

$$1/T = (\partial S/\partial U)_{V,N_i}, \tag{2.170}$$

$$\mu_i = -T(\partial S/\partial N_i)_{U,V,N_{j\neq i}}, \tag{2.171}$$

we can find the Lagrange multipliers by carrying out the partial derivatives of entropy with respect to internal energy and the number of particles:

$$\beta = 1/(k_B T), \tag{2.172}$$

$$\gamma_i = -\beta\mu_i. \tag{2.173}$$

Finally, the multiplier α is determined from the normalization condition, Eq. (2.166),

$$\alpha = \ln \Xi - 1 \tag{2.174}$$

where

$$\Xi \equiv \sum_{\nu} \exp\left[-\beta\left(E_{\nu} - \sum_{i} \mu_i N_{i,\nu}\right)\right] \tag{2.175}$$

is called the grand partition function. The final expression for the probability distribution is obtained by substituting Eqs. (2.173)–(2.175) into Eq. (2.168):

$$p_{\nu} = \frac{1}{\Xi} \exp\left[-\beta\left(E_{\nu} - \sum_{i} \mu_i N_{i,\nu}\right)\right]. \tag{2.176}$$

Eq. (2.176) indicates that, in the grand canonical ensemble, the microstate probability depends on the instantaneous energy and the number of particles in the system.

2.9.2 Grand Potential

All thermodynamic properties can be expressed in terms of either the microstate probability or the grand partition function. From p_{ν}, we can derive the entropy from the Gibbs equation, Eq. (2.163). Substituting p_{ν} with Eq. (2.176) leads to

$$S = U/T - \sum_{i} \mu_i \langle N_i \rangle / T - \Omega/T = (U - G - \Omega)/T \tag{2.177}$$

where

$$\Omega \equiv -k_B T \ln \Xi = U - G - TS. \tag{2.178}$$

The quantity defined in Eq. (2.178) is called the grand potential, a free energy equivalent to the Helmholtz energy for a closed system at constant temperature and volume. From the viewpoint of classical thermodynamics, the grand potential may be understood as a Legendre transformation of the Helmholtz energy such that it employs T, V, and μ_i as independent variables.

Like the Helmholtz energy for a canonical ensemble, the grand potential provides a good starting point to derive the thermodynamic properties of an open system. For a uniform system such as a bulk gas or liquid, the Legendre transformation from the Helmholtz energy to the grand potential yields a fundamental equation

$$d\Omega = -SdT - PdV - \sum_{i} N_i d\mu_i. \tag{2.179}$$

Eq. (2.179) provides a general link between the grand potential and other thermodynamic properties:

$$S = -\left(\frac{\partial \Omega}{\partial T}\right)_{V,\mu_i}, \tag{2.180}$$

$$P = -\left(\frac{\partial \Omega}{\partial V}\right)_{T,\mu_i}, \tag{2.181}$$

$$N_i = -\left(\frac{\partial \Omega}{\partial \mu_i}\right)_{V,T,\mu_{j\neq i}}. \tag{2.182}$$

From Eq. (2.178), we have $U - G - TS = -PV$ thus $\Omega = -PV$, indicating that, in a uniform system, pressure may be understood as the negative of the grand potential density (or the density of a thermodynamic potential). Substituting $\Omega = -PV$ into Eq. (2.179) leads to the familiar Gibbs–Duhem equation

$$0 = -SdT + VdP - \sum_i N_i d\mu_i. \tag{2.183}$$

Other thermodynamic properties can also be derived from the grand partition function. For example, we can express the internal energy and the average number of particles i in terms of the grand potential:

$$U = \sum_\nu E_\nu p_\nu = -(\partial \ln \Xi/\partial\beta)_{V,\beta\mu_{i=1,m}} = (\partial\beta\Omega/\partial\beta)_{V,\beta\mu_{i=1,m}}, \tag{2.184}$$

$$N_i = \sum_\nu N_{i,\nu} p_\nu = (\partial \ln \Xi/\partial\beta\mu_i)_{V,\beta} = -(\partial\Omega/\partial\mu_i)_{V,\beta}. \tag{2.185}$$

Eq. (2.184) is equivalent to the Gibbs–Helmholtz equation in classical thermodynamics. Unlike those derived from the fundamental equation, the relations between thermodynamic properties and the grand partition function are applicable to both uniform and inhomogeneous systems.

2.9.3 Mass and Energy Fluctuations

The grand partition function can be utilized to derive the mean-square deviations for the number of particles i and the total energy

$$\langle \delta N_i^2 \rangle \equiv \langle N_i^2 - \langle N_i \rangle^2 \rangle = (\partial \langle N_i \rangle/\partial\beta\mu_i)_{V,\beta,\mu_{j\neq i}}, \tag{2.186}$$

$$\langle \delta E^2 \rangle \equiv \langle E^2 - \langle E \rangle^2 \rangle = -(\partial U/\partial\beta)_{V,\beta\,\mu_{i=1,m}}. \tag{2.187}$$

Because $\beta = 1/(k_B T)$ is an intensive variable, Eq. (2.186) suggests that, in an open system, the relative deviation of the number of particles i, $\sqrt{\langle \delta N_i^2 \rangle}/\langle N_i \rangle$, scales as $1/\sqrt{N_i}$. For a typical thermodynamic system, the average number of particles of species i is on the order of $N_i \sim 10^{23}$. As a result, we expect that the fluctuation of the number of particles is extremely small (except at the critical point of phase transitions). Similarly, Eq. (2.187) suggests that the relative deviation of the total energy is extremely small for a macroscopic system.

Because the system volume is fixed in the grand canonical ensemble, the mean-square deviation for the number of particles i can be alternatively written as

$$\langle \delta N_i^2 \rangle = k_B T V (\partial \rho_i/\partial\mu_i)_{T,\mu_{j\neq i}} \tag{2.188}$$

where $\rho_i \equiv \langle N_i \rangle / V$ stands for the average number density of species i. As the left side of Eq. (2.188) is non-negative, it predicts that, in an open system at constant temperature and constant chemical potentials of all other species in the system, the average density of a chemical species always increases with its own chemical potential. Like non-negative heat capacity or non-negative isothermal compressibility for equilibrium systems, the non-negative variation of the number density with respect to the chemical potential is another requirement of thermodynamic stability.

For a one-component system, Eq. (2.188) can be written as

$$\frac{\langle \delta N^2 \rangle}{\langle N \rangle^2} = \frac{k_B T V}{\langle N \rangle^2}(\partial\rho/\partial\mu)_T = \frac{k_B T V}{\langle N \rangle^2}\left(\frac{\partial \rho}{\partial P}\right)_T \left(\frac{\partial P}{\partial \mu}\right)_T = \frac{k_B T}{V}\kappa_T \tag{2.189}$$

where $\kappa_T \equiv (\partial\rho/\partial P)_T/\rho$ is the isothermal compressibility. Eq. (2.189) is known as the fluctuation-compressibility theorem, which is often used in studying the relation between fluctuation and thermodynamic behavior near the critical point of phase transition.

2.9.4 Summary

This section introduces the grand partition function, Eq. (2.175), for describing the probability distribution of microstates in an open system, Eq. (2.176). In addition, we introduce grand potential Ω as a free energy convenient for the application of thermodynamics to open systems. For a system at a fixed temperature, chemical potential, and total volume, it follows from Eq. (2.177) that a maximum entropy subject to the macroscopic constraints corresponds to a minimum in the grand potential. In other words, the grand potential is minimized for an open system at equilibrium, which may be understood as another statement of the second law of thermodynamics.

2.10 Transformation Between Ensembles

Statistical ensembles are mental copies of a thermodynamic system that can be specified by different sets of macroscopic variables, i.e., constraints on microstate distributions. We expect that, from a macroscopic viewpoint, different ensembles are equivalent in terms of their predictions of thermodynamic properties. However, different ensembles are distinguishable from a microscopic perspective, as manifested in different microstate distributions. In particular, for small systems (viz., systems having size comparable to molecular dimensionalities), thermodynamic properties may depend on the choice of a specific ensemble. This is because different microstate distributions lead to different fluctuations of thermodynamic quantities. Different ensembles are equivalent only in the thermodynamic limit, i.e., when the system size is infinitely large and the boundary effects are negligible.

In this section, we first elucidate how the concept of ensemble can be generalized by following the Gibbs phase rule. Next, we discuss the connections between different ensembles in the context of the Legendre transformation and demonstrate their equivalency in the thermodynamic limit. The ensemble dependence of mean quantities and the fluctuation effects play a significant role in the thermodynamics of small systems and have practical implications for the applications of molecular simulation methods.

2.10.1 Thermodynamic Constraints

The Gibbs phase rule asserts that, to specify a thermodynamic system at equilibrium, the number of intensive variables depends on the number of chemical species, the number of coexisting phases, and the number of independent chemical reactions. Besides, one additional variable is needed to define the system size. Once these variables are given, the thermodynamic system is well defined from a macroscopic perspective. From a microscopic perspective, however, different ways to specify a thermodynamic system macroscopically lead to different microstate distributions and, subsequently, different ensembles.

The equilibrium distribution of microstates is determined by maximizing the Gibbs entropy subject to a specific set of constraints associated with the thermodynamic variables to specify a macroscopic system. For example, in a microcanonical ensemble, the only constraint on the microstate probability is the normalization condition. In this case, all microstates have identical *a priori* probabilities. In a canonical ensemble, an additional constraint is applied to the system energy, leading to the Boltzmann distribution of microstates, and a grand canonical ensemble imposes one more constraint related to the average number of particles. Other ways are also possible to define a thermodynamic system, such as constant temperature, pressure, and number of particles (viz., NPT ensemble), or constant temperature, volume, number of solute molecules, and the chemical potential of the solvent (viz., semi-grand canonical ensemble). Indeed, the number of ways to define a thermodynamic system is numerous, resulting in many kinds of ensembles!

2.10.2 Conjugate Ensembles

From a macroscopic perspective, thermodynamic properties predicted by different ensembles are expected to be equivalent. To a certain degree, different ways to specify the constraints of a thermodynamic system are analogous to different independent variables used in Legendre transformations, i.e., the switching of independent thermodynamic variables between the conjugate pairs as appeared in the fundamental equation of thermodynamics. For example, the microcanonical and canonical ensembles are linked by replacing the total energy E with its conjugate variable, $\beta = 1/(k_B T)$ or T, as the macroscopic constraint. The change of independent variables corresponds to the Legendre transformation from entropy, $S = S(E, V, N)$, to the Helmholtz energy, $F = F(T, V, N)$

$$\beta F = -S/k_B + \beta U. \tag{2.190}$$

While the Legendre transformation converts the thermodynamic potential, it also transforms the microcanonical to canonical ensemble in the statistical representation of a thermodynamic system.

To unveil the implications of the Legendre transformation from a microscopic perspective, consider in general an ensemble for a thermodynamic system specified by an extensive property X, which is in dimensionless form for convenience, and a few other thermodynamic variables that are required to define the equilibrium state. Let $\Phi(X)$ be the free energy, also in proper dimensionless form, when the set of thermodynamic properties, including X are used as independent variables. For simplicity, the unspecified thermodynamic variables are not shown in function $\Phi(X)$ because they will not be altered in our discussion.

The Legendre transformation from variable X to its conjugate thermodynamic variable ξ, which is intensive because of the conjugate relation, results in a different free energy,[43]

$$\Psi \equiv \Phi + \xi X \tag{2.191}$$

and a new ensemble with ξ being fixed while X is allowed to fluctuate. According to the fundamental equation of thermodynamics, the conjugate variables and the corresponding free energies are related by

$$\xi = -\left(\frac{\partial \Phi}{\partial X}\right), \tag{2.192}$$

$$X = \left(\frac{\partial \Psi}{\partial \xi}\right). \tag{2.193}$$

Again, all variables are dimensionless. It is also understood that some other thermodynamic variables are fixed in the partial derivatives.

Let p_ν^X be the probability of microstate ν in the ensemble with X fixed as a constant. In a general way, we can write p_ν^X in terms of

$$p_\nu^X = \frac{W_\nu^X}{W_X} \tag{2.194}$$

---

43 We adopt a positive sign in Eq. (2.191) so that the Legendre transformations starting from entropy $-S/k_B$ correspond to the addition of constraints when the microcanonical ensemble is transformed into other ensembles.

where W_ν^X is the weight function of microstate ν in the ensemble with fixed X, and $W_X = \sum_\nu W_\nu^X$ is the partition function. The latter is associated with free energy $\Phi(X)$ according to

$$\Phi(X) = -\ln W_X. \tag{2.195}$$

Note that, for the microcanonical ensemble, $\Phi = -S/k_B$. Therefore, any difference between $\Phi(X)$ and $-S/k_B$ can be attributed to the Legendre transformations of entropy, or equivalently, additional constraints on the distribution of microstates introduced in defining the thermodynamic system in a macroscopic way (e.g., $\beta F = -S/k_B + \beta U$ arises from the additional constraint of a constant temperature or β).

In the ensemble with ξ being fixed, the probability density of the system at the same microstate but having a macroscopic quantity X is then given by

$$p_\nu(X) = p(X) p_\nu^X \tag{2.196}$$

where $p(X)$ is the probability density of the system with X. We may find $p(X)$ by maximizing the Gibbs entropy subject to all constraints *inherited* from the original ensemble, plus the normalization condition

$$\int dXp(X) = 1, \tag{2.197}$$

as well as a constant mean value of X

$$\int dXp(X)X = \langle X \rangle_\xi \tag{2.198}$$

where $\langle \cdots \rangle_\xi$ denotes an average in the ensemble with ξ being fixed. Using the Lagrange multiplier method, we obtain

$$p(X) = W_X e^{-\lambda X}/\alpha \tag{2.199}$$

where λ is the Lagrange multiplier affiliated with Eq. (2.198), and α is a normalization constant that can be determined from Eq. (2.197)

$$\alpha = \int dXW_X e^{-\lambda X}. \tag{2.200}$$

Substituting Eqs. (2.194) and (2.199) into (2.196) gives

$$p_\nu(X) = W_\nu^X e^{-\lambda X}/\alpha. \tag{2.201}$$

Accordingly, the probability of the system being in microstate ν is

$$p_\nu^\xi = \int dXp_\nu(X) = \int dXW_\nu^X e^{-\lambda X}/\alpha \equiv \frac{W_\nu^\xi}{W_\xi} \tag{2.202}$$

where

$$W_\nu^\xi \equiv \int dXW_\nu^X e^{-\lambda X}, \tag{2.203}$$

$$W_\xi \equiv \sum_\nu W_\nu^\xi = \int dXW_X e^{-\lambda X} = \alpha. \tag{2.204}$$

Eq. (2.202) suggests that W_ν^ξ represents the weight function of microstate ν for the ensemble with ξ being fixed, and W_ξ is the corresponding the partition function. The latter defines a dimensionless free energy corresponding to the Legendre transformation from X to ξ,

$$\Psi = -\ln W_\xi. \tag{2.205}$$

From Eqs. (2.203)–(2.205), we have

$$\frac{\partial \Psi}{\partial \lambda} = -\frac{1}{W_\xi}\frac{\partial}{\partial \lambda}\sum_\nu \int dXW_\nu^X e^{-\lambda X} = \int dXp(X)X = \langle X \rangle_\xi. \tag{2.206}$$

A comparison of Eq. (2.206) with the classical thermodynamic relation, Eq. (2.193), indicates $\lambda = \xi$, i.e., the Lagrange multiplier is related to the intensive thermodynamic variable in conjugate with extensive variable X. Accordingly, Eq. (2.200) can be rewritten as

$$W_\xi = \int dXW_X e^{-\xi X}. \tag{2.207}$$

Eq. (2.207) provides an exact relation between the partition functions of conjugate ensembles. The integrand is maximized when

$$d(W_X e^{-\xi X}) = W_X e^{-\xi X}(d\ln W_X - \xi dX) = 0. \tag{2.208}$$

From Eqs. (2.195) and (2.192), we have

$$d\ln W_X/dX = -d\Phi/dX = \xi \tag{2.209}$$

which indicates that, according to Eq.(2.208), the maximum contribution to W_ξ comes from the average value of $X = \langle X \rangle_\xi$. As to be discussed below, in the thermodynamic limit, the fluctuation of X makes negligible contributions to the ensemble average of any dynamic quantity. As a result, the integration in Eq. (2.207) may be approximated only by the maximum term

$$W_\xi \approx W_X e^{-\xi X}. \tag{2.210}$$

Interestingly, Eq. (2.210) reproduces the relation between free energies Φ and Ψ, Eq. (2.191), as predicted by the Legendre transformation. Therefore, the *maximum-term approximation* becomes exact in the thermodynamic limit! Indeed, it can be shown that the probability density of the system to have extensive variable X can be approximated by the Gaussian distribution

$$p(X) = \frac{1}{\sqrt{2\pi\sigma_X^2}}\exp\left[-\frac{(X-\overline{X})^2}{2\sigma_X^2}\right] \tag{2.211}$$

where $\sigma_X^2 \equiv \langle (X-\overline{X})^2 \rangle_\xi$. Because both X and σ_X^2 scale with the system size ($\sim N$), Eq. (2.211) indicates that the relative width of the Gaussian distribution diminishes as N increases. In the thermodynamic limit ($N \to \infty$), $p(X) \to \delta(X-\overline{X})$, the Dirac delta function.

From an experimental perspective, the probability of observing any appreciable deviation of X from the mean value is negligibly small for macroscopic systems. In other words, a macroscopic system specified by a constant value of X, along with other independent variables fixed, is thermodynamically equivalent to that specified by its corresponding conjugate variable ξ.

2.10.3 Ensemble Dependence of Statistical Quantities

Armed with the connection between different ensembles through the Legendre transformation, we are in a position to analyze dynamic quantities predicted by different ensembles. Here, we follow a procedure reported by Lebowitz et al.[44]

44 Lebowitz J. L., Percus J. K., and Verlet L., "Ensemble dependence of fluctuations with application to machine computations", *Phys. Rev.* 250–254, 1673 (1967).

Suppose that A is an arbitrary dynamic quantity of a macroscopic system that is specified by two conjugate ensembles, one with constant X, an extensive variable, and the other with constant ξ, an intensive variable conjugate to X. In general, other thermodynamic variables are needed to fully specify the system, but they are irrelevant to our discussion. For convenience, we refer to the two ensembles as ensemble-X and ensemble-ξ, respectively.

In ensemble-X, the average of A is given by

$$\langle A \rangle_X = \sum_\nu A_\nu W_\nu^X / W_X \tag{2.212}$$

where A_ν denotes the instantaneous value of A in microstate ν, which depends only on microscopic variables. When the ensemble is transformed due to switching the fixed variable from X to ξ (with all other variables unchanged), the ensemble average of A becomes

$$\langle A \rangle_\xi = \sum_\nu A_\nu W_\nu^\xi / W_\xi. \tag{2.213}$$

In Eqs. (2.212) and (2.213), the weight functions and partition functions are related to each other through Eqs. (2.203) and (2.204).

To find out the difference between $\langle A \rangle_\xi$ and $\langle A \rangle_X$, we consider the dependence of $\langle A \rangle_X$ on X by taking a Taylor expansion

$$\langle A \rangle_X = \langle A \rangle_{\overline{X}} + \Delta X \left. \frac{\partial \langle A \rangle_X}{\partial X} \right|_{X=\overline{X}} + \frac{1}{2} \Delta X^2 \left. \frac{\partial^2 \langle A \rangle_X}{\partial X^2} \right|_{X=\overline{X}} + \cdots. \tag{2.214}$$

where $\overline{X} = \langle X \rangle_\xi$ is the expected value of X in ensemble-ξ, and $\Delta \overline{X} = X - \overline{X}$. Applying ensemble average on both sides of Eq. (2.214) leads to

$$\langle A \rangle_\xi = \langle A \rangle_{\overline{X}} + \frac{1}{2} \langle \Delta X^2 \rangle_\xi \left. \frac{\partial^2 \langle A \rangle_X}{\partial X^2} \right|_{X=\overline{X}} + \cdots. \tag{2.215}$$

In Eq. (2.215), the linear terms vanish because $\langle \Delta X \rangle_\xi = 0$, and the mean-square fluctuation is given by

$$\langle \Delta X^2 \rangle_\xi = -\frac{\partial \overline{X}}{\partial \xi}. \tag{2.216}$$

The relation between $\langle A \rangle_\xi$ and $\langle A \rangle_X$ is more informative in its inverse form

$$\begin{aligned} \langle A \rangle_{\overline{X}} &= \langle A \rangle_\xi - \frac{1}{2} \langle \Delta X^2 \rangle_\xi \frac{\partial^2}{\partial \overline{X}^2} \left\{ \langle A \rangle_\xi - \frac{1}{2} \langle \Delta X^2 \rangle_\xi \frac{\partial^2}{\partial \overline{X}^2} \{\cdots\} \right\} + \cdots \\ &= \langle A \rangle_\xi + \frac{1}{2} \frac{\partial \overline{X}}{\partial \xi} \frac{\partial^2 \langle A \rangle_\xi}{\partial \overline{X}^2} + \cdots = \langle A \rangle_\xi + \frac{1}{2} \frac{\partial^2 \langle A \rangle_\xi}{\partial \xi \partial \overline{X}} + \cdots. \end{aligned} \tag{2.217}$$

Because both $\langle A \rangle_\xi$ and $\overline{X}$ are extensive variables and ξ is intensive, Eq. (2.217) indicates that the leading order in the difference between $\langle A \rangle_\xi$ and $\langle A \rangle_X$ is independent of the system size. In the thermodynamic limit, the extensive properties approach infinity while the difference between the averages of any dynamic quantity in the conjugate ensembles remains finite. As a result, this difference becomes negligible (except under special conditions when the fluctuation term diverges). In analogous to Legendre transformations, different ensembles are formally related by the variations of the conjugate variables. Therefore, in the thermodynamic limit, different ensembles are equivalent for predicting the mean values of dynamic quantities.

We may also apply a similar procedure to find the ensemble dependence of the fluctuations of arbitrary dynamic variables. Suppose that A and B are two dynamic quantities of the system discussed above. Their correlated fluctuations in ensembles ξ and X are given by

$$\langle \delta A \delta B \rangle_X \equiv \langle AB \rangle_X - \langle A \rangle_X \langle B \rangle_X, \tag{2.218}$$

$$\langle \delta A \delta B \rangle_\xi \equiv \langle AB \rangle_\xi - \langle A \rangle_\xi \langle B \rangle_\xi. \tag{2.219}$$

According to Eq. (2.217), $\langle AB \rangle_X$ and $\langle A \rangle_X \langle B \rangle_X$ can be expressed to the leading order as

$$\langle AB \rangle_X = \langle AB \rangle_\xi + \frac{1}{2} \frac{\partial^2 \langle AB \rangle_\xi}{\partial \xi \partial \overline{X}} + \cdots. \tag{2.220}$$

$$\langle A \rangle_X \langle B \rangle_X = \langle A \rangle_\xi \langle B \rangle_\xi + \frac{1}{2} \langle A \rangle_\xi \frac{\partial^2 \langle B \rangle_\xi}{\partial \xi \partial \overline{X}} + \frac{1}{2} \langle B \rangle_\xi \frac{\partial^2 \langle A \rangle_\xi}{\partial \xi \partial \overline{X}} + \cdots. \tag{2.221}$$

Substituting Eqs. (2.220) and (2.221) leads to

$$\langle \delta A \delta B \rangle_X \equiv \langle \delta A \delta B \rangle_\xi + \frac{\partial \xi}{\partial \overline{X}} \frac{\partial \langle A \rangle_X}{\partial \xi} \frac{\partial \langle B \rangle_X}{\partial \xi} + \cdots. \tag{2.222}$$

Note that the leading term in Eq. (2.222) scales linearly with the system size, which is the same as the mean-square fluctuations of the dynamic quantities in different ensembles. As a result, the fluctuations of dynamic variables are always ensemble-dependent.

2.10.4 Summary

In closing, transformations between different ensembles may be understood as an extension of Legendre transformations in classical thermodynamics. Whereas thermodynamic potentials are considered equivalent when an independent variable is replaced by its conjugate counterpart, the variable change leads to a separate set of macroscopic constraints and thus a different distribution of microstates. As a result, different ensembles predict different fluctuations in dynamic quantities. The average properties predicted by different ensembles are equivalent only when the system size is infinitely large. This is, of course, never the case in molecular simulation. Eqs. (2.217) and (2.222) provide the leading order corrections for the conversions of numerical results from different ensembles.

2.11 Generalized Ensembles

In statistical mechanics, ensembles are introduced to analyze the probability distribution of microstates so that the properties of a macroscopic system can be predicted from the microscopic behavior of individual particles. Traditionally, the statistical analysis is accomplished mostly by using various mean-field approximations or correlation functions to describe the fluctuations of different dynamic quantities. Modern applications of statistical mechanics are often based on simulation methods that provide numerical procedures for the systematic sampling of microstates according to certain statistical distributions. Toward that end, Monte Carlo (MC) and molecular dynamics (MD) methods are frequently used to generate microstates by the progressive changes of the dynamic variables of individual particles following deterministic or stochastic equations of motion. In principle, simulation methods can generate microstates in any ensemble. However, a perennial problem is computational efficiency, which often deteriorates for complex systems where the simulation trajectory can easily be trapped into so-called local-minimum-energy states (viz., long-lived, or metastable states). In this section, we discuss some basic strategies that are commonly adopted to enhance sampling in the context of generalized ensembles. Some of these strategies will be discussed in more detail in Section 5.12 within the context of MC simulation. More extensive discussions of this important topic are available in the literature.[45]

2.11.1 Expanded Ensemble

While a regular ensemble is concerned with a single thermodynamic state, an expanded ensemble deals with microstates over a range of thermodynamic conditions. The original idea was proposed for more efficient sampling of microstates at low temperatures with Monte Carlo simulation.[46] In a canonical ensemble, the probability of a system at low-energy microstates increases as temperature falls. To avoid being trapped in or excessively sampling low-energy states, the expanded ensemble is devised to generate microstates over a range of temperatures such that microstates can be uniformly sampled in the entire energy space. The concept is akin to that adopted by multicanonical ensemble.[47] These methods have profound implications for later developments of virtually all enhanced sampling methods (Section 5.12).

To make the above ideas concrete, consider a thermodynamic system containing N particles in volume V over a range of temperatures, $T_1, T_2, \ldots, T_M$. At each temperature, we have a canonical partition function

$$Q_m = \sum_\nu e^{-\beta_m E_\nu} \tag{2.223}$$

where $\beta_m = 1/(k_B T_m)$. In the expanded ensemble, the partition function is defined in terms of the microstates at all temperatures with a weighted probability:

$$Q \equiv \sum_m Q_m e^{q_m} \tag{2.224}$$

where q_m is a weight factor to be determined by simulation such that microstates at different temperatures can be sampled with equal probability.

According to Eq. (2.224), the probability of the system at any specific temperature is

$$p_m = Q_m e^{q_m}/Q. \tag{2.225}$$

To ensure a uniform distribution of microstates, an obvious choice of the weight factor is $q_m = -\ln Q_m = \beta_m F_m$ such that $p_m = 1/Q$. In other words, the weight factor is related to the Helmholtz energy F_m for the system at temperature T_m.

---

45 For example, Sugita Y., Mitsutake A. and Okamoto Y., "Generalized-ensemble algorithms for protein folding simulations", in *Lecture notes in physics*, Edited by W. Janke, Springer-Verlag, Berlin, 2008, 369–407; Berg B. A., "A brief history of the introduction of generalized ensembles to Markov chain Monte Carlo simulations", *Eur. Phys. J.* 226, 551–565 (2017).

46 Lyubartsev A. P., et al., "New approach to Monte Carlo calculation of the free energy: method of expanded ensembles", *J. Chem. Phys.*, 96, 1776 (1992).

47 Berg B. A. and Neuhaus T., "Multicanonical algorithms for first order phase transitions", *Phys. Lett. B* 267, 249–253 (1991).

In general, the reduced Helmholtz energy, $\beta_m F_m$, is unknown before the simulation. However, the problem may be circumvented by implementing an iterative procedure with a reasonable initial guess for each q_m. According to Eq. (2.225), the probability ratio between different energy states is given by

$$\frac{p_m}{p_k} = \frac{Q_m}{Q_k} e^{q_m - q_k} = \exp(-\beta_m F_m + \beta_k F_k + q_m - q_k). \tag{2.226}$$

Eq. (2.226) allows for the estimation of the relative reduced Helmholtz energy from the simulation results, i.e., $\beta_m F_m$ can be determined from the probabilities p_m and p_k calculated from simulation relative to that corresponding to an arbitrarily pre-selected reference state $\beta_k F_k$ (e.g., the state with the highest temperature). The approximate free energy provides a new estimate, for q_m and procedure is iterated until satisfactory results are obtained for both the non-Boltzmann distribution of microstates and the relative Helmholtz energies.[48]

2.11.2 Multicanonical Ensemble

Like the expanded ensemble methods, a generalized ensemble is referred to as a broad class of simulation methods that intend to sample microstates without being trapped in local-minimum-energy states. The essential idea is to generate a uniform distribution of microstates over a range of thermodynamic conditions of practical interest. The procedure can also be utilized to calculate thermodynamic properties, including both mean variables and free energy, through the density of states.

To elucidate the idea of sampling microstates with equal probability, consider the canonical partition function

$$Q \equiv \sum_{\nu} e^{-\beta E_\nu} = \int dE w(E) e^{-\beta E} \tag{2.227}$$

where $w(E)$ stands for the density of states. As discussed in the previous section, the probability density for the system to have energy E follows the Gaussian distribution

$$p(E) = w(E)e^{-\beta E} \approx \frac{1}{\sqrt{2\pi\sigma_E^2}} \exp\left(-\frac{(E - \overline{E})^2}{2\sigma_E^2}\right) \tag{2.228}$$

where $\overline{E} = \langle E \rangle$ stands for the mean energy (viz., internal energy), and $\sigma_E^2 = \langle (E - \overline{E})^2 \rangle$ is the mean-square deviation of the total energy. In the thermodynamic limit, microstates in the canonical ensemble are narrowly distributed around the mean energy.

In the multicanonical ensemble (MU)[49] – one of the most popular and earliest generalized ensemble methods, microstates are sampled according to a uniform probability distribution

$$p_{MU}(E) \sim w(E) w_{MU}(E) = \text{constant} \tag{2.229}$$

where $w_{MU}(E)$ denotes a weight factor. Like that used in the enhanced ensemble, the weight factor is unknown *a priori* but can be iteratively calculated from simulation. According to Eq. (2.228), at any given temperature, the probability density is given by

$$p(E, T) \sim \frac{p_{MU}(E)}{w_{MU}(E)} e^{-\beta E}. \tag{2.230}$$

---

48 The iterative procedure to determine the weight factor is also known as simulated tempering. Marinari E. and Parisi G., "Simulated tempering: a new Monte Carlo scheme", *Europhys. Lett.* 19, 451 (1992).

49 Berg B. and Neuhaus T., "Multicanonical ensemble: a new approach to simulate first-order phase transitions", *Phys. Rev. Lett.* 68 (1), 9–12 (1992).

With $p_{MU}(E)/w_{MU}(E)$ computed from other techniques like Wang–Landau sampling (Section 5.12), Eq. (2.230) can be used to predict the ensemble averages for the system over a range of temperatures (viz., multicanonical)

$$\langle M \rangle_T = \frac{\int dEM(E)p(E,T)}{\int dEp(E,T)} \tag{2.231}$$

where $M(E)$ represents any dynamic quantity. In simulation, the latter can be calculated from the corresponding values of the quantity at microstates with energy E.

2.11.3 Multidimensional Generalized Ensembles

Armed with the analogy between transforming among different ensembles and Legendre transformations, it is straightforward to extend the multicanonical ensemble methods to multidimensional generalized ensembles (MG). In a generic way, we may write the partition function for an arbitrary ensemble

$$Z \equiv \sum_{\nu} \exp\left[-\sum_k \xi_k X_{k,\nu}\right] \tag{2.232}$$

where ξ_k is an intensive variable that is held as a constant for the thermodynamic system under consideration, and X_k is an extensive variable conjugate with ξ_k. For simplicity, both ξ_k and X_k are in dimensionless forms.

In terms of a multidimensional density of states $w(\mathbf{X})$, Eq. (2.232) can be rewritten as

$$Z = \int d\mathbf{X}\, w(\mathbf{X}) \exp[-\mathfrak{I} \cdot \mathbf{X}] \tag{2.233}$$

where $\mathfrak{I} \equiv (\xi_1, \xi_1, \ldots)$ and $\mathbf{X} \equiv (X_1, X_2, \ldots)$ are short notations for the conjugate pairs, and $\mathfrak{I} \cdot \mathbf{X}$ stands for a generalized energy function. If microstates can be generated with a uniform probability distribution, as in the microcanonical ensemble, the probability density satisfies

$$p_{MG}(\mathbf{X}) \sim w(\mathbf{X}) w_{MG}(\mathbf{X}) = \text{constant} \tag{2.234}$$

where $w_{MG}(\mathbf{X})$ is a weight function to be determined by simulation. Over a range of thermodynamic conditions, the probability density can be calculated from

$$p(\mathfrak{I}, \mathbf{X}) \sim \frac{p_{MG}(\mathbf{X})}{w_{MG}(\mathbf{X})} e^{-\mathfrak{I}\cdot\mathbf{X}}. \tag{2.235}$$

Accordingly, any physical quantity can be calculated from the ensemble average

$$\langle M \rangle_T = \frac{\int d\mathbf{X} M(\mathbf{X}) p(\mathfrak{I}, \mathbf{X})}{\int d\mathbf{X} p(\mathfrak{I}, \mathbf{X})}. \tag{2.236}$$

Like the multicanonical ensemble method, the weight factor can be found from either molecular dynamics or Monte Carlo simulations.[50]

A similar procedure can be adopted to generalize the expanded ensemble method. In comparison with the one-dimensional generalized ensemble, which extends the thermodynamic condition by changing only a single thermodynamic variable, the multidimensional generalized ensemble method accounts for variations of all intensive variables to define a macroscopic system. Consequently, this approach enables sampling microstates over a broader range of the parameter space. Understandably, the numerical procedure is also more complicated due to the increase in dimensionality.

50 Mitsutake A. and Okamoto Y., "Multidimensional generalized-ensemble algorithms for complex systems", *J. Chem. Phys.* 130, 214105 (2009); Chodera J. D. and Shirts M. R., "Replica exchange and expanded ensemble simulations as Gibbs sampling: simple improvements for enhanced mixing", *J. Chem. Phys.* 135, 194110 (2011).

2.11.4 Summary

Generalized ensemble refers to a broad variety of strategies in molecular simulation to enhance microstate sampling. It extends the notion of a conventional ensemble by accounting for the microstate distribution over a range of thermodynamic conditions. The numerical procedures are applicable to both molecular dynamics and Monte Carlo simulations. In comparison with regular simulation methods, simulation in the generalized ensemble allows the system to avoid barriers between low-energy states, thus enhancing sampling efficiency.

2.12 Chapter Summary

Statistical mechanics describes the microstates of a thermodynamic system in the context of ensembles. A key hypothesis is ergodicity, which asserts the equivalence of the expected values of dynamic quantities derived from the statistical distribution of microstates and the corresponding time averages of the system under investigation. Another fundamental hypothesis of statistical mechanics is the principle of equal *a priori* probabilities. For equilibrium systems, the latter is fully consistent with the second law of thermodynamics. With these two hypotheses, statistical mechanics allows us to derive microstate distributions and predict macroscopic properties based on the dynamic behavior of individual particles. In particular, the fluctuation–dissipation theorem also allows us to predict the kinetic and transport coefficients that are used in phenomenological models of time-dependent processes.

An ensemble refers to the mental copies of a particular macroscopic system under investigation; it entails all possible microstates of a thermodynamic system. In analogy to the different forms of free energy and the fundamental equation of thermodynamics introduced by the Legendre transformations, different ensembles can be constructed for thermodynamic systems specified by different sets of macroscopic variables. Accordingly, the thermodynamic potentials in various ensembles and the associated partition functions are intrinsically related by the corresponding Legendre transformations. Whereas all ensembles are equivalent in terms of the extensive thermodynamic variables of macroscopic systems, they generate different distributions of microstates, leading to variations in the fluctuations of dynamic variables. As the number of macroscopic ways to prepare a thermodynamic system is numerous, a wide range of ensembles can be constructed accordingly; microcanonical, canonical, and grand canonical ensembles are only a few common examples. Which ensemble shall we use in practical applications? The answer depends on how we specify an equilibrium system and what specific properties are of practical interest. That specification, in turn, depends on what information is available concerning the system under consideration.

Molecular dynamics provides a powerful procedure to predict the properties of matter either from first principles or from various forms of molecular models. Mathematically, the procedure is exact and applicable to the computation of both thermodynamic and transport properties of equilibrium systems. However, the atomistic approach is not directly applicable to macroscopic systems or their interactions with an environment. In molecular simulation, the environmental effects are implemented through various mathematical strategies (e.g., thermostats and barostats) to modify the equations of motion such that the ensemble distributions of microstates can be faithfully reproduced. To make molecular simulation computationally more efficient, generalized ensemble methods provide numerical strategies for efficiently sampling a broad range of microstates without being trapped in local energy minima.

2.A Virial Theorem and Virial Equation

The virial theorem was introduced by Rudolf Clausius in 1870. It asserts that, for a system containing a large number of particles, the average total kinetic energy K is linearly proportional to the summation of the average particle position multiplied by particle force

$$\overline{K} = -\frac{1}{2}\sum_i \overline{\mathbf{r}_i \cdot \mathbf{F}_i} \tag{2.A.1}$$

where the overbar stands for average over time, $\mathbf{r}_i$ and $\mathbf{F}_i$ are the instantaneous position and force of particle i. Clausius defined the right side of Eq. (2.A.1) as virial, a word derived from Latin meaning force or energy.

The virial theorem was originally derived in the context of classical mechanics. It was shown later by Vladimir Fock that the same equation holds true for quantum systems.[51] In this appendix, we discuss the virial theorem in the context of ergodic hypothesis, using a one-component simple fluid as an example. We will also elucidate an application leading to the virial equation.

2.A.1 Virial Theorem

Virial theorem is applicable to both classical and quantum systems. The basic idea is that the total kinetic energy of a system is linearly related to the potential energy. We may elucidate this idea with a system of classical particles.

Consider a thermodynamic system consisting of a large number of spherical particles. For any particle i, Newton's equation predicts

$$m_i \ddot{\mathbf{r}}_i = \mathbf{F}_i \tag{2.A.2}$$

where m_i stands for the particle mass, $\mathbf{F}_i$ is the instantaneous force on particle i, and $\ddot{\mathbf{r}}_i \equiv d^2\mathbf{r}_i/dt^2$ is acceleration, i.e., the second-order derivative of the particle position $\mathbf{r}_i$ with respect to time t.

Multiplying Eq. (2.A.2) by $\mathbf{r}_i$ and then applying the ensemble average on both sides leads to

$$m_i\langle \mathbf{r}_i \cdot \ddot{\mathbf{r}}_i\rangle = \langle \mathbf{r}_i \cdot \mathbf{F}_i\rangle. \tag{2.A.3}$$

Rearranging the left side of Eq. (2.A.3) gives

$$\langle \mathbf{r}_i \cdot \ddot{\mathbf{r}}\rangle_i = d\langle \mathbf{r}_i \cdot \dot{\mathbf{r}}_i\rangle/dt - \left\langle \dot{\mathbf{r}}_i^{\,2}\right\rangle. \tag{2.A.4}$$

Using the ergodic hypothesis, we can evaluate the ensemble average in terms of the time average for the first term on the right side of Eq. (2.A.4)

$$\langle \mathbf{r}_i \cdot \dot{\mathbf{r}}\rangle_i = \lim_{\tau\to\infty}\frac{1}{\tau}\int_0^\tau \mathbf{r}_i \cdot \dot{\mathbf{r}}_i dt = \lim_{\tau\to\infty}\frac{\mathbf{r}_i(\tau)^2 - \mathbf{r}_i(0)^2}{2\tau}. \tag{2.A.5}$$

Eq. (2.A.5) suggests that $\langle \mathbf{r}_i \cdot \dot{\mathbf{r}}_i\rangle$ is independent of time, i.e.,

$$d\langle \mathbf{r}_i \cdot \dot{\mathbf{r}}_i\rangle/dt = 0. \tag{2.A.6}$$

With the help of Eq. (2.A.6), substituting Eq. (2.A.4) into (2.A.3) gives

$$-\left\langle \mathbf{p}_i^2/m_i\right\rangle = \langle \mathbf{r}_i \cdot \mathbf{F}_i\rangle \tag{2.A.7}$$

where $\mathbf{p}_i = m_i\dot{\mathbf{r}}_i$ is the instantaneous momentum of particle i. Summing both sides of Eq. (2.A.7) over all particles leads to the virial theorem

$$\langle K\rangle = -\frac{1}{2}\sum_i \langle \mathbf{r}_i \cdot \mathbf{F}_i\rangle \tag{2.A.8}$$

where $K = \sum_i \mathbf{p}_i^2/(2m_i)$ is the total kinetic energy.

51 Fock V., "Bemerkung zum Virialsatz", *Z. Phys. A* 63 (11), 855–858 (1930).

2.A.2 Virial Equation

For a uniform system of spherical particles, the pressure may be understood as the average force per unit area exerted by the particles on an imaginary boundary (viz., container). For each particle near the boundary surface, the total force $\mathbf{F}_i^{total}$ includes a contribution due to all other particles in the system $\mathbf{F}_i$, and that due to its interaction with the container $\mathbf{F}_i{}'$, i.e.,

$$\mathbf{F}_i^{total} = \mathbf{F}_i + \mathbf{F}_i{}'. \tag{2.A.9}$$

In the thermodynamic limit, the system pressure is immaterial to the microscopic details of particle–boundary interactions. For simplicity, we may assume that the interaction between each particle and the boundary can be described by an elastic collision at the surface. Accordingly, the boundary force $\mathbf{F}_i{}'$ takes place only at the surface, and the ensemble average of the total force on the surface due to all particles in the system is related to the pressure.

With the boundary effect described by elastic collisions, the contribution to the virial due to the force from the container is related to surface integration

$$\sum_i \langle \mathbf{r}_i \cdot \mathbf{F}_i{}' \rangle = -\oint \mathbf{r} \cdot \mathbf{n} P dA \tag{2.A.10}$$

where dA represents a surface element with a normal in the direction of the unit vector $\mathbf{n}$, $\oint dA$ means integration over the system boundary, $\mathbf{r}$ stands for the position of the surface element dA, and the negative sign indicates that the force from the container is in the opposite direction of that from the particle. According to the divergence theorem, the surface integration can be replaced by

$$-\oint \mathbf{r} \cdot \mathbf{n} P dA = -P \int \nabla \cdot \mathbf{r} dV = -3PV \tag{2.A.11}$$

In deriving the last equality in Eq. (2.A.10), we have used the identity

$$\nabla \cdot \mathbf{r} = (\partial/\partial x + \partial/\partial y + \partial/\partial z) \cdot (x, y, z)^T = 3. \tag{2.A.12}$$

The combination of Eqs. (2.A.8)–(2.A.10) leads to

$$P = \frac{1}{3V} \sum_i \langle \mathbf{p}_i^2/m_i + \mathbf{r}_i \cdot \mathbf{F}_i \rangle. \tag{2.A.13}$$

Eq. (2.A.13) corresponds to a special application of the virial theorem to a thermodynamic system with an imaginary boundary that exerts elastic collisions with the particles.

For classical particles, the positions and momenta are independent variables. We may thus scale the system coordinates without changing the particle momenta (i.e., $\mathbf{r} \to \lambda \mathbf{r}$ where λ represents a scaling factor). The inter-particle interactions become negligible in the limit of $\lambda \to \infty$ while the total number of particles N is fixed. As a result, the dilation of coordinates leads to the ideal gas law

$$P = N k_B T / V. \tag{2.A.14}$$

A comparison of Eqs. (2.A.13) and (2.A.14) indicates

$$\sum_i \langle \mathbf{p}_i^2 / 2m_i \rangle = 3 N k_B T / 2. \tag{2.A.15}$$

Because the momenta of individual particles are independent, Eq. (2.A.15) predicts

$$\langle \mathbf{p}_i^2 / 2m_i \rangle = 3 k_B T / 2. \tag{2.A.16}$$

Eq. (2.A.16) indicates that the mean kinetic energy is the same for all particles; each dimension has an average of $k_B T/2$. As discussed in Chapter 3, the same average kinetic energy in each degree of freedom for particle motion is known as the equipartition theorem.

The virial equation is obtained by substituting Eq. (2.A.15) into (2.A.13)

$$P = \rho k_B T + \frac{1}{3V}\sum_i \langle \mathbf{r}_i \cdot \mathbf{F}_i \rangle \tag{2.A.17}$$

where $\rho = N/V$ is the particle density.

2.B Nosé's Thermostat

In this Appendix, we follow Tuckerman[52] to show that Nosé's Hamiltonian leads to the probability density in the phase space consistent with that given by the canonical ensemble. While Nosé's thermostat follows Hamiltonian mechanics, which generates microstates corresponding to a microcanonical ensemble, it introduces an imaginary particle that regulates the velocities of real particles in such a way that the average over the extra degrees of freedom associated with dimensionless parameter s and the fictitious momentum p_s transforms the microcanonical distribution into the canonical distribution.

To elucidate the essential ideas, consider the microcanonical partition function (viz., the continuous limit of the total number of microstates times a dimensional constant) according to Nosé's Hamiltonian

$$W = \int d\mathbf{r}^N \int d\mathbf{p}^N \int ds \int dp_s \delta[\mathcal{H}(\mathbf{r}^N, \mathbf{p}^N, s, p_s) - E] \tag{2.B.1}$$

where E is the total energy, and $\delta(x)$ stands for a one-dimensional Dirac delta function. Let $\mathbf{p}_i' = \mathbf{p}_i/s$, then $d\mathbf{p}^N = d\mathbf{p}'^N s^{3N}$ so Eq. (2.B.1) becomes

$$W = \int d\mathbf{r}^N \int d\mathbf{p}'^N \int ds \int dp_s s^{3N} \delta[\mathcal{H}'(\mathbf{r}^N, \mathbf{p}'^N, s, p_s) - E] \tag{2.B.2}$$

where

$$\mathcal{H}'(\mathbf{r}^N, \mathbf{p}'^N, s, p_s) = \sum_{i=1}^{N} \frac{|\mathbf{p}_i'|^2}{2m} + \Phi(\mathbf{r}^N) + \frac{p_s^2}{2\varpi} + (3N+1)k_B T \ln s. \tag{2.B.3}$$

In Eq. (2.B.2), the rescaled momenta are independent variables and can thus be rewritten as $\mathbf{p}$ in the following derivation. After relabeling, we see that the first two terms on the right side of Eq. (2.B.3) correspond to the total energy of N classical particles.

In Eq. (2.B.2), the integrations over parameters s and p_s can be carried out analytically with the help of the mathematical identity

$$\delta(f(s)) = \frac{\delta(s - s_0)}{f'(s_0)} \tag{2.B.4}$$

where $f(s)$ is an analytical function with a single zero at s_0. Substituting in Eq. (2.B.2) with

$$f(s) = \mathcal{H}'(\mathbf{r}^N, \mathbf{p}^N, s, p_s) - E = \sum_{i=1}^{N} \frac{|\mathbf{p}_i|^2}{2m} + \Phi(\mathbf{r}^N) + \frac{p_s^2}{2\varpi} + (3N+1)k_B T \ln s - E \tag{2.B.5}$$

leads to

$$W = \int d\mathbf{r}^N \int d\mathbf{p}^N \int ds \int dp_s s^{3N} \frac{\delta(s - s_0)}{f'(s_0)} \tag{2.B.6}$$

---

52 Tuckerman M. E., *Statistical mechanics: theory and molecular simulation*. Oxford University Press, 2010, p. 180.

where s_0 is obtained from $f(s) = 0$

$$s_0 = \exp\left[\left(E - \sum_{i=1}^{N} \frac{|\mathbf{p}_i|^2}{2m} - \Phi(\mathbf{r}^N) - \frac{p_s^2}{2\varpi}\right) \Big/ (3N+1)k_B T\right]. \tag{2.B.7}$$

From Eq. (2.B.5), one can find

$$f'(s_0) = \frac{(3N+1)k_B T}{s_0}. \tag{2.B.8}$$

With the analytical expression for $f'(s_0)$, Eq. (2.B.6) can then be evaluated analytically

$$\begin{aligned} W &= \int d\mathbf{r}^N \int d\mathbf{p}^N \int dp_s \frac{{s_0}^{3N+1}}{(3N+1)k_B T} \\ &= \frac{1}{(3N+1)k_B T} \int d\mathbf{r}^N \int d\mathbf{p}^N \int dp_s \exp\left[\left(E - \sum_{i=1}^{N} \frac{|\mathbf{p}_i|^2}{2m} - \Phi(\mathbf{r}^N) - \frac{p_s^2}{2\varpi}\right) \Big/ k_B T\right] \\ &= \frac{e^{E/k_B T}\sqrt{2\pi\varpi k_B T}}{(3N+1)k_B T} \int d\mathbf{r}^N \int d\mathbf{p}^N \exp\left\{-\left[\sum_{i=1}^{N} \frac{|\mathbf{p}_i|^2}{2m} + \Phi(\mathbf{r}^N)\right] \Big/ k_B T\right\} \end{aligned} \tag{2.B.9}$$

Eq. (2.B.7) predicts that the probability density of the particles in the phase space is indeed given by the canonical distribution

$$p(\mathbf{r}^N, \mathbf{p}^N) = \frac{1}{Q} \exp\left\{-\beta\left[\sum_{i=1}^{N} \frac{|\mathbf{p}_i|^2}{2m} + \Phi(\mathbf{r}^N)\right]\right\} \tag{2.B.10}$$

where $\beta = 1/(k_B T)$, and the canonical partition function Q can be derived by considering an ideal gas of non-interacting particles (see Chapter 3)

$$Q = \frac{1}{N!h^{3N}} \int d\mathbf{r}^N \int d\mathbf{p}^N \exp\left\{-\beta\left[\sum_{i=1}^{N} \frac{|\mathbf{p}_i|^2}{2m} + \Phi(\mathbf{r}^N)\right]\right\}. \tag{2.B.11}$$

Further Readings

Alder B. J. and Wainwright T. E., "Molecular motions", *Sci. Am.* 201 (4), 113–127 (1959).

Allen M. P. and Tildesley D. J., *Computer simulation of liquids* (2nd Edition). Oxford University Press. Chapter 3, 2017.

Bian X., Kim C. and Karniadakis G. E., "111 years of Brownian motion", *Soft Matter* 12, 6331–6346 (2016).

Callen H. B., *Thermodynamics and an introduction to thermostatistics.* Wiley. Chapters 15–17, 1985.

Chandler D., *Introduction to modern statistical mechanics.* Oxford University Press. Chapters 3 and 8, 1986.

Frenkel D. and Smit B., *Understanding molecular simulation* (2nd Edition). Academic Press. Chapters 4 and 6, 2002.

Gubbins K. E. and Moore J. D., "Molecular modeling of matter: impact and prospects in engineering", *Ind. Eng. Chem. Res.* 49, 3026–3046 (2010).

Hill T.L., *An introduction to statistical thermodynamics.* Dover Publications. Chapter 2, 1986.

van Gunsteren W. F., "Validation of molecular simulation: an overview of issues", *Angew. Chem. Int. Eds.* 57, 884–902 (2018).

Problems

2.1 Consider an isolated system with N distinguishable but noninteracting particles. Each particle can be in one of two possible energy states: one at ϵ, and the other at $-\epsilon$. Let E be the total energy of the system

(i) Derive equations for the number of particles at each energy state;
(ii) Derive the system entropy;
(iii) Derive the system temperature;
(iv) Verify the prediction of the third law of thermodynamics, i.e., $S \to 0$ when $T \to 0$ K;
(v) Show that at some values of E and N, the absolute temperature can be negative.

2.2 Suppose that the model system considered in Problem 2.1 is immersed in a thermal bath at temperature T.

(i) Derive equations for the number of particles at each energy state, $< N_+ >$ and $< N_- >$;
(ii) Show that the system entropy can be expressed as

$$S = -k_B \left[< N_+ > \ln(< N_+ > /N) + < N_- > \ln(< N_- > /N) \right].$$

(iii) Derive an expression for the internal energy;
(iv) Verify the prediction of the third law of thermodynamics, i.e., $S \to 0$ when $T \to 0$ K;
(v) In absolute scale, can the system temperature be negative?

2.3 Proposed by Loup Verlet in 1967, the Verlet algorithm is one of the most popular integration schemes used in molecular dynamics (MD) simulation. The numerical method can be derived from the Taylor expansions of the particle position at two consecutive time steps, i.e., from t_{n-1} to t_n and from t_n to t_{n+1}, such that the time reversal symmetry of Newton's equation of motion is preserved:

$$\mathbf{r}_{n+1} = \mathbf{r}_n + \mathbf{v}_n \delta t + \frac{1}{2!}\mathbf{a}_n \delta t^2 + \frac{1}{3!}\dot{\mathbf{a}}_n \delta t^3 + O(\delta t^4), \tag{A}$$

$$\mathbf{r}_{n-1} = \mathbf{r}_n - \mathbf{v}_n \delta t + \frac{1}{2!}\mathbf{a}_n \delta t^2 - \frac{1}{3!}\dot{\mathbf{a}}_n \delta t^3 + O(\delta t^4), \tag{B}$$

where $\mathbf{v}_n = d\mathbf{r}_n/dt$, $\mathbf{a}_n = d\mathbf{v}_n/dt$, and $\dot{\mathbf{a}}_n = d\mathbf{a}_n/dt$. Adding Eqs. (A) and (B) leads to

$$\mathbf{r}_{n+1} = 2\mathbf{r}_n - \mathbf{r}_{n-1} + \frac{\mathbf{f}_n}{m}\delta t^2 + O(\delta t^4), \tag{C}$$

where $\mathbf{f}_n$ is the net force on the particle at t_n, and m is the particle mass. On the other hand, subtracting Eq. (A) with (B) gives the velocity equation:

$$\mathbf{v}_n = \frac{\mathbf{r}_{n+1} - \mathbf{r}_{n-1}}{2\delta t} + O(\delta t^3). \tag{D}$$

(i) What are the advantages of the Verlet algorithm in comparison with Euler's method

$$\mathbf{r}_{n+1} = \mathbf{r}_n + \mathbf{v}_n \delta t + \frac{\mathbf{a}_n}{2!}\delta t^2 + O(\delta t^3), \qquad \mathbf{v}_{n+1} = \mathbf{v}_n + \mathbf{a}_n \delta t + O(\delta t^2)?$$

(ii) The three terms on the left side of Eq. (C) involves two large numbers (viz., $2\mathbf{r}_n$ and $\mathbf{r}_{n-1}$) and one small number, $(\mathbf{f}_n/m)\delta t^2$. The numerical issue can be avoided in the so-called leapfrog algorithm, a slight modification of Verlet's algorithm using the position at time step n and the velocity at time step $n - 1/2$ to update both position and velocity at the next step:

$$\mathbf{r}_{n+1} = \mathbf{r}_n + \mathbf{v}_{n+1/2}\delta t, \tag{E}$$

$$\mathbf{v}_{n+1/2} = \mathbf{v}_{n-1/2} + \frac{\mathbf{f}_n}{m}\delta t. \tag{F}$$

Show that the leapfrog algorithm are accurate up to the third order (δt^3) for both position and velocity.

(iii) Show that Eqs. (A) and (B) are equivalent to those from the velocity Verlet algorithm

$$\mathbf{r}_{n+1} = \mathbf{r}_n + \mathbf{v}_n\delta t + \frac{\mathbf{f}_n}{2m}\delta t^2, \tag{G}$$

$$\mathbf{v}_{n+1} = \mathbf{v}_n + \frac{(\mathbf{f}_n + \mathbf{f}_{n+1})}{2m}\delta t. \tag{H}$$

Does the velocity Verlet algorithm satisfy the time reversal symmetry of Newton's equation?

2.4 Exactly solvable models are often used in statistical mechanics for testing theoretical methods and simulation algorithms. In this exercise, we use the one-dimensional (1D–) harmonic oscillator as a benchmark to test the velocity Verlet algorithm. Consider a particle of mass m subject to a harmonic potential $\varphi(x) = kx^2/2$ where k represents the force constant, and x is the particle position. The equation of motion for the particle, $m\frac{d^2x}{dt^2} = -kx$, has the following exact solutions for the position and velocity:

$$x(t) = A\cos(\omega t + \phi), \qquad v(t) = -A\omega\sin(\omega t + \phi),$$

where $\omega = \sqrt{k/m}$ is the angular frequency, A is the oscillation amplitude, and ϕ denotes the phase. While k and m are intrinsic properties of the system, A and ϕ are set by the initial conditions. For simplicity, we adopt $1/\omega$ as the unit of time, A as the unit of length, $A\omega$ as the unit of velocity, kA as the unit of force, and kA^2 as the unit of energy. The initial condition is set as $x_0 = 1$ and $v_0 = 0$ at $t = 0$. Prepare a compute program to implement the velocity Verlet algorithm

$$\mathbf{r}_{n+1} + \mathbf{r}_{n-1} = 2\mathbf{r}_n + \frac{\mathbf{f}_n}{m}\delta t^2 \qquad \mathbf{v}_{n+1} = \mathbf{v}_n + \frac{(\mathbf{f}_n + \mathbf{f}_{n+1})}{2m}\delta t,$$

for the 1D–harmonic oscillator and study the following:

(i) Compare the simulation and exact results for the particle position and velocity using integration time $\delta t = 0.1$ for the first 100 steps;

(ii) Suppose that the accuracy of the simulation can be measured by the mean-square error (MSE) of the particle position

$$MSE = \frac{1}{n_{step}}\sum_{n=1}^{n_{step}}[x_n - x(t_n)]^2$$

where n_{step} is the number of integration steps, and $x(t_n)$ stands for the particle position at time t_n predicted by the exact result. Plot MSE versus δt for $n_{step} = 1000$ and discuss how the integration time should be selected in MD simulation.

(iii) Set $\delta t = 0.1$, how does the simulation error propagate with time in terms of the MSE?

(iv) Set $\delta t = 0.1$, how does the system energy propagate with time? Compare the result with that from Euler's algorithm:

$$\mathbf{r}_{n+1} = \mathbf{r}_n + \mathbf{v}_n \delta t, \qquad \mathbf{v}_{n+1} = \mathbf{v}_n + \frac{\mathbf{f}_n}{m} \delta t.$$

2.5 The characteristic time of particle motions (τ) is often used to estimate the time step (δt) in MD simulation. A good rule of thumb is $\delta t \approx 0.1\tau$. Suppose that MD simulation is used to calculate the properties of liquid argon near its triple point ($T_t = 83.804$ K, $\rho_L = 35.475$ mol/L, $P_t = 0.06895$ MPa). Using the Lennard–Jones model ($\epsilon/k_B = 119.8$ K, $\sigma = 3.405$ Å), estimate the characteristic time of particle motions and proper time step in both absolute and appropriate dimensionless units. What is a proper duration of simulation?

2.6 In Andersen's method for keeping a constant temperature during MD simulation, the frequency of velocity reassignment (viz., the probability of particle collision at each MD step) can be estimated by considering a simulation cell with volume V surrounded by a much larger heat bath at temperature T. When there is a temperature fluctuation in the simulation cell so that its average temperature is $T + \Delta T$, it will gain or lose energy at a rate proportional to the temperature difference ΔT.

(i) Show that the rate at which energy is absorbed/released by the system during time step Δt can be written as

$$\frac{\Delta E}{\Delta t} = -a\kappa \Delta T V^{1/3},$$

where κ is the thermal conductivity, and a is a dimensionless constant that depends on the shape of the simulation cell.

(ii) Show that the change in the total kinetic energy of the system due to the heat transfer can be compensated by rescaling the average velocity of particles

$$\frac{\Delta E}{\Delta t} = \frac{3}{2} \nu N k_B \Delta T,$$

where ν stands for the frequency of velocity reassignment, and N is the number of particles in the system.

(iii) Derive the frequency of velocity reassignment

$$\nu = \frac{2a\kappa}{3k_B \rho^{1/3} N^{2/3}},$$

where $\rho = N/V$.

(iv) Show that the frequency of velocity reassignment will be much smaller than the frequency of particle collisions when N is sufficiently large.

2.7 There is an increasing interest in the practical applications of so-called low-dimensional materials, for example, in studies of electrodes for energy storage, ion channels in biological membranes, and molecules in catalytic environments. Computer simulation is useful for understanding the thermodynamic properties and dynamical behavior of such systems which can be qualitatively different from those displayed in bulk systems. In this exercise, we use the one-dimensional Lennard–Jones (1D-LJ) model to simulate the structure and dynamic properties of spherical particles in a narrow pore.

(i) Prepare a MD program for simulating 1D-LJ fluids in the canonical ensemble using the velocity Verlet algorithm and Anderson thermostat; Use dimensionless units for all physical quantities, i.e., distance in σ, energy in ϵ, temperature in k_BT/ϵ, velocity in $\sqrt{k_BT/m}$, the particle number density in $\rho\sigma$, and time in $\sigma\sqrt{m/\epsilon}$. Show the simulation results at $k_BT/\epsilon = 1$ and $\rho\sigma = 0.7$.

(ii) Testing the simulation results with the Maxwell–Boltzmann velocity distribution for the particle velocity;

(iii) Calculate the radial distribution function of the 1D system, estimate the correlation length ξ, and discuss how ξ may vary with the system temperature and density.

(iv) Calculate the velocity autocorrelation function, estimate the correlation time τ, and discuss how τ may vary with the system temperature and density.

(v) Calculate the mean-square displacement and discuss how the diffusive behavior of the one-dimensional fluid differ from that corresponding to a bulk system.

2.8 Show that Nosé's equations of motion

$$\frac{d\mathbf{r}_i}{dt} = \frac{\mathbf{p}_i}{ms^2}, \qquad \frac{d\mathbf{p}_i}{dt} = \mathbf{F}_i, \qquad \frac{ds}{dt} = \frac{p_s}{\varpi}, \qquad \frac{dp_s}{dt} = \sum_i \frac{p_i^2}{ms^3} - \frac{(3N+1)k_BT}{s},$$

can be transformed into the Nosé–Hoover equations

$$\frac{d\mathbf{r}_i}{dt'} = \frac{\mathbf{p}'_i}{m}, \qquad \frac{d\mathbf{p}'_i}{dt'} = \mathbf{F}_i - \xi\mathbf{p}'_i, \qquad \frac{ds}{dt'} = \xi s, \qquad \frac{d\xi}{dt'} = \frac{1}{\varpi}\left[\sum_i \frac{p'^2_i}{m} - (3N+1)k_BT\right],$$

with $\xi = p_s/\varpi$ by rescaling momentum and time: $\mathbf{p}'_i = \mathbf{p}_i/s$ and $t' = t/s$.

2.9 A general form of the fluctuation-dissipation theorem may be derived by considering an arbitrary dynamic property of a thermodynamic system in response to the change in the total energy. Suppose that a system is prepared at a *nonequilibrium* condition by applying an external force f conjugated with dynamic quantity A such that the system energy is given by

$$\bar{E}_\nu = E_\nu + \Delta E_\nu,$$

where E_ν is the system energy without the perturbation, $\Delta E_\nu = -fA_\nu$, and ν denotes microstates. At time $t = 0$, the external force is turned off and the system is relaxing to equilibrium. Show that the dynamic quantity averaged over the initial nonequilibrium condition

$$\bar{A}(t) \equiv \frac{\sum_\nu A_\nu(t)e^{-\beta\bar{E}_\nu}}{\sum_\nu e^{-\beta\bar{E}_\nu}},$$

reaches to the equilibrium value linearly proportional to its fluctuation at the equilibrium condition, i.e.,

$$\Delta\bar{A}(t) = \beta f < \delta A(0)\delta A(t) >, \qquad \text{(J)}$$

where $\Delta\bar{A}(t) \equiv \bar{A}(t) - < A >$, $\delta A(0) \equiv A_\nu(0) - < A >$, and $\delta A(t) \equiv A_\nu(t) - < A >$. Eq. (J) is a general form of the fluctuation-dissipation theorem.

2.10 One of the simplest models to explain the fluctuation-dissipation theorem is provided by a one-dimensional harmonic system that contains only a single particle. Suppose that the potential energy is given by $\varphi(x) = \frac{1}{2}kx^2$ where k is a constant and x represents the particle coordinate. At temperature T, the probability density of the system taking the value of x is described by the Boltzmann equation

$$p(x) = \frac{1}{\mathcal{N}}e^{-\beta kx^2/2},$$

where $\beta = 1/(k_B T)$, and $\mathcal{N}$ is a normalization constant. Now consider how the system responds to a weak external field that is linearly dependent on the particle coordinate, $-fx$, where f is a constant.

(i) What is the probability density under the additional potential?
(ii) How does the additional potential change the mean position of the particle?
(iii) How does the additional potential change the fluctuation of the particle position?
(iv) Show that the deviation of the mean position is linearly proportional to the particle fluctuation in the original system.

Hint: Under the additional potential, the probability density is still described by the Boltzmann equation.

2.11 Lars Onsager won the 1968 Nobel Prize in Chemistry for his theoretical contributions to the thermodynamics of irreversible processes. A cornerstone of his pioneering work was the regression hypothesis, which states that the relaxation of a macroscopic quantity from non-equilibrium conditions is governed by the same equation describing microscopic fluctuations. Discuss Onsager's regression hypothesis in the context of the fluctuation-dissipation theorem and explain why it is applicable only to relaxation of dynamic quantities near equilibrium conditions.

2.12 The Green–Kubo equation provides a general formula for computing transport coefficients through MD simulation

$$\gamma = \int_0^\infty dt < \dot{\xi}(t)\dot{\xi}(0) >,$$

where γ represents a transport coefficient (within a multiplicative constant), ξ is the mechanical variable associated with the particular transport property under consideration, and $\dot{\xi}(t)$ signifies a time derivative. Show that the above equation can be equivalently written as

$$\gamma = \frac{1}{2}\lim_{t\to\infty}\frac{d}{dt} < (\xi(t) - \xi(0))^2 > .$$

Which equation is more convenient to use in MD simulation?

2.13 The Langevin equation is numerically more complicated to implement in MD simulation than Newton's equation because the change in velocity is related to a stochastic variable, which makes it not differentiable in conventional sense. Show that

(i) Integration of the Langevin equation over the time step from t_n to t_{n+1} leads to

$$\mathbf{v}_{n+1} - \mathbf{v}_n = \int_{t_n}^{t_{n+1}} dt'(\mathbf{f}/m) - \gamma_0(\mathbf{r}_{n+1} - \mathbf{r}_n) + \omega_{n+1}/m. \tag{K}$$

where $\mathbf{f} = -\nabla\varphi(\mathbf{r})$ is the conservative force, and

$$\omega_{n+1} \equiv \int_{t_n}^{t_{n+1}} dt'\xi(t'),$$

is a Gaussian random number with $< \omega_n > = 0$ and $< \omega_n \cdot \omega_l > = 6\gamma_0 m k_B T \delta t \delta_{nl}$, $\delta t = t_{n+1} - t_n$ is the step length, and δ_{nl} is the Kronecker-delta function.

(ii) Up to the second-order accuracy in the step length, the numerical integration of $d\mathbf{r}/dt = \mathbf{v}$ leads to

$$\mathbf{r}_{n+1} = \mathbf{r}_n + b\delta t\mathbf{v}_n + \frac{b\delta t^2}{2m}\mathbf{f}_n + \frac{b\delta t}{2m}\omega_{n+1}, \tag{L}$$

where

$$b \equiv \frac{1}{1+\gamma_0\delta t/2},$$

(iii) Up to the second-order accuracy in the step length, Eq. (K) can be rewritten as

$$\mathbf{v}_{n+1} = \mathbf{v}_n + \frac{\delta t}{2m}(\mathbf{f}_n + \mathbf{f}_{n+1}) - \gamma_0(\mathbf{r}_{n+1} - \mathbf{r}_n) + \omega_{n+1}/m. \tag{M}$$

(iv) Eqs. (L) and (M) reduce to the velocity-Verlet algorithm when $\gamma_0 = 0$.
(v) How is ω_n, a contribution to the velocity change from the stochastic force, implemented in MD simulation?

2.14 Prepare a computer program for the Brownian motion of a single particle in one dimension based on the results obtained from Problem 2.13. Validate your code by comparing the simulation results with the analytical solutions for

(i) the probability density for the particle velocity

$$p(v) = \sqrt{\frac{a}{\pi}}e^{-av^2},$$

where $a \equiv m/2k_BT(1-e^{-2\gamma_0 t})$;

(ii) the mean-square displacement

$$R^2(t) = <|x(t)-x(0)|^2> = \frac{2k_BT}{m\gamma_0}\left(t - 1/\gamma_0 + e^{-\gamma_0 t}/\gamma_0\right);$$

(iii) the velocity autocorrelation function

$$c(t) = <v(t)v(0)> = \frac{k_BT}{m}e^{-\gamma_0 t}.$$

Hint: Identify characteristic time and length scales and use dimensionless units for particle position $x(t)$ and velocity $v(t)$.

2.15 In the canonical ensemble, the magnitude of energy fluctuations is related to the constant-volume heat capacity

$$<\delta E^2> = k_BT^2C_V.$$

Starting from this equation, verify that, for a classical system of N spherical particles at constant T and V, fluctuations in the total kinetic and potential energies are given by, respectively,

$$<\delta K^2> = 3N(k_BT)^2/2,$$
$$<\delta\Phi^2> = k_BT^2(C_V - 3Nk_B/2).$$

2.16 Cross fluctuations are important for describing the mutual influences of dynamic variables in thermodynamic systems.

(i) Show that, in the canonical ensemble, the cross fluctuations of any dynamic quantity A and the system energy E can be calculated from

$$<\delta A\delta E> = -\left(\frac{\partial <A>}{\partial\beta}\right)_V.$$

(ii) Considering a classical system of N spherical particles at constant T and V, show that the cross fluctuations of the total potential energy Φ and the system virial $\mathcal{V}$ are given by

$$< \delta\Phi\delta\mathcal{V} > = k_B T^2(V\gamma_V - Nk_B),$$

where $\gamma_V \equiv (\partial P/\partial T)_V$ is the thermal pressure coefficient.

2.17 For a classical system of N spherical particles at constant volume V and temperature T, the instantaneous pressure may be defined as

$$\mathcal{P} = 2K/3V + \mathcal{V}/V$$

where K is the kinetic energy of translational motions, and $\mathcal{V}$ is the system virial given by

$$\mathcal{V} = -\frac{1}{3}\sum_i^N \mathbf{r}_i \cdot \nabla_{\mathbf{r}_i}\Phi.$$

Verify the following equations:

(i) The ensemble average of the instantaneous pressure can be written as

$$< \beta\mathcal{P} > = \frac{\partial \ln Q}{\partial V},$$

where $\beta = 1/(k_B T)$, and Q is the canonical partition function.

(ii) The fluctuations of instantaneous pressure is given by

$$< \delta\mathcal{P}^2 > = \frac{k_B T}{V}\left[\frac{2 < \mathcal{P} >}{3} - \frac{1}{\kappa_T} + \frac{Nk_B T}{V} + \frac{1}{9V} < \mathbf{r}^N \cdot \nabla\nabla\Phi \cdot \mathbf{r}^N >\right],$$

where $\kappa_T \equiv -V^{-1}(\partial V/\partial P)_T$ is the isothermal compressibility, $\mathbf{r}^N \equiv \{\mathbf{r}_1, \mathbf{r}_2, \ldots, \mathbf{r}_N\}$ and $\nabla\nabla\Phi$ denotes a matrix of second-order gradients.

(iii) The above equation for $< \delta\mathcal{P}^2 >$ allows for the computation of the isothermal compressibility through molecular simulation. Comment its computational efficiency in comparison to a "simpler" compressibility equation

$$\kappa_T = \frac{V}{Nk_B T} + \frac{1}{k_B T}\int_0^\infty dr 4\pi r^2[g(r) - 1],$$

where $g(r)$ stands for the radial distribution function.

(iv) The pressure-energy fluctuations satisfy

$$< \delta\Phi\delta\mathcal{P} > = k_B T^2(\gamma_V - \rho k_B)$$

where $\rho = N/V$ and $\gamma_V \equiv (\partial P/\partial T)_V$. In molecular simulation, this equation is often used to calculate thermal pressure coefficient γ_V.

2.18 The Jahn–Teller effect is related to the geometrical distortion of molecules or ions in a crystal that results from certain electron configurations. In a coarse-grained representation of such an effect, the distortion of each lattice site may be represented by a particle that may exist in one of three energy states

$$\epsilon_1 = \epsilon_2 = bx^2 - \frac{1}{2}cx,$$

$$\epsilon_3 = bx^2 + cx,$$

where $b > 0$ and $c > 0$ are constants, and $x \geq 0$ is a parameter that characterizes the uniform strain of the material.

(i) Assume that the lattice sites are independent to each other, find the canonical partition function for this system;
(ii) What is the reduced Helmholtz energy per lattice site, $a = \beta F/N$, as a function of temperature and x?
(iii) If parameter x is a small value but can change at constant T, what is its equilibrium value, $\overline{x}$? Assume that $a(x)$ can be expanded in a power series in x to the cubic oder, x^3.
(iv) Show that the system has a critical temperature T_c beyond which the three energy states become identical.

2.19 Show

$$\left(\frac{\partial \beta G}{\partial \beta}\right)_{N,\beta P} = \left(\frac{\partial \beta G}{\partial \beta}\right)_{N,P} - P\left(\frac{\partial G}{\partial P}\right)_{N,\beta}.$$

2.20 Suppose that a perfect crystal have N lattice sites and M interstitial locations. The Frenkel defects correspond to the migration of atoms from lattice sites to interstitial locations. Because a large positive energy ϵ is associated with each Frenkel defect, the number of displaced atoms n is typically much smaller than N or M.
(i) How many ways are there to remove n atoms from N sites and displace them at M interstitial locations?
(ii) Assume that the total volume is fixed and that the atoms have no degrees of freedom other than their locations, derive the canonical partition function for the crystal.
(iii) What is the average number of displaced atoms at temperature T?
(iv) Assume $N = M$ and $\epsilon = 1$ eV, what is the concentration of the displaced atoms at $T = 1000$ K?

2.21 Consider the grand canonical ensemble for a two-component system, show that the numbers of molecules for different species are correlated through

$$\langle N_1 N_2\rangle - \langle N_1\rangle\langle N_2\rangle = k_B T\left(\frac{\partial \langle N_1\rangle}{\partial \mu_2}\right)_{V,T,\mu_1} = k_B T\left(\frac{\partial \langle N_2\rangle}{\partial \mu_1}\right)_{V,T,\mu_2}.$$

2.22 Based on the fundamental equation for the grand potential, show the following useful relations:
(i)

$$\left(\frac{\partial P}{\partial \mu_i}\right)_{T,\mu_{j\neq i}} = \left(\frac{\partial \langle N_i\rangle}{\partial V}\right)_{T,\mu_i}.$$

(ii)

$$\rho_i = \left(\frac{\partial P}{\partial \mu_i}\right)_{T,\mu_{j\neq i}},$$

where $\rho_i = \langle N_i\rangle/V$.
(iii) In an ideal-gas mixture, the chemical potential for each species is given by

$$\mu_i = k_B T \ln(\rho_i \Lambda_i^3),$$

where Λ_i is a constant depending on temperature and the identify of chemical species i.

(iv) The fluctuation in the number of molecules in an open system is

$$\langle \delta N_i^2 \rangle = k_B T V \left(\frac{\partial^2 P}{\partial \mu_i^2} \right)_{V,T,\mu_{j \neq i}}.$$

(v) The cross fluctuation in the number of molecules is

$$\langle \delta N_i \delta N_j \rangle = k_B T \left(\frac{\partial \langle N_i \rangle}{\partial \mu_j} \right)_{V,T,\mu_{k \neq j}} = k_B T \left(\frac{\partial \langle N_j \rangle}{\partial \mu_i} \right)_{V,T,\mu_{k \neq i}}.$$

2.23 The following exercise is helpful to understand how different ensembles affect the averages and fluctuations of thermodynamic variables. Suppose that we know the expected value of a dynamic quantity A in an ensemble specified by extensive variable X as well as other parameters that will not be altered. The ensemble average of A is given by

$$< A >_X = \sum_{\nu} A_\nu W_\nu^X / W_X,$$

where W_ν^X is the appropriate statistical weight for microstate ν, and $W_X \equiv \sum_\nu W_\nu^X$ is the corresponding partition function. Let ξ be the conjugate intensive variable of X. From the viewpoint of classical thermodynamics, the variable change from X to ξ constitutes a Legendre transformation. In the new ensemble with X in the original ensemble replaced by ξ that is fixed as a constant, the partition function can be calculated from the Laplace transform

$$W_\xi = \int dX e^{-X\xi} W_X.$$

Show the following identifies in the ensemble with ξ being fixed:

(i) The expected value of A in the new ensemble

$$< A >_\xi = \int dX < A >_X W_X e^{-X\xi} / W_\xi.$$

(ii) Let $\Psi(\xi) \equiv -\ln W_\xi$, then we have

$$< X >_\xi = \partial \Psi(\xi) / \partial \xi,$$

and the mean-square fluctuation of X is

$$< \delta X^2 >_\xi = -\frac{\partial < X >_\xi}{\partial \xi} = -\partial^2 \Psi(\xi) / \partial \xi^2.$$

The relation between the fluctuations in X and the linear response of its mean $< X >$ to conjugate intensive variable ξ is known as the linear response theory.

(iii) To the leading order of the system size, the expected values of A from the two ensembles are related by

$$< A >_{\overline{X}} \approx < A >_\xi + \frac{1}{2} \frac{\partial^2 < A >_\xi}{\partial \xi \partial \overline{X}},$$

where $\overline{X} = < X >_\xi$.

(iv) In the thermodynamic limit, $N \to \infty$, the two ensembles predict the same expected values of A per particle, i.e.,

$$< a >_\xi = < a >_X,$$

where $a = A/N$.

(v) The mean-square fluctuations of A from the two ensembles are related by

$$< \delta A^2 >_X \approx < \delta A^2 >_\xi + \frac{\partial \xi}{\partial X}\left(\frac{\partial < A >_\xi}{\partial \xi}\right)^2.$$

(vi) For a thermodynamic system with N classical particles represented by a microcanonical ensemble, the mean-square fluctuation of the translational kinetic energy K is given by

$$\frac{< \delta K^2 >}{< K >} = k_B T\left(1 - \frac{3Nk_B}{2C_V}\right),$$

where T is the absolute temperature, and C_V is the constant-volume heat capacity.

2.24 The partition functions of conjugate ensembles are related by the Laplace transform

$$W_\xi = \int dX e^{-X\xi} W_X,$$

where X is an extensive variable, and ξ is the conjugate intensive variable of X. Accordingly, at a fixed value of ξ, the probability density of the system as a function of X is

$$p(X) = W_X e^{-X\xi}/W_\xi.$$

Both X and ξ are expressed in dimensionless forms. Show that, in the ensemble with ξ being fixed,

(i) The probability density $p(X)$ exhibits a maximum at $X = < X >_\xi$;

(ii) When the system is sufficiently large, $p(X)$ can be approximated by the Gaussian distribution

$$p(X) = p(\overline{X}) \exp\left[-\frac{(X - \overline{X})^2}{2\sigma_X^2}\right],$$

where $\overline{X} = < X >_\xi$ and $\sigma_X^2 \equiv < (X - \overline{X})^2 >_\xi$.

(iii) $p(\overline{X}) = 1/\sqrt{2\pi\sigma_X^2}$;

(iv) In the thermodynamic limit, the probability density becomes a Dirac-delta function, i.e.,

$$\lim_{N\to\infty} p(X) = \delta(X - \overline{X}).$$

3

Ideal Gases and Single-Molecule Thermodynamics

Historically, predicting the thermodynamic properties of noninteracting molecular systems represents one of the most successful applications of quantum and statistical mechanics. These applications remain relevant today not only because the ideal-gas model provides a good representation of the properties of real gases at low density or low pressure, but it is also important for understanding chemical reactions and kinetics from molecular perspectives. Besides, accurate prediction of the thermodynamic characteristics of individual molecules is essential for creating novel chemicals with unique capabilities because fundamental properties such as Gibbs free energy, enthalpy, heat capacity, and standard entropy play a critical role in determining the stability and reaction energies, thus enabling informed design decisions.

In this chapter, we will derive the thermodynamic properties of noninteracting molecular systems using the canonical and grand canonical ensembles. For representing microstates and molecular energy of relatively simple molecules, we adopt single-particle quantum-mechanical models to describe the translational, rotational, and vibrational motions. The procedure will be illustrated first with ideal monatomic gases (e.g., argon) through which we establish the connection between the translational partition function and thermodynamic properties. The statistical-thermodynamic analysis is then applied to diatomic and polyatomic systems in which the rotational and vibrational motions are described in terms of the rigid-rotor model and the harmonic-oscillator model, respectively. We will elucidate some applications of the ideal-gas models by considering the thermodynamics of gas adsorption and hydrate formation.

In this chapter, we will also discuss coarse-grained models to represent polymer conformations. We will exemplify the applications of such models to single-molecule thermodynamics, i.e., the use of thermodynamic principles to understand the dynamic behavior and properties of individual molecules, such as their interaction energy, entropy, and free energy landscape. The study of single-molecule thermodynamics is a key area of research in many fields, including chemistry, physics, and biology. It has led to new insights into the behavior of complex molecular systems, such as proteins and DNA, and has enabled researchers to design new materials and devices with tailored properties.

3.1 Noninteracting Molecular Systems

Noninteracting molecular systems often serve as a useful reference to quantify the thermodynamic properties of gases and liquid mixtures of practical interest. The connection can be easily established by considering a hypothetical process that varies the molecular density (e.g., by reducing

Fundamentals and Practice in Statistical Thermodynamics, First Edition. Jianzhong Wu and John M. Prausnitz.

Companion website: www.wiley.com/go/jianzhongwu/fundamentalsandpractice

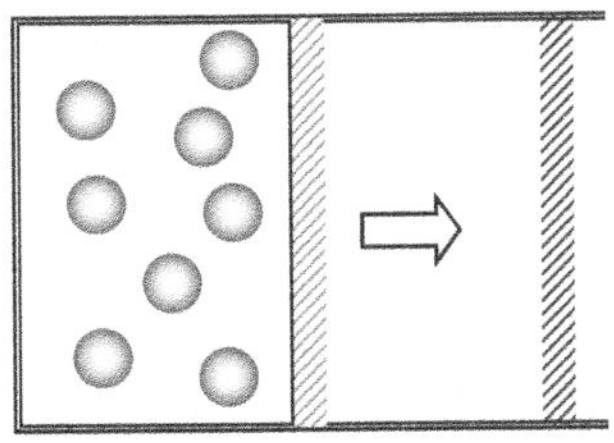

Figure 3.1 The ideal-gas state can be reached by the isothermal expansion of a real gas (or a liquid) to $P = 0$ or $V = \infty$ at constant temperature T and the total number of molecules N.

the pressure or increasing the volume of a real gas with the number of molecules fixed). Similar strategies are commonly used in classical thermodynamics to evaluate the difference between the properties of an interacting system at different thermodynamic conditions.

Figure 3.1 shows schematically the isothermal expansion of a real gas (or a liquid) to zero pressure (or infinite volume) with the system temperature T and the total number of molecules N fixed as constants. As the number density of molecules approaches zero, the separation between molecules becomes infinitely large so that the intermolecular interactions become negligible. As well documented in classical thermodynamics, we can evaluate the changes in thermodynamic properties based on an equation of state that relates the system volume to temperature T, pressure P, and the number of molecules for each species N_i, which is often expressed as $V = V(T, P, N_i)$ or $P = P(T, V, N_i)$. With such connections, we can predict, for example, the chemical potential of any specie in a real gas (or liquid) relative to that of the same species as a pure ideal gas at system temperature and pressure $P = 1$ atm

$$\mu_i = \mu_i^0 + k_B T \ln(P y_i) + \int_0^P \left(\overline{v}_i - \frac{k_B T}{P}\right) dP \tag{3.1}$$

where $\mu_i{}^0$ is the chemical potential of species i at the ideal-gas state (*viz.*, the reference state), and $\overline{v}_i$ stands for the partial molecular volume. With an analytical expression $V = V(T, P, N_i)$, we can derive the partial molecular volume from

$$\overline{v}_i = \left(\frac{\partial V}{\partial N_i}\right)_{T,P,N_{j\neq i}} \tag{3.2}$$

and integrate the right side of Eq. (3.1) to obtain the chemical potential of the real system. According to the ideal gas law, $PV=NRT$, the last term on the right side of Eq.(3.1) disappears for an ideal gas.

In chemical thermodynamics, the chemical potential is often expressed in terms of fugacity f_i

$$\mu_i = \mu_i^0 + k_B T \ln\left(f_i/f_i^0\right). \tag{3.3}$$

Because the fugacity of species i in the reference state is exactly known, $f_i^0 = 1$ atm, a comparison of Eqs. (3.3) and (3.1) indicates

$$f_i = P y_i \phi_i \tag{3.4}$$

where ϕ_i is called the fugacity coefficient, a dimensionless quantity that can be calculated from

$$k_B T \ln \phi_i = \int_0^P \left(\overline{v}_i - \frac{k_B T}{P}\right) dP. \tag{3.5}$$

Eq. (3.3) reveals that, in any real gas system, the chemical potential of a molecule, an abstract thermodynamic variable, is intrinsically related to that in a noninteracting reference system $\mu_i{}^0$, as well as to the volumetric properties that are accessible through experiment or predicted by an equation of state.

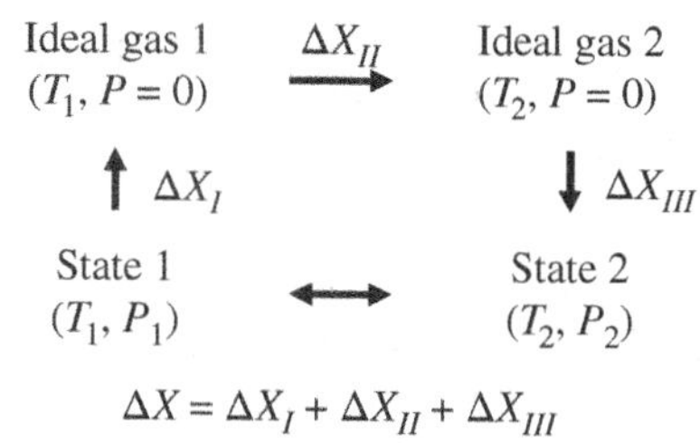

Figure 3.2 The difference in thermodynamic quantity X between two equilibrium states referred to as State 1 and State 2, can be computed by employing a thermodynamic cycle that connects them to their corresponding ideal gas states.

The connection between noninteracting and interacting systems is also useful for calculating the changes of thermodynamic properties between arbitrary equilibrium states. As shown schematically in Figure 3.2, we can connect thermodynamic states at different temperatures and pressures with an imaginary process that consists of three reversible steps:

(I) expansion of the system from the initial state (State 1) to an ideal-gas state (IG 1) at constant temperature T_1;
(II) changing the temperature of the ideal-gas system from T_1 to T_2;
(III) and the isothermal compression of the ideal-gas system (IG 2) to the final state of the real system (State 2).

Because variations in thermodynamic properties are independent of the imaginary path, we can predict the change of any thermodynamic quantity from

$$\Delta X = \Delta X_I + \Delta X_{II} + \Delta X_{III}. \tag{3.6}$$

Eq. (3.6) shows a general relation between the thermodynamic properties of a real system and those corresponding to an ideal-gas system.

We may elucidate the application of the above procedure using enthalpy as an example

$$\Delta H = \Delta H_I + \Delta H_{II} + \Delta H_{III}. \tag{3.7}$$

The processes (steps I and III) related to the isothermal expansion and compression lead to

$$\Delta H_I + \Delta H_{III} = -\int_0^{P_1} \left(\frac{\partial H}{\partial P}\right)_{T_1} dP + \int_0^{P_2} \left(\frac{\partial H}{\partial P}\right)_{T_2} dP \tag{3.8}$$

where

$$(\partial H/\partial P)_T = V - T(\partial V/\partial T)_P \tag{3.9}$$

can be calculated from an equation of state. Meanwhile, the change in the enthalpy of an ideal gas can be calculated from the ideal-gas heat capacity $C_p^{IG}(T)$

$$\Delta H_{II} = \int_{T_1}^{T_2} \left(\frac{\partial H}{\partial T}\right)_{P=0} dT = \int_{T_1}^{T_2} C_p^{IG} dT. \tag{3.10}$$

With analytical expressions for the equation of state and the ideal-gas heat capacity, Eqs. (3.7)–(3.10) allow us to predict the enthalpy of a thermodynamic system at arbitrary conditions.

3.1.1 Summary

This introductory section serves as a reminder that the ideal-gas model is a useful approximation for low-pressure gases. Additionally, it is often employed as a reference to predict the thermodynamic properties of nonideal systems. As discussed in later sections, statistical-thermodynamic models of ideal-gas systems are not only valuable for understanding thermodynamic quantities from an atomistic viewpoint but also for modeling equilibrium constants as well as kinetic and transport properties.

3.2 Monatomic Ideal Gases

A monatomic ideal gas consists of individual atoms not interacting with each other (e.g., low-density argon or krypton). From the viewpoint of statistical mechanics, each atom can be considered as an independent subsystem with the microscopic degrees of freedom defined by the quantum states of electrons and nuclei, i.e., the quantum states of individual atoms.

For a system with N noninteracting atoms, each microstate can be expressed as

$$\nu = (\nu_1, \nu_2, \cdots, \nu_N) \tag{3.11}$$

where $\nu_{k=1,2\cdots N}$ represents the quantum state of atom k. If the N atoms are identical, the canonical partition function is

$$Q = \sum_{\nu} \exp(-\beta E_{\nu}) = \frac{1}{N!}\sum_{\nu_1}\sum_{\nu_2}\cdots\sum_{\nu_N} \exp(-\beta E_{\nu}) \tag{3.12}$$

where $\beta = 1/(k_B T)$, k_B stands for the Boltzmann constant, T is the absolute temperature, and $N!$ is introduced to account for the indistinguishability of identical atoms that occupy the same space.[1]

In an ideal gas, the molecules are assumed not to interact with each other. As a result, the total energy at each microstate, E_{ν}, is given by the summation of the energies of individual atoms

$$E_{\nu} = \varepsilon_{\nu_1} + \varepsilon_{\nu_2} + \ldots + \varepsilon_{\nu_N} \tag{3.13}$$

where ε_{ν_k} represents the energy of atom k at microstate ν_k. Substituting Eq. (3.13) into (3.12) leads to

$$Q = \frac{1}{N!}\sum_{\nu_1} \exp(-\beta\varepsilon_{\nu_1})\sum_{\nu_2} \exp(-\beta\varepsilon_{\nu_2}) \ \ldots \ \sum_{\nu_N} \exp(-\beta\varepsilon_{\nu_N}) = \frac{q^N}{N!} \tag{3.14}$$

where q is called the single-molecule partition function.

Like a regular canonical partition function, the single-molecule partition function is defined as a summation of the Boltzmann factor over all microstates of a single molecule

$$q \equiv \sum_{n} \exp(-\beta\varepsilon_n). \tag{3.15}$$

Although our discussion in this section is concerned with monatomic molecules, Eqs. (3.11)–(3.15) are similarly applicable to all noninteracting molecular systems.

3.2.1 The Born–Oppenheimer Approximation

The Born–Oppenheimer approximation assumes that, in a molecular system, the degrees of freedom of atomic nuclei can be decoupled from those corresponding to electrons. The assumption can be justified by the fact that nuclear mass is much larger than that of electrons: the rest mass of a proton, the smallest nucleus, is more than 1836 times the effective mass of an electron.

---

1 As to be discussed in Chapter 4, the correction for indistinguishability can also be derived as the classical limit of quantum mechanics.

According to the Born–Oppenheimer approximation, the energy of a molecular system can be divided into two components, i.e., the energy of nuclei and that of electrons. The energy of atomic nuclei can be further decomposed into kinetic energy due to the atomic motions and the nuclear energy of protons and neutrons. The latter is constant under typical thermodynamic conditions thus can often be ignored unless unclear reactions are of concern. Meanwhile, electronic energy depends on the electron–electron and electron–nucleus interactions, while the atomic positions are fixed.

In an atomic ideal gas, the energy for each atom can be written as

$$\varepsilon = \varepsilon_e + \varepsilon_{trans} \tag{3.16}$$

where ε_e represents the energy of electrons in each atom, and ε_{trans} denotes the kinetic energy due to the translational motion of the entire atom. With the assumption that the electronic degrees of freedom and the translational motions of individual atoms are independent to each other, we can fractionate the single-molecule partition function into contributions from electronic and translational microstates

$$q = q_e q_{trans}. \tag{3.17}$$

As discussed in the following, we can determine q_e and q_{trans} independently from the quantum states of electrons and those corresponding to the translational motions of a single atom, respectively.

3.2.2 Electronic Partition Function

Electrons in an isolated atom (viz., no chemical bonding or interaction with other atoms) may exist in multiple quantum states with the same or different energy levels. The number of quantum states with the same electronic energy is called degeneracy.

For an isolated atom, we can predict the electronic energies based on standard quantum-mechanical methods. Alternatively, electronic energies can be measured with atomic spectroscopy. Based on the electronic energy levels, we can calculate the electronic partition function

$$q_e = \sum_k g_k e^{-\beta \varepsilon_{e,k}} = g_0 + g_1 e^{-\beta \varepsilon_{e,1}} + \cdots \tag{3.18}$$

where $g_k = 2J_k + 1$ is the degeneracy of energy state $\varepsilon_{e,k}$, with J_k being the quantum number for the total angular momentum of all electrons. In Eq. (3.18), subscript 0 denotes the ground state of the electrons, subscript 1 represents the first excited state, and so on for other energy levels. By convention, the electronic energy is set to zero at the ground state.

For most atoms in the periodic table, their electronic energies are available from standard physicochemical handbooks or various online libraries (e.g., the NIST Atomic Spectra Database).[2] Table 3.1 lists the electronic configuration, the term symbol, and the quantum number for the total electronic angular momentum for the first few energy levels of an argon atom. In quantum mechanics, the term symbol is an abbreviated description of the quantum state of an electronic system.

In practical applications, we may express the electronic energy in temperature units, $\theta_{e,k} \equiv \varepsilon_{e,k}/k_B$. In terms of the "electronic temperatures", the electronic partition function becomes

$$q_e = g_0 + g_1 e^{-\theta_{e,1}/T} + g_2 e^{-\theta_{e,2}/T} + \cdots \tag{3.19}$$

One advantage of using Eq. (3.19) over (3.18) is that θ_e provides a direct indication of the temperature beyond which the excited states of electrons must be considered in evaluating

2 https://www.nist.gov/pml/atomic-spectra-database.

Table 3.1 Microstates of electrons in an isolated argon atom.

Configuration	Term	*J*	Energy (cm)	θ_e (× 10^5K)
$3s^23p^6$	^1S	0	0.0000	—
$3s^23p^5(^2p^o{}_{3/2})4s$	$^2[3/2]^o$	2	93 143.7600	1.3398
—	—	1	93 750.5978	1.3486
$3s^23p^5(^2p^o{}_{1/2})4s$	$^2[1/2]^o$	0	94 553.6652	1.3601
—	—	1	95 399.8276	1.3723
$3s^23p^5(^2p^o{}_{3/2})4p$	$^2[1/2]$	1	104 102.0990	1.4975
—	—	0	107 054.2720	1.5399
...	...	...	...	...

In spectroscopy measurements, the energy is conventionally reported in the units of cm, which can be converted to Joule using the Plank constant h and the speed of light c, i.e., $E(\text{J}) = E(\text{cm}) \times 10^2\ hc = E(\text{cm}) \times 1.986 \times 10^{-21}$.
Source: Data from Kramida A., Ralchenko Yu., Reader J. and NIST ASD Team. NIST Atomic Spectra Database (ver. 5.9), National Institute of Standards and Technology (NIST), Gaithersburg, MD, 2021, https://doi.org/10.18434/T4W30F.

the electronic partition function. For example, Table 3.1 suggests that the excited states make negligible contribution to the electronic partition function when the temperature is less than about 10^5 K. Above this temperature, the excited states make significant contributions to the electronic partition function.

Figure 3.3 presents the electronic partition function for argon over a broad range of temperatures. When $T < \theta_{e,1} \approx 10^5$ K, the electronic partition function is the same as the degeneracy of the electrons at the ground state, i.e., $q_e \approx g_0 = 1$. In this case, the excited states make no contribution to the thermodynamic properties of the monatomic ideal gas. Because thermodynamic systems of practical interest rarely reach such an elevated temperature, the electronic partition function is often not explicitly considered unless one is concerned with chemical reactions. As the excitation energy of an atom is typically much higher than k_BT, the statement holds true for most noninteracting molecular systems.

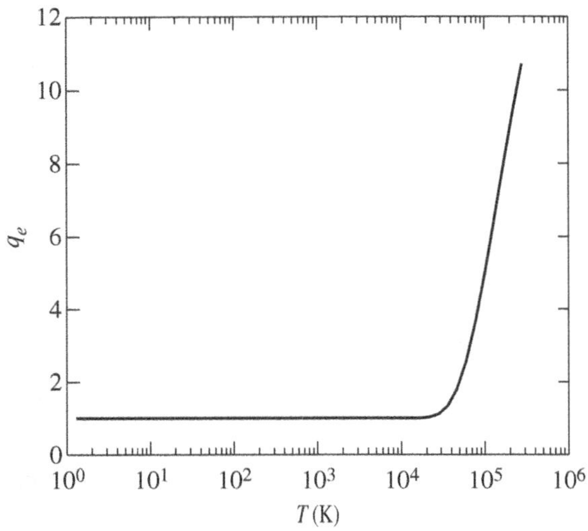

Figure 3.3 The electronic partition function for argon is predicted according to the energy levels presented in Table 3.1.

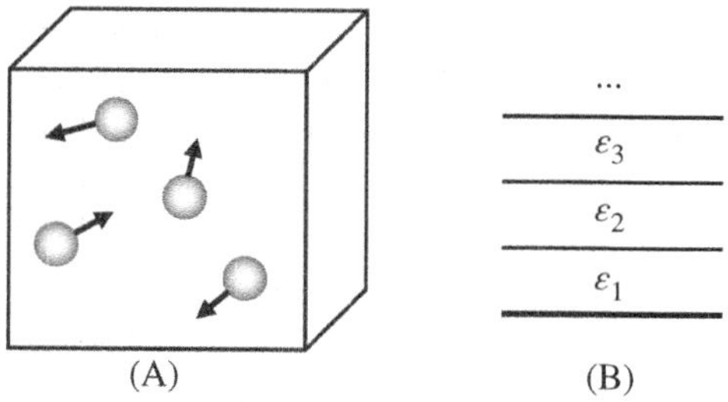

Figure 3.4 (A) Schematic of a cubic box contains N noninteracting atoms moving in different directions. (B) The energy levels (ε_i) for the translational motion of each atom.

3.2.3 Translational Partition Function

To identify the microstates related to atomic motions, i.e., the microstates corresponding to the translational motions of atomic nuclei, consider N noninteracting particles in a cubic box of macroscopic volume $V = a^3$ as shown schematically in Figure 3.4. Because the translational motion is only concerned with the nuclear position, the particle structure plays no role in determining the microstates. In other words, the particle size or shape is immaterial to the translational degrees of freedom.

According to the Schrödinger equation (Problem 3.1), a particle in a box may exist in many quantum states. At each quantum state, the translational motion is associated with a kinetic energy

$$\varepsilon_i = \frac{h^2}{8ma^2}\left(n_x^2 + n_y^2 + n_z^2\right), \qquad \mathbf{n} = (n_x, n_y, n_z) \tag{3.20}$$

where subscript i denotes a quantum state corresponding to the translational motion, $h = 6.626 \times 10^{-34}$ J·s is the Planck constant, m is the particle mass, $a = V^{1/3}$ is the box dimension, and $n_x, n_y, n_z = 1, 2, 3, \cdots$ are quantum numbers related to the translational motion of the particle in x, y, and z directions, respectively. In deriving Eq. (3.20), we assume that the box exerts no force on the particles other than maintaining the system sie and temperature.

The translational partition function is defined as a summation of the Boltzmann factors of all microstates corresponding to the translational motion

$$\begin{aligned} q_{trans} &\equiv \sum_i \exp(-\beta\varepsilon_i) \\ &= \sum_{n_x=1}^{\infty}\sum_{n_y=1}^{\infty}\sum_{n_z=1}^{\infty} \exp\left[-\frac{\beta h^2}{8ma^2}\left(n_x^2+n_y^2+n_z^2\right)\right] = \left(\sum_{n=1}^{\infty}\exp\left[-\frac{\beta h^2 n^2}{8ma^2}\right]\right)^3. \end{aligned} \tag{3.21}$$

In Eq. (3.21), the first equal sign follows from the fact that each microstate is defined by a set of quantum numbers $n_x, n_y, n_z = 1, 2, 3, \cdots$; and the second equal sign follows the equivalency of x, y, and z directions for a particle in a cubic box.

At ambient conditions, the dimensionless quantity in Eq. (3.21), $\alpha = \beta h^2/(8ma^2)$, is extremely small (e.g., $\alpha \approx 2 \times 10^{-22}$ for an argon gas at $T = 300$ K in 1 m^3 box). As a result, the summation over n may be replaced by an integration with respect to a continuous variable $x \equiv \sqrt{\alpha}n$

$$\sum_{n=1}^{\infty} \exp[-\alpha n^2] \approx \frac{1}{\sqrt{\alpha}}\int_0^{\infty} e^{-x^2}\, dx = \frac{1}{\sqrt{\alpha}}\frac{\sqrt{\pi}}{2} = \sqrt{\frac{2\pi m a^2}{\beta h^2}} = \frac{a}{\Lambda} \tag{3.22}$$

where $\Lambda \equiv \sqrt{\beta h^2/2\pi m}$ is called the de Broglie thermal wavelength or the thermal wavelength.

Substituting Eq. (3.22) into (3.21) yields

$$q_{trans} = \frac{V}{\Lambda^3} \tag{3.23}$$

where $V = a^3$ is the volume of the cubic box. Eq. (3.23) provides a simple interpretation of the translational partition function, i.e., q_{trans} can be understood as the number of ways to place a particle on a lattice with V/Λ^3 cells. Each quantum state corresponding to the translational motion occupies a single cell of volume Λ^3. For typical atomic systems at room temperature, the de Broglie thermal wavelength is about a few angstroms (10^{-10} m). As the system size is macroscopic, the translational partition function is a vast number.

3.2.4 Thermodynamic Properties of Monatomic Ideal Gases

Based on the electronic and translational partition functions given by Eqs. (3.18) and (3.23), respectively, we can now calculate the canonical partition function for a monatomic ideal gas with N noninteracting atoms

$$Q = \frac{q_e^N q_{trans}^N}{N!}. \tag{3.24}$$

Eq. (3.24) provides a starting point to derive the thermodynamic properties of the ideal gas system. For example, the Helmholtz energy is given by

$$F = -k_B T \ln Q = -k_B T[N \ln(q_e V/\Lambda^3) - \ln N!]. \tag{3.25}$$

Using the Stirling approximation, $\ln N! \approx N \ln N - N$, we can rewrite Eq. (3.25) as

$$\beta F/N = -\ln q_e + \ln(\rho\Lambda^3) - 1 \tag{3.26}$$

where $\rho \equiv N/V$ represents the number density of the ideal-gas molecules. At ambient conditions, the electronic partition function may be approximated by the degeneracy of the electronic energy at the ground state, $q_e \approx g_0$. For systems without chemical reactions, the electronic partition function is often irrelevant to practical applications.

From Helmholtz energy F, we can readily derive analytical expressions for pressure P, internal energy U, and chemical potential μ:

$$P = -\left(\frac{\partial F}{\partial V}\right)_{N,T} = -Nk_B T\left(\frac{\partial \ln \rho}{\partial V}\right)_{N,T} = \frac{Nk_B T}{V}, \tag{3.27}$$

$$U = \left(\frac{\partial \beta F}{\partial \beta}\right)_{N,V} = U_e + 3N\left(\frac{\partial \ln \Lambda}{\partial \beta}\right)_{N,V} = U_e + 3\,Nk_B T/2, \tag{3.28}$$

$$\mu = \left(\frac{\partial F}{\partial N}\right)_{T,V} = k_B T \ln(\rho\Lambda^3/q_e). \tag{3.29}$$

Eq. (3.27) is the familiar ideal gas law. Other quantities can also be derived following standard thermodynamic relations. If the electrons stay at the ground state, the electronic energy U_e is often set as zero.

The entropy of a monatomic ideal gas is related to the difference between the internal energy and the Helmholtz energy

$$S/(Nk_B) = \beta(U - F)/N = -\ln(\rho\Lambda^3) + 5/2 + S_e/(Nk_B) \tag{3.30}$$

where S_e denotes the entropy of electrons. Eq. (3.30) is known as the Sackur–Tetrode equation, named after two physicists who derived it independently in the early 20th century.[3] The Sackur–Tetrode equation provides an accurate description of the entropy related to the translational motion of molecules. For example, Figure 3.5 compares the theoretical predictions with

3 Grimus W., "100th anniversary of the Sackur-Tetrode equation", *Ann. Phys. (Berlin)* 525, A32–A35 (2013).

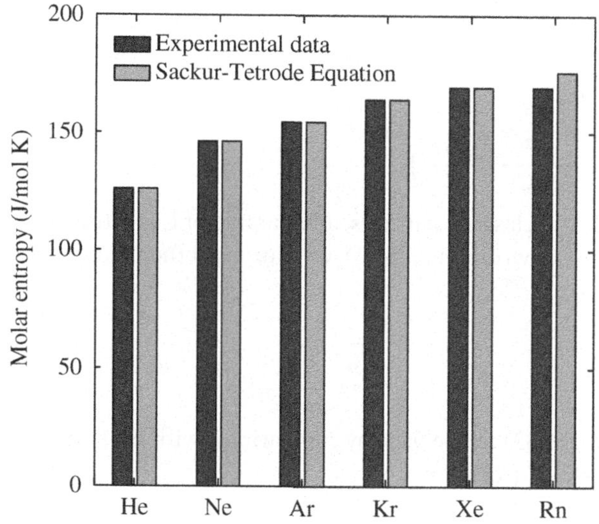

Figure 3.5 The molar entropies of noble gases at the standard state (viz., 25 °C and 1 atm) from the Sackur–Tetrode equation, viz. Eq. (3.30), and from calorimetry measurements.

experimental data for the entropies of five noble gases at the standard state (*viz.*, 25 °C and 1 atm). The experimental data are determined from calorimetry measurements based on the 3rd law of thermodynamics.

For convenience, the translational entropy can be expressed as a function of temperature T (in K), pressure P (in atm), and the molecular weight W (in g/mol) of the ideal-gas molecule:

$$S_{trans}/(Nk_B) = \ln(T^{5/2}W^{3/2}/P) + S_0/(Nk_B) \tag{3.31}$$

where $S_0/(Nk_B) \approx -1.16486$ is known as the Sackur–Tetrode constant. At 300 K and 1 atm, Eq. (3.31) predicts that the reduced entropy for argon ($W = 39.95$) is $S/(Nk_B) = 18.6$, which corresponds to a molar entropy of 155 J/(K·mol). This number agrees perfectly with the experimental value.

It should be noted that the Sackur–Tetrode equation is accurate only at high temperature, i.e., at conditions when the summation of the quantum states can be represented by an integration, Eq. (3.22). The approximation breaks down at low temperature. Because the thermal wavelength diverges at low temperature, the Sackur–Tetrode equation does not predict zero entropy at $T = 0$ K.

3.2.5 Continuous Microstates

According to Newtonian mechanics, the microstate of a monatomic ideal gas is defined by the positions and momenta of individual atoms in continuous space. To derive the partition function, consider a canonical system containing N spherical particles. Without inter-particle interactions, the total energy depends only on the particle momenta $\{\mathbf{p}_{i=1,2,\cdots N}\}$

$$E(\mathbf{r}^N, \mathbf{p}^N) = \sum_{i=1}^{N} \frac{p_i^2}{2m} \tag{3.32}$$

where $\mathbf{r}^N = (\mathbf{r}_1, \mathbf{r}_2, \cdots, \mathbf{r}_N)$, $\mathbf{p}^N = (\mathbf{p}_1, \mathbf{p}_2, \cdots, \mathbf{p}_N)$, m is the particle mass, and $p_i^2 = |\mathbf{p}_i|^2$. Because the positions and momenta of classical particles are continuous variables, the canonical partition function is defined by the integration over the phase space

$$Q = \frac{1}{N!C} \int d\mathbf{p}^N \int d\mathbf{r}^N \exp\left[-\beta \sum_{i=1}^{N} \frac{p_i^2}{2m}\right]$$
$$= \frac{V^N}{N!C} \left[\int_{-\infty}^{\infty} \exp(-\beta p^2/2m) dp\right]^{3N} = \frac{V^N}{N!C^N} \left(\frac{2m\pi}{\beta}\right)^{3N/2} \tag{3.33}$$

where $N!$ accounts for the indistinguishability of classical particles, and parameter C is introduced to make the partition function dimensionless. In writing Eq. (3.33), we have used the mathematic identity

$$\int_{-\infty}^{\infty} dx e^{-x^2/b} = \sqrt{\pi b} \tag{3.34}$$

where $b > 0$ is a positive constant.

For monatomic systems, parameter C in Eq. (3.33) can be fixed by comparing it with the partition function from the quantum-mechanical model, Eq. (3.24)

$$C = h^3/q_e. \tag{3.35}$$

Eq. (3.35) suggests that, in evaluating the partition function for a system of N atoms represented by classical particles, we may simply replace the summation over microstates by integration over the phase space

$$\sum_{\nu} \to \frac{q_e^N}{N!h^{3N}} \int d\mathbf{p}^N \int d\mathbf{r}^N. \tag{3.36}$$

For each particle with the kinetic energy less than $\varepsilon = p^2/2m$, Eq. (3.36) can be written as

$$\sum_{\nu_1} \to \frac{1}{h^3} \int d\mathbf{p} \int d\mathbf{r} = \frac{V}{h^3}\frac{4\pi p^3}{3} = \frac{V}{h^3}\frac{4\pi}{3}(2m\varepsilon)^{3/2} \tag{3.37}$$

where ν_1 denotes all possible microstates. If we express the summation in terms of an integration with respect to ε, Eq. (3.37) becomes

$$\sum_{\nu_1} \to \int g(\varepsilon) d\varepsilon = \frac{V}{h^3}\frac{4\pi}{3}(2m\varepsilon)^{3/2} \tag{3.38}$$

where $g(\varepsilon)$ is called the density of states, which has an explicit form of

$$g(\varepsilon) = \frac{V}{h^3} 2\pi (2m)^{3/2} \varepsilon^{1/2}. \tag{3.39}$$

Although Eq. (3.39) is derived from a monatomic ideal gas as a classical system, the expression is generically applicable to the continuous representation of other quantum systems.

3.2.6 The Maxwell–Boltzmann Distribution

According to Newtonian mechanics, the kinetic energy for the translational motion of a classical particle is related to the particle mass m and velocity $\mathbf{v}$

$$\varepsilon = mv^2/2 \tag{3.40}$$

where $v = |\mathbf{v}|$. Because the kinetic and the potential energies are independent of each other, the canonical partition function predicts that the particle velocity follows the Boltzmann distribution

$$p(\mathbf{v}) \sim \exp(-\beta m v^2/2). \tag{3.41}$$

From the normalization condition $\int p(\mathbf{v})d\mathbf{v} = 1$, we can fix the proportionality constant in Eq. (3.41), leading to

$$p(\mathbf{v}) = \left(\frac{m}{2\pi k_B T}\right)^{3/2} \exp\left[-\frac{mv^2}{2k_B T}\right]. \tag{3.42}$$

Eq. (3.42) is known as the *Maxwell–Boltzmann distribution* for the velocities of classical particles. It is applicable to the translational motions of all classical systems, either ideal gas or nonideal gas. Because the motions in (x, y, z) directions are independent and $v^2 = v_x^2 + v_y^2 + v_z^2$, a similar equation can be written for each component of the particle velocity

$$p(v_\alpha) = \left(\frac{m}{2\pi k_B T}\right)^{1/2} \exp\left[-\frac{mv_\alpha^2}{2k_B T}\right] \tag{3.43}$$

where $\alpha = x, y$, or z, and $p(v_x)p(v_y)p(v_z) = p(\mathbf{v})$.

The Maxwell–Boltzmann distribution provides insights on the physical meanings of temperature and pressure of an ideal gas from a molecular perspective. For the translational motion of a classical particle, the average kinetic energy in each direction of motion is proportional to the absolute temperature

$$K_\alpha = \int_{-\infty}^{\infty} dv_\alpha \sqrt{\frac{\beta m}{2\pi}} e^{-\beta m v_\alpha^2/2} \left(mv_\alpha^2/2\right) \overset{x=\sqrt{\beta m/2}v_\alpha}{=} \frac{k_B T}{\sqrt{\pi}} \int_{-\infty}^{\infty} x^2 e^{-x^2} dx = \frac{k_B T}{2}. \tag{3.44}$$

Eq. (3.44) indicates that, other than a universal constant $k_B/2$, the absolute temperature is equivalent to the translational kinetic energy of individual particles. The ideal gas pressure can be derived from the collision of gas molecules on a testing surface. As shown in Figure 3.6, the elastic collision of each gas molecule results in a momentum transfer of $2mv_z^+$, where v_z^+ represents the velocity moving toward the surface. The gas pressure corresponds to the average force per unit area of the imaginary testing surface due to particle collisions

$$\begin{aligned} P &= \int_0^{\infty} dv_z p(v_z) \frac{(2mv_z/dt)}{A} (\rho A v_z dt) \\ &= 2\rho \int_{-\infty}^{\infty} dv_z \sqrt{\frac{\beta m}{2\pi}} e^{-\beta m v_z^2/2} \left(mv_z^2/2\right) = \rho k_B T. \end{aligned} \tag{3.45}$$

In Eq. (3.45), $2mv_z/dt$ corresponds to the force due to the momentum transfer of a colliding particle with velocity v_z^+ over an infinitesimal time dt, $\rho A v_z dt$ is the number of particles that collide with the surface in dt, and the second equal sign follows from Eq. (3.44). Eq. (3.45) is again the familiar

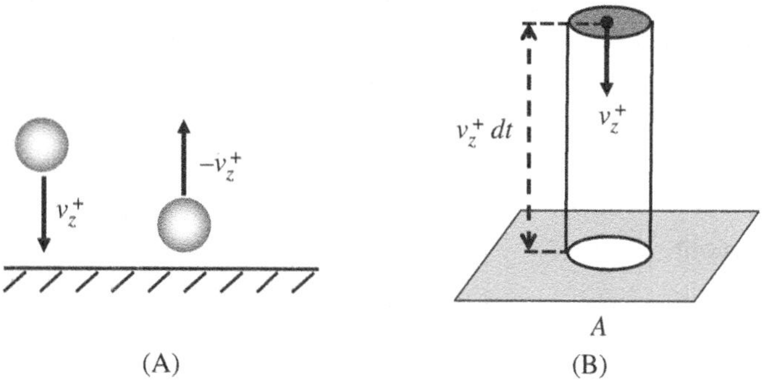

Figure 3.6 (A) Schematic of elastic collision between an ideal-gas molecule and an imaginary testing surface. (B) The swept volume in differential time *dt* through surface area *A* by ideal-gas molecules of velocity v_z^+.

ideal-gas law; its connection with molecular motions provides a mechanical interpretation of the ideal-gas pressure.

The Maxwell–Boltzmann distribution plays a key role in establishing the kinetic theory of gases. For example, the interaction of ideal-gas molecules with a surface is related to gas effusion. As shown in Figure 3.6(B), the flux of gas molecules through a hole at the boundary is the same as the number of molecular collisions per surface area. Accordingly, the gas flux is given by

$$J = \int_0^\infty dv_z p(v_z) \frac{\rho A v_z dt}{A dt} = \rho \int_0^\infty dv_z \sqrt{\frac{\beta m}{2\pi}} e^{-\beta m v_z^2/2} v_z = \frac{P}{\sqrt{2\pi m k_B T}}. \quad (3.46)$$

Eq. (3.46) is known as the Hertz–Knudsen equation, commonly used in describing the kinetics of catalytic reactions, gas effusion, and evaporation processes.

3.2.7 Summary

The Born–Oppenheimer approximation posits that the microstates of a molecular system can be represented by the quantum states of atomic nuclei and electrons. In the case of a monatomic ideal gas, it predicts that the single-molecule partition function can be expressed as the product of the translational and electronic partition functions. Under ambient conditions, the former can be described by the particle-in-a-box model, whereas the electronic partition functions can often be approximated by the degeneracy of the electrons at the ground state. By utilizing the single-molecule partition function, we can predict all thermodynamic properties of an ideal gas following standard statistical-mechanical equations. Alternatively, the single-molecule partition function can be obtained from a classical perspective. The thermodynamic model provides a foundation to derive the Maxwell–Boltzmann distribution for the velocities of classical particles and the kinetic theory of gases.

3.3 Diatomic Molecules

In this and the next few sections, we discuss statistical-mechanical methods for predicting the thermodynamic properties of ideal-gas systems containing small molecules, which consist of only a few atoms. For such systems, we assume that each molecule adopts a single conformation characterized by a minimum energy, and that the fluctuation in the molecular structure primarily arise from thermal energy ($\sim k_B T$). The microstates due to the atomic motions can be described by simple quantum-mechanical models with translational, rotational, and vibrational degrees of freedom. Meanwhile, electronic microstates can be described using the ground-state approximation. Various modifications of the so-called rigid-rotor-harmonic-oscillator model have been developed for the analyses of molecular structure and electronic energies of polyatomic systems. In practice, such statistical thermodynamic analyses are readily available from standard quantum-chemistry software.

3.3.1 The Internal Degrees of Freedom

A diatomic molecule consists of two atomic nuclei surrounded by electrons. According to the Born–Oppenheimer approximation, electronic energies can be decoupled from those corresponding to atomic motions. The microstates of a diatomic molecule are thus defined by the translational motions related to the position of the molecular center-of-mass, and by variables that characterize the electronic structure, molecular orientation, and the distance between the two nuclei. While the

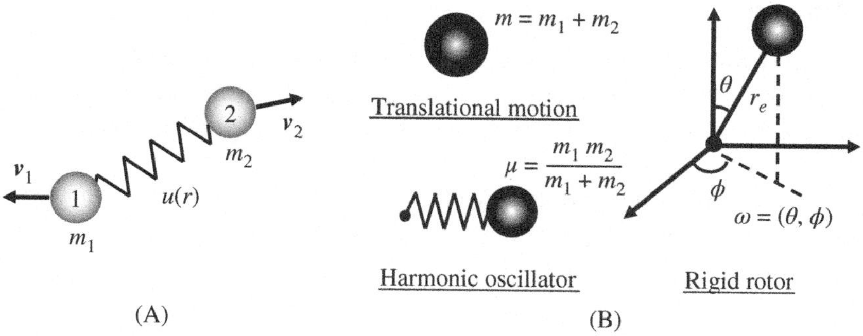

Figure 3.7 Equivalent representations for the dynamics of two particles connected by a bond potential $u(r)$. The particle motions can be described in terms of (A) the positions and velocities of individual particles or equivalently, (B) the translational motion for the center of mass, bond oscillation, and rotation with an effective mass $\mu = m_1 m_2/(m_1 + m_2)$.

translational motion of the entire molecule depends on the system volume, those variables describing the electronic states, bond vibration, and rotation are independent of the system size and thus are commonly referred to as the molecular internal degrees of freedom.

Relative to the energies of two isolated atoms at their respective ground states, the total energy of a diatomic molecule includes an attractive energy responsible for the bond formation $-D_e$, an electronic energy ε_e above that corresponding to the ground-state energy of the molecule, and additional contributions associated with atomic motions. As shown schematically in Figure 3.7,[4] the atomic motions of a diatomic molecule can be equivalently described in terms of the translocation of the center of mass plus bond vibration and rotation (Problem 3.8).

With the atomic degrees of freedom represented by translational, rotational, and vibrational motions, we can divide the total energy of a diatomic molecule into 5 mutually independent components.

$$\varepsilon_i = -D_e + \varepsilon_e + \varepsilon_{trans} + \varepsilon_{rot} + \varepsilon_{vib} \tag{3.47}$$

where the first two terms on the right represent the total electronic energy, ε_{trans} is the kinetic energy corresponding to the translational motion of the entire molecule (viz., the displacement of the molecular center of mass), ε_{rot} denotes the rotational energy relative to the center of mass, and ε_{vib} is the vibrational energy due to bond stretching. Correspondingly, we may write the single-molecule partition function as a product of the partition functions related to different degrees of freedom

$$q = q_n q_e q_{trans} q_{rot} q_{vib} \tag{3.48}$$

where $q_n = e^{\beta D_e}$ denotes the nuclear partition function.

For a thermodynamic system of N identical diatomic molecules not interacting with each other, the canonical partition function is given by

$$Q = \frac{q^N}{N!} \tag{3.49}$$

where $N!$ accounts for the fact that the molecules are indistinguishable.

---

4 Here, μ refers to an effective mass or reduced mass, which is a conventional notation and should not be confused with chemical potential.

3.3.2 Interatomic Potential

According to the Hellman–Feynman theorem from quantum mechanics, the interaction between atomic nuclei is fully determined by the electronic structure or the local density of electrons. For each nuclear configuration, the electronic structure can be predicted from quantum-mechanical calculations.

The interatomic potential describes the total energy of the electrons as a function of nuclear distance. The interatomic potential can be represented using a semi-empirical function

$$v_B(r) = -D_e + D_e\left[1 - e^{-k(r-r_e)}\right]^2 \tag{3.50}$$

where r is the interatomic distance, D_e is the minimum potential energy, r_e is the equilibrium distance (i.e., the bond length), and parameter k determines the width of attraction. Eq. (3.50) is called the Morse potential, which provides a convenient description of bond formation in terms of inter-atomic distance. Table 3.2 shows the Morse parameters for three representative diatomic molecules.

As shown in Figure 3.8, the bond dissociation energy D_0 corresponds to the difference between the minimum electronic energy D_e and the zero-point energy of the nuclei. The latter is a reminiscence of the uncertainty principle, i.e., the nuclear positions cannot be exactly fixed. With the assumption that the bond vibration can be described in terms of a harmonic oscillator, the dissociation energy is given by

$$D_0 = D_e - h\nu/2 \tag{3.51}$$

Table 3.2 Parameters in the Morse potential for three diatomic molecules.

	r_e (Å)	D_e (kJ/mol)	k (Å^{-1})
O_2	1.208	502.7	2.66
B_2	1.590	283.4	1.88
F_2	1.411	159.6	2.85

Source: Adapted from Qadeer S., et al., "Vibrational levels of a generalized Morse potential", *J. Chem. Phys.* 157 (14), 144104 (2022).

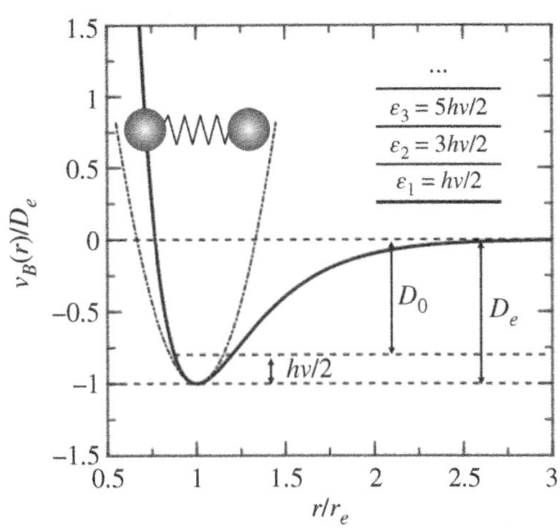

Figure 3.8 The electronic energy versus internuclear distance represented by the Morse potential (solid line) and by the harmonic oscillator model (dotted line). The inset shows the energy levels predicted by the harmonic-oscillator model. Here, ν represents the vibrational frequency. The dissociation energy D_0 corresponds to the difference between the minimum energy D_e and the zero-point energy at the equilibrium bond length r_e.

where ν denotes the bond vibrational frequency. Unlike the minimum potential energy $-D_e$, the bond dissociation energy of a diatomic molecule D_0 can be measured experimentally with spectroscopic or calorimetric methods. It corresponds to the difference in ground-state energy between a diatomic molecule and two isolated atoms.

3.3.3 Electronic and Translational Partition Functions

Like those for monatomic systems, the electronic and translational components of the single-molecule partition function are given by, respectively,

$$q_e = \sum_i g_i e^{-\beta\varepsilon_{e,i}} = g_0 + g_1 e^{-\beta\varepsilon_{e,1}} + \cdots \tag{3.52}$$

$$q_{trans} = \frac{V}{\Lambda^3} \tag{3.53}$$

where g_i is the degeneracy of the electronic state with energy $\varepsilon_{e,i}$, and $\Lambda = \sqrt{h^2/(2\pi m k_B T)}$ is the de Broglie thermal wavelength of the diatomic molecule.

As mentioned above, electronic energy $\varepsilon_{e,i}$ is defined relative to D_e with the atomic positions fixed and the ground-state energy set to zero. At ambient conditions, the electronic partition function can be approximated by $q_e \approx g_0$ because the excitation energy is typically much larger than $k_B T$. In practical applications, electronic degeneracy is often given by the output files of quantum-chemistry calculations.

It should be noted that the microstates and kinetic energies related to the translational motion of a diatomic molecule are the same as those for an atom except that the atomic mass is replaced by the molecular mass, $m = m_1 + m_2$. As discussed earlier for monatomic systems, the translational partition function can be interpreted as the number of ways to place a diatomic molecule on a lattice with V/Λ^3 cells such that each translational quantum state occupies a volume of Λ^3.

3.3.4 Rotational Partition Function

The microstates and energies corresponding to the rotation of a diatomic molecule can be solved from the Schrödinger equation for a rigid rotor (Problem 3.10)

$$\varepsilon_{rot} = \frac{n_R(n_R+1)h^2}{8\pi^2 I} \tag{3.54}$$

where $n_R = 0, 1, 2, \cdots$ are the quantum numbers, and I stands for the moment of inertia. The latter is related to the bond length r_e and the atomic masses of both atoms

$$I = m_1 m_2 r_e^2/(m_1 + m_2). \tag{3.55}$$

The rotational partition function is given by a summation over all rotational quantum states

$$q_{rot} = \frac{1}{\sigma}\sum_{n_R=0}^{\infty}(2n_R+1)\exp\left[-\frac{n_R(n_R+1)h^2}{8\pi^2 I k_B T}\right] \tag{3.56}$$

where $2n_R + 1$ accounts for the degeneracy at the rotational energy level ε_{rot}, and σ is called the rotational symmetry number, i.e., the number of equivalent orientations for the diatomic molecule. Intuitively, the symmetry number accounts for the atomic indistinguishability when the two nuclei are identical. It is $\sigma = 2$ for homonuclear molecules such as O_2 and N_2, and $\sigma = 1$ for heteronuclear molecules such as CO and HF.

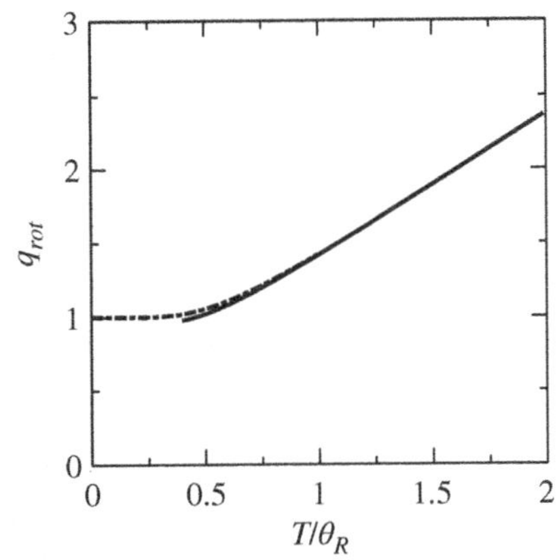

Figure 3.9 The rotational partition function predicted by Eqs. (3.57) and (3.58), respectively, for diatomic molecules at low (dashed line) and high (solid line) temperatures.

At low temperature ($T < \theta_R$), we may approximate the rotational partition function by the first few terms on the right side of Eq. (3.56)

$$q_{rot} = \frac{1}{\sigma}\left[1 + 3e^{-2\theta_R/T} + 5e^{-6\theta_R/T} + 7e^{-12\theta_R/T}\ldots\right] \tag{3.57}$$

where $\theta_R \equiv h^2/(8\pi^2 k_B I)$ is called the rotational temperature. At high temperature ($T > \theta_R$), the summation can be evaluated using the Euler–Maclaurin formula

$$q_{rot} = \frac{T}{\sigma\theta_R}\left[1 + \frac{1}{3}(\theta_R/T) + \frac{1}{15}(\theta_R/T)^2 + \frac{4}{315}(\theta_R/T)^3 + \cdots\right]. \tag{3.58}$$

Figure 3.9 shows the predictions of Eqs. (3.57) and (3.58) for a heteronuclear diatomic molecule ($\sigma = 1$). The low- and high-temperature predictions cross at $T = \theta_R$, which provides a simple criterion for the selection of Eqs. (3.57) or (3.58). For typical diatomic molecules, θ_R is on the order of 10 K (see Table 3.3). Therefore, at ambient conditions, the rotational partition function may be approximated by

$$q_{rot} \approx \frac{T}{\sigma\theta_R}. \tag{3.59}$$

Note that Eq. (3.59) neglects all corrections from the power series of θ_R/T.

3.3.5 Vibrational Partition Function

With the bond vibration represented by *a harmonic oscillator*, the interatomic potential becomes

$$v_B(r) = -D_e + \frac{1}{2}k_s(r - r_e)^2 \tag{3.60}$$

where k_s is the spring constant. The quantum states and corresponding energies can be derived from the Schrödinger equation for the one-dimensional (1D) oscillation of a single particle with an effective mass (Problem 3.13)

$$\varepsilon_{vib} = (n_v + 1/2)h\upsilon \tag{3.61}$$

where $n_v = 0, 1, 2\cdots$ are the vibrational quantum numbers, and

$$\upsilon = \frac{1}{2\pi}\sqrt{\frac{k_s(m_1 + m_2)}{m_1 m_2}} \tag{3.62}$$

stands for the bond vibrational frequency. Note that $h\upsilon/2$ corresponds to the zero-point energy of the nuclei, which does not disappear even at $T = 0$ K.

Table 3.3 Parameters for some diatomic molecules.

	Electronic state	θ_v (K)	θ_R (K)	D_0 (kJ/mol)	r_e (Å)
B_2	$^3\Sigma_g^-$	1512.3	1.74	290.0	1.590
Br_2	$^1\Sigma_g^+$	468.0	0.12	139.9	2.2811
CO	$^1\Sigma^+$	3121.3	2.78	1070.1[a)]	1.12832
Cl_2	$^1\Sigma_g^+$	805.2	0.35	242.9	1.9872
D_2	$^1\Sigma_g^+$	4481.6	43.79	443.3	0.74152
F_2	$^1\Sigma_g^+$	1319.0	1.28	158.2	1.41264
HBr	$^1\Sigma^+$	3810.5	12.18	366.2	1.41444
HCl	$^1\Sigma^+$	4302.4	15.24	431.4	1.27456
HF	$^1\Sigma^+$	5953.1	30.14	569.7	0.91685
H_2	$^1\Sigma_g^+$	6331.1	87.54	430.554[b)]	0.74144
I_2	$^1\Sigma_g^+$	308.6	0.05	152.3	2.666
NO	$^2\Pi_{1/2}$	2739.2	2.41	630.6	1.15077
N_2	$^1\Sigma_g^+$	3392.8	2.87	944.9	1.09769
O_2	$^3\Sigma_g^-$	2273.1	2.08	498.5	1.20752
S_2	$^3\Sigma_g^-$	1043.9	0.42	430.0	1.88941

Here D_0 stands for the bond dissociation energy. The upper-corner number of the electronic state symbol stands for spin multiplicity (viz., the electronic degeneracy).

a) Saberi M., Vlemmings W. H. T. and De Beck E. "Photodissociation of CO in the outflow of evolved stars", *A&A*, 625, A81 (2019).

b) Cheng C. et al., "Dissociation energy of the hydrogen molecule at 10^{-9} accuracy", *Phys. Rev. Lett.* 121, 013001 (2018).

Source: From *CRC Handbook of Chemistry and Physics* (103rd Edition). 2022.

The vibrational partition function is obtained by a summation of over the quantum states

$$q_{vib} = \sum_{n_v=0}^{\infty} e^{-\beta(n_v+1/2)h\upsilon} = \frac{e^{-\theta_v/2T}}{1 - e^{-\theta_v/T}} \tag{3.63}$$

where $\theta_v \equiv h\upsilon/k_B$ is called the vibrational temperature. At elevated temperature ($T > \theta_v$), Eq. (3.63) predicts $q_{vib} \approx \theta_v/T$. At low temperature ($T < \theta_v$), $q_{vib} \approx e^{-\theta_v/2T}$, i.e., the vibrational partition function is dominated by the ground state of the harmonic oscillator or the zero-point energy. It is worth noting that the harmonic oscillator model breaks down when the vibrational frequency is exceedingly small. As $\theta_v \to 0$, Eq. (3.63) predicts $q_{vib} \to \infty$, which is not physically meaningful.

3.3.6 Thermodynamic Properties of Diatomic Ideal Gases

Substituting the electronic, translational, vibrational, and rotational partition functions into Eq. (3.49) gives the canonical partition function of an ideal gas of diatomic molecules

$$Q = \frac{(q_n q_e q_{trans} q_{rot} q_{vib})^N}{N!}. \tag{3.64}$$

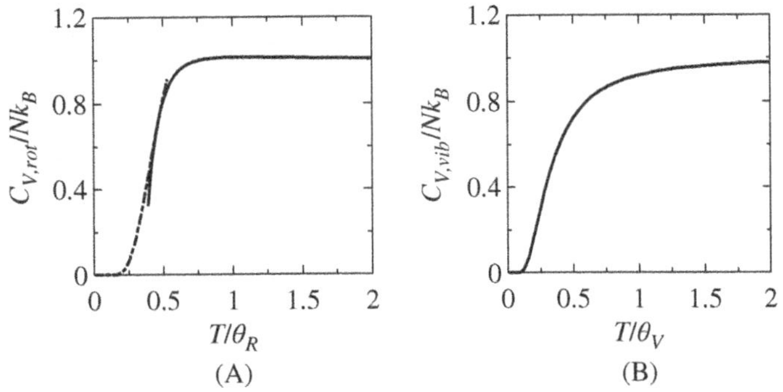

Figure 3.10 The rotational and vibrational contributions to the heat capacity of a diatomic ideal gas. (A) Predictions of the rigid rotor model. The dashed and solid lines are calculated from Eqs. (3.57) and (3.58), respectively. (B) The vibrational heat capacity is predicted by the harmonic oscillator model.

For convenience, $e^{-\theta_v/2T}$ from the vibrational partition function is often combined with the nuclear partition function $q_n = e^{\beta D_e}$. In that case, Eq. (3.64) can be rewritten as

$$Q = \frac{\left(q_e q_{trans} q_{rot} q_{vib}^* e^{\beta D_0}\right)^N}{N!} \tag{3.65}$$

where $D_0 = D_e - h\nu/2$, and the modified vibrational partition function is defined as

$$q_{vib}^* = \frac{1}{1 - e^{-\theta_v/T}}. \tag{3.66}$$

From Eq. (3.65), we can derive the internal energy, entropy, and heat capacity following the standard statistical-thermodynamic relations:

$$\beta U/N \approx \frac{5}{2} + \frac{\theta_v/T}{e^{\theta_v/T} - 1} - \frac{D_0}{k_B T}, \tag{3.67}$$

$$S/(Nk_B) \approx -\ln(\rho\Lambda^3) + 7/2 + \ln(T/\sigma\theta_R) + \frac{\theta_v/T}{e^{\theta_v/T} - 1} - \ln\left(1 - e^{-\theta_v/T}\right) + \ln g_0, \tag{3.68}$$

$$\frac{C_V}{Nk_B} \approx \frac{5}{2} + \left(\frac{\theta_v}{T}\right)^2 \frac{e^{\theta_v/T}}{\left(e^{\theta_v/T} - 1\right)^2}. \tag{3.69}$$

In the above equations, we assume $q_e \approx g_0$ and $q_{rot} \approx T/\sigma\theta_R$.

Figure 3.10 shows the contributions to the heat capacity of an ideal gas of diatomic molecules due to the rotational and vibrational motions. Note that the low-temperature approximation for the rotational partition function predicts an accurate heat capacity when $T < \theta_R/2$. In terms of translational motions, the kinetic energy of a diatomic molecule is the same as that of a monatomic molecule. In other words, this component of the internal energy (viz., $3k_BT/2$ per molecule) is independent of the molecular structure. From the classical perspective, the Maxwell–Boltzmann distribution for molecular velocity is directly applicable to the translational motions of diatomic (and polyatomic) molecules. A similar probability distribution can also be derived for the angular velocity of the molecular rotation (Problem 3.12).

Table 3.3 presents the vibrational and rotational temperatures, θ_v and θ_R, along with the rotational symmetry number σ, bond length r_e, and dissociation energy D_0 for a few diatomic molecules. With the molecular parameters from spectroscopy measurements or quantum-chemistry calculations, the single-molecule partition function can be used to predict all thermodynamic properties including the various forms of free energies and chemical potential.

We observe from Table 3.3 that, except iodine and bromide, the vibrational temperature of a diatomic molecule is much higher than room temperature. Therefore, the vibrational energy can often be approximated by $k_B\theta_v/2$ (viz., the zero-point energy of a harmonic oscillator). As shown in Figure 3.10, neither rotation nor vibration makes a significant contribution to the heat capacity as the temperature approaches absolute zero. When $T \gg \theta_v$, the vibrational energy approaches k_BT, which includes equal contributions from the kinetic and potential energy as predicted by the equipartition theorem (Box 3.1).

3.3.7 Summary

Simple quantum-mechanical models can be used to accurately describe the thermodynamic properties of a diatomic ideal gas by considering the translational, rotational, and vibrational degrees of freedom. While these models are relatively simple, they provide important insights into fundamental concepts such as microstates, the equal partition theorem, and the additivity and relativity of extensive thermodynamic variables. By applying these models, we can predict the behavior of diatomic ideal gases and gain a deeper understanding of their thermodynamic properties.

Box 3.1 Equipartition Theorem

Classical physics asserts that both translational and rotational motions of an object result in a kinetic energy proportional to the velocity squared in each dimension. Using a general form of kinetic energy $k\omega^2$, where k is a constant and ω denotes velocity, we can calculate the average kinetic energy according to the Boltzmann distribution

$$\langle k\omega^2 \rangle = \frac{\int_{-\infty}^{\infty} k\omega^2 e^{-\beta k\omega^2} d\omega}{\int_{-\infty}^{\infty} e^{-\beta k\omega^2} d\omega}$$
$$= -\frac{d \ln \int_{-\infty}^{\infty} e^{-\beta k\omega^2} d\omega}{d\beta} = -\frac{d\left[\ln (\beta k)^{-1/2} \int_{-\infty}^{\infty} e^{-x^2} dx\right]}{d\beta} = \frac{1}{2\beta} = \frac{k_BT}{2} \tag{3.70}$$

where $x = \sqrt{\beta k}\omega$. A similar procedure can be applied to any energy in a quadratic form (e.g., the potential energy $kx^2/2$ in the harmonic oscillator model).

Eq. (3.70) indicates that the translational and rotational motions make an equal contribution (viz., $k_BT/2$ per degree of freedom or dimension) to the internal energy. In statistical mechanics, the uniform distribution of the kinetic average over the translational and rotational degrees of freedom is known as *the equipartition theorem* (also called the *law of equipartition* or *equipartition principle*). At a fixed bond length, a diatomic molecule has 5 degrees of freedom, i.e., 3 associated with the translational motions in x, y, z directions, and 2 with the rotation, the average kinetic energy is $5k_BT/2$, corresponding to the translational and rotational motions of the entire molecule. If the bond vibration is represented by a harmonic oscillator, both kinetic and potential energy contribute $k_BT/2$ to the internal energy per molecule.

3.4 Polyatomic Molecules

A typical molecule consists of two or more atoms bonded together with a potential energy much larger than the average kinetic energy of individual atoms ($\sim k_BT$). While there are only slightly over 100 elements in the periodic table, the types of molecules are unlimited. It has been estimated that the number of carbon-based compounds with a molecular weight of less than 500 Daltons is on the order of 10^{63}, the atomic total mass for a single copy of these molecules would be larger than that of the known universe.[5] Indeed, understanding the properties of a single molecule is not an easy task.

The properties of polyatomic molecules are much more complex than those of atomic or diatomic molecules. This complexity arises from several factors, such as multibody interactions in the bond potential, multiple low-energy conformations, and molecular flexibility. As a result, predicting the properties of polyatomic molecules requires advanced statistical-thermodynamic methods, such as *ab initio* molecular dynamic simulations or coarse-grained models. However, if a polyatomic molecule contains only a few atoms (e.g., H_2O, CO_2, NH_3), we may predict its single-molecule partition function and thermodynamic properties using the rigid rotor-harmonic oscillator approximation, as discussed in the previous section.

3.4.1 Molecular Structure

The 3-dimensional structure of a polyatomic molecule can be determined from crystallographic data or quantum-mechanical calculations. In the latter approach, the procedure is like that for determining the bond length of a diatomic molecule except that the mathematical complexity rises rapidly as the number of atoms in the molecule increases. Instead of calculating the potential energy as a function of the interatomic distance, the potential energy of a polyatomic molecule depends on the positions of all atoms, $\mathbf{R} = (\mathbf{r}_1, \mathbf{r}_2, \cdots, \mathbf{r}_n)$. The molecular configuration is determined by minimizing the potential energy $v_B(\mathbf{R})$ in a multidimensional space. The calculation leads to the minimum energy of a polyatomic molecule relative to the ground-state energies of individual atoms at rest

$$D_e = -\min\{v_B(\mathbf{R})\}. \tag{3.71}$$

As the potential energy of a polyatomic molecule does not vary with the molecular position, the quantum-mechanical calculation is typically set up with the origin of the coordinate system coinciding with the molecular center of mass

$$\mathbf{r}_{cm} \equiv \frac{1}{m}\sum_{i=1}^{n} \mathbf{r}_i m_i \tag{3.72}$$

where $m = \sum_{i=1}^{n} m_i$. The atomic positions are then expressed within the molecular frame according to the CIF (Crystallographic Information File) format.[6]

3.4.2 Single-Molecule Partition Function

Following the Born–Oppenheimer approximation, we can define the microstates of a polyatomic molecule in terms of the nuclear and electronic degrees of freedom. If the molecule has a rigid structure, the microstates due to the atomic motions can be further divided into those corresponding to the translational, rotational, and vibrational motions.

According to the rigid rotor-harmonic oscillator approximation, the total energy of a polyatomic molecule consists of five mutually independent contributions

$$\varepsilon_i = -D_e + \varepsilon_e + \varepsilon_{trans} + \varepsilon_{rot} + \varepsilon_{vib} \tag{3.73}$$

---

5 Bohacek R. S., McMartin C. and Guida W. C., "The art and practice of structure-based drug design: a molecular modelling perspective", *Med. Res. Rev.* 16, 3–50 (1996).

6 https://www.ccdc.cam.ac.uk/Community/depositastructure/cifsyntax/.

where the electronic energy ε_e is defined relative to the minimum energy of the molecule $-D_e$. Eq. (3.73) is similar to that corresponding to a diatomic molecule. Because the different components of the molecular energy are independent of each other, the single-molecule partition function can be written as a product of the partition function related to different modes of atomic motions

$$q = q_n q_e q_{trans} q_{rot} q_{vib} \tag{3.74}$$

where $q_n = e^{\beta D_e}$ denotes the nuclear partition function. Like that for a diatomic molecule, the single-molecule partition function of a polyatomic molecule includes contributions from the electronic degrees of freedom q_e, the molecular translational motion q_{trans}, and the rotational and vibrational motions, q_{rot} and q_{vib}, respectively.

3.4.2.1 Electronic and Translational Partition Functions

Like that for a monatomic or diatomic molecule, the electronic partition function can be approximated by the degeneracy at the ground state, $q_e \approx g_0$. The assumption is justified because, for most molecules that are chemically stable, the electronic excitation energy is much larger than $k_B T$. Meanwhile, the translational partition function is given by $q_{trans} = V/\Lambda^3$, where the thermal wavelength Λ is calculated from the total mass of the molecule, $m = \sum_{i=1}^{n} m_i$. Because the pressure of an ideal gas is exclusively determined by the translational degrees of freedom, $q_{trans} = V/\Lambda^3$ implies that the ideal-gas law is universally applicable to all molecular systems.

3.4.2.2 Rotational Partition Function

For a polyatomic molecule with a linear structure, such as C_2H_2 and CO_2, the rotational motion can be described in terms of the zenith and azimuth angles, $\omega = (\theta, \phi)$. The rotational partition function is the same as that for a diatomic molecule except that the moment of inertia is defined as

$$I = \sum_{i=1}^{n} m_i d_i^2 \tag{3.75}$$

where d_i is the distance of the nucleus i to the molecular center of mass. As discussed in Section 3.3, at elevated temperature, the rotational partition function is related to the rotational temperature θ_R and symmetry number σ

$$q_{rot}^{L} \approx \frac{T}{\sigma \theta_R} \tag{3.76}$$

where $\theta_R \equiv h^2/(8\pi^2 k_B I)$. The symmetry number σ accounts for the number of different ways in which the molecule can achieve the same orientation in space by rotation.

Different that for a diatomic molecule, the rotational temperature θ_R for a nonlinear molecule is determined by three principal moments of inertia I_A, I_B, and I_C (Problem 3.11)

$$\theta_R = \frac{h^2}{8\pi^2 k_B}(I_A I_B I_C)^{-1/3}. \tag{3.77}$$

At high temperature, the rotational partition function is given by

$$q_{rot}^{NL} = \frac{\pi^{1/2}(T/\theta_R)^{3/2}}{\sigma}. \tag{3.78}$$

3.4.2.3 Vibrational Partition Function

A polyatomic molecule with n atoms has $3n$ degrees of freedom to specify the atomic motions. If the molecule has a linear structure, the translational motion of the entire molecule can be described by

3 coordinates (x, y, z), and the rotational motion around the center of mass is described by 2 polar angles (θ, ϕ), the remaining $3n-5$ degrees of freedom define atomic vibrations. If the polyatomic molecule is nonlinear, the rotational motion entails 3 Euler angles, and the vibrational motions have $3n$–6 degrees of freedom.[7]

Each independent way of vibration is commonly referred to as a vibration mode. According to the harmonic oscillator model (Problem 3.13), a summation over all quantum states of the vibration leads to a vibrational partition function

$$q_{vib,i} = \frac{e^{-\theta_{v,i}/2T}}{1 - e^{-\theta_{v,i}/T}} \tag{3.79}$$

where $\theta_{v,i} = h\nu_{v,i}/k_B$. The vibrational partition function for the entire molecule accounts for the quantum states for all vibration modes and is thus given by

$$q_{vib} = \prod_{i=1}^{3n-5 \text{ or } 3n-6} \frac{e^{-\theta_{v,i}/2T}}{1 - e^{-\theta_{v,i}/T}} \tag{3.80}$$

where the number of terms $3n-5$ or $3n-6$ depends on the number of vibrational modes, i.e., whether the molecule has a linear or nonlinear structure. If the contributions from the zero-point energy of vibration are combined with the nuclear partition function, the modified vibrational partition functions become

$$q^*_{vib} = \prod_{i=1}^{3n-5 \text{ or } 3n-6} \frac{1}{1 - e^{-\theta_{v,i}/T}}. \tag{3.81}$$

In the application of the harmonic-oscillator model to a polyatomic molecule, we may establish a connection between the vibrational frequencies and the intramolecular interactions through the normal coordinate analysis. The mathematical procedure is quite involved and can be found in specialized texts.[8] Briefly, the vibrational frequencies for a polyatomic molecule with n atoms are calculated from

$$\nu_i = \frac{1}{2\pi}\sqrt{\lambda_i/c}. \tag{3.82}$$

where $i = 1, 2, \cdots, 3n-5$ or $3n-6$ for a linear or a nonlinear molecule, respectively, c is the speed of light, and λ_i are the eigenvalues of a Hessian matrix defined by the intramolecular potential. In Supplementary Materials, the mathematical procedure is elucidated in detail for calculating ν_i of CO_2, one of the simplest polyatomic molecules.

3.4.2.4 Thermodynamic Properties of Polyatomic Ideal Gases

From the single-molecule partition function, we can readily derive internal energy, entropy, heat capacity, and other thermodynamic properties of a polyatomic ideal gas. For a system containing N identical noninteracting polyatomic molecules with a linear structure (e.g., CO_2), the equations for internal energy, heat capacity, and entropy are given by

$$\beta U/N \approx \frac{5}{2} + \sum_{i=1}^{3n-5} \frac{\theta_{v,i}/T}{e^{\theta_{v,i}/T} - 1} - \beta D_0, \tag{3.83}$$

---

7 For a polyatomic molecule at the transition state of chemical reaction, one of the eigenvalues is negative, which leads to two "imaginary frequency" corresponding to the square roots of a negative number. Therefore, a polyatomic molecule has $3n$–6 (linear) or $3n$–7 (nonlinear) vibrational modes at the transition state.

8 See, e.g., Wilson E. B. Jr,., Decius J. C. and Cross P. C., *Molecular vibrations: the theory of infrared and Raman vibrational spectra*. Dover Publications, 1980.

$$\frac{C_V}{Nk_B} \approx \frac{5}{2} + \sum_{i=1}^{3n-5} \left(\frac{\theta_{v,i}}{T}\right)^2 \frac{e^{\theta_{v,i}/T}}{\left(e^{\theta_{v,i}/T} - 1\right)^2}, \tag{3.84}$$

$$S/(Nk_B) \approx -\ln(\rho\Lambda^3) + 7/2 + \ln(T/\sigma\theta_R) + \sum_{i=1}^{3n-5} \left[\frac{\theta_{v,i}/T}{e^{\theta_{v,i}/T} - 1} - \ln\left(1 - e^{-\theta_{v,i}/T}\right)\right] + \ln g_0 \tag{3.85}$$

where $D_0 = D_e - \sum_{i=1}^{3n-5} h\nu_{v,i}/2$ stands for the molecular dissociation energy. For nonlinear polyatomic molecules, the statistical-thermodynamic equations are

$$\beta U/N \approx 3 + \sum_{i=1}^{3n-6} \frac{\theta_{v,i}/T}{e^{\theta_{v,i}/T} - 1} - \beta D_0, \tag{3.86}$$

$$\frac{C_V}{Nk_B} \approx 3 + \sum_{i=1}^{3n-6} \left(\frac{\theta_{v,i}}{T}\right)^2 \frac{e^{\theta_{v,i}/T}}{\left(e^{\theta_{v,i}/T} - 1\right)^2}, \tag{3.87}$$

$$S^{IG}/(Nk_B) \approx -\ln(\rho\Lambda^3) + 4 + \ln\left(\frac{1}{\sigma}\sqrt{\frac{\pi T^3}{\theta_R^3}}\right) + \sum_{i=1}^{3n-6} \left[\frac{\theta_{v,i}/T}{e^{\theta_{v,i}/T} - 1} - \ln\left(1 - e^{-\theta_{v,i}/T}\right)\right] + \ln g_0 \tag{3.88}$$

where $D_0 = D_e - \sum_{i=1}^{3n-6} h\nu_{v,i}/2$. The above equations are accurate for ideal gases of small polyatomic molecules at the classical or high-temperature limit.

Table 3.4 presents the symmetry numbers and vibrational temperatures for a few polyatomic molecules. With the molecular structure and vibrational frequencies obtained from quantum-chemistry calculations or spectroscopy, the statistical-thermodynamic equations allow us to estimate the thermodynamic properties of polyatomic ideal gases (see Problems 3.17 and 3.18 for specific examples).

Table 3.4 The symmetry number, rotational temperature, vibrational temperature, and dissociation energy for some polyatomic molecules.

	σ	θ_{rot} (K)	θ_{vib} (K)	D_0 (kJ/mol)
CO_2	2	0.561	3360, 954(2), 1890	1596.2
H_2O	2	40.1, 20.9, 13.4	2290, 5160, 5360	917.6
NH_3	3	13.6, 13.6, 8.92	4800, 1360, 4880(2), 2330(2)	1158.1
ClO_2	2	2.50, 0.478, 0.400	1360, 640, 1600	378.2
SO_2	2	2.92, 0.495, 0.422	1660, 750, 1960	1062.7
N_2O	2	0.603	3200, 850(2), 1840	1103.7
NO_2	2	11.5, 0.624, 0.590	1900, 1980, 2330	928.0
CH_4	12	7.54, 7.54, 7.54	4170, 2180(2), 4320(3), 1870(3)	1640.5
CH_3Cl	3	7.32, 0.637, 0.637	4270, 1950, 1050, 4380(2), 2140(2) 1460(2)	1551.0
CCl_4	12	0.0823, 0.0823, 0.0823	660, 310(2), 1120(3), 450(3)	1292.0

The numbers in the parentheses denote degeneracy.
Source: Adapted from McQuarrie D. A. *Statistical mechanics*. University Science Books, 2000.

3.4.3 Summary

We conclude this section by noting that the thermodynamic properties of an ideal gas can, in theory, be predicted by combining quantum and statistical mechanical calculations. While this approach is conceptually simple and effective for small molecules, it can become complex when dealing with larger molecules. Nevertheless, with the aid of quantum chemistry software,[9] such calculations can be performed routinely for practical applications.

3.5 Chemical Equilibrium in Ideal-Gas Mixtures

The accurate prediction of the thermodynamic properties of noninteracting molecular systems highlights the power and versatile utility of statistical thermodynamics. On the one hand, the idealized molecular systems provide a useful platform for understanding abstract concepts of statistical mechanics such as microstates, canonical partition function, and additivity and relativity of thermodynamic quantities in terms of microscopic variables related to electronic quantum states and atomic motions. On the other hand, analytical equations derived from statistical thermodynamics are directly applicable to realistic chemical systems of practical interest.

In this section, we will discuss how statistical thermodynamics can be used to predict the equilibrium constants of gas-phase reactions. To keep things simple, we will only provide numerical values for a single reaction between a few small molecules. However, with the help of quantum-chemistry software, similar calculations can be performed for more complex systems, including those involving condensed phases and heterogeneous reactions. Thanks to the increasing power of computers, *ab initio* thermodynamics, which involves predicting thermodynamic properties from the first principles of quantum mechanics, has emerged as one of the most important applications of statistical mechanics.

3.5.1 Non-Reacting Ideal-Gas Mixtures

Molecules in an ideal gas do not interact with each other. As a result, the statistical-thermodynamic equations derived for single-component systems can be readily extended to mixtures containing multiple molecular species.

As the microstates of individual molecules are independent of each other, the canonical partition function for a multicomponent ideal gas can be written as

$$Q = \prod_i \frac{q_i^{N_i}}{N_i!} \tag{3.89}$$

where N_i and q_i are the number of molecules of species i and its single-molecule partition function, respectively. Like for a pure ideal gas, the canonical partition function provides a starting point to predict the thermodynamic properties of the mixture.

For example, the reduced Helmholtz energy for an ideal-gas mixture of monatomic species is given by

$$\beta F = -\ln Q = \sum_i N_i \left[\ln\left(\rho_i \Lambda_i^3\right) - 1 - \ln q_{e,i}\right] \tag{3.90}$$

9 For example, Zheng J. et al., "MSTor: a program for calculating partition functions, free energies, enthalpies, entropies, and heat capacities of complex molecules including torsional anharmonicity", *Comput. Phys. Commun.* 183, 1803–1812 (2012).

where $\rho_i \equiv N_i/V$ is the molecular number density of species i, $\Lambda_i = \sqrt{\beta h^2/2\pi m_i}$ is the thermal wavelength, and $q_{e,i}$ is the electronic partition function. From the Helmholtz energy F, we can derive the ideal-gas pressure,

$$P = -\left(\frac{\partial F}{\partial V}\right)_T = -k_B T \sum_i N_i \left(\frac{\partial \ln \rho_i}{\partial V}\right)_T = \frac{N k_B T}{V} \tag{3.91}$$

where $N = \sum_i N_i$. As expected, the ideal-gas law is applicable to pure as well as mixed ideal gases. Similarly, the internal energy, entropy, and the chemical potential of each species are given by

$$U = \left(\frac{\partial \beta F}{\partial \beta}\right)_{V,N_i} = U_e + \sum_i N_i \frac{\partial \ln \Lambda_i^3}{\partial \beta} = U_e + \frac{3 N k_B T}{2}, \tag{3.92}$$

$$S/k_B = \beta U - \beta F = S_e/k_B - \sum_i N_i \left[\ln\left(\rho_i \Lambda_i^3\right) - 5/2\right], \tag{3.93}$$

$$\beta \mu_i = \left(\frac{\partial \beta F}{\partial N_i}\right)_{T,N_{j\neq i}} = -\left(\frac{\partial \ln Q}{\partial N_i}\right)_{T,N_{j\neq i}} = -\ln(q_i/N_i) \tag{3.94}$$

where U_e and S_e denote electronic energy and entropy, respectively. The electronic properties are determined by the quantum states of electrons within individual atoms. Eq. (3.94) follows from Eq. (3.89) and the Stirling approximation, $\ln N_i! \approx N_i \ln N_i - N_i$. The generic expression is valid for monatomic as well as polyatomic ideal gases.

Similar equations can be derived from ideal-gas mixtures of diatomic and polyatomic molecules. Interestingly, Eq. (3.93) predicts that the entropy of chemical species i in the ideal-gas mixture is the same as that of a pure ideal gas at the same density ρ_i and temperature. Because the molecules do not interact with each other in an ideal-gas system, the total entropy is equal to the summation of those for individual species.

3.5.2 Predicting Chemical Equilibrium Constant

The single-molecule partition functions and ideal-gas properties are instrumental for understanding chemical reactions and kinetics from molecular perspectives. The practical application may be exemplified by the theoretical prediction of the equilibrium constants of gas-phase reactions. Consider a generic expression for a single homogeneous reaction

$$\sum_i \nu_i A_i = 0 \tag{3.95}$$

where ν_i denotes the stoichiometric coefficient, and A_i stands for a chemical species. By convention, $\nu_i > 0$ means A_i is a product, $\nu_i < 0$ for a reactant, and $\nu_i = 0$ for any inert species in the system. As discussed in the standard texts of chemical thermodynamics, the condition of chemical equilibrium follows from the second law of thermodynamics. For a system at a given temperature and pressure, a chemical reaction reaches equilibrium when the total Gibbs energy is minimized. Accordingly, we can find a linear relation among the chemical potentials of the reacting species

$$\sum_i \nu_i \mu_i = 0. \tag{3.96}$$

To find the equilibrium composition, it is conventional to express the chemical potential for each species in terms of that corresponding to a reference state μ_i^0 and activity a_i

$$\mu_i = \mu_i^0 + k_B T \ln a_i. \tag{3.97}$$

For gas-phase reactions, the reference state is typically selected as a pure ideal gas of the same chemical species at system temperature and unit pressure ($P_0 = 1$ atm). Meanwhile, the activity of chemical species i in the gas phase is related to the system pressure P, the mole fraction y_i, and the fugacity coefficient ϕ_i

$$a_i = Py_i\phi_i/P_0. \tag{3.98}$$

In an ideal-gas mixture, $\phi_i = 1$; for a nonideal gas (or liquid mixture), $\phi_i = \phi_i(T, P, y_i)$ can be calculated from an equation of state as discussed in Section 3.1.

Based on the canonical partition function given in Eq. (3.89), we can predict the chemical potential of each chemical species in the reference state from the single-molecule partition function q_i. For a pure ideal gas at system temperature T and reference pressure P_0, the reference chemical potential is given by

$$\mu_i^0 = -k_BT\ln(q_ik_BT/P_0V). \tag{3.99}$$

For convenience, we define the *reference* single-molecule partition function

$$q_i^0 \equiv q_ik_BT/P_0V = \frac{k_BTq_i^*}{P_0\Lambda_i^3} \tag{3.100}$$

where $q_i^* \equiv q_i\Lambda_i^3/V$. Note that q_i^* excludes the contribution due to the translational degree of freedom and that, and for each chemical species, q_i^0 depends only on temperature. As expected, the reference chemical potential, $\mu_i^0 = -k_BT\ln q_i^0$, is independent of the system pressure and composition.

With the chemical potentials expressed in terms of μ_i^0 and a_i, Eq. (3.96) becomes

$$K = \prod_i a_i^{\nu_i} \tag{3.101}$$

where K is called the equilibrium constant, given by

$$K \equiv \exp\left(-\sum_i \nu_i\beta\mu_i^0\right) = \prod_i (q_i^0)^{\nu_i}. \tag{3.102}$$

Note that all quantities in the above equations are dimensionless. Eq. (3.101) is known as the law of mass action. With q_i^0 prediction from the single-molecule partition function for each chemical species, we can calculate equilibrium constant K without experimental input.

It is instructive to elucidate the above procedure with some numerical values. Due to concerns over climate change, significant efforts have been made to convert CO_2 into valuable chemicals and fuels. One of the elementary reactions in CO_2 hydrogenation is

$$CO_2 + H_2 \rightleftharpoons CO + H_2O. \tag{3.103}$$

According to Eq. (3.102), the equilibrium constant can be calculated from the single-molecule partition functions of the underlying chemical species

$$K = \frac{q_{CO}^0 q_{H_2O}^0}{q_{CO_2}^0 q_{H_2}^0} \tag{3.104}$$

where

$$q_{CO/H_2}^0 \approx (g_0 + g_1e^{-\beta\varepsilon_{ne,1}} + \cdots) \times \left(\frac{k_BT}{P_0\Lambda^3}\right) \times \left(\frac{T}{\sigma\theta_R}\right) \times \left(\frac{1}{1-e^{-\theta_v/T}}\right) \times e^{\beta D_0}, \tag{3.105}$$

$$q_{CO_2}^0 \approx (g_0 + g_1e^{-\beta\varepsilon_{ne,1}} + \cdots) \times \left(\frac{k_BT}{P_0\Lambda^3}\right) \times \left(\frac{T}{\sigma\theta_R}\right) \times \prod_{i=1}^{4}\left(\frac{1}{1-e^{-\theta_{v,i}/T}}\right) \times e^{\beta D_0}, \tag{3.106}$$

$$q_{H_2O}^0 \approx (g_0 + g_1e^{-\beta\varepsilon_{ne,1}} + \cdots) \times \left(\frac{k_BT}{P_0\Lambda^3}\right) \times \left(\frac{\pi^{1/2}T^{3/2}}{\sigma(\theta_A\theta_B\theta_C)^{1/2}}\right) \times \prod_{i=1}^{4}\left(\frac{1}{1-e^{-\theta_{v,i}/T}}\right) \times e^{\beta D_0}. \tag{3.107}$$

Table 3.5 Molecular parameters for predicting the equilibrium constant of gas-phase reaction $CO_2 + H_2 \rightleftharpoons CO + H_2O$.

	W	g_0	θ_R (K)	σ	θ_V (K)	D_0 (kJ/mol)
H_2	2	1	87.54	2	6331.1	430.554
CO_2	44	1	0.561	2	3360, 954, 954, 1890	1596.2
H_2O	18	1	40.1, 20.9, 13.4	2	5360, 5160, 2290	917.6
CO	28	1	2.78	1	3121.3	1070.1

In the above equations, we omit distinctions in the molecular parameters among different chemical species for clarity. It is understood that, for example, the thermal wavelengths for different chemical species are not the same, even though the same symbol Λ is used on the right side of the above equations.

With the inputs of parameters characterizing different electronic states (g_i, ε_i), molecular mass (m_i), rotational temperatures (θ_R), and symmetry number (σ), vibrational temperatures (θ_v), we can readily calculate the equilibrium constant from Eq. (3.104). Table 3.5 presents the molecular parameters for the hydrogenation reaction, and Figure 3.11 shows the Van't Hoff plot for the equilibrium constant versus the inverse temperature. It predicts that, at 298 K, the equilibrium constant is $K \approx 1.0405 \times 10^{-5}$, which agrees well with the experimental value of 9.67×10^{-6}. The discrepancy between the theoretical result and experimental data primarily stems from the calculation of the bond energies. Hydrogenation of CO_2 to CO via the reverse water–gas shift reaction has been recognized as one of the most promising processes for CO_2 utilization because CO can be used in downstream Fischer–Tropsch reaction and methanol synthesis, among other industrial applications.[10]

3.5.3 Summary

Statistical thermodynamics provides a powerful tool for predicting the thermodynamic properties of chemical systems with minimal or no experimental data. Although the example discussed here is elementary, similar calculations can be applied to more complex molecular systems with the assistance of modern computers and quantum-chemistry software. By extending these methods to

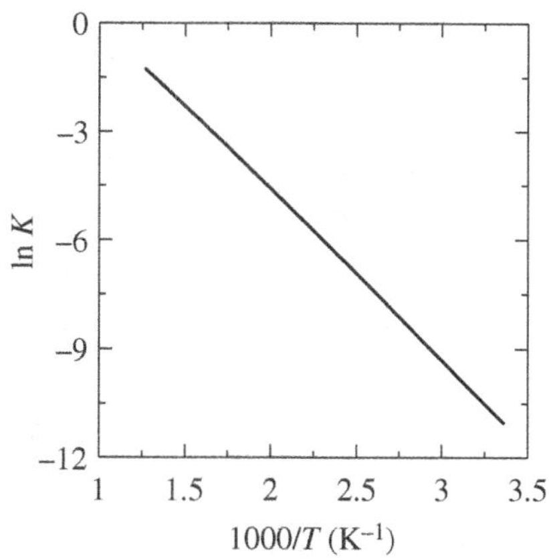

Figure 3.11 The Van't Hoff plot of the equilibrium constant for the hydrogenation of CO_2 to CO and H_2O predicted from Eq. (3.104) with the parameters shown in Table 3.5. The negative slope is a characteristic of endothermic reactions.

10 Jia C. et al., "The thermodynamics analysis and experimental validation for complicated systems in CO_2 hydrogenation process", *J. Energy Chem.* 25, 1027–1037 (2016).

account for solvent effects, we can also predict reduction potentials for electron transfer and other electrochemical reactions and half-reactions in both aqueous and nonaqueous solutions.[11]

3.6 Thermodynamics of Gas Adsorption

Thermodynamic models of gas adsorption play a crucial role in the engineering design of separation and purification processes, as well as in characterizing the surface properties of porous materials through gas adsorption. In practical applications, the equilibrium between adsorbate (gas molecules) and adsorbent (solid surface or substrate) is typically described by adsorption isotherms, which express the amount of adsorption per unit surface area as a function of the gas pressure at a constant temperature. In this section, we demonstrate several applications of the ideal-gas partition functions in the grand canonical ensemble, which are instrumental in deriving adsorption isotherms.

3.6.1 The Langmuir Isotherm

The Langmuir model is named after Irving Langmuir, a surface chemist famous for his pioneering work on the adsorption of gas molecules on solid surfaces. The Langmuir isotherm is commonly applied to gas adsorption on a planar surface at low pressure such that the gas phase behaviors like an ideal gas and the interaction among gas molecules at the surface is negligible. The Langmuir isotherm is also applicable for describing solute adsorption from liquid solutions.

To derive the Langmuir isotherm for gas adsorption, consider the adsorption of a low-pressure gas on a chemically uniform solid surface. Figure 3.12 shows schematically a statistical–mechanical

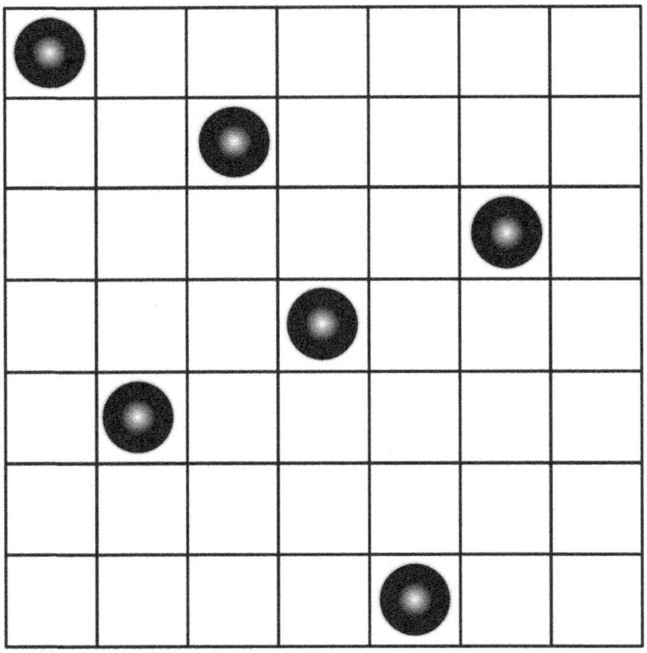

Figure 3.12 Schematic of the Langmuir model of gas adsorption. The gas molecules on the surface are depicted as spherical particles filling a two-dimensional lattice.

11 Marenich A. V. et al., "Computational electrochemistry: prediction of liquid-phase reduction potentials", *Phys. Chem. Chem. Phys.* 16, 15068–15106 (2014).

representation of the gas adsorption. For simplicity, we assume that gas molecules do not interact with each other such that the bulk phase is an ideal gas, and that the solid surface can be described in terms of a two-dimensional lattice with independent binding sites. Each lattice site can bind with only one gas molecule, i.e., the adsorption of gas molecules leads to a monolayer coverage on a planar surface. With these assumptions, gas molecules adsorbed on the surface constitute an open system of noninteracting molecules that can be described with a grand canonical ensemble. Each microstate is specified by the occupancy of the lattice sites along with the quantum state of the gas molecules

$$\nu = (n_1\nu_1, n_2\nu_2, \cdots n_{N_s}\nu_{N_s}) \tag{3.108}$$

where N_s stands for the total number of lattice sites, $n_i = 0, 1$ denotes the number of occupancies of lattice site i, and ν_i is the quantum state of the gas molecule occupying site i.

If $n_i = 0$, there are no gas molecules binding with the surface at the lattice site i; if $n_i = 1$, the lattice site is occupied by a gas molecule. The total number of gas molecules on the surface is thus related to the occupancy of the lattice sites

$$N_\nu = \sum_{i=1}^{N_s} n_i. \tag{3.109}$$

As the gas molecules on the surface are assumed not interacting with each other, the total energy of the system can be written as

$$E_\nu = \sum_{i=1}^{N_s} n_i \varepsilon_{\nu_i} \tag{3.110}$$

where ε_{ν_i} represents the energy of a gas molecule at quantum state ν_i. In writing Eq. (3.110), we assume that the energy of the substrate is a constant invariant with gas adsorption. In other words, the energy of each adsorbed gas molecule is defined relative to a fixed surface energy.

Assuming that gas molecules are immobile at the solid surface (e.g., by strong binding with the surface functional groups) such that they are *distinguishable* by their lattice locations, we can write the grand canonical partition function as

$$\Xi = \sum_\nu e^{-\beta(E_\nu + N_\nu \mu)} = \prod_{i=1}^{N_s} \sum_{n_i=0,1} \sum_{\nu_i} e^{-\beta(\varepsilon_{\nu_i} - \mu)n_i} = \prod_{i=1}^{N_s} (1 + q_s e^{\beta\mu}) = (1 + q_s e^{\beta\mu})^{N_s} \tag{3.111}$$

where $q_s \equiv \sum_{\nu_i} e^{-\beta\varepsilon_{\nu_i}}$ is the single-molecule partition function for an adsorbed molecule, and μ is the chemical potential. The exponent N_s in Eq. (3.111) results from the assumption that the lattice sites are independent of each other. Note that q_s is the same for all lattice sites because the surface is assumed to be uniform. For a heterogeneous surface, a single-molecule partition function may be defined for each type of lattice site. For gas molecules with a rigid structure, q_s can be estimated from quantum-mechanical models like those used for describing the single-molecule partition function (Problem 3.28).

From the grand partition function, we obtain the average number of gas molecules in the system (viz., the average number of adsorbed gas molecules)

$$<N> = \left(\frac{\partial \ln \Xi}{\partial \beta\mu}\right)_\beta = N_s \frac{q_s e^{\beta\mu}}{1 + q_s e^{\beta\mu}}. \tag{3.112}$$

At low pressure, the chemical potential of a pure gas can be approximated by that of an ideal gas

$$\beta\mu = \beta\mu_0 + \ln P \tag{3.113}$$

where μ_0 is the chemical potential of an ideal gas at system temperature and unit pressure. Substituting Eq. (3.113) into (3.112) yields the fraction of surface sites occupied by gas molecules

$$\phi = \frac{<N>}{N_s} = \frac{bP}{1 + bP} \tag{3.114}$$

where $b \equiv q_s \exp(\beta\mu_0) = q_s/q_i^0$ is a parameter that depends only on temperature. As discussed in Section 3.5, $q_i^0 = q_i k_B T/P_0 V$ depends on temperature and the internal degrees of freedom of the gas molecule.

Eq. (3.114) is known as the Langmuir adsorption isotherm. The original equation was derived with arguments much simpler than those discussed above (viz., the kinetics of adsorption/desorption) for understanding the adsorption of gas molecules on metal wires in incandescent light balls. Although the Langmuir isotherm can be interpreted from different perspectives, the statistical-thermodynamic derivation provides microscopic insights into the physics of gas adsorption and a theoretical basis to predict parameter b from first principles. More importantly, the statistical-thermodynamic model offers guidelines for the systematic improvement of the Langmuir isotherm by relaxing some of its assumptions.

The Langmuir isotherm can be used to represent a variety of experimental data for gas adsorption at low pressure (e.g., adsorption of ethyl chloride on charcoal, nitrogen on TiO_2, and O_2 or CO on silica). For example, Figure 3.13 shows that the Langmuir isotherm agrees well with the experimental results for the adsorption of nitrogen and nitrous oxide on a steamed-charcoal surface at 20 °C.[12] When parameter b is obtained from fitting the adsorption data, one can estimate the total surface area of the adsorbent based on the projected area of each gas molecule sitting on the surface. In principle, the method can be used to characterize the surface areas of porous materials such as catalysts or catalyst supports. However, because of its severe simplifying assumptions and the use of adjustable parameter, the Langmuir isotherm is often considered as an empirical tool for representing adsorption data. For physical (as opposed to chemical) adsorption, a much better adsorption isotherm is provided by allowing adsorbing molecules to form more than one adsorbed layer.

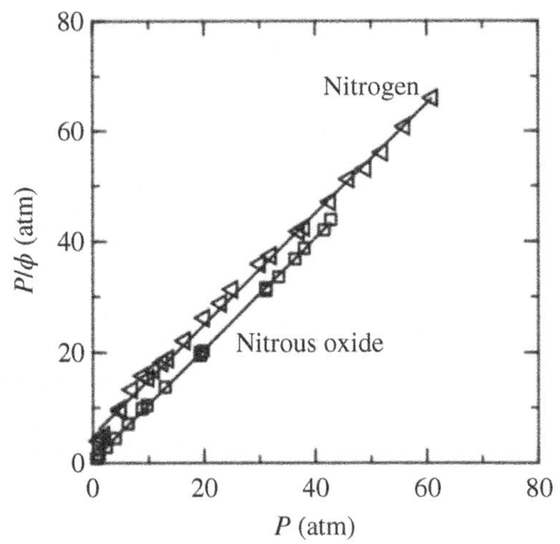

Figure 3.13 Correlation of experimental data (symbols) with that of the Langmuir isotherm (lines) for adsorption of nitrogen and nitrous oxide on a steamed-charcoal surface at 20 °C.

12 Choi W. and Leu M. T., "Nitric acid uptake and decomposition on black carbon (soot) surfaces: its implications for the upper troposphere and lower stratosphere", *J. Phys. Chem. A* 102 (39), 7618–7630 (1998).

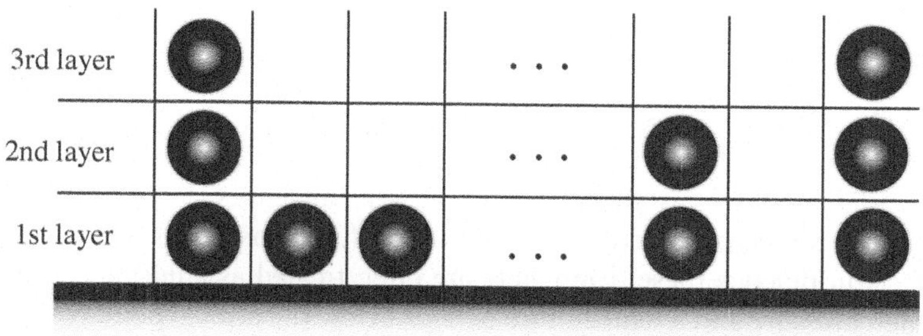

Figure 3.14 Schematic of multilayer gas adsorption at a planar surface (side view). Here gas molecules are depicted as spheres; and each square represents a lattice site near the surface (the shaded area) that can accommodate at most a single gas molecule.

3.6.2 The Brunauer–Emmett–Teller (BET) Isotherm

In deriving the Langmuir isotherm, we assume that gas molecules form a monolayer once they are adsorbed on a solid surface. This assumption is relaxed in the Brunauer–Emmett–Teller (BET) isotherm, which also assumes independent surface binding but allows for the further interaction of the surface sites with more gas molecules, i.e., it accounts for multilayer gas adsorption.

Figure 3.14 shows schematically the multilayer adsorption model. In contrast to the Langmuir model, here a surface site may adsorb a vertical column of gas molecules: the gas molecule closest to the surface has an energy ε_0, determined by the direct interaction of the gas molecule with the surface; and all successive gas molecules in the vertical column have an adsorption energy ε_1, determined primarily by the interactions between neighboring gas molecules. As in the Langmuir model, we assume that the surface is uniform, and all lattice sites are identical and independent of each other. Besides, we assume that gas molecules within the same layer do not interact with each other. In other words, we consider vertical interactions but neglect horizontal interactions between adsorbed molecules.

Like the Langmuir model, gas molecules on the surface constitute a grand canonical ensemble defined by the system temperature, the gas chemical potential, and the surface area of the substrate. Each microstate is specified by the occupation numbers of the lattice sites. The grand partition function is thus given by

$$\Xi = \sum_{i=1}^{N_s} \sum_{n_i=0}^{\infty} q_0 q_1^{n_i-1} e^{\beta\mu n_i} = \left(1 + q_0 e^{\beta\mu} + q_0 q_1 e^{2\beta\mu} + q_0 q_1^2 e^{3\beta\mu} + \cdots\right)^{N_s} \tag{3.115}$$

where q_0 and q_1 stand for the single-molecule partition function for a gas molecule directly binding with the surface and that of a molecule in the second and higher layers of adsorption, and N_s is the number of surface sites. The average number of gas molecules on the surface can be obtained from the partitional derivative of the grand partition function

$$<N>= \left(\frac{\partial \ln \Xi}{\partial \beta\mu}\right)_\beta = N_s q_0 \lambda \frac{1 + 2q_1\lambda + 3(q_1\lambda)^2 + \cdots}{1 + q_0\lambda\left[1 + q_1\lambda + (q_1\lambda)^2 + \cdots\right]} = N_S \frac{q_0\lambda}{(1 - q_1\lambda + q_0\lambda)(1 - q_1\lambda)} \tag{3.116}$$

where $\lambda \equiv e^{\beta\mu}$.

With the chemical potential approximated by that for an ideal gas, Eq. (3.113), the surface coverage of the adsorbed molecules is given by

$$\phi = \frac{\langle N\rangle}{N_S} = \frac{q_0\lambda}{(1 - q_1\lambda + q_0\lambda)(1 - q_1\lambda)} = \frac{k_0 P}{(1 - k_1 P + k_0 P)(1 - k_1 P)} \tag{3.117}$$

where

$$k_0 = q_0 e^{\beta \mu_0}, \tag{3.118}$$

$$k_1 = q_1 e^{\beta \mu_0}. \tag{3.119}$$

Eq. (3.117) is known as the BET adsorption isotherm, named after the inventors.[13]

Whereas in principle k_0 and k_1 can be predicted from the partition functions of gas molecules in the bulk and at the surface, in practice these parameters are often treated as adjustable that vary with temperature. At high pressure and a temperature near or below the critical point of the bulk vapor–liquid transition, the gas adsorption on a solid surface is more likely to follow the BET isotherm than the Langmuir isotherm. When $k_1 P \ll 1$ (low pressure), and $-\varepsilon_0 \gg -\varepsilon_1$ such that adsorption in the first layer is strongly favored relative to adsorption of higher layers, the BET equation reduces to the Langmuir form, Eq. (3.114), because, in that event, $\phi = k_0 P/(1 - k_0 P)$.

To fit experimental data, the BET isotherm has two adjustable parameters, one parameter more than the Langmuir isotherm. However, parameter k_1 can often be identified as the reciprocal of the saturation pressure because, as the gas pressure approaches saturation, a bulk liquid begins to emerge on the adsorbing surface,[14] leading to $\phi \to \infty$. According to Eq. (3.117), $\phi \to \infty$ is satisfied when $k_1 P_S \to 1$, where P_S is the saturation pressure. With the assumption $k_1 = 1/P_S$, we may rewrite the BET isotherm as

$$\phi = \frac{k_0 P/P_S}{(1 - P/P_S + k_0 P/P_S)(1 - P/P_S)}. \tag{3.120}$$

Rearranging Eq. (3.120) gives a linear relation between the "surface coverage" and gas pressure

$$\frac{P}{\phi(P_S - P)} = 1/k_0 + (1 - 1/k_0) P/P_S. \tag{3.121}$$

Eq. (3.121) provides a simple criterion to test whether experimental data follow the BET isotherm.

Figure 3.15 shows a successful application of the BET equation to nitrogen adsorption on nonporous silica at 77 K, below the critical temperature of nitrogen (126.2 K). In deriving the

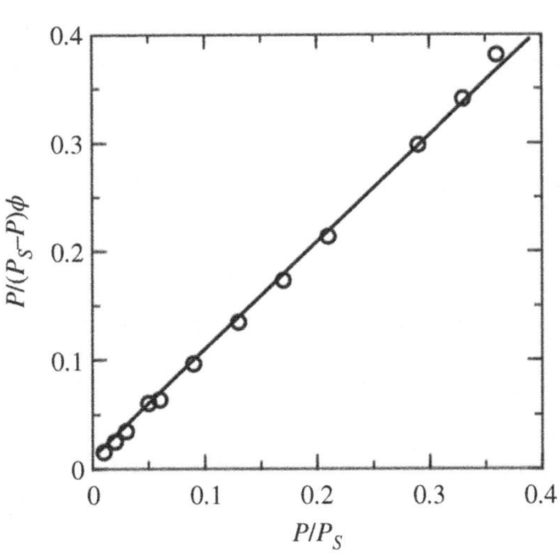

Figure 3.15 Nitrogen adsorption on nonporous silica at 77 K predicted by the linearized form of the BET isotherm (line). The symbols are experimental data. Here ϕ is the fraction of surface sites covered by nitrogen molecules, P is the gas pressure, and P_S is the vapor pressure of nitrogen at 77 K. Source: Adapted from Everett D. H. et al.[15]

13 Brunauer S., Emmett P. H. and Teller E. "Adsorption of gases in multimolecular layers", *J. Am. Chem. Soc.* 60 (2), 309–319 (1938).

14 Assuming that wetting transition takes place at the surface, i.e., the number of adsorption layers diverges as the gas pressure approaches the saturation point.

15 Everett D. H. et al. "The SCI/IUPAC/NPL project on surface area standards", *J. Appl. Chem.* 24, 199–219 (1974).

BET adsorption isotherm, we assume a uniform surface and no horizontal interactions between adsorbed molecules. These assumptions are often not justified for practical systems of gas adsorption. Nevertheless, compared with experimental results, the BET equation gives a good representation of experimental data provided that, empirically, the range of reduced pressure is restricted to $0.05 < P/P_S < 0.35$. At reduced pressures below 0.05, the BET equation underestimates adsorption because of surface heterogeneity; gas molecules have a strong affinity for the few high-energy surface sites during the initial stages of adsorption. At reduced pressures above 0.35, the BET equation often overestimates adsorption because the binding energy falls as the number of adsorption layers further increases.

3.6.3 Gas Adsorption in Porous Materials

Before closing this section, we discuss another example that is concerned with gas adsorption in porous materials. Consider a low-pressure gas in equilibrium with a porous medium that contains pores (or cages) where each pore can contain up to n molecules. Figure 3.16 presents a schematic picture of cages and resident molecules. For simplicity, we assume that all cages are identical and independent of each other, i.e., gas molecules in neighboring cages do not interact with each other. For gas molecules in N_c independent cages, the grand partition function is thus given by

$$\Xi = \xi^{N_c} \tag{3.122}$$

where ξ stands for the grand partition function of a single cage.

To find the single-cage partition function, we assume further that the interaction between each gas molecule and the cage can be represented by an "average" ε. Because a cage of volume v may contain a maximum of n molecules, the grand partition function for gas molecules in a single cage can be approximated as

$$\xi = 1 + q_{in}e^{-\beta(\varepsilon-\mu)}(v/\Lambda^3) + \sum_{i=2}^{n} \frac{q_{in}^i e^{-i\beta(\varepsilon-\mu)}}{i!\Lambda^{3i}}[v - (i-1)v_0]^i \tag{3.123}$$

where Λ is the thermal wavelength, q_{in} is the single-molecule partition function, and v_0 is the excluded volume of each gas molecule. Intuitively, $[v-(i-1)v_0]/\Lambda^3$ represents the number of ways to insert an additional gas molecule in a cage that has already been occupied by $(i-1)$ molecules. The first term on the right side of Eq. (3.123) corresponds to the empty cage; the second term corresponds to that for a single molecule in the cage, and the summation corresponds to contributions

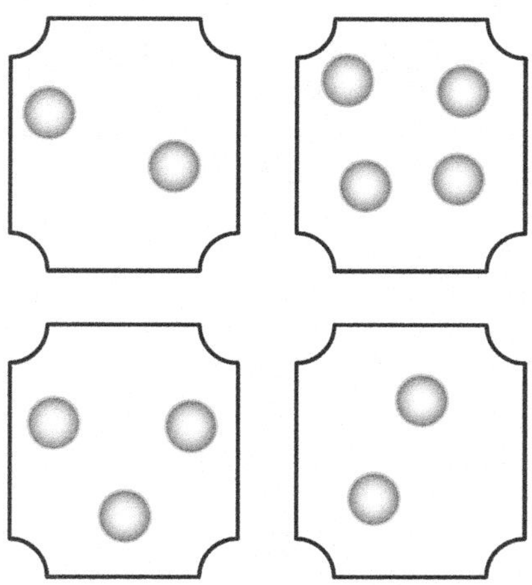

Figure 3.16 Schematic of gas adsorption in a porous medium. In this picture, the porous material has four cages (pores) and each can accommodate up to 4 gas molecules.

due to the adsorption of more gas molecules. The maximum number of molecules occupying a cage, n, can be estimated from the largest integer less than v/v_0.

Like the BET adsorption isotherm discussed above, we can derive the average number of gas molecules in the porous medium from the grand canonical partition function. For convenience, we define a temperature-dependent constant

$$K \equiv q_{in} e^{-\beta(\epsilon-\mu_0)}(v/\Lambda^3). \tag{3.124}$$

At low pressure, the gas chemical potential can be approximated by that of an ideal gas, i.e., $\beta\mu = \beta\mu_0 + \ln P$. The number of gas molecules per cage is then given by

$$\frac{<N>}{N_c} = \frac{1}{N_c}\frac{\partial \ln \Xi}{\partial \beta\mu} = \frac{KP + \sum_{i=2}^{n}\{KP[1-(i-1)v_0/v]\}^i/(i-1)!}{1 + KP + \sum_{i=2}^{n}\{KP[1-(i-1)v_0/v]\}^i/i!}. \tag{3.125}$$

Eq. (3.125) contains two model parameters, i.e., a temperature-dependent parameter K and the molecular excluded volumes v_0. These parameters can be obtained by fitting to the experimental results.

Figure 3.17 shows gas adsorption (e.g., O_2, N_2, CO) in zeolites over a wide range of temperatures. Table 3.6 presents the pertinent parameters. Although the simple statistical-mechanical model neglects gas–gas interactions except the excluded volume effects, Figure 3.17 suggests that it

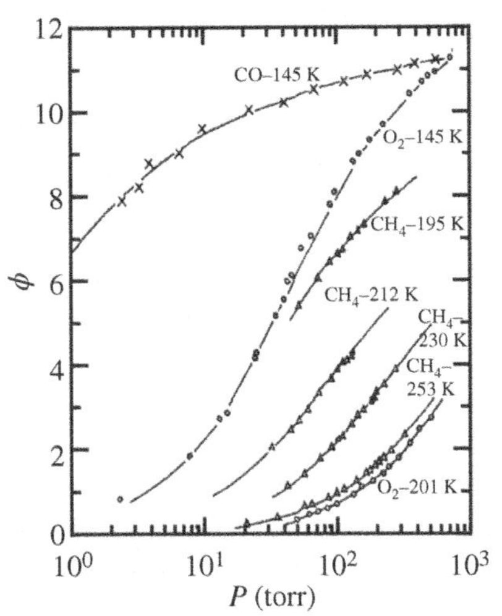

Figure 3.17 Equilibrium isotherms for adsorption of O_2, CO, and CH_4 in 5A zeolite at different temperatures. Symbols are experimental data, and the curves are from Eq. (3.125) with parameters shown in Table 3.6. Source: Adapted from Ruthven.[15]

Table 3.6 Molecular parameters in Eq. (3.125) for correlating single-gas adsorption data on a 5A Zeolite.

Gases	T (K)	K (Torr^{-1})	v_0 (Å^3)
O_2	145	0.31	46
—	201	0.007	46
CO	145	50	59
CH_4	195	—	64.5
—	212	0.083	64.5
—	230	0.033	64.5
—	253	0.011	64.5

provides an excellent fitting of the gas adsorption data. The model has been successfully extended to predict equilibrium sorption of binary gas mixtures.[16]

3.6.4 Summary

In this section, we have observed that the grand-canonical ensemble is a suitable method for characterizing thermodynamic systems in contact with a bulk gas phase. Such systems are frequently encountered in various engineering applications, including gas separation, energy storage, and heterogeneous chemical reactions. Additionally, adsorption isotherms are commonly employed to characterize porous materials, such as measuring specific surface area. Statistical thermodynamic models can be beneficial in all these applications since they provide quantitative correlations and a microscopic understanding of the physical phenomena governing gas adsorption processes.

3.7 Thermodynamics of Gas Hydrates

In offshore wells or cold climates, such as Alaska, moist natural gas can transform into gas hydrates, which may have a major impact on fluid flow during recovery or transfer through pipelines. The formation of hydrates is influenced by factors such as gas composition, temperature, pressure, and water content. A reliable thermodynamic model is valuable for predicting gas–solid equilibrium, where natural gas (mostly consisting of methane with lesser amounts of light hydrocarbons, nitrogen, carbon dioxide, and hydrogen sulfide) is in equilibrium with a gas hydrate.

In this section, we discuss the van der Waals and Platteeuw model for describing the thermodynamics of gas hydrate formation. Since its publication in 1959, the statistical thermodynamic model has been widely used in the oil and gas industries for flow assurance analysis and CO_2 sequestration.[17] Some researchers in these engineering communities consider the van der Waals and Platteeuw model as "one of the most successful applications of statistical thermodynamics in common engineering practice".

3.7.1 Gas Hydrates

Gas hydrates, also known as clathrate hydrates,[18] are crystalline solids containing "guest" molecules (e.g., methane or CO_2) in the cages of "host" water molecules. Large deposits of methane hydrates have been discovered in Canada, the Gulf of Mexico, and Siberia under the seabed and under permafrost. It is estimated that the mass of natural gas trapped in hydrates is more than twice that of carbon in all other known fossil fuels on the Earth.

Figure 3.18 presents three common clathrate structures of natural gas hydrates.[19] Structure I (sI) hydrates contain small gas molecules such as methane or ethane. The unit cell consists of 46 water molecules forming 2 small dodecahedral cavities (cages) and 6 large tetradecahedral cavities. Each cage can accommodate one guest molecule. Structure II (sII) hydrates are formed by

16 Ruthven D. M., *Principles of adsorption and adsorption processes*. Wiley, 1984.
17 de Azevedo-Medeiros F. et al., "Sixty years of the van der Waals and Platteeuw model for Clathrate hydrates—a critical review from its statistical thermodynamic basis to its extensions and applications", *Chem. Rev.* 120, 13349–13381 (2020).
18 The Greek root of clathrate means claw.
19 Hassanpouryouzband A. et al., "Gas hydrates in sustainable chemistry", *Chem. Soc. Rev.* 49, 5225–5309 (2020).

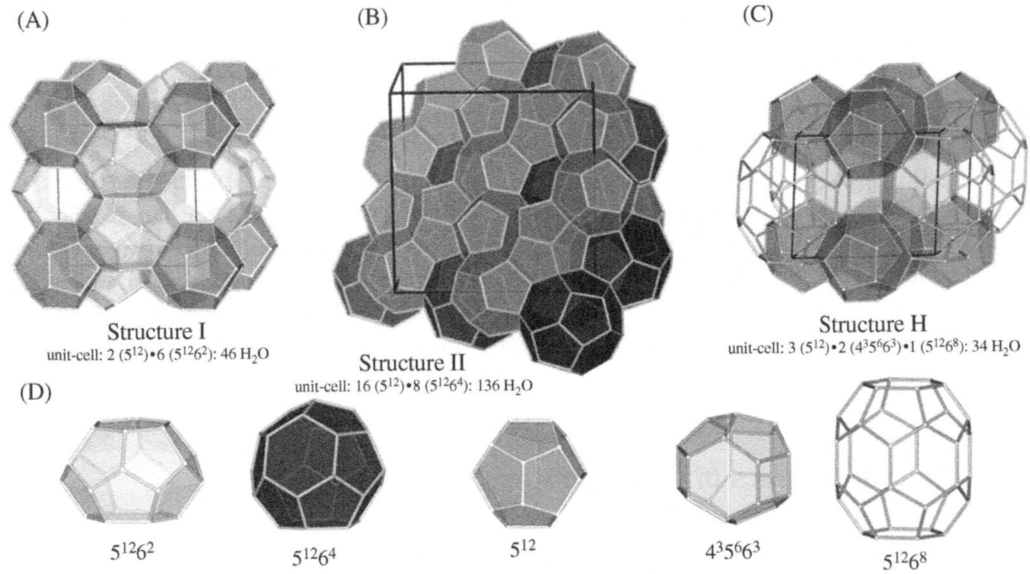

Figure 3.18 Typical unit cells of gas hydrates: (A) structure I, (B) structure II, and (C) structure H. (D) Five different host water cages. Here, n^m stands for a polyhedron with m faces and n edges on each face. Source: Reproduced from Hassanpouryouzband et al.[18]

larger guest molecules such as propane. In this case, each unit cell is made of 136 water molecules along with 24 guest molecules. These water molecules form 16 small dodecahedral cavities and 8 large hexakaidecahedral cavities. Finally, structure H hydrates (sH) have cavities for both small and large guest molecules. Each unit cell consists of 34 water molecules and 6 guest molecules. The large cavities of sH hydrates can hold gas molecules like isopentane and ethylcyclohexane.

3.7.2 The van der Waals and Platteeuw Model

Conceptually, the van der Waals and Platteeuw model is similar to the Langmuir model for gas adsorption in multicomponent systems. To elucidate the basic ideas, consider a clathrate hydrate in equilibrium with an m-component gas mixture at a given temperature, pressure, humidity, and dry gas composition. To understand hydrate formation and the composition of various gas compounds in the gas hydrate, we assume that the crystal structure can be represented by n types of cages and that each cage can accommodate at most one guest molecule. The guest molecules cannot diffuse from one cage to another within the hydrate phase, and interaction between guest molecules from different cages is negligible. With these assumptions, the hydrate phase is equivalent to a set of independent open subsystems; each subsystem is a water cage in equilibrium with an m-component gas mixture.

Like that established for describing gas adsorption in a porous material, the grand canonical ensemble can be used to find the amount and composition of natural-gas molecules in various cages of the gas hydrate. For each subsystem (cage), a microstate may be specified by the occupancy number (viz., 0 and 1) along with the quantum states of the guest and host molecules. Accordingly, the grand canonical partition function for cage i can be written as

$$\Xi_i = 1 + \sum_{j=1}^{m} q_{i,j} e^{\beta\mu_j} \tag{3.126}$$

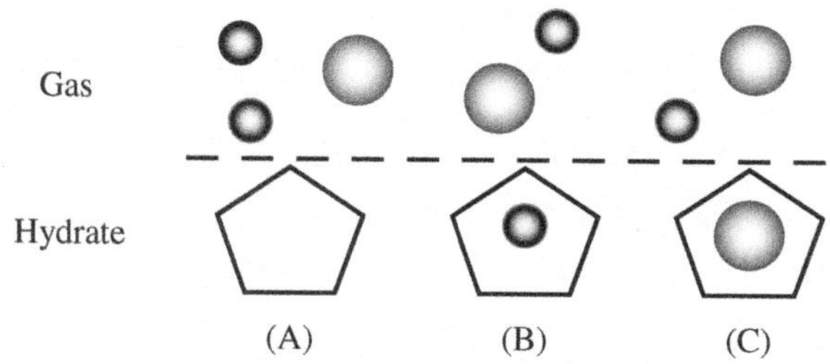

Figure 3.19 The number of ways that a hydrate cage (shown as a pentagon) can be filled with one gas molecule from a binary mixture. The cage accommodates at most one gas molecule: (A) the cage is unoccupied; (B) the cage contains a gas molecule of type 1; and (C) the cage contains a gas molecule of type 2.

where m is the number of gas components in the system, subscripts i and j denote the type of cage and the type of guest molecule, respectively; $q_{i,j}$ is the single-molecule partition function of molecule j in cage i. Following the usual notation, μ_j is the chemical potential of molecule j and $\beta = 1/(k_B T)$. In writing Eq. (3.126), we have assumed that the microstate of the host molecules are fixed.

The first term (unity) on the right side of Eq. (3.126) corresponds to the microstate where the cage is empty; and $q_{i,j}e^{\beta\mu_j}$ corresponds to the contribution from all microstates where cage i is occupied by a gas molecule of type j. For a cage in equilibrium with an m-component gas mixture such that it contains at most one gas molecule, there are $m + 1$ situations including an empty cage or a cage containing one of the m different gas molecules. Figure 3.19 illustrates the number of ways that a hydrate cage can be in equilibrium with a two-component gas mixture.

As the subsystems (cages) are independent of each other, the partition function of the hydrate system can be written as the product of the partition functions of individual cages,[20]

$$\Xi = \prod_{i=1}^{n} \left(1 + \sum_{j=1}^{m} q_{i,j} e^{\beta\mu_j}\right)^{\upsilon_i N_W} \tag{3.127}$$

where N_W is the total number of water (ice) molecules, and υ_i is the number of cages of type i per water molecule. From the partition function, we can derive the thermodynamic properties of practical interest. For example, the average number of guest molecules of type j in the hydrate phase, $<N_j>$, can be determined by differentiating $\ln\Xi$ with respect to $\beta\mu_j$

$$<N_j> = \frac{\partial \ln \Xi}{\partial \beta\mu_j} = \sum_{i=1}^{n} \frac{\upsilon_i N_W q_{i,j} e^{\beta\mu_j}}{1 + \sum_{j=1}^{m} q_{i,j} e^{\beta\mu_j}} \tag{3.128}$$

where $<N_j>$ is the number of guest molecules of type j in all types of cages. Because $<N_j>$ is a linear homogeneous function[21] of the total number of cages, Eq. (3.128) allows us to find the number of molecules of type j in cages of type i

$$<N_{i,j}> = \frac{\upsilon_i N_W q_{i,j} e^{\beta\mu_j}}{1 + \sum_{l=1}^{m} q_{i,l} e^{\beta\mu_l}}. \tag{3.129}$$

It follows that, $x_{i,j}$, the fraction of cages of type i occupied by gas molecules of type j is

$$x_{i,j} = \frac{<N_{i,j}>}{\upsilon_i N_W} = \frac{q_{i,j} e^{\beta\mu_j}}{1 + \sum_{l=1}^{m} q_{i,l} e^{\beta\mu_l}}. \tag{3.130}$$

20 This assumption is equivalent to saying that the grand potential of the hydrate system is given by the sum of the those of the individual cavities.

21 A linear homogeneous function satisfies $f(\alpha\mathbf{x}) = \alpha f(\mathbf{x})$ where α is an arbitrary scalar parameter and $\mathbf{x}$ represents a set of independent variables. In thermodynamics, all extensive properties are linear homogeneous functions of the system size.

Given an equation of state for the bulk gas phase, we can obtain the reduced chemical potential for each gas compound in terms of fugacity f_j or activity a_j

$$\beta\mu_j = \beta\mu_j^0 + \ln\left(f_j/f_j^0\right) = \beta\mu_j^0 + \ln a_j \tag{3.131}$$

where μ_j^0 and f_j^0 are, respectively, the chemical potential and the fugacity of gas compound j in the reference state, and $a_j = f_j/f_j^0$ denotes the activity of compound j. For the gas phase, the reference state is conventionally defined by pure ideal gas j at the system temperature and unit pressure, i.e., $P_0 = 1$ atm or 1 bar. Substituting Eq. (3.131) into (3.130) gives the fraction of cages i occupied by gas molecules of type j

$$x_{i,j} = \frac{C_{i,j}a_j}{1 + \sum_{l=1}^{m} C_{i,l}a_l} \tag{3.132}$$

where constants $C_{i,j} \equiv q_{i,j}e^{\beta\mu_j^0}$ are called the Langmuir coefficients. Accordingly, the fraction of empty cages of type i is

$$x_{i,0} = 1 - \sum_{j=1}^{m} x_{i,j} = 1\bigg/\left(1 + \sum_{l=1}^{m} C_{i,l}a_l\right). \tag{3.133}$$

Eqs. (3.132) and (3.133) represent a key result of the van der Waals and Platteeuw model for predicting the composition of a natural gas hydrate.

As mentioned above, hydrate formation affects both recovery and transportation processes for natural gas and oil industries. To avoid detrimental effects, it is desirable to forecast the incipient conditions and identify strategies to avoid hydrate formation (for CO_2 sequestration, to promote hydrate formation).[22] Toward that end, we need information on the chemical potential of water molecules in the hydrate phase, μ_w^h, and the incipient conditions can be predicted by comparing it with that in the aqueous solution, μ_w^a. While the latter can be calculated from an equation of state (or an excess Gibbs energy model),[23] the water chemical potential in the hydrate phase is typically expressed in terms of that of an empty hydrate lattice μ_w^{h*}, and an excess due to water interacting with the guest molecule $\Delta\mu_w^h$

$$\mu_w^h = \mu_w^{h*} + \Delta\mu_w^h. \tag{3.134}$$

Because an empty hydrate is not thermodynamically stable, the reference state is hypothetical, and the chemical potential of water molecules depends on the hydrate structure as well as temperature and pressure. This quantity can be predicted from conventional statistical-thermodynamic procedures to predict the thermodynamic properties of a crystalline solid (Section 6.9). Meanwhile, the excess chemical potential due to water–gas interactions can be calculated from the grand partition function.

From Eq. (3.127), we find that the reduced excess chemical potential of water is

$$\beta\Delta\mu_w^h = -\frac{\partial \ln \Xi}{\partial N_W} = -\sum_{i=1}^{n} \upsilon_i \ln\left(1 + \sum_{j=1}^{m} q_{i,j}e^{\beta\mu_j}\right) = -\sum_{i=1}^{n} \upsilon_i \ln\left(1 + \sum_{j=1}^{m} C_{i,j}a_j\right). \tag{3.135}$$

From Eqs. (3.133) to (3.135), we obtain a simple expression for the chemical potential of water in the hydrate phase

$$\mu_w^h = \mu_w^{h*} + \sum_{i=1}^{n} \upsilon_i \ln x_{i,0}. \tag{3.136}$$

22 Zheng et al., "Carbon dioxide sequestration via gas hydrates: a potential pathway toward decarbonization", *Energy Fuels* 34 (9), 10529–10546 (2020).

23 Prausnitz J. M., de Azevedo E. G., Lichtenthaler R., *Molecular thermodynamics of fluid-phase equilibria* (3rd Edition). Pearson Press, 1998.

Because $x_{i,0}$ is a small number, Eq. (3.136) indicates that the incorporation of guest molecules reduces the water chemical potential, making the hydrate phase thermodynamically more stable. Eq. (3.136) also explains why gas hydrates are formed at temperatures above the freezing point of pure water, and why increasing the gas pressure promotes hydrate formation.

Like the equilibrium constant in the Langmuir isotherm, C_{ij} can be determined either from quantum-mechanical calculations or classical models of guest–host interactions. The theoretical procedure is discussed briefly in Box 3.2. In general, these coefficients depend on the microscopic details of gas molecules interacting with the hydrate cages. In practical applications, the Langmuir coefficients are often treated as adjustable parameters, obtained by fitting Eq. (3.132) with experimental data (e.g., for a hydrate in equilibrium with a pure gas). Once C_{ij} is fixed, Eq. (3.132) can be used to predict the gas composition and conditions for hydrate formation at different thermodynamic conditions. For example, Figure 3.20 shows the pressure versus temperature phase diagram for hydrate formation in the binary mixtures of ethane and methane. The van der Waals and Platteeuw model fits the experimental data extremely well. It predicts the lower structural transition point at 274.2 K within 1% of the experimentally determined value. The theoretical predictions at and above incipient hydrate formation conditions reveal unusual phenomena such as minimum pressure azeotropes and solid-solid phase equilibria.[24]

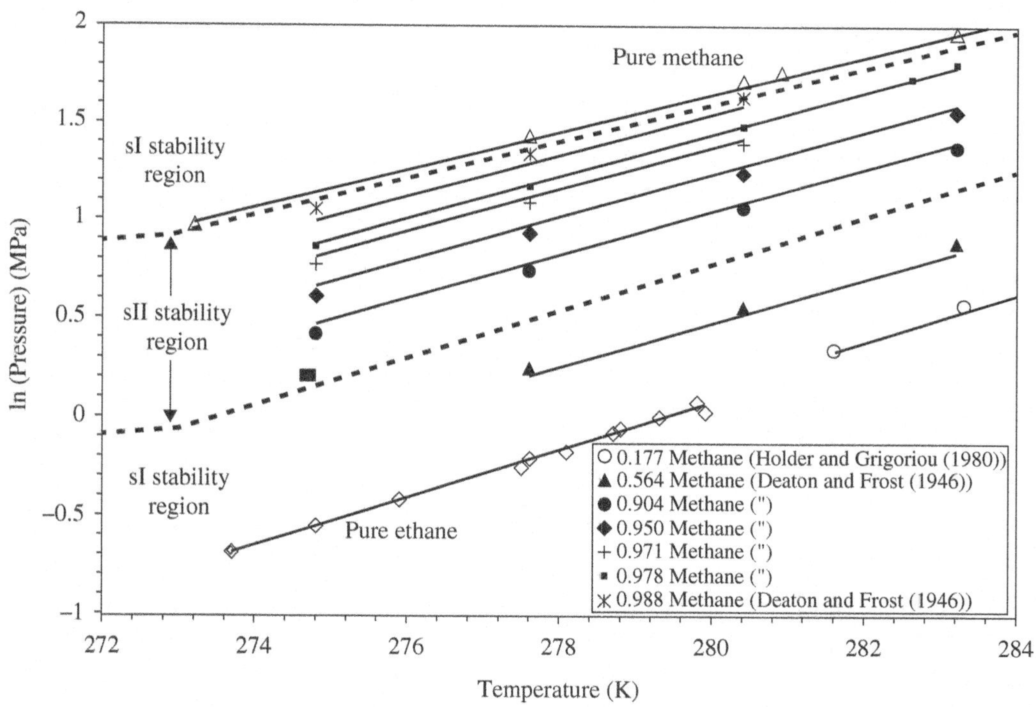

Figure 3.20 Incipient hydrate equilibrium data for the methane–ethane–water system (lines are predictions). The thick, dashed lines represent the sII incipient hydrate stability boundary. Source: Adapted from Ballard and Sloan.[23]

24 Ballard A. L. and Sloan E. D., "Hydrate phase diagrams for methane + ethane + propane mixtures", *Chem. Eng. Sci.* 55, 5773–5782 (2000).

box 3.2 Modeling the Langmuir Coefficients

The Langmuir coefficient, $C_{i,j} \equiv q_{i,j} e^{\beta\mu_j^0}$, depends on $q_{i,j}$, the single-molecule partition function of the guest molecule j in cage i of the hydrate, and $\beta\mu_j^0$, the reduced chemical potential of guest molecule j at the ideal-gas reference state. The latter may also be expressed in terms of single-molecule partition function for the gas molecule at the ideal-gas reference state (q_j^0)

$$\beta\mu_j^0 = -\ln\left(q_j^0/N\right) = -\ln\left(q_j^{\text{int}}/\rho_0\Lambda_j^3\right). \tag{3.137}$$

where q_j^{int} represents the contribution due to the internal degrees of freedom (e.g., rotation and vibration) of molecule j, $\rho_0 = P_0/k_BT$ is the number density of the gas molecules in the ideal-gas reference state, and Λ_j is the thermal wavelength.

With the assumption that the confinement has negligible influence on the intramolecular motions, the single-molecule partition function of guest molecule j in cage i can be expressed as

$$q_{i,j} = \frac{q_j^{\text{int}}}{\Lambda_j^3} \int d\mathbf{r} e^{-\beta u_{i,j}(\mathbf{r})} \tag{3.138}$$

where $u_{i,j}(\mathbf{r})$ represents the potential energy of guest molecule j in cage i. Substituting Eqs. (3.137) and (3.138) into the definition of the Langmuir coefficients leads to

$$C_{i,j} \equiv q_{i,j} e^{\beta\mu_j^0} = \rho_0\Lambda_j^3 \frac{q_{i,j}}{q_j^{\text{int}}} = \frac{P_0}{k_BT} \int d\mathbf{r} e^{-\beta u_{i,j}(\mathbf{r})}. \tag{3.139}$$

With further assumptions that each cage can be represented by a spherical cavity of radius R_i, and that $u_{i,j}(\mathbf{r})$ depends only on the radial distance, we have $d\mathbf{r} = 4\pi r^2 dr$, and Eq. (3.139) becomes

$$C_{i,j} = \frac{4\pi}{k_BT} \int_0^{R_i} e^{-\beta u_{i,j}(r)} r^2 dr \tag{3.140}$$

where the pressure of the reference state, P_0= 1 atm (or 1 bar), is omitted. Eq. (3.140) is equivalent to that used in the Lennard–Jones–Devonshire model to describe the thermodynamic properties of liquids or gases with the cell model.[25]

3.7.3 Summary

The van der Waals and Platteeuw model serves as an excellent example of the application of statistical thermodynamics in industrial practice. Despite its simplicity, the thermodynamic model performs well over a broad range of pressures and temperatures that are crucial for industrial applications. It can be easily applied to clathrate hydrates with multiple guests and to conditions other than those used to estimate the model parameters. For further discussion on this topic, we refer the reader to the outstanding monograph authored by Sloan and Koh.[26]

---

25 Lennard-Jones J. E., Devonshire A. F., "Critical phenomena in gases – I", *Proc. R. Soc. Lond. A* 163, 53 (1937); "Critical phenomena in gases. II. Vapour pressures and boiling points", ibid A165, 1 (1938).
26 Sloan E. D. and Koh C., *Clathrate hydrates of natural gases*. CRC Press, New York, 2007.

3.8 Ideal Polymer Chains

Unlike small molecules, a polymer chain is intrinsically flexible, particularly at the length scale comparable to its size. Polymer conformation refers to the three-dimensional arrangement of the atoms or atomic units due to the rotation around individual chemical bonds.[27] Just like atomic structure is essential for predicting the thermodynamic properties of small molecules, polymer conformation plays a significant role in understanding the functionality and physical properties of polymeric materials such as elasticity and phase behavior.

In this and the next few sections, we discuss some of the simplest statistical-thermodynamic models for describing the conformation of linear polymer chains. These models are often used as a reference for studying the properties of polymeric materials, just like how the ideal-gas model is used to study real gases and liquids. For a more in-depth discussion on this topic, we recommend interested readers to specialized texts.[28]

3.8.1 Conformation of Polymer Chains

As the word polymer (poly = many and mer = unit or segment) implies, polymer is a macromolecule with many repetitive monomers or segments. Such molecules are abundant in nature: polysaccharides, polypeptides, DNA, and RNA are essential components of life. Additional examples of natural polymers include starch, glycogen, and polyesters that commonly appear in bacteria, plants, and animals for storing energy and carbon. Many synthetic materials used in daily life, such as artificial rubber, nylon, Kevlar fibers, and various forms of plastics, are also polymers. Typically, such materials are synthesized from fossil fuels and may persist in the environment for hundreds of years. While their importance in modern life is undeniable, the pressing environmental concerns on climate change and plastic pollution are calling for alternative production routes with renewable resources and for more efficient recycling.

The number of repeating segments in a polymer chain is called the degree of polymerization, which typically ranges from a few hundred to several thousand for synthetic polymers and can be much larger for natural polymers. Because of the large molecular weight, polymers may vary in size from a few nm to μm or even longer. For example, human X-chromosome, a single DNA chain with over 150 million base pairs, has a length of about 50 mm when it is fully extended. Because of the enormous number of segments in a polymer chain, its structure is much more complicated than those corresponding to small molecules.

The conformation of a polymer chain depends on several factors such as the chemical structure of the polymer, the solvent conditions, and temperature. As the number of ways that a polymer chain may exist (viz., the number of conformers) increases exponentially with the chain length (e.g., a polymer chain of 100 segments would have $3^{100} \approx 5 \times 10^{47}$ conformers if each segment takes 3 rotational states), a first-principles approach to describe the microstates of a polymer chain in a solvent is computationally demanding even with powerful supercomputers and often unnecessary from the practical perspective. Different from the quantum-mechanical models of atomic motions in small molecules, coarse-grained models are commonly used to describe the statistics of polymer conformations.

---

27 Both configuration and conformation describe molecular structure but the two concepts should not be conflated. While the former refers to the spatial arrangement of atoms including bond connectivity, the latter concerns only with the relative orientations and positions of individual atoms within a specific bond connectivity.
28 For example, Yamakawa H., *Modern theory of polymer solutions*. Harper & Row, 1971.

3.8.2 Freely Jointed Chain Model

A polymer chain is considered *ideal* if the effects due to short-range repulsion among polymer segments are perfectly balanced by longer-range attractions. Such situations may arise for polymers in either a dilute or concentrated solution[29] or even in polymer melt.

The freely jointed chain (FJC) model is used to represent ideal polymer chains that are able to assume any conformation with equal probability. As illustrated in Figure 3.21, an FJC with m segments consists of $(m-1)$ bonds of equal length b. The segments are assumed to be structureless and have no excluded volume, and each bond is *free* to take any orientation in the three-dimensional space. Because there is no interaction among the polymer segments other than the bond connectivity, the bond orientations are random. Accordingly, the FJC model is also known as the random flight model, and the polymer conformation is equivalent to the trajectory of a 3-dimensional random walk (Section 3.12).

3.8.2.1 End-to-End Distance

The end-to-end distance of a linear polymer chain refers to the separation between the polymer ends. It is a crucial parameter commonly used in describing the statistical properties of polymer conformation. As shown in Figure 3.21, polymer segments in an ideal chain are linearly connected by rigid bonds with random orientations. The end-to-end distance depends on the bond orientations, and it is typically much smaller than the length of a fully stretched chain $L = (m-1)b$.

To derive the probability distribution for the end-to-end distance of an ideal chain, consider a FJC with m segments linearly connected by bonds of equal length b. We place the first segment at the origin of a Cartesian coordinate, and let the other segments take a "random flight", i.e., the second segment is placed on the surface of a sphere of radius b with the center at the origin; the third segment takes another random flight of distance b from the second, and so on for all other segments. Let $\mathbf{r}_i$ be the position of the ith segment and $\mathbf{b}_i$ the vector connecting the ith segment and the $(i+1)^{th}$ segment. Thus, the position of the first segment is $\mathbf{r}_1 = \mathbf{0}$, and the position of the ith segment, $i \geq 2$, can be expressed as the sum of the bond vectors

$$\mathbf{r}_i = \sum_{k=2}^{i} \mathbf{b}_{k-1}. \tag{3.141}$$

Because $\mathbf{b}_i$ has equal probability in all directions, its average overall bond orientation gives

$$<\mathbf{b}_i> = 0. \tag{3.142}$$

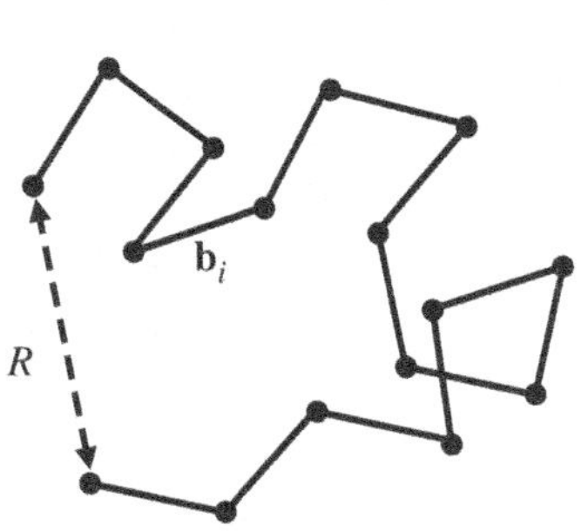

Figure 3.21 The freely jointed chain (FJC) model for ideal polymer chains. Here R represents the end-to-end distance, and $\mathbf{b}_i$ is a vector connecting the ith and $(i+1)^{th}$ structureless segments (dots).

29 The solution condition that makes a polymer become an ideal chain is called theta point; the corresponding solvent is referred to as the theta solvent. For example, at room temperature, 1-dodecanol may serve as a theta solvent for polyethylene, a synthetic polymer made of CH_2 units.

Meanwhile, the orientations of different bonds are uncorrelated, implying

$$<\mathbf{b}_i \cdot \mathbf{b}_j> = b^2\delta_{ij} \tag{3.143}$$

where δ_{ij} denotes the Kronecker delta function, i.e., $\delta_{ij} = 1$ if $i = j$ and 0 otherwise. From the viewpoint of statistical mechanics, $<\cdots>$ is equivalent to the ensemble average over all microstates of the polymer chain. Here, microstates are referred to as different polymer conformations.[30]

With the first polymer segment fixed at the origin, Eq. (3.142) predicts that the average position of the end segment $\mathbf{r}_m$, and the average of the end-to-end vector $\mathbf{R} = \mathbf{r}_m - \mathbf{r}_1$, should be zero

$$<\mathbf{R}> = <\mathbf{r}_m> = <\sum_{k=1}^{m-1}\mathbf{b}_k> = 0. \tag{3.144}$$

According to Eq. (3.143), the square of the end-to-end vector is

$$<\mathbf{R}^2> = \left\langle \sum_{i=1}^{m-1}\sum_{j=1}^{m-1}\mathbf{b}_i \cdot \mathbf{b}_j \right\rangle = \sum_{i=1}^{m-1}\langle \mathbf{b}_i \cdot \mathbf{b}_i\rangle + 2\sum_{i=1}^{m-2}\sum_{j>i}^{m-1}\langle \mathbf{b}_i \cdot \mathbf{b}_j\rangle = (m-1)b^2. \tag{3.145}$$

Because $m >> 1$ for a typical polymer, the number of polymer segments is often treated as equal to the number of bonds, i.e., $m - 1 \approx m$. Thus, the end-to-end distance is given by

$$R = \sqrt{<\mathbf{R}^2>} \approx b\sqrt{m}. \tag{3.146}$$

As discussed above, R is significantly smaller than the contour length of the polymer chain, $L = b(m-1) \approx bm$. Because $m \gg 1$, the average end-to-end distance is still much larger than the length scale of individual segments, b. Eq. (3.146) also suggests that the space occupied by a polymer chain is mostly empty, i.e., the average segment density of the polymer chain within a sphere defined by the end-to-end distance

$$\rho_{avg} = \frac{m}{4\pi R^3/3} \sim 1/\sqrt{m} \tag{3.147}$$

is smaller than that of a polymer melt ($\sim 1/b^3$) by orders of magnitude.

It should be noted that the "segments" and "bonds" in the FJC model should not be identified as monomers and chemical bonds in a real polymer. The FLC segments may be understood as a short sequence of chemical units along the polymer backbone such that the effective interactions between such sequences are nullified by the cancellation of the attraction and repulsion components of interactions. Accordingly, each bond represents the connection of "renormalized" units without chemical details. Such a coarse-grained view of long polymer chains was first conceived by Werner Kuhn, who developed the first theoretical model of the viscosity of polymer solutions using statistical mechanics. To his credit, bond length b in the FJC model is often referred to as the Kuhn length. Because both the polymer contour length and the end-to-end distance are physical variables, the Kuhn length may be estimated from the ratio of these two quantities according to the formula provided by the FJC model. From $R^2 = (m-1)b^2$ and $L = b(m-1)$, we obtain

$$b = R^2/L. \tag{3.148}$$

Eq. (3.148) allows us to estimate the Kuhn length of real polymer chains based on the end-to-end distance and polymer contour length. For example, theoretical calculations and X-ray-scattering experiments indicate that the carbon–carbon bond in polyethylene has a length of about $b_0 \approx 1.53$ Å

30 Mazars M., "Statistical physics of the freely jointed chain", *Phys. Rev. E* 53 (6), 6297–6319 (1995).

and the bond angle among three consecutive carbon atoms is $\theta \approx 113°$.[31] Assuming that the carbon atoms take a zig-zag structure in the fully extended state, we may estimate the contour length from

$$L = m_0 b_0 \sin(\theta/2) = 0.834 m_0 b_0 \tag{3.149}$$

where m_0 is the degree of polymerization, i.e., the number of CH_2 units in each polyethylene chain. Neutron scattering experiments for the polymer melt indicate the mean-square end-to-end distance of polyethylene chains can be expressed as[32]

$$R^2 \approx 5.7 m b_0^2 \tag{3.150}$$

where the coefficient is obtained from fitting to the experimental data. According to Eq. (3.148), the Kuhn length is $b = R^2/L \approx 6.83 b_0 \approx 10.4$ Å. Thus, the number of coarse-grained segments in the FJC model is related to the degree of polymerization through $m = L/b \approx 0.122 m_0$. In other words, each segment in the FJC model corresponds to about 8.2 methylene groups.

Polyethylene has a simple monomeric structure, making it one of the easiest polymers to comprehend. However, it also holds great significance in both technological and commercial applications. Due to these attributes, polyethylene is commonly utilized as a benchmark for both experimental and theoretical studies of polymer systems.

3.8.2.2 Radius of Gyration

Another important measure of polymer conformation is the radius of gyration, R_g, defined as the average distance of the segments from the molecular center of mass, $\mathbf{r}_{cm} = \frac{1}{m}\sum_{i=1}^{m} \mathbf{r}_i$. Following this definition, the radius of gyration for a FJC with segments is given by

$$R_g^2 = \frac{1}{m}\sum_{i=1}^{m} <(\mathbf{r}_i - \mathbf{r}_{cm})^2> . \tag{3.151}$$

Because

$$\sum_{i=1}^{m}(\mathbf{r}_i - \mathbf{r}_{cm})^2 = \sum_{i=1}^{m}\mathbf{r}_i^2 - 2\sum_{i=1}^{m}\mathbf{r}_i \cdot \mathbf{r}_{cm} + m\mathbf{r}_{cm}^2 = \sum_{i=1}^{m}\mathbf{r}_i^2 - \frac{1}{m}\sum_{i=1}^{m}\sum_{j=1}^{m}\mathbf{r}_i \cdot \mathbf{r}_j \tag{3.152}$$

and

$$\mathbf{r}_i \cdot \mathbf{r}_j = \left(\mathbf{r}_i^2 + \mathbf{r}_j^2 - \mathbf{r}_{ij}^2\right)/2 \tag{3.153}$$

where $\mathbf{r}_{ij} = \mathbf{r}_i - \mathbf{r}_j$, we can rewrite Eq. (3.151) as

$$R_g^2 = \frac{1}{2m^2}\sum_{i=1}^{m}\sum_{j=1}^{m} <\mathbf{r}_{ij}^2> . \tag{3.154}$$

According to Eq. (3.145), $<\mathbf{r}_{ij}^2> = | i - j | b^2$ and thus Eq. (3.154) becomes

$$\begin{aligned} R_g^2 &= \frac{b^2}{m^2}\sum_{i=1}^{m}\sum_{j=1}^{i-1}(i-j) = \frac{b^2}{m^2}\sum_{i=1}^{m}\frac{(i-1)\cdot i}{2} \\ &= \frac{b^2}{2m^2}\left[\frac{m(m+1)(2m+1)}{6} - \frac{m(m+1)}{2}\right] = \frac{b^2}{6m}(m^2 - 1). \end{aligned} \tag{3.155}$$

31 Miao M. S. et al., "Conformation and electronic structure of polyethylene: a density-functional approach", *Phys. Rev. B* 54 (15), 10430–10435 (1996).

32 Wu S., "Predicting chain conformation and entanglement of polymers from chemical structure", *Polym. Eng. Sci.* 32, 823–830 (1992).

For a long polymer chain, $m >> 1$ and $m^2 - 1 \approx m^2$, Eq. (3.155) leads to the well-known expression for the radius of gyration of an ideal chain

$$R_g = b\sqrt{m/6}. \tag{3.156}$$

A comparison of Eqs. (3.145) and (3.156) gives $R_g^2 = R^2/6$. According to the FJC model, both the end-to-end distance and the radius of gyration are proportional to $m^{1/2}$.

3.8.3 Statistics of Polymer Conformations

Whereas the ensemble methods are applicable to the statistics of polymer conformations, the physical properties of polymer systems (e.g., viscosity and elasticity) are often more conveniently described in terms of the probability distribution for the end-to-end distance or radius of gyration. For a FJC with m segments, the probability distribution of the end-to-end vector can be evaluated exactly. Because the derivation is quite lengthy, here we reproduce only the final result[28]

$$p(\mathbf{R}) = \frac{1}{2^m(m-3)!\pi b^2 R} \sum_{k=0}^{k\le(m-1-R/b)/2} (-1)^k \binom{m-1}{k} (m-1-2k-R/b)^{m-3}. \tag{3.157}$$

As $p(\mathbf{R})$ stands for the probability density of finding the end segment at position $\mathbf{R}$ with the first segment fixed at the origin, it satisfies the normalization condition

$$\int d\mathbf{R}p(\mathbf{R}) = \int dR 4\pi R^2 p(\mathbf{R}) = 1. \tag{3.158}$$

In terms of the end-to-end distance $R = |\mathbf{R}|$, the probability distribution is

$$\begin{aligned} p(R) &= 4\pi R^2\, p(\mathbf{R}) \\ &= \frac{R}{2^{m-2}(m-3)!b^2} \sum_{k=0}^{k\le(m-1-R/b)/2} (-1)^k \binom{m-1}{k} (m-1-2k-R/b)^{m-3}. \end{aligned} \tag{3.159}$$

As shown in Eq. (3.141), the end position of a FJC can be written as a summation of bond vectors. When the number of segments is sufficiently large, we can derive an approximate expression for $p(\mathbf{R})$ using the central limit theorem.[33] Based on the mean and the second-order moment of the end-to-end vector given by Eqs. (3.144) and (3.145), respectively, the central limit theorem asserts that the probability distribution for the end segment (viz., the end-to-end vector) is given by the Gaussian distribution

$$p(\mathbf{R}) = \left(\frac{3}{2\pi\, mb^2}\right)^{3/2} \exp\left[-\frac{3R^2}{2\, mb^2}\right]. \tag{3.160}$$

Figure 3.22 compares the probability distribution for the end-to-end distance predicted from the exact and approximate methods. While the central limit theorem is strictly valid for an infinite series of random numbers, the Gaussian approximation provides a satisfactory description of the end-segment distribution even for relatively short chains ($m \sim 11$). A major drawback of the Gaussian approximation is that, as shown in Figure 3.22, it erroneously gives a finite probability even when the end-to-end distance is longer than its contour length. The error is introduced due to the finite length of polymer chains and is significant only when the chain length is extremely small.

---

33 The central limit theorem asserts that the summation of an infinite series of independent random variables satisfies the Gaussian distribution.

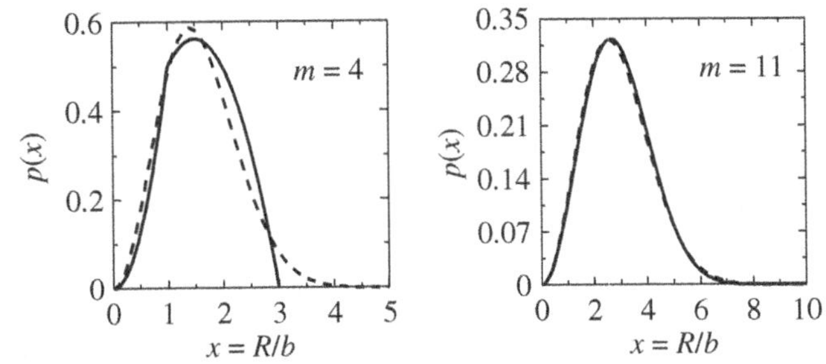

Figure 3.22 Approximately, the probability density for the end-to-end distance of a freely jointed chain (solid lines) follows the Gaussian distribution (dashed lines) when ($m > 10$). Here $p(x) = 4\pi bR^2 p(\mathbf{R})$ is dimensionless.

3.8.4 Thermodynamic Properties of a Single Chain

Just like the distribution of microstates in a canonical ensemble, the conformational probability of a polymer chain is related to its thermodynamic properties. For example, the FJC model predicts that, for a polymer with a given end-to-end distance R, the Helmholtz energy is given by

$$F(R) = -k_B T \ln p(R) = k_B T \left[\frac{3R^2}{2mb^2} - 2\ln(R/b) \right]. \tag{3.161}$$

Eq. (3.161) is obtained from the Boltzmann distribution for the end-to-end distance

$$p(R) \sim e^{-\beta F(R)} \tag{3.162}$$

neglecting the terms independent of R. Because the FJC model ignores segment–segment interactions other than the bond connectivity, the entropy of a single polymer chain varies with the end-to-end distance

$$S(R)/k_B = -F(R)/(k_B T) = -\frac{3R^2}{2mb^2} + 2\ln(R/b). \tag{3.163}$$

The above thermodynamic equations are useful for understanding polymer conformation and mechanical behavior. For example, minimizing the Helmholtz energy with respect to R

$$\frac{\partial F}{\partial R} = k_B T[3R/mb^2 - 2/R] = 0. \tag{3.164}$$

gives the most probable end-to-end distance

$$R_0 = \sqrt{2mb^2/3}. \tag{3.165}$$

Eq. (3.165) predicts that the polymer size scales with the square root of the chain length, which is consistent with that predicted by the mean-square end-to-end distance or the radius of gyration.

3.8.5 Two Illustrative Applications

Before closing this section, we discuss two examples on the application of the FJC model. We may understand the elasticity of a polymer chain in terms of the variation of the end-to-end vector **R** in response to an external force **f** in the same direction. As discussed above, the Helmholtz energy of an ideal chain with a given end-to-end vector **R** is related to the probability distribution

$$F(\mathbf{R}) = -k_B T \ln p(\mathbf{R}) = k_B T \left(\frac{3R^2}{2\,mb^2} \right) \tag{3.166}$$

where terms independent of **R** have been neglected. Note that the Helmholtz energy for fixing the position of the end segment is different from that for fixing the end-to-end distance. In the latter case, the polymer chain has angular degrees of freedom thus a lower free energy.

Now suppose that a polymer is extended from R to αR along the direction of the end-to-end vector **R**, Eq. (3.166) predicts that the polymer would experience a mechanical force in the opposite direction of **R**

$$\mathbf{f} = -\nabla F(\mathbf{R}) = -\frac{3k_BT}{mb^2}\mathbf{R}. \tag{3.167}$$

Eq. (3.167) indicates that the elastic modulus of a polymeric chain is inversely proportional to the chain length, and that the dimensions of a long polymer chain can be more easily influenced by the application of an external force than a short polymer. In addition, Eq. (3.167) predicts that the elasticity of a polymer increases with temperature as commonly observed in experiments.

Another application of the FJC model is concerned with the cyclization of polymer chains as described by the Jacobson–Stockmayer theory.[34] One special case for the conformational probability is $R = 0$, which corresponds to the probability of the two ends of a linear polymer chain coming together. From the practical perspective, the cyclization probability is important for understanding the rate of ring formation during polymer synthesis, and it is also relevant to the functionality of biopolymers including DNA and RNA. For example, DNA looping is frequently observed in living systems and plays a critical role in many cellular processes, including gene expression and regulation, DNA replication and recombination, DNA repair, and chromosome segregation.

According to Eq. (3.160), the probability of polymer ends within a distance b such that cyclization reactions may take place during polymer synthesis is given by

$$\begin{aligned} p_{cycl} &= \int_0^b dR4\pi R^2 p(\mathbf{R}) \\ &= \left(\frac{3}{2\pi\, mb^2}\right)^{3/2} \int_0^b dR4\pi R^2 \exp\left[-\frac{3R^2}{2\, mb^2}\right]. \end{aligned} \tag{3.168}$$

Because $R < b$ and for typical polymers (3.168) $m > > 1$, we have $e^{-3R^2/2\, mb^2} \approx 1$ and Eq. (3.168) becomes

$$p_{cycl} \approx \left(\frac{3}{2\pi\, mb^2}\right)^{3/2} \int_0^b dR4\pi R^2 = \left(\frac{3}{2\pi\, mb^2}\right)^{3/2} \frac{4\pi b^3}{3} = \left(\frac{6}{\pi}\right)^{1/2} m^{-3/2}. \tag{3.169}$$

Eq. (3.169) predicts that the cyclization probability decreases with the chain length according to the power law $p_{cycl} \sim m^{-3/2}$. The probability of ring closure provides a basis to analyze the composition of molecular species in reversible polycondensation reactions.

3.8.6 Summary

The FJC model has gained widespread use in polymer physics and biophysics due to its ability to capture some essential features of polymer chains with minimal assumptions. It is commonly employed to describe various phenomena such as rubber elasticity, DNA conformation, protein folding, and single-molecule force spectroscopy. Nevertheless, it is important to recognize that the FJC model has significant limitations. Specifically, it neglects the excluded volume effects of segments, solvent quality, chain stiffness, and specific interactions among monomers. More sophisticated models are required to accurately describe polymers in different conditions and environments.

---

34 Jacobson H. and Stockmayer W. H., "Intramolecular reaction in polycondensations. I. The theory of linear systems", *J. Chem. Phys.* 18, 1600 (1950).

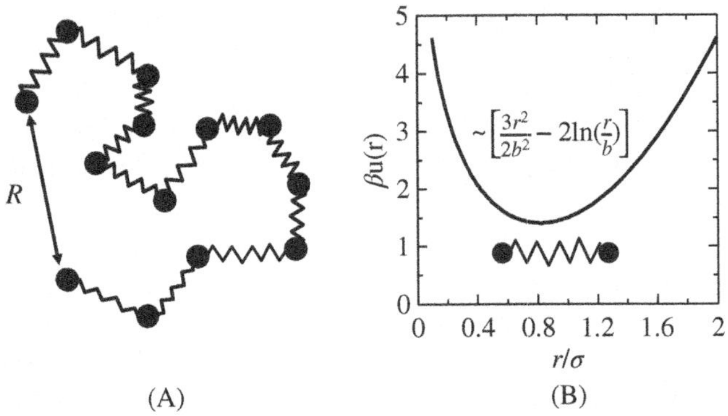

Figure 3.23 (A) In a Gaussian chain, polymer segments are noninteracting particles, and the bond vectors satisfy the 3-dimensional Gaussian distribution. (B) The average bond potential between polymer segments according to the Gaussian chain model.

3.9 Gaussian Chains

The Gaussian-chain model is a variation of the FJC model for ideal polymer chains. It is often used as a reference for the continuous representation of polymer conformations as described in the polymer field theory (Section 8.7).

3.9.1 The Gaussian-Chain Model

As shown schematically in Figure 3.23, the polymer segments in a Gaussian chain are represented by noninteracting particles, and the bond vectors are assumed uncorrelated. Different from the FJC model, the Gaussian-chain model assumes that the bond lengths are not fixed. Instead, the bond vector connecting segments i and $i+1$ follows the 3-dimensional Gaussian distribution

$$p(\mathbf{r}_i) = \left(\frac{3}{2\pi b^2}\right)^{3/2} \exp\left(-\frac{3\mathbf{r}_i^2}{2b^2}\right) \tag{3.170}$$

where b represents the standard deviation, i.e., the square root of the variance. Because the bond vectors are free to take any direction by random, the angular average of the bond vector is zero, $<\mathbf{r}_i> = 0$, and the variance of the bond vector is $<|\mathbf{r}_i|^2> = b^2$. The standard deviation provides a characteristic length for the chain connectivity. Like that for the end-to-end vector of a FJC, the most probable value of the bond length occurs at $b_0 = \sqrt{2/3}b$, which is directly derived from Eq. (3.165).

Intuitively, the bond-vector probability in a Gaussian chain may be interpreted in terms of the canonical distribution for the polymer bonds represented by classical harmonic oscillators[35]

$$p(\mathbf{r}) \sim \exp[-\beta k_s r^2/2] \tag{3.171}$$

where $\beta = 1/(k_B T)$, and k_s is the spring constant. A comparison of Eqs. (3.170) and (3.171) indicates

$$k_s = 3k_B T/b^2. \tag{3.172}$$

According to Eq. (3.171), the bond potentials in the Gaussian chain are represented by

$$v_B(r) = k_B T\left[\frac{3r^2}{2b^2} - 2\ln(r/b)\right]. \tag{3.173}$$

35 In a harmonic oscillator, the potential to displace the system from its equilibrium position is proportional to the square of the displacement.

Eq. (3.173) neglects a constant that does not affect the probability distribution of the bond vectors. The mathematical form is identical to the free energy of a free joint chain discussed in the previous section, Eq. (161).

3.9.2 Characteristic Size of a Gaussian Chain

Despite the difference in chain connectivity, the Gaussian-chain model and the FJC model yield the same equations for the end-to-end distance and the radius of gyration. In addition, these two models give the same end-to-end vector distribution function for a long polymer chain.

The equivalence of the FJC and Gaussian-chain models for the end-to-end vector follows because, in a Gaussian chain, we also have $<\mathbf{r}_i> = <\mathbf{r}_i \cdot \mathbf{r}_{j\neq i}> = 0$ and $<|\mathbf{r}_i|^2> = b^2$. As the derivation for the radius gyration of an FJC is equally applicable to a Gaussian chain, $R_g = b\sqrt{m/6}$ is valid for both models. Furthermore, according to the central limit theorem, the end-to-end vector of a long polymer chain satisfies the Gaussian distribution independent of the details of chain connectivity (as long as it is represented by uncorrelated random variables). Therefore, the Gaussian distribution is equally applicable to the end-to-end distribution of a Gaussian chain

$$p(\mathbf{R}) = \left(\frac{3}{2\pi\, mb^2}\right)^{3/2} \exp\left[-\frac{3R^2}{2\, mb^2}\right] \tag{3.174}$$

where $\mathbf{R} \equiv \mathbf{r}_m - \mathbf{r}_1$ stands for the end-to-end vector. The fact that FJC and Gaussian-chain models predict the same chain dimensionality and end-to-end distribution implies that the microscopic details of chain connectivity may not significantly impact the overall properties of long polymer chains.

3.9.3 Scale Invariance

Scale invariance represents a fundamental characteristic of polymeric systems and plays a key role in developing polymer theory and scaling laws. Intuitively, it refers to a phenomenon observed in polymer systems, where the physical properties of polymers remain invariant across different length scales. The scale invariance implies that the behavior of polymer chains is similar at all length scales, from the molecular level up to the macroscopic level.

The Gaussian-chain model provides a simple understanding of scale invariance in polymeric systems. To elucidate, consider two segments i and j in a Gaussian chain with $j > i$. The probability of these two segments separated by a vector $\mathbf{r}$ is given by

$$p_{ij}(\mathbf{r}) \sim \langle \delta((\mathbf{x}_j - \mathbf{x}_i) - \mathbf{r}) \rangle \tag{3.175}$$

where $\mathbf{x}_i$ and $\mathbf{x}_j$ represent the segment positions, $\delta(\mathbf{r})$ is the three-dimensional Dirac delta function, $<\cdots>$ represents the average over all possible bond connections between segments i and j, and the proportionality constant can be fixed by normalization of the probability density.

An analytical expression for $p_{ij}(\mathbf{r})$ can be found by first considering Eq. (3.175) in the Fourier space

$$\begin{aligned}\hat{p}_{ij}(\mathbf{k}) \sim \langle \exp[-i\mathbf{k} \cdot (\mathbf{x}_j - \mathbf{x}_i)] \rangle &= \prod_{n=i}^{j-1} \langle \exp(-i\mathbf{k} \cdot \mathbf{r}_n) \rangle \\ &= \prod_{n=i}^{j-1} \exp\left(-k^2 <\mathbf{r}_n^2> /6\right) = \exp[-k^2 (j-i) b^2/6]\end{aligned} \tag{3.176}$$

where $\mathbf{r}_i=\mathbf{x}_{i+1}-\mathbf{x}_i$. Applying the inverse Fourier transform to the above equation gives

$$p_{ij}(\mathbf{r}) \sim \exp\left[-\frac{3r^2}{2(j-i)b^2}\right]. \tag{3.177}$$

In the derivation of Eq. (3.176), we have used the following formula for the Gaussian average (Appendix 8B)

$$< e^{\mathbf{k}\cdot\mathbf{r}} >= \exp(< \mathbf{r}^2 > k^2/6). \tag{3.178}$$

where $\mathbf{r}$ is a Gaussian variable and $\mathbf{k}$ is an arbitrary vector. By setting i and j as the end segments, we find that the end-to-end distribution function predicted by Eq. (3.177) is equivalent to that for a FJC discussed in Section 3.8. Note that the central limit theorem is not used in the derivation of Eq. (3.177).

A comparison of Eqs. (3.170) and (3.177) suggests that the probability distribution for the separation between any two points in a Gaussian chain is invariant with the length scales. In other words, the relations between different variables (e.g., $R_g \sim m^{1/2}$) do not change when both quantities are scaled with a common factor. If we group λ segments together as a new unit, the number of grouped segments is m/λ and each new segment has the length of $b/\sqrt{\lambda}$. Therefore, the relation for the radius of gyration $R_g^2 = (b/\sqrt{\lambda})^2(m/\lambda)/6 = b^2m/6$ remains the same.

3.9.4 Intra-Chain Correlation Functions

Although there is no correlation between segments from different ideal chains, the chain connectivity results in strong correlations among segments from the same polymer. The intrachain correlation functions provide useful insights into the internal structure of the polymer conformation.

To derive the segment–segment correlation function for a Gaussian chain, we may place an arbitrary segment, i, at the origin of a Cartesian coordinate and consider the radial distribution of all other segments around i. The average density of the other segments around segment i is given by

$$\rho_i(\mathbf{r}) = \sum_{j=1}^{m} < \delta[\mathbf{r} - (\mathbf{r}_j - \mathbf{r}_i)] > \tag{3.179}$$

where $<\cdots>$ denotes the ensemble average over all conformations of the polymer chain, and $\delta(\mathbf{r})$ is the 3-dimensional Dirac delta function. Since segment i is selected arbitrarily, the correlated density of the polymer should be represented by an average over all segments

$$\rho(\mathbf{r}) = \frac{1}{m}\sum_{i=1}^{m}\rho_i(\mathbf{r}) = \frac{1}{m}\sum_{i=1}^{m}\sum_{j=1}^{m} < \delta[\mathbf{r} - (\mathbf{x}_j - \mathbf{x}_i)] > \tag{3.180}$$

where $\rho(\mathbf{r})$ stands for the probability of two segments in the Gaussian chain separated by vector $\mathbf{r}$.

To evaluate the ensemble average in Eq. (3.180), we first express $\rho(\mathbf{r})$ in the Fourier space

$$\hat{\rho}(\mathbf{k}) = \int d\mathbf{r}\, e^{i\mathbf{k}\cdot\mathbf{r}}\rho(\mathbf{r}) = \frac{1}{m}\sum_{i=1}^{m}\sum_{j=1}^{m}\langle \exp[i\,\mathbf{k}\cdot(\mathbf{x}_j - \mathbf{x}_i)]\rangle. \tag{3.181}$$

Following (3.178), we can rewrite Eq. (3.181) as

$$\hat{\rho}(\mathbf{k}) = \frac{1}{m}\sum_{i=1}^{m}\sum_{j=1}^{m}\exp\left[-k^2\left\langle(\mathbf{x}_j - \mathbf{x}_i)^2\right\rangle/6\right]. \tag{3.182}$$

Because the separation between any two segments satisfies the Gaussian distribution, we have

$$\left\langle(\mathbf{x}_j - \mathbf{x}_i)^2\right\rangle = |j - i|b. \tag{3.183}$$

Substitution of Eq. (3.183) into (3.182) gives the final expression for the correlated density

$$\hat{\rho}(\mathbf{k}) = \frac{1}{m}\sum_{i=1}^{m}\sum_{j=1}^{m} \exp[-\mid j-i \mid k^2b^2/6] \approx mf(kR_g) \tag{3.184}$$

with $x = kR_g$ and

$$f(x) = \frac{2}{x^4}[e^{-x^2} - 1 + x^2]. \tag{3.185}$$

Eq. (3.184) [36] was first derived by Peter Debye and, to his credit, $\hat{\rho}(\mathbf{k})$ is often called the Debye function.

For small value of $x, f(x) \approx 1 - x^2/3$, the Debye function reduces to

$$\hat{\rho}(\mathbf{k}) = m\left(1 - k^2R_g^2/3\right). \tag{3.186}$$

Conversely, for $x >> 1, f(x) \approx 2/x^2$ and

$$\hat{\rho}(\mathbf{k}) = 2m/\left(k^2R_g^2\right). \tag{3.187}$$

Eqs. (3.186) and (3.187) give the asymptotic expressions of $\hat{\rho}(\mathbf{k})$ at small and large values of the wave vector $\mathbf{k}$, respectively. An interpolation of these two equations gives

$$\hat{\rho}(\mathbf{k}) \approx \frac{m}{1 + k^2R_g^2/2}. \tag{3.188}$$

Figure 3.24 compares the correlated densities predicted by Eqs. (3.184) and (3.188). Except at small value of $x = kR_g$, the difference between these two equations is unsubstantial. Unlike Eq. (3.184), however, Eq. (3.188) provides an analytical expression for the inverse Fourier transform of the local density of polymer segments

$$\rho(\mathbf{r}) = \frac{m}{2\pi R_g^2}\frac{e^{-\sqrt{2}r/R_g}}{r}. \tag{3.189}$$

Eq. (3.189) indicates that, for a Gaussian chain, the radius of gyration R_g may be interpreted as the correlation length (within a proportional constant $\sqrt{2}$) for the distribution of polymer segments. Because the radius of gyration increases with the chain length $R_g \sim \sqrt{m}$, the intra-chain correlation diverges as $m \to \infty$. The divergence of the correlation length resembles the critical behavior of phase transition as predicted by a mean-field theory (Chapter 5).

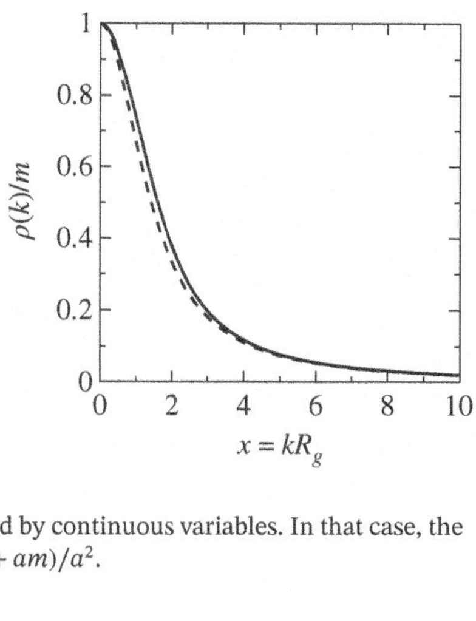

Figure 3.24 Correlated density of an ideal chain in the Fourier space. The solid line represents the reduced density calculated from the Debye function, and the dashed line is from Eq. (3.188). Here, m is the number of segments per chain, $x = kR_g$ where R_g is the radius of gyration and $k = |\mathbf{k}|$ is the magnitude of wave vector $\mathbf{k}$.

---

36 This equation is evaluated by assuming that i and j can be represented by continuous variables. In that case, the double summation becomes $\int_0^m dx \int_0^m dy \exp(-a \mid x-y \mid) = 2(e^{-am} - 1 + am)/a^2$.

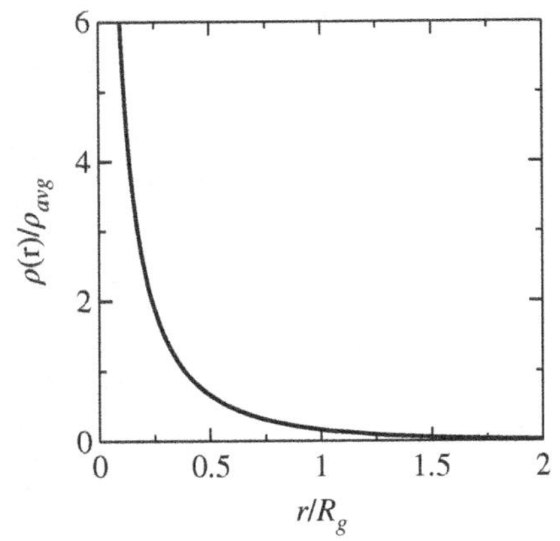

Figure 3.25 The reduced correlated density of an ideal chain in real space. Here, R_g is the radius of gyration, **r** is the end-to-end vector, and $\rho_{avg} = 3m/\left(4\pi R_g^3\right)$ is the average density of polymer segments within a sphere of radius R_g.

Figure 3.25 shows the correlated density of polymer segments predicted by Eq. (3.189). We see that the segment density falls sharply when $r > R_g$, suggesting that most segments are located within a spherical coil of radius R_g, with the density falling monotonically from the center. As predicted by the FJC model, the average density of polymer segments within the coil scales, $\rho_{avg} = 3m/\left(4\pi R_g^3\right) \sim m^{-1/2}$, is extremely small when $m > > 1$.

The Gaussian-chain model allows us to construct a continuous representation of polymer chains commonly used in the polymer field theory. Toward that end, consider a Gaussian chain of finite contour length (L) and Kuhn length (b). In the limit that the average bond length is infinitely small, $s \to 0$, the number of segments approaches $m \to \infty$ in order to keep a finite contour length $L = (m-1)s$. Because $L = ms$ is fixed, the scale invariance suggests that probability distribution for the bond vector connecting the $(i-1)^{th}$ and i^{th} segments of the Gaussian chain can be written as

$$p(\mathbf{r}_i) = \left(\frac{3}{2\pi bs}\right)^{3/2} \exp\left(-\frac{3r_i^2}{2bs}\right) \tag{3.190}$$

According to Eq. (3.190), the probability that the chain has a conformation with the bond vectors given by $\{\mathbf{r}_{i=1,2\cdots}\}$ is

$$p[\{\mathbf{r}_i\}] \sim \prod_{i=1}^{m-1} \exp\left[-\frac{3r_i^2}{2bs}\right] = \exp\left[-\sum_{i=1}^{m-1} \frac{3(\mathbf{R}_{i+1} - \mathbf{R}_i)^2}{2bs}\right] \tag{3.191}$$

where $\mathbf{R}_i$ represents the position of segment i. As $s \to 0$, the summation in Eq. (3.191) can be expressed as an integration

$$\sum_{i=1}^{m-1} (\mathbf{R}_{i+1} - \mathbf{R}_i)^2/s = \sum_{i=1}^{m-1} \left(\frac{\mathbf{R}_{i+1} - \mathbf{R}_i}{s}\right)^2 s \to \int_0^L ds \left(\frac{\partial \mathbf{R}}{\partial s}\right)^2 \tag{3.192}$$

where the function $\mathbf{R}(s)$ specifies the local position of the polymer backbone (viz., a contour path), and $\partial\mathbf{R}/\partial s$ stands for the local bond orientation. Substitution of Eq. (3.192) into Eq. (3.191) gives

$$p[\mathbf{R}(s)] \sim \exp\left[-\frac{3}{2b}\int_0^L ds \left(\frac{\partial \mathbf{R}}{\partial s}\right)^2\right]. \tag{3.193}$$

Eq. (3.193) gives the probability of a continuous polymer to have a contour path $\mathbf{R}(s)$. Because for a Gaussian chain, $Lb = 6R_g^2$, Eq. (3.193) can be re-expressed as

$$p[\mathbf{R}(n)] \sim \exp\left[-\frac{1}{4R_g^2}\int_0^1 dn \left(\frac{\partial \mathbf{R}}{\partial n}\right)^2\right] \tag{3.194}$$

where $n \equiv s/L$ is dimensionless. We will return to Eq. (3.193) in Section 8.7 to extend the statistical mechanics of simple fluids to that of continuous polymers.

3.9.5 Summary

In this section, we introduce the discrete and continuous representations of idealized polymer chains using Gaussian statistics. The scale invariance ensures that both particle-based and continuous models are equivalent in describing the statistical properties of polymer chains. The idealized model can be extended to describe nonideal polymer systems. Gaussian statistics plays a fundamental role in applying field-theoretical tools in polymer physics. Some of these applications will be further discussed in Chapter 8.

3.10 Statistics of Copolymer Chains

A copolymer chain contains two or more different types of monomers. Depending on the arrangement of the repeat units, copolymers can be classified into various categories, including statistical, block, graft, star copolymers, and gels. The statistical analysis of copolymer chains is important for understanding the behavior of these materials, such as their thermodynamic properties and their response to external factors such as temperature and pressure.

Schematically, Figure 3.26 shows the molecular architectures of a few representative copolymers that contain two chemically distinct repeat units, A and B. Statistical copolymers are made of monomers connected in a random sequence along the polymer backbone. A simple example of a statistical copolymer is provided by a heteropolymer where the different repeat units are randomly distributed along the chain.[37] Block copolymers have repeat units that exist in long sequences, or blocks, of the same type. In most cases (but not always), block and statistical copolymers are linear. Graft copolymers, or comb polymers, contain branched chains that have a chemical structure different from that of the polymer backbone. The properties of statistical copolymers are normally between those of the corresponding homopolymers. However, block and graft copolymers often preserve the characteristics of individual polymer blocks. As a result, block or graft copolymers may be used to combine the desirable properties of the homopolymers into a single material with improved mechanical, flow, optical, barrier, or other desired physiochemical properties.

3.10.1 Statistics of an Ideal Block Copolymer

As for homopolymers, polymer chains in a melt of flexible block copolymers take conformations like those for an ideal chain. Different from a homopolymer, however, a block-copolymer chain consists of segments with different identities; correlations between different segments from the same chain are closely related to the polymer architecture.

To illustrate the difference between the structure of a homopolymer and that of a block copolymer, consider a diblock copolymer with m_A segments in block A and m_B segments in block B. A similar procedure can be applied to multi-block copolymers. We assume that each polymer block can be represented by a Gaussian chain. The single-chain density correlation function (viz., the intra-chain correlation function) is defined as

$$\omega_{ij}(\mathbf{r}) = \frac{1}{m_i m_j} \sum_{\alpha=1}^{m_i} \sum_{\beta=1}^{m_j} \langle \delta[\mathbf{r} - (\mathbf{x}_\alpha - \mathbf{x}_\beta)] \rangle \tag{3.195}$$

37 A copolymer with a random distribution of repeat units is called a random copolymer.

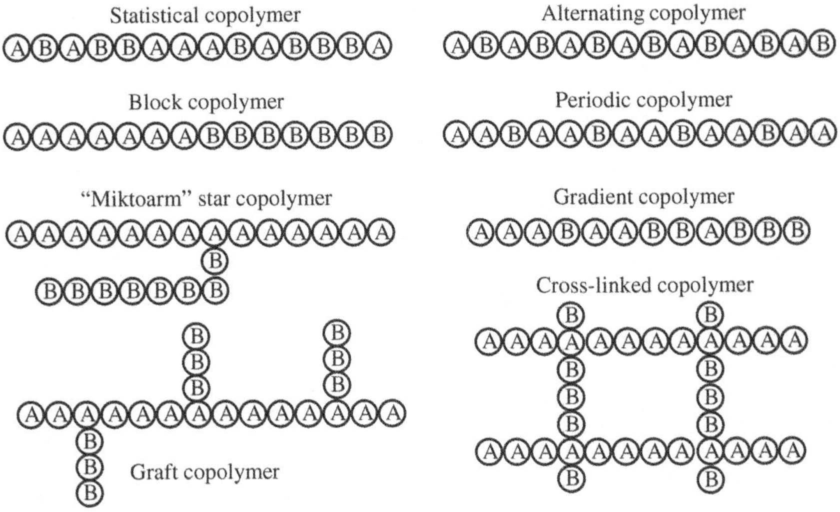

Figure 3.26 Schematic of the molecular architectures of various copolymers composed of two chemically distinct repeat units A and B.

where $\langle \cdots \rangle$ denotes an average over all conformations of the polymer chain, and $i, j = A$ or B.

Following the mathematical procedure discussed in Section 3.9, we can obtain an analytic expression for the single-chain density correlation function in the Fourier space

$$\hat{\omega}_{ij}(k) = \begin{cases} 2\left(e^{-x_i} - 1 + x_i\right)/x_i^2 & i = j \\ \left(e^{-x_i} - 1\right)\left(e^{-x_j} - 1\right)/(x_i x_j) & i \neq j \end{cases} \tag{3.196}$$

where $x_i \equiv k^2 R_{g,i}^2$ with $R_{g,i} = \sqrt{m_i b_i^2/6}$ being the radius of gyration for an ideal chain with m_i Kuhn segments. In Eq. (3.196), the first line corresponds to the Debye function; and the second line is derived from the end-to-end distance for a composite Gaussian chain with n segments from block A and m segment from block B. The latter is given by $\sqrt{nb_A^2 + mb_B^2}$, where b_A and b_B are the Kuhn lengths of blocks A and B. (Problem 3.33)

Figure 3.27 shows single-chain density correlation functions predicted by Eq. (3.196) for a diblock copolymer of A and B. All three correlation functions decay monotonically with the reduced wave vector $x = kR_g$; the rate of decay reflects the correlation length. As expected, the correlation between different blocks is always longer than that between segments within the same block.

As discussed in Section 3.9, analytical expressions for the single-chain density correlation functions can be found by an asymptotic analysis. In the limit of small k (viz., $k \to 0$), we can expand Eq. (3.196) in a power series of k and truncate the series after the second-order term,

$$\hat{\omega}_{ij}(k) = \begin{cases} 1 - R_{g,i}^2 k^2/3 & i = j, \\ 1 - \left(R_{g,i}^2 + R_{g,j}^2\right) k^2/2 & i \neq j. \end{cases} \tag{3.197}$$

For large k (viz., $k \to \infty$), Eq. (3.196) becomes

$$\hat{\omega}_{ij}(k) = \begin{cases} 2/\left(R_{g,i}^2 k^2\right) & i = j, \\ 1/\left(R_{g,i}^2 R_{g,j}^2 k^4\right) & i \neq j \end{cases} \tag{3.198}$$

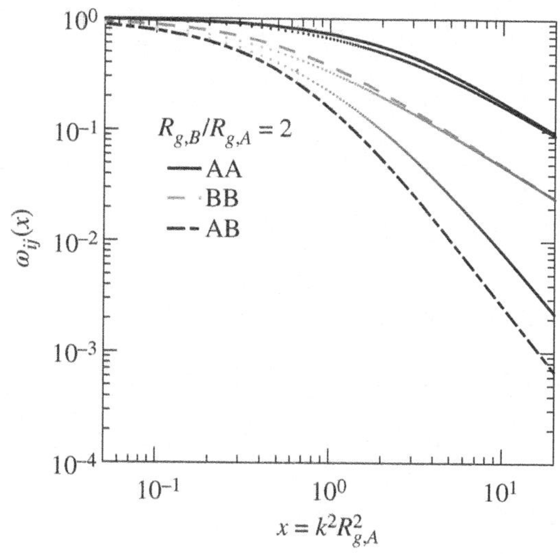

Figure 3.27 The single-chain density correlation function for a diblock copolymer where R_g is the radius of gyration for the corresponding polymer. Here, the radius of gyration for block B is twice that for block A. The bold lines correspond to the predictions of Eq. (3.196), while the dotted lines represent the predictions of Eq. (3.199).

A simple interpolation of Eqs. (3.197) and (3.198) is given by

$$\hat{\omega}_{ij}(k) = \begin{cases} \left(1 + R_{g,i}^2 k^2/2\right)^{-1} & i = j, \\ \left(1 + R_{g,i}^2 k^2/2\right)^{-1} \cdot \left(1 + R_{g,j}^2 k^2/2\right)^{-1} & i \neq j. \end{cases} \tag{3.199}$$

Eq. (3.199) indicates that the asymptotic behavior of the single-chain density correlation function for segments that belong to the same block is identical to that for a homopolymer. Furthermore, it shows that the single-chain density correlation function between segments from different blocks in the same polymer chain can be written as a convolution of individual correlation functions.

3.10.2 Summary

Copolymers are composed of two or more types of monomers and can exhibit unique properties that are not observed in homopolymers. The Gaussian-chain model can be used to describe the intrachain correlation functions similar to that used for homopolymers. As explained in Section 8.7, the arrangement of copolymer chains plays a crucial role in comprehending and formulating copolymers with specific properties and applications.

3.11 Semi-Flexible Chains

Both the freely jointed chain (FJC) and Gaussian-chain models assume that a polymer chain is highly flexible, where the bond orientations are uncorrelated. However, this assumption does not apply to semi-flexible polymers, especially in the case of biomacromolecules such as double-stranded DNA, actins, microtubules, and polymers that exhibit liquid crystalline order at high concentrations. In such cases, the freely rotating chain (FRC) model provides a more realistic description of the polymer conformations. The worm-like chain (WLC) model may be considered as a continuous extension of the FRC model.[38] The semi-flexible chain models have applications in various fields, such as biophysics, polymer science, and materials science.

38 The term wormlike chain is used because the continuous chain can be envisioned to bend only gradually and smoothly, somewhat like a worm.

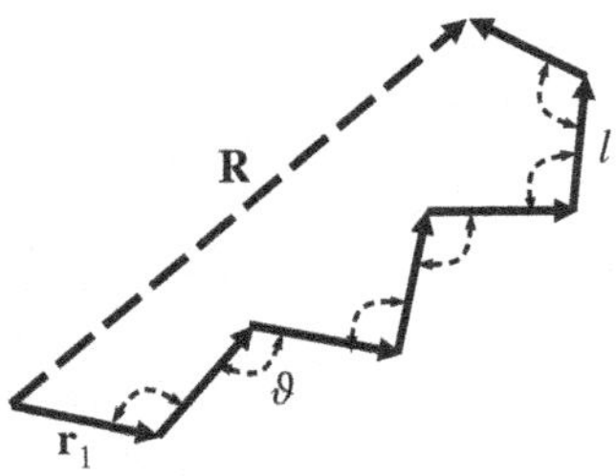

Figure 3.28 A freely rotating chain (FRC) is similar to a freely jointed chain (FJC) but with a fixed bond length l and bond angle ϑ.

3.11.1 Freely Rotating Chains

As shown in Figure 3.28, a FRC consists of noninteracting segments with rigid bonds of length l similar to those in a FJC.[39] However, in FRC the bond angles are fixed to a constant value $90° < \vartheta < 180°$. Accordingly, the orientations of neighboring bonds are strongly correlated, i.e.,

$$<\mathbf{r}_i \cdot \mathbf{r}_{i+1}> = l^2 \cos(\pi - \vartheta). \tag{3.200}$$

If $\vartheta = 180°$, the chain becomes a straight line, and the polymer is rigid and has only a single conformation. In the following discussion, we assume $\vartheta \neq 180°$.

The persistence length is a characteristic length scale that describes the degree of stiffness or flexibility of a semi-flexible polymer chain. According to the FRC model, the internal stiffness or the rigidity of a polymer chain is reflected in the alignment of the first bond vector $\mathbf{r}_1$ with the end-to-end vector, $\mathbf{R} = \sum_{i=1}^{m-1} \mathbf{r}_i$. Quantitatively, the alignment can be measured in terms of the vector product

$$\begin{aligned}<\mathbf{R} \cdot \mathbf{r}_1> &= \sum_{i=1}^{m-1} <\mathbf{r}_i \cdot \mathbf{r}_1> = \sum_{i=1}^{m-1} \frac{<\mathbf{r}_i \cdot \mathbf{r}_{i-1}> \cdot <\mathbf{r}_{i-1} \cdot \mathbf{r}_{i-2}> \cdot \cdots \cdot <\mathbf{r}_2 \cdot \mathbf{r}_1>}{l^{2(i-2)}} \\ &= l^2 \sum_{i=1}^{m-1} \cos^{i-1}(\pi - \vartheta) = l^2 \frac{1 - \cos^{m-1}(\pi - \vartheta)}{1 + \cos(\vartheta)}.\end{aligned} \tag{3.201}$$

As the bond angle increases, the polymer chain becomes more rigid, as reflected by a larger value of $<\mathbf{R} \cdot \mathbf{r}_1>$. Therefore, we may define the persistence length by the projection of the end-to-end vector $\mathbf{R}$ in the direction of the first bond vector $\mathbf{r}_1$

$$\xi_p \equiv \lim_{m\to\infty} \frac{<\mathbf{R} \cdot \mathbf{r}_1>}{l} = \frac{l}{1 + \cos\vartheta}. \tag{3.202}$$

In Eq. (3.202), we take the limit of infinite chain length such that the persistence length reflects the characteristics of the polymer backbone. Unlike $<\mathbf{R} \cdot \mathbf{r}_1>$, the persistence length is fully determined by the chemical characteristics of the bond connectivity. Intuitively, it describes the flexibility of a chain or a length scale on which the polymer maintains its orientation in the initial bond direction. For example, the persistence length of an FJC is one segment length and that of a rod-like polymer is the same as the polymer contour length.

Based on Eq. (3.200), we can evaluate the end-to-end distance of an FRC analytically. Note that for $i \geq j$

$$<\mathbf{r}_i \cdot \mathbf{r}_j> = \frac{<\mathbf{r}_i \cdot \mathbf{r}_{i-1}> \cdot <\mathbf{r}_{i-1} \cdot \mathbf{r}_{i-2}> \cdot \cdots \cdot <\mathbf{r}_{j+1} \cdot \mathbf{r}_j>}{l^{2(i-j-1)}} = l^2 \cos^{i-j}(\pi - \vartheta). \tag{3.203}$$

---

39 The bond length in the FRC model l is not necessarily identical to the Kuhn length in the FJC model. Similarly, the bond angle ϑ should not be identified as the angle of real chemical bonds.

Thus, the ensemble average of the square of the end-to-end vector can be written as

$$<\mathbf{R}^2> = \sum_{i=1}^{m-1}\sum_{j=1}^{m-1} <\mathbf{r}_i \cdot \mathbf{r}_j> = \sum_{i=1}^{m-1} \left[<\mathbf{r}_i \cdot \mathbf{r}_i> + 2\sum_{j=1}^{i-1} <\mathbf{r}_i \cdot \mathbf{r}_j> \right]$$
$$= l^2(m-1) + 2l^2 \sum_{i=1}^{m-1}\sum_{j=1}^{i-1} \cos^{(i-j)}(\pi - \vartheta). \tag{3.204}$$

Using the algebraic relation for $x = \cos(\pi - \vartheta) < 1$

$$\sum_{i=1}^{m-1}\sum_{j=1}^{i-1} x^{(i-j)} = \sum_{i=1}^{m-1} (x^{i-1} + x^{i-2} + \cdots + x) = \sum_{i=1}^{m-1} \left(\frac{1-x^i}{1-x} - 1 \right)$$
$$= \frac{x(m-1)}{1-x} - \frac{1}{1-x}\sum_{i=1}^{m-1} x^i = \frac{x(m-1)}{1-x} - \frac{x - x^m}{(1-x)^2}, \tag{3.205}$$

we may evaluate the double summation in Eq. (3.204)

$$<\mathbf{R}^2> = (m-1)l^2 + 2l^2 \frac{\cos(\pi-\vartheta)(m-1)}{1-\cos(\pi-\vartheta)} - 2l^2 \frac{\cos(\pi-\vartheta) - \cos^m(\pi-\vartheta)}{[1-\cos(\pi-\vartheta)]^2}$$
$$= (m-1)l^2 \left[\frac{1-\cos\vartheta}{1+\cos\vartheta} + \frac{2}{m-1} \frac{\cos\vartheta + \cos^m(\pi-\vartheta)}{(1+\cos\vartheta)^2} \right] \tag{3.206}$$

Figure 3.29 shows the end-to-end distance for a FRC of different chain lengths as a function of the persistence length. As expected, R increases with the persistence length and, in the asymptotic limit (as $\vartheta \to \pi$), it approaches the polymer contour length.

3.11.2 The Flory Characteristic Ratio

The Flory characteristic ratio provides an alternative way to characterize the flexibility of a polymer chain. The dimensionless parameter is defined as the ratio of the mean-square end-to-end distance of a polymer chain to the product of the number of bonds and the square of the bond length. In the context of FRC model, the Flory characteristic ratio is given by

$$C_F \equiv \frac{<\mathbf{R}^2>}{(m-1)l^2} = \frac{1-\cos\vartheta}{1+\cos\vartheta} + \frac{2}{m-1} \frac{\cos\vartheta + \cos^m(\pi-\vartheta)}{(1+\cos\vartheta)^2}. \tag{3.207}$$

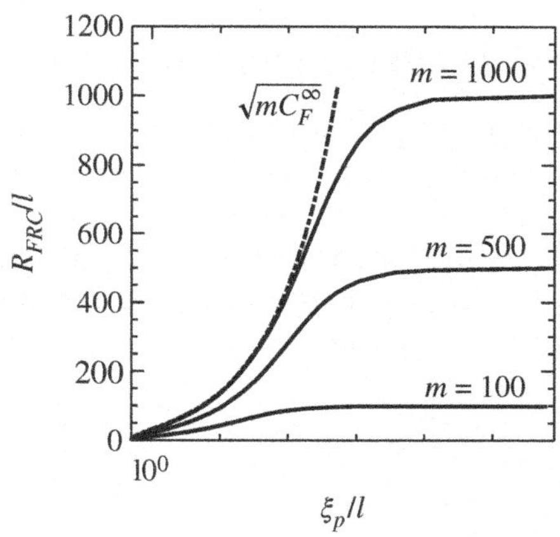

Figure 3.29 The end-to-end distance versus the persistence length of a freely rotating chain (FRC). The dashed line shows the prediction of Eq. (3.209). Here, m stands for the number of polymer segments per chain.

Table 3.7 Single-chain parameters for some polymer melts.

Compound	l (Å)	ϑ	C_F^{∞}	ξ_p (Å)
Polyethylene (PE)	1.53	135°	5.7	5.12
Polycarbonate (PC)	6.61	114°	2.4	11.24
Poly(vinyl chloride) (PVC)	1.53	140°	7.6	6.58
Poly(methyl methacrylate) (PMMA)	1.53	141°	7.9	6.81
Polydimethylsiloxane (PDMS)	1.64	138°	6.8	6.40
Polyoxymethylene (POM)	1.43	140°	7.5	6.07
Polycaprolactam (Nylon 6)	1.71	136°	6.2	6.15
Polystyrene (PS)	1.53	146°	10.8	9.03
Poly(ethylene terephthalate) (PET)	2.68	127°	4.10	6.82

Source: Adapted from Wu S., "Predicting chain conformation and entanglement of polymers from chemical structure", *Polym. Eng. Sci.* 32, 823–830 (1992).

Because $<\mathbf{R}^2>$ is an experimentally measurable quantity (e.g., through light scattering), the Flory characteristic ratio provides a convenient route to estimate the parameters in the FRC model (Table 3.7).[40] For $m \gg 1$, Eq. (3.207) reduces to

$$C_F^{\infty} \approx \frac{1-\cos\vartheta}{1+\cos\vartheta}. \tag{3.208}$$

Accordingly, the end-to-end distance can be expressed as

$$R = \sqrt{<\mathbf{R}^2>} \approx l\sqrt{m\frac{1-\cos\vartheta}{1+\cos\vartheta}} = l\sqrt{mC_F^{\infty}}. \tag{3.209}$$

In Eq. (3.209), the approximation holds only when the persistence length is not too large (see Figure 3.29).

Eq. (3.209) indicates that, in the long-chain limit, the end-to-end distance of an FRC is proportional to the chain length, $\sqrt{m}$, similar to that for an FJC or a Gaussian chain. Intuitively, a long FRC can be considered as an FJC but with an effective Kuhn length depending on the Flory characteristic ratio. Because $90° < \vartheta < 180°$, the end-to-end distance of an FRC is significantly larger than that of an FJC or a Gaussian chain with the same bond length and number of segments.

3.11.3 The Worm-Like Chain Model

The WLC model may be understood as a continuous extension of the FRC model for semi-flexible polymers. This model is commonly used to represent semi-flexible polymers including DNAs and polypeptides. The WLC model also provides a good description of the end-to-end distance distribution of protein loops in comparison with experimental data.[41]

In the continuous extension of the FRC model, we let the bond length $l \to 0$, the number of segments $m \to \infty$, and the bond angle $\vartheta \to 180°$ while the contour length $L = ml$ and the persistence

40 Wu S., "Predicting chain conformation and entanglement of polymers from chemical structure", *Polym. Eng. Sci.* 32, 823–830 (1992).

41 Zhou H. X., "Polymer models of protein stability, folding, and interactions", *Biochemistry* 43 (8), 2141–2154 (2004).

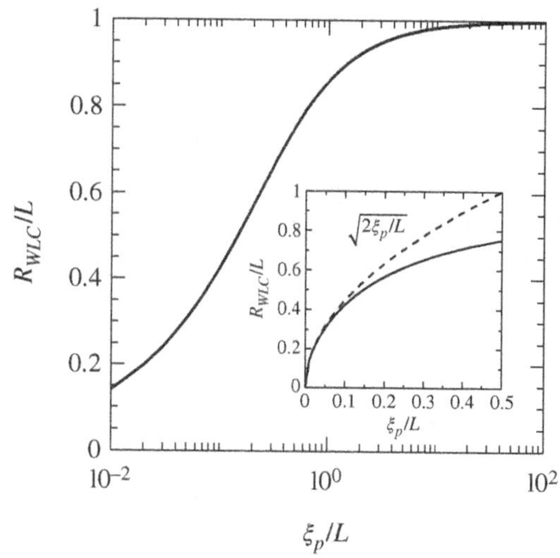

Figure 3.30 The end-to-end distance for a worm-like chain (WLC) versus the persistence length.

length $\xi_p = l/([1 + \cos(\vartheta)]$ remain fixed. Accordingly, the mean square end-to-end distance for a WLC can be derived from that for an FRC, Eq. (3.206),

$$<\mathbf{R}^2_{WLC}> = \lim_{\vartheta \to 180^\circ} \left\{ \frac{l^2 m[1 - \cos(\vartheta)]}{[1 + \cos(\vartheta)]} + \frac{2l^2[\cos^m(\pi - \vartheta) + \cos(\vartheta)]}{[1 + \cos(\vartheta)]^2} \right\}$$
$$= \lim_{\vartheta \to 180^\circ} \left\{ 2L\xi_p + 2\xi_p^2[\cos^m(\pi - \vartheta) - 1] \right\}. \tag{3.210}$$

As $\vartheta \to 180^\circ, x \equiv \pi - \vartheta \to 0$, $\cos^m x \approx (1 - x^2/2)^m \approx e^{-mx^2/2}$, and $\xi_p = l/(1 - \cos x) \approx 2l/x^2$. Accordingly, Eq. (3.210) can be evaluated analytically

$$<\mathbf{R}^2_{WLC}> = 2L\xi_p + 2\xi_p^2(e^{-mx^2/2} - 1) = 2L\xi_p + 2\xi_p^2 \left(e^{-L/\xi_p} - 1\right). \tag{3.211}$$

In terms of the average end-to-end distance, Eq. (3.211) can be written as

$$R_{WLC} = \sqrt{< \mathbf{R}^2_{WLC} >} = \sqrt{2L\xi_p + 2\xi_p^2 \left(e^{-L/\xi_p} - 1\right)}. \tag{3.212}$$

Figure 3.30 shows the end-to-end distance of a WLC versus the persistence length. As $\xi_p \to \infty$, i.e., when the polymer chain becomes rigid, Eq. (3.212) reduces to

$$R_{WLC} = L - L^2/(6\xi_P) + \cdots. \tag{3.213}$$

As expected, Eq. (3.213) predicts $R_{WLC} = L$ for a rigid rod. Conversely, the polymer becomes highly flexible as $L/\xi_p \to \infty$. In that case, $R_{WLC} \approx \sqrt{2L\xi_p}$ which corresponds to the end-to-end distance for a Gaussian chain with Kuhn length $b = 2\xi_p$ and the number of segments $m = L/b$. Therefore, the Gaussian chain and rigid-rod models may be considered as two extreme limits of the WLC model.

In the continuous limit, a combination of Eqs. (3.201) and (3.202) leads to the projection of the end-to-end vector in the direction of the first bond

$$\frac{<\mathbf{R} \cdot \mathbf{r}_1>}{l} = \xi_p \left[1 - \lim_{l \to 0}(1 - l/\xi_p)^{L/l}\right] = \xi_p \left(1 - e^{-L/\xi_p}\right). \tag{3.214}$$

As $L \to \infty$, Eq. (3.214) predicts that the persistence length equals to ξ_p. With the polymer backbone represented by a continuous curve, $\mathbf{r}(s)$, the radius of gyration for a WLC is given by

$$R_g^2 = \frac{1}{2m^2} \int_0^m \int_0^m dsds' <[\mathbf{r}(s) - \mathbf{r}(s')]^2> = \frac{1}{m^2} \int_0^m ds(m - s) <\mathbf{R}^2(s)> \tag{3.215}$$

where $\mathbf{R}(s) = \mathbf{r}(s) - \mathbf{r}(0)$. Using the analytical expression for the end-to-end distance of a WLC given by Eq. (3.211), we can evaluate the above equation analytically (Problem 3.34)

$$R_g^2 = \frac{L\xi_p}{3} - \xi_p^2 + \frac{2\xi_p^3}{L} - \frac{2\xi_p^4}{L^2}\left(1 - e^{-L/\xi_p}\right). \tag{3.216}$$

Like the end-to-end distance, the radius of gyration of a WLC reduces to that for a Gaussian chain and to that for a rigid rod at extreme limits. As the persistence length goes to zero, Eq. (3.216) predicts

$$\lim_{\xi_p \to 0} R_g^2 = \frac{L\xi_p}{3} \tag{3.217}$$

which corresponds to that for a Gaussian chain with Kuhn length $b = 2\xi_p$. As the persistence length goes to infinity, we have from Eq. (3.216)[42]

$$\lim_{\xi_p \to \infty} R_g^2 = \frac{L^2}{12} \tag{3.218}$$

which corresponds to that for a rigid rod.

3.11.4 Summary

The FRC model and the WLC model are two complementary theoretical methods that are commonly used to describe the conformation of flexible polymer chains. The FRC model characterizes a polymer chain as a series of rigid bonds that are linked by freely rotating joints, while the WLC model describes the polymer as a continuous curve possessing a finite persistence length. Both models can be used to determine the average end-to-end distance, the persistence length, and the force-extension curve of semi-flexible polymers at different solution conditions.

3.12 Random-Walk Models

The random-walk model is one of the simplest models in statistical mechanics and is relevant to various natural phenomena, including diffusion, polymer physics (such as the FJC model discussed in Section 3.8), and the movements of microorganisms or cells in a liquid medium. The statistical–mechanical description of a random walk can be applied to particle motion in a microfluidic device and molecular transport in protein channels or microporous materials, such as zeolites. In this section, we discuss a few simple random-walk models and their applications.

3.12.1 One-Dimensional Random Walk

The 1D random walk is also known as the "drunkard's walk", where the steps taken by the walker are entirely random and unpredictable. Imagine a drunkard walking in a narrow lane such that he takes a random step in either a forward or a backward direction at a fixed pace. Here by random, we mean that the direction of a particular step is completely independent of any previous step. Therefore, the drunkard's location after each step taken depends exclusively on his location after the previous step. We ask how the drunkard's position varies with time (or the number of steps), $x(t)$, given the original position is $x(0) = 0$.

To study the random-walk problem from a statistical–mechanical perspective, we may employ a 1D lattice shown in Figure 3.31. For simplicity, we assume that both the step length and the duration

42 $R_g^2/L^2 = \frac{1}{3x} - \frac{1}{x^2} + \frac{2}{x^3} - \frac{2}{x^4}(1 - e^{-x}) = \frac{1}{12} - \frac{x}{60} + \cdots$ as $x \equiv L/\xi_p \to 0$.

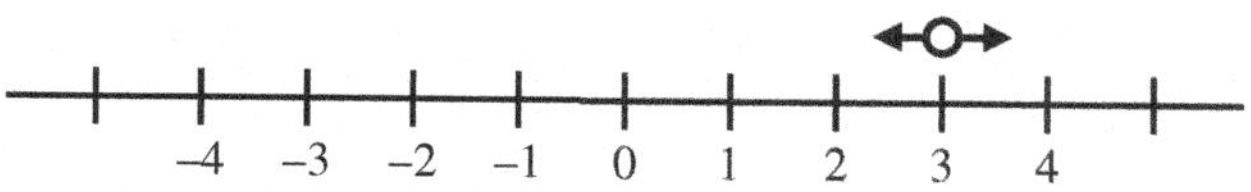

Figure 3.31 A drunkard takes a random walk on a one-dimensional lattice. Here, the drunkard is in position 3 at a given moment. After his next step, he will be at either position 2 or 4 with equal probability.

of each step are fixed. The system contains only one "particle", i.e., the drunkard. At any moment, the "microstate" is defined by an integer, k, that specifies the drunkard's position on the lattice

$$x = k\Delta \tag{3.219}$$

where Δ stands for the step length. After a long walk, the drunkard could end up in many positions, i.e., many microstates. The probability that the drunkard is at a given position corresponds to the probability distribution of microstates in an "ensemble" of drunkards walking in one dimension at the same pace and with the same step length.

Suppose that each drunkard in the ensemble starts from the origin at time $t = 0$. At time t, the position depends on the numbers of forward and backward steps taken, designated as n_F and n_B, respectively. Let the total number of steps be $n = t/\tau$, where τ stands for the duration per step. After n steps, the location of a particular drunkard is related to an integer

$$k = (n_F - n_B) \tag{3.220}$$

where n_F and n_B satisfy

$$n = n_F + n_B. \tag{3.221}$$

Rearrangement of Eqs. (3.220) and (3.221) gives the numbers of the forward and backward steps, respectively:

$$n_F = (n + k)/2, \tag{3.222}$$

$$n_B = (n - k)/2. \tag{3.223}$$

At each step, a drunkard moves forward or backward by one lattice site. After n steps, the total number of possible paths is 2^n. Because the number of drunkards in the ensemble can be arbitrarily large, all paths will be fully explored. The probability of a drunkard at position $x = k\Delta$ is given by the number of paths that lead to that position, $n!/(n_F!\,n_B!)$, divided by the total number of paths for n steps:

$$p(k, n) = \frac{n!}{2^n n_F! n_B!}. \tag{3.224}$$

Plugging Eqs. (3.222) and (3.223) into Eq. (3.224) for n_F and n_B, and using Stirling's formula,[43] we obtain

$$p(k, n) = \frac{1}{\sqrt{2\pi n}} \exp\left(-\frac{k^2}{2n}\right). \tag{3.225}$$

Eq. (3.225) can be written in terms of continuous variables (x, t). In that case, we find that the probability density, i.e., the probability that the drunkard's position is between $(x, x + \Delta)$ with $x = k\Delta$ and $t = n\tau$, follows the Gaussian distribution

$$p(x, t) = \frac{p(k, n)}{\Delta} = \frac{1}{\sqrt{4\pi Dt}} \exp\left(-\frac{x^2}{4Dt}\right) \tag{3.226}$$

43 $\ln n! \approx n \ln n - n + \ln\sqrt{2\pi n}$ for $n >> 1$. Stirling's approximation is remarkably accurate even for moderate n. For example, it approximates $\ln(10!) = 15.104$ by 15.096; the percentage error is 0.1%.

where $D \equiv \Delta^2/2\tau$ is a diffusion coefficient (or diffusivity). As required, $p(x, t)$ satisfies the normalization condition

$$\int_{-\infty}^{\infty} p(x,t)dx = 1. \tag{3.227}$$

Mathematically, Eq. (3.226) conforms to the central limit theorem, which states that the sum of many random steps leads to the Gaussian distribution.

We can identify the physical significance of parameter D by considering the balance equation for calculating the probability that a drunkard is at position k after $n+1$ steps:

$$p(k,n+1) = p(k-1,n)/2 + p(k+1,n+1)/2 \tag{3.228}$$

Eq. (3.228) indicates that at the nth step, the drunkard must be at either position $(k-1)$ or $(k+1)$ with equal probability. Subtracting both sides of Eq. (3.228) by $p(k, n)$ and replacing the finite differences with differentials yield

$$\frac{\partial p}{\partial n} \approx p(k,n+1) - p(k,n), \tag{3.229}$$

$$\frac{\partial^2 p}{\partial k^2} \approx \frac{p(k-1,n) + p(k+1,n+1) - 2p(k,n)}{4}. \tag{3.230}$$

A combination of Eqs. (3.228)–(3.230) gives

$$\frac{\partial p}{\partial n} = \frac{1}{2}\frac{\partial^2 p}{\partial k^2}. \tag{3.231}$$

In terms of continuous variables (x, t), Eq. (3.231) can be expressed as the familiar 1D diffusion equation:

$$\frac{\partial p}{\partial t} = \frac{\Delta^2}{2\tau}\frac{\partial^2 p}{\partial x^2} = D\frac{\partial^2 p}{\partial x^2} \tag{3.232}$$

where $D \equiv \Delta^2/2\tau$ can be identified with the Fickian diffusion coefficient.[44]

Figure 3.32 shows the probability density for the drunkard's position at three representative times. At $t = 0$, all drunkards in the ensemble are at the origin; the initial probability density is thus given by the Dirac delta function

$$p(x,0) = \delta(0). \tag{3.233}$$

As time elapses, the probability density becomes more uniform, and eventually, it is evenly distributed. To a good approximation, the mean-square deviation (MSD), $W(t)$, provides a good measure of the distribution range:

$$W(t) = \langle x^2 \rangle = \int_{-\infty}^{\infty} p(x,t)x^2 dx = 2Dt. \tag{3.234}$$

While, on average, the position for an ensemble of drunkards is always at the origin, the probability of any drunkard returning to the origin is about $1/\sqrt{2\pi n}$, which is extremely small after a long duration.

The drunkard's walk discussed above provides a suitable model for a variety of practical systems including proteins moving along double-stranded (ds) DNA or RNA, diffusion of particles in a microfluidic device, and molecular transport in small pores. For example, single-molecule experimental studies indicate that protein diffusion on a nucleic acid chain closely follows the 1D

44 This equation was first reported by Albert Einstein early in the 20th century using similar arguments.

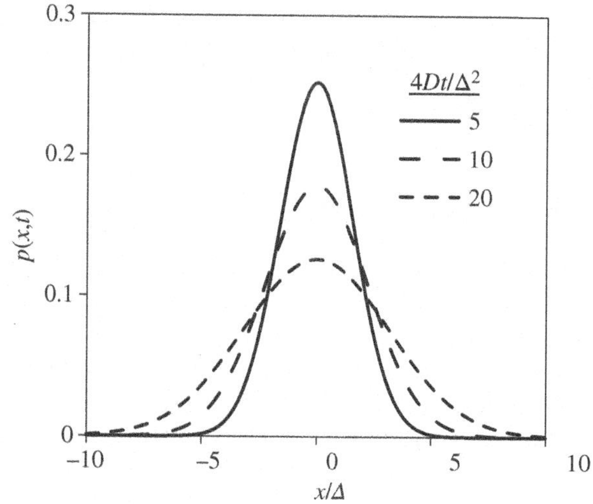

Figure 3.32 The probability density of a random walk becomes more uniform as the (reduced) time progresses.

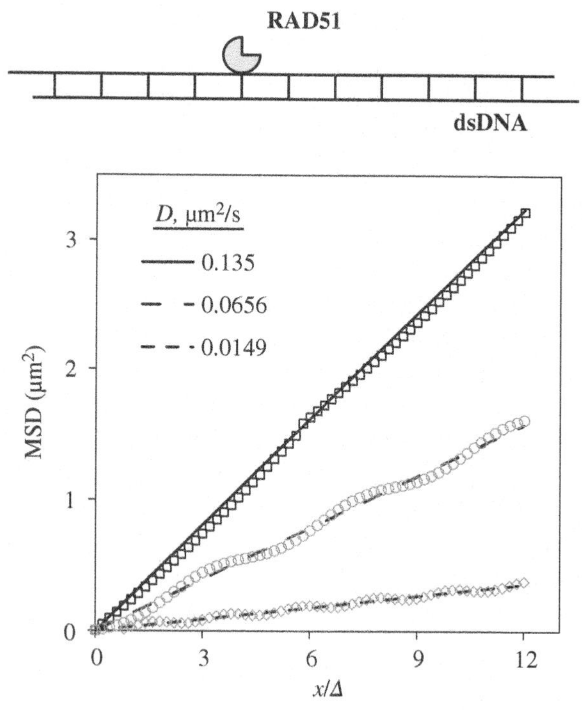

Figure 3.33 The mean-squared deviation for the thermal diffusion of RAD51 along the backbone of an extended DNA molecule. The solid lines are fitted with Eq. (3.234), and the points are from experiments. Source: Adapted from Granéli A. et al.[44]

random walk. To illustrate, Figure 3.33 compares the MSD predicted by Eq. (3.234) with experiments for RAD51 and two mutants diffusing along the helical axis of a long dsDNA chain. Given that the experimental time interval $\tau = 0.124$ s, we can estimate the step length from the experimental diffusion coefficient $\Delta = \sqrt{2D\tau}$, that is about 0.1 μm (≈300–400 base pairs).[45] RAD51 is the primary eukaryotic recombinase responsible for repairing damaged DNA and for initiating

45 Granéli A. et al., "Long-distance lateral diffusion of human Rad51 on double-stranded DNA", *PNAS* 103 (5), 1221–1226 (2006).

DNA-strand exchange during homologous recombination.[46] The random-walk model describes faithfully the 1D diffusion of all three DNA-bound proteins and thus may enable the recombinase to scan DNA for regions in need of repair.

3.12.2 Single-File Diffusion

Single-file diffusion (SFD) is a term used to describe the one-dimensional movement of interacting particles within a narrow pore, where the particles are unable to pass each other. This exclusion of mutual passage results in a unique diffusion process that can be modeled as a random walk of many drunkards in a narrow lane. The SFD process is particularly relevant to zeolites, which are used for molecular sieving and catalysis, as it can help to better understand how molecules move within these porous materials. The SFD phenomenon has significant implications for various fields, including materials science, physics, and chemistry.

The trajectory of each particle in SFD is not influenced by the presence of other particles at a time scale shorter than the mean collision time. In that case, the probability distribution is the same as that for a random-walk model discussed above. At the long-time limit, the probability density becomes[47]

$$p(x,t) = \frac{1}{\sqrt{4\pi M t^{1/2}}} \exp\left(-\frac{x^2}{4Mt^{1/2}}\right) \tag{3.235}$$

where M represents the single-file mobility. Parameter M depends on the particle occupation fraction, the particle diffusivity, and the step length.

To illustrate the applications of the SFD model to a real system, Figure 3.34 shows the probability density of paramagnetic colloidal spheres confined in a narrow circular trench fabricated by photolithography.[48] The diameter of a colloidal sphere is a few micrometers. The excellent representation of the experimental data by Eq. (3.235) indicates that the colloidal particles in narrow microchannels follow closely the SFD process. While both the short- and long-time limits of the probability density of the SFD process are remarkably similar to those for the random-walk model, the collision between particles significantly slows the 1D diffusion process. According to Eqs. (3.226) and (3.235), the MSD for the short- and long-time limits are $<x^2> = 2Dt$ and $2M\sqrt{t}$, respectively. An interpolation of the two limits gives a satisfactory approximation for the MSD of the entire SFD process[49]

$$\frac{1}{<x^2>} = \frac{1}{2Dt} + \frac{1}{2M\sqrt{t}}. \tag{3.236}$$

Figure 3.35 shows that Eq. (3.236) is satisfactory for the SFD process at both short and long-time scales.

46 Homologous recombination is a type of genetic recombination in which nucleotide sequences are exchanged between two similar or identical molecules of DNA. It is most widely used by cells to accurately repair harmful breaks that occur on both strands of DNA, known as double-strand breaks.
47 For a short derivation, see Käger J., "Straightforward derivation of the long-time limit of the mean-square displacement in one-dimensional diffusion", *Phys. Rev. A* 45, 4173 (1992).
48 Wei Q. H. et al., "Single-file diffusion of colloids in one-dimensional channels", *Science* 287, 625–627 (2000).
49 Lin B. H. et al., "From random walk to single-file diffusion", *Phys. Rev. Lett.*, 94, 216001 (2005).

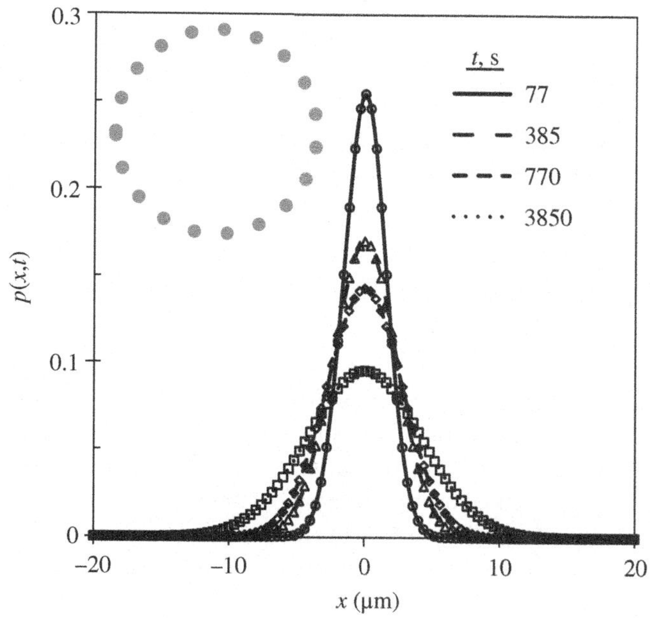

Figure 3.34 The probability density of particle displacement evolving with time. The solid lines are fits of the experimental data to Eq. (3.235) with $M = 0.14\ \mu m^2/s^{1/2}$. The inset shows image of colloidal particles trapped by a scanning laser beam. The trap itself was not shown. The colloidal particles follow a modified single-file diffusion. Source: Adapted from Wei Q. H. et al.[48]

Figure 3.35 The log–log plot of the mean-square displacement (MSD) versus time at different particle number densities. As predicted by Eq. (3.236), MSD scales linearly with time at the short-time scale and the square root of time at the long-lime scale. Source: From Lutz C. et al. [50]

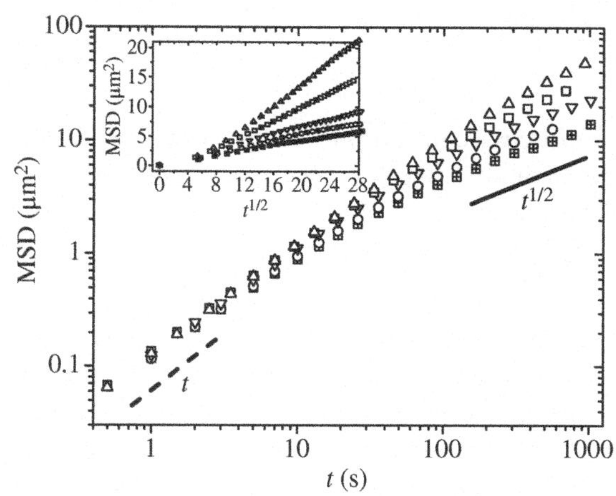

3.12.3 High-Dimensional Random Walk

The 1D random-walk model can be readily extended to two or three dimensions. For the 3-dimensional case, the random-walk model is identical to the FJC model of ideal chains.

To understand the random walk from the perspective of diffusion, imagine that an astronaut takes a random spacewalk in outer space. Given the original position at $t = 0$, the probability density at position $\mathbf{r} = (x, y, z)$ is the product of three 1D distribution functions because the motions in

50 Lutz C. et al. "Single-file diffusion of colloids in one-dimensional channels", *Phys. Rev. Lett.* 93, 026001 (2004).

different directions are independent:

$$p(\mathbf{r}, t) = \frac{1}{(4\pi Dt)^{3/2}} \exp\left(-\frac{r^2}{4Dt}\right) \tag{3.237}$$

where $r^2 = x^2 + y^2 + z^2$. Eq. (3.237) assumes that the motion is the same in directions x, y, and z. Like the 1D case, the probability density satisfies the diffusion equation

$$\frac{\partial p(\mathbf{r}, t)}{\partial t} = D\nabla^2 p(\mathbf{r}, t) \tag{3.238}$$

where $\nabla^2 = \partial^2/\partial x^2 + \partial^2/\partial y^2 + \partial^2/\partial z^2$, and diffusion coefficient D is proportional to the MSD

$$<r^2> = \int p(\mathbf{r}, t)\, r^2 d\mathbf{r} = 6Dt \tag{3.239}$$

Eq. (3.239) is equivalent to that for the polymer end-to-end distance derived from the FJC model. As discussed in Section 3.8, at a length scale much larger than the monomer size, the conformation of a polymer chain in a highly dilute solution can be described in terms of a three-dimensional random walk, i.e., the polymer backbone is represented by a connection of n random vectors, each of length $b = \Delta$. Using a continuous representation of the position but a discrete representation for the chain length, the random-walk model gives the end-to-end distance by a Gaussian distribution:

$$p(\mathbf{r}, n) = \frac{1}{(2\pi nb^2/3)^{3/2}} \exp\left(-\frac{3r^2}{2nb^2}\right) \tag{3.240}$$

Eq. (3.240) is equivalent to the probability density of three independent 1D random walks in x, y, and z directions each with step length $\sqrt{b^2/3}$.

The probability density for the end-to-end distance is also related to the Helmholtz energy of the polymer chain, $F(\mathbf{r}, n)$, according to the Boltzmann distribution

$$p(\mathbf{r}, n) \sim \exp\left[-\frac{F(\mathbf{r}, n)}{k_B T}\right] \tag{3.241}$$

As discussed in Section 3.8, Eqs. (3.240) and (3.241) suggest that the elastic behavior of a polymer chain follows Hooke's law, i.e., when a polymer is stretched, the retraction force is proportional to the extension

$$\mathbf{f}_{elastic} = -\nabla F(\mathbf{r}, n) \approx -\frac{3k_B T}{nb^2}\mathbf{r}. \tag{3.242}$$

While it is generally challenging to measure the elasticity of a single polymer chain, atomic force microscopy proves useful for measuring the mechanical properties of large dsDNA molecules. For example, Figure 3.36 shows how the stretching force is related to the extension for a 97 kilobase (kb) λ-DNA in an aqueous solution at room temperature.[51] Hooke's law is valid only at small extensions. Because the end-to-end distance must be smaller than the polymer contour length, $L = nb$, the random-walk model is not valid at large extensions of the dsDNA chain. An interpolation of the asymptotic limits of small and large extensions leads to the following analytical function for the force-extension relationship:

$$\frac{f_{elastic}\xi_p}{k_B T} = \frac{r}{L} - \frac{1}{4} + \frac{1}{4(1 - r/L)^2} \tag{3.243}$$

where ξ_p stands for the polymer persistence length. Figure 3.36 shows that Eq. (3.243) provides an excellent fit of the experimental data.

---

51 Bustamante C. et al., "Entropic elasticity of lambda-phage DNA", *Science* 265, 1599–1600 (1994).

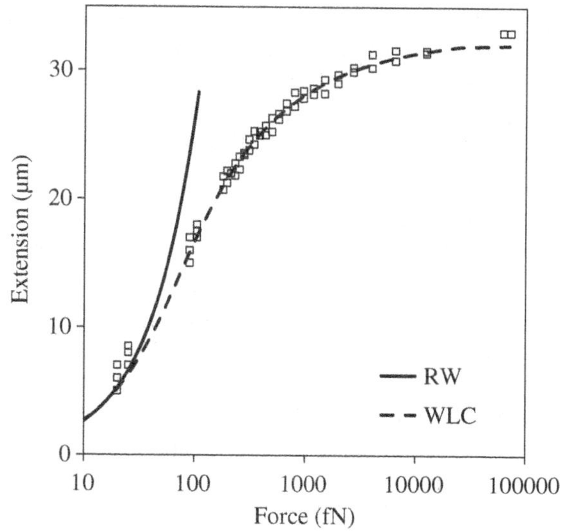

Figure 3.36 Stretching force, in units of femtonewton (fN), versus extension for a long dsDNA chain. The points are experimental data from atomic-force microscopy; the solid line is from the random walk (RW) model ($L = 32.7$ μm, $b = 100$ nm), and the dashed line is from Eq. (3.243) with $L = 32.8$ μm and $\xi_p = 53.4$ nm. Source: Adapted from Bustamante C. et al.[51]

3.12.4 Summary

The random-walk model is a fundamental concept in statistical mechanics that has broad implications in understanding the behavior of physical and biological systems. It provides a simple yet powerful framework for modeling the motion of particles in a variety of environments, from the diffusion of molecules in solution to the movement of cells in tissue. By treating the motion of particles as a series of random steps, the random-walk model enables us to calculate probabilities and predict the behavior of complex systems. The model has many applications in a wide range of fields, including condensed matter physics, materials science, chemistry, biology, and mathematics.[52] Its simplicity and versatility make the random-walk model a useful tool for scientists and engineers seeking to understand and design complex systems.

3.13 Chapter Summary

An ideal gas is an idealized model of a real gas where intermolecular interactions are neglected. The lack of intermolecular interactions implies that its thermodynamic properties can be predicted according to the microscopic behavior of individual molecules (viz., subsystems). If each molecule consists of only a single atom, we can determine the entropy and internal energy based on the particle-in-a-box model. For ideal-gas systems consisting of polyatomic molecules, we need to use additional quantum-mechanical models to describe molecular rotation and the vibrational motions of individual atoms. The rigid-rotor-harmonic-oscillator model provides a systematic way to predict the single-molecule partition function and thermodynamic properties of ideal-gas systems consisting of small molecules with rigid molecular structures.

The assumptions of harmonic oscillators and rigid rotors are not accurate for systems containing complex molecules. Harmonic treatments often fail to describe large amplitude vibrations. Besides, complex polyatomic molecules typically have more than one minimum energy state (i.e., more

52 Rudnick J. and Gaspari G., *Elements of the random walk: An introduction for advanced students and researchers*, Cambridge University Press, 2010.

than one stable configuration). In practical applications, the statistical-thermodynamic models are thus often modified to account for multiple molecular structures and the anharmonicity of atomic vibrations. The theoretical predictions of the partition function and thermodynamic properties of individual molecules are instrumental for understanding chemical equilibrium, reaction kinetics, as well as transport phenomena including both the equilibrium and kinetic constants of gas-phase reactions.

The statistical–mechanical models of uniform ideal-gas systems can be extended to inhomogeneous conditions using the grand canonical ensemble. Such methods are commonly used in the development of thermodynamic models for describing gas adsorption in nanoporous materials (e.g., zeolites) and hydrate formation. In general, the grand-canonical ensemble methods are useful for open systems that allow for the exchanges of mass and energy with the surroundings.

Quantum-mechanical models of small molecules are not suitable for predicting the microstates of polymer chains. Instead, for systems containing polymers or macromolecules, we rely on coarse-grained models and describe the microstate distribution in terms of the statistics of polymer conformations. The physical properties of polymer systems are often insensitive to the microscopic details of individual segments due to scale invariance, which is not observed in typical small-molecule systems. Coarse-grained models provide a simplified way to describe the behavior of polymer chains, making it easier to predict their properties under different conditions.

We can describe the conformations of polymer chains from either discrete or continuous viewpoints. In the case of long polymer chains, the end-to-end distance of a FJC or a Gaussian chain can be derived using the central limit theorem. The internal stiffness or rigidity of polymer chains can be represented by the FRC or WLC models. The scaling relation between the polymer size and chain length, $R \sim \sqrt{m}$, is applicable to long ideal chains regardless of the microscopic details of the bond potential. Because the intra-chain correlation length increases with the polymer size and diverges in the limit of infinite chain length, the properties of polymer systems are intrinsically scaling invariant thus suitable for scaling analysis (to be discussed in Section 5.10). Although coarse-grained models provide no atomic details for polymer chains, they are adequate at the length scale comparable to the polymer size (e.g., the radius of gyration). By quantifying the conformations of polymer chains with statistical mechanics, we can better understand the behavior of complex polymer systems, which find applications in materials science and biophysics.

Further Readings

Dill K. A. and Bromberg S., *Molecular driving forces*. Garland Science, Chapters 11, 13, 33, 2011.

Hill T. L., *An introduction to statistical thermodynamics*. Dover Publications, New York, Chapters 4, 6, 7, 8, 9, 10, 10, 12, 13, 1986.

Masud N., Porter M. A. and Lambiotte R., "Random walks and diffusion on networks", *Phys. Rep.* 716–717, 1–58 (2017).

McQuarrie D. A., *Statistical mechanics*. University Science Books, Chapters 5, 6, 8, 9, 2000.

Wang Z. G., "Polymer conformation—a pedagogical review", *Macromolecules* 50, 23, 9073–9114 (2017).

Ye J. and Truhlar D. G., "Simple approximation for the ideal reference state of gases adsorbed on solid-state surfaces", *J. Am. Chem. Soc.* 144, 12850–12860 (2022).

Problems

3.1 The "particle in a box" model is commonly used to represent the microstates affiliated with the translational degrees of freedom for particle motions. To identify these microstates, consider a particle moving freely in x direction over the range $0 \le x \le L$. Assume that the wave function for the single-particle motion is described by the Schrödinger equation

$$-\frac{\hbar^2}{2m}\frac{d^2\psi(x)}{dx^2} = \epsilon_T \psi(x), \tag{A}$$

where $\hbar = h/2\pi$ is the reduced Planck's constant, m is the particle mass, and ϵ_T stands for the kinetic energy due to the one-dimensional translational motion.

(i) Solve Eq. (A) with the boundary condition $\psi(0) = \psi(L) = 0$.

(ii) Show that the energy levels are given by

$$\epsilon_T = \frac{n^2h^2}{8mL^2},$$

with $n = 1, 2, 3, \ldots$.

(iii) Generalize the results for the translational motion of a single particle in three dimensions.

3.2 Predict the entropy, Helmholtz energy, chemical potential, and heat capacity of argon gas at 25 °C and 1 atm using the statistical-thermodynamic equations for a monatomic ideal gas. Compare your result with the entropy of argon obtained from calorimetric experiment, 155 J/(mol K).

3.3 In evaluating the translational partition function, the continuous approximation becomes problematic when $\rho\Lambda^3 \sim 1$.

(i) Estimate the temperature below which the Sackur–Tetrode equation breaks down for predicting the entropy of argon as an ideal gas at 1 atm.

(ii) What are the number density and the thermal wavelength of the argon gas at this temperature? and how are these numbers compared with those at 300 K?

3.4 The electronic energy of atoms and ions can be determined from atomic spectroscopy and the data have been well tabulated (e.g., the National Institute of Standards and Technology (NIST) Atomic Spectra Database). The table below shows the first few electronic states of Li in terms of electron configuration, term symbol for the electronic state, degeneracy and energy level.

(i) Based on the first few electronic energy levels, calculate the electronic partition function q_e for a Li atom at 300 and 10^6 K. How are the numbers compared with the single-particle translational partition function q_T with volume $V = 1\,\text{m}^3$ at these temperatures?

(ii) How does the electronic partition function contribute to the entropy for Li as an ideal gas at 300 K and 1 atm and at 10^6 K and 1 atm?

Configuration	Term	Degeneracy (g_i)	Energy level (ϵ_i, cm^{-1})
$1s^2 2s$	$^2S_{1/2}$	2	0.00
$1s^2 2p$	$^2P_{1/2}$	2	14903.66
	$^2P_{3/2}$	4	14904.00
$1s^2 3s$	$^2S_{1/2}$	2	27206.12
$1s^2 3p$	$^2P_{3/2}$	4	30925.38
	$^2P_{1/2}$	2	30925.38
$1s^2 3d$	$^2D_{3/2}$	4	31283.08
	$^2D_{3/2}$	6	31283.12

3.5 Consider gas molecules passing through a small hole as shown schematically in Figure P3.5. Assume that the gas molecules do not interact with each other and that the planar wall has negligible thickness. Based on the Maxwell–Boltzmann distribution for the molecular velocity, show that the flux of gas molecules, i.e., the number of gas molecules passing through the hole per unit area and unit time, is given by the Knudsen equation

$$J = \rho\sqrt{\frac{k_B T}{2\pi m}}$$

where ρ is the number density, k_B is the Boltzmann constant, T is the absolute temperature, and m is the mass of a gas molecule.

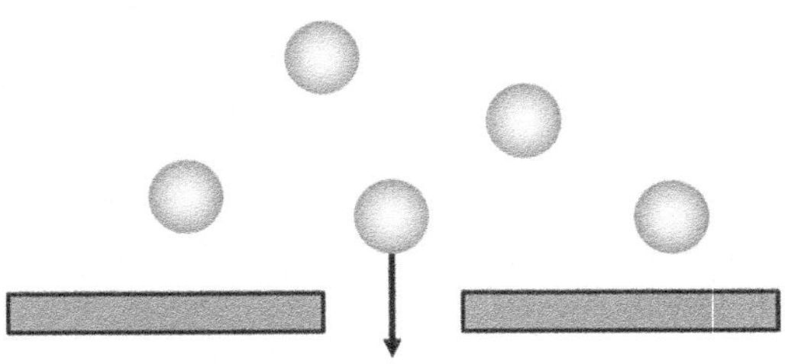

Figure P3.5 Effusion of gas molecules through a hole of negligible thickness.

3.6 Liquid evaporation is a nonequilibrium process in which there is a net transport of liquid molecules across a liquid–vapor interface into the vapor phase. Approximately, the rate of condensation (number of molecules per unit area per unit time) can be estimated from the Hertz–Knudsen (HK) equation, which is related to the number of the gas molecules impinging on the liquid surface multiplied by an efficiency factor α

$$J = \alpha\rho\sqrt{\frac{k_B T}{2\pi m}},$$

where ρ is the number density of the gas molecules, and m is the molecular mass. At equilibrium, the rate of gas molecule condensation is the same as the rate of evaporation. Assume $\alpha = 0.01$, estimate how long it takes to dry up a pond of $A = 100$ m^2 in surface area and $h = 10$ m in depth open to 50 °C dry air. At 50 °C, the saturation pressure for liquid water is $p^{sat} = 1.227 \times 10^4$ Pa. The water density is about $d \approx 1000$ kg/m^3, and molecular weight is $W = 18$ g/mol.

3.7 Historically, effusion was considered for the separation of 235U and 238U isotopes in the early development of atom bombs. To use the effusion method, the uranium isotopes are first converted to uranium hexafluoride (UF_6) gas, which is forced to diffuse through porous barriers repeatedly, each time becoming a little more enriched in slightly lighter 235U isotope. The 235U to 238U ratio in a typical uranium ore is 1:140. Based on the Knudsen equation, estimate how many separation stages are required to obtain 90% 235U.

3.8 Consider two particles with mass m_1 and m_2 linked together by a bond potential $\upsilon_B(r)$, where r is the center-to-center distance.

(i) Prove that the relative motions of the two particles can be represented by that of one particle with an effective mass $\mu = m_1 m_2/(m_1 + m_2)$.

(ii) Assume that the inter-particle potential is described by

$$\upsilon_B(r) = -D_e + \frac{1}{2}k(r - r_e)^2.$$

where $-D_e$ stands for the bond formation energy. Show that the relative motion of the two particles can be represented by a harmonic oscillator with frequency

$$\nu = \frac{1}{2\pi}\sqrt{\frac{k}{\mu}}.$$

(iii) Show that, at fixed separation $r = r_e$, the rotation of the two particles is equivalent to that of one particle with effective mass μ.

3.9 Consider an ideal gas mixture with two types of molecules 1 and 2. Show that the relative velocity between a type 1 molecule and a type 2 molecule satisfies the Maxwell–Boltzmann distribution

$$p(\mathbf{v}) = \left(\frac{\mu}{2\pi k_B T}\right)^{3/2} \exp\left(-\frac{\mu \upsilon^2}{2k_B T}\right),$$

where $\mu = m_1 m_2/(m_1 + m_2)$. What is the average speed of the relative velocity? Hint: Express the velocities of the two molecules in terms of the relative velocity $\mathbf{v_r} = \mathbf{v}_2 - \mathbf{v}_1$ and the center of mass velocity $\mathbf{v}_c = (m_1\mathbf{v}_1 + m_2\mathbf{v}_2)/(m_1 + m_2)$.

3.10 The rotational partition function for diatomic and linear polyatomic molecules can be derived by applying the Schrödinger equation to a rigid rotor, i.e., a particle of mass μ freely rotating around the origin with fixed distance r.

(i) Starting from the Schrödinger equation

$$-\frac{\hbar^2}{2\mu}\nabla^2\psi = \epsilon_R \psi,$$

where $\hbar = h/2\pi$ is the reduced Planck's constant, and ϵ_R is the kinetic energy of rotation, show that the wave function of a rigid rotor is given by the spherical harmonics

$$\psi(\theta, \phi) = Y_n^m(\theta, \phi) = (-1)^m \sqrt{\frac{2n+1}{4\pi}\frac{(n-m)!}{(n+m)!}} P_n^m(\cos\theta) e^{im\phi},$$

where $n = 0, 1, 2, \ldots,$ $m = -n, -n+1, \ldots, n-1, n$, θ is the zenith angle, ϕ is the azimuthal angle, and $P_n^m(x)$ are the associated Legendre functions.

(ii) Show that the energy levels of a rigid rotor are

$$\epsilon_R = \frac{n(n+1)h^2}{8\pi^2 I}$$

where $n = 0, 1, 2, \ldots$ is a rotational quantum number. What is the degeneracy of each rotational energy level?

(iii) Show that, at low temperature ($T \ll \theta_R$), the rotational partition function has the leading terms

$$q_{rot} = 1 + 3e^{-2\theta_R/T} + 5e^{-6\theta_R/T} + 7e^{-12\theta_R/T} + \cdots,$$

where $\theta_R \equiv (h^2/8\pi^2 I k_B)$ is the rotational temperature.

(iv) Using the Euler–Maclaurin formula

$$\sum_{n=L}^{M} f(n) = \int_L^M f(x)dx + \frac{1}{2}[f(L) + f(M)] + \sum_{k=1}^{\infty} \frac{B_{2k}}{(2k)!}[f^{(2k-1)}(M) - f^{(2k-1)}(L)],$$

where n is integer between $[L, M]$, $f(x)$ is an analytical function with $f^{(k)}$ being the kth derivative of f, and B_k is the kth Bernoulli number, show that the rotational partition function can be expressed as a power series of the inverse temperature

$$q_{rot} = \frac{T}{\theta_R}\left[1 + \frac{1}{3}\frac{\theta_R}{T} + \frac{1}{15}\left(\frac{\theta_R}{T}\right)^2 + \frac{4}{315}\left(\frac{\theta_R}{T}\right)^3 + \cdots\right].$$

The first few Bernoulli numbers are $B_2 = 1/6$, $B_4 = -1/30$, $B_6 = 1/42$, ...

3.11 The rotational motions of a nonlinear polyatomic molecule are considerably more complicated than that corresponding to a diatomic or linear polyatomic molecule. In general, the moment of inertia is described by a 3×3 tensor that depends on the mass and position of multiple atoms:

$$\mathbf{I} = \begin{bmatrix} I_{xx} & I_{xy} & I_{xz} \\ I_{yx} & I_{yy} & I_{yz} \\ I_{zx} & I_{zy} & I_{zz} \end{bmatrix} = \begin{bmatrix} \sum_k m_k(y_k^2 + z_k^2) & -\sum_k m_k x_k y_k & -\sum_k m_k x_k z_k \\ -\sum_k m_k y_k x_k & \sum_k m_k(x_k^2 + z_k^2) & -\sum_k m_k y_k z_k \\ -\sum_k m_k z_k x_k & -\sum_k m_k z_k y_k & -\sum_k m_k(x_k^2 + y_k^2) \end{bmatrix},$$

where m_k stands for atomic mass, (x_k, y_k, z_z) is the atomic position relative the molecular center of mass, and the summation over index k applies to all atoms from the polyatomic molecule. The simplest scenario is called a spherical top, which applies to molecules such as CH_4 and C_{60} where the three principal moments of inertia (i.e., the eigenvalues of the moment tensor) are identical $I_A = I_B = I_C = I$. In this case, the Schrödinger equation predicts that the rotation has quantified energy levels ϵ_R and degeneracy g_n

$$\epsilon_R = \frac{n(n+1)\hbar^2}{2I} \quad \text{and} \quad g_n = (2n+1)^2.$$

Verify that, at high temperature, the rotational partition function can be approximated by

$$q_{rot} \approx \frac{\sqrt{\pi}}{\sigma}\left(\frac{T}{\theta_R}\right)^{3/2}, \tag{G}$$

where $\theta_R \equiv \hbar^2/(2Ik_B)$, and σ stands for the molecular symmetry number, i.e., the number of ways in which the molecule can achieve through rotation to reach the same orientation in space (e.g., $\sigma = 12$ for methane, corresponding to the threefold symmetry about each of the

4 C–H bonds). Eq. (G) can be generalized for nonlinear molecules either with a symmetric top ($I_A = I_B \neq I_C$) or with an asymmetric top ($I_A \neq I_B \neq I_C$)

$$q_{rot} \approx \frac{\sqrt{\pi}}{\sigma}\left(\frac{T^3}{\theta_{R,A}\theta_{R,B}\theta_{R,C}}\right)^{1/2},$$

with $\theta_{R,i} \equiv \hbar^2/(2I_i k_B)$ and $i = A, B$, or C.

3.12 According to the classical physics, the kinetic energy of a rigid rotor is determined by the three principal moments of inertia $\mathbf{I} = (I_A, I_B, I_C)$ and corresponding angular velocity $\boldsymbol{\omega} = (\omega_A + \omega_B + \omega_C)$:

$$K = \frac{1}{2}(I_A\omega_A^2 + I_B\omega_B^2 + I_C\omega_C^2).$$

Show that the angular velocity ω follows the Maxwell–Boltzmann distribution

$$p(\boldsymbol{\omega}) = \sqrt{\frac{I_A I_B I_C}{(2\pi k_B T)^3}}\exp\left[-\frac{1}{2k_B T}(I_A\omega_A^2 + I_B\omega_B^2 + I_C\omega_C^2)\right],$$

and that each component of the angular velocity has an average kinetic energy of $k_B T/2$.

3.13 The microstates associated with bond vibration in a diatomic molecule is conventionally represented by the harmonic oscillator model, i.e., a particle of effective mass μ subject to a harmonic potential

$$v(x) = \frac{1}{2}k_s x^2,$$

where k_s is the spring constant, x is the deviation from the equilibrium bond length, and the effective mass is related to those of the two bonded atoms $\mu = m_1 m_2/(m_1 + m_2)$. The microstate energy ϵ_V is defined by the solutions of the Schrödinger equation

$$-\frac{\hbar^2}{2m}\frac{d^2\psi(x)}{dx^2} + \frac{1}{2}k_s x^2\psi(x) = \epsilon_V\psi(x),$$

where $\hbar = h/2\pi$ is the reduced Planck's constant.

(i) Show that the wave function of the harmonic oscillator is given by the Hermite functions

$$\psi(x) = \frac{1}{\sqrt{2^n n!}}\left(\frac{\mu\omega}{\pi\hbar}\right)^{1/4} e^{-\frac{m\omega x^2}{2\hbar}} H_n\left(\sqrt{\frac{\mu\omega}{\hbar}}x\right),$$

where $n = 0, 1, 2, \ldots$, $\omega = \sqrt{k_s/\mu}$ is the angular frequency of the oscillator, and $H_n(x)$ are called the Hermite polynomials.

(ii) Show that the vibrational energy levels are given by

$$\epsilon_V = (n + 1/2)\hbar\omega.$$

What is the degeneracy of each vibrational energy level? Why does not the vibrational energy vanish at the ground state?

(iii) Plot the wave function for the four lowest energy levels $n = 0, 1, 2$, and 3 and discuss their significance.

(iv) Show that the vibrational partition function is

$$q_{vib} = \frac{1}{2\sinh(\theta_v/2T)},$$

where $\theta_v = \hbar\omega/k_B$ is the vibrational temperature.

(v) Discuss experimental/computational ways to determine the angular frequency.

3.14 Estimate the fraction of iodine molecules in an ideal gas at the excited vibrational states ($n > 0$) at room temperature ($T = 298.15$ K). The vibrational temperature for iodine is $\theta_v = 310$ K.

3.15 Predict the molecular partition function, molar entropy, and constant-pressure molar heat capacity for HF at 25 °C and 1 atm based on the following parameters:

	W, g/mol	Electronic state	θ_v, K	θ_R, K	D_0, kJ/mol	r_e, Å
HF	20.006	1Σ+	5953.1	30.14	569.7	0.91685

Compare the theoretical results with experimental values of $S = 173.78$ J/(mol K) and $C_P = 29.14$ J/(mol K).

3.16 Plot the constant-pressure molar heat capacity for HF as an ideal gas from 0 to 10^5 K. Assume that the electrons remain unexcited at all temperatures.

3.17 Predict the entropy of methyl radical, CH_3*, as an ideal gas at 298.15 K and 1 atm based on the following data for the molecular weight, structure, electronic and vibrational energy levels:

(1) Molecular weight: $W = 15.0345$,
(2) CH bond length: $b = 1.0767$ Å,
(3) Electronic energy levels (cm^{-1}): 46239, 59972, 66536, 69853.44 72508,
(4) Vibrational energy levels (cm^{-1}): 3004.43, 606.453, 3160.821,1396, 3160.821,1396.
Data from the NIST Chemistry WebBook (https://webbook.nist.gov/cgi/inchi?ID=C2229074&Mask=800).

3.18 Predict the single-molecule partition function for water at 298.15 K and 1 atm based on the following parameters

	W, g/mol	g_0	θ_R, K	σ	θ_v, K	D_0
H_2O	18.0153	1	40.1, 20.9, 13.4	2	5360, 5160, 2290	917.6, kJ/mol

3.19 Predict the equilibrium constant for the hydrogenation reactions of CO_2 to CH_4 in the gas phase at 300 and 500 K.

$$CO_2 + 4H_2 \rightleftharpoons CH_4 + 2H_2O.$$

Compare the theoretical results with experimental value at 298 K, $K = 7.79 \times 10^{19}$.

	W, g/mol	g_0	θ_R, K	σ	θ_v, K	D_0, kJ/mol
H_2	2	1	87.54	2	6331.1	430.554
CO_2	44	1	0.561	2	3360, 954(2), 1890	1596.2
H_2O	18	1	40.1, 20.9, 13.4	2	5360, 5160, 2290	917.6
CH_4	16	1	7.54, 7.54, 7.54	12	4170, 2180(2), 4320(3), 1870(3)	1640.5

3.20 Predicting the thermodynamic properties of chemical systems such as equilibrium constants from first principles is extremely sensitive to the input from quantum-mechanical calculations. Fortunately, the results from quantum-mechanical predictions have been improving over the years, which makes *ab initio* thermodynamics become more and more reliable. In this exercise, predict the various contributions to the molecular partition function of CO at 298.15 K and 1 atm based on the following four sets of "well-known" parameters:

	W, g/mol	g_0	θ_R, K	σ	θ_υ, K	D_0
CRC103*	28.01	1	2.78	1	3121.3	1076.6, kJ/mol
Dill and Bromberg	28.01	1	2.77	1	3122	1070.1, kJ/mol
McQuarrie	28.01	1	2.77	1	3103	255.8, kcal/mol
Hill	28.01	1	2.77	1	3070	9.14, eV

*CRC Handbook of Chemistry and Physics 103rd Edition, 2022; Dill K. A. and Bromberg S., Molecular Driving Forces, 2011, Garland Science; McQuarrie D.A., Statistical Mechanics. 2000, University Science Books; Hill T.L., An Introduction to Statistical Thermodynamics. 1986, New York: Dover Publications.

3.21 Predict the molar entropy and chemical potential of hydrogen gas at 298.15 K and 1 atm based on the following parameters

	W, g/mol	g_0	θ_R, K	σ	θ_υ, K	D_0, kJ/mol
H_2	2	1	87.54	2	6331.1	430.554

3.22 Consider the ionization of hydrogen atoms in the gas phase

$$H \rightleftharpoons H^+ + e^-.$$

(i) What is the equilibrium constant at $T = 10000$ K?
(ii) When the above system is in equilibrium at 10000 K and 0.5 atm, what is the fraction hydrogen atoms ionized?
(iii) What is the Gibbs energy of reaction ΔG^0 for the ionization of hydrogen atoms at 25 °C?

Assume that the translational partition function of electrons can be described by the classical approximation, i.e., it has a form the same as that for classical particles. The ionization energy of the hydrogen atom is $D_0 = 13.59844$ eV.

3.23 Predict the equilibrium constant and the fraction of hydrogen molecules dissociated at 3000 K and 5 atm for hydrogen dissociation reaction:

$$H_2 \rightleftharpoons 2H.$$

What is the Gibbs energy of reaction ΔG^0 for hydrogen dissociation at 25 °C? The dissociation energy of a hydrogen molecule is $D_0 = 430.554$ kJ/mol.

3.24 The standard hydrogen electrode (SHE) is defined as a platinum (Pt) metal submerged in an aqueous solution with 1 M H^+ concentration in equilibrium with a hydrogen gas at

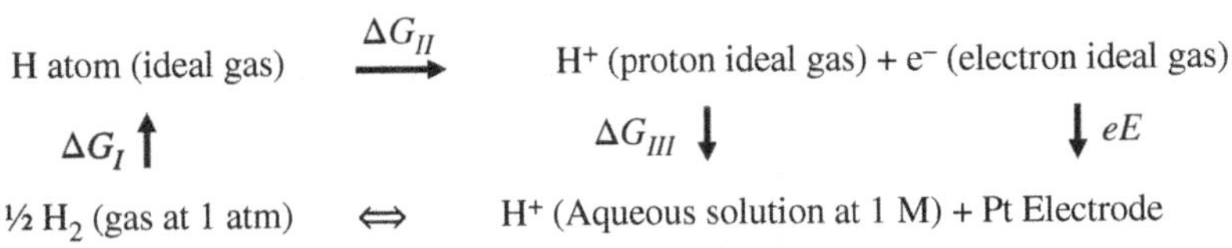

Figure P3.24 The Born–Haber cycle for determining the electric potential of the standard hydrogen electrode (SHE).

T = 25 °C and P = 1 atm. Based on the Born–Haber cycle as shown in Figure P3.24, predict the "absolute" electric potential

$$E \equiv (\mu_e - \mu_e^0)/e,$$

where e stands for the unit charge, μ_e is the electron chemical potential in the Pt metal, and μ_e^0 is the chemical potential of noninteracting electrons (viz., ideal gas) at $T = 25\,°\text{C}$ and $P = 1$ atm. Compare the theoretical result with the common value used in the literature $E_{SHE} = -4.44 \pm 0.02$ V.
(Hint: The dissociation free energy of a hydrogen molecule and the ionization energy of a hydrogen atom can be derived from the single-molecule partition functions discussed in Problems 3.23 and 3.22. According to first-principles solvation-included electronic structure calculations, the hydration free energy of proton is $G_{\text{sol}} = -1097.9$ kJ/mol.[53]

3.25 Consider the canonical ensemble for a one-component ideal gas in a uniform electric field $\mathbf{E}_0$ introduced by two parallel electrodes with opposite charges as shown schematically in Figure P3.25. Each gas molecule has polarizability α and permanent electric dipole moment d. Assume that the number density of the ideal-gas molecules is extremely low such that they do not interact with the charged surfaces and that the gas molecules are independently polarized.

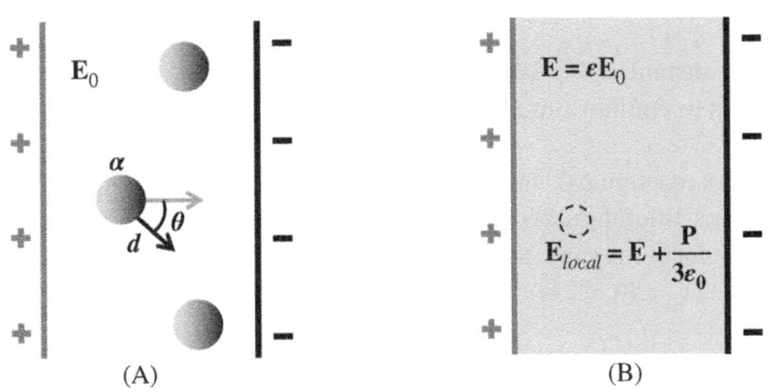

Figure P3.25 Molecular (A) and continuous (B) representations of an ideal gas in an external electric field $\mathbf{E}_0$. Each gas molecule experiences an external potential u^{ext} that depends on the local electric field $\mathbf{E}_{local}$, molecular polarizability α, permanent dipole moment $\mathbf{d}$, and the dipole orientation θ. If the ideal gas is described as continuum with dielectric constant ϵ, the local electric field $\mathbf{E}_{local}$ is related to the electric field in the medium $\mathbf{E}$ and the polarization density $\mathbf{P}$ through the Lorentz equation $\mathbf{E}_{local} = \mathbf{E} + \mathbf{P}/3\epsilon_0$.

53 Zhan C. G. and Dixon D. A., "Absolute hydration free energy of the proton from first-principles electronic structure calculations", *J. Phys. Chem. A.* 105 (51), 11534–11540 (2001).

(i) Show that the external potential for each gas molecule is given by

$$u^{ext} = -\frac{1}{2}\alpha E_0^2 - d\cos\theta E_0,$$

where $E_0 = |\mathbf{E}|$ and $d = |\mathbf{d}|$.

(ii) Show that the single-molecule partition function for the ideal gas in external electric field $\mathbf{E_0}$ can be written as

$$q = q_0 e^{\alpha\beta E_0^2/2}\left(\frac{\sinh y}{y}\right),$$

where q_0 is the single-molecule partition function without the external potential, $\beta = 1/(k_B T)$, and $y \equiv \beta d E_0$.

(iii) Verify that the ideal-gas law remains valid in the presence of the electric field, i.e., $P = \rho k_B T$ where $\rho = N/V$ is the number density of the ideal-gas molecules in the system.

(iv) Show that the mean dipole moment of each ideal-gas molecule is given by

$$\mathfrak{m} = \alpha\mathbf{E_0} + \mathbf{d}\mathcal{L}(y),$$

where $\mathcal{L}(y) = \coth y - 1/y$ is the Langevin function.

(v) In the continuous model, the dielectric constant is defined as $\epsilon \equiv E/E_0$. Show that the assumption of independent polarizations of the ideal-gas molecules leads to

$$\epsilon - 1 = \frac{\rho}{\epsilon_0}\left(\alpha + \frac{d^2}{3k_B T}\right).$$

(vi) The polarization of each ideal-gas molecule introduces its own electric field which subsequently affects the polarization of other molecules in the system. In classical electrostatics, the polarization effect is described by the Lorentz relation

$$\mathbf{E}_{local} = \mathbf{E} + \frac{\mathbf{P}}{3\epsilon_0},$$

where $\mathbf{E}_{local}$ stands for the electric field at the position of a particular molecule. Based on the Lorentz relation and the ideal-gas model for the average polarization density, $\mathbf{P} = \rho\mathfrak{m}$, show that the dielectric constant of the system can be described by the Debye–Langevin equation:

$$\frac{\epsilon - 1}{\epsilon + 2} = \frac{\rho}{3\epsilon_0}\left(\alpha + \frac{d^2}{3k_B T}\right).$$

3.26 Estimate the dielectric constant of water vapor at 100 °C and 1 atm using the Debye–Langevin equation. The polarizability volume of a water molecule is $\alpha' = 1.48 \times 10^{-24}$ cm^3, and the dipole moment is $d = 1.84$ Debye. Note that $\alpha = 4\pi\epsilon_0\alpha'$ and 1 Debye = $3.335640952 \times 10^{-30}$ Coulomb×meter. Compare your result with the experimental value $\epsilon = 1.00589$.

3.27 Consider chemisorption of an ideal gas on a surface with N_s independent binding sites. Assume that the adsorption does not change the molecular flexibility and that molecules binding to the surface can be described by the harmonic oscillator model. Show that parameter b in the Langmuir isotherm can be predicted from

$$b = \frac{q_e^s q_{vib}^s e^{-\beta\epsilon}}{q_e q_{rot}}\left(\frac{P_0\Lambda^3}{k_B T}\right).$$

In the above equation, q_e^s is the electronic partition function of an adsorbed molecule; q_{vib}^s is the vibrational partition function for the surface binding *without* the zero-point-energy contribution

$$q_{vib}^s = [(1 - e^{-\theta_x/T})(1 - e^{-\theta_y/T})(1 - e^{-\theta_z/T})]^{-1},$$

where $\theta_{x,y,z}$ are the surface vibrational temperatures; ϵ is the surface binding energy; Λ is the thermal wavelength of the gas molecules; $P_0 = 1$ atm; q_e and q_{rot} are the electronic and rotational partition functions in the gas phase. How would you predict the surface vibrational temperatures?

3.28 The Langmuir model of gas adsorption assumes that gas molecules are localized once they are attached to a planar surface. This assumption is reasonable if the adsorption energy ϵ is significantly larger than $k_B T$. Otherwise, the adsorbed molecules will have two-dimensional translational degrees of freedom (Figure P3.28).

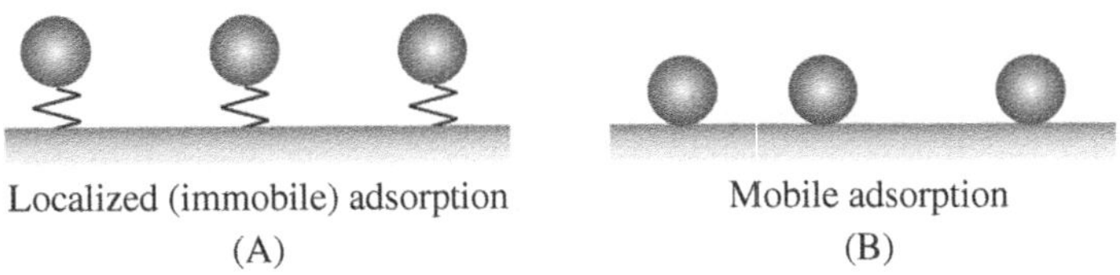

Figure P3.28 Two extreme cases of gas adsorption. A. In the localized adsorption model, the adsorbed molecules are tethered to surface sites. B. In the mobile adsorption model, the adsorbed molecules can move freely on the surface.

(i) Derive the single-molecule canonical partition function of a localized molecule on the surface using the harmonic oscillator model for surface tethering;
(ii) Derive the single-molecule canonical partition function of a mobile molecule on the surface using the harmonic oscillator model for surface tethering and the particle-in-a-box model for the two-dimensional translational motions on the surface;
(iii) Derive the grand partition function for the adsorption of mobile gas molecules on the surface;
(iv) Compare the adsorption isotherms derived from the localized and mobile models. Which model predicts more adsorption?

3.29 The Langmuir model is also used to describe monolayer adsorption in dilute solutions. Image a solid surface in contact with a dilute solution at temperature T. The surface can bind with the solute molecules in the solution with some binding sites independent of each other. To account for the molecular excluded volume effect, each surface site can bind at most one solute molecule with energy ϵ. For simplicity, we assume that the solute molecules have no internal degree of freedom either in the bulk solution or at the surface.

(i) Derive internal energy, entropy, and the number of adsorbed molecules using the grand partition function;
(ii) Suppose that the canonical ensemble is used to describe the same system, how is the equation for entropy compared with that derived from the grand canonical ensemble?
(iii) In a dilution solution, the solute chemical potential can be written as $\mu = \mu^0 + k_B T \ln C$, where μ^0 is the chemical potential at a reference state at unit concentration, and C is

the solute concentration, verify the Langmuir adsorption isotherm

$$\theta \equiv N/n = \frac{KC}{1+KC},$$

where $K \equiv e^{-\beta(\epsilon-\mu^0)}$, N is the average number of adsorbed molecules, and n is the total number of binding sites.

(iv) If each solute molecule at the surface can vibrate relative to its binding site, how would you modify the adsorption isotherm?

3.30 Suppose that a surface with n binding sites is in contact with a bulk system containing two types of particles with chemical potentials μ_A and μ_B. Assume that each surface site can bind at most one particle and that the surface sites are independent to each other. Let ϵ_A is the energy when a surface site is occupied by particle A, and ϵ_B is the energy when a surface site is occupied by particle B.

(i) Derive an expression for the grand partition function of the surface as an open system;
(ii) How are the surface coverages of A and B dependent on temperature?

3.31 The Langmuir-type model is applicable to ligand binding in certain biological systems (e.g., binding tryptophan to Trp aporepressor). Suppose that a protein has n independent binding sites for a ligand with molar concentration C_L in the solution. Assume that all protein binding sites have the same binding affinity.

(i) Derive an expression for the grand partition function of the protein binding with the ligand;
(ii) Predict the average number of ligand molecules bound with the protein, N, based on the grand partition function;
(iii) The Scatchard plot refers to a linear correlation between N/C_L and N. Discuss the physical significance of the slope and interception.

3.32 The Scatchard plot (discussed in Problem 3.31) is applied to ligand binding to a protein with identical and fully independent sites. It is an idealized model because in reality binding with the protein sites are often correlated with each other. Another idealized model is that ligand binding is fully cooperative, i.e., the number of ligand molecule associated with a protein is either zero or n, the total number of binding sites available. Let C_L be the ligand concentration in the solution, and ΔE be the total binding energy when all sites are occupied.

(i) Derive an expression for the grand partition function of the fully cooperative ligand binding;
(ii) Predict the average number of ligand molecules associated with the protein per binding, θ, based on the grand partition function;
(iii) The Hill plot is referred to as a linear correlation between $\ln[\theta/(1-\theta)]$ and $\ln C_L$. Discuss the physical significance of the slope and interception.

3.33 For a diblock copolymer of Gaussian chains A and B, the intra-chain correlation functions are defined as

$$\omega_{i,j}(r) = \frac{1}{m_A m_B}\sum_{n=1}^{m_A}\sum_{m=1}^{m_B}\langle\delta[\mathbf{r}-(\mathbf{r}_n-\mathbf{r}_m)]\rangle,$$

where i, j denotes block A or B, m_A and m_B are the number of Kuhn segments in blocks A and B, respectively. Show that the three-dimensional Fourier transform of $\omega_{i,j}(r)$ is given by

$$\hat{\omega}_{i,j}(k) = \begin{cases} 2(e^{-x_i} - 1 + x_i)/x_i^2 & i = j \\ (e^{-x_i} - 1)(e^{-x_j} - 1)/(x_i x_j) & i \neq j \end{cases},$$

where $x_i \equiv k^2 R_{g,i}^2$, and $R_{g,i}$ is the radius of gyration for block i.

3.34 Verify that the square of the radius of gyration for a worm-like chain (WLC) of persistence length ξ_p and contour length L is given by

$$R_g^2 = \frac{L\xi_p}{3} - \xi_p^2 + \frac{2\xi_p^3}{L} - \frac{2\xi_p^4}{L^2}\left(1 - e^{-L/\xi_p}\right).$$

3.35 The elasticity of a single polymer chain can be described by using a constant temperature and constant tension ensemble as shown schematically in Figure P3.35. Consider a freely-jointed chain (FJC) with one end fixed and the other end subject to a constant force f in the direction of the end-to-end vector, $\mathbf{R} = \mathbf{r}_m - \mathbf{r}_1$, m is the number of polymer segments. Because the bond lengths are fixed, the polymer configuration is uniquely defined by the orientations of individual bonds, i.e., the polar and azimuthal angles for each bond, $\{\theta_i, \phi_i\}$, $i = 1, 2, \ldots, m-1$.

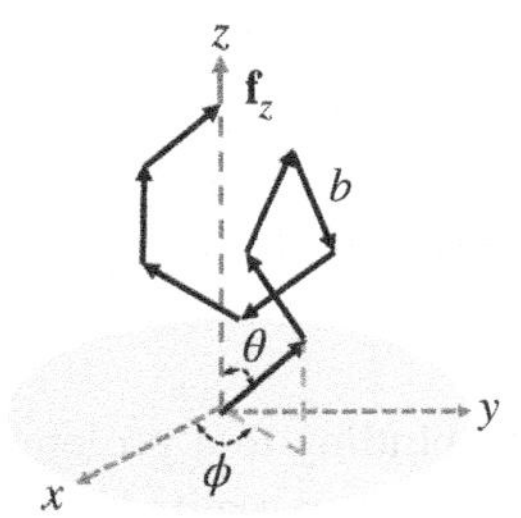

Figure P3.35 A freely-jointed chain (FJC) with one end fixed and the other end subject to a constant force $\mathbf{f}_z$ along the direction of the end-to-end vector (viz., z-direction).

(i) Show that the end-to-end distance is given by

$$R = b\cos\theta_1 + b\cos\theta_2 + \cdots + b\cos\theta_{m-1},$$

where b is the bond length.

(ii) Assume that the polymer energy is linearly proportional to the force $u(R) = -fR$, evaluate the canonical partition function

$$Z = \sum_{\nu} \exp\left[-\frac{u(R)}{k_B T}\right],$$

where ν stands for the microstates of the system.

(iii) Show that the average end-to-end distance due to the external force is given by

$$\langle R \rangle = \left(\frac{\partial \ln Z}{\partial \beta f}\right)_T = L\left[\coth\left(\frac{fb}{k_B T}\right) - \frac{k_B T}{fb}\right],$$

where $\beta = 1/(k_B T)$, and $L = (m-1)b$ is the contour length.

(iv) Plot the reduced force (βfb) versus the relative extension distance (R/L). Show that for small elongations, the polymer elasticity follows Hooke's law

$$f = \frac{3k_BT}{b}\frac{\langle R\rangle}{L},$$

and that, for strong stretching (i.e., $R \sim L$),

$$f = \frac{k_BT}{b}\frac{L}{L - \langle R\rangle}.$$

(v) What is the relative fluctuation of the chain length in the absence of the external force? Hint: $< R^2 > -\langle R\rangle^2 = (\partial\langle R\rangle/\partial\beta f)$.

3.36 The elasticity of rubbery materials can be understood in terms of the conformation of individual polymer chains. To illustrate, consider the elongation of a rubber from its original length L_0 at relaxation to a stretched length L under an external force τ. Approximately, the elongation does not change the total volume of the rubbery material. To account for the additional degree of freedom due to the anisotropic extension, the fundamental equation of thermodynamics for an elastic material can be written as

$$dF = -TdS - pdV + \tau dL, \tag{R}$$

where τ is shown as a scalar with the understanding that dL occurs in the same direction of the force. Eq. (R) provides a starting point to describe the elastic behavior of rubbery materials with statistical thermodynamics.

(i) Assume that the rubber elongation from L_0 to L takes place in the z direction and that the cross section of the rubber is uniform. How are the dimensions in the x and y directions, $L_x^0 \times L_y^0$ at the relaxation state, changed by the elongation in the z direction if the rubber volume remains unchanged.

(ii) How does the rubber elongation affect the end-to-end distance of the polymer chain?

(iii) Assume that the rubber can be represented by an affine network (Figure P3.36) of freely-joined chains (FJC) each with m segments of equal bond length b. What is the change in the Helmholtz energy for each chain if they are randomly orientated in the polymer network?

(iv) How is the elastic force of the rubber related to the characteristics of individual chains?

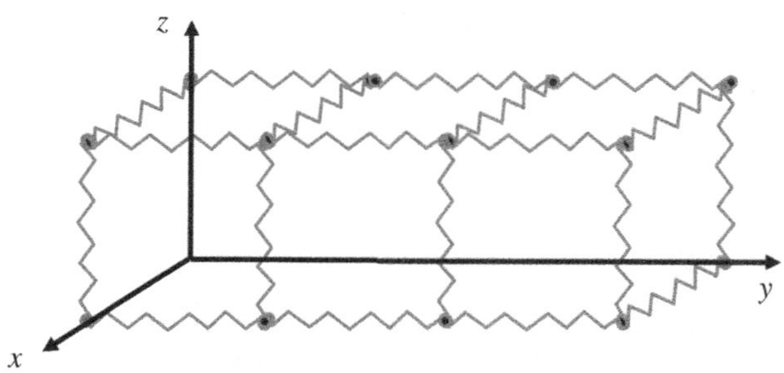

Figure P3.36 A rubbery material represented by an affine network where polymer chains are cross-linked with a cubic structure.

3.37 Smith et al.[54] reported a modified freely-jointed chain (MFJC) model for the extension and force relation of single polymer chains

$$\langle R\rangle = L\left[\coth(\beta fb) - \frac{1}{\beta fb}\right](1 + f/k), \tag{S}$$

where k stands for the stretch modulus, accounting for the variation of the bond length in response to the external force. It was found that the MFJC model provides an excellent description of the elastic behavior of single-stranded (ss) DNA chains in comparison with experimental data at 25 °C with parameters $b = 15$ Å, $L = 27\ \mu$m, and $k = 800$ pN.

(i) Show that the modified FJC equation is consistent with Hooke's law of elasticity in the limits of weak and strong stretching.
(ii) Estimate the reversible work to stretch a ssDNA chain from the fully relaxed state to 50% of its contour length.
(iii) Estimate the reversible work to stretch a ssDNA chain from the fully relaxed state to 100% of its contour length.
(iv) Assume that the stretch modulus is a constant, how does the reversible work of ssDNA stretching change as the temperature increases?

3.38 Shokri et al. reported a quantitative model to describe the force-induced meting of a double-stranded (ds) DNA into single-stranded (ss) DNA.[55] The essential idea is that the force–extension curve for a partially melted DNA is given by a linear combination of those corresponding to ssDNA and dsDNA chains, i.e., the DNA length per base pair is given by

$$b = (1 - f_a) \times b_{dsDNA} + f_a \times b_{ssDNA},$$

where f_a denotes the fractional DNA melted, b_{dsDNA} and b_{ssDNA} are the per base pair length for dsDNA and ssDNA, respectively. Assume that the force–extension relation for dsDNA is described by the WLC model

$$\frac{f\xi_p}{k_BT} = \frac{b_{dsDNA}}{b^{max}_{dsNDA}} - \frac{1}{4} + \frac{1}{4(1 - b_{dsDNA}/b^{max}_{dsDNA})^2},$$

and that for the ssDNA is described by the modified freely-jointed chain (MFJC) model

$$b_{ssDNA} = b^{max}_{ssDNA}\left[\coth\left(\frac{fb_0}{k_BT}\right) - \frac{k_BT}{fb_0}\right](1 + f/k),$$

plot the force–extension curve (i.e., b versus f) for a partially melted DNA with $f_a = 0.5$ and compare the results with those for pure ssDNA and pure dsDNA. The following parameters are available from fitting to the experimental data at $T = 298.15$ K: for dsDNA, $\xi_p = 48$ nm, $b^{max}_{dsDNA} = 0.34$ nm; for ssDNA, $b_0 = 1.5$ nm, $k = 800$ pN, and $b^{max}_{ssDNA} = 0.56$ nm. What was the main conclusion of the single-molecule experiment?

54 Smith S. B., et al., "Overstretching B-DNA: The elastic response of individual double-stranded and single-stranded DNA molecules", *Science*, 271, 795–799 (1996).
55 Shokri et al., "DNA overstretching in the presence of glyoxal: Structural evidence of force-induced DNA melting", *Biophys. J.* 95, 1248–1255 (2008).

4

Thermodynamics of Photons, Electrons, and Phonons

Photons and electrons are the key players of diverse photochemical, photoelectrical, and electrochemical processes important for a wide variety of technological applications. Because both photons and electrons are quantum particles, their thermodynamic properties are drastically different from those corresponding to a system of classical particles such as a noble gas. Phonons are also quantum particles, more precisely, quasiparticles representing various vibrational modes of atoms in condensed matter. While the importance of real particles such as electrons and photons are self-evident for their broad engineering applications, one might be wondering the usefulness of quasiparticles. As we will see from this chapter, the statistical thermodynamics of phonons plays a fundamental role in understanding diverse properties of condensed matter, especially the collective behavior of solids and some liquids, such as heat capacity, thermal conductivity, and electrical conductivity.

According to quantum mechanics, particles (and quasiparticles) exist in quantum states that can be described in terms of wave functions. At each quantum state, the properties of a many-particle system are manifested as the expectation values of observables (i.e., measurable physical quantities). Because the microscopic properties can be uniquely determined by the wave functions, quantum states provide a natural starting point to study the thermodynamic properties of many-particle systems using statistical mechanics.

In this chapter, we start with some basic concepts to define quantum states for many-particle systems. Next, we proceed with the standard statistical-mechanical procedures to derive the thermodynamic properties of ideal quantum systems, namely those comprising noninteracting quantum particles or quasiparticles. Like the ideal-gas model for gases and fluids, these idealized models provide a starting point and a useful reference to describe the properties of photons, electrons, and phonons, as well as other more complicated quantum systems.

4.1 Quantum Particles

In classical and quantum systems alike, identical particles have the same intrinsic properties such as the rest mass and electric charge; there is no experimental method or physical feature by which we would be able to distinguish one from another. From the viewpoint of statistical mechanics, particle identity affects microstate counting, and consequently, the entropy of a thermodynamic system.

For quantum systems, particle identity is also associated with the permutation symmetry of many-body wave functions that are not directly detectable by experimental means. The so-called exchange effect has no classical analog yet has profound implications on the statistical distribution

Fundamentals and Practice in Statistical Thermodynamics, First Edition. Jianzhong Wu and John M. Prausnitz.

Companion website: www.wiley.com/go/jianzhongwu/fundamentalsandpractice

of quantum states. According to quantum mechanics, the permutation symmetry of a many-body wave function is responsible for diverse properties of matter including the volume of any object and quantum entanglement (i.e., the quantum state of multiple particles cannot be separated into those corresponding to individual particles even when they are far apart).

4.1.1 The Gibbs Paradox

Before delving into quantum effects, it is instructive to discuss the thermodynamic consequence of particle identity from a classical perspective. In his celebrated *Statistical Mechanics* book published in 1902,[1] J. W. Gibbs presented a puzzling analysis for the entropy of mixing between two ideal gases. Because the result appears inconsistent with the 2nd law of thermodynamics, the paradox has provoked much debate.

Consider a box containing N_1 molecules of ideal gas 1 and N_2 molecules of ideal gas 2 at the same temperature T and pressure P. As shown schematically in Figure 4.1, the two gas systems are initially separated with a diaphragm such that gas 1 has volume V_1 and gas 2 has volume V_2. Now imagine that the diaphragm is removed allowing the two gases to mix with each other and reach a new equilibrium state without changing the temperature and pressure. According to the classical thermodynamic relations, the entropy of mixing can be calculated from

$$\frac{\Delta S}{(Nk_B)} = x_1 \ln x_1 + x_2 \ln x_2 \tag{4.1}$$

where $N = N_1 + N_2, x_1 = N_1/N$, and $x_2 = N_2/N$. As $0 < x_1 < 1$ and $0 < x_2 < 1$, Eq. (4.1) predicts that the entropy of mixing is always positive. The entropy increase is easily understandable because the gas molecules gain more accessible volume after mixing.

In an ideal gas, molecules do no interact with each other. As a result, the increase in entropy due to the removal of the diaphragm is expected to be independent of the intrinsic properties of the gas molecules. In other words, gas molecules have larger accessible volume without the diaphragm regardless of their identities. However, classical thermodynamics asserts that the total entropy should remain unchanged if we mix two *identical* gases at the same temperature and pressure. This apparent inconsistency is known as the Gibbs paradox.

Why does not more accessible volume for each molecule introduced by mixing two identical gases at the same temperature and pressure result in an increase in the total entropy? The answer to this question lies in the *distinguishability* of the gas molecules, i.e., microstates differing only by permutations of the positions of identical particles should not be counted as distinct. The Gibbs paradox would be resolved if we considered particles in volumes 1 and 2 identical even *before* the mixing. In other words, there would be no entropy of mixing if we do not identify molecules by the subsystems before the mixing. The nonzero entropy of mixing ensues only if the molecules are distinguished by certain identifiable features, for example, the belonging of particles to a certain subsystem.

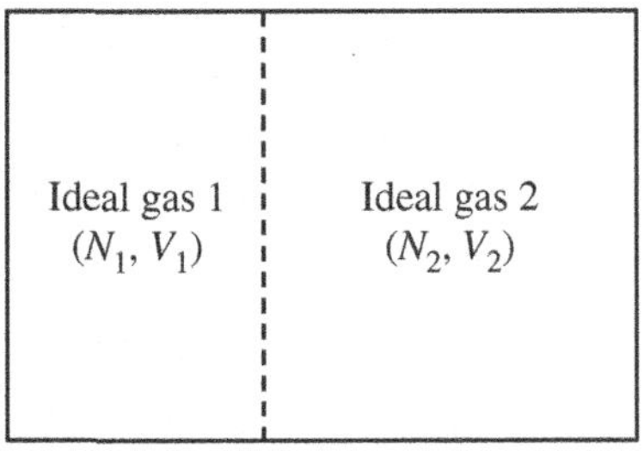

Figure 4.1 The entropy of mixing for two gases of the same temperature and pressure depends on the distinguishability of the gas molecules.

1 Gibbs J. W., *Elementary principles in statistical mechanics*. Dover Publications, Reprint edition (2014).

4.1.2 Permutation Symmetry

While classical particles may be "tagged" with their coordinates, quantum particles do not possess definite positions. In other words, identical quantum particles cannot be distinguished by their positions or the trajectory of motion. As a result, the quantum indistinguishability results in additional consequences that are not shown in classical systems.

To illustrate the implication of indistinguishability from the quantum-mechanical point of view, consider a system with N identical quantum particles. For simplicity, we assume that all particles are in the same spin state.[2] At each microstate (viz., quantum state), the properties of the system are uniquely determined by wave function $\Psi(\mathbf{r}_1, \mathbf{r}_2, \ldots \mathbf{r}_N)$, where $\mathbf{r}_{i=1,2,\ldots N}$ refers to the particle coordinates. According to the statistical interpretation of quantum mechanics, the wave function defines the probability density of finding the particles at various positions in the system

$$p(\mathbf{r}_1, \mathbf{r}_2, \ldots \mathbf{r}_N) \sim |\Psi(\mathbf{r}_1, \mathbf{r}_2, \ldots \mathbf{r}_N)|^2 \tag{4.2}$$

where the proportionality constant may be fixed by the normalization condition. Wave functions $\Psi(\mathbf{r}_1, \mathbf{r}_2, \ldots \mathbf{r}_N)$ and $-\Psi(\mathbf{r}_1, \mathbf{r}_2, \ldots \mathbf{r}_N)$ describe the same quantum state because both satisfy the Schrödinger equation and yield the same probability density $p(\mathbf{r}_1, \mathbf{r}_2, \ldots \mathbf{r}_N)$.

Now imagine that the coordinates of two identical particles in the system, say i and j, are exchanged.[3] The indistinguishability of these particles means that the system remains at the same quantum state. However, we would have two choices to assign the wave function after the particle exchange. The wave function may remain unchanged (symmetric) or switch the sign (anti-symmetric). Particles with the same wave function after the permutation are called bosons, named after theoretical physicist Satyendra N. Bose. Photons and phonons are two common examples of bosons. Conversely, particles with the antisymmetric wave function are called fermions, named after Enrico Fermi. Ordinary quantum particles are either bosons or fermions, more or less like the two genders in the animal kingdom.[4] The standard model of elementary particles includes 24 different types of fermions, including electrons.

4.1.3 Occupation Numbers

The symmetric or anti-symmetric nature of wave functions in response to particle permutation leads to drastic different properties due to their effects on the particle occupation of quantum states. The permutation effect can be elucidated by considering a pair of *non-interacting* particles 1 and 2. As shown schematically in Figure 4.2, suppose $f(\mathbf{r})$ and $g(\mathbf{r})$ are one-particle wave functions for a particle at position A and another particle at position B. To ensure no interaction, we may assume that these particles are far away from each other such that the wave functions do not overlap. If these particles are distinguishable, the system would have two possible quantum states with wave functions given by the product of the single-particle wave functions, $f(\mathbf{r}_1)g(\mathbf{r}_2)$ and $f(\mathbf{r}_2)g(\mathbf{r}_1)$, which correspond to particle 1 at position A and particle 2 at position B, and vice versa.

---

2 Spin is associated with the intrinsic angular momentum of quantum particles. In the presence of an external magnetic field, it takes half-integer ($S_{1/2+n}$) or integer (S_n) values of $\hbar = h/(2\pi)$, where h is the Planck constant and $n > 0$ is an integer. Because quasiparticles represent the collective motions of a many-particle system (e.g., vibrational excitation of a crystal lattice), they carry no intrinsic angular momentum (thus do not have spin).

3 An alternative and more rigorous approach has been proposed without the introduction of particle indices. See Leinass J. M. and Myrheim J. "On the theory of identical particles," *Il Nouvo Cimento* 37 (1), 1–23 (1977).

4 In one- or two-dimensional systems, there exists a continuum of intermediate cases between bosons and fermions.

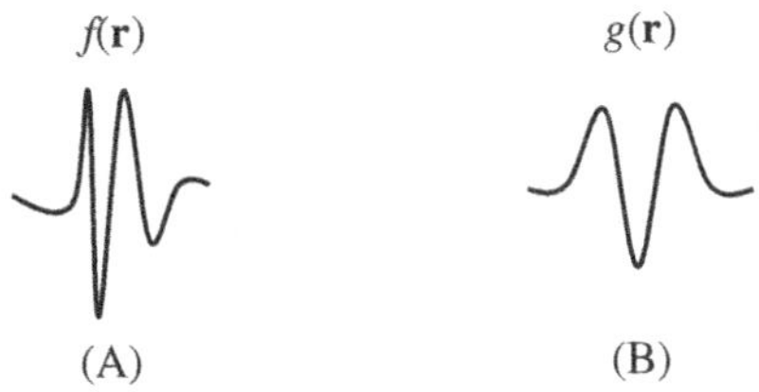

Figure 4.2 A schematic representation for the single-particle wave function for a particle at position A and that for another identical particle at position B.

If these particles are identical, the two quantum states become the same, i.e., there is only a single quantum state.

In order to satisfy the symmetric or anti-symmetric properties due to the particle exchange, the wave function for the two-particle system can be constructed as

$$\Psi(\mathbf{r}_1, \mathbf{r}_2) = f(\mathbf{r}_1)g(\mathbf{r}_2) \pm f(\mathbf{r}_2)g(\mathbf{r}_1) \tag{4.3}$$

where the positive sign applies to bosons (symmetric) and the negative sign to fermions (anti-symmetric). When the two single-particle states are identical, Eq. (4.3) predicts that $\Psi(\mathbf{r}_1, \mathbf{r}_2)$ vanishes for fermions but remains finite for bosons. Because a wave function that is zero everywhere is considered unphysical, the vanishing wavefunction is a manifestation of the Pauli's exclusion principle: no two fermions can occupy the same single-particle quantum state. By contrast, there is no such restriction on the number of bosons to be in the same quantum state.

The above analysis for the particle occupation of quantum states can be extended to a system with N identical quantum particles that do not interact with each other. From the viewpoint of statistical mechanics, each microstate ν (viz., quantum state) can be specified by wave function $\Psi_\nu(\mathbf{r}_1, \mathbf{r}_2, \ldots \mathbf{r}_N)$, which corresponds to a solution of the N-body Schrödinger equation. If the particles do not interact with each other, the N-particle wave function can be expressed in terms of a complete orthonormal set of single-particle wave functions, $\{\varphi_i(\mathbf{r})\}_{i=1,2,\ldots}$,

$$\int d\mathbf{r}\ \varphi_i^*(\mathbf{r})\varphi_j(\mathbf{r}) = \delta_{ij} \tag{4.4}$$

where δ_{ij} is the Kronecker delta function.

To satisfy the permutation symmetry of particle coordinates, the N-particle wave function for non-interacting bosons can be written as

$$\Psi(\mathbf{r}_1, \mathbf{r}_2, \ldots \mathbf{r}_N) = \frac{1}{\sqrt{N!}} \sum_P P[\varphi_{\alpha 1}(\mathbf{r}_1)\varphi_{\alpha 2}(\mathbf{r}_2) \ldots \varphi_{\alpha N}(\mathbf{r}_N)] \tag{4.5}$$

where $P[\cdots]$ denotes the exchange of the coordinates for a pair of particles, the summation applies to all possible permutations of the index set $\{\alpha_1, \alpha_2, \ldots, \alpha_N\}$, and the prefactor $1/\sqrt{N!}$ arises from the normalization condition.

For bosons, the number of particles in each single-particle state is unrestricted, i.e., the occupation number can be any non-negative integer $n_\alpha = 0, 1, 2 \ldots$. Subsequently, the microstate of the system is completely defined by the occupation number in each single-particle state

$$\nu = \{n_{\alpha 1}, n_{\alpha 2}, \ldots\}. \tag{4.6}$$

The total number of particles in the system is given by a summation of the occupation number in all single-particle states

$$N = \sum_{\alpha_i=1}^{\infty} n_{\alpha_i}. \tag{4.7}$$

For fermions, the N-particle wave function satisfies the permutation anti-symmetry. In this case, the combination of the single-particle wave functions leads to the Slater determinant, named after John C. Slater, a theoretical physicist who made major contributions to the theory of the electronic structure of atoms, molecules, and solids,

$$\Psi(\mathbf{r}_1, \mathbf{r}_2, \ldots \mathbf{r}_N) = \frac{1}{\sqrt{N!}} \begin{vmatrix} \varphi_1(\mathbf{r}_1) & \varphi_1(\mathbf{r}_2) & \cdots & \varphi_1(\mathbf{r}_N) \\ \varphi_2(\mathbf{r}_1) & \varphi_2(\mathbf{r}_2) & \cdots & \varphi_2(\mathbf{r}_N) \\ \vdots & \vdots & \vdots & \vdots \\ \varphi_N(\mathbf{r}_1) & \varphi_N(\mathbf{r}_2) & \cdots & \varphi_N(\mathbf{r}_N) \end{vmatrix} \tag{4.8}$$

where the prefactor $1/\sqrt{N!}$ is again required by the normalization condition. Because the determinant vanishes whenever two single-particle wave functions are identical, Eq. (4.8) predicts that, as stated in the Pauli exclusion principle, no two fermions can be in the same quantum state. Accordingly, the occupation number for each single-particle state is either $n_\alpha = 0$ or 1. As for bosons, the microstate of the system is completely defined by the occupation number in each single-particle state, and the total number of particles in the system is given by a summation of the occupation numbers.

Before closing this section, it is instructive to consider a simple system with three identical particles that may occupy three single-particle quantum states. As illustrated in Figure 4.3, the system would have 10 distinctive microstates if the particles are bosons. For three identical fermions, however, there is only a single microstate, i.e., one fermion for each single-particle state, in order to satisfy the Pauli exclusion principle.

4.1.4 Summary

In summary, the distinguishability of particles in a thermodynamic system is an important concept in statistical mechanics because it is related to microstate counting thus entropy. For classical systems, the significance of particle identity may be illustrated with the Gibbs paradox. Without defining the particle identity, the entropy of mixing would disappear. For quantum systems, the particle identity is intrinsically associated with the permutation symmetry of the many-body wave function. The symmetry or anti-symmetry nature of particle permutation leads to different occupation numbers for bosons and fermions. The occupation numbers can be used to define the microstates of a quantum system from which we can derive the partition functions and thermodynamic properties following standard statistical-thermodynamic equations.

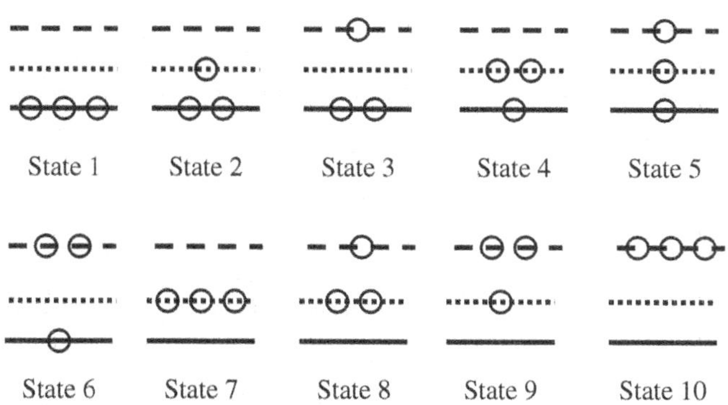

Figure 4.3 Microstates for three identical bosons in three single-particle quantum states.

4.2 Quantum Statistics

In this section, we discuss the statistical distributions of microstates for thermodynamic systems containing noninteracting quantum particles, viz., bosons and fermions. The grand canonical ensemble will be used to derive the average occupation number of the non-interacting particles at different single-particle states. Applications of the statistical-mechanical equations to systems of practical interests will be discussed in subsequent sections.

4.2.1 Single-Particle States

As discussed in Section 4.1, the microstates for a system of noninteracting quantum particles can be described in terms of the single-particle wave functions.[5] For both bosons and fermions, the single-particle states are associated with the solutions to the Schrödinger equation[6]

$$-\frac{\hbar^2}{2m}\nabla^2\psi(\mathbf{r}) = \varepsilon\psi(\mathbf{r}) \tag{4.9}$$

where $\hbar = h/(2\pi)$ with h being the Planck constant, m is the particle mass, and ε stands for the particle energy. The linear partial differential equation may be solved with the normalization condition

$$\int d\mathbf{r}|\psi(\mathbf{r})|^2 = 1 \tag{4.10}$$

plus appropriate boundary conditions.

With a boundary condition such that the quantum particle is confined in a cubic box of length L (viz., a particle in a box), we may find a general solution for the single-particle wave function from the Schrödinger equation:

$$\psi(\mathbf{r}) = (2/L)^{3/2}\sin(k_x x)\sin(k_y y)\sin(k_z z) \tag{4.11}$$

where $\mathbf{r} = (x, y, z)$ stands for particle position, and $\mathbf{k} = (k_x, k_y, k_z)$ is the wavevector taking discrete values[7]

$$\mathbf{k} = \frac{\pi}{L}(n_x, n_y, n_z),\ n_{x,y,z} = 1, 2, \ldots. \tag{4.12}$$

Each wavevector represents a single-particle quantum state, which has an energy of

$$\varepsilon_{\mathbf{k}} = \frac{\hbar^2 k^2}{2m} \tag{4.13}$$

where $k = |\mathbf{k}|$. Comparing Eq. (4.13) with that for the kinetic energy of a classical particle, $\varepsilon = |\mathbf{p}|^2/2m$, we may associate wavevector $\mathbf{k}$ with the particle momentum $\mathbf{p} = \hbar\mathbf{k}$, or with the particle velocity $\mathbf{v} = \hbar\mathbf{k}/m$. Similar to that for a classical particle, the kinetic energy of a quantum particle is nonnegative.

---

5 For simplicity, we do not consider degeneracy arising from particle spin. In general, fermions have half-integer spins such as 1/2, 3/2, 5/2, …, and bosons have integer spins 1, 2, …. For a system of non-interacting quantum particles with spin S, the single-particle wave function is associated with $2S + 1$ quantum states of the same energy, which can be described by a degeneracy of $g = 2S + 1$.

6 Strictly speaking, the Schrodinger equation is not applicable to relativistic particles such as photons, which are massless, relativistic spin-1 particles.

7 The negative values of k_x, k_y and k_z change only the sign of the wave function thus result in the same single-particle quantum state. $n_{x,y,z} = 0$ is not allowed because that would lead to zero particle density.

4.2.2 The Bose–Einstein Statistics

Consider now the grand canonical ensemble for a system of identical bosons that do not interact with each other. The grand partition function is defined as

$$\Xi = \sum_{\nu} e^{-\beta(E_\nu - \mu N_\nu)}. \tag{4.14}$$

where $\beta = 1/(k_B T)$, μ is the chemical potential, E_ν and N_ν stand for the total energy and the number of particles at microstate state ν, respectively.

At each microstate ν, the system energy and the number of particles can be expressed in terms of the occupation numbers of single-particle states, $n_\alpha = 0, 1, 2, \ldots$,

$$E_\nu = \sum_\alpha n_\alpha \varepsilon_\alpha, \tag{4.15}$$

$$N_\nu = \sum_\alpha n_\alpha, \tag{4.16}$$

where ε_α represents the energy of the single-particle state α as determined by the Schrödinger equation, Eq. (4.13).

Because each microstate is fully specified by the occupation numbers of the single-particle states, we may replace the summation over the microstates in Eq. (4.14) with that for all possible occupation numbers:

$$\Xi = \sum_{\nu=\{n_1,n_2,\ldots\}} \exp\left[-\beta \sum_\alpha n_\alpha(\varepsilon_\alpha - \mu)\right] = \sum_{\nu=\{n_1,n_2,\ldots\}} \prod_\alpha \exp[-\beta(\varepsilon_\alpha - \mu)n_\alpha]. \tag{4.17}$$

Switching the sequence of summation and product in Eq. (4.17) leads to

$$\Xi = \prod_\alpha \sum_{n_\alpha=0}^{\infty} \exp[-\beta(\varepsilon_\alpha - \mu)n_\alpha] = \prod_\alpha \frac{1}{1 - e^{\beta(\mu - \varepsilon_\alpha)}}. \tag{4.18}$$

Accordingly, the grand potential is given by

$$\Omega = -k_B T \ln \Xi = k_B T \sum_\alpha \ln\left[1 - z e^{-\beta \varepsilon_\alpha}\right] \tag{4.19}$$

where $z \equiv e^{\beta\mu}$ is called fugacity.[8]

For noninteracting quantum particles in a cubic box of length L, the single-particle states are defined by wavevectors $\mathbf{k} = \frac{\pi}{L}(n_x, n_y, n_z)$, as shown in Eq. (4.12). Because L is a macroscopic quantity, π/L approaches zero in the thermodynamic limit ($L \to \infty$). Thus, we can replace the summation over the microstates with integration

$$\sum_\alpha = \sum_{n_x=1}^{\infty}\sum_{n_y=1}^{\infty}\sum_{n_z=1}^{\infty} \approx \frac{L^3}{\pi^3}\int_0^\infty dk_x \int_0^\infty dk_y \int_0^\infty dk_z. \tag{4.20}$$

Using Eq. (4.20) and $\varepsilon_\alpha = \hbar^2 k^2/2m$, Eq. (4.19) becomes

$$\begin{aligned}\Omega &= k_B T \frac{L^3}{\pi^3}\int_0^\infty dk_x \int_0^\infty dk_y \int_0^\infty dk_z \ln[1 - z e^{-\beta\hbar^2 k^2/2m}] \\ &= \frac{k_B T V}{8\pi^3}\int_0^\infty dk 4\pi k^2 \ln[1 - z e^{-\beta\hbar^2 k^2/2m}]\end{aligned} \tag{4.21}$$

8 It should not be confused with the fugacity f in chemical thermodynamics. The latter is defined as $\mu = \mu_0 + k_B T \ln(f/f_0)$ where subscript 0 stands for a reference state. While in Eq. (4.19) z is dimensionless, f has the units of pressure.

where $V = L^3$ is the system volume, and a factor of 1/8 arises from the change in the integration boundaries from the positive octant to the entire space.

Upon replacement $x = \beta\hbar^2k^2/2m$ and integration by parts, Eq. (4.21) can be alternatively expressed as[9]

$$\Omega = -\frac{4k_BTV}{3\sqrt{\pi}\Lambda^3}\int_0^\infty dx\frac{x^{3/2}}{e^x/z-1} = -\frac{k_BTV}{\Lambda^3}\mathrm{Li}_{5/2}(z) \tag{4.22}$$

where $\Lambda \equiv h/\sqrt{2\pi mk_BT}$ is called the De Broglie thermal wavelength, and $\mathrm{Li}_{5/2}(z)$ is a polylogarithm function.

Following the standard statistical-mechanical equations for the grand canonical ensemble, we may derive thermodynamic properties from the grand partition function or the grand potential. For example, the average number of particles in each single-particle state is given by

$$\langle n_\alpha\rangle = \frac{1}{\Xi}\prod_\alpha\sum_{n_\alpha=1}^\infty n_\alpha\exp[-\beta(\varepsilon_\alpha-\mu)n_\alpha] = -\frac{\partial\ln\Xi}{\partial\beta\varepsilon_\alpha}. \tag{4.23}$$

Substituting Eq. (4.18) into (4.23) gives

$$\langle n_\alpha\rangle = \frac{1}{e^{\beta(\varepsilon_\alpha-\mu)}-1}. \tag{4.24}$$

Eq. (4.24) is known as the Bose–Einstein distribution. It predicts the number of bosons in each single-particle state.

Figure 4.4 illustrates the dependence of the occupation number $\langle n_\alpha\rangle$ on the single particle energy ε_α for three values of the reduced chemical potential $\beta\mu$. At a fixed chemical potential, the occupation number increases monotonically as temperature rises (β becomes smaller). Because both β and $\langle n_\alpha\rangle$ are nonnegative, Eq. (4.24) indicates that, in a stable system, the chemical potential of bosons must be smaller than the ground-state energy, i.e., $\mu < \varepsilon_0$. When the chemical potential is the same as the ground-state energy $\mu = \varepsilon_0$, Eq. (4.24) predicts $\langle n_0\rangle = \infty$, implying that an infinitely large number of bosons will simultaneously occupy the same single-particle state. This quantum phenomenon was first predicted by Albert Einstein in 1925 who extended the statistical theory of photons developed by Satyendra N. Bose in 1924. The Bose–Einstein

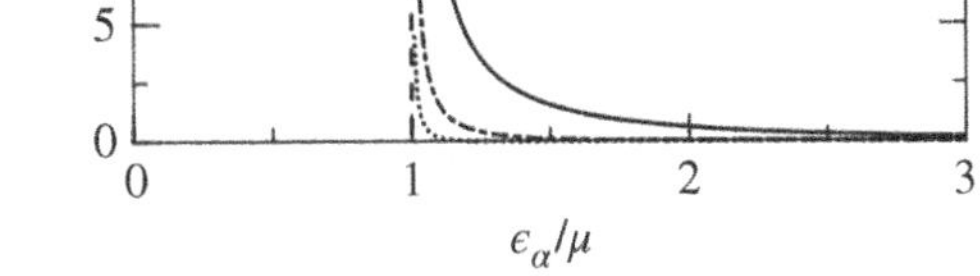

Figure 4.4 The Bose–Einstein distribution for the occupation number of the single-particle energy levels for three values of reduced chemical potential $\beta\mu = 1.0$, 5.0 and 20. The Bose-Einstein condensation takes place when the chemical potential is the same as the ground state energy, $\mu = \epsilon_0$. In that case, an infinite number of bosons simultaneously occupy the ground state. For illustration purposes, here we assume that the chemical potential is always positive.

9 $\int_0^\infty dkk^2\ln[1-ze^{-\beta\hbar^2k^2/2m}] = \frac{k^3}{3}\ln[1-ze^{-\beta\hbar^2k^2/2m}]_0^\infty - \frac{1}{3}\int_0^\infty dkk^3\frac{z(\beta\hbar^2k^2/m)e^{-\beta\hbar^2k^2/2m}}{1-ze^{-\beta\hbar^2k^2/2m}} = -\frac{1}{3}\left(\frac{2m}{\beta\hbar^2}\right)^{3/2}\int_0^\infty dx\frac{x^{3/2}}{e^x/z-1}$ where the first term vanishes because $\lim_{k\to\infty}k^3\ln(1-ze^{-\beta\hbar^2k^2/2m}) = -\lim_{k\to\infty}zk^3e^{-\beta\hbar^2k^2/2m} = 0$. The polylogarithm function is defined as $\mathrm{Li}_n(z) = \frac{1}{\Gamma(n)}\int_0^\infty dx\frac{x^{n-1}}{e^x/z-1}$ where $\Gamma(n)$ is gamma function. $\Gamma(3/2) = \pi^{1/2}/2$ and $\Gamma(5/2) = 3\pi^{1/2}/4$.

condensate (BEC), a state of matter that is typically formed when a low density boson gas is cooled to temperatures very close to absolute zero, was discovered experimentally about 70 years later by Eric A. Cornell, Wolfgang Ketterle and Carl E. Wieman, three physicists who shared the Nobel prize in 2001. While BEC was born out of merely theoretical curiosity, the unconventional state of matter is potentially useful for a wide range of practical applications including novel optoelectronic devices such as atom lasers, ultrafast optical switches, and quantum computers.

4.2.3 The Fermi–Dirac Statistics

We may repeat the above procedure for a system of noninteracting fermions. In this case, the occupation number for each single-particle state is restricted to $n_\alpha = 0$ or 1. Accordingly, the grand partition function is given by

$$\Xi = \prod_\alpha \sum_{n_\alpha=0,1} \exp[-\beta(\varepsilon_\alpha - \mu)n_\alpha] = \prod_\alpha \left[1 + e^{-\beta(\varepsilon_\alpha-\mu)}\right]. \tag{4.25}$$

Using Eq. (4.20), we can derive the grand potential similar to that for bosons

$$\Omega = -k_BT\ln\Xi = -k_BT\sum_\alpha \ln\left[1 + ze^{-\beta\varepsilon_\alpha}\right]$$

$$= -\frac{4k_BTV}{3\sqrt{\pi}\Lambda^3}\int_0^\infty dx\frac{x^{3/2}}{e^x/z+1} = \frac{k_BTV}{\Lambda^3}\mathrm{Li}_{5/2}(-z). \tag{4.26}$$

The average number of particles in single-particle state α is thus given by

$$\langle n_\alpha\rangle = \frac{1}{\Xi}\prod_\alpha \sum_{n_\alpha=0,1} n_\alpha \exp[-\beta(\varepsilon_\alpha - \mu)n_\alpha] = -\frac{\partial \ln\Xi}{\partial\beta\varepsilon_\alpha}. \tag{4.27}$$

Substituting Eq. (4.25) into (4.27) yields the Fermi–Dirac distribution:

$$\langle n_\alpha\rangle = \frac{1}{e^{\beta(\varepsilon_\alpha-\mu)}+1}. \tag{4.28}$$

Figure 4.5 shows the average occupation number for a single-particle state as a function of the energy level. As the exponential term is nonnegative, Eq. (4.28) predicts that the average occupation number for any single-particle state is confined between 0 and 1. If the energy of a single-particle state is much smaller than the chemical potential, it is most likely occupied by one fermion because the chemical potential of the quantum particle is typically much larger than the thermal energy at ambient conditions (i.e., $\beta\mu \gg 1$). When the temperature is absolute zero ($\beta\mu = \infty$), the Fermi–Dirac distribution becomes a step function, i.e., a single-particle state is occupied if $\epsilon_\alpha < \mu$, and unoccupied otherwise. In that case, the chemical potential is often referred to as the Fermi energy.

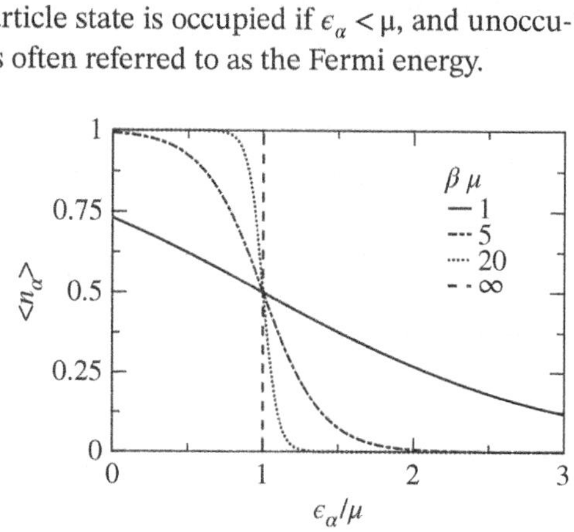

Figure 4.5 The Fermi–Dirac distribution for the occupation number of single-particle quantum states as a function of the energy. For illustration purposes, here we assume that the chemical potential is always positive.

4.2.4 The Classical Limit

At high temperature and low particle density, we expect that the quantum effects become less important so that the particle distributions would reduce to the classical limit. To verify this conjecture, we may write the average number of particles at a single-particle state as

$$\langle n_\alpha \rangle = \frac{1}{e^{\beta(\varepsilon_\alpha - \mu)} \pm 1} \tag{4.29}$$

where the positive sign applies to fermions and the negative sign to bosons.

When the temperature is sufficiently high while the particle density is low, the particles are accessible to a very large number of energy states such that, on average, the occupation number at any particular state is extremely small, $\langle n_\alpha \rangle \ll 1$. Accordingly, Eq. (4.29) predicts that the exponential term must be much larger than 1. Because the tiny occupation number should be independent of any specific value of ε_α, which is positive for both ideal fermions and bosons, $e^{\beta(\varepsilon_\alpha - \mu)} \gg 1$ implies that $e^{-\beta\mu} \gg 1$ or $z \to 0$ (which corresponds high temperature and low particle density). Under this circumstance, both the Fermi–Dirac distribution and the Bose–Einstein distribution can be approximated by

$$\langle n_\alpha \rangle \approx e^{-\beta(\varepsilon_\alpha - \mu)}. \tag{4.30}$$

The total number of particles in the system is thus given by

$$N = \sum_\alpha \langle n_\alpha \rangle \approx e^{\beta\mu} \sum_\alpha e^{-\beta\varepsilon_\alpha}. \tag{4.31}$$

Replacing $e^{\beta\mu}$ in Eq. (4.30) with the relation given by Eq. (4.31) results in the familiar Boltzmann equation for the distribution of noninteracting classical particles

$$\frac{\langle n_\alpha \rangle}{N} \approx \frac{e^{-\beta\varepsilon_\alpha}}{\sum_i e^{-\beta\varepsilon_i}}. \tag{4.32}$$

As shown in Figure 4.6, the classical limit is approached when the single-particle energy ε_α surpasses the chemical potential μ with a magnitude significantly larger than $k_B T$.

In the classical limit, the grand potential of fermions, Eq. (4.26), becomes identical to that for bosons, Eq. (4.19), and both can be expressed as

$$\Omega = \pm k_B T \sum_\alpha \ln\left[1 \mp z e^{-\beta\varepsilon_\alpha}\right] \approx -k_B T e^{\beta\mu} \sum_\alpha e^{-\beta\varepsilon_\alpha} = -N k_B T. \tag{4.33}$$

Recalling from Section 2.9 that $\Omega = -PV$ for a uniform system, Eq. (4.33) predicts the ideal-gas law for a thermodynamic system containing noninteracting classical particles (viz., the classical limit of quantum particles)

$$P = \rho k_B T \tag{4.34}$$

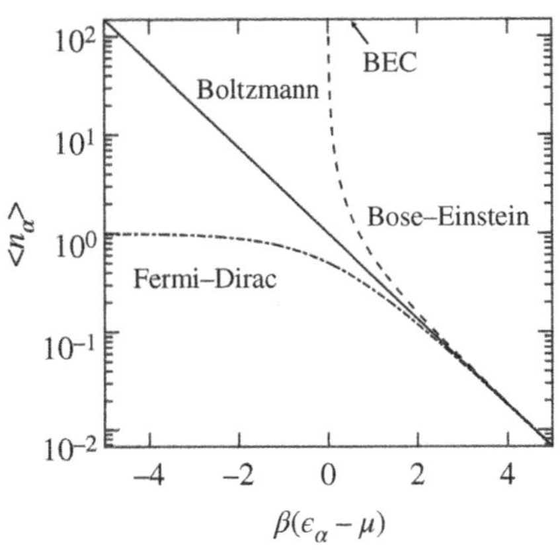

Figure 4.6 The distributions of noninteracting quantum particles reduce to the classical Boltzmann limit. The arrow marks the condition leading to the Bose–Einstein condensate (BEC).

where $\rho = N/V$ is the average number density. Because $-\mathrm{Li}_{5/2}(z) \approx \mathrm{Li}_{5/2}(-z) \approx -z$ as $z \to 0$, Eqs. (4.26) and (4.22) also predict that the grand potential of classical particles is given by

$$\Omega = -\frac{k_B TV}{\Lambda^3} z = -\frac{k_B TV}{\Lambda^3} e^{\beta\mu}. \tag{4.35}$$

Substituting $\Omega = -PV = -\rho k_B TV$ into Eq. (4.35) leads to a simple expression for the reduced chemical potential of a classical ideal gas system

$$\beta\mu = \ln(\rho\Lambda^3). \tag{4.36}$$

In addition, we may obtain from Eq. (4.35) analytical expressions for the internal energy

$$U = \left(\frac{\partial \beta\Omega}{\partial \beta}\right)_{V,\beta\mu} = -\left(\frac{\partial V e^{\beta\mu}/\Lambda^3}{\partial \beta}\right)_{V,\beta\mu} = -\frac{3}{2\beta}\beta\Omega = \frac{3N}{2} k_B T \tag{4.37}$$

and for the ideal-gas entropy

$$S = -\left(\frac{\partial \Omega}{\partial T}\right)_{V,\mu} = \left(\frac{\partial}{\partial T}\frac{k_B TV}{\Lambda^3} e^{\beta\mu}\right)_{V,\mu} = -\frac{5\Omega}{2T} + \frac{\mu}{k_B T}\frac{\Omega}{T}$$
$$= -Nk_B[\ln(\rho\Lambda^3) - 5/2] \tag{4.38}$$

where we have used $\Omega/T = -Nk_B$ and $\beta\mu = \ln(\rho\Lambda^3)$. As expected, the internal energy given by Eq. (4.37) is the same as the average kinetic energy of classical particles predicted by the Maxwell–Boltzmann equation. As discussed in Section 3.2, Eq. (4.38) is known as the Sackur–Tetrode equation. Interestingly, the factor of $N!$ is not utilized in quantum statistics, suggesting that its presence in the partition functions of classical systems is merely a remnant of quantum phenomena. Because the Sackur–Tetrode equation is derived from the classical limit, it is not consistent with the third law of thermodynamics (viz., $S = 0$ at $T = 0$ K).

4.2.5 Summary

The preceding statistical-thermodynamic analysis reveals that noninteracting quantum particles are distributed into the single-particle states according to either the Bose–Einstein distribution or the Fermi–Dirac distribution, depending on the intrinsic characteristics of the quantum particles, i.e., whether the particles are bosons or fermions, respectively. In the classical limit, the Boltzmann distribution is recovered from both distributions, leading to the conventional ideal-gas equations. For noninteracting fermions at absolute zero temperature, a single-particle state is fully occupied if its energy level is lower than the Fermi energy, and it is unoccupied otherwise. Unlike noninteracting fermions or classical particles, bosons have the unique ability to form Bose–Einstein condensate (BEC) when the ground state energy is lower than the chemical potential. This results in an unusual state of matter that has proven useful in exploring a wide range of questions in fundamental physics as well as technological applications.

4.3 Thermodynamics of Light

Light consists of photons,[10] a type of elementary particle first introduced by Albert Einstein to represent the discrete units of radiation energy. By treating light as a quantum fluid, i.e., a statistical-mechanical system of noninteracting photons, we will be able to understand not only its thermodynamic properties such as internal energy and entropy but also thermal radiation, a ubiquitous property of matter at finite temperature. Besides, the statistical-thermodynamic

10 Photon is derived from the Greek word *phos* that means light. The term was coined by Gilbert N. Lewis in 1926.

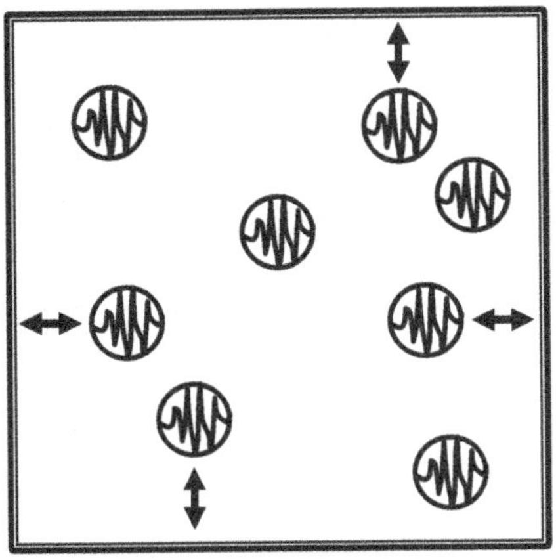

Figure 4.7 Schematic of photons confined within a container (viz., cavity) that can emit and adsorb light without any energy loss.

description of light is useful for a broad spectrum of engineering applications, in particular for quantifying the thermodynamic efficiency of solar energy conversion including photoelectronic devices and photosynthesis.

4.3.1 Photon Gas

Photons are bosons, i.e., multiple photons can occupy the same single-particle state. Different from conventional bosons, however, photons possess no rest mass and thus are not subject to the conservation of mass or the number of particles.[11] Even in a closed system at thermal equilibrium, the total number of phonons is subject to fluctuations due to light emission and absorption from the surrounding (*viz.*, a container).

To derive the thermodynamic properties for a system of noninteracting photons (also known as photon gas), consider as illustrated in Figure 4.7 a cubic container of side length L at temperature T. We assume that light emission and absorption from the inner surface of the container are balanced without any energy loss. For photons inside the container, each microstate is completely specified by a set of occupation number n_1, n_2, ... in different single-particle states denoted, respectively, by energy levels $\varepsilon_1, \varepsilon_2, \ldots$. Because photons are bosons, the possible occupation number for each single-particle state is $n_\alpha = 0, 1, 2, \ldots$.

With the assumption that photons do not interact with each other, the total energy of the system is the sum of the energies of individual photons

$$E_\nu = n_1\varepsilon_1 + n_2\varepsilon_2 + \cdots. \tag{4.39}$$

Given temperature T and accessible volume V defined by the container, the canonical partition function for the photon gas is[12]

$$Q = \sum_\nu \exp(-\beta E_\nu) = \sum_{n_{i=1,2,\ldots}=0}^{\infty} \exp[-\beta(n_1\varepsilon_1 + n_2\varepsilon_2 + \cdots)]. \tag{4.40}$$

As noninteracting photons are independent of each other, the summation in Eq. (4.40) may be evaluated analytically

$$Q = \sum_{n_1=0}^{\infty} \exp(-\beta n_1\varepsilon_1) \sum_{n_2=0}^{\infty} \exp(-\beta n_2\varepsilon_2) \cdots = \prod_{\alpha=1}^{\infty} \frac{1}{1 - e^{-\beta\varepsilon_\alpha}}. \tag{4.41}$$

11 According to Einstein's special theory of relativity, the mass of an object increases with its velocity. When an object is at rest (relative to the observer), it has the usual mass that we call the "rest mass."

12 Strictly speaking, the statistical-mechanical description corresponds to that for a grand canonical ensemble of zero chemical potential instead of the canonical ensemble.

Accordingly, the Helmholtz energy of photons is given by

$$F = -k_B T \ln Q = k_B T \sum_{\alpha=1}^{\infty} \ln\left[1 - e^{-\beta\varepsilon_\alpha}\right]. \tag{4.42}$$

From the canonical partition function, Eq. (4.41); we can derive the internal energy, i.e., the total energy of photons on average,

$$U = -\left(\frac{\partial \ln Q}{\partial \beta}\right)_V = \sum_{\alpha=1}^{\infty} \frac{\varepsilon_\alpha}{e^{\beta\varepsilon_\alpha} - 1}. \tag{4.43}$$

Other thermodynamic properties may also be obtained from the partition function or the Helmholtz energy.

The average number of photons at any single-particle energy state α is given by ensemble average

$$\langle n_\alpha \rangle = \frac{\sum_\nu n_\alpha e^{-\beta E_\nu}}{\sum_\nu e^{-\beta E_\nu}} = \frac{\sum_\nu -\left(\partial e^{-\beta E_\nu}/\partial \beta\varepsilon_\alpha\right)}{\sum_\nu e^{-\beta E_\nu}} = -\left(\frac{\partial \ln Q}{\partial \beta\varepsilon_\alpha}\right) = \frac{1}{e^{\beta\varepsilon_\alpha} - 1}. \tag{4.44}$$

Eq. (4.44) is known as the Planck distribution. In comparison with the occupation number of a single-particle state predicted by the Bose–Einstein distribution, Eq. (4.44) suggests that $\mu = 0$, i.e., photons in equilibrium with its container has a zero chemical potential.[13]

4.3.2 Photon Density of States

The density of states (DOS) is a useful concept to describe microstate distribution for any thermodynamic system. Here, we will use DOS to evaluate the summation over single-particle states and the spectral properties of light, which are often expressed in terms of wavevector, frequency, or wavelength.

Imagine that, as described by wave-particle duality, each photon can be represented by a standing wave contained in a cubic box of length L. In *each* dimension, the photon wavelength λ is related to the box length by

$$L = n\lambda/2 \tag{4.45}$$

where $n = 1, 2, \ldots$ is a positive integer. According to Planck's hypothesis, the photon energy is proportional to the frequency of the electromagnetic wave υ

$$\varepsilon = h\upsilon \tag{4.46}$$

where h is Planck's constant. The photon frequency is related to the wavelength and the speed of light $c = 2.9979 \times 10^8$ m/s

$$\upsilon = \frac{c}{\lambda} = \frac{cn}{2L}. \tag{4.47}$$

In each dimension, the energy levels for the single-particle states of photons must satisfy

$$\varepsilon_\alpha = h\upsilon_\alpha = hcn/(2L). \tag{4.48}$$

Because L is a macroscopic quantity and Planck's constant is extremely small, Eq. (4.48) suggests that the summation over the different energy levels can be replaced by an integration with

---

13 The chemical potential of photons is not zero if light emission and absorption from the container are not balanced. For example, when photons are generated from a light emitting diode (LED), the chemical potential of light is determined by the chemical potentials of electrons and holes.

the frequency treated as a continuous variable from 0 to ∞. For a three-dimensional system, the summation over all single-particle states corresponds to a product to those in three-independent dimensions[14]:

$$\sum_{\alpha} = 2 \times \sum_{n_x=0}^{\infty} \sum_{n_y=0}^{\infty} \sum_{n_z=0}^{\infty} = 2 \times \frac{(2L)^3}{c^3} \int_0^{\infty} dv_x \int_0^{\infty} dv_y \int_0^{\infty} dv_z = \frac{2V}{c^3} \int_0^{\infty} 4\pi v^2 dv \tag{4.49}$$

where $V = L^3$ is the system volume, and a factor of 2 arises from the spin states of photons (or intuitively, the factor of 2 accounts for photon polarization because light can be electrically polarized in both clockwise and counter-clockwise directions).

From Eq. (4.49), we find that the density of states (DOS) for noninteracting photons is given by

$$g(v) = \frac{8\pi v^2 V}{c^3}. \tag{4.50}$$

Eq. (4.50) indicates that DOS is a quadratic function of the photon frequency. As discussed in the following, the divergence of DOS at high frequency has profound implications in the early development of quantum mechanics.

4.3.3 Thermodynamic Properties of Photons

Armed with an analytical expression for DOS, we can now derive the thermodynamic properties of noninteracting photons in closed forms. Replacing the summation in Eq. (4.43) with an integration over the frequency according to Eq. (4.49) leads to[15]

$$U = \int_0^{\infty} \frac{hv}{e^{\beta hv} - 1} g(v) dv = \frac{8\pi V}{c^3 \beta^4 h^3} \int_0^{\infty} \frac{x^3 dx}{e^x - 1} = \frac{8\pi^5 V}{15(hc)^3 \beta^4}. \tag{4.51}$$

Eq. (4.51) predicts that the internal energy density of photons is proportional to the fourth power of the absolute temperature, i.e.,

$$U/V = \sigma T^4 \tag{4.52}$$

where $\sigma = 8\pi^5 k_B^4/[15(hc)^3] \approx 7.5676 \times 10^{-16}$ J/(m^3 K^4) is a universal constant. From Eq. (4.52), we can readily obtain the heat capacity per unit volume

$$C_V/V = (\partial U/\partial T)_V/V = 4\sigma T^3. \tag{4.53}$$

Eq. (4.53) indicates that the heat capacity of photon gas is also proportional to the third power of the absolute temperature and vanishes at 0 K.

We can also evaluate the Helmholtz energy of photons by replacing the summation in Eq. (4.42) with integration[16]

$$\begin{aligned} F &= k_B T \int_0^{\infty} \ln[1 - e^{-\beta hv}] g(v) dv \\ &= \frac{8\pi k_B T V}{c^3 \beta^3 h^3} \int_0^{\infty} \ln[1 - e^{-x}]\, x^2 dx = -\frac{8\pi^5 V}{45(hc)^3 \beta^4} = -\frac{\sigma}{3} T^4 V. \end{aligned} \tag{4.54}$$

---

14 Alternatively, we can obtain Eq. (4.49) using the relation from Section 4.2, $\sum_\alpha = 2 \times \frac{L^3}{\pi^3} \int_0^\infty dk_x \int_0^\infty dk_y \int_0^\infty dk_z$, and $\mathbf{k} = (2\pi/c)v$ is the wavevector. The latter is from de Broglie's equation $\varepsilon = |\mathbf{p}|c = hv$ where $\mathbf{p} = \hbar\mathbf{k}$ is the photon momentum.

15 In deriving Eq. (4.51), we use the formula $\int_0^\infty \frac{x^3 dx}{e^x - 1} = \frac{\pi^4}{15}$.

16 $\int_0^\infty \ln[1 - e^{-x}] x^2 dx = \frac{1}{3} \int_0^\infty \ln[1 - e^{-x}] dx^3 = \frac{1}{3} \ln[1 - e^{-x}] x^3 \Big|_0^\infty - \frac{1}{3} \int_0^\infty \frac{x^3 e^{-x} dx}{1 - e^{-x}} = -\frac{1}{3} \int_0^\infty \frac{x^3 dx}{e^x - 1} = -\frac{\pi^4}{45}$.

Accordingly, the pressure of photons can be derived from the thermodynamic relation

$$P = -\left(\frac{\partial F}{\partial V}\right)_T = \frac{\sigma}{3}T^4. \tag{4.55}$$

For noninteracting photons in thermal equilibrium with a container, the Gibbs energy vanishes, i.e., $G = F + PV = 0$. The result is expected from $\mu = (\partial F/\partial N)_{T,V} = 0$.

We may obtain other thermodynamic properties of photons following standard thermodynamic relations. For example, the specific entropy of noninteracting photons can be derived from Eqs. (4.52) and (4.54)

$$S/V = (U - F)/(TV) = 4\sigma T^3/3. \tag{4.56}$$

Eq. (4.56) conforms to the third law of thermodynamics, i.e., $S = 0$ at $T = 0$ K. What may not be intuitive is that the entropy of noninteracting photons is linearly proportional to the volume. At a fixed temperature, the constant entropy density of a photon gas is in stark contrast to the strong dependence of the entropy density of an ideal gas on the number density, $S^{IG}/V \sim -\rho \ln \rho$.

The average number of photons in the container can be obtained from the summation of the occupation numbers over all energy levels. Replacing the summation in Eq. (4.44) by integration over the photon frequency leads to

$$N = \sum_\alpha \langle n_\alpha \rangle = \int_0^\infty \frac{1}{e^{\beta h\upsilon} - 1} g(\upsilon) d\upsilon$$
$$= \frac{8\pi V}{c^3\beta^3h^3} \int_0^\infty \frac{x^2}{e^x - 1} dx = \frac{16\pi\zeta(3)k_B^3 V}{c^3h^3} T^3 \tag{4.57}$$

where $\zeta(3) \approx 1.2021$ according to the Riemann zeta function.[17] Eq. (4.57) indicates that the number density of photons is completely determined by the absolute temperature

$$\rho = N/V = \frac{16\pi\zeta(3)k_B^3}{c^3h^3} T^3 \approx 2.0287 \times 10^7 T^3 \tag{4.58}$$

where ρ has the units of $1/\mathrm{m}^3$ and T is in kelvin. From the photon density and the specific internal energy, we can obtain the average energy per photon

$$U/N = \frac{\sigma c^3 h^3}{16\pi\zeta(3)k_B^3} T \approx 2.7012 k_B T. \tag{4.59}$$

Interestingly, the average internal energy per photon is about 1.8 times the kinetic energy of a classical particle at the same temperature, $3k_BT/2$.

4.3.4 Spectral Energy of Photons

The discussion above shows that, based on statistical mechanics, we can deduce the thermodynamic properties of light. However, practical applications such as thermal radiation or solar energy conversion are often concerned with the properties of light with a specific range of frequencies. The spectral properties can be conveniently described in terms of the density of states (DOS).

According to Eq. (4.51), the internal energy includes contributions from photons over the entire range of frequencies. For photons with frequency between ν and $\nu + d\nu$, the internal energy per

---

17 The Riemann zeta function is defined as $\zeta(s) = \frac{1}{\Gamma(s)} \int_0^\infty x^{s-1}/(e^x - 1)dx$ where $\Gamma(s)$ stands for the Gamma function. For any positive integer n, $\Gamma(n) = (n - 1)!$.

unit volume u_υ can be deduced from the energy per photon, the probability density and the density of states

$$U/V = \frac{1}{V}\int_0^\infty \frac{h\upsilon}{e^{\beta h\upsilon}-1}g(\upsilon)d\upsilon \equiv \int_0^\infty u_\upsilon d\upsilon \tag{4.60}$$

which gives

$$u_\upsilon = \frac{8\pi}{c^3}\frac{h\upsilon^3}{e^{\beta h\upsilon}-1}. \tag{4.61}$$

In terms of the angular frequency $\omega = 2\pi\upsilon$ or wavelength $\lambda = c/\upsilon$, Eq. (4.60) may be expressed as, respectively,

$$u_\omega = \frac{\hbar\omega^3}{\pi^2c^3}\frac{1}{e^{\beta\hbar\omega}-1}, \tag{4.62}$$

$$u_\lambda = \frac{8\pi hc}{\lambda^5}\frac{1}{e^{\beta hc/\lambda}-1} \tag{4.63}$$

where we have used $d\upsilon = d\omega/2\pi = -(c/\lambda^2)d\lambda$ and $\hbar = h/2\pi$. Eqs. (4.61)–(4.63) are different forms of the Planck law for the distribution of the photon energy. It should be noted that u_λ, u_ω, and u_υ are different functions and have different units. While u_λ has the units of energy per volume per unit wavelength, u_ω and u_υ have the units of energy time per unit volume.

At low frequency (or high temperature), $\beta h\upsilon \ll 1$, the exponential term in Eq. (4.61) may be approximated by $e^{\beta h\upsilon} \approx 1 + \beta h\upsilon$ and thus

$$u_\upsilon \approx \frac{8\pi\upsilon^2}{\beta c^3}. \tag{4.64}$$

Eq. (4.64) is known as the Rayleigh–Jeans law, which was originally established from the classical laws of physics.[18]

If Eq. (4.64) is integrated with respect to υ, the internal energy density would diverge, i.e.,

$$U/V = \int_0^\infty u_\upsilon d\upsilon = \frac{8\pi}{\beta c^3}\int_0^\infty \upsilon^2 d\upsilon = \infty. \tag{4.65}$$

Apparently, the infinite internal energy is unphysical because, at any finite temperature, the system energy must be finite. Historically, the contribution to the internal energy from high frequencies was known as the *ultraviolet catastrophe*,[19] which played an important role in the early development of quantum mechanics. The divergence of the internal energy density is avoided by the quantization of the photon energy, a seminal concept proposed by Planck in the beginning of twentieth century.

Figure 4.8 shows the spectral energy density predicted by Eq. (4.62) at three temperatures. At each temperature, u_λ vanishes at both low and high wavelengths while exhibits a maximum at an intermediate wavelength. The wavelength corresponding to the maximum internal energy density can be obtained from $du_\lambda/d\lambda = 0$. Note that

$$u_\lambda = \frac{8\pi}{\beta^5h^4c^4}\frac{(\beta hc/\lambda)^5}{e^{\beta hc/\lambda}-1} = \frac{8\pi}{\beta^5h^4c^4}\frac{x^5}{e^x-1} \tag{4.66}$$

---

18 Eq. (4.64) can be directly obtained from the equal partition theorem for classical systems, i.e., the number of vibration modes with the frequency between υ and $\upsilon + d\upsilon$, is $8\pi\upsilon^2 d\upsilon/c^3$ according to Eq. (4.51), and the energy for each mode is $k_BT = \beta^{-1}$.

19 The phrase refers to the fact that, when applied to thermal radiation, the Rayleigh–Jeans law predicts experimental results accurately at low radiative frequencies, but it begins to depart from empirical observations as the frequency reaches the ultraviolet region of the electromagnetic spectrum.

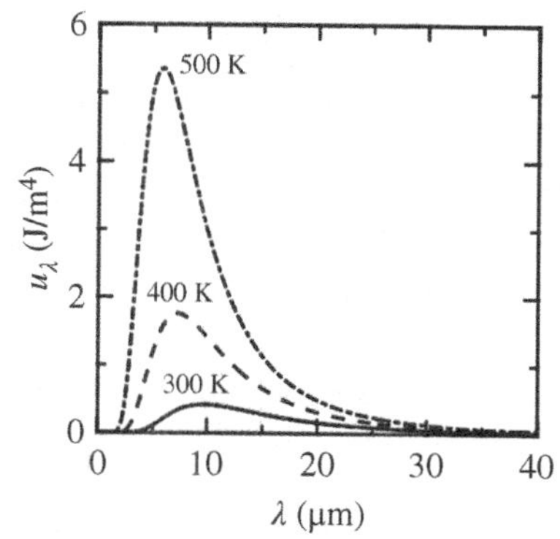

Figure 4.8 The spectrum of thermal radiation according to Eq. (4.62). As predicted by Wien's law, the peak of emission occurs at a wavelength inversely proportional to absolute temperature (in kelvins), i.e., $\lambda_u^{max} \approx 2.8978 \times 10^{-3}/T$ (in meters).

where $x \equiv \beta hc/\lambda$. The maximum u_λ occurs when

$$\frac{d}{dx}\left(\frac{x^5}{e^x - 1}\right) = 0. \tag{4.67}$$

We may solve Eq. (4.67) numerically to find a single solution of $x \approx 4.9651$. Accordingly, the wavelength at maximum u_λ is

$$\lambda_u^{\max} = \beta hc/x \approx 2.8978 \times 10^{-3}/T. \tag{4.68}$$

where $\lambda_u^{\max}$ has the units of meters, and temperature T is in K. Eq. (4.68) is known as Wien's displacement law, discovered experimentally before the advent of quantum mechanics.

4.3.5 Summary

We see from this section that, much like a regular gas, a photon gas has thermodynamic properties. Different from an ideal gas of classical particles, however, light has no mass and thus its thermodynamic properties are quite unique. For example, noninteracting photons in thermal equilibrium with the surroundings have a zero chemical potential. The specific internal energy, specific Helmholtz energy, and pressure are all proportional to the fourth power of the absolute temperature, while the entropy and the number density of photons are proportional to the third power of the absolute temperature. As discussed in Section 4.4, the thermodynamic properties of photons play an important role in many engineering applications including solar energy conversion, light emitting diodes, and photosynthesis.

4.4 Radiation and Solar Energy Conversion

The thermodynamic approach to radiation has been routinely used in atmosphere sciences, clinical applications, and diverse fields of radiative energy transfer, i.e., the conversion of the electromagnetic energy into electrical energy or chemicals. Because of recent interests in the development of renewable energy from solar power, the literature on the thermodynamics of solar energy conversion has been vast and keeps on growing.[20] In this section, we discuss radiation from thermodynamic perspective with a focus on its connection with the equations derived from statistical mechanics.

20 For an introductory overview, see for example, Crabtree G. W. and Lewis N. S. "Solar energy conversion", *Phys. Today* 60 (3), 37 (2007).

4.4.1 Thermal Radiation

Thermal radiation is a universal property of all materials. In general, the emission and absorption of light depends on temperature, the radiation frequency as well as the microscopic details of the material-photon interactions. While some materials emit or absorb electromagnetic waves more efficiently than others, the blackbody model assumes that the radiation efficiency is 100% for electromagnetic waves of arbitrary frequency or wavelength. For a real material, the emissivity ε_λ is defined as the ratio of emitted energy to the amount that would be radiated if the material were a blackbody at wavelength λ. Conversely, the absorptivity α_λ is defined as the ratio of the energy absorbed by the material to the energy of the incident radiation at a specific wavelength. The Kirchhoff's law of thermal radiation asserts $\varepsilon_\lambda = \alpha_\lambda$ for all materials at thermodynamic equilibrium, i.e., the absorption and emission of photons are balanced at any wavelength.

Although the emission and absorption efficiency of light is far from perfect for real materials, the blackbody model provides a convenient starting point to understand thermal radiation without entailing atomic details. To establish a connection between radiation and the thermodynamic properties of photons, consider a differential area dA at a blackbody surface as shown schematically in Figure 4.9. At equilibrium, the emission and absorption of photons are balanced, i.e., photons are absorbed and emitted from the surface at the same rate without any energy loss. For photons of arbitrary wavelength, the flux of the radiation energy from differential area dA over time dt is the same as the energy of photons adsorbed by the same surface

$$I(\lambda)dAdt = \frac{u_\lambda}{4\pi}\int_0^{2\pi} d\phi \int_0^{\frac{\pi}{2}} \sin\theta d\theta cdt \cos\theta dA \tag{4.69}$$

where $I(\lambda)$ stands for the radiation intensity or the spectral irradiance, i.e., the spectral radiation energy per unit area per unit time for photons with wavelength λ, u_λ is the internal energy density of the photons, c is the speed of light, and cdt is the distance traveled by the photons. The integration over the solid angles ϕ and θ accounts for emission over all directions leaving the surface, i.e., $0 \le \phi \le 2\pi$ and $0 \le \theta \le \pi/2$, and 4π is a normalization constant for the solid angles. In writing Eq. (4.69), we assume that the radiation form the blackbody surface is uniform in all directions.

Upon substituting the spectral energy density derived from Section 4.3, Eq. (4.63), into Eq. (4.69) and integrating off the solid angles, we obtain an analytical expression for the spectral irradiance[21]

$$I(\lambda) = \frac{c}{4}u_\lambda = \frac{2\pi hc^2}{\lambda^5}\frac{1}{e^{\beta hc/\lambda} - 1}. \tag{4.70}$$

Eq. (4.70) is known as Planck's law of blackbody radiation. It indicates that the spectral radiation intensity is exclusively determined by temperature and the photon wavelength.

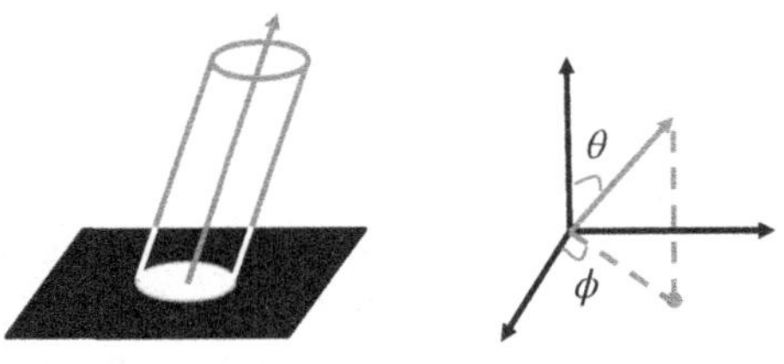

Figure 4.9 Radiation from a differential area dA of a blackbody surface. In the direction specified by solid angles (θ, ϕ), the emission energy over time dt is given by the energy of photons contained in a cylinder with volume $cdt \cos\theta dA$, where c is the speed of light. For uniform radiation, $d\phi \sin\theta d\theta/4\pi$ represents the probability of radiation for the solid angle within the range of $[\theta, \theta + d\theta]$ and $[\phi, \phi + d\phi]$.

21 Without the factor of π, the equation would define the radiance (or surface brightness), which has the units of power per unit area per steradian (the unit of solid angle).

Integration of the spectral radiation intensity over the entire range of the wavelength leads to the Stefan-Boltzmann law for thermal radiation:

$$I = \int_0^{\infty} I(\lambda)d\lambda = \frac{c}{4} \cdot \frac{8\pi^5}{15(hc)^3\beta^4} = \Theta T^4 \tag{4.71}$$

where $\Theta = 2\pi^5 k_B^4/(15h^3c^2) = 5.67 \times 10^{-8}$ W/(m^2 K^4) is the Stefan–Boltzmann constant. While Θ was originally determined from experiment, the statistical-mechanical derivation provides physical insights into its microscopic origin. More importantly, the theoretical analysis reveals the quantum nature of photons. Historically, the validation of the Planck's law for photon energy with the radiation energy opened a gateway to the development of quantum mechanics early in the twentieth century.

From a practical perspective, one of the most important applications of Eq. (4.70) is that it allows us to understand the spectrum of light from solar radiation, which empowers virtually everything on the Earth. As shown in Figure 4.10, the blackbody model provides an excellent representation of solar radiation by using an effective surface temperature of $T = 5778$ K. Interestingly, the irradiance peaks within the range of the visible wavelengths, which has no doubt played an important role in the biological evolutions of photosynthesis and animal visions. The difference between the solar spectrum above Earth's atmosphere and that at the sea level reflects light absorption by gas molecules in the air (and by reflection or albedo effect). For example, the missing radiation at the ultraviolet end of the spectrum (i.e., irradiance below 300 nm) is attributed to the absorption by the ozone (O_3) layer. Because ozone depletion would lead to enhanced ultraviolet light radiation that is harmful to plants and animals, the United Nations General Assembly (UNGA) designated September 16 as the International Day for the Preservation of the Ozone Layer.

4.4.2 Thermodynamic Limits of Solar Energy Conversion

Like any energy conversion process, the thermodynamic laws set a physical limit to the efficiency of solar energy conversion. To identify such a generic constraint, Figure 4.11 shows a schematic representation of a solar energy conversion device where the light irradiance I_{in} is balanced by a net radiative energy flux I_{out}, plus heat flux q and power generation w per unit area of radiation.

In a steady-state operation, the first law predicts that change in the radiation energy is balanced by the rate of heat and work per unit area

$$I_{in} = I_{out} + q + w. \tag{4.72}$$

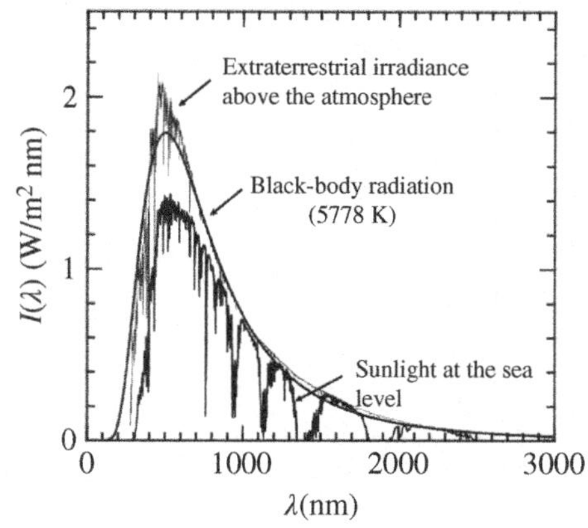

Figure 4.10 The spectrum of solar radiation at the sea level (black, ASTM G-173-03 data from NREL) and that at the top of the Earth's atmosphere (grey line). The solid line is predicted from Eq. (4.70) for the blackbody radiation at $T = 5778$ K.

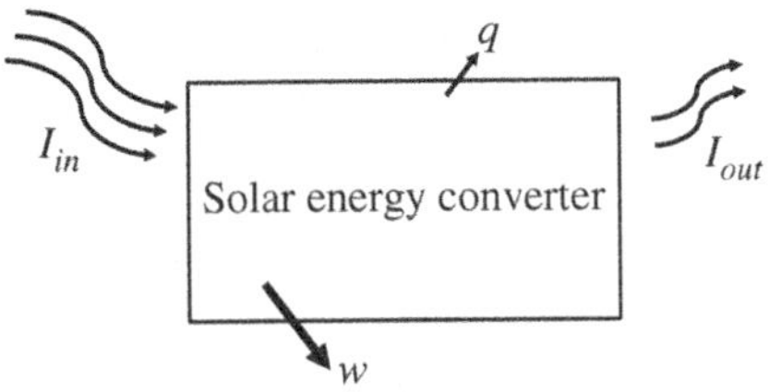

Figure 4.11 The spectrum of exergy from solar radiation at the sea level and the thermodynamic efficiency of solar energy conversion.

The second law states that the entropy generation must be nonnegative, i.e., the rate of entropy flux must include a nonnegative entropy production

$$S_{in} = S_{out} + q/T_0 - S_{gen} \tag{4.73}$$

where S stands for the entropy flux associated with the photons, $S_{gen} \geq 0$ is the rate of entropy generation per unit irradiation area of the solar energy conversion device, and T_0 is the temperature of the surroundings. Combining the expressions for the first and second laws leads to an upper limit for the available work from the solar energy conversion

$$w \leq (I_{in} - I_{out}) - T_0(S_{in} - S_{out}) \equiv \vartheta \tag{4.74}$$

where ϑ is called exergy. Eq. (4.74) is applicable to any steady-state process; it provides a generic limit on the maximum work available from the solar energy.

We can obtain the spectral and total radiation entropies per unit area per unit time following a procedure similar to that used for the internal energy. Similar to the spectral irradiance given by Eq. (4.70), the spectral entropy of radiation is given by (Problem 4.16)

$$S(\lambda) = \frac{c}{4} s_\lambda = \frac{2\pi k_B c}{\lambda^4} \left[\frac{\beta hc/\lambda}{e^{\beta hc/\lambda} - 1} - \ln(1 - e^{-\beta hc/\lambda}) \right], \tag{4.75}$$

and the integration of $S(\lambda)$ with respect to the wavelength gives the total radiation entropy

$$S = \frac{c}{4} \times (S/V) = 4\Theta T^3/3. \tag{4.76}$$

Assuming that I_{out} corresponds to the radiative energy flux of a blackbody at T_0, we can readily calculate the maximum available work as a function of the photon frequency.

Figure 4.12 presents the spectral exergy and the thermodynamic efficiency based on the blackbody model for solar radiation. Here, the thermodynamic efficiency is defined as the ration of the

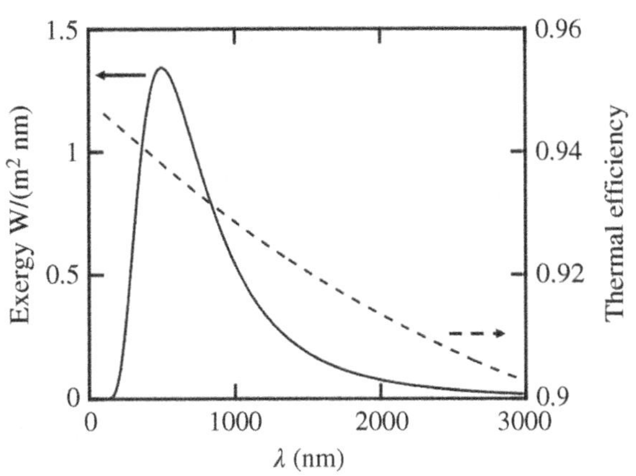

Figure 4.12 The spectrum of exergy from solar radiation at the sea level and the thermodynamic efficiency of solar energy conversion.

exergy $\vartheta(\lambda)$ and the inlet radiation energy $I_{in}(\lambda)$

$$\eta_T(\lambda) = \frac{\vartheta(\lambda)}{I_{in}(\lambda)}. \tag{4.77}$$

Interestingly, the exergy profile virtually coincides with the solar irradiance. While the thermodynamic efficiency for the solar energy conversion is larger than 90% over the entire range of frequency, it declines noticeably as the wavelength increases. The efficiency is highest for photons with shortest wavelengths, indicating that the solar energy is almost completely convertible into other forms of useful energy.

Using $I = \Theta T^4$ and $S = 4\Theta T^3/3$, we may also evaluate the thermodynamic efficiency for the entire spectrum of the solar energy (Problem 4.22)

$$\eta_T = \frac{\vartheta}{I_{in}} = 1 - \frac{4}{3}\left(\frac{T_0}{T}\right) + \frac{1}{3}\left(\frac{T_0}{T}\right)^4. \tag{4.78}$$

Eq. (4.78) is known as Petela's formula.[22] With the Earth's surface temperature set as $T_0 = 288$ K, Petela's formula predicts that thermodynamic efficiency of solar energy conversion is about 93%.

4.4.3 Spectrum Loss

The above analysis indicates that the solar energy can be nearly entirely converted into other forms of useful energy. However, the maximum efficiency of solar energy conversion under terrestrial conditions is significantly smaller, typically less than 31%. The disparity between the generic thermodynamic analysis and experimental observations primarily stems from spectrum loss, i.e., there exists a threshold wavelength λ_g beyond which the energy of photons is unable to drive electronic processes of energy conversion or photoreactions. The concept of spectrum loss was first introduced by Trivich and Flinn,[23] and independently by Shockley and Queisser,[24] in the context of the thermodynamic efficiency of p–n junction solar cells. The thermodynamic analysis is equally applicable to virtually all solar energy conversion processes, including photosynthesis and photovoltaics.

For a p–n junction with the bandgap energy ε_g, the threshold wavelength of photons is

$$\lambda_g = hc/\varepsilon_g. \tag{4.79}$$

According to the so-called *ultimate efficiency hypothesis*,[25] each photon with energy greater than ε_g is able to produce one electronic charge. Otherwise, the photon energy cannot be converted by the photoelectronic device. Therefore, the maximum electrical power (P_{max}) generated by the solar radiation, which is often referred to as the Shockley–Queisser (SQ) limit, is equal to the product of bandgap energy ε_g and the flux of photons with wavelength $\lambda \leq \lambda_g$:

$$P_{max} = \varepsilon_g J_g \tag{4.80}$$

where the photon flux can be calculated from the spectral solar energy radiation

$$J_g = \int_0^{\lambda_g} \frac{\lambda I(\lambda)}{hc} d\lambda. \tag{4.81}$$

22 Petela R. "Exergy of undiluted thermal radiation", *Solar Energy* 74 (6), 469–488 (2003).
23 Trivich D. and Flinn P. "Maximum efficiency of solar energy conversion by quantum processes", *Solar energy research*, edited by Daniels F. and Duffie J., London: Thames and Hudson, p. 143 (1955).
24 Shockley W. and Queisser H. J., "Detailed balance limit of efficiency of p-n junction solar cells", *J. Appl. Phys.* 32 (3), 510–519 (1960).
25 The name "ultimate efficiency" is probably somewhat misleading, as it takes into account only the first law of thermodynamics. Here the 2nd law efficiency as given by Eq. (4.77) was not included. Because the thermodynamic efficiency is nearly perfect, the 2nd law efficiency has relatively minor effect on the overall energy conversion.

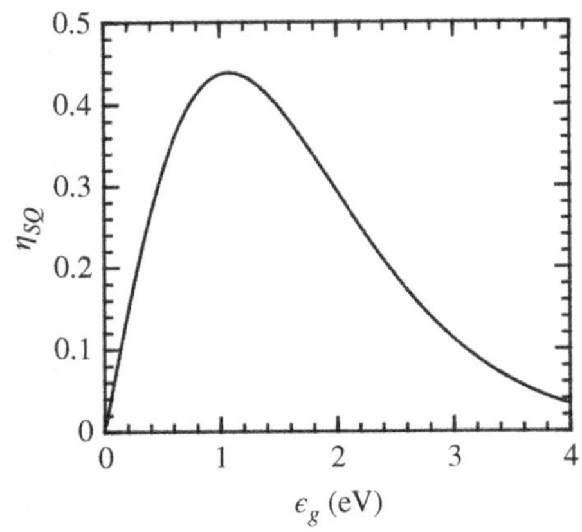

Figure 4.13 The ultimate efficiency for solar energy conversion by a single threshold solar cell versus the bandgap energy ε_g.

If $I(\lambda)$ is approximated by that corresponding to a blackbody radiation of temperature T, Eq. (4.81) can be integrated analytically to give[26]

$$J_g = \int_0^{\lambda_g} \frac{2\pi c}{\lambda^4} \frac{1}{e^{\beta hc/\lambda} - 1} d\lambda = \frac{2\pi}{\beta^3 h^3 c^2} f_g(\beta hc/\lambda_g) \tag{4.82}$$

where $f_g(x) = x^2 \mathrm{Li}_1(e^{-x}) + 2x\mathrm{Li}_2(e^{-x}) + 2\mathrm{Li}_3(e^{-x})$ and $x \equiv \beta hc/\lambda_g$.

Figure 4.13 shows the ultimate efficiency of solar energy conversion (Problem 4.19)

$$\eta_{SQ} \equiv P_{\max}/I = 15xf_g(x)/\pi^4, \tag{4.83}$$

as a function of the bandgap energy. Here, the flux of photons is calculated from Eq. (4.82) as predicted by the blackbody model for solar radiation ($T = 5778$ K). The maximum efficiency $\eta_{\max} \approx 0.4388$ takes place when the bandgap energy is $\varepsilon_g \approx 1.08$ eV, which is in excellent agreement with experimental observations.[27]

The existence of a threshold wavelength is equally applicable to other solar energy conversion processes, including photosynthesis. For both natural and artificial solar syntheses, the threshold wavelength can be predicted from the Gibbs energy of reaction associated with the solar energy conversion (ΔG°), the number of photons (n) involved in the chemical process, and the energy loss per photon (U_{loss})[28]

$$\lambda_g = \frac{hc}{\Delta G^\circ/n + U_{loss}}. \tag{4.84}$$

Accordingly, the efficiency of solar energy conversion may be estimated from

$$\eta = \frac{J_g \cdot (\Delta G^\circ/n)}{I_{in}} \eta_Q \tag{4.85}$$

where J_g and I_{in} represent the flux of solar photons with a wavelength below λ_g and the flux of the total solar energy, respectively; and η_Q stands for the quantum yield for converting the adsorbed solar energy into product. In writing Eq. (4.85), we assume that each photon contributes a free energy of $\Delta G^\circ/n$ to the photosynthesis. Like the SQ efficiency, Eq. (4.85) does not account for the second law efficiency.

The thermodynamic analysis is useful for the optimization of the solar energy conversion by tuning the pertinent physical parameters. To illustrate, Figure 4.14 shows the ideal maximum

---

26 $\int_x^\infty \frac{x^2}{e^x - 1} dx = x^2\mathrm{Li}_1(e^{-x}) + 2x\mathrm{Li}_2(e^{-x}) + 2\mathrm{Li}_3(e^{-x})$.
27 Ehrler B., et al., "Photovoltaics reaching for the Shockley–Queisser limit", *ACS Energy Lett.* 5, 3029–3033 (2020).
28 Bolton J. R., Haught A. F., and Ross R. T., "Chapter 11 – Photochemical energy storage: an analysis of limits", in *Photochemical conversion and storage of solar energy*, edited by Connolly J. S., Academic (1981).

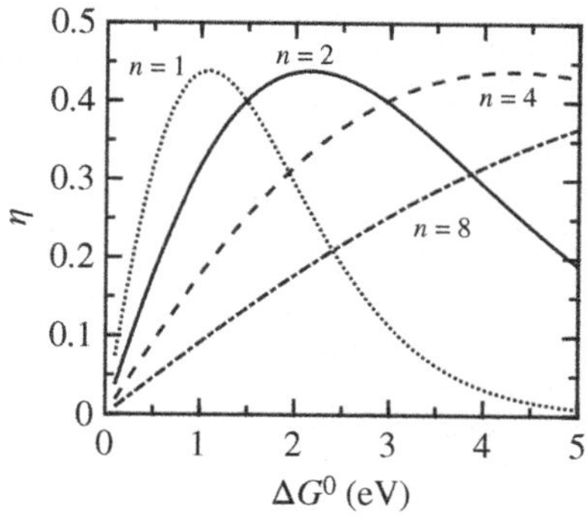

Figure 4.14 The efficiency of solar syntheses versus the Gibbs energy of reaction $\Delta G°$. Here n stands for the number of photons involved in the solar energy conversion.

efficiency predicted by Eq. (4.85) (viz., with the assumptions of $\eta_Q = 1$, $U_{loss} = 0$, and blackbody radiation for J_g). The numerical values may be discussed in the context of the chemical reaction for natural photosynthesis

$$H_2O(l) + CO_2(g) \xrightarrow{8h\nu} (CHOH) + O_2(g) \tag{4.86}$$

where (CHOH) represents one carbon equivalent of carbohydrate. According to Eq. (4.86), the reaction involves eight protons and the Gibbs energy of reaction is $\Delta G° \approx 468.8$ kJ ≈ 4.86 eV. Figure 4.14 predicts a maximum efficiency of 0.360, which is virtually identical to that obtained from much more sophisticated theoretical analysis.[29] Figure 4.14 indicates that the maximum efficiency for a single-threshold solar conversion process is $\eta_{max} \approx 0.4388$, which is close to the number (~0.46) calculated from solar irradiance data at terrestrial conditions. The optimal $\Delta G°$ depends on the number of photons involved in the chemical reaction, which is determined by the mechanistic details of the photosynthesis.

4.4.4 Thermodynamic Limits of Solar Fuel

Before closing this section, let's discuss another interesting application on the thermodynamics of thermal radiation. In light of the practical importance of solar fuels as the world is shifting away from fossil fuels to renewable energies, it is often desirable to examine the thermodynamic limits on the production of energy-rich compounds by the photochemical conversion and storage of solar energy. Among many reactions proposed for solar energy conversion, none has generated as much attention as the photolysis of liquid water

$$H_2O(l) \xrightarrow{h\nu} H_2(g) + \frac{1}{2}O_2(g). \tag{4.87}$$

Efficient and economic water splitting may have revolutionary impacts on chemical and energy industries to empower a green hydrogen economy.[30]

At the standard state, the Gibbs energy of reaction for water splitting is $\Delta G° \approx 237.14$ kJ/mol, which is the same as the Gibbs energy of formation for liquid water (but with an opposite sign). The formation of each hydrogen molecule requires the transfer of two electrons, implying that an electronic energy of $E° = \Delta G°/2F \approx 1.229$ eV, where $F = 96.485$ kJ/mol is the Faraday constant, must be supplied for the electron transfer. If the Gibbs energy of reaction is powered by a single photon, its wavelength must be no less than $\lambda_g = hc/\Delta G° \approx 504.4$ nm, assuming there is no photon energy loss.

29 Hill R. and Rich P.R., "A physical interpretation for the natural photosynthetic process", *PNAS* 80 (4), 978–982 (1983).
30 Bolton J. R., Strickler S. J., and Connolly J. S., "Limiting and realizable efficiencies of solar photolysis of water", *Nature* 316 (6028), 495–500 (1985).

Table 4.1 Solar conversion efficiencies and other related data for water photolysis with different photons per formation for each hydrogen molecule.

Scheme	n	U_{loss} (eV)	λ_g (nm)	η (%)
S1	1	0.49	420	5.3 (9.98)
S2	2	0.37	775	30.7 (29.1)
S4	4	0.31	1340	30.6 (28.7)

All numbers are from the literature[27] except those in the parentheses predicted from Eq. (4.85) with the blackbody radiation model for the incident light.

If the reaction involves two photons for the formation of each hydrogen molecule, the threshold wavelength would be $\lambda_g = 2hc/\Delta G^\circ \approx 1008.8$ nm. These values must be modified if we consider the photon energy loss. The unused energy per photon is estimated to be $U_{loss} \approx 0.37$ eV. Accordingly, Eq. (4.84) predicts that the threshold wavelength $\lambda_g \approx 438.4$ and 775 nm for the single-photon and 2-photon water splitting reactions, respectively.

Table 4.1 presents the efficiencies of water photolysis for various schemes (with the assumption of ideal quantum yield $\eta_Q = 1$). The thermodynamic analysis is helpful for understanding the mechanisms of catalytic reactions and the design of efficient chemical processes for solar fuel synthesis.

4.4.5 Summary

This brief discussion of radiation and solar energy conversion illustrates both the general principles and specific applications of thermodynamics. With the assumption of perfect emission and absorption efficiency, the blackbody model is universally applicable to radiation processes for any materials. Remarkably, both radiation spectrum and its conversion to other forms of energy can be analyzed quantitatively in terms of the thermodynamic properties of noninteracting photons (viz., photon gas). While the thermodynamics of light seems exceedingly abstract, statistical mechanics plays an essential role not only in predicting the energy of thermal radiation but also in the analysis of the spectral energy and entropy densities that are intrinsically associated with the microstates of photons. In this section, we also see that statistical thermodynamics is useful for the analysis of specific chemical and energy conversion processes including plant photosynthesis and water splitting. Although the thermodynamic analysis does not account for many chemical details thus provides only the first-order approximation, it nonetheless offers valuable guidance for the engineering development of solar technology.

4.5 The Free-Electron Model of Metals

The free-electron model of metals, proposed by Arnold Sommerfeld in 1927,[31] has been instrumental in advancing our understanding of the electronic properties of metallic systems. The

31 Arnold Sommerfeld pioneered the developments of atomic and quantum physics. He is also known as the PhD or postdoctoral supervisor for a number of Nobel Prize winners including Werner Heisenberg, Wolfgang Pauli, Peter Debye, and Hans Bethe.

Sommerfeld theory neglects electron–electron interactions but incorporates the Pauli exclusion principle, which states that no two electrons can occupy the same quantum state. Additionally, it approximates the interaction of electrons with the ionic lattice of a metallic material by a uniform background, thereby allowing free electrons to be described as a Fermi gas. Despite its simplicity, the Sommerfeld theory provides surprisingly accurate descriptions of the electronic properties of certain metallic systems, especially sp-bonded monovalent metals like sodium and potassium. Moreover, the model provides a nonintuitive understanding of the electronic properties of metallic systems and serves as a useful reference for describing solid materials important for practical applications.

4.5.1 The Density of Free Electrons

We may estimate the average density of free electrons in a metal from its mass density and the number of valence electrons for each atom. For example, at ambient conditions, the mass density of copper is around 8.92×10^3 kg/m^3. Assuming one free electron for each copper atom, we find the average number density of free electrons from its atomic weight 63.55 and the atomic mass constant 1.66×10^{-27} kg

$$\rho = \frac{1 \times 8.92 \times 10^3}{63.55 \times 1.66 \times 10^{-27}} = 8.46 \times 10^{28} \ \text{m}^{-3}. \tag{4.88}$$

Eq. (4.88) indicates that the number density of free electrons in a metal is a few thousand times greater than that corresponding to a conventional ideal gas at $P = 1$ atm and $T = 298.15$ K, $\rho = P/k_BT \approx 2.45 \times 10^{25}$ m^{-3}.

For a metallic system, the number density of free electrons is often expressed in terms of the Wigner-Seitz radius, i.e., the radius of a sphere whose volume is the same as the volume per electron in the macroscopic system

$$r_s \equiv \left(\frac{3}{4\pi\rho}\right)^{1/3}. \tag{4.89}$$

Because the Wigner–Seitz radius is related to the density of the crystal, it is often used to calculate the lattice constant or vice versa. The Wigner–Seitz radius is also used in the calculation of the electronic properties of crystals, such as the band structure and the density of states. For the electron density given by Eq. (4.88), the Wigner–Seitz radius is $r_s = 1.39$ Å. The number is only slightly larger than the atomic radius of copper (~1.35 Å), suggesting that the metal atoms are tightly packed.

Figure 4.15 shows the Wigner–Seitz radii for a number of metals.[32] Here, the numerical values are presented in terms of the Bohr radius $a_0 = 0.529$ Å, which is often adopted as a unit length in atomic physics. For the same group of metals in the periodic table, the density of free electrons increases with the atomic size. Interestingly, those metals with the largest valence electrons per atom (e.g., Bi and Sb) have relatively small free electron density. For all metals shown in Figure 4.15, the Wigner–Seitz radius varies from about two to six times the Bohr radius, suggesting that the number density of free electrons is about the same order of magnitude in all bulk metals.

32 The numerical results are obtained from Ashcroft N. W. and Mermin N. D., *Solid state physics*. Saunders, 1976.

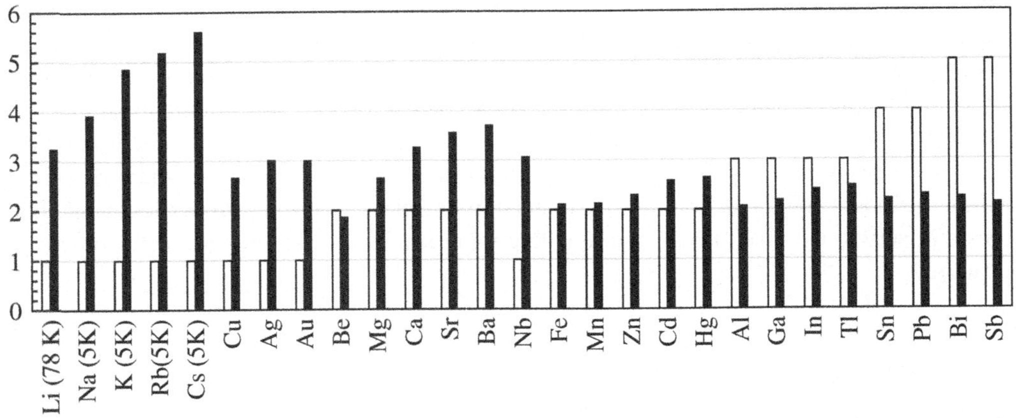

Figure 4.15 The Wigner–Seitz radius (in units of Bohr radius $a_0 = 0.529$ Å, solid bars) and the number of valence electrons per atom (open bars) for some selected metals near room temperature (or otherwise specified).

4.5.2 Translational Symmetry

Unlike gases and liquids, a metal has a crystalline structure manifested in terms of the periodic arrangement of atoms on a three-dimensional (3D) lattice. Although the atomic structure is inconsequential to the free-electron model, it is instructive to understand the implication of the crystalline symmetry on the electronic structure and thermodynamic properties before applying the Fermi–Dirac statistics for the microstate distribution of free electrons.

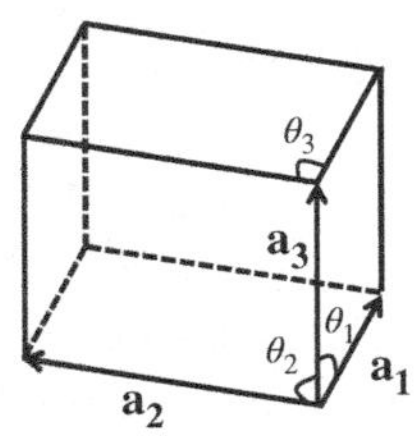

Figure 4.16 In a 3D crystal, the primitive vectors, $\mathbf{a}_1$, $\mathbf{a}_2$, and $\mathbf{a}_3$, are defined by three shortest arrowed lines connecting equivalent atoms. The space spanned by these vectors consists of a primitive unit cell (PUC) which has a volume of $V_{PUC} = (\mathbf{a}_1 \times \mathbf{a}_2) \cdot \mathbf{a}_3$. PUC is able to reproduce the entire structure of a crystal by only translational displacements. The PUC volume depends on the lengths and inclination angles of the primitive vectors, θ_1, θ_2, and θ_3.

Regardless of the complexity of atomic details, all crystals adhere to translational symmetry. Mathematically, the translational symmetry in the atomic coordinates of a crystal can be described in terms of the Bravais lattice,[33] i.e., a set of points in space defined by three primitive vectors $\mathbf{a}_1$, $\mathbf{a}_2$, and $\mathbf{a}_3$

$$\mathbf{R} = n_1\mathbf{a}_1 + n_2\mathbf{a}_2 + n_3\mathbf{a}_3 \tag{4.90}$$

where n_1, n_2, and n_3 are integers. As shown in Figure 4.16, the primitive vectors are associated with the primitive unit cell (PUC), i.e., the smallest space unit that can be used to reproduce the entire crystalline structure only by translational displacements.

The translational periodicity of a crystal implies that any local property of the system must satisfy the periodic boundary conditions

$$f(\mathbf{r} + \mathbf{R}) = f(\mathbf{r}). \tag{4.91}$$

Taking the Fourier transform on both sides of Eq. (4.91) gives

$$\begin{aligned} \int d\mathbf{r} e^{-i\mathbf{k}\cdot\mathbf{r}} f(\mathbf{r} + \mathbf{R}) &= \int d\mathbf{r} e^{-i\mathbf{k}\cdot(\mathbf{r}-\mathbf{R})} f(\mathbf{r}) \\ &= \int d\mathbf{r} e^{-i\mathbf{k}\cdot\mathbf{r}} f(\mathbf{r}) \equiv \hat{f}(\mathbf{k}). \end{aligned} \tag{4.92}$$

33 It is named after A. Bravais, a nineteenth century mathematical physicist who defined 14 unique crystalline lattices for three-dimensional systems.

Rearranging Eq. (4.92) leads to

$$\hat{f}(\mathbf{k})(e^{i\mathbf{k}\cdot\mathbf{R}} - 1) = 0 \tag{4.93}$$

which indicates that $\hat{f}(\mathbf{k})$ is nonzero *only* at those values of $\mathbf{k}$ that satisfy

$$e^{i\mathbf{k}\cdot\mathbf{R}} = 1. \tag{4.94}$$

As vector $\mathbf{R}$ is defined by the primitive vectors, Eq. (4.90), those special values of $\mathbf{k}$ satisfying Eq. (4.94) can be used to define a reciprocal Bravais lattice

$$\mathbf{G} = m_1\mathbf{b}_1 + m_2\mathbf{b}_2 + m_3\mathbf{b}_3 \tag{4.95}$$

where m_1, m_2, and m_3 are also integers, $\mathbf{b}_1$, $\mathbf{b}_2$, and $\mathbf{b}_3$ are reciprocal primary vectors given by

$$\mathbf{b}_1 = \frac{2\pi(\mathbf{a}_2 \times \mathbf{a}_3)}{\mathbf{a}_1 \cdot (\mathbf{a}_2 \times \mathbf{a}_3)}, \mathbf{b}_2 = \frac{2\pi(\mathbf{a}_3 \times \mathbf{a}_1)}{\mathbf{a}_2 \cdot (\mathbf{a}_3 \times \mathbf{a}_1)}, \text{and } \mathbf{b}_3 = \frac{2\pi(\mathbf{a}_1 \times \mathbf{a}_2)}{\mathbf{a}_3 \cdot (\mathbf{a}_1 \times \mathbf{a}_2)}. \tag{4.96}$$

The primary vectors in the real space and those in the reciprocal space satisfy the mathematical relation

$$\mathbf{a}_i \cdot \mathbf{b}_j = 2\pi\delta_{ij} \tag{4.97}$$

where $i, j = 1, 2, 3$, and δ_{ij} is the Kronecker-delta function. It is easy to show the dot product of vectors $\mathbf{R}$ and $\mathbf{G}$ satisfies

$$\mathbf{R} \cdot \mathbf{G} = 2\pi I \tag{4.98}$$

with $I = n_1m_1 + n_2m_2 + n_3m_3$ is an integer. Accordingly, vectors from the real and reciprocal Bravais lattices are related by

$$e^{i\mathbf{G}\cdot\mathbf{R}} = 1. \tag{4.99}$$

Therefore, Eq. (4.98) provides the desired solution to Eq. (4.94). Because function $\hat{f}(\mathbf{k})$ is nonzero only at the positions defined by the reciprocal lattice, the inverse Fourier transform of $\hat{f}(\mathbf{k})$ leads to

$$f(\mathbf{r}) = \frac{1}{V}\sum_{\mathbf{G}} \hat{f}(\mathbf{G})\, e^{i\mathbf{G}\cdot\mathbf{r}}. \tag{4.100}$$

Eq. (4.100) suggests that any local property of a crystal can be uniquely determined by its Fourier transform at the reciprocal Bravais lattice.

4.5.3 The Wave Function of Free Electrons

In the free-electron model, we use the single-particle Schrödinger equation to describe the quantum states of individual electrons

$$-\frac{\hbar^2}{2m}\nabla^2\psi(\mathbf{r}) = \varepsilon\psi(\mathbf{r}). \tag{4.101}$$

where m represents the electron rest mass,[34] $\hbar = h/2\pi$ is the reduced Planck constant, and ε is the electronic energy. Different from that for "a particle in a box," the single-electron wave function for a crystalline system satisfies the periodic boundary condition

$$\psi(\mathbf{r} + \mathbf{R}) = \psi(\mathbf{r}). \tag{4.102}$$

34 In some physics texts, the electron mass is denoted as m_e^*, the effective mass of one electron, to emphasis that a free electron never exists.

As we see from the following, Eq. (4.102) leads to a solution of the Schrödinger equation that is quite different from that for a particle in a box.

Applying the Fourier transform to both sides of Eq. (4.101) yields

$$\frac{\hbar^2 k^2}{2m}\widehat{\psi}(\mathbf{k}) = \varepsilon\widehat{\psi}(\mathbf{k}) \tag{4.103}$$

where $\mathbf{k}$ is the wavevector, and $k = |\mathbf{k}|$. Rearranging Eq. (4.103) gives

$$\left(\frac{\hbar^2 k^2}{2m} - \varepsilon\right)\widehat{\psi}(\mathbf{k}) = 0, \tag{4.104}$$

which suggests that $\widehat{\psi}(\mathbf{k})$ is nonvanishing only when the single-particle energy is

$$\varepsilon(\mathbf{k}) = \frac{\hbar^2 k^2}{2m}. \tag{4.105}$$

Eq. (4.104) suggests that $\widehat{\psi}(\mathbf{k})$ is proportional to the Dirac-delta function $\delta(\mathbf{k})$. Its inverse Fourier transform can be written as

$$\psi(\mathbf{r}) = \frac{e^{i\mathbf{k}\cdot\mathbf{r}}}{\sqrt{V}} \tag{4.106}$$

where the proportionality constant $1/\sqrt{V}$ is obtained by imposing the normalization condition

$$\int d\mathbf{r}|\psi(\mathbf{r})|^2 = 1. \tag{4.107}$$

Eq. (4.107) follows the fact that the probability of finding an electron in volume V is unity.

To fix the possible values of $\mathbf{k}$ in the single-particle wave function, we need to use the periodic boundary condition, Eq. (4.102). Substituting Eq. (4.106) into (4.102) yields

$$e^{i\mathbf{k}\cdot(\mathbf{r}+\mathbf{R})} = e^{i\mathbf{k}\cdot\mathbf{r}} \tag{4.108}$$

which is equivalent to

$$e^{i\mathbf{k}\cdot\mathbf{R}} = 1. \tag{4.109}$$

Eq. (4.109) indicates that wavevector $\mathbf{k}$ coincides with the vectors defined by the reciprocal Bravais lattice. The smallest nonzero value of $\mathbf{k}$ is determined by the maximum of vector $\mathbf{R}$. A combination of Eqs. (4.90) and (4.109) predicts

$$\mathbf{k} = \left(\frac{2\pi n_x}{L_x}, \frac{2\pi n_y}{L_y}, \frac{2\pi n_z}{L_z}\right), \quad n_{x,y,z} = 0, \pm 1, \pm 2, \ldots. \tag{4.110}$$

where L_x, L_y, and L_z are the dimensionalities of the entire system along the three directions of the primitive translational vectors $\mathbf{a}_1$, $\mathbf{a}_2$, and $\mathbf{a}_3$ as shown in Figure 4.16.

We may compare the wave function that satisfies the periodic boundary condition with that corresponding to a particle in a box by considering a one-dimensional (1D) system of length L_x. With the periodic boundary condition, k_x takes discrete values of $2\pi n_x/L_x$ with $n_x = 0, \pm 1, \pm 2, \ldots$. For each k_x, the wave function $\varphi(x) = e^{ik_x x}/\sqrt{L_x}$ predicts a uniform electron density of $1/L_x$. For an electron confined in a one-dimensional box, however, the possible values of k_x become $\pi n_x/L_x$ with $n_x = 1, 2, \ldots$. In this case, the wave function is given by $\varphi(x) = \sqrt{2/L}\sin(k_x x)$, which yields a nonuniform density in the range $0 \leq x \leq L_x$, as shown in Figure 4.17. With the periodic boundary condition, n_x can be any integer including zero and negative values. However, for an electron confined in a one-dimensional box, n_x takes only positive integers because a negative n_x only reverses the sign

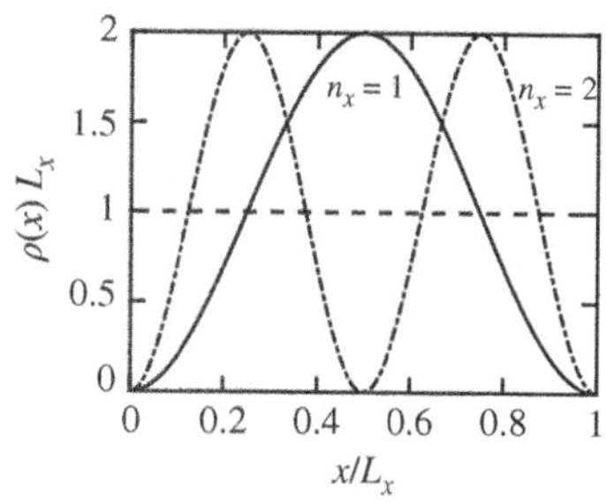

Figure 4.17 The local density of a free electron with periodic boundary condition (red solid line) and that confined in a 1D-box with the wavevector corresponding to $n_x = 1$ (black dashed line) and 2 (blue dash-dotted line).

of the wave function without changing the quantum state. Importantly, $n_x = 0$ is allowed with the periodic boundary condition but should be excluded for the case of a particle in a box because the latter would lead to $\varphi(x) = \sqrt{2/L}\sin(k_x x) = 0$. If a wave function is zero everywhere, it is not physically meaningful.

4.5.4 Thermodynamic Properties of Free Electrons

Based on the single-particle states predicted by the Schrödinger equation, we can derive the thermodynamic properties of free electrons following a procedure similar to that discussed in Section 4.2.3. For free electrons in a metal, each single-particle state is specified by the wavevector **k** given by Eq. (4.110) and the direction of electron spins. Because an electron has two possible values for its intrinsic angular momentum (spin ½), each energy level obtained from the Schrödinger equation can accommodate up to *two* electrons, one with angular momentum $\hbar/2$ and another $-\hbar/2$ as conventionally referred to as spin "up" and "down." Accordingly, the grand partition function for the entire system is given by

$$\Xi = \prod_{\alpha} [1 + e^{-\beta(\varepsilon_\alpha - \mu)}]^2. \tag{4.111}$$

where the summation applies to all possible energy states as specified by Eq. (4.105).

Similar to that for a fermion gas, the grand potential is given by[35]

$$\Omega = -2k_B T \sum_\alpha \ln\left[1 + z e^{-\beta\varepsilon_\alpha}\right] = -\frac{8k_B TV}{3\sqrt{\pi}\Lambda^3}\int_0^\infty dx \frac{x^{3/2}}{e^x/z+1} = \frac{2k_B TV}{\Lambda^3}\mathrm{Li}_{5/2}(-z). \tag{4.112}$$

In writing Eq. (4.112), we take the continuous limit of the wavevectors

$$\sum_\alpha = \sum_{n_x=-\infty}^{\infty}\sum_{n_y=-\infty}^{\infty}\sum_{n_z=-\infty}^{\infty} \approx \int_{-\infty}^{\infty}\frac{L_x}{2\pi}dk_x \int_{-\infty}^{\infty}\frac{L_y}{2\pi}dk_y \int_{-\infty}^{\infty}\frac{L_z}{2\pi}dk_z = \frac{V}{\vartheta\pi^3}\int_0^\infty dk_x \int_0^\infty dk_y \int_0^\infty dk_z \tag{4.113}$$

where $\vartheta = (1 - \cos^2\theta_1 - \cos^2\theta_2 - \cos^2\theta_3) + 2[\cos\theta_1 \cos\theta_2 \cos\theta_3]^{1/2}$ is a constant related to the inclination angles of the primitive vectors as shown in Figure 4.16. For a cubic lattice, the inclination angles are $\theta_1 = \theta_2 = \theta_3 = \pi/2$ so that $\vartheta = 1$. For simplicity, ϑ is dropped from Eq. (4.112) and in the following discussion because it has no effects on the thermodynamic properties of the system. As shown in Figure 4.18, the polylogarithm function varies smoothly with fugacity $z \equiv e^{\beta\mu}$

35 The integrals in Eq. (4.112) and in following equations are written as $\int_0^\infty dx \frac{x^{n-1}}{e^x/z+1} = -\Gamma(n)\mathrm{Li}_n(-z)$ where $\Gamma(n)$ is gamma function, and $\mathrm{Li}_n(-z)$ is polylogarithm function. These special functions can be conveniently calculated with standard math programs such Matlab.

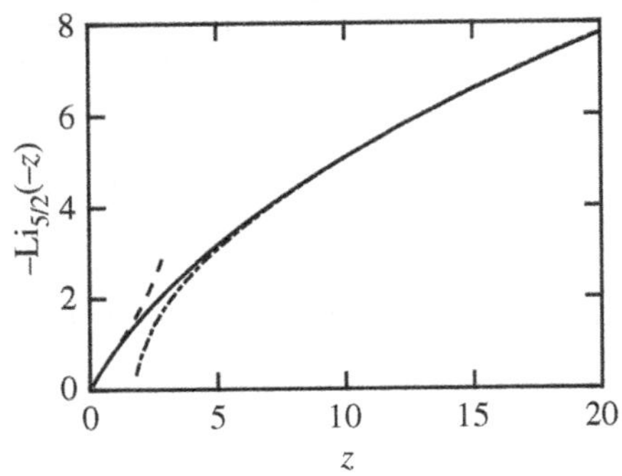

Figure 4.18 A polylogarithm function (solid line) and its approximations at small (dashed line) and large values of the independent variable (dash-dotted line) as given by Eq. (4.114).

and can be approximated by conventional functions at small (classical limit) and large values (low-temperature limit) of z:

$$-\text{Li}_s(-z) = \begin{cases} z - \dfrac{z^2}{2^s} + \dfrac{z^3}{3^s} - \cdots & z \to 0 \\ \dfrac{(\ln z)^s}{\Gamma(s+1)} + \dfrac{\pi^2}{6}\dfrac{(\ln z)^{s-2}}{\Gamma(s-1)} + \dfrac{7\pi^4}{360}\dfrac{(\ln z)^{s-4}}{\Gamma(s-3)} + \cdots & z \gg 1 \end{cases}. \tag{4.114}$$

From the grand potential, we can derive other thermodynamic functions following the standard equations of statistical mechanics. For a uniform system of free electrons, the pressure is related to the grand potential by

$$P = -\frac{\Omega}{V} = -\frac{2k_BT}{\Lambda^3}\text{Li}_{5/2}(-z). \tag{4.115}$$

A partial derivative of the reduced grand potential with respect to β gives the internal energy

$$U = \left(\frac{\partial \beta\Omega}{\partial \beta}\right)_{V,\beta\mu} = 2V\text{Li}_{5/2}(-z)\left(\frac{\partial \Lambda^{-3}}{\partial \beta}\right) = -\frac{3k_BTV}{\Lambda^3}\text{Li}_{5/2}(-z). \tag{4.116}$$

In deriving Eq. (4.116), we utilize the definitions of fugacity $z \equiv e^{\beta\mu}$ and the thermal wavelength $\Lambda \equiv h/\sqrt{2\pi m k_B T}$, which gives $(\partial \ln \Lambda^3/\partial\beta) = 3/(2\beta)$. A comparison of Eqs. (4.115) and (4.116) yields a linear relation between the pressure and the internal energy density

$$P = \frac{2U}{3V}. \tag{4.117}$$

Interestingly, Eq. (4.117) is valid for both classical and quantum systems of noninteracting particles.

Noting that $d\text{Li}_s(z)/dz = \text{Li}_{s-1}(z)/z$ and $(\partial z/\partial \mu)_\beta = \beta z$, we can find the ensemble average for the number electrons in the system

$$\begin{aligned} N &= -\left(\frac{\partial \Omega}{\partial \mu}\right)_{V,\beta} = \frac{2k_BTV}{\Lambda^3}\left(\frac{\partial \text{Li}_{5/2}(-z)}{\partial(-z)}\right)\left(\frac{\partial z}{\partial \mu}\right)_\beta \\ &= \frac{2k_BTV}{\Lambda^3}\left(\frac{1}{-z}\right)\text{Li}_{3/2}(-z)(\beta z) = -\frac{2V}{\Lambda^3}\text{Li}_{3/2}(-z). \end{aligned} \tag{4.118}$$

Thus, the average electron density is given by

$$\rho\Lambda^3 = -2\text{Li}_{3/2}(-z). \tag{4.119}$$

Finally, we can derive the system entropy from

$$S = (U - \Omega - N\mu)/T = \frac{k_BV}{\Lambda^3}[-5\text{Li}_{5/2}(-z) + (2\ln z)\text{Li}_{3/2}(-z)]. \tag{4.120}$$

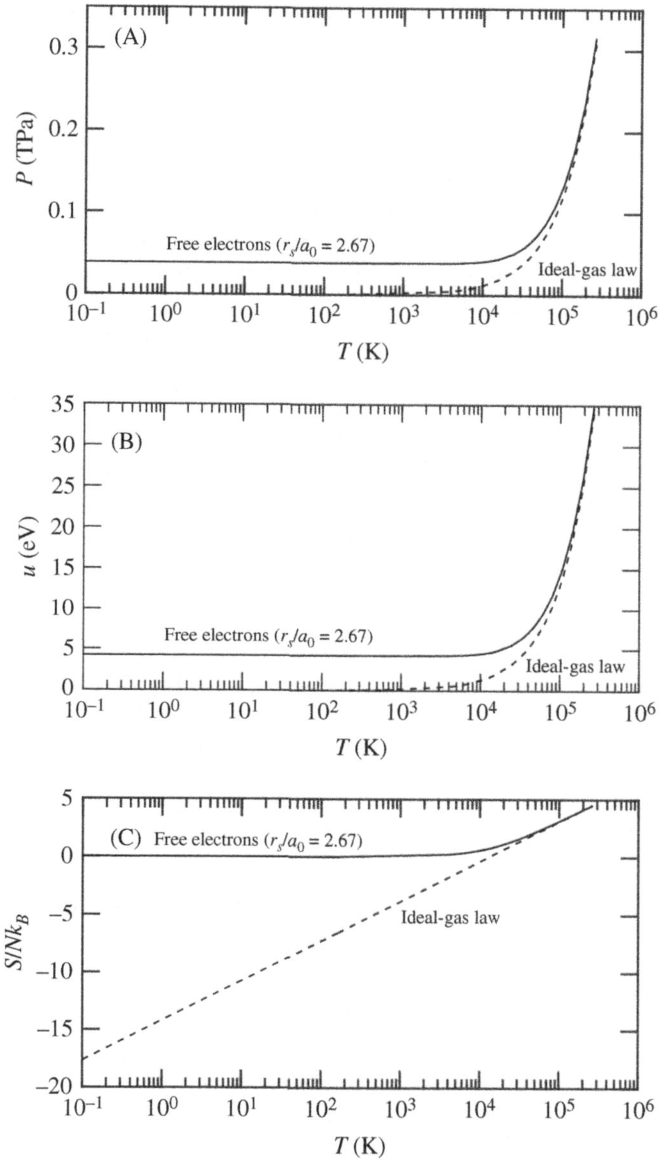

Figure 4.19 Thermodynamic properties of free electrons versus the absolute temperature predicted by the Sommerfeld free-electron theory. A. Pressure P; B. Internal energy per electron ($u = U/N$); and C. Reduced entropy S/Nk_B. Approximately, the reduced Wigner–Seitz radius $r_s/a_0 = 2.67$ corresponds to that of free electrons in copper. From comparison, the dashed lines present the results corresponding to a system of noninteracting classical particles (viz., ideal gas).

As shown in Figure 4.19, Eq. (4.120) predicts $S = 0$ at 0 K, consistent with the 3rd law of thermodynamics.

In the classical limit ($z \to 0$), $\mathrm{Li}_{5/2}(-z) = -z$. Because $\ln z = \beta\mu$, Eqs. (4.115)–(4.120) can be reduced to the familiar thermodynamic equations

$$P = \rho k_B T, \tag{4.121}$$

$$\beta\mu = \ln(\rho\Lambda^3/2), \tag{4.122}$$

$$S = -Nk_B[\ln(\rho\Lambda^3/2) - 5/2]. \tag{4.123}$$

Because of each electron can exist in two spin states, Eqs. (4.121)–(4.123) correspond to the pressure, chemical potential, and entropy for a system containing *two* types of noninteracting classical particles (viz., spin up and spin down) of the same density $\rho/2$ and thermal wavelength Λ.

While the mathematic procedure discussed above is quite straightforward, the thermodynamic properties from the grand potential are not very informative because they are expressed in terms of fugacity $z \equiv e^{\beta\mu}$. Nevertheless, with a modern computer, all thermodynamic properties can be readily calculated in term of temperature and the average electron density $\rho = N/V$. For example, Figures 4.19 and 4.20 present representative thermodynamic properties of free electrons near 0 K and in the classical limit, respectively. Over a broad range of temperatures (from 0 K up to about 10^4 K), the thermodynamic properties of the free electrons are virtually identical to those corresponding to 0 K, implying that electronic interactions are mostly dominated by quantum effects. As expected, the classical model (ideal-gas law) fails at low temperature. As discussed in Section 4.2.4, the classical results are recovered at high temperature and low density, i.e., when z or $\rho\Lambda^3 \to 0$. Because $\rho r_s^3 = 3/(4\pi)$, Figure 4.20 suggests that the quantum effects become significant when the Wigner–Seitz diameter ($2r_s$) is smaller than the thermal wavelength Λ.

4.5.5 Properties of Metals at Low Temperature

To understand the electronic properties at low temperature, we may express the thermodynamic properties of free electrons in terms of the density of states (DOS). The number of quantum states between ε and $\varepsilon + d\varepsilon$ can be deduced from the electron energy $\varepsilon = \hbar^2k^2/2m$ and the discrete wavevectors given by Eq. (4.110):

$$g(\varepsilon)d\varepsilon = 2 \times \frac{dk_x}{(2\pi/L_x)}\frac{dk_y}{(2\pi/L_y)}\frac{dk_z}{(2\pi/L_z)} = \frac{8\pi k^2 dk}{(2\pi)^3/V} = \frac{V}{2\pi^2}\left(\frac{2m}{\hbar^2}\right)^{3/2}\varepsilon^{1/2}d\varepsilon. \tag{4.124}$$

where a factor of 2 accounts for two single-electron states at each energy level (viz., spin up and spin down). By combing the density of states

$$g(\varepsilon) = \frac{V}{2\pi^2}\left(\frac{2m}{\hbar^2}\right)^{3/2}\varepsilon^{1/2} = 4\pi V\varepsilon^{1/2}\left(\frac{2m}{h^2}\right)^{3/2} \tag{4.125}$$

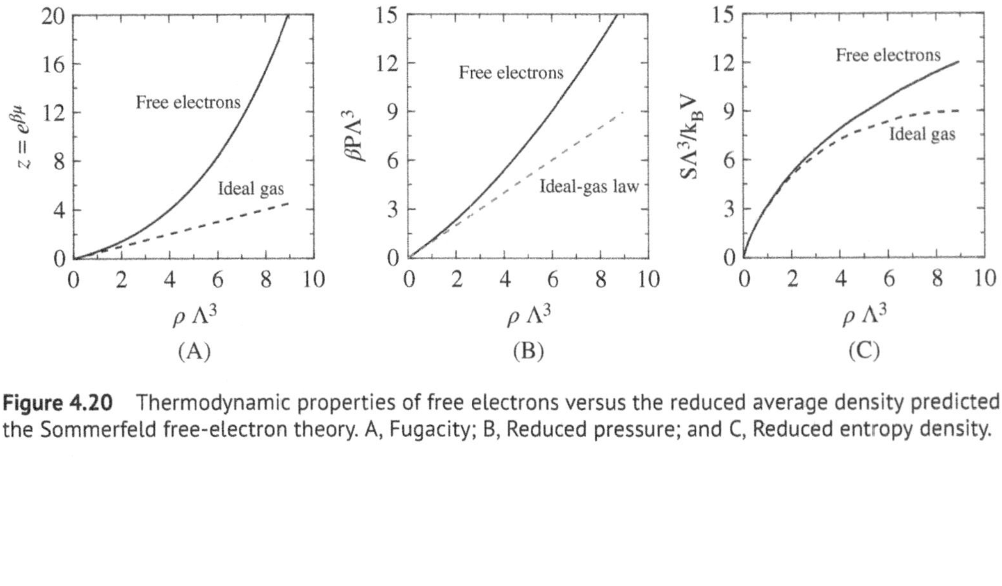

Figure 4.20 Thermodynamic properties of free electrons versus the reduced average density predicted by the Sommerfeld free-electron theory. A, Fugacity; B, Reduced pressure; and C, Reduced entropy density.

and the Fermi–Dirac distribution function $n(\varepsilon) = 1/[e^{\beta(\varepsilon-\mu)} + 1]$, we can obtain the following expressions for the number of electrons in the system and the internal energy

$$N = \int d\varepsilon g(\varepsilon) n(\varepsilon) = \int d\varepsilon \frac{g(\varepsilon)}{e^{\beta(\varepsilon-\mu)} + 1}, \tag{4.126}$$

$$U = \int d\varepsilon g(\varepsilon) \varepsilon n(\varepsilon) = \int d\varepsilon \frac{\varepsilon g(\varepsilon)}{e^{\beta(\varepsilon-\mu)} + 1}. \tag{4.127}$$

Because the asymptotic behavior of $n(\varepsilon)$ is well known, Eqs. (4.126) and (4.127) allow us to derive the average electron density and the average kinetic energy near zero absolute temperature.

At $T = 0$ K, the Fermi–Dirac distribution function becomes a step function

$$n(\varepsilon) = \begin{cases} 1 & \varepsilon \le \mu, \\ 0 & \varepsilon > \mu. \end{cases} \tag{4.128}$$

In this case, we can integrate Eq. (4.126) analytically

$$N = \int_0^\mu d\varepsilon g(\varepsilon) = 4\pi V \left(\frac{2m}{h^2}\right)^{3/2} \int_0^\mu d\varepsilon \varepsilon^{1/2} = \frac{8\pi V}{3} \left(\frac{2m\mu}{h^2}\right)^{3/2}. \tag{4.129}$$

Rearranging Eq. (4.129) provides a direct relation between the chemical potential and the average electron density

$$\mu\ (T = 0\ \mathrm{K}) = \left(\frac{h^2}{2m}\right) \left(\frac{3\rho}{8\pi}\right)^{2/3}. \tag{4.130}$$

In quantum mechanics, μ ($T = 0$ K) is commonly referred to as the Fermi energy, ε_F. Because at $T = 0$ K, the Fermi–Dirac distribution function reduces to a step function, the Fermi energy represents the difference between the highest and the lowest energies of the single-particle states occupied by the electrons.

Typically, the Fermi energy of a metallic system is positive and has the numerical value on the order of a few electron volts (eV). For example, substituting the electron density for copper into Eq. (4.130) gives

$$\varepsilon_F = \left(\frac{(6.626 \times 10^{-34})^2}{2 \times 9.109 \times 10^{-31}}\right) \left(\frac{3 \times 8.46 \times 10^{28}}{8\pi}\right)^{2/3} \approx 1.126 \times 10^{-18}\ \mathrm{J} \approx 7.025\ \mathrm{eV}. \tag{4.131}$$

In practical applications, the chemical potential at 0 K is conveniently represented by the Fermi temperature $T_F = \varepsilon_F/k_B$, or the Fermi wavevector $k_F = \sqrt{2m\varepsilon_F}/\hbar = (3\pi^2\rho)^{1/3}$. For copper, $T_F \approx 8.18 \times 10^4$ K, which amounts to $\beta\mu = T_F/T \approx 274$ at room temperature ($T = 298.15$ K). Accordingly, the fugacity of electrons, $z \approx e^{274}$, is an astronomic number. Because the density of free electrons is of the same order of magnitude for most metals (Figure 4.15), the Sommerfeld theory suggests that, at ambient conditions, the electronic properties of a metallic system are similar to those at 0 K.

Using Eq. (4.130), we can express the Fermi wavevector in terms of the electron density or the Wigner–Seitz radius

$$k_F = (3\pi^2\rho)^{1/3} = \frac{(9\pi/4)^{1/3}}{r_s} \approx \frac{3.63}{(r_s/a_0)}\ \text{Å}^{-1}. \tag{4.132}$$

Using a typical value for the Wigner-Seitz radius ($r_s/a_0 \sim 1$), Eq. (4.132) predicts that the Fermi velocity, $v_F = \hbar k_F/m$, is on the order of 10^6 m^2/s. From a classical perspective, such a high velocity at $T = 0$ K is rather astonishing!

After some algebra,[36] we can also obtain the asymptotic behavior of the internal energy in the limit of $k_B T/\varepsilon_F \to 0$

$$U = \frac{3N\varepsilon_F}{5}\left[1 + \frac{5\pi^2}{12}\left(\frac{k_B T}{\varepsilon_F}\right)^2 + \cdots\right]. \tag{4.133}$$

For a typical metal, the Fermi energy is on the order of a few electron volts. Therefore, even near room temperature, $k_B T/\varepsilon_F \sim 10^{-2}$, Eq. (4.133) affirms that that the internal energy of a metallic system is essentially the same as that at 0 K.

From Eq. (4.133), we can derive the heat capacity

$$C_V/(Nk_B) = \frac{\pi^2}{2}\frac{k_B T}{\varepsilon_F} + \cdots. \tag{4.134}$$

In comparison with that for a system of noninteracting classical particles ($C_V = 3Nk_B/2$), the heat capacity of free electrons is negligibly small.

4.5.6 Bulk Modulus and Electrical Conductivity of Metallic Materials

Before closing this section, it might be instructive to elucidate two relatively simple applications of the Sommerfeld theory. Based on the equations derived above, we may deduce the bulk modulus of a metallic material at 0 K[37]

$$B \equiv -V\left(\frac{\partial P}{\partial V}\right)_T = \frac{2\rho\varepsilon_F}{3} \approx \left(\frac{6.13}{r_s/a_0}\right)^5 \text{ GPa}. \tag{4.135}$$

For the reasons discussed above, even though Eq. (4.135) is derived from 0 K, it is applicable to metallic systems at ambient conditions. Table 4.2 compares the theoretical bulk moduli with experimental data for several metals.[32] Although the agreement is far from perfect, it is rather remarkable that the simply theory of free electrons yields the right order of magnitude, even without any consideration of Coulomb interactions. Because free electrons do not interact with each other, the good agreement suggests that the stiffness of metallic materials is largely determined by the Pauli exclusion principle.

The Sommerfeld theory is often used to describe the electrical conductivity of metallic materials. Assuming that, in the presence of an external electrical field, the directed electron motion can be described by Newton's equation, we can show that the electrical conductivity is given by[38]

$$\sigma = \frac{\rho e^2 \lambda}{m v_F} \tag{4.136}$$

Table 4.2 Bulk moduli in GPa for some selected metals.

Metal	Li	Na	K	Rb	Cs	Cu	Ag	Al
Theory	23.9	9.23	3.19	2.28	1.54	63.8	34.5	228
Exp.	11.5	6.42	2.81	1.92	1.43	134.3	99.9	76.0

Source: Adapted from Ashcroft N. W. and Mermin N. D., *Solid state physics*. Saunders, 1976.

36 See Problem 4.27.
37 See Problem 4.29.
38 See Problem 4.31.

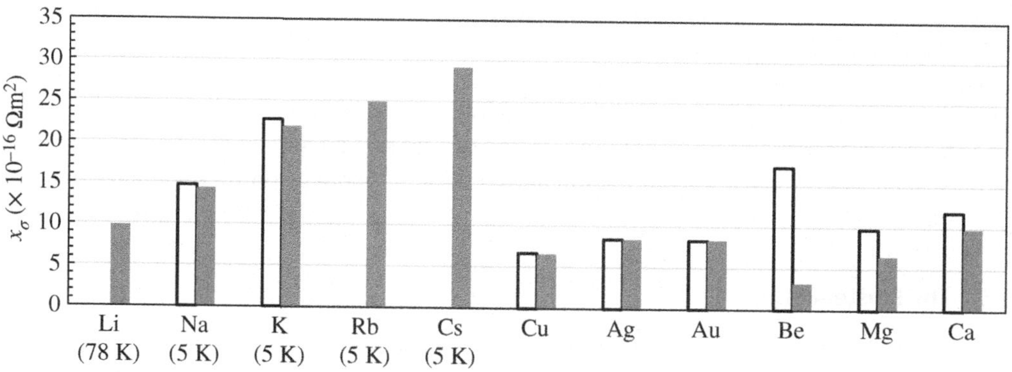

Figure 4.21 The numerical value of $x_\sigma \equiv mv_F/\rho e^2$ predicted by the Sommerfeld theory (shaded bars) in comparison with the results from DFT calculations (open bars). Source: DFT data from Gall.[39]

where v_F is the Fermi velocity, and λ represents the electron mean-free path, i.e., the average distance that electrons can travel between two successive collisions with other particles. Based on the electrical conductivity measured from experiments, Eq. (4.136) predicts that the mean-free path of free electrons in a metal is much larger than inter-atomic distances, typically on the order of a few to tens nanometers. The large mean-free path implies that electron motion is not impeded by the collision with the atomic nuclei of a metallic system.[39]

Eq. (4.136) can be used to estimate the electrical conductivity of metal wires. Suppose that d is the relevant length scale (e.g., the wire width or the grain size) of a nanowire, it has been shown the electrical conductivity can be approximated by[39]

$$\sigma_d = \frac{\rho e^2 d}{m v_F} = d/x_\sigma. \tag{4.137}$$

where $x_\sigma = mv_F/\rho e^2$. Eq. (4.137) predicts that the electrical conductivity of metal wires is linear proportional to the wire thickness d. Figure 4.21 shows the numerical values of x_σ for certain metals at room temperature in comparison with those obtained from highly sophisticated DFT calculations.[38] The results are surprisingly good for many metals, suggesting that the free-electron model provides a useful starting point for the computational screening of narrow wires with high conductivity.

4.5.7 Summary

The Sommerfeld theory is based on the Fermi–Dirac statistics for non-interacting electrons. It offers a comprehensive explanation of the electronic properties of metallic systems across all temperatures. Despite its simplicity, the Sommerfeld theory offers nearly quantitative predictions for the thermodynamic and transport properties of numerous metallic systems.

4.6 Ideal Solids and Phonons

As we have seen in Section 4.5, under ambient conditions, the electrons in a metal mostly exist in the ground state, meaning that the electronic properties are not significantly different from

39 Gall D., "Electron mean free path in elemental metals", *J. Appl. Phys.* 119, 085101 (2016).

those of the same material at 0 K. In this section, we will discuss two simple models of solids that account for the atomic degrees of freedom, specifically the microstates associated with the vibrational motions of atoms within a lattice. These collective vibrational modes are known as phonons, a type of quasiparticle. Representing atomic vibrational motions as quasiparticles is convenient from a theoretical standpoint, as it enables us to describe thermodynamic properties of crystalline materials, such as heat capacity, and the quantum effects of phonons on thermal and electrical conductivity.

4.6.1 The Einstein Model

In 1907, Albert Einstein put forward an idealized model for the atomic vibrations that occur within a crystalline lattice.[40] According to this model, the vibrations of each atom can be described as a harmonic oscillator with quantized energies.[41] Like the free-electron model for metals or the ideal-gas model for real fluids, the Einstein model serves as a useful starting point for understanding the thermodynamic properties of crystalline solids, particularly heat capacity. Due to its simplicity, the Einstein model is often referred to as the ideal-solid model.

To demonstrate how the properties of an ideal solid can be derived from statistical thermodynamics, let us consider a monatomic crystal consisting of only a single type of atom. We assume that, relative to some ground-state energy, the kinetic energy of each atom can be described by that corresponding to a three-dimensional (3D) harmonic oscillator:

$$E_i = \left(n_{i,x} + \frac{1}{2}\right) h\upsilon_{i,x} + \left(n_{i,y} + \frac{1}{2}\right) h\upsilon_{i,y} + \left(n_{i,z} + \frac{1}{2}\right) h\upsilon_{i,z}, \quad n_{i,x}\ n_{i,y}\ n_{i,z} = 0, 1, 2, \ldots \tag{4.138}$$

where $h = 6.626 \times 10^{-34}$ J s is Planck's constant, $i = 1, 2, \ldots N$ is the atomic indices, $n_{i,x}$, $n_{i,y}$ and $n_{i,z}$ are nonnegative integers (*viz.*, quantum numbers), and $\upsilon_{i,x}$, $\upsilon_{i,y}$, $\upsilon_{i,z}$ are the characteristic frequencies of a harmonic oscillator representing the vibrational motion of atom i in the 3D space.

In an ideal solid, we assume that all phonons are independent of each other and have the same vibration frequency. Accordingly, a solid with N atoms amounts to a system of $3N$ noninteracting phonons (viz., a phonon gas). The total energy is given by the sum of the energies of individual phonons:

$$E_\nu = \sum_{i=1}^{N} E_i = \sum_{j=1}^{3N} \left(n_j + \frac{1}{2}\right) h\upsilon_j \tag{4.139}$$

where subscript ν denotes a microstate, which is defined by a set of occupation numbers $n_j = 0, 1, 2, \ldots$. For a system of $3N$ identical phonons, each microstate can be specified by a set of quantum numbers $\{n_j\}_{j=1,2,\ldots N}$, representing the $3N$ energy levels of harmonic oscillators.

The thermodynamic properties of phonons can be readily derived using the canonical ensemble. As the phonons are assumed independent to each other, the canonical partition function is given by

$$Q = \sum_\nu e^{-\beta E_\nu} = \prod_{j=1}^{3N} \sum_{n_j=0,1,\ldots} e^{-\beta\left(n_j+\frac{1}{2}\right)h\upsilon_j} = \prod_{j=1}^{3N} \frac{e^{-\beta h\upsilon_j/2}}{1 - e^{-\beta h\upsilon_j}}. \tag{4.140}$$

---

40 Einstein A., "Theory of radiation and the theory of specific heats", *Annalen der Physik (German)*, 4, 180–190 (1907).

41 A harmonic oscillator is a dynamic system of a single particle where the restoring force from the equilibrium position is proportional to displacement, i.e., $F = -kx$. According to the Schrödinger equation, a harmonic oscillator takes discrete energy levels, $(n + 1/2)h\upsilon$, $n = 0, 1, 2, \ldots$, with the characteristic frequency given by $\upsilon = (1/2\pi)(k/m)^2$ where m stands for the particle mass.

Thus, the Helmholtz energy of the system is

$$F = -k_B T \ln Q = k_B T \sum_{i=1}^{3N} \ln \left(e^{\beta h\upsilon_i/2} - e^{-\beta h\upsilon_i/2} \right). \tag{4.141}$$

From Eq. (4.141), we can derive other thermodynamic properties following standard equations of statistical mechanics.

To evaluate the summation in Eq. (4.141), we need information on the frequencies of the harmonic oscillators. In general, these frequencies depend on the microscopic details of atomic interactions and the crystalline structure. To the zeroth-order approximation, we may assume that all harmonic oscillators have the same vibrational frequency υ. Accordingly, Eq. (4.141) is simplified to

$$F = 3Nk_B T \ln \left(e^{\theta_E/2T} - e^{-\theta_E/2T} \right) \tag{4.142}$$

where $\theta_E \equiv h\upsilon/k_B$ is called the Einstein temperature. For typical metals such as copper, the Einstein temperature is around 200–300 K.[42] For diamond, the value is substantially higher, on the order of 1300 K.

From Eq. (4.142), we can derive the following expressions for the internal energy U and the constant-volume heat capacity C_V:

$$U = \left(\frac{\partial \beta F}{\partial \beta} \right)_V = \frac{3N\theta_E k_B}{2} + \frac{3Nk_B\theta_E}{e^{\theta_E/T} - 1}, \tag{4.143}$$

$$C_V = \left(\frac{\partial U}{\partial T} \right)_V = 3Nk_B \left(\frac{\theta_E}{T} \right)^2 \frac{e^{\theta_E/T}}{\left(e^{\theta_E/T} - 1 \right)^2}. \tag{4.144}$$

Because the energy for each atom is relative to some reference state (e.g., electronic energy of the system at 0 K), we understand that Eq. (4.143) only accounts for the internal energy due to atomic vibrations. Interestingly, Eq. (4.144) suggests that $C_V/(3Nk_B)$ is a universal function of reduced temperature, T/θ_E, much like the prediction of the corresponding-state principle in classical thermodynamics.

Figure 4.22 shows the heat capacity of an ideal solid predicted by the Einstein model. At sufficiently high temperature, $\theta_E/T \ll 1$ so that

$$(\theta_E/T)e^{\theta_E/2T} / \left(e^{\theta_E/2T} - 1 \right) \sim 1. \tag{4.145}$$

In that case, Eq. (4.144) reduces to the Dulong–Petit law, first proposed by Pierre Louis Dulong and Alexis Thérèse Petit in 1819 based on experimental observations,

$$\tilde{C}_V \simeq 3R \tag{4.146}$$

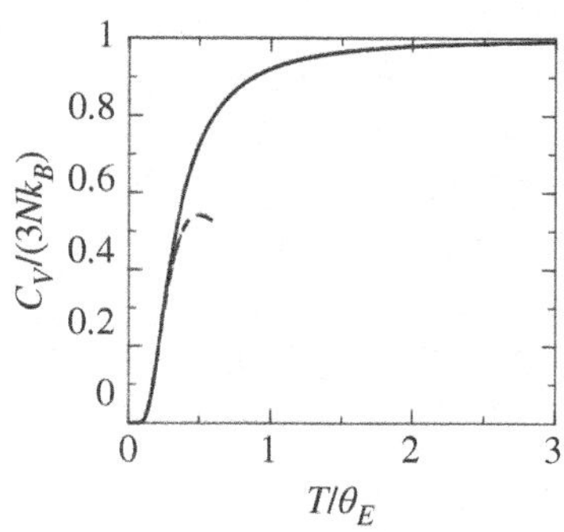

Figure 4.22 Heat capacity of solids predicted by the Einstein model (solid line). The dashed line is the prediction from the asymptotic expression, Eq. (4.147).

42 Tabor D., *Gases, liquids and solids and other states of matter*. Cambridge University Press. Chapter 9, 1991.

where $\widetilde{C}_V$ represents the molar heat capacity, and R is the gas constant. The Dulong–Petit law agrees reasonably well with experiment data for the heat capacities of a large number of metals at ambient temperature (about 25 J/(mol K)). For example, at 298 K, the experimental heat capacity of Cu is 25.5 J/(mol K), very close to that of Al, 24.4 J/(mol K). The empirical observation provides a good estimate for the heat capacity of monatomic crystals.

At the low-temperature limit, the Einstein model predicts that the heat capacity of a monatomic solid approaches

$$C_V \to 3Nk_B\left(\frac{\theta_E}{T}\right)^2 e^{-\theta_E/T}\ (T \to 0). \tag{4.147}$$

Although Eq. (4.147) is inaccurate in comparison with experimental data, which indicate $C_V \sim T^3$ as $T \to 0$, the Einstein model correctly predicts that the heat capacity is a strong function of temperature near absolute zero and approaches a constant at high temperature as predicted by the Dulong-Petit law. Furthermore, Eq. (4.147) indicates that C_V vanishes at 0 K consistent with the third law of thermodynamics.

In the early twentieth century, Einstein's work on ideal solids provided strong support for the quantum theory by showing that the energy of a harmonic oscillator can only take on discrete values. This was a revolutionary idea at the time because it contradicted classical physics, which predicted that energy could take on any value. Einstein's work on ideal solids demonstrated that the energy of a solid's atomic vibrations is quantized, meaning it can only exist in discrete packets of energy called quanta. This discovery provided strong evidence for the validity of the quantum theory and was a major milestone in the development of modern physics.

4.6.2 The Debye Model

While Einstein's model of ideal solids assumed that all phonons have the same energy, a more realistic approximation was proposed by Peter Debye to account for the distribution of vibrational frequencies of atoms in a solid.[43] Debye's model has been shown to accurately predict the heat capacity of solids at a wide range of temperatures, from very low temperatures up to temperatures near the melting point of the material. This has made it a valuable tool for understanding the thermodynamic properties of solids and for developing new materials for practical applications.

In Debye's model, phonons are represented by standing waves. Similar to those solved from the Schrödinger equation for a particle in a box (see Problem 3.1), the wavevectors take only discrete values

$$\mathbf{k} = \frac{\pi}{L}\mathbf{n} \tag{4.148}$$

where $\mathbf{n} = (n_x, n_y, n_z)$, $n_x, n_y, n_z = 1, 2, \ldots$ are positive integers. Figure 4.23 illustrates schematically the wave functions corresponding to some possible wavevectors for a one-dimensional crystal of length L.

According to Eq. (4.148), the wavenumber, i.e., the magnitude of wavevector $\mathbf{k}$, is given by

$$k^2 = \left(\frac{\pi}{L}\right)^2 \left(n_x^2 + n_y^2 + n_z^2\right). \tag{4.149}$$

The number of possible standing waves with the wavenumber less than k^2 is equal to the number of all possible combinations of n_x, n_y, n_z that satisfy

$$k^2 \le \left(\frac{\pi}{L}\right)^2 \left(n_x^2 + n_y^2 + n_z^2\right). \tag{4.150}$$

43 Debye P., "On the theory of specific heat", *Annalen der Physik (German)*, 4, 789–839 (1912).

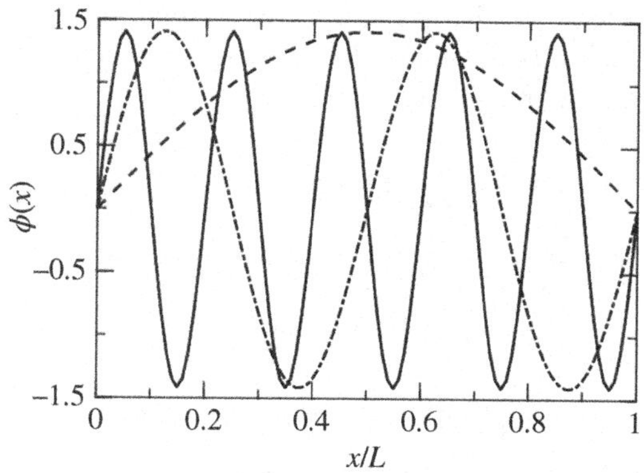

Figure 4.23 In the Debye model, phonons are represented by standing waves (represented by different curves) as shown here for a one-dimensional crystal lattice of total width L. The wavelength for each standing wave satisfies $\lambda_n = 2L/n$ where $n = 1, 2, \ldots$, is a positive integer.

Eq. (4.150) indicates that all possible quantum numbers $\mathbf{n} = (n_x, n_y, n_z)$ must be located within one-eighth of a sphere of radius kL/π where all three coordinates are positive. In the continuous limit, the number of quantum states is thus equal to one-eighth of the volume of the sphere enclosed by radius kL/π:

$$n(k) = \frac{1}{8} \times \frac{4\pi}{3}\left(\frac{Lk}{\pi}\right)^3 = \frac{\pi}{6}\left(\frac{Lk}{\pi}\right)^3. \tag{4.151}$$

Accordingly, the number of standing waves with the magnitude of wavevector between k and $k + dk$ is given by

$$dn(k) = \frac{Vk^2 dk}{2\pi^2} = g(k)dk \tag{4.152}$$

where $V = L^3$ is the system volume, and $g(k) = Vk^2/2\pi^2$ represents the density of states (DOS) for phonons.

The frequency of a standing wave is related to the wavenumber through the relation

$$k = 2\pi\upsilon/c \tag{4.153}$$

where c represents the sound velocity. From Eqs. (4.152) and (4.153), we find an analytic expression for the DOS in terms of the frequency

$$g(\upsilon) = g(k)\frac{dk}{d\upsilon} = \frac{Vk^2}{2\pi^2}\frac{2\pi}{c} = \frac{4\pi V\upsilon^2}{c^3}. \tag{4.154}$$

Because each phonon is associated with one standing wave, the integration of $g(\upsilon)$ from zero to a maximum frequency υ_D equals the total number of phonons

$$\int_0^{\upsilon_D} g(\upsilon)d\upsilon = 3N. \tag{4.155}$$

Substitution of Eq. (4.154) into (4.155) gives

$$\upsilon_D = \left(\frac{9N}{4\pi V}\right)^{1/3} c. \tag{4.156}$$

This maximum frequency υ_D is called the Debye frequency. In terms of υ_D, the DOS of phonons in form of the frequency distribution $g(\upsilon)$ becomes

$$g(\upsilon) = \begin{cases} \dfrac{9N\upsilon^2}{\upsilon_D^3} & 0 \le \upsilon \le \upsilon_D, \\ 0 & \upsilon > \upsilon_D. \end{cases} \tag{4.157}$$

Based on the frequency distribution derived from Debye's model, we can now evaluate the Helmholtz energy of the system by replacing the summation in Eq. (4.141) with an integration over the frequency

$$F = k_B T \sum_{i=1}^{3N} \ln\left(e^{\beta h\upsilon_i/2} - e^{-\beta h\upsilon_i/2}\right) = -k_B T \int_0^{\infty} g(\upsilon) \ln\left(\frac{e^{-\beta h\upsilon/2}}{1 - e^{-\beta h\upsilon}}\right) d\upsilon. \tag{4.158}$$

Substituting Eq. (4.157) into (4.158) gives an analytical expression for the Helmholtz energy

$$F = -\frac{9Nk_B T}{\upsilon_D^3} \int_0^{\upsilon_D} \upsilon^2 \ln\left(\frac{e^{-\beta h\upsilon/2}}{1 - e^{-\beta h\upsilon}}\right) d\upsilon. \tag{4.159}$$

Accordingly, the internal energy is given by

$$U = \left(\frac{\partial \beta F}{\partial \beta}\right)_{N,V} = \frac{9Nk_B T}{\upsilon_D^3} \int_0^{\upsilon_D} \left(\frac{h\upsilon}{2k_B T} + \frac{h\upsilon/(k_B T)}{e^{h\upsilon/(k_B T)} - 1}\right) \upsilon^2 d\upsilon, \tag{4.160}$$

and the constant-volume heat capacity is

$$C_V/(Nk_B) = \frac{1}{Nk_B}\left(\frac{\partial U}{\partial T}\right)_V = 9\left(\frac{T}{\theta_D}\right)^3 \int_0^{\theta_D/T} \frac{x^4 e^x}{(e^x - 1)^2} dx \tag{4.161}$$

where $\theta_D \equiv h\upsilon_D/k_B$ is called the Debye temperature. While in principle the Debye temperature may be estimated from the elastic constant, it is mostly obtained by direct fitting with experimental data for C_V at low temperature. Table 4.3 gives the Debye temperatures for some atomic crystals.

Figure 4.24 compares the heat capacities predicted by the Debye theory and by the Einstein theory. Similar to the Einstein theory, the Debye theory also predicts that, in dimensionless units, the heat capacity of a solid is a universal function of temperature. While both theories predict the convergence of the heat capacity to the Dulong–Petit law at high temperature, they differ significantly

Table 4.3 The Debye temperatures (K) for some atomic crystals.

Ag	225	Fe	470	Na	158	Si	645
Al	428	Ga	320	Nb	275	Sn	200
Ar	92	Gd	200	Ne	75	Sr	147
As	282	Ge	374	Ni	450	Ta	240
Au	165	Hf	252	Os	500	Te	153
Ba	110	Hg	71.9	Pb	105	Th	163
Be	1440	In	108	Pd	274	Ti	420
Bi	119	Ir	420	Pt	240	Tl	78.5
C	2230	K	91	Rb	56	U	207
Ca	230	Kr	72	Re	430	V	380
Cd	209	La	142	Rh	480	W	400
Co	445	Li	344	Rn	64	Xe	64
Cr	630	Lu	210	Ru	600	Y	280
Cs	38	Mg	400	Sb	211	Yb	120
Cu	343	Mn	410	Sc	360	Zn	327
Dy	210	Mo	450	Se	90	Zr	291

Source: Adapted from Kittel C. *Introduction to solid state physics*. Wiley (7th Edition), 1995.

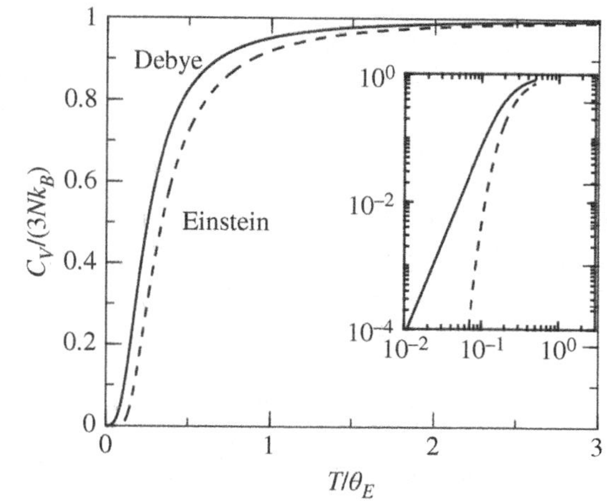

Figure 4.24 The reduced heat capacity of a monatomic crystal according to the theories of Einstein and Debye. Here, θ is either the Einstein temperature or the Debye temperature. The inset shows theoretical predictions at low temperature.

at low temperature. The Einstein theory predicts that, near $T = 0$ K, the heat capacity scales as

$$C_V/Nk_B \to 3\left(\frac{\theta_E}{T}\right)^2 e^{-\theta_E/T} \tag{4.162}$$

whereas the Debye theory predicts that the heat capacity follows the third power of temperature, similar to that for photons (Eq. 4.53):

$$C_V/Nk_B \to \frac{12\pi^4}{5}\left(\frac{T}{\theta_D}\right)^3. \tag{4.163}$$

The low-temperature limit of the Debye theory was found in good agreement with experiment. For the Einstein theory, the low-temperature heat capacity falls to zero more rapidly than T^3.

The Debye theory of atomic crystals has been remarkably successful for describing the experimental observations of the heat capacities for a wide range of solid materials. When compared with experiment, however, the frequency distribution is not at all accurate. As shown schematically in Figure 4.25, serious errors are noticeable at high frequencies, even though it agrees reasonably well with experiment in the low-frequency end. Fortunately, for macroscopic thermodynamic quantities, such as heat capacity, errors at high frequencies are often less important. At high temperature, the heat capacity is determined by an average over the entire frequency range with the classical limit of $C_V/N = 3k_B$ as $T/\theta_D \to \infty$. As a result, to a good approximation, errors in the Debye theory for the high-frequency contributions tend to cancel. At low temperature, the low-frequency part of the vibrational modes is well approximated by the continuous theory of elasticity. The Debye cut-off is expected because each phonon is associated with one standing wave and high frequency implies many oscillation cycles.

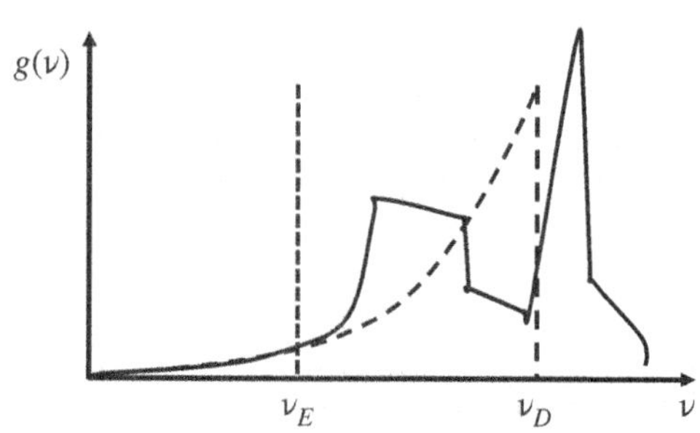

Figure 4.25 Schematic of frequency distribution $g(\nu)$ for a monatomic crystal. The solid curve represents typical experimental results while the bold dashed curve represents Debye's approximation. Einstein's model uses only a single frequency as represented by a Dirac delta function (the perpendicular line), i.e., $g(\nu) = \delta(\nu - \nu_E)$.

4.6.3 Summary

The quantized vibrational modes of a crystal lattice are referred to as phonons, which are a type of pseudoparticle commonly used in solid-state physics to describe the atomic vibrational properties of materials. The statistical-thermodynamic theories proposed by Einstein and Debye rely on the assumption of independent phonons, represented by the harmonic oscillator model. This idealized model is particularly useful in explaining the heat capacity of solids at low temperatures. Specifically, the Debye model is precise at low frequencies (or long wavelengths), where the atomic nature of the solid is negligible. Under these conditions, a perfect crystal can be viewed as a continuous elastic body, and the harmonic oscillators (i.e., phonons) are standing waves with frequencies that are independent of the atomic details. At high temperatures, the theories of Einstein and Debye reduce to the Dulong-Petit law. These models, along with further improvements accounting for various sources of phonon–phonon and electron–phonon interactions, can facilitate the rational design of new materials for a broad range of energy technologies.[44]

4.7 Chapter Summary

The thermodynamic properties of photons, electrons, and phonons have significant relevance in many engineering applications, such as solar energy conversion, photosynthesis, and electrochemical reactions that are crucial for modern technological advancements. Although the highly idealized models discussed in this chapter may not fully capture the complex behavior of photoelectronic interactions in practical applications involving complex chemical systems and condensed materials, they serve to illustrate how even the basic principles of statistical thermodynamics, without considering particle-particle interactions, can aid in our understanding and prediction of the unique properties of macroscopic systems, which arise from the quantum nature of the individual constituents, often in a quantitative manner.

This chapter also exemplifies "Unreasonable Effectiveness of Mathematics."[45] The single-particle states of non-interacting electrons, photons and phonons can all be described in terms of the Schrödinger equation.[46] For photons and phonons, we may solve the differential equation with the boundary condition of a particle in a cubic box of length L, which gives a wavevector taking only discrete values, $\mathbf{k} = \pi\mathbf{n}/L$, where $\mathbf{n} = (n_x, n_y, n_z)$, $n_x, n_y, n_z = 1, 2, \ldots$. The solution is slightly different for free electrons in a metal because we use the periodic boundary condition, yielding $\mathbf{k} = 2\pi\mathbf{n}/L$, with $n_{x,y,z} = 0, \pm 1, \pm 2, \ldots$. In both cases, the energy for each single-particle state can be expressed as $\varepsilon(\mathbf{k}) = \hbar^2k^2/(2m)$ with $m = h\upsilon/c$ standing for an effective mass for photons or phonons and c for the particle velocity. Based on the single-particle states and their corresponding energies, statistical mechanics allows us to evaluate the partition function and, subsequently, the microstate distribution and all thermodynamic properties. What is truly remarkable is that the same mathematical procedure can be extended to virtually any system by using the density functional theory (DFT) with an effective one-body potential to account for multi-body interactions. In that context, this chapter may be best summarized by quoting Bertrand Russell,[47] "Physics is mathematical not because we know so much about the physical world, but because we know so little; it is only its mathematical properties that we can discover."

---

44 Fultz B., "Vibrational thermodynamics of materials", *Prog. Mater. Sci.* 55, 247–352 (2010).

45 Wigner, E. P., "The unreasonable effectiveness of mathematics in the natural sciences", *Commun. Pure Appl. Math.* 13, 1–14 (1960).

46 Formally, the Schrödinger equation may be applied to photons and phonons by using an effective mass $h\upsilon/c$ and wavevector $\mathbf{k} = 2\pi(\upsilon_x, \upsilon_y, \upsilon_z)$ where c is the particle velocity.

47 Russell B., "An outline of philosophy", *The nature of our knowledge of physics*. George Allen and Unwin, 1927.

Further Readings

Ashcroft N. W. and Mermin N. D., *Solid state physics.* Harcourt: Orlando. Chapter 2, 1976.

Blundell S.J and Blundell K. M., *Concepts in thermal physics.* Oxford University Press. Chapters 29 and 30, 2006.

Coridan R. H. et al., "Methods for comparing the performance of energy-conversion systems for use in solar fuels and solar electricity generation", *Energy Environ. Sci.*, 8, 2886–2901 (2015).

Huang K., *Introduction to statistical physics.* CRC Press. Chapters 14–18, 2010.

Liu Z. K., "Computational thermodynamics and its applications", *Acta Mater.*, 200, 745–792 (2020).

Mizutani U., *Introduction to the electron theory of metals.* Cambridge University Press. Chapters 3 and 4, 2009.

Ritz E. T., Li S. J. and Benedek N. A. "Thermal expansion in insulating solids from first principles", *J. Appl. Phys.* 126, 171102 (2019).

Singh M.R, Clark E. L. and Bell A. T., "Thermodynamic and achievable efficiencies for solar-driven electrochemical reduction of carbon dioxide to transportation fuels", *PNAS*, 112 (45), E6111–E6118 (2015).

Widom B., *Statistical mechanics.* Cambridge University Press. Chapter 8, 2009.

Problems

4.1 Consider a system with two identical particles that each may exist in one of two single-particle quantum states. How many microstates that the two particles may exist if they are (a) classical particles, (b) bosons, and (c) fermions?

4.2 How many microstates may the two protons in a hydrogen molecule exist? What are the compositions of the spin isomers if these microstates are randomly distributed?

4.3 In a thermodynamic system of many particles, the quantum effects on the translational motion become significant when the thermal de Broglie wavelength Λ is comparable to the average inter-particle spacing. The condition $\rho\Lambda^3 = 1$ thus defines a so-called "degeneracy temperature"

$$T_0 = \frac{h^2}{2\pi m k_B}\rho^{2/3}$$

beyond which the quantum effects may be neglected. Assume that the molar volume of the system is 22.4 L/mol, estimate the degeneracy temperature for (a) hydrogen gas; (b) nitrogen gas; and (c) electron gas.

4.4 Based on the grand potential for noninteracting bosons derived in the text (Eq. (4.22))

$$\Omega = -\frac{k_B T V}{\Lambda^3}\mathrm{Li}_{5/2}(z),$$

where $z = e^{\beta\mu}$ represents fugacity, and Λ is the thermal de Broglie wavelength, show the following expressions for

(i) pressure

$$P = \frac{k_B T}{\Lambda^3}\mathrm{Li}_{5/2}(z),$$

(ii) internal energy

$$U = \frac{3k_BTV}{2\Lambda^3}\mathrm{Li}_{5/2}(z),$$

(iii) the average number of boson particles

$$N = \frac{V}{\Lambda^3}\mathrm{Li}_{3/2}(z),$$

(iv) entropy

$$S = \frac{k_BV}{\Lambda^3}\left[\frac{5}{2}\mathrm{Li}_{5/2}(z) - (\ln z)\mathrm{Li}_{3/2}(z)\right].$$

4.5 Show that, in the classical limit, the thermodynamic properties of noninteracting bosons are given by:

(i) chemical potential

$$\mu = k_BT\ln(\rho\Lambda^3),$$

(ii) grand potential

$$\Omega = -Nk_BT,$$

(iii) pressure

$$P = \frac{Nk_BT}{V},$$

(iv) internal energy

$$U = \frac{3}{2}Nk_BT,$$

(v) entropy

$$S = -Nk_B[\ln\rho\Lambda^3 - 5/2].$$

4.6 Based on the grand potential for noninteracting fermions

$$\Omega = \frac{k_BTV}{\Lambda^3}\mathrm{Li}_{5/2}(-z),$$

where $z = e^{\beta\mu}$, show the following expressions for

(i) pressure

$$P = -\frac{k_BT}{\Lambda^3}\mathrm{Li}_{5/2}(-z),$$

(ii) internal energy

$$U = -\frac{3k_BTV}{2\Lambda^3}\mathrm{Li}_{5/2}(-z),$$

(iii) the average number of particles

$$N = -\frac{V}{\Lambda^3}\mathrm{Li}_{3/2}(-z),$$

(iv) entropy

$$S = -\frac{k_BV}{\Lambda^3}\left[\frac{5}{2}\mathrm{Li}_{5/2}(-z) - (\ln z)\mathrm{Li}_{3/2}(-z)\right].$$

4.7 Show that, in the classical limit, the thermodynamic properties of noninteracting fermions are the same as those corresponding to bosons, i.e., both are reduced to the equations corresponding to an ideal gas of spherical particles:

(i) chemical potential

$$\mu = k_B T \ln(\rho\Lambda^3),$$

(ii) grand potential

$$\Omega = -Nk_B T,$$

(iii) pressure

$$P = \frac{Nk_B T}{V},$$

(iv) internal energy

$$U = \frac{3}{2}Nk_B T,$$

(v) entropy

$$S = -Nk_B[\ln \rho\Lambda^3 - 5/2].$$

4.8 What is the change in entropy when one mole of noninteracting bosons expand isothermally to double the volume at $z_0 = 0.5$. How does the entropy change vary with the initial fugacity z_0?

4.9 What is the change in entropy when one mole of noninteracting fermions expand isothermally to double the volume at $z_0 = 0.5$. How does the entropy change with the initial fugacity z_0?

4.10 Calculate the entropy of mixing at constant temperature and pressure for equal moles of

(i) two boson gases;
(ii) two fermion gases;
(iii) two classical ideal gases.

In all cases, assume that the fugacity of both species before the mixing is $z_0 = 0.5$.

4.11 Show that the Gibbs paradox can be resolved by using the analytical expressions of entropy of non-interacting quantum particles. Does the paradox hold in the classical limit?

4.12 Consider the density of a noninteracting boson gas as a function of the fugacity z

$$\rho\Lambda^3 = \mathrm{Li}_{3/2}(z).$$

As shown in Figure P4.12, the polylogarithm function $\mathrm{Li}_{3/2}(z)$ has real values only when $z \leq 1$. Therefore, there exists a maximum number density beyond which the bosons will condense into a single-particle state with zero momentum (viz., the ground state with zero energy). This phenomenon is known as the Bose–Einstein condensation.

(i) Verify that $(\partial\rho/\partial\mu)_T$ diverges at $z = 1$;
(ii) At a given temperature, the Bose–Einstein condensation takes place at density

$$\rho_c = \zeta(3/2)/\Lambda^3,$$

where $\zeta(3/2) \approx 2.612$ is the Riemann zeta function.

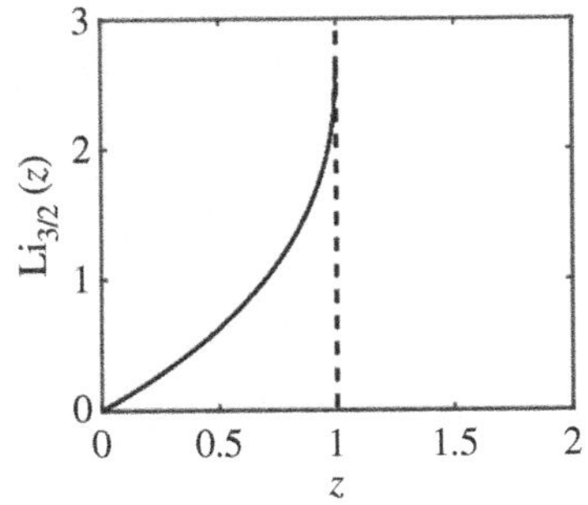

Figure P4.12 The reduced density of a boson gas, $\rho\Lambda^3 = \mathrm{Li}_{3/2}(z)$, versus the fugacity z.

(iii) At a given density, the Bose–Einstein condensation takes place at temperature

$$T_c = \frac{h^2}{2\pi m k_B}\left[\rho/\zeta(3/2)\right]^{2/3}.$$

(iv) Sketch the $\rho \sim T$ phase diagram for the Bose–Einstein condensation.

(v) At $T < T_c$, the fraction of the Bose–Einstein condensate (BEC) is given by

$$N_0/N = 1 - (T/T_c)^{3/2}.$$

4.13 Estimate the number of photons in an oven at 500 K with the volume of 0.5 m^3. What is the molar internal energy of the photons?

4.14 Consider a photon gas as an open system of volume V, temperature T, and chemical potential μ.

(i) Write an expression for the grand canonical partition function;

(ii) Show that the average number of photons derived from the grand canonical ensemble is the same as that from the canonical ensemble

$$\langle N\rangle = 16\zeta(3)\pi V\left(\frac{k_BT}{hc}\right)^3,$$

where $\zeta(3) \approx 1.2021$ is the Riemann zeta function.

(iii) Show that the relative variance in the number of photons is a constant

$$\frac{\langle N^2\rangle - \langle N\rangle^2}{\langle N\rangle} = \frac{\pi^2}{6\zeta(3)}.$$

(iv) Explain why the photon gas does not satisfy the fluctuation-compressibility theorem

$$\frac{\langle N^2\rangle - \langle N\rangle^2}{\langle N\rangle^2} = \frac{k_BT}{V}\kappa_T,$$

where κ_T stands for the isothermal compressibility.

4.15 Suppose that a photon have an effective mass of $m = h\upsilon/c^2$, derive the pressure of a photon gas from the kinetic theory

$$P = \frac{1}{3}\rho m\langle \upsilon^2\rangle.$$

Show that the above result is equivalent to that derived from the thermodynamic route discussed in Section 4.3.

(Hint: For photons, the average velocity is given by the speed of light, i.e., $\langle \upsilon^2\rangle = c^2$, and $mc^2 = h\upsilon$ corresponds to the energy per photon.)

4.16 The entropy analysis plays a major role in understanding the thermodynamic efficiency of solar energy conversion including photosynthesis and biological evolutions.

(i) Verify the following equation for the spectral entropy density for a system of noninteracting photons

$$s_\lambda = \frac{8\pi k_B}{\lambda^4}\left[\frac{\beta hc/\lambda}{e^{\beta hc/\lambda}-1} - \ln(1-e^{-\beta hc/\lambda})\right].$$

(ii) Show that the wavelength at the maximum spectral entropy density is given by

$$\lambda_s^{max} \approx 3.0029 \times 10^{-3}/T,$$

where λ_s^{max} has the units of meter and T is in kelvin.

4.17 Plot the spectral irradiance of solar radiation above Earth's atmosphere based on black-body radiation. What is the intensity of solar energy at the top of the atmosphere? Could you estimate the solar energy intensity at the sea level?

4.18 The photon flux is important in determining the number of electrons and hence the current produced from a solar cell.

(i) Show that the flux of photons from the surface of a blackbody at temperature T is given by

$$J = \frac{4\pi\zeta(3)k_B^3}{c^2h^3}T^3.$$

(ii) Estimate the photon flux at the sea level from the solar radiation.

(iii) Plot the spectral photon flux from the Sun at the sea level.

4.19 Verify that, for a single p-n junction solar cell, the Shockley–Queisser (SQ) efficiency (i.e., the ultimate efficiency) is given by

$$\eta_{SQ} = \frac{15}{\pi^4}x_g f_g(x_g),$$

where $x_g = \beta hc/\lambda_g$ with λ_g being the threshold photon wavelength, and

$$f_g(x) = x^2\mathrm{Li}_1(e^{-x}) + 2x\mathrm{Li}_2(e^{-x}) + 2\mathrm{Li}_3(e^{-x}).$$

4.20 The global warming may be understood in terms of the enhanced greenhouse effect, i.e., the increase of the concentrations of greenhouse gases such as carbon dioxide, methane, nitrous oxide due to human activities – particularly burning fossil fuels (coal, oil, and natural gas). Because the Earth temperature is regulated by the exchange of energy with the universe mainly through radiations, Planck's law may shed useful light on the problem.

(i) Estimate the Earth temperature if there is no atmosphere; Assume that 30% of the Sun light would be reflected into space.

(ii) Plot the spectral energy density due the Earth's radiation by assuming it as a blackbody with an average temperature of $T = 255$ K.

(iii) Compare the spectrum with that of the sunlight.

4.21 The chemical potential of photons is nonzero if light is emitted from the excited states of matter. For example, in a light emitting diode (LED), an electron (e), and a hole (h) recombine to produce a photon (γ), which has a chemical potential

$$\mu_\gamma = \mu_e + \mu_h,$$

where μ_e and μ_h are the chemical potentials of the electron and hole, respectively.

(i) Show that the spectral energy density for the LED light is given by

$$I(\omega) = \frac{\hbar\omega^3}{\pi^2 c^3}\frac{1}{e^{\beta(\hbar\omega-\mu_\gamma)} - 1},$$

where ω stands for the angular frequency of the light.

(ii) Assume that photons cannot be emitted by a LED device if the energy is below the bandgap, i.e., $\hbar\omega < \epsilon_g$, and the emissivity is perfect otherwise. Use $\mu_\gamma = \epsilon_g = 1$ eV and $T = 300$ K, plot the spectral radiation energy density of the LED light and compare it with that from a blackbody radiation at the same temperature.

4.22 Verify Petela's equation for the thermodynamic efficiency for the whole spectrum of the solar energy

$$\eta_T \equiv \frac{\vartheta}{I_{in}} = 1 - \frac{4}{3}\left(\frac{T_0}{T}\right) + \frac{1}{3}\left(\frac{T_0}{T}\right)^4,$$

where ϑ stands for exergy, I_{in} is total solar radiation, $T = 5778$ K, and T_0 is ambient temperature.

4.23 The photosynthetically active radiation (PAR) region is defined as the spectral range of solar radiation that can be utilized by photosynthetic organisms (or devices) during photosynthesis.

(i) The efficiency of this radiation is usually defined as

$$\eta_{PAR} = \frac{\int_{\lambda_1}^{\lambda_2} I(\lambda)d\lambda}{\int_0^\infty I(\lambda)d\lambda}.$$

What is the value of η_{PAR} if the spectral energy of solar radiation can be represented by that of a blackbody at $T = 5778$ K and that the photosynthesis takes place in the region from $\lambda_1 = 400$ to $\lambda_2 = 700$ nm?

(ii) As a consequence of the entropy increase, not all the radiation reaching the Earth's surface is useful to produce work. The second law PAR efficiency is defined as

$$\eta_{PAR}^{2nd} = \frac{\int_{\lambda_1}^{\lambda_2} \vartheta(\lambda)d\lambda}{\int_0^\infty I(\lambda)d\lambda}.$$

The spectral distribution of the exergy of radiation is defined as

$$\vartheta(\lambda) \equiv I(\lambda, T) - I(\lambda, T_0) - T_0[S(\lambda, T) - S(\lambda, T_0)],$$

where $T_0 = 298.15$ K stands for the ambient temperature, and $S(\lambda, T)$ is the spectral entropy density. What is the value of η_{PAR}^{2nd} under the condition the same as that in part (i)?

4.24 It has been shown that the upper limit of the solar energy conversion efficiency is achieved when the energy loss per photon is about $U_{loss} \sim 0.3$–0.4 eV, the exact value depending on the threshold wavelength λ_g. Estimate the threshold wavelength and the efficiency of solar energy conversion for the following fuel generation reactions by photosynthesis for both one-photo and two-photo systems. For simplicity, assume quantum yield $\eta_Q = 1$ and $U_{loss} = 0.37$ eV for all the photosynthesis reactions. Here, n_e is the number of electrons which should be transferred in an electrochemical reaction for the reaction as written.

	Reaction	n_e	ΔG^o (kJ/mol)
(1)	$H_2O(l) \rightarrow H_2(g) + \frac{1}{2}O_2(g)$	2	237
(2)	$CO_2(g) \rightarrow CO(g) + \frac{1}{2}O_2(g)$	2	257
(3)	$CO_2(g) + H_2O(l) \rightarrow HCOOH(l) + \frac{1}{2}O_2(g)$	2	286
(4)	$CO_2(g) + H_2O(l) \rightarrow HCOH(g) + O_2(g)$	4	522
(5)	$CO_2(g) + 2H_2O(l) \rightarrow CH_3OH(l) + \frac{3}{2}O_2(g)$	6	703
(6)	$CO_2(g) + 2H_2O(l) \rightarrow CH_4(g) + 2O_2(g)$	8	818
(7)	$N_2 + 3H_2O(l) \rightarrow 2NH_3(g) + \frac{3}{2}O_2(g)$	6	678
(8)	$CO_2(g) + H_2O(l) \rightarrow \frac{1}{6}C_6H_{12}O_6(l) + O_2(g)$	4	480

4.25 Verify the following expression for the velocity distribution of free electrons

$$p(\mathbf{v}) = \frac{(m/\hbar)^3}{4\pi^3\rho}\frac{1}{\exp[\beta(mv^2/2 - \mu)] + 1},$$

and show that it reduces to the Maxwell–Boltzmann distribution at high temperature

$$p(\mathbf{v}) = \left(\frac{m}{2\pi k_B T}\right)^{3/2} \exp[-\beta m v^2/2].$$

4.26 For free electrons at 0 K, show that the internal energy can be expressed as:

$$U = \frac{3N}{5}\epsilon_F,$$

where ϵ_F is the Fermi energy.

4.27 Show that, for free electrons at low temperature,

(i) the chemical potential can be written as

$$\mu/\epsilon_F = 1 - \frac{\pi^2}{12}\left(\frac{k_B T}{\epsilon_F}\right)^2 + \cdots,$$

(ii) the internal energy is given by

$$\frac{U}{N\epsilon_F} = \frac{3}{5}\left[1 + \frac{5\pi^2}{12}\left(\frac{k_B T}{\epsilon_F}\right) + \cdots\right],$$

where ϵ_F is the Fermi energy.

4.28 Show that, according to the free-electron model, the reduced entropy per free electron at low temperature is the same as the reduced heat capacity:

$$\frac{S}{Nk_B} = \frac{C_V}{Nk_B} = \frac{\pi^2}{2}\frac{k_B T}{\epsilon_F}.$$

4.29 Show that, at 0 K, the bulk modulus of a metal can be estimated from

$$B \equiv -V\left(\frac{\partial P}{\partial V}\right)_T = \frac{2\rho\epsilon_F}{3},$$

where is ρ is the number density of free electrons, and ϵ_F is the Fermi energy.

4.30 The metallic bonds can be largely attributed to electron delocalization over macroscopic dimensions. To elucidate the microscopic origin of metallic bonds, we may estimate the cohesive energy of a metal from the difference between the internal energy of free electrons at 0 K and that of each free electron confined in confined in a cube of length $a = (1/\rho)^{1/3}$. Show that the free-electron model predicts the cohesive energy per free electron is given by

$$\Delta u = \left[\frac{(3\pi^2)^{2/3}}{5\pi^2} - 1\right] \times \frac{3h^2}{8ma^2} \approx -0.302 \times \frac{h^2}{ma^2}.$$

Compare the theoretical prediction with experimental data for some real metals and discuss the discrepancies.

4.31 The free-electron model is often used as a starting point to interpret the electrical conductivity of metallic materials. Image that an external electric field $\mathbf{E}$ is applied to a metal of free electron density ρ, resulting in an electrical current $\mathbf{j}$. The electrical conductivity is defined as

$$\sigma \equiv |\mathbf{j}|/|\mathbf{E}|.$$

The current may be attributed to all free electrons moving with an average velocity $\mathbf{v}$, i.e.,

$$\mathbf{j} = -\rho e\mathbf{v}.$$

Assuming that the directed motion of the free electrons is gained from interacting with the external field over the relaxation time τ (i.e., a duration determined by the consecutive collisions of the electron with other particles in the background), we may estimate the average velocity from Newton's equation

$$\mathbf{v} = -e\mathbf{E}\tau/m.$$

Using these expressions for $\mathbf{v}$ and $\mathbf{j}$, we deduce the Drude equation for the electrical conductivity

$$\sigma = \frac{\rho e^2\tau}{m}.$$

(i) Estimate the relaxation time of free electrons in copper under ambient condition based on its free electron density and electrical conductivity;
(ii) Based on the relaxation time obtained above, predict the mean free path $\lambda = \tau|\mathbf{v}|$ from $|\mathbf{v}| = \sqrt{3k_BT/m}$, i.e., the electron velocity derived from the classical limit;
(iii) Repeat the above procedure using $|\mathbf{v}| = v_F = \hbar k_F/m$, i.e., the low temperature limit of the electron velocity;
(iv) Discuss the physical significance of the mean free path and why the classical velocity adopted by the classical Drude model may lead to misleading conjectures.

4.32 The Wiedemann–Franz law is an empirical correlation between the thermal conductivity (κ) and the electrical conductivity (σ) for metallic materials. It states that the ratio of κ and σ is linearly proportional to the absolute temperature (T), i.e.,

$$\frac{\kappa}{\sigma T} = A,$$

where the proportionality constant A has a numerical value between $2.1 - 3.4 \times 10^{-8}$ WΩK^{-2} for common metals. We may use the free-electron model to justify/understand this empirical relation.

(i) Derive the following expression for the thermal conductivity

$$\kappa = \frac{1}{3}\rho v^2 \tau c_v,$$

where v is the mean-square electron velocity, τ is the electron relaxation time (see Problem 4.31), and c_v represents the molar constant-volume heat capacity of electrons. (Hint: Equating the heat flux in the x-direction, $\mathbf{j}_{x,q} = -\kappa(\partial T/\partial x)$ with the convection of energy in the same direction $-\rho \mathbf{v}_x \cdot (\mathbf{v}_x \tau)(\partial u/\partial x)$, where ρ is the electron density, and u is the molar internal energy.)

(ii) Derive the ratio of κ and σ based on the classical model (i.e., the Maxwell–Boltzmann distribution) for the average velocity

$$\frac{\kappa}{\sigma T} = \frac{3}{2}\left(\frac{k_B}{e}\right)^2.$$

(iii) Derive the ratio of κ and σ based on the results from the Sommerfeld theory at the low temperature limit

$$\frac{\kappa}{\sigma T} = \frac{\pi^2}{3}\left(\frac{k_B}{e}\right)^2.$$

(iv) Discuss the theoretical predictions of the classical and quantum models.

4.33 The thermal conductivity of a solid material consists of contributions due to electronic motions and lattice vibration. As discussed in Problem 4.32, the electronic component is well described by the Wiedemann–Franz law. To a first-order approximation, the lattice contribution may be described in terms of phonons as the energy carriers. Accordingly, the thermal conductivity is given by

$$\kappa = \frac{1}{3}\rho v^2 \tau c_v,$$

where ρ and c_v now correspond to the number density and molar heat capacity of phonons, and τ is the phonon relaxation time. The phonon velocity may be simply approximated by the low-frequency speed of sound

$$v = \sqrt{\frac{B + 4C/3}{\hat{\rho}}},$$

where B and C stand for the bulk modulus and shear modulus of the elastic materials, respectively, and $\hat{\rho}$ is the mass density. These equations provide useful guidelines to design novel materials with low thermal conductivity.[48]

---

48 See, e.g., Toberer E. S. et al, "Phonon engineering through crystal chemistry", *J. Mater. Chem.*, 21, 15843–15852 (2011).

(i) Discuss specific strategies that might be taken to reduce the lattice thermal conductivity. Can you name some specific materials that meet the criteria?
(ii) Estimate the low-frequency speed of sound for a typical steel alloy with $B = 170$ GPa, $C = 80$ GPa and $\hat{\rho} = 7{,}700$ kg/m^3.
(iii) Assume that the Debye temperature for an alloy is $\theta_D = 470$ K and the molecular weight is 60 g/mol, predict the specific heat capacity of the material at $T = 300$ K based on the Debye model.
(iv) Using the lattice thermal conductivity $\kappa = 60$ W/(m K), estimate the phonon relaxation time and mean free path and discuss their significance.

4.34 The Debye function is defined as

$$D(x) \equiv \frac{3}{x^3}\int_0^x \frac{t^3}{e^t - 1}dt.$$

Show that
(i) the molar entropy of a monatomic solid predicted by the Debye model is given by

$$S/R = 4D(\theta_D/T) - 3\ln(1 - e^{-\theta_D/T}),$$

where R is the gas constant, and $\theta_D \equiv h\upsilon_D/k_B$.
(ii) at high temperature, the above equation can be approximated by

$$S/R = 4 + 3\ln(T/\theta_D).$$

(iii) at low temperature, the entropy is given by

$$S/R = \frac{4\pi^4}{5}(T/\theta_D)^3.$$

4.35 It has been demonstrated that the Debye model can be used to estimate the entropy at the melting point for a number of monatomic compounds (e.g., Li, Ga, Na, Hg, Al, Mg, In, K, Zn, Cu, Rb, Ag, Si, Cs, Pb, and Au).[49] The Debye temperature was correlated with the ratio of the densities of the solid and liquid phases

$$\frac{\theta_D^S}{\theta_D^L} \approx \left(\frac{\rho^S}{\rho^L}\right)^\gamma,$$

where γ is known as the Grüneisen parameter. For example, at the melting point ($T = 1356$ K), the densities of copper at the solid and liquid states are 8.2572 g/cm^3 and 7.992 g/cm^3 respectively, and the Grüneisen parameter is $\gamma \approx 1.01$. Estimate the entropy of fusion and compare the result with Richard's melting rule, an empirical observation showing that the normal fusion entropy is approximately a universal constant, $\Delta S_f/R \approx 1.1$.

4.36 Show that the Debye model for the heat capacity in the low- and high-temperature limits:

$$C_V/(3Nk_B) = \begin{cases} \frac{4\pi^4}{5}(T/\theta_D)^3, & T \to 0\,\text{K}, \\ 1, & T \to \infty. \end{cases}$$

49 Lilley D., Jain A., and Prasher R., "A simple model for the entropy of melting ofmonatomic liquids", *Appl. Phys. Lett.* 118, 083902 (2021).

4.37 Experimental data suggest that, at low temperature, the heat capacity of many metals can be written in the form

$$C_V/R = AT + BT^3,$$

where R is gas constant, A and B are empirical coefficients that depend on the specific metal. Can you identify the physical meanings of parameters A and B? How are the two contributions to C_V compared with each other for gold at 20 K? Please compare the heat capacity predicted by this equation with that predicted by the Debye model. The Debye temperature for gold is 165 K, and the Fermi temperature is about 6.39×10^4 K.

5

Cooperative Phenomena and Phase Transitions

In order to acquire a deeper understanding of macroscopic properties resulting from interactions among a vast number of particles, it is advantageous to analyze simple physical models. Once these models are comprehended, similar techniques can be employed to represent more complicated real-world systems. The Ising model embodies such attributes, rendering it an invaluable framework for investigating cooperative phenomena and phase transitions taking place in both natural and industrial systems.

The Ising model was originally proposed to explain spontaneous phase changes observed in certain metallic materials, where they shift from a paramagnetic state to a ferromagnetic state. When a thermodynamic system is described by this model, each particle can exist in one of two states: "spin up" or "spin down." Despite its simplicity, this two-state model can be utilized to represent a wide range of cooperative phenomena, including ligand binding with proteins, the formation of helical structures in polypeptides or DNA, and the ionization of macromolecules.

The Ising model is particularly valuable for understanding the universal aspects of phase transitions and critical phenomena. In addition, it is frequently employed to elucidate fundamental Figure 5.4. The Helmholtz concepts in statistical mechanics, such as correlation length, mean-field approximation, transfer-matrix techniques, order parameter, broken symmetry, and universality. Furthermore, the Ising model finds practical use in data analysis within the fields of machine learning and statistical inference. Its application in these domains allows for the exploration and interpretation of complex datasets. By leveraging the Ising model, researchers can gain insights into the underlying patterns, dependencies, and statistical properties of the data they analyze.

In this chapter, our focus will be on establishing the partition functions for one- and two-dimensional Ising models. While the exact partition function for the three-dimensional Ising model is not currently known, it can be calculated by using mean-field approximation or simulation method (to be discussed in Chapter 6). We will explore the thermodynamic properties and correlation functions obtained from these Ising models, showcasing their relevance in understanding and analyzing various cooperative phenomena that have practical significance.

Additionally, this chapter will introduce theoretical procedures for describing phase transitions and critical phenomena. We will elucidate why different physical systems, such as gases or magnets, exhibit similar critical behavior. By delving into these concepts, we will gain a deeper understanding of the underlying principles governing phase transitions and critical phenomena in diverse thermodynamic systems.

Fundamentals and Practice in Statistical Thermodynamics, First Edition. Jianzhong Wu and John M. Prausnitz.

Companion website: www.wiley.com/go/jianzhongwu/fundamentalsandpractice

5.1 Spins and Ferromagnetism

Ferromagnetic materials, such as iron, cobalt, and nickel, find wide application in electrical and electromechanical devices. Below a certain temperature, these materials undergo a spontaneous transition from a paramagnetic state, where atomic magnetic moments are randomly oriented, to a ferromagnetic state, characterized by nearly complete alignment of the atomic magnetic moments. The study of this ferromagnetic transition was pioneered by Wilhelm Lenz and his PhD student Ernst Ising in 1925. They introduced a simple mathematical model, later known as the Ising model, which holds a prominent place in statistical mechanics, both in terms of its fundamental principles and practical applications.

Microscopically, ferromagnetism can be attributed to the alignment of the spin states associated with unpaired electrons of individual atoms in a magnetic material. As discussed in Chapter 4, spin is a quantum degree of freedom associated with the angular momentum of electrons. While the quantum-mechanical property does not have a classical counterpart, it may be loosely thought of as the intrinsic angular momentum of a spherical particle, as schematically illustrated in Figure 5.1.[1] The electron spin behaves like a tiny magnet and can be aligned either parallel or antiparallel to an external magnetic field **B**. When it aligns with the external magnetic field, the spin has a negative energy of $-h \equiv -\mu_e |\mathbf{B}|$, where μ_e stands for intrinsic magnetic moment. In this case, the electronic state is commonly referred to as "spin up." Conversely, the spin takes a positive energy of $+h \equiv \mu_e |\mathbf{B}|$ when its magnetic dipole is opposite to the direction of the external field, which is called "spin down."

In a magnetic material, the interactions between neighboring spins are responsible for the phase transition in a ferromagnetic material from a paramagnetic state to a ferromagnetic state. Approximately, the spin–spin interactions can be represented by a lattice model where each lattice site accomodates a spin with two microscopic states, i.e., spin up and spin down. The ferromagnetic interaction favors two neighboring spins aligned in the same direction (i.e., with a negative energy). Conversely, antiferromagnetic interaction yields a positive energy when two neighboring spins are pointing to the same direction. Because the energy is minimized by arranging the spins in a regular pattern while entropy favors the random distribution of spin orientations, a phase transition takes place below a critical temperature. For the transition from a paramagnetic state to a ferromagnetic state, the critical temperature is called the Curie temperature, named after Pierre Curie,

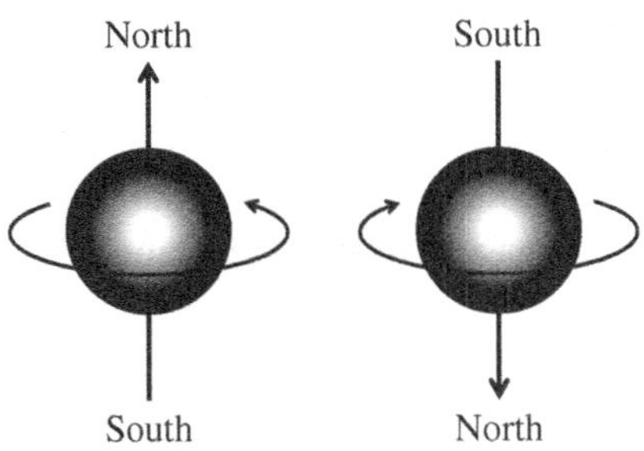

Figure 5.1 The self-rotation of a spherical particle in clockwise and counterclockwise directions.

1 Spin is a quantum mechanical property of elementary particles that is distinct from their translational and rotational motion. While spin has no classical analog, it can be loosely understood as the intrinsic angular momentum of a particle. In this sense, one may think of spin as the particle "spinning" around its own axis in either a clockwise or counterclockwise direction, though this should not be taken too literally as the concept of spin is fundamentally quantum mechanical and cannot be fully understood in classical terms.

who received the Nobel Prize jointly with his wife Marie Curie in 1903. Similarly, antiferromagnetic interactions can also lead to order-disorder transition. In this case, the critical temperature is known as the Néel temperature, named after Louis Néel, who first identified antiferromagnetism (ordering with neighboring spins aligned in opposite directions) and won the Nobel Prize in 1970 for his studies of the magnetic properties of solids.

Lenz and Ising initially used the one-dimensional version of the two-state model, subsequently known as the Ising chain, to describe the magnetic phase transition of ferromagnetic materials. However, in the thermodynamic limit, the Ising chain does not exhibit spontaneous magnetization at a finite temperature. The phase transition was not identified until 1944, when Lars Onsager published an exact solution for the two-dimensional Ising model.

The Ising model has been extended to describe phase transitions and critical phenomena in a wide range of thermodynamic systems. In the case of its application to nonmagnetic systems like ternary mixtures of liquid water, oil, and amphiphiles as elucidated in Section 5.7, each lattice site, or "spin," can assume multiple values to represent the spatial distribution of different chemical species. By employing the Ising model to analyze this system, it becomes feasible to investigate the influence of interactions among various chemical species on the microscopic structure and phase behavior of microemulsions. This type of extension underscores the significant utility of the Ising model in describing complex many-body systems.

5.1.1 Summary

The Ising model was originally developed to explain magnetic phase transition in ferromagnetic materials. Over time, its applications in statistical mechanics have been extended, making it a versatile theoretical tool to represent various properties of many-body systems. In later applications, the term "spin" is used for the sake of simplicity to describe the microscopic states of individual elements in a macroscopic system. In that regard, a spin is a variable capable of assuming two or more values and is not necessarily linked to any magnetic or electronic properties. This broader interpretation of "spin" will be adopted throughout the remaining sections of this chapter.

5.2 The Ising Chain Model

The one-dimensional (1D) Ising model, commonly referred to as the Ising chain, stands as a fundamental statistical–mechanical model for elucidating cooperative phenomena arising from interactions among individual entities in a thermodynamic system. These interactions encompass physical, chemical, or biological aspects and may give rise to cooperative behavior, i.e., novel properties that are not displayed by the entities in isolation. Because the partition function can be derived analytically, the 1D-Ising model provides a nontrivial platform to elucidate the collective behavior or coordination of individual elements.

Figure 5.2 illustrates a simple representation of an Ising chain with N spins. We assume that each spin can only be in one of the two states, as depicted by the up and down arrows. For an Ising chain with ferromagnetic interactions, two immediate-neighboring spins acquire a negative energy ($-\varepsilon < 0$) when they are oriented in the same direction (aligned); and the energy is positive otherwise ($+\varepsilon > 0$). In the presence of an external magnetic field, each spin has a negative energy $-h < 0$ if it is aligned with the direction of the external field (i.e., spin up), and a positive energy $+h > 0$ if it is oriented opposite to the magnetic field (i.e., spin down).[2]

2 See Section 5.1 for the relation between spin energy and magnetic field.

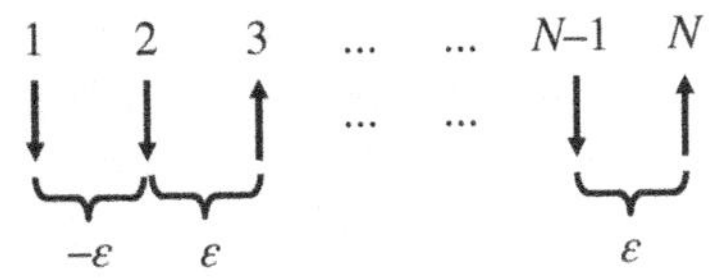

Figure 5.2 1D-Ising model for a chain of N spins with nearest-neighboring interactions.

5.2.1 The Partition Function for an Ising Chain

For an Ising chain with N spins, each microstate ν is specified by the spin orientations (i.e., ups and downs for all spins)

$$\nu = \{s_1, s_2, s_i, \dots, s_N\}. \tag{5.1}$$

For numerical convenience, we may designate $s_i = +1$ when spin i is up, and $s_i = -1$ when spin i is down. Accordingly, we can write the system energy as

$$E_\upsilon = -(s_1 + s_2 + \dots + s_N)h - (s_1 s_2 + s_2 s_3 + \dots + s_{N-1}s_N)\varepsilon \tag{5.2}$$

where $-s_i h$ stands for the energy of spin i due to the presence of the external field; $-s_i s_{i+1}\varepsilon$ stands for the coupling energy between spin i and its nearest neighbor $i+1$ (Figure 5.3).

The canonical partition function,[3] Q, is defined as a summation of the Boltzmann factors for all microstates

$$Q = \sum_\nu e^{-\beta E_\nu} = \prod_{i=1}^{N} \sum_{s_i=\pm 1} \exp\left[\beta h \sum_{i=1}^{N} s_i + \beta\varepsilon \sum_{i=1}^{N-1} s_i s_{i+1}\right] \tag{5.3}$$

where $\beta = 1/(k_B T)$, and the summation over all microstates is written as $\prod_{i=1}^{N}\sum_{s_i=\pm 1}$. This summation has 2^N terms, encompassing all possible orientations of N spins in the Ising chain.

Without the external field ($h = 0$), we can evaluate the partition function analytically by replacing the summation over the spin orientations with that for parameter $b_i = s_i s_{i+1} = \pm 1$, i.e.,

$$Q = 2\prod_{i=1}^{N-1} \sum_{b_i=\pm 1} \exp\left(\beta\varepsilon \sum_{i=1}^{N-1} b_i\right) = 2(e^{\beta\varepsilon} + e^{-\beta\varepsilon})^{N-1} = 2^N \cosh^{N-1}(\beta\varepsilon) \tag{5.4}$$

where a factor of 2 accounts for two possible orientations for one of the end spins. Intuitively, parameter b_i reflects the alignments of neighboring spins, or bond formation because of the difference in energy due to different alignments.

According to Eq. (5.4), the Helmholtz energy of the Ising chain is

$$F = -k_B T \ln Q = -k_B T N\left[\ln 2 + \frac{N-1}{N} \ln\cosh\beta\varepsilon\right]. \tag{5.5}$$

Spin pairs	Energy
↑↓ ↓↑	$\varepsilon > 0$
↓↓ ↑↑	$-\varepsilon < 0$

Figure 5.3 Interaction of two immediate-neighboring spins in a ferromagnetic system. If neighboring spins are aligned in parallel, the pair energy is $-\varepsilon$; if neighboring spins are antiparallel, the pair energy is $+\varepsilon$. Here, ε is a positive number with the units of energy.

3 In the context of the lattice model (Section 5.7) where the spin state is associated with the occupancy by a molecule, h would have the physical meaning of chemical potential. In that case, Eq. (5.3) would represent the grand canonical partition function.

From the Helmholtz energy, we can derive other thermodynamic properties. If the spins do not interact with each other, $\varepsilon = 0$ and $\cosh \beta\varepsilon = 1$; Eq. (5.5) leads to the entropy of a randomly oriented spins as predicted by the Boltzmann equation, $S = k_B \ln 2^N$.

In the presence of an external field, the partition function can also be evaluated analytically. However, the algebra becomes more involved (see Appendix 5.A). In the limit of large N, the partition function can be approximated by

$$Q \approx \lambda_+^N \tag{5.6}$$

with

$$\lambda_+ = e^{\beta\varepsilon}\left[\cosh(\beta h) + \sqrt{\sinh^2(\beta h) + e^{-4\beta\varepsilon}}\right]. \tag{5.7}$$

Eq. (5.6) becomes exact when $N = \infty$. For $h = 0$, $\lambda_+ = 2\cosh(\beta\varepsilon)$, and Eq. (5.4) is recovered is recovered for $N = \infty$.

From the partition function, we can readily obtain the Helmholtz energy of the Ising chain in the presence of an external field.

$$F = -k_B T \ln Q = -Nk_B T \ln \lambda_+. \tag{5.8}$$

In dimensionless units, the Helmholtz energy per spin is given by

$$\beta F/N = -\beta\varepsilon - \ln\left[\cosh(\beta h) + \sqrt{\sinh^2(\beta h) + e^{-4\beta\varepsilon}}\right]. \tag{5.9}$$

Based on Eq. (5.9), we may derive other thermodynamic properties by various differentiations of the Helmholtz energy.

Figure 5.4 shows the Helmholtz energy as a function of the reduced external potential and coupling energy. At given external field (βh = constant), the Helmholtz energy decreases monotonically when the reduced coupling energy ($\beta\varepsilon$) increases, indicating that the alignment of spins becomes increasingly preferred over random orientation. In other words, the individual elements become more cooperative as the nearest-neighbor energy increases. When the reduced coupling energy is fixed ($\beta\varepsilon$ = constant), the Helmholtz energy is a symmetric function of βh; and it decreases monotonically as the external field increases its magnitude. In the presence of a strong external field ($|\beta h| \gg 1$), Eq. (5.9) may be approximated as

$$\beta F/N \approx -\beta\varepsilon - \beta h. \tag{5.10}$$

Eq. (5.10) becomes exact when all spins are aligned in the direction of the external field. In that case, the entropy of the Ising chain disappears, and the internal energy is identical to the Helmholtz energy.

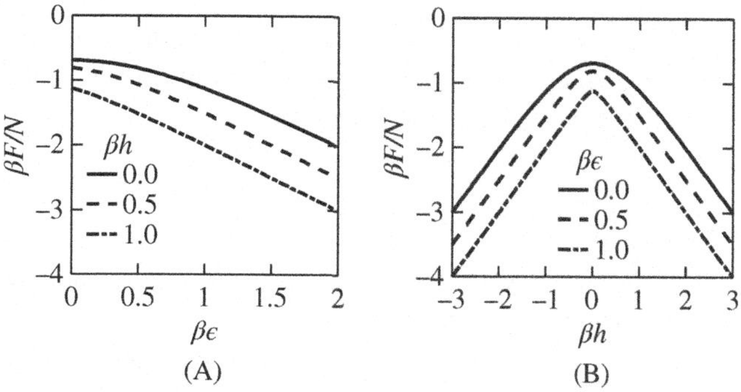

Figure 5.4 The Helmholtz energy per spin for an Ising chain depends on the coupling energy $\beta\varepsilon$ (A) and external magnetic field βh (B). All quantities are in dimensionless units.

5.2.2 Thermodynamic Properties of an Ising Chain at Zero Field

When $h = 0$ (no external field), Eq. (5.5) predicts that the Helmholtz energy of an Ising chain with a large number of spins simplifies to

$$\beta F/N = -\ln[2\cosh(\beta\varepsilon)]. \tag{5.11}$$

From Eq. (5.11), it is straightforward to derive the internal energy per spin

$$U/N = \partial(\beta F/N)/\partial\beta = -\varepsilon\tanh(\beta\varepsilon), \tag{5.12}$$

the reduced entropy

$$S/Nk_B = -\beta\varepsilon\tanh(\beta\varepsilon) + \ln[2\cosh(\beta\varepsilon)]. \tag{5.13}$$

From Eq. (5.12), we can also derive the heat capacity

$$C_V = \left(\frac{\partial U}{\partial T}\right)_N = \frac{(\beta\varepsilon)^2 Nk_B}{\cosh^2(\beta\varepsilon)}. \tag{5.14}$$

As discussed in Section 2.3.3, C_V reflects the fluctuation of the system energy and, in dimensionless units, is given by

$$C_V/k_B = <(\beta E)^2> - <\beta E>^2. \tag{5.15}$$

Similar expressions can be derived from the Helmholtz energy under the influence of an external field, as given by Eq. (5.9); however, these equations are considerably more complicated.

Figure 5.5 shows the Helmholtz energy, the internal energy, the entropy, and the heat capacity of an Ising chain without an external field. All these thermodynamic properties are smooth functions of the reduced reciprocal temperature $\beta\varepsilon$. As temperature approaches zero ($\beta\varepsilon \to \infty$), the spins are in the lowest energy state where they are perfectly aligned. In this case, the heat capacity vanishes because, when $k_B T$ is negligible compared to the coupling energy ε, and a small rise in temperature does not change the spin orientations. As temperature rises further, the internal energy becomes less negative because of the presence of the spins of opposite orientations. At high temperature ($\beta\varepsilon \to 0$), the spin orientations are nearly randomly distributed. In this case, the Helmholtz energy is dominated by the entropy contribution. Both the internal energy and the heat capacity approach zero because of the random spin distribution. Whereas the Helmholtz energy, the entropy, and the internal energy of an Ising chain are all monotonic functions of $\beta\varepsilon$, the heat capacity exhibits a maximum at $\beta\varepsilon \approx 1.2$, reflecting a competition between the energy and the entropy effects.

5.2.3 Magnetization

In the presence of an external field, the spins are biased toward the direction of the external field in order to attain the lowest free energy. The average magnetization per spin is given by an ensemble average of the overall spin orientations, which can be derived from Eq. (5.6),

$$m = \frac{1}{N}\left\langle \sum_{i=1}^{N} s_i \right\rangle = \frac{1}{N}\frac{\partial \ln Q}{\partial(\beta h)} = \frac{\sinh(\beta h)}{\sqrt{\sinh^2(\beta h) + e^{-4\beta\varepsilon}}}. \tag{5.16}$$

Eq. (5.16) predicts $m = 0$ when $h = 0$, meaning that the spins have an equal likelihood of being in an up or down state, there is no spontaneous magnetism at any temperature.

Figure 5.6 shows that, for large N, spin magnetization m is a smooth function of the reduced external energy βh and the reduced coupling parameter $\beta\varepsilon$. At large βh, $\beta h \gg 1$ and $\sinh^2(\beta h) \gg \exp[-4\beta\varepsilon]$ so that $m \approx 1$, meaning that all spins are aligned in the direction of the

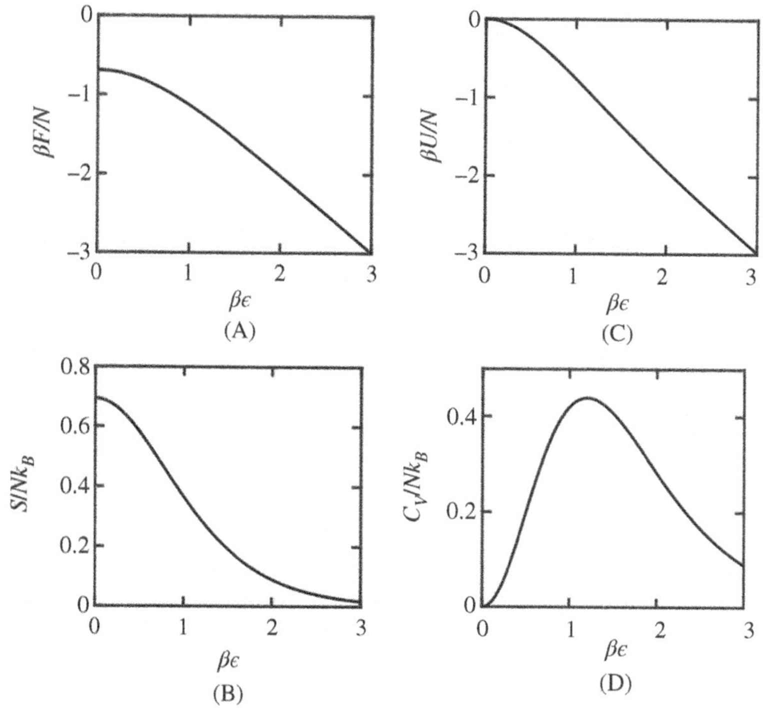

Figure 5.5 Thermodynamic properties of an Ising chain in the absence of an external field. (A) Helmholtz energy; (B) internal energy; (C) entropy; (D) heat capacity. All quantities are in dimensionless units.

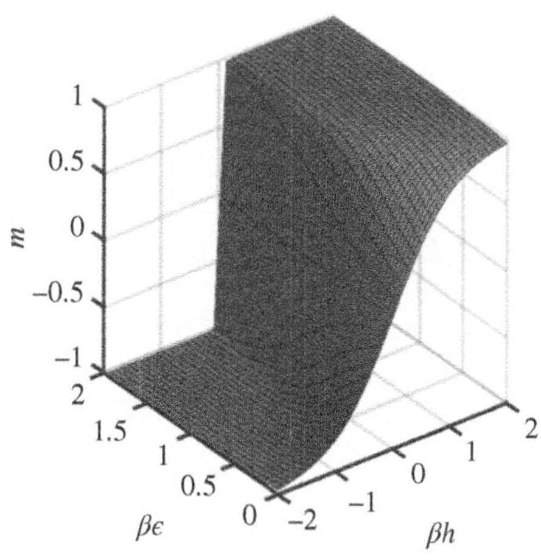

Figure 5.6 Magnetization m as a function of the reduced external field βh, and the reduced interaction energy $\beta \varepsilon$.

external field. As temperature T approaches zero, the system reaches magnetic saturation for any nonvanishing external field.[4]

The response of the spin magnetization to an applied magnetic field is characterized by magnetic susceptibility, which can be derived from the second-order derivative of the partition function

4 At saturation, all spins are aligned, i.e., $m = 1$ or -1.

(or the first-order derivative of magnetization) with respect to βh:

$$\chi = -\frac{1}{N} \cdot \frac{\partial^2 \ln Q}{\partial(\beta h)^2} = \frac{\partial m}{\partial(\beta h)} = \frac{e^{-4\beta\varepsilon} \cosh(\beta h)}{[\sinh^2(\beta h) + e^{-4\beta\varepsilon}]^{3/2}}. \tag{5.17}$$

At zero external field ($\beta h = 0$), Eq. (5.17) reduces to $\chi = e^{2\beta\varepsilon}$, suggesting that the magnetic susceptibility increases exponentially with the coupling energy. In other words, the Ising chain becomes easily magnetized by a magnetic field as the coupling energy increases.

By carrying out the derivatives of the partition function, we see that the magnetic susceptibility is related to the fluctuations of the magnetization around the average value

$$\chi = \frac{1}{N}\left[\left\langle \sum_{i=1}^{N}\sum_{j=1}^{N} s_i s_j \right\rangle - \left\langle \sum_{i=1}^{N} s_i \right\rangle^2\right]. \tag{5.18}$$

Because the terms inside the square brackets can be rewritten as

$$\left\langle \sum_{i=1}^{N}\sum_{j=1}^{N} s_i s_j \right\rangle - \left\langle \sum_{i=1}^{N} s_i \right\rangle^2 = \left\langle \sum_{i=1}^{N} (s_i - \langle s_i \rangle) \right\rangle^2, \tag{5.19}$$

susceptibility is always nonnegative. Intuitively, it provides a measure of the correlation between the orientations of neighboring spins. If the spin orientations are totally independent, $\chi = 0$, i.e., there is no correlation among the spins. On the other hand, the susceptibility takes a positive value and increases if the spin orientations are more strongly correlated with each other.

Figure 5.7 visualize the magnetic susceptibility of an Ising chain as a function of βh and $\beta\varepsilon$. As shown in Figure 5.7A and B, the maximum of χ occurs at zero external field ($\beta h = 0$), indicating

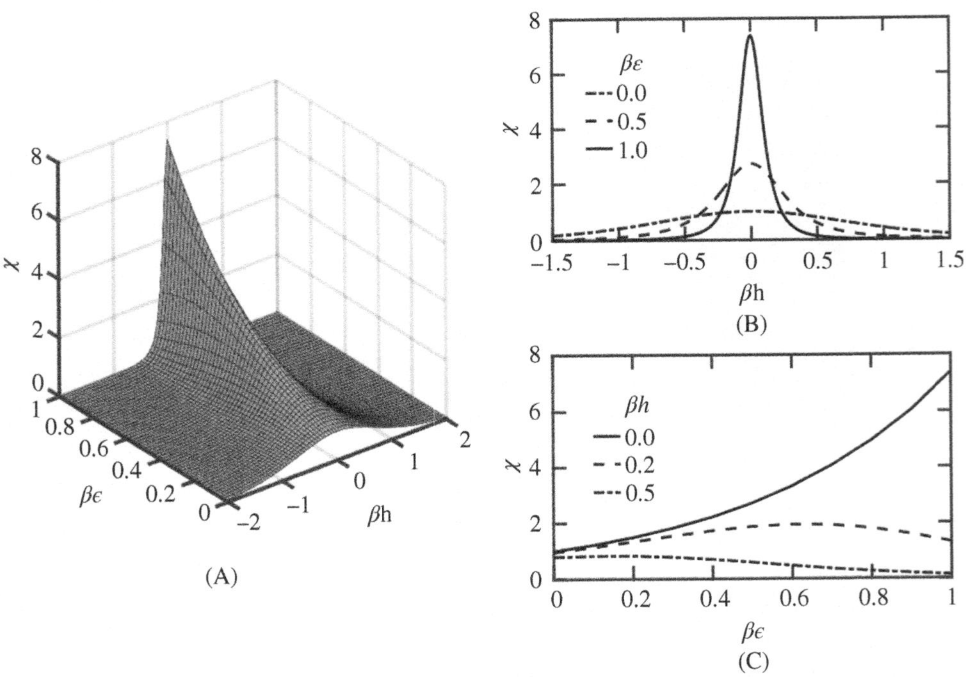

Figure 5.7 Susceptibility of a long Ising chain as a function of βh and $\beta\varepsilon$. (A) $\chi(\beta h, \beta\varepsilon)$ predicted by Eq. (5.17). (B) Variation of the susceptibility with the (reduced) external field, βh. (C) Variation of the susceptibility with the (reduced) coupling energy, $\beta\varepsilon$.

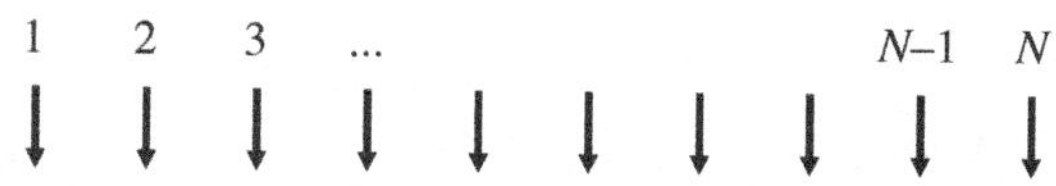

Figure 5.8 An ordered Ising chain refers to the state when all spins are aligned, i.e., all up or all down. In this picture, all spins are down.

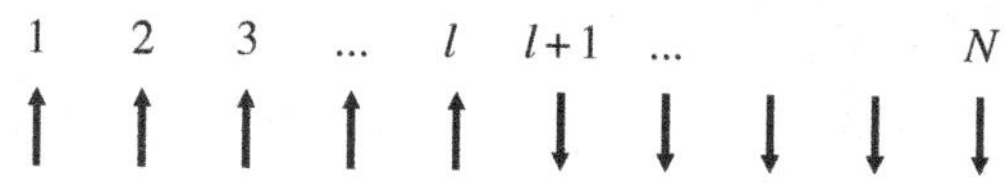

Figure 5.9 An Ising chain where $s_i = 1$ for 1 to l, and $s_i = -1$ for $i = l+1$ to N.

that the external field suppresses fluctuations in the spin orientations. In this case, the susceptibility rises monotonically with the coupling energy $\beta\varepsilon$. As $\beta\varepsilon$ increases, the neighboring spins become more correlated, i.e., the orientation of one spin strongly affects those of neighboring spins. In the presence of a strong external field (viz., a large valuate of βh), the susceptibility decreases monotonically with $\beta\varepsilon$, as illustrated by Figure 5.7C. The opposite trends suggest that, in the presence of an external field, χ is maximized at an intermediate value of the coupling energy.

The absence of spontaneous magnetization in an Ising chain is understandable because the energy reduction due to the alignment of spins is insufficient to overcome the entropy loss relative to the state where the spin orientations are randomly distributed. To illustrate the interplay of entropy and internal energy, we may consider an Ising chain where all spins are down, as shown in Figure 5.8. In this case, the spins are in an ordered state ($m = -1$) with the total energy of $E_o = (N-1)\varepsilon$, where subscript "o" stands for an ordered state. Now consider a set of disordered states, all with the same energy. One of such disordered states is shown in Figure 5.9. Suppose that for all positions from 1 to l, the spin is up and for all subsequent positions, the spin is down. There are $N-1$ such disordered states with the same energy (because l may vary from 2 to N). Regardless of the value of l, the system energy is $E = E_o + 2\varepsilon$.

At finite temperature T, the Helmholtz energy difference, ΔF, between the two states shown in Figures 5.8 and 5.9 is approximately given by[5]

$$\Delta F = 2\varepsilon - k_B T \ln(N-1) \tag{5.20}$$

where the first term on the right accounts for the change in energy, and the second term for the change in entropy. In the limit $N \to \infty$, ΔF is always negative for all $T > 0$, indicating that disordered states are always more likely in the presence of thermal fluctuation. In other words, with only nearest-neighbor interactions, spontaneous magnetization does not take place at any temperature in an Ising chain of infinite length.[6]

5.2.4 Spin–Spin Correlation Functions

As discussed above, the interaction between neighboring spins leads to spin–spin correlations. Quantitatively, we can describe the cooperative behavior in terms of correlation functions.

Specifically, the pair distribution function (PDF) describes how a pair of spins, i and j, are correlated in terms of their orientations in the same microstate,

$$g_{ij} \equiv \langle s_i s_j \rangle. \tag{5.21}$$

5 Equation (5.20) is approximate because the difference in energy ΔU and the difference in entropy ΔS are evaluated independently, with approximations $\Delta U = 2\varepsilon$ and $\Delta S = k_B \ln(N-1)$.

6 At temperatures below $k_B T/\varepsilon \approx 2/\ln(N-1)$, the free energy attains a minimum value at a finite value of magnetization ($m \neq 0$), indicating that an Ising chain will have a preference toward spin alignments when the chain length is finite. However, spontaneous magnetization or symmetry breaking does not occur because the free energy barrier for switching between spin up and spin down states remains finite.

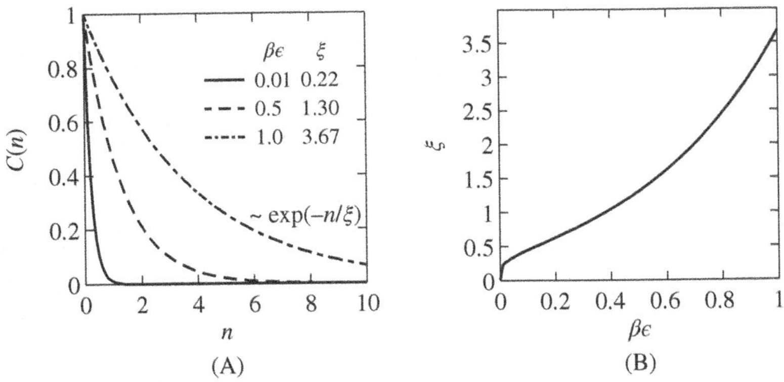

Figure 5.10 (A) The pair correlation functions for spin distributions in a field-free Ising chain. (B) The correlation length versus the reduced nearest-neighbor energy at $\beta h = 0$.

For a cyclic or long Ising chain, there are no end effects, or the end effects are insignificant. In that case, g_{ij} depends only on the separation between the two spins, $n = |i - j|$. If $n = 0$, $g_{ii} = 1$, i.e., the spin is self-correlated. On the other hand, when $n \gg 1$, $g_{ij} = 0$, we say that the orientations of the two spins are independent to each other, i.e., they are uncorrelated.

Alternatively, the correlation between neighboring spins can be described by the pair correlation function (PCF), which is defined as the correlated fluctuation of spins deviating from their mean orientations:

$$C_{ij} \equiv \langle (s_i - \langle s_i \rangle)(s_j - \langle s_j \rangle) \rangle = \langle s_i s_j \rangle - \langle s_i \rangle \langle s_j \rangle. \tag{5.22}$$

In the absence of external field, $\langle s_i \rangle = 0$ so that $C_{ij} = g_{ij}$. For a cyclic or long Ising chain, PCF also depend only on the distance between spins i and j, $n = |j - i|$.

As discussed in Appendix 5.A, we can find an analytical expression for the spin–spin correlation function. For a cyclic or long Ising chain without the external field, we have

$$C(n) = \tanh^n(\beta\varepsilon) \equiv e^{-n/\xi}. \tag{5.23}$$

In Eq. (5.23), $\xi = 1/\ln[\coth(\beta\varepsilon)]$ defines the correlation length of this system. As discussed in Section 2.2.3, here the correlation length provides a measure of the distance over which the orientations of two spins in an Ising chain are correlated with each other. When the distance between spins is greater than the correlation length, the correlation effect can be considered insignificant, and thus the spins can be treated as independent.

Figure 5.10 illustrates the PCF $C(n) = C_{ij}$ and the correlation length ξ of an Ising chain in the absence of an external field at three reciprocal temperatures. At zero separation, $n = 0$ and $C(0) = 1$ because $s_i s_i = 1$ and $\langle s_i \rangle = 0$. In this case, we say that the spins are self-correlated. At large separations, $C(n) \to 0$ and the spin orientations become independent of each other. Figure 5.10B shows that, in the absence of the external field, the correlation length grows monotonically with the coupling energy. As $T \to 0$, $\xi \approx e^{2\beta\varepsilon}/2$ diverges, suggesting that all spins are correlated, i.e., all spins in the Ising chain align in the same directions at 0 K.

5.2.5 Summary

By studying the Ising chain model, we can gain valuable insights into the emergence of cooperative phenomena resulting from the interactions between individual elements within a thermodynamic system. In particular, analytical results derived from the 1D-Ising model are useful to comprehend abstract statistical-mechanical concepts such as PDFs, PCFs, and correlation length, which are absent in classical thermodynamics. These concepts are crucial for understanding diverse cooperative phenomena that emerge from the complex interactions among a large number of elements. In the following sections (Sections 5.3 and 5.4), we will elucidate applications of the 1D-Ising model to problems of practical interests.

5.3 Ionization of Weak Polyelectrolytes

The Ising model provides a general theoretical framework to describe cooperative phenomena in chemical and biological systems. Before more extensive discussions of this generic model, it is instructive to illustrate how a system of "spins" can be related to the microstates of specific chemical systems and their corresponding thermodynamic properties. In this section, we consider one such example, i.e., the ionization of pH-responsive polymers. This example elucidates the usefulness of a general statistical-mechanical model to represent seemingly unrelated phenomena of practical interest.

Weak polyelectrolytes are pH-responsive polymers consisting of ionizable monomers such as carboxylic acids (—COOH) and amines (—NH). Unlike their monomeric counterparts, the electrostatic behavior of a weak polyelectrolyte depends not only on the solution pH but also on interactions between neighboring segments. Because of the prominent effects of electrostatic interactions, ionization often plays a pivotal role in various applications of pH-responsive polymers such as drug delivery, gene delivery, sensors, membranes, and chromatography.[7]

Application of the 1D-Ising model to acid-base equilibria of weak polyelectrolytes was pioneered by Marcus, Steiner and others.[8] The so-called site-binding model provides a simple framework to rationalize the ionization properties of pH-responsive polymers. In addition to describing the ionization of weak polyelectrolytes, the site-binding model is applicable to a large class of adsorption processes including ions or ligands binding with macromolecules.[9]

As an illustrative example, consider the titration of a polybase, such as a single polyethylenimine (PEI) chain, in an electrolyte solution. Each repeating unit of this cationic polymer contains an amine (NH) group and two methylene groups (CH_2CH_2). Depending on the salt concentration and solution pH, the monomer can be in one of two charge states, i.e., the amine group is either positively charged or neutral. For a dilute polymer solution at high electrolyte concentration, we may neglect interactions between different polyelectrolyte chains, and the ionization behavior can be understood based on the extended structure of an isolated polymer chain. As shown schematically in Figure 5.11, the charge distribution along the polymer backbone can be described

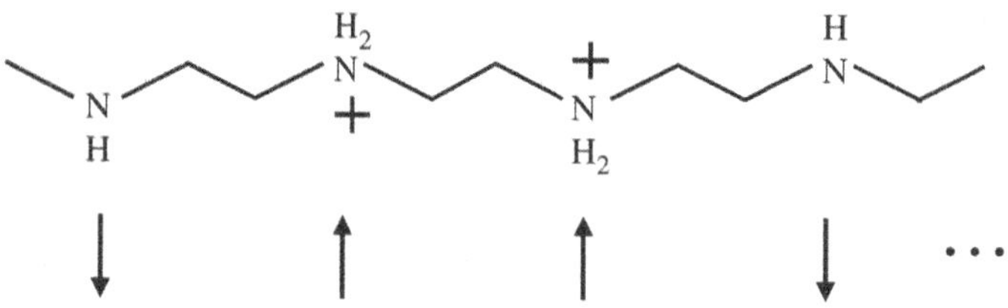

Figure 5.11 Protonation of a pH-responsive polymer such as polyethyleneimine (top) can be described by an Ising model (bottom). Here, the spin orientation denotes the electrostatic charge of each monomer ($-CH_2CH_2NH-$).

7 Kocak G., Tuncer C. and Butun V., "pH-responsive polymers", *Polym. Chem.* 8 (1), 144–176 (2017).
8 Marcus R. A., "Titration of polyelectrolytes at higher ionic strengths", *J. Phys. Chem.* 58 (8), 621–623 (1954), Steiner R. F., "Some aspects of pair interactions for a linear array of sites, as applied to adsorption problems", *J. Chem. Phys.* 22(8), 1458–1459 (1954); Katchalsky A., Mazur J. and Spitnik P., "Polybase properties of polyvinylamine", *J. Polym. Sci.* 23(104), 513–532 (1957).
9 Lavrinenko I. A. et al., "Cooperative oxygen binding with hemoglobin as a general model in molecular biophysics", *Biophysics* 67, 327–337 (2022).

in terms of the spin variable, i.e., "spin up" if the ionizable group (NH) is protonated, and "spin down" if it is neutral. At high electrolyte concentration, the electrostatic interaction is short-ranged (see Chapter 9). Consequently, protonation primarily affects the interaction among the nearest-neighboring segments along the polymer chain, while the polymer conformation, which mainly influences long-range intra-chain interactions, remains decoupled from its ionization behavior.[10]

As far as PEI protonation is concerned, we can describe the microstates of the polymer chain by various distributions of electrostatic charges along the polymer backbone. The spin variable, $s_i = \pm 1$, can be utilized to represent the protonation status of each amine group

$$s_i = 2n_i - 1 = \begin{cases} -1 & n_i = 0 \\ 1 & n_i = 1 \end{cases} \tag{5.24}$$

where $n_i = 0$ means that the amine group in monomer i is neutral, and $n_i = 1$ means that it is protonated with one additional hydrogen atom. The system energy includes a contribution due to the protonation of individual monomers. To account for the cooperative effect, it includes an additional term to account for interactions among ionized groups. As illustrated in Box 5.1, we can estimate the protonation energy from the dissociation equilibrium of individual monomers. To the first-order approximation, the cooperative effect can be captured with only the nearest-neighbor interactions.

Box 5.1 Estimation of the Protonation Energy

Proton dissociation from each polymer segment can be written as chemical reaction

$$BH^+ \rightleftharpoons B + H^+. \tag{5.25}$$

For polyethyleneimine, B stands for its repeating unit, $-CH_2CH_2NH-$. For each monomer, the Henderson–Hasselbalch equation relates the concentration ratio of the charged and the neutral segments to solution pH and the *intrinsic* dissociation constant K

$$\text{pH} = \text{pK} + \log[\text{B}]/[\text{BH}^+] \tag{5.26}$$

where the square brackets stand for the monomeric concentration. Approximately, proton dissociation from $-CH_2CH_2NH-$ is the same as that from $CH_3CH_2NH_2$, which has a pK value around 10.

Alternatively, the concentration ratio may be considered as the relative probability of a polymer segment in the protonated state to that in the neutral state. Without interactions with other amine groups along the polymer chain, the relative probability satisfies the Boltzmann distribution

$$\frac{[\text{B}]}{[\text{BH}^+]} = \exp(-\beta\Delta\Gamma) \tag{5.27}$$

where $\Delta\Gamma$ stands for the deprotonation energy (i.e., the energy difference between the neutral and the protonated groups). A comparison of Eqs. (5.27) and (5.26) leads to the protonation energy

$$\beta\Delta\Gamma = \ln[\text{BH}^+] - \ln[\text{B}] = \ln 10(\text{pK} - \text{pH}). \tag{5.28}$$

In application of the 1D-Ising model to the ionization of weak polyelectrolytes, each microstate is defined by the charge status of ionizable monomers, $n_i = 0, 1, i = 1, 2, \ldots, N$, where N represents

10 Gallagos A. and Wu J., "Hierarchical model of weak polyelectrolytes with ionization and conformation consistency", *Macromolecules* 56(12), 4760–4772 (2023).

the total number of ionizable sites in the polymer chain. Accordingly, the system energy is given by

$$E_v = \sum_{ij}{}' n_i n_j u - \sum_i n_i \Delta\Gamma \tag{5.29}$$

where subscript v represents microstates of the system, i and j are the indices of neighboring monomers, u stands for the interaction energy between nearest-neighboring segments. With the assumption that all ionizable sites are equivalent, we can derive the one-body energy, $\Delta\Gamma$, from the solution pH and the intrinsic dissociation constant of each protonated amine group (see Box 5.1). Intuitively, the one-body energy represents the proton chemical potential or the difference between the energy of a repeating unit in the neutral state and that in the protonated state.[11] In Eq. (5.29), the primed sum extends only to the nearest-neighbor segments.

With identities $n_i = (s_i + 1)/2$ and $n_i n_j = (s_i s_j + s_i + s_j + 1)/4$, we can rewrite the total energy in terms of the spin variable

$$E_v = \sum_{ij}{}' s_i s_j u/4 - \sum_i s_i(\Delta\Gamma - u)/2 + \Theta \tag{5.30}$$

where Θ is a constant independent of the microstate of the system. The partition function is given by

$$Q = \prod_{i=1}^{N} \sum_{s_i=\pm 1} \exp\left[-\beta \sum_{ij}{}' s_i s_j u/4 + \beta \sum_i s_i(\Delta\Gamma - u)/2\right] \tag{5.31}$$

where $\beta = 1/(k_B T)$, k_B is the Boltzmann constant, and T is the absolute temperature. Using parameters $\varepsilon \equiv -u/4$ and $h \equiv (\Delta\Gamma - u)/2$, we see that Eq. (5.31) is identical to the partition function of the 1D-Ising model. Therefore, all equations derived from the Ising model can be directly adopted to describe the ionization of weak polyelectrolytes.

For pH-responsible polymers, a property of central interest is the variation of the polymer charge in response to pH changes in the solution. The charge distribution along the polymer backbone can be specified by the probability of protonation for individual amine groups

$$\theta_i = \sum_v n_i e^{-\beta E_v}/Q. \tag{5.32}$$

For a short PEI chain, we can evaluate the partition function Q as well as the degree of ionization θ_i by directly enumerating the microstates. The degree of protonation for the entire polymer chain can be obtained by averaging over all protonation segments

$$\theta = \frac{1}{N}\sum_{i=1}^{N} \theta_i. \tag{5.33}$$

For the ionization of a single amine group, $N = 1$ and $Q = 1 + e^{\beta\Delta\Gamma}$. Using $\beta\Delta\Gamma = \ln 10(\text{pK} - \text{pH})$ discussed in Box 5.1, we can simplify Eq. (5.33) to

$$\theta = \frac{1}{1 + e^{-\beta\Delta\Gamma}} = \frac{1}{1 + K/a_H} \tag{5.34}$$

where K denotes the dissociation constant of the protonated amine, and a_H is the proton activity. Rearrangement of Eq. (5.34) reduces to the familiar Langmuir equation for dissociation equilibrium

$$K = \frac{(1 - \theta)a_H}{\theta}. \tag{5.35}$$

---

11 Strictly speaking, the relative probability is related to the difference between the free energy of the monomer in the protonated state and that in the deprotonated state. Here it is referred to as the energy difference because the quantity is independent of the microstates of the 1D-Ising model.

For $N = 2$, $Q = 1 + 2e^{\beta\Delta\Gamma} + e^{2\beta\Delta\Gamma - \beta u}$, and Eq. (5.33) becomes

$$\theta = \frac{1}{2} \cdot \frac{a_H/K_1 + 2a_H^2/K_2}{1 + a_H/K_1 + a_H^2/K_2} \tag{5.36}$$

where $K_1 = K/2$ and $K_2 = K^2 e^{\beta u}$. When the chain length is sufficiently long, the polymer end effects are negligible. In that case, we can obtain an analytical expression from Eq. (5.16)

$$\theta = \frac{\sinh[\beta(\Delta\Gamma - u)/2]}{2\sqrt{\sinh^2[\beta(\Delta\Gamma - u)/2] + e^{\beta u}}} + \frac{1}{2}. \tag{5.37}$$

Figure 5.12 shows the degree of protonation as a function of pH for a long ionizable polyelectrolyte (e.g., PEI with pK = 10). Without interaction among ionizable groups ($\beta u = 0$), the titration curve is identical to that for the corresponding monomeric species. The nearest-neighbor interaction introduces an energy barrier for the ionization of consecutive amine groups, as manifested by a plateau in the titration curve at $\theta = 0.5$. The titration curve broadens as the repulsion energy increases. Qualitatively, the predictions of the Ising model are in excellent agreement with experimental observations.

While the discussion above is focused on linear polymers, the site-binding model can be extended to ionization of branched and dendrimeric polyelectrolytes. From the modeling perspective, the main difference between linear and branched polymers lies in the evaluation of the partition function. For branched polymers, analytical expressions are no more available for either the partition function or the degree of protonation. Nevertheless, if the polymer does not contain too many ionizable groups, we can still evaluate the probability of ionization by direct enumeration of the microstates.[12]

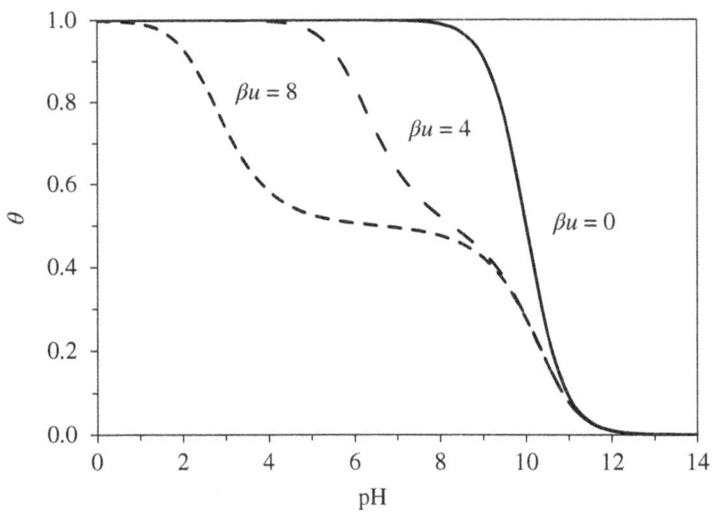

Figure 5.12 Degree of protonation for a weakly ionizable polyelectrolyte (e.g., PEI) at different levels of nearest-neighbor repulsion represented by βu. For $\beta u = 0$, the titration curve is the same as that for the monomer.

12 For large polymers, the numerical issue can be resolved with the help of Monte Carlo simulation, which will be discussed in Chapter 6.

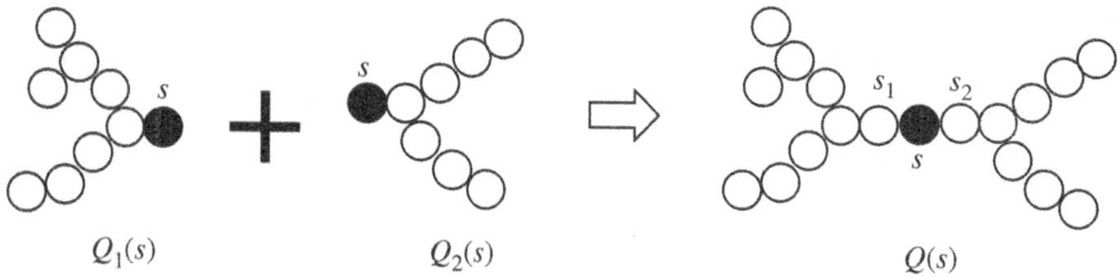

Figure 5.13 Joining two tree-like structures together leads to a new tree-like structure. Here each cycle represents an ionizable monomer, and the filled circle represents a root segment with its charge state denoted by *s*. Source: Adapted from Borkovec and Koper[13].

For a dendrimeric or comb-like polyelectrolyte, the polymer backbone exhibits a tree-like structure, as shown schematically in Figure 5.13.[13] A recursion relation can be established to calculate the restricted partition function, i.e., the partition function for a fraction of an ionizable polymer with a predefined charge state for the root segment

$$Q(s) = \sum_{s_{i \neq r}=\pm 1} \delta_{ss_r} e^{-\beta E_v} \tag{5.38}$$

where E_v is given by Eq. (5.30), δ_{ss_r} stands for the Kronecker delta function (i.e., $\delta_{ss_r} = 1$ if $s = s_r$ and $\delta_{ss_r} = 0$ otherwise), and subscript r denotes the root segment. When two polymer fragments are joined together by adding a new monomer as shown schematically in Figure 5.13, the restricted partition function for the merged polymer fragment becomes

$$Q(s) = \sum_{s_1=\pm 1} \sum_{s_2=\pm 1} Q_1(s_1) Q_2(s_2) \tag{5.39}$$

where $Q_1(s)$ and $Q_2(s)$ are the restricted partition functions for fragments 1 and 2, respectively. After merging two polymer fragments, the new monomer becomes the root segment, and the charge states of the original root segments, s_1 and s_2, are accounted for explicitly by the summations in Eq. (5.39). With the restricted partition function for the entire polymer obtained from the above recursion relation, we can calculate the partition function for the entire system

$$Q = \sum_{s=\pm 1} Q(s). \tag{5.40}$$

Subsequently, the degree of ionization can be numerically calculated from

$$\theta = \frac{1}{N} \frac{\partial \ln Q}{\partial \beta h}. \tag{5.41}$$

To illustrate, Figure 5.14 shows the titration curves for two branched polyelectrolytes (e.g., PEI in dendrimeric and comb-like structures).[14] In analogy to a linear chain, protonation of the dendrimeric polymer proceeds in two steps: the first step occurs when the solution pH is about equal to pK, similar to the protonation of individual monomers; and the next step takes place around a reduced pH that allows the proton to overcome the repulsion from neighboring protonated sites. The intermediate state lies at $\theta = 2/3$, corresponding to a condition when only the odd shells of

13 Borkovec M. and Koper, G. J. M. "Proton binding characteristics of branched polyelectrolytes", *Macromolecules* 30 (7), 2151–2158 (1997).
14 Koper G. J. M. and Borkovec M. "Proton binding by linear, branched, and hyperbranched polyelectrolytes", *Polymer* 51 (24), 5649–5662 (2010).

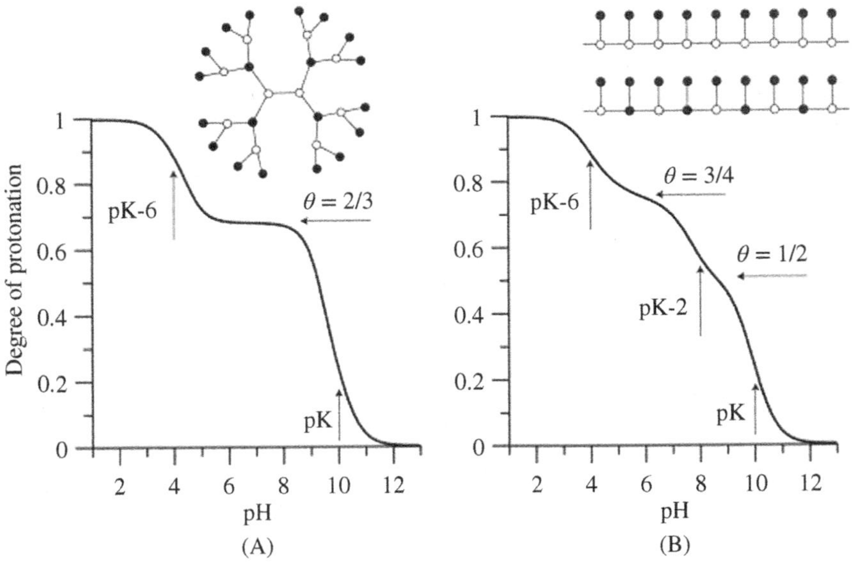

Figure 5.14 Titration curves for (A) dendritic and (B) comb-like weak polyelectrolytes. Here $pK = 10$ is for all ionizable sites, and the reduced nearest-neighbor energy is $\beta u \approx 4.6$. The insets show the intermediate states of ionization. Source: Reproduced from Koper and Borkovec 2010/with permission of Elsevier.

the dendrimer are protonated. The protonation behavior of the comb-like polymer is slightly more complicated. In the basic region, each site protonates independently around $pH = pK$. When every second site becomes protonated, the nearest-neighbor interactions can be avoided by protonating the singly coordinated sites, leading to an intermediate plateau at $\theta = 1/2$. The polymer protonates further until every second site of the backbone is filled up, corresponding to the second intermediate state at $\theta = 3/4$. Further protonation is possible only at sites that have three protonated neighboring sites.

5.3.1 Summary

The examples discussed in this section highlight the application of the Ising model in understanding the ionization behavior of weak polyelectrolytes. With the binding constant and nearest-neighbor energy treated as adjustable parameters, the site-binding model provides a generic framework to describe the properties of weak polyelectrolytes in concentrated salt solutions. As the nearest-neighbor model ignores the long-range interactions, its predictive capability is compromised for several reasons. Firstly, the ionizable groups within the polymer chain may have different intrinsic dissociation constants. Secondly, the nearest-neighbor model is inadequate to capture electrostatic interactions between ionizable groups at low salt concentration. Lastly, the lattice model overlooks the coupling between ionization states and the conformations of the macromolecules. For better performance, it is crucial to consider additional factors such as incorporating the polymer's conformation, accounting for differences between various ionizable groups along the polymer backbone (such as primary, secondary, and tertiary amines), and considering interactions between protonated groups beyond nearest neighbors.[10] By taking these factors into account, we can achieve a more comprehensive understanding of the ionization behavior of weak polyelectrolytes.

5.4 The Zimm–Bragg Model of Helix-Coil Transition

Proteins are linear polymers, more specifically, long-chain polypeptides in which amino acid residues[15] are linked together by peptide bonds. Unlike typical linear synthetic polymers, proteins often adopt a well-defined three-dimensional conformation known as the native structure, which is crucial for their biological functions. In this section, we introduce a simple statistical-mechanical model that describes the transition between α-helix and coil conformations in polypeptides.[16] The model was initially proposed by Zimm and Bragg,[17] building upon an extension of the one-dimensional Ising model.

5.4.1 α-Helix/Coil Transition in Biopolymers

The α-helix/coil transition arises from the coexistence of two conformations within a polypeptide chain, i.e., a spring-like helical structure and a random-coil structure. To construct a statistical-mechanical model for representing the α-helix/coil transition, the Zimm–Bragg model assumes that each amino acid residue (viz., segment) can be in one of two microstates analogous to spin up and spin down in an Ising chain. Schematically, Figure 5.15 illustrates the microstate of a polypeptide chain in terms of the helical (h) or coil (c) states of individual segments. For a segment in the helical state, it forms a hydrogen bond with another segment in the polypeptide chain. The hydrogen bonding is determined by the chemistry of the amino acid residues, specifically through the formation of a hydrogen bond between the carbonyl oxygen of amino acid residue i and the amide hydrogen of residue $i+3$. Here, i and $i+3$ refer to the positions of amino-acid residues in the protein sequence starting from the C-terminus.[18] It is important to note that due to the span of each hydrogen bond across four amino acid residues along the backbone, the first three carbonyl oxygen atoms at the C-terminus and the last three amide hydrogen atoms at the N-terminus are unable to adopt the helical conformations.

Intuitively, the formation of a helical structure resembles the coordination seen in military drills between a "sergeant" and "soldiers". Once a helical state is initiated by a starting helical segment (the "sergeant"), additional helical states follow in a sequential manner, akin to the orderly

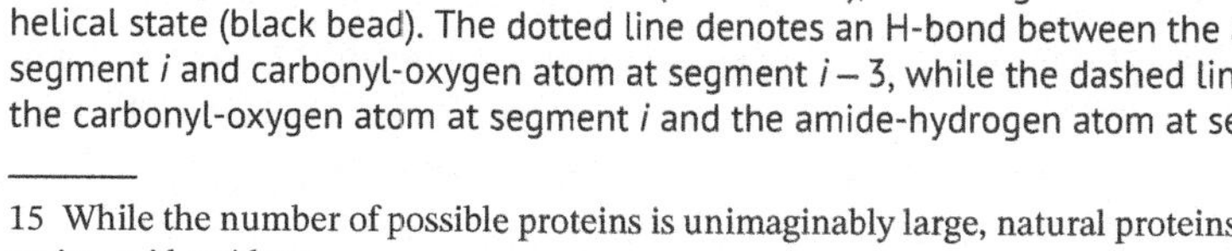

Figure 5.15 The modified 1D-Ising model for representing the helical state and coil state of individual segments in polypeptide conformations. The microstates are described in terms of the helical and coil states of segments: a random coil state (white bead), a starting helical state (gray bead), or a propagating helical state (black bead). The dotted line denotes an H-bond between the amide-hydrogen atom from segment i and carbonyl-oxygen atom at segment $i-3$, while the dashed line denotes the H-bond between the carbonyl-oxygen atom at segment i and the amide-hydrogen atom at segment $i+3$.

15 While the number of possible proteins is unimaginably large, natural proteins consist of only twenty types of amino acid residues.

16 The nomenclatures α-helix and β-sheet were introduced by William T. Astbury who made pioneering X-ray diffraction studies of biological molecules.

17 Zimm B. H. and Bragg J. K. "Theory of the phase transition between helix and random coil in polypeptide chains", *J. Chem. Phys.* 31, 526–535 (1959).

18 The C-terminus (or C-terminal end) is the end of an amino-acid chain terminated by a carboxyl group (—COOH). The N-terminus (or N-terminal end) refers to the end of a protein or polypeptide chain terminated by an amino group (—NH2).

progression of "soldiers" in a queue.[19] This is because in a polypeptide chain, the formation of the first hydrogen bond requires ordering of three intervening residues. As a result, segments in the starting helical state ("sergeants") possess a free-energy barrier arising from ordering of 3 consecutive amino acid residues. After the first hydrogen bond is established, formation of an additional hydrogen bond or propagating at the end of a preexisting helix requires fixing only one amino acid residue ("solider") in the helical orientation. Consequently, the free-energy cost is much smaller for the formation of a second, third, and consecutive hydrogen bonds in comparison with that for formation of the first hydrogen bond.

The Zimm–Bragg model described above resembles that for an Ising chain except that each segment may exist in one of 3 different states, i.e., one coil state and two helical states in starting or propagating positions.[20] Clearly, the microscopic model is concerned only with the thermodynamics associated with the transition of a polymer chain from the random coil to the helical conformation; it does not describe any atomic details of individual segments or their spatial arrangements. In other words, the Ising-like model is designed to determine how many residues in a polypeptide chain are in helical or coil structures at a fixed solution condition.

5.4.2 Partition Function for a Modified Ising Model

To describe the helical distribution of amino acid residues in a polypeptide chain, the Zimm–Bragg model uses the coil state as a reference and assumes that its free energy is zero. For a segment in the helical state, its free energy is denoted as u_s if it is at the starting position of an α-helix, or as u_a if it propagates a preexisting α-helix (i.e., its preceding residue is also in the helical state). The free energy difference between different amino acid residues can be understood as the free energy change when they are switched from one state to another. Near the transition, this free energy is expected to be on the order of $k_B T$.

To derive the statistical-mechanical equations, consider first a long polypeptide such that the end effects can be ignored. For an m-residue polypeptide, the Zimm–Bragg model predicts that the partition function is given by

$$Q = \prod_{i=1}^{m} \sum_{r_i = h_s, h_a, c} \exp\left[-\beta \sum_{i=1}^{m} u_i\right] \tag{5.42}$$

where r_i represents the conformational state of residue i ($r_i = h_s$, h_a, or c, where h stands for helical and c for coil), and u_i denotes the corresponding free energy. For an amino-acid residue in a coil state, $r_i = c$ and $u_i = 0$; for those in a helical state, $r_i = h_s$ or h_a and $u_i = u_s$ or u_a, depending on the conformational state of the preceding residue, i.e., whether it starts a helical structure or propagates a preexisting structure.

Eq. (5.42) resembles the partition function of an Ising chain except that each residue can take 3 states as denoted by variable $r_i = h_s$, h_a, or c. In addition, the system energy is specified by the sequence instead of the nearest-neighbor energy. Similar to that for an Ising chain, the partition function can be evaluated by using a modified transfer-matrix method (Appendix 5.B):

$$Q \approx \lambda_+^m \tag{5.43}$$

---

19 Green M. M. et al. "The macromolecular route to chiral amplification", *Angew. Chem. Int. Ed.* 38, 3139–3154 (1999).

20 Generalization of the Ising model to account for multiple states for each spin is called the Potts model, named after Renfrey Potts, who proposed the model during his Ph.D. work in mathematics.

where

$$\lambda_+ = \left[\left(1 + e^{-\beta u_a}\right) + \sqrt{\left(1 - e^{-\beta u_a}\right)^2 + 4e^{-\beta u_s}}\right] /2. \tag{5.44}$$

Like that for the Ising chain, Eq. (5.43) is valid for large m and becomes exact as $m \to \infty$.

From the partition function given by Eq. (5.43), we can readily calculate the average number of propagating residues ("soldiers") in a helical structure

$$\langle n_a \rangle = -\left(\frac{\partial \ln Q}{\partial \beta u_a}\right)_{T,m,u_a} = \frac{m}{\lambda_+}\left[\frac{e^{-\beta u_a}}{2} + \frac{e^{-\beta u_a}\left(e^{-\beta u_a} - 1\right)}{2\sqrt{\left(1 - e^{-\beta u_a}\right)^2 + 4e^{-\beta u_s}}}\right], \tag{5.45}$$

and the average number of starting helical residues ("sergeants")

$$\langle n_s \rangle = -\left(\frac{\partial \ln Q}{\partial \beta u_s}\right)_{T,m,u_a} = \frac{m}{\lambda_+}\left[\frac{e^{-\beta u_s}}{\sqrt{(1 - e^{-\beta u_a})^2 + 4e^{-\beta u_s}}}\right]. \tag{5.46}$$

The total number of residues in the helical structure is thus given by the sum of the two types of helical residues

$$<n_h> = <n_a> + <n_s> = \frac{me^{-\beta u_a}}{2\lambda_+}\left\{1 + \frac{2e^{-\beta(u_s - u_a)} + e^{-\beta u_a} - 1}{\sqrt{\left(1 - e^{-\beta u_a}\right)^2 + 4e^{-\beta u_s}}}\right\}. \tag{5.47}$$

From the practical perspective, a property of great interest in the helix-coil transition is the average fraction of residues in the helical conformation, or the helical fraction, denoted by θ. From Eq. (5.47), we have

$$\theta = \frac{<n_h>}{m} = \frac{s}{2\lambda_+}\left\{1 + \frac{2\sigma + s - 1}{\sqrt{(1 - s)^2 + 4s\sigma}}\right\} \tag{5.48}$$

where $s \equiv e^{-\beta u_a}$ and $\sigma \equiv e^{-\beta(u_s - u_a)}$. Recalling that u_a represents the change in free energy when a segment is transformed from a coil state to a helical state, we may consider s as the helix propagation parameter as it is related to the statistics of adding individual segments to an existing helix. The other parameter in the model, σ may be understood as the helix nucleation parameter because it is related to additional cost of free energy to generate a leading helical segment. In terms of the dimensionless parameters s and σ, parameter λ_+ is given by

$$\lambda_+ = \frac{1}{2}\left[(1 + s) + \sqrt{(1 - s)^2 + 4s\sigma}\right]. \tag{5.49}$$

Figure 5.16 shows variation of the helical fraction versus the propagation parameter for some representative values of the nucleation parameter. The average fraction of residues in the helical state exhibits a typical cooperative behavior, i.e., the transition from coil to helical state becomes steeper as the individual elements become more cooperative.

In general, both parameters s and σ vary with temperature. Because the nucleation parameter σ reflects the conformational factors associated with the helix initiation, it is relatively insensitive to temperature changes. To find the dependence of parameter s on temperature, we recall that u_a corresponds to the free energy difference between individual segments in the coil and helix states. The free energy can be expressed in terms of an energetic parameter by using the Gibbs–Helmholtz relation

$$\frac{\partial \beta\, u_a}{\partial \beta} = \varepsilon_a \tag{5.50}$$

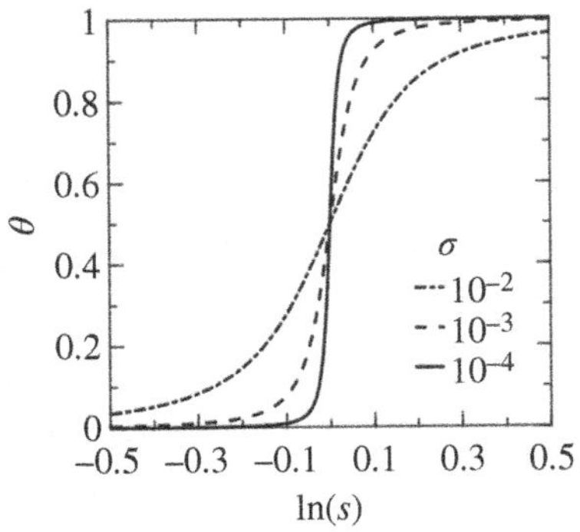

Figure 5.16 The helical fraction versus the helix propagation parameter s for several values of nucleation parameter σ.

where ε_a is the energy change associated with extending one helical unit onto a preexisting helical structure. Integration of Eq. (5.50) yields

$$\ln(s) = -\beta u_a = \frac{\varepsilon_a}{k_B}(1/T_m - 1/T) \tag{5.51}$$

where T_m be the system temperature when $s = 1$ (or $u_a = 0$). In practical applications, we may assume ε_a and σ to be independent of temperature; both can be obtained by the best fit of the theory to some experimental data for helix-coil transition.

If all segments are totally uncorrelated, i.e., the transition is noncooperative, $\sigma = 1$ and Eq. (5.49) predicts

$$\lambda_+ = 1 + s. \tag{5.52}$$

From Eq. (5.48), we find

$$\theta = \frac{s}{1+s} = \frac{e^{-\beta u_a}}{1 + e^{-\beta u_a}} \tag{5.53}$$

Eq. (5.53) corresponds to the distribution of independent segments in the helical and coil states. On the other hand, $\sigma = 0$ corresponds to polypeptides with an extremely high nucleation barrier for the helical formation. In that case, we have from Eq. (5.49)

$$\lambda_+ = \frac{1}{2}[(1 + s) + |1 - s|]. \tag{5.54}$$

For $s > 1$, we have from Eq. (5.54) $\lambda_+ = s$. Accordingly, Eq. (5.48) predicts $\theta = 1$, i.e., all segments are in the helical states. For $s < 1$, Eq. (5.48) predicts $\lambda_1 = 1$ and $\theta = 0$, i.e., all segments are in the coil states. The "all" or "nothing" transition occurs at $s = 1$, indicating that the individual segments are strongly correlated in the helical-coil transition.

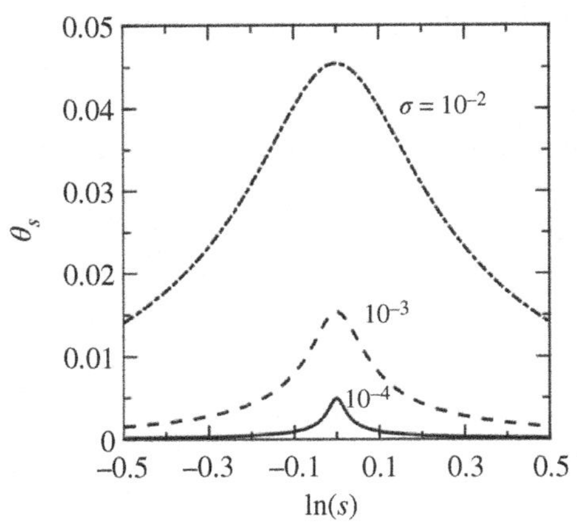

Figure 5.17 The fraction of starting helical segments versus the helix propagation parameter s for several values of nucleation parameter σ.

As shown in Figure 5.17, the fraction of starting helical segments, $\theta_s \equiv <n_s>/m$, shows a maximum at $s = 1$. The maximum number of helical regions is $<n_s>_{max} = m\sigma^{1/2}/2$, which occurs at the middle point of helix-coil transition. The average number of starting helical residues is identical to the average number of helical regions in a polypeptide chain. For a typical polypeptide, $\sigma \sim 10^{-4}$, the maximum number of helical regions is thus on the order of $m/200$, suggesting that a separated helical region occurs every 200 residues on average. Because $\theta = 0.5$ at $s = 1$, the Zimm–Bragg model predicts that the average length of a helical sequence is about ~100, which is in good agreement with experimental observation.

5.4.3 Short Polypeptide Chains

The Zimm–Bragg model is also applicable to short polypeptide chains. In this case, it is reasonable to consider only one helical region. Because of the end effects, the maximum number of hydrogen bonds is $m-3$. The partition function for a short chain is thus given by

$$Q = 1 + \sigma \sum_{i=1}^{m-3} (m - i - 2)s^i \tag{5.55}$$

where $(m-i-2)$ corresponds to the number of ways to arrange an i-residue helical region in an m-residue polypeptide chain. The helical fraction θ can be obtained from the partition function in Eq. (5.55) by

$$\theta = \frac{1}{m}\frac{d\ln Q}{d\ln s} = \frac{\sum_{i=1}^{m-3}(i+3)(m-i-2)s^i}{m\left[\sigma^{-1} + \sum_{i=1}^{m-3}(m-i-2)s^i\right]}$$
$$= \frac{m(s-1) - 2 - s^{-m+2}[3(m-2)s^2 - (7m-16)s + 4(m-3)]}{m(s-1)\{1 + \sigma^{-1}(s-1)^2 s^{-m+1} - [(m-3)(s-1)+s]s^{-m+2}\}}. \tag{5.56}$$

The analysis of helix-coil transition has been tested using experimental data. For example, Figure 5.18 compares theoretical and experimental helical fractions for poly-γ-benzyl-L-glutamate chains with average length 1500, 46, or 26 residues.[21] The solvent is 70:30 (by weight) dichloroacetic acid and ethylene dichloride. The long-chain model performs well for the polypeptide with 1500 residues, and the helical fractions for the short chains can be described quantitatively by

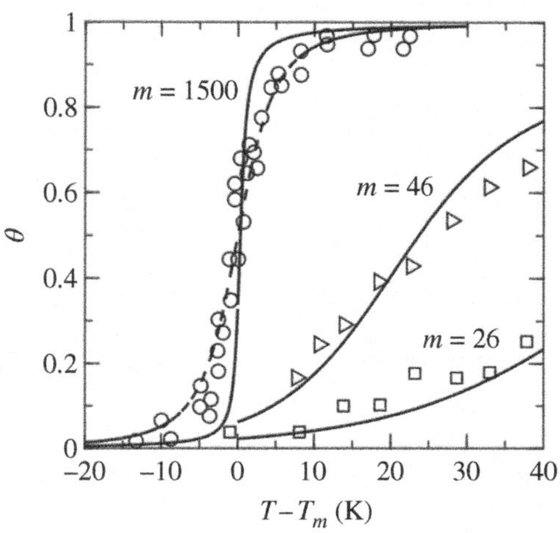

Figure 5.18 Theoretical and experimental results for the helical fraction θ as a function of temperature for poly-γ-benzyl-L-glutamate chains of three different lengths, denoted by m. The points represent experimental data from Zimm et al.[21], solid lines are calculated from Eq. (5.56), and the dashed line is from Eq. (5.48).

21 Zimm B. H., Doty P. and Iso K. "Determination of the parameters for helix formation in poly-gamma-benzyl-L-glutamate", *Proc. Natl. Acad. Sci.*, 45, 1601–1604 (1959).

Eq. (5.56). In correlation of the experimental results, we assume that $\sigma = 2 \times 10^{-4}$ independent of temperature. Using $\varepsilon_a = 0.89$ kcal/mol and $T_m = 285$ K obtained by best fitting, we see that the agreement between calculated and experimental data is quite reasonable, suggesting that the simple theory captures the essential physics of the α-helix/coil transition.

5.4.4 Summary

In summary, the simple model discussed in this section intends to capture the essential features of the α-helix/coil transition in proteins as a cooperative process. It explains that the helical regions in a protein do not grow indefinitely and that the average length of a helical region in proteins is independent of the total number of residues. Qualitatively, these predictions are in good agreement with experimental observations.

The Zimm–Bragg model can be similarly used to describe the helix/coil transition in DNA and cooperativity in racemizing supramolecular systems[19]. Similar to the Ising model, it assumes that every repeating unit can be either in helix or in coil state. While each spin may exist in one of two states in the Ising model, the Zimm–Bragg model further distinguishes between helical segments in the starting and propagating positions ("sergeant and soldiers"). However, it takes no explicit consideration of interactions between the side groups. Besides, it assumes that the helical structures are either all right- or all left- handed conformations. Despite these drastic assumptions, the theoretical predictions seem to be in accord with experiment observation of helical structures of natural biopolymers.[22]

5.5 Two-Dimensional Ising Model

The Ising model may be constructed on various two-dimensional (2D) lattices, as illustrated in Figure 5.19. In contrast to the 1D-Ising model, the 2D-Ising model is capable of describing a phase transition from a disordered state to an ordered state when the system temperature drops below a critical value. Because exact results for both the partition function and thermodynamic properties are available for certain special cases, the 2D-Ising model is a useful tool for capturing the general behavior of phase transitions in two-dimensional systems, including the universality of the critical phenomena and the associated power laws (Section 5.11). In addition, the 2D-Ising model is directly applicable to magnetic thin films, gas adsorption at a planar surface, and monolayers of phospholipids.

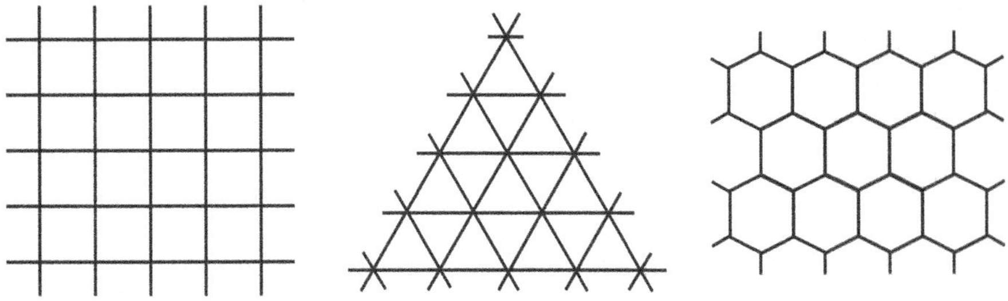

(A) (B) (C)

Figure 5.19 Two-dimensional square (A), triangular (B), and honey-comb (C) lattices.

22 Poland D. and Scheraga H. A., *Theory of helix–coil transitions in biopolymers*. Academic Press, 1970.

5.5.1 Onsager's Solution

In 1944, Lars Onsager first solved the square-lattice 2D-Ising model at zero field.[23] While similar derivations can be extended to triangular and hexagonal lattices, no analytical solution has yet been found for any two-dimensional model in the presence of an external field. The mathematical details of the derivations can be found in specialized texts.[24] Here we introduce only the definition of the 2D square-lattice Ising model and discuss various thermodynamic properties and physical concepts commonly used for describing phase transitions.

Schematically, Figure 5.20 shows spins on a 2D-square lattice where each spin interacts with four nearest neighbors. Similar to that for the 1D-Ising model, the total energy of the system is given by

$$E_\nu = -h\sum_i\sum_j s_{i,j} - \varepsilon\sum_i\sum_j (s_{i,j}s_{i+1,j} + s_{i,j}s_{i,j+1}) \tag{5.57}$$

where microstate ν is defined by the spin orientations $\{s_{i,j} = \pm 1\}$, subscripts i and j designate the spin index (*viz*., row and column on the square lattice), h stands for the interaction energy between a spin and an external potential (viz., external field), and ε is a coupling parameter describing the interaction energy between nearest neighboring spins. For a ferromagnetic system, the coupling energy is positive ($\varepsilon > 0$); and it is negative for an antiferromagnetic system.

For the square-lattice 2D-Ising model, Onsager was able to derive the partition function at zero external field ($h = 0$):

$$Q = \sum_\nu e^{-\beta E_\nu} = \left[2\cosh(2\beta\varepsilon)e^I\right]^N \tag{5.58}$$

where N is the total number of spins,

$$I = \frac{1}{\pi}\cdot\int_0^{\pi/2} \ln\frac{1+\sqrt{1-q^2\sin^2 x}}{2}dx \tag{5.59}$$

and

$$q \equiv 2\sinh(2\beta\varepsilon)/\cosh^2(2\beta\varepsilon). \tag{5.60}$$

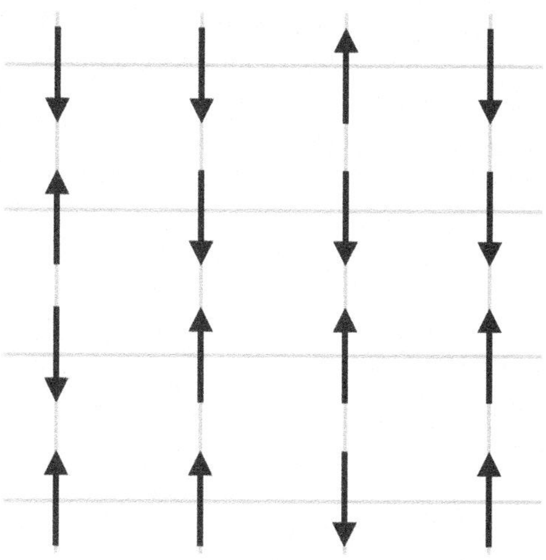

Figure 5.20 The 2D-Ising model on a square lattice where each spin interacts with four nearest neighbors.

23 Onsager L. "Crystal statistics I. A two-dimensional model with an order-disorder transition", *Phys. Rev.* 65, 117–149 (1944).

24 McCoy B. M. and Wu T. T., *The two-dimensional Ising model*. Dover Publications, 2014.

Eq. (5.58) is exact in the thermodynamic limit, i.e., when the number of spins in the system approaches infinity ($N \to \infty$).

From the partition function, it is straightforward to obtain thermodynamic properties. In dimensionless units, the Helmholtz energy, the internal energy, and the heat capacity are given by

$$\beta F/N = -\ln[2\cosh(2\beta\varepsilon)] - I, \tag{5.61}$$

$$\beta U/N = -\beta\varepsilon\coth(2\beta\varepsilon)\left\{1 + \frac{2K_1}{\pi}[2\tanh^2(2\beta\varepsilon) - 1]\right\}, \tag{5.62}$$

$$C_V/(Nk_B) = \frac{4}{\pi}(\beta\varepsilon\coth(2\beta\varepsilon))^2$$
$$\{K_2 - (1 - \tanh^2(2\beta\varepsilon))\,[\pi/2 + K_1(2\tanh^2(2\beta\varepsilon) - 1)\,]\,\} \tag{5.63}$$

where

$$K_1 = \int_0^{\pi/2} \frac{1}{\sqrt{1 - q^2\sin^2 x}}dx, \tag{5.64}$$

$$K_2 = \int_0^{\pi/2} \frac{q^2\sin^2 x}{\sqrt{1 - q^2\sin^2 x}}dx. \tag{5.65}$$

We plot in Figure 5.21 functions q, K_1, and K_2 in terms of the dimensionless reciprocal temperature, $\beta\varepsilon$. Function q exhibits a maximum of $q = 1$ at $\beta\varepsilon = -\frac{1}{2}\ln(\sqrt{2} - 1) \approx 0.4407$, and it vanishes at both zero and infinite temperatures. Functions K_1 and K_2 are related to elliptic integrals; both functions show a singularity at $q = 1$. As discussed latter, this condition is associated with the singularity of some thermodynamic properties and the phase transition in the 2D-Ising model.

Figure 5.22 presents the Helmholtz energy, the internal energy, the entropy, and the heat capacity versus the reduced reciprocal temperature, $\beta\varepsilon$. At high temperature ($\beta\varepsilon \to 0$), the spins are randomly oriented as evidenced by the disappearance of the average internal energy per spin, $\beta U/N = 0$. In this case, the entropy per spin is equivalent to the entropy of mixing for an equal molar mixture of a two-component ideal gas, $S/Nk_B = \ln 2 \approx 0.6931$. At the low-temperature limit ($\beta\varepsilon \to \infty$), all spins are aligned in the same direction. In that case, the entropy disappears ($S/Nk_B \to 0$), and the Helmholtz energy and the internal energy become identical (the energy per spin is -2ε).

One of the most remarkable predictions of Onsager's derivation is that the 2D-Ising model exhibits a disorder-to-order transition when the temperature is below a critical value, determined

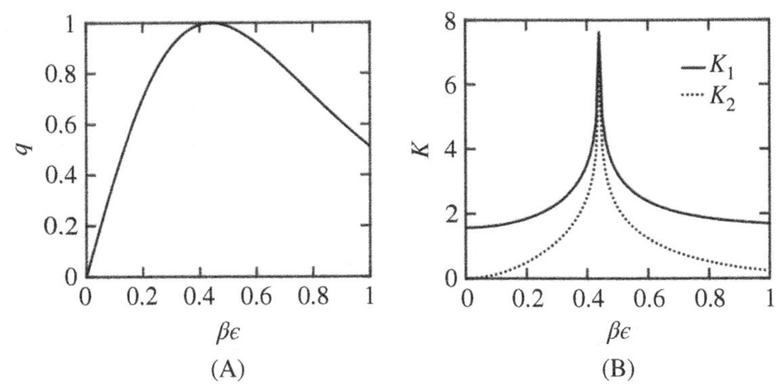

Figure 5.21 Functions q (A), K_1 and K_2 (B) used in the analytical solution of the 2D-Ising model by Onsager.

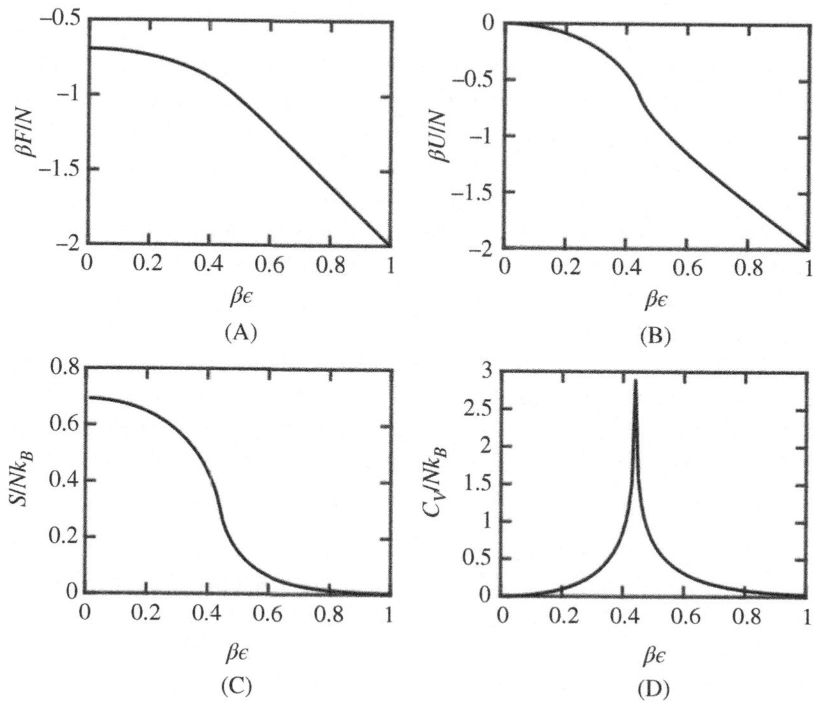

Figure 5.22 Thermodynamic properties of the square-lattice 2D-Ising model at zero external field. (A) Helmholtz energy; (B) Internal energy; (C) Entropy; (D) Heat capacity. The heat capacity shows singularity at the critical temperature ($\beta_c\varepsilon \approx 0.4407$ or $k_BT_c/\varepsilon \approx 2.269$).

from the following equation

$$\sinh\left(\frac{2\varepsilon}{k_BT_c}\right) = 1. \tag{5.66}$$

The solution to Eq. (5.66), $k_BT_c/\varepsilon = 2/\ln(1+\sqrt{2}) \approx 2.269185\ldots$, coincides to the condition for $q = 1$ and the divergence of functions K_1 and K_2 discussed above. Here, T_c is called the critical temperature (or the Curie temperature when the Ising model is applied to magnetic systems).[25]

Above the critical temperature, the average magnetization is zero, i.e., $m = 0$. In this case, the system is disordered because the spins have the same probability for the up and down orientations. Below the critical temperature ($T \leq T_c$), the system is ordered with a finite average magnetization given by

$$m = [1 - \sinh(2\beta\varepsilon)^{-4}]^{1/8}. \tag{5.67}$$

The analytical expression for the average magnetization was first published by C. N. Yang.[26]

As the system approaches the critical temperature, the magnetic susceptibility diverges due to long-range spin–spin correlations. While concise expressions for such properties are not attainable, it has been shown that the magnetic susceptibility satisfies the following asymptotic result[24]

$$k_BT\chi(T) \approx C_{0\pm}|1 - T_c/T|^{-7/4} \tag{5.68}$$

where $C_{0+} \approx 0.963$ and $C_{0-} \approx 0.026$ are coefficients for $T > T_c$ and $T < T_c$, respectively.

25 The Curie temperature is the maximum temperature where the material is ferromagnetic. At any temperature above the Curie temperature, there is no spontaneous magnetization.
26 Yang C. N. "The spontaneous magnetization of a 2-dimensional ising model", *Phys. Rev.* 85(5), 808–815 (1952).

5.5.2 Broken Symmetry

In the absence of an external field, the Ising model is fully symmetric in terms of the orientations of individual spins. While this symmetry should be preserved at any temperature, statistical thermodynamics predicts that the system exists in only one of the two possible macroscopic states below the critical temperature. In statistical mechanics, the phase transition in a macroscopic system from a symmetric state to an asymmetric state is often referred to as "*broken symmetry*." The concept is fundamental in understanding various physical phenomena, such as ferromagnetism, superconductivity, and phase transitions in thermodynamic systems. This term highlights the change from a state where all possible outcomes are equally likely (symmetric) to a state where one specific outcome is favored over others (asymmetric).

As illustrated schematically in Figure 5.23, the 2D-Ising model exhibits a free energy landscape that displays two minima. This indicates that the system can exist in either one of two states: the state with the majority of spins up or the state with the majority of spins down. The free energy barrier separating these two macroscopic states scales with the system size L according to

$$\Delta F \sim L^{d-1} \tag{5.69}$$

where d stands for the spatial dimensionality, i.e., $d = 2$ for the 2D-Ising model. In the thermodynamic limit, $L \to \infty$, meaning that the transition from one state to the other is impossible. In other words, the macroscopic system will exist in either the state with the majority of spins up or the state with the majority of spins down.

Figure 5.24 shows the average magnetization, $m = \langle s_i \rangle$, as a function of the reduced reciprocal temperature, $\beta\varepsilon$. Below the critical temperature ($\beta\varepsilon > 0.4407$), the majority of the spins are aligned in a single direction (here m is greater than zero) even without an external field. In this case, the system is "ordered" in the sense that one type of spin orientation (here spin ups) dominates. Above the critical temperature, however, the spins are randomly oriented, and the average magnetization is zero. Because m reflects the status of "order" or alignment of spins in the Ising model, the average magnetization is called the *order parameter*. Intuitively, the order parameter can be understood as a coordinate to describe transition between ordered and disordered phases.

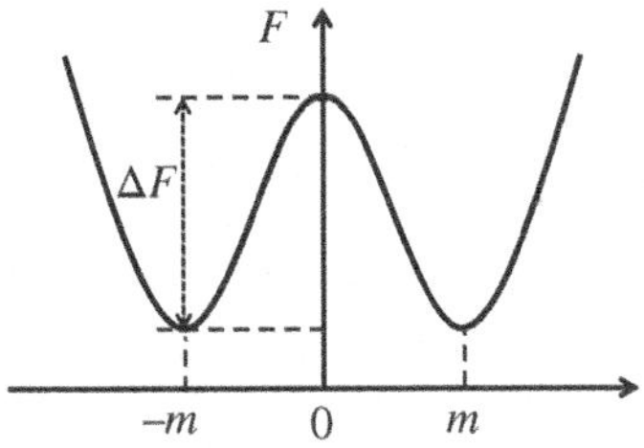

Figure 5.23 Below the critical temperature, the 2D-Ising model predicts equal Helmholtz energy for two macroscopic states, one with the majority of the spins up and the other with the majority of the spins down. The free energy barrier (ΔF) separating the two thermodynamic states diverges in the thermodynamic limit ($N \to \infty$) such that the system can exist in only one of the two states.

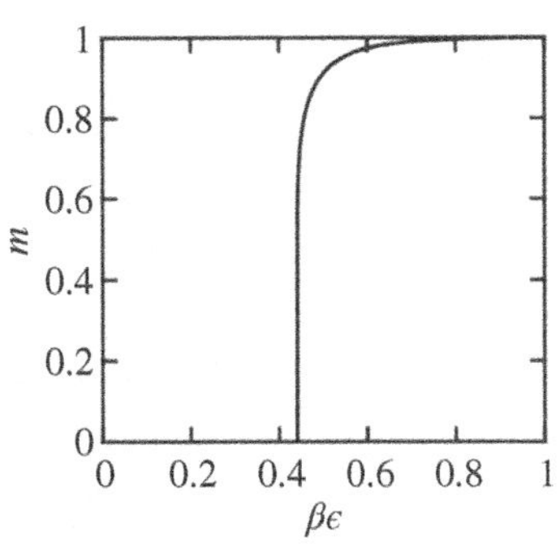

Figure 5.24 The average magnetization per spin versus the reduced reciprocal temperature for the 2D-Ising model without an external field. The phase transition occurs at $\beta_c\varepsilon \approx 0.4407$ or $k_BT_c/\varepsilon \approx 2.269$.

5.5.3 Critical Lipidomics

Before concluding this section, it is informative to highlight an application of the 2D-Ising model in analyzing the phase behavior of lipid bilayers.[6] In most living cells, the lipid membranes are composed of multiple components including lipids of different molecular weights, cholesterol, and proteins. At high temperature, all components in the bilayer mix uniformly in a single uniform phase. As the temperature is lowered, a transition may occur and the lipids separate laterally into coexisting domains enriched in cholesterol and sphingolipids (known as "lipid rafts").

There is growing evidence that, in living cells, the lipid composition is regulated to maintain a certain distance to the critical point of the phase transition that can be described by the 2D-Ising model.[27] To illustrate, we show in Figure 5.25 that the structure factor[28] for a biomimetic lipid bilayer consists of 25 : 20 : 55 mol% of dipalmitoylphosphatidylcholine (DPPC), diphytanoyl-phosphatidylcholine (diPhyPC), and cholesterol at different temperatures.[29] Experimentally, the structure factor was obtained from the Fourier transform of the two-point correlation function determined from fluorescence microscopy of vesicles.

According to the 2D-Ising model, the structure factor varies with temperature following the asymptotic relation

$$\hat{S}(k)k^{7/4} \sim \left[\frac{\xi k}{1+(\xi k)^2}\right]^{7/4} \tag{5.70}$$

where $\xi = \xi_0/(T/T_c - 1)$ is the correlation length. As shown in Figure 5.25B, the experimental data agree perfectly with the prediction of the 2D-Ising model. It has been reported that vesicles isolated from the plasma membranes of living rat basophilic leukemia (RBL-2H3) mast cells and other cell types also display critical behavior similar to the model lipid bilayer.[30]

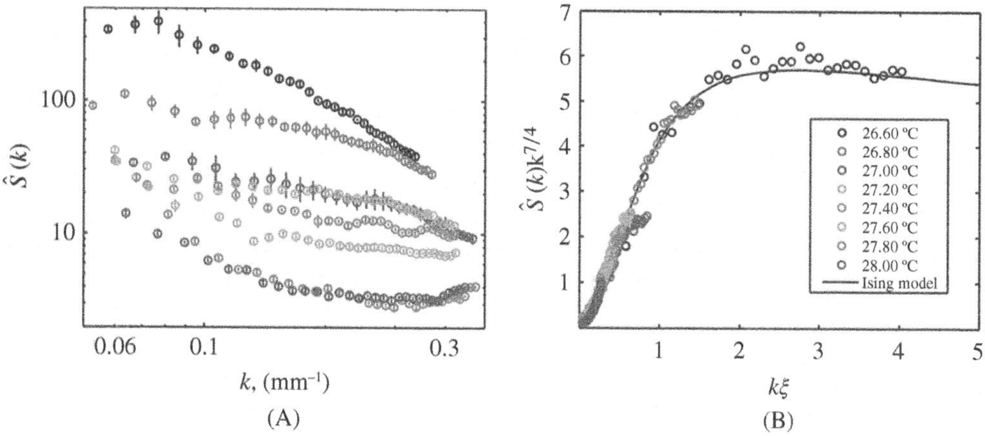

Figure 5.25 The structure factor for a biomimetic lipid bilayer at eight temperatures obtained from experiment (A) and fitted with the 2D-Ising model (B). The lowest temperature shown is close to the critical temperature, $T_c = 26.4\,°C$. Source: Reproduced from Honerkamp-Smith et al.[29]

27 Veatch S. L. and Cicuta P. "Critical lipidomics: the consequences of lipid miscibility in biological membranes", in *Physics of biological membranes*. Bassereau P. and Sens P. eds., Switzerland AG Springer Nature. 141–168, 2018.
28 See Section 7.2.
29 Honerkamp-Smith A. R. et al., "Line tensions, correlation lengths, and critical exponents in lipid membranes near critical points", *Biophys. J.* 95 (1), 236–246 (2008).
30 Levental I., Grzybek M. and Simons K. "Raft domains of variable properties and compositions in plasma membrane vesicles", *PNAS* 108 (28), 11411–11416 (2011).

5.5.4 Summary

We have discussed in this section that the partition function and the free energy can be analytically derived for the square-lattice 2D-Ising model. The 2D system exhibits a phase transition from a disordered phase to an ordered phase at a critical temperature accompanied by the divergence of the magnetization and the correlation length. The mathematical model can be utilized to elucidate important concepts such as broken symmetry and order parameter commonly used to describe phase transitions and critical phenomena. From a practical perspective, the 2D-Ising model is useful for understanding lipid miscibility in biological membranes and other 2D phase transitions.

5.6 Mean-Field Methods

In statistical mechanics, exact results are rarely obtainable for nonideal systems, i.e., for systems with a large number of interacting elements. Therefore, practical applications of statistical mechanics are often based on various mean-field approximations that neglect or only partially account for the correlation effects. In this section, we discuss three common mean-field procedures within the context of the three-dimensional (3D) Ising model. Similar ideas are applicable to more realistic thermodynamic models. Unlike its lower-dimensional counterparts, the 3D-Ising model lacks exact expressions for the partition function or the thermodynamic properties. However, numerical results can be obtained through Monte Carlo simulation (Chapter 6), which serves as a reliable benchmark for quantitatively evaluating approximate methods.

5.6.1 The Weiss Molecular Field Theory

Weiss molecular field theory, also known as the Weiss mean field theory, was proposed by Pierre Weiss in 1907 as an extension of the molecular field theory developed by Pierre Curie for describing the phase behavior of magnetic materials. The mean-field approximation asserts that the individual magnetic moments in a material interact with each other through an effective molecular field, which is the sum of the applied external field and the average field produced by all other moments in the material. This effective field acts on each magnetic moment, causing it to align or anti-align with the field.

Within the context of the Ising model, each spin has two microstates, described by $s = \pm 1$. With the neighboring spins represented by average magnetization m, the energy for an arbitrary spin i is approximated by

$$E_i = -hs_i - \varepsilon \sum_{j=1}^{Z} s_i s_j \approx -hs_i - \varepsilon \sum_{j=1}^{Z} s_i m \tag{5.71}$$

where h represents the external field, ε is the coupling energy, and Z stands for the number of nearest neighbors for spin i (*viz.*, the coordination number of the lattice; $Z = 6$ for a simple cubic (SC) lattice). In writing Eq. (5.71), we neglect the fluctuation of neighboring spins, i.e., $s_j = m$. Accordingly, the total energy of the system is simplified to[31]

$$E_\nu = \sum_{\{s_i\}} \left\{ -hs_i - \frac{\varepsilon}{2} \sum_{j=1}^{Z} s_i s_j \right\} \approx -\sum_{\{s_i\}} h_a s_i + Zm^2 \varepsilon / 2 \tag{5.72}$$

31 In writing this equation, we approximate $s_i s_j = [m + (s_i - m)] \cdot [m + (s_j - m)] \approx (s_i + s_j)m - m^2$ by neglecting the correlated spin fluctuation $(s_i - m)(s_j - m) = 0$. Accordingly, the mean-field energy is given by $\sum_{\{s_i\}}(-hs_i - Zm\varepsilon s_i) + Z\varepsilon m^2/2$, and the effective external field is given by $h_a = h + Zm\varepsilon$.

where a factor of 2 takes in to account that the pair potential involves two neighboring spins, and the effective external field for each spin is given by

$$h_a = h + Zm\varepsilon. \tag{5.73}$$

The Weiss molecular field theory neglects the fluctuations of the microscopic states of neighboring spins. In other words, it considers only the average energy due to spin–spin interactions, i.e., the mean-field energy.

Within the aforementioned mean-field approximation, the energy of the Ising model can be treated as equivalent to that of a set of noninteracting spins. Consequently, we are able to analytically evaluate the partition function

$$\begin{aligned} Q &= \sum_{v=\{s_i\}} \exp(-\beta E_v) \approx \prod_{i=1}^{N} \exp\left\{ \sum_{s_i=\pm 1} (\beta h_a s_i) \right\} \exp(-\beta Z m^2 \varepsilon/2) \\ &= [2\cosh(\beta h_a)]^N \exp(-\beta N Z m^2 \varepsilon/2). \end{aligned} \tag{5.74}$$

The Helmholtz free energy is given by

$$\beta F/N = -\frac{\ln Q}{N} \approx -\ln\{2\cosh[\beta(h + Z\varepsilon m)]\} + \beta Z m^2 \varepsilon/2. \tag{5.75}$$

Eq. (5.74) also allows us to determine the average magnetization self-consistently

$$m = \frac{1}{N}\frac{\partial \ln Q}{\partial \beta h} \approx \tanh[\beta(h + Z\varepsilon m)]. \tag{5.76}$$

While in general m must be solved numerically, some useful results can be readily obtained without the numerical solution of Eq. (5.76). For $\varepsilon = 0$, $m = \tanh(\beta h)$ corresponds to the average magnetization of noninteracting spins. In that case, there is no interaction between the spins and the mean-field approximation reproduces the exact result. For $h = 0$, $m = 0$ is always a solution.

A plot of $y = x$ and $y = \tanh(x)$, as shown schematically by Figure 5.26, indicates that, without the external field ($h = 0$), Eq. (5.76) yields nonzero solutions only when $\beta Z\varepsilon > 1$. The opposite values of the average magnetization ($\pm m$) can be attributed to a phase transition. A critical temperature can be identified from $\beta Z\varepsilon = 1$. Near the critical temperature $k_B T_c/\varepsilon = Z$, m is small. As a result, we can make a Taylor expansion of the hyperbolic tangent function

$$m = \tanh(\beta Z\varepsilon m) \approx \beta Z\varepsilon m - (\beta Z\varepsilon m)^3/3 + \cdots. \tag{5.77}$$

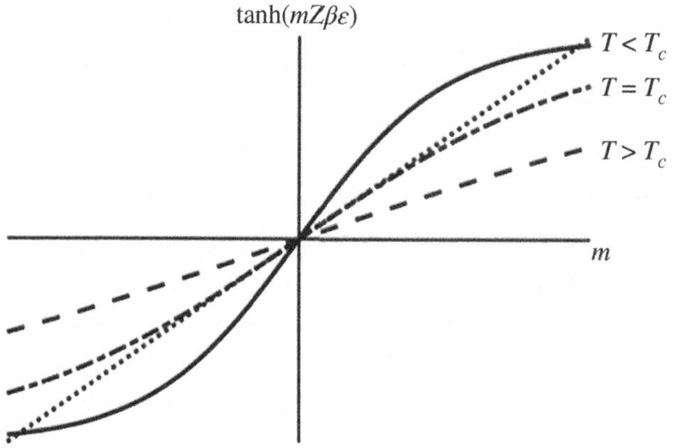

Figure 5.26 Solutions from Eq. (5.76) for the average magnetization when $h = 0$.

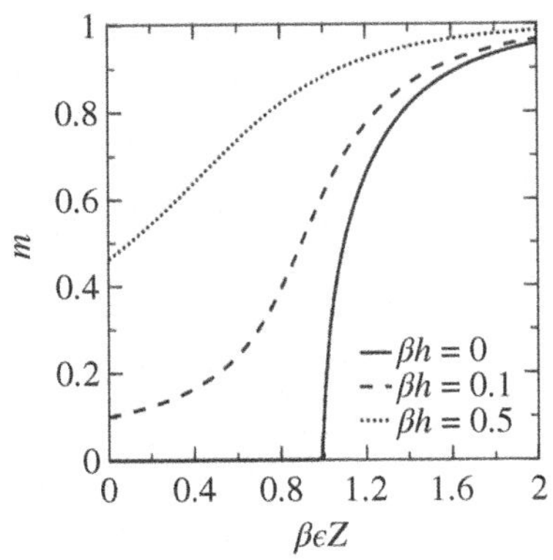

Figure 5.27 The average magnetization of the Ising model predicted by the mean-field approximation.

Neglecting high-order terms in Eq. (5.77) leads to

$$m(T) \approx \frac{1}{\sqrt{3}\alpha^{3/2}}(\alpha - 1)^{1/2} \tag{5.78}$$

where $\alpha = \beta Z\varepsilon = T_c/T$. Eq. (5.78) indicates that, as the system approaches the phase transition from low temperature, the average magnetization (viz., the order parameter) diminishes following a power law relationship

$$m(T) \sim (T_c/T - 1)^{1/2}. \tag{5.79}$$

As discussed further below, the mean-field theory predicts a universal exponent; however, when compared to the exact result, the predicted exponent is incorrect.

Figure 5.27 shows the mean-field prediction of the average magnetization for three values of βh. When $h = 0$, the mean-field theory predicts a phase transition at $\beta\varepsilon Z = 1$ as discussed above; nonzero solution exists only when $\beta Z\varepsilon > 1$. For any finite value of h, the average magnetization is a smooth function of $\beta\varepsilon$, i.e., m increases monotonically with the coupling energy. In this case, the system does not undergo a phase transition.

The Weiss molecular field theory may be similarly applied to 1D and 2D-Ising models. For the Ising chain, the coordination number is $Z = 2$; the mean-field theory predicts an erroneous phase transition at $k_BT_c/\varepsilon = 1$ due to the neglect of fluctuation effects. For the square-lattice Ising model, $Z = 4$ and the mean-field theory predicts critical temperature at $k_BT_c/\varepsilon = 4$, which is significantly larger than the Onsager solution $k_BT_c/\varepsilon \approx 2.269$. For the 3D-Ising model on a SC lattice ($Z = 6$), Monte Carlo simulation shows a critical temperature at $k_BT_c/\varepsilon \approx 4.5115$,[32] while the Weiss molecular field theory predicts $k_BT_c/\varepsilon = 6$. Because the mean-field theory neglects fluctuations that stabilize the disordered phase, it yields a critical temperature much larger than the exact result.

5.6.2 The Gibbs–Bogoliubov Variational Principle

The central idea behind mean-field theory is to approximate correlated interactions in a thermodynamic system using an effective one-body potential or mean field. Mathematically, the problem can be reformulated by finding a noninteracting reference system with the one-body potential that accurately reproduces the exact partition function or, equivalently, the exact Helmholtz energy. This alternative approach is commonly known as the Gibbs–Bogoliubov variational principle.

32 Ferrenberg A. M., Xu, J. and Landau D. P. "Pushing the limits of Monte Carlo simulations for the three-dimensional Ising model", *Phys. Rev. E* 97 (4), 043301 (2018).

While the variational principle is applicable to a wide range of thermodynamic systems, here we discuss its basic concepts within the context of the Ising model. Consider first the Hamiltonian for a noninteracting reference system given by

$$E_\nu^0 = -h_0 \sum_{i=1}^{N} s_i \tag{5.80}$$

where h_0 defines the one-body energy of a spin in the reference system. Its partition function is given by

$$Q_0 = \sum_{\{s_i\}} \exp\left(-\beta E_\nu^0\right) = \prod_{i=1}^{N} \sum_{s_i=\pm 1} \exp(\beta h_0 s_i) = [2\cosh(\beta h_0)]^N. \tag{5.81}$$

For the noninteracting system, the spin–spin correlation can be written as the square of the magnetization

$$< s_i s_j >_0 =< s_i >_0 < s_j >_0 = m_0^2. \tag{5.82}$$

This relationship arises from the absence of interactions between spins in the reference system. The Helmholtz energy and the average magnetization of the reference system are, respectively, given by

$$\beta F_0 = -\ln Q_0 = -N \ln(2\cosh \beta h_0), \tag{5.83}$$

$$m_0 =< s >_0 = -\frac{1}{N}\frac{\partial F_0}{\partial h_0} = \tanh(\beta h_0). \tag{5.84}$$

We now express the partition function of the real system relative to that of the reference system as

$$\frac{Q}{Q_0} = \frac{\sum_{\{s_i\}} \exp(-\beta \Delta E_\nu) \exp\left(-\beta E_\nu^0\right)}{\sum_{\{s_i\}} \exp\left(-\beta E_\nu^0\right)} = \langle \exp(-\beta \Delta E_\nu)\rangle_0 \tag{5.85}$$

where $<\cdots>_0$ represents an average in the reference ensemble, and

$$\Delta E_\nu = E_\nu - E_\nu^0 = \sum_{\{s_i\}} \left\{ -(h - h_0)s_i - \frac{\varepsilon}{2}\sum_{j=1}^{Z} s_i s_j \right\}. \tag{5.86}$$

Following the Jensen inequality,[33] we have

$$\langle \exp(-\beta \Delta E_\nu)\rangle_0 \geq \exp(-\langle \beta \Delta E_\nu\rangle_0). \tag{5.87}$$

Accordingly, Eq. (5.85) predicts that the Helmholtz energy of the interacting system satisfies the inequality relation

$$\beta F \leq \beta F_0 + \langle \beta \Delta E_\nu \rangle_0 = N\Phi_0 \tag{5.88}$$

where Φ_0 represents the reduced mean-field free energy per spin. With the expressions for βF_0 and ΔE_ν given by Eqs. (5.83) and (5.86), we obtain

$$\Phi_0 \equiv -\ln(2\cosh \beta h_0) + \beta(h_0 - h)\tanh(\beta h_0) - Z\beta\varepsilon \tanh^2(\beta h_0)/2. \tag{5.89}$$

In writing Eq. (5.89), we have used the expressions $m_0 = \tanh(\beta h_0)$ and $< s_i s_j >_0 = m_0^2$ discussed above for the noninteracting system.

---

33 In probability theory, Jensen's inequality indicates that the expectation of any convex function $f(x)$ satisfies $f[E(x)] \leq E[f(x)]$ where E stands for expectation and x is a random variable. In statistical mechanics, Eq. (5.87) is often referred to as the Gibbs-Bogoliubov inequality.

Eq. (5.88) indicates that $N\Phi_0$ represents an upper bound for the reduced Helmholtz energy βF. To make the mean-field approximation most accurate, $\beta F \approx N\Phi_0$, it is thus reasonable to seek an effective one-body potential that minimizes the right side of Eq. (5.88). The first-order derivative of Φ_0 with respect to the effective one-body potential

$$\frac{\partial \Phi_0}{\partial \beta h_0} = -\frac{\sinh(\beta h_0)}{\cosh(\beta h_0)} + \frac{\beta(h_0 - h)}{\cosh^2(\beta h_0)} + \tanh(\beta h_0) - \frac{Z\varepsilon \tanh(\beta h_0)}{\cosh^2(\beta h_0)} = 0 \tag{5.90}$$

leads to

$$h_0 = h + Z\varepsilon \tanh(\beta h_0). \tag{5.91}$$

Using Eq. (5.84), $m_0 = \tanh(\beta h_0)$, we can find the optimal one-body potential

$$h_0 = h + Z\varepsilon m_0. \tag{5.92}$$

Note that the optimal one-body potential given by Eq. (5.92) is the same as the mean-field potential used in the Weiss molecular field theory, Eq. (5.73). In other words, the one-body potential employed in the Weiss molecular field theory corresponds to the one that minimizes the mean-field free energy.

By substituting Eq. (5.91) into Eq. (5.89), we obtain the reduced mean-field free energy corresponding to h_0

$$\beta F/N \approx -\ln(2\cosh \beta h_0) + \beta(h_0 - h)^2/2Z\varepsilon. \tag{5.93}$$

Subsequently, we can determine the average magnetization

$$m = -\frac{1}{N}\frac{\partial \beta F}{\partial \beta h} = \frac{h_0 - h}{Z\varepsilon}. \tag{5.94}$$

A comparison of Eqs. (5.92) and (5.94) indicates that the variational method predicts a self-consistent description of the average magnetization, i.e., $m = m_0$.

Substituting Eq. (5.91) into Eq. (5.94) yields

$$m = \tanh(\beta h + Z\beta\varepsilon m). \tag{5.95}$$

Eq. (5.95) is identical to the self-consistent equation derived from the Weiss molecular field theory, Eq. (5.76). Besides, the two mean-field methods yield the same Helmholtz energy. For $h = 0$, both methods predict the same critical temperature of $T_c = \varepsilon Z/k_B$. It should be noted that, although the Gibbs–Bogoliubov variational principle and the Weiss molecular field theory generate the same results for the Ising model, these two methods are not interchangeable. While the variational principle can be used to derive the mean-field results, it can also provide more accurate solutions by incorporating higher-order correlations beyond the mean-field approximation.

5.6.3 The Bragg–Williams Theory

In addition to the self-consistent approaches discussed above, the mean-field free energy can be formulated in terms of the order parameter, or more specifically for the Ising models, average magnetization m. This procedure to derive the Helmholtz energy of a lattice system is known as the Bragg–Williams approximation.

Consider again the Ising model with N spins arranged on a 3D lattice. The total energy of the system depends on the arrangement of the nearest neighboring spins and, on average, it is given by

$$U = -h(N_+ - N_-) - \varepsilon(N_{++} + N_{--} - N_{+-}) \tag{5.96}$$

where N_+ and N_- represent the average number of spins up and the average number of spins down, respective; N_{++} is the average number of nearest neighbors with both spins up, N_{--} is the average number of nearest neighbors with both spins down, and $N_{+-} = N_{-+}$ is the average number of nearest neighbors with one spin up and one spin down. These quantities satisfy the conservation relations for the total number of spins and the total number of possible neighbors

$$N = N_+ + N_- \tag{5.97}$$

$$N_+ Z = 2N_{++} + N_{+-} \tag{5.98}$$

$$N_- Z = 2N_{--} + N_{-+} \tag{5.99}$$

where Z is the number of nearest neighbors. The average magnetization per site is given by

$$m = (N_+ - N_-)/N. \tag{5.100}$$

From Eqs. (5.97) and (5.98), we can express N_+ and N_- in terms of m

$$N_\pm = N(1 \pm m)/2. \tag{5.101}$$

With the assumption that the local spin composition near each spin is the same as the overall composition,[34] we can write the average number for each type of nearest neighboring bonds:

$$N_{++} = (N_+ Z/2) \cdot N_+/N = ZN_+^2/2N \tag{5.102}$$

$$N_{--} = (N_- Z/2) \cdot N_-/N = ZN_-^2/2N \tag{5.103}$$

$$N_{+-} = (N_+ Z/2) \cdot N_-/N = ZN_+ N_-/2N \tag{5.104}$$

Substituting Eqs. (5.101)–(5.104) into Eq. (5.96) leads to the internal energy per spin obtained from the mean-field approximation

$$U/N \approx -Z\varepsilon m^2/2 - mh. \tag{5.105}$$

Eq. (5.105) is the same as the average energy derived from the Weiss mean-field theory.

The local composition approximation implies that the spins of different orientations are randomly mixed. As a result, the reduced per spin entropy is the same as that corresponding to the entropy of mixing between two ideal gases

$$\begin{aligned} S/Nk_B &\approx -(N_+/N)\ln(N_+/N) - (N_-/N)\ln(N_-/N) \\ &= -\frac{1+m}{2}\ln\frac{1+m}{2} - \frac{1-m}{2}\ln\frac{1-m}{2}. \end{aligned} \tag{5.106}$$

The reduced Helmholtz energy per spin is thus derived by the combination of Eqs. (5.105) and (5.106)

$$f(m) \equiv F/Nk_BT = \frac{1+m}{2}\ln\frac{1+m}{2} + \frac{1-m}{2}\ln\frac{1-m}{2} - Z\beta\varepsilon m^2/2 - m\beta h. \tag{5.107}$$

Figure 5.28 illustrates the reduced Helmholtz energy per spin $f(m)$ as a function of the order parameter m for the 3D-Ising model on a cubic lattice ($Z = 6$) without the external field ($h = 0$). At high temperature (e.g., $k_BT/\varepsilon = 10$), $f(m)$ exhibits a concave shape with a minimum at $m = 0$. At low temperature (e.g., $k_BT/\varepsilon = 5$), it shows a maximum at $m = 0$ and two symmetric minima when the order parameter satisfies

$$\frac{\partial f(m)}{\partial m} = 0 = \frac{1}{2}\ln\frac{1+m}{1-m} - Z\beta\varepsilon m. \tag{5.108}$$

Using the mathematical identity $\text{arctanh}(x) = \frac{1}{2}\ln\frac{1+x}{1-x}$, we can verify that Eq. (5.108) is identical to the expression for m derived from the Gibbs–Bogoliubov variational principle, Eq. (5.95). At the

34 This is also known as the local composition approximation.

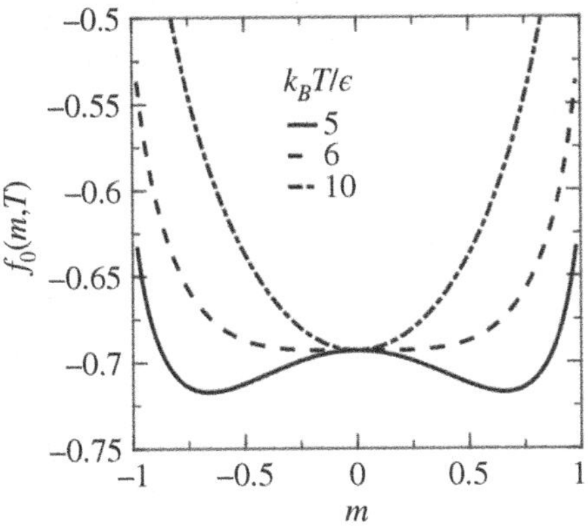

Figure 5.28 The reduced free energy per spin as a function of the order parameter for the field-free 3D-Ising model on a cubic lattice predicted by the Bragg–Williams theory.

critical temperature $k_BT_C/\varepsilon = 6, f(m)$ exhibits an inflection point, i.e., the three lowest derivatives of $f(m)$ disappear at $m = 0$ because the minimum and maximum points are coming together

$$\frac{\partial f(m)}{\partial m} = \frac{\partial^2 f(m)}{\partial m^2} = \frac{\partial^3 f(m)}{\partial m^3} = 0. \tag{5.109}$$

The Bragg–Williams theory suggests that, at a fixed temperature, the free energy is a unique function of order parameter m. Near the critical point, the order parameter is small such that $f(m)$ can be expanded in terms of a Taylor series

$$f(m) = -\ln 2 + \frac{1}{2}(1 - Z\beta\varepsilon)m^2 + \frac{1}{12}m^4 + \cdots. \tag{5.110}$$

Eq. (5.110) is a simple form of the Landau theory for phase transitions that will be discussed in Section 5.9. Because of the symmetry under reflection $m \to -m$, we have only even terms in Eq. (5.110).

5.6.4 Summary

Mean-field methods can be understood through the use of an effective one-body potential or an average structure. In the Weiss mean-field theory and the Gibbs–Bogoliubov variational principle, an effective one-body potential is introduced to map the properties of the interacting system into those of a noninteracting system. In the Bragg–Williams approximation, the mean-field theory is formulated in terms of the average structure or order parameter. While different mean-field approaches produce identical results for the Ising model, it is important to note that they may not be equivalent when applied to more complex thermodynamic systems. Besides, the mean-field methods neglect fluctuations or correlation effects that may be significant in certain thermodynamic systems. Their predictions are most accurate in systems where these effects are relatively weak or can be accounted for separately. For systems with strong fluctuations or strong correlation effects, more advanced techniques, such as Monte Carlo simulations or renormalization group (RG) methods, are often employed to account for these important factors and provide a more accurate description of the system.

5.7 Lattice Models

In this section, we will demonstrate how the Ising model can be transformed into statistical-mechanical theories of real fluids by reinterpreting the physical significance of the spin variables.

Specifically, we can utilize the spin orientations to represent the occupancy of lattice sites by actual molecules. While the lattice model does not capture the microscopic intricacies of intermolecular forces, it offers a general framework to understand, and sometimes even predict, the phase behavior of various thermodynamic systems with practical significance. We will illustrate such isomorphic transformations of the Ising model in the context of vapor-liquid equilibrium for simple fluids, followed by demixing in multicomponent liquid mixtures, and finally the phase behavior of complex fluids such as microemulsions.

5.7.1 Lattice-Gas Models

Consider a lattice model for a one-component fluid such that each cell contains at most one molecule. Let n_i= 0 or 1 be the occupation number of the ith cell. With the assumption that the intermolecular interactions are short-ranged, the total potential energy of the system can be represented by a pairwise summation of the nearest-neighbor energies. Other than a constant related to the molecular kinetic energy that does not influence the phase behavior of the thermodynamic system, we can express the grand partition function as

$$\Xi = \sum_{\{n_i=0,1\}} \exp\left\{\frac{\beta w}{2}\sum_{i=1}^{N}\sum_{j=1}^{Z} n_i n_j + \beta\mu\sum_{i=1}^{N} n_i\right\} \tag{5.111}$$

where $w > 0$ stands for the magnitude of the contact energy between two neighboring molecules $(-w)$, Z is the coordination number of the lattice, and μ denotes the fluid chemical potential.

By expressing occupation number n_i in terms of spin variable $s_i = 2n_i - 1 = \pm 1$, we can rewrite Eq. (5.111) as

$$\Xi = \sum_{\{s_i=\pm 1\}} \exp\left\{\frac{\beta w}{8}\sum_{i=1}^{N}\sum_{j=1}^{Z} s_i s_j + \beta(Zw/4 + \mu/2)\sum_{i=1}^{N} s_i + \beta N(wZ/8 + \mu/2)\right\} \tag{5.112}$$

Because the last term on the right side of Eq. (5.112) is independent of the system configuration, the partition function of the lattice gas is identical to that for an Ising model with parameters $\varepsilon = w/4$ and $h = Zw/4 + \mu/2$. Accordingly, all results derived from the Ising model can be directly applied to the fluid system. In the context of the lattice-gas model, "spin up" or $s_i = 1$ means that lattice site i is occupied by one molecule ($n_i = 1$), and spin down or $s_i = -1$ means that lattice site i is empty.

Eq. (5.112) allows us to relate chemical potential μ, number density ρ, and grand potential Ω to standard variables in the Ising model:

$$\mu = 2(h - Z\varepsilon), \tag{5.113}$$

$$\rho = \frac{1}{V}\sum_{i=1}^{N} < n_i >= \frac{N(1+m)}{2V}, \tag{5.114}$$

$$\beta\Omega = -\ln\Xi = \beta F_{\text{Ising}}(\varepsilon, h) - \beta N(h - Z\varepsilon/2) \tag{5.115}$$

where V stands for the total volume, and $<n> = (m+1)/2$ is the average occupation number of the lattice. Because the lattice-gas model describes a uniform system, the grand potential is directly proportional to the system pressure

$$\beta PV = -\beta\Omega. \tag{5.116}$$

To attained an analytical expression for the Helmholtz energy, we may use the Gibbs–Bogoliubov variational principle (Section 5.6):

$$\beta F_{\text{Ising}}/N \approx -\ln\{2\cosh[\beta(h + Z\varepsilon m)]\} + Z\beta\varepsilon m^2/2 \tag{5.117}$$

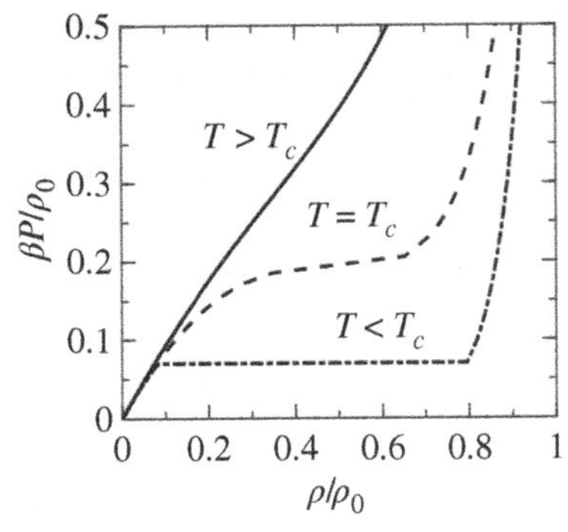

Figure 5.29 Isotherms of a one-component fluid predicted by the lattice-gas model. The pressure versus molecular density is continuous above the critical temperature ($T > T_C$), exhibits an inflection point at the critical temperature ($T = T_C$), and shows a vapor-liquid transition below the critical temperature ($T < T_C$). Here, $\rho_0 = N/V$ represents the number density of the lattice cells.

where

$$m = \tanh(\beta h + Z\beta\varepsilon m). \tag{5.118}$$

Figure 5.29 shows the reduced pressure of the lattice gas as a function of the reduced number density predicted from the mean-field theory. As expected, the system exhibits a critical temperature at $k_B T_C/\varepsilon = Z$ where Z is the coordination number of the lattice. Above the critical temperature ($T > T_C$), the pressure varies smoothly with the molecular density, while below the critical temperature ($T < T_C$), the system exhibits instability within a certain range of molecular densities that the system is unstable. Qualitatively, the isotherms are similar to those predicted from a typical equation of state for simple fluids (e.g., the van der Waals theory).

Although the lattice-gas model is rarely used to describe the phase behavior of one-component fluids, it does capture the qualitative behavior of compressibility effects and vapor–liquid-like phase transition. Importantly, it helps us to understand why the Ising model captures the universality of the thermodynamic properties of real systems near the critical point of transitions (Section 5.11).

5.7.2 Liquid–Liquid Demixing

The isomorphic transformation of the Ising model can be similarly applied to liquid mixtures that display demixing or liquid–liquid phase separation. To elucidate this application, consider a lattice model for a binary mixture of pure liquids A and B. Similar to the lattice-gas model discussed above, we assume that the liquid is incompressible such that each cell is occupied by either one molecule of species A or by one molecule of species B.

Given N_A molecules of species A and N_B molecules of species B on the lattice, the microstates of the system are defined by the spatial arrangements of A and B molecules. Assuming that the intermolecular interactions are short-ranged, we can write the total potential energy of the system in terms of the nearest-neighbor interactions similar to that used in the Bragg–Williams theory (Section 5.6)

$$E_v = N_{AA}\varepsilon_{AA} + N_{AB}\varepsilon_{AB} + N_{BB}\varepsilon_{BB} \tag{5.119}$$

where N_{AA}, N_{AB}, and N_{BB} are the numbers of different contacts between A and B molecules, and ε_{AA}, ε_{AB} and ε_{BB} are the corresponding contact energies, respectively. A similar expression can be written for the potential energy of pure liquids

$$E_0 = Z(N_A\varepsilon_{AA} + N_B\varepsilon_{BB})/2 \tag{5.120}$$

where Z is the coordination number of the lattice, and a factor of 2 accounts for the involvement of two molecules for each pair interaction.

As discussed in Section 5.6, the total number of nearest neighbors for species A and that for species B on the lattice are given by

$$N_A Z = 2N_{AA} + N_{AB} \tag{5.121}$$

$$N_B Z = 2N_{BB} + N_{BA} \tag{5.122}$$

where $N_{AB} = N_{BA}$. Based on Eqs. (5.120)–(5.122), we can rewrite the total energy of the mixture in terms of that for the pure liquids plus the mixing energy

$$E_v = E_0 + N_{AB}w/Z \tag{5.123}$$

where $w \equiv Z(\varepsilon_{AB} - \varepsilon_{AA}/2 - \varepsilon_{BB}/2)$. Intuitively, w may be understood as "the exchange energy," i.e., the change in the total energy when a molecule from pure liquid A is exchanged with a molecule from pure liquid B, which leads to $\Delta E = Z(\varepsilon_{AB} - \varepsilon_{AA})/2 + Z(\varepsilon_{AB} - \varepsilon_{BB})/2 = w$.

To relate the system energy with the microstates, we designate $n_i = 0$ when lattice site i is taken by a molecule of type A, and $n_i = 1$ when it is occupied by a molecule of type B. The interaction energy ε_{ij} between adjacent sites i and j can then be expressed as

$$\varepsilon_{ij} = (n_i + n_j - 2n_i n_j)w/Z. \tag{5.124}$$

It is easy to verify that $\varepsilon_{ij} = 0$ if i and j sites are occupied by the same species, and $\varepsilon_{ij} = w/Z$ if one site is occupied by A and the other by B. Therefore, the total potential energy can be rewritten as

$$E_v = E_0 + \sum_{i=1}^{N}\sum_{j=1}^{Z}(n_i + n_j - 2n_i n_j)w/Z. \tag{5.125}$$

To make a connection with the Ising model, we now consider the grand partition function for the binary mixture

$$\Xi = \sum_{\{n_i\}} \exp\left\{-\beta E_0 - \frac{\beta w}{Z}\sum_{i=1}^{N}\sum_{j=1}^{Z}(n_i + n_j - 2n_i n_j) + \beta\mu_A \sum_{i=1}^{N} n_i + \beta\mu_B\left(N - \sum_{i=1}^{N} n_i\right)\right\} \tag{5.126}$$

Similar to the lattice-gas model for one-component fluids discussed above, we can express the occupation number terms of the spin variable s_i

$$s_i = 2n_i - 1. \tag{5.127}$$

Substituting Eq. (5.127) into Eq. (5.126) yields

$$\Xi = \sum_{\{s_i=\pm 1\}} \exp\left\{\frac{\beta\varepsilon}{2}\sum_{i=1}^{N}\sum_{j=1}^{Z} s_i s_j + \beta h \sum_{i=1}^{N} s_i - \beta E_0'\right\} \tag{5.128}$$

where $\varepsilon = w/Z$, $h = (\mu_A - \mu_B)/2$, and $E_0' = E_0 + N(w - \mu_A - \mu_B)/2$. Except for a constant (E_0') that is irrelevant for the describing the phase behavior of the system, Eq. (5.128) is identical to the partition function of an Ising model. Accordingly, all thermodynamic relations derived from the Ising model are directly applicable to the lattice model for binary liquid mixtures.

Whereas discussions above are concerned with one or two-component fluids, the lattice model can be extended to multicomponent systems by increasing the number of states per lattice site, i.e., from two states such as occupied/unoccupied or A/B occupations as discussed above to multiple states. In statistical mechanics, the multistate generalization of the Ising model is known as the n-component spin model. For example, Figure 5.30 shows the phase diagrams of ternary systems predicted by a 3-state lattice model.[35] With the free energy of the lattice system evaluated

---

35 Yang J. Y., et al. "A new molecular thermodynamic model for multicomponent Ising lattice", *J. Chem. Phys.* 125 (16), 164506 (2006).

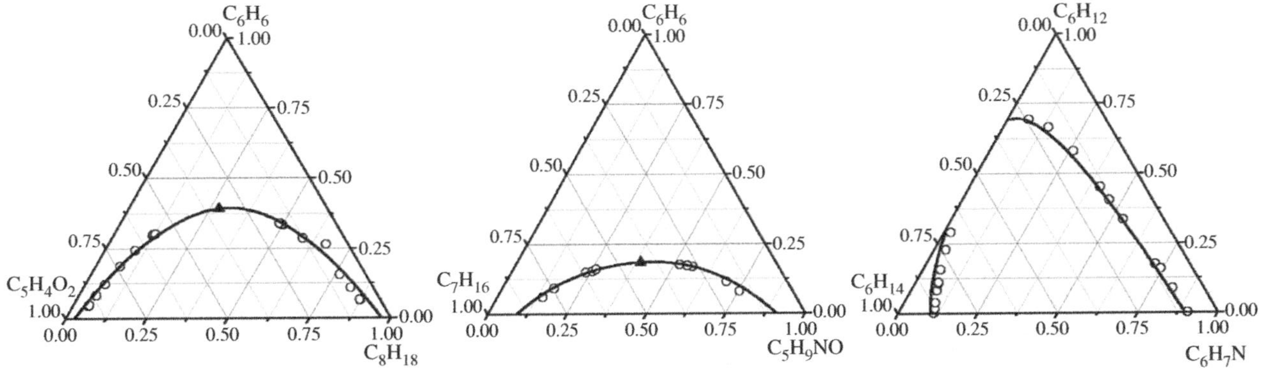

Figure 5.30 Liquid–liquid phase diagrams for, from left to right, furfural/2,2,4-trimethyl-pentane/benzene at 298.15 K, heptane/1-methyl-2-pyrrolidone/benzene at 298.15 K, and aniline/hexane/methyl-cyclopentane at 307.65 K. The symbols are from experimental data and the lines are correlations with a three-state lattice-gas model. Source: Yang et al.[35]

from Monte Carlo simulation, the simple model was able to reproduce experimental data near quantitatively.

5.7.3 Microemulsions

The lattice model provides a valuable tool for describing phase diagrams in systems that involve liquid water, oil, and surfactants (or amphiphiles, which possess dual affinities for both water and oil). Unlike conventional fluids and mixtures discussed earlier, these systems can exhibit not only macroscopic phase separations but also mesoscopic structures, including micelles, microemulsions, and lamellae, resulting from the self-assembly of amphiphilic molecules. Theoretical understanding of the phase behavior and mesoscopic structures plays a crucial role in harnessing the diverse functionalities of surfactants and other amphiphilic systems in applications such as cosmetics, drug delivery, enhanced oil recovery, and more.

A number of lattice models have been proposed to describe the phase behavior of water–oil–amphiphile systems. In the following, we discuss a relatively simple model proposed by Gompper and Schick,[40] which is most relevant to the Ising model. Consider a lattice model for a ternary system containing liquid water (W), oil (O), and amphiphiles (A). Similar to the discussions above for simple fluids, let $n_i^\alpha = 1$ if lattice site i is taken by a molecule of species $\alpha =$ W, O, or A, and $n_i^\alpha = 0$ otherwise. Each lattice site is occupied by only one molecule, and vacancy is not allowed. With the assumption of pairwise additivity and nearest-neighbor interactions, the grand canonical ensemble of the ternary system can be expressed as

$$\Xi = \sum_{\{n_i\}} \exp\left\{\beta \sum_{\alpha,\beta} \sum_{i,j}{}' \varepsilon_{\alpha\beta} n_i^\alpha n_j^\beta + \beta \sum_{\alpha} \sum_{i=1}^{N} n_i^\alpha \mu_\alpha - \beta H_{AMP}\right\} \tag{5.129}$$

where the prime sign in the first term denotes summation over nearest-neighbor interactions, $\varepsilon_{\alpha\beta} > 0$ represents the magnitude of the contact energy between molecules α and β, μ_α is the chemical potential for species α, and H_{AMP} accounts for the amphiphilic nature of amphiphiles, i.e., they are energetically favored to be positioned between water and oil molecules. Approximately, the amphiphilic interaction can be represented in term of three-body interactions

$$H_{AMP} = L\sum_{i,j,k}{}' \left(n_i^W n_j^A n_k^W + n_i^O n_j^A n_k^O - n_i^W n_j^A n_k^O - n_i^O n_j^A n_k^W\right) \tag{5.130}$$

where the prime sign denotes summation over three adjacent sites in a straight line, $L > 0$ represents a 3-body energy due to an amphiphilic molecule. Eq. (5.130) captures the essential physics for the interaction of each amphiphilic molecule with solvent molecules, i.e., it introduces a positive energy of $L > 0$ if an amphiphilic molecule is sandwiched between two water (WAW) or two oil molecules (OAO), a negative energy of $-L$ if it is between a water molecule and an oil molecule (WAO and OAW). For example, a negative energy of $-L$ is introduced when site j is taken by an amphiphilic molecule meanwhile its immediate neighboring sites i and k are occupied by water and oil molecules.

To transform the lattice model for the three-component mixture into a modified Ising model, we define spin variable for the ith lattice site via

$$n_i^W = s_i(1 + s_i)/2, \tag{5.131}$$

$$n_i^O = -s_i(1 - s_i)/2, \tag{5.132}$$

$$n_i^A = 1 - s_i^2. \tag{5.133}$$

Accordingly, $s_i = 1$, -1 and 0 correspond to site i occupied by a molecule of type W, O, and A, respectively. In terms of the spins, we can rewrite the Hamiltonian for the ternary system as[36]

$$\Xi = \sum_{\{s_i\}} \exp\left\{ \beta \sum_{i,j}{}' \left[J\, s_i s_j + K s_i^2 s_j^2 + C\left(s_i s_j^2 + s_j s_i^2 \right) \right] \right.$$
$$\left. + \beta \sum_i \left(H s_i - \Delta s_i^2 \right) - \beta L \sum_{i,j,k}{}' s_i \left(1 - s_j^2 \right) s_k \right\}. \tag{5.134}$$

In Eq. (5.134), the indices i, j, and k denote lattice sites. The first sum represents energy related to interactions over all direct contacts, the second term sums over all individual sites, and the third term sums over all amphiphile-mediated interactions; a term independent of spin variable s_i is omitted because it has no influence on the phase behavior.

A comparison of Eqs. (5.129) and (5.134) indicates that parameters J, K, C are related to the contact energies

$$4J = \varepsilon_{WW} + \varepsilon_{OO} - 2\varepsilon_{WO}, \tag{5.135}$$

$$4C = 2(\varepsilon_{OA} - \varepsilon_{WA}) - (\varepsilon_{OO} - \varepsilon_{WW}), \tag{5.136}$$

$$J + K - 2C = \varepsilon_{OO} + \varepsilon_{AA} - 2\varepsilon_{OA}, \tag{5.137}$$

and that parameters H and Δ are related to the contact energies as well as the chemical potentials of three species

$$H = (\mu_W - \mu_O)/2 + 3(\varepsilon_{WA} - \varepsilon_{OA}), \tag{5.138}$$

$$\Delta = \mu_A - (\mu_W + \mu_O)/2 - 3(\varepsilon_{WA} + \varepsilon_{OA} - 2\varepsilon_{AA}). \tag{5.139}$$

In writing the above equations, we assume that the coordination number of the three-dimensional lattice is $Z = 6$.

We can evaluate the partition function given in Eq. (5.129) or (5.134) using the mean-field approximation. Without considering the fluctuations of the spin variables, Eq. (5.134) predicts that the internal energy of the lattice system is given by

$$U = -\sum_{i,j}{}' [J\, M_i M_j + K Q_i Q_j + C(M_i Q_j + M_j Q_i)]$$
$$- \sum_i (H M_i - \Delta Q_i) + L \sum_{i,j,k}{}' M_i (1 - Q_j) M_k \tag{5.140}$$

where $M_i = \langle s_i \rangle_0$ and $Q_i = \langle s_i^2 \rangle_0$ represent order parameters,[37] and $\langle \cdots \rangle_0$ represents ensemble average within a mean-field Hamiltonian. Using Eqs. (5.131)–(5.133), we can express the average compositions of the three components in the system in terms of the order parameters

$$x_i^W \equiv \langle n_i^W \rangle = (M_i + Q_i)/2, \tag{5.141}$$

$$x_i^O \equiv \langle n_i^O \rangle = (Q_i - M_i)/2, \tag{5.142}$$

$$x_i^A \equiv \langle n_i^A \rangle = (1 - Q_i). \tag{5.143}$$

Accordingly, the entropy of mixing is given by

$$S/k_B = -\sum_i \left[x_i^W \ln x_i^W + x_i^O \ln x_i^O + x_i^A \ln x_i^A \right]. \tag{5.144}$$

---

36 Without the last term, the Hamiltonian is known as the Blume–Emery–Griffiths model for spin-1 systems, which represents one of the simplest extensions of the spin-1/2 Ising model.

37 We will discuss the definition and physical significance of order parameters in Section 5.8.

Combing Eqs. (5.140) and (5.144) leads to the Helmholtz energy of the entire system

$$F = -\sum_{i,j}{}' [J M_i M_j + K Q_i Q_j + C(M_i Q_j + M_j Q_i)] - \sum_i (H M_i - \Delta Q_i)$$
$$+ L \sum_{i,j,k}{}' M_i (1 - Q_j) M_k + k_B T \sum_i \left[x_i^W \ln x_i^W + x_i^O \ln x_i^O + x_i^A \ln x_i^A\right]. \quad (5.145)$$

The phase behavior of the ternary system depends on, in addition to temperature, the contact energies and chemical potentials of individual species as shown in Eqs. (5.135)–(5.139). Given a set of model parameters J, K, C, H and Δ, we can calculate the spatial distributions of order parameters $\{M_i\}$ and $\{Q_i\}$ by minimizing the Helmholtz energy given by Eq. (5.145). Subsequently, we can determine the phase diagram by comparing the free energies of different phases. For any particular system, the molecular properties and the contact energies are fixed. As a result, parameters J, K, C are constants, and the phase diagram depends only on temperature T, and parameters H and Δ. While parameter H contrasts the energetic difference between water and oil molecules, Δ is related to the relative composition of the amphiphiles to water and oil concentrations.

For macroscopic phases, order parameters $\{M_i\}$ and $\{Q_i\}$ are the same for all lattice sites. In that case, Eq. (5.145) can be simplified by using $M_0 = M_i$ and $Q_0 = Q_i$, and $x_\alpha = x_i^\alpha$:

$$F_0/N = -Z\left(J M_0^2 + K Q_0^2 + 2C M_0 Q_0\right) - (H M_0 - \Delta Q_0)$$
$$+ Z L M_0^2 (1 - Q_0)/2 + k_B T (x_W \ln x_W + x_O \ln x_O + x_A \ln x_A) \quad (5.146)$$

where subscript "0" denotes uniform systems. Similar to that used in the Bragg–Williams theory (Section 5.6.3), order parameters M and Q can be solved by the minimization of the Helmholtz energy. Subsequently, the overall composition of the system can be calculated from Eqs. (5.141) to (5.143).

Despite its simplicity, the lattice model proposed by Gompper and Schick is able to reproduce many essential features of the phase behavior observed in water–oil–amphiphile systems. As shown in Figure 5.31, in the absence of an amphiphile or at a low concentration of amphiphiles, a system with equal concentrations of oil and water undergoes phase separation, resulting in the formation of two distinct phases: one enriched in water and the other enriched in oil. As the amphiphile concentration increases, microemulsion takes place in the water phase at low temperatures (o/w or the Winsor type I microemulsion, commonly denoted as "$\underline{2}$"), in the oil phase (w/o or the Winsor type II microemulsion, denoted as "$\overline{2}$") at high temperatures, and in a bicontinuous phase (Winsor type III microemulsion) coexisting with the bulk phases at intermediate temperatures. Winsor type III microemulsion is an isotropic solution with a microstructure consisting of a multiply connected sponge-like 3D bilayer. The lattice model is able to capture the characteristic progression from two-phase to three-phase and to two-phase coexistence ($\underline{2} \rightarrow 3 \rightarrow \overline{2}$) by varying the temperature.[38] At high amphiphile concentrations, the system may exist as a lamellar phase (L_α),[39] i.e., a sheet-like two-dimensional structure of amphiphilic molecules.

Figure 5.32 presents a phase diagram for the water–oil–amphiphile system at constant temperature ($k_B T/J = 4.45$) predicted by the Gompper–Schick model.[40] Unlike the phase diagram shown in Figure 5.30, the ternary system exhibits seven phases: an oil-rich (OR), a water-rich (WR), and disordered fluid phase (DIS), a symmetric lamellar phase (LAM), two asymmetric lamellar phases (LAM'), and an amphiphile-rich phase (AR). Three-phase coexistence triangles and tie lines are shown schematically only on one half of the diagram. Having a qualitative understanding of the

38 Gradzielski M., et al. "Using microemulsions: formulation based on knowledge of their mesostructure", *Chem. Rev.* 121 (10), 5671–5740 (2021).

39 L_α is also known as lamellar smectic phase in which the constituent molecules arrange themselves into layers, akin to sheets or planes, with each layer having long-range positional order.

40 Gompper G. and Schick M. "Lattice model of microemulsions", *Phys. Rev. B* 41 (13), 9148–9162 (1990).

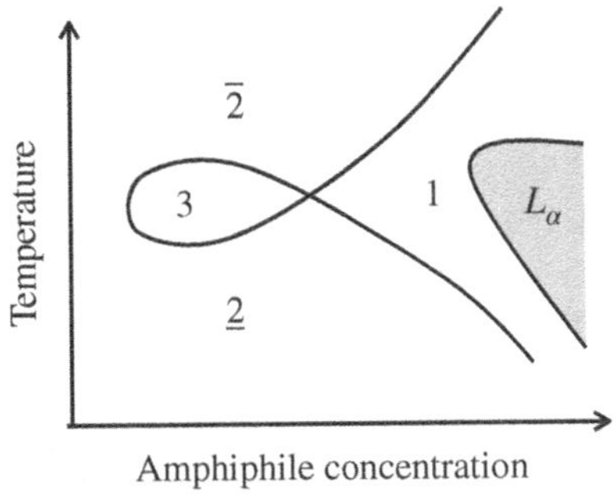

Figure 5.31 Schematic phase diagram for a water–oil–amphiphile system at equal oil and water concentrations. Here 1, 2, and 3 stand for the number of phases, L_α represents a lamellar phase.

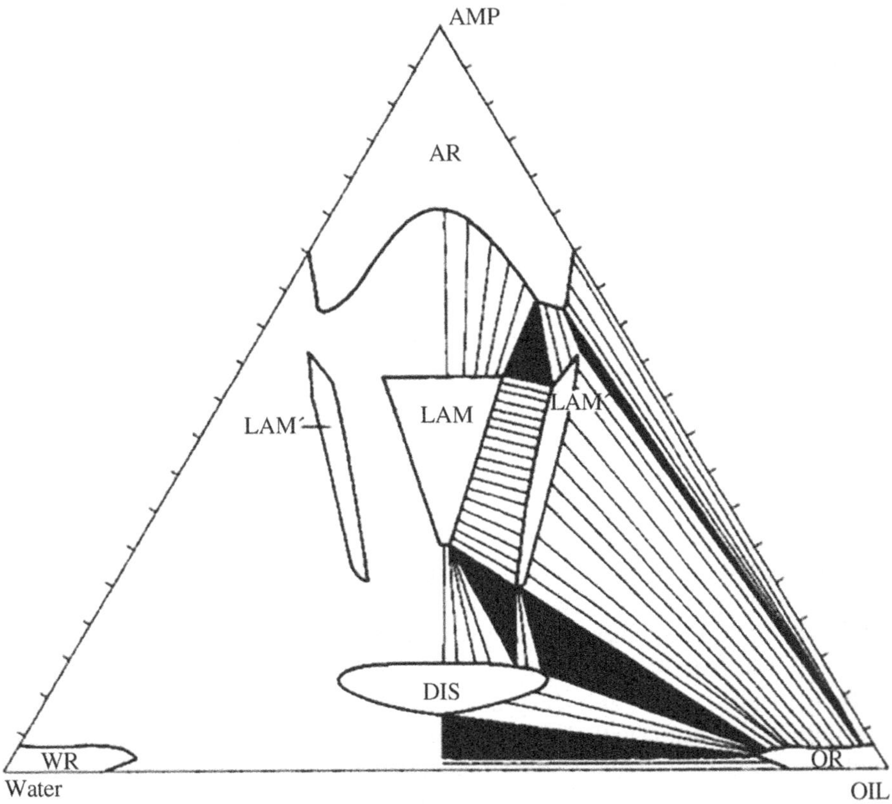

Figure 5.32 The phase diagram predicted by a lattice model for a water–oil–amphiphile system that exhibits seven phases: an oil-rich (OR), a water-rich (WR), and disordered fluid phase (DIS), all at low amphiphile concentration, a symmetric lamellar phase (LAM), two asymmetric lamellar phases (LAM) which are related to one another by interchange of oil and water, and an amphiphile-rich phase (AR). Three-phase coexistence triangles and tie lines (schematic) are shown on one half of the diagram. Reproduced from Gompper and Schick.[40]

transitions between different phases is highly valuable for formulating various microemulsion phases suitable for different applications.

5.7.4 Summary

Lattice model is a simple but powerful tool to describe the thermodynamic behavior of a wide range of physical systems, including gases, liquids, and microemulsions. While it is drastically simplified from a molecular perspective, lattice model is able to capture the essential phase behavior and many mesoscopic structures observed in experiments because physical systems often exhibit similar phase behavior at mesoscopic and macroscopic scales, despite their differences underlying the microscopic details. Furthermore, by adjusting the parameters, a lattice model is possible to achieve quantitative agreement between theoretical predictions and experimental observations, making it a powerful tool for understanding the behavior of physical systems at a mesoscopic scale.

5.8 Order Parameters and Phase Transitions

A thermodynamic system may exist in various macroscopic and mesoscopic phases each characterized by distinct microscopic structures. The transition between these phases often entails significant changes in thermodynamic properties. A key objective of statistical mechanics is to comprehend how systems undergo phase transitions in response to external parameters like temperature, pressure, or chemical composition. In this section, we establish the terminology commonly used to describe phase transitions. The subsequent section will delve into the applications of these concepts, specifically within the framework of the Landau theory—a phenomenological approach that captures the general behavior of a broad range of phase transitions.

5.8.1 Order Parameters

Order parameter is a concept instrumental for describing the states of thermodynamic systems. Intuitively, it may be understood as a quantity that provides a measure of the degree of order associated with different phases (e.g., the degree of magnetization in the Ising model or the difference between the densities of liquid and vapor phases at consistence). While the definition of order parameter appears intuitive for simple phase transitions, its generalization may not be an easy task for more complicated systems.

For any thermodynamic system, the equilibrium properties are uniquely determined by the microscopic structure, which are often quantified in terms of the spatial distribution of the constituent particles or the atomic density profiles.[41] If a thermodynamic system consists of only one type of particles, the density profile can be written as a 3-dimensional (3D) function, $\rho(\mathbf{r})$. If different kinds of particles exist in the system, the microscopic structure is defined by multiple density profiles, $\rho_1(\mathbf{r}), \rho_2(\mathbf{r}), \ldots$, such that each kind of particle has its own density profile. Because an equilibrium system in each phase has a unique microscopic structure, the density profiles provide the ultimate details for characterizing phase changes.

Describing different phases in terms of their 3D density profiles is theoretically rigorous but practically inconvenient due to its mathematical complexity. While the density profiles are intrinsically related to the microscopic structures of individual components, *order parameters* are introduced to characterize the structure at macroscopic or mesoscopic levels. From the mathematic point of view, an order parameter may be understood as a simplified representation of the microscopic structure or certain averaged properties;[42] it intends to capture the most significant features of the thermodynamic state in terms of either macroscopic property or microscopic structure, such as average density or spin orientation, or various forms of symmetry in the density profiles.[43] For example, Figure 5.33 illustrates three basic forms of symmetry commonly observed in the atomic distributions of molecular systems.

41 In the density functional theory (DFT), the Hohenberg–Kohn theorem asserts that the thermodynamic properties of an equilibrium system are uniquely determined by the density profiles of individual particles.

42 M. Fisher explains that order parameters are "certain general, overall properties especially as regards locality and symmetry: those then serve to govern the most characteristic behavior on scales greater than atomic", *Rev. Mod. Phys.* 70(2), 653–681 (1998).

43 Under certain circumstances, the density profile itself is referred to as the order parameter, or more precisely, the *local* order parameter.

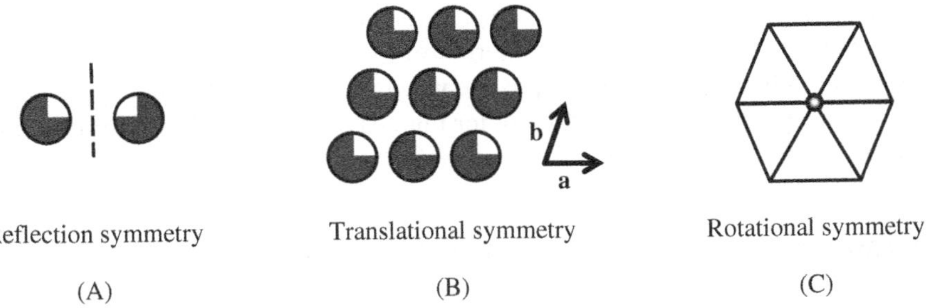

Figure 5.33 Possible symmetries in the microscopic structure of molecular systems. (A) Reflection symmetry. (B) Translational symmetry. (C) Rotational symmetry.

Without invoking any mathematical details on the density profiles, it is instructive to consider the relation between symmetry and order parameters in some relatively simple systems. For example, the field-free Ising model displays the reflection symmetry in the sense that lattice sites are occupied by spins with equal probabilities of up and down orientations. Consequently, the average magnetization serves as the order parameter to investigate phase transitions between disordered and ordered phases. On the other hand, a simple fluid like argon is isotropic, meaning that its properties remain invariant under rotation-symmetry operations. In such cases, the bulk density can be utilized as an order parameter to describe the vapor-liquid transition. Conversely, a liquid crystal may exhibit various orientational and translational orders due to the nonspherical shapes of its constituent molecules. For phase transitions in these systems, order parameters are often defined in terms of the leading coefficients of the molecular density profile, expanded with certain basis functions that adhere to the intrinsic symmetry of the underlying phases.

We may elucidate the connection between order parameters and density profiles by considering a crystalline phase that shows a discrete, lattice translational symmetry relative to the positions of individual atoms. The presence of translational symmetry implies that the density profile conforms to discrete lattice translational symmetry

$$\rho(\mathbf{r}) = \rho(\mathbf{r} + \mathbf{T_n}) \tag{5.147}$$

where $\mathbf{T_n} = (n_1\mathbf{a}_1 + n_2\mathbf{a}_2 + n_1\mathbf{a}_3)$ is known as *the Bravais lattice* of the crystal, $n_{i=1,2,3} = 0, \pm1, \pm2\cdots$ are integers, and $\mathbf{a}_{i=1,2,3}$ are the primitive translational vectors of the unit cell, as illustrated in Figure 5.34.

The order parameters of the crystalline phase are often defined in terms of the Fourier expansion of the atomic density profile

$$\rho(\mathbf{r}) = \sum_{\lambda} \hat{\rho}(\Gamma_\lambda) \exp(i\Gamma_\lambda \cdot \mathbf{r}) \tag{5.148}$$

where $\Gamma_\lambda = \lambda_1\mathbf{b}_1 + \lambda_2\mathbf{b}_2 + \lambda_3\mathbf{b}_3$ represents the reciprocal lattice vectors, $\mathbf{b}_{i=1,2,3}$ are the primitive translational vectors of the reciprocal lattice, $\lambda = (\lambda_1, \lambda_1, \lambda_3)$ denote an integer vector with

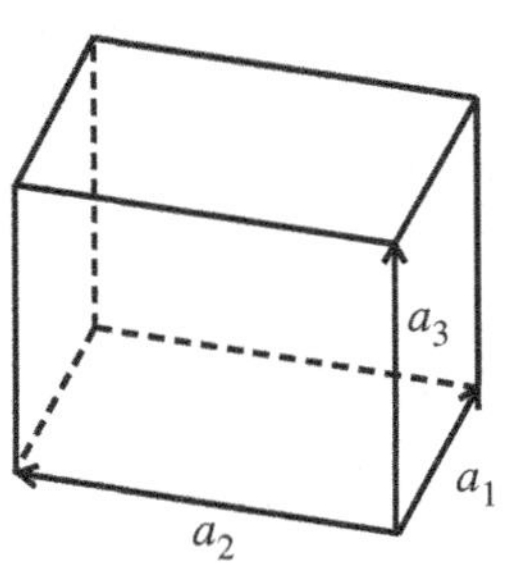

Figure 5.34 A unit cell of a crystal lattice consists of three primitive translational vectors, $\mathbf{a}_{i=1,2,3}$. The unit cell volume is given by $V_0 = \mathbf{a}_1 \cdot (\mathbf{a}_2 \times \mathbf{a}_3)$.

$\lambda_{i=1,2,3} = 0, \pm 1, \pm 2 \cdots$, and $\hat{\rho}(\Gamma_\lambda)$ are the Fourier coefficients of the expansion. The reciprocal lattice vectors can be calculated from the primitive translational vectors of the unit cell through the reciprocal lattice vector formula

$$\mathbf{b}_i = 2\pi \frac{\mathbf{a}_j \times \mathbf{a}_k}{\mathbf{a}_i \cdot (\mathbf{a}_j \times \mathbf{a}_k)}. \tag{5.149}$$

The 3-dimensional Fourier transform of the density profile within a unit cell is defined as

$$\hat{\rho}(\mathbf{k}) = \frac{1}{v_0} \int_{v_0} d\mathbf{r} \rho(\mathbf{r}) \exp(i\mathbf{k} \cdot \mathbf{r}). \tag{5.150}$$

Because $\Gamma_\lambda \cdot \mathbf{T_n} = 2\pi(n_1\lambda_1 + n_2\lambda_2 + n_3\lambda_3)$, basis functions $\exp(i\Gamma_\lambda \cdot \mathbf{r})$ satisfy the translational symmetry of the crystalline phase. When the system undergoes a transition from an isotropic phase to the crystalline phase, $\hat{\rho}(\Gamma_\lambda)$ are manifested in terms of the magnitudes of the Bragg peaks in scattering experiments. To have a complete characterization of the translational order, we need in principle all coefficients in Eq. (5.148). Because there is no periodicity in the density profile of a disordered phase, a few leading terms are often sufficient to serve as the order parameters for modeling the crystallization process.

The aforementioned approach can be similarly used to introduce *order parameters* that reflect the orientational orders of molecular distributions observed in different phases of liquid crystals. For instance, in a thermodynamic system composed of cylindrically shaped molecules, an isotropic-to-nematic phase transition occurs as the temperature decreases or the molecular density increases. The nematic phase, which exhibits fluid-like properties, holds significant technological importance as many electro-optical applications are based on nematic liquid crystals. In the isotropic phase, the molecular orientations are nearly random, while in the nematic phase, the nonspherical molecules align preferentially in a particular direction. This orientational ordering enables us to define a *director*, which is a unit vector indicating the direction of molecular distributions in the nematic phase.

We can describe the change in the microscopic structure of a liquid crystal under the isotropic-nematic phase transition in terms of the probability distribution of the molecular orientations, $p(\theta, \varphi)$, where (θ, φ) stands for the polar and azimuth angles of each molecule, as illustrated in Figure 5.35. The angular order parameters are often referred to as the coefficients in a spherical harmonic expansion of the probability distribution for the molecular orientations

$$p(\theta, \varphi) = \sum_{l=0}^{\infty} \sum_{k=-l}^{l} p_{lk} Y_{lk}(\theta, \varphi) \tag{5.151}$$

where $Y_{lk}(\theta, \varphi)$ stands for the normalized spherical harmonic functions, and p_{lk} are coefficients determined by the orthogonality relations

$$p_{lk} = \int d\theta \int d\varphi p(\theta, \varphi) Y_{lk}^*(\theta, \varphi) = \langle Y_{lk}^*(\theta, \varphi) \rangle \tag{5.152}$$

where the asterisk denotes complex conjugation, and <$\cdots$> stands for the average over the distribution of the molecular orientations.

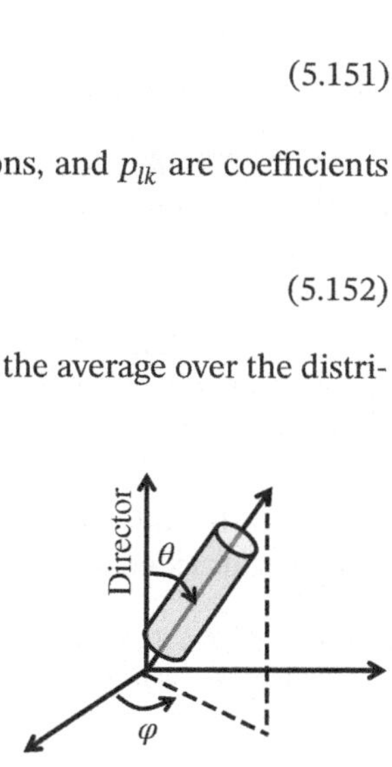

Figure 5.35 Orientation of a cylindrically shaped molecule relative to a given direction, i.e., a director, can be described with polar angle θ and azimuth angle φ.

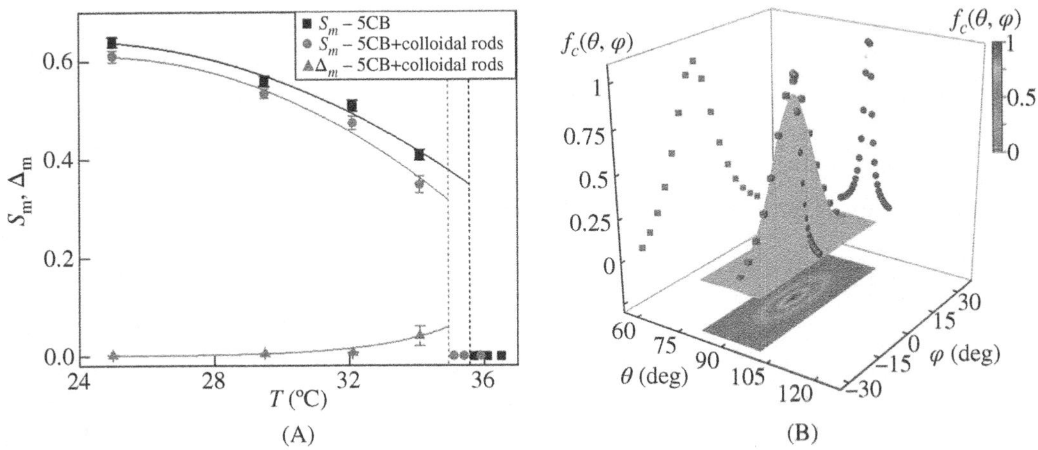

Figure 5.36 Order parameters for molecular orientations (A) and the orientational distributions of nanorods (B) for cylindrical colloidal particles dispersed in a liquid crystal consisting of pentyl-cyanobiphenyl (5CB) molecules. Panel B shows the biaxial distribution for the orientation of nanorods at $T = 27.5$ °C. A phase diagram can be constructed according to different orientations of colloids and molecules (denoted by "c" and "m"). Source: Adapted from Mundoor et al.[45]

For a liquid crystal consisting of cylindrical particles, nontrivial coefficients in the spherical harmonic expansion, Eq. (5.151), appear only after the second-rank level.[44] As a result, the first two lowest coefficients are commonly used to characterize isotropic to nematic phase transition:

$$S_m = \sqrt{\frac{4\pi}{5}} \langle Y_{20}^*(\theta, \varphi) \rangle = \frac{1}{2} \langle 3\cos^2\theta - 1 \rangle, \tag{5.153}$$

$$\Delta_m = \sqrt{\frac{4\pi}{15}} \langle Y_{22}(\theta, \varphi) + Y_{2-2}(\theta, \varphi) \rangle = \langle \sin^2\theta \cos 2\varphi \rangle. \tag{5.154}$$

Figure 5.36 shows the nematic order parameters and orientational distribution of nanorods in a hybrid system of molecular-colloidal liquid crystals.[45] In the isotropic phase, the molecules are uniformly oriented and $S_m = 0$. If all molecules align exactly with the director, $\cos^2\theta = 1$ and $S_m = 1$. Order parameter Δ_m signals the alignment of the molecular orientations in two directions, i.e., a nonzero value of Δ_m indicates the existence of a biaxial nematic phase.

5.8.2 Classification of Phase Transitions

A system undergoing phase transition leads to the variation of its microscopic structure and, subsequently, all pertinent thermodynamic properties. Depending on how the thermodynamic properties change at the transition point, phase transitions are conventionally classified into the first-order, the second-order, and higher orders. As shown schematically in Figure 5.37, a first-order phase transition exhibits abrupt changes in the order parameter and extensive thermodynamic variables such as entropy, internal energy, and the number of particles. These properties can be obtained from the first-order derivatives of the free energy with respect to their conjugate fields (e.g., $S = -(\partial F/\partial T)_{V,N}$, $U = (\partial \beta F/\partial \beta)_{V,N}$, $N = -(\partial \Omega/\partial \mu)_{V,T}$). In contrast, a second-order phase transition, also known as a continuous phase transition, does not exhibit discontinuity in the order parameter or extensive thermodynamic variables. However, at a second-order phase transition,

44 Turzi S. S. "On the Cartesian definition of orientational order parameters", *J. Math. Phys.* 52, 053517 (2011).
45 Mundoor H., et al. "Hybrid molecular-colloidal liquid crystals", *Science* 360, 768–771 (2018).

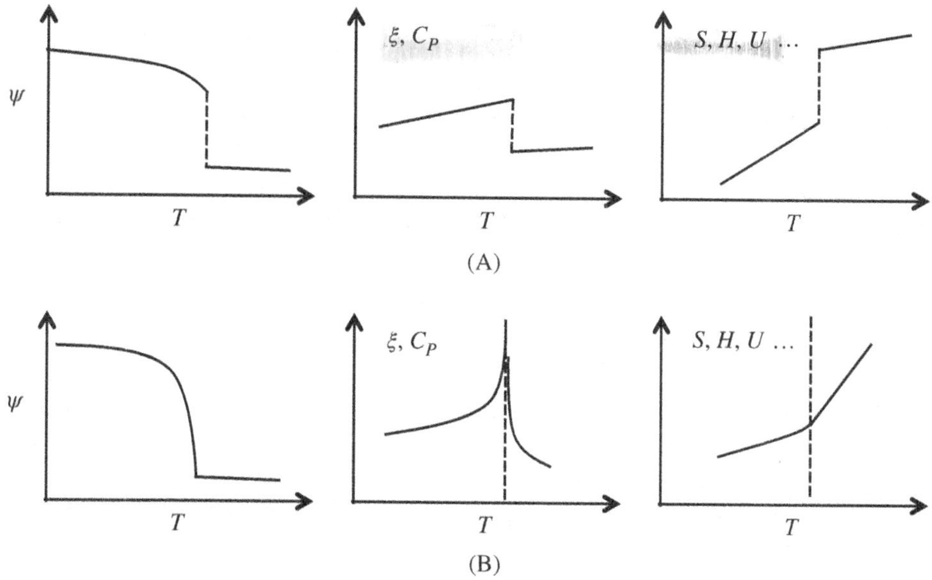

Figure 5.37 Properties of a thermodynamic system vary with temperature near a first-order phase transition (A) and near a second-order phase transition (B). Displayed above are schematic changes for order parameter ψ, correlation length ξ, and extensive thermodynamic variables such as heat capacity C_p, entropy S, enthalpy H, internal U, etc.

the second-order derivatives of the free energy, such as heat capacity, compressibility, and thermal expansion coefficients, experience discontinuities. In principle, it is possible to define third and higher-order transitions by considering additional derivatives of the free energy. However, these higher-order transitions are rarely used or studied in practice because they occur less frequently and often involve complex interactions that are challenging to analyze and characterize.

Familiar examples of first-order phase transitions include the evaporation of a liquid or a solid and its reversible processes, transitions among various solids and liquid crystal phases, and diverse structural changes in surfactant or amphiphilic systems. These transitions typically occur with the absorption or release of latent heat and are often accompanied with metastability, which can manifest as metastable phenomena like overheating or overcooling. Second-order and high-order transitions are less common. In addition to vapor–liquid or liquid–liquid equilibria at the critical point, examples often cited as second-order phase transitions include ferromagnetic (ferroelectric) transitions, structural transitions among various solid states, and transitions related to superconductivity or superfluids. Second-order phase transitions are typically manifested in terms of a divergent susceptibility, an infinite correlation length, and a power-law decay of correlations near the critical point.

5.8.3 Summary

This section introduces two important concepts for describing phase transitions. Order parameter reflects the characteristic properties of different macroscopic phases or the symmetry of microscopic structure; this concept is closely related to the density profiles of individual particles in a thermodynamic system. Classification of phase transitions is based on the continuity of order parameter and thermodynamic variables when the system changes from one phase to another phase. During a first-order phase transition, the order parameter undergoes a discontinuous jump, while a second-order phase transition features a continuous change in the order parameter. Additionally, there is a distinct difference in the behavior of thermodynamic variables between these two types of phase transitions. In a first-order transition, thermodynamic variables often exhibit discontinuities or jumps at the transition point. In a second-order transition, thermodynamic variables do not display such abrupt changes or discontinuities at the critical point. Instead, they typically exhibit continuous variations with discontinuities in the second-order derivatives of the free energy, such as heat capacity, compressibility, and thermal expansion coefficients, as the system undergoes the transition.

5.9 The Landau Theory of Phase Transitions

The Landau theory, named after the renowned physicist Lev Landau, offers a comprehensive framework for describing phase transitions across a broad range of thermodynamic systems. Initially devised for second-order phase transitions, this phenomenological approach proves applicable also to diverse types of first-order phase transitions, encompassing structural and magnetic transitions, superfluid transitions, liquid–solid and liquid–vapor transitions, as well as the self-assembly phenomena observed in liquid crystals and complex fluids. In this section, we introduce briefly the essential concepts of the Landau theory. For more comprehensive discussions on this important subject, we refer to specialized texts.[46]

5.9.1 Second-Order Phase Transition

As discussed in Section 5.8, a thermodynamic system undergoing a second-order phase transition may be characterized by the continuous increase of an order parameter, i.e., zero for the disordered phase and a finite value for the ordered phase. Near the critical point, the order parameter is small such that the free energy density may be expressed in terms of the Taylor expansion[47]

$$f(m) = f_0 + a_1 m + a_2 m^2/2 + a_3 m^3/3 + a_4 m^3/4 + a_5 m^3/5 + \cdots \tag{5.155}$$

where m denotes an appropriate order parameter associated with the phase transition, f_0 corresponds to the free energy density when the system is in the disordered phase, $a_{i=1,2,\ldots}$ are phenomenological coefficients that account for the change in the free energy density due to the emergence of ordering, and the fractional coefficients are introduced for the convenience of later discussion.

Eq. (5.155) is known as the Landau expansion of the free energy density. Both f_0 and a_i are dependent on macroscopic thermodynamic variables such as temperature T, pressure P, and the composition of chemical species. In principle, the dependence of f_0 and a_i on thermodynamic and material properties can be predicted from a microscopic theory. Alternatively, these phenomenological parameters can be obtained by empirical correlation with certain experimental results. At equilibrium, the free energy density attains a minimum with respect to the order parameter

$$\partial f/\partial m = 0. \tag{5.156}$$

46 Jean-Claude T. and Pierre T., *The Landau theory of phase transitions: application to structural, incommensurate, magnetic and liquid crystal systems*. World Scientific Press, 1987; Hohenberg P. C. and Krekhov A. P., "An introduction to the Ginzburg-Landau theory of phase transitions and nonequilibrium patterns", *Phys. Rep.* 572, 1–42 (2015).

47 It should be noted that the order parameter is typically not a scalar; it can take the form of a scalar, vector, or tensor, as discussed in the previous section. Besides, the free energy density (or the free energy per particle) is not analytical at the critical point.

Eq. (5.156) allows for determining the order parameter and, subsequently, variations of thermodynamic properties due to the formation of the ordered phase.

Near a second-order phase transition, the ordered phase is thermodynamically stable below the critical temperature ($T < T_c$). At equilibrium, the order parameter takes $m = 0$ for $T \geq T_c$ and $m \neq 0$ for $T < T_c$. If the free energy density of the system is invariant to the sign of the order parameter, i.e., $f(m) = f(-m)$, all odd terms on the right side of Eq. (5.155) must be absent. Figure 5.38 illustrates the free energy density curve, showing a minimum at $m = 0$ for $T \geq T_c$, and a local maximum at $m = 0$ along with two symmetric minima at $\pm m \neq 0$ for $T < T_c$. To capture this behavior, the simplest form of the Landau expansion is given by

$$f(m) = f_0 + a_2 m^2/2 + a_4 m^4/4 \tag{5.157}$$

where $a_4 > 0$ is required in order to have a lower bound for the free energy density.

Above the critical temperature, the disordered phase is thermodynamically stable, while below the critical temperature, it becomes unstable. Because the thermodynamic stability of the phase depends on the critical temperature, a_2 must change the sign at the critical point. Near the critical temperature, we can thus express a_2 as

$$a_2 = a_2^0(T - T_c) \tag{5.158}$$

where $a_2^0 > 0$ is independent of temperature.

Minimizing the free energy density given by Eq. (5.157) leads to

$$\partial f/\partial m = a_2 m + a_4 m^3 = 0. \tag{5.159}$$

Therefore, for $T < T_c$, the order parameter satisfies

$$m = \pm\sqrt{a_2^0(T_c - T)/a_4}. \tag{5.160}$$

Eq. (5.160) agrees with the prediction of mean-field methods discussed in Section 5.5. Because the Landau expansion aims to describe the phase transition in the vicinity of the critical point, it is often assumed that the parameter a_4 is independent of temperature.

Plugging Eq. (5.160) into Eq. (5.157) yields the free energy density of the ordered phase

$$f = f_0 - \frac{\left[a_2^0(T - T_c)\right]^2}{4a_4} \tag{5.161}$$

From the free energy density, we can readily derive other thermodynamic properties. With the assumption that a_4 is independent of temperature, the entropy density and the heat capacity per

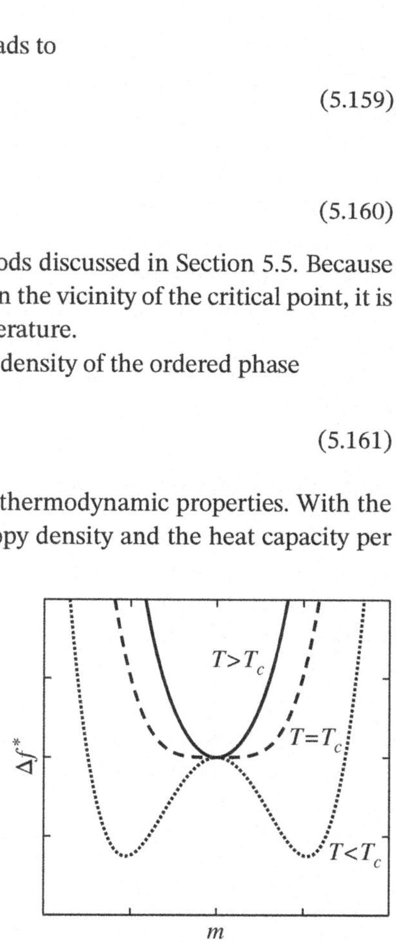

Figure 5.38 The free energy density versus the order parameter near a second-order phase transition. Here $\Delta f^* \equiv (f - f_0)/a_0 T_c$.

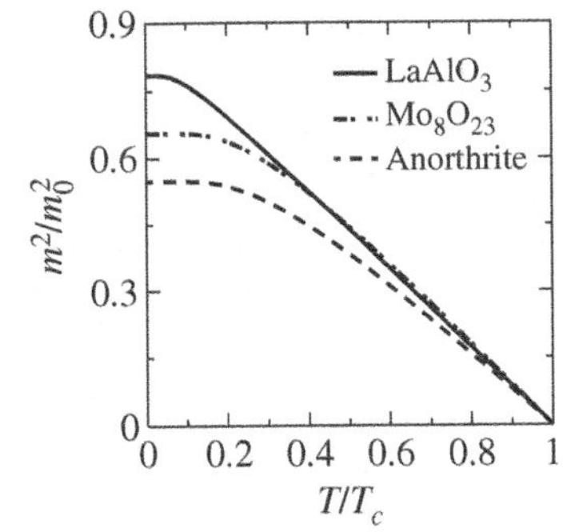

Figure 5.39 Theoretical prediction of the (normalized) order parameter for structural transitions in minerals. Source: Adapted from Salje.[49]

volume of the ordered phase are given by

$$s = -(\partial f/\partial T)_m = s_0 + \frac{\left(a_2^0\right)^2}{2a_4}(T - T_c), \tag{5.162}$$

$$c_v = T(\partial s/\partial T)_m = c_{v,0} + \left(a_2^0\right)^2 T/\left(2a_4\right), \tag{5.163}$$

where s_0 and $c_{v,0}$ correspond to the entropy and heat capacity per volume corresponding to the disordered phase at the same temperature, respectively.

The Landau theory of second-order phase transitions finds wide application in various fields, including magnetic transitions, metal-to-superconductor transitions, and superfluid transitions.[48] Second-order phase transitions are also frequently observed in various crystalline minerals such as perovskites. For example, Figure 5.39 illustrates the temperature dependence of the order parameter for three specific minerals.[49] By incorporating modifications to account for quantum effects at low temperatures, the Landau theory can accurately capture structural phase transitions, demonstrating excellent agreement with experimental results.

During a second-order phase transition, the free energy and its first-order derivatives, such as entropy, exhibit continuity. However, the second-order derivatives of the free energy, including heat capacity, isothermal compressibility, and thermal expansion coefficient, become discontinuous at the transition point. As a consequence, the continuous transition can lead to significant changes in the physical properties of a material, such as acoustic velocity, thermal expansion, and electrical conductivity.

5.9.2 First-Order Phase Transition

While the preceding discussion strictly applies to second-order phase transitions, a similar approach can be used to describe first-order phase transitions. However, in the case of first-order transitions, both the first derivatives of the free energy and the order parameter exhibit discontinuities at the transition point. The presence of a latent heat and the coexistence of two phases make the mathematical treatment significantly more complex.

To elucidate the Landau theory of first-order phase transition, consider again a system with positive-negative symmetry in terms of the order parameter.[50] Because $f(m) = f(-m)$, no odd terms are absent in the landau expansion. Therefore, the simplest way for describing a first-order phase

48 Chaikin P. M. and Lubensky T. C., *Principles of condensed matter physics*. Cambridge University Press, 1995.
49 Salje E. K. H. "Application of Landau theory for the analysis of phase-transitions in minerals", *Phys. Rep.* 215 (2), 49–99 (1992).
50 Problem 5.22 discusses the Landau theory for systems that do not satisfy the invariance condition $f(T, m) = f(T, -m)$ so that odd terms are allowed in the Taylor expansion of the free energy density.

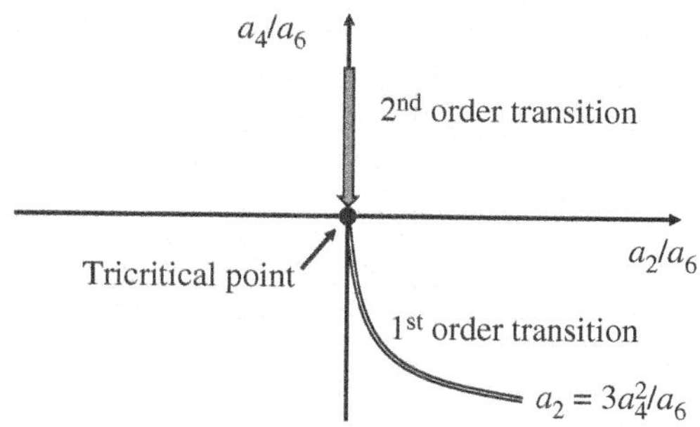

Figure 5.40 Phase diagram predicted by the free energy density given by Eq. (5.164).

transition is by retaining the first three lowest-order even terms in Eq. (5.155)

$$f(m) = f_0 + a_2m^2/2 + a_4m^4/4 + a_6m^6/6. \tag{5.164}$$

As discussed above, the coefficient for the highest-order term must be positive in order to maintain the system stability, i.e., $a_6 > 0$.

When a first-order transition takes place, we have the same free energy for the ordered and disordered phases

$$\Delta f \equiv f(m) - f_0 = a_2m^2/2 + a_4m^4/4 + a_6m^6/6 = 0. \tag{5.165}$$

At equilibrium, the order parameter for the ordered phase can be identified by minimizing the free energy density

$$\partial f/\partial m = a_2m + a_4m^3 + a_6m^5 = 0. \tag{5.166}$$

From Eqs. (5.165) and (5.166), we can obtain the order parameter at the transition point

$$m^2 = -3a_4/4a_6. \tag{5.167}$$

Substituting Eq. (5.167) into Eq. (5.166) yields

$$a_2 = 3a_4^2/16a_6. \tag{5.168}$$

Because $a_6 > 0$, Eq. (5.168) indicates that a_2 must also be positive near the first-order phase transition.

In Figure 5.40, the double solid line illustrates the phase diagram of the first-order transition. When $a_4 < 0$, Eq. (5.167) yields two nonzero order parameters. In this case, the first-order phase transition takes place, leading to the coexistence of three phases, i.e., the disordered phase with $m = 0$ in equilibrium with two ordered phases with $m = \pm\sqrt{-3a_4/4a_6}$. The three-phase coexistence terminates at $a_4 = 0$. For $a_4 > 0$, Eq. (5.167) does not yield any real values of the order parameter, indicating that a first-order phase transition is not possible.

When the order parameter is sufficiently small, the sixth-order term in Eq. (5.164) becomes negligible. Therefore, for $a_4 > 0$, we expect that the system will undergo a second-order phase transition when the temperature is sufficiently low, similar to that predicted by Eq. (5.157). At $a_4 = 0$, Eq. (5.164) predicts an additional second-order phase transition and, near its transition point, the order parameter varies with temperature with critical exponent of 1/4 rather than 1/2

$$m = \pm[|a_2|/a_6]^{1/4}. \tag{5.169}$$

Because the first-order phase transition also terminates at $a_4 = 0$, the thermodynamic condition where both a_2 and a_4 are zero leads to a tricritical point.[51] As shown in Figure 5.40, at this condition,

51 The term "tricritical point" is used to describe a point at which the second-order phase transition curve intersects with the first-order phase transition curve. Coined by Lev Landau in 1937, the concept of a tricritical point was introduced to characterize the critical point of a continuous transition.

the first-order and second-order phase transition curves join together. This kind of phase behavior has been observed in certain thermodynamic systems, including water–oil–nonionic amphiphile mixtures.[52] At the tricritical point, the distinction between three phases becomes less pronounced and the properties of the substance, such as density, heat capacity, and compressibility, may exhibit unusual behavior.

When the temperature is sufficiently low, the ordered phase becomes more stable than the disordered phase. To satisfy such a condition, we may write a_2 as

$$a_2 = a_2^0(T - T_{c2}) \tag{5.170}$$

where $a_2^0 > 0$ is independent of temperature, and T_{c2} is the critical temperature for the occurrence of the second-order phase transition. When $T < T_{c2}$, Eq. (5.168) is no more valid, implying that the first-order phase transition ceases to exist. Schematically, Figure 5.41 presents the free energy density predicted by Eq. (5.164). The first-order phase transition occurs at $T = T_{tr}$. At a certain temperature, here designated as $T_{c2} \leq T_{tr}$, a second-order transition phase transition may also take place. Below this temperature ($T < T_{c2}$), only the ordered phase is thermodynamically stable. Above the first-order transition temperature, a critical temperature of metastability can be identified such that the free energy density displays a local minimum at a finite value of the ordered parameter. This temperature, designated as T_{c1}, can be found from the inflection point for the free-energy-density curve

$$T_{c1} = T_{c2} + \frac{a_4^2}{4a_2^0 a_6}. \tag{5.171}$$

At temperature $T > T_{c1}$, the free energy shows only one minimum at $m = 0$, i.e., the system exists in a disordered state. For temperatures between T_{c1} and T_{tr}, the free energy shows two minimum values, one at $m = 0$ and an additional one at $m \neq 0$ corresponding to a metastable state.

The polymorphic transformation of quartz provides one simple application of the Landau theory of the first-order transition. Crystalline quartz exists as the α phase at low temperature and the β phase at high temperature. Figure 5.42 shows the crystalline structures of these two phases and the temperature-dependent of the order parameter, which is related to the distortion of atomic coordinates.[53] A first-order phase transition occurs at temperature $T_{tr} = 847$ K, arising from the rotation of SiO_4 tetrahedra with additional small distortions of the bond lengths and bond angles. With only few parameters, the Landau theory is able to describe the phase transition and various

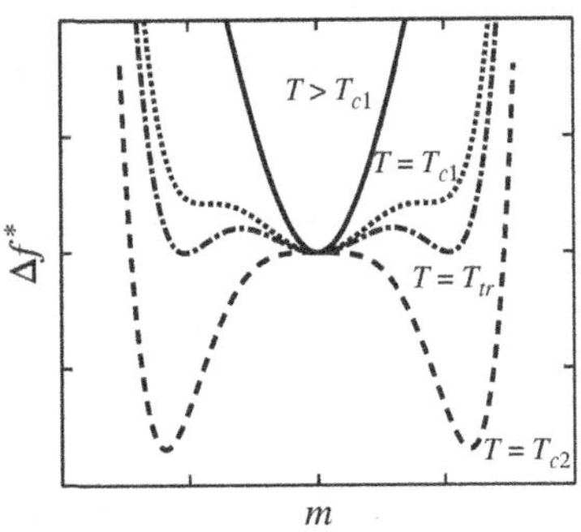

Figure 5.41 The reduced free energy density versus the order parameter for a system exhibiting a first-order phase transition at T_{tr}. Other lines correspond to conditions when only the disordered phase is stable ($T > T_{c1}$), a metastable state starts to emerge ($T = T_{c1}$), and the second-order phase transition takes place at $T = T_{c2}$. The reduced free energy density is defined as $\Delta f^* \equiv \Delta f / a_2^0 T_{c2}$.

52 Indekeu J. O. and Koga K. "Wetting and nonwetting near a tricritical point", *Phys. Rev. Lett.* 129 (22), 224501 (2022).

53 Antao S. M. "Quartz: structural and thermodynamic analyses across the alpha–beta transition with origin of negative thermal expansion (NTE) in beta quartz and calcite", *Acta Crystallogr. B* 72, 249–262 (2016).

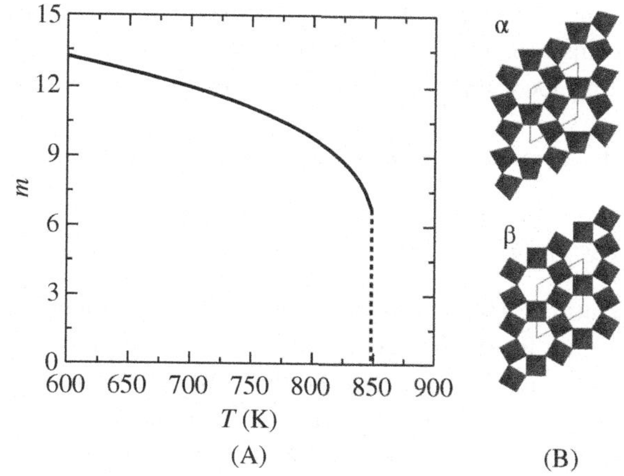

Figure 5.42 (A) Variation of the order parameter with temperature for quartz near the α-β phase transition. (B) Two polymorphs of quarts with the $(SiO_4)^{-4}$ tetrahedra linked along corners to four other tetrahedra. Source: Adapted from Antao.[53]

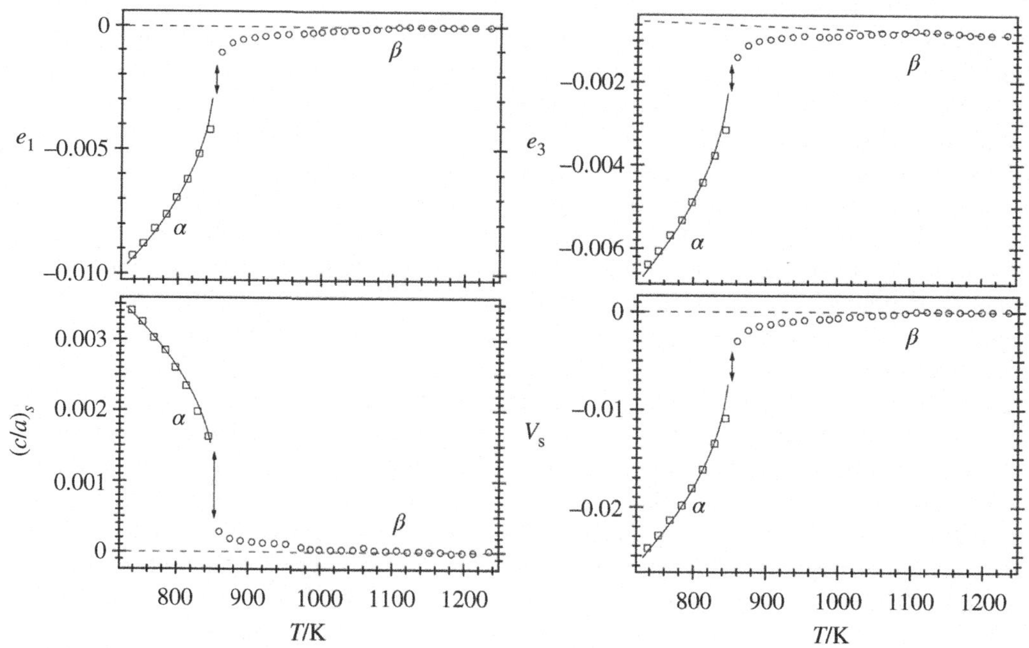

Figure 5.43 Mechanical properties of quartz predicted by the Landau theory in comparison with experimental data. Source: Adapted from Antao.[53]

properties of the ordered phase (α phase) in quantitative agreement with experimental measurements (Figure 5.43).

5.9.3 The Ginzburg–Landau Theory

In the previous discussions, we assumed that phase transitions involve a single scalar order parameter independent of position. However, this assumption is often an oversimplification.

As discussed in Section 5.8, order parameters can have diverse forms and may exhibit spatial variation throughout the system. To address this broader scope, the Ginzburg–Landau theory extends the Landau expansion by incorporating local order parameters to capture the spatial variations.

To incorporate the spatial variation of an order parameter, we may express the total free energy as a functional of $m(\mathbf{r})$

$$F[m(\mathbf{r})] = \int d\mathbf{r} \left\{ f(m) + \frac{c_2}{2}(\nabla m)^2 + \frac{c_4}{2}(\nabla^2 m)^2 \right\} \tag{5.172}$$

where $m(\mathbf{r})$ stands for the order parameter at position $\mathbf{r}$, $f(m)$ corresponds to the free energy density of a uniform phase, and the gradient terms account for the free energy arising from the inhomogeneity of the system. Like that in the original Landau theory, the local order parameter $m(\mathbf{r})$ can be obtained by minimizing the total free energy (see Appendix 8A for a brief introduction to the calculus of variations). Subsequently, we can derive both structure and thermodynamic properties of the system in an inhomogeneous phase.

For systems invariant to the sign of the order parameter, we can write the free energy density corresponding to a uniform phase in terms of a power series as given by Eq. (5.164)

$$f(m) = f_0 + a_2 m^2/2 + a_4 m^4/4 + a_6 m^6/6 + \cdots. \tag{5.173}$$

Alternatively, the local free energy may be derived from a microscopic theory of the bulk phase as originally proposed in the van der Waals theory of capillarity.[54] A similar approach was adopted in the Swift–Hohenberg equation for describing the dynamics and pattern formation in systems with instabilities and spatial modulations, the Cahn–Hilliard theory, as well as the Flory–Huggins-de Gennes theory, for the kinetics of phase transitions.[55]

To ensure thermodynamic stability, the phenomenological coefficients in Eq. (5.172) are typically selected such that $c_2 < 0$ and $c_4 > 0$. The coefficient for the square gradient of the order parameter $c_2 < 0$ means that the system may be stabilized in an inhomogeneous state or favors the spontaneous formation of interfacial areas. Meanwhile, the 4th-order gradient term is included to make the system stable without unlimited growth of the interfacial area. Without the 4th-order gradient term, the coefficient for the square gradient must be positive.

Intuitively, the Ginzburg–Landau theory can be understood as a gradient expansion of the local free energy density relative to that of a uniform system. Because the Landau expansion of the free energy invokes no specific knowledge on the microscopic details or correlation functions, Eq. (5.172) is applicable to a wide variety of phase transitions involving spatially modulated phases as well as self-assembly processes. Mathematically, the gradient expansion is adequate only for weakly nonuniform systems where the order parameter is small and varies smoothly over space. However, in practice, the functional form derived from the Ginzburg–Landau theory is often employed beyond the so-called "weak segregation" limit, indicating its usefulness in a broader range of scenarios. In addition to phase transitions near critical points, the Ginzburg–Landau theory provides valuable insights into the interfacial properties of coexisting phases where the order parameter exhibits spatial variation and can be used to describe interfacial phenomena such as wetting transitions.[56]

---

54 van der Waals J. D., "The thermodynamic theory of capillarity under the hypothesis of a continuous variation of density", *J. Stat. Phys.* 20, 197–200 (1979).

55 Swift J. and Hohenberg P.C., "Hydrodynamic fluctuations at the convective instability", *Phys. Rev. A* 15(1), 319–328 (1977). Cahn J. W. and Hilliard J. E., "Free energy of a nonuniform system. I. Interfacial free energy", *J. Chem. Phys* 28(2), 258–267 (1958). de Gennes P. G. "Dynamics of fluctuations and spinodal decomposition in polymer blends", *J. Chem. Phys.* 72, 4756 (1980).

56 de Gennes P. G., Brochard-Wyart F., Quéré D., *Capillarity and wetting phenomena*. Springer, Berlin, 2003.

5.9.4 The Ginzburg Criterion

One important assumption in the Ginzburg–Landau theory is that the correlation between local dynamic variables is negligible compared to their mean values. This assumption can be assessed quantitatively using the Ginzburg criterion, which measures the impact of fluctuations on the free energy. The Ginzburg criterion allows for an evaluation of the relative importance of fluctuation effects and helps determine whether the mean-field approximation is valid for a given system. If the fluctuations have a significant influence on the free energy, indicating a breakdown of the assumption, more sophisticated approaches beyond the mean-field theory may be necessary to accurately describe the system.

At a coarse-grained level, the effects of fluctuations can be quantified by measuring the variation in the local order parameter

$$\delta m(\mathbf{r}) = \widetilde{m}(\mathbf{r}) - m(\mathbf{r}) \tag{5.174}$$

where $\widetilde{m}(\mathbf{r})$ represents the instantaneous value associated with $m(\mathbf{r})$. Imagine that the order parameter is coupled with an external field $\varphi(\mathbf{r})$, we can evaluate the fluctuation effect using the partition function

$$Q = \sum_{\widetilde{m}(\mathbf{r})} \exp\left\{-\beta F[\widetilde{m}(\mathbf{r})] - \int d\mathbf{r}\widetilde{m}(\mathbf{r})\beta\varphi(\mathbf{r})\right\} \tag{5.175}$$

where $F[\widetilde{m}(\mathbf{r})]$ is given by Eq. (5.172). The summation overall all possible forms of $\widetilde{m}(\mathbf{r})$ is often referred to as the functional integration, which can be understood as a summation of all possible microstates but at the coarse-grained level. Apparently, the free energy corresponding to the Ginzburg–Landau theory (*viz.*, the mean-field theory) is recovered if the local order parameter is fixed as $\widetilde{m}(\mathbf{r}) = m(\mathbf{r})$.

According to the self-consistent mean-field approximation, the local order parameter minimizes the free energy of the inhomogeneous system

$$\frac{\delta F}{\delta m(\mathbf{r})} + \varphi(\mathbf{r}) = 0. \tag{5.176}$$

Meanwhile, the correlated fluctuation of the local order parameter can be evaluated in terms of the density–density correlation function

$$\chi(\mathbf{r}, \mathbf{r}') \equiv \langle\delta\widetilde{m}(\mathbf{r})\delta\widetilde{m}(\mathbf{r}')\rangle = -\frac{\delta m(\mathbf{r})}{\delta\beta\varphi(\mathbf{r}')}. \tag{5.177}$$

Substituting the mean-field free energy, Eq. (5.172), into Eq. (5.176) yields

$$\frac{\partial f}{\partial m} - c_2\nabla^2 m + c_4\nabla^4 m + \varphi = 0. \tag{5.178}$$

We can obtain an analytical expression for $\chi(\mathbf{r}, \mathbf{r}')$ by applying a functional derivative with respective to $-\beta\varphi(\mathbf{r}')$ on both sides of Eq. (5.178)

$$\left[\frac{\partial^2 f}{\partial m^2} - c_2\nabla^2 + c_4\nabla^4\right]\chi(\mathbf{r}, \mathbf{r}') - \delta(\mathbf{r} - \mathbf{r}')k_B T = 0 \tag{5.179}$$

where $\delta(\mathbf{r} - \mathbf{r}')$ stands for the Dirac function, and function f is given by Eq. (5.173).

For a uniform system without an external field, the order parameter is independent of the position (i.e., uniform throughout the system). In that case, the density–density correlation function depends only on distance $r = |\mathbf{r} - \mathbf{r}'|$ and can be obtained by taking the Fourier transform on both sides of Eq. (5.179)

$$\beta\hat{\chi}(k) = \frac{1}{f''(m) + c_2k^2 + c_4k^4}. \tag{5.180}$$

If the Landau free energy is taken a quadratic form, i.e., $f''(m) \approx a_2 > 0$, Eq. (5.180) reduces to the density–density correlation function used in the Teubner–Strey theory of microemulsions (Section 5.10).

As $k \to 0$, the long-range component of the density–density correlation function is dominated by the leading order in the denominator of Eq. (5.180). In that case, we can obtain the density–density correlation function in the real space by taking the inverse Fourier transform

$$\chi(r) \approx \int \frac{d\mathbf{k}}{(2\pi)^3} \frac{k_B T e^{-i\mathbf{k}\cdot\mathbf{r}}}{f''(m) + c_2 k^2} = -\frac{k_B T}{4\pi c_2} \frac{\exp(-r/\xi)}{r} \tag{5.181}$$

where $c_2 < 0$, and $\xi = [-c_2/f''(m)]^{1/2} \approx (-c_2/a_2)^{1/2}$ represents the correlation length, i.e., it dictates how fast the correlation decays at long distance.[57]

Equation (5.181) indicates that the Ginzburg–Landau theory accounts for correlation at the length scale comparable to ξ. If a system displays substantial fluctuations beyond the correlation length, it is expected that the mean-field approximation breaks down. The Ginzburg criterion assesses the condition when the mean-field approximation becomes problematic. A key parameter in this assessment is the relative mean fluctuation of the order parameter across the coherence volume of $V_\xi \approx 4\pi\xi^3/3$, which is defined as

$$N_G \equiv \frac{1}{V_\xi} \int_{V_\xi} d\mathbf{r} \frac{<\delta m(0)\delta m(\mathbf{r})>}{m^2} = \frac{4\pi}{m^2 V_\xi} \int_0^\xi r^2 dr \cdot \chi(r) = -\frac{3(1-2/e)}{4\pi} \frac{k_B T}{m^2 \xi c_2} \tag{5.182}$$

where we have evaluated the integral using the density–density correlation function given by Eq. (5.181). Meanwhile, we can estimate the order parameter from

$$m^2 \approx -a_2/a_4 \approx -c_2/(\xi^2 a_4) \tag{5.183}$$

where the second equation follows the relation $\xi^2 \approx -c_2/a_2$, and $a_4 > 0$. Substituting Eq. (5.183) Eq. (5.182) yields

$$N_G = \frac{3(1-2/e)}{4\pi} \frac{k_B T \xi a_4}{c_2^2} \approx 0.063 \frac{k_B T \xi a_4}{c_2^2}. \tag{5.184}$$

Equation (5.184) predicts that the fluctuation ignored by the mean-field approximation becomes significant when $N_G \approx 1$, i.e., when the correlation length of the system is on the order of

$$\xi_G = \frac{4\pi}{3(1-2/e)} \frac{c_2^2}{k_B T a_4} \approx 15.8 \frac{c_2^2}{k_B T a_4} \tag{5.185}$$

where ξ_G is known as the Ginzburg length.[58] This criterion facilitates the determination of the range of applicability of the mean-field approximation and aids in identifying situations where the fluctuations are too pronounced for using the mean-field approach. If $N_G \ll 1$, it is reasonable to assume that the fluctuation effects are negligible beyond the coherence volume, and thus the mean-field approximation is justified.

Whereas the Ginzburg length is finite, the correlation length diverges when the system approaches to the critical temperature. As a result, we expect that the mean-field approximation fails beyond certain temperature. Using $a_2 = a_2^0|T - T_c|$ and $\xi^2 \approx -c_2/a_2$, we find the correlation length diverges with temperature according to

$$\xi \approx \xi_0 |T - T_c|^{-1/2} \tag{5.186}$$

---

57 The three-dimensional Fourier transform is given by $\int d\mathbf{r} \frac{e^{-br}}{4\pi r} e^{i\mathbf{k}\cdot\mathbf{r}} = \frac{1}{4\pi} \int_0^\infty r dr \int_0^{2\pi} d\varphi \int_0^\pi \sin\theta d\theta e^{-br+ikr\cos\theta} = \frac{1}{2} \int_0^\infty r \exp(-br) dr \int_{-1}^1 dt e^{ikrt} = \frac{1}{2} \int_0^\infty \frac{\sin(kr) e^{-br}}{k} dr = \frac{1}{k^2+b^2}$.

58 Alternatively, the Ginzburg length can be defined as the distance at which the fluctuation of the order parameter becomes comparable to its mean, i.e., when $\chi(\xi_G) \approx m^2 k_B T$ or $\xi_G \approx 4\pi e c_2^2/(k_B T a_4)$.

where $\xi_0^2 = -c_2/a_2^0$. Accordingly, the temperature beyond which the mean-field approximation breaks down can be estimated from

$$N_G \approx 0.063\frac{k_B T\xi_0 a_4}{c_2^2}|T - T_c|^{-1/2}. \tag{5.187}$$

Theoretically, the above procedure may be extended to systems of any dimensionality. In general, the dependence of the relative mean fluctuation as defined in Eq. (5.182), on temperature takes the form of

$$N_G \sim |T - T_c|^{d/2-2}. \tag{5.188}$$

where d stands for dimensionality. When $d=3$, Eq. (5.188) reduced to Eq. (5.187), as expected. It predicts that the relative mean fluctuation diverges the same as the correlation length $\xi \sim |T - T_c|^{-1/2}$. Interestingly, N_G remains finite at the critical point when $d \geq 4$, implying that the mean-field approximation is adequate even near the critical temperature.

5.9.5 Summary

The Landau theory of phase transitions is a powerful theoretical framework for understanding the thermodynamic behavior of materials near phase transition. Unlike conventional statistical-mechanical methods, the Landau theory assumes that the free energy of a thermodynamic system can be expressed as a function of the order parameters and their gradients. These order parameters provide information about the degree of order or symmetry in the thermodynamic system and can take many different forms depending on the specific phase transition being studied. In the next section, we will elaborate on the applications of the Landau theory in describing the phase behavior of microemulsions and liquid crystals.

5.10 Microemulsion and Liquid Crystals

As discussed in Section 5.9, the Landau theory provides a powerful framework for analyzing and predicting the phase behavior of thermodynamic systems. It allows us to describe the equilibrium states and phase transitions without delving into the microscopic details. In this section, we explore the applications of the Landau theory for understanding the structure and phase behavior of complex fluids, with a specific focus on two key aspects: the interpretation of microemulsion structures and the utilization of fundamental equations to describe the phase behavior of liquid crystals.

5.10.1 The Teubner–Strey Theory of Microemulsions

The Teubner–Strey theory of microemulsions provides a relatively simple example for application of the Landau–Ginzburg theory to self-assembly systems.[59] Here, the order parameter is represented by the relative concentration of water or oil in a ternary mixture containing amphiphilic molecules.

By retaining only the quadratic term in the local the free energy density, the Landau–Ginzburg theory predicts that the Helmholtz energy of a microemulsion can be written as

$$F[m(\mathbf{r})] = \int d\mathbf{r}\left\{a_2 m^2 + \frac{c_2}{2}(\nabla m)^2 + \frac{c_4}{2}(\nabla^2 m)^2\right\}. \tag{5.189}$$

59 Teubner M. and Strey R. "Origin of the scattering peak in microemulsions", *J. Chem. Phys.* 87 (5), 3195–3200 (1987).

As discussed in Section 5.9, Eq. (5.189) yields an analytical expression for the density–density correlation function Eq. (5.180):

$$\beta \hat{\chi}(k) = \frac{1}{a_2 + c_2 k^2 + c_4 k^4} \tag{5.190}$$

where $a_2 > 0$, $c_2 < 0$, $c_4 > 0$, and the stability condition requires

$$4a_2 c_4 - c_2^2 > 0. \tag{5.191}$$

As the density–density correlation function is directly related to the spectrum of neutron or X-ray scattering, Eq. (5.190) provides a theoretical basis for interpreting experimental data and for understanding formation of microemulsions in amphiphilic systems. Specifically, small angle scattering experiments provide valuable information on the long-range limit of the density–density correlation function, which may be transformed into the real space as

$$\chi(r) \sim \frac{de^{-r/\xi}}{2\pi r} \sin\left(\frac{2\pi r}{d}\right) \tag{5.192}$$

where $r/\xi >> 1$ and

$$\xi = \left[\frac{1}{2}\left(\frac{a_2}{c_4}\right)^{1/2} + \frac{1}{4}\frac{c_2}{c_4}\right]^{-1/2}, \tag{5.193}$$

$$d = 2\pi \left[\frac{1}{2}\left(\frac{a_2}{c_4}\right)^{1/2} - \frac{1}{4}\frac{c_2}{c_4}\right]^{-1/2}. \tag{5.194}$$

Equation (5.192) suggests that d is related to a characteristic length associated with the domain size of microemulsion, and ξ may be interpreted as the correlation length for the density fluctuations. For both water in oil (W/O) and oil in water (O/W) microemulsions, the domain size corresponds to the diameter of spherical droplets for the dispersed phase, and the correlation length reflects density fluctuation within each spherical domain. For bicontinuous microemulsions, the sinusoidal term accounts for the alternating domains of oil and water phases with an average periodicity of d, and the exponential term is related to the short-range correlation within the water or oil domain.

In dimensionless form, the density–density correlation function can be written as[60]

$$\frac{\hat{\chi}(k)}{\hat{\chi}(0)} = \frac{1}{1 + \lambda \tilde{k}^2 + \tilde{k}^4/4} \tag{5.195}$$

where $\tilde{k} = k\xi_L$, $\xi_L = (4c_4/a_2)^{1/4}$, and $\lambda = c_2/(\sqrt{4a_2c_4})$. Because $4a_2c_4 - c_2^2 > 0$, we have $|\lambda| < 1$. Eq. (5.195) suggests that, in reduced units, we can characterize the scattering intensity by using a single parameter λ. From Eqs. (5.193) and (5.194), we can show that λ is related to the ratio of the correlation length and domain size by

$$\frac{\xi}{d} = \sqrt{\frac{1-\lambda}{1+\lambda}}. \tag{5.196}$$

Figure 5.44 presents representative scattering spectra of microemulsion systems as predicted by Eq. (5.195). For strongly structured microemulsions, the domains are correlated over a distance larger than their size, $\xi/d >> 1$ and $\lambda \approx -1$. When $\lambda \to -1$, the correlation length $\xi/\xi_L = 1/\sqrt{1+\lambda}$ diverges, signaling the instability of the microemulsion phase. In this case, the domain size remains finite, $d/\xi_L = 1/\sqrt{1-\lambda} = 1/\sqrt{2}$, which corresponds to the period of a long-range ordered lamellar phase. When $\lambda = 0$, the correlation length is the same as the domain size. The condition is often

60 Ciach A. and Gozdz W. T. "Nonelectrolyte solutions exhibiting structure on the nanoscale", *Annu. Rep. Prog. Chem. Sect. C: Phys. Chem.* 97, 269–314 (2001).

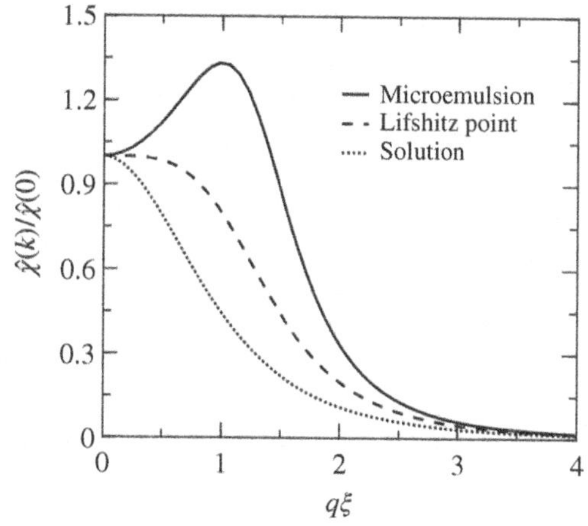

Figure 5.44 The Teubner–Strey theory captures a universal behavior for the structure of microemulsions.

referred to as *the Lifshitz point*, which is manifested in the disappearance of the maximum in the scattering intensity. Beyond this condition, the density–density correlation function decays without oscillations, i.e., there are no well-defined domains of self-assembly. When $\lambda \to 1$, the correlation length vanishes and the system has a bicontinuous structure with totally random interfaces.

The Teubner–Strey theory has been routinely used to describe the scattering spectra of a wide variety of microemulsions. The domain size and correlation length obtained from the fit parameters provide insights into the microscopic structure. For example, Figure 5.45 shows the small angle neutron scatting (SANS) spectra for nonionic microemulsions containing water, a surfactant (Brij 96), ethyl oleate, and hexanol.[61] The system is relevant for a variety of pharmaceutical and cosmetic formulations, particularly those involving emulsions and liquid crystal-based products such as creams, lotions, and drug delivery systems.

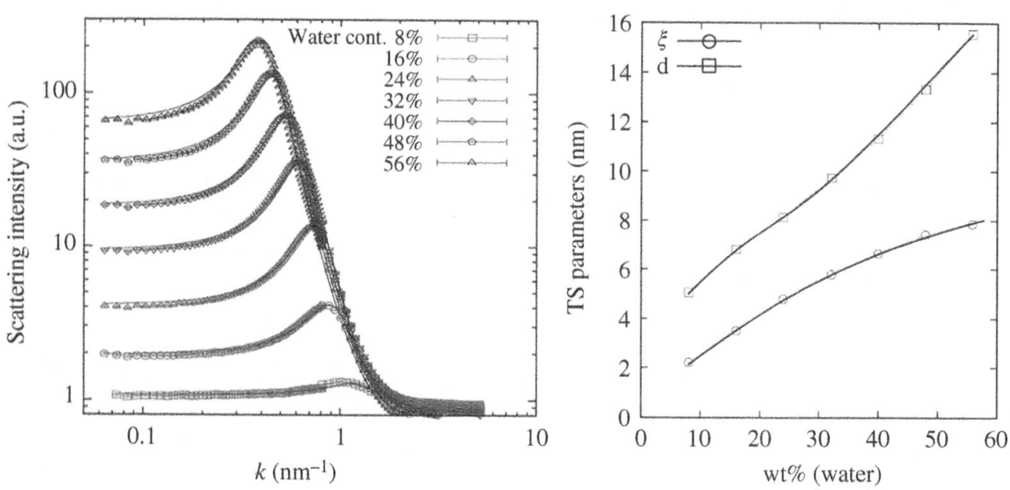

Figure 5.45 Small angle neutron scatting (SNAS) spectra for nonionic microemulsions containing water, Brij 96 surfactant, ethyl oleate, and hexanol. The solid lines represent correlations with the Teubner–Strey (TS) theory with two fit parameters ξ and d changing with the water percent.

61 Kaur G., et al., "Probing the microstructure of nonionic microemulsions with ethyl oleate by viscosity, ROESY, DLS, SANS, and cyclic voltammetry", *Langmuir* 28(29), 10640–10652 (2012).

We see that the Teubner–Strey theory reproduce the scatting intensity curves well. By fitting the Landau theory with the scattering spectra, one can determine the domain size and correlation length at different water contents. As expected, the domain size is linearly increasing with the amount of water contained, implying that the microscopic structure swells proportionally upon the addition of water. This example illustrates how the scattering experiments shed lights for the formulation of microemulsions with desired microscopic structures. As the microscopic structure of a microemulsion cannot be observed directly, the statistical-mechanical model is indispensable for the interpretation of the scattering spectra that provide only indirect information.

5.10.2 The Landau-de Gennes Theory

The de Gennes theory of liquid crystals provides another well-known example for applications of the generalized Landau theory.[62] For liquid crystal phases, de Gennes formulated the Landau free energy in terms of a Q-tensor order parameter

$$F[\hat{\mathbf{Q}}(\mathbf{r})] = \int d\mathbf{r} \left\{ f_B(\hat{\mathbf{Q}}) + f_E(\hat{\mathbf{Q}}, \nabla\hat{\mathbf{Q}}) \right\} \tag{5.197}$$

where $\hat{\mathbf{Q}}$ is a second-order tensor related to the distribution of molecular orientations, f_B corresponds to the free energy of a bulk uniform system, and f_E accounts for elastic energy arising from the spatial inhomogeneity of the system.

The Q-tensor is defined as

$$\hat{\mathbf{Q}} = \int d\varpi p(\varpi)\, \varpi \otimes \varpi - \hat{I}/3 \tag{5.198}$$

where $p(\varpi)$ is the probability distribution for molecular orientation ϖ, $\hat{\mathbf{I}}$ is a unit vector. The order parameter provides a measure of the orientational order relative to the random distribution molecular orientations ($\hat{\mathbf{Q}} = 0$).

It can be shown that the Q-tensor is symmetric ($\hat{\mathbf{Q}} = \hat{\mathbf{Q}}^T$) and traceless ($\mathrm{Tr}\hat{\mathbf{Q}} = 0$). The three eigenvalues of $\hat{\mathbf{Q}}$, λ_1, λ_2, and $\lambda_3 = -\lambda_1 - \lambda_2$, are related to the first two lowest coefficients in the spherical harmonic expansion of the probability distribution for molecular orientations as discussed in Section 5.8.1:

$$S_m = 2\lambda_1 + \lambda_2, \tag{5.199}$$

$$\Delta_m = \lambda_1 + 2\lambda_2. \tag{5.200}$$

The nematic phase is uniaxial if $\lambda_1 = \lambda_2$, which implies that the liquid-crystal molecules are aligned in a single direction. Otherwise, if $\lambda_1 \neq \lambda_2$, it is referred to as biaxial because the liquid crystal molecules are aligned in two directions.

The bulk free energy can be expressed in terms of a polynomial function of the eigenvalues of the Q-tensor

$$f_B(\hat{\mathbf{Q}}) = f_0 + a_2 \mathrm{Tr}\hat{\mathbf{Q}}^2/2 - a_3 \mathrm{Tr}\hat{\mathbf{Q}}^3/3 + a_4(\mathrm{Tr}\hat{\mathbf{Q}}^2)^2/4 + \cdots \tag{5.201}$$

where f_0 is the free energy of an isotropic phase, $a_2 = a_2^0(T - T_{tr})$ such that it changes the sign at the isotropic-nematic transition temperature T_{tr}, a_2^0, a_3, and a_4 are system-specific parameters that are often assumed independent of temperature, $\mathrm{Tr}\hat{\mathbf{Q}}^2 = \sum_{i=1}^{3} \lambda_i^2$ and $\mathrm{Tr}\hat{\mathbf{Q}}^3 = \sum_{i=1}^{3} \lambda_i^3$. In Eq. (5.201), we retain the third-order terms to describe first-order transition because the free energy density may depend on the sign of the order parameter.

---

62 de Gennes P. G. and Prost J., *The physics of liquid crystals* (2nd Edition). Oxford University Press, 1993.

Similar to the gradient term in the generalized Landau expansion, the elastic energy density is typically written as a quadratic function of $\nabla\hat{\mathbf{Q}}$

$$f_E(\hat{\mathbf{Q}}) = \frac{1}{2}\sum_{i,j,k=1}^{3}\left\{L_1(\nabla_{\mathbf{r}_k}\hat{\mathbf{Q}}_{ij})^2 + L_2(\nabla_{\mathbf{r}_j}\hat{\mathbf{Q}}_{ij})(\nabla_{\mathbf{r}_k}\hat{\mathbf{Q}}_{ik}) + L_3(\nabla_{\mathbf{r}_k}\hat{\mathbf{Q}}_{ij})(\nabla_{\mathbf{r}_j}\hat{\mathbf{Q}}_{ik})\right\}. \tag{5.202}$$

where L_1, L_2, and L_3 are elastic constants satisfying the constraints

$$L_1 > 0; -L_1 < L_3 < 2L_1; \quad 10L_2 > -6L_1 - L_3. \tag{5.203}$$

de Gennes' theory is often employed to interpret experimental data as well as to classify possible topologies of the phase diagrams of liquid crystals. It can be applied to phase transitions from an isotropic liquid to a nematic liquid crystal and, after some modifications, other forms of liquid crystal phases.[63]

5.10.3 Summary

The Landau theory has proven to be highly effective in explaining a diverse array of phase transitions across different types of materials and thermodynamic systems. In this section, we highlight only two specific examples that are most relevant for understanding the structure and phase behavior of complex fluids. The Teubner–Strey theory plays a crucial role in determining the equilibrium structure of microemulsions. It complements to the lattice model discussed in Section 5.7, enabling quantitative descriptions of droplet size, interfacial area, and phase compositions within microemulsions. On the other hand, the de Gennes theory provides a systematic framework for characterizing the different phases of liquid crystals, including the nematic, smectic, and cholesteric phases. It finds applications in various fields, including display technologies, optics, and materials science.

5.11 Critical Phenomena and Universality

Critical phenomena are associated with the remarkable behavior exhibited by a thermodynamic system as it approaches the critical point of a phase transition. A classic example is the vapor–liquid transition in a simple fluid, which only occurs below a specific temperature known as the critical temperature. Above the critical temperature, the system exists as a single phase. However, in close proximity to the critical point, the system displays extraordinary thermophysical properties, such as the divergence of the heat capacity and the disappearance of the vapor-liquid boundary. This critical behavior can also manifest in more complex systems, including magnetic materials, liquid crystals, membranes, superfluids, and various solid states. Despite the distinct microscopic characteristics of these systems, they exhibit remarkable similarity in the critical behavior. Due to the fundamental role of critical phenomena in macroscopic phase transitions, there is a wealth of specialized literature dedicated to exploring this fascinating subject.[64]

In this section, we discuss the fundamental concepts of criticality using the Ising model as an example. We will explore the relationship between critical properties exhibited by thermodynamic

---

63 Majumdar A., and Lewis A., A theoretician's approach to nematic liquid crystals and their applications. In *Variational methods in molecular modeling*, Wu J., Ed. Springer, 2016; pp. 223–254.

64 (a) Ma S., *Modern theory of critical phenomena*. Perseus Pub., 2000; (b) Nishimori H. and Ortiz G., *Elements of phase transitions and critical phenomena*. Oxford University Press, 2011; (c) Stanley H. E., *Introduction to phase transitions and critical phenomena*. Oxford University Press, 1987.

systems under phase transition and delve into mathematical descriptions like scale invariance, self-similarity, and power-law scaling. These mathematical tools provide valuable means for characterizing the critical behavior of various natural and artificial systems.[65] By understanding these connections, we can gain insights into critical behavior across a wide range of phenomena. In the subsequent section, we will present a theoretical procedure for describing critical phenomena through the application of the renormalization group (RG) theory. The RG theory provides a computational framework for quantifying the behavior of thermodynamic systems near their critical points and for elucidating the underlying principles governing critical phenomena.

5.11.1 Singular Behavior

Near the critical point, a macroscopic system displays singular behavior characterized by the divergence of various thermodynamic variables such as heat capacity, magnetic susceptibility, or isothermal compressibility. This singularity emerges due to the large-scale fluctuations of local dynamic variables, as evidenced by the rapid growth of the correlation length as the system approaches the critical point. At the critical point, fluctuations are correlated across all length scales, spanning from microscopic to macroscopic levels. Because there are no identifiable natural length scales to describe the thermodynamic system, the singular behavior remains invariant regardless of the specific length scale used to characterize the correlation effects. The absence of characteristic length scales is known as *scale-free* or *scale invariance*. As illustrated in Box 5.2, any scale invariant quantity can be mathematically described in terms of a power-law relation.

Box 5.2 Scale Invariance, Power Law, and Homogeneous Function

A one-dimensional scalar function $f(x)$ is called scale invariant if its relative value depends *only* on the ratio of the independent variable, i.e.,

$$f(x)/f(y) = \phi(x/y) \tag{5.204}$$

where $\phi(x)$ is any analytical function. Eq. (5.204) suggests that function $f(x)$ varies in the same way at all scales, i.e., its relative value does not change with the scale of its independent variable x.

It is straightforward to show that only power functions are scale invariant. To see this, we take a derivative with respective to x on both sides of Eq. (5.204) and then let $y = x$

$$f'(x)/f(x) = \phi'(1)/x. \tag{5.205}$$

Integration of Eq. (5.205) with respect to x yields

$$f(x) = kx^a \tag{5.206}$$

where k is the integration constant, and $a = \phi'(1)$ is known as the "dimension" of function$f(x)$.

If we scale the independent variable x by a factor λ, a positive number, Eq. (5.206) becomes

$$f(\lambda x) = \lambda^a f(x), \tag{5.207}$$

or equivalently

$$f(\lambda^{1/a} x) = \lambda f(x). \tag{5.208}$$

65 Khaluf Y., et al., "Scale invariance in natural and artificial collective systems: a review", *J. R. Soc. Interface* 14(136), 0662 (2017).

Equation (5.208) indicates that function $f(x)$ is invariant under the change of scale from $x \to \lambda^{1/a}x$, provided that its value is scaled by a factor λ. Therefore, we say that function $f(x)$ satisfies the *power law scaling*.

The definition of scale invariance can be easily extended to the multidimensional space. If function $f(x_1, x_2, \cdots, x_n)$ is scale-invariant in each dimension, then it must satisfy the power law relation

$$f\left(\lambda^{1/a_1}x_1, \lambda^{1/a_2}x_2, \cdots \lambda^{1/a_n}x_n\right) = \lambda f(x_1, x_2, \cdots x_n) \tag{5.209}$$

where $a_1, a_2 \ldots, a_n$ are constants, and λ is an arbitrary positive number. If $a_{i=1,2,\cdots,n} = 1$, $f(x_1, x_2, \cdots, x_n)$ is called a homogeneous function. Otherwise, it is referred to as a generalized homogeneous function (GHF).

The scale invariance of a macroscopic system at the critical point explains why the divergence of thermodynamic variables must follow power-law descriptions. As a system approaches the critical point, the *mathematical forms* of thermodynamic functions do not change when the independent variables are *scaled* with certain factors. The latter can be represented by an effective distance from the critical temperature, $t \equiv |1 - T/T_C|$. Accordingly, the thermodynamic variables are subject to the power-law description

$$X \sim t^{\lambda} \tag{5.210}$$

where X stands for the deviation of a particular thermodynamic variable from its critical value, and λ is called the critical exponent corresponding to property X.

In Eq. (5.210), the proportionality constant is system-specific, depending on microscopic details. However, the mathematical forms are determined only by the spatial dimensionality of the thermodynamic system, invariant with microscopic details. Therefore, the mathematical form, including the exponent, is generic and referred to as the *universality* of the critical behavior, meaning that it is independent of the system details.

5.11.2 Scaling Relations Predicted by the Ising Model

We may use the Ising model to illustrate the aforementioned general relations (also known as scaling laws or scaling relations) underlying the critical behavior of thermodynamic properties near the critical point. Both 2D and 3D-Ising models predict a second-order phase transition as the system approaches the critical point. Qualitatively, the critical behavior similar to that underlying the vapor-liquid transformation of a simple fluid or any other continuous phase transition in general.

According to the 2D-Ising model, the critical temperature depends on the spin arrangements. For example, $k_B T_c/\varepsilon \approx 2.26919$, 3.64096, and 1.51865 have been obtained for square (S), triangular (T) and hexagonal (H) lattices, respectively.[66] While the critical temperature depends on the lattice structure and coupling parameter ε, the spontaneous magnetization near the critical point follows similar power-law relations:

$$m^S \approx 1.22241\left(1 - T/T_C^S\right)^{1/8}, \tag{5.211}$$

$$m^T \approx 1.20327\left(1 - T/T_C^T\right)^{1/8}, \tag{5.212}$$

$$m^H \approx 1.25318\left(1 - T/T_C^H\right)^{1/8}, \tag{5.213}$$

66 McCoy B. M. and Wu T. T., *The two-dimensional Ising model*. Harvard University Press, Cambridge, Mass., 1973.

where superscripts S, T, and H denote the lattice types. The universality is manifested as the identical power-law exponent, 1/8, applicable to all 2D-Ising models. This critical exponent is determined by the symmetry of the Ising model (viz., spin up/spin down) and the dimensionality of the system.

Similar to the 2D-Ising model, the 3D-Ising model also predicts a phase transition below a critical temperature. Likewise, the exact value of the critical temperature depends on the microscopic details of spin–spin interactions, including the lattice type. While for 3D-Ising model, the analytical solutions are no more available, the critical temperature can be calculated from Monte Carlo simulation.[67] The numerical results indicate that the reduced critical temperature is $k_B T_C/\varepsilon \approx 4.52$, 6.44, and 9.84 for SC body-centered-cubic (BCC), and face-centered-cubic (FCC) 3D lattice structures, respectively. Again, the correlation and thermodynamic properties near the critical point follow the power-law descriptions, with the exponent independent of microscopic details. For example, spontaneous magnetization can be described by the scaling relation

$$m \sim (1 - T/T_C)^{\beta} \tag{5.214}$$

where $\beta \approx 0.3263$ is applicable to all lattice types. Even for the same thermodynamic quantity, the exponent predicted by the 3D-Ising model is different from that by the 2D-Ising model. This difference can be attributed to the variation of dimensionality.

As discussed in Section 5.2, in the Ising model, the correlation function is defined as:

$$C(r) \equiv \langle\sigma_i \sigma_j\rangle - \langle\sigma_i\rangle\langle\sigma_j\rangle \tag{5.215}$$

where r represents the distance between spins i and j. At the asymptotic limit, i.e., when $r \to \infty$, the correlation function can be expressed as

$$C(r) \sim \frac{e^{-r/\xi}}{r^{d-2+\eta}} \tag{5.216}$$

where ξ denotes the correlation length, and d stands for the dimensionality of the system (e.g., $d = 2$ for 2D-Ising models), η is a constant depending only on dimensionality ($\eta = 1/4$ and $\eta \approx 0.03631$ for $d = 2$ and 3, respectively). At the critical point, the correlation function is scale-invariant thus becomes a power function

$$C(r) \sim \frac{1}{r^{d-2+\eta}}. \tag{5.217}$$

A comparison of Eqs. (5.216) and (5.217) indicates that the correlation length diverges, i.e., ξ approaches infinity at the critical point. According to the 2D-Ising model, the divergence of the correlation length follows the power-law scaling

$$\xi^S \approx 0.321825\left|1 - T/T_C^S\right|^{-1}, \tag{5.218}$$

$$\xi^T \approx 0.184119\left|1 - T/T_C^T\right|^{-1}, \tag{5.219}$$

$$\xi^H \approx 0.0960959\left|1 - T/T_C^H\right|^{-1} \tag{5.220}$$

for the square, triangular, and hexagonal lattices, respectively. Again, the critical exponent is *universal*, depending only on the system symmetry and dimensionality, immaterial to the lattice types.

Table 5.1 summarizes the scaling relations and the corresponding critical exponents for various thermodynamic properties predicted by the 2D and 3D-Ising models.[68] Because the phase rule

67 Ferrenberg A. M., Xu, J. and Landau D. P. "Pushing the limits of Monte Carlo simulations for the three-dimensional Ising model", *Phys. Rev. E* 97(4), 043301 (2018).

68 For four-dimensional (4D) systems, critical exponents depend on the specific model and universality class of the phase transition.

Table 5.1 Critical scaling and exponents for the second-order phase transition.

	Scaling	Relations	2D	3D
α	$C_v \sim \lvert t \rvert^{-\alpha}$	$\alpha = 2 - \nu d$	0	0.11003
β	$m \sim (^{-}t)^{\beta}$	$\alpha + 2\beta + \gamma = 2$	1/8	0.32630
γ	$\chi \sim \lvert t \rvert^{-\gamma}$	$\gamma = \nu(2-\eta)$	7/4	1.23708
δ	$m \sim \lvert h \rvert^{1/\delta}$	$\gamma = \beta(\delta - 1)$	15	4.79123
ν	$\xi \sim \lvert t \rvert^{-\nu}$	—	1	0.62993
η	$C(r) \sim 1/r^{d-2+\eta}$	—	1/4	0.03631

Here, d stands for the system dimensionality, and the power law for heat capacity does not apply to the 2D-Ising model. The exponents for the 3D-Ising model are from simulation.
Source: Adapted from Ferrenberg et al. (2018).
Lundow P. H. and Campbell I. A., "The Ising universality class in dimension three: corrections to scaling", *Physica A* 511, 40–53 (2018).
Source: Adapted from Ferrenberg et al. (2018).

remains valid near the critical point, the thermodynamic variables as well as the critical exponents are interrelated as discovered first by Widom (see Problem 5.23).[69] For the Ising models, we have only two independent critical exponents, implying that 4 relations can be identified among the six exponents listed in this table. The various exponents in the Widom relations are satisfied independent of the universality class.[70] One can verify that these relations hold true for the 2D-Ising models and remain valid when $D = 3$ within the limits of numerical precision. It should be noted that, for the 2D-Ising model, the heat capacity does not follow the power law.[71] It diverges near the critical point according to a logarithm relation

$$C_V/k_B N \approx -\frac{8}{\pi}\left(\frac{\varepsilon}{k_B T_c}\right)^2 \ln\left(\left|1 - \frac{T}{T_c}\right|\right) + \text{constant}. \tag{5.221}$$

5.11.3 Summary

Critical phenomena are commonly observed in thermodynamic systems that undergo second-order phase transitions. These transitions are characterized by unusual behavior, such as the divergence of thermodynamic functions and correlation length, which can be attributed to the invariance of length scales. Universality is a key aspect of critical phenomena, as it dictates the power-law scaling relations. These scaling relations have been extensively validated using experimental data from various systems.

The concept of universality is particularly valuable for practical applications because it enables the transfer of experimental results for the thermodynamic properties of one system near its critical point to other systems within the same universality class. This approach is frequently employed when analyzing complex biological phenomena. While considering microscopic details can make

69 Widom B. "Equation of state in neighborhood of critical point", *J. Chem. Phys.* 43(11), 3898–3905 (1965).
70 The universality class refers to a group of physical systems that exhibit the same critical behavior near a phase transition, regardless of their microscopic details. Systems with the same spatial and order dimensionalities are said to belong to the same universality class.
71 Exceptions to universality were discussed by R. J. Baxter in the context of the eight-vertex lattice model *Ann. Phys. New York* 70 (1), 193–228 (1972).

certain natural phenomena computationally challenging, the behavior near the critical point can often be captured effectively using simplified models.[72]

5.12 Renormalization Group (RG) Theory

The renormalization group (RG) theory provides a general theoretical framework for understanding critical phenomena in a wide range of physical systems, from magnets and superconductors to liquid crystals and polymers. The theoretical procedure was first introduced by Leo Kardanoff in the early 1960s and further developed by Kenneth G. Wilson[73] for describing continuous phase transitions. In this method, the partition function of a thermodynamic system that exhibits scale-invariant properties is evaluated by the RG transformation, i.e., the evaluation of the partition function by iterative grouping of microstates. By accounting for large-scale fluctuations "piece by piece," the RG transformation leads to a quantitative description of universality and critical exponents.

5.12.1 RG Transformation for an Ising Chain

We may elucidate the basic ideas of RG transformation using the 1D-Ising model. As discussed in Section 5.2, the canonical partition function for the Ising chain is given by

$$Q(K,N) = \sum_{s_1=\pm 1}\sum_{s_2=\pm 1}\cdots\sum_{s_N=\pm 1} \exp[K(s_1 s_2 + s_2 s_3 + \cdots + s_{N-1}s_N)] \tag{5.222}$$

where $K = \beta\varepsilon > 0$ stands for the coupling parameter. For $N \gg 1$, we can evaluate Eq. (5.222) analytically with the method of transfer matrix

$$Q(K,N) = \mathrm{Tr}\begin{pmatrix} e^{K} & e^{-K} \\ e^{-K} & e^{K} \end{pmatrix}^N = [2\cosh K]^N \tag{5.223}$$

where Tr stands for trace, i.e., the sum of the diagonal elements of a square matrix. From the partition function, we can find the reduced free energy per spin

$$f \equiv \beta F/N = -(1/N)\ln Q(K,N) = -\ln(2\cosh K). \tag{5.224}$$

In the RG transformation, we evaluate the partition function iteratively by spin decimation, i.e., by considering the degrees of freedom for a block of spins at a time. The procedure is illustrated schematically in Figure 5.46, where each block consists of two neighboring spins. By writing out the orientations of the even-numbered spins explicitly, we can express the partition function as

$$\begin{aligned} Q(K,N) = &\sum_{s_1=\pm 1}\sum_{s_3=\pm 1}\cdots\{\exp[K(s_1+s_3)] + \exp[-K(s_1+s_3)]\} \\ &\times\{\exp[K(s_3+s_5)] + \exp[-K(s_3+s_5)]\}\cdots \end{aligned} \tag{5.225}$$

On the right side of Eq. (5.225), the summation in each curly bracket can be formulated as

$$\exp[K(s+s')] + \exp[-K(s+s')] = A(K')\exp[K'ss'] \tag{5.226}$$

---

72 (a) Veatch S. L., et al. "Critical fluctuations in plasma membrane vesicles", *ACS Chem. Biol.* 3 (5), 287–293 (2008); (b) Munoz M. A. "Colloquium: criticality and dynamical scaling in living systems", *Rev. Mod. Phys.* 90(3), 031001 (2018).

73 Wilson won the 1982 Nobel Prize in Physics for his theoretical contributions to critical phenomena in connection with phase transitions.

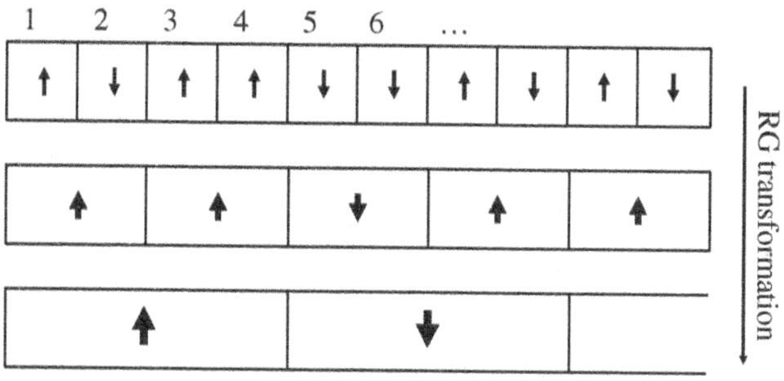

Figure 5.46 RG transformation for 1D-Ising model. The degree of freedom is integrated out systematically by grouping the pairs of nearest spins.

where $s' = \pm 1$ and

$$K' = (1/2)\ln\cosh 2K, \tag{5.227}$$

$$A(K') = 2e^{K'}. \tag{5.228}$$

After the transformation, the partition function for blocks of spins becomes self-similar to that of the original system, i.e.,

$$Q(K,N) = A(K')^{N/2}\sum_{s_1=\pm1}\sum_{s_3=\pm1}\cdots\{\exp[K'(s_1s_3+s_3s_5+\cdots]\} = A(K')^{N/2}Q(K',N/2). \tag{5.229}$$

Eq. (5.229) transforms the partition function for N spins to that for$N/2$ spins. Such a procedure is known as the Kadanoff transformation, which reduces the microscopic degrees of freedom by spin decimation or coarse-graining.

Eq. (5.229) indicates that, except for the prefactor $A(K')^{N/2}$, the partition function preserves its functional form after the RG transformation. To obtain thermodynamic properties, we can rewrite Eq. (5.229) in terms of the reduced free energy per spin

$$f(K,N) = -\frac{1}{2}\ln A(K') + \frac{1}{2}f(K',N/2). \tag{5.230}$$

As mentioned above, application of the RG transformation to the Ising chain reduces the degrees of freedom by half. After the $(i+1)^{\text{th}}$ transformation, the reduced free energy per spin is given by

$$f(K_i,N_i) = -\frac{1}{2}\ln A(K_{i+1}) + \frac{1}{2}f(K_{i+1},N_{i+1}) \tag{5.231}$$

where $N_{i+1} = N_i/2$ and $K_{i+1} = (1/2)\ln\cosh 2K_i$. By a continuous application of the RG transformations, the recursive relation, Eq. (5.231), allows us to evaluate the free energy of the Ising chain in the thermodynamic limit ($N \to \infty$).

As shown in Figure 5.47, the coupling parameter K declines quickly with the RG transformation and approaches zero in the asymptotic limit. For the results shown in this figure, we set the initial coupling parameter arbitrarily to $K = 10$, which corresponds to a correlation length of $\xi = 1/\ln[\coth K] \approx 2.4 \times 10^8$. Remarkably, the strong coupling between neighboring spins vanishes after about 30 RG iterations. When the coupling parameter is sufficiently small, the correlation between neighboring spins becomes unimportant and thus the free energy can be estimated from $f(K,N) \approx -\ln 2$. For a finite value of K, the free energy can be calculated from Eq. (5.231), starting from a small value of K.

Figure 5.48 shows the free energy for the 1D-Ising model calculated from RG iterations in comparison with that from the exact result, Eq. (5.224). Because the numerical error for $K \approx 0$

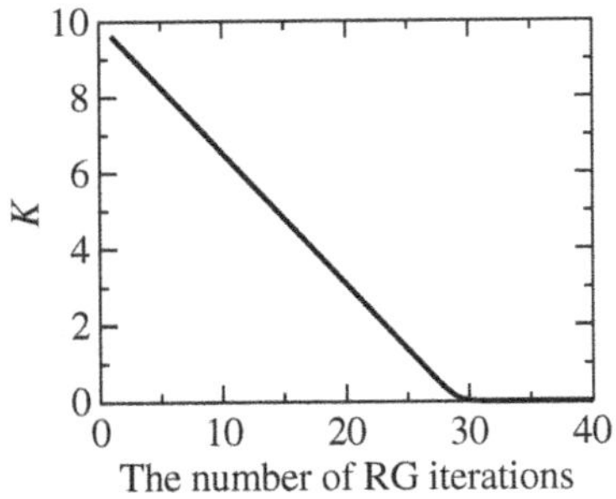

Figure 5.47 Variation of the coupling parameter K with the number of the RG iterations.

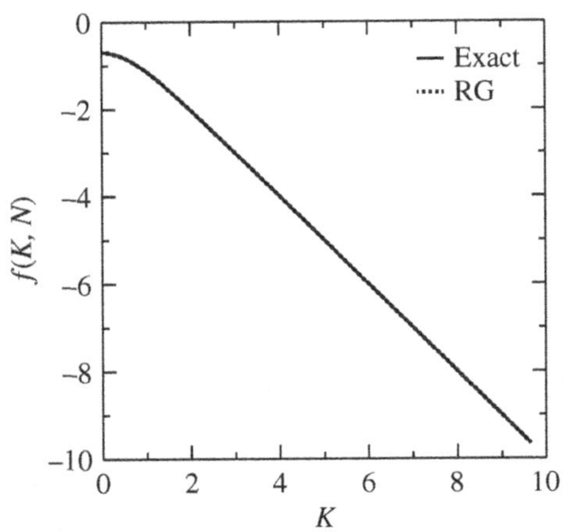

Figure 5.48 The reduced free energy per spin calculated from the RG iterations in comparison with the exact result calculated from Eq. (5.224).

(weak coupling) can be controlled to an arbitrarily small value, the results obtained from the RG transformation are indistinguishable from the exact solution.

Regrettably, for most nontrivial systems, the functional form of the partition function is not preserved after partially integrating certain degrees of freedom for the microstates. Nevertheless, application of the RG transformation to the 1D-Ising model provides a key insight into the emergence of singularity and self-similarity for thermodynamic systems near the critical point of a second-order phase transition. It helps uncover the underlying mechanisms and behaviors that drive critical phenomena, shedding light on the universal aspects of phase transitions.

5.12.2 General RG Transformation

To comprehend the critical behavior of a second-order phase transition, consider a general form of the reduced Hamiltonian for a large number of spins on a d-dimensional lattice ($N \to \infty$)

$$-\beta E[\mathbf{K}, \{s_i\}, N] = K_0 + K_1 \sum_i s_i + K_2 \sum_{i,j}^{[1]} s_i s_j + K_3 \sum_{i,j}^{[2]} s_i s_j + K_4 \sum_{i,j,k}^{[3]} s_i s_j s_k + K_5 \sum_{i,j,k,l}^{[4]} s_i s_j s_k s_l + \cdots \tag{5.232}$$

where $\mathbf{K} = (K_0, K_1, K_2 \cdots)$ denotes an infinite-dimensional vector that contains all coupling constants for multi-body interactions among the spins, the summations are applied to the orientations of individual spins, to all pairs of nearest neighbors ([1]), to all pairs of next nearest neighbors ([2]), to all triplets of nearest neighbors ([3]), and so on. The Hamiltonian for a conventional Ising model is recovered if we assign $K_1 = -\beta h$, $K_2 = -\beta\varepsilon$, and $K_0 = K_3 = K_4 = \cdots = 0$ where h and ε are the one-body and pair interaction energies.

Now apply the RG transformation by the decimation procedure for blocks of spins. Without loss of generality, we assume that each block contains L^d spins before the decimation where $L > 1$ represents a step length, and the microstates for each block of spins have been integrated except a single spin at the center of the block. The overall reduced energy for the system of block spins becomes

$$-\beta E[\mathbf{K}', \{s_i\}, N/L^d] = K_0' + K_1' \sum_i s_i + K_2' \sum_{i,j}^{[1]} s_i s_j + K_3' \sum_{i,j}^{[2]} s_i s_j + K_4' \sum_{i,j,k}^{[3]} s_i s_j s_k \cdots \tag{5.233}$$

where d stands for the dimensionality of the system.

Formally, the coupling constants for individual spins and blocks of spins can be written as

$$\mathbf{K}' = \mathbb{R}(\mathbf{K}) \tag{5.234}$$

where $\mathbb{R}$ is a function of vector $\mathbf{K}$ that is yet to be determined. The condition of $\mathbf{K}^* = \mathbf{K}' = \mathbf{K}$ represents a fixed point of the RG iteration. Apparently, $\mathbf{K}^* = \mathbf{0}$ satisfies Eq. (5.234) because, in this case, all spins are uncorrelated regardless of the specific details of "blocking." In other words, the uncorrelated system represents a fixed point of the RG transformation. Another trivial fixed point takes place at $\mathbf{K}^* = \infty$, which corresponds to the condition of complete alignment of all spins due to the infinitely strong spin–spin coupling.

A second-order phase transition occurs if Eq. (5.234) has a nontrivial fixed point. At such a condition, we have $\mathbf{K}^* = \mathbf{K}_C$, and the continuous application of the RG transformation keeps all properties of the system unchanged. In other words, the repeated application of the RG transformation results in a set of thermodynamic properties that exhibit the same functional form as the original Hamiltonian. The term "renormalization group" is derived from the mathematical notion that the RG transformation satisfies the group behavior, i.e., the successive application of the RG transformation resembles the composition of mathematical groups

$$\mathbb{R}^{n+m} = \mathbb{R}^n \cdot \mathbb{R}^m \tag{5.235}$$

where n and m are positive integers. While a reverse RG transformation can be defined for simple systems like the Ising chain, it is important to note that the procedure is not generally reversible for more complex systems. In this sense, the RG transformations form not a full group, but rather a semi-group. A semi-group refers to a mathematical structure where the composition of elements is associative, but the existence of inverses is not guaranteed for every element. In the context of RG transformations, this means that while forward transformations can be applied iteratively, reversing the transformations to retrieve the original system may not be possible in general.[74]

As the Hamiltonian given by Eq. (5.233) accounts for all conceivable spin–spin interactions, we expect that the RG transformation preserves the partition function in its original functional form, i.e.,

$$Q(\mathbf{K}, N) = Q(\mathbf{K}', N/L^d). \tag{5.236}$$

Eq. (5.236) predicts that the reduced free energy per spin satisfies

$$f(\mathbf{K}, N) = -\frac{1}{N} \ln Q(\mathbf{K}, N) = \frac{1}{L^d} f(\mathbf{K}', N/L^d). \tag{5.237}$$

Eq. (5.237) thus provides a general recursive relation for the free energy. It may be understood as a generalization of the recursive relation for the 1D-Ising model given by Eq. (5.231).

74 In mathematics, a group is an algebraic structure consisting of a set and rules for binary operations. While the operations in a group are both associative and invertible, a semigroup does not allow invertible operations.

Similar to the 1D-Ising model, the RG transformation allows us to account for long-range correlations because the interaction between the spins now takes place in the units of larger and larger blocks. After n iterations, the coarse-grained length scale is L^n, and the coupling parameters vary from $\mathbf{K}^{(0)}$ at the initial condition to $\mathbf{K}^{(n)}$. Intuitively, the RG iteration mimics a dynamic process in a high-dimensional parameter space $\mathbf{K}$ that moves from the initial thermodynamic condition toward a critical point with fixed values of the coupling parameters. The RG flow reduces the range of the correlated fluctuations because the interactions are "screened" by more and more neighboring spins.

As each RG iteration reduces the relative length scale by a factor L, the correlation length of the system varies accordingly with the coupling constants

$$\xi(\mathbf{K}') = \xi(\mathbf{K})/L. \tag{5.238}$$

Eq. (5.238) indicates that the correlation length is reduced after each RG transformation. At the fixed point, we have

$$\xi(\mathbf{K}^*) = \xi(\mathbf{K}^*)/L. \tag{5.239}$$

Eq. (5.239) can be satisfied only when the correlation length is either zero or infinity. The former corresponds to a trivial fixed point, and the latter occurs at the critical point of second-order phase transition.

5.12.3 RG for the 2D-Ising Model

To make the above ideas more concrete, let us apply the RG transformation to the 2D-Ising model on a square lattice without an external magnetic field. As discussed in Section 5.5, the partition function for this system is given by

$$Q(\mathbf{K}, N) = \sum_{\{s_i=\pm 1\}} \exp\left[K_2 \sum_{ij}^{[1]} s_i s_j\right]. \tag{5.240}$$

In comparison with Eq. (5.232), the coupling vector in the original Hamiltonian is given by

$$\mathbf{K} = (0,\ 0,\ K_2,\ 0, \cdots). \tag{5.241}$$

Following a spin-decimation procedure similar to that used for the 1D-Ising model, we may carry out the RG transformation by summation over half of the spins. As shown schematically in Figure 5.49, each spin on a square lattice interacts with its 4 nearest neighbors. Summation over the up and down orientations for every other nearest-neighboring spin leads to

$$Q(\mathbf{K}, N) = \sum_{\{s_{i'}=\pm 1\}} \prod_i \{\exp[K_2(s_{i-x} + s_{i+x} + s_{i-y} + s_{i+y})] + \exp[-K_2(s_{i-x} + s_{i+x} + s_{i-y} + s_{i+y})]\} \tag{5.242}$$

where subscripts i and i' represent spins summed and not yet summed, respectively.

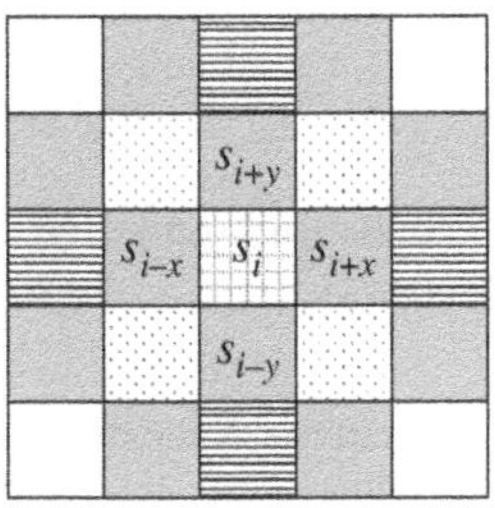

Figure 5.49 Each spin interacts (e.g., spin i at the center) with 4 nearest spins ($i\pm x$; $i\pm y$) on a 2-dimensional square lattice. After summation over the microstates for every other spins, the center spin has 4 nearest neighbors (the dotted sites) and 4 next nearest neighbors (the lined sites).

The partition function given by Eq. (5.242) does not have the same functional form of the original 2D-Ising model. However, we can cast it into that corresponding to the generic Hamiltonian given by Eq. (5.232). Because the terms inside the curved brackets involve 4 spin variables, we need a minimum of 4 parameters to transform the reduced energy into the general form. The simplest possibility is to express the summation in terms of the lowest orders of spin–spin interactions

$$\begin{aligned}&\exp[K_2(s_{i-x} + s_{i+x} + s_{i-y} + s_{i+y})] + \exp[-K_2(s_{i-x} + s_{i+x} + s_{i-y} + s_{i+y})]\\&\quad = \exp[K_0' + (1/2)K_2'(s_{i-x}s_{i+y} + s_{i+y}s_{i+x} + s_{i+x}s_{i-y} + s_{i-y}s_{i-x})\\&\qquad + K_3'(s_{i-x}s_{i+x} + s_{i-y}s_{i+y}) + K_5' s_{i-x}s_{i+x}s_{i-y}s_{i+y}]\end{aligned} \tag{5.243}$$

where the coefficients can be identified by ensuring the correctness of Eq. (5.243) under all values of the pertinent spins

$$K_0' = \ln 2 + (1/2)\ln\cosh 2K_2 + (1/8)\ln\cosh 4K_2, \tag{5.244}$$

$$K_2' = (1/4)\ln\cosh 4K_2, \tag{5.245}$$

$$K_3' = (1/8)\ln\cosh 4K_2, \tag{5.246}$$

$$K_5' = (1/8)\ln\cosh 4K_2 - (1/2)\ln\cosh 2K_2. \tag{5.247}$$

Substituting Eq. (5.243) into Eq. (5.242) leads to a partially summed partition function

$$Q(\mathbf{K}, N) = \sum_{\{s_{i'}=\pm 1\}} \exp\left[NK_0'/2 + K_2'\sum_{i',j'}^{[1]} s_{i'}s_{j'} + K_3'\sum_{i',j'}^{[2]} s_{i'}s_{j'} + K_5'\sum_{i',j',k',l'}^{[4]} s_{i'}s_{j'}s_{k'}s_{l'}\right]. \tag{5.248}$$

Although the original model entails only nearest-neighbor interactions, the RG transformation results in multiple spin interactions due to the removal of certain degrees of freedom. In terms of the general expression for the partition function, summation over half of the spins leads to a new coupling vector

$$\mathbf{K}' = \left(K_0',\ 0,\ K_2',\ K_3',\ 0,\ K_5',\ 0, \cdots\right). \tag{5.249}$$

From a theoretical perspective, we may understand the transformed Hamiltonian as a free energy, i.e., a multi-body potential of mean force arising from partial integration of the degrees of freedom for the microstates. Further application of the RG transformation leads to more nonzero terms in the coupling vector. To circumvent cumbersome mathematics, we may assume that four-spin interactions are negligible (i.e., $K_5' \approx 0$), and that the energy due to the nearest and the next nearest-neighbor interactions can be approximated by that between an effective nearest-neighbor interaction

$$K_2'\sum_{i',j'}^{[1]} s_{i'}s_{j'} + K_3'\sum_{i',k'}^{[2]} s_{i'}s_{k'} \approx \left(K_2' + K_3'\right)\sum_{i',j'}^{[1]} s_{i'}s_{j'}. \tag{5.250}$$

As shown in Figure 5.49, the number of the nearest and the next nearest neighbor pairs for each filled lattice site (unsummed) are exactly the same. Eq. (5.250) would be precise if the nearest and the next nearest spins take the same orientation.

With these assumptions, the partially summed partition function retains the original form for the 2D-Ising model

$$Q(K_2, N) = \exp\left(NK_0'/2\right)\ Q\left[\left(K_2' + K_3'\right), N/2\right]. \tag{5.251}$$

Subsequently, the free energy per spin becomes

$$f(K_2, N) = \left[K_0' + f\left(K_{2+3}', N/2\right)\right]/2 \tag{5.252}$$

where

$$K'_{2+3} \equiv K'_2 + K'_3 = (3/8)\ln\cosh 4K_2. \tag{5.253}$$

Eq. (5.251) indicates that, similar to that for 1D-Ising chain, the RG transformation leads to a reduction of the number of spin variables by half with a new coupling parameter K'_{2+3}.

Figure 5.50 shows how the coupling parameter K_2 varies with the RG transformation. Depending on its initial value, K_2 converges to either zero or infinity as the RG iteration progresses. As predicted by Eq. (5.253), the bifurcation takes place at a fixed point, i.e., when the coupling parameter satisfies

$$K_2^C = (3/8)\ln\cosh 4K_2^C. \tag{5.254}$$

Eq. (5.254) can be solved numerically, and an approximate solution is given by $K_2^C \approx 0.506981$. From the numerical point of view, the fixed point is unstable because the RG iteration always moves the system away from the condition corresponding to K_2^C. In particular, the RG flow leads to singular behavior at K_2^C because the final state of the system bifucates at this particular point.

Unlike that in the 1D-Ising model, the fixed point in the 2D model has a clear physical significance; it corresponds to the critical point of the magnetization phase transition. Even with drastic approximations discussed above, the critical temperature predicted by the RG transformation, $K_2^C \approx 0.506981$, is close to the exact value of $K_2^C = (1/2)\ln(\sqrt{2}+1) \approx 0.440687$, and it is much better than that predicted by the mean-field theory, $K_2^C = 0.25$.

To evaluate thermodynamic properties using the RG approach, we can begin with either a very small or a very large value of the coupling parameter K_2. Under these extreme conditions, the thermodynamic properties are precisely known because all spins are either randomly oriented or aligned in the same direction. Consequently, the reduced Helmholtz energy per spin can be expressed as follows:

$$f = \begin{cases} -\ln 2 & K_2 \to 0 \\ -2K_2 & K_2 \to \infty \end{cases}. \tag{5.255}$$

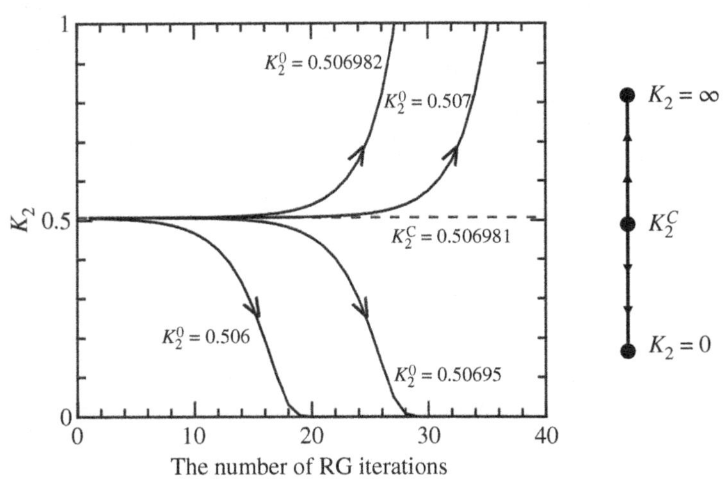

Figure 5.50 Variation of the coupling parameter K_2 with the number of the RG iterations for a 2D-Ising model. The coupling constant approaches 0 or ∞ depending on the initial value of K_2^0. The fixed point, $K_2 = K_2^C$, corresponds to the critical point of the phase transition.

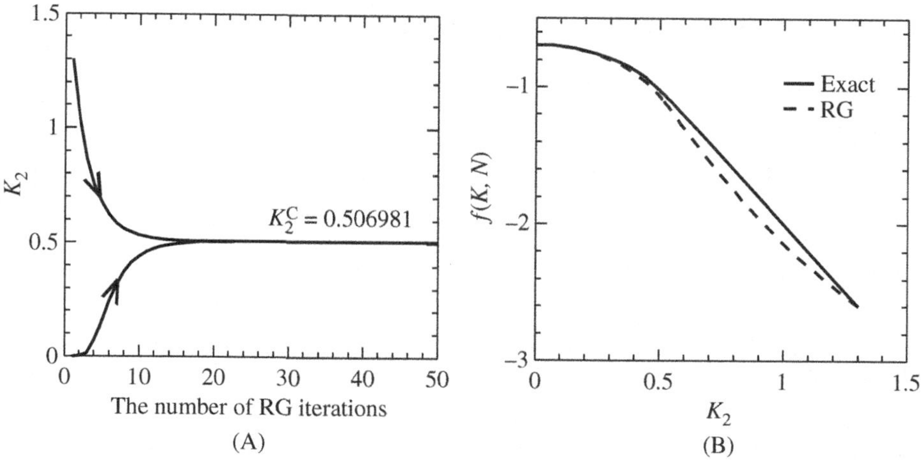

Figure 5.51 (A) Variation of the coupling parameter K_2 with the number of steps in the reversed RG iteration for the 2D-Ising model. (B) The RG free energy in comparison with the exact result.

The free energy between the two extreme values can be calculated using Eq. (5.252), where the coupling parameter is obtained from the backward RG iteration

$$K_2 = (1/4)\cosh^{-1}\left(e^{8K_2'/3}\right). \tag{5.256}$$

Figure 5.51A depicts the variation of the coupling parameter K_2 in the RG transformation when starting from small and large values. As anticipated, K_2 converges to the unstable fixed point after a few reverse RG iterations, $K_2^C \approx 0.506981$. Figure 5.51B displays the reduced free energy per spin over the entire range of K_2. Despite using crude approximations, the RG transformation predicts a free-energy profile that closely matches the exact results. The excellent agreement highlights the effectiveness of the RG approach in capturing the critical behavior, even when simplifications are made in the calculations.

5.12.4 RG Transformation Near the Critical Point

For the general form of the Hamiltonian given by Eq. (5.233), application of the RG transformation should not change the coupling parameters significantly near the critical point, i.e., $\mathbf{K}$ should be similar to that corresponding to the fixed point, $\mathbf{K}_C$. Accordingly, we may express $\mathbf{K}$ in term of a small deviation relative to the critical value

$$\mathbf{K} = \mathbf{K}_C + \delta\mathbf{K}. \tag{5.257}$$

After another step of RG transformation, we can express the new set of coupling parameters as

$$\mathbf{K}' = \mathbf{K}_C + \delta\mathbf{K}' = \mathbf{K}'(\mathbf{K}) \tag{5.258}$$

where function $\mathbf{K}'(\mathbf{K})$ can be represented by the Taylor series

$$\mathbf{K}'(\mathbf{K}_C + \delta\mathbf{K}) = \mathbf{K}_C + \hat{\mathbf{A}} \cdot \delta\mathbf{K} + \cdots \tag{5.259}$$

with each term in the coefficient matrix defined as $\hat{\mathbf{A}}_{ij} = \left(\partial K_i'/\partial K_j\right)\Big|_{\mathbf{K}=\mathbf{K}_C}$. Because $\delta\mathbf{K}$ is a small quantity, we may retain only the linear term in the Taylor expansion. As a result, the RG iteration leads to a linear variation of the coupling parameters

$$\delta\mathbf{K}' = \hat{\mathbf{A}} \cdot \delta\mathbf{K}. \tag{5.260}$$

Note that matrix $\hat{\mathbf{A}}$ is determined by the properties of the system at the critical point and by the step length L. If we assume that matrix $\hat{\mathbf{A}}$ can be diagonalized,[75] there exists a set of orthonormal eigen vectors $\{\mathbf{e}_i\}$ satisfying

$$\hat{\mathbf{A}} \cdot \mathbf{e}_i = \Lambda_i \mathbf{e}_i \tag{5.261}$$

where $\{\Lambda_i\}$ stands for the eigenvalues. Because the RG transformation follows the group properties for different step lengths L and L',

$$\hat{\mathbf{A}}(L) \cdot \hat{\mathbf{A}}(L') = \hat{\mathbf{A}}(L \cdot L'), \tag{5.262}$$

the corresponding eigenvalues must satisfy

$$\Lambda_i(L)\Lambda_i(L') = \Lambda_i(L \cdot L'). \tag{5.263}$$

Apparently, Eq. (5.263) has the boundary condition $\Lambda_i(1) = 1$.

We now take logarithm on both sides of Eq. (5.263) and differentiate the terms with respect to L'

$$\frac{d \ln \Lambda_i(L')}{dL'} = \frac{d \ln \Lambda_i(L \cdot L')}{d(LL')} \frac{d(LL')}{dL'}, \tag{5.264}$$

By setting $L' = 1$, Eq. (5.264) leads to

$$d \ln \Lambda_i(L)/d \ln L = [d\Lambda_i(L')/dL']_{L'=1} \equiv \lambda_i. \tag{5.265}$$

where λ_i is a constant independent of L. Solving the above differential equation gives

$$\Lambda_i(L) = L^{\lambda_i}. \tag{5.266}$$

The vectors corresponding to the coupling parameters before and after RG transformation can be written as

$$\delta\mathbf{K} = \sum_i k_i \mathbf{e}_i \tag{5.267}$$

$$\delta\mathbf{K}' = \sum_i k_i' \mathbf{e}_i \tag{5.268}$$

where k_i and k_i' are respectively the projections of $\delta\mathbf{K}$ and $\delta\mathbf{K}'$ in the direction of $\mathbf{e}_i$. Substituting above identities into Eq. (5.260) yields

$$\sum_i k_i' \mathbf{e}_i = \sum_i \Lambda_i k_i \mathbf{e}_i \tag{5.269}$$

or

$$k_i' = \Lambda_i k_i. \tag{5.270}$$

Eq. (5.270) indicates that, when the coupling parameters are expressed in terms of the eigenvectors, those components of $\delta\mathbf{K}$ with $|\Lambda_i| > 1$ grow with the RG flow, while those with $|\Lambda_i| < 1$ shrinks with the RG flow, and those for $|\Lambda_i| = 1$ do not change. Correspondingly, $\{k_i\}$ can be classified in terms of relevant, irrelevant, and marginal variables. Starting from a condition near the fixed point, the RG flow moves away from the fixed point in the directions of eigenvectors corresponding to the relevant variables, but converges to the fixed point in the directions related to the irrelevant variables. The situation is somewhat similar to that discussed for the 2D-Ising model as shown in

---

75 In general, $\hat{\mathbf{A}}$ is not necessarily symmetric and its eigenvalues may not be real. Nevertheless, a similar procedure can be established after minor modifications for asymmetric matrices.

Figure 5.50, where K_2 is a relevant coupling constant and no irrelevant or marginal variables ones are considered.

With the coupling parameters expressed in terms of the eigenvectors of $\hat{\mathbf{A}}$, the recursive relation the free energy per spin becomes

$$\Delta f(k_1, k_2, \cdots) = \frac{1}{L^d}\Delta f(\Lambda_1 k_1, \Lambda_2 k_2, \cdots) \tag{5.271}$$

where $\Delta f = f - f_c$ represents the *singular* part of the free energy. Eq. (5.271) indicates that the free energy is a generalized homogeneous function (GHF) of variables $\{k_i\}$. Near the critical point, we expect that the free energy density should be independent of step length L. To make it evident, we consider the *singular* part of the free energy after nth RG iterations

$$\Delta f(k_1, k_2, \ldots) = \frac{1}{L^{nd}}\Delta f(L^{n\lambda 1} k_1, L^{n\lambda 2} k_2, \ldots) \tag{5.272}$$

where we have expressed the eigenvalues in terms of $\Lambda_i = L^{\lambda_i}$. Because Δf is a GHF, we can rewrite Eq. (5.272) as

$$\Delta f(k_1, k_2, \ldots) = \left(k_1/k_1^0\right)^{d/\lambda_1} \Delta f\left[k_1^0, \left(k_1/k_1^0\right)^{-\lambda_2/\lambda_1} k_2, \ldots\right] \tag{5.273}$$

where $k_1^0 = k_1 L^{n\lambda_1}$ represents the value of k_1 at a reference condition. Note that the scaling factor is absent in Eq. (5.273), suggesting that it has no influence on the final free energy.

Within the framework of the Ising model, the thermodynamic properties are fully determined by two parameters, temperature (T) and magnetic field (h). These parameters are considered mutually independent, allowing us to define a thermal scaling variable, denoted by k_t with eigenvalue Λ_t, and a magnetic scaling variable k_h with eigenvalue Λ_h. For a uniform Ising system without the magnetic field ($h = 0$), $\Delta f = 0$ at the critical point. As a result, both k_t and k_h approach zero and have the form of a linear dependence on the distance from the critical point

$$k_t = t/t_0 + \cdots, \quad k_h = h/h_0 + \cdots \tag{5.274}$$

where $t = (T - T_c)/T$, t_0 and h_0 are scaling parameters, and $\cdots$ represents higher-order terms. As a result, we can rewrite the *singular* part of the free energy given by Eq. (5.273) as

$$\Delta f(k_t, k_h, \ldots) = (t/t_0)^{d/\lambda_t} \Phi\left[(h/h_0) \cdot (t/t_0)^{-\lambda_h/\lambda_t}\right] \tag{5.275}$$

where Φ represents a scaling function, λ_t and λ_h are exponents related to Λ_t and Λ_h. Eq. (5.275) suggests that the thermodynamic properties of the Ising system must follow power-law scaling. Although an exact form of the free energy remains unknown, we can derive the relations between different critical exponents. Using thermodynamic relations for the specific heat capacity,

$$c_V \sim \frac{\partial^2 f}{\partial t^2} \sim |t|^{-\alpha} \tag{5.276}$$

Carrying out the partial derivative of f at $h = 0$, we can obtain from Eq. (5.275)

$$\alpha = 2 - \frac{d}{\lambda_t}. \tag{5.277}$$

Similarly, we can find the scaling relations for the average magnetization and susceptibility

$$m = (\partial f/\partial h)|_{h=0} \sim |t|^{\beta}, \tag{5.278}$$

$$\chi = (\partial^2 f/\partial h^2)\Big|_{h=0} \sim |t|^{-\gamma} \tag{5.279}$$

with the critical exponents

$$\beta = \frac{d - \lambda_h}{\lambda_t}, \quad \gamma = \frac{2\lambda_h - d}{\lambda_t}. \tag{5.280}$$

Under RG transformation, the correlation length and reduced temperature scale as

$$\xi' = \xi/L, \tag{5.281}$$

$$t' = \Lambda_t t = L^{\lambda_t} t. \tag{5.282}$$

The above equations indicate $\xi^{\lambda_t} t = \xi'^{\lambda_t} t'$, i.e., invariant with the RG iteration. As a result, we have the scaling relation for the correlation length

$$\xi \sim |t|^{\nu} \tag{5.283}$$

with $\nu = 1/\lambda_t$.

5.12.5 Summary

The RG theory is a valuable tool for understanding the power-law scaling and critical behavior of second-order or continuous phase transitions. By employing the RG transformation, we can approximate the critical exponents and their relationships. While the RG process can be applied to systems exhibiting scale invariance, it does not provide a definitive confirmation of universality, which remains an open question in statistical mechanics. Scale invariance suggests that the critical behavior of phase transitions is insensitive to the specific microscopic interactions, as these interactions change under the RG transformation. The classification of coupling parameters into relevant, irrelevant, and marginal variables defines a universality class, representing a collection of systems that converge toward the same fixed point.

5.13 Generalized Ising Models

The Ising model, originally formulated with binary spin states, has been generalized to describe a wide range of cooperative phenomena and phase transitions. These extensions primarily focus on augmenting the lattice site either with additional spin states or higher-dimensional spin variables. For example, the Potts model introduces multiple spin states per lattice site, while the n-vector model involves increasing the dimensionality of the spin variables. These generalized models are useful to advance our comprehension of phase transitions occurring in different universality classes. By allowing for a more diverse range of spin states or higher-dimensional spin variables, the generalized Ising model provide a comprehensive framework for capturing the critical behaviors and the fundamental mechanisms of phase transitions and cooperative phenomena.

5.13.1 The Potts Model

The Potts model extends the standard Ising model for each spin to m possible states.[76] Similar to the Ising model, it accounts for only nearest-neighbor interactions. We assign a coupling energy of ε if they are in different states, and $-(m-1)\varepsilon$ if neighboring spins i and j are in the same states. Accordingly, the total energy of the system is thus given by

$$E_\nu = -\varepsilon \sum_{\{i,j\}} [m\delta_{s_i s_j} - 1] \tag{5.284}$$

76 Wu F. Y. "The Potts model", *Rev. Mod. Phys.* 54(1), 235–268 (1982).

where $\delta_{s_i s_j}$ is the Kronecker delta function, and $\{i, j\}$ represents all possible nearest neighbors on the lattice. When $m = 2$, the Potts model reduces to the Ising model. If $\varepsilon > 0$, the spin alignment is favored at low temperature, i.e., the system becomes ferromagnetically ordered.

The Potts model has found applications in diverse fields, encompassing machine learning (e.g., neural networks), condensed-matter physics (e.g., percolation and foam behavior), biology (e.g., protein folding and biological membranes), medicine (e.g., cancer cell development and beating heart cells), and sociology (e.g., flocking birds and social behavior). By allowing each element to exist in multiple states, the Potts model effectively captures the complexity of landscapes with multiple local minima, such as those observed in spin glasses. The remarkable versatility of the Potts model has facilitated the derivation of exact results in specific cases, serving as benchmarks for validating numerical simulations or approximations, and offering guidance for further theoretical advancements.[77]

5.13.2 *n*-Vector Model

In the n-vector model, each spin is characterized by an n-dimensional unit vector

$$\mathbf{s}_i = \{s_1, s_2, \dots, s_n\} \tag{5.285}$$

where subscript i denotes a lattice site in a d-dimensional space. The n components of the spin vector take continuous variables and satisfy

$$\sum_{\alpha=1}^{n} s_\alpha^2 = 1 \tag{5.286}$$

with $\alpha = 1, 2, \cdots, n$ indexing for different dimensions of the spin vector $\mathbf{s}_i$. At a given microstate, the system energy is determined by the interaction between nearest neighboring spins

$$E_\nu = -\varepsilon \sum_{\{i,j\}} \mathbf{s}_i \cdot \mathbf{s}_j \tag{5.287}$$

where ε stands for the coupling energy, and the sum runs over all pairs of nearest neighbors on the lattice.

When $n = 1$, the n-vector model is equivalent to the regular Ising model. For $n = 2$, it is referred to as the XY model. In this scenario, the spin vector can be represented as $\mathbf{s}_i = \{\cos\theta_i, \sin\theta_i\}$. The XY model is commonly employed in the study of phase transitions occurring in hexatic liquid crystals and the Berezinskii–Kosterlitz–Thouless (BKT) quantum phase transition.[78] Furthermore, when n equals 3, the n-vector model is known as the Heisenberg model, which proves valuable in describing long-range ordering observed in magnetic systems.[79]

de Gennes demonstrated that a self-avoiding random walk, which is often used to describe polymer excluded volume effect, can be mapped into an n-vector model near its critical point at the limit of $n = 0$.[80] Based on the scaling relations for the spin–spin correlation function established for the "magnetic" system,[81] de Gennes derived a power law relationship between the radius of

77 Baxter R. J., *Exactly solved models in statistical mechanics* (Dover Edition). Dover Publications, 2007.
78 José J. V., *40 years of Berezinskii-Kosterlitz-Thouless theory*. World Scientific, 2013.
79 de Albuquerque D. F. and de Arruda A. S., "Heisenberg model in a random field: phase diagram and tricritical behavior", *Physica A* 316(1–4), 13–18 (2002).
80 de Gennes P. G., "Exponents for the excluded volume problem as derived by the Wilson method", *Phys. Lett. A* 38, 229 (1972). While $n = 0$ does not make any sense from a physical perspective, it can be shown that the limit of $n \to 0$ is well defined mathematically.
81 De Gennes P. G., *Scaling concepts in polymer physics*. Cornell University Press, 1979.

gyration R_g and the degree of polymerization N of a polymer chain in a good solvent at the infinite dilute limit[82]

$$R_g \sim N^{\nu} \tag{5.288}$$

where ν is the same as the critical exponent for the divergence of the correlation length in the n-vector system, $\xi \sim |1 - T/T_C|^{-\nu}$. For the 3-dimensional "magnetic" system ($d = 3$ and $n = 0$ in the n-vector model), the RG theory predicts $\nu \approx 0.592$, which is in excellent agreement with numerical results from Monte Carlo simulation for a self-avoiding random walk and with the empirical fitting of experimental data for a dilute solution of polymers in a good solvent (Figure 5.52).[83]

The polymer-magnet isomorphism for an isolated polymer in a solution was generalized by des Cloizeaux for polymer solutions at intermediate concentrations.[84] In the presence of a magnetic

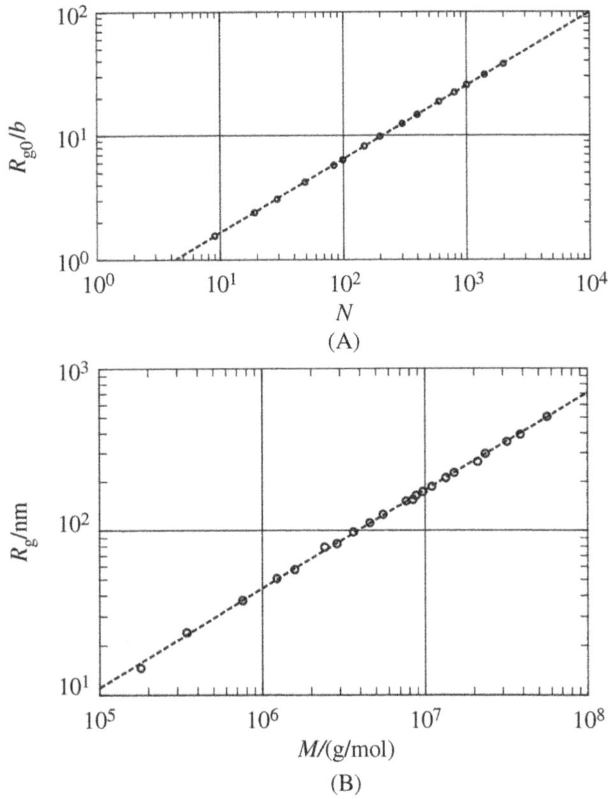

Figure 5.52 (A) Radius of gyration (R_{g0}) versus the number N of the steps in a self-avoiding random walk on a cubic lattice of lattice unit (b). The symbols are results from Monte Carlo simulation, the dashed line, $R_{g0}/b = 0.4205 \times N^{0.5934}$, is obtained by the best fit of the simulation data. (B) Radius of gyration (R_g) as a function of the molecular weight (M) of polystyrene in a binary mixture of toluene and benzene. The symbols are experimental data and the dashed line is the optimal fit with R_g/[nm] = 0.01234 × (M/[g/mol])$^{0.5936}$. Source: Reproduced from Teraoka.[83]

82 A similar relation, $R_g \approx bN^{3/5}$, was obtained by Paul Flory for a self-avoiding random walk of step length b. While this expression is remarkably close to that by de Gennes, Flory reached the good final result due to cancellation of errors.

83 Teraoka I., *Polymer solutions: an introduction to physical properties*. Wiley, 2002.

84 des Cloizeaux J., "The Lagrangian theory of polymer solutions at intermediate concentrations", *J. Phys. (Paris)* 36, 281 (1975).

Figure 5.53 The reduced osmotic pressure versus the reduced concentration for poly(α-methyl styrene) of different molecular weights in liquid toluene at 25 °C. The ideal osmotic pressure is $\Pi_{\text{ideal}} = cN_A k_B T/M$, where c in g/m^3 is the polymer concentration, M is the molecular weight (g/mol), and N_A is the Avogadro number. The overlap concentration is calculated from $c^* = (M/N_A)/\left(4\pi R_g^3/3\right)$ where R_g the radius of gyration for individual polymer chains at infinite dilution. Reproduced from Teraoka.[83]

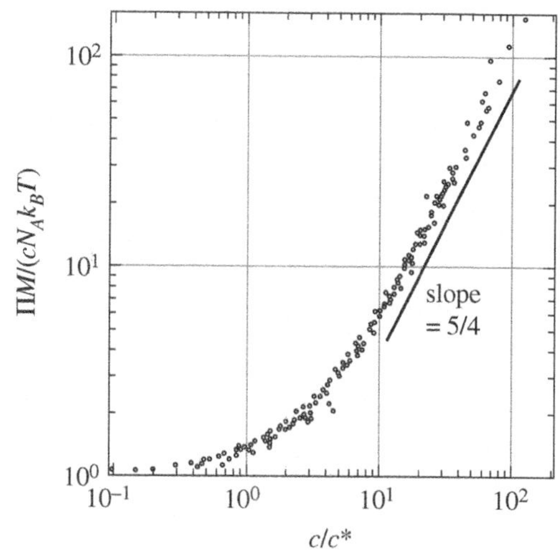

field, the partition function of the n-vector model ($n = 0$) is analogous to the grand canonical partition function of a polymer system with excluded volume interactions. By applying the linear relationship between magnetization and the external field near the critical point, des Cloizeaux obtained the first-order correction of the osmotic pressure due to interaction between polymer chains

$$\Pi \sim \rho^{3\nu/(3\nu-1)} \tag{5.289}$$

where ρ stands for the number density of polymer segments, and $\nu \approx 3/5$ is again the critical exponent for the magnetic system. The inter-chain interaction becomes significant at a crossover density when the osmotic pressure predicted from Eq. (5.289) becomes comparable to that for an ideal polymer solution ($\Pi_{\text{ideal}} = \rho k_B T/N$)

$$\rho^* \sim N^{1-3\nu}. \tag{5.290}$$

As expected intuitively, the crossover density occurs when the polymer chains occupy the entire space $N\rho^* R_g^3 \sim 1$. As shown in Figure 5.53, the transition from the dilute to semi-dilute regimes predicted by the des Cloizeaux theory agrees well with the experimental results for the osmotic pressure of real polymer systems in a good solvent.

It is worth highlighting that the calculations based on the RG approach can be directly applied to polymer systems without relying on the polymer-magnet analogy.[85] In addition to considering the polymer structure, which includes parameters such as the mean-square end-to-end distance, a closed-form expression can be derived for the osmotic pressure in both the dilute and semi-dilute regimes[86]

$$\Pi/\Pi_{\text{real}} = 1 + \frac{X}{2}\exp\{[X^{-1} + (1 - X^{-2})\ln(1 + X)]/4\} \tag{5.291}$$

where $X = 3.49\rho/\rho^*$.

85 Schäfer L., *Excluded volume effects in polymer solutions as explained by the renormalization group*, Springer, 1999.
86 Ohta T. and Oono Y., "Conformation space renormalization theory of semi-dilute polymer-solutions", *Phys. Lett. A* 89(9), 460–464 (1982).

5.13.3 Summary

Scientific progress is often nonlinear and can arise from seemingly unrelated fields. The extension of the n-vector model by de Gennes exemplifies this phenomenon. Originally developed for studying phase transitions and critical phenomena, the generalized Ising model has had significant implications for understanding the properties of polymer systems. This unexpected connection between two distinct areas of research highlights how insights from one field can have profound impacts on another, leading to new understandings and breakthroughs. It underscores the importance of interdisciplinary approaches and the potential for cross-pollination of ideas in advancing scientific knowledge, as advocated by this text.

5.14 Chapter Summary

In the realm of statistical mechanics, valuable insights can be obtained by investigating abstract models that display cooperative behavior due to interactions among multiple elements. The knowledge gained from studying these abstract models can be generalized to solve seemingly unrelated real-world problems. The Ising model serves as an exemplary illustration of such a model. Originally devised to elucidate the properties of ferromagnetic materials, the Ising model simplifies complex interactions among particles in thermodynamic systems, making it applicable to cooperative phenomena across different scales. At the molecular level, the Ising model has been employed to analyze cooperative ligand binding, ionization of weak polyelectrolytes, and the formation of biological structures. At the thermodynamic limit, the model explains the occurrence of phase transitions and the emergence of universality in critical phenomena. The simplicity of the Ising model allows for analytical treatment, rendering it broad applications in diverse scientific fields.

5.A The Partition Function of an Ising Chain

In this appendix, we present the mathematical details for driving the partition function and average magnetization of an Ising chain in an external field. From the partition function, we can also obtain the spin–spin correlation functions.

5.A.1 Direct Enumeration of the Partition Function

For an Ising chain with N spins under the influence of an external field, the partition function can be expressed as

$$Q = \sum_{s_1=\pm 1} \sum_{s_2=\pm 1} \cdots \sum_{s_N=\pm 1} \exp[\beta h s_1 + \beta \varepsilon s_1 s_2 + \beta h s_2 + \beta \varepsilon s_2 s_3 + \cdots + \beta \varepsilon s_{N-1} s_N + \beta h s_N] \quad (5.A.1)$$

where the notation is the same as that used in Section 5.2.

If N is a small number, we can evaluate Eq. (5.A.1) by directly enumerating all possible combinations of up and down orientations for each spin represented by variable $s_i = \pm 1$. For example, $N = 2$ gives $2 \times 2 = 4$ microstates, corresponding to the 4 combinations of spin orientations:

$$Q = \sum_{s_1=\pm 1} \sum_{s_2=\pm 1} \exp[\beta h s_1 + \beta \varepsilon s_1 s_2 + \beta h s_2] = e^{2\beta h + \beta \varepsilon} + e^{-\beta \varepsilon} + e^{-\beta \varepsilon} + e^{-2\beta h + \beta \varepsilon} \quad (5.A.2)$$

When N is a large number, we can obtain an analytical expression for Q using vector and matrix operations. For example, it is straightforward to verify that Eq. (5.A.2) can be written as

$$Q = \sum_{\mathbf{s}_1}\sum_{\mathbf{s}_2} \mathbf{s}_1 \cdot \begin{pmatrix} e^{2\beta h+\beta\varepsilon} & e^{-\beta\varepsilon} \\ e^{-\beta\varepsilon} & e^{-2\beta h+\beta\varepsilon} \end{pmatrix} \cdot \mathbf{s}_2^T \tag{5.A.3}$$

where unit vector $\mathbf{s}_i = (1, 0)$ represents $s_i = 1$, and $\mathbf{s}_i = (0, 1)$ represents $s_i = -1$; superscript T stands for matrix transpose. The summation over all possible choices for vectors $\mathbf{s}_1$ and $\mathbf{s}_2$ lead to the four Boltzmann factors in Eq. (5.A.2). As discussed in the following, the matrix expression can be easily extended to systems with a large number of spins.

5.A.2 End Effects

For an Ising chain with a large number of spins ($N \gg 1$), the total energy may be approximated by summing the pair interaction energies

$$E_v = -h\sum_{i=1}^{N} s_i - \varepsilon\sum_{i=1}^{N-1} s_i s_{i+1} \approx -\sum_{i=1}^{N} [h(s_i + s_{i+1})/2 + \varepsilon s_i s_{i+1}] \tag{5.A.4}$$

where $s_{N+1} = s_1$. In Eq. (5.A.4), we have added an extra term, $-\varepsilon s_N s_{N+1} - h(s_1 - s_{N+1})/2$, to the total energy of the Ising chain for easier mathematical operation.[87] The approximation introduces a superficial interaction between the end spins, similar to that in a cyclic Ising chain (or an Ising ring). In the case of a long linear chain with open ends, this extra energy is insignificant in comparison to the total energy of the system.

5.A.3 The Transfer Matrix Method

The transfer-matrix method is a mathematical technique useful to evaluate the partition function of 1D and 2D-Ising models. According to Eq. (5.A.4), the energy of an Ising chain can be written in terms of pair interactions. For each nearest-neighboring pair, the summation over all possible values of the Boltzmann factor can be expressed in terms of a 2×2 matrix:

$$\sum_{s_i=\pm 1}\sum_{s_{i+1}=\pm 1} \exp[\beta h(s_i/2 + s_{i+1}/2) + \beta\varepsilon s_{i+1}] = \sum_{\mathbf{s}_i}\sum_{\mathbf{s}_j} \mathbf{s}_i \cdot \begin{pmatrix} e^{\beta h+\beta\varepsilon} & e^{-\beta\varepsilon} \\ e^{-\beta\varepsilon} & e^{-\beta h+\beta\varepsilon} \end{pmatrix} \cdot \mathbf{s}_j^T. \tag{5.A.5}$$

Accordingly, the partition function for the entire chain becomes

$$Q = \sum_{v} e^{-\beta E_v} = \sum_{\{\mathbf{s}_i\}} \left(\mathbf{s}_1 \cdot \hat{\mathbf{M}} \cdot \mathbf{s}_2^T\right) \cdot \left(\mathbf{s}_2 \cdot \hat{\mathbf{M}} \cdot \mathbf{s}_3^T\right) \cdot \quad (\cdots) \quad \cdot \left(\mathbf{s}_N \cdot \hat{\mathbf{M}} \cdot \mathbf{s}_{N+1}^T\right) \tag{5.A.6}$$

where $\mathbf{s}_{N+1} = \mathbf{s}_1$ and

$$\hat{\mathbf{M}} = \begin{pmatrix} e^{\beta(\varepsilon+h)} & e^{-\beta\varepsilon} \\ e^{-\beta\varepsilon} & e^{\beta(\varepsilon-h)} \end{pmatrix} \tag{5.A.7}$$

is called the transfer matrix.

Noting that

$$\sum_{\mathbf{s}_i} \mathbf{s}_i^T \cdot \mathbf{s}_i = \begin{pmatrix} 1 \\ 0 \end{pmatrix} (1\ 0) + \begin{pmatrix} 0 \\ 1 \end{pmatrix} (0\ 1)$$
$$= \begin{pmatrix} 1 & 0 \\ 0 & 0 \end{pmatrix} + \begin{pmatrix} 0 & 0 \\ 0 & 1 \end{pmatrix} = \begin{pmatrix} 1 & 0 \\ 0 & 1 \end{pmatrix} = \hat{\mathbf{I}} \tag{5.A.8}$$

87 The assumption would be exact if we consider an Ising ring where the first and last spins are next to each other.

where $\hat{\mathbf{I}}$ denotes the 2×2 unit matrix, we can rewrite the partition function as

$$Q = \sum_{\mathbf{s}_1} \mathbf{s}_1 \cdot \hat{\mathbf{M}}^N \cdot \mathbf{s}_1^T = \mathrm{Tr}(\hat{\mathbf{M}}^N) \tag{5.A.9}$$

where $\mathrm{Tr}(\hat{\mathbf{M}}) = \sum_i \hat{\mathbf{M}}_{ii}$ stands for the matrix trace.

Because $\hat{\mathbf{M}}$ is a symmetric matrix taking a positive value for every element, the finite-dimensional spectral theorem states that it can be diagonalized with a unitary matrix $\hat{\mathbf{U}}$

$$\hat{\mathbf{U}}^T \cdot \hat{\mathbf{M}} \cdot \hat{\mathbf{U}} = \begin{pmatrix} \lambda_+ & 0 \\ 0 & \lambda_- \end{pmatrix} \tag{5.A.10}$$

where λ_+ and λ_- are two eigenvalues of matrix $\hat{\mathbf{M}}$, and $\lambda_+ > \lambda_- > 0$. Using the identity $\hat{\mathbf{U}}^T \cdot \hat{\mathbf{U}} = \hat{\mathbf{I}}$ and the trace invariance of a matrix under cyclic permutations, we can evaluate Eq. (5.A.9) analytically

$$\begin{aligned} Q &= \mathrm{Tr}(\hat{\mathbf{M}}^N) = \mathrm{Tr}(\hat{\mathbf{U}} \cdot \hat{\mathbf{U}}^T \hat{\mathbf{M}} \hat{\mathbf{U}} \cdot \hat{\mathbf{U}}^T \hat{\mathbf{M}} \hat{\mathbf{U}} \cdots \hat{\mathbf{U}}^T \hat{\mathbf{M}} \hat{\mathbf{U}} \cdot \hat{\mathbf{U}}^T) \\ &= \mathrm{Tr}\left\{ \hat{\mathbf{U}} \begin{pmatrix} \lambda_+^N & 0 \\ 0 & \lambda_-^N \end{pmatrix} \hat{\mathbf{U}}^T \right\} = \mathrm{Tr}\left\{ \begin{pmatrix} \lambda_+^N & 0 \\ 0 & \lambda_-^N \end{pmatrix} \hat{\mathbf{U}}^T \hat{\mathbf{U}} \right\} = \lambda_+^N + \lambda_-^N \end{aligned} \tag{5.A.11}$$

If N is a large number, we have $(\lambda_+/\lambda_-)^N >> 1$ so that the partition function can be approximated by

$$Q \approx \lambda_+{}^N. \tag{5.A.12}$$

The eigenvalues of transfer matrix $\hat{\mathbf{M}}$ can be found from

$$\mathrm{Det}(\hat{\mathbf{M}} - \lambda \hat{\mathbf{I}}) = \begin{vmatrix} e^{\beta(\varepsilon+h)} - \lambda & e^{-\beta\varepsilon} \\ e^{-\beta\varepsilon} & e^{\beta(\varepsilon-h)} - \lambda \end{vmatrix} = 0 \tag{5.A.13}$$

where Det stands for matrix determinant. Solution of Eq. (5.A.13) for λ gives two eigenvalues

$$\lambda_\pm = e^{\beta\varepsilon}[\cosh(\beta h) \pm \sqrt{\sinh^2(\beta h) + e^{-4\beta\varepsilon}}] \tag{5.A.14}$$

Without the external field, we have $h = 0$, $\cosh(\beta h) = 1$ and $\sinh(\beta h) = 0$ so that the eigenvalues become

$$\lambda_\pm = e^{\beta\varepsilon} \pm e^{-\beta\varepsilon}. \tag{5.A.15}$$

From the eigenvalues, we can find the eigenvectors and derive the unitary matrix (see Problems 5.4–5.5)

$$\hat{\mathbf{U}} = \begin{pmatrix} \cos\phi & -\sin\phi \\ \sin\phi & \cos\phi \end{pmatrix} \tag{5.A.16}$$

where $\cot(2\phi) = e^{2\beta\varepsilon} \sinh(\beta h)$ and $0 < \phi < \pi/2$.

5.A.4 Average Magnetization

The average magnetization for any particular spin is given by the ensemble average

$$<s_i> = \frac{1}{Q} \sum_\nu s_i e^{-\beta E_\nu} = \frac{1}{Q} \sum_{\{\mathbf{s}_k\}} (\mathbf{s}_1 \cdot \hat{\mathbf{M}} \cdot \mathbf{s}_2{}^T) \cdot (\mathbf{s}_2 \cdot \hat{\mathbf{M}} \cdots \hat{\mathbf{M}} \cdot \mathbf{s}_i{}^T)\ s_i\ (\mathbf{s}_i \quad \cdots \hat{\mathbf{M}} \cdot \mathbf{s}_N{}^T). \tag{5.A.17}$$

We may rewrite the summation related to spin i as

$$\sum_{\mathbf{s}_i} \mathbf{s}_i^T \cdot s_i \cdot \mathbf{s}_i = \begin{pmatrix}1\\0\end{pmatrix}(1\ 0) - \begin{pmatrix}0\\1\end{pmatrix}(0\ 1)$$
$$= \begin{pmatrix}1 & 0\\0 & 0\end{pmatrix} - \begin{pmatrix}0 & 0\\0 & 1\end{pmatrix} = \begin{pmatrix}1 & 0\\0 & -1.\end{pmatrix}. \tag{5.A.18}$$

With the identity $\sum_{\mathbf{s}_{k\neq i}} \mathbf{s}_k^T \cdot \mathbf{s}_k = \hat{\mathbf{I}}$, the average magnetization of spin i is then given by

$$<s_i> = \frac{1}{Q}\sum_{v} s_i e^{-\beta E_v} = \frac{1}{Q}\sum_{\bar{s}_1} \mathbf{s}_1 \cdot \hat{\mathbf{M}}^{i-1} \cdot \begin{pmatrix}1 & 0\\0 & -1\end{pmatrix} \cdot \hat{\mathbf{M}}^{N-i+1} \mathbf{s}_1^T$$
$$= \frac{1}{Q}\mathrm{Tr}\left\{\hat{\mathbf{M}}^N \cdot \begin{pmatrix}1 & 0\\0 & -1\end{pmatrix}\right\} = \frac{1}{Q}\mathrm{Tr}\left\{\hat{\mathbf{U}} \cdot \begin{pmatrix}\lambda_+^N & 0\\0 & \lambda_-^N\end{pmatrix} \cdot \hat{\mathbf{U}}^T \cdot \begin{pmatrix}1 & 0\\0 & -1\end{pmatrix}\right\}$$
$$= \frac{1}{Q}\mathrm{Tr}\left\{\begin{pmatrix}\lambda_+^N & 0\\0 & \lambda_-^N\end{pmatrix} \cdot \begin{pmatrix}\cos\phi & \sin\phi\\-\sin\phi & \cos\phi\end{pmatrix} \cdot \begin{pmatrix}1 & 0\\0 & -1\end{pmatrix} \cdot \begin{pmatrix}\cos\phi & -\sin\phi\\\sin\phi & \cos\phi\end{pmatrix}\right\}$$
$$= \cos(2\phi)\frac{\lambda_+^N - \lambda_-^N}{\lambda_+^N + \lambda_-^N} \approx \cos(2\phi). \tag{5.A.19}$$

where the last expression is valid for $N \gg 1$. When $h=0$, $\cot(2\phi)=e^{2\beta\varepsilon}\sinh(\beta h)=0$ and, as expected, $<s_i>=\cos(2\phi)=0$.

5.A.5 Spin–Spin Pair Correlation Functions

The above procedure can be similarly used to find the correlation function between two arbitrary spins in the Ising chain. With the help of Eq. (5.A.18), the spin–spin correlation function can be expressed as

$$<s_i s_{i+n}> = \frac{1}{Q}\mathrm{Tr}\left\{\hat{\mathbf{M}}^{N-n} \cdot \begin{pmatrix}1 & 0\\0 & -1\end{pmatrix} \cdot \hat{\mathbf{M}}^{N-n} \cdot \begin{pmatrix}1 & 0\\0 & -1\end{pmatrix}\right\}$$
$$= \frac{1}{Q}\mathrm{Tr}\left\{\begin{pmatrix}\lambda_+^{N-n} & 0\\0 & \lambda_-^{N-n}\end{pmatrix} \cdot \begin{pmatrix}0 & 1\\1 & 0\end{pmatrix} \cdot \begin{pmatrix}\lambda_+^n & 0\\0 & \lambda_-^n\end{pmatrix} \cdot \begin{pmatrix}0 & 1\\1 & 0\end{pmatrix}\right\}$$
$$= \frac{\lambda_+^{N-n}\lambda_-^n + \lambda_-^{N-n}\lambda_+^n}{\lambda_+^N + \lambda_-^N}. \tag{5.A.20}$$

For $N>>1$ and $h=0$, $\lambda_+^N \gg \lambda_-^N$, the spin–spin correlation function becomes

$$<s_i s_{i+n}> \approx (\lambda_-/\lambda_+)^n = \tanh^n(\beta\varepsilon) \tag{5.A.21}$$

where λ_+ and λ_- are given by Eq. (5.A.15). As discussed in Section 5.A.2, Eq. (5.A.21) provides a basis for the definition of the correlation length.

5.B The Partition Function in the Zimm–Bragg Model of Helix/Coil Transition

In this appendix, we outline briefly the derivation of the partition function in the Zimm–Bragg model using the transfer-matrix method discussed in Appendix 5.A. The notation for the helix/coil transition is defined in Section 5.B.4 of the main text.

5.B.1 Recursive Relation for the Partition Function

Let Q_m^c represent the partition function of an m-residue polypeptide whose final residue is in a coil state, and Q_m^h represent the partition function when the final residue is in the helical state. The partition function of an m-residue polypeptide can thus be expressed as

$$Q = Q_m^c + Q_m^h. \tag{5.B.1}$$

Eq. (5.B.1) can be written as the inner product of two vectors

$$Q = \left(Q_m^c, Q_m^h\right)\begin{pmatrix}1\\0\end{pmatrix} + \left(Q_m^c, Q_m^h\right)\begin{pmatrix}0\\1\end{pmatrix} = \left(Q_m^c, Q_m^h\right)\begin{pmatrix}1\\1\end{pmatrix} \tag{5.B.2}$$

where the unit vectors, $\begin{pmatrix}1\\0\end{pmatrix}$ and $\begin{pmatrix}0\\1\end{pmatrix}$ denote the coil and helical states of the residues, respectively.

If the final residue is in a helical state, we can relate Q_m^h to the partition functions of an $(m-1)$-residue polypeptide

$$Q_m^h = Q_{m-1}^c e^{-\beta u_s} + Q_{m-1}^h e^{-\beta u_a}. \tag{5.B.3}$$

In writing Eq. (5.B.3), we consider the possibility that the final residue may exist in one of two energy states: either as an initial segment or as an additional segment within the helical conformation. Similarly, Q_m^c can be written as

$$Q_m^c = Q_{m-1}^c + Q_{m-1}^h. \tag{5.B.4}$$

Thus, the vector $\left(Q_m^c, Q_m^h\right)$ can be calculated from

$$\left(Q_m^c, Q_m^h\right) = \left(Q_{m-1}^c, Q_{m-1}^h\right) \cdot \hat{\mathbf{M}} \tag{5.B.5}$$

where the transfer matrix is defined as

$$\hat{\mathbf{M}} = \begin{pmatrix}1 & e^{-\beta u_s}\\ 1 & e^{-\beta u_a}\end{pmatrix}. \tag{5.B.6}$$

The above procedure can be repeated iteratively until it reaches the first residue, which can only exist in either a coil state or an initiating helical state

$$\left(Q_1^c, Q_1^h\right) = (1, 0)\begin{pmatrix}1 & e^{-\beta u_s}\\ 1 & e^{-\beta u_a}\end{pmatrix}. \tag{5.B.7}$$

Using the recursive relation, we obtain the desired expression for the partition function of an m-residue polypeptide,

$$Q = (1, 0)\begin{pmatrix}1 & e^{-\beta u_s}\\ 1 & e^{-\beta u_a}\end{pmatrix}^m \begin{pmatrix}1\\1\end{pmatrix}. \tag{5.B.8}$$

5.B.2 Diagonalization of the Transfer Matrix

We can evaluate the mth-power matrix in Eq. (5.B.8) using a similarity transformation

$$\hat{\mathbf{T}}^{-1} \cdot \hat{\mathbf{M}} \cdot \hat{\mathbf{T}} = \begin{pmatrix}\lambda_+ & 0\\ 0 & \lambda_-\end{pmatrix} \tag{5.B.9}$$

where $\hat{\mathbf{T}}$ is the transformation matrix, $\hat{\mathbf{T}}^{-1}$ is the matrix inverse of $\hat{\mathbf{T}}$, and $\lambda_\pm$ are the eigenvalues of the transfer matrix $\hat{\mathbf{M}}$. The latter can be calculated from

$$\begin{vmatrix}1-\lambda & e^{-\beta u_s}\\ 1 & e^{-\beta u_a}-\lambda\end{vmatrix} = 0 \tag{5.B.10}$$

which gives

$$\lambda_{\pm} = \left[\left(1 + e^{-\beta u_a}\right) \pm \sqrt{\left(1 - e^{-\beta u_a}\right)^2 + 4e^{-\beta u_s}} \right] /2. \tag{5.B.11}$$

The transformation matrix $\hat{\mathbf{T}}$ is derived from the corresponding eigenvectors

$$\hat{\mathbf{T}} = \begin{pmatrix} 1-\lambda_- & 1-\lambda_+ \\ 1 & 1 \end{pmatrix}, \tag{5.B.12}$$

and its inverse is given by

$$\hat{\mathbf{T}}^{-1} = \frac{1}{\lambda_+ - \lambda_-} \begin{pmatrix} 1 & \lambda_+ - 1 \\ -1 & 1-\lambda_- \end{pmatrix}. \tag{5.B.13}$$

After the diagonalization of $\hat{\mathbf{M}}$, we can evaluate Eq. (5.B.8) analytically

$$\begin{aligned} Q &= (1,0)\hat{\mathbf{T}} \begin{pmatrix} \lambda_+^m & 0 \\ 0 & \lambda_-^m \end{pmatrix} \hat{\mathbf{T}}^{-1} \begin{pmatrix} 1 \\ 1 \end{pmatrix} \\ &= (1-\lambda_-,\ 1-\lambda_+) \begin{pmatrix} \lambda_+^m & 0 \\ 0 & \lambda_-^m \end{pmatrix} \begin{pmatrix} \lambda_+/(\lambda_+ - \lambda_-) \\ -\lambda_-/(\lambda_+ - \lambda_-) \end{pmatrix} \\ &= \left[\lambda_+^{m+1}(1-\lambda_-) - \lambda_-^{m+1}(1-\lambda_+) \right] /(\lambda_+ - \lambda_-). \end{aligned} \tag{5.B.14}$$

Since $\lambda_+ > \lambda_-$, Eq. (5.B.14) can be approximated by

$$Q \approx \frac{\lambda_+^{m+1}(1-\lambda_-)}{\lambda_+ - \lambda_-} \approx \lambda_+^m. \tag{5.B.15}$$

The error of this approximation becomes negligibly small for $m \gg 1$.

Further Readings

Aksel T. Barrick D. "Analysis of repeat-protein folding using nearest-neighbor statistical mechanical models", *Methods Enzymol.* 455, 95–125 (2009).

Binney J. J., Dowrick N. J., Fisher A. J., and Newman M. E. J., *The theory of critical phenomena: an introduction to the renormalization group*. Clarendon Press, Oxford, 1992.

Chandler D., *Introduction to modern statistical mechanics*. Oxford University Press, Chapter 5, 1987.

Ising T., Folk R., Kenna R., Berche B. and Holovatch Y., "The fate of Ernst Ising and the fate of his model", *J. Phys. Stud.* 21 (3), 3002, (2017).

McCoy, B. M., *Advanced statistical mechanics*. Oxford University Press, Chapter 10, 1987.

Problems

5.1 Find the eigenvalues and eigenvectors of the following matrices

$$\text{(i)}\,\hat{\mathbf{A}} = \begin{pmatrix} a & c \\ b & d \end{pmatrix}, \qquad \text{(ii)}\,\hat{\mathbf{B}} = \begin{pmatrix} a & c \\ c & d \end{pmatrix}.$$

Here, $a \neq b \neq c \neq d \neq 0$ are real numbers.

5.2 Show that the eigenvectors for matrix $\hat{\mathbf{B}}$ in Problem 5.1 are orthogonal. How about the eigenvectors for matrix $\hat{\mathbf{A}}$?

5.3 Show that matrix $\hat{\mathbf{B}}$ in Problem 5.1 can be diagonalized with a unitary matrix $\hat{\mathbf{U}}$, i.e.,

$$\hat{\mathbf{U}}^T \cdot \begin{pmatrix} a & c \\ c & d \end{pmatrix} \cdot \hat{\mathbf{U}} = \begin{pmatrix} \lambda_1 & 0 \\ 0 & \lambda_2 \end{pmatrix}.$$

5.4 Show that the eigenvalues of the transfer matrix for the 1D-Ising model

$$\hat{\mathbf{M}} = \begin{pmatrix} e^{\beta h+\beta\epsilon} & e^{-\beta\epsilon} \\ e^{-\beta\epsilon} & e^{-\beta h+\beta\epsilon} \end{pmatrix},$$

are

$$\lambda_{1,2} = e^{\beta\epsilon}\left[\cosh(\beta h) \pm \sqrt{\sinh^2(\beta h) + e^{-4\beta\epsilon}}\right], \tag{C}$$

and the eigenvectors are

$$v_{1,2} = \begin{pmatrix} 1 \\ -e^{2\beta\epsilon}\left[\sinh(\beta h) \mp \sqrt{\sinh^2(\beta h) + e^{-4\beta\epsilon}}\right] \end{pmatrix}. \tag{D}$$

5.5 Consider the transfer matrix for the 1D-Ising model

$$\hat{\mathbf{M}} = \begin{pmatrix} e^{\beta h+\beta\epsilon} & e^{-\beta\epsilon} \\ e^{-\beta\epsilon} & e^{-\beta h+\beta\epsilon} \end{pmatrix}.$$

(i) Verify that the eigenvectors are orthogonal;
(ii) Find the corresponding unitary matrix $\hat{\mathbf{U}}$ that diagonalizes the transfer matrix.

5.6 Consider a cyclic Ising chain with N spins at zero field. Verify the following equations for the partition function and thermodynamic properties:

(i) Partition function

$$Q = 2^N(\cosh^N K + \sinh^N K),$$

where $K \equiv \beta\epsilon$.

(ii) Helmholtz energy

$$\beta F = -N \ln 2 - \ln[\cosh^N K + \sinh^N K].$$

(iii) Internal energy

$$\frac{U}{N\epsilon} = -\tanh K\left(\frac{1+\tanh^{N-2} K}{1+\tanh^N K}\right).$$

(iv) Heat capacity

$$\frac{C_V}{Nk_B} = \frac{K^2}{\cosh^2 K}\frac{1+(N-1)(\tanh^{N-2} K - \tanh^N K) - \tanh^{2N-2} K}{(1+\tanh^N K)^2}.$$

(v) The average squared magnetization

$$\overline{M^2} \equiv \left\langle \left(\sum_{i=1}^{N} s_i\right)^2 \right\rangle = Ne^{2K}\left(\frac{1-\tanh^N K}{1+\tanh^N K}\right). \tag{E}$$

5.7 A cyclic Ising chain consisting of N spins displays a paramagnetic to ferromagnetic transition at a finite temperature. The transition is not characterized by the mean magnetization per spin m, which is always zero due to the symmetry, but by a drastic increase of the average squared magnetization $\overline{M^2}$.

(i) Show how $\chi^2 \equiv \overline{M^2}/N^2$ varies with temperature for $N = 10, 20$, and 100;

(ii) The transition temperature (viz., the Curie temperature) can be identified from the second-order derivative of the average squared magnetization

$$\frac{\partial^2 \overline{M^2}}{\partial T^2} = 0.$$

Show how the transition temperature varies with the number of spins and compare your numerical results with the prediction of Eq. (5.20), $k_B T/\epsilon = 2 / \ln(N-1)$.

5.8 The partition function for an open-ended Ising chain with N spins was reported by Altenberger and Dahler[88]:

$$Q(K,H) = A_+\lambda_+^{N-1} + A_-\lambda_-^{N-1} \tag{H}$$

where

$$\lambda_\pm = e^K\left[\cosh H \pm \sqrt{\sinh^2 H + e^{-4K}}\right]$$

$$A_\pm = \cosh H \pm [\sinh^2 H + e^{-2K}][\sinh^2 H + e^{-4K}]^{-1/2},$$

with $K \equiv \beta\epsilon$ and $H \equiv \beta h$. From $Q(K,H)$, we can derive the average squared magnetization

$$\overline{M^2} \equiv \left\langle \left(\sum_{i=1}^{N} s_i \right)^2 \right\rangle = Ne^{2K}[1 + \sinh(2K)(\tanh^N K - 1)/N].$$

Show how the paramagnetic to ferromagnetic transition temperature varies with the number of spins and compare the numerical results with the prediction of Eq. (5.20), $k_B T/\epsilon = 2 / \ln(N-1)$.

5.9 For an Ising chain consisting of N spins at zero field, the partition function may be expressed in terms of bonding between neighboring spins

$$Q = 2\prod_{i=1}^{N-1}\sum_{b_i=\pm 1} \exp\left(\beta\epsilon\sum_{i=1}^{N-1} b_i\right),$$

where $b_i = s_i s_{i+1}$, and a factor of two accounts for two possible orientations for one of the end spins. Show that

(i) $< b_k > = \tanh(\beta\epsilon)$;

(ii) $< s_k s_{k+n} > = \tanh^n(\beta\epsilon)$.

5.10 Cooperative phenomena are commonplace in chemical and biological systems. The essential concept may be elucidated with a two-state model. For the simplest case, imagine that ligand molecules of finite concentrations are able to bind with a substrate (e.g., a protein or polymer) with two identical binding sites at infinite dilution. When each binding site is occupied by a ligand molecule, the free energy of the system is changed by ΔG_b. Cooperativity is introduced by an additional energy w when both sites are occupied.

---

88 Altenberger A. and Dahler J.,"One-dimensional Ising model for spin systems of finite size", *Adv. Chem. Phys.* 112, 337–356 (2007).

(i) Suppose that the chemical reaction for ligand binding of each site may be expressed as

$$\mathrm{L} + \mathrm{S} \rightarrow \mathrm{LS},$$

where L and S represent the ligand and the substrate, respectively. Show that, at infinite dilution of the substrate, ΔG_b can be calculated from the intrinsic binding constant K_b and ligand activity [L]

$$\beta \Delta G_b = -\ln K_b - \ln[\mathrm{L}].$$

(ii) Derive the following expression for the average occupation number of each binding site

$$\alpha = \frac{K_b[\mathrm{L}] + K_b^2[\mathrm{L}]^2 e^{-\beta w}}{1 + 2K_b[\mathrm{L}] + K_b^2[\mathrm{L}]^2 e^{-\beta w}}.$$

(iii) Show that, when $w = 0$, the average occupation number per site is given by the Langmuir binding equation:

$$\alpha = \frac{K_b[\mathrm{L}]}{1 + K_b[\mathrm{L}]}.$$

(iv) Show that the probability of double site occupation is given by

$$\alpha_{12} = \frac{K_b^2[\mathrm{L}]^2 e^{-\beta w}}{1 + 2K_b[\mathrm{L}] + K_b^2[\mathrm{L}]^2 e^{-\beta w}}.$$

(v) Supposed that the intrinsic binding constant for each site is $K_b = 10$, plot the binding curve in terms of the average number of ligand molecules binding with each site as a function of the ligand activity for $\beta w = 0, 1.0, -1.0$.

5.11 The 1D-Ising model provides a systematic way to represent cooperative phenomena such as ligand binding.

(i) Show that ligand binding with a substrate discussed in Problem 5.10 can be represented by the Ising model.

(ii) Evaluate the average occupation number per site as a function of the ligand activity if the substrate is able to bind a large number of ligand molecules.

(iii) Compare the binding curve for substrates with two and that with a large number of sites using $K_b = 10$ and $\beta w = -3$ and discuss the implications of cooperativity.

5.12 The Ising model has been used to analyze the extent of genetic markers shared by siblings.[89] In the so-called sib pair analysis, each data point for an allele (viz., any of the alternative forms of a gene) being shared and not shared by a sib pair may be represented by $s = +1$ or -1, with the linkage between adjacent genetic markers represented by nearest-neighbor interactions, and the effect of a disease gene by an external energy. For a chromosome with N markers, the total energy is thus given by

$$E_\nu = -\sum_{i=1}^{N-1} \epsilon_i s_i s_{i+1} - \sum_{i}^{N} h_i s_i,$$

where $\nu = \{s_i\}_{i=1,2,\ldots,N}$ stands for an identity-by-descent (IBD) sharing configuration (viz., microstate), "genetic parameters" ϵ_i and h_i are dimensionless, equivalent to the coupling

89 Majewsk J. et al.,"The Ising model in physics and statistical genetics", *Am. J. Hum Genet.* 69(4), 853–862 (2001).

and external energies in a regular Ising model divided by the absolute temperature and the Boltzmann constant. The genetic parameters can be obtained by fitting the Ising model with experimental data for the statistics of gene expression.

(i) Verify that the following iterative relations may be used to evaluate the partition function

$$Q_N = Q_N^+ + Q_N^-,$$

where $Q_N^+ = e^{h_N}[Q_{N-1}^+ e^{\epsilon_{N-1}} + Q_{N-1}^- e^{-\epsilon_{N-1}}]$ is the value of the partition function for a chromosome with N markers, given that the last (Nth) position is positive ($s_N = +1$), and $Q_N^- = e^{-h_N}[Q_{N-1}^+ e^{-\epsilon_{N-1}} + Q_{N-1}^- e^{\epsilon_{N-1}}]$ is the value of the partition function for the same chromosome, given that the last position is negative ($s_N = -1$).

(ii) The genetic parameters, ϵ_i and h_i, can be estimated by using the maximum likelihood method, i.e., by maximizing the probability of observing a particular dataset

$$\mathrm{L} = \prod_{i=1}^{N} p_i,$$

where p_i stands for the probability of observing IBD at marker i. Find an expression for the probability of observing IBD at marker i using the Ising model.

(iii) Under the simplest hypothesis that there is only one marker k affiliated with a disease such that the influence parameter $h_k \neq 0$ and $h_{i\neq k} = 0$. Find an expression for the LOD (logarithm of the odds) score

$$\mathrm{LOD} = \log(L/L_0),$$

where L_0 stands for the likelihood with $h_i = 0$ for all markers.

5.13 An extended one-dimensional Ising model has been utilized to quantify the folding cooperativity of *de novo*-designed helical repeat proteins (DHRs).[90] In this model, the conformations of individual repeats in a modular helix–loop–helix–loop protein are represented as either folded or unfolded, resulting in $2n$ possible conformations for an n-repeat superhelical array. For each conformation, the energy is determined by the intrinsic folding energy of each repeat (ΔG_i) as well as the coupling ("interfacial") free energies (ϵ) between adjacent folded repeats. Because the sequences of the N- and C-terminal capping repeats differ from the sequence of central repeats, three intrinsic energies (ΔG_N, ΔG_R, and ΔG_C) were used to represent the intrinsic folding energies of folded repeats at N-terminal, center, and C-terminal, respectively.

(i) Show that the canonical partition function for this system is given by

$$Q = \begin{pmatrix} 0 & 1 \end{pmatrix} \begin{pmatrix} \kappa_N\tau & 1 \\ \kappa_N & 1 \end{pmatrix} \begin{pmatrix} \kappa_R\tau & 1 \\ \kappa_R & 1 \end{pmatrix}^{n-2} \begin{pmatrix} \kappa_C\tau & 1 \\ \kappa_C & 1 \end{pmatrix} \begin{pmatrix} 1 \\ 1 \end{pmatrix},$$

where $\kappa_N = \exp(-\beta\Delta G_N)$, $\kappa_R = \exp(-\beta\Delta G_R)$, $\kappa_C = \exp(-\beta\Delta G_C)$, and $\tau = \exp(-\beta\epsilon)$.

(ii) What is the fraction of folded repeats in the superhelical array?

5.14 The one-dimensional Ising model has been applied to understand the cooperativity of histone H1 in binding with a chromatin.[91] The cooperative process results in a long-range

---

90 Barrick et al., "Extreme stability in de novo-designed repeat arrays is determined by unusually stable short-range interactions", *PNAS*, 115(29), 7539–7544 (2018).

91 Ishii H., "A statistical-mechanical model for regulation of long-range chromatin structure and gene expression", *J. Theor. Biol.* 203, 215–228 (2000).

correlation of histone H1 distribution along the nucleosome array, which explains several aspects of gene regulation and chromatin structure. In this mode, the chromatin structure is represented by a one-dimensional lattice such that each site represents a nucleosome. Each microstate of the model system (viz., chromatin conformation) is defined by the status of histone binding, i.e., a set of numbers $\{n_1, n_2, \dots, n_N\}$ where $n_i = 0$ or 1 stands for the occupation number for nucleosome i, and N is the total number of binding sites. Accordingly, the system energy is given by

$$E_\nu = -a\sum_i n_i n_{i+1} - (b+\mu)\sum n_i,$$

where a is the free energy associated with the cooperative interaction of histone molecules, b stands for the binding energy, and μ is the chemical potential of histone H1.

(i) Verify that the canonical partition function can be written as

$$Q = \lambda_1^N + \lambda_2^N,$$

where $\lambda_{1,2} = [AB + 1 \pm \sqrt{(AB-1)^2 + 4B}]/2$ with $A = \exp(a/k_BT)$ and $B = \exp[(b+\mu)/k_BT]$.

(ii) What is the probability of nucleosome i occupied by a histone H1 molecule?

(iii) Ishii found that the joint probability for the simultaneous occupancy of nucleosomes i and j is given by

$$< n_i n_j > = \cos^4\phi + e^{|j-i|/\xi_c}\cos^2\phi\sin^2\phi,$$

where $\cot\phi = [AB - 1 + \sqrt{(AB-1)^2 + 4B}]/(2\sqrt{B})$, and $\xi_c = 1/\ln(\lambda_1/\lambda_2) > 0$ stands for the correlation length. Show that the joint probability that nucleosome j is occupied while the nucleosome i is not occupied is given by

$$\frac{< (1-n_i)n_j >}{< 1 - n_i >} = < n > (1 - e^{|j-i|/\xi_c}).$$

(iv) Plot ξ_c as a function of $b + \mu$ for $T = 300$ K and $a \approx 7.5$ kcal/mol, and discuss how gene regulatory DNA sequences can affect the chromatin structure over a long distance.

5.15 For the Ising model on the square lattice, the critical point may be identified from the Kramers–Wannier duality, i.e., there is a one-to-one correspondence between terms in the high- and low-temperature series expansions of the partition function. According to the duality relation, the critical temperate satisfies $\tanh(K_C) = e^{-2K_C}$, where $K_C = \beta_C\epsilon$. Show

(i) $\sinh(2K_C) = 1$;

(ii) $K_C = [\ln(1+\sqrt{2})]/2$.

5.16 The Weiss molecular field theory may be alternatively formulated by neglecting the correlated spin fluctuations, i.e., by assuming $(s_i - m)(s_j - m) \approx 0$.

(i) Show that the mean-field energy is given by

$$E_\nu^0 = \sum_i \left[-(h + Zm\epsilon)s_i + Z\epsilon m^2/2\right];$$

(ii) Derive a self-consistent equation for determining the average magnetization;

(iii) Compare the mean-field Helmholtz energy with those obtained from the Gibbs–Bogoliubov variational principle and the Bragg–Williams theory.

5.17 Show that the free energy for the Ising model derived from the Bragg–Williams theory

$$\beta F/N = \frac{1+m}{2}\ln\frac{1+m}{2} + \frac{1-m}{2}\ln\frac{1-m}{2} - Z\beta\epsilon m^2/2 - m\beta h,$$

is identical to that from the Gibbs–Bogoliubov variational principle

$$\beta F/N = -\ln\{2\cosh[\beta(h+Zm)]\} + Z\beta\epsilon m^2/2.$$

5.18 The Gibbs–Bogoliubov variational principle may be applied to a generalized Ising model where the external field is dependent on the identity of spins, i.e., the total energy for each microstate is given by

$$E_\nu = \sum_i -h_i s_i - \frac{\epsilon}{2}\sum_{i,j}{}' s_i s_j,$$

where $'$ means that spins i and j are nearest neighbors. Show that the effective mean field for each spin is given by

$$h_i^0 = h_i + \epsilon\sum_j{}' m_j.$$

5.19 Kornyshev[92] proposed a lattice-gas model to describe the capacitance for the electric double layer (EDL) of ionic liquids and molten salts in the presence of an electric field (e.g., introduced by an electrode). Similar to the Bragg–Williams theory, the lattice-gas mode starts with a mean-field free energy for a system of N_+ monovalent cations and N_- monovalent anions placed on a lattice

$$F = e\psi(N_+ - N_-) + B_+N_+^2 + B_-N_-^2 + B_\pm N_+N_- - k_BT\ln W,$$

where e stands for unit charge, ψ is the electric potential, B_+, B_-, and $B_\pm$ are binary energy parameters, W is the number of ways to place cations and anions on the lattice, and N is the total number of lattice sites. With the assumption that cations and anions are randomly mixed with each other and with the unoccupied lattice sites, W is given by

$$W = \frac{N!}{(N-N_+-N_-)!N_+!N_-!}.$$

(i) Supposed that the chemical potentials of cations and anions are given by μ_+ and μ_-, what is the grand potential of the lattice system?

(ii) Assuming that $B_+ = B_- = B_\pm = 0$, show that the ion concentrations can be related to their corresponding values in the bulk (viz., when $\psi = 0$)
(Hint: by minimizing the grand potential)

$$c_+ = c_0\frac{\exp(-\beta e\psi)}{1-\gamma+\gamma\cosh(\beta e\psi)}$$
$$c_- = c_0\frac{\exp(\beta e\psi)}{1-\gamma+\gamma\cosh(\beta e\psi)},$$

where $c_+ = N_+/v$, $c_- = N_-/v$ are ion concentrations for the lattice system, and v is the lattice volume. $\gamma \equiv 2N_0/N$ is called the lattice-saturation parameter, which provides a measure of the bulk ion concentration. Within the lattice model, N_0 corresponds to the number of cations or anions in the bulk (viz., when $\psi = 0$), and $c_0 = N_0/v$ is the concentration of cations (or anions).

92 Kornyshev A., "Double-layer in ionic liquids: paradigm change?", *J. Phys. Chem. B.*, 111, 5545–5557 (2007).

(iii) Consider a planar electrode placed at position $x = 0$ with surface potential ψ_s. The electric potential disappears in the bulk, i.e., $\psi = 0$ at $x = \infty$. Assuming that the above relation between ion concentration and electric potential is valid at any position in space (viz., local density approximation), find an expression for the electric potential as a function of the perpendicular distance x from the surface based on the Poisson equation

$$\frac{\partial^2 \psi}{\partial x^2} = -\frac{4\pi \rho_e}{D},$$

where $D \equiv 4\pi\epsilon_0\epsilon_D$, $\epsilon_0 = 8.854 \times 10^{-12}$ F/m is the vacuum permittivity, ϵ_D stands for the dielectric constant of the ionic fluid, and ρ_e is the local electrical charge density

$$\rho_e = e(c_+ - c_-) = -\frac{2ec_0 \sinh(\beta e\psi)}{1 + 2\gamma \sinh^2(\beta e\psi/2)}.$$

(iv) The differential capacitance per unit surface area of the electrode is defined as

$$C_d \equiv \frac{\partial Q}{\partial \psi_s},$$

where Q is the surface charge density. Derive an expression for capacitance C_d as a function of the surface voltage.

(v) Plot the capacitance versus the electric potential for several values of γ and discuss how the differential capacitance varies with the ion concentration in the bulk.

5.20 Small angle x-ray scattering (SAXS) and small-angle neutron scattering (SANS) are experimental techniques commonly used for quantifying the microscopic structure of matter at atomic dimensions. The intensity and angular distribution of the scattered radiation depend on the fluctuations of electron density in the case of x-rays or nuclear density in the case of neutrons. In both cases, the scattered intensity I can be represented by the Fourier transform of the density–density correlation function $\chi(r)$

$$I(\mathbf{k}) = \langle \eta^2 \rangle \int d\mathbf{r}\chi(r)e^{i\mathbf{k}\cdot\mathbf{r}}, \tag{M}$$

where $\langle \eta^2 \rangle$ is the mean square fluctuation of the scattering density, and the wave vector $\mathbf{q}$ depends on the scattering angle θ and the wave length of the incident beam.

(i) Based on the density–density correlation function from the Teubner–Strey theory of microemulsions

$$\chi(r) = \frac{e^{-r/\xi}}{\kappa r} \sin(\kappa r),$$

show that the scattering intensity can be expressed as

$$I(\mathbf{k}) = \frac{8\pi \langle \eta^2 \rangle c_4}{\xi(a_2 + c_2k^2 + c_4k^4)}, \tag{N}$$

where $k = |\mathbf{k}|$, and

$$\xi = \left[\frac{1}{2}\left(\frac{a_2}{c_4}\right) + \frac{1}{4}\frac{c_2}{c_4}\right]^{-1/2}, \qquad 1/\kappa = \left[\frac{1}{2}\left(\frac{a_2}{c_4}\right) - \frac{1}{4}\frac{c_2}{c_4}\right]^{-1/2}.$$

(ii) Consider an inhomogeneous medium consisting of two types of scattering domains 1 and 2 (e.g., oil and water in a microemulsion). Assume that the fluctuation of the scattering density can be represented by the random distribution of these two scattering regions on a lattice, show that the mean square fluctuation is given by

$$\langle \eta^2 \rangle = \Phi_1 \Phi_2 \delta^2,$$

where $\Phi_1 = N_1/N$ and $\Phi_2 = N_2/N$ are the fractions of lattice sites occupied by 1 and 2, respectively, $N = N_1 + N_2$ is the total number of lattice sites, and δ is the difference in the amplitude of scatterings from the two scattering regions.
Hint: $\langle \eta^2 \rangle = \frac{\delta^2}{N} \sum_{i,j=1}^{N} [\langle n_i n_j \rangle - \langle n_i \rangle \langle n_j \rangle]$ where $n_i = 1$ is the lattice site is occupied by 1 and $n_i = 0$ otherwise.

5.21 The Maier–Saupe theory for predicting phase transition in a nematic liquid crystal is based on the mean-field approximation for van der Waals interaction between rod-like molecules. Specifically, the pair potential between two anisotropic molecules averaged over a random distribution for the center of mass is given by

$$u_{12} = -\epsilon \left(\frac{3}{2} \cos^2 \theta_{12} - \frac{1}{2} \right),$$

where $\epsilon > 0$ is an energy parameter, θ_{12} is the angle between the long axes of the two molecules. In the mean-field approximation, the two-body potential is replaced by an effective one-body field

$$u_1 = -\epsilon S \left(\frac{3}{2} \cos^2 \theta - \frac{1}{2} \right),$$

where θ is the angle between the long axis of a molecule and the z-axis (i.e., the preferred direction of molecular alignment), and S is the orientational order parameter determined self-consistently from the Boltzmann equation

$$S = < \frac{3}{2} \cos^2 \theta - \frac{1}{2} > = \frac{\int_0^\pi \sin \theta d\theta (\frac{3}{2} \cos^2 \theta - \frac{1}{2}) e^{-\beta u_1}}{\int_0^\pi \sin \theta d\theta e^{-\beta u_1}}. \quad \text{(P)}$$

(i) Show that $S = 0$, which corresponds to the system in an isotropic phase, satisfies the self-consistent equation, Eq. (P);

(ii) Derive the reduced Helmholtz energy of the nematic phase according to the mean-field approximation:

$$\beta F/N = -\frac{1}{N} \ln Q^{IG} - \ln 2 - \frac{3}{2} \ln(\pi) + \frac{\alpha S}{3} - \ln \frac{\text{erfi}(\sqrt{\alpha S})}{\sqrt{\alpha S}}.$$

where $\alpha = 3\beta\epsilon/2$, Q^{IG} stands for the partition function for a system of noninteracting particles, and erfi(x) is the imaginary error function.

(iii) Find nontrivial solution of S numerically and plot it as a function of the reduced temperature $k_B T/\epsilon$;

(iv) Plot the reduced Helmholtz energy per molecule in the nematic phase relative to that in the isotropic phase as a function of reduced temperature $k_B T/\epsilon$;

(v) What are the critical temperature and the orientational order parameter at the isotropic–nematic phase transition? Is it a first-order or second-order phase transition at the critical point?

5.22 When the free energy density (f) of a system does not satisfy the invariance condition $f(m) = f(-m)$ for the order parameter m, a first-order phase transition can be described with the

Landau theory by using a fourth-order polynomial that contains both even and odd terms

$$f(m) = f_0 + a_2 m^2/2 + a_3 m^3/3 + a_4 m^4/4,$$

where f_0 is the free energy density for the disordered state ($m = 0$), $a_2 = a_2^0(T - T_0)$, $a_3 < 0$, and $a_4 > 0$. With parameter $a_2^0 > 0$, T_0 controls the sign of a_2 in response to temperature changes.

(i) Show that the disordered phase is always stable when $T > T_A \equiv T_0 + a_3^2/(4a_2^0 a_4)$.

(ii) A metastable phase with $m \neq 0$ emerges when the temperature is in the range of $T_C \equiv T_0 + 2a_3^2/(9a_0 a_4) < T < T_A$.

(iii) The system undergoes a first-order phase transition at T_C, i.e., the disordered phase is in equilibrium with an ordered phase.

(iv) When $T_0 < T < T_C$, the ordered phase is stable while the disordered phase is metastable.

(v) When $T < T_0$, only the ordered phase is stable, and the disordered phase is unstable.

(vi) Plot the Landau free energy f as a function of m at $T = T_0, T_C$ and T_A and prepare a phase diagram of the system with parameters $f_0 = 0$ and $T_0 = a_2^0 = -a_3 = a_4 = 1$. Label the transition temperatures.

5.23 The interconnections between different critical exponents were established by B. Widom.[93] The assumption was that, near the critical point, the Gibbs energy per particle, $g \equiv G/N$, includes two contributions, one is referred to as a regular part g_r that has an analytical form, and the other is called a scaling part g_c that has the singular behavior. For a one-component system, the free energy density can be written as

$$g(T, \xi) = g_r(T, \xi) + g_s(\Delta T, \Delta \xi),$$

where T is temperature, ξ (e.g., pressure) is an intensive variable of the system that is conjugated with a dynamic variable x (e.g., particle density or magnetization), and Δ stands for the deviation from the critical condition. The critical behavior implies that the scaling part of the free energy is a generalized homogeneous function, i.e., it satisfies

$$g_s(\lambda^p \Delta T, \lambda^q \Delta \xi) = \lambda g_s(\Delta T, \Delta \xi), \tag{U}$$

where $\lambda > 0$ denotes the scaling factor, p and q are parameters.

(i) Using the thermodynamic relation $x = (\partial g/\partial \xi)_T$, show

$$\beta = \frac{1-q}{p},$$

where β is the exponent describing the difference between the two coexisting phases in dynamic variable x, i.e., $\Delta x \sim \Delta T^{\beta}$.

(ii) For the scaling relation, $\Delta x \sim \Delta \xi^{1/\delta}$, exponent δ satisfies

$$\delta = \frac{q}{1-q}.$$

(iii) According to the linear response theory, the derivative of x with respect to ξ (viz., susceptibility) is related to the fluctuation

$$< \delta x^2 > = -\left(\frac{\partial x}{\partial \xi}\right)_T.$$

---

93 Widom B., "Equation of state in neighborhood of critical point". *J. Chem. Phys.* 43, 3898–3905 (1965).

Show that the critical exponent for susceptibility is given by

$$\gamma = \frac{2q-1}{p}.$$

(iv) Based on the thermodynamic relation between heat capacity and reduced free energy

$$C_\xi/N = -T\left(\frac{\partial^2 g}{\partial T^2}\right)_\xi,$$

show that the critical exponent for heat capacity, $C_\xi \sim (1/\Delta T)^\alpha$, is given by

$$\alpha = 2 - 1/p.$$

6

Monte Carlo Simulation

Since its inception in the early 1950s, molecular simulation has revolutionized the field of statistical thermodynamics by providing a powerful tool to investigate and understand both the macroscopic properties and microscopic behavior of complex fluids, materials, and biological systems. With the rapid advancement of computational power and algorithms, molecular simulation has gained increasing significance as a valuable alternative to experimental measurements and analytical theories. It enables the determination of both the microscopic structure and thermodynamic properties for virtually any thermodynamic system. While experimental measurements often face limitations in terms of cost, time, or the ability to access extreme thermodynamic conditions, molecular simulation offers a virtual laboratory that enables probing and manipulating any system under precisely controlled conditions.

Two complementary approaches are commonly employed in molecular simulation, i.e., by tracking the dynamic motions of individual particles or by statistical sampling of microstates. The first approach is called molecular dynamics (MD) simulation, which traces the trajectories of individual particles using the equations of motion derived from quantum or classical mechanics. As described in Chapter 2, MD simulation provides insights into the dynamic behavior and temporal evolution of the system. The second approach is Monte Carlo (MC)[1] simulation, which utilizes stochastic methods[2] to sample the microstates of a thermodynamic system and analyze the resulting ensemble averages. MC simulation focuses on exploring thermodynamic properties and equilibrium behavior by proposing random changes in configuration and accepting or rejecting them based on certain predefined criteria to control the microstate distribution.

Both MD and MC simulations have their merits and limitations, and their selection depends on the specific properties of the system being studied. While MD simulation provides a versatile tool for predicting both thermodynamic (e.g., equation of state, heat capacity, and free energy) and transport properties (e.g., diffusivity, viscosity, and thermal conductivity), MC simulation has its own advantages; it is particularly valuable in studying the equilibrium structure and phase behavior of complex thermodynamic systems, such as polymers, polymer gels, composites, and colloidal dispersions. In MC simulation, microstates are sampled according to the equilibrium distribution. Compared to MD simulation, MC simulation often has better numerical efficiency due to the absence of force calculations. Moreover, MC simulation can be tailored to sample microstates through unphysical evolutions, thereby enhancing its efficiency in sampling a diverse range of microstates.

1 Monte Carlo is the capital city of Monaco, a tiny country near the French-Italian border, famous for its gambling casinos whose operation is based on the statistics of random numbers.
2 Stochastic means non-deterministic, characterized by randomness or probability.

Fundamentals and Practice in Statistical Thermodynamics, First Edition. Jianzhong Wu and John M. Prausnitz.

Companion website: www.wiley.com/go/jianzhongwu/fundamentalsandpractice

This chapter serves as an introduction to the fundamental concepts of MC simulation, exploring its applications in obtaining structural and thermodynamic properties. Our discussion primarily focuses on relatively simple systems, where simulation methods can be easily implemented and compared against analytical results. Several MATLAB codes are available from the companion website of this text for practice and for understanding the essential ingredients of MC simulation programs. For advanced simulation techniques and the practical use of professional software capable of handling complex systems, we refer readers to specialized texts.

6.1 Importance Sampling

Any nontrivial macroscopic system contains an enormous number of microstates. As a result, it is virtually impossible to enumerate *ALL* microstates and calculate ensemble averages. In MC simulation, the ensemble averages are evaluated in a statistical manner, i.e., by sampling the microstates according to an appropriate probability distribution. To a certain degree, the procedure is analogous to that practiced by a pollster who wants to predict the outcome of a political election. While the poll for a conventional election is often conducted by random sampling such that the opinion from every voter is equally weighted, MC simulation follows a biased sampling process, i.e., microstates making more contributions to the ensemble average are sampled with higher frequency. The biased sampling scheme is justified because, in a political poll for the election of a public officer, every voter who is polled is "worth" as much as any other polled voter. However, for a thermodynamic system, the ensemble average is evaluated with a non-uniform probability distribution of microstates, i.e., some microstates are "worth" more than others.

Statistical thermodynamics presumes that, in a macroscopic system, the probabilities of microstates are dictated by the system energy as well as the constraints defining the equilibrium state. Different thermodynamic constraints, such as fixed temperature, pressure or the chemical potentials of certain components, lead to different statistical ensembles and probability distributions. To ensure that the ensemble averages are properly evaluated, MC sampling is typically biased in favor of *statistically important* microstates, i.e., those states that make more contributions to the ensemble average. Such a biased sampling scheme is called *importance sampling*.

A close analogy of importance sampling is provided by the media prediction of the outcome of the U.S. Senate vote for an important political decision. A credible reporter will focus on the opinions from the members of the Senate instead of those from ordinary people in the street. In other words, the reporter, using the language of MC simulation, is practicing the method of *importance sampling*.

6.1.1 Microstate Probability

To make the ideas concrete, imagine that MC simulation is applied to a canonical system of volume V containing N particles at temperature T. The internal energy of the system U is given by an ensemble average of the total energy of microstates E_ν

$$U = \frac{1}{Q}\sum_{\nu} E_\nu e^{-\beta E_\nu} \tag{6.1}$$

where $Q = \sum_\nu e^{-\beta E_\nu}$ is the canonical partition function, $\beta = 1/(k_B T)$, and the summation applies to all accessible microstates. When the chemical nature of particles is specified, we can in principle calculate E_ν by using either quantum or classical mechanics that describe the kinetic and potential energies of the system at each microstate.

If a thermodynamic system has a relatively small number of microstates (e.g., a short Ising chain), we may evaluate the summation in Eq. (6.1) explicitly. However, as explained in the following, direct enumeration is prohibitive in time and computational cost for most realistic systems. To avoid direct enumeration, Eq. (6.1) can be alternatively written as

$$U = \sum_{v} p_v E_v \tag{6.2}$$

where $p_v = e^{-\beta E_v}/Q$ represents the probability of the canonical system in microstate v. Eq. (6.2) suggests that not all microstates contribute equally to the ensemble average: the microstates of high values of p_v contribute more than those with low values of p_v. For efficient evaluation of the ensemble average, we may sample those states with high values of p_v, i.e., the important microstates while ignoring those with negligible probability.

In comparison with Eq. (6.1), Eq. (6.2) has the advantage of not being explicitly dependent on the canonical partition function Q, which is unknown for most systems of practical concern. More important, Eq. (6.2) indicates that the internal energy is simply the average of the total energies of the system if the microstates are sampled according to the probability distribution p_v. Because calculation of this average does not require enumeration of all microstates in the system, the ensemble average can be *estimated* by generating a sufficiently large number of microstates that represent the probability distribution p_v. Using importance sampling, we do not calculate the total energy E_v for every microstate. Instead, we evaluate E_v for selected (important) microstates that make most significant contributions to the ensemble average.

6.1.2 Biased Sampling

The purpose of importance sampling is to keep the number of samples as small as possible without serious loss of accuracy. Toward that end, we desire generating a relatively small number of microstates with a prescribed microstate distribution p_v^0. Based on these microstates, the (canonical) ensemble average for an arbitrary dynamic quantity X can be calculated from

$$\langle X \rangle \approx \frac{\sum_{v=1}^{n} X_v e^{-\beta E_v}/p_v^0}{\sum_{v=1}^{n} e^{-\beta E_v}/p_v^0} \tag{6.3}$$

where n is the total number of *sampled microstates*. Each term in the summation is divided by p_v^0 to take into consideration that the samples are generated with the *biased* probability distribution p_v^0 and that identical microstates may be generated during the sampling. Contrary to the summation in Eq. (6.1) (that includes *ALL* accessible microstates), the summation in Eq. (6.3) is controlled by sampling microstates with *a priori* probability.

Eq. (6.3) is approximate because the summation does not encompass all microstates. However, with a judicious selection of the representative microstates, the error introduced by this approximation can be made arbitrarily small. If the samples (microstates) are randomly generated, p_v^0 is a constant. In that case, Eq. (6.3) reduces to

$$\langle X \rangle \approx \frac{\sum_{v=1}^{n} X_v e^{-\beta E_v}}{\sum_{v=1}^{n} e^{-\beta E_v}}. \tag{6.4}$$

Eq. (6.4) amounts to the direct evaluation of the ensemble average provided that the summation applies to all microstates, in other words, if the number of samples n becomes extremely large. One subtle difference between the summations in Eq. (6.4) and that in the definition of the partition function or ensemble average is that the microstates can be repeatedly sampled in MC simulation.

As to be discussed in later sections, through *importance sampling*, MC simulation is able to predict macroscopic properties using relatively modest computer resources, much less than what would be required if we used Eq. (6.1), rather than Eq. (6.3), to find property <X>.

6.1.3 An Illustrative Example

To facilitate a comparison between importance sampling and random sampling, we consider a simple system that has only 5 possible microstates; one of these has a reduced energy $\beta E_1 = -5$, significantly lower than those of other 4 microstates with a reduced energy $\beta E_\nu = -1$, $\nu = 2, 3, 4, 5$. Eq. (6.5) gives the canonical ensemble average of the reduced energy:

$$<\beta E> = \frac{\sum_{\nu=1}^{5} \beta E_\nu \exp(-\beta E_\nu)}{\sum_{\nu=1}^{5} \exp(-\beta E_\nu)} = \frac{-5 \cdot e^5 - e - e - e - e}{e^5 + e + e + e + e} = -4.727 \tag{6.5}$$

where $e = 2.71828\cdots$. Note that -4.727 is close to -5, the first microstate $[\beta E_1 = -5]$ makes a dominant contribution to $<\beta E>$. Accordingly, this state is referred to as an *important state*.

Now calculate the ensemble average following two sampling methods, one with random sampling and the other with importance sampling. This simple system has five microstates. Now, instead of sampling all five states, we take only three samples. In the first method, the samples are taken at random. As listed in Table 6.1, $p_\nu^0 = 1/5$ is the chance of choosing the low-energy state $(\beta E_\nu = -5)$, and $p_\nu^0 = 4/5$ is the chance of choosing a high-energy state $(\beta E_\nu = -1)$. The probabilities of having low and high-energy states remain the same for the second and third samples. When all three samples are taken, there are four possible energy distributions, as given in Table 6.2. In the first possible distribution, all three samples are in the high-energy state, and the probability for this case is $(4/5)^3 = 51.2\%$. In the second case, one sample has high energy, and two samples have low energy; the probability is $3 \cdot (1/5) \cdot (4/5)^2 = 38.4\%$. In the third case, two samples have high energy and one sample has low energy; the probability is $3 \cdot (1/5)^2 \cdot (4/5) = 9.6\%$. In the last case, all samples have low energy and the probability is $(1/5)^3 = 0.8\%$.

We now calculate the average reduced energy according to the approximate ensemble average

$$<\beta E> \approx \frac{\sum_{\nu=1}^{3} \beta E_\nu \exp(-\beta E_\nu)/p_\nu^0}{\sum_{\nu=1}^{3} \exp(-\beta E_\nu)/p_\nu^0}. \tag{6.6}$$

Eq. (6.6) is approximate because only 3 accessible microstates are used to evaluate the ensemble average. For random sampling, the sampling probability p_ν^0 is a constant (here 1/5),

Table 6.1 A comparison of random sampling and importance sampling.

Microstates ν	Reduced energy βE_ν	Probability by random sampling p_ν^0	Probability by importance sampling p_ν^0
1	−5	1/5	0.932
2	−1	1/5	0.017
3	−1	1/5	0.017
4	−1	1/5	0.017
5	−1	1/5	0.017

Table 6.2 There are four possible distributions (I, II, III, IV) of reduced energy when 3 samples are taken from 2 energy states.

		I	II	III	IV
	Sample 1	−1	−5	−5	−5
	Sample 2	−1	−1	−5	−5
	Sample 3	−1	−1	−1	−5
Random Sampling	Overall probability	$\left(\frac{4}{5}\right)^3$	$3\cdot\left(\frac{4}{5}\right)^2\cdot\left(\frac{1}{5}\right)$	$3\cdot\left(\frac{4}{5}\right)\cdot\left(\frac{1}{5}\right)^2$	$\left(\frac{1}{5}\right)^3$
		=51.2%	=38.4%	=9.6%	=0.8%
	Average energy	−1	−4.859	−4.964	−5
Importance Sampling	Overall probability	$\left(\frac{4e}{e^5+4e}\right)^3$	$3\cdot\left(\frac{4e}{e^5+4e}\right)^2\cdot\left(\frac{e^5}{e^5+4e}\right)$	$3\cdot\left(\frac{4e}{e^5+4e}\right)\cdot\left(\frac{e^5}{e^5+4e}\right)^2$	$\left[\frac{e^5}{(e^5+4e)}\right]^3$
		≈0.03%	≈1.3%	≈17.8%	≈80.9%
	Average energy	−1	−2.333	−3.667	−5

independent of v. Thus, the denominator in Eq. (6.6) can be understood as a normalization factor.[3] Using random sampling, Eq. (6.6) predicts that the average reduced energies $\langle\beta E\rangle$ for the above four different combinations of samples are, −1, −4.859, −4.964, and −5, respectively (Table 6.2). In other words, there is a 51.2% chance that the "simulation" provides a reduced internal energy ($\beta E_v = -1$) that is significantly different from the exact result. Because random sampling does not distinguish between low and high-energy microstates, the situation becomes even worse when the number of high-energy (unimportant) states is very large, as in a typical realistic system.

Now we follow the second method, where we use importance sampling. In choosing microstates, we give preference to those having lower energy according to the Boltzmann distribution, that is, $p_v^0 \propto e^{-\beta E_v}$. We again take three samples. The probability of each microstate sampled is also given in Table 6.1. In this case, whenever a sample is taken, there is $\frac{e^5}{e^5+4e} \approx 93.2\%$ chance that the sample is in the low-energy state, and a chance of $\frac{4e}{(e^5+4e)} \approx 6.8\%$ that the sample is in a high-energy state. The simulation gives an estimated reduced energy with $0.068^3 \approx 0.03\%$ chance that $\langle\beta E\rangle$ is equal to −1; $3\times0.932\times0.068^2 \approx 1.3\%$ chance that it is equal to −2.333; $3\times0.932^2\times0.068 \approx 17.8\%$ chance that it is equal to −3.667; $0.932^3 \approx 80.9\%$ chance that it is equal to −5 (Table 6.2). Compared with random sampling, where we have a 51.2% chance of a poor estimate, $<\beta E> = -1$, the importance sampling method is much more efficient because it gives an 80.9% chance to yield $<\beta E> = -5$, which is close to the exact result $<\beta E> = -4.727$. This example illustrates that importance sampling significantly improves the simulation efficiency.

3 The concept "normalization" factor is a necessary part of the concept "average." Suppose we measure the heights of ten trees and we want to calculate the average height. We add the ten measured heights and then divide by 10. In this calculation the normalization factor is 10.

6.1.4 Sampling with the Ensemble Probability

We see from the above example that the choice of sampling scheme has a dominating influence on the efficiency of MC simulation. In implementing importance sampling, one may use various schemes to generate microstates that are significant for evaluating an ensemble average. By far the most popular selection of p_ν^0 is to make it identical to the microstate probability distribution in the ensemble under investigation. For example, in a canonical ensemble, p_ν^0 is specified as the Boltzmann distribution

$$p_\nu^0 \propto e^{-\beta E_\nu}. \tag{6.7}$$

Substitution of Eq. (6.7) into Eq. (6.3) yields the familiar result

$$\langle X \rangle = \sum_{\nu=1}^{n} X_\nu / n. \tag{6.8}$$

Eq. (6.8) indicates that if the samples are generated according to the ensemble distribution, the ensemble average becomes a simple arithmetic average of X over the set of n microstates sampled by MC simulation.

Similar selection of p_ν^0 is applicable to other ensembles. In the grand canonical ensemble for identical classical particles where μ, T, V are fixed, the ensemble sampling probability is

$$p_\nu^0 \propto \frac{V^{N_\nu}}{\Lambda^{3N_\nu} N_\nu!} \exp(-\beta E_\nu + \beta\mu N_\nu). \tag{6.9}$$

Similarly, in the isobaric–isothermal ensemble of identical particles where N, T, P are fixed, the ensemble sampling probability is given by

$$p_\nu^0 \propto V_\nu^N \exp[-\beta(E_\nu + PV_\nu)]. \tag{6.10}$$

With the sampling scheme described by Eqs. (6.9) and (6.10), the ensemble averages $<X>$ in the grand canonical and in the isobaric–isothermal ensemble, respectively, are also calculated using the simple arithmetic average, Eq. (6.8).

Because we cannot include all accessible microstates in a simulation, we must select (i.e., generate or create) those microstates that we choose to include. The desired sequence of microstates in MC simulation is generated sequentially by a computer, following a randomly selected initial configuration. How can we create a sequence of microstates such that this sequence satisfies a pre-specified distribution, p_ν^0, as given in Eq. (6.7), (6.9), or (6.10)? To address this question, we will discuss in the next section some basic properties of the Markov chain stochastic process. This process provides the mathematical basis for creating a sequence of microstates that satisfy a pre-specified probability distribution.

6.1.5 Summary

This section explores the prediction of ensemble averages through MC sampling, with a specific focus on the significance of importance sampling. Importance sampling enables the efficient evaluation of ensemble averages, requiring a relatively small number of microstates. This concept is crucial in virtually all MC methods. By designing the importance sampling to align with the microstate distribution, thermodynamic properties can be calculated using straightforward arithmetic averages. In the next section, we will discuss how to generate a sequence of statistically significant microstates that adhere to a predetermined distribution of microstates.

6.2 Monte Carlo Sampling

As discussed in Section 6.1, MC simulation focuses on computing ensemble averages through the sampling of significant microstates, namely those that occur most frequently within the ensemble. This section addresses the implementation of importance sampling and explores the methodology behind it. Specifically, we introduce the use of a Markov process[4] for generating microstates that satisfy an *a priori* specified statistical distribution.

6.2.1 Monte Carlo Moves

In MC simulation, we sample the microstates through MC moves, i.e., generating microstates according to a certain probability distribution. For example, the microstates of an Ising chain can be generated by a sequential or random variation of the spin orientations. Similarly, the configuration of a molecular system can be sampled by changing the atomic positions.

Typically, a computer is utilized to perform MC calculations and to store the physical details of each sampled microstate. Such information includes, with molecular systems as an example, the positions of individual atoms as well as the total and pair potential energies. Because the memory of any computer is limited, and because an ensemble average encompasses a huge number of microstates, we can rarely save all numerical details of sampled microstates. Instead, we calculate and save the overall properties of a sequence of microstates generated by MC sampling. Here, "sampling" means first, creation of new microstates and second, justification of whether a newly generated microstate, which is known as the trial state, is accepted as a valid sample for calculating ensemble averages.

In the process of sampling microstates, the most recently generated microstate that contributes to the ensemble averages is called the *current state*. This current state is updated according to a probability when a new microstate is sampled. The transition from the current state to a new microstate is called a *MC step* or a *MC move*. Such transition requires two steps: the trial move and a test for acceptance. For systems containing only spherical particles, a trial microstate can be achieved by a random perturbation of the current state via moving the position of a particle. Additions or removals of particles, or volume scaling, can also be implemented to generate a trial microstate. The trial state is accepted or rejected as an additional valid microstate dictated by importance sampling, i.e., by an *a priori* specified sampling probability.

Figure 6.1 illustrates a few possible MC moves for updating the microstate of a molecular system. More complex algorithms are possible for polymer systems. At any given MC step, the computer stores the microscopic details of the current and newly generated microstates.

6.2.2 Monte Carlo Cycles

MC simulation typically consists of two stages: the first stage is called equilibrium, and the second stage is called production. In each stage, microstates are generated by a cyclic process updating the current state.

Figure 6.2 illustrates a schematic flow chart of a typical MC simulation program. During the *equilibration stage*, the microstates are sequentially generated following a stochastic process called

---

4 A Markov process, or Markov chain, is a stochastic process in which the outcome of the next step depends solely on the current state. For example, consider a shooting game where the results can be either a hit or a miss. We specify that the shooter hits a target with an 80% probability *if* the previous result is a hit, and 40% probability *if* the previous result is a miss. This shooting game is a Markov process because the outcome of each shot, hit or miss, depends only on the outcome of the previous shot. See Appendix 6.A for further details.

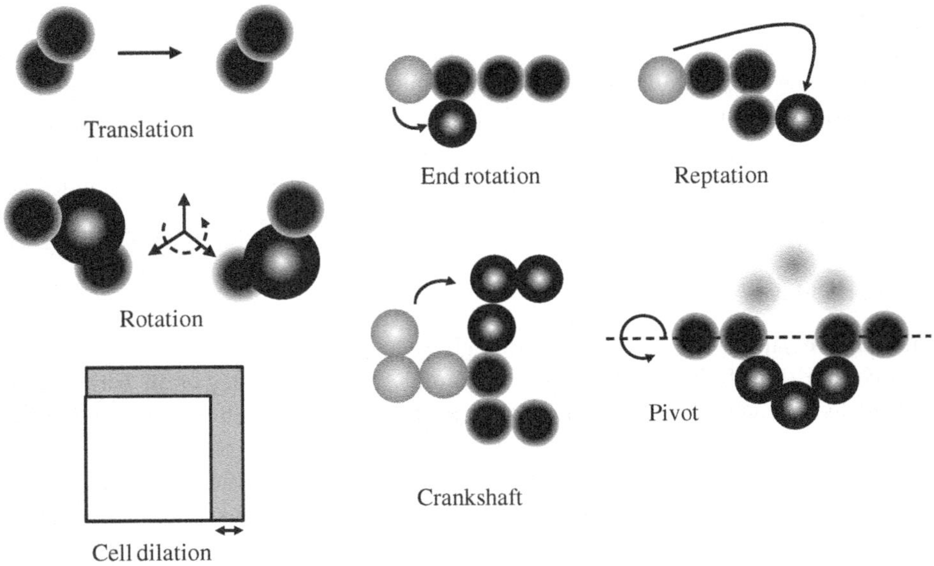

Figure 6.1 Possible MC moves to generate new microstates in a molecular system.

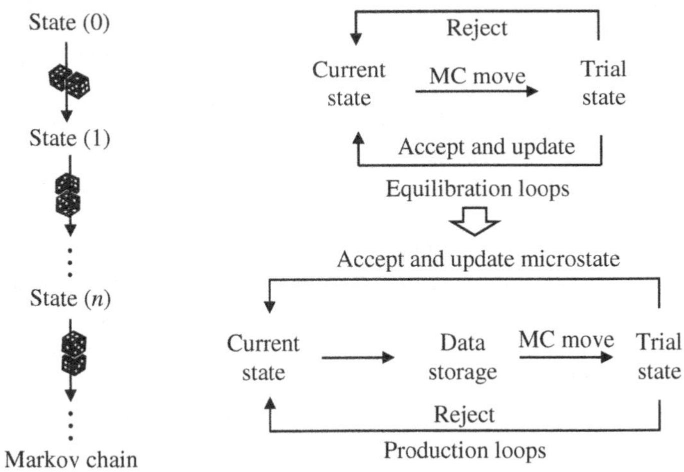

Figure 6.2 Schematic of a Markov process and the flow chart for Monte Carlo (MC) simulation. In the Markov process, the transition among different states is dictated by a probability.

a Markov chain. Equilibrium is established if newly generated microstates are uncorrelated with the initial state, i.e., the statistics of the new microstates are independent of the microstate used in the initial simulation. At this point, the successive microstates generated by the Markov process will follow an equilibrium distribution, independent of any bias that might be introduced in the initial state. As discussed in Appendix 6.A, the equilibration stage ensures that the generated microstates follow an *a priori*-specified distribution as required in importance sampling. During the production stage, the properties of interest are calculated for each newly generated microstate.

In terms of the computation, a key difference between the equilibration stage and the production stage arises from the calculation and storage of system properties. During the equilibration stage of a simulation, the focus is to relax the system and reach an equilibrium state. The properties of the system, such as energy, pressure, and temperature, are typically calculated and monitored during this stage to ensure that they have converged to their desired values. However, these properties are often not stored or recorded for further analysis. In other words, the equilibration stage is concerned only with those properties that are related to the sampling probability because the microstates sampled in this stage do not contribute to the ensemble averages. In contrast, during the production stage, the objective is to generate microstates that are statistically significant and accurately represent the equilibrium behavior of the system. In this stage, the properties of the system are not only calculated but also stored for subsequent analysis. These properties can include thermodynamic properties such as internal energy and pressure, correlation functions such as the radial distribution functions (RDF), or any other relevant quantities of interest. The number of microstates used to calculate the ensemble averages is related to the relaxation time, which dictates the number of MC steps in the production stage.

When calculating ensemble averages in a simulation, it is important to keep in mind that the samples should include not only the newly accepted microstates but also the original states when a trial state is rejected. Including the original states for the rejected MC moves ensures that the statistical distribution of microstates is properly represented and that accurate results can be obtained through continuing the sampling process. If a trial state is discarded, the creation of this state and any related calculations do not make any contribution to the simulation results. On the other hand, for a given number of calculations, the trial states determine the scope of sampling. Although the probability of acceptance may be high, the simulation can be inefficient if the samples represent only a small part of all possible configurations.

In general, the accuracy of the simulation results improves with an increased number of sampled microstates. By sampling a larger number of microstates, a more comprehensive exploration of the phase space can be achieved, resulting in more precise estimates of ensemble averages and thermodynamic properties. Therefore, to obtain reliable and accurate simulation results, it is always beneficial to increase the number of sampled microstates as much as possible within the limitations of computational resources and time constraints.

6.2.3 Balance Condition

Toward understanding the theoretical basis of MC simulation, it is imperative to consider the following questions:

1. First, how do the criteria for acceptance of trial microstates lead to a pre-specified importance sampling scheme? What is the relation between the update criterion and the distribution of sampled microstates? A related question is, how is the initial microstate selected? Does the choice of the initial microstate affect the final result?
2. Second, what is the most efficient way to create the trial states? How can we efficiently generate a large number of important samples?

By addressing these questions, one can develop a more comprehensive understanding of the theoretical foundations that underpin the practical implementation of MC simulation.

The answer to the first question is known exactly. The mathematical basis for addressing the first question is a stochastic theorem for the Markov chain process (Appendix 6.A). The stochastic process defines a probability for transition from one microstate to another which is called the update criterion. The Markov chain process ensures that, under specified conditions, the microstates sampled by MC simulation satisfy an *a priori* equilibrium distribution. The answer to the second question is more subtle. We will address this question in later sections.

To elucidate the creation of a set of microstates that satisfies a pre-specified probability distribution, consider an initial microstate that is generated according to an ARBITRARY probability distribution $\mathbf{p}^{(0)} = \left\{p_1^{(0)}, p_2^{(0)}, \cdots, p_i^{(0)}, \cdots\right\}^T$. Here, index i specifies a microstate, $p_i^{(0)}$ is the probability of microstate i is generated in initialization, and superscript T stands for vector transpose. The initial probability distribution is normalized, i.e.,

$$\sum_i p_i^{(0)} = 1 \tag{6.11}$$

where the summation applies to *ALL* microstates of the system.

Suppose that each new microstate is subsequently generated following a Markov chain process (i.e., the probability of selecting a new microstate depends only on the current state). After a single Markov step, a new microstate is created according to the transition matrix $\widehat{\mathbf{M}}$, which specifies the probability of transition from one microstate to another (see Appendix 6.A for the detail). At the first step, the microstate probability distribution becomes

$$\mathbf{p}^{(1)} = \widehat{\mathbf{M}} \cdot \mathbf{p}^{(0)}. \tag{6.12}$$

Similarly, the probability distribution after k Markov steps is

$$\mathbf{p}^{(k)} = \widehat{\mathbf{M}} \cdot \mathbf{p}^{(k-1)} = \widehat{\mathbf{M}}^k \cdot \mathbf{p}^{(0)}. \tag{6.13}$$

MC simulation requires that, as k becomes very large, the limiting probability distribution of the Markov chain process

$$\lim_{k\to\infty} \mathbf{p}^{(k)} = \widehat{\mathbf{M}} \cdot \mathbf{p} = \mathbf{p} \tag{6.14}$$

becomes identical to the desired probability (such as the one required by importance sampling). This limiting probability should be independent of the arbitrary initial probability distribution $\mathbf{p}^{(0)}$.

Eq. (6.14) holds true ONLY if the Markov chain process is *regular*, i.e., after a finite number of steps m, there is a non-zero probability of transition from any one microstate to any other and *vice versa*. In other words, the transition matrix of a *regular* Markov chain process satisfies

$$\widehat{\mathbf{M}}_{ij}^m > 0, \tag{6.15}$$

where m is an arbitrary integer, i and j represent two arbitrary microstates.

Eq. (6.14) also indicates that the transition matrix $\widehat{\mathbf{M}}$ of a regular Markov chain process has a unit eigenvalue,[5] and that the corresponding eigenvector $\mathbf{p}$ is the limiting probability distribution. Further, it implies that, after a sufficiently large number of steps, the probability distribution $\mathbf{p}$ converges to the eigenvector of the transition matrix, independent of the initial distribution. Therefore, we may select a transition matrix such that its eigenvector corresponds to the desired probability distribution of microstates. Since the limiting probability distribution (or the eigenvector of the transition matrix) is independent of the initial distribution $\mathbf{p}^{(0)}$, a typical MC simulation may start with an arbitrarily selected initial configuration. As the limiting probability distribution converges to the eigenvector of the transition matrix, we can devise a Markov chain process that yields a desired statistical distribution.

Eq. (6.14) suggests that once a limiting distribution is reached, it becomes invariant upon further Markov transitions. In other words, additional MC moves will not destroy the equilibrium distribution. In this case, the summation of all transition probabilities from an arbitrary microstate i to any

5 An eigenvalue λ of a matrix $\widehat{\mathbf{M}}$ is defined as $\widehat{\mathbf{M}} \cdot \mathbf{p} = \lambda\mathbf{p}$, where the vector $\mathbf{p}$ is the eigenvector corresponding to eigenvalue λ.

other microstate j must be exactly the same as the summation obtained from any other microstate to microstate i,

$$\sum_j \mathbf{p}_i \cdot \widehat{\mathbf{M}}_{ij} = \sum_j \mathbf{p}_j \cdot \widehat{\mathbf{M}}_{ji}. \tag{6.16}$$

Because $\sum_j \widehat{\mathbf{M}}_{ij} = 1$, we have an equivalent expression,

$$\mathbf{p}_i = \sum_j \mathbf{p}_j \cdot \widehat{\mathbf{M}}_{ji}. \tag{6.17}$$

Eq. (6.16) is called the *balance condition*; it provides a key relation between the desired probability distribution $\mathbf{p}$ for importance sampling and the transition matrix $\widehat{\mathbf{M}}$ of a Markov chain process. To ensure the convergence of statistical sampling, the balance condition is a prerequisite for all MC simulation algorithms.

In MC simulation, the balance condition can be intuitively understood as "material balance" in a steady-state process. Suppose that an equilibrium distribution of microstates is established by a Markov process. Upon further generation of microstates, the distribution of microstates remains unchanged because, as illustrated in Figure 6.3, the probability of any microstate becomes constant when the transitions from and to this microstate are balanced.

In most practical algorithms for MC simulation, the balance condition, Eq. (6.16), is satisfied by imposing a significantly stricter criterion

$$\mathbf{p}_i \cdot \widehat{\mathbf{M}}_{ij} = \mathbf{p}_j \cdot \widehat{\mathbf{M}}_{ji}. \tag{6.18}$$

Eq. (6.18) is sufficient (but not necessary) to fulfill the balance condition. This over-specified condition is called *detailed balance* or *microscopic reversibility*. It requires that the stochastic transitions between any two microstates are balanced, i.e., the "flux" or probability of transition from state i to state j is the same as that from state j to state i.

It should be noted that the (detailed) balance condition alone is insufficient to determine all elements of the transition matrix. Actually, there could be a number of transition matrices that satisfy the balance condition because for a system with n microstates, the matrix $\widehat{\mathbf{M}}$ has n^2 degrees of freedom. However, for a given probability distribution $\mathbf{p}$, there are only n balance equations, plus n normalization conditions. Even if we impose the regular condition $\widehat{\mathbf{M}}_{ij}^m > 0$, the definition of the transition matrix $\widehat{\mathbf{M}}$ from the limiting distribution $\mathbf{p}$ is not unique. As a consequence, there are multiple approaches to selecting a transition matrix that satisfies a specific probability distribution. Naturally, different choices of the transition matrix can result in varying levels of sampling efficiency.

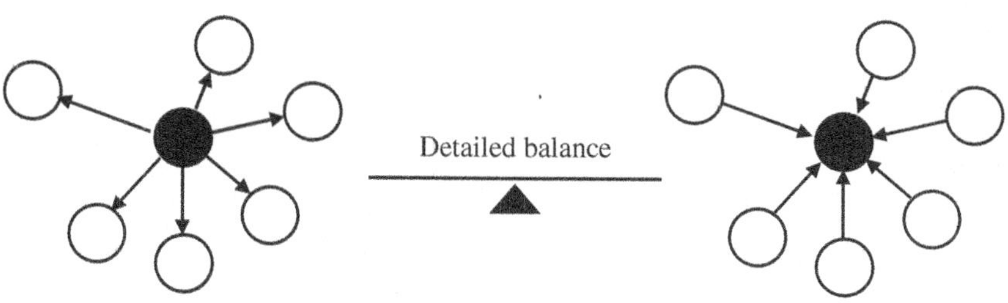

Figure 6.3 Balance condition guarantees that, at equilibrium, the probability distribution of microstates will not change due to transition among microstates created by the Markov process, i.e., the rate of transition out of any particular microstate is the same as the overall rate of transition from all other microstates into this microstate. In detailed balance, the probability of transition from state i to state j is the same as that from state j to state i.

6.2.4 Summary

The Markov chain process serves as a mathematical framework for generating microstates that adhere to a predetermined probability distribution. Importance sampling can be achieved by enforcing detailed balance or microscopic reversibility during the MC moves. In the upcoming section, we will delve into the Metropolis–Hastings algorithm, which offers a straightforward yet robust method for sampling microstates in different ensembles.

6.3 The Metropolis–Hastings Algorithm

In the previous section, we indicate that microstates satisfying an *a priori* probability distribution can be generated by following a stochastic process. For a regular Markov process, the limiting probability is determined by the principal eigenvector of the transition matrix. In general, there are many possible choices of the transition matrix that give the same limiting distribution of microstate probability. One simple yet elegant method for the selection of an efficient transition matrix was devised by Metropolis, Rosenbluth, Rosenbluth, Teller and Teller in 1953.[6] This method, often referred to as the Metropolis or MR^{2}T^2 method or the Metropolis–Hastings algorithm,[7] has been used extensively in practical applications of MC simulation.

6.3.1 Detailed Balance Condition

In the Metropolis method, each MC move (i.e., transition from microstate i to another microstate j) consists of two independent steps. As shown schematically in Figure 6.4, in the first step, we propose a new microstate j as a target microstate (to move to) according to a trial probability given by t_{ij}. In the second step, the proposed microstate j is accepted (or rejected) according to a probability given by a_{ij}. The overall probability of transition from microstate i to microstate j is thus given by

$$\widehat{\mathbf{M}}_{ij} = t_{ij}a_{ij}. \tag{6.19}$$

To ensure that the Markov process has a limiting distribution, a detailed balance condition is imposed on the transition matrix $\widehat{\mathbf{M}}$

$$p_i\widehat{\mathbf{M}}_{ij} = p_j\widehat{\mathbf{M}}_{ji}. \tag{6.20}$$

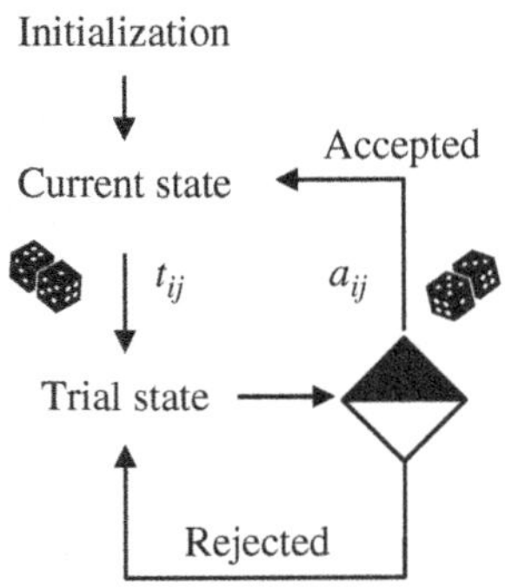

Figure 6.4 Schematic of the Metropolis–Hastings algorithm for MC moves. The probability of sampling of a new microstate is dictated by the trial probability t_{ij} multiplied by the acceptance probability a_{ij}.

6 Metropolis N., Rosenbluth A. W., Rosenbluth M. N., Teller A. H. and Teller E., "Equation of state calculations by fast computing machines", *J. Chem. Phys.* 21 (6), 1087–1092 (1953).

7 A more general procedure was proposed by Hastings W. K., "Monte Carlo sampling methods using Markov chains and their applications", *Biometrika* 57 (1), 97–109 (1970).

Eq. (6.20) indicates that the probability of direct transitions from microstate i to j is the same as that of reverse transitions, i.e., from microstate j to i. Substitution of Eq. (6.19) into (6.20) gives

$$p_i t_{ij} a_{ij} = p_j t_{ji} a_{ji}. \tag{6.21}$$

For simple systems (containing uniform spherical particles), a trial move is typically generated by a (small) random change in the configuration of the current state. In this case, the probability of the trial moves is symmetric, meaning that

$$t_{ij} = t_{ji}. \tag{6.22}$$

As a result, the trial probability does not show explicitly in the simulation because it cancels on both sides of Eq. (6.21). Hence, t_{ij} is often not specified explicitly. Substituting Eq. (6.22) into (6.21) leads to

$$\frac{a_{ij}}{a_{ji}} = \frac{p_j}{p_i}. \tag{6.23}$$

In most MC simulation programs, the probability distribution for microstates is *a priori* specified by the statistical ensemble for the particular thermodynamic system under consideration. To satisfy Eq. (6.23), we may define the acceptance probability a_{ij} as

$$a_{ij} = \min(1,\ \ p_j/p_i) \tag{6.24}$$

where 'min' stands for the minimum of 1 and p_j/p_i. If $p_j/p_i > 1$, $a_{ij} = 1$, i.e., the proposed microstate is accepted. If $p_j/p_i < 1$, the proposed microstate is accepted only with probability p_j/p_i. This simple selection of a_{ij} satisfies the *detailed balance* condition, Eq. (6.20). As a result, the probability distribution p_i must be the limiting distribution of the Markov process. In other words, the ensemble distribution can be recovered when the transition matrix is determined by the Metropolis–Hastings algorithm, i.e., Eqs. (6.22) and (6.24).

The Metropolis–Hastings algorithm plays a central role in early implementation of MC simulation. Remarkably, it suggests that importance sampling can be achieved by knowing only the *relative distribution* (p_j/p_i) of microstates. Eq. (6.24) determines maximal absolute acceptances a_{ij} compatible with given ratio p_j/p_i. This feature presents a key advantage because, in most cases, we do not know the partition function or the absolute values of microstate probabilities. However, for most thermodynamic systems of practical interest, the relative probability can be easily calculated from the properties of individual microstates. As long as the detailed balance condition is satisfied (sufficient but not necessary), the microstates sampled in the course of MC simulation satisfy the *a priori* specified ensemble distribution.

6.3.2 Acceptance Probability in Various Ensembles

We now present the acceptance probability a_{ij} in various ensembles when the trial probability t_{ij} is symmetric. In a canonical ensemble, where the number of particles N, temperature T, and volume V are specified, the microstate probability is proportional to the Boltzmann factor, $p_i \propto e^{-\beta E_i}$. According to the Metropolis method, Eq. (6.24), the acceptance probability a_{ij} is[8]

$$a_{ij} = \min(1,\ \ e^{-\beta \Delta E}) \tag{6.25}$$

where $\Delta E = E_j - E_i$ represents the change in the total energy when we move from microstate i to microstate j. If the trial microstate j has an energy lower than that for microstate i, ΔE is negative so

8 The pre-exponential factor in the Boltzmann distribution for p_i need not be specified because it cancels when we consider the ratio p_i/p_j.

that $a_{ij} = 1$ and microstate j is accepted. Otherwise, if ΔE is positive, a random number between 0 and 1 is generated by the computer; the trial state is accepted only if the random number is less than $e^{-\beta\Delta E}$. The extensive use of random numbers in this procedure (reminiscent of gambling), which involves the generation of random trial states and the evaluation of their acceptance or rejection, is responsible for the name "Monte Carlo", derived from a city famous for its elegant gambling casinos. The stochastic nature of the simulation aptly mirrors the element of chance inherent in both gambling and MC simulation techniques.

In an isobaric–isothermal ensemble with fixed temperature T, number of particles N, and pressure P, the probability that the system exists in a microstate i with total energy E_i and volume V_i is given by

$$p_i \propto V_i^N \exp[-\beta(E_i + PV_i)]. \tag{6.26}$$

It follows from Eq. (6.24) that the Metropolis method for acceptance probability a_{ij} is

$$a_{ij} = \min\left\{1, \left(\frac{V_j}{V_i}\right) \exp[-\beta(\Delta E + P\Delta V)]\right\} \tag{6.27}$$

where j stands for a selected trial microstate, and $\Delta V = V_j - V_i$. In an isobaric–isothermal ensemble, a trial MC move can be selected either by a change in configuration (as in canonical-ensemble simulation) or by a change of the system volume. When T, P and N are fixed, the system volume is a dynamic variable that fluctuates around its equilibrium value.

Similarly, in a grand-canonical ensemble, where T, V, and chemical potential μ are fixed, the configurational probability of a system with total potential energy E_i and number of particles N_i is given by

$$p_i \propto \frac{V^N}{\Lambda^{3N} N!} \exp(-\beta E_i + \beta\mu N_i) \tag{6.28}$$

where Λ is the de Broglie thermal wavelength ($\Lambda = h/\sqrt{2\pi m k_B T}$, m is the particle mass, and h is Planck's constant). In that case, the Metropolis acceptance probability a_{ij} is given by

$$a_{ij} = \min\left\{1, \frac{V^{\Delta N} N_j!}{\Lambda^{3\Delta N} N_i!} \exp(-\beta\Delta E + \beta\mu\Delta N)\right\} \tag{6.29}$$

where ΔN is usually either 0 or ± 1. In the grand-canonical ensemble, a trial MC move can be a change in the particle position ($\Delta N = 0$), or insertion ($\Delta N = 1$) or deletion ($\Delta N = -1$) of a particle consistent with the constraints of constant T, V and μ.

6.3.3 Summary

The publication of the Metropolis method signifies the inception of contemporary MC simulation, underpinning the crucial foundation for sampling microstates without the precise knowledge of the probability or partition function. Although various alternatives for the transition matrix have been proposed in the literature, the Metropolis method remains the most commonly adopted in practical MC simulation programs. Interestingly, to date, no mathematical proof has emerged to establish the superiority of the Metropolis–Hastings algorithm as the optimal choice for achieving a predefined probability distribution. Nonetheless, the wide adoption and success of the Metropolis method in various applications attest to its effectiveness and reliability in generating samples that conform to desired probability distributions.

6.4 Monte Carlo Simulation for an Ising Chain

In this section, we elucidate the numerical details of MC simulation using an Ising chain as an example. Due to its simplicity, simulation of a short Ising chain allows us to understand multiple aspects of a MC simulation program, including the preparation of the initial microstate, implementation of MC moves with an acceptance/rejection criterion, relaxation from the initial configuration, and calculation of ensemble averages.

6.4.1 Essential Ingredients of MC Simulation

As discussed in Section 6.2, MC simulation typically involves two stages: equilibration and production. The equilibration stage begins with an arbitrarily chosen initial configuration (microstate) that is consistent with the macroscopic constraints of the system (e.g., in a canonical ensemble we fix temperature, the system volume, and the total number of particles). The initial configuration is then updated by a sequence of steps (viz., MC moves) according to a Markov chain process (e.g., the Metropolis method) until a limiting probability distribution is achieved. Upon equilibrium, further updating of microstates does not change the probability distribution. This concludes the equilibration stage.

Once equilibrium has been attained, the trajectory of the microstate evolution evolves into the production stage where a large number of microstates (typically $\sim 10^5$ MC cycles per particle) are generated following the same Markov process. Meanwhile, properties of interest (structural and thermodynamic properties[9]) are calculated with appropriate ensemble averages. We conclude the production stage when the calculated thermodynamic property (or structure) shows no significant change with the number of samples.

6.4.2 MC Moves for an Ising Chain

As discussed in Chapter 5, the thermodynamic properties of an Ising chain are exactly known. For example, in the absence of an external field ($h = 0$), the internal energy per spin is given by

$$\frac{<E>}{N\varepsilon} = (1/N - 1)\tanh \beta\varepsilon \tag{6.30}$$

and the constant volume heat capacity is

$$\frac{C_v}{Nk_B} = \frac{<\delta E^2>}{Nk_B^2 T^2} = \left(1 - \frac{1}{N}\right)\left[\frac{\beta\varepsilon}{\cosh \beta\varepsilon}\right]^2 \tag{6.31}$$

where $\delta E = E - <E>$ represents energy fluctuation. In Eqs. (6.30) and (6.31), N is the number of spins in the Ising chain, and ε represents the coupling parameter, i.e., negative of the energy due to interaction between a pair of nearest neighbors.

At a given temperature T, the microstates of the Ising chain follow the canonical distribution

$$p_\nu \propto e^{-\beta E_\nu} \tag{6.32}$$

where ν stands for a microstate. To initiate MC simulation, we first create a microstate by randomly assigning the spin orientations, $\{s_i = \pm 1\}_{i=1,2,\cdots N}$. The initial configuration can be generated by

9 Structural properties are the average structure and correlation of the system. Some examples are the density distribution of molecules in a non-uniform system, the pair correlation function, and the configurations of large polyatomic molecules such as polymers or biomacromolecules.

using random numbers: For each spin, we take a random number between 0 and 1. If the random number is larger than 0.5, we assign $s_i = +1$ (spin up); otherwise, $s_i = -1$ (spin down). Next, we use the Metropolis algorithm to sample important microstates according to the Boltzmann distribution given in Eq. (6.32). We may propose the trial state by flipping a randomly selected spin, e.g., by changing the orientation of spin i from s_i=−1 to +1 or vice versa. According to the Metropolis method, the acceptance probability for the trial move is given by

$$a = \min\left\{1, e^{-\beta\Delta E}\right\}. \tag{6.33}$$

where ΔE is the energy of the trial state minus that of the current state. The MC move is accepted if a is larger than another random number between 0 and 1 generated Independently by the computer; otherwise, it is rejected. In the former case, we replace the current microstate with the trial state. The rejection means no change in the current microstate. The process continues until the system reaches equilibrium, i.e., until the generation of new microstates satisfies the equilibrium canonical distribution. That concludes the equilibration stage.

In the production stage, the MC move continues. However, the internal energy of the system at the current microstate is recorded during the course of simulation. Typically, the sample is collected not after every MC move but after a certain number of MC steps such that the samples are not strongly correlated with each other. To calculate the internal energy and heat capacity, we make the ensemble averages of E and E^2 using arithmetic averages after a sufficient number of samples has been collected according to the Metropolis algorithm.

To obtain numerical results, suppose that we simulate an Ising chain with 10 spins as shown schematically by Figure 6.5. The interaction energy between neighboring spins i and $i+1$ is given by

$$\varepsilon_{i,i+1} = -s_i s_{i+1}\varepsilon \tag{6.34}$$

where $s_i = 1$ stands for spin up, and $s_i = -1$ for spin down. For simplicity, let $\beta\varepsilon = 1$. The total energy of the Ising chain for a given configuration (some spins are up, and some spins are down) is given by

$$E = \sum_{i=1}^{9} \varepsilon_{i,i+1} \tag{6.35}$$

Suppose now that the initial configuration generated is given by Figure 6.5. In this configuration, the system has a total (reduced) energy of $\beta E = -1$.

To start the Markov chain process, we propose a trial configuration by randomly flipping a randomly selected spin. The acceptance or rejection of the trial move is determined by the Metropolis algorithm, Eq. (6.33). Figure 6.6 elucidates the first five steps in the Markov process with the initial configuration given in Figure 6.5. In step 1, the first spin is randomly selected; a flip of this spin causes a change in reduced energy (CRE) of 2. In this case, the trial configuration is rejected because the Boltzmann factor (BF), e^{-2}, is less than random number 0.44 generated by the computer. Similarly, steps 2 and 3 are also rejected according to the Metropolis algorithm. Steps 4 and 5 are accepted because in each case, the Boltzmann factor is larger than the according random numbers. Note that the generation of the random number is unnecessary for MC moves leading to an energy reduction (or no change in energy); with BF ≥ 1, these moves (e.g., steps 4 and 5 in Figure 6.6) are automatically accepted. The process proceeds until the configuration of the Ising

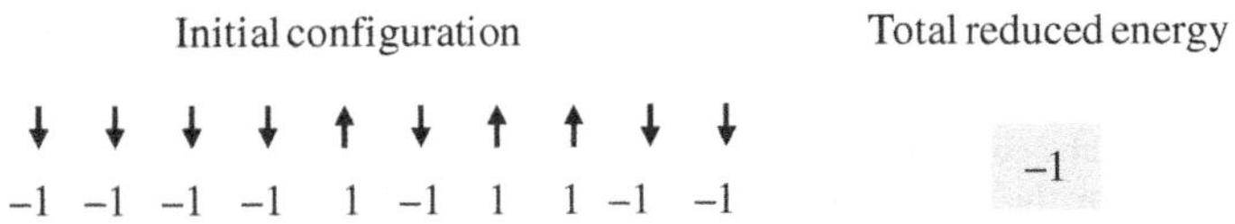

Figure 6.5 An Ising chain with 10 spins with a reduced energy $\beta E_v = -1$.

	Initial configuration				Total reduced energy
	–1 –1 –1 –1 1 –1 1 1 –1 –1				–1
Step: 1	Spin flipped: –1	CRE: 2	BF: .14	Random number: .44	Rejected
	–1 –1 –1 –1 1 –1 1 1 –1 –1				–1
Step: 2	Spin flipped: 3	CRE: 4	BF: .02	Random number: .43	Rejected
	–1 –1 **–1** –1 1 –1 1 1 –1 –1				–1
Step: 3	Spin flipped: 10	CRE: 2	BF: .14	Random number: .78	Rejected
	–1 –1 –1 –1 1 –1 1 1 –1 **–1**				–1
Step: 4	Spin flipped: 7	CRE: 0	BF: 1.00	Random number: .02	Accepted
	–1 –1 –1 –1 1 –1 **–1** 1 –1 –1				–1
Step: 5	Spin flipped: 5	CRE:4	BF: 54.60	Random number: .13	Accepted
	–1 –1 –1 –1 **–1** –1 –1 1 –1 –1				–5

Figure 6.6 A sample Markov chain process for Monte Carlo simulation of an Ising chain containing 10 spins. CRE, change in reduced energy; BF, Boltzmann factor.

chain is uncorrelated with the initial state, i.e., the microscopic details of new microstates (here spins up or down) are independent of that arbitrarily selected for the initial state. For the production stage, we record the total energy of each accepted microstate and calculate the average total energy and average heat capacity by arithmetic averages. In this production calculation, the microstates used in calculating the ensemble averages may be repeated. For example, when a trial microstate is rejected, the original microstate should be counted as a valid sample in calculating ensemble averages.

We may compare the results from MC simulation for internal energy and heat capacity at different temperatures with the exact results from Eqs. (6.30) and (6.31). Figure 6.7 presents the

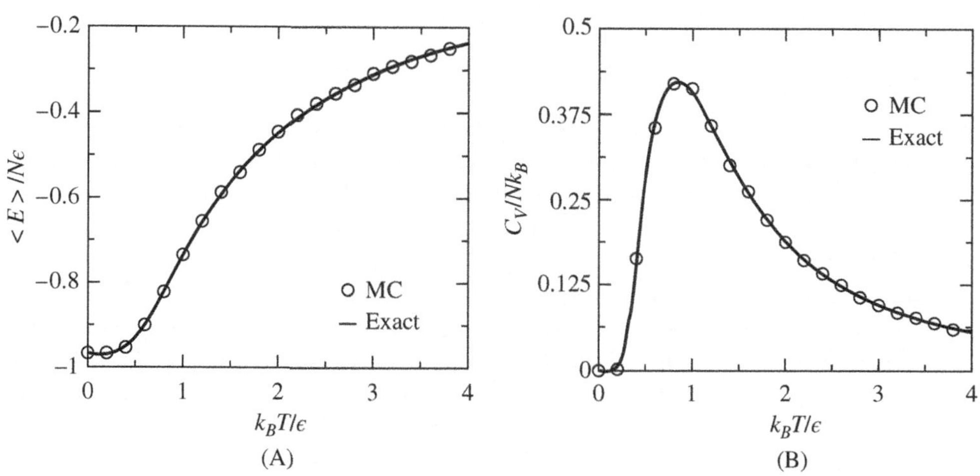

Figure 6.7 Reduced internal energy (A) and reduced heat capacity (B) for an Ising chain with $N = 30$ spins. The lines are calculated from the analytical theory (exact) and the points are from Monte Carlo (MC) simulation.

internal energy and heat capacity calculated from simulation after 200,000 MC steps for both the equilibrium stage and the production stage. Even with a relatively small number of samples, MC simulation shows excellent agreement with the exact results.

6.4.3 Summary

The 1D-Ising model serves as a simple framework for illustrating the numerical details of MC simulation. Its simplicity allows for a clear exploration of various aspects of a simulation program including the system representation, energy calculation, MC moves, acceptance/rejection criteria, equilibration, and sampling techniques. Conversely, the MC program can be utilized to compute the properties of any system that can be represented by the 1D Ising model.

6.5 Simulation Size

Despite the considerable computational power of modern computers, it is not feasible to apply MC simulation directly to macroscopic systems that contain an enormous number of particles. This is primarily due to limitations in computer memory and computation time. Moreover, conducting simulations of macroscopic systems is unnecessary, as the dynamic quantities are typically correlated at the microscopic length scales. As a result, molecular simulation can be performed using smaller systems. In this section, we discuss fluctuation and boundary effects in MC simulation, two fundamental challenges that arise when simulating a macroscopic system with a relatively small number of particles. We will explore strategies for overcoming these obstacles and demonstrate their application to MC simulation of gas adsorption using a two-dimensional lattice model.

6.5.1 Fluctuation Effect

Fluctuations are particularly important for small systems. As discussed in Chapter 2, the relative root-mean-square deviation of an extensive thermodynamic quantity is proportional to $1/\sqrt{N}$, where N is the total number of particles in the system. For example, in a canonical system, the relative root-mean-square deviation of the internal energy is given by

$$\sqrt{\langle(\delta E)^2\rangle}/\langle E\rangle = \sqrt{k_B T^2 C_V/\langle E\rangle} \propto 1/\sqrt{N}. \tag{6.36}$$

The proportionality in Eq. (6.36) follows because both the heat capacity C_V and the internal energy $<E>$ are extensive properties proportional to N. When the system is remote from the critical point of a second-order phase transition, the relative energy fluctuation is negligibly small in a typical macroscopic system. For example, the relative fluctuation in a macroscopic system with 10^{23} particles is on the order of 10^{-12}, which is insignificant in most circumstances. However, the relative fluctuation is about 0.03 for a small system of 1000 particles.

In MC simulation, the number of particles simulated depends on several factors, including the complexity of molecules in the system, the thermodynamic properties of interest, the required accuracy, and the availability of computer power. In the case of a simple fluid far from the critical condition, the correlation length is of the same order of magnitude as the particle diameter. In such cases, simulating a few hundred particles is often sufficient for accurate calculation of the microscopic structure and thermodynamic properties. Near the critical point, however, the dynamic properties are correlated over long distances. In that case, the simulation results are highly sensitive to system size. This sensitivity can be evaluated by comparing simulation outcomes for different particle numbers while keeping the intensive properties of the system constant, such as temperature or density. Due to the large correlation lengths near the critical condition, it becomes challenging to calculate critical properties using MC simulation (discussed in Section 6.11). Although periodic boundary conditions (PBC) can minimize the boundary effect, they are unable to eliminate the finite-size effect on the fluctuations of thermodynamic properties. This limitation arises because fluctuations in the image boxes mirror those in the simulation box. The finite-size effect becomes most significant when simulating fluid properties in the critical region.

6.5.2 Boundary Effect

Unlike a large system, a small system has a significant fraction of particles exposed to the surface. Approximately, the ratio of the number of particles at the surface to that inside a bulk system is proportional to $1/N^{1/3}$. Due to the variation in interfacial properties compared to those in the bulk, the behavior of a small system differs significantly from that of a large system. A standard procedure for eliminating the boundary effect is by the use of PBC, i.e., replicating the simulation box throughout the space to form an infinitely large system.

Figure 6.8 shows schematically the PBC for a two-dimensional system. Here, the central box, i.e., the system for simulation calculations, is highlighted with a shaded background. The image boxes (delineated by the dashed lines) contain particles in exactly the same relative positions as those in the central box. In MC simulation, we consider the interaction of each particle with all other

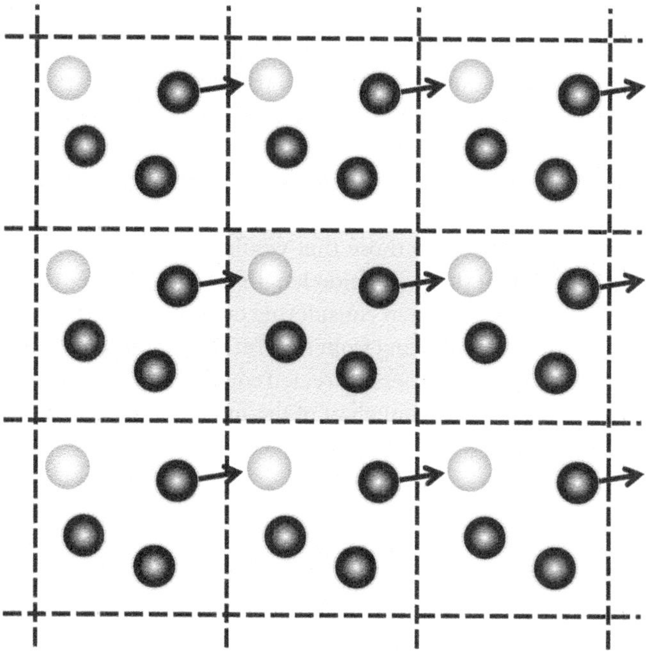

Figure 6.8 A schematic representation of the periodic boundary conditions for a two-dimensional system. Here, the center (shaded) represents the simulation box that contains three particles (darker colored), and those with the dashed boundaries are its periodic images i.e., the image boxes contain exactly the same number of particles with identical relative positions. In MC simulation, we consider the interactions between particles in the simulation box and the interactions between each simulated particle and all image particles generated by the periodic boundary conditions. When a particle leaves the simulation box (depicted as the sphere with an arrow), an identical particle (depicted as the lighter colored sphere) enters the simulation box from the opposite direction.

particles within the central simulation box and, in addition, with the images of all particles in the periodic boxes. If every particle in the "real" box interacts with every other particle in ALL the other boxes, we virtually have an infinite system, as desired for the calculation of thermodynamic properties of a macroscopic system. But how does MC simulation handle an "infinite" system?

When we move the position of a particle in the simulation box (shaded), an identical move is applied to all its images generated by PBC. As shown in Figure 6.8, as a particle exists the simulation box (labeled by an arrow), one of its images enters from the opposite direction (depicted as a lighter colored sphere). In terms of the particle coordinates, the application of PBC leads to a new position for the particle within the simulation cell

$$\mathbf{r}_{PBC} = \mathbf{r} + \mathbf{L} \tag{6.37}$$

where $\mathbf{L} = (L_x, L_y, L_z)$ represents the dimensionality of the simulation box.

By using PBC, we do not need to keep track of the absolute positions for an infinite number of particles. Further, we may truncate the potential energy if the correlation is dominated by short-range particle-particle interactions.[10] In that case, we consider only the interaction of each particle with other particles within a small distance, instead of the interaction between all particles in an infinite system. While the truncation works remarkably well for systems with short-ranged forces, it may result in substantial error for systems with long-range forces, such as those between charged particles or ions. In those cases, special techniques such as the Ewald sum method must be used to calculate the energy of a virtually "infinite" system.[11]

In MC simulation, the central box only provides a convenient coordinate frame for measuring the relative positions of particles under simulation. With PBC, the surface effect disappears because the simulation system is not affected by the particular location of the boundaries. As long as the size of the simulation box is much larger than the correlation length[12] of the real macroscopic system under investigation, MC simulation provides a reliable method for calculating the thermodynamic properties.

By considering interactions between each particle within the simulation box and all other particles in the same box and those in the periodic images, MC simulation for particles in a finite-size box provides equilibrium properties essentially identical to those that would be obtained by the simulation of an infinitely-large system, provided that the correlation length is much smaller than the box size. It should be noted that a subtle point arises when considering energy calculations in periodic systems. When evaluating the energy change resulting from trial MC moves, interactions within the central box and those extending across its boundaries are treated equally. However, when determining the *prorated* energy of the central box system, only half of the interactions extending beyond the boundaries contribute to the calculation.

6.5.3 Gas Adsorption on a Planar Surface

In the remainder of this section, we discuss the size effects in MC simulation by considering a simple example, i.e., application of MC simulation to a two-dimensional system representing gas adsorption. We will demonstrate the application of PBC for macroscopic systems and the use of simulation results to validate theoretical predictions.

10 As discussed in Section 6.6, the truncation of particle-particle interactions introduces a small error in the simulation outputs which can be estimated by using a mean-field theory.
11 See, for example, Frenkel D. and Smit B., *Understanding molecular simulation*. Academic Press, Chapter 12, 2002.
12 As discussed in Section 2.2, correlation length refers to the maximum separation between two particles where one particle "feels" the "existence" of the other. The presence of particle A will not affect the behavior of particle B if the distance between A and B exceeds the correlation length.

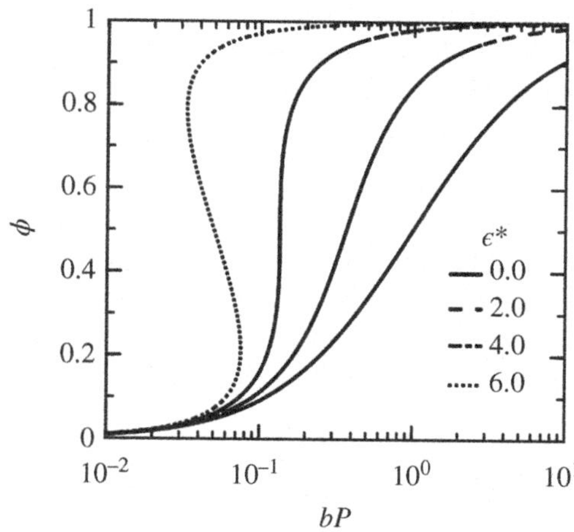

Figure 6.9 Adsorption isotherms according to the Langmuir theory ($\varepsilon^* = 0$) and the Fowler–Guggenheim mean-field theory. Here $\varepsilon^* = \beta Z\varepsilon$, and ϕ stands for the surface coverage. According to the Fowler–Guggenheim theory, the two-dimensional system exhibits a vapor–liquid-like phase transition when $\varepsilon^* \geq 4$.

Suppose that we want to find the adsorption of a low-density gas on a solid surface as discussed in Section 3.6. For simplicity, we assume that the surface can be represented by a two-dimensional lattice where each site adsorbs no more than one gas molecule. If we assume that the gas phase is ideal and that the adsorbed gas particles do not interact with each other, we have the Langmuir-type adsorption isotherm[13]

$$\frac{\phi}{1-\phi} = bP \tag{6.38}$$

where ϕ is the fraction of lattice sites covered by adsorbed gas molecules, and P is the gas pressure. As discussed in Section 3.6, parameter b depends on the system temperature and the microscopic details of the solid–gas interactions.

To make the model more realistic, we need to consider interactions between neighboring adsorbed molecules on the lattice. For this case, it is difficult to deduce a rigorous analytical expression for the adsorption isotherm. Nevertheless, we may use the Weiss molecular field theory (Section 5.6) to derive an approximate adsorption isotherm

$$\frac{\phi}{1-\phi} = be^{\beta Z\varepsilon\varphi}P \tag{6.39}$$

where Z is the number of nearest neighbors in the two-dimensional lattice, and ε is negative of the interaction energy between a pair of adjacent adsorbed gas molecules. Eq. (6.39) is known as the Fowler–Guggenheim theory of gas adsorption.

When $\varepsilon = 0$, Eq. (6.39) reduces to the Langmuir isotherm. When the force between adjacent adsorbed molecules is attractive, $\varepsilon > 0$; otherwise, $\varepsilon < 0$. The exponential term in Eq. (6.39) takes into account the additional adsorption energy (viz., cooperative effects) due to the attraction between adsorbed molecules. The Fowler–Guggenheim theory can be readily extended to adsorption of nonideal gases by replacing the pressure with the fugacity.

As shown in Figure 6.9, the adsorption isotherm from Eq. (6.39) resembles the Langmuir isotherm at low pressure, i.e., the surface coverage increases monotonically with pressure. However, the adsorption isotherm from the Fowler–Guggenheim theory becomes drastically different from the Langmuir isotherm as $\varepsilon^* \equiv \beta Z\varepsilon$ increases. According to the Fowler–Guggenheim theory, the two-dimensional system exhibits a vapor–liquid-like phase transition when $\varepsilon^* \geq 4$.

13 An isothermal plot of ϕ versus P (or fugacity) is called an adsorption isotherm.

To validate the aforementioned theoretical analysis through MC simulation, we will utilize a two-dimensional lattice model. In this model, the adsorption or desorption of gas molecules is depicted by the occupation of lattice sites. For this system, the grand partition function is

$$\Xi = \sum_{\nu} q_s^{N_\nu} \exp(-\beta E_\nu + \beta N_\nu \mu - \beta N_\nu \varepsilon_s) \tag{6.40}$$

where microstate ν is determined by the occupancy status of individual lattice sites, q_s is the intrinsic partition function for each gas molecule adsorbed on the surface, E_ν is the total energy due to interaction among N_ν adsorbed gas molecules at microstate ν; ε_s is the external energy due to the surface attraction, and μ is the chemical potential of the gas molecules. For a pure ideal gas, the chemical potential can be expressed as

$$\mu = \mu_0 + k_B T \ln P \tag{6.41}$$

where μ_0 is the chemical potential at the reference state, i.e., the ideal gas at system temperature and unit pressure (e.g., 1 bar). For a nonideal gas, the pressure should be replaced with the gas fugacity.

We may simplify Eq. (6.40) by combining certain constants

$$\Xi = \sum_{\nu} \exp[-\beta E_\nu + N_\nu \ln Pb] \tag{6.42}$$

where $b = q_s \exp(-\beta\mu_0 - \beta\varepsilon_s)$ is the same as the phenomenological parameter used in the Langmuir adsorption isotherm. If the gas molecules do not interact with each other at the surface, $E_\nu = 0$, and

$$\Xi = \sum_{\nu} (Pb)^{N_\nu} = \prod_{i=1}^{N_s} (1 + Pb) = (1 + Pb)^{N_s} \tag{6.43}$$

where N_s stands for the total number of surface sites. In this case, the grand potential leads to the Langmuir adsorption isotherm, Eq. (6.38).

Assume that the two-dimensional simulation box is a square lattice with L sites in each dimension. The total number of surface sites is then given by $N_s = L \times L$. With the PBC in both horizontal and vertical directions, we can simulate the fraction of surface covered by gas molecules and the average energy per site. To determine the appropriate system size, we consider in the following the simulation results for $L = 4, 6, 8, 10, 12, 15$.

We may generate an initial configuration for the simulation system by a random assignment of each site as empty or occupied by a gas molecule. For this simple system, the MC moves can be achieved by adding a gas molecule to an empty site or by deleting an adsorbed molecule from an occupied site. The probability that the surface has total surface energy E_ν [14] and number of adsorbed particles N_ν is given by the grand-canonical distribution

$$P_\nu \propto e^{-\beta E_\nu + N_\nu \ln Pb}. \tag{6.44}$$

According to the Metropolis algorithm, the probability of acceptance for each MC move is given by

$$a = \min\{1, \exp(-\beta\Delta E + \Delta N \ln Pb)\}. \tag{6.45}$$

where ΔE is the change in surface energy, and ΔN is the change in the number of adsorbed gas molecules. If a gas molecule is *added* to an empty lattice site of index (i, j), the change in the intermolecular energy is

$$\Delta E = -\varepsilon(s_{i+1,j} + s_{i,j+1} + s_{i-1,j} + s_{i,j-1}) \tag{6.46}$$

14 The interaction energy between a gas molecule and the surface is accounted for by parameter b.

where $s_{i,j} = 1$ means that the site (i, j) is occupied, and $s_{i,j} = 0$ means that the site is empty. With the periodic boundary condition, we have $s_{L+1,j} = s_{i,j}$, and $s_{1,j} = s_{L,j}$, where L stands for the box length. Similar relations apply to the j index.

As a gas molecule is added to an empty site, the change in the number of adsorbed molecules is $\Delta N = 1$. Eq. (6.45) predicts the new configuration is accepted if $\exp(-\beta\Delta E + \ln Pb)$ is larger than a random number between 0 and 1; otherwise, the original configuration is retained. Conversely, if an adsorbed molecule is *deleted* from an occupied site, the change in the intermolecular energy is

$$\Delta E = \varepsilon(s_{i+1,j} + s_{i,j+1} + s_{i-1,j} + s_{i,j-1}), \tag{6.47}$$

and the change in the number of molecules $\Delta N = -1$. In this case, if $\exp(-\beta\Delta E - \ln Pb)$ is larger than a random number between 0 and 1, the deletion is accepted; otherwise, the original configuration is retained.

The system reaches equilibrium after about 10^4 MC cycles, i.e., after each lattice site has been tried by addition/removal of a gas molecule 10^4 times. In the production stage (activated upon completion of the equilibration stage), both the number of molecules adsorbed on the lattice and the system energy are recorded for each MC cycle. The ensemble averages are then calculated after 10^5 cycles. For this simple system, further sampling does not change the results significantly.

Figure 6.10 shows the surface coverages and total intermolecular energies for the systems with $L = 4, 6, 8, 10, 12$, and 15, respectively. Here, the reduced interaction energy is $\varepsilon^* = \beta Z\varepsilon = 6$, above the critical reduced energy ($\varepsilon^* = 4$) predicted by the Fowler–Guggenheim theory. The simulation results indicate that the system-size effect becomes insignificant when $L > 8$. Thus, we may assume that $L = 12$ is a safe choice for the size of the two-dimensional simulation box[15] for this system.

Figure 6.11 compares the simulation results with the predictions of the theoretical methods. When $\varepsilon^* = 0$, the surface coverage predicted by the Langmuir isotherm is exact in the context of the adsorption model considered here. It shows perfect agreement between simulation and (Langmuir)

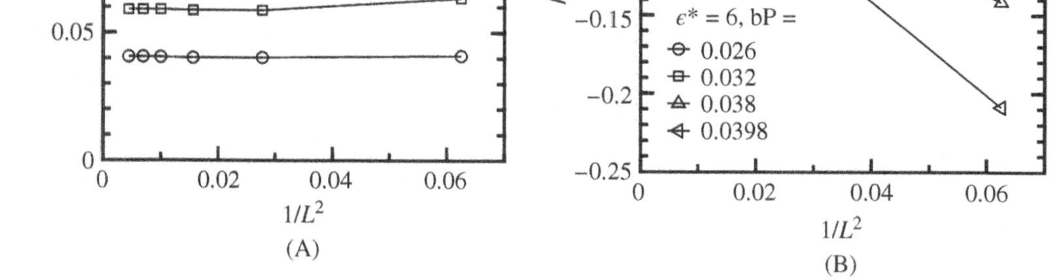

Figure 6.10 The effect of simulation size (L^2 = number of lattice sizes) on the surface coverage (A) and the reduced average energy per molecule (B) calculated from MC simulation. Points are simulation results for $L = 4, 6, 8, 10, 12, 15$, and the lines are to guide the eye.

15 Near the critical point of the two-dimensional vapor–liquid equilibrium, the system-size effect becomes more prominent.

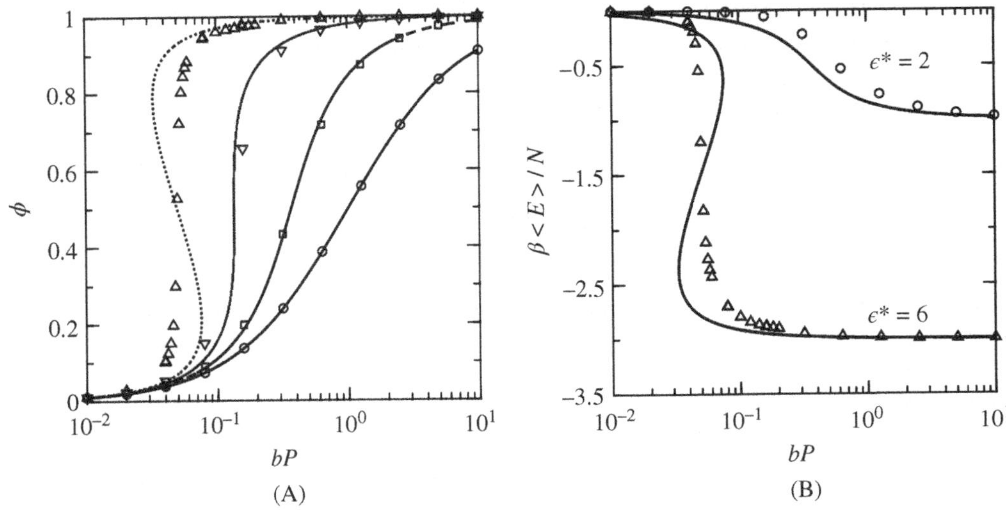

Figure 6.11 Adsorption isotherms (A) and the reduced average surface energy per molecule (B) for the adsorption of a low-density gas on a two-dimensional lattice with lateral interaction energy $-\varepsilon$ the same as those shown in Figure 6.10. The simulation data (symbols) are added for comparison with the predictions of the Fowler–Guggenheim theory. Agreement between theory and simulation is perfect when $\varepsilon^* = 0$ but declines as ε^* increases. While the mean-field theory predicts a phase transition at $\varepsilon^* = 6$, the phase transition was not confirmed by simulation at this condition.

theoretical results for $\varepsilon^* = 0$. However, when $\varepsilon^* = 6$, the Fowler–Guggenheim theory and simulation are not in agreement. The discrepancy is introduced by the mean-field approximation.

6.5.4 Summary

By utilizing PBC, MC simulation can produce highly accurate results for macroscopic systems. Although it is introduced as an artifact, the presence of periodicity does not compromise the precision of the simulation, provided that the size of the simulated system is sufficiently large in comparison with the correlation length. Selecting an appropriate simulation size is crucial for ensuring accurate outcomes. As a general rule, the simulation box size should be at least several times larger than the correlation length.

This section also elucidates the numerical procedure involved in applying MC simulation to a two-dimensional system. While it is often difficult to obtain analytical solutions in statistical thermodynamics, numerical results can be readily obtained by running a simple simulation. Nonetheless, analytical methods should be considered valuable for understanding the underlying physics and for analyzing the simulation data.

6.6 MC Simulation for Simple Fluids

In this section, we discuss MC simulation for studying the structure and thermodynamic properties of simple fluids such as argon or methane, typically represented by the Lennard-Jones (LJ) model. By employing the Metropolis algorithm, we can compute the radial distribution function (RDF), internal energy, and pressure as they vary with temperature and density. The LJ model plays a crucial role in establishing modern liquid-state theories in both bulk and confined environments, and the MC simulation results serve as a benchmark for testing the theoretical advancements.

6.6.1 Configurational Averages

Consider a one-component fluid with volume V, absolute temperature T, and N spherical particles. According to the LJ model, the pair potential is given by

$$u(r) = 4\varepsilon\left[\left(\frac{\sigma}{r}\right)^{12} - \left(\frac{\sigma}{r}\right)^{6}\right] \tag{6.48}$$

where r is the center-to-center distance, parameter ε represents the maximum magnitude of attraction, and parameter σ characterizes the particle diameter. Due to its simplicity, the LJ model holds significant appeal for a broad range of practical applications in statistical mechanics. In particular, the LJ model is well-suited for elucidating the fundamental concepts of molecular simulation methods, as well as for investigating interfacial phenomena and phase transitions.

For a uniform LJ fluid, the canonical partition function can be expressed as (Section 7.1)

$$Q = \frac{V^N}{N!\Lambda^{3N}} \int d\mathbf{r}^N \exp[-\beta\Phi(\mathbf{r}^N)] \tag{6.49}$$

where Λ stands for the thermal wavelength, $\Phi(\mathbf{r}^N)$ is the total potential energy, i.e., the energy due to the interaction among N particles at configuration $\mathbf{r}^N = (\mathbf{r}_1, \mathbf{r}_2, \cdots \mathbf{r}_N)$. Assuming pairwise additivity, we may express the total potential energy as

$$\Phi(\mathbf{r}^N) = \sum_{i=1}^{N-1}\sum_{j=i+1}^{N} u(r_{ij}) \tag{6.50}$$

where r_{ij} is the separation between particles i and j.

The internal energy of a LJ fluid contains two parts, one from the kinetic energy of the particles, and the other from the intermolecular potential. For a system of spherical particles at temperature T, the kinetic energy can be determined precisely; it is directly proportional to the average energy resulting from the translational motion of each particle, $3k_BT/2$, as predicted by the Maxwell–Boltzmann equation. However, the potential energy must be evaluated through the ensemble average

$$U = \frac{3}{2}Nk_BT + \langle\Phi(\mathbf{r}^N)\rangle \tag{6.51}$$

where $<\cdots>$ stands for the average over all possible configurations of the system. At a given particle configuration $\mathbf{r}^N$, we can calculate the total potential energy using Eq. (6.50). Because the potential energy is uncorrelated with the particle motions, the ensemble average can be calculated from

$$\langle\Phi(\mathbf{r}^N)\rangle = \frac{\int d\mathbf{r}^N \Phi(\mathbf{r}^N) e^{-\beta\Phi(\mathbf{r}^N)}}{\int d\mathbf{r}^N e^{-\beta\Phi(\mathbf{r}^N)}} = \langle\Phi(\mathbf{r}^N)\rangle_C \tag{6.52}$$

where $<\cdots>_C$ stands for configurational average, i.e., the average of a dynamic quantity according to the configurational probability density

$$p(\mathbf{r}^N) \sim e^{-\beta\Phi(\mathbf{r}^N)}. \tag{6.53}$$

A common procedure for calculating the system pressure is by using the virial equation.[16] For a one-component fluid of argon-like particles, the virial equation is given by

$$P = \rho k_BT + \frac{1}{6V}\left\langle \sum_{i\neq j}^{N} (\mathbf{r}_{ij} \cdot \mathbf{f}_{ij}) \right\rangle_C \tag{6.54}$$

16 See Appendix 2.A. The virial equation, used here, should not be confused with the virial equation of state.

where $\rho = N/V$ is the number density of the uniform system, $\mathbf{r}_{ij} = \mathbf{r}_j - \mathbf{r}_i$ is the vector connecting the positions of two particles given by $\mathbf{r}_i$ and $\mathbf{r}_j$, and $\mathbf{f}_{ij}$ is the force between particles i and j. For a pair of spherical particles, the force is related to the two-body potential

$$\mathbf{f}_{ij} = -\frac{\partial u(r_{ij})}{\partial r_{ij}} \frac{\mathbf{r}_{ij}}{r_{ij}}. \tag{6.55}$$

where $r_{ij} = |\mathbf{r}_{ij}|$ is the center-to-center distance.

The RDF, $g(r)$, is defined as the average density $\rho(r)$ at a distance r from a tagged particle divided by the bulk density ρ [17]

$$g(r) = \frac{\rho(r)}{\rho}. \tag{6.56}$$

Because of the symmetry of spherical particles, at a fixed temperature and density, the RDF depends only on the radial distance.

Figure 6.12 presents a schematic picture of the RDF and the numerical values for a LJ fluid at a liquid-like condition. As the LJ particles experience both repulsive and attractive interactions, the local density $\rho(r)$ differs from the average bulk density when r is small or of the order of molecular diameter σ. When r is smaller than σ, the local density vanishes due to the harsh repulsion and $g(r)$ approaches zero. At high density, the oscillatory density profile is mainly determined by the particle excluded volume,[18] little influenced by the inter-particle attraction. At large separations, the local density is not affected by the presence of the tagged particle at the origin and therefore, when $r \gg \sigma$, $g(r)$ approaches unity. In terms of an ensemble average, Eq. (6.56) can be expressed as

$$g(r) = \frac{1}{\rho N} \sum_{i=1}^{N} \left\langle \frac{dn_i(r)}{4\pi r^2 dr} \right\rangle_C \tag{6.57}$$

where $dn_i(r)$ is the number of particles around particle i in the spherical shell between r and $r + dr$, and $4\pi r^2 dr$ is the shell volume.

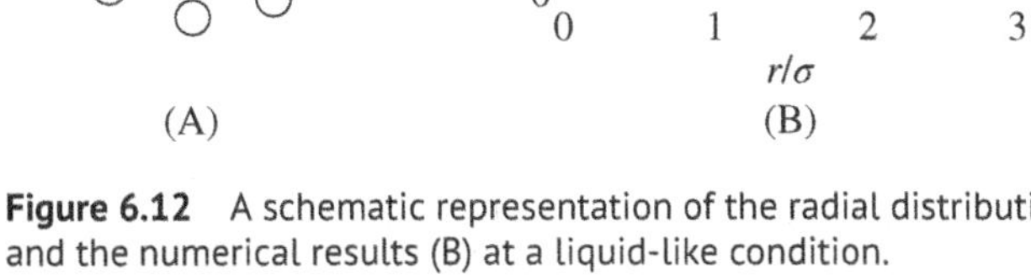

Figure 6.12 A schematic representation of the radial distribution function $g(r)$ of a Lennard-Jones fluid (A) and the numerical results (B) at a liquid-like condition.

---

17 This abbreviated notation of $g(r)$ is incomplete. Radial distribution function g depends not only on r but also on bulk density and temperature.

18 Excluded-volume effects arise because particles cannot occupy the same space at the same time.

6.6.2 MC Moves in the Configurational Space

The above analysis indicates that thermodynamic properties of a LJ fluid can be calculated by using MC simulation in the configuration space. In the following, we discuss the evaluation of the internal energy, pressure, and RDF using the Metropolis algorithm.

Typically, MC simulation is carried out with a certain number of particles placed in a cubic simulation box with PBC. The selections of box size and the number of particles depend on the average density and the correlation length.[19] In general, the simulation box should be much larger than the correlation length such that the periodic boundary condition does not introduce artificial effects. When the temperature and density are not close to those at the vapor–liquid critical point, satisfactory results can be obtained by using only a few hundred particles.

To start the simulation, the particles are randomly placed within the simulation box (or alternatively, at high density, placed uniformly in a cubic lattice). MC moves are carried out by making random displacements of individual particles. Specifically, a particle in the simulation box is selected at random. The trial move is achieved by a small displacement of this particle in a random direction:

$$x_{trial} = x + (2R - 1)\delta r_{\max} \tag{6.58}$$

$$y_{trial} = y + (2R - 1)\delta r_{\max} \tag{6.59}$$

$$z_{trial} = z + (2R - 1)\delta r_{\max} \tag{6.60}$$

where R represents a random number in the range from 0 to 1, and $\delta r_{\max}$ is the maximum length of displacement, i.e., the maximum distance a particle moves in each step. A conventional selection for $\delta r_{\max}$ is such that the probability of acceptance for the move is about 50%.

The change in total potential energy due to the displacement of the trial particle, $\Delta E = \Phi\left(\mathbf{r}^N_{trial}\right) - \Phi(\mathbf{r}^N)$, is calculated by using Eq. (6.50). If $\Delta E < 0$, the system has a lower potential energy in the trial configuration and the trial position is accepted. Otherwise, a new random number R between 0 and 1 is generated, and the trial move is accepted only when $R < e^{-\beta\Delta E}$. If the displacement is rejected, the original Cartesian coordinates are retained until the next trial. The process is repeated until the system reaches equilibrium, that is, when further trials do not significantly change the total energy of the system. At this stage, the distribution of microstates satisfies the Boltzmann distribution. The production stage closely resembles the equilibration stage, with the main difference being the calculation of microscopic properties and local densities. These microscopic properties are then stored for the purpose of computing configurational averages. Alternatively, to avoid the storing step, all running averages can be updated before moving on to the next MC move.

Figure 6.13 shows the simulation results for the average total potential and compressibility factor. Figure 6.14 gives the RDFs at 3 representative conditions. In the MC simulation, it is common practice to truncate the interaction potential between particles beyond a certain distance (typically $r_c \approx 4 - 5\sigma$), which is called the truncation radius or cutoff distance. The truncation of the potential greatly reduces the computational cost and can be justified not only because the interaction between particles i and j declines rapidly with r_{ij} but because the correlation length is typical finite. However, the choice of the cutoff distance should be made with caution. Setting the cutoff too small may introduce artifacts due to the abrupt truncation of the potential. On the other hand, selecting a cutoff distance that is too large may result in unnecessary computational overhead.

19 As discussed in Section 6.2.2, correlation length is the separation between particles beyond which the radial distribution function is approximately unity. It is the minimum distance between two particles where one particle does not "feel" the presence of the other.

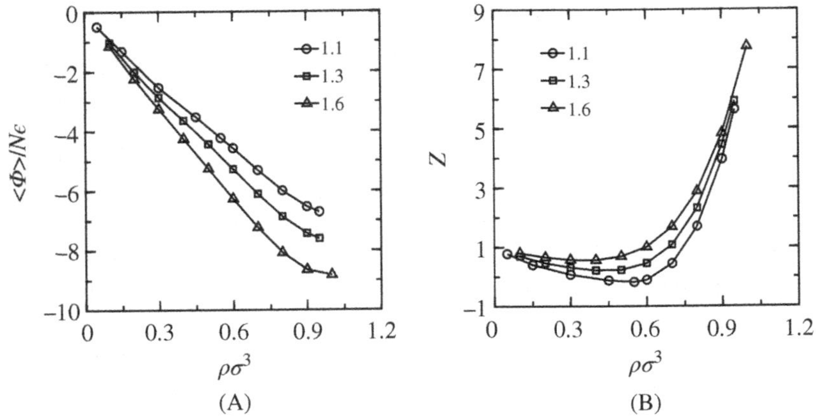

Figure 6.13 The average reduced total potential energy (A) and compressibility factor (B) of a Lennard–Jones fluid versus the reduced density ($\rho\sigma^3$) for three values of the reduced temperature k_BT/ε = 1.1, 1.3, and 1.6. Points are from MC simulation and lines are to guide the eye. N is the total number of particles; ε and σ are the Lennard–Jones parameters.

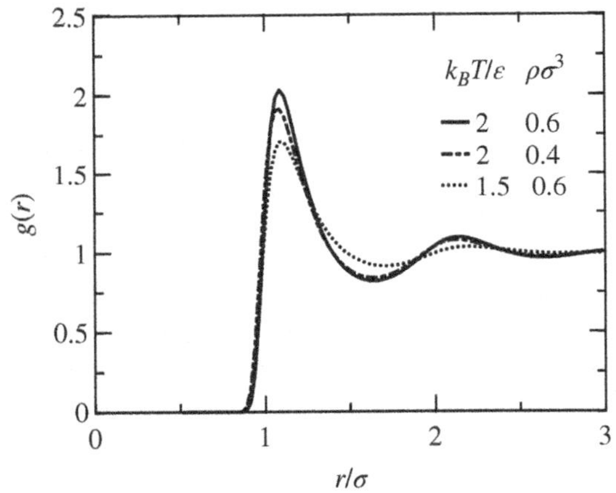

Figure 6.14 Radial distribution functions of a Lennard–Jones fluid calculated from Monte Carlo simulation. The magnitude of the first peak rises with increasing density but falls slightly with increasing temperature.

We can estimate the corrections for pressure (ΔP) and internal energy $<\Delta E>$ due to the potential cutoff using the mean-field approximation. With the assumption that $g(r) = 1$ beyond the cutoff distance (viz., $r > r_c$), Eqs. (6.52) and (6.54) predict

$$<\Delta E> = 2\pi\rho \int_{r_c}^{\infty} dr\, r^2 u(r) = \frac{8\pi\varepsilon}{9}\rho\sigma^3 \left[\left(\frac{\sigma}{r_c}\right)^9 - 3\left(\frac{\sigma}{r_c}\right)^3\right], \tag{6.61}$$

$$\Delta P = -\frac{2\pi\rho^2}{3}\int_{r_c}^{\infty} dr r^3 \frac{\partial u(r)}{\partial r} = \frac{16\pi\varepsilon}{3}\rho^2\sigma^6\left[\frac{2}{3}\left(\frac{\sigma}{r_c}\right)^9 - \left(\frac{\sigma}{r_c}\right)^3\right]. \tag{6.62}$$

For a typical case, the corrections for energy and pressure are less than a few percent. Because of the longer-ranged attractive term in the LJ potential, Eq. (6.48), both corrections are entirely dominated by the attractive contributions, decreasing with the cutoff distance as r_c^{-3}.

Figure 6.15 shows how the average total potential and the compressibility factor change with the number of microstates used in calculating the ensemble averages. With about 10^5 samples generated by importance sampling, the total potential and pressure of the LJ fluid can be calculated with good accuracy. In this calculation, the box size is about 6.3σ, and the number of particles in the box is 150.

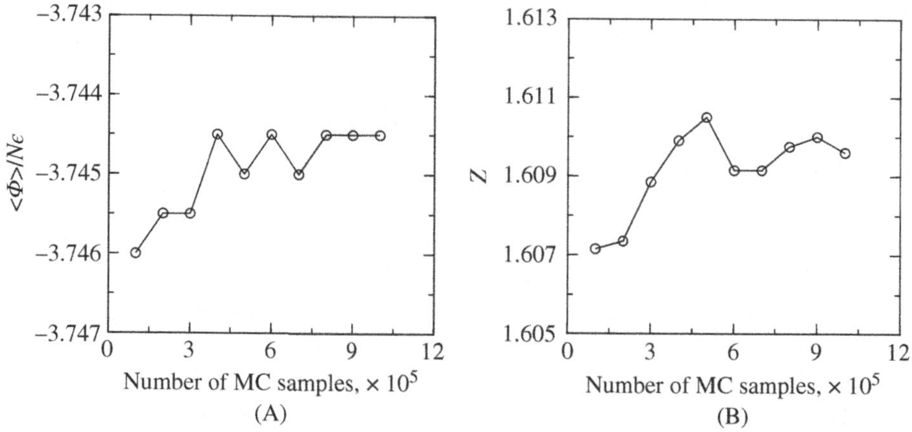

Figure 6.15 Average reduced total potential energy (A) and compressibility factor (B) versus the number of MC samples. Here $k_B T/\varepsilon = 2$ and $\rho\sigma^3 = 0.6$. The lines become essentially flat when the number of trials exceeds (about) 8×10^5.

6.6.3 Summary

It is evident that the Metropolis algorithm is highly versatile and can be equally applied to lattice models as well as continuous thermodynamic models. When dealing with systems that are far from the critical point, it is relatively straightforward to carry out MC simulation to calculate both structural and thermodynamic properties. The simulation program can be easily implemented with a personal computer. In principle, this procedure can be directly extended to LJ mixtures and more complicated molecular systems. In subsequent sections, we will explore advanced simulation methods to improve the efficiency of MC sampling including those used for the simulation of phase diagrams and critical properties.

6.7 Biased MC Sampling Methods

In the conventional Metropolis algorithm, microstates are sampled by sequentially modifying a single variable associated with the dynamics of a thermodynamic system. This can involve small displacements of individual particles or flipping the orientation of a spin, for example. The MC move, to some extent, resembles how microstates are updated in MD simulation, as the transition between microstates occurs on the timescale of individual particles. While this sampling approach is suitable for relatively simple systems, it becomes problematic for complex systems like polymers, where the distribution of microstates reflects collective motions of a large number of particles (e.g., many atoms in a polymer chain). In such systems, important microstates are not easily reachable from one another. If trial moves are solely generated through the displacements of individual particles, an excessively large number of MC steps would be required to sample all significant microstates, just like replicating the dynamics of a polymer system through atomistic motions would necessitate an exceptionally long MD simulation.

In comparison to MD simulation, one major advantage of MC simulation is that microstates can be sampled through non-physical moves, i.e., the transition among important microstates is not constrained by the system dynamics. MC simulation provides an accurate estimation of the ensemble average as long as the sampled microstates follow the correct statistical distribution. In this section, we discuss biased sampling as a generic strategy to promote transitions among important microstates. Section 6.12 provides a more systematic description of enhanced sampling through alternative procedures.

6.7.1 The Generalized Metropolis Algorithm

In the symmetric sampling scheme, microstates are explored by consecutive small changes in a low-dimensional dynamic variable. The probability of a trial move from microstate i to a neighboring microstate j is the same as that from j to i, i.e., $t_{ij} = t_{ji}$. Because the trial probability cancels in the balance condition, there is no need to specify the probability of sampling explicitly. The detailed balance condition is automatically satisfied when the probability of acceptance for the MC move is selected as the ratio of the predefined Boltzmann factors.

Asymmetric trial moves enable to accelerate MC sampling by taking bigger MC steps, i.e., more changes in the microstate in each MC move. In MC simulation, bigger MC steps are often desirable because they lead to microstates less correlated with each other and thus more representative in evaluating the ensemble averages. However, the probability of successful move typically falls as the step length increases. The acceptance rate can be improved by taking biased trial moves in favor of those microstates that yield higher probability of transition between important microstates. In that case, the trial probability is selected in an asymmetric fashion, i.e., $t_{ij} \neq t_{ji}$; those microstates with higher accepting probability are sampled more frequently.

Similar to the conventional Metropolis method, the acceptance probability in asymmetric sampling is selected according to the detailed balance condition, i.e., the probability of transition from microstate i to j is the same as that from j to i

$$p_i a_{ij} t_{ij} = p_j a_{ji} t_{ji} \tag{6.63}$$

where p_i and p_j stand for the equilibrium probabilities of microstates i to j, respectively. The probability acceptance is thus given by

$$a_{ij} = \min\left(1, \frac{p_j t_{ji}}{p_i t_{ij}}\right). \tag{6.64}$$

Eq. (6.64) indicates that any biased sampling leads to the correct microstate distribution when the acceptance probability is modified in such a way that the bias is removed from the sampling scheme.

6.7.2 Orientational Bias Monte Carlo

We may elucidate the general idea of biased sampling by considering the orientational distribution of a point dipole in a uniform electrical field. As shown schematically in Figure 6.16A, the orientation of a point dipole depends on the potential energy

$$u_d = -dD\cos\theta \tag{6.65}$$

where $D = |\mathbf{D}|$ is the magnitude of the electric field, and θ is the angle between the direction of the field and the axis of the permanent dipole d. At given temperature T, the orientational distribution is exactly known (see Problem 6.9). The probability density of observing the dipole angle between θ and$\theta + d\theta$ is

$$p(\theta) = \sin\theta e^{y\cos\theta}/q_0 \tag{6.66}$$

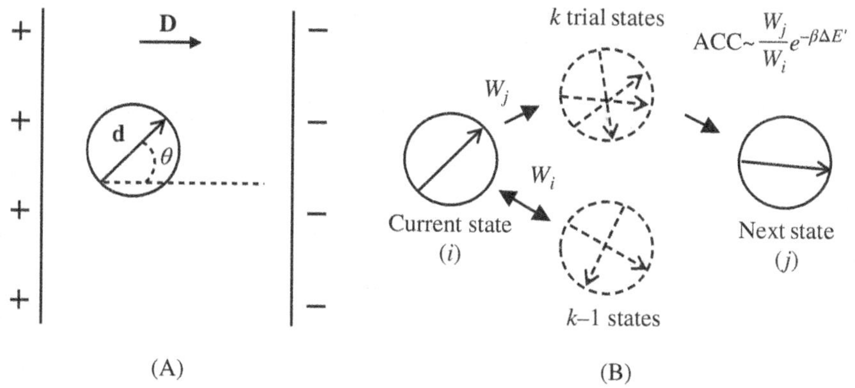

Figure 6.16 Orientational bias MC simulation. (A) A point dipole **d** with angle θ relative to a uniform electric field **D**. (B) The MC move is selected from one of the k trial orientations according to the Rosenbluth weight (W_j and W_i) and a Boltzmann factor ($e^{-\beta\Delta E'}$) related to other degrees of freedom. For an ideal gas of point dipoles, $\Delta E' = 0$.

where $y = \beta dD$ and $q_0 = 2\sin y/y$. For a permanent dipole with $d = 1$ Debye near an electrode about 1 V, the electric field is on the order of $D \sim 10^9$ volts/m. Thus, parameter y is in the range of [0, 10] near room temperature.

Now suppose we are interested to calculate the orientational distribution of the point dipole by running MC simulation. For this simple system, each microstate may be understood as a particular value for the orientational angle θ. The probability distribution can be calculated with either the Metropolis algorithm or a biased sampling scheme. In the former case, angle θ is sampled by generating a series of random numbers between $[0, \pi]$. The probability of acceptance for a transition from θ_i to θ_j is determined by

$$a_{ij} = \min(1,\ p_j/p_i) = \min\left(1,\ \frac{\sin\theta_j e^{y\cos\theta_j}}{\sin\theta_i e^{y\cos\theta_i}}\right). \tag{6.67}$$

In practical simulations, the use of biasing factor $\sin\theta$ is typically avoided by replacing angle sampling between 0 and π with the generation of $\cos\theta$ uniformly between −1 and 1. As discussed above, the probability of the trial move is not explicitly shown in the Metropolis algorithm because it cancels out in the detailed balance condition.

Figure 6.17 shows the probability density of θ for three values of the reduced electrical energy, $y = 1$, 5, and 10. In a weak electric field ($y \to 0$), Eq. (6.66) predicts $p(\theta) \to \sin\theta/2$. As the electric field increases, the maximum probability takes place at a smaller value of θ because the energy is minimized when the dipole is aligned in the direction of the external field. As expected, the Metropolis algorithm reproduces the exact result provided sufficient MC steps are taken.

Because the external field has a strong influence on the orientational distribution of the point dipole, it is reasonable to expect that the simulation will be more efficient if the sampling is biased toward low energy states, i.e., toward the direction of the external field. As illustrated by Figure 6.17B, an intuitive way to implement such a biased scheme is to generate several angles at each MC step and select one according to their probabilities:

$$t_{ij} = p_j \Big/ \sum_{n=1}^{k} p_n \tag{6.68}$$

where subscript j represents the selected state from k attempted orientations. To identity the acceptance criterion using Eq. (6.64), we need to know the trial probability for the reverse process of

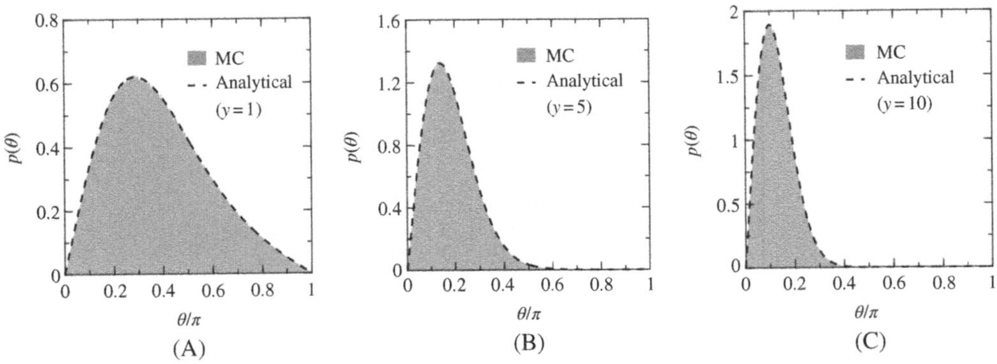

Figure 6.17 The angular distribution for a point dipole in an electric field. The reduced electrical energy has the values of $y = 1$, 5 and 10 in panels A, B and C, respectively. The shaded areas are calculated from MC simulation, and the dashed lines are exact results. Averaged over 10^7 MC cycles, the Metropolis method and the orientational bias algorithm (with $k = 10$ trial moves) yield virtually identical results.

transferring from state j to i. It can be shown that this probability can also be calculated from the Boltzmann distribution[20]

$$t_{ji} = \frac{p_i}{p_i + \sum_{m=1}^{k-1} p_m} \tag{6.69}$$

Eq. (6.69) indicates that the reverse probability amounts to selecting state i along with additional $k - 1$ angles generated *independently* according to the same protocol in the trial move. Substituting Eqs. (6.68) and (6.69) into (6.64) gives the probability of acceptance for biased sampling

$$a_{ij} = \min\left(1, \frac{p_j t_{ji}}{p_i t_{ij}}\right) = \min\left(1, \frac{p_j}{p_i} \times \frac{p_i / \left(p_i + \sum_{n=1}^{k-1} p_n\right)}{p_j / \sum_{m=1}^{k} p_m}\right) = \min\left(1, \frac{W_j}{W_i}\right) \tag{6.70}$$

where $W_j = \sum_{n=1}^{k} p_n$ and $W_i = p_i + \sum_{m=1}^{k-1} p_m$ are weight factors, and subscripts n and m are referred to different sets of trial states.

The detailed balance condition ensures that the biased sampling leads to the probability distribution identical to that from the Metropolis algorithm. However, the efficiency is not the same. As shown in Figure 6.18, with the same number of MC states, the orientational bias simulation is significantly more accurate than the Metropolis algorithm. As expected, the biased sampling increases the rate of acceptance (ACC) for the MC moves, which is particularly important for MC sampling with big steps. Understandably, the biased sampling scheme is not necessary for calculating the angle distribution of a point dipole in a uniform electrical field. Nevertheless, this simple system helps to illustrate how biased sampling can be implemented beyond the Metropolis algorithm. From the practical perspective, orientational bias may find applications in sampling the microstates of thermodynamic systems containing non-spherical particles or polar molecules (e.g., liquid crystals or electrolyte solutions). For such systems, the intermolecular interactions depend not only on the particle positions but also on particle orientations. Similar to sampling the orientational distribution of a point dipole, MC moves favoring low-energy orientations will significantly improve the sampling efficiency.

20 See Frenkel D. and Smit B., *Understanding molecular simulation* (2nd Edition). Academic Press, pp. 327–329.

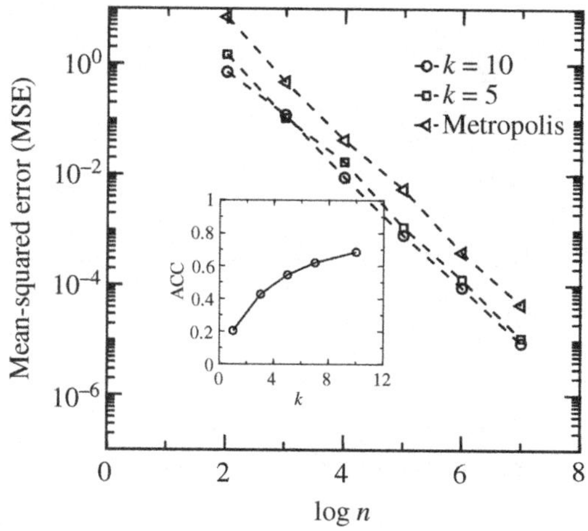

Figure 6.18 The mean squared error (MSE) versus the number of MC steps (n) for the angular distribution of a point dipole in a uniform electric field ($y = 10$) calculated according to the Metropolis and orientational-bias sampling methods ($k = 5$ and 10). The inset plots the dependence of the average acceptance rate (ACC) on the number of attempted orientations k.

6.7.3 Configurational Bias Monte Carlo

The configurational bias (CB) method was originally proposed by Rosenbluth and Rosenbluch for sampling the conformations of a single polymer chain on a lattice in the early history of MC simulation.[21] However, a rigorous foundation for generating polymer conformations according to the Boltzmann statistics was not established until several decades later.[22] Today, the asymmetric sampling method represents perhaps the most popular scheme for sampling the configurational space of polymeric systems.

The essential idea of CB sampling is to make efficient MC moves at the level of polymer chains instead of individual segments. While straightforward molecular moves (e.g., by translations or rotations of the entire chain) are possible in a dilute system, the probability of acceptance for taking big MC steps diminishes as the polymer concentration increases and vanishes in concentrated polymer systems or polymer melts. In CB sampling, we generate polymer conformations segment-by-segment such that the position of each segment is selected from a number of trial moves according to a Boltzmann weight. The biased scheme enhances conformational sampling by avoiding the overlap of the trial moves with other molecules in the system.

More specifically, Figure 6.19 illustrates how we update the conformation of a polymer chain with l spherical segments in the CB simulation. Suppose that one polymer chain is selected by random in a particular MC move at step i. To create a new polymer conformation for the next microstate, we start by selecting a new position to place the first bead (1) of the polymer chain from k trial positions according to the Boltzmann distribution

$$t_{i1} = \frac{e^{-\beta\varepsilon_1}}{w_j(1)} \tag{6.71}$$

where subscript j denotes the polymer conformation to be sampled, ε_1 is the interaction energy between the first bead and all *other* molecules in the system, $w_j(1) = \sum_{n=1}^{k} e^{-\beta\varepsilon_n}$ is the normalization constant for the Boltzmann probability, and subscript n denotes the index of the k trial positions.

21 Rosenbluth M. N. and Rosenbluth A. W., "Monte Carlo calculation of the average extension of molecular chains", *J. Chem. Phys.* 23, 356 (1955).
22 Siepmann J. I. and Frenkel D., "Configurational bias Monte Carlo: a new sampling scheme for flexible chains", *Mol. Phys.* 75, 59–70 (1992); De Pablo J. J., Laso M. and Suter U. W., "Simulation of polyethylene above and below the melting point", *J. Chem. Phys.* 96, 2395–2403 (1992).

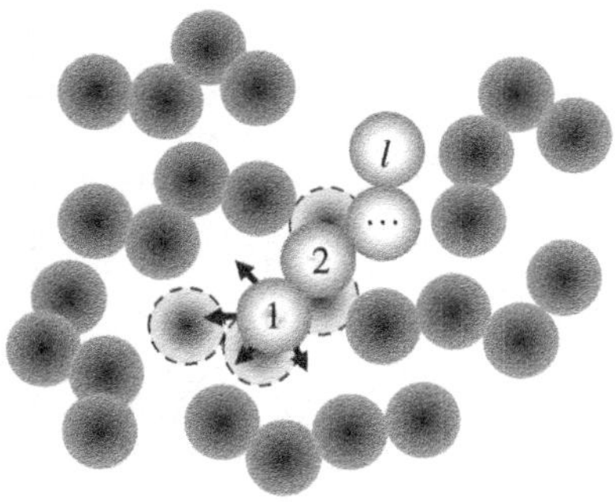

Figure 6.19 In configurational bias sampling, a polymer conformation is built up segment by segment in each MC step where the position of each segment is selected from k possible attempts with a probability dictated by the Boltzmann distribution. The dashed lined circles represent polymer segments at the current chain conformation, the numbered spheres represent the segments at the new polymer conformation, and the darker spheres represent polymer segments in the background.

Next, we select the position for the second bead of the polymer chain in the new conformation also according to the Boltzmann distribution

$$t_{i2} = \frac{e^{-\beta\varepsilon_2 - \beta\Delta\varepsilon_B}}{w_j(2)} \tag{6.72}$$

where ε_2 is the energy of the second bead due to its interaction with all *other* chains in the system, $\Delta\varepsilon_B$ presents an additional energy arising from the interaction between the first and second beads (including the bond energy), and $w_j(2) = \sum_{n=1}^{k} e^{-\beta\varepsilon_n - \beta\Delta\varepsilon_{B,n}}$ is the normalization constant for selecting the second bead out of k trial positions. The process continues until the position for the last segment l has been selected.

The overall probability of generating a new polymer conformation is given by

$$t_{ij} = t_{i1} \times t_{i2} \times \cdots \times t_{il} = \prod_{\alpha=1}^{l} \frac{e^{-\beta\varepsilon_\alpha - \beta\Delta\varepsilon_{B,\alpha}}}{w_j(\alpha)} = \frac{e^{-\beta E_j}}{W_j} \tag{6.73}$$

where $E_j = \sum_{\alpha=1}^{l}(\varepsilon_\alpha + \Delta\varepsilon_{B,\alpha})$ represents the total energy of the polymer chain in conformation j, and $W_j = \prod_{\alpha=1}^{l} w_j(\alpha)$ is called the Rosenbluth factor, i.e., the normalization constant for the probability of conformation sampling. It should be noted that E_j includes contributions from both inter- and intramolecular interactions (viz., both bonding and nonbonding energies of the polymer). As the position of each polymer segment is selected according to the Boltzmann distribution, the biased sampling scheme excludes overlapping among polymer chains thus increasing the probability of generating a trial conformation.

For the reverse process from configuration j to i, $n - 1$ trial positions are generated for each polymer segment and the probability of original position is given by

$$t_{j\alpha} = \frac{e^{-\beta\varepsilon_\alpha}}{w_i(\alpha)} \tag{6.74}$$

where $w_i(\alpha) = e^{-\beta\varepsilon_\alpha} + \sum_{m=1}^{k-1} e^{-\beta\varepsilon_{\alpha,m}}$ is the normalization factor, and ε_α includes both bonding and nonbonding energies. Accordingly, the reverse trial probability for the entire chain is given by

$$t_{ji} = \frac{e^{-\beta E_i}}{W_i} \tag{6.75}$$

where E_i represents the total energy of the polymer chain in conformation i, and $W_i = \prod_{\alpha=1}^{l} w_i(\alpha)$ is the Rosenbluth factor in the original polymer conformation. In the canonical ensemble, the probability of polymer in different conformations is proportional to the Boltzmann factor

$$p_j/p_i = e^{-\beta(E_j - E_i)}. \tag{6.76}$$

Substituting Eqs. (6.73) to (6.76) into Eq. (6.64) leads to the probability of acceptance for changing the entire polymer conformation[23]

$$a_{ij} = \min(1, W_j/W_i). \tag{6.77}$$

Not surprisingly, Eq. (6.77) is identical to that for the orientational biased sampling for a point dipole in an electric field, viz., Eq. (6.70). Indeed, the criterion for the biased sampling is the same, both are based on the Boltzmann distribution of microstates.

As mentioned above, CB sampling is commonly used in MC simulation of polymer systems. Because the polymer conformations are built up segment-by-segment, it allows for the MC moves to any portion of a polymer chain or portions of several chains.[24] The flexibility in updating polymer conformations allows for optimizing the efficiency of simulation by balancing the size of MC steps and the acceptance rate.

One drawback with the segment-by-segment generation of polymer conformations is that it may lead to dead ends, i.e., the probability of finding segment positions diminishes as the chain length increases. To avoid such situation, we may use extended CB strategies such as the recoil-growth scheme[25] or the Pruned-Enriched Rosenbluth Method (PERM).[26] The basic idea behind these extended CB methods is to avoid investing computational effort in generating trial moves that are eventually rejected. The recoil-growth scheme is designed to escape from the dead ends by "recoiling back" a few monomers and retrying the growth process using another trial orientation. Unlike the conventional CB method that generates a polymer conformation based on the energy of a single segment at each step, the recoil growth scheme allows for the evaluation of the statistical weight for several segments before a segment is irrevocably added to the trial conformation.

In PERM, we make multiple copies of partially grown chains instead of a single chain in order to improve the successful rate for generating an entirely new chain conformation. The rationale behind pruning is that it is not useful to spend much computer time on the generation of conformations that have a low Rosenbluth weight. Therefore, it is advantageous to discard ("prune") such irrelevant conformations at an early stage. The idea behind enrichment is to make multiple copies of partially grown chains that have a large statistical weight.

6.7.4 Summary

Considerable efforts have been made to improve the efficiency of MC simulations. These endeavors are often focused on generating multiple trial states and implementing "smart" MC moves, allowing for sampling microstates with statistical significance. In contrast to the equations of motion in MD simulation, MC sampling offers a notable advantage by incorporating non-physical moves, which expedite sampling and circumvent energy barriers. Biased sampling methods are especially valuable in navigating the configurational space of complex thermodynamic systems, such as polymer melts and liquid crystals.

---

23 Because the segment position is a continuous variable, the probability density is normalized with a differential volume that depends on specific coordinate system, i.e., $\int p(\mathbf{r})d\mathbf{r} = 1$. Eq. (6.77) must include an additional term if different coordinates are used in t_{ij} and t_{ji}. For details, see Dodd L. R., Boone T. D. and Theodorou D. N., "A concerted rotation algorithm for atomistic Monte Carlo simulation of polymer melts and glasses", *Mol. Phys.*, 78, 961 (1993).

24 Mavrantzas V. G., "Using Monte Carlo to simulate complex polymer systems: recent progress and outlook", *Front. Phys.* 9, 661367 (2021).

25 Consta S. et al., "Recoil growth algorithm for chain molecules with continuous interactions", *Mol. Phys.* 97, 1243–1254 (1999).

26 Grassberger P., "Pruned-enriched Rosenbluth method: Simulations of theta polymers of chain length up to 1,000,000", *Phys. Rev. E* 56, 3682–3693 (1997).

6.8 Free-Energy Calculation Methods

MC simulation is efficient to calculate configuration averages such as those entailed in the internal energy, pressure, and radial distribution function (RDF). The procedure is relatively straightforward because configurational properties can be directly obtained from the positions of individual particles and the underlying intermolecular forces. However, the calculation of free energy (or entropy) is more challenging because they are inherently linked to the absolute values of microstate probabilities or the partition function. Determining the precise values of these quantities requires additional computational techniques beyond the straightforward analysis of particle positions and intermolecular forces.

To see why a direct free-energy calculation is problematic, consider the connection between Helmholtz energy F and canonical partition function Q

$$F = -k_B T \ln Q \tag{6.78}$$

where k_B is the Boltzmann constant, and T is the absolute temperature. Calculation of the partition function requires integrations over all degrees of freedom in the system (e.g., the positions and momenta of classical particles). Because of the high dimensionality, Q cannot be calculated directly for most nonideal systems. This difficulty, however, can be circumvented by considering ΔF, i.e., the difference in Helmholtz energy between two thermodynamic states. By using standard thermodynamic relations or well-established statistical–mechanical equations, we can evaluate the Helmholtz energy of a system indirectly.

In this section, we will explore strategies for calculating the free energy and chemical potential using MC simulation. These calculations play a crucial role in describing a wide range of chemical and biochemical phenomena, including chemical reaction equilibrium, molecular solvation, molecular association, macromolecular stability, and enzyme catalysis.[27] By determining the free energy and chemical potential, we gain valuable insights into the thermodynamics and energetics of these systems, enabling a deeper understanding of their behavior and properties.

6.8.1 Thermodynamic Integration

Thermodynamic integration refers to a class of simulation methods that can be used to predict relative free energies, i.e., the free energy of a system relative to that of a reference state. These methods involve the numerical integration of thermodynamic derivatives through simulations conducted under different conditions.

One of the most straightforward methods to obtain relative free energy is by integrating its derivative with respect to pressure or inverse temperature. For example, consider the partial derivatives of the Helmholtz energy with respect to volume (V) and with respect to inverse temperature, $\beta = 1/(k_B T)$, for a one-component system containing N classical particles

$$(\partial F/\partial V)_{N,T} = -P, \tag{6.79}$$

$$(\partial \beta F/\partial \beta)_{N,V} = U. \tag{6.80}$$

Eq. (6.79) can be derived from the fundamental equation of thermodynamics, and Eq. (6.80) is the familiar Gibbs–Helmholtz equation. By selecting a reference state (e.g., ideal gas or harmonic

27 Mey A. S. J. S. et al., "Best practices for alchemical free energy calculations", *Living J. Comput. Mol. Sci.* 2 (1), 18378 (2020).

crystal) with a known Helmholtz energy, integration of Eq. (6.79) or (6.80) provides a thermodynamic route connecting the reference state with the system under investigation:

$$F(V) = F(V_0) - \int_{V_0}^{V} P(V')dV', \tag{6.81}$$

$$\beta F = (\beta F)_0 - \int_{T_0}^{T} \frac{U(T')}{k_B T'^2} dT', \tag{6.82}$$

where subscript 0 denotes a reference system. By performing MC simulation for pressure P and internal energy U at a series of conditions along the thermodynamic path, Eqs. (6.81) and (6.82) allow us to estimate the difference in the Helmholtz energy.

In molecular simulation, calculation of Helmholtz energy is not limited to integration with respect to physical variables like volume or reciprocal temperature. For example, the Helmholtz energy can also be calculated from perturbation, i.e., from the variation of the Helmholtz energy with respect to the intermolecular potential.[28] The numerical procedure is known as coupling-parameter method. For a system of spherical particles, the integration can be carried out through simulation over a set of hypothetical systems defined by the pair potential

$$u_\lambda(\mathbf{r}_1, \mathbf{r}_2) = (1-\lambda)u_0(\mathbf{r}_1, \mathbf{r}_2) + \lambda u(\mathbf{r}_1, \mathbf{r}_2) \tag{6.83}$$

where $0 \le \lambda \le 1$ is called the coupling parameter. When $\lambda = 1$, Eq. (6.83) gives $u(\mathbf{r}_1, \mathbf{r}_2)$, the potential energy for two real particles sitting at positions $\mathbf{r}_1$ and $\mathbf{r}_2$; when $\lambda = 0$, Eq. (6.83) reduces to the reference pair potential, $u_0(\mathbf{r}_1, \mathbf{r}_2)$. Integration of the variational energy with respect to λ provides a way to obtain the difference in the Helmholtz energy[29]

$$F = F_0 + \frac{\rho^2}{2} \int_0^1 d\lambda \int d\mathbf{r}_1 \int d\mathbf{r}_2 g(\mathbf{r}_1, \mathbf{r}_2, u_\lambda) \Delta u(\mathbf{r}_1, \mathbf{r}_2) \tag{6.84}$$

where subscript 0 denotes a reference system at temperature, volume, and number density $\rho = N/V$ identical to those of the system of interest, and $\Delta u(\mathbf{r}_1, \mathbf{r}_2) = u_1(\mathbf{r}_1, \mathbf{r}_2) - u_0(\mathbf{r}_1, \mathbf{r}_2)$. In this reference system, the pairwise-additive intermolecular potential is given by $u_0(\mathbf{r}_1, \mathbf{r}_2)$. The RDF $g(\mathbf{r}_1, \mathbf{r}_2, u_\lambda)$ corresponds to that for a system whose two-body potential is given by Eq. (6.83).

In Eq. (6.84), the integration with respect to particle positions can be evaluated with MC simulation using the perturbation potential at coupling parameter λ. Consequently, the thermodynamic integration becomes

$$F = F_0 + \frac{1}{2} \int_0^1 d\lambda < \sum_{i=1}^{N-1} \sum_{j \neq i}^{N} \Delta u(\mathbf{r}_i, \mathbf{r}_j) >_\lambda. \tag{6.85}$$

In writing Eq. (6.85), we have replaced the radial distribution with a double summation over indices i and j, i.e., overall particle pairs.

Numerical implementation of the thermodynamic-integration methods is rather straightforward. However, there are practical issues that one should keep in mind. Firstly, these methods can be computationally time-demanding because they require running multiple computer simulations for the numerical integration. Each simulation needs to be sufficiently long to minimize statistical errors and achieve high accuracy, resulting in a high computational burden. Secondly, it is crucial to carefully choose the integration path to avoid unwanted phase transitions. Selecting an appropriate path ensures that the system remains in a well-behaved state throughout the integration, preventing any abrupt changes or discontinuities in the calculated free energy values.

28 This is known as the functional derivative in the calculus of variation, i.e., $\delta F/\delta u(\mathbf{r}_1, \mathbf{r}_2) = g(\mathbf{r}_1, \mathbf{r}_2)$.

29 Equation (6.84) corresponds to the functional integration, a generalization of conventional integration to the functional space.

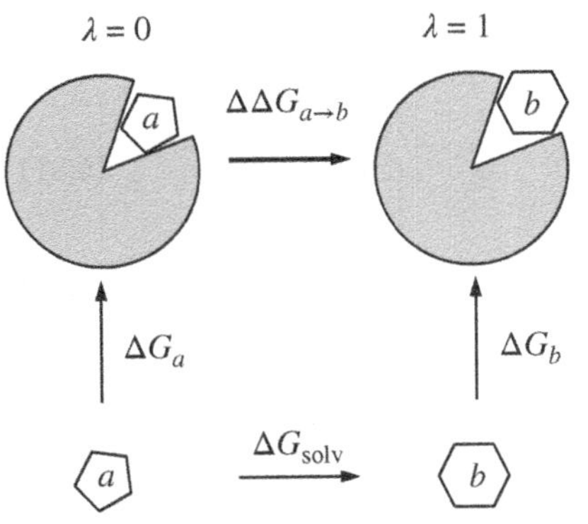

Figure 6.20 Schematic of a thermodynamic cycle to calculate the difference in the free energy of binding ΔG between ligands a and b with a protein. Both $\Delta\Delta G_{a \to b}$ and ΔG_{solv} can be calculated with the thermodynamic integration method.

The coupling-parameter method is not restricted to systems with pairwise additive intermolecular potentials. It can be similarly applied to a wide range of problems involving non-covalent binding. For example, we can use the coupling-parameter method to evaluate the binding free energy for ligand-protein interactions as shown schematically in Figure 6.20. Relative to that of a reference process with the binding free energy ΔG_a for ligand a, the free energy of binding for the target ligand b can be calculated through the coupling-parameter method for the difference in solvation free energy ΔG_{solv}, and for the replacement free energy $\Delta\Delta G_{a \to b}$. These types of calculations play a crucial role in computational-assisted product design, particularly in the fields of drug discovery and personalized medicine.

6.8.2 Sampling Chemical Potential

For a canonical system of classical particles at constant temperature (T) and volume (V), the particle chemical potential can be related to the change in the Helmholtz energy F when a particle is added to (or removed from) the system:

$$\mu = F(N+1) - F(N)|_{T,V}. \tag{6.86}$$

Eq. (6.86) is exact in the limit of large N, corresponding to a macroscopic system.

In Eq. (6.86), the change in Helmholtz energy can be evaluated by considering a system containing N+1 particles and a fictitious system that has N particles plus a "ghost" particle. Here, the "ghost" particle is identical to a real particle except that it does not interact with any other particles in the system. Therefore, the contribution of the "ghost" particle to the Helmholtz energy of the fictitious system is the same as the chemical potential of an ideal gas at the system temperature T, volume V, and total number of particles N,

$$\mu^{id} = F(N,1) - F(N)\Big|_{T,V} \tag{6.87}$$

where $F(N, 1)$ denotes the Helmholtz energy of the fictitious system. By comparing Eqs. (6.86) and (6.87), we find that the excess chemical potential can be expressed as the difference in the Helmholtz energy

$$\mu^{ex} \equiv \mu - \mu^{id} = F(N+1) - F(N,1). \tag{6.88}$$

Eq. (6.88) provides a starting point to evaluate the chemical potential of classical particles using MC simulation.

Using the relation between the Helmholtz energy and the canonical partition function, Eq. (6.78), we obtain

$$e^{-\beta F} = Q = \frac{Q^{id}}{V^N} \int d\mathbf{r}^N e^{-\beta\Phi(\mathbf{r}^N)} \tag{6.89}$$

where Q^{id} stands for the ideal-gas partition function. Plugging Eq. (6.89) into (6.88) gives

$$-\beta\mu^{ex} = \ln Q(N+1) - \ln Q(N,1) = \ln \frac{\int d\mathbf{r}^N e^{-\beta\Phi(N+1)}}{\int d\mathbf{r}^N e^{-\beta\Phi(N,1)}}. \tag{6.90}$$

The ideal-gas part of the partition function (Q^{id}/V^N) cancels because the real and fictitious systems have the same number of particles.

Eq. (6.90) can be expressed in terms of an ensemble average

$$e^{-\beta\mu^{ex}} = \frac{\int d\mathbf{r}^N e^{-\beta\Phi(N,1)} e^{-\beta\Delta\Phi}}{\int d\mathbf{r}^N e^{-\beta\Phi(N,1)}} \equiv \langle e^{-\beta\Delta\Phi}\rangle_+ \tag{6.91}$$

where the bracket $< \ldots >_+$ denotes an ensemble average for a system that contains N real particles plus one "ghost" particle, and $\Delta\Phi = \Phi(N+1) - \Phi(N, 1)$, i.e., the difference between the potential energy of the real system with $N+1$ particles and that of the fictitious system (with N real particle and one ghost particle). Eq. (6.91) relates the excess chemical potential μ^{ex}, a property that has no direct microscopic counterpart, to an ensemble average.

In calculation of Eq. (6.91) with simulation, the ensemble average can be evaluated by inserting a "test" particle, as shown schematically in Figure 6.21, to a system containing N real particles

$$e^{-\beta\mu^{ex}} = \frac{1}{n_s}\sum_{i=1}^{n_s} e^{-\beta\Delta\Phi(i)} \tag{6.92}$$

where n_s is the total number of insertions (viz., MC samples), and $\Delta\Phi(i)$ stands for the potential energy due to the interaction of the inserted particle with N real particles in the system. This potential is the same as the difference in the total potential energy between a system containing $N+1$ particles and a fictitious system containing N real particles and one "ghost" particle. Because the ensemble average applies to the fictitious system, the "test" particle (i.e., the "ghost" particle) is used only for the calculation of $\Delta\Phi(i)$ at a particular configuration of N particles and at the given position of the "ghost" particle, i.e., the "ghost" particle is not involved in the MC moves.

In molecular simulation, the idea of using a test particle (viz., "ghost" particle) to calculate the excess chemical potential is called Widom's insertion method,[30] named after Benjamin Widom who first derived Eq. (6.91). In principle, the excess chemical potential can be evaluated by either adding or removing a particle from the system. In the latter case, the excess chemical potential is related to the ensemble average of $e^{\beta\Delta\Phi}$ in the system containing N particles

$$e^{-\beta\mu^{ex}} = \frac{\int d\mathbf{r}^N e^{-\beta\Phi(N)}}{\int d\mathbf{r}^N e^{-\beta\Phi(N)} e^{\beta\Delta\Phi}} \equiv 1/\langle e^{\beta\Delta\Phi}\rangle. \tag{6.93}$$

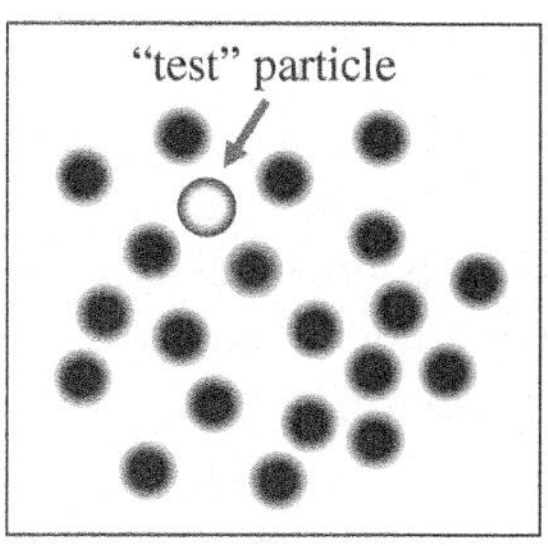

Figure 6.21 In the particle-insertion method, a "test" particle is placed in the simulation box at a random position; the average interaction energy between this test particle and all other particles is related to the excess chemical potential as given by Eq. (6.91).

30 Widom B., "Some topics in the theory of fluids", *J. Chem. Phys.*, 39 (11), 2808–2812 (1963).

Unlike Eq. (6.91), here the ensemble average entails no "ghost" particle. While both Eqs. (6.91) and (6.93) are mathematically rigorous, they differ in numerical efficiency when evaluated by MC simulation. In Eq. (6.93), the large values are associated with the small probabilities of microstates, which are dictated by the Boltzmann factor $e^{-\beta\Delta\Phi}$. As a result, the numerical efficiency of this alternative method is often low. This is because the occurrence of rare events, corresponding to large changes in energy, requires a significant amount of computational effort to accurately sample and calculate the associated probabilities. In Widom's insertion method, by contrast, large values of $e^{-\beta\Delta\Phi}$ coincide with the most probable distribution of the equilibrium configurations. Because of its poor numerical efficiency, Eq. (6.93) is rarely used in MC simulation of the excess chemical potential.

It is worth noting that the thermodynamic integration method, as discussed earlier, offers a reliable approach for calculating the difference in the Helmholtz energy given by Eq. (6.88). Here, $\lambda = 0$ represents an uncoupled (or ghost) particle, while $\lambda = 1$ signifies the complete coupling of the test particle with the surrounding environment. In contrast to Widom's insertion method, the alternative procedure is more time-consuming because the progress variable λ gradually increases from 0 to 1.

Widom's insertion method has been widely used in computational chemistry and statistical mechanics to estimate the chemical potential of a substance. Regrettably, the insertion becomes inefficient at high particle densities because of the molecular excluded-volume effect. As a system becomes dense, it is difficult to find a "hole" large enough to accommodate the test particle. The excluded-volume effect is particularly pronounced when we are concerned with a system containing multiple types of particles or polymers where the probability of inserting a test particle at a random position is extremely small. Various strategies have been proposed to improve the sampling efficiency. For example, at high densities, the insertion probability can be extrapolated from a series of calculations for insertion of smaller particles. For polymeric systems, the sampling efficiency can be improved following configuration-biased methods.

As mentioned above, the correct value of the excess chemical potential is obtained in the limit of an infinitely large system ($N \to \infty$). However, the system size is finite in practical simulations. As a result, a finite-size correction to μ^{ex} is often required when N is not sufficiently large. A simple way to do that is by carrying out simulations for different values of N and extrapolating μ^{ex} with respect to $1/N$. If an equation of state is known for the system under investigation, the leading correction can be estimated from[31]

$$\Delta\mu^{ex}(N) = \frac{1}{2N}\left(\frac{\partial P}{\partial \rho}\right)_T \left\{1 - k_B T\left(\frac{\partial \rho}{\partial P}\right)_T - \rho k_B T \frac{(\partial^2 P/\partial \rho^2)_T}{(\partial P/\partial \rho)_T}\right\}. \tag{6.94}$$

Eq. (6.94) can be obtained by a comparison of the reversible work to insert a particle using the canonical and grand canonical ensembles at the same volume and temperature.

6.8.3 Summary

Thermodynamic integration and particle insertion are two common strategies to calculate free energy using MC simulation. Thermodynamic integration focuses on the variations of free energy in response to changes in thermodynamic conditions or model parameters. By simulating the system under a range of conditions, we can determine the free energy difference between the system of interest and a reference state (or reference system). Conversely, the particle insertion method is able to evaluate the excess chemical potential at a specific condition of interest. However, this

31 Siepmann J. I., McDonald I. R. and Frenkel D., "Finite-size corrections to the chemical potential", *J. Phys. Condensed Matter* 4, 679 (1992).

method becomes less efficient at high particle densities due to the increased likelihood of overlap or steric repulsion. In addition to these approaches, more sophisticated methods, such as multiple histogram analysis and enhanced sampling techniques, will be discussed in subsequent sections. These methods offer advanced strategies for accurately estimating free energy in MC simulation, addressing challenges associated with complex systems and improving sampling efficiency.

6.9 Simulation of Crystalline Solids

Many industrial products, especially pharmaceuticals, agrochemicals, and crystalline materials, are manufactured and utilized in their solid forms. The thermodynamic properties of these systems play a crucial role in industrial design and manufacturing processes, as they help determine solubility and phase behavior during fluid–solid and solid–solid transitions. In this section, we present fundamental concepts for predicting the free energy of solids using MC (or MD) simulation.

6.9.1 Permutation Symmetry

A solid distinguishes itself from a liquid or gas in terms of structural rigidity, i.e., the localization of molecules or atoms near their equilibrium positions. While an amorphous solid lack any regular pattern in the microscopic structure, molecules are self-organized in a crystalline phase with lattice-like order and periodicity.[32] As discussed in the following, the ordered packing of atomic species has profound implications on the thermodynamic properties of a crystalline solid.

From a microscopic perspective, one of the fundamental characteristics of crystalline solids is the presence of permutation symmetry in the statistical behavior of individual particles located on the lattice sites. We may elucidate this unique feature of crystals by considering a system of spherical particles as shown schematically in Figure 6.22. Because classical particles are indistinguishable from each other, the thermodynamic properties of the system will not be changed by switching the particle positions. The invariance with respect to the spatial arrangement of particles is referred to as the permutation symmetry.

We can express the canonical partition function for a system of N indistinguishable classical particles as follows (Section 6.6.1)

$$Q = \frac{1}{N!\Lambda^{3N}} \int d\mathbf{r}^N e^{-\beta\Phi(\mathbf{r}^N)} \tag{6.95}$$

where $N!$ accounts for the permutation symmetry, Λ is the thermal wavelength, and $\Phi(\mathbf{r}^N)$ represents the potential energy. In principle, Eq. (6.95) is equally applicable to both fluid and solid

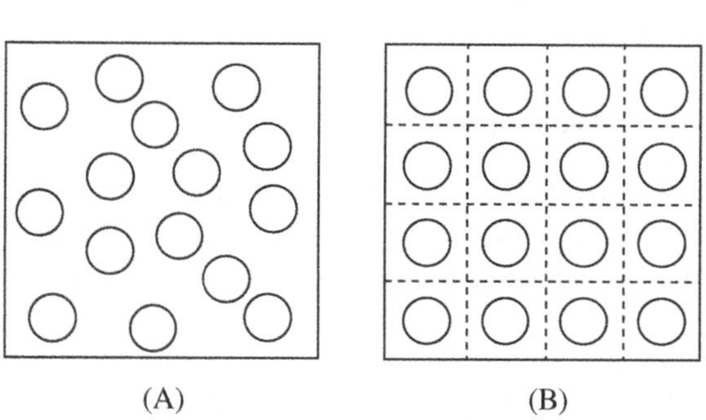

Figure 6.22 Identical particles in a fluid phase can switch their positions through translational motions (A). However, the position change is not allowed in a solid phase because particles are localized (B).

32 A quasicrystal is a solid with ordered structure but no periodicity.

phases, given that the temperature is sufficiently high to render quantum effects negligible. While application of the canonical partition function to a fluid system is rather straightforward, a subtle issue arises as the exchange of positions rarely takes place among particles in the solid phase. In fact, a perfect crystal exhibits non-ergodic behavior due to the division of its configurational space into $N!$ equivalent subdomains that are inaccessible to each other. As a result, the "real" entropy of the system is never observed within the relevant time scale for practical applications.

The issue with the permutation symmetry can be resolved if we consider the inherent rigidity of a solid phase, i.e., particles are confined within a single configurational sub-space such that they are distinguishable by their positions (viz., by considering "a specific solid state" instead of "a generic solid state").[33] This idea of distinguishability of identical particles in a solid phase is rather intuitive from a classical perspective, and the application of quantum statistics to atomic crystals does not make any difference. Since particles in the solid state can be distinguished by their positions, the canonical partition function for a crystal is then given by

$$Q_S = \frac{1}{\Lambda^{3N}} \int d\mathbf{r}^N e^{-\beta\Phi(\mathbf{r}^N)} \tag{6.96}$$

where subscript S denotes a single configurational sub-space. In comparison with Eq. (6.95), Eq. (6.96) does not include the factorial $N!$ because the permutation symmetry is not applied to a *specific* permutation of particles.

It is worth noting that both the *specific* and *generic* interpretations of a crystalline solid yield the same relative thermodynamic quantities. In the case of a *generic* solid, we incorporate a factor of $1/N!$ in the partition function, analogous to the normal canonical system of identical particles. In that case, we must also consider $N!$ mutually inaccessible sub-domains in evaluation of the configurational integration, leading to the cancellation of the factorial term.

Eq. (6.96) can be readily extended to crystalline solids with different particles such as alloys or solid solutions. If the lattice sites are occupied by different particles, the canonical partition function can be similarly written as

$$Q_S = \left(\prod_{i=1}^{N} \frac{1}{\Lambda_i^3}\right) \int d\mathbf{r}^N e^{-\beta\Phi(\mathbf{r}^N)}. \tag{6.97}$$

Eq. (6.97) indicates that, different from that for a gas or liquid mixture, the entropy of mixing is not materialized in a multicomponent alloy or solid solution.[34] Although Eq. (6.97) is not strictly valid at low temperature, the absence of the factorial term is fully consistent with the third law of thermodynamics, i.e., the entropy of a perfect crystal vanishes at 0 K. In other words, the third law is equally applicable to perfect crystalline solids at absolute zero temperature, either with only a single type of atoms (e.g., metals) or with multiple types of atoms (e.g., alloys, ceramics, or molecular crystals).

6.9.2 The Einstein Crystal

Einstein crystal provides one of the simplest representations of classical particles in a solid state.[35] As the thermodynamic properties of an Einstein crystal can be readily derived analytically, the simple model is often used as a reference for simulating the free energy of real crystals.

---

33 Alexander S., "What is a solid?", *Physica A* 249, 266–275 (1998).

34 Schneider J. M., "How high is the entropy in high entropy ceramics?", *J. Appl. Phys.* 130, 150903 (2021).

35 It should not be confused with the Einstein model of phonons discussed in Section 4.6. In that case, a solid state is represented by noninteracting phonons.

As illustrated schematically in Figure 6.22B, an Einstein crystal is represented by a lattice with N independent cells of the same volume, $v = V/N$, with each cell accommodating only a single particle. The total potential of the system is given by the summation of the harmonic energies of individual particles

$$\Phi_0(\mathbf{r}^N) = \sum_{i=1}^{N} \frac{k}{2} |\mathbf{r}_i - \mathbf{r}_{0,i}|^2 \tag{6.98}$$

where k represents the spring constant, $\mathbf{r}_i$ is the position of particle i, and $\mathbf{r}_{0,\,i}$ corresponds to the position of a lattice site affiliated with particle i. The harmonic potential restricts the position of each particle near a lattice site.

We may evaluate the canonical partition function of the Einstein crystal by integrating the harmonic potential

$$Q_0 = \frac{1}{\Lambda^{3N}} \int d\mathbf{r}^N e^{-\beta\Phi_0(\mathbf{r}^N)} = \frac{1}{\Lambda^{3N}} \left\{ \int_{-\infty}^{\infty} dx \exp\left[-\frac{\beta k}{2}(x - x_0)^2 \right] \right\}^{3N} = \left(\frac{2\pi}{\beta k \Lambda^2} \right)^{3N/2} \tag{6.99}$$

where Λ represents the thermal wavelength, $\beta = 1/(k_B T)$, x denotes the translational degree of freedom for the particle motion in a particular dimension, and $3N$ is the total degrees of freedom for all particles. In contrast to the partition function of a conventional canonical system with N identical particles, Eq. (6.99) does not include the factorial term, $1/N!$. Again, this is because the particles are labeled by the lattice sites, i.e., the affiliation between the potential energy and lattice positions makes the particles in the Einstein crystal distinguishable.

From Eq. (6.99), we can derive the Helmholtz energy

$$F_0 = -k_B T \ln Q_0 = \frac{3Nk_B T}{2} \ln\left(\frac{\beta k \Lambda^2}{2\pi} \right) \tag{6.100}$$

and all other thermodynamic quantities. Particularly, Eq. (6.100) allows us to derive the entropy

$$S_0 = -\left(\frac{\partial F_0}{\partial T} \right)_{V,N} = -3Nk_B \ln\left(\frac{\beta h}{2\pi} \sqrt{k/m} \right) \tag{6.101}$$

where h is Planck's constant, and m is the particle mass. Note that Eq. (6.101) is not consistent with the third law because the Einstein crystal entails classical approximations that are not valid at $T = 0$ K. As expected, both F_0 and S_0 are linearly scaled with the number of particles N. Apparently, the extensiveness of the thermodynamic quantities is also preserved if we consider the Einstein crystal in *a generic* solid state. In that case, the factorial term $1/N!$ in the partition function cancels with $N!$ mutually inaccessible configurational sub-domains for the solid state.

If the Einstein crystal consists of particles with different spring constants and masses, Eq. (6.100) can be written as

$$F_0 = \frac{3k_B T}{2} \sum_{i=1}^{N} \ln\left(\frac{\beta k_i \Lambda_i^2}{2\pi} \right) \tag{6.102}$$

where $\Lambda_i = h/\sqrt{2\pi m_i k_B T}$ is the thermal wavelength for particle i. Eq. (6.102) follows from the assumption that particles in the Einstein crystal are independent of each other. As expected, Eq. (6.102) reduces to Eq. (6.100) when all particles are identical.

6.9.3 The Frenkel–Ladd Method

The coupling-parameter method is often used to calculate the free energy of a solid system. The idea was proposed first by Frenkel and Ladd for atomic crystals, i.e., crystalline solids consisting of a single type of spherical particles.[36] As discussed later, the procedure can be modified for calculating the free energies of molecular crystals.

In the Frenkel–Ladd method, the reference potential was selected to be the same as that of an Einstein crystal as discussed above. To make a connection between the partition functions of the reference and real crystals, we define a coupling potential

$$\Phi_\lambda(\mathbf{r}^N) = \Phi\left(\mathbf{r}_0^N\right) + \lambda\left[\Phi(\mathbf{r}^N) - \Phi\left(\mathbf{r}_0^N\right)\right] + (1-\lambda)\Phi_0(\mathbf{r}^N) \tag{6.103}$$

where $\Phi(\mathbf{r}^N)$ and $\Phi_0(\mathbf{r}^N)$ are the potential energies of the real and reference systems, respectively, $\Phi\left(\mathbf{r}_0^N\right)$ is the potential energy of the real crystal when all particles are at their equilibrium positions, and $0 \leq \lambda \leq 1$ stands for a coupling parameter connecting the real and reference systems. For each value of λ, we can write the partition function corresponding to the coupling potential $\Phi_\lambda(\mathbf{r}^N)$

$$Q_\lambda \equiv \frac{1}{\Lambda^{3N}} \int d\mathbf{r}^N e^{-\beta\Phi_\lambda(\mathbf{r}^N)}. \tag{6.104}$$

When $\lambda = 0$, Q_0 corresponds to the partition function of the Einstein crystal multiplied a constant $\exp\left[-\beta\Phi\left(\mathbf{r}_0^N\right)\right]$; for $\lambda = 1$, Q_1 is the partition function of the real system.

Note that

$$\frac{\partial \ln Q_\lambda}{\partial \lambda} = -\frac{\int d\mathbf{r}^N (\partial\beta\Phi_\lambda/\partial\lambda) e^{-\beta\Phi_\lambda(\mathbf{r}^N)}}{\int d\mathbf{r}^N e^{-\beta\Phi_\lambda(\mathbf{r}^N)}} = -\beta <\Delta\Phi(\mathbf{r}^N)>_\lambda \tag{6.105}$$

where $\Delta\Phi(\mathbf{r}^N) = \Phi(\mathbf{r}^N) - \Phi_0(\mathbf{r}^N) - \Phi\left(\mathbf{r}_0^N\right)$, and $<\cdots>_\lambda$ represents an ensemble average in the canonical system of N particles with the coupling potential $\Phi_\lambda(\mathbf{r}^N)$. Integration of Eq. (6.105) with respect to λ leads to

$$F_1 = F_0 + \int_0^1 d\lambda <\Delta\Phi(\mathbf{r}^N)>_\lambda \tag{6.106}$$

where $F_0 = -k_BT \ln Q_0$ and $F_1 = -k_BT \ln Q_1$. The free energy of the real crystal is thus given by

$$F = F_0 + \int_0^1 d\lambda <\Delta\Phi(\mathbf{r}^N)>_\lambda. \tag{6.107}$$

In Eq. (6.107), the integration with respect to λ can be accomplished numerically with, for example, the Gauss–Legendre quadrature method.

6.9.4 Center of Mass Constraints and Finite-Size Effects

In evaluating the ensemble average of the potential energy difference at different values of λ with MC (or MD) simulation, it is customary to fix the center of mass (COM) for all particles simulated in order to avoid artificial effects due to the net motion of the entire system.[37] Specifically, the free energy of the crystal is calculated through a thermodynamic cycle shown in Figure 6.23. The quantity of interest corresponds to the free energy per particle for a real crystal (i.e., F/N as $N \to \infty$), which can be calculated with molecular simulation (step 2) plus corrections due to the finite size effects (viz., steps 1 and 3).

---

36 Frenkel D. and Ladd A. J., "New Monte Carlo method to compute the free energy of arbitrary solids: application to the fcc and hcp phases of hard spheres", *J. Chem. Phys.* 81, 3188–3193 (1984).

37 Alternatively, the artifact can be avoided by fixing a single particle in the simulation box. See Vega C. and Noya E. G., "Revisiting the Frenkel-Ladd method to compute the free energy of solids: the Einstein molecule approach", *J. Chem. Phys.* 127, 154113 (2007) and Vega C. et al., "Determination of phase diagrams via computer simulation: methodology and applications to water, electrolytes and proteins", *J. Phys.: Condens. Matter* 20, 153101 (2008).

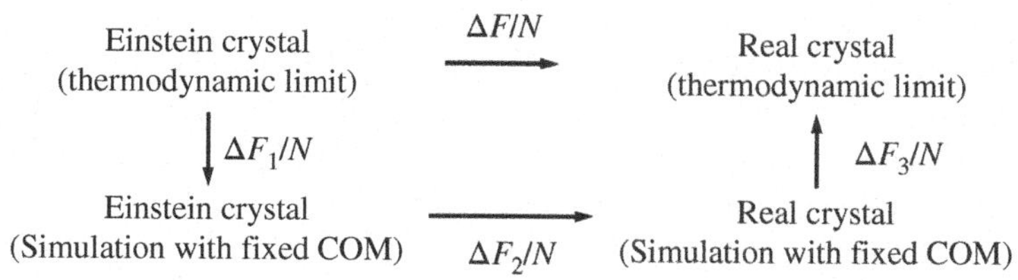

Figure 6.23 The Frenkel–Ladd method for calculating the free energy of an atomic crystal with identical particles. The difference between the free energy per particle of the real system and that of the reference system, $\Delta F/N$, includes contributions due to fixing the center of mass (COM) for the particles simulated, $\Delta F_1/N$ and $\Delta F_3/N$, in addition to that introduced by the change of the potential energy.

As the simulation is applied to a system with N particles with fixed COM, the difference between the free energy per particle of the real system and that of the reference includes three contributions

$$\Delta F/N = (F - F_0)/N = \Delta F_1/N + \Delta F_2/N + \Delta F_3/N. \tag{6.108}$$

In Eq. (6.108), the first term on the right is affiliated with fixing the particle COM in the Einstein crystal; the second term accounts for the variation of the potential energy as appeared in regular thermodynamic integration

$$\Delta F_2 = \int_0^1 d\lambda < \Phi(\mathbf{r}^N) - \Phi_0(\mathbf{r}^N) >_{\lambda,\mathrm{COM}}; \tag{6.109}$$

and the third term is related to the release of the COM constraint for the real crystal.

In the canonical ensemble, the position of each particle in the Einstein crystal follows a Gaussian distribution with the probability density in each dimension given by

$$p(x) = \left(\frac{\beta k}{2\pi}\right)^{1/2} \exp\left[-\frac{\beta k}{2}(x - x_0)^2\right] \tag{6.110}$$

where x_0 is the equilibrium coordinate of the particle. Because the harmonic potentials are independent of each other, the COM position for N particles, $x_{\mathrm{COM}} = \frac{1}{N}\sum_{i=1}^{N} x_i$, also satisfies the Gaussian distribution. In each dimension, the probability density is given by

$$p(x_{\mathrm{COM}}) = \frac{1}{\sqrt{2\pi}\sigma} \exp\left[-\frac{(x_{\mathrm{COM}} - x_{\mathrm{COM},0})^2}{2\sigma^2}\right] \tag{6.111}$$

where the variance is obtained from

$$\sigma^2 \equiv < (x_{\mathrm{COM}} - x_{\mathrm{COM},0})^2 > = \frac{1}{N^2}\sum_i^N < (x_i - x_{i,0})^2 > = \frac{1}{N\beta k}. \tag{6.112}$$

Fixing the particle COM amounts to reducing the degrees of freedom affiliated with x_{COM} thereby resulting in a change in the free energy

$$\Delta F_1 = 3k_BT \ln(\sqrt{2\pi}\sigma) = -3k_BT \ln\left(\frac{N\beta k}{2\pi}\right)^{1/2} \tag{6.113}$$

where a factor of 3 arises from the 3 coordinates for describing the COM position. In Eq. (6.113), the effects of COM on particle momenta are not considered because they contribute only to the ideal-gas part of the free energy. Such effect, along with the dimensionality of the logarithm term in Eq. (6.113), cancels with that appeared in step 3 when the constraint of COM position is removed for the real crystal.

Upon the removal of the COM constraint for the real crystal (viz., step 3 in Figure 6.23), the change in the free energy is also affiliated with the COM degree of freedom. Because of periodicity, the COM position is confined within a Wigner–Seitz (WS) cell, i.e., a primitive cell of the crystal structure that contains only a single lattice site. Fixing the COM position leads to the reduction of the accessible volume, and thereby leading to a free-energy change

$$\Delta F_3 = -k_B T \ln v_{WS} \tag{6.114}$$

where v_{WS} is the volume of a WS cell. For an atomic system with N identical particles, each particle occupies a single WS cell, and thus $v_{WS} = V/N$. Similar to the derivation of Eq. (6.113), Eq. (6.114) does not include the effect on particle momenta due to fixing the COM position; the particle momenta do not contribute to the difference between the free energy of the real and reference crystals.

The above procedure can be directly extended to systems containing different types of particles. When the particles in a solid structure are chemically distinguishable, i.e., each particle on the lattice has its own spring constant k_i and mass m_i, the COM position is then defined as

$$\mathbf{r}_{\text{COM}} = \sum_i^N m_i \mathbf{r}_i / M \tag{6.115}$$

where $M = \sum_i^N m_i$ is the total mass of the system. In this case, the COM position for the particles on the Einstein crystal still satisfies the Gaussian distribution. In each dimension, the probability density for the COM position is given by

$$p(x_{\text{COM}}) = \frac{1}{\sqrt{2\pi}\sigma} \exp\left[-\frac{(x_{\text{COM}} - x_{\text{COM},0})^2}{2\sigma^2} \right] \tag{6.116}$$

where the variance is obtained from

$$\sigma^2 = <(x_{\text{COM}} - x_{\text{COM},0})^2> = \sum_{i=1}^{N} <(x_i - x_{i,0})^2> (m_i/M)^2 = \sum_{i=1}^{N} \frac{m_i^2}{M^2 \beta k_i}. \tag{6.117}$$

Accordingly, fixing the COM position results in a change of the Helmholtz energy

$$\Delta F_1 = 3k_B T \ln(\sqrt{2\pi\sigma^2}) = \frac{3k_B T}{2} \ln\left(\sum_{i=1}^{N} \frac{2\pi m_i^2}{M^2 \beta k_i} \right). \tag{6.118}$$

The particle identify has no influence on the thermodynamic integration (ΔF_2) or the effects due to fixing of the COM position (ΔF_3).

Collecting all terms in the thermodynamic cycle shown in Figure 6.23, we obtain the reduced free energy for an atomic crystal of different particles

$$\beta F = \beta\Phi\left(\mathbf{r}_0^N\right) + \frac{3}{2}\sum_{i=1}^{N} \ln\left(\frac{\beta k_i \Lambda_i^2}{2\pi} \right) + \beta \int_0^1 d\lambda <\Delta\Phi(\mathbf{r}^N)>_{\lambda,\text{COM}} + \frac{3}{2} \ln\left(\sum_{i=1}^{N} \frac{2\pi m_i^2}{M^2 \beta k_i v_{WS}^{2/3}} \right). \tag{6.119}$$

On the right side of Eq. (6.119), the first term corresponds to the reduced static or lattice energy, which can be calculated from first-principles methods. The second term represents a generic expression for the reduced free energy of the Einstein crystal where all particles have different spring constants and different masses; the third term accounts for the change in the potential energy from the reference to the real crystal; and the last term, which is now properly dimensionless in the logarithm, arises from a correction due to fixing COM position. Note that the last term on the right side of Eq. (6.119) depends on the system size; it scales as $\sim \ln N$ and disappears in the thermodynamic limit, because $(\ln N)/N \to 0$ as $N \to \infty$. The scaling relation allows for the extrapolation of the simulation results to obtain the free energy of the real crystal.

6.9.5 Free Energies of Molecular Crystals[38]

So far, we assume that each WS cell contains one spherical particle. The assumption is valid only for relatively simple crystals such as those made of noble gas molecules or certain colloidal crystals. In general, the WS cell may contain multiple atoms that are clustered together with covalent or ionic bonding. The complexity in the internal degrees of freedom is responsible for the rich structures of crystalline materials of practical interest, and it also makes the free-energy calculation much more challenging.

Although the Einstein crystal can still be used as a reference system, the direct application of the thermodynamic cycle shown in Figure 6.23 to molecular crystals is often plagued by the divergence of the integrand in Eq. (6.109). An alternative method is by using the so-called lattice-coupling expansion method.[39] Suppose that each WS cell contains one molecule with multiple atoms. As shown schematically in Figure 6.24, we may first consider molecules in the solid phase with the COM position for each molecule coupled with the center of a WS cell using a harmonic potential. The variation in the potential energy can be implemented by using a coupling parameter $0 \leq \lambda \leq 1$ like that in the thermodynamic integration

$$\Phi_\lambda(\mathbf{r}^N, \varpi^N) = \Phi(\mathbf{r}^N, \varpi^N) + \lambda \Phi_0(\mathbf{r}^N) \tag{6.120}$$

where $\mathbf{r}_i$ and ϖ_i represent the molecular COM position and other degree of freedoms affiliated with molecule i, respectively. When $\lambda = 0$, Φ_λ reduces to the potential energy of the real system; and for $\lambda = 1$ it corresponds to that of an "interacting Einstein crystal." The change in the free energy due to coupling with the lattice can be calculated from molecular simulation

$$F = F_1 - \int_0^1 d\lambda < \Phi_0(\mathbf{r}^N)>_\lambda. \tag{6.121}$$

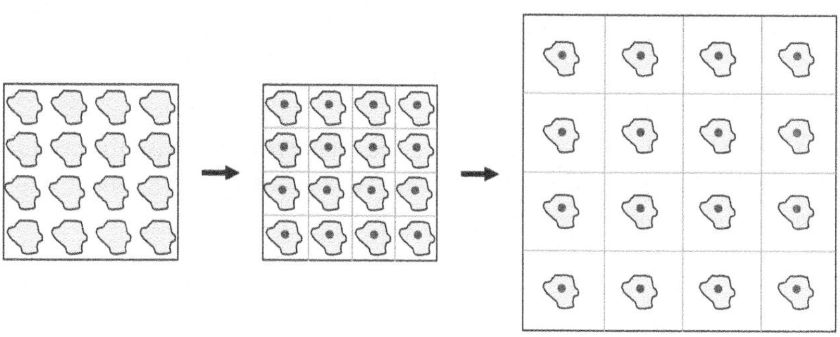

Figure 6.24 In calculation of the free energy with the lattice-coupling expansion method, the molecules in a crystalline solid are first tied up to a lattice by applying a harmonic potential to the center of mass. The lattice is then uniformly expanded with the molecules constrained within individual cells.

---

38 This subsection follows Frenkel D. and Smit B., *Understanding molecular simulation (10.3)*. Academic Press, 2002.

39 Meijer E. J. et al., "Location of melting point at 300 K of nitrogen by Monte Carlo simulation", *J. Chem. Phys.* 92, 7570 (1990).

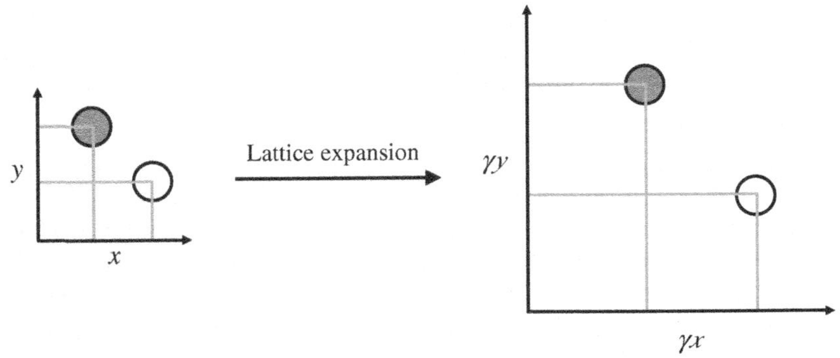

Figure 6.25 Schematic of the coordinate changes after a uniform expansion of the lattice.

Because molecules are already localized in the solid phase, it is reasonable to expect that the integrand in Eq. (6.121) is a smooth function of λ.

As the molecular COM positions are tied to the WS cells by harmonic potentials, the second step in the free-energy calculation is to expand the lattice uniformly until the intermolecular interactions become insignificant. Figure 6.25 illustrates schematically the uniform expansion of particle coordinates for a two-dimensional system. The change in the position of each atom can be calculated by the linear scaling $\mathbf{r} \to \gamma\mathbf{r}$, where $\gamma \geq 1$ is a scaling parameter. The free energy change due to the lattice expansion can also be calculated from the thermodynamic integration method

$$F_1 = F_\infty - \int_1^\infty d\gamma < \sum_k \mathbf{r}_k \cdot \nabla_{\mathbf{r}_k} \Phi_1(\mathbf{r}^N, \varpi^N, \gamma) >_\gamma \tag{6.122}$$

where F_∞ corresponds to the free energy of a single molecule in vacuum, $\Phi_1(\mathbf{r}^N, \varpi^N, \gamma)$ represents the potential energy of the interacting Einstein crystal after the coordinate expansion, and subscript k denotes the index of each atom. Due to the coupling between the molecular COM positions and the WS cells, the crystal structure is not disrupted upon the lattice expansion.

The simulation methods discussed above require as an input both the crystal structure and the spring constants for individual particles. For each vibrational mode, the spring constant can be fixed by reproducing the mean-square displacement of the constrained variable

$$\sigma_x^2 \equiv <(x_{\text{COM}} - x_{\text{COM},0})^2> = \frac{1}{\beta k_x} \tag{6.123}$$

where the ensemble average can be calculated by simulating the real crystal. Because the crystal structure is generally anisotropic, the spring constants in different directions of the coordinate may be different even for the same particle (or molecular COM).

6.9.6 Summary

Molecular simulation is commonly used to study the thermodynamic properties of crystalline solids. In principle, we can determine the crystal structure by comparing the free energies of all feasible polymorphs. The most thermodynamically stable structure corresponds to the one with the lowest free energy. While the process is conceptually straightforward, predicting the thermodynamically stable crystal structure poses computational challenges, primarily due to the limited availability of a complete set of potential crystal structures. Additionally, the free energy of a solid varies with different thermodynamic conditions such as temperature and pressure.

In practical applications, the crystal structure of a solid material is often predicted by minimizing the lattice energy at 0 Kelvin and 0 bar pressure, which can be calculated using first-principles methods. The candidate crystal structures obtained from these calculations usually exhibit lattice energies within a certain range (e.g., approximately 10 kJ/mol) of the global minimum energy. Further refinement and evaluation, such as through experimental validation or more advanced computational techniques, may be necessary to confirm the thermodynamically stable crystal structure under specific conditions.[40]

6.10 Monte Carlo Simulation of Fluid Phase Equilibria

Phase-equilibrium calculations are essential, among other applications, for the computational design of industrial separation processes and for the assessment of the stability of thermodynamic systems. In general, direct simulation of coexisting phases can be challenging because of the interfacial phenomena and finite-size effects. Special techniques are often required to construct phase diagrams through molecular simulation. In this section, we discuss the Gibbs-ensemble method and the Gibbs–Duhem integration (GDI) method, two commonly used strategies for obtaining the compositions of coexisting phases. In the next section, we will explore the multiple-histogram reweighting method, another technique useful for phase-equilibrium calculations.

6.10.1 The Gibbs-Ensemble Method

In MC simulation, we employ a small number of particles to represent the microstates of a thermodynamic system by utilizing periodic boundary conditions (PBC). However, phase transition is a collective phenomenon depending on interactions among a large number of particles. Because of the interfacial effects and long-range correlations, the properties of coexisting phases are not directly accessible through the simulation of small systems. Consider, for example, vapor–liquid equilibrium for a one-component fluid. In a macroscopic system with two coexisting phases, the interfacial energy is negligible compared to the total energies of the bulk systems. In a small system, however, the energy of the vapor–liquid interface may be comparable to the total energy. This is due to the fact that, unlike that in a macroscopic system, a significant fraction of molecules or particles may reside at the interface. Besides, in a small system, this interfacial energy results in a kinetic barrier preventing the formation of two phases within a single simulation box. Because of this energy barrier, direct simulation often yields metastability over a range of conditions much broader than that for a macroscopic system. Consequently, simulation over a small system is likely to persist in either one of these metastable phases without "knowing" that the vapor has condensed or that the liquid has evaporated. Furthermore, the simulated density of the vapor phase may be significantly larger, and that for the liquid phase may be substantially lower, than those corresponding to macroscopic equilibrium.

The Gibbs-ensemble method for phase-equilibrium calculations was developed by Panagiotopoulos.[41] The central idea is to calculate the properties of coexisting phases by avoiding the phase boundary or the interfacial region. As shown schematically in Figure 6.26, the bulk properties of two coexisting phases are simulated in two separated cells using methods similar to

40 Bowskill D. H. et al., "Crystal structure prediction methods for organic molecules: state of the art", *Annu. Rev. Chem. Biomol. Eng.* 12, 593–623 (2021).

41 Panagiotopoulos A., "Direct determination of phase coexistence properties of fluids by Monte Carlo simulation in a new ensemble", *Mol. Phys.* 61, 813–826 (1987).

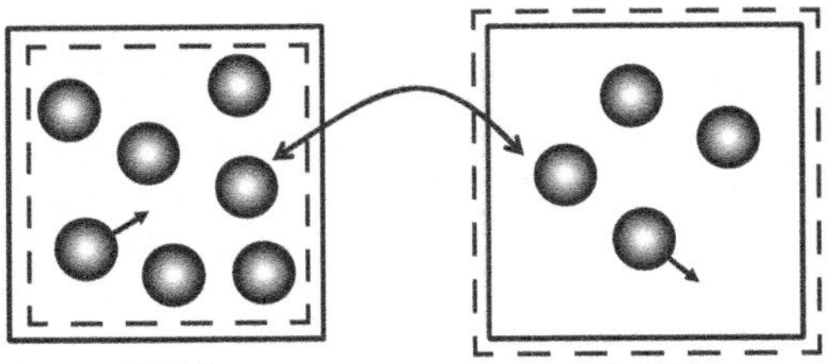

Figure 6.26 Monte Carlo moves in the Gibbs-ensemble method: particle displacement (single arrow), volume changes (dashed lines), and particle transfer (double arrows).

those for a standard MC simulation with PBC. The phase coexistence requires that each cell should reach the equilibrium condition, and that the temperature, pressure and the chemical potential of each chemical species in one simulation cell are, respectively, equal to those in the other. With the temperature set to be the same for both cells, the internal equilibrium is achieved by the displacements of particles within each cell as in standard MC simulation. However, the equalities of pressure and individual chemical potentials must be achieved by performing additional MC "moves" between the two sub-systems; the equality of pressures is satisfied by changes in the volumes of the two simulation cells while the total volume remains constant. For each chemical species, the equality of chemical potential is achieved through the exchange of molecules between the two cells, while fixing the total number of molecules in the system. Accordingly, the system reaches a state of equilibrium where the chemical potentials of all components are equalized.

To establish a clear understanding of the MC moves, consider vapor–liquid equilibrium in a one-component system containing N spherical particles in total volume V and at constant temperature T. As shown in Figure 6.26, the system is represented by two cells (viz., phase regions) with volumes V_I and V_{II} such that $V_I + V_{II} = V$; the corresponding number of particles are N_I and N_{II} with $N_I + N_{II} = N$. Because these two regions are *NOT* directly in contact with each other, the particles in one region do not interact with those in the other. As a result, the total potential energy of the entire system is given by $\Phi_I + \Phi_{II} = \Phi$ where Φ_I and Φ_{II} represent the potential energies of the subsystems.

For this composite canonical system (constant T, V, and N with internal constraints), the partition function is similar to that of a normal canonical ensemble except that the particles are divided into two simulation cells

$$Q_{NTV} = \frac{1}{N!}\frac{1}{V}\int_0^V dV_I \sum_{N_I=0}^{N} \left(\frac{N!}{N_I!N_{II}!}\right) \frac{V_I^{N_I} V_{II}^{N_{II}}}{\Lambda^{3N}} \int d\mathbf{x}_I^{N_I}\, e^{-\beta\Phi_I} \int d\mathbf{x}_{II}^{N_{II}}\, e^{-\beta\Phi_{II}} \tag{6.124}$$

where Λ is the thermal wavelength, and $\beta = 1/k_BT$; $N!$ accounts for the particle identity, which is the same as that appeared in the partition function of N identical classical particles as in a normal canonical ensemble; the integration with respect to V_I represents different ways for partitioning the total volume; and the binomial coefficient $N!/(N_I!N_{II}!)$ corresponds to the number of ways to distribute N particles into two cells for each value of N_I. Here, the particle coordinates are expressed in terms of the dimensionless scale such that the integration boundaries are independent of volume, i.e., $d\mathbf{x}_I^{N_I} = d\mathbf{r}_I^{N_I}/V_I^{N_I}$ and $d\mathbf{x}_{II}^{N_{II}} = d\mathbf{r}_{II}^{N_{II}}/V_{II}^{N_{II}}$, with $\mathbf{r}_I^{N_I} = \left(\mathbf{r}_I^{(1)}, \mathbf{r}_I^{(2)}, \cdots, \mathbf{r}_I^{(N_I)}\right)$ and $\mathbf{r}_{II}^{N_{II}} = \left(\mathbf{r}_{II}^{(1)}, \mathbf{r}_{II}^{(2)}, \cdots, \mathbf{r}_{II}^{(N_{II})}\right)$ standing for the configurations of subsystems I and II, respectively.

In this internally constrained canonical system, particles in region I do no interact with particles in region II and vice versa. Therefore, the interfacial energy is not considered in the simulation. At a given configuration of the particles in each region, the probability density that region I contains N_I particles with a volume between V_I and $V_I + dV_I$ is proportional to

$$p(N_I, V_I; N, V, T) \sim \frac{V_I^{N_I}}{N_I!}\frac{V_{II}^{N_{II}}}{N_{II}!} e^{-\beta(\Phi_I+\Phi_{II})}. \tag{6.125}$$

Different from a normal canonical ensemble, the probability density depends on, in addition to the Boltzmann factor, how the total volume and particles are divided between the two subsystems.

Suppose that the trial MC moves for particle displacements, volume changes and particle transfers are all symmetric, i.e., the probability of taking a trial move from a configuration i to configuration j is identical to that of the reverse process, from configuration j to configuration i. In that case, the detailed balance condition is satisfied by setting the acceptance criterion for each type of move according to the Metropolis algorithm. For a displacement step within region I, the probability of acceptance a_{ij} is the same as that in a conventional canonical ensemble

$$a_{ij} = \min\left(1, e^{-\beta\Delta\Phi_I}\right). \tag{6.126}$$

A similar relation holds for displacement moves in region II. In Eq. (6.126), $\Delta\Phi$ denotes the change of potential energy resulting from a particle displacement in a subsystem. If the energy change is negative, $\Delta\Phi < 0$. Eq. (6.126), predicts $a_{ij} = 1$, and thus the displacement is accepted. Otherwise, the acceptance probability is determined by taking a random number between 0 and 1, i.e., the displacement is accepted only if the random number is smaller than $e^{-\beta\Delta\Phi}$.

The second type of trial move concerns the variation of cell volumes. Typically, the volume change is achieved with $\ln(V_I/V_{II})$ treated as a random variable within the range from zero to a prescribed maximum. For a small volume change ΔV in region I with a corresponding change $-\Delta V$ in region II, the Metropolis algorithm give the acceptance probability

$$a_{ij} = \min\left[1, \exp\left\{-\beta\Delta\Phi_I - \beta\Delta\Phi_{II} + (N_I+1)\ln\frac{V_I+\Delta V}{V_I} + (N_{II}+1)\ln\frac{V_{II}-\Delta V}{V_{II}}\right\}\right]. \tag{6.127}$$

As for the displacement move, the volume change is accepted if the exponential term in Eq. (6.127) is larger than unity; otherwise, the MC move is accepted with the probability equal to a_{ij}.

Finally, the acceptance criterion for particle transfer, say for the transfer of a particle from region II to region I, is given by

$$a_{ij} = \min\left[1, \frac{N_{II}V_I}{(N_I+1)V_{II}}\exp(-\beta\Delta\Phi_I - \beta\Delta\Phi_{II})\right]. \tag{6.128}$$

Eq. (6.128) implies that there is a finite probability to transfer the last particle out of a region (I or II). Nonetheless, the probability of transferring a particle out of an empty region is always zero because in that event, $a_{ij} = 0$.

The acceptance probabilities for volume and particle exchange, as described by Eqs. (6.127) and (6.128), respectively, can be interpreted as the product of the probabilities of concurrent MC moves in the isobaric and grand canonical simulation of the two subsystems. In these moves, the attempted changes in volume and the number of particles in one system are oppositely equal to those in the other system. This restriction leads to the cancellation of pressure and chemical potential-dependent Boltzmann factors in the acceptance equations.

The Gibbs phase rule predicts that only one intensive variable (e.g., temperature) can be specified for a one-component system with two coexisting phases. At conditions remote from the critical point[42] of the vapor–liquid phase transition, the equilibrium densities of the vapor and liquid phases are determined by the corresponding ensemble averages in the two simulation cells. Meanwhile, the saturation vapor pressure can be obtained from the virial equation as

42 Near the critical point, we must consider large fluctuations in density; these fluctuations are suppressed due to the finite box size.

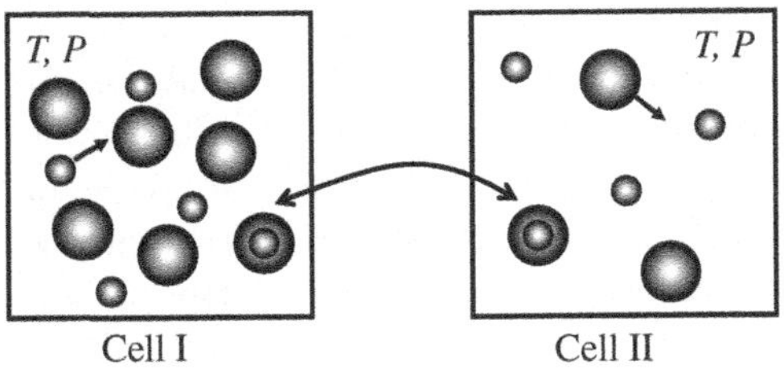

Figure 6.27 A sketch of identity exchange for a Gibbs ensemble simulation of a binary system (represented by core-shell spheres). An exchange of molecule identity from a large molecule in region I is accompanied by a simultaneous change in the identity of a small molecule in region II.

discussed in Section 6.6.1. For phase equilibrium in a binary system, both temperature and pressure must be specified in advance. In that case, the phase coexistence is most conveniently calculated using a constrained isothermal–isobaric ensemble, i.e., the simulation is performed at constant temperature and pressure with fixed overall composition and total number of particles partitioning between two "coexistence" regions. The volume changes in the two regions are now made independently; we no longer have $\Delta V_I = -\Delta V_{II}$ because now pressure for each region, not total volume, is kept constant. Equality of chemical potential for each component is achieved through particle transfers following an acceptance probability similar to that given by Eq. (6.128) with $N_{I,i}$ and $N_{II,i}$ representing the number of particles of component i in the two simulation cells.

For a binary system containing particles that differ greatly in size, the transfer of the large particles between two regions is often inefficient because of the small probability for successful insertion of a large particle at a randomly selected position in a condensed phase. To improve sampling efficiency,[43] we may use the identity-exchange method as illustrated in Figure 6.27. In this method, only small particles are transferred between the two regions, and the large particles are transferred indirectly, by changing a small particle into a larger one in one of the two regions concomitant to a simultaneous reverse change in the other region. That is, a trial move consists of a random selection of a small (large) particle i in region I and a large (small) particle j in region II followed by a switch in their identities. According to the Metropolis algorithm, the acceptance probability for an identity exchange can be expressed as

$$a_{ij} = \min\left(1, \frac{N_{I,i}N_{II,j}}{(N_{I,j}+1)(N_{II,i}+1)} \exp(-\beta\Delta\Phi_I - \beta\Delta\Phi_{II})\right) \tag{6.129}$$

where $N_{I,i}$ stands for the number of particles of type i in region I, and a similar notation is used for $N_{I,j}$, $N_{II,i}$ and $N_{II,j}$.

Particle-identity exchange does not alter the overall composition of the system that includes both regions I and II. The particle transfer and exchange ensure the equality of chemical potential for the small particles (i) and the equality of relative chemical potential for the large particles (j):

$$\mu_{I,i} = \mu_{II,i} \text{ (from the transfer step for a small particle)}, \tag{6.130}$$

$$\mu_{I,j} - \mu_{I,i} = \mu_{II,j} - \mu_{II,i} \text{ (from the exchange step for a large particle)}. \tag{6.131}$$

Eqs. (6.130) and (6.131) indicate that the chemical potentials are equal for both small and large particles in the two coexisting phases.

The Gibbs-ensemble method has found extensive applications in MC simulation of fluid-phase equilibria. When accurate molecular force fields are available to describe intermolecular interactions, the Gibbs-ensemble method can provide reliable phase-coexistence properties, even for

43 Good sampling efficiency is achieved when successful (accepted) moves are more frequent than those that are unsuccessful.

systems containing hydrogen-bonding species that exhibit high nonideality. However, as the density of the liquid phase rises, the computational efficiency of the Gibbs-ensemble method diminishes. This is primarily due to the low acceptance probability associated with the particle-transfer step, similar to what is observed in the particle insertion method discussed in Section 6.8. Unfortunately, the Gibbs-ensemble method cannot be directly employed to study solid–fluid equilibria. This is because exchanging particles between the two phases would require the addition or removal of molecules within an otherwise perfect crystal, resulting in the introduction of artificial point defects through MC moves. Additionally, near the critical point, the Gibbs-ensemble simulation becomes unstable as the MC moves can inadvertently lead to the presence of two or more phases within each region. In such cases, histogram-reweighting techniques offer a more suitable alternative, especially when determining the equilibrium conditions across a multitude of temperatures, pressures, and compositions.

6.10.2 The Gibbs–Duhem Integration

The Gibbs–Duhem integration (GDI) method was proposed by Kofke.[44] The basic idea is to determine the coexistence line (viz., the phase boundary) by utilizing the Gibbs–Duhem equation, which, in its most general form, can be written as

$$0 = \sum_i X_i d\xi_i \tag{6.132}$$

where X_i and ξ_i are a pair of conjugate thermodynamic variables with X_i being extensive and ξ_i intensive. The intensive variables, such as enthalpy, volume, and the number of molecules for each species, can be directly calculated from ensemble averages. By contrast, the intensive variables are either specified as constraints (e.g., temperature and pressure in the *NPT* ensemble) or calculated with integration methods (e.g., chemical potential).

For two bulk phases α and γ in equilibrium, the intensive variables satisfy

$$\xi_i^\alpha = \xi_i^\gamma. \tag{6.133}$$

Based on Eqs. (6.132) and (6.133), we may identify an intensive variable, ξ_0, as a coordinate to predict the variation of other intensive variables, $\xi_{i=1,2,\ldots}$, along the coexistence curve

$$\left(\frac{\partial \xi_i}{\partial \xi_0}\right)_{\xi_j \neq \xi_i \text{ or } 0} = -\frac{\Delta X_0}{\Delta X_i} \tag{6.134}$$

where ΔX_i stands for the difference in the extensive variable between the coexisting phases. Eq. (6.134) may be recognized as the generalized Clausius–Clapeyron equation.

To elucidate the essential concepts of the GDI method, consider the vapor–liquid–solid coexistence lines in a one-component system. Figure 6.28 shows schematically the phase diagram. Along the vapor–liquid coexistence line, the saturation pressure varies with temperature according to the Clausius–Clapeyron equation

$$\frac{dP}{dT} = \frac{\Delta H}{T\Delta V} \tag{6.135}$$

where ΔH and ΔV stand for the changes in enthalpy and volume due to the vapor-to-liquid phase transition. We may express Eq. (6.135) in a slightly modified form

$$P' = -\frac{\Delta H}{\beta P \Delta V} \tag{6.136}$$

---

44 Kofke D. A., "Direct evaluation of phase coexistence by molecular simulation via integration along the saturation line", *J. Chem. Phys.* 98, 4149 (1993).

where $P' \equiv d \ln P/d\beta$. Within a small range of temperature, P' does not vary as significantly as dP/dT, thereby improving the accuracy of numerical integration.

Suppose that we have identified a particular point (P_0 and derivative P_0') on the vapor–liquid coexistence curve using some other method (e.g., the Gibbs-ensemble method), the change in pressure from P_0 to P_1, corresponding to $\Delta\beta$, a small variation in β, can be estimated from the trapezoidal rule

$$\ln P_1 \approx \ln P_0 + \Delta\beta \cdot \left(P_0' + P_1'\right)/2 \tag{6.137}$$

where P_1' is obtained from Eq. (6.136) using a conventional isothermal–isobaric simulation for ΔH and ΔV at (T_1, P_1). Because P_1 is initially unknown, to be determined from Eq. (6.137), we may use a predictor–corrector iteration algorithm to calculate P_1'. Specifically, P_1' is first set equal to P_0'. This first estimate of P_1, called the predictor, is calculated from Eq. (6.137). With the predictor pressure P_1, P_1' is then obtained from a conventional isothermal–isobaric simulation at (N, P_1, T); Eq. (6.137) provides an improved estimate for P_1, which is called the corrector. The calculation is repeated until P_1' converges, that is, until further iterations produce no significant change in P_1'. The entire procedure is then repeated for the next state point on the coexistence line in the phase diagram. The efficiency of integration can be improved when it is combined with the histogram-reweighting method (Section 6.11) where extensive MC simulations are performed at similar conditions.

The GDI method can be applied to simulation of both fluid–fluid and fluid–solid equilibria including multicomponent systems. For a one-component system, the vapor–liquid–solid phase diagram, as shown in Figure 6.28, can be calculated by performing four simulation steps: (i) to locate one point on the vapor-pressure line using the Gibbs-ensemble method; (ii) to calculate the entire vapor–liquid coexistence line by the GDI; (iii) to simulate the solid–fluid coexistence using the Einstein-crystal model for the solid phase and the GDI method; (iv) to determine the solid–liquid–vapor triple point by finding the intercept of vapor–liquid and solid–liquid coexistence curves; and (v) to calculate the solid-vapor line starting with the triple point, again using GDI.

Application of the GDI method to multicomponent systems is slightly more complicated because it entails more intensive variables. For a binary system with two coexisting phases, we need to specify two out of the four intensive variables, T, P, μ_1 and μ_2, that must satisfy the equilibrium condition. Because the chemical potential cannot be directly evaluated from the ensemble average, temperature and the fugacity fraction provide a convenient choice as the independent variables for the simulation. In this case, the generalized Clausius–Clapeyron equation can be written as (see Problem 6.21)

$$\left(\frac{\partial\beta}{\partial z_2}\right)_P = \frac{x_2^\alpha - x_2^\gamma}{z_1 z_2 (h^\alpha - h^\gamma)} \tag{6.138}$$

where $0 \le z_2 \equiv f_2/(f_1 + f_2) \le 1$ stands for the fugacity fraction of species 2, and the fugacity follows its conventional definition $\mu_i = \mu_i^0 + k_B T \ln\left(f_i/f_i^0\right)$, superscript '0' stands for a reference state.

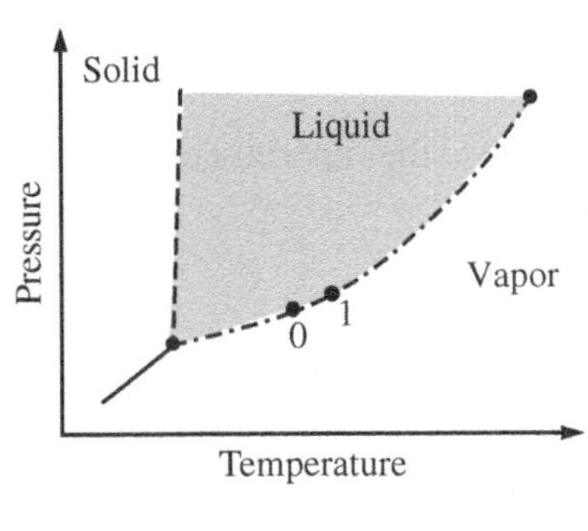

Figure 6.28 When applying the Gibbs–Duhem integration method to determine vapor–liquid equilibrium in a one-component fluid, the phase boundary is computed using the Clapeyron equation. The alterations in enthalpy and volume resulting from phase transitions can be obtained through MC simulations. The gradient of the liquid–vapor coexistence curve, for example ranging from "0" to "1", can be determined by conducting isothermal–isobaric ensemble simulations utilizing a predictor-corrector algorithm.

In Eq. (6.138), we use the fugacity fraction instead of chemical potential because this quantity is bounded between 0 and 1. The extensive variables on the right side of Eq. (6.138) can be calculated using a semi-grand canonical ensemble (at constant N, P, T and z_2). Toward that end, MC simulation is applied to a composite system with two regions similar to those in the Gibbs ensemble method. The temperature and pressure are fixed for each subsystem as in normal NPT simulation. The constant fugacity fraction is achieved by particle identity exchanges, i.e., by the random selection of a particle within each cell and switching the identity from species 1 to 2 or vice versa with the probability of acceptance specified by the Metropolis method

$$a_{ij} = \min\left(1, \left(\frac{z_2}{1-z_2}\right)^{\pm 1} \exp(-\beta\Delta\Phi)\right) \tag{6.139}$$

where the positive and negative signs "±" apply to the identity exchange from 1 to 2 and from 2 to 1, respectively. Using the saturation temperatures of pure species (i.e., $z_2 = 0$ and 1) as the boundary conditions, we can numerically integrate Eq. (6.138) with respect to z_2 and determine the coexistence curves.

One appealing feature of the Gibbs–Duhem-integration method is that there is no need to calculate chemical potentials of different species explicitly, thereby avoiding the difficulties encountered with particle insertions or particle transfers. One caveat is that the accuracy of the GDI is limited by the accumulation of errors in numerical integration. Toward minimizing that limitation, multiple-histogram reweighting provides an excellent procedure for increasing the accuracy of GDI for phase-diagram calculations.

Before closing this section, we present an application of GDI for simulating the phase diagram for a one-component system containing buckminsterfullerenes[45] (C_{60}). As shown in Figure 6.29, a buckminsterfullerene molecule is composed of 60 carbon atoms interconnected to create a polygonal structure that closely approximates a sphere. The bonds in a C_{60} molecule are arranged as the panels on a soccer ball, joined by hexagons and pentagons. One of the most striking properties of C_{60} is that small molecules can be introduced inside the cage of carbon atoms. Another striking property is provided by metallic compounds of buckminsterfullerene (notably K_3C_{60}, i.e., three potassium atoms per C_{60} molecule); these compounds are superconducting at low temperatures. Further, some derivatives of buckminsterfullerene show biological activity; they have been used to attack cancer or other diseases because the soccer-ball-shaped molecules can enter the active site of an enzyme and block its catalytic action.

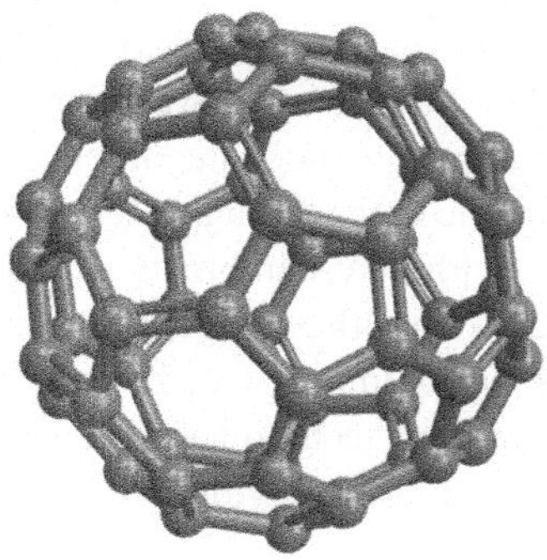

Figure 6.29 Molecular structure of buckminsterfullerene. Each sphere represents a carbon atom.

45 The name Buckminsterfullerene follows from resemblance of the molecule to the geometric structures invented by the American architect Richard Buckminster Fuller.

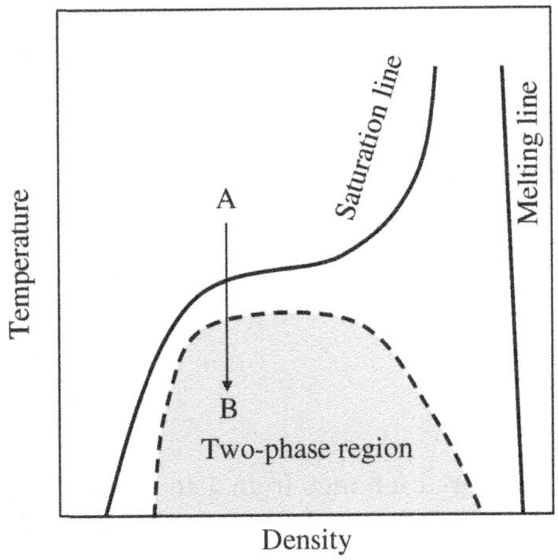

Figure 6.30 Schematic of the phase diagram for C_{60} calculated from the GDI method for fluid–solid coexistence (solid line) and from the Gibbs-ensemble method for fluid–fluid coexistence (dashed line). The solid phase has an fcc structure. See text for explanation of process A→B.

The phase diagram for a thermodynamic system containing C_{60} particles was predicted by the combination of the Gibbs-ensemble and the Gibbs–Duhem-integration methods.[46] The pairwise additive potential between C_{60} molecules was constructed by assuming that the carbon atoms on two different C_{60} molecules interact through the Lennard–Jones (LJ) potential. Schematically, Figure 6.30 shows the vapor–liquid and fluid–solid coexistence curves on a plot of temperature versus density. Unlike simple fluids, such as argon or methane, C_{60} does not have a stable liquid phase because the vapor–liquid coexistence curve for C_{60} lies underneath that for fluid–solid equilibria. As a result, the liquid phase is metastable relative to the solid with a face-centered-cubic (fcc) structure. This phase diagram suggests that rapid cooling from the vapor phase produces an amorphous C_{60} solid (i.e., a liquid or glass). Suppose C_{60} is cooled from A to B. If cooling is rapid, at the sublimation temperature B, C_{60} does not crystallize but, instead, upon further cooling, forms metastable liquid-like aggregates; these aggregates resemble an amorphous (not crystalline) solid. To obtain crystalline solid of C_{60} from the vapor, a possible method is isothermally to compress the vapor slowly to a temperature well above 1800 K until the density reaches the freezing line. Figure 6.30 explains why, as experimentally observed, upon rapid cooling, high-temperature synthesis of C_{60} in the gas phase often gives an amorphous soot-like deposit.

6.10.3 Summary

The Gibbs ensemble and thermodynamic integration methods are commonly used to calculate phase diagrams through MC simulation. The Gibbs ensemble method is particularly effective for determining liquid–vapor phase diagrams and properties of liquid mixtures. On the other hand, thermodynamic integration methods are based on the principle of connecting different thermodynamic states through a series of intermediate states. These techniques require running simulations for different combinations of control variables such as temperature and pressure. By collecting data from these simulations, we can construct phase diagrams and obtain thermodynamic properties of interest. As discussed in later sections, other methods and algorithms, such as advanced sampling techniques and histogram reweighting, are also available for calculating phase diagrams from molecular simulation.

---

46 Hagen M. H. J. et al., "Does C_{60} have a liquid phase?", *Nature* 365, 425–426 (1993).

6.11 Histogram Reweighting Analysis

The practical application of molecular modeling often involves studying the variations of thermodynamic variables, such as temperature, pressure, density, and composition, across a wide range of conditions. However, conventional simulation methods mostly focus on one specific thermodynamic state at a time, similar to experiments. Predicting the variations in the properties of an equilibrium system with different thermodynamic variables can be a time-intensive process. From a practical perspective, it would be desirable to carry out simulation at one condition in such a way that the sampled microstates can be used to predict not just the thermodynamic properties of that specific state, but also those at neighboring conditions. In this section, we discuss the optimal utilization of simulation data based on histogram reweighting for the distribution of collective variables. The numerical procedure aims to extract thermodynamic properties across a range of conditions using a limited set of sampled microstates. In other words, it maximizes the utilization of simulation data for predicting thermodynamic properties more efficiently.

6.11.1 Single Histogram Reweighting

It was recognized early in the history of molecular simulation that microstates generated from a single simulation (i.e., simulation at a particular thermodynamic condition) could be utilized to extrapolate the properties of the same thermodynamic system at neighboring conditions. The method for making this extrapolation is called histogram reweighting.[47] While the procedure is not directly useful for practical applications, it elucidates the important concepts underlying multiple histogram reweighting and inspires the late development of a variety of non-Boltzmann sampling methods.

To understand how thermodynamic properties at different conditions are interrelated, consider MC simulation in a canonical ensemble. For example, the internal energy U can be calculated from the ensemble average

$$U = <E_\nu> = \int E \cdot p(E)\, dE \tag{6.140}$$

where E_ν represents the total energy at microstate ν, and $p(E)$ is the probability density of the system with energy E. The latter can be calculated by counting the number of microstates $n(E)$ with total energy between E and $E + dE$[48]

$$p(E)dE = n(E)/n_T \tag{6.141}$$

where n_T denotes the total number of configurations sampled by the MC simulation. It is important to note that the quantities on both sides of Eq. (6.141) are dimensionless.

The essential idea of histogram reweighting is to predict the thermodynamic properties of a system by utilizing the probability density obtained at other conditions. In a canonical ensemble, $p(E)$

---

47 A histogram is a graphical display of tabulated distribution of a certain valuable; reweighting means calculation of the distribution at a different condition.

48 Moe precisely, here dE should be written as ΔE, a finite energy difference used in calculating $n(E)$ by MC simulation. Unless inside an integral, $p(E)dE$ should not be confused as an infinitesimal quantity.

is related to the Boltzmann factor $e^{-\beta E}$ and to the number of microstates $W(E)$ with energy between E and $E + dE$

$$p(E)dE = W(E)e^{-\beta E}/Q \tag{6.142}$$

where Q is the canonical partition function. From Eqs. (6.141) and (6.142), we can determine $W(E)$ from $n(E)/n_T$ for a range of energy values within an unknown constant Q. Note that $W(E)$ is also dimensionless, corresponding to the microcanonical ensemble partition function for the system with the total energy between E and $E + dE$.

Now suppose that we run MC simulation at temperature T_0 and have calculated the energy distribution function $p_0(E)$ from the histogram analysis. We want to extrapolate the internal energy of the same system at a neighboring temperature T. Toward that end, we apply Eq. (6.142) to both temperatures, T and T_0. Because the microcanonical partition function $W(E)$ is independent of temperature, the energy distribution function at T is related to that at T_0 by

$$p(E)/p_0(E) = e^{-(\beta-\beta_0)E}Q_0/Q \tag{6.143}$$

where Q_0 is the canonical partition function at T_0. Using the normalization condition

$$\int p(E)dE = 1 \tag{6.144}$$

where the limits of the integral are determined by the range of all possible energies, we can obtain $p(E)$ by combining Eqs. (6.143) and (6.144)

$$p(E) = \frac{e^{-(\beta-\beta_0)E}p_0(E)}{\int e^{-(\beta-\beta_0)E}p_0(E)\,dE}. \tag{6.145}$$

In writing Eq. (6.145), we have replaced the unknown constant Q/Q_0 by the normalization condition for $p(E)$. The right-hand side of Eq. (6.145) can be evaluated through numerical integration. With $p(E)$ calculated from Eq. (6.145), we can subsequently predict the internal energy at T using Eq. (6.140). While for simplicity our discussion is focused on the internal energy, the procedure can be similarly applied to other thermodynamic properties. For example, the heat capacity can be calculated from the fluctuation of the system energy

$$C_V(T) = \frac{<E_v^2> - <E_v>^2}{k_BT} = \frac{1}{k_BT}\left\{\int E^2 \cdot p(E)\,dE - \left[\int E \cdot p(E)\,dE\right]^2\right\}. \tag{6.146}$$

The histogram reweighting method is directly applicable to analyzing simulation data from other ensembles. For example, in a grand canonical ensemble of a one-component fluid at temperature T volume V and chemical potential μ_0, the probability that the system contains N molecules with a total energy between E and $E + dE$ is given by

$$p_0(N, E)dE = W(N, E)e^{-\beta E+\beta N\mu_0}/\Xi_0 \tag{6.147}$$

where $W(N, E)$ represents the number of microstates of the system at fixed V, and Ξ_0 is the grand canonical partition function at state 0. Once $p_0(N, E)$ is evaluated by a regular grand canonical simulation at T, V, and μ_0, we can obtain the distribution of N and E for the system at the same T and V but at a different μ, provided that μ is a chemical potential not far from μ_0. Following the same procedure as that used to obtain Eq. (6.145), we can derive the new energy distribution function

$$p(N, E) = \frac{e^{\beta N(\mu-\mu_0)}p_0(N, E)}{\sum_{N=0}^{\infty}\int e^{\beta N(\mu-\mu_0)}p_0(N, E)\,dE}. \tag{6.148}$$

Subsequently, the distribution function $p(N, E)$ can be used to calculate the average number of particles $<N>$ and the internal energy U at T, V, and μ

$$<N>= \sum_{N=0}^{\infty} \int N \cdot p(N, E)\, dE, \tag{6.149}$$

$$U = \sum_{N=0}^{\infty} \int E \cdot p(N, E)\, dE, \tag{6.150}$$

where the limits of the integrals are determined by the range of all possible energies. From the average number of particles $<N>$ and system volume V, we can calculate the average density $\rho = <N>/V$. Therefore, the histogram reweighting analysis yields a relation between chemical potential μ and number density ρ at some fixed temperature; this relation is applicable for phase-equilibrium calculations. A similar extrapolation method may be applied to a small variation in temperature while keeping V and N constant.

While the single-histogram reweighting method is conceptually appealing, the numerical procedure is not directly useful in practice because, at equilibrium, the thermodynamic properties are typically distributed around their mean values with only small fluctuations. In other words, the distribution function such as $p(N, E)$ obtained from a single MC simulation is typically restricted to a small range of thermodynamic conditions. To make more reliable predictions, we need to use the multiple-histogram-reweighting method, which generalizes the histogram-reweighting analysis for microstates sampled at a single thermodynamic state to those at multiple conditions. The advantage of multiple-histogram-reweighting analysis is that it covers a broad range of conditions. Numerically, it hinges on interpolation rather than extrapolation as used in single-histogram reweighting.

6.11.2 Multiple Histogram Reweighting

The multiple-histogram-reweighting analysis was proposed by Ferrenberg and Swendsen.[49] The procedure is useful not just for predicting ensemble averages at different thermodynamic conditions, but also for free-energy and phase-diagram calculations. As discussed in the subsequent section, the mathematical procedure can be applied in a similar manner to umbrella sampling.

To elucidate the essential ideas of multiple-histogram reweighting, imagine that simulation data are available from canonical-ensemble MC simulations at a set of temperatures $\{T_i\}$. We wish to find the optimal prediction of the energy distribution $p(E)$ as a function of temperature T. As discussed above for the single-histogram reweighting method, the simulation at each temperature provides an estimate of $p_i(E)$. The probability density that the system has a total energy between E and $E + dE$ at temperature T_i is given by

$$p_i(E)dE = n_i(E)/n_T^{(i)} \tag{6.151}$$

where $n_i(E)$ stands for the number of configurations at T_i with the energy between E and $E + dE$, and $n_T^{(i)}$ is the total number of configurations sampled at this temperature.

To find the best estimate of $p(E)$ at some temperature T from a series of energy distributions $\{p_i(E)\}$, we may first calculate $p(E)$ using the single-histogram reweighting method. From any one of the energy distributions $p_i(E)$, we have

$$p(E) = p_i(E)e^{-(\beta-\beta_i)E}Q_i/Q \tag{6.152}$$

49 Ferrenberg A.M. and Swendsen R. H., "New Monte Carlo technique for studying phase transitions", *Phys. Rev. Lett.* 61, 2635 (1988); "Optimized Monte Carlo data analysis", *ibid.*, 63, 1195 (1989).

where Q_i stands for the canonical partition function at T_i, and Q stands for the canonical partition function at T. To provide an improved prediction of $p(E)$ from a series of simulations at different temperatures, we construct a linear combination of the individual estimates, Eq. (6.152), using a weighted average

$$p(E) = \sum_i w_i(E) p_i(E) e^{-(\beta-\beta_i)E} Q_i/Q \tag{6.153}$$

where the weight function $w_i(E)$ is yet to be determined.

In the so-called weighted histogram analysis method (WHAM),[50] we seek to minimize the statistical error, i.e., minimizing the variance of $p(E)$ with respect to the weight function $w_i(E)$ [51]

$$\delta^2 p(E) \equiv \overline{p^2(E)} - \overline{p(E)}^2 \tag{6.154}$$

where the quantities under the bars stand for the expected values if multiple simulations were performed at the same set of parameters (i.e., T, N, V). To a good approximation, we may assume that the simulation results for the microstate distributions at different conditions are uncorrelated. In that case, the variance of $p(E)$ can be written as

$$\delta^2 p(E) \approx \sum_i \left[w_i(E) e^{-(\beta-\beta_i)E} Q_i/Q\right]^2 \delta^2 p_i(E) \tag{6.155}$$

where $\delta^2 p_i(E)$ stands for the variance of $p_i(E)$.

With an explicit expression for the variance of $p(E)$, we can determine weight function $w_i(E)$ using the Lagrangian-multiplier method. Specifically, minimizing $\delta^2 p(E)$ from Eq. (6.155) with respect to $w_i(E)$, subject to the normalization condition $\sum_i w_i(E) = 1$, yields

$$2w_i(E)\left[e^{-(\beta-\beta_i)E} Q_i/Q\right]^2 \delta^2 p_i(E) = \alpha \tag{6.156}$$

where α is the Lagrange multiplier to be determined from the normalization condition. Eq. (6.156) indicates that, to minimize the statistical error in $p(E)$, the weighting function $w_i(E)$ should be inversely proportional to $\delta^2 p_i(E)$, the variance of the individual estimates for $p(E)$.

In a typical simulation, the microstates are generated following the Markov-chain process. In order to combine multiple histograms, a reasonable amount of overlap must be present between neighboring histograms. The variance of the energy distribution at each temperature $\delta^2 p_i(E)$ can be obtained by assuming that $n_i(E)$, the number of microstates with energy between E and $E + dE$, follows the Poisson distribution.[52] Accordingly, the variance of $n_i(E)$ is given by[53]

$$\overline{n_i^2(E)} - (\overline{n_i(E)})^2 \approx \overline{n_i(E)} \tag{6.157}$$

As the total number of configurations $n_T^{(i)}$ for calculating the histograms is fixed for each simulation run, Eq. (6.151) predicts that the variance of $p_i(E)$ is subsequently given by

$$\delta^2 p_i(E)(dE)^2 = \overline{n_i(E)}/\left[n_T^{(i)}\right]^2 = W(E)e^{-\beta_i E}/\left[Q_i n_T^{(i)}\right] \tag{6.158}$$

---

50 Kumar S. et al., "The weighted histogram analysis method for free-energy calculations on biomolecules", *J. Comput. Chem.* 13, 1011 (1992).

51 In statistics, variance provides a measure of the dispersion of a set of data points around their mean value. A small variance means a small statistical error.

52 The Poisson distribution describes the number of events that occur randomly and independently in a specific time period or region. For example, the number of people visiting San Francisco per year follows approximately the Poisson distribution. The probability that the number of visitors in San Francisco per year is n, is given by $p(n) = s^n e^{-\overline{n}}/n!$ where $\overline{n}$ is the average of n. For a Poisson distribution, the average is also equal to the variance.

53 Here $n_i(E)$ follows the Poisson distribution only approximately because at a fixed condition, two consecutive MC moves are not truly independent.

In writing the second equality in Eq. (6.158), we have used the identity following Eq. (6.151)

$$\overline{n_i(E)} = n_i(E) = [p_i(E)dE]\, n_T^{(i)} = W(E)e^{-\beta_i E} n_T^{(i)} / Q_i. \tag{6.159}$$

Substitution of Eq. (6.158) into (6.156) gives the weight function

$$w_i(E) = \frac{\alpha(dE)^2 Q^2 e^{2\beta E}}{2W(E)} \cdot \frac{n_T^{(i)} e^{-\beta_i E}}{Q_i}. \tag{6.160}$$

To eliminate the Lagrange multiplier, we use the normalization condition $\sum_i w_i(E) = 1$; Eq. (6.160) then becomes

$$w_i(E) = \frac{n_T^{(i)} e^{-\beta_i E}}{Q_i} \Big/ \sum_j \frac{n_T^{(j)} e^{-\beta_j E}}{Q_j}. \tag{6.161}$$

Substituting Eq. (6.161) into (6.153) gives the final expression for the optimal estimate of the energy distribution $p(E, T)$ at temperature T

$$p(E, T) = \frac{e^{-\beta E}}{Q} \frac{\sum_i n_T^{(i)} p_i(E)}{\sum_j n_T^{(j)} e^{-\beta_j E} / Q_j}. \tag{6.162}$$

In Eq. (6.162), both Q and Q_j remain unknown. However, their ratio can be obtained self-consistently from

$$Q_j / Q = \frac{\int d\mathbf{r}^N e^{-\beta_j E}}{\int d\mathbf{r}^N e^{-\beta E}} = \int p(E) e^{-(\beta_j - \beta)E} dE. \tag{6.163}$$

Once $p(E, T)$ is obtained from Eq. (6.162), we can then predict internal energy and related thermodynamic properties (e.g., the Helmholtz energy difference between two temperatures) over a broad range of temperatures.

As an example of the multiple-histogram-reweighting method, consider the 2-dimensional Ising model discussed in Chapter 5. This model provides a simple platform for examining the efficiency of a simulation technique because the partition function is exactly known. At a given microstate or spin configuration, the total energy is given by

$$E_\nu = -\varepsilon \sum_{i,j}{}' s_{ij} \tag{6.164}$$

where spins s_i and s_j are either 1 or −1, and ε is the coupling parameter, i.e., the energy of interaction between two identical spins. The primed sum $\sum'$ denotes that only interactions between nearest neighbors contribute to the total energy.

Figure 6.31 presents the energy distributions for a 20 × 20 square-lattice at five different temperatures. Here, the simulation results were calculated from the Metropolis algorithm with 10^5 MC cycles (microstates per spin). At zero field, the reduced critical temperature of the two-dimensional Ising model is $k_B T_c / \varepsilon \approx 2.692$. Due to the strong thermal fluctuations, the energy distribution is broadest at the critical point; it is appreciably sharper at temperatures either lower or higher than the critical temperature.

Figure 6.32 compares the energy distribution function at $T/T_c = 1.05$ obtained from the single and multiple histogram reweighting methods. Although the single-histogram analysis works reasonably well when the temperature is close to that used as the reference, the multiple-histogram method yields more reliable and accurate predictions over a broad range of conditions. This example illustrates that, once we have obtained the energy distribution functions for a few temperatures, the multiple-histogram reweighting method allows us to predict thermodynamic properties (internal energy in this example) at other temperatures.

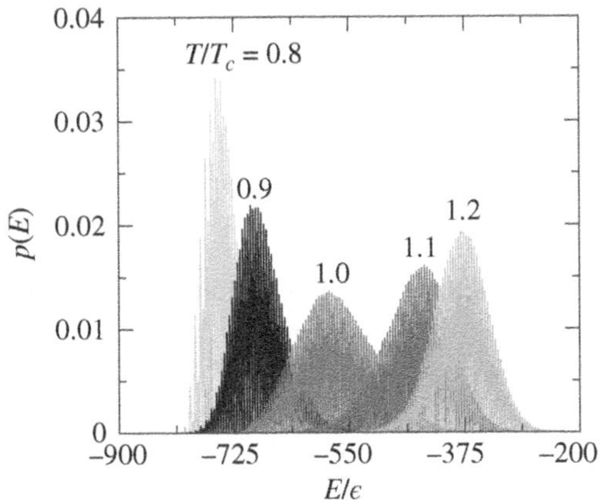

Figure 6.31 The energy distribution function of a 20×20 Ising lattice at five reduced temperatures. The energy distribution is broadest near the critical temperature.

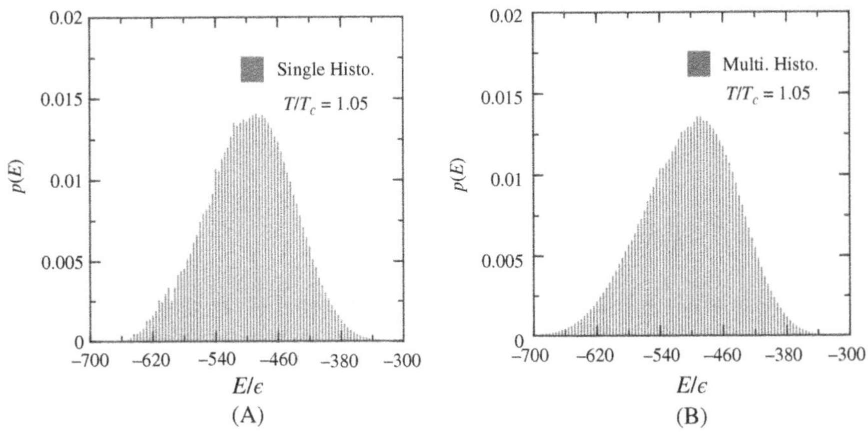

Figure 6.32 The energy distribution function at $T/T_c = 1.05$ based on single (A) and multiple (B) histogram reweighting methods. In panel A, $p(E)$ is predicted from the energy distribution function at a single temperature $T/T_c = 1$. In panel B, $p(E)$ is predicted from the energy distribution functions at 5 temperatures as shown in Figure 6.31.

One of the most valuable information attained from multiple-histogram reweighting is the variation of free energy. Based on the ratio of partition functions obtained from Eq. (6.163), we can readily calculate the difference in the reduced Helmholtz energy

$$\Delta(\beta F) = -\ln(Q_j/Q_i). \tag{6.165}$$

Figure 6.33 compares the numerical results from the histogram analysis with Onsager's expression for the partition function of 2D-Ising model. The free energy is typically not accessible through direct simulations.

The multiple-histogram reweighting method can also be applied to MC simulation in the grand canonical ensemble at different temperatures and different chemical potentials. Suppose we know $p_i(N, E)$ for a one-component system at a set of μ_i and T_i. The optimal prediction of $p(N, E)$ at temperature T and chemical potential μ can be derived following the same procedure that leads to Eq. (6.162):

$$p(N, E) = \frac{e^{-\beta(E-N\mu)}}{\Xi} \frac{\sum_i n_T^{(i)} p_i(N, E)}{\sum_j n_T^{(j)} e^{-\beta_j(E-N\mu_j)}/\Xi_j} \tag{6.166}$$

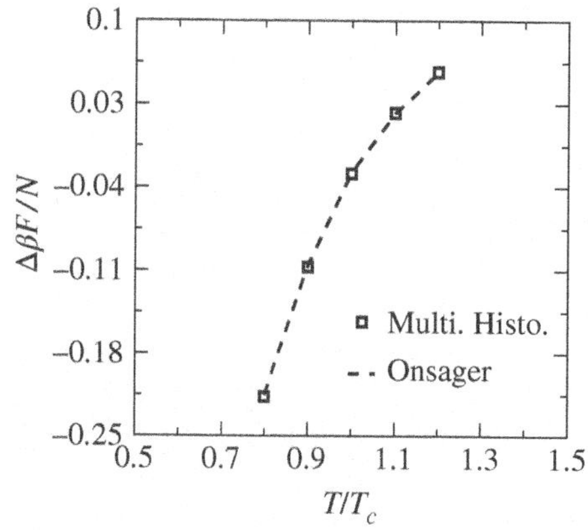

Figure 6.33 The relative Helmholtz energy of a two-dimensional Ising model calculated from the multiple-histogram reweighting (symbols) and from Onsager's equation.

where Ξ is the grand partition function. For every thermodynamic state i, the ratio of grand partition functions can be obtained self-consistently either from

$$\Xi_j/\Xi = \sum_{N=0}^{\infty} \int dE\, p(N,E)e^{-(\beta_j-\beta)E+(\beta_j\mu_j-\beta\mu)N} \tag{6.167}$$

or from the normalization condition

$$\sum_{N=0}^{\infty} \int dE \frac{e^{-\beta_k(E-N\mu_k)}\sum_i n_T^{(i)} p_i(N,E)}{\sum_j n_T^{(j)} e^{-\beta_j(E-N\mu_j)}/\Xi_j} = \Xi_k. \tag{6.168}$$

Similar to the canonical case, Eq. (6.168) is a special form of Eq. (6.167) where T and μ have been set equal to T_k and μ_k, respectively.

Once we have obtained $p(N, E)$, the distribution of energy and the number of molecules for a fixed T and μ, all thermodynamic properties of the system related to E and N can be predicted over a broad range of temperatures and chemical potentials covered by the simulation. For example, the internal energy can be calculated from

$$U(T,\mu) = \sum_N \int dE \;\; p(N,E)\cdot E \tag{6.169}$$

and the average density is given by

$$\rho(T,\mu) = \frac{1}{V}\sum_N \int dE \;\; p(N,E)\cdot N. \tag{6.170}$$

Because $PV = k_BT \ln \Xi$ for a bulk system, the difference in pressure between isothermal states 1 and 2 is related to the ratio of the corresponding grand potentials

$$(P_1 - P_2)V = k_BT \ln(\Xi_1/\Xi_2) \tag{6.171}$$

where the ratio of the grand partition functions can be calculated from Eq. (6.167).

6.11.3 Histogram Reweighting for Phase-Equilibrium Calculations

Multiple histogram reweighting provides quantitative information on the relative free energy of a system at different thermodynamic states. If the system exhibits phase transition, the free energy can be directly utilized to locate the coexistence curves if the grand-canonical MC simulation covers thermodynamic states on both sides of the phase boundary.

To elucidate the numerical procedure, suppose that we intend to predict the vapor–liquid coexistence curve for a one-component fluid. Based on grand-canonical MC simulations at a set of

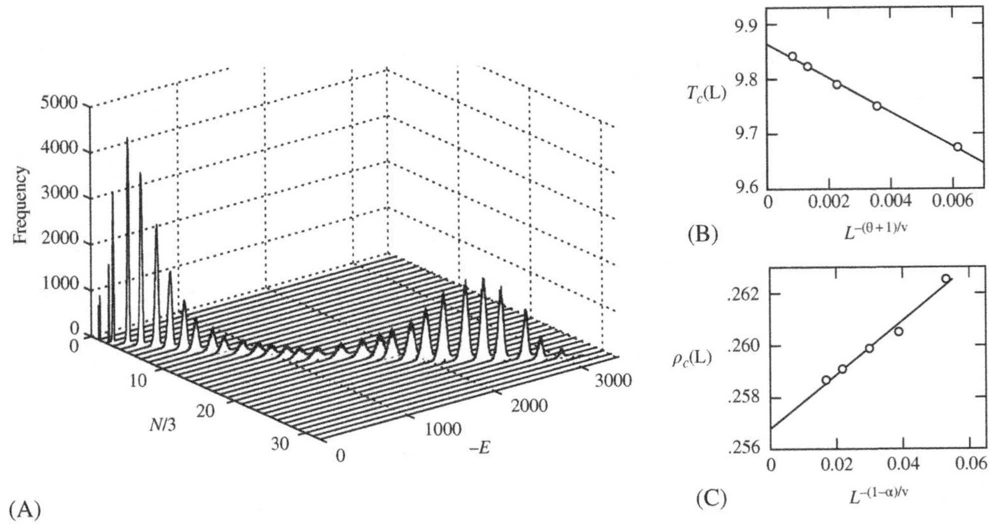

Figure 6.34 (A) The frequency of observing total energy E and number of particles N for an octane-like system represented by a $10 \times 10 \times 10$ cubic lattice model with PBC. The data were collected with grand-canonical simulation at reduced temperature $T^* = 11.5$ and chemical potential $\mu^* = -60.4$ following the notation of the original publication.[54]

temperatures and chemical potentials, we may use the multiple histogram reweighting method, Eq. (6.166), to construct the distributions of the number of particles and the total energy for both liquid and vapor sides of the phase diagram. For example, Figure 6.34 illustrates the simulation data for an octane-like system represented by a lattice model.[54] The probability distribution peaks at two thermodynamic states, corresponding to the two coexisting phases at the vapor side and liquid side of the phase diagram. At a fixed temperature, the vapor pressure can be identified by adjusting the chemical potential such that it yields an equal grand potential for the vapor and liquid states.[55] According to Eqs. (6.171) and (6.167), the equality of grand potential can be satisfied by equating the integral under the probability distribution function $p(N, E)$. From the temperature and chemical potential, the coexisting densities are obtained by locating the number of particles corresponding to the two maxima of the combined histogram.

Figure 6.35 compares the vapor–liquid coexistence curves for several n-alkanes calculated from MC simulation with experimental data.[56] The MC simulation was based on a united-atom model, as shown schematically in Figure 6.35A, where each spherical bead represents a functional group such as methyl or methylene (viz., united atom). The spherical particles interact with each other through the Mie potential

$$u(r_{ij}) = c_n \varepsilon_{ij} \left[\left(\frac{\sigma_{ij}}{r_{ij}} \right)^n - \left(\frac{\sigma_{ij}}{r_{ij}} \right)^m \right] \tag{6.172}$$

54 Panagiotopoulos A. Z., "Monte Carlo methods for phase equilibria of fluids", *J. Phys.: Condens. Matter* 12, R25–R52 (2000).
55 Because the system volume is fixed, the equality of the grand partition functions is equivalent to the equality of the pressures.
56 Potoff J. J. and Bernard-Brunel D. A., "Mie potentials for phase equilibria calculations: application to alkanes and perfluoroalkanes", *J. Phys. Chem. B* 113, 14725–14731 (2009).

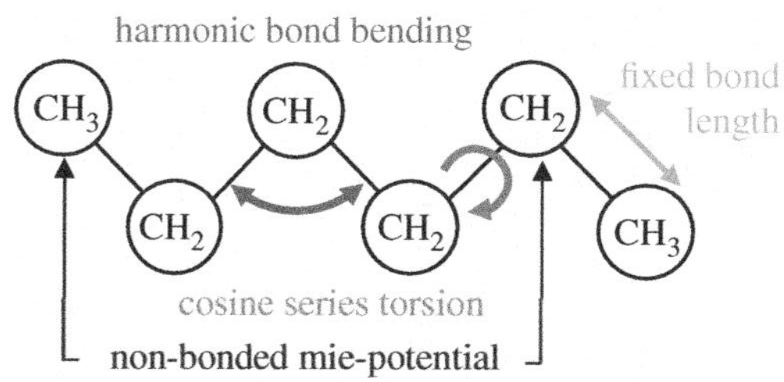

(A)

Group	ε (K)	σ (Å)	*n*
CH_4	161.0	3.74	14
CH_3	121.25	3.783	16
CH_2	61.0	3.99	16

(B)

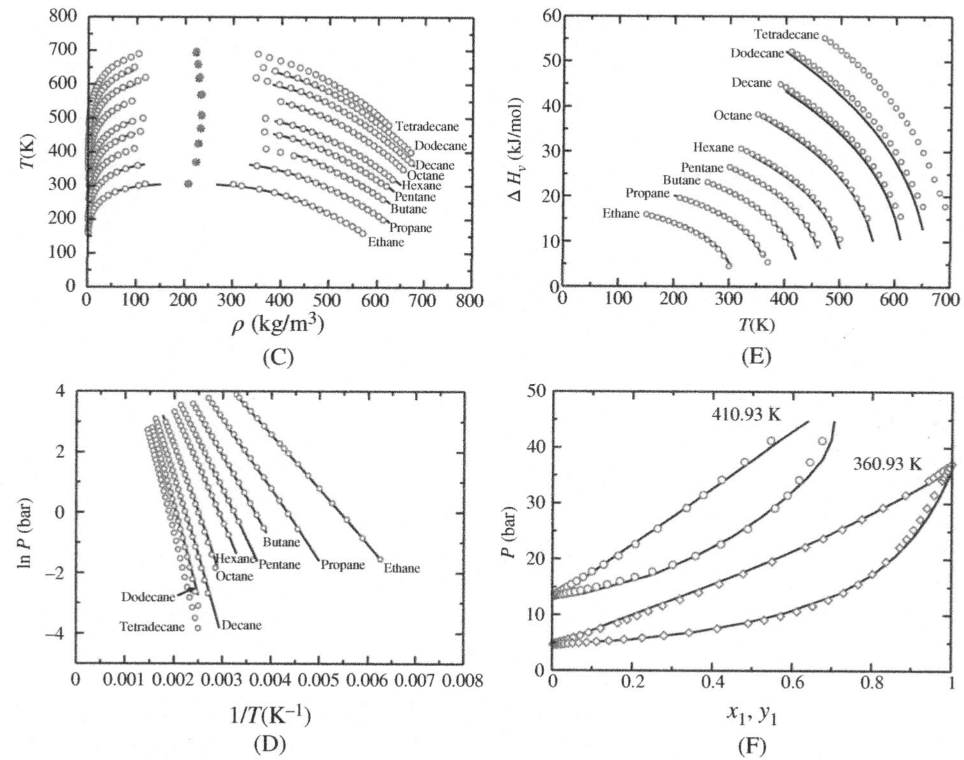

Figure 6.35 Monte Carlo simulation of vapor–liquid equilibrium. (A) In the united atom model, the nonbonded interactions are represented by the Mie potential, and the intramolecular interactions are described in terms of a fixed bond length, a harmonic potential for the bending energy, and a cosine series of the dihedral angle for the torsional potential. (B) The model parameters. Simulation results for the vapor–liquid coexistence curves (C), vapor pressure (D), and heat of vaporization (E) for n-alkanes. The lines represent experimental data; open circles are from Monte Carlo simulation with multiple histogram reweighting. Stars and filled circles correspond to critical points determined from experiment and simulation, respectively. (F) Phase diagram for a binary mixture of propane and n-pentane. Source: Reproduced from reference.[56]

where r_{ij}, ε_{ij}, and σ_{ij} are the separation, well depth, and collision diameter for the pair of united atoms i and j, respectively, and $c_n = (n/m)^{m/(n-m)} n/(n-m)$ is used to make the minimum of the potential equal to $-\varepsilon_{ij}$. Eq. (6.172) can be considered as a generalization of the LJ potential, which adopts $n = 12$, $m = 6$, and $c_n = 4$. For the simulation results reported in this work, $m = 6$ is used for the inter-particle attractions. The parameters for interaction between different particles are expressed in terms of the Lorentz–Berthelot combining rules

$$\varepsilon_{ij} = \sqrt{\varepsilon_{ii}\varepsilon_{jj}} \text{ and } \sigma_{ij} = (\sigma_{ii} + \sigma_{jj})/2. \tag{6.173}$$

Meanwhile, an arithmetic average is used to determine the repulsion exponents for the cross interactions

$$n_{ij} = (n_{ii} + n_{jj})/2. \tag{6.174}$$

Panel B of Figure 6.35 lists the parameters for nonbonded interactions among different types of united atoms. The parameters for methylene and methyl groups were obtained by best fitting the vapor pressure and saturated liquid densities of ethane and n-decane. For intramolecular interactions, the bond length between different pseudo-atoms is fixed at 1.54 Å, and the bond bending potential is represented by a harmonic function

$$u_{bend}(\theta)/k_B = \frac{k_\theta}{2}(\theta - \theta_0)^2 \tag{6.175}$$

with $\theta_0 = 114.0^\circ$ and $k_\theta = 62.5$ K/rad^2. The rotational potential is expressed as a cosine series of the dihedral angle ϕ:

$$u_{tors}(\phi)/k_B = \alpha_1(1 + \cos\phi) + \alpha_2(1 - \cos 2\phi) + \alpha_3(1 + \cos 3\phi) \tag{6.176}$$

with $\alpha_1 = 355.03$, $\alpha_2 = -68.19$ and $\alpha_3 = 791.32$, in units of K/mol, taken from the OPLS force field.

In Figure 6.35, the vapor–liquid coexistence curves, vapor pressures, and heats of vaporization for various alkanes were calculated from multiple histogram reweighting discussed above. In all cases, the saturated liquid densities are predicted with less that 1% error, and vapor pressures within 2–3% in comparison with experimental data. Figure 6.35C shows the heat of vaporization for each compound from the normal boiling point to within 20 K of the critical temperature. Here, the simulation results were obtained from the histogram data for the enthalpies of both the gas and liquid phases. For n-alkanes from C_2H_6 to C_5H_{12}, the theoretical predictions agree closely with experimental data. However, we see more significant deviations for longer alkanes, with the error increasing with the chain length. This is due in part to small errors in the coexistence densities and vapor pressures, which increase slowly with the chain length. Application of histogram reweighting method for phase equilibrium calculations is not limited to one-component systems. Essentially the same procedure can be used for calculating phase separation in mixtures. For example, Figure 6.35F shows the pressure-composition diagram for n-propane and n-pentane at two temperatures. In both cases, the simulation and experimental results are in excellent agreement, validating the transferability of the model parameters.

6.11.4 Finite-Size Scaling Near the Critical Point

To locate the critical point from multiple histogram reweighting, Wilding and Bruce proposed the mixed-field finite-size scaling method based on the similarities between different systems for the scaling fields and corresponding extensive variables in the same Ising universality class.[57]

57 Wilding N. B. and Bruce A. D., "Density fluctuations and field mixing in the critical fluid", *J. Phys.:Condens. Matter* 4, 3087 (1992).

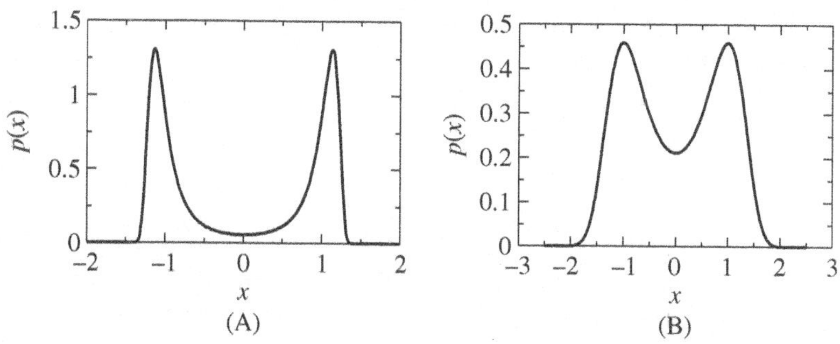

Figure 6.36 The universal probability distribution functions of the order parameter x for the 2D-Ising (A) and 3D-Ising (B) universality classes.

The universality *ansatz* allows for the prediction of the critical temperatures and densities from the histogram analysis of finite systems.

The main idea of the mixed-field finite-size scaling method is that, for a finite system with sufficiently large correlation length ξ near the critical point, the fluctuations of thermodynamic variables are universal and depend only on specific combinations of the system size L and the relevant scaling fields that measure deviations from criticality. At the critical point, the probability distribution $p(x)$ of the order parameter x has a universal form.[58] For the Ising model, the order parameter x is simply defined by the average magnetization.

Figure 6.36 shows the probability distributions for the 2D-Ising and 3D-Ising universality classes that are obtained from MC simulations. Approximately, these functions can be represented analytically by[59]

$$p(x) = \begin{cases} \dfrac{1}{A}\exp(-ax^{16} - bx^2 - c\,|\,x\,|\, - d) & \text{2D-Ising class} \\ \dfrac{1}{B}\exp[-(x^2-1)^2(kx^2+t)] & \text{3D-Ising class} \end{cases} \tag{6.177}$$

where $A = 0.9995$, $a = 0.058137$, $b = -2.68949$, $c = -0.11235$, and $d = 2.884162$; and for the 3D case, $B = 2.0549$, $k = 0.158$, and $t = 0.776$.

For a one-component fluid, the order parameter is proportional to a linear combination of the number of particles N and the total potential energy Φ

$$x \sim N - s\Phi \tag{6.178}$$

where s is called the field mixing parameter. For multicomponent systems, we use an extra field mixing parameter for each added component. For a binary system, the order parameter is given by

$$x \sim N_1 - s\Phi - qN_2 \tag{6.179}$$

where q is the field mixing parameter for the number of particles of component 2. By running grand canonical MC simulations for different system sizes near the critical condition, we can combine the histograms from the different simulation runs and estimate the critical parameters at a given system size by fitting the probability distribution of the mixed-field order parameter with the appropriate universal form (e.g., 2D or 3D Ising universality classes).

58 See Section 5.8 for the definition of order parameter.

59 Liu Y., Panagiotopoulos A. Z. and Debenedetti P. G., "Finite-size scaling study of the vapor–liquid critical properties of confined fluids: Crossover from three dimensions to two dimensions", *J. Chem. Phys.* 132, 144107 (2010); Tsypin M. M. and Blöte H. W. J., "Probability distribution of the order parameter for the three-dimensional Ising-model universality class: A high-precision Monte Carlo study", *Phys. Rev. E* 62, 73 (2000).

Once we have determined the critical parameters for systems of different sizes, the asymptotical behavior is described by the finite-size scaling laws

$$T_c(\infty) - T_c(L) \sim L^{-(\theta+1)/\nu}, \tag{6.180}$$

$$\rho_c(\infty) - \rho_c(L) \sim L^{-(\alpha+1)/\nu}, \tag{6.181}$$

where θ, α, and ν are parameters depending on the universality classes of the phase transition. The values of these parameters are (0.54, 0.629, 0.119) and (2, 1, 0) for the 3D- and 2D-Ising universality classes, respectively. As shown in Figure 6.34B and C, Eqs. (6.180) and (6.181) allow us to identify the critical conditions for infinite system size by scaling those corresponding to the finite systems.

The histogram reweighting analysis enables the accurate determination of the vapor pressure and vapor density. It has also been utilized to construct the phase diagrams of polymer solutions of various chain length. Based on simulation of polymer solutions on a lattice model, scaling relations were confirmed for the chain length dependence of the critical temperature and critical volume fraction of polymer solutions.

6.11.5 Summary

Histogram reweighting analysis is a powerful technique for estimating the free energy of a system at various thermodynamic states, leveraging the distribution of collective variables and statistical analysis. In principle, this method can be employed to predict equilibrium properties based on simulations conducted at a single condition. In the next section, we will discuss the procedures involved in utilizing such techniques. Our focus in this section is on trajectories obtained from importance sampling, where the microstates are biased toward those crucial for calculating the properties of a particular thermodynamic state. Because the simulation trajectories may not cover microstates that are important at other conditions, the use of multiple-histogram reweighting analysis helps overcome this limitation. By employing this method, we can address deficiencies in the simulation data by effectively incorporating information from multiple thermodynamic states.

6.12 Enhanced Sampling Methods

The ultimate goal of molecular simulation is to effectively sample microstates of thermodynamic systems, allowing for reliable calculation of ensemble averages while maintaining reasonable computational costs. Efficient sampling is an essential component of all simulation methods, regardless of MC, MD, or other techniques. Over the years, numerous enhanced sampling methods have been proposed, including replica exchange molecular dynamics (REMD), metadynamics, accelerated molecular dynamics (AMD), bias-exchange metadynamics (BEM), and many others. In this section, we discuss two basic strategies to enhance the efficiency of microstate sampling. The first strategy involves sampling through a generalized ensemble, which encompasses a wide range of thermodynamic conditions. The second strategy is known as umbrella sampling, which utilizes bias potentials to facilitate transitions between low-energy states. Given the continuous advancements in computational resources, algorithmic innovations, and methodological insights, further developments in enhanced sampling methods are anticipated. For a more in-depth discussion of various enhanced sampling techniques and their applications, interested readers should refer to the literature.[60]

---

60 See for example, Hénin J. et al., "Enhanced sampling methods for molecular dynamics simulations", *Living J. Comput. Mol. Sci.* 4 (1), 1583 (2022).

6.12.1 The Quasi-Ergodic Problem

As discussed in Section 6.3, the Metropolis–Hastings algorithm is biased toward MC moves from high-energy microstates to those of lower energy. When a new microstate is explored with an energy lower than that of its immediately preceding state, the MC move is always accepted. In other words, the transition probability from high energy to low energy is unity. For a MC move that leads to a microstate with a higher energy, the new microstate is accepted with the probability

$$p(\Delta E > 0) = e^{-\beta \Delta E}. \tag{6.182}$$

Eq. (6.182) indicates that the probability of a successful transition from a low-energy microstate to a high-energy microstate falls exponentially as the reduced energy gap $\beta\Delta E$ increases.

The importance sampling strategy as implemented in the Metropolis–Hastings algorithm is efficient for a thermodynamic system where the total energy fluctuates near its equilibrium value over the entire configurational space. While it is commonly used for sampling microstates due to thermal fluctuations, this method becomes problematic for systems that exhibit multiple regions of low-energy configurations (viz., multiple metastable states). When metastable states are separated by large energy barriers, the MC moves toward high-energy microstates are mostly rejected, and thus the microstates sampled by the Metropolis algorithm will be trapped in a certain region of the configurational space. The quasi-ergodic problem is encountered in diverse thermodynamic systems, including polymer melts, glassy materials, and biological systems of practical interest. Apparently, a proper representation of the entire configurational space is critically important for predicting the structural and thermodynamic properties of such systems.

The sampling problem may arise in simulating systems as simple as an ideal gas of n-butane. A n-butane molecule may exist in the trans (a.k.a., anti) or gauche states as defined by the relative positions of the two methyl groups. Figure 6.37 shows the variation of the molecular energy as a function of the dihedral angle ϕ. Infrared spectrum measurements indicate that the energy of the trans conformer, a metastable state from the perspective of statistical thermodynamics, is lower than those of the gauche conformers, other metastable states, by about 2.8 kJ/mol.[61] While the difference between the energies of trans and gauche states is not too much larger than the thermal energy at room temperature (k_BT~2.5 kJ/mol), these metastable states are separated by energy barriers that are substantially higher than the thermal energy, with the trans-to-gauche, the gauche-to-gauche, and the gauche-to-trans transitions determined to be 15.1, 13.7 and 12.3 kJ/mol, respectively. These energy barriers prohibit the easy transition between different conformer states

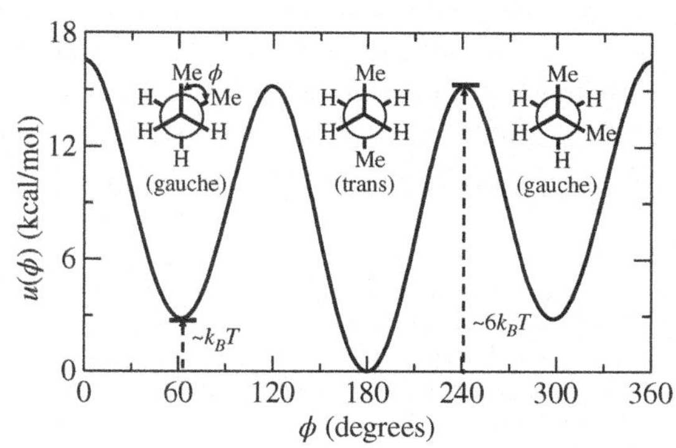

Figure 6.37 The torsional energy of a n-butane molecule as a function of the dihedral angle. The potential energy is calculated according to a semi-empirical fitting of the experimental data from infrared spectrum measurements. Source: Adapted from Herrebout W. A. et al.[61]

61 Herrebout W. A. et al., "Enthalpy difference between conformers of n-butane and the potential function governing conformational interchange", *J. Phys. Chem.* 99, 578–585 (1995).

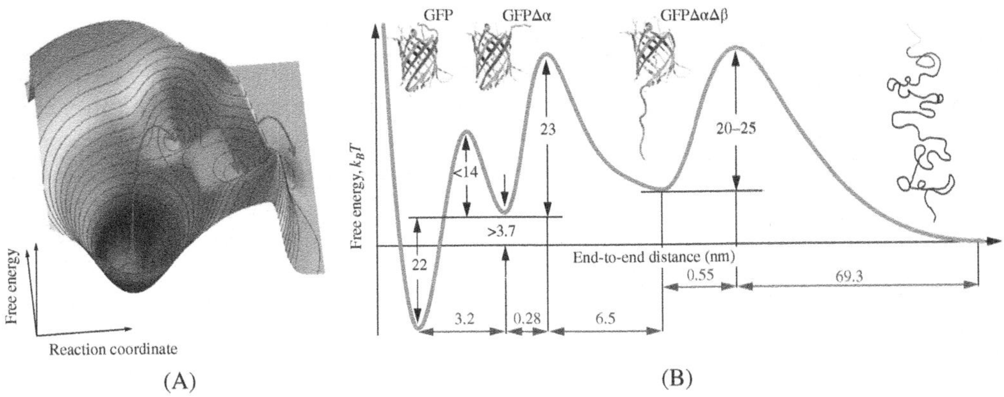

Figure 6.38 Free-energy landscape of a green fluorescent protein (GFP) derived from single-molecule force spectroscopy. (A) Cartoon of the multidimensional energy landscape. The arrows indicate the course of the mechanical unfolding pathway. (B) Projection of the energy landscape along the unfolding pathway onto one reaction coordinate. Source: Reproduced from reference.[63]

with conventional sampling schemes like the Metropolis method or simple MD simulation. At low temperature, importance sampling would render information relevant only to microstates near one of the conformer states.

Understandably, polymers and biomolecules are much more complicated than n-butane. Because of the complexity in inter- and intra-molecular interactions, such systems typically exhibit a large number of low-energy states. For example, Figure 6.38 illustrates the energy landscape[62] of a green fluorescent protein (GFP) derived from single-molecule force spectroscopy. When a mechanical force is applied between the N- and C-terminals of a GFP molecule, the protein transforms from its native state into the completely unfolded state through a complex energy landscape.

Figure 6.38 shows that, under the mechanical stress, the protein unfolds through two intermediate states, labelled as GFPΔα to GFPΔαΔβ.[63] The transition from the native state to GFPΔα is characterized by the detachment of a seven-residue N-terminal α-helix from the beta barrel structure of the protein. The free-energy barrier is about $36k_BT$ per molecule at room temperature, prohibiting the straightforward simulation of the transition through the Metropolis sampling ($e^{-36} \sim 10^{-16}$) or MD simulation. Similar free energy barriers exist for the transition from GFPΔα to GFPΔαΔβ and from the GFPΔαΔβ state into the completely denatured state. It is worth mentioning that GFP is among the most important proteins used in biotechnology.[64] A detailed knowledge of the energy landscape is important for diverse applications, ranging from the studies of protein localization in living cells to development of fluorescent pH or ion sensors.

62 An energy landscape is referred to the variation of the total energy in terms of few collective variables (CVs), i.e., certain features of microstates introduced to identify the metastable states of a specific thermodynamic system (e.g., the dihedral angle to define different conformers of n-butane). Because the microstates are integrated in expressing the system energy in terms of CVs, an energy landscape is also called the free-energy landscape. CVs are also known as reaction coordinates for chemical systems and order parameters in the physics literature.
63 Dietz H. and Rief M., "Exploring the energy landscape of GFP by single-molecule mechanical experiments", *PNAS* 101 (46), 16192–16197 (2004).
64 Valeur B. and Beerberan-Santos M. N., *Molecular fluorescence: principles and applications*. John Wiley & Sons, 2012.

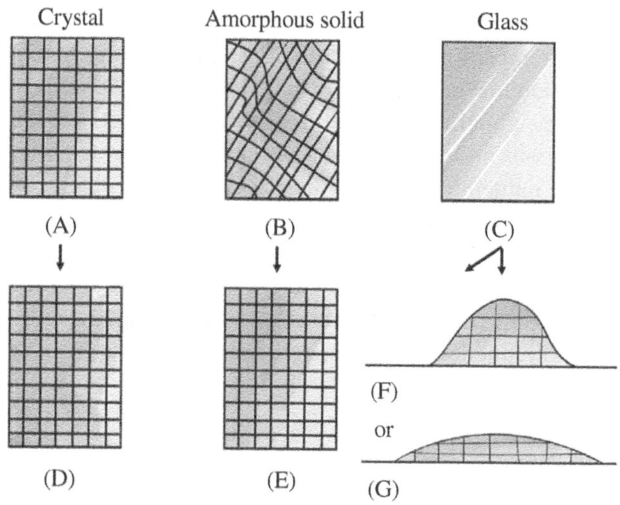

Figure 6.39 While a crystalline material (crystal) keeps its internal structure and external configuration at infinitely long-time scales (A, D), an amorphous solid will recrystallize via atomic diffusion at sufficiently long time (B, E), and a glass will ultimately relax and flow until an equilibrium state is reached (E, F or G), where the shape of the final state depends on the kinetics of crystallization and interfacial relaxation. Source: Reproduced from reference.[65]

A "rugged" energy landscape is commonplace not only in systems containing macromolecules but also in glassy systems that consist of seemingly simple objects such as spherical particles at high packing density. Unlike its low-density counterparts, a glassy system exhibits solid-like behavior at short times but liquid-like at long times (Figure 6.39)[65]. In this case, the sampling issue arises from the large gap between the macroscopic timescales and the actual time that can be explored in standard molecular simulation. From the point of view of thermodynamics, a glassy system can be characterized by a unique set of metastable states that are mutually inaccessible through thermal fluctuations. A macroscopic state of the system must be described by a superposition of these metastable states with certain probability distribution. Because the transition among different glass states cannot be realized through short-time dynamic processes, efficient sampling of the configuration space is crucial for understanding the properties of glassy materials, supercooled liquids, and liquid-to-glass transitions.

6.12.2 Generalized Ensemble Methods

In generalized ensemble methods, we simulate a thermodynamic system over a broad range of conditions simultaneously. At each condition, the microstates are sampled according to a conventional simulation protocol, and a stochastic process is designed to swap microstates between neighboring thermodynamic states.

As discussed in Section 2.11, the microstates encompassing different thermodynamic conditions constitute a generalized ensemble, which has its own partition function and microstate probability distribution. The stochastic hopping between different thermodynamic states can be implemented by using a procedure similar to the Metropolis method. We may elucidate this idea with the parallel

---

65 Zanotto E. D. and Mauro J. C., "The glassy state of matter: its definition and ultimate fate", *J. Non Cryst. Solids* 471, 490–495 (2017).

tempering method, which is also known as replica-exchange simulation.[66] In this method, we run simulation for a thermodynamic system at the temperature of interest along with the same system at higher temperatures (viz., *replicas*). In addition to conventional MC moves within each replica, parallel tempering swaps microstates between neighboring temperatures (viz., *tempering*) with the acceptance probability specified by the Metropolis criterion. The microstate swapping facilitates transition between metastable states either by reducing the energy barriers due to the change of thermodynamic conditions and/or by circumventing the energy barriers due to the disruption of correlations among microstates in the configurational space.

Figure 6.40 shows schematically the application of *parallel* simulations to n replicas with the same sampling scheme. Except microstate swapping between neighboring temperatures, the replicas do not interact with each other. The partition function for the generalized ensemble is given by

$$Q_E = \prod_{i=1}^{n} Q(T_i) = \prod_{i=1}^{n} \sum_{\nu=1} e^{-\beta_i E_\nu} \tag{6.183}$$

where $\beta_i = 1/k_B T_i$. Similar to that in a regular canonical ensemble, the configurational probability of a microstate ν in replica i is determined by its configuration $\mathbf{x}$ and temperature T_i

$$p_\nu = e^{-\beta_i E(\mathbf{x}_\nu)}/Q_E. \tag{6.184}$$

With symmetric trial moves for microstate swapping, the Metropolis criterion predicts that the acceptance probability for the exchange of a microstate in configuration $\mathbf{x}_i$ at temperature T_i with that in configuration $\mathbf{x}_j$ at temperature T_j is given by

$$a_{ij} = \min\left\{1, \frac{e^{-\beta_i E(\mathbf{x}_j) - \beta_j E(\mathbf{x}_i)}}{e^{-\beta_i E(\mathbf{x}_i) - \beta_j E(\mathbf{x}_j)}}\right\} = \min\{1, e^{\Delta\beta\Delta E}\ \} \tag{6.185}$$

where $\Delta\beta = \beta_j - \beta_i$ and $\Delta E = E(\mathbf{x}_j) - E(\mathbf{x}_i)$. The microstate swapping resembles a random walk in the temperature space that allows the low-energy states to wander from low temperatures to high temperatures. While the energy barriers cause inefficient sampling with regular simulation methods at low temperature, MC moves at high temperatures make transitions among metastable states possible even for thermodynamic systems with complex energy landscapes.

Figure 6.41 illustrates a simple application of the parallel tempering method for sampling the dihedral angle of n-butane at 100 K. While the Metropolis method is not able to reproduce the exact result even after 5 million MC states, parallel tempering with 3 temperatures (100, 200, and 300 K) provides accurate probability distribution for the dihedral angle.

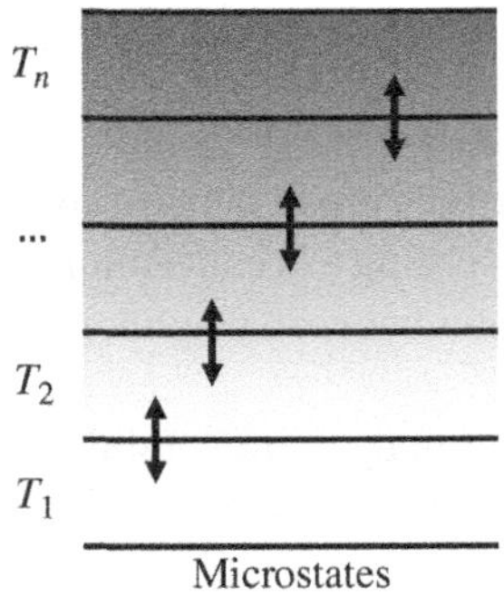

Figure 6.40 Schematic representation of parallel tempering simulation. A thermodynamic system is simulated along with multiple replicas at higher temperatures. In general, $T_1 < T_2 < \cdots T_n$ where T_1 is normally the temperature of interest, and T_n should be sufficiently high in order to overcome the free-energy barriers. In addition to regular MC moves within each replica, the microstates are swapped between adjacent temperatures according to the Metropolis criterion.

66 Earl D. J. and Deem M. W., "Parallel tempering: theory, applications, and new perspectives", *Phys. Chem. Chem. Phys.* 7, 3910 (2005).

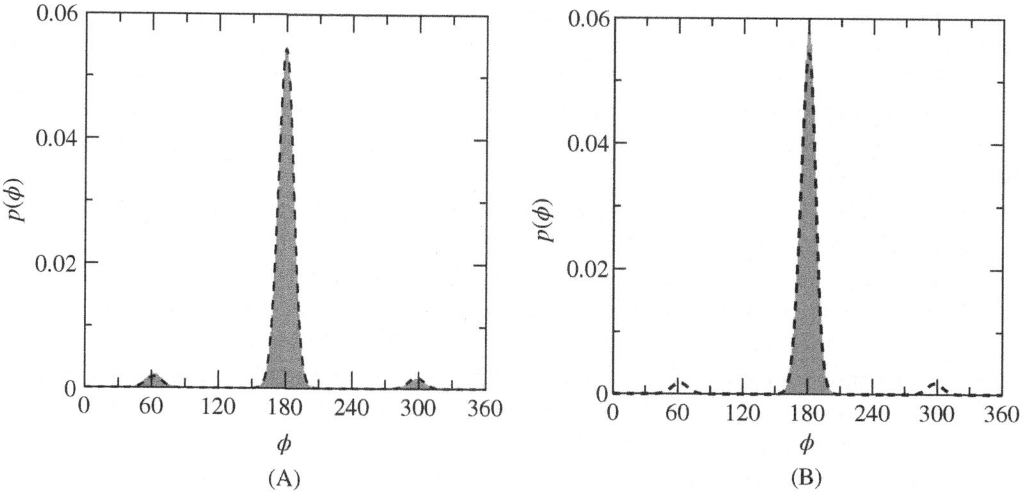

Figure 6.41 The probability distribution for the dihedral angle of n-butane at 100 K calculated from parallel tempering (A) and the Metropolis-Hastings algorithm (B). The dashed lines represent the exact results, and the shaded areas represent results from MC simulation.

A key hypothesis of the generalized ensemble methods is that energy barriers dividing different local minimum-energy states can be effectively conquered by changing the thermodynamic variables (e.g., increasing the system temperature or reducing the particle density) and/or by microstate swapping between neighboring states. The general concept is not limited to replicas at different temperatures. Alternative replicas with different thermodynamic variables, or with modified force fields, can also be used to reduce the energy barriers. Like the original Metropolis-Hastings algorithm, parallel tempering satisfies the detailed balance condition. Because the same sampling method is used within each replica, the enhanced sampling strategy is fully compatible with any simulation methods that are conventionally used for sampling microstates at a specific thermodynamic condition.

In numerical implementation of parallel tempering, we need to determine the temperature range and the number of replicas simulated. In general, the lowest temperature is selected according to the condition of practical interest. On the other hand, the highest temperature depends on the energy barriers dividing various metastable states of the system, i.e., it must be sufficiently elevated to overcome the energy barriers leading to trapped states. The number of replicas can be determined either by maximizing the overall probability of replica exchange between two extremal temperatures. A semi-empirical rule is that the temperatures should be set such that about 20% of all the swap attempts are accepted.[67]

6.12.3 Umbrella Sampling

Umbrella sampling was initially introduced by Torrie and Valleau.[68] Conceptually, the simulation technique resembles the multiple histogram reweighting method, where the thermodynamic properties of a system are predicted by utilizing ensemble averages from several reference systems.

---

67 Kone A. and Kofke D.A., "Selection of temperature intervals for parallel-tempering simulations", *J. Chem. Phys.* 122, 206101 (2005).

68 Torrie G. M. and Valleau J. P., "Nonphysical sampling distributions in Monte Carlo free-energy estimation: umbrella sampling", *J. Comput. Phys.* 23, 187–199 (1977).

Instead of running simulation at different thermodynamic conditions, such as varying temperature or chemical potential, umbrella sampling involves performing multiple simulations under a series of bias potentials, often referred to as "umbrellas." These bias potentials partition the configurational space of the system into smaller regions, known as "windows," which exhibit smooth energy landscapes. By applying bias potentials, umbrella sampling allows for enhanced sampling of regions that would otherwise be less frequently explored in a conventional simulation. Each window represents a specific range of the bias potential, and the collective data from all the windows are subsequently combined to obtain the unbiased thermodynamic properties of the system.

In umbrella sampling, the bias potentials are typically formulated in terms of collective variables (CVs), also known as reaction coordinates or order parameters. These parameters reflect certain features of a thermodynamic system that can be utilized to differentiate different low-energy regions of the configurational space. For example, in the case of n-butane conformers discussed above, the dihedral angle provides a one-dimensional CV that effectively captures the relevant degrees of freedom associated with the conformational changes. Similarly, the end-to-end distance of GFP reflects the molecular extension under a mechanical force. It is important to note that CVs can be either one-dimensional or multi-dimensional, depending on the complexity of the system and the specific phenomena of interest; they can be different even for the same system under consideration. For example, while the end-to-end distance is adequate for understanding the mechanical denaturation of proteins, multiple reaction coordinates, such as energy, the radius of gyration, the fraction of native contacts, and the similarity of the configuration of natively contacting residues are needed to describe protein folding or denaturation.[69]

The selection of a proper set of CVs is critically important in umbrella sampling because it determines the efficiency of simulation. Regardless of their complexity, each CV or reaction coordinate can be understood as a "coarse-grained" parameter that can be obtained by integration of the microstates

$$\bar{\xi} = \int d\mathbf{x}\, \delta[\bar{\xi} - \xi(\mathbf{x})] \tag{6.186}$$

where $\mathbf{x}$ represents the system configuration, δ stands for the Dirac-delta function, and $\xi(\mathbf{x})$ maps the microstates into some physical properties of the metastable states. Once these parameters are identified, the range of reaction coordinates is split into a number of windows using a set of bias potentials, as shown schematically in Figure 6.42A. While these potentials may take arbitrary forms, harmonic functions are commonly employed to confine the microstate sampling within

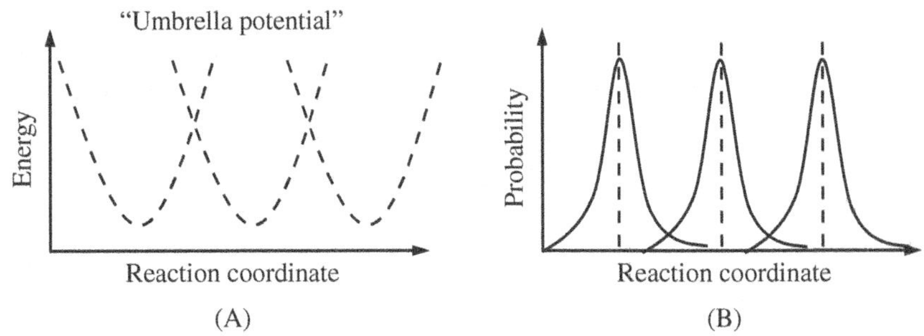

Figure 6.42 The harmonic bias potentials for umbrella sampling (A) and the probability distribution within each sampling window (B).

69 Cho S.S., Levy Y. and Wolynes P. G., "P versus Q: structural reaction coordinates capture protein folding on smooth landscapes", *PNAS* 103, 586-591 (2006).

a particular region of the configurational space. For systems that entail only a one-dimensional reaction coordinate, the bias potential is given by

$$w_i(\xi) = \frac{k}{2}(\xi - \xi_i)^2 \tag{6.187}$$

where parameter k determines the strength of the bias sampling, and ξ_i is a specific value of the reaction coordinate associated with window i. The parameter k is selected before the simulation; to ensure efficient sampling, this parameter should be sufficiently large such that the sampling is confined to a specific sampling window yet not too large to avoid too many windows that would be needed to cover the entire configurational space.

Given a set of bias potentials, as defined by Eq. (6.187), we can carry out conventional simulations as in the multiple histogram reweighting method. From simulation under each bias potential, we can predict the probability distribution in terms of the coarse-grained reaction coordinate $\bar{\xi}$

$$p_i^w(\bar{\xi}) = \frac{\int d\mathbf{x}\ e^{-\beta[E(\mathbf{x})+w_i(\xi(\mathbf{x}))]}\delta[\xi(\mathbf{x}) - \bar{\xi}]}{\int d\mathbf{x}\ e^{-\beta[E(\mathbf{x})+w_i(\xi(\mathbf{x}))]}} \tag{6.188}$$

where $E(\mathbf{x})$ represents the total energy of the original system. Eq. (6.188) describes the probability to observe $\bar{\xi}$ after the system is biased by adding potential $w_i(\xi(\mathbf{x}))$.

Similar to the single-histogram reweighting analysis, $p_i^w(\bar{\xi})$ can be utilized to predict the unbiased probability distribution for $\bar{\xi}$

$$\begin{aligned} p_i(\bar{\xi}) &= \frac{\int d\mathbf{x}\ e^{-\beta E(\mathbf{x})}\delta[\xi(\mathbf{x}) - \bar{\xi}]}{\int d\mathbf{x}\ e^{-\beta E(\mathbf{x})}} \\ &= \frac{\int d\mathbf{x}\ e^{\beta w_i(\xi)}e^{-\beta[E(\mathbf{x})+w_i(\xi)]}\delta[\xi(\mathbf{x}) - \bar{\xi}]}{\int d\mathbf{x}\ e^{-\beta[E(\mathbf{x})+w_i(\xi(\mathbf{x}))]}} \times \frac{\int d\mathbf{x}\ e^{-\beta[E(\mathbf{x})+w_i(\xi(\mathbf{x}))]}}{\int d\mathbf{x}\ e^{-\beta E(\mathbf{x})}} \\ &= p_i^w(\bar{\xi})e^{\beta w_i(\bar{\xi})} < e^{-\beta w_i(\xi(\mathbf{x}))} > \end{aligned} \tag{6.189}$$

where $<\cdots>$ represent the ensemble average in the original (unbiased) system. From Eq. (6.189), we can calculate the free energy along the coarse-grained reaction coordinate

$$F_i(\bar{\xi}) \equiv -k_BT\ln p_i(\bar{\xi}) = -k_BT\ln p_i^w(\bar{\xi}) - w_i(\bar{\xi}) - k_BT\ln < e^{-\beta w_i(\xi(\mathbf{x}))} > . \tag{6.190}$$

$F_i(\bar{\xi})$ is also known as the potential of mean force (PMF), i.e., the reversible work associated with an average force when the system is changed along a specific reaction coordinate. In Eq. (6.190), the last term on the right can be formally written as

$$< e^{-\beta w_i(\xi(\mathbf{x}))} > = \int p(\bar{\xi})e^{-\beta w_j(\bar{\xi})}d\bar{\xi} \tag{6.191}$$

where $p(\bar{\xi})$ stands for the probability distribution of $\bar{\xi}$ in the original system. Eq. (6.191) indicates that $\left\langle e^{-\beta w_i(\xi(\mathbf{x}))} \right\rangle$ is independent of $\bar{\xi}$. As a result, differentiating both sides of Eq. (6.190) with respect to $\bar{\xi}$ leads to

$$\frac{dF_i(\bar{\xi})}{d\bar{\xi}} = -k_BT\frac{d\ln p_i^w(\bar{\xi})}{d\bar{\xi}} - \frac{dw_i(\bar{\xi})}{d\bar{\xi}}. \tag{6.192}$$

Eq. (6.192) is the key equation for calculating the potential of mean force by umbrella sampling. By evaluating the free-energy derivative at a set of ξ_i and integrating the numerical results, we can calculate the free energy of the original system.

Alternatively, we can predict the free energy using the weighted histogram analysis method (WHAM). As discussed in Section 6.11, we can derive an optimal probability density from multiple simulations

$$p(\bar{\xi}) = \frac{1}{Q}\frac{\sum_i n_T^{(i)} p_i(\bar{\xi})}{\sum_j n_T^{(j)} e^{-\beta w_j(\bar{\xi})}/Q_j} \tag{6.193}$$

where the ratio of partition functions Q and Q_j is obtained self-consistently from

$$Q_j/Q = \int p(\bar{\xi}) e^{-\beta w_j(\bar{\xi})} d\bar{\xi} = < e^{-\beta w_j(\xi(\mathbf{x}))} > . \tag{6.194}$$

Eq. (6.193) is obtained by minimizing the statistical error of $p(\bar{\xi})$. From $p(\bar{\xi})$, we can finally calculate the potential of mean force

$$F(\bar{\xi}) = -k_B T \ln p(\bar{\xi}). \tag{6.195}$$

Unlike Eq. (6.192), WHAM avoids the numerical integration with respect to the reaction coordinate, and thus often yields more accurate predictions.

Umbrella sampling facilitates the generation of configurations corresponding to high-energy states, which would be excluded in conventional molecular simulation. This method is particularly useful for systems where the energy landscape or the energy barriers dividing different metastable states can be identified *à priori*. A trivial example is an ideal gas of n-butane where the total energy is primarily determined by the torsional potential of individual molecules. In this case, the energy landscape can be expressed as a function of the dihedral angle.

6.12.4 Thermodynamic-Scaling Method

The thermodynamic-scaling method may be considered as an extension of the umbrella sampling.[70] It is particularly useful for phase-equilibrium calculations.

We may illustrate the basic ideas of the thermodynamic-scaling simulation by considering vapor–liquid equilibria for a one-component fluid where the number of particles N is fixed while the volume V may vary. By representing the positions of all particles $\mathbf{r}^N$ in units of the system size,[71] the canonical partition function can be expressed as

$$Q = \frac{(V/\Lambda^3)^N}{N!} \int d\mathbf{x}^N e^{-\beta \Phi(\mathbf{x}^N, V)} \tag{6.196}$$

where Λ stands for the thermal wavelength, $\mathbf{x}^N$ is the position of all particles in dimensionless units, and $\mathbf{\Phi}$ represents the intermolecular potential. The microstates over a substantial range of temperatures and densities can be generated simultaneously by following a generalized probability distribution $p_0(\mathbf{x}^N)$ that, as explained later, is constructed self-consistently. When the number of particles N is fixed, we can calculate the ensemble average of an arbitrary dynamic quantity M at a particular thermodynamic state (T, V) according to

$$<M> = \frac{<M(\mathbf{x}^N) e^{-\beta \Phi(\mathbf{x}^N)}/p_0(\mathbf{x}^N)>_0}{<e^{-\beta \Phi(\mathbf{x}^N)}/p_0(\mathbf{x}^N)>_0} \tag{6.197}$$

where $<\cdots>$ denotes an average over the sampling distribution function $p_0(\mathbf{x}^N)$. The difference in Helmholtz energy F between thermodynamic states (T_1, V_1) and (T_2, V_2) can be obtained from

$$\beta_2 F_2 - \beta_1 F_1 = N \ln \frac{V_1 \Lambda_2^3}{V_2 \Lambda_1^3} + \ln \left\{ \frac{<e^{-\beta_1 \Phi(\mathbf{x}^N, V_1)}/p_0(\mathbf{x}^N)>_0}{<e^{-\beta_2 \Phi(\mathbf{x}^N, V_2)}/p_0(\mathbf{x}^N)>_0} \right\}. \tag{6.198}$$

---

70 Valleau J. P., "Density-scaling: A new monte carlo technique in statistical mechanics", *J. Comput. Phys.*, 96 (1), 193–216 (1991).

71 $\mathbf{x} = (x, y, z)/V^{1/3}$ where stands for the position of a particle and is the system volume.

The selection of an efficient probability distribution function $p_0(\mathbf{x}^N)$ is critical for successful implementation of the thermodynamic-scaling method. Ideally, $p_0(\mathbf{x}^N)$ should sample all configurations of the system that are important for the set of targeted thermodynamic states with roughly comparable weights. A general procedure to construct $p_0(\mathbf{x}^N)$ can be obtained by a linear combination of the Boltzmann weights at discrete thermodynamic states of interest

$$p_0(\mathbf{x}^N) = \sum_i \sum_j w_{ij} e^{-\beta_i \Phi(\mathbf{x}^N, V_j)} \tag{6.199}$$

where the weight functions $\{w_{ij}\}$ for all thermodynamic states over a grid of $\beta_i = 1/k_B T_i$ and volumes remain to be specified. To ensure a uniform sampling of all microstates, a logical choice for w_{ij} is given by

$$w_{ij} = e^{\beta_i F_{ij}^{ex}} \tag{6.200}$$

where F_{ij}^{ex} represents the excess Helmholtz energy[72] of the system at (T_i, V_j). Because the excess free energy is related to the summation of Boltzmann factors at various microstates

$$e^{-\beta_i F_{ij}^{ex}} = \int d\mathbf{x}^N e^{-\beta_i \Phi(\mathbf{x}^N, V_j)}, \tag{6.201}$$

Eq. (6.200) ensures that every term in Eq. (6.199) yields the same contribution to the configuration space for calculating the average according to $p_0(\mathbf{x}^N)$.

Although F_{ij}^{ex} is not known in advance, estimates for F_{ij}^{ex} from a simplified mean-field theory may be used as an initial guess. Alternatively, carrying out simulation on small systems may provide adequate estimates because they are not sensitive to the particular choice of w_{ij}. More accurate estimates of w_{ij} can be obtained by histogram weighting over energy distributions. The procedure is iterated until convergence is achieved for the input and output excess Helmholtz energies.

Once we know the Helmholtz energies at different temperatures and volumes, from Eq. (6.198), the vapor–liquid coexisting densities can be identified by constructing a double tangent line on the isothermal Helmholtz energy curve as a function of system volume for each temperature. Schematically, Figure 6.43 illustrates the Helmholtz energy as a function of system volume. Because $P = -(\partial F/\partial V)_T$, the two states corresponding to the two interception points in the

Figure 6.43 For a one-component system of fixed N and T, the vapor–liquid coexisting densities ($\rho_L = N/V_L$ and $\rho_V = N/V_V$) can be identified by constructing a common tangent line on the curve relating Helmholtz energy to volume.

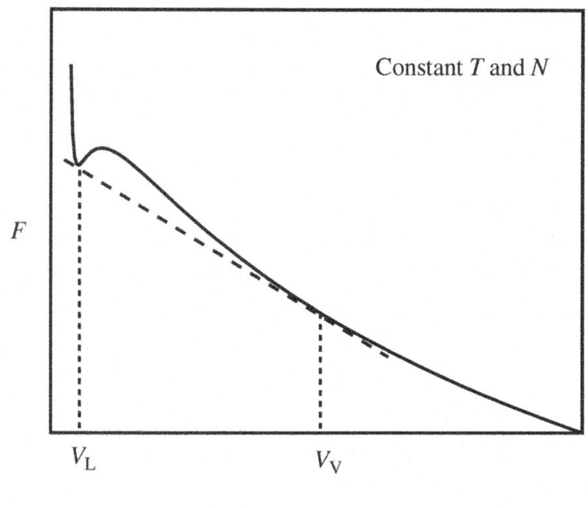

72 Here, the excess Helmholtz energy refers to the Helmholtz energy of the system in excess of that of an ideal gas at the same temperature and particle density.

common tangent line lead to the same pressure. The difference in the Helmholtz energy between these two states is given by

$$\Delta F = -(\partial F/\partial V)_T \cdot \Delta V = -P\Delta V \tag{6.202}$$

which leads to

$$\Delta G = 0. \tag{6.203}$$

Eq. (6.203) indicates that the common tangent line yields two phases with the same pressure and chemical potential.

The thermodynamic-scaling method eliminates the energy barriers for simulating the vapor–liquid equilibrium. Compared with the original umbrella sampling method, one disadvantage of thermodynamic scaling is that it involves some of the sampled thermodynamic states that are not of practical interest, thereby reducing the computational efficiency. Besides, numerical problems may arise in calculating the potential energies associated with volume scaling due to the molecular excluded-volume effects. Nevertheless, as an alternative to the multiple-histogram method, the thermodynamic-scaling method can be generalized for MC simulations covering a broad range of thermodynamic states.

6.12.5 The Wang–Landau Algorithm

The Wang–Landau algorithm[73] takes random walks in the energy space such that each MC move is accepted with a probability proportional to the reciprocal of the density of states. The entropic sampling leads to a simulation where the energy barriers become invisible. As a result, the Wang–Landau algorithm is able to sample all accessible microstates (favorable and less favorable) much faster than the Metropolis algorithm. The method has been successfully used in many problems of statistical physics, biophysics, and others. It is particularly efficient in simulating the first-order phase transitions, where simple canonical simulations have difficulty in overcoming the tunneling barrier between coexisting phases at the transition temperature.

As discussed in Section 6.11, we can express the canonical partition function as an integration of the density of states $w(E)$

$$Q = \int dE \;\; w(E)e^{-\beta E} \tag{6.204}$$

where $w(E)dE \equiv g(E)$ corresponds to the number of microstates with the total energy between E to $E + dE$. In terms of the system energy, the probability density is accordingly given by

$$p(E, T)dE = g(E)e^{-\beta E}/Q. \tag{6.205}$$

As the density of states $w(E)$ is independent of temperature, we would be able to calculate both the partition function Q and the probability density $p(E, T)$ at any temperature *if* we could simulate $g(E)$ for all possible energies. Subsequently, the thermodynamic properties of the system can be calculated at all temperatures. For example, the internal energy is given by

$$U = \langle E \rangle = \int dE \;\; p(E, T)E \tag{6.206}$$

---

73 Wang F. and Landau D. P., "Efficient, multiple-range random walk algorithm to calculate the density of states", *Phys. Rev. Lett.* 86 2050 (2001).

and the specific heat capacity can be determined from the fluctuation of the total energy

$$C_V = \frac{1}{k_B T^2}\{<E^2> - <E>^2\}. \tag{6.207}$$

The Helmholtz energy can be calculated directly from the partition function

$$F = -k_B T \ln Q = -k_B T \ln \int dE \ w(E) e^{-\beta E} \tag{6.208}$$

and the entropy can then be easily calculated from

$$S = (U - F)/T. \tag{6.209}$$

Regrettably, direct calculation of $g(E)$ is not possible for most systems of practical interest. This is because the task would be equivalent to the evaluation of the microcanonical partition function or the total number of microstates. The Wang–Landau algorithm circumvents this issue by using a self-correcting procedure and is able to estimate the density of states efficiently in an iterative manner. The idea may be most easily understood within the framework of the multicanonical ensemble (MU) discussed in Section 2.11. In the MU method, microstates are sampled according to a uniform probability distribution, i.e.,

$$p_{MU}(E) \sim g(E) w_{MU}(E) = \text{constant}. \tag{6.210}$$

Eq. (6.210) requires that the weight factor is inversely proportional to the density of states

$$w_{MU}(E) \sim 1/g(E) = e^{-S/k_B} \tag{6.211}$$

where $S = k_B \ln g(E)$ is recognized as the entropy of the system with energy E in an isolated state. At given temperature T, the Metropolis sampling would lead to the canonical distribution

$$p(E, T) \sim g(E) e^{-\beta E}. \tag{6.212}$$

A comparison of Eqs. (6.210) and (6.212) suggests that a uniform distribution in the energy space can be achieved when the microstates are sampled according to entropy, i.e.,

$$p(E_o \to E_n) = \min\{g(E_o)/g(E_n),\ 1\} = \min\{e^{-\Delta S/k_B},\ 1\} \tag{6.213}$$

where E_o and E_n represent respectively the energies of the current and trial microstates in the MC move, and $\Delta S = k_B \ln[g(E_n)/g(E_o)]$ is the corresponding entropy change. Eq. (6.213) implies that a uniform energy distribution can be attained if the MC moves leading to the reduction of entropy are always accepted. To a certain degree, the updates for $g(E)$ provide a self-correcting scheme; if a microstate is visited more often than it would be under equal sampling of all microstates, its weight decreases accordingly, resulting in fewer visits to that state. Because the acceptance criterion is determined by the system entropy, we call MC moves according to Eq. (6.213) as entropic sampling.

It is instructive to elucidate how entropy sampling leads to a uniform energetic probability distribution using an N-spin cyclic Ising chain as a model system. As discussed in Section 6.4.2, the partition function of a cyclic Ising chain is exactly known

$$Q = (e^{-\beta\varepsilon} + e^{\beta\varepsilon})^N = \sum_{n=0}^{N} \binom{N}{n} e^{-\beta E_n} \tag{6.214}$$

where n represents the number of nearest neighboring spins aligned in parallel, ε is the coupling energy, and $E_n = (N - 2n)\varepsilon$ is the system energy corresponding to n. Eq. (6.214) indicates that the density of states for the cyclic Ising chain is

$$g(E) = \binom{N}{n} = \frac{N!}{n!(N-n)!}. \tag{6.215}$$

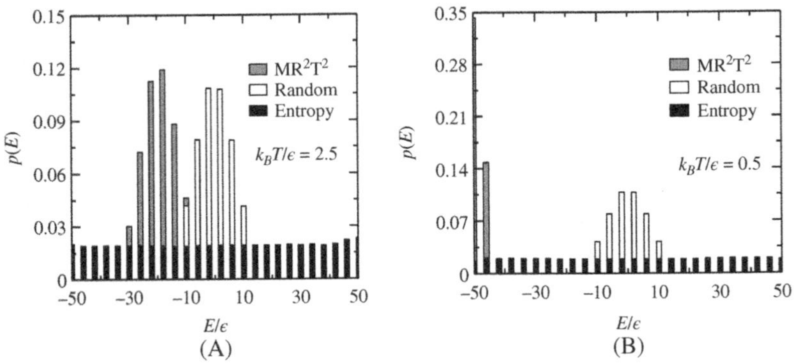

Figure 6.44 The probability distribution as a function of energy for a cyclic Ising chain with 50 spins. The results are generated from the Metropolis algorithm (MR^2T^2), random sampling, and the Wang–Landau algorithm at two reduced temperatures 2.5 and 0.5.

With an analytical expression for $g(E)$, we can carry out MC simulation using Eq. (6.213) as the sampling scheme. Figure 6.44 shows the probability density as a function of energy for $N = 50$. As expected, the entropy sampling generates microstates with a uniform distribution of energy. In comparison, the Metropolis algorithm leads to microstates with the energy near the equilibrium value (viz., canonical distribution), and the random flipping of the spin orientation favors microstates near zero energy. While the entropic sampling leads to uniform probability distribution in the energy space, both Metropolis and random sampling are biased toward a particular energy value.

As mentioned above, $g(E)$ is unknown for most practical systems. A simple yet elegant scheme used in the Wang–Landau algorithm is to calculate the density of states self-consistently through iterations. Specifically, $g(E) = 1$ is set for all possible energies of the system at the beginning of simulation. A randomly selected trial microstate with energy E_n is accepted if $g(E_n) \leq g(E_o)$. If $g(E_n) > g(E_o)$, the MC move is accepted with a probability $g(E_o)/g(E_n)$. To avoid repeated visiting of this energy state, the existing value of $g(E_n)$ is modified by multiplying a factor $f > 1$ (typically $f = e^1 \approx 2.7183$), i.e., $g(E_n) \to f \times g(E_n)$. If the trial move is rejected, the current density of states $g(E_o)$ is modified by the same modification factor, $g(E_o) \to f \times g(E_o)$. The process continues until the random walk yields a flat energy distribution. Subsequently, the modification factor is reduced by $f \to \sqrt{f}$ (or by other functions leading to $f \to 1$). The procedure repeats until $f \approx 1$. At the end of simulation, we have the desired density of states that corresponds to a uniform probability distribution in the energy space.

Figure 6.45 shows the density of states calculated from the Wang-Landau algorithm for the cyclic Ising chain discussed above. The simulation starts with a random assignment of spin orientations as the initial configuration and $f = e^1 \approx 2.7183$ is used as the initial value. For this simple system, the energy histogram becomes nearly flat after 10^5 MC moves according to Eq. (6.213), with $g(E)$ calculated along with simulation. The iteration quick converges after a few iterations by reducing the modification factor according to $f \to \sqrt{f}$. Although the probability density for the energy distribution is not exactly uniform at the end of the simulation, it yields the density of states in excellent agreement with the exact results (Figure 6.45). From $g(E)$, we can determine thermodynamic properties using equations as given by Eqs. (6.206)–(6.209). Figure 6.46 compares that the numerical values for Helmholtz energy, internal energy, and heat capacity with the exact results for the one-dimensional Ising model. Again, the agreement is almost perfect!

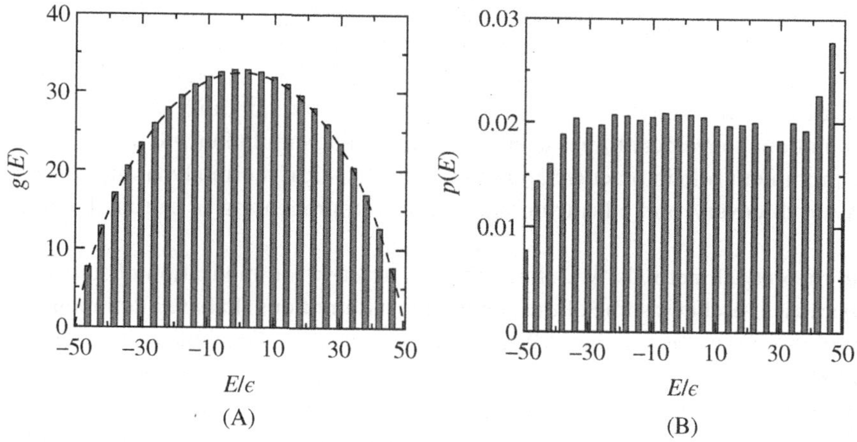

Figure 6.45 The density of states and the probability density of a cyclic Ising chain with 50 spins calculated from the Wang–Landau algorithm.

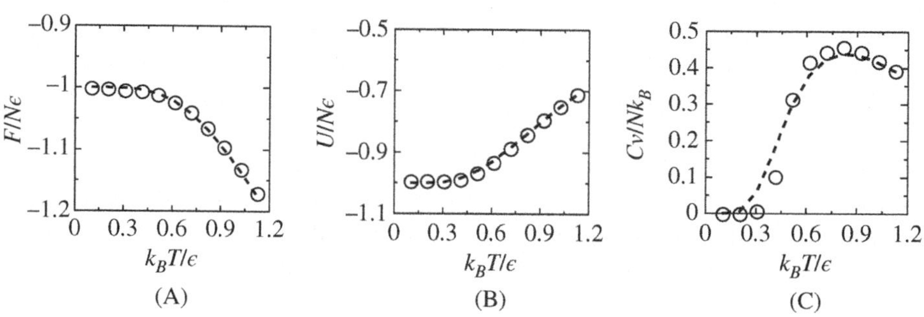

Figure 6.46 Comparison of thermodynamic properties calculated from the Wang–Landau sampling (symbols) with exact results (dashed lines) for a cyclic Ising chain with 50 spins.

6.12.6 Summary

Conventional MC and MD techniques are not directly applicable to sampling transformation among different metastable states in complex systems. The objective of enhanced sampling is to solve the quasi-ergodicity problem by applying bias potentials or sampling over a broad range of thermodynamic conditions. In either case, one can obtain the ensemble averages of physical quantities as functions of model parameters or thermodynamic conditions by the single-histogram and/or multiple-histogram reweighting techniques. While these two families of enhanced sampling methods are often considered as complementary, recent developments indicate that they can be used in tandem as a unified approach to improve the sampling efficiency.[74]

6.13 Chapter Summary

Molecular simulation is a powerful computational tool in statistical thermodynamics that has been extensively used to investigate and comprehend the microscopic behavior of complex fluids,

74 Invernizzi M., Piaggi P. M. and Parrinello M. "Unified approach to enhanced sampling", *Phys. Rev. X* 10 (4), 041034 (2020).

materials, and biological systems. With its ability to provide detailed insights into both structural and thermodynamic properties, molecular simulation complements experimental measurements and analytical theories, offering a valuable alternative for studying the phase transitions and critical behavior of diverse thermodynamic systems including both fluid–fluid and fluid–solid equilibria. As computational power continues to advance, molecular simulation will undoubtedly play an even more prominent role in practical applications of statistical thermodynamics such as materials design, drug discovery, and advancing our overall comprehension of thermodynamic phenomena.

In comparison to theoretical methods, MC simulation is attractive for application of statistical thermodynamics because it eschews some of the simplifying assumptions that are often required to obtain analytic solutions. On the one hand, MC simulation provides benchmark data for the calibration of analytical statistical-thermodynamic theories including, for example, those for fluid-phase equilibria, self-assembly of surfactants or block copolymers, and for folding and aggregation of proteins. For systems where the interactions between molecules (i.e., force fields) are well established, simulation results are now sufficiently reliable to replace experimental data. Unlike experiments, however, molecular simulation provides not only the physical properties of a molecular system as ensemble averages but in addition, it renders microscopic details that are often not directly accessible from experiments. On the other hand, MC simulation is useful for obtaining properties of those systems where experimental studies are difficult or impossible, in particular those that are toxic, explosive, or subjected to extreme conditions, e.g., chemical or biological warfare agents or systems at very high temperature or pressure.

It is important to note that numerical results obtained from molecular simulations are accurate within the framework of a thermodynamic model used to represent the energy of the system, but they are never exact. Calculating an ensemble average exactly would require considering all possible microstates, which is generally infeasible for most practical systems. In principle, molecular simulations can generate microstates according to the relative probability distribution for a given statistical ensemble, thereby enabling accurate predictions of ensemble averages within a specific molecular model. Consequently, the accuracy of simulation results is primarily constrained by the semi-empirical models employed to represent intermolecular and intramolecular interactions. As these models often rely on experimental inputs for parameter calibration, the reliability of molecular simulation may diminish when applied outside the conditions for which the semi-empirical models are applicable. Additionally, the accuracy of simulation results is influenced by statistical errors, which depend on the number of microstates used to estimate the ensemble average and the methodology employed for microstate generation.

6.A Stochastic Processes and Markov Chains

This appendix provides a concise introduction to stochastic processes, which serve as the fundamental mathematical foundation for MC simulation. For a more comprehensive understanding and application of stochastic processes, further study and exploration of the topic are recommended.

6.A.1 Stochastic Processes

A stochastic process refers to a series of events in which the transition from one event to the next is determined by probability. Consequently, a stochastic process consists of a collection of random variables, which each represents the outcome of the event at a given time.

To illustrate, the act of repeatedly flipping a coin can be viewed as a simple stochastic process, wherein each flip has a 50% chance of resulting in heads or tails (assuming a fair coin). Here, "heads" and "tails" are random variables, representing the two states of the coin (viz., the system). Another example is a random walk on a two-dimensional square lattice. Suppose that the walker starts from an initial position and takes one of the four adjacent neighbors by random, i.e., each direction has an equal probability of 1/4 for the next step. In this case, the random variables correspond to the positions of the lattice sites.

6.A.2 The Markov Chains

A Markov process, also known as the Markov chain, is a stochastic process in which the probability of each event depends solely on the outcome of its immediate precedent event. In other words, the future states of the Markov process are determined solely by its present state and are not affected by how the system arrived at its current state.

A Markov process is finite if the number of possible outcomes (viz., states) is limited. In this case, each event can be understood as a transition from any one state to any other state with the probability specified by a transition matrix, denoted by $\widehat{\mathbf{M}}$. Each element of the transition matrix, M_{ij}, defines the probability of the transition from a state i to state j. Because the probability of transition from one state to another is nonnegative and normalized, all elements in the transition matrix satisfy

$$M_{ij} \geq 0, \tag{6.A.1}$$

and the normalization condition

$$\sum_j M_{ij} = 1. \tag{6.A.2}$$

To elucidate a Markov process, imagine a shooting game with only two outcomes, hit or miss for each shot. Suppose that a shooter hits the target with 80% probability if the previous shot was a hit, and 40% probability if the previous shot was a miss. Therefore, the transition matrix is given by

$$\widehat{\mathbf{M}} = \begin{pmatrix} 0.8 & 0.2 \\ 0.4 & 0.6 \end{pmatrix}. \tag{6.A.3}$$

According to Eq. (6.A.3), $M_{11} = 0.8$, and $M_{12} = 1 - 0.8 = 0.2$ is determined from the normalization of probability; $M_{21} = 0.4$, and similarly $M_{22} = 1 - 0.4 = 0.6$. If "1" stands for a hit and "2" for a miss, then $M_{12} = 0.2$, for example, means that if the shooter hits the target at one time, the probability that he/she will miss on the next shot is 0.20. The process is stochastic because the outcome depends on a probability, and it is a Markov process because the probability for the outcome of the next step is solely dependent on the current state.

A Markov process is *ergodic* if any two states are mutually accessible, that is, after a finite number of steps, there is a non-zero probability of access from one state to the other and vice versa. In an ergodic process, the transition matrix satisfies

$$(\widehat{\mathbf{M}}^m)_{ij} > 0. \tag{6.A.4}$$

where $m > 0$ is a finite integer, which may depend on i and j. A Markov process is *regular* if Eq. (6.A.4) holds true for some fixed m, independent of i and j.

6.A.3 The Perron–Frobenius Theorem

The Perron–Frobenius theorem is central to understanding the mathematical basis of MC simulation:

> Let $\widehat{\mathbf{M}}$ be the transition matrix of a regular Markov process. Then $\lim_{k\to\infty}\widehat{\mathbf{M}}^k = \widehat{\mathbf{M}}^\infty$ exists and the matrix $\widehat{\mathbf{M}}^\infty$ has identical rows corresponding to the limiting probability distribution of the states.

A proof of this theorem can be found in Bapat and Raghavan.[75]

To understand the Perron–Frobenius theorem, consider again the shooting game discussed earlier. Suppose that the shooter hits the target at the first shot, thus the probability that he/she hits again at the second shot is 0.8, which can be calculated from

$$(1,\ 0)\begin{pmatrix} 0.8 & 0.2 \\ 0.4 & 0.6 \end{pmatrix} = (0.8,\ 0.2) \tag{6.A.5}$$

where the vector (1, 0) specifies the initial state, and (0.8, 0.2) specifies the second state, i.e., 0.8 chance of a hit and 0.2 chance of a miss. The probability that the shooter hits at the third state is given by

$$(1,\ 0)\begin{pmatrix} 0.8 & 0.2 \\ 0.4 & 0.6 \end{pmatrix}^2 = (0.72,\ 0.28). \tag{6.A.6}$$

One may repeat the above calculation and find that the chance of a hit after m shots is specified by the probability vector

$$(1,\ 0)\begin{pmatrix} 0.8 & 0.2 \\ 0.4 & 0.6 \end{pmatrix}^m \tag{6.A.7}$$

which approaches the limiting value of (2/3, 1/3) as $m \to \infty$. Now suppose that the first shot is a miss, the chance of hit or miss after m shots is given by

$$(0,\ 1)\begin{pmatrix} 0.8 & 0.2 \\ 0.4 & 0.6 \end{pmatrix}^m \tag{6.A.8}$$

which also approaches the limiting value (2/3, 1/3) as $m \to \infty$. In other words, after a large number of shots, the probability distribution becomes *independent of the initial state*, and approaches a limiting value. This result follows directly from the Perron–Frobenius theorem, which indicates that the limiting probability distribution is determined by

$$\lim_{m\to\infty}\begin{pmatrix} 0.8 & 0.2 \\ 0.4 & 0.6 \end{pmatrix}^m = \begin{pmatrix} 2/3 & 1/3 \\ 2/3 & 1/3 \end{pmatrix}. \tag{6.A.9}$$

Because the two rows of the limiting matrix are identical, the final probability distribution is independent of the initial state.

In MC simulation, the microstates are updated based on regular Markov processes. Consequently, the limiting distribution, which represents the long-term behavior of the system, is solely determined by the transition matrix and is independent of the initial configuration. The microstates

75 Bapat R. B. and Raghavan T. E. S., *Nonnegative matrices and applications.* Cambridge University Press, 40–50 (1997).

created by a regular Markov process are *ergodic*, meaning that the time average and ensemble average yield equivalent results. Ergodicity is crucial for obtaining statistically representative equilibrium properties of a system through MC simulation.

Further Readings

Allen M. P. and Tildesley D. J., *Computer simulation of liquids* (2nd Edition). Oxford University Press, Chapters 4, 9 and 10, 2017.

Chandler D., *Introduction to modern statistical mechanics*. Oxford University Press, New York, Chapter 6, 1987.

Frenkel, D. and Smit B., *Understanding molecular simulation* (2nd Edition). Academic Press, Chapters 3, 5 and 7, 2002.

Landau D. and Binder K., *A guide to Monte Carlo simulations in statistical physics* (5th Edition). Cambridge University Press, 2021.

Price S. L., "Control and prediction of the organic solid state: a challenge to theory and experiment", *Proc. Roy. Soc. A: Math. Phys. Eng. Sci.* 474 (2217), 20180351 (2018).

Problems

6.1 Generate a sequence of random numbers between 0 and 1 and calculate their mean and standard deviation versus the number of samples. Discuss how your numerical results are compared with the exact values of the mean and standard deviation.

6.2 In probability theory, the universality of the uniform refers to the concept that if we have a continuous random variable denoted as x, with a probability density function $p(x)$, then its cumulative distribution function (CDF)

$$F(x) = \int_{-\infty}^{x} f(x')dx'$$

corresponds to a random variable that follows a uniform distribution in the interval [0, 1]. The universality of the uniform is easily proved by the change of variables, $y = F(x)$, which leads to

$$1 = \int_{-\infty}^{\infty} f(x)dx = \int_{0}^{1} f(x)\frac{dx}{dy}dy = \int_{0}^{1} \frac{f(x)}{f(x)}dy = \int_{0}^{1} dy.$$

The above equation indicates that y is a random variable that has a uniform probability distribution in the interval [0, 1]. Based on the universality of the uniform, show that the exponential distribution

$$f(x) = \begin{cases} \frac{1}{\langle x \rangle} \exp\left(-\frac{x}{\langle x \rangle}\right), & 0 < x < \infty, \\ 0, & \text{otherwise,} \end{cases}$$

can be generated from $x = -\langle x \rangle \ln y$ where $\langle x \rangle$ represents the mean, and y is a random number with a uniform probability distribution in [0, 1].

6.3 Prepare a Monte Carlo (MC) program to estimate the value of π and discuss how the numerical error varies with the sample size.

6.4 Suppose that x_1 and x_2 are uniformly and independently distributed random numbers between 0 and 1. The Box–Muller transformation, defined as

$$y_1 = \sqrt{-2\ln x_1}\cos(2\pi x_2), \qquad y_2 = \sqrt{-2\ln x_1}\sin(2\pi x_2),$$

yields two independent standard normal distributions, i.e., variables distributed between $-\infty$ and ∞ with the probability density of $p(y) = e^{-y^2}/\sqrt{\pi}$. Prepare a MC program to generate 1000 sample points following the Gaussian distribution

$$p(x) = \frac{1}{\sqrt{2\pi\sigma^2}}\exp\left[-\frac{(x-x_0)^2}{2\sigma^2}\right],$$

where $x_0 = 2$ and $\sigma = 0.5$. Verify the results by checking the first few moments of the probability distribution:

$$\langle x\rangle = x_0; \qquad \langle x^2\rangle = x_0^2 + \sigma^2;$$
$$\langle x^3\rangle = x_0^3 + 3x_0\sigma^2; \qquad \langle x^4\rangle = x_0^4 + 6x_0^2\sigma^2 + 3\sigma^4.$$

6.5 For a system of classical particles at absolute temperature T, the velocity of translational motions follows the Maxwell–Boltzmann distribution

$$p(\mathbf{v}) = \sqrt{\frac{m}{2\pi k_B T}}\exp\left[-\frac{mv^2}{2k_B T}\right].$$

Prepare a program to assign the velocities of $N = 1000$ particles according to the Maxwell–Boltzmann distribution and verify your result. Assume that all particle have the same mass, $m = 6\times 10^{-26}$ kg, and that the system temperature is $T = 298.15$ K.

6.6 MC simulations often require generating variables within a finite range $[a, b]$ with a given probability density function $p(x)$. This task can be accomplished by rejection sampling, which is also known as the acceptance–rejection algorithm. The basic idea can be illustrated with Figure P6.6, which demonstrates the generation of variables with probability density $p(x)$ by uniformly sampling the independent variable and retaining the samples falling under the region defined by the graph of the probability density function. Based on the rejection sampling, prepare a program to generate 10^6 numbers from -1 to 1 with a probability satisfying the Gaussian distribution, $p(x) \sim e^{-x^2/2}$.

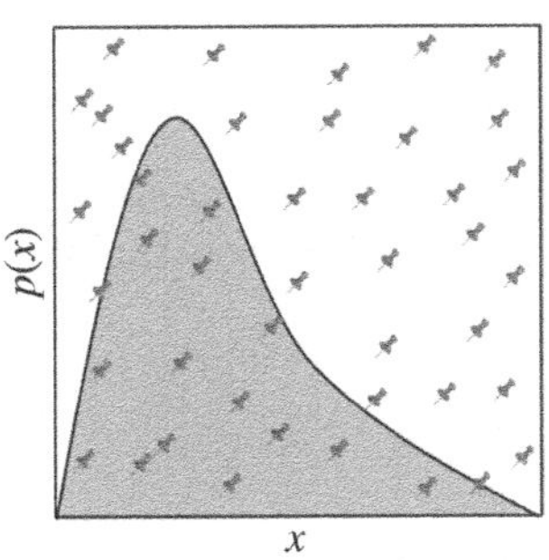

Figure P6.6 Rejection sampling keeps the random samples in the region under the graph of its probability density function, $p(x)$.

6.7 Consider a shooting game as discussed in Appendix 6A. Suppose that a shooter hits a target with an 80% probability if the previous result is a hit, and 40% probability if the previous result is a miss. Assume that the shooter made the first shot, and consecutive shooting follows a Markov chain process. What's the probably that the shooter hit the target in 5th shot? How is it compared with the equilibrium probability.

6.8 Fahidy demonstrated that the Markov chain model is a convenient tool for representing the kinetics of multicomponent linear chemical reactions.[76] In doing so, he related the compositions of chemical species with a probability vector and the rate constants with the transitional matrix. Suppose that three chemical species A, B, C react in a mixture with the rate equations

$$\frac{d[A]}{dt} = -0.3[A] + 0.1[B],$$

$$\frac{d[B]}{dt} = 0.2[A] - 0.3[B] + 0.1[C],$$

$$\frac{d[C]}{dt} = 0.1[A] + 0.2[B] - 0.1[C],$$

where $[\cdots]$ stands for molar concentration. How many independent reactions are there in this system? If the initial composition is $x_A = 0.4$, $x_B = 0.3$, and $x_C = 0.3$, what is the equilibrium composition?

6.9 Consider a one-component gas in an external electric field D as depicted schematically in Figure P6.9.
Assume that each gas molecule has polarizability α and permanent dipole $\mathbf{d}$. When the gas density is sufficiently low, the intermolecular interactions can be neglected and, due to the presence of an electric field, each molecule experiences an external energy of

$$u_{ext} = -\frac{1}{2}\alpha D^2 - dD\cos\theta,$$

where $u_{ind} = -\frac{1}{2}\alpha D^2$ represents the energy due to the molecular polarization, and $u_d = -dD\cos\theta$ is the energy due to the permanent dipole $\mathbf{d}$.

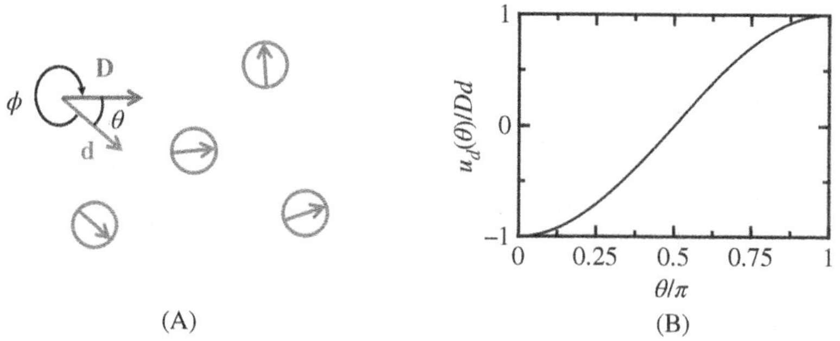

Figure P6.9 A. An ideal gas of point dipole **d** in an electric field **D**. B. The reduced potential energy due to the permanent dipole **d** as a function of the angle θ relative to the direction of the electric field **D**. Here ϕ represents the azimuthal angle around the direction of the electric field.

76 Fahidy T. Z., "Solving chemical kinetics problems by the Markov-chain approach", *Chem. Eng. Educ.*, Winter, 42–43 (1993).

(i) Show that the canonical partition function of the system can be written as

$$Q = \frac{1}{N!}\left\{\frac{q_{int}V}{4\pi\Lambda^3}\int_0^{2\pi} d\phi \int_0^{\pi} d\theta \sin\theta e^{-\beta u_{ext}}\right\}^N,$$

where N is the number of gas molecules in the system, Λ is the thermal wave length, q_{int} represents the intrinsic partition function related to the rotational, vibrational, and electronic degrees of freedom of each gas molecule, and V is the system volume.

(ii) Show that the probability density of angle θ is given by

$$p(\theta) = \frac{e^{-\beta u_d(\theta)}\sin\theta}{q_0},$$

where $u_d(\theta) = -dD\cos\theta$, $q_0 = (2/y)\sinh y$, and $y = \beta dD$.

(iii) Plot $p(\theta)$ for $y = 1,5$ and 10.

(iv) Why isn't $p(\theta)$ maximized at $\theta = 0$?

(v) Prepare a MC program for sampling the distribution of θ using the Metropolis algorithm and compare the simulation results with the analytic prediction for $y = 1, 5$, and 10.

6.10 Implement the Metropolis algorithm for MC simulation of a cyclic Ising chain with $N = 50$ spins and carry out the following analyses:

(i) Consider $\beta\epsilon = 0.5$ and $\beta h = 0$, validate your code by comparing the internal energy and heat capacity with the analytical solutions:

$$\beta U/N = -\beta\epsilon \tanh(\beta\epsilon),$$

$$C_V/Nk_B = \left[\frac{\beta\epsilon}{\cosh(\beta\epsilon)}\right]^2.$$

Why does not the simulation yield an exactly zero magnetization at zero field?

(ii) Consider $\beta\epsilon = 0.5$ and $\beta h = 0.5$, compare your simulation result with the average magnetization per spin predicted by the equation:

$$m = \frac{\sinh(\beta h)}{\sqrt{\sinh^2(\beta h) + e^{-4\beta\epsilon}}}.$$

(iii) Discuss how the simulation results vary with the number of MC steps for $\beta\epsilon = 0.5$ and $\beta h = 0$.

(iv) Can you simulate the reduced Helmholtz energy βF at $\beta\epsilon = 0.5$ and $\beta h = 0.5$?

6.11 In molecular simulation, the density of states (DOS) is often used both in analyzing the simulation data and in constructing advanced algorithms to speed up microstate sampling. In a canonical ensemble, the DOS is formally related to the partition function through

$$Q = \int dE w(E) e^{-\beta E},$$

where $w(E)dE$ represents the number of microstates with the total energy between E and $E + dE$. Derive an expression for $w(E)$ for a cyclic Ising chain with N spins at zero field and verify the analytical result using MC simulation.

6.12 Very often molecular simulation is limited by the system size and computer time. Suppose that your Ising chain simulation can run only up to 10^4 MC steps per spin.

(i) Compare your simulation results with the exact values for the internal energy, the mean magnetization, and the heat capacity at zero field and three reduced temperatures, $k_BT/\epsilon = 10,1$, and 0.5.

(ii) Construct the density of states (DOS) at these temperatures and compare with the exact results.

(iii) How does the accuracy of your simulation results change with temperature?

6.13 Implement the Metropolis algorithm for MC simulation of an open-end Ising chain with $N = 50$ spins and carry out the following studies:

(i) Consider $\beta\epsilon = 0.2$ and $\beta h = 0$, validate your results by comparing the internal energy and heat capacity with the exact results

$$\beta U/N = -(1 - 1/N)\beta\epsilon \tanh(\beta\epsilon),$$

$$C_V/Nk_B = (1 - 1/N)\left[\frac{\beta\epsilon}{\cosh(\beta\epsilon)}\right]^2.$$

(ii) Consider $\beta\epsilon = 0.5$ and $\beta h = 0.5$, compare your simulation result with the average magnetization per spin predicted by the analytical equation:

$$m = \frac{\sinh(\beta h)}{\sqrt{\sinh^2(\beta h) + e^{-4\beta\epsilon}}}.$$

(iii) Show how the simulation results vary with the number of MC steps for $\beta\epsilon = 0.2$ and $\beta h = 0$.

(iv) How would you calculate the Helmholtz energy?

6.14 Carry out MC simulation for the following thermodynamic properties versus temperature for a cyclic Ising chain with $N = 50$ spins without external field (i.e., $h = 0$). Compare the simulation results with the analytic solutions:

(i) The reduced internal energy

$$\frac{U}{N\epsilon} = -\tanh K\left(\frac{1 + \tanh^{N-2} K}{1 + \tanh^N K}\right).$$

(ii) The mean magnetization per spin

$$m = 0.$$

(iii) The reduced heat capacity

$$\frac{C_V}{Nk_B} = \frac{K^2}{\cosh^2(K)}\frac{1 + (N-1)(\tanh^{N-2} K - \tanh^N K) - \tanh^{2N-2} K}{(1 + \tanh^N K)^2}.$$

(iv) The normalized average squared magnetization

$$\chi^2 \equiv \left\langle\left(\sum_{i=1}^{N} s_i\right)^2\right\rangle \Big/ N^2 = \frac{e^{2K}}{N}\frac{1 - \tanh^N K}{1 + \tanh^N K}.$$

6.15 Carry out MC simulation for the following thermodynamic properties versus temperature for an open-end Ising chain with $N = 50$ spins without external field (i.e., $h = 0$). Compare the simulation results with the analytic solutions:

(i) The reduced internal energy

$$\frac{U}{N\epsilon} = -(1 - 1/N)\tanh K.$$

(ii) The mean magnetization per spin

$$m = 0.$$

(iii) The reduced heat capacity

$$\frac{C_V}{Nk_B} = (1 - 1/N)\frac{K^2}{\cosh^2(K)}.$$

(iv) The normalized average squared magnetization

$$\chi^2 == \left\langle \left(\sum_{i=1}^{N} s_i \right)^2 \right\rangle \Big/ N^2 = \frac{e^{2K}}{N}[1 + \sinh(2K)(\tanh^N K - 1)/N].$$

6.16 An Ising chain of finite length exhibits an order-disorder transition at temperature $k_B T_c/\epsilon \approx 2/\ln(N-1)$. To obtain the degree of magnetization below T_c, we may perform MC simulation in the presence of an external field and calculate the desired result through non-Boltzmann sampling

$$m = \frac{1}{N}\frac{\langle \sum_\nu \sum_i s_i \exp(\beta \Delta E_\nu) \rangle}{\langle \exp(\beta \Delta E_\nu) \rangle},$$

where $\Delta E_\nu \equiv E_\nu - E_{0,\nu}$, subscript "0" denotes the system at zero field, and $\langle \cdots \rangle$ stands for the ensemble average with total energy E_ν. Implement the non-Boltzmann sampling with the non-zero field, calculate the magnetization of a cyclic Ising chain with $N = 50$ spins below and above T_c, and compare the simulation results with those obtained from Problem 6.14.

6.17 Prepare a computer program for the MC simulation of the 2D-Ising model with the square lattice and zero field using the Metropolis method. Investigate the dependence of the total energy, heat capacity, and magnetization on temperature near the Curie temperature. Compare the simulation results with Onsager's theory discussed in Section 5.5 and explore how the accuracy of numerical results vary with temperature.

6.18 Investigate the size effects on the simulation results using a MC simulation program for the 2D-Ising model. Report the simulation results at two reduced temperatures, $T/T_c = 2.0$ and 0.5, for the internal energy per spin, heat capacity, magnetization, and susceptibility with different lattice sites in each dimension, L = 5,10, 20,30, 50. Discuss the accuracy of the simulation results versus the system size.

6.19 For the square-lattice Ising model, the correlation function is defined by neighboring spins either within the same row or the same column

$$C(n) = \langle s_{i,j} s_{i+n,j} \rangle - \langle s_{i,j} \rangle \langle s_{i+n,j} \rangle = \langle s_{i,j} s_{i,j+n} \rangle - \langle s_{i,j} \rangle \langle s_{i,j+n} \rangle,$$

where (i, j) are spin indexes. Perform MC simulation to calculate $c(n)$ using a 20×20 lattice with the periodic boundary conditions at a few different temperatures, and demonstrate the divergence of the correlation length at the critical point.

6.20 Due to the divergence of the correlation length, thermodynamic properties calculated from MC simulation become inaccurate near the critical point. Nevertheless, the simulation data can be utilized to extrapolate accurate results using the scaling relations derived from the renormalization group theory (Section 5.12). To demonstrate this, suppose that the heat capacity and susceptibility have been simulated with a 20 × 20 lattice Ising model shown in Figure P6.20.

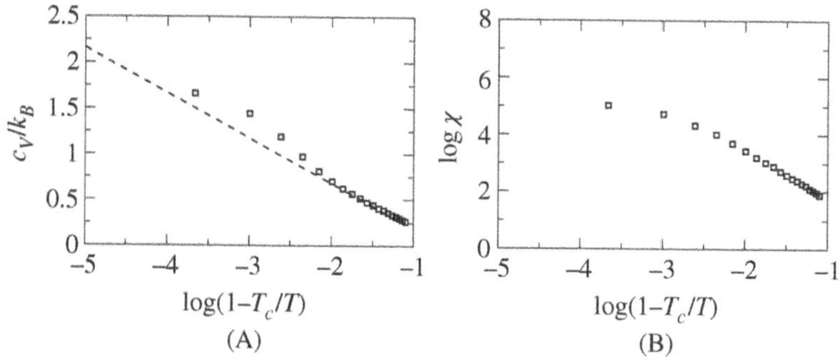

Figure P6.20 The heat capacity (A) and susceptibility (B) of the 2D-Ising model obtained from finite size MC simulation (Critical2Ddata.m).

Extrapolate the simulation data to the critical point using appropriate scaling relations (Critical2Ddata.m).

6.21 Consider a binary mixture containing components 1 and 2 at constant pressure. Show that the Clapeyron equation for equilibrium between two phases α and γ can be written as

$$\left(\frac{\partial \beta}{\partial z_2}\right)_P = \frac{x_2^\alpha - x_2^\gamma}{z_1 z_2 (h^\alpha - h^\gamma)},$$

where $x_{i=1,2}$ represents the mole fractions, $z_{i=1,2} = f_i/f_T$ is the fugacity fraction of component i, and h stands for the molar enthalpy. Here, the fugacity is defined in terms of chemical potential

$$\mu_i = \mu_i^0 + k_B T \ln(f_i/f_i^0),$$

where superscript "0" denotes a reference state. The total fugacity is $f_T = f_1 + f_2$. Hint: For a binary system, $d(\beta G) = H d\beta + \beta V dP + \beta\mu_1 dN_1 + \beta\mu_2 dN_2$.

6.22 Perform the Wang–Landau sampling for the 2-dimensional Ising model using a 20 × 20 square lattice with the periodic boundary conditions. Use the density of states (DOS) from MC simulation, predict the Helmholtz energy, internal energy and heat capacity per spin over the entire range of temperature.

6.23 The torsional potential for n-butane can be expressed as a cosine-function series[77]:

$$u(\phi) = \frac{1}{2}\sum_{i=1}^{6} A_i(1 - \cos i\phi),$$

77 Herrebout W. A., et al., "Enthalpy difference between conformers of n-butane and the potential function governing conformational interchange", *J. Phys. Chem.*, 99, 578–585 (1995).

where ϕ is the dihedral angle (or torsion angle), the coefficients from A_1 to A_6 have the units of kJ/mol and are given by

$$A = [2.8948, 0.5144, 13.7323, 0.4785, -0.0718, -0.4306].$$

The torsional potential predicts three energy minima corresponding to trans ($\phi = 180°$), gauche− ($\phi = 60°$) and gauche+ ($\phi = 300°$) conformers of n-butane.

(i) Derive an analytical expression for predicting the probability distribution of the dihedral angle as a function of temperature. Assume that the kinetic energy of the n-butane atoms has no effect on the distribution of the dihedral angle.
(ii) Develop a MC scheme for sampling the probability distribution of the dihedral angle using the Metropolis method.
(iii) Compare the analytical and simulation results at $T = 100$ and 300 K and discuss how the efficiency of the sampling varies with temperature.
(iv) Estimate the equilibrium constant for the transformation between the trans ($120° < \phi < 240°$) and gauche states ($0° < \phi < 120°$ or $240° < \phi < 360°$) of n-butane at 300 K. What is the equilibrium composition of different conformers of n-butane in the gas phase?

6.24 Develop a parallel tempering scheme for sampling the probability distribution of the dihedral angle of n-butane using three replicas at temperature T = 100, 200, and 300 K. Show that the probability density conforms to the Boltzmann distribution for each replica.

7

Simple Fluids and Colloidal Dispersions

To understand the thermodynamic properties of a macroscopic system, it is helpful to start with its microscopic structure, i.e., the relative arrangements of particles such as atoms or molecules at different conditions. From an atomic perspective, statistical mechanics describes the microscopic structure of chemical systems in terms of the local atomic or particle densities and correlation functions. These functions can be defined by the ensemble averages of particle positions or the response of the free energy to variations in the one-body particle potential and particle–particle interactions. In a nonideal thermodynamic system, the particle positions are highly correlated with each other due to particle–particle interactions.

In this chapter, we introduce molecular theories to predict the microscopic structure and thermodynamic properties of simple fluids, i.e., macroscopic systems consisting of classical particles of spherical shape. By classical, we mean that the particle motions and potential energy can be described within the realm of Newtonian mechanics. While our discussion is mostly focused on one-component systems with a pairwise additive potential, similar concepts, such as direct and total correlation functions, potential of mean force, and cavity correlation function, can be established for mixtures and more complex thermodynamics systems including polymeric fluids and aqueous solutions (to be discussed in Chapters 8 and 9, respectively). The theoretical methods discussed in this Chapter can be directly used to predict the microscopic structure and thermodynamic properties of a wide range of practical systems including not just gases or liquids of nonpolar small molecules but metallic alloys, colloidal dispersions, aqueous solutions nanoparticles, and globular proteins. For example, the one-dimensional model of classical particles serves as a useful framework for understanding steric effects in the statistical arrangement of nucleosomes in chromatins. This model, which shares similarities with the Kornberg–Stryer model in genomics, provides insights into the spatial organization and interactions of nucleosomes within chromatin structures.

7.1 Microstates in the Phase Space

For chemical systems represented by a large number of classical particles, the microstates are associated with the positions and momenta of individual particles. Due to the continuous nature of these variables, we need to expand the formal statistical–mechanical equations from summation over discrete microstates to integration over the phase space. As discussed in Chapter 3, the phase space for a system of classical particles is defined by a set of continuous coordinates that specify the positions and momenta of all particles. In evaluating the partition functions of statistical ensembles and the ensemble averages of various dynamic variables, we need to describe the microstate distribution in terms of the probability density in the phase space instead of the probabilities of individual microstates.

Fundamentals and Practice in Statistical Thermodynamics, First Edition. Jianzhong Wu and John M. Prausnitz.

Companion website: www.wiley.com/go/jianzhongwu/fundamentalsandpractice

7.1.1 Continuous Microstates

According to classical mechanics, the translational motion of a spherical particle can be described with six continuous variables, i.e., three for position $\mathbf{r} = (x, y, z)$ and three for momenta $\mathbf{p} = (p_x, p_y, p_z)$. For a system containing N spherical particles, we need $6N$ continuous variables to define the positions and momenta of all particles, $(\mathbf{r}_i, \mathbf{p}_i)$, $i = 1, 2, \cdots, N$. These variables constitute a $6N$-dimensional hyperspace known as the phase space. For short notation, we designate each point in the phase space as $\nu \equiv (\mathbf{r}^N, \mathbf{p}^N)$ with $\mathbf{p}^N \equiv \{\mathbf{p}_1, \mathbf{p}_2, \cdots, \mathbf{p}_N\}$ and $\mathbf{r}^N \equiv \{\mathbf{r}_1, \mathbf{r}_2, \cdots, \mathbf{r}_N\}$.

Schematically, Figure 7.1 shows the continuous variables describing the position and momentum of a spherical particle and its extension to coordinates in the phase space for describing a many-particle system. Because both $\mathbf{p}$ and $\mathbf{r}$ are three-dimensional variables, the phase space for a system of N spherical particles has $6N$ coordinates, i.e., $6N$ degrees of freedom. At any point in the pase space, the system has a total energy that includes two contributions: a kinetic energy due to particle motion, and a potential energy due to particle interactions. According to classical mechanics, the former is determined by the particle momenta or velocities; and the latter depends only on the particle positions. At any moment, the force on each particle is defined by the relative positions of all particles in the system, which is also known as the system configuration.

7.1.2 The Uncertainty Principle

To define the partition function of a classical system with continuous variables, we may utilize an elementary concept from quantum mechanics.[1] According to the Heisenberg uncertainty principle, the position $\mathbf{r}_i$ and momentum $\mathbf{p}_i$ of any particle i are subject to the inequality

$$d\mathbf{r}_i \cdot d\mathbf{p}_i \geq \left(\frac{h}{4\pi}\right)^3 \tag{7.1}$$

where $d\mathbf{r}_i$ and $d\mathbf{p}_i$ stand for, respectively, infinitesimal changes in position and momentum; $h = 6.6261 \cdot 10^{-34}$ J s is Planck's constant. Heisenberg's uncertainty principle asserts that we cannot specify both the position and momentum of a particle simultaneously.

The uncertainty principle implies that the number of quantum states for an infinitesimal volume[2] $d\nu$ in phase space must be a finite, not zero number. As the phase space is a $6N$-dimensional hyperspace of real variables, the differential volume must obey the inequality

$$d\nu \equiv d\mathbf{r}_1 \cdot d\mathbf{r}_2 \cdots d\mathbf{r}_N \cdot d\mathbf{p}_1 \cdot d\mathbf{p}_2 \cdots d\mathbf{p}_N \geq \left(\frac{h}{4\pi}\right)^{3N}. \tag{7.2}$$

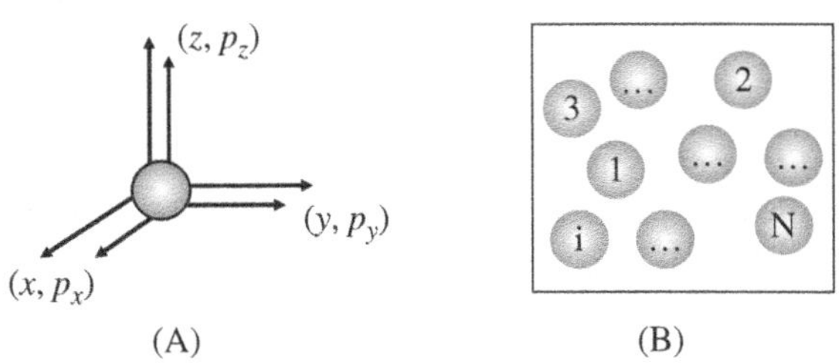

Figure 7.1 (A) According to classical mechanics, a spherical particle has 6 degrees of freedom (x, y, z, p_x, p_y, p_z), where $\mathbf{r} = (x, y, z)$ specifies the particle position, and $\mathbf{p} = (p_x, p_y, p_z)$ for the momentum; (B) for a system consisting of N spherical particles, each microstate is fully defined by the positions and momenta of all particles.

1 Historically, statistical mechanics was established many years before quantum mechanics. Here, we discuss the "extension" to ensure that the quantum and classical descriptions would result in identical thermodynamic properties.
2 Here, volume is not defined in the usual three-dimensional space but in the phase space with $6N$ dimensions.

Because each microstate occupies a minimum "volume" of $(h/4\pi)^{3N}$, the number of microstates $n(v)$ in phase-space volume dv is proportional to

$$n(v) \propto dv/(h^{3N}N!). \tag{7.3}$$

In Eq. (7.3), the proportionality constant is yet to be determined, and $N!$ is introduced to account for the indistinguishability of classical particles.[3] Eq. (7.3) provides a simple connection between an infinitesimal volume dv in the phase space and the number of possible microstates. In the following, we formulate the partition functions of classical systems utilizing this connection.

7.1.3 Classical Partition Functions

To extend the definitions of partition functions to systems consisting of classical particles, consider first the canonical ensemble for a system consisting of N spherical particles at temperature T and volume V. Let $E(v)$ be the total energy of the system with the particle positions and momenta specified by an infinitesimal volume, $dv = d\mathbf{r}^N d\mathbf{p}^N$, and $p(v)$ represent the probability density of microstates. As mentioned above, the number of microstates in dv is proportional to $1/(h^{3N}N!)$. Therefore, we can write the probability density of microstates in the phase space in terms of the degeneracy and the Boltzmann factor

$$p(v)dv \propto \frac{dv}{N!h^{3N}} e^{-\beta E(v)} \tag{7.4}$$

where $\beta = 1/(k_B T)$, and k_B is the Boltzmann constant. Eq. (7.4) is equivalent to the Boltzmann distribution for the discrete microstates of a canonical ensemble discussed in Section 2.2.3.

The proportionality constant in Eq. (7.4) can be determined from the normalization condition for the probability density

$$\int p(v)dv = 1. \tag{7.5}$$

Substituting Eq. (7.4) into (7.5) yields

$$p(v)dv = \frac{1}{Q} \cdot \frac{dv}{N!h^{3N}} e^{-\beta E(v)} \tag{7.6}$$

where Q is recognized as the canonical partition function

$$Q = \frac{1}{N!h^{3N}} \int dv e^{-\beta E(v)}. \tag{7.7}$$

The multidimensional integration in Eq. (7.7) is commonly known as the phase integral.

Eq. (7.7) often serves as a starting point to describe the thermodynamic properties of classical systems. The procedure can be similarly used to derive the partition functions (and the microstate probability densities) in microcanonical, grand canonical, and isothermal–isobaric ensembles. Table 7.1 summarizes the final results. These expressions are similar to those corresponding to discrete systems except that the summation of microstates must be replaced by integration over the continuous variables of the phase space.

3 As discussed in Chapter 4, indistinguishability means that if the positions and momenta of two molecules are exchanged, the system remains in the same microstate.

Table 7.1 Partition functions and phase-space probability densities for thermodynamic systems of classical particles in conventional ensembles.

Ensemble	Partition function	Probability density
Microcanonical	$W(N,V,E) = \int_{E,V=const.} dv \frac{1}{N!h^{3N}}$	$p(v) = \frac{1}{W} \cdot \frac{1}{N!h^{3N}}$
Canonical	$Q(N,V,T) = \int dv \frac{e^{-\beta E(v)}}{N!h^{3N}}$	$p(v) = \frac{1}{Q} \cdot \frac{1}{N!h^{3N}} e^{-\beta E(v)}$
Grand canonical	$\Xi(\mu,V,T) = \sum_N \int dv \frac{e^{-\beta[E(v)-N\mu]}}{N!h^{3N}}$	$p(v) = \frac{1}{\Xi} \cdot \frac{1}{N!h^{3N}} e^{-\beta[E(v)-N\mu]}$
Isobaric-isothermal	$Y(N,P,T) = \sum_N \int dv \frac{e^{-\beta[E(v)+PV(v)]}}{N!h^{3N}}$	$p(v) = \frac{1}{Y} \cdot \frac{1}{N!h^{3N}} e^{-\beta[E(v)+PV(v)]}$

7.1.4 The Extended Gibbs Entropy

In Section 1.4.2, we define the Gibbs entropy in terms of the summation over discrete microstates. To extend the definition to systems of classical particles, we must replace the summation over discrete microstates with integration in the phase space. The procedure is similar to that used for the continuous extensions of the partition functions.

When the microstates of a thermodynamic system are described by discrete variables, the Gibbs entropy is given by

$$S \equiv -k_B \sum_{\nu} p_{\nu} \ln p_{\nu} \tag{7.8}$$

where p_ν stands for the probability of the system in microstate ν. For a system of classical particles, the microstates are described in terms of continuous variables. Because the probability density must be normalized, Eq. (7.8) can be formally written as

$$S = -k_B \int dv p(v) \ln p_{\nu}. \tag{7.9}$$

In Eq. (7.9), p_ν stands for the probability of discrete microstate ν with the particle positions and momenta in an infinitesimal phase-space volume dv, and $p(\nu)$ represents the *probability density*, i.e., $p(\nu)dv$ is the *probability* of a classical system with the microstates associated with the differential volume dv in the phase space. From the perspective of discrete microstates, the system has a fixed energy $E(\nu)$ within infinitesimal phase-space volume dv. Accordingly, all microstates associated with dv have the same probability p_ν.

As shown in Eq. (7.3), an infinitesimal volume in phase space dv contains $n(\nu)$ microstates. Because all microstates in dv have the same total energy, the probability of each microstate ν is related to the probability density in the phase space

$$p_{\nu} = p(v)N!h^{3N} \tag{7.10}$$

where $N!h^{3N}$ comes from Eq. (7.6), accounting for the degeneracy of microstates in dv.

Substituting Eq. (7.10) into (7.9) yields

$$S = -k_B \int dv p(v) \ln[p(v)N!h^{3N}]. \tag{7.11}$$

Using the expression for $p(\nu)$ from Eq. (7.6), we get

$$S = \int dv \frac{E(\nu)}{T} p(v) + k_B \ln Q = U/T + k_B \ln Q \tag{7.12}$$

where U is the internal energy of the system given by

$$U = \int d\upsilon \ E(\upsilon) \cdot p(\upsilon). \tag{7.13}$$

Apparently, Eq. (7.12) is consistent with the thermodynamic relation

$$S = (U - F)/T \tag{7.14}$$

where $F = -k_B T \ln Q$ is the Helmholtz energy.

7.1.5 Configurational Integral

According to classical mechanics, the total energy for a system of N spherical particles can be divided into kinetic energy K and potential Φ

$$E(\mathbf{r}^N, \mathbf{p}^N) = K(\mathbf{p}^N) + \Phi(\mathbf{r}^N) \tag{7.15}$$

As previously mentioned, $\mathbf{p}^N = \{\mathbf{p}_1, \mathbf{p}_2, \cdots, \mathbf{p}_N\}$ is a multidimensional vector that specifies the particle momenta, and $\mathbf{r}^N = \{\mathbf{r}_1, \mathbf{r}_2, \cdots, \mathbf{r}_N\}$ specifies the particle positions, i.e., a configuration of the N-particle system.

For a system of N identical spherical particles, the kinetic energy is fully determined by the translational motions of all particles

$$K(\mathbf{p}^N) = \sum_{i=1}^{N} \mathbf{p}_i^2/(2m) \tag{7.16}$$

where m represents the particle mass. Meanwhile, the total potential Φ depends on the particle positions, $\mathbf{r}^N = \{\mathbf{r}_1, \mathbf{r}_2, \cdots, \mathbf{r}_N\}$. The potential energy is often represented by the approximation of pairwise additivity[4]

$$\Phi(\mathbf{r}^N) \approx \sum_{i=1}^{N}\sum_{j>i}^{N} u(|\,\mathbf{r}_i - \mathbf{r}_j\,|) = \frac{1}{2}\sum_{i \neq j} u(r_{ij}) \tag{7.17}$$

where $u(r)$ stands for the pair potential, and $r_{ij} \equiv |\,\mathbf{r}_i - \mathbf{r}_j|$ is the center-to-center distance between two spherical particles. In Eq. (7.17), the summation includes all ij pairs, and the factor ½ prevents double accounting.

Substituting Eq. (7.15) into Eq. (7.7) leads to a simplified expression for the canonical partition function of the classical system

$$Q = \frac{1}{N!h^3} \int d\mathbf{p}^N \int d\mathbf{r}^N \exp[-\beta K(\mathbf{p}^N) - \beta\Phi(\mathbf{r}^N)] = \frac{1}{N!\Lambda^{3N}} \int d\mathbf{r}^N e^{-\beta\Phi(\mathbf{r}^N)} \tag{7.18}$$

where $\Lambda = h/\sqrt{2\pi m k_B T}$ is thermal de Broglie wave length. In Eq. (7.18), the kinetic term K has been integrated analytically. A comparison of Eq. (7.18) with the partition function for a system of noninteracting particles (i.e., an ideal gas) $Q^{IG} = V^N/(N!\Lambda^{3N})$ results in

$$Q/Q^{IG} = \frac{1}{V^N} \int d\mathbf{r}^N \exp[-\beta\Phi(\mathbf{r}^N)] \tag{7.19}$$

where the integration over the particle positions is called the configurational integral.

4 In principle, we need to consider three-body interactions, four-body interactions, and so forth. However, calculation of many-body interactions is a challenging task. The pairwise approximation takes into account the key contribution to the total interaction energy. For practical applications, effective pair potentials are often used in order to make the calculated thermodynamic properties agree with experimental measurements.

Eq. (7.19) suggests that the main task in application of statistical mechanics to classical systems is to evaluate the configurational integral. Because of the high dimensionality (on the order of 10^{23} for macroscopic systems), the direct integration is not practical from a numerical perspective. Therefore, alternate methods must be used in order to predict macroscopic properties analytically. Toward that end, we introduce in Section 7.2 the concepts of correlation functions and explore their connections with thermodynamic properties.

7.1.6 Summary

For a thermodynamic system consisting of classical particles, the microstates are represented by the phase-space variables, i.e., by the positions and momenta of individual particles. Because these variables are continuous, summation over discrete microstates must be replaced with integration in the phase space. As the variables related to the particle momenta can be evaluated analytically, development of efficient methods for direct or indirect calculation of the configurational integral is one central task in applying statistical mechanics to classical systems.

7.2 Radial Distribution Function and Structure Factor

A bulk fluid of spherical particles appears structureless from a macroscopic point of view, i.e., the particle density is spatially invariant. From a microscopic perspective, however, the particles are not randomly distributed around each other, and their relative positions are strongly influenced by the particle–particle interactions. As the potential energy depends on the system configuration, the positions of neighboring particles are strongly correlated with each other. For example, two particles cannot take the same position because of finite size. Meanwhile, each particle is more likely surrounded by other particles interacting with an attractive potential than those with a repulsive energy.

In statistical mechanics, the correlated distributions of particles are described in terms of correlation functions. Figure 7.2 illustrates some intuitive examples on how particles in a macroscopic system exhibit correlations regarding their positions. For a thermodynamic system containing classical particles, the total potential energy is often expressed in terms of pairwise additive potentials. Typically, the pair potential includes a steep repulsion at short distance and an attraction beyond the collision diameter. For spherical particles, the spatial correlation functions can be described using either the radial distribution function (RDF) or the structure factor. These two concepts are complementary. The RDF is physically intuitive but difficult to measure directly. Conversely, the structure factor is abstract but can be measured with scattering experiments. In this section, we discuss these concepts and demonstrate their interrelation. The connection of the correlation functions with thermodynamic properties will be discussed in subsequent sections.

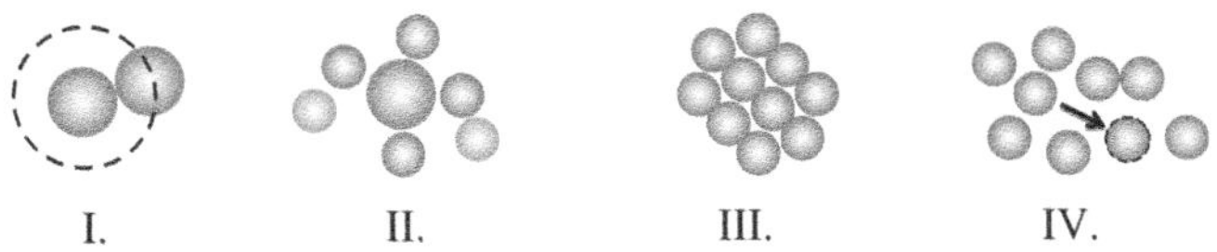

Figure 7.2 A few examples of inter-particle correlations. (I) The closest distance between two particles is dependent on their physical sizes; (II) a particle is more likely surrounded by other particles with stronger attractive energy; (III) in a crystalline phase, particle positions are coordinated according to a lattice structure; (IV) particle motion from one position to another depends on interactions with a number of other particles in the surrounding.

7.2.1 Radial Distribution Function (RDF)

The RDF, often designated as $g(r)$, specifies the nonrandom distribution of particles at radial distance r from an arbitrarily tagged particle. As shown schematically in Figure 7.3, RDF can be intuitively interpreted as the normalized local density, i.e., the local density $\rho(r)$ divided by the average density in the bulk, ρ_b. For a condensed system such as a liquid near the triple point, RDF depends strongly on bulk density ρ_b but weakly on temperature T. It is worth noting that the notation $g(r)$ is slightly misleading because it depends not only on r; a more precise but cumbersome notation would be $g(r, \rho_b, T)$.

To express RDF in terms of an ensemble average, consider a uniform system consisting of N spherical particles within volume V at temperature T. Imagine particle (1) is tagged and positioned at the coordinate center, i.e., $\mathbf{r}_1 = (0, 0, 0)$. As mentioned above, $g(r)$ is related to the local density of the $N - 1$ other particles in the system around the tagged particle. Note that the probability of finding another particle (say 2) at position $\mathbf{r}$ is described by the Dirac delta function $\delta(\mathbf{r}_2 - \mathbf{r})$. Because there are $N - 1$ free particles in the system, the average particle density at position $\mathbf{r}$ is given by

$$\rho(\mathbf{r}) = \sum_{i=2}^{N} < \delta(\mathbf{r}_i - \mathbf{r}) > = \frac{(N-1)\int d\mathbf{r}_2 d\mathbf{r}_3 \cdots d\mathbf{r}_N \delta(\mathbf{r}_2 - \mathbf{r}) \exp\{-\beta\Phi(\mathbf{r}^N)\}}{\int d\mathbf{r}_2 d\mathbf{r}_3 \cdots d\mathbf{r}_N \exp[-\beta\Phi(\mathbf{r}^N)]}$$
$$= \frac{(N-1)\int d\mathbf{r}_3 \cdots d\mathbf{r}_N \exp\{-\beta\Phi(\mathbf{r}^N)\}}{\int d\mathbf{r}_2 d\mathbf{r}_3 \cdots d\mathbf{r}_N \exp[-\beta\Phi(\mathbf{r}^N)]}. \tag{7.20}$$

In above equation, $\Phi(\mathbf{r}^N)$ stands for the total potential energy, and $N - 1$ emerges because the particle at position $\mathbf{r}$ can be any one of the $N - 1$ free particles in the system. Because the local density satisfies spherical symmetry, i.e., $\rho(\mathbf{r}) = \rho(r)$ where $r = |\mathbf{r}|$ represents the radial distance, the normalized density, $g(r) \equiv \rho(r)/\rho_b$, is thus called the radial distribution function.

With the assumption of pairwise additivity, the total potential energy is given by

$$\Phi(\mathbf{r}^N) = \sum_{i=1}^{N}\sum_{j>i}^{N} u(r_{ij}) \tag{7.21}$$

where $u(r)$ denotes the pair potential, and $r_{ij} \equiv |\mathbf{r}_i - \mathbf{r}_j|$ is the center-to-center distance. With one particle fixed at the origin and another at position $\mathbf{r}$, we can rewrite Eq. (7.21) as

$$\Phi(\mathbf{r}^N) = u(r) + \Phi'(\mathbf{r}^{N-2}) \tag{7.22}$$

where $\Phi'(\mathbf{r}^{N-2}) = \sum_{i=3}^{N}[u(r_i) + u(|\mathbf{r} - \mathbf{r}_i|)] + \sum_{i=3}^{N-1}\sum_{j=i+1}^{N} u(r_{ij})$ accounts for the interaction of the two particles at the origin and position $\mathbf{r}$ with $N - 2$ other particles in the system, as well as the interactions of the N–2 other particles.

Substituting Eq. (7.22) into Eq. (7.20) gives

$$\rho(r) = (N-1)e^{-\beta u(r)} \frac{\int d\mathbf{r}_3 d\mathbf{r}_4 \cdots d\mathbf{r}_N \exp\{-\beta\Phi'(\mathbf{r}^{N-2})\}}{\int d\mathbf{r}_2 d\mathbf{r}_3 \cdots d\mathbf{r}_N \exp[-\beta\Phi(\mathbf{r}^N)]}. \tag{7.23}$$

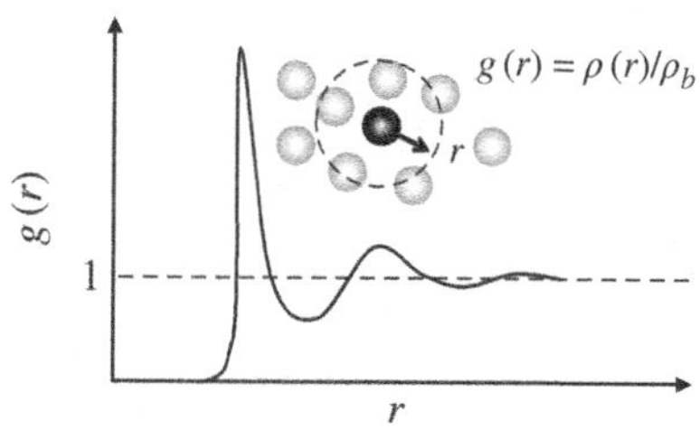

Figure 7.3 Radial distribution function $g(r)$ represents the normalized local density near an arbitrarily tagged particle in a uniform system. It vanishes when the distance is smaller than the particle size and approaches unity at large separation.

For a macroscopic system, N is extremely large such that $N - 1$ may be replaced by N. Thus, dividing both sides of Eq. (7.23) by the average number density in the bulk, $\rho_b = N/V \approx (N-1)/V$ leads to

$$g(r) = \frac{\rho(r)}{\rho_b} = e^{-\beta u(r)} \frac{V \int d\mathbf{r}_3 d\mathbf{r}_4 \cdots d\mathbf{r}_N \exp\{-\beta\Phi'(\mathbf{r}^{N-2})\}}{\int d\mathbf{r}_2 d\mathbf{r}_3 \cdots d\mathbf{r}_N \exp[-\beta\Phi(\mathbf{r}^N)]}. \tag{7.24}$$

Eq. (7.24) suggests that RDF includes a contribution from the direct particle–particle interaction (viz. the pair potential) and an indirect contribution due to the presence of other particles. In a system with low particle density, the average separation between any pair of particles far exceeds the range of interaction. In that case, the system behaviors virtually like an ideal gas. It follows from Eq. (7.24) that

$$g(r) \approx e^{-\beta u(r)} \tag{7.25}$$

where volume V cancels with integration $\int d\mathbf{r}_2 = V$. Eq. (7.25) indicates that, at low density, the radial distribution function depends only on the direct particle–particle interaction. In other words, the particle distribution relative to a tagged particle depends only on the reduced pair potential $\beta u(r)$.

At high density, the particles are close to each other. Because the steep repulsion prohibits particles to overlap, we expect that the local structure is mainly determined by the excluded volume effects, much like the close packing of hard spheres. In that case, the attraction between particles would play a minor role in determining the microscopic structure. For molecular systems, the similarity between the structure of a simple liquid near the triple point and that of a hard-sphere (HS) system was first recognized by J. D. van der Waals (vdW). However, the quantitative verification was reported much later, not until that the radial distribution functions of a realistic liquid could be accurately determined from molecular simulation.[5]

Figure 7.4 shows the radial distribution function of a Lennard–Jones (LJ) fluid at two representative densities, one at low particle density and another at high density. Here the exact results,

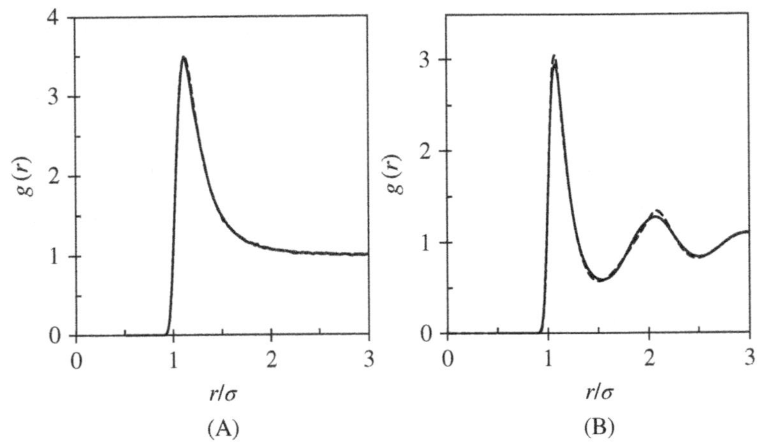

Figure 7.4 Radial distribution function, $g(r)$, for a Lennard–Jones fluid in a dilute gas (A) and in a liquid (B). The reduced density in the gas phase is $\rho\sigma^3 = 0.001$ and that in the liquid is $\rho\sigma^3 = 0.85$. In both panels, the reduced temperature is $k_BT/\varepsilon = 0.8$. The solid lines are from Monte Carlo simulation; the dashed line in A is calculated from Eq. (7.25) and that in B is from an effective hard-sphere model.

5 Verlet L., "Computer experiments on classical fluids", *Phys. Rev.* 165 (1), 201–214 (1968).

calculated from Monte Carlo (MC) simulations, are compared with predictions of Eq. (7.25) for the low-density case and, for the high-density case, with the radial distribution function of a HS system. As mentioned above, the radial distribution of particles at low density follows the Boltzmann equation, viz., Eq. (7.25). At high density, the radial distribution function can be quantitatively represented by that of a HS fluid as defined by the WCA theory (to be discussed in Section 7.10).

7.2.2 Potential of Mean Force

The potential of mean force, $W(r) \equiv -k_B T \ln g(r)$, represents the average potential between a pair of particles surrounded by all other particles in the system.[6] This potential may be understood as the reversible work to bring two particles from infinitely apart to distance r when the two particles are subject to a mean force due to particle–particle interactions. In the limit of low density, the potential of mean force is identical to the two-body potential.

The connection between $-k_B T \ln g(r)$ and a mean potential becomes evident by considering its gradient. From Eq. (7.24), we get

$$\begin{aligned} -\nabla \ln g(r) &= -\nabla \ln e^{-\beta u(r)} - \nabla \ln \int d\mathbf{r}_3 \cdots d\mathbf{r}_N \exp[-\beta \Phi'(\mathbf{r}^{N-2})] \\ &= \nabla \beta u(r) + \frac{\int d\mathbf{r}_3 \cdots d\mathbf{r}_N \nabla \beta u(|\mathbf{r}-\mathbf{r}_i|) \exp[-\beta \Phi'(\mathbf{r}^{N-2})]}{\int d\mathbf{r}_3 \cdots d\mathbf{r}_N \exp[-\beta \Phi'(\mathbf{r}^{N-2})]} \\ &= \nabla \beta u(r) + \left\langle \sum_{i=3}^{N} \nabla \beta u(|\mathbf{r}-\mathbf{r}_i|) \right\rangle_{N-2} \end{aligned} \tag{7.26}$$

where $<\cdots>_{N-2}$ stands for the ensemble average over all other $N-2$ particles surrounding the particle at the origin and another particle at position $\mathbf{r}$. Eq. (7.26) indicates that the gradient of $W(r)$ consists of a direct force between two particles $\nabla u(r)$, and an averaged force due to the interaction of the particle at position $\mathbf{r}$ with all other particles in the system. Because the gradient of a potential gives the (negative) force, $-\nabla W(r)$ stands for a "mean force," and $W(r)$ is thus referred to as the potential corresponding to this mean force.

As mentioned above, the radial distribution function includes two contributions, one due to the pair potential as described by the Boltzmann equation, Eq. (7.25), and the other due to the interaction of a pair of particles at distance r with all other particles in the system. The latter contribution is called the cavity correlation function

$$y(r) = g(r) e^{\beta u(r)}. \tag{7.27}$$

Substituting Eq. (7.27) into (7.26) gives

$$-\nabla \ln y(r) = \left\langle \sum_{i=3}^{N} \nabla \beta u(|\mathbf{r}-\mathbf{r}_i|) \right\rangle_{N-2}. \tag{7.28}$$

Eq. (7.28) suggests that the cavity correlation function is associated with the interaction between two "cavities," referring to two particles at distance r that do not directly interact with each other but with all other particles in the system.

6 The connection between the potential of mean force and the radial distribution function is also known as the reversible work theorem.

7.2.3 Scattering Experiments

The microscopic structure of a thermodynamic system can be detected through diffraction analysis, e.g., the scattering of X-ray or neutron beams by a large number of particles. To understand the essential ideas, Figure 7.5 shows schematically a scattering beam of wavevector $\mathbf{k}_0$ interacting with a one-component system of spherical particles. In the presence of a particle at position $\mathbf{r}_\alpha$, the Born approximation, a scattering theory proposed by Max Born in the early days of quantum theory development, predicts that the scattered wave with a wavevector $\mathbf{k}_1$ has an amplitude proportional to $e^{i(\mathbf{k}_1-\mathbf{k}_0)\cdot\mathbf{r}_\alpha}$, where the proportionality constant is dependent on the microscopic details of the beam-particle interactions. A superposition of the scattered waves by all particles gives the total wave amplitude at the detector

$$\text{total wave amplitude at the detector} \sim \sum_\alpha e^{i\mathbf{k}\cdot\mathbf{r}_\alpha} \tag{7.29}$$

where $\mathbf{k} \equiv \mathbf{k}_1 - \mathbf{k}_0$ is the scattering wavevector.

The intensity of the scattered wave observed at the detector is proportional to the square of the amplitude of the total scattered wave

$$\text{intensity} \sim \left\langle \left| \sum_\alpha e^{i\mathbf{k}\cdot\mathbf{r}_\alpha} \right|^2 \right\rangle \tag{7.30}$$

where an ensemble average is added to account for all possible particle positions. Note that the summation in Eq. (7.30) can be written in terms of the instantaneous particle density $\widetilde{\rho}(\mathbf{r}) = \sum_\alpha \delta(\mathbf{r}-\mathbf{r}_\alpha)$, where $\delta(\mathbf{r})$ stands for the Dirac function,

$$\sum_\alpha e^{i\mathbf{k}\cdot\mathbf{r}_\alpha} = \int d\mathbf{r} \sum_\alpha \delta(\mathbf{r}-\mathbf{r}_\alpha) e^{i\mathbf{k}\cdot\mathbf{r}} = \int d\mathbf{r}\widetilde{\rho}(\mathbf{r}) e^{i\mathbf{k}\cdot\mathbf{r}}. \tag{7.31}$$

Because the scattering from a uniform background makes no contribution to the system structure, we may subtract the average density ρ_b from the instantaneous particle density $\widetilde{\rho}(\mathbf{r})$ in Eq. (7.31). In terms of the density fluctuations, the scattering intensity thus becomes

$$I(\mathbf{k}) \sim \int d\mathbf{r} \int d\mathbf{r}' \langle \Delta\widetilde{\rho}(\mathbf{r}) \Delta\widetilde{\rho}(\mathbf{r}') \rangle e^{i\mathbf{k}\cdot(\mathbf{r}-\mathbf{r}')} \tag{7.32}$$

where $\Delta\widetilde{\rho}(\mathbf{r}) = \widetilde{\rho}(\mathbf{r}) - \rho_b$, and $\chi(\mathbf{r}, \mathbf{r}') \equiv \langle \Delta\widetilde{\rho}(\mathbf{r}) \Delta\widetilde{\rho}(\mathbf{r}') \rangle$ is called the density–density correlation function. As discussed below, for a uniform system of spherical particles, $\chi(\mathbf{r}, \mathbf{r}')$ depends only on the distance $|\mathbf{r}-\mathbf{r}'|$. In that case, Eq. (7.32) indicates that the scattering intensity is proportional to the three-dimensional Fourier transform of $\chi(r)$ [7]

$$I(k) \sim \hat{\chi}(k). \tag{7.33}$$

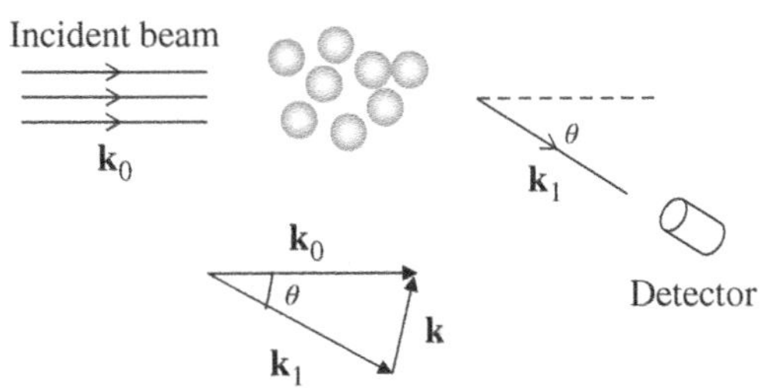

Figure 7.5 Schematic of neutron or X-ray scatting (diffraction) experiment. Here $\mathbf{k}_0$ represents the wavevector of an incident beam, and $\mathbf{k}_1$ is the wavevector of the scattered beam. In elastic scattering, the magnitude of the wavevector is unchanged due to particle scattering, i.e., $|\mathbf{k}| = |\mathbf{k}_1| = |\mathbf{k}_0|$.

7 The three-dimensional (3D) Fourier transform of a function $f(\mathbf{r})$ is defined as $\hat{f}(\mathbf{k}) = \int d\mathbf{r} f(\mathbf{r}) \exp[i\mathbf{k}\cdot\mathbf{r}]$ where $i = \sqrt{-1}$. The inverse Fourier transform is defined as $f(\mathbf{r}) = (2\pi)^{-3} \int d\mathbf{k}\hat{f}(\mathbf{k}) \exp[-i\mathbf{k}\cdot\mathbf{r}]$. For a scalar function like RDF, the 3D Fourier transform becomes $\hat{f}(k) = (1/k) \int_0^\infty dr 4\pi r \sin(kr) f(r)$.

The density–density correlation function and the radial distribution function are directly related to each other. Because the ensemble average of the instantaneous particle density is the same as the average density $<\widetilde{\rho}(\mathbf{r})>=\rho_b$, we may express the density-density correlation function as

$$\chi(\mathbf{r},\mathbf{r}')=<\widetilde{\rho}(\mathbf{r})\widetilde{\rho}(\mathbf{r}')>-\rho_b^2=<\sum_{\alpha}\delta(\mathbf{r}-\mathbf{r}_\alpha)\sum_{\alpha'}\delta\left(\mathbf{r}'-\mathbf{r}_{\alpha'}\right)>-\rho_b^2. \tag{7.34}$$

The first term on the right side of Eq. (7.32) includes two contributions corresponding to $\alpha=\alpha'$ and $\alpha\neq\alpha'$. In the former case, the ensemble average is given by

$$<\sum_{\alpha}\delta(\mathbf{r}-\mathbf{r}_\alpha)\delta(\mathbf{r}'-\mathbf{r}_\alpha)>=<\sum_{\alpha}\delta(\mathbf{r}-\mathbf{r}_\alpha)\delta(\mathbf{r}'-\mathbf{r})>=\rho_b\delta(\mathbf{r}'-\mathbf{r}). \tag{7.35}$$

In the case of $\alpha\neq\alpha'$, the ensemble average defines the two-body density function

$$\rho(\mathbf{r},\mathbf{r}')=<\sum_{\alpha}\delta(\mathbf{r}-\mathbf{r}_\alpha)\sum_{\alpha'\neq\alpha}\delta\left(\mathbf{r}'-\mathbf{r}_{\alpha'}\right)> \tag{7.36}$$

Intuitively, $\rho(\mathbf{r},\mathbf{r}')$ describes the probability of finding a particle at $\mathbf{r}$ and another particle at $\mathbf{r}'$. The local densities are uncorrelated when $\mathbf{r}$ and $\mathbf{r}'$ are far apart, thus $\rho(\mathbf{r},\mathbf{r}')=\rho_b^2$. When positions $\mathbf{r}$ and $\mathbf{r}'$ are close to each other, the two-body density function is related to the radial distribution function

$$\rho(\mathbf{r},\mathbf{r}')=\rho_b^2 g(|\,\mathbf{r}-\mathbf{r}'\,|) \tag{7.37}$$

The equivalence of Eqs. (7.37) and (7.24) is evident because the two-body density may be understood in terms of ρ_b, the average particle density at $\mathbf{r}$, and $\rho_b g(r)$, the correlated particle density at $\mathbf{r}'$. Inserting Eqs. (7.35) and (7.37) into (7.34) yields

$$\chi(\mathbf{r},\mathbf{r}')=\rho_b^2 g(|\,\mathbf{r}-\mathbf{r}'\,|)+\rho_b\delta(\mathbf{r}'-\mathbf{r})-\rho_b^2. \tag{7.38}$$

From Eq. (7.38) and (7.32), we obtain a direct connection between the radial distribution function and scattering intensity

$$\begin{aligned} I(k) &\sim \int d\mathbf{r}\int d\mathbf{r}'\left[\rho_b^2 g(|\,\mathbf{r}-\mathbf{r}'\,|)+\rho_b\delta(\mathbf{r}-\mathbf{r}')-\rho_b^2\right]e^{i\mathbf{k}\cdot(\mathbf{r}-\mathbf{r}')} \\ &= \rho_b V\left\{1+\rho_b\int d\mathbf{r}\,[g(r)-1]e^{i\mathbf{k}\cdot\mathbf{r}}\right\} \end{aligned} \tag{7.39}$$

where V is the system volume. Eq. (7.39) provides the theoretical basis for analyzing scattering measurements. In principle, the radial distribution function can be derived from the inverse Fourier transform of the scattering intensity at different values of the scattering wavevector.

7.2.4 Static Structure Factor

The static structure factor, or simply the structure factor, is a concept mostly associated with the analysis of scattering experiments. Formally, the structure factor is defined as

$$\widehat{S}(k)\equiv 1+\rho_b\int d\mathbf{r}[g(r)-1]\,e^{i\mathbf{k}\cdot\mathbf{r}}=1+\rho_b\widehat{h}(k) \tag{7.40}$$

where $\widehat{h}(k)$ stands for the Fourier transform of the total correlation function, $h(r)\equiv g(r)-1$. Substituting Eq. (7.40) into (7.39) indicates that the intensity of the scattered wave is proportional to the structure factor and the number of scattering points (viz., particles) in the system[8]

$$I(k) \sim N\hat{S}(k) \tag{7.41}$$

with $N = \rho_b V$. A comparison of Eq. (7.41) and (7.33) shows

$$\hat{S}(k) = \hat{\chi}(k)/\rho_b. \tag{7.42}$$

At low density, $\rho_b \approx 0$ so that Eq. (7.40) predicts $\hat{S}(k) \approx 1$, suggesting no correlation between particles. For a concentrated system, $\hat{S}(k)$ often displays a strong peak due to nearest-neighbor interactions. From the first peak position k_m, we can estimate the average distance between the neighboring particles, $2\pi/k_m$.

Figure 7.6 shows the total correlation functions of several noble gases in both the Fourier space and in the real space.[9] We see that, within experimental uncertainties, the structure factor, which is related to $\hat{h}(k)$ as defined in Eq. (7.40), can be faithfully reproduced by the Lennard-Jones (LJ) model. As expected, the total correlation function shows a strong peak at $r = \sigma$, where σ is the particle diameter. This peak arises from the excluded-volume interactions. Correspondingly, both $\hat{h}(k)$ and $\hat{S}(k)$ show the first peak at $k_m\sigma \approx 2\pi$.

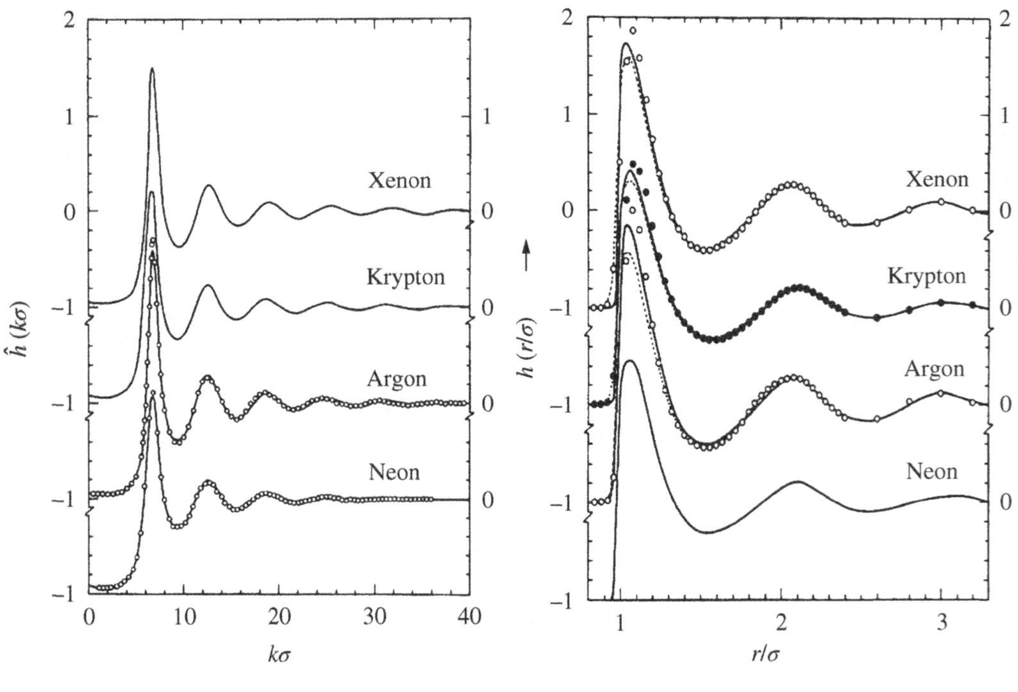

Figure 7.6 Total correlation function in the Fourier space (left panel) and in the real space (right panel) for various noble gases in the liquid state from neutron scattering experiment (points) and from theoretical prediction. The LJ parameters and thermodynamic conditions corresponding to each noble gas can be found in the original publication. Source: Adapted from Narten et al.[9]

8 Here the scattering points are referred to individual atoms. For colloidal or macromolecular systems, each particle consists of a large number of atoms and the collective behavior is represented by the form factor. While the structure factor accounts for inter-particle correlations, the form factor reflects atomic distribution within each particle. If the particle is rigid, the form factor can be calculated from the atomic density and particle shape (viz., form).

9 Narten A. H., Blum L., Fowler R. H., "Mean spherical model for structure of Lennard–Jones fluids", *J. Chem. Phys.* 60 (9), 3378–3381 (1974).

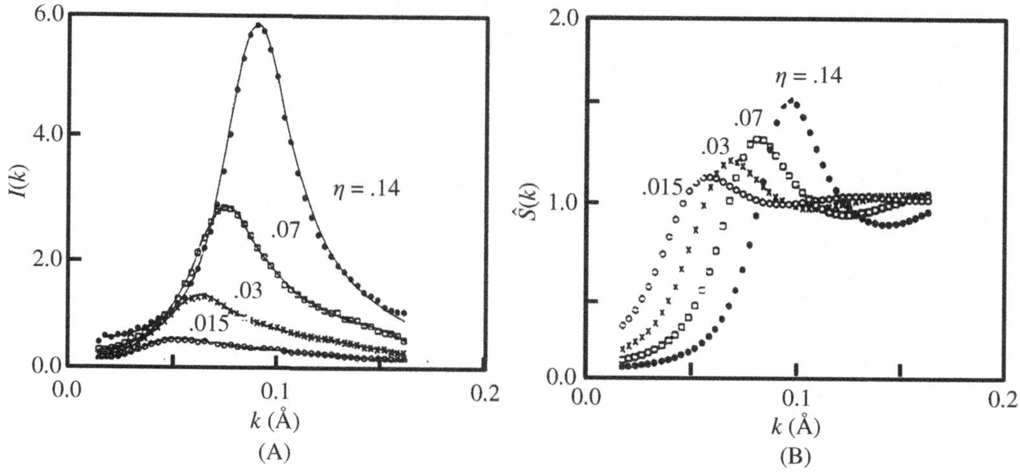

Figure 7.7 (A) The neutron-scattering intensity for aqueous micellar solutions of ionic detergent lithium dodecyl sulfate (LDS) at different concentrations. Here symbols are from experimental data, and the solid lines are fitted with the micelles represented by spherical particles. η stands for the volume fraction of micelles, the temperature is 37 °C. (B) The structure factor of micelles according to neutron scattering. Source: Reproduced from Bendedouch et al.[10]

The concepts discussed above are applicable not just to simple fluids such as noble gases, they can also be used to describe the microscopic structures of colloidal dispersions and aqueous solutions of globular proteins. For example, Figure 7.7 shows the scattering intensity and the structural factor for an aqueous solution of spherical micelles measured from neutron-scattering experiments.[10] The structure factor can be perfectly fitted with that for a system of spherical particles, suggesting that the microscopic structure and thermodynamic properties of the spherical micelles are similar to those corresponding to a simple fluid. From the peak positions in the scattering intensity or the structure factor, we see that the micelle size becomes smaller as the surfactant volume fraction increases.

7.2.5 Summary

The radial distribution function (RDF), the density fluctuation function, and the structure factor are three complementary concepts in statistical mechanics used to quantify the microscopic structure of thermodynamic systems consisting of spherical particles. These functions are linearly related to each other in the real space or through the 3D Fourier transformation. In the next section, we will discuss the applications of these correlation functions for determining thermodynamics properties.

7.3 Structure–Property Relations

In comparison with classical thermodynamics, one major advantage of statistical thermodynamics is that it deals with both macroscopic properties and microscopic structure. For fluid systems, the latter is typically described by various correlation functions. In this section, we introduce some

---

10 Bendedouch D., Chen S. H. and Koehler W. C., "Determination of inter-particle structure factors in ionic micellar solutions by small-angle neutron-scattering", *J. Phys. Chem.* 87 (14), 2621–2628 (1983).

basic relations between correlation functions and internal energy, pressure, and chemical potential. These structure–property relations allow us to calculate other thermodynamic properties using the standard thermodynamic relations.

7.3.1 Energy Equation

For a closed system containing N spherical particles at temperature T and volume V, the internal energy can be formally calculated from the partial derivatives of the canonical partition function Q with respect to the inverse temperature

$$U = -\left(\frac{\partial \ln Q}{\partial \beta}\right)_{N,V}. \tag{7.43}$$

As discussed in Section 7.1, we may write Q in terms of the configuration integral

$$Q = \frac{1}{N!\Lambda^{3N}} \int d\mathbf{r}^N e^{-\beta\Phi(\mathbf{r}^N)} \tag{7.44}$$

where $\Lambda = \sqrt{\beta h^2/2\pi m}$ represents the thermal wavelength. Substituting Eq. (7.44) into (7.43) leads to

$$U = \frac{3N}{2\beta} + < \Phi(\mathbf{r}^N) > \tag{7.45}$$

where the kinetic energy is obtained from $(\partial \ln \Lambda^{3N}/\partial\beta)$, and $\Phi(\mathbf{r}^N)$ represents the total potential energy due to particle–particle interactions.

In Eq. (7.45), the second term on the right corresponds to the total potential energy averaged over the configurational space $\mathbf{r}^N = (\mathbf{r}_1, \mathbf{r}_2, \ldots, \mathbf{r}_N)$, i.e.,

$$< \Phi(\mathbf{r}^N) > = \frac{\int d\mathbf{r}^N \Phi(\mathbf{r}^N) e^{-\beta\Phi(\mathbf{r}^N)}}{\int d\mathbf{r}^N e^{-\beta\Phi(\mathbf{r}^N)}} = \int d\mathbf{r}^N \Phi(\mathbf{r}^N)\rho(\mathbf{r}^N) \tag{7.46}$$

where $\rho(\mathbf{r}^N) = \langle\delta(\mathbf{r}^N)\rangle$ stands for the N-body density function. In practical applications, the potential energy is often approximated as pairwise additive

$$\Phi(\mathbf{r}^N) = \sum_{i=1}^{N}\sum_{j>i}^{N} u(|\, \mathbf{r}_i - \mathbf{r}_j \,|) \tag{7.47}$$

where $u(r)$ represents the pair potential. Accordingly, we can evaluate the average potential energy based on the radial distribution function $g(r)$

$$\begin{aligned} < \Phi(\mathbf{r}^N) > &= \sum_{i=1}^{N}\sum_{j>i}^{N} \int d\mathbf{r}d\mathbf{r}' u(|\, \mathbf{r} - \mathbf{r}' \,|) < \delta(\mathbf{r} - \mathbf{r}_i)\delta(\mathbf{r} - \mathbf{r}_j) > \\ &= \rho_b^2 \int d\mathbf{r}d\mathbf{r}' u(|\, \mathbf{r} - \mathbf{r}' \,|) g(|\, \mathbf{r} - \mathbf{r}' \,|) = 2\pi\rho_b^2 V \int dr r^2 u(r) g(r). \end{aligned} \tag{7.48}$$

In writing Eq. (7.48), we have used the relation between the density–density correlation function $\rho(\mathbf{r}, \mathbf{r}')$ and the radial distribution function (Section 7.2)

$$\rho(\mathbf{r}, \mathbf{r}') =< \sum_i \delta(\mathbf{r} - \mathbf{r}_i) \sum_{j\neq i} \delta(\mathbf{r}' - \mathbf{r}_j) > = \rho_b^2 g(|\, \mathbf{r} - \mathbf{r}' \,|). \tag{7.49}$$

Combination of Eqs. (7.48) and (7.45) yields

$$U = \frac{3N}{2\beta} + 2\pi\rho_b \int dr\, r^2 u(r) g(r). \tag{7.50}$$

Eq. (7.50) is known as the energy equation. It suggests that, for a uniform system of spherical particles, the average kinetic energy per particles is proportional to temperature with a universal proportionality constant (1.5 k_B). This linear relation provides a microscopic interpretation of temperature from a classical perspective, i.e., temperature may serve as a measure of the kinetic energy due to the translational motion of classical particles. As the kinetic energy is nonnegative, the absolute temperature is always positive and the zero value is achieved only when the particles are at rest.

The relation between the internal energy and radial distribution function is intuitive upon considering, as shown in Figure 7.8, an arbitrarily tagged particle interacting with all other particles in the system. The number of particles with the center of mass located in a spherical shell of radius r and thickness dr is $4\pi r^2 dr\, \rho_b g(r)$, and the potential energy due to the interaction of the tagged particle with the free particles from this shell is $4\pi r^2 dr\, \rho_b g(r)u(r)$. Integration with respect to radius r gives the total potential energy of the tagged particle. To avoid the double counting of pair potentials, we need a factor ½.

Eq. (7.50) allows us to calculate the internal energy of a uniform system without the evaluation of the configurational integral. The numerical procedure is rather straightforward provided that we have analytical expressions for both the radial distribution function $g(r)$ and the pair potential $u(r)$. Regrettably, neither $g(r)$ nor $u(r)$ is exactly known for most systems of practical interest. With a molecular model for $u(r)$, we can in principle calculate $g(r)$ using molecular dynamics or Monte Carlo simulations. Alternatively, $g(r)$ can be predicted from the integral-equation theories that will be discussed in Section 7.4. In both molecular simulation and theoretical predictions, we need a reliable expression for the pair potential $u(r)$. While the pair potential can be obtained from first-principles calculations, it may not yield accurate thermodynamic properties at high density because Eq. (7.50) ignores multi-body interactions. Conversely, the pair potential that reproduces experimental data may not correspond to an accurate pair potential for two isolated molecules. To best match experimental data, the pair potential $u(r)$ must vary with the average density of the system.

To illustrate the effects of multi-body interactions, Figure 7.9 compares the heat of vaporization for liquid argon predicted from first-principles calculations of pair and three-body potentials with those from effective pair potentials and experimental data.[11] An accurate description of the two-body or even three-body interactions is often insufficient to reproduce the total potential energy of the liquid phase. In particular, the pair potential from first-principles calculations fails to reproduce the coexistence curve of the vapor–liquid equilibrium (VLE); it overestimates the critical temperature while underestimating the vapor pressure. These deficiencies can be partially rectified by the addition of a density-dependent three-body term. Conversely, the optimized Mie

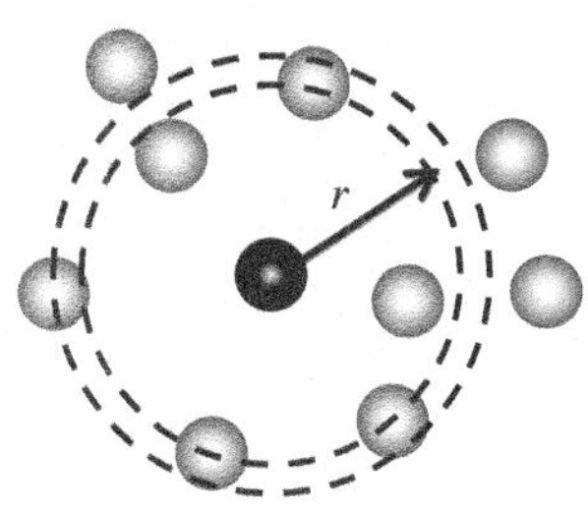

Figure 7.8 Distribution of particles around an arbitrarily tagged particle (shaded) in a uniform system of many identical particles.

11 Deiters U. K. and Sadus R. J., "Fully a priori prediction of the vapor–liquid equilibria of Ar, Kr, and Xe from ab initio two-body plus three-body interatomic potentials", *J. Chem. Phys.* 151, 034509 (2019).

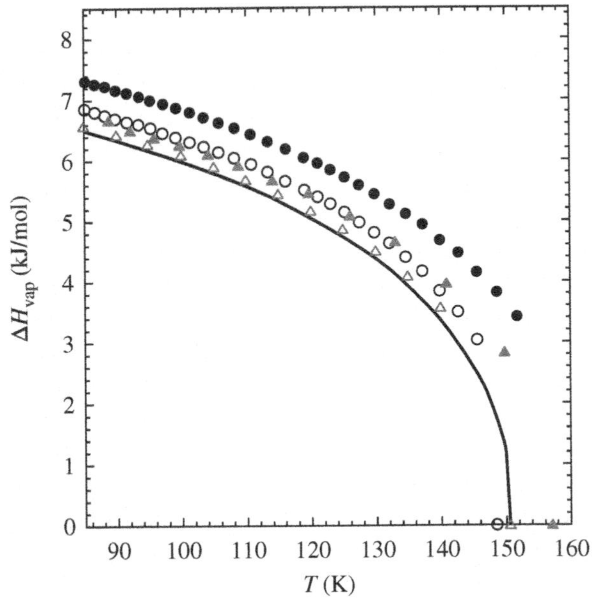

Figure 7.9 Comparison of the heat of evaporation for argon from experimental data (solid line) with that from Monte Carlo simulation based on a two-body potential (solid circles), a three-body potential (empty circles), the Lennard–Jones potential (solid triangles), and the optimized Mie potential (empty triangles). Source: Reproduced from Deiters and Sadus.[11]

potential[12] leads to an accurate prediction of the heat of evaporation while the effective two-body potential does not represent interaction between two argon molecules in a vacuum. Because the parameters are obtained by fitting with the VLE data, its performance is likely less satisfactory for other thermodynamic properties.

7.3.2 Compressibility Equation

For a uniform system of spherical particles, we can predict the pressure either from the compressibility equation

$$(\partial \rho_b / \partial \beta P)_T = 1 + \rho_b \int d\mathbf{r} h(r) = \hat{S}(0) \tag{7.51}$$

or from the virial equation[13]

$$P = k_B T \rho_b - \frac{\rho_b^2}{6} \int d\mathbf{r} \frac{\partial u(r)}{\partial r} r g(r). \tag{7.52}$$

While a pairwise additive potential is implicitly assumed in deriving the virial equation, the compressibility equation does not require any explicit knowledge about the potential energy.

The compressibility equation links the pressure of a thermodynamic system to density fluctuations. To find the connection between the bulk pressure and the fluctuation of particle density, consider the grand canonical ensemble at temperature T and chemical potential μ. The grand partition function is

$$\Xi = \sum_{N=0}^{\infty} \frac{1}{N! \Lambda^{3N}} \int d\mathbf{r}^N \exp[-\beta \Phi(\mathbf{r}^N) + N \beta \mu]. \tag{7.53}$$

12 The optimized Mie (OMie) potential adjusts the powers in the repulsive and attractive branches of the Lennard–Jones potential.
13 The concept of virial was introduced by Rudolf Clausius; in Latin, virial means force. It should be noted that the virial equation is not the same as the virial expansion.

For an open system, the particle number N is a dynamic variable. As discussed in Chapter 2, its average is related to the partial derivative of the grand partition function with respect to the reduced chemical potential

$$\langle N\rangle =< \int d\mathbf{r}\sum_{i=1}^{N}\delta(\mathbf{r}-\mathbf{r}_i) > = \partial \ln \Xi/\partial\beta\mu \tag{7.54}$$

where $<\cdots>$ denotes the ensemble average, and the second-order partial derivative yields the mean-square deviation in N

$$\langle N^2\rangle - \langle N\rangle^2 = \partial\langle N\rangle/\partial\beta\mu. \tag{7.55}$$

Eq. (7.54) indicates that the ensemble average of N^2 is related to the density-density correlation function $\rho(\mathbf{r},\mathbf{r}')$

$$\begin{aligned}< N^2 > &= \int d\mathbf{r}\int d\mathbf{r}' < \sum_{i=1}^{N}\delta(\mathbf{r}-\mathbf{r}_i)\sum_{j=1}^{N}\delta(\mathbf{r}'-\mathbf{r}_j) > \\ &= \int d\mathbf{r}\int d\mathbf{r}'[\rho(\mathbf{r})\delta(\mathbf{r}-\mathbf{r}')+\rho\left(\mathbf{r},\mathbf{r}'\right)] =< N > \left[1+\rho_b\int d\mathbf{r}g(r)\right].\end{aligned} \tag{7.56}$$

For a bulk system, the average particle density is uniform, i.e., $\rho(\mathbf{r}) = \rho_b = <N>/V$. Using the identity $g(r) = h(r) + 1$ and $\rho_b \int d\mathbf{r} = <N>$, we can rewrite the left side of Eq. (7.55) as

$$< N^2 > - < N>^2 =< N > \left[1+\rho_b\int d\mathbf{r}h(r)\right]. \tag{7.57}$$

Because V is fixed in the grand canonical ensemble, the right side of Eq. (7.55) can be expressed in terms of the bulk density

$$\partial\langle N\rangle/\partial\beta\mu = V(\partial\rho_b/\partial\beta\mu). \tag{7.58}$$

For a one-component system at fixed temperature T, the Gibbs–Duhem equation predicts

$$\partial\beta P/\partial\rho_b = \rho_b(\partial\beta\mu/\partial\rho_b). \tag{7.59}$$

Substituting Eq. (7.59) into (7.58) gives

$$\partial\langle N\rangle/\partial\beta\mu =< N > (\partial\rho_b/\partial\beta P). \tag{7.60}$$

Now replacing the left and right sides of Eq. (7.55) with the expressions given by Eq. (7.57) and (7.60), respectively, we get the compressibility equation

$$(\partial\rho_b/\partial\beta P)_T = 1+\rho_b\int d\mathbf{r}h(r). \tag{7.61}$$

Eq. (7.61) is so-named because the partial derivative is related to the isothermal compressibility

$$\kappa_T = \frac{1}{\rho_b}\left(\frac{\partial\rho_b}{\partial P}\right)_T. \tag{7.62}$$

Note that the integral on the right side of Eq. (7.61) amounts to the Fourier transform of the total correlation function at wavevector $q = 0$. As a result, the compressibility equation can also be expressed in terms of the structure factor

$$\rho_b k_B T\kappa_T = \widehat{S}(0) \tag{7.63}$$

where $\widehat{S}(0) = 1+\rho_b\widehat{h}(0)$. Eq. (7.63) follows from $\widehat{S}(k) = 1+\rho_b\widehat{h}(k)$ discussed in Section 7.2. It predicts that, for an ideal gas, $\kappa_T^{IG} = (\rho_b k_B T)^{-1}$, $\widehat{S}^{IG}(0) = 1$. For a near incompressible system (e.g., a liquid near the triple point), $\kappa_T \approx 0$, and thus $\hat{S}(0)$ would be exceedingly small. Near the critical point of the second-order phase transition, $(\partial\rho_b/\partial P)_T \to \infty$, implying that both the compressibility and the structure factor diverge.

7.3.3 Virial Equation

Another route to connect the pressure of a classical fluid with the microscopic structure is by using the virial theorem. As discussed in Appendix 2A, the virial theorem predicts

$$P = \rho_b k_B T + \frac{1}{3V}\sum_i < \mathbf{r}_i \cdot \mathbf{F}_i > \tag{7.64}$$

where $\mathbf{r}_i$ stands for the particle position and $\mathbf{F}_i$ for the force. With the pairwise additive approximation, we may write the force on particle i due to other particles in the system as a summation of all pair interactions

$$\mathbf{F}_i = \sum_{j\neq i} \mathbf{F}_{ij} = -\sum_{j\neq i} \partial u(r_{ij})/\partial r_{ij} \tag{7.65}$$

where $r_{ij} = |\mathbf{r}_i - \mathbf{r}_j|$. Substituting Eq. (7.65) into (7.64) gives

$$\begin{aligned}\sum_i \mathbf{F}_i \cdot \mathbf{r}_i &= \sum_i \sum_{j\neq i} \mathbf{F}_{ij} \cdot \mathbf{r}_i = \frac{1}{2}\sum_i \sum_{j\neq i} (\mathbf{F}_{ij} \cdot \mathbf{r}_i + \mathbf{F}_{ji} \cdot \mathbf{r}_j) \\ &= \frac{1}{2}\sum_i \sum_{j\neq i} (\mathbf{F}_{ij} \cdot \mathbf{r}_i - \mathbf{F}_{ij} \cdot \mathbf{r}_j) = \frac{1}{2}\sum_i \sum_{j\neq i} \mathbf{F}_{ij} \cdot \mathbf{r}_{ij}\end{aligned} \tag{7.66}$$

where $\mathbf{r}_{ij} = \mathbf{r}_i - \mathbf{r}_j$, the second equality is obtained from the equivalence of indices i and j, and the third equality follows from Newton's third law, $\mathbf{F}_{ij} = -\mathbf{F}_{ji}$. In terms of the radial distribution function, the virial equation is given by

$$P = k_B T \rho_b + \frac{1}{6V} < \sum_i \sum_{j\neq i} \mathbf{F}_{ij} \cdot \mathbf{r}_{ij} > = k_B T \rho_b - \frac{\rho_b^2}{6}\int d\mathbf{r} \frac{\partial u(r)}{\partial r} r g(r). \tag{7.67}$$

The virial equation is convenient for practical applications when both the radial distribution function and the intermolecular potential are available. At low density, $g(r) \approx e^{-\beta u(r)}$ so that Eq. (7.67) can be written as

$$\begin{aligned}P &= k_B T \rho_b + \frac{k_B T \rho_b^2}{6}\int dr 4\pi r^3 \left[\frac{\partial(e^{-\beta u(r)} - 1)}{\partial r}\right] \\ &= k_B T \rho_b - 2\pi k_B T \rho_b^2 \int dr r^2 [e^{-\beta u(r)} - 1] = k_B T \rho_b (1 + B\rho_b)\end{aligned} \tag{7.68}$$

where $B \equiv -2\pi \int dr r^2 [e^{-\beta u(r)} - 1]$ is the second virial coefficient. Eq. (7.68) corresponds to the lowest-order virial expansion of the gas pressure.

In comparison with simulation results, the pressure calculated from the virial equation is often less accurate than that from the compressibility equation. Similar trend is observed in comparison with experimental data. With the same radial distribution function, the virial equation often introduces a larger error because any discrepancy in the structure is maximized due to integration along with the inter-particle force at small separations. By contrast, the compressibility equation does not involve the pair potential. In practical applications, we may use the virial and compressibility equations together to improve the theoretical predictions by imposing the thermodynamic consistency, i.e., by requiring the two equations to yield the same result.

7.3.4 Chemical Potential Equation

We can calculate the chemical potential from the compressibility equation discussed above. Specifically, a combination of Eqs. (7.60) and (7.61) gives

$$\rho_b(\partial\beta\mu/\partial\rho_b)_T = [1 + \rho_b\hat{h}(0)]^{-1}. \tag{7.69}$$

For an ideal gas, the total correlation function is everywhere zero so that $\hat{h}(0) = 0$. Thus, integration of Eq. (7.69) gives

$$\beta\mu^{IG} = \ln(\rho_b\Lambda^3) \tag{7.70}$$

where the integration constant ($\ln \Lambda^3$) can be fixed by comparing the continuous and discrete representations of microstates as discussed in Section 7.1.

Alternatively, we can derive the chemical potential from the Helmholtz energy

$$F = F^{IG} + F^{ex} \tag{7.71}$$

where F^{IG} denotes the Helmholtz energy of an ideal gas at the same (N, T, V), and F^{ex} represents the excess Helmholtz energy due to inter-particle interactions. Accordingly, the chemical potential can be written as

$$\mu = \mu^{IG} + \mu^{ex} \tag{7.72}$$

where μ^{IG} is the chemical potential of the ideal gas, and μ^{ex} represents the excess chemical potential. As discussed in Section 6.6, we can calculate the excess chemical potential μ^{ex} using Windom's insertion method, i.e., by differentiation of the excess Helmholtz energy with respect to N

$$\mu^{ex} = \left(\frac{\partial F^{ex}}{\partial N}\right)_{T,V} = F^{ex}(N, T, V) - F^{ex}(N-1, T, V). \tag{7.73}$$

In molecular simulation, Eq. (7.73) may be implemented with a coupling-parameter method, i.e., by using a coupling parameter for the interaction between a tagged particle and all other particles in the system. At any given value of the coupling parameter $0 \le \lambda \le 1$, the tagged particle interacts with any other particle in the system with a reduced potential $\lambda u(r)$, where $u(r)$ is the normal pair potential. When $\lambda = 0$, the tagged particle does not interact with other particles, i.e., it is uncoupled from the system; when the interaction between the tagged particle and any other particle is the same as that between any two untagged particles, it is fully coupled. The excess chemical potential μ^{ex} thus represents the reversible work to transfer one particle into a canonical system from an ideal-gas state. The work of "insertion" can be obtained by integrating with respect to the coupling parameter from $\lambda = 0$ to $\lambda = 1$.

To put the above description in terms of equations, let $\rho(r, \lambda) = \rho_b g(r, \lambda)$ represent the average number density of other particles at radial distance r from the tagged particle. Due to the interactions of the tagged particle with other particles located in a spherical shell between r and $r + dr$, the reversible work to increase coupling parameter λ by $d\lambda$ is

$$dW = u(r)d\lambda \cdot \rho(r, \lambda) \cdot 4\pi r^2 dr. \tag{7.74}$$

The excess chemical potential corresponds to the reversible work when λ is increased from $\lambda = 0$ to $\lambda = 1$ for all possible values of radial distance

$$\mu^{ex} = \rho_b \int_0^1 d\lambda \int u(r)g(r, \lambda) \cdot 4\pi r^2 dr. \tag{7.75}$$

Substituting Eq. (7.75) into Eq. (7.72) gives

$$\mu = k_B T \ln(\rho_b \Lambda^3) + \rho_b \int_0^1 d\lambda \int u(r) g(r, \lambda) \cdot 4\pi r^2 dr. \tag{7.76}$$

Eq. (7.76) indicates that, to calculate the chemical potential, we need to know the radial distribution function $g(r, \lambda)$ for the entire range of the coupling parameter $0 \leq \lambda \leq 1$. In principle, $g(r, \lambda)$ can be calculated either from integral-equation theories (Section 7.4) or through molecular simulation.

7.3.5 Summary

With analytic expressions for the pair potential and the radial distribution function, we can in principle calculate all thermodynamic properties of thermodynamic systems comprising spherical particles. Whereas the virial equation is derived with the assumption of pairwise additivity for the inter-particle potential, this approximation is not necessary in deriving the virial theorem or the compressibility equation. Using the same radial distribution function or the same total correlation function, the pressure calculated from the compressibility equation is often more reliable than that from the virial equation. We will discuss in the next few sections on the theoretical prediction of the radial distribution function and other correlation functions.

7.4 Integral Equation Theories

Practical applications of statistical mechanics often hinge on analytical methods that are able to rapidly predict thermodynamic properties of various systems over a broad range of conditions. For example, an iterative evaluation of chemical potentials of many species is often required in multicomponent phase-equilibrium calculations. In this and next few sections, we introduce conventional liquid-state theories to predict the correlation functions and bulk properties of simple fluids. Extension of similar methods to molecular systems and mixtures will be discussed in Chapter 8.

7.4.1 The Ornstein–Zernike Equation

Analytical methods to calculate correlation functions in fluid systems are mostly based on the Ornstein–Zernike (OZ) equation.[14] The integral equation can be derived from diagrammatic expansion or, more straightforwardly, from functional analysis.[15] Alternatively, the OZ equation may serve as a definition of the direct correlation function.

To minimize mathematical exposure, we introduce the OZ equation following a heuristic approach. As discussed in the previous section, the total correlation function, defined as $h(r) \equiv g(r) - 1 = [\rho(r) - \rho_b]/\rho_b$, corresponds to the change in the local particle density $\rho(r)$ at radial distance r (normalized by bulk density ρ_b) due to the presence of another particle at the origin. If $h(r) = 0$, the local density is unaffected by the tagged particle. The particle positions are uncorrelated when the separation is sufficiently far away or if the particles do not interact with each other. In the latter case, the local particle density is independent of the position, i.e., $\rho(r)$ is not affected by the presence of a particle at the origin, and thus the local density is the same as the bulk density ρ_b. For a system with interacting particles, the total correlation function is finite

14 Ornstein L. S. and Zernike F., "Accidental deviations of density and opalescence at the critical point of a single substance", *Proc. R. Netherlands Acad. Arts Sci.* 17, 793–806 (1914).
15 Hansen J.-P. and McDonald I. R., *Theory of simple liquids: with applications to soft matter*. Academic Press, 2013.

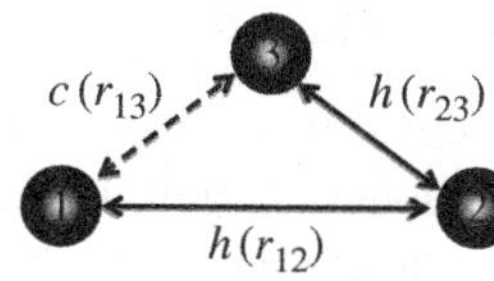

Figure 7.10 The OZ equation decomposes the total correlation function $h(r_{12})$ into a direct contribution due to interactions between particles at positions 1 and 2, i.e., the direct correlation function $c(r_{12})$, and an indirect contribution due to interaction of the particles at positions 1 and 2 with all other particles in the system, i.e., an integration $c(r_{13})h(r_{23})$ over position 3.

within the range of the correlation length. If $h(r) > 0$, the presence of a particle at the origin makes the local density larger than ρ_b; and vice versa if $h(r) < 0$, the local density is depressed.

In a many-particle system, the total correlation function between two particles at positions 1 and 2 can be intuitively divided into direct and indirect contributions. As illustrated in Figure 7.10, the direct contribution, here denoted by $c(r_{12})$, describes how the change in the local particle density at position 2 is correlated with the change in local particle density at position 1 (or more precisely, how the system Helmholtz energy responds to the changes in the local particle densities at positions 1 and 2). The indirect contribution accounts for the correlations in particle density at all other positions.

In the limit $\rho_b \to 0$, the indirect contribution disappears. In this case, we would have

$$h(r_{12}) = c(r_{12}). \tag{7.77}$$

At finite particle density, the indirect contribution arises due to the interaction of particles at positions 1 and 2 with other particles in the system. Such contributions depend on the position of the other particles and can be subsequently decomposed into direct and indirect contributions. As a result, the total correlation function can be expressed as an integration of the possible positions of all other particles overall the entire space

$$h(r_{12}) = c(r_{12}) + \rho_b \int d\mathbf{r}_3 c(r_{13})h(r_{23}). \tag{7.78}$$

Eq. (7.78) is known as the OZ equation. As our derivation is heuristic, it may also serve as the definition of direct correlation function $c(r_{12})$.

In the Fourier space, the OZ equation can be expressed as

$$\hat{h}(k) = \hat{c}(k) + \rho_b \hat{h}(k)\hat{c}(k) \tag{7.79}$$

where the hat expression ^ denotes the Fourier transform. Recalling that the structure factor is proportional to the total correlation function, we may obtain a simple algebraic relation

$$\hat{S}(k) = 1 + \rho_b \hat{h}(k) = \frac{1}{1 - \rho_b \hat{c}(k)}. \tag{7.80}$$

Using the geometric series $(1 - x)^{-1} = 1 + x + x^2 + \cdots$, we can rearrange the right side of Eq. (7.80) to obtain

$$\hat{h}(k) = \hat{c}(k)[1 + \rho_b \hat{c}(k) + \cdots]. \tag{7.81}$$

As discussed above, the total correlation function becomes the same as the direct correlation function when the number density ρ_b is sufficiently small.

Although the OZ equation looks rather abstract, reasonable approximations can be readily established for both the total and direct correlation functions. On the one hand, we expect that $h(r) \approx -1$ when r is smaller than the molecular collision diameter σ. This is because, as classical particles do not overlap, $g(r) \approx 0$ when $r < \sigma$. On the other hand, intermolecular interactions become insignificant at sufficiently large separation. In that limit, the direct correlation function may be represented by $c(r) \approx -\beta u(r)$, i.e., the variation of the free energy to local density changes is directly related to the pair potential. Based on these approximations, we can find the radial distribution function from the OZ equation and predict thermodynamic properties subsequently.

7.4.2 Closure

To solve the OZ equation (i.e., to obtain the total and direct correlation functions), we need an additional relation linking the direct and total correlation functions. This additional relation is commonly known as the *closure*. Formally, an exact closure can be derived from diagrammatic expansion or from the classical density functional theory (cDFT)[16]

$$h(r) = c(r) + \ln y(r) + B(r) \tag{7.82}$$

where $y(r) \equiv g(r)\exp[\beta u(r)]$ represents the cavity correlation function, and $B(r)$ is the bridge function. In the diagrammatic expansion, $B(r)$ is associated with the "bridge diagrams," thereby leading to the name.

Intuitively, we may understand the three terms on the right side of Eq. (7.82) as follows. The first term arises from direct correlation; the second term accounts for indirect two-body contributions to $h(r)$; and the third term accounts for all contributions to $h(r)$ due to multi-body correlations. With an explicit expression for bridge function $B(r)$, we can solve for $h(r)$ and $c(r)$ from Eq. (7.82) and the OZ equation (Eq. (7.78)) using appropriate numerical schemes.[17]

7.4.3 Hypernetted-Chain (HNC) Approximation

The bridge function is relatively insignificant in determining the total and direct correlation functions because it is nonzero mostly in the range of r where $u(r)$ diverges. If we ignore $B(r)$ in Eq. (7.82) completely, the closure for the OZ equation becomes

$$h(r) \approx c(r) + \ln y(r). \tag{7.83}$$

Eq. (7.83) is known as the hypernetted-chain (HNC) approximation, a name originally introduced in the cluster expansion of the Helmholtz energy due to intermolecular interactions.[18] In cDFT, Eq. (7.83) can be readily derived from a quadratic expansion of the excess Helmholtz energy with respect to the variation of the density profile $\rho(r)$.[16]

Extensive results from molecular simulation suggest that, for a system of classical particles far from the vapor–liquid critical point, the bridge function is not sensitive to the specific form of the pair potential. In other words, at the same temperature and bulk density, the bridge function determined for one system can be similarly applied to other systems of the same temperature and average density
(but with different inter-particle potentials). The *quasi-universality* of the bridge function has inspired substantial theoretical efforts to improve the performance of the HNC closure.[19] For example, the reference hypernetted chain (RHNC) approximation replaces the bridge function by that of a reference system at the same temperature and density[20]

$$h(r) - h_0(r) = c(r) - c_0(r) + \ln[y(r)/y_0(r)] \tag{7.84}$$

16 Wu J., Density functional theory for liquid structure and thermodynamics, In "*Molecular thermodynamics of complex systems*", Lu X., Hu Y., Eds. Springer, Berlin, 2009.
17 See, e.g., Peplow A.T., Beardmore R. E., and Bresme F., "Algorithms for the computation of solutions of the Ornstein-Zernike equation", *Phys. Rev. E* 74, 046705 (2006).
18 Morita T., "Theory of classical fluids-hyper-netted chain approximation. 1. Formulation for a one-component system", *Prog. Theor. Phys.* 20 (6), 920–938 (1958).
19 Rosenfeld Y., "Universality of bridge functions and its relation to variational perturbation theory and additivity of equations of state", *Phys. Rev. A* 29, 2877 (1984).
20 Lado F., Foiles S. M. and Ashcroft N. W., "Solutions of the reference-hypernetted-chain equation with minimized free energy", *Phys. Rev. A* 28, 2374 (1983).

where subscript "0" denotes the reference system. The parameters of the reference system can be determined by minimizing the Helmholtz free energy.

In practical applications, the hard-sphere (HS) model is often adopted to derive the reference total and direct correlation functions. The idea is similar to that of the modified hypernetted-chain (MHNC) approximation, which utilizes the bridge function of a HS fluid at the same particle density[21]

$$h(r) = c(r) + \ln[y(r)] + B_{HS}(r). \tag{7.85}$$

The HS diameter can be determined from thermodynamic consistency such that constraining the pressure from the virial equation is the same as that from the compressibility equation. Alternatively, it can be obtained by the extremum condition for the Helmholtz energy, or by equating the second density derivative of the free energy of the actual system to that of the reference system.[22]

For a uniform HS fluid, the bridge function has been accurately determined from computer simulation. According to the simulation results, the HS bridge function can be approximated by the semi-empirical Verlet modified (MV) equation[23]

$$B_{HS}(r) \approx \frac{1}{2} \frac{[h_{HS}(r) - c_{HS}(r)]^2}{1 + a\,[h_{HS}(r) - c_{HS}(r)]} \tag{7.86}$$

with

$$a = 1.1 - \rho_b \sigma_{HS}^3/3. \tag{7.87}$$

Figure 7.11 shows $B_{HS}(r)$ at different particle densities predicted by the MV equation. We see that the bridge function vanishes at low density and becomes more significant as the density increases. Importantly, $B_{HS}(r)$ is most pronounced at short separation (viz., $r/\sigma_{HS} < 1$).

The MHNC closure agrees well with simulation results for a number of model systems. It is also able to describe the correlation functions of realistic systems such as alloys. For example, Figure 7.12 compares the structure factor and radial distribution function of liquid gallium predicted by MHNC with those obtained from X-ray scattering measurements.[24] Here, the effective pair potential is derived from quantum-mechanical calculations. The agreement between the theoretical results and experimental data is reasonably good, considering the simplicity of the physical model for the metallic liquid and numerical errors introduced in the Fourier transformation.

HNC and its modifications provide accurate results for a variety of thermodynamic systems including not just simple fluids but also electrolyte solutions, molten salts, and various colloidal dispersions. One caveat in application of HNC is that the radial distribution function (and subsequently thermodynamic properties) must be calculated with numerical methods. Although the computational cost remains insignificant in comparison with molecular simulation, the main issue is that, at certain conditions, the OZ equation may not even have a numerical solution. For practical purposes, it would be convenient to use analytical (but often less accurate) solutions of the OZ equation for obtaining the radial distribution function and thermodynamic properties.

---

21 Rosenfeld Y. and Ashcroft N. W., "Theory of simple classical fluids: Universality in the short-range structure", *Phys. Rev. A* 20, 1208 (1979).
22 Rosenfeld Y. and Blum L., "Fluids in contact with a hard surface: Universality of the bridge functions for the density profile", *J. Chem. Phys.* 85, 2197 (1986).
23 Verlet L., "Integral equations for classical fluids", *Mol. Phys.* 41, 183 (1980); Labík S., Malijevský A. and Smith W. R., "An accurate integral equation for molecular fluids", *Mol. Phys.* 73, 87 (1991).
24 Hoshino K. "Dynamical structure of the liquid Ge by the viscoelastic theory", *J. Phys. Soc. Jpn.* 71 (10), 2466–2470 (2002).

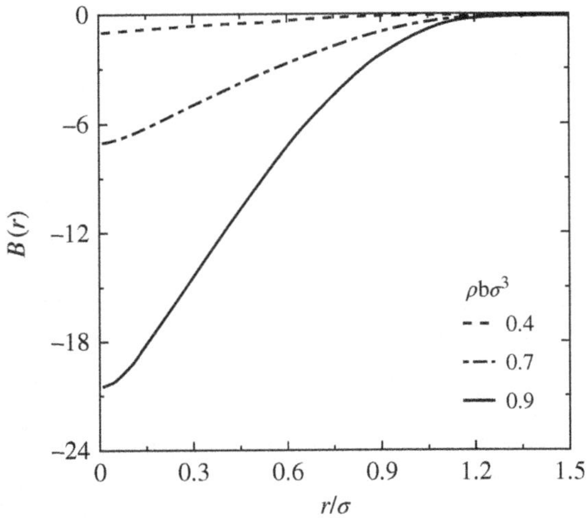

Figure 7.11 Bridge functions for hard-sphere systems at different reduced densities. In the right panel, each high-density curve has been shifted upward by one unit.

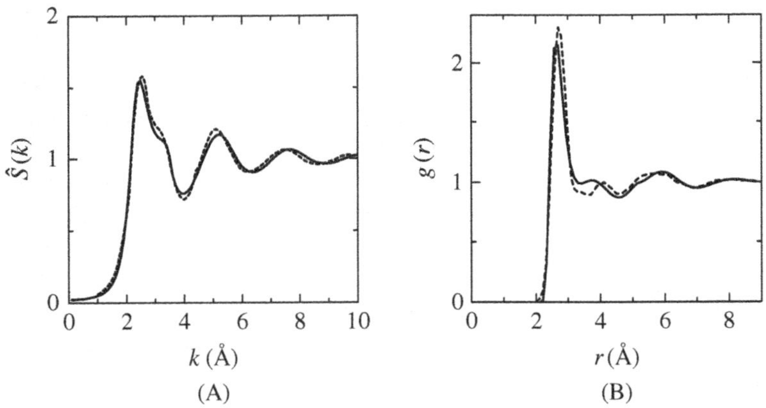

Figure 7.12 The structure factor (A) and radial distribution function (B) of liquid gallium at 1253 K and 1 atm. The solid lines are obtained from the MHNC predictions and the dashed lines are from the X-ray experimental results.

7.4.4 The Percus–Yevick (PY) Equation

The Percus–Yevick (PY) equation assumes that radial distribution function $g(r)$ can be expressed as direct correlation function $c(r)$ plus "indirect" cavity correlation function $y(r)$

$$g(r) = c(r) + y(r). \tag{7.88}$$

Formally, Eq. (7.88) may be obtained from the HNC closure, Eq. (7.83), by the linear expansion

$$\exp[h(r) - c(r)] \approx 1 + h(r) - c(r). \tag{7.89}$$

Alternatively, the PY closure may be understood as an approximation of the bridge function

$$B(r) = h(r) - c(r) - \ln y(r) \approx y(r) - 1 - \ln y(r). \tag{7.90}$$

In comparison with other closures for the OZ equation, one major advantage of the PY closure is that, as to be discussed in the next two sections, it yields analytical solutions for HS systems including sticky hard spheres and HS mixtures.

The PY closure predicts the direct and total correlation functions and thermodynamic properties with good accuracy for systems with short-range potentials such as those describing a simple fluid. As it can be obtained by a linear expansion of the HNC closure, one may expect that the HNC closure is superior to the PY closure. Although this expectation holds for most systems, in particular for those containing charged particles, the thermodynamic properties of HS systems predicted by the PY closure are surprisingly more accurate than those from the HNC closure. To a certain degree, the adoption of the PY bridge function in Eq. (7.90) explains the good performance. In the next section, we discuss the properties of hard spheres and sticky hard spheres predicted by the PY theory.

7.4.5 Mean-Spherical Approximation (MSA)

For thermodynamic systems consisting of particles with a HS-like core, the radial distribution function $g(r)$ is zero when the distance r is smaller than the hard-core diameter σ, i.e.,

$$h(r) = -1, \ r < \sigma. \tag{7.91}$$

At large separations, $g(r) \approx 1$, and thus $h(r) \approx 0$. Accordingly, we may use the approximation $h(r) \approx \ln[1 + h(r)]$ and write the HNC approximation as

$$c(r) \approx -\beta u(r), \ r \geq \sigma. \tag{7.92}$$

Eqs. (7.91) and (7.92) are known as the mean-spherical approximation (MSA), which was originally introduced in applying the lattice-gas model to a continuous fluid with spherical hard-core interactions.[25]

Using the MSA closure, Blum and others obtained the analytical solutions of the OZ equation for ionic systems and for systems with the hard-core multiple Yukawa potential[26]

$$\beta u(r) = \begin{cases} \infty & r < \sigma \\ -\sum_{n=1}^{M} K_n e^{-z_n(r-\sigma)}/r & r \geq \sigma \end{cases} \tag{7.93}$$

where M stands for the number of Yukawa terms, K_n and z_n are Yukawa parameters. The Yukawa model[27] is applicable to a wide range of systems including simple liquids, electrolytes, molten salts, liquid metals, dense plasma, colloidal dispersions, micelles, and microemulsions. While analytical solutions to the MSA equation often require lengthy derivations, the final equations are in closed form thus convenient for practical use. We will discuss their applications to electrolyte solutions in Chapter 9.

---

25 Lebowitz J. L. and Percus J. K., "Mean Spherical model for lattice gases with extended hard cores and continuum fluids", *Phys. Rev.* 144 (1), 251 (1966).

26 Blum L., "Mean spherical model for asymmetric electrolytes. 1. Method of solution", *Mol. Phys.* 30 (5), 1529–1535 (1975); Blum L., Vericat F., Herrerapacheco J. N., "On the analytical solution of the Ornstein-Zernike equation with Yukawa closure", *J. Stat. Phys.* 66 (1–2), 249–262 (1992).

27 The Yukawa model, named after physicist Hideki Yukawa, was originally introduced to describe the nuclear force between nucleons in particle physics.

Because the analytical solutions are attainable, the Yukawa potential can also be used for a self-consistent improvement of the MSA closure by adding one additional Yukawa term in the direct correlation function

$$c(r) \approx -\beta u(r) + K^* e^{-z^*(r-\sigma)}/r, \; r \geq \sigma \tag{7.94}$$

where parameters K^* and z^* are fixed from thermodynamic consistency or from known theoretical or experimental results. The self-consistent procedure is called the generalized MSA (GMSA).[28] Similar ideas have been implemented in the context of the self-consistent Ornstein–Zernike approximation (SCOZA).[29]

Figure 7.13 shows, as an example, that GMSA is able to describe the structure factors of colloidal dispersions represented by the two-Yukawa model.[30] Apparently, the structure factor predicted by GMSA agrees well with Monte Carlo (MC) simulation over a broad range of conditions. Here, the colloidal force includes contributions from a short-range attraction and long-range screened Coulomb repulsion (see Section 9.3.4). The MC simulation identifies four distinctive microstructures: monomer-dominated states, clustered, clustered percolated, and random percolated system. When the concentration is increased, the magnitude of the first peak of the structure factor is an indication of the transition from a monomeric fluid to a cluster fluid and then to a cluster-percolated state (shown in the left panel of Figure 7.13 from bottom to top, respectively). When the concentration increases at higher temperature, the system goes from a monomer-dominated state to random percolated states (shown in the right panel of Figure 7.13 from bottom to top, respectively).

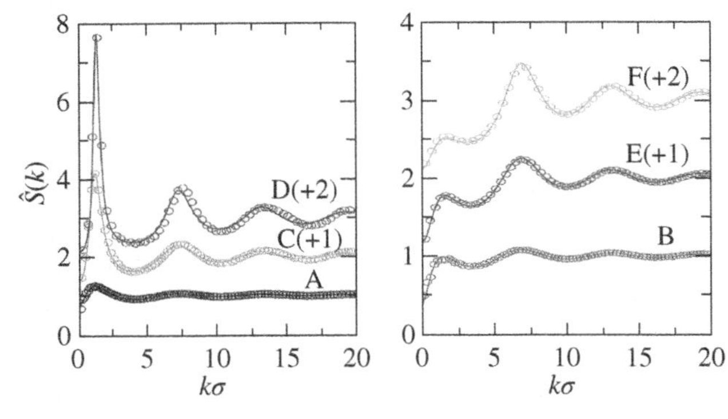

Figure 7.13 Structure factor of a two-Yukawa fluid obtained from MC simulation (open circles) and GMSA (solid lines). In the left panel, $T^* = 0.25$ and the volume fractions of colloids are $\eta = 0.01$, 0.05, and 0.15, represented by dark-colored solid (A), dashed (C), and dotted (D) lines, respectively. In the right panel, $T^* = 0.46$ and the volume fractions are $\eta = 0.05$, 0.15, 0.25, represented by gray-colored solid (B), dashed (E), and dotted (F) lines, respectively. The reduced temperature T^* represents the relative magnitude of the potential energy well depth to thermal energy. For clarity, the curves at higher densities have been shifted upward.

28 Henderson D. and Blum L., "Generalized mean spherical approximation for hard spheres", *Mol. Phys.* 32 (6), 1627–1635 (1976); Høye J. S., Lebowitz J. L. and Stell G., "Generalized mean spherical approximations for polar and ionic fluids", *J. Chem. Phys.* 61, 3253–3260 (1974).
29 Pini D., Stell G. and Høye J. S., "Self-consistent approximation for fluids and lattice gases", *Int. J. Thermophys.* 19, 1029-1038 (1998); Pini D., Stell G. and Wilding N. B., "A liquid-state theory that remains successful in the critical region", *Mol. Phys.* 95, 483–494 (1998).
30 Vazquez-Rodriguez O. and Ruiz-Estrada H., "An explicit expression for the static structure factor for a multi-Yukawa fluid: the one-component case", *Phys. Chem. Liq.* 54 (5), 632–646 (2016).

7.4.6 Summary

The Ornstein-Zernike (OZ) equation represents a class of integral-equation theories that deal with the direct and total correlation functions of many-body systems. Coupled with a suitable two-body potential function and a closure such as HNC, PY, and MSA, the OZ equation has experienced much success in describing both correlation functions and thermodynamic properties of thermodynamic systems including not just simple fluids but also electrolyte solutions, molten salts, liquid metals, dense plasma, colloidal dispersions, micelles, and microemulsions.

It should be noted that the integral-equation theories are not exact; thermodynamic properties obtained from different routes are often different from each other (viz., thermodynamically inconsistent). The discrepancy, commonly known as thermodynamic inconsistency, is introduced in the closure. Empirically, it is often found that thermodynamic properties predicted by the energy equation are better than those by compressibility or virial equations. To a certain degree, the good performance for the energy equation explains why reasonable results can often obtained from the mean-filed approximation that ignores the radial distribution function completely.

7.5 Hard-Sphere Model

The hard-sphere (HS) model provides a simple representation of molecular excluded-volume effects. It is universally applicable to chemical systems and materials including those containing atoms, small molecules, as well as mesoscopic particles such as globular proteins and colloidal particles. While an attractive potential is not included, the HS model captures the strong repulsion arising from the finite size of classical particles. Because of its simplicity, the HS model has been extensively adopted as a useful reference for modeling diverse systems of practical interest. In this section, we discuss analytical results obtained from the PY theory, i.e., the analytical solution of the OZ equation with the PY closure. While the PY theory offers analytical solutions for both one-component hard spheres and HS mixtures,[31] our discussion here is focused on single-component HS fluids.

7.5.1 The PY Theory for Hard-Sphere Fluids

As discussed in Section 7.4, the OZ equation for a one-component system can be expressed as

$$\hat{h}(k) = \hat{c}(k) + \rho_b \hat{h}(k)\hat{c}(k) \tag{7.95}$$

where $\hat{h}(k)$ and $\hat{c}(k)$ represent, respectively, total and direct correlation functions in the Fourier space, and ρ_b is the number density in the bulk. The PY closure is given by

$$c(r) = g(r) - y(r) = g(r)[1 - e^{\beta u(r)}]. \tag{7.96}$$

For a HS fluid, $g(r) = 0$ for $r < \sigma$, and $u(r) = 0$ for $r \geq \sigma$ vanishes. Thus, the PY closure can be expressed in terms of two partially known functions

$$\begin{cases} c(r) = 0 & r \geq \sigma, \\ h(r) = -1 & r < \sigma. \end{cases} \tag{7.97}$$

31 Santos A. et al., "Structural and thermodynamic properties of hard-sphere fluids", *J. Chem. Phys.* 153, 120901 (2020).

The analytical solution of the OZ equation that satisfies Eq. (7.97) was obtained first by Michael Wertheim[32] and, independently, by Everett Thiele.[33] For mathematical details, tutorial derivations were given by Douglas Henderson.[34] It is worth noting that, for HS systems, the PY closure is indistinguishable from the mean-spherical approximation (MSA).

Box 7.1 Hard-sphere Model

The HS model has a long history. Ancient Greek philosopher Democritus believed that the world was made of atoms so hard that would never break. Throughout the history, much interest has been devoted to the efficiency of hard-sphere packing, i.e., the arrangement of hard spheres such that they occupy a minimal volume. Astronomer Johannes Kepler attempted to solve the problem using geometry. He argued that the most efficient way to pack uniform hard spheres is by arranging them in a face-centered-cubic (FCC) lattice, very much like that often seen in grocery stores for displaying oranges or apples. Kepler did not prove his conjecture; indeed, a mathematical proof is so difficult that the conjecture remains unsolved for nearly four centuries. Packing of hard spheres was of great interest a few hundred years ago for efficient loading of cannon balls in a warship. Despite its simplicity, the HS model provides a fair representation of typical colloids because the range of interaction between colloidal particles is often much smaller than the particle diameter. To avoid coagulation, the dispersing medium (i.e., the solvent) can be adjusted such that the attraction between colloidal particles is minimized. Material chemists employ HS-like nanoparticles for the fabrication of nanostructured materials.[35]

7.5.2 Correlation Functions

For a uniform HS fluid, the PY theory predicts that the direct correlation function is given by

$$c(r) = \begin{cases} \lambda_1 + \lambda_2(r/\sigma) + \lambda_3(r/\sigma)^3 & r \leq \sigma \\ 0 & r > \sigma \end{cases} \tag{7.98}$$

where

$$\lambda_1 = -\frac{(1+2\eta)^2}{(1-\eta)^4} \tag{7.99}$$

$$\lambda_2 = \frac{3\eta(2+\eta)^2}{2(1-\eta)^4} \tag{7.100}$$

$$\lambda_3 = \frac{\eta\lambda_1}{2} \tag{7.101}$$

and $\eta = \pi\rho\sigma^3/6$ stands for the packing fraction or volume fraction. Eq. (7.98) can be obtained from the Laplace transforms of the OZ equation along with the PY closure.[36] From the direct correlation function, one can readily obtain the total correlation function, the cavity correlation function, and

32 Wertheim M., "Exact solution of the Percus-Yevick integral equation for hard spheres", *Phys. Rev. Lett.* 10, 321–323 (1963).
33 Thiele E., "Equation of state for hard spheres", *J. Chem. Phys.* 39 (2), 474–479 (1963).
34 Henderson D., "Analytic methods for the Percus-Yevick hard sphere correlation functions", *Condens. Matter Phys.* 12 (2), 127–135 (2009).
35 Smirnov A. V. et al., "Hard-sphere close-packing models: possible applications for developing promising ceramic and refractory materials", *Glass Ceram.* 75, 345–351 (2019).
36 The Laplace transform of a function $f(r)$ with $r > 0$ is defined by $\hat{f}(s) = \int_0^\infty f(r)e^{-rs}dr$.

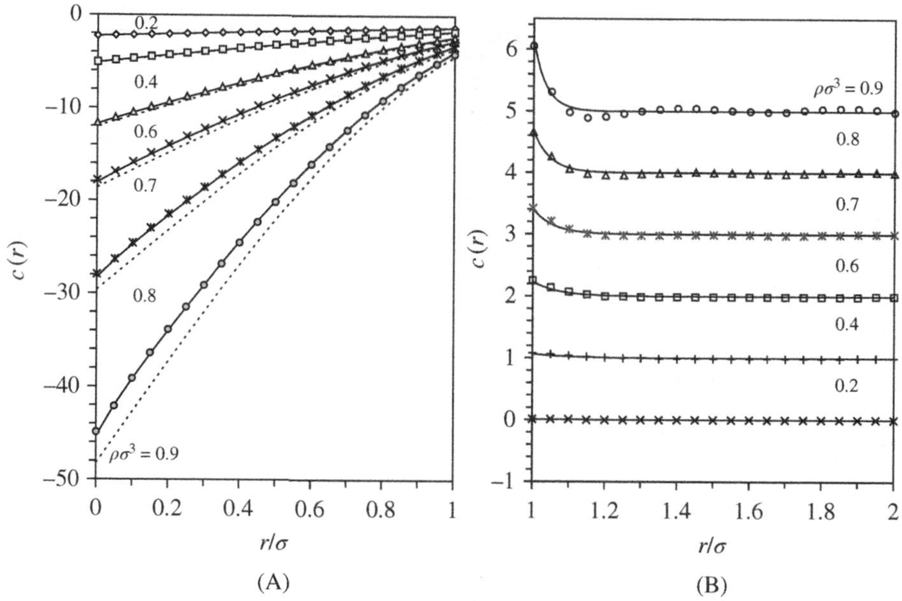

Figure 7.14 The direct correlation functions of hard-sphere fluids calculated from the PY theory (dashed lines), GMSA (solid lines), and MC simulation (symbols). The curves for higher densities have been shifted up by one unit each for clarity. The simulation results are from Groot et al.[38]

the bridge function by numerical means. An analytical expression is also available for the radial distribution function in the Laplace transform.[37]

Figure 7.14 presents the direct correlation functions of several HS fluids predicted by the OZ equation with the PY closure and with the GMSA. For comparison, the symbols are calculated from MC simulation.[38] A similar comparison is given in Figure 7.15 for the first peak of the radial distribution function for three HS systems.[39] While the PY theory becomes less satisfactory as the volume fraction increases, GMSA shows near perfect agreement with MC at all densities. As discussed in Section 7.4, GMSA imposes the thermodynamic consistency by adding a Yukawa-type function to approximate the long-range behavior of the direct correlation function. Alternatively, the thermodynamic consistency can be imposed by using the Rogers–Young (RY) closure[40]

$$g(r) = e^{-\beta u(r)}\{\exp[\gamma(r)f(r)] + f(r) - 1\}/f(r) \tag{7.102}$$

where $\gamma(r) = h(r) - c(r), f(r) = 1 - e^{-\alpha r}$, and parameter α is fixed by satisfying thermodynamic consistency, i.e., the bulk pressure calculated from the virial equation is the same as that calculated from the compressibility equation. While imposing thermodynamic consistency improves numerical performance, the procedure is less convenient for practical applications because the parameters must be adjusted at every thermodynamic state.

---

37 Lebowitz J. L., "Exact solution of generalized Percus-Yevick equation for a mixture of hard spheres", *Phys. Rev.* 133 (4A), A895–A899 (1964).

38 Groot R. D., van der Eerden J. P., and Faber N. M., "The direct correlation function in hard sphere fluids", *J. Chem. Phys.* 87, 2263-2270 (1987).

39 Barker J. A. and Henderson D., "Monte Carlo values for the radial distribution function of a system of fluid hard spheres", *Mol. Phys.* 21, 187 (1971).

40 Rogers F. J. and Young D. A., "New, thermodynamically consistent, integral-equation for simple fluids", *Phys. Rev. A* 30 (2), 999–1007 (1984).

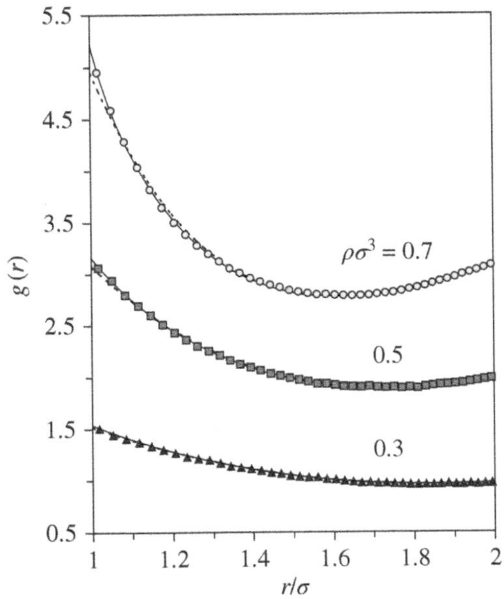

Figure 7.15 The first peak of radial distribution function of bulk hard spheres calculated from the PY theory (dashed lines), GMSA (solid lines), and MC simulation (symbols). The pair correlation functions for the reduced densities $\rho\sigma^3 = 0.5$ and 0.7 have been shifted up by one and two units, respectively.

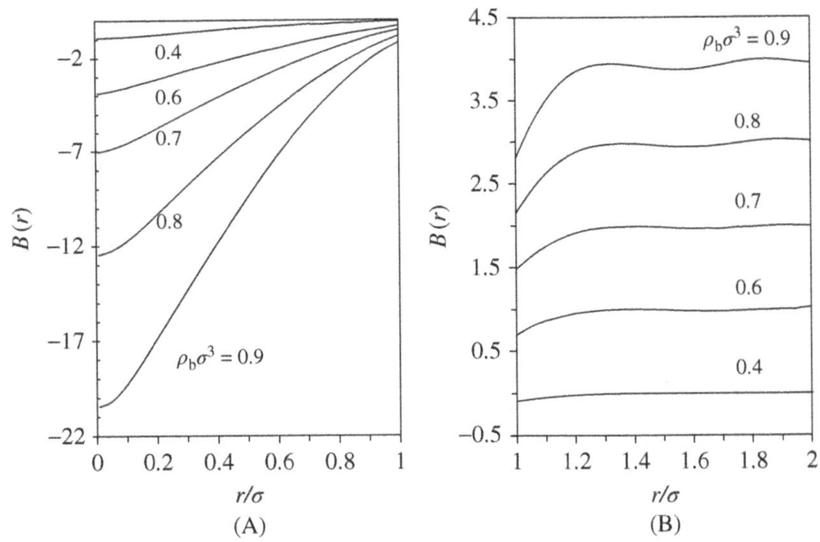

Figure 7.16 The bridge functions of several hard-sphere fluids predicted by GMSA. In panel B, the curves for higher densities have been shifted up by one unit each for clarity.

Figure 7.16 shows the bridge functions of hard-sphere fluids at five bulk densities predicted by GMSA. Clearly, $B(r)$ is most significant in the small range of r and its magnitude increases with the bulk density. Because $B(r)$ coincides with the pair potential $u(r)$ in the exact closure, it has relatively small effect in predicting the total and direct correlation functions from the OZ equation.

7.5.3 Hard-Sphere Equation of State

The equation of state for hard-sphere fluids can be derived from either the virial equation or the compressibility equation. As discussed in Section 7.4, the virial equation is given by

$$\beta P/\rho_b = 1 - \frac{\beta\rho_b}{6}\int d\mathbf{r}\frac{du(r)}{dr}rg(r). \tag{7.103}$$

In Eq. (7.103), the discontinuity in the pair correlation function can be avoided by using the cavity correlation function $y(r) = g(r)e^{\beta u(r)}$

$$\beta P/\rho_b = 1 + \frac{2\pi\rho_b}{3}\int drr^3y(r)\frac{de^{-\beta u(r)}}{dr}. \tag{7.104}$$

Figure 7.17 shows the cavity correlation functions of three hard-sphere systems calculated from GMSA in comparison with MC simulation.[41] Unlike the direct and total correlation functions, the cavity correlation function is a continuous function of r even when the pair potential is discontinuous.

As shown in Figure 7.18, the Boltzmann factor for a hard-sphere potential is a step function. The first-order derivative of this function results in a one-dimensional Dirac-delta function

$$\frac{de^{-\beta u(r)}}{dr} = \delta(r - \sigma). \tag{7.105}$$

Substituting Eq. (7.105) into Eq. (7.104) gives

$$\beta P/\rho_b = 1 + \frac{2\pi}{3}\rho_b\sigma^3y(\sigma) = 1 + 4\eta y(\sigma). \tag{7.106}$$

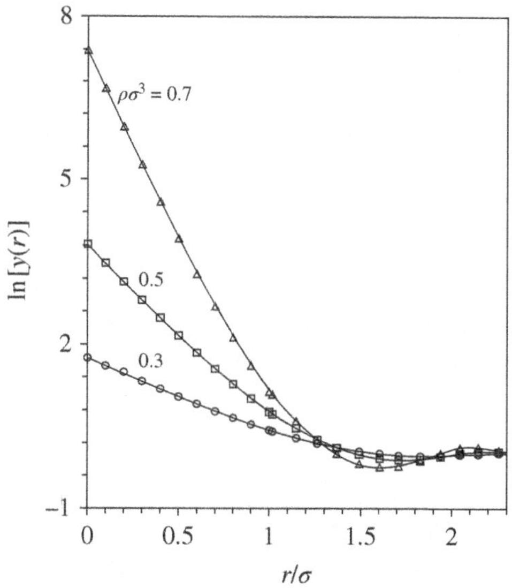

Figure 7.17 The cavity correlation functions of uniform hard spheres calculated from GMSA (solid lines) and from Monte Carlo simulation (symbols). Source: Adapted from Labik and Malijevsky.[41]

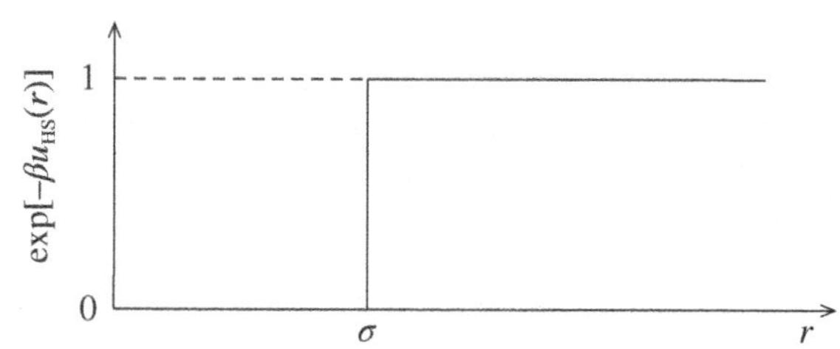

Figure 7.18 The Boltzmann factor corresponding to the hard-sphere potential.

41 Labik S. and Malijevsky A., "Monte-Carlo simulation of the background correlation-function of non-spherical hard body-fluids. 1. General-method and illustrative results", *Mol. Phys.* 53, 381 (1984).

According to the PY theory, the cavity correlation function at contact ($r = \sigma$) is given by

$$y(\sigma) = \frac{1+\eta/2}{(1-\eta)^2}. \tag{7.107}$$

Using Eq. (7.107), we obtain the virial equation for one-component hard-sphere fluids

$$Z^V = \frac{1+2\eta+3\eta^2}{(1-\eta)^2} \tag{7.108}$$

where $Z = P/(\rho_b k_B T)$ is the compressibility factor, and superscript V denotes that the result is derived from the virial equation.

Alternatively, we can derive the equation of state from the compressibility equation

$$\beta\left(\frac{\partial P}{\partial \rho_b}\right)_T = \left[1 + \rho_b \int d\mathbf{r} h(r)\right]^{-1} = [1 + \rho_b \hat{h}(0)]^{-1}. \tag{7.109}$$

By rearranging the OZ equation, Eq. (7.95), we get

$$[1 + \rho\hat{h}(q)]^{-1} = [1 - \rho\hat{c}(q)]. \tag{7.110}$$

Thus, in terms of the direct correlation function, the compressibility equation can be expressed as

$$\beta\left(\frac{\partial P}{\partial \rho_b}\right)_T = 1 - \rho_b\hat{c}(0). \tag{7.111}$$

According to Eq. (7.98), the right-hand-side of Eq. (7.111), after some algebra, is

$$1 - \rho_b\hat{c}(0) = 1 - 4\pi\rho_b \int_0^\sigma r^2 c(r)dr = \frac{(1+2\eta)^2}{(1-\eta)^4}. \tag{7.112}$$

Because the pressure vanishes as $\rho_b \to 0$, integration of Eq. (7.111) with respect to η leads to

$$P^* \equiv \frac{\pi\beta P\sigma^3}{6} = \int_0^\eta \frac{(1+2\eta)^2}{(1-\eta)^4} d\eta = \frac{\eta(1+\eta+\eta^2)}{(1-\eta)^3}. \tag{7.113}$$

Eq. (7.113) is known as the Thiele equation, named after Everett Thiele who first derived this relation. The compressibility factor, $Z = P/(\rho_b k_B T)$, is obtained by rearranging Eq. (7.113)

$$Z^C = \frac{1+\eta+\eta^2}{(1-\eta)^3} \tag{7.114}$$

where superscript C denotes results from the compressibility equation, Eq. (7.111).

Regrettably, the virial equation and compressibility equation do not yield the same compressibility factor. This discrepancy arises from the thermodynamic inconsistency of the PY closure. As shown in Figure 7.15, the PY theory underestimates the contact value of the radial distribution function, especially at high densities. The equation of state from the virial equation is slightly inferior to that from the compressibility equation. By a careful comparison of the compressibility factors calculated from both methods and those from molecular simulation, Carnahan and Starling (CS) found that an empirical combination of Z^C and Z^V provides a more satisfactory equation of state for hard-sphere fluids[42]

$$Z^{CS} = \frac{2Z^c}{3} + \frac{Z^v}{3} = \frac{1+\eta+\eta^2-\eta^3}{(1-\eta)^3}. \tag{7.115}$$

---

42 Carnahan N. F. and Starling K. E., "Equation of state for nonattracting rigid spheres", *J. Chem. Phys.* 51 (2), 635–636 (1969).

Other semi-empirical modifications have been proposed, such as the Kolafa (K) equation[43]

$$Z^K = \frac{1 + \eta + \eta^2 - 2(\eta^3 + \eta^4)/3}{(1 - \eta)^3}, \tag{7.116}$$

the Yelash–Kraska (YK) equation[44]

$$Z^{YK} = \frac{3 + 8\eta + 14\eta^2 + 14\eta^3 + 40\eta^4/3}{3 - 4\eta}, \tag{7.117}$$

and the Speedy (S) equation for the metastable region $(0.494 \leq \eta \leq 0.545)$[45]

$$Z^S = \frac{2.67}{(1 - 1.543\eta)}. \tag{7.118}$$

Figure 7.19 compares the compressibility factor calculated from molecular simulations and those from various hard-sphere equations of states.[46] In general, the agreement between theory and simulation is excellent for hard spheres in the fluid state ($\eta \leq 0.494$). The Kolafa, Eu-Ohr,[47] and YK equations are slightly more accurate than the CS equation but the improvement in accuracy is relatively insignificant. Beyond the freezing density ($\eta > 0.494$), the Speedy equation is notably more accurate than other equations.

As mentioned above, the hard-sphere model is commonly used as a reference to describe the structure and thermodynamic properties of chemical systems of practical interest, in particular for colloidal dispersions and aqueous solutions of globular proteins. As shown in Figure 7.20, at certain conditions, the properties of some uncharged colloidal suspensions are remarkably close

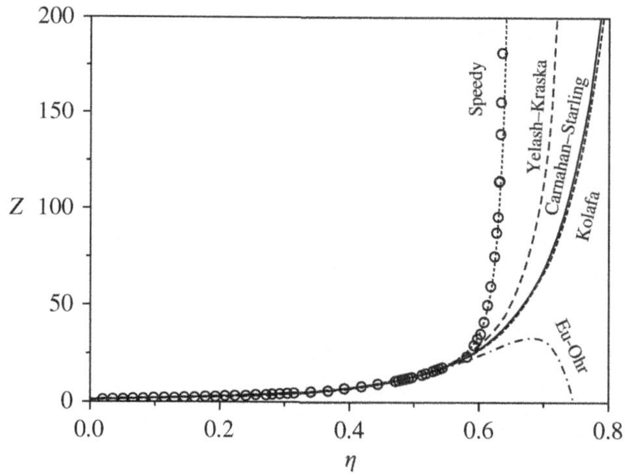

Figure 7.19 Equation of state for a hard-sphere fluid. Points are results from Monte Carlo simulation, and the lines are from various equation of state. The freezing packing fraction is $\eta = 0.494$. Source: Adapted from Wu and Sadus.[46]

---

43 Kolafa J., Labik S. and Malijevsky A., "Accurate equation of state of the hard sphere fluid in stable and metastable regions", *Phys. Chem. Chem. Phys.* 6 (9), 2335–2340 (2004).

44 Yelash L. V. and Kraska T., "A generic equation of state for the hard-sphere fluid incorporating the high-density limit", *Phys. Chem. Chem. Phys.* 3 (15), 3114–3118 (2001).

45 Speedy R. J., "The equation of state for the hard-sphere fluid at high-density – the glass-transition", *Phys. B and C* 121 (1–2), 153–161 (1983).

46 Wu G. W. and Sadus R. J., "Hard sphere compressibility factors for equation of state development", *AIChE J.* 51 (1), 309–313 (2005).

47 Eu B. C. and Ohr Y. G., "Thermodynamically consistent equation of state of hard sphere fluids, *J. Chem. Phys.* 118 (5), 2264–2269 (2003).

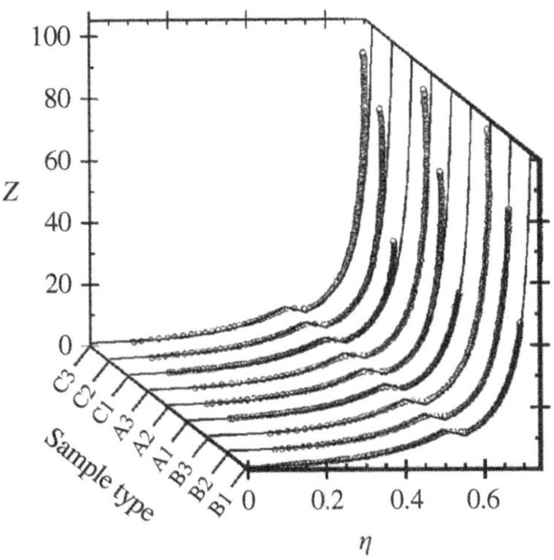

Figure 7.20 Osmotic compressibility factors (Z) for nine polystyrene suspensions (sample type labeled with letters and numbers for different particle sizes and preparation procedures). The solid-lines are predicted from the CS equation (liquid branch), and a vdW type equation of state for solids (see Section 7.8), $Z^{solid} = 3/1 - \eta/\eta_c)$ where $\eta_c = 0.74$ for a face-centered-cubic crystal (solid branch); the points are experimental results. Source: Adapted from Rutger et al.[48]

to those corresponding to hard-sphere systems.[48] The CS equation yields excellent agreement with experimental compressibility factors for colloids of different sizes and solution conditions.

7.5.4 Free Energy and Chemical Potential for Hard-Sphere Fluids

Following standard thermodynamic relations, we can readily derive other thermodynamic properties of hard-sphere systems from the equation of state. For example, the excess Helmholtz energy, which is defined relative that of an ideal gas at system temperature and density, can be calculated from the thermodynamic relation

$$\left(\frac{\partial \beta F^{ex}/N}{\partial \eta}\right)_{T,N} = \frac{Z}{\eta}. \tag{7.119}$$

Substituting the CS equation into Eq. (7.119) gives, after integration,

$$\beta F^{ex}/N = \int_0^{\eta} d\eta \frac{1+\eta+\eta^2-\eta^3}{\eta(1-\eta)^3} = \frac{4\eta - 3\eta^2}{(1-\eta)^2}. \tag{7.120}$$

In deriving Eq. (7.120), we have used the boundary condition that, as the particle density goes to zero, the excess Helmholtz energy vanishes. For a one-component system, the excess chemical potential can be obtained from the Helmholtz energy and compressibility factor

$$\beta\mu^{ex} = \beta F^{ex}/N + Z - 1 = \frac{\eta(8 - 9\eta + 3\eta^2)}{(1-\eta)^3}. \tag{7.121}$$

Other thermodynamic properties can be calculated from pressure and chemical potential.

7.5.5 Summary

The hard-sphere model plays an important role in the development of liquid-state theories for predicting the structure and thermodynamic properties of diverse condensed-matter systems spanning from simple and complex fluids to soft materials. The PY solution of the OZ equation for hard-sphere fluids serves not only as a useful reference for simple fluids but also as a theoretical

48 Rutgers M. A. et al., "Measurement of the hard-sphere equation of state using screened charged polystyrene colloids", *Phys. Rev. B* 53 (9), 5043–5046 (1996).

basis for calculating the thermodynamic properties of virtually all condensed-matter systems where atomic excluded volume plays a role.

7.6 The Sticky Hard-Sphere Model of Colloids and Globular Proteins

The interaction between colloidal particles is often dominated by various forms of short-range attraction that are significant over a length scale much smaller than the particle size. Therefore, colloidal dispersions and the liquid solutions of macromolecules (e.g., globular proteins) or micelles may be represented by the sticky hard-sphere (SHS) model. According to this model, the pair potential is given by a square-well-type function

$$\beta u(r) = \begin{cases} \infty & r < \sigma \\ \ln[12\tau d/(\sigma + d)] & \sigma \le r \le \sigma + d\,, \\ 0 & r > \sigma + d \end{cases} \tag{7.122}$$

taken in the limit of $d \to 0$ while fixing the second virial coefficient. In Eq. (7.122), τ is a dimensionless parameter that typically increases with temperature.

Figure 7.21 shows the sticky potential at a finite value of the width parameter d. It includes a hard-sphere component and a square-well attraction at short distance. As d approaches zero, the logarithm term becomes negative infinity while the second virial coefficient remains unchanged (see Problem 7.14).

7.6.1 Predictions of the Percus–Yevick (PY) Theory

For sticky hard spheres, the OZ equation with the PY closure was first solved by Rodney J. Baxter using the Fourier transform method similar to that for hard-sphere systems.[49] Based on the correlation functions derived from the PY closure, several methods can be used to derive the thermodynamic properties. These include the compressibility equation (c), the virial equation (v), the energy equation (e), the zero-separation theorem (ZS),[50] and three routes for calculating the chemical potential.

For example, the equation of state obtained from the compressibility route is given by

$$\beta P/\rho_b = \frac{1+\eta+\eta^2}{(1-\eta)^3} - \eta\lambda\frac{18(2+\eta)-\eta\lambda^2}{36(1-\eta)^3} \tag{7.123}$$

where λ is related to parameter τ and packing fraction η

$$\lambda = 6\,\{1-\tau+\tau\eta^{-1} - [(1-\tau+\tau\eta^{-1})^2 - (1+2\eta^{-1})/6]^{1/2}\}. \tag{7.124}$$

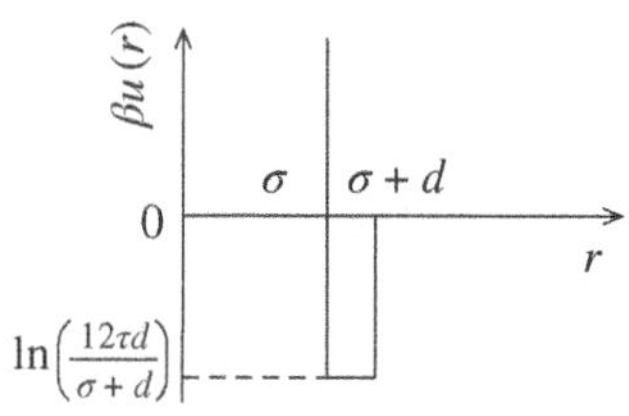

Figure 7.21 The sticky hard-sphere potential can be understood as a square-well potential in the limit of the well width $d \to 0$ while keeping the second virial coefficient as a constant. Here σ stands for the hard-sphere diameter.

49 Baxter R. J, "Percus-Yevick equation for hard spheres with surface adhesion", *J. Chem. Phys.* 49 (6), 2770–2774 (1968).

50 The zero-separation theorem states that the excess chemical potential of a sticky hard-sphere fluid is related to the cavity correlation function $y(r)$ by $\beta\mu^{ex} = \ln y(0)$. See Problem 7.9.

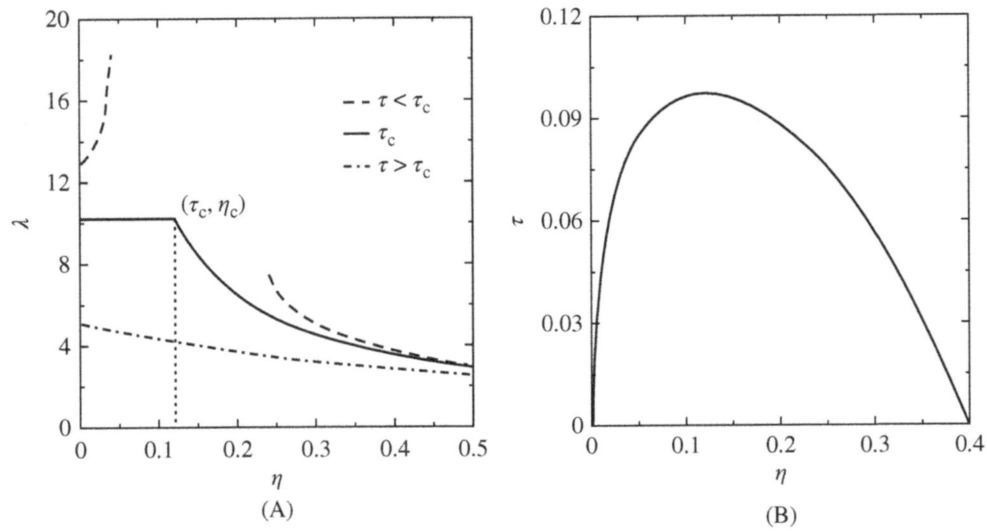

Figure 7.22 (A) Variation of λ with η, the packing fraction of sticky hard spheres, at three values of the sticky parameter τ. (B) Plot of τ versus η under which Eq. (7.124) has no real solution.

The first term on the right side of Eq. (7.123) corresponds to the compressibility factor for a hard-sphere fluid, and the second term accounts for the sticky attraction. The latter disappears when τ becomes large ($\lambda \to 0$).

Figure 7.22A shows the dependence of λ with the packing fraction of sticky hard spheres for 3 values of τ. When τ is greater than $\tau_c = (2 - \sqrt{2})/6$, we can calculate λ from Eq. (7.124) for any reasonable value of packing fraction η. In this case, λ decreases monotonically with the packing fraction and approaches 0 as $\tau \to \infty$. When $\tau = \tau_c$, $\lambda = 1/\tau_c$ for $\eta < \eta_c = (3\sqrt{2} - 4)/2$; and beyond this packing fraction λ declines monotonically with η. For $\tau < \tau_c$, there is a range of η within which Eq. (7.124) has no real value for λ.

The lack of a real solution from the PY theory is sometimes interpreted as that the system is unstable and undergoes a discontinuous vapor–liquid phase transition. Figure 7.22B shows the boundary in the $\tau - \eta$ plane that separates the conditions with a real value of λ (above the curve) and that without a real value of λ (beneath the curve). While the boundary of real solutions resembles the coexistence curve of VLE, the two densities at the same τ, which are solved from $(1 - \tau + \tau\eta^{-1})^2 = (1 + 2\eta^{-1})/6$, do not lead to an equal chemical potential or the same pressure as required by the equilibrium between two coexisting phases.

From Eq. (7.123) and the Gibbs–Duhem equation,[51]

$$\rho_b \left(\frac{\partial \mu}{\partial \rho_b} \right)_T = \left(\frac{\partial P}{\partial \rho_b} \right)_T, \tag{7.125}$$

we can find the chemical potential

$$\beta\mu = \ln(\rho_b \Lambda^3) - \ln(1 - \eta) + \frac{3\eta(4 - \eta)}{2(1 - \eta)^2} + \beta P v_0 + J \tag{7.126}$$

where Λ is the thermal wave length, $v_0 = \pi\sigma^3/6$ is the particle volume, and parameter J is determined from

51 Equation (7.125) follows from $d\mu = vdP$ at constant temperature where $v = 1/\rho$ is molar volume.

$$
\begin{aligned}
J = & \frac{3\lambda\eta\,(\lambda\eta - 8\eta - 2)}{2(1-\eta)^2} + \frac{6\eta(2+\eta)}{(1-\eta)^2} - \frac{18\tau\eta}{1-\eta} \\
& + \frac{\tau_c(3\tau\tau_c - 1)^2}{(1-\tau_c)} \ln\left|\frac{\lambda - 3\tau_c}{\tau^{-1} - 3\tau_c}\right| - \frac{(6\tau - \tau_c)^2}{\tau_c(1-\tau_c)} \ln\left|\frac{\lambda - 6\tau_c^{-1}}{\tau^{-1} - 6\tau_c^{-1}}\right|
\end{aligned}
\tag{7.127}
$$

As expected, the chemical potential reduces to that of an ideal gas in the limit of low density.

Based on the PY closure, thermodynamic properties derived from different routes are not identical. To illustrate the thermodynamic inconsistency, Figure 7.23 shows the reduced pressure ($\eta Z = \pi\beta P\sigma^3/6$) and the vapor-liquid phase diagram of the SHS model predicted from 6 different thermodynamic routes.[52] In comparison with MC simulation, the results from the chemical potential route are most accurate, and those from the ZS are the worst. It should be noted the chemical potential route depends on an arbitrary thermodynamic pathway connecting the SHS system to a reference without the sticky potential. The results from three chemical potential routes (μ_A, μ_B and μ_C) are also different. Because of the approximations in the PY closure, thermodynamic inconsistency is inevitable. Fortunately, because different routes lead to similar thermodynamic properties, the inconsistency has little consequence for practical applications when the model parameters are fitted with experimental data.

As mentioned above, the sticky-hard-sphere model provides a convenient representation of the thermodynamic properties and phase behavior of colloidal dispersions and protein solutions. Such properties are relevant for understanding the onset of crystallization and amorphous-aggregation processes. For example, the osmotic pressures of egg-white lysozyme solutions can be successfully represented with Eq. (7.123) at a variety of solution conditions. Figure 7.24 presents a comparison of theoretical results and experimental data.[53] Although the sticky-hard-sphere model does not describe the atomic details, it offers a convenient way to correlate experimental data and yields useful insights into the phase behavior of globular proteins.

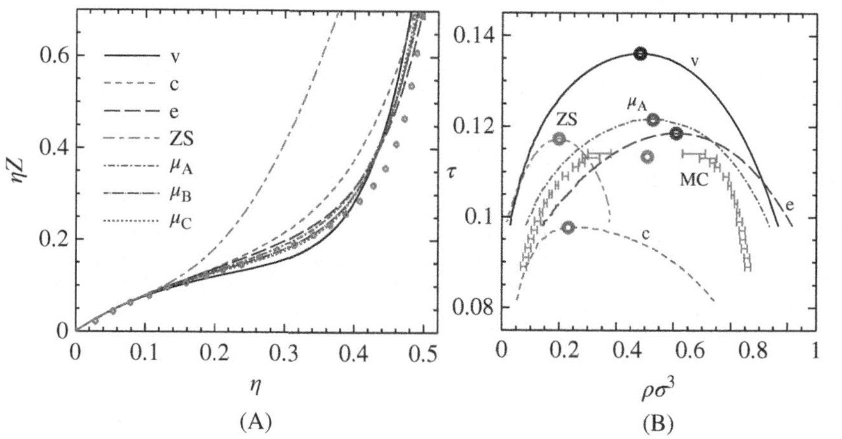

Figure 7.23 The reduced pressure ($\eta Z = \pi\beta P\sigma^3/6$) and coexistence curve (solid line) for the sticky hard-sphere model predicted from different thermodynamic routes (see the main text for the meaning of the legends). Source: Adapted from Rohrmann and Santos.[52]

52 Rohrmann R. D. and Santos A., "Equation of state of sticky-hard-sphere fluids in the chemical-potential route", *Phys. Rev. E* 89 (4), 042121 (2014).

53 Piazza R., "Interactions in protein solutions near crystallisation: a colloid physics approach", *J. Cryst. Growth* 196 (2–4), 415–423 (1999).

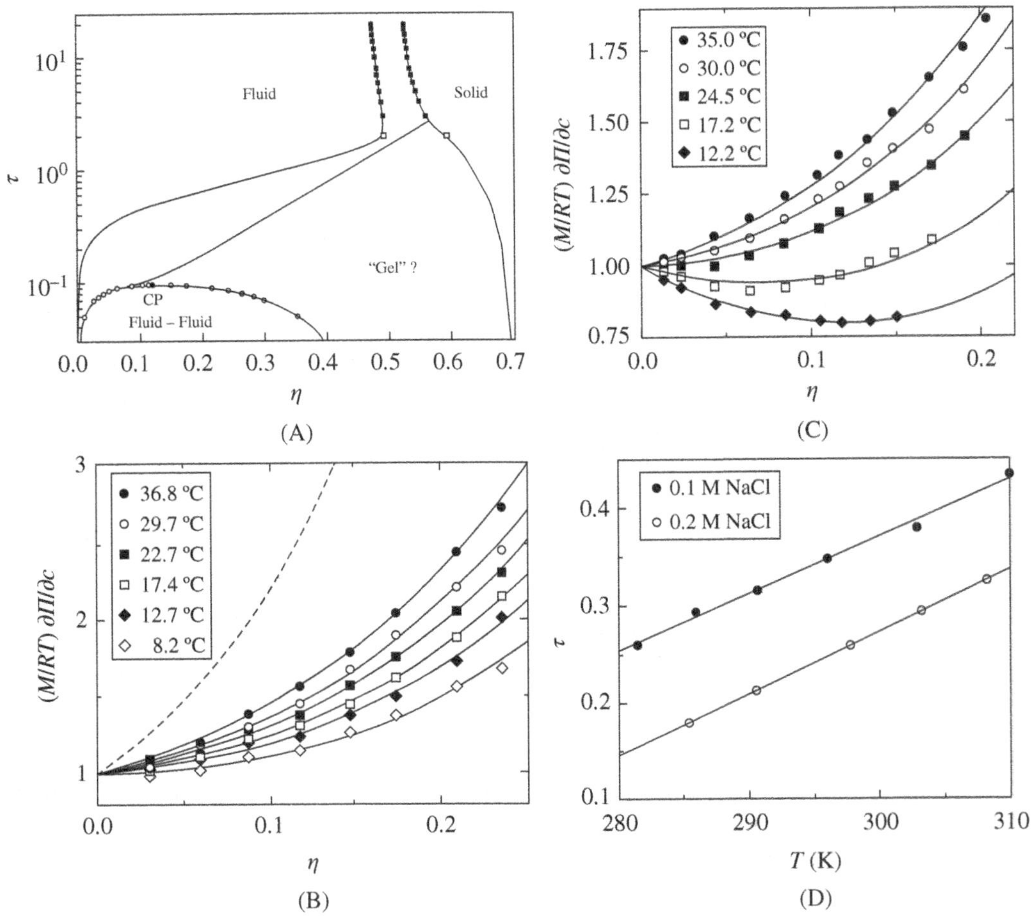

Figure 7.24 (A) Phase diagram for a system of sticky hard spheres calculated from the Baxter theory for the fluid phase and a generalized effective liquid-density theory for the solid phase. The solid squares are calculated results for the fluid–solid equilibrium, and open points represent the spinodal line. The line underneath the fluid-solid coexistence stands for the gelation boundary, i.e., the condition leading to formation of protein gels. (B) Reduced osmotic compressibility of hen egg-white lysozyme in NaAcO buffer + 0.1 M NaCl at pH = 3.7. Points refer to experimental data and the lines are fitted to the sticky hard-sphere model; (C) Same as (B) for samples containing 0.2 M NaCl; (D) temperature dependence of the sticky parameter obtained from the fits to the two sets of measurement shown in (B) and (C). In this plot, Π represents osmotic pressure, M is the protein molecular weight, and R is the gas constant, c and η are, respectively, the concentration and packing fraction of proteins. Source: Adapted from Piazza.[53]

7.6.2 Summary

The sticky hard-sphere (SHS) model may be understood as an extension of the hard-sphere model by including a surface adhesion mimicking attraction between classical particles. Because in a liquid solution the attraction between neutral particles is mostly dominated by short-range forces, the sticky-hard sphere model often provides a reasonable approximation for colloidal dispersions and aqueous solutions of globular proteins or nanoparticles.

7.7 The van der Waals Theory

Developed in the 1870s, the van der Waals theory of bulk fluids provides a remarkable success in the early applications of molecular thermodynamics.[54] Its main hypothesis, i.e., the repulsive and attractive components of intermolecular interactions play distinct roles in determining thermodynamic properties, has profound influences on later development of liquid-state theories. The van der Waals theory indicates that, in a liquid state, the excess entropy is mainly determined by repulsive interactions or excluded volume effects, while the attractive forces are responsible for the excess internal energy in comparison with that of an ideal gas at the same temperature and molecular density.

7.7.1 Mean-Field Potential

To elucidate the salient features of the van der Waals theory in the context of statistical thermodynamics, consider a canonical system containing N spherical particles at temperature T and volume V. As discussed in Section 7.1, the canonical partition function is given by

$$Q = \sum_{\upsilon} e^{-\beta E_{\upsilon}} = \frac{1}{N!\Lambda^{3N}} \int d\mathbf{r}^N e^{-\beta\Phi(\mathbf{r}^N)} \tag{7.128}$$

where $d\mathbf{r}^N = d\mathbf{r}_1 d\mathbf{r}_2 \cdots d\mathbf{r}_N$ represents an infinitesimal volume in a $3N$ dimensional space, $\mathbf{r}_i$ designates the position of particle i, Λ is the de Broglie thermal wave length, and $\Phi(\mathbf{r}^N)$ stands for the total potential energy.

In van der Waals's theory, the configurational integral in Eq. (7.128) is evaluated with the mean-field assumption, i.e., the interaction of any single particle i with all other particles in the system is approximated by an *average* potential $\varphi(\mathbf{r}_i)$ (viz., "mean-field" potential). The total potential energy of the system is thus given by

$$\Phi(\mathbf{r}^N) \approx \sum_{i=1}^{N} \varphi(\mathbf{r}_i). \tag{7.129}$$

Substituting Eq. (7.129) into (7.128) leads to

$$Q \approx \frac{1}{N!\Lambda^{3N}} \left\{ \int d\mathbf{r} \exp[-\beta\varphi(\mathbf{r})] \right\}^N. \tag{7.130}$$

In general, $\varphi(\mathbf{r})$ depends on the local properties of the system with a particle fixed at position $\mathbf{r}$. For a uniform system, the local properties are independent of the particle position.

In contrast to the exact result, the mean-field approximation reduces the multidimensional configurational integral to an integration with respect to the position of a single particle. As a result, the thermodynamic properties can be calculated analytically by considering only the *average* interaction of a single particle with all the other particles in the system. As shown schematically in Figure 7.25, the single-particle potential $\varphi(\mathbf{r})$ may be divided into two parts. The short-range repulsion accounts for the molecular size as represented by a hard sphere of diameter σ; and the van der Waals attraction accounts for the mutual influence of the interacting molecules at longer separations (see Supplementary Materials IV for intermolecular potentials). The repulsive part of

54 van der Waals J. D., On the continuity of the gaseous and liquid states, Universiteit Leiden (1873).

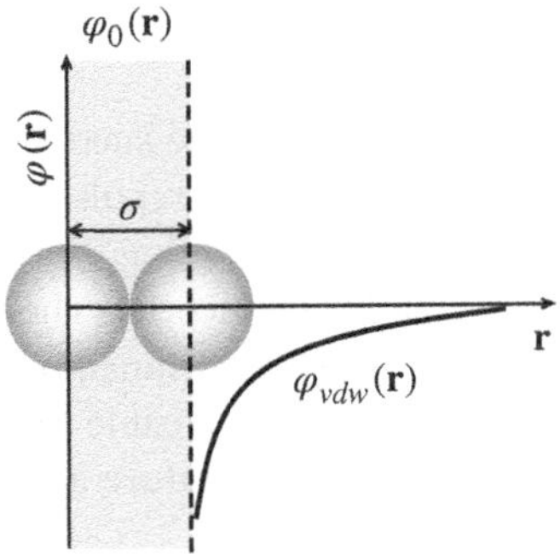

Figure 7.25 Schematic of the mean-field potential $\varphi(\mathbf{r})$ and its decomposition into excluded volume effect and long-range vdw attraction, $\varphi_0(\mathbf{r})$ and $\varphi_{vdw}(\mathbf{r})$.

the mean-field potential prevents the overlap of classical particles, i.e., any two particles cannot occupy the same position simultaneously. Meanwhile, the attractive potential is responsible for the vapor–liquid transition.

Corresponding to the two contributions of the mean-field potential, the partition function can be written as

$$Q \approx \frac{1}{N!\Lambda^{3N}}\left\{\int d\mathbf{r}\exp[-\beta\varphi_0(\mathbf{r}) - \beta\varphi_{vdw}(\mathbf{r})]\right\}^N = \frac{q_0^N q_{vdw}^N}{N!\Lambda^{3N}} \tag{7.131}$$

where q_0 and q_{vdw} are the integrations of the mean-field potential over different domains, i.e.,

$$q_0 = \int_0 d\mathbf{r}\exp[-\beta\varphi_0(\mathbf{r})], \tag{7.132}$$

$$q_{vdw} = \frac{\int_0 d\mathbf{r}\exp[-\beta\varphi_0(\mathbf{r}) - \beta\varphi_{vdw}(\mathbf{r})]}{\int_0 d\mathbf{r}\exp[-\beta\varphi_0(\mathbf{r})]} = \langle\exp[-\beta\varphi_{vdw}(\mathbf{r})]\rangle_0. \tag{7.133}$$

In above equations, subscripts 0 and *vdw* denote the space dictated by short-range and long-range interactions, respectively; and $<\cdots>_0$ stands for average over the short-range potential $\varphi_0(\mathbf{r})$.

7.7.2 Excluded Volume Approximation

We may evaluate the short-range component of the partition function using the excluded-volume approximation. As shown in Figure 7.26, the excluded volume of a spherical particle represents the space due to its occupancy not accessible to the center of other particles.[55] Around any given sphere, a spherical volume of diameter σ is inaccessible to the center of other spheres of the same diameter, which creates an inaccessible volume of $4\pi\sigma^3/3$. Because the interaction involves two particles, the excluded volume per particle is $2\pi\sigma^3/3$. For a system containing N spherical particles, the total excluded volume is $2\pi\sigma^3 N/3$. Therefore, we may estimate the short-range component of the partition function from the total volume accessible to each particle (viz., the free volume)

$$q_0 = \int_0 d\mathbf{r}\exp[-\beta\varphi_0(\mathbf{r})] \approx \int d\mathbf{r} - \int_{\text{excluded}} d\mathbf{r} = V - \frac{2\pi\sigma^3 N}{3}. \tag{7.134}$$

7.7.3 The Attractive Energy of a Tagged Particle

To calculate the mean-field potential due to van der Waals' attraction, $\varphi_{vdw}(\mathbf{r})$, consider an arbitrarily chosen particle (viz., a tagged particle) interacting with all the other particles in the system

55 The excluded volume of a molecule is different from its own volume; it varies with the size of the molecule with which it interacts.

Figure 7.26 The excluded volume due to interaction between two spherical particles of the same diameter σ is $4\pi\sigma^3/3$, shown schematically by the dashed circle. Because the exclusion involves two particles, the excluded volume for each particle is $2\pi\sigma^3/3$.

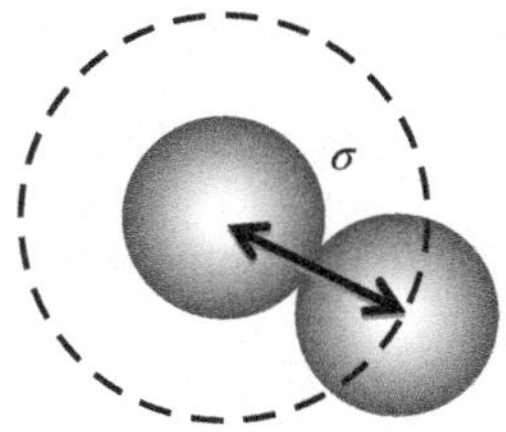

Figure 7.27 A spherical shell of thickness dr at distance r from a tagged particle at the center contains $4\pi r^2 \rho(r)dr$ particles, where $\rho(r)$ represents the average number density at position **r** with the tagged particle placed at the origin. The total potential experienced by the tagged particle is then given by Eq. (7.135).

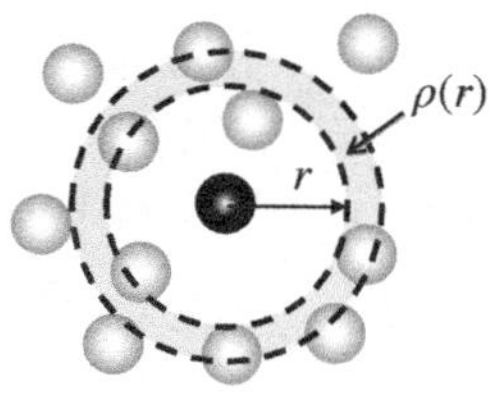

as shown in Figure 7.27. Let $u_A(r)$ be the pair potential between any two nonoverlapping particles separated by distance r. The potential energy contributed by the tagged particle is equal to the sum of all pair potentials

$$\varphi_{vdw} = \frac{1}{2}\int_\sigma^\infty dr 4\pi r^2 \rho(r) u_A(r) \tag{7.135}$$

where $\rho(r)$ represents the average density of particles at distance r, and ½ accounts for the fact that pair potential $u_A(r)$ involves two particles.

For a uniform system, $\varphi_{vdw}(\mathbf{r})$ is independent of the position of the tagged particle. With the further assumption that the local density is the same as the bulk density, $\rho(r) \approx \rho_b = N/V$, Eq. (7.135) becomes

$$\varphi_{vdw} = 2\pi\rho_b \int_\sigma^\infty dr r^2 u_A(r). \tag{7.136}$$

Substituting Eq. (7.136) into (7.133) gives the long-range component of the partition function

$$q_{vdw} = \langle \exp[-\beta\varphi_{vdw}(\mathbf{r})]\rangle_0 = e^{-\beta\varphi_{vdw}}. \tag{7.137}$$

7.7.4 Thermodynamic Properties

From Eqs. (7.131), (7.134), and (7.137), we obtain the partition function corresponding to the van der Waals theory

$$Q = \frac{(V - 2\pi\sigma^3 N/3)^N}{N!\Lambda^{3N}} \exp(-N\beta\varphi_{vdw}). \tag{7.138}$$

Accordingly, the reduced Helmholtz energy is given by

$$\beta F = -\ln Q = \beta F^{IG} + N\ln\left(\frac{V}{V - 2\pi\sigma^3 N/3}\right) + N\beta\varphi_{vdw} \tag{7.139}$$

where $\beta F^{IG} = N\ln(\rho_b\Lambda^3) - N$ is to the reduced Helmholtz energy for a system of noninteracting particles.

In the conventional notation, the excluded volume is expressed in terms of parameter

$$b \equiv 2\pi\sigma^3/3, \tag{7.140}$$

and the average attractive energy is expressed as $\varphi_{vdw} = -a\rho_b$ where

$$a \equiv -2\pi \int_{\sigma}^{\infty} dr r^2 u_A(r). \tag{7.141}$$

With this nomenclature, the reduced Helmholtz energy per particle is given by

$$\beta F/N = \ln \rho\Lambda^3 - 1 + \ln\left(\frac{V}{V - Nb}\right) - \frac{N\beta a}{V}. \tag{7.142}$$

From the Helmholtz energy, we obtain the van der Waals equation of state

$$P = -\left(\frac{\partial F}{\partial V}\right)_{T,N} = \frac{Nk_BT}{V - Nb} - \frac{N^2 a}{V^2} = \frac{k_BT}{\upsilon - b} - \frac{a}{\upsilon^2} \tag{7.143}$$

where $v = V/N = 1/\rho_b$ represents the per particle volume. The reduced chemical potential is given by

$$\beta\mu = \left(\frac{\partial \beta F}{\partial N}\right)_{T,V} = \ln\left(\frac{\rho_b\Lambda^3}{1 - b\rho_b}\right) + \frac{\rho_b b}{1 - \rho_b b} - 2\rho_b\beta a. \tag{7.144}$$

In the van der Waals equation of state, parameters a and b have intuitive physical meanings. The former is related to an overall attractive energy, i.e., the integration of the attractive potential over the entire space; and the latter represents the molecular excluded volume. In van der Waals's original work, both parameters are assumed independent of temperature. Accordingly, it predicts that the excess internal energy of the system, $U^{ex} \equiv U - U^{IG}(T, N, V)$, is attributed to the attractive energy

$$U^{ex} = \left(\frac{\partial \beta F^{ex}}{\partial \beta}\right)_{V,N} = -\frac{N^2 a}{V}. \tag{7.145}$$

Meanwhile, the excess entropy, $S^{ex} \equiv S - S^{IG}(T, N, V)$, determined by

$$S^{ex} = -\left(\frac{\partial F^{ex}}{\partial T}\right)_{V,N} = N \ln\left(\frac{V - Nb}{V}\right), \tag{7.146}$$

is dependent only on the excluded-volume effects. Although the long-range and short-range contributions to the internal energy and entropy are not entirely correct, similar ideas were adopted in later development of liquid-state theories (Section 7.11).[56]

7.7.5 Vapor–Liquid Transition

The van der Waals equation of state predicts that the pressure is a cubic function of the number density or molar volume. As shown in Figure 7.28, beyond the critical temperature $T_c = 8a/(27k_Bb)$, Eq. (7.143) is satisfied by only one density or molar volume for the entire range of pressure. But for $T < T_c$, it may yield three molar volumes in a certain range of the pressure, signaling a phase transition. At the saturation pressure, the smallest root corresponds to a saturated liquid, the largest root corresponds to the coexisting vapor, and the middle root has no physical significance.

To calculate the coexistence densities at a fixed temperature, the pressure and chemical potential of the vapor phase (superscript V) must be the same as those for the liquid phase (superscript L)

$$P^V = P^L, \tag{7.147}$$

$$\mu^V = \mu^L. \tag{7.148}$$

56 Chandler D., Weeks J. D., and Andersen H. C., "Van der Waals picture of liquids, solids, and phase transformations", *Science* 220, 787–794 (1983).

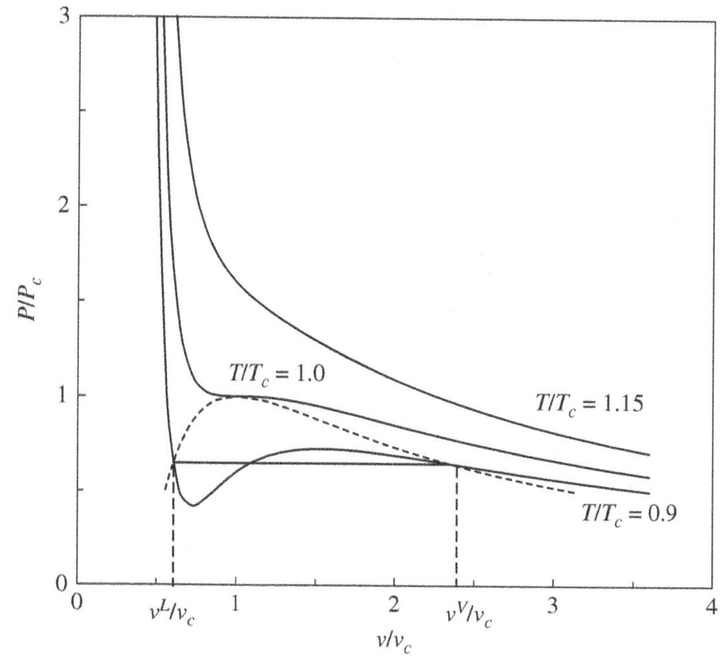

Figure 7.28 Isothermal lines and the coexistence curve predicted by the van der Waals theory. Each solid line represents an isotherm; the dotted line shows the vapor–liquid coexistence curve, and the horizontal dashed line shows the coexistence between a liquid and a vapor. Here subscript "c" denotes the critical point.

When the system temperature is below the critical temperature T_c, the solution of Eqs. (7.147) and (7.148), along with an equation of state, yields the saturation pressure as well as the coexistent vapor and liquid densities (or molecules per unit volume). The phase transition of a one-component fluid is characterized by a density difference in the coexisting phases. As temperature rises, this difference declines continuously until it vanishes at the critical temperature T_c.

Figure 7.28 shows the vapor–liquid coexistence curve and isothermal lines predicted by the van der Waals equation of state. For any fixed temperature higher than the critical temperature T_c, the van der Waals equation of state predicts that the pressure is a monotonically decreasing function of density. Below the critical temperature, it gives a negative isothermal compressibility, $\kappa = -\frac{1}{V}\left(\frac{\partial V}{\partial P}\right)_T$, for a range of densities that are thermodynamically unstable. The isotherm inside the vapor–liquid coexistence curve is known as the van der Waals loop. Qualitatively, the phase diagram predicted by the van der Waals equation of state captures the essential features of the vapor-liquid phase behavior of simple fluids.

7.7.6 Principle of Corresponding States

At the critical point, the three roots of the van der Waals equation of state become identical. In this case, the isothermal line exhibits a point of inflection

$$\left(\frac{\partial P}{\partial V}\right)_{N,T_c} = \left(\frac{\partial^2 P}{\partial V^2}\right)_{N,T_c} = 0. \tag{7.149}$$

Based on Eq. (7.149), we can express constants a and b in terms of critical temperature T_c and critical pressure P_c

$$a = \frac{27(k_B T_c)^2}{64P_c}, \tag{7.150}$$

$$b = \frac{k_B T_c}{8P_c}. \tag{7.151}$$

Accordingly, the critical volume per molecule v_c is given by

$$v_c = \frac{Z_c k_B T_c}{P_c}. \tag{7.152}$$

The van der Waals theory predicts that, in dimensionless units of $T_r \equiv T/T_c$, $P_r \equiv P/P_c$, and $v_r \equiv v/v_c$, the equation of state satisfies a universal relation

$$P_r = \frac{8T_r}{3v_r - 1} - \frac{3}{v_r^2}. \tag{7.153}$$

The van der Waals equation predicts that, at the critical point, the compressibility factor is Z_c=3/8, which is substantially higher than the experimental results for simple fluids, $Z_c \approx 0.27 - 0.29$. Therefore, if the van der Waals theory is used to calculate v_c with experimental P_c and T_c as the input, the result would be too large. Although Eq. (7.153) is not accurate in comparison with experimental observations, nearly universal relations can be established among reduced pressure, reduced temperature, and reduced volume for a large number of simple fluids. In classical thermodynamics, such quasi-universal relations are known as the principle of corresponding states.

7.7.7 An Improved Model for the Excluded Volume Effects

Whereas the van der Waals equation of state captures the essential features of vapor–liquid equilibria, its performance is not satisfactory from a quantitative perspective. In particular, the van der Waals theory predicts too small a value for the liquid density. The poor performance is partially related to the inadequate description of the molecular excluded volume. The repulsive part of the van der Waals theory requires that $v > b = 2\pi\sigma^3/3$, implying that the packing fraction η

$$\eta \equiv \pi\rho_b\sigma^3/6 < 1/4. \tag{7.154}$$

This upper limit is well below that of a typical liquid near the triple point.

To avoid the unreasonable small value of the packing fraction, we may replace the repulsive terms in Eqs. (7.143) and (7.144) with those from the CS equation of state for hard spheres (Section 7.5)

$$\frac{P}{\rho_b k_B T} = \frac{1 + \eta + \eta^2 - \eta^3}{(1 - \eta)^3} - \frac{a\rho_b}{k_B T}, \tag{7.155}$$

$$\beta\mu = \ln(\rho_b\Lambda^3) + \frac{\eta(8 - 9\eta + 3\eta^2)}{(1 - \eta)^3} - 2\beta\rho_b a. \tag{7.156}$$

While the van der Waals theory assumes that the excluded volume is additive, the CS equation accounts for the correlation effects nearly quantitatively within the hard-sphere model. The modified van der Waals equation of state [viz., Eq. (7.155)] also predicts a universal packing fraction and a universal reduced temperature at the critical point

$$\eta_c \approx 0.130444, \tag{7.157}$$

$$a/(k_B T_c \sigma^3) \approx 5.55079. \tag{7.158}$$

According to MC simulation for the LJ fluids, the critical packing fraction is $\eta_c^{LJ} = 0.165$, and the critical temperature is $k_B T/(4\varepsilon^{LJ}) = 0.328$. These numbers are in reasonable agreement with the predictions of the modified van der Waals theory.[57]

7.7.8 Cubic Equations of State

The van der Waals theory and its numerous modifications predict that, when temperature is expressed as T/T_c and density as ρ_b/ρ_c or $\rho\sigma^3$, the vapor–liquid coexistence curve is universally applicable to many simple fluids, independent of the details of intermolecular interactions. As discussed above, the generalized phase diagram provides a theoretical foundation for the principle of corresponding states that is useful for early applications of liquid-state theories.

Many variations of the van der Waals equation of state have appeared in the literature.[58] For applications to VLE calculations, agreement with experiment can be much improved when parameter a is treated as a function of temperature

$$a = \alpha(T)\frac{(k_B T_c)^2}{P_c} \tag{7.159}$$

where $\alpha(T)$ is determined by fitting experimental data for the saturation pressure and liquid-density of pure species. At given temperature and pressure, the van der Waals theory yields a cubic equation for the molar volume or compressibility factor. Accordingly, its empirical modifications are conventionally known as the cubic equations of state.

One great success of the cubic equations of state is that they are directly applicable to fluid mixtures. It is customary to express parameters a and b as functions of composition. The usual mixing rules, also known as van der Waals mixing rules, are

$$a = \sum_i \sum_j x_i x_j a_{ij} \tag{7.160}$$

$$b = \sum_i x_i b_i \tag{7.161}$$

where x_i represents the mole fraction of component i. The cross-coefficient is often approximated by

$$a_{ij} = (1 - k_{ij})\sqrt{a_i a_j} \tag{7.162}$$

where k_{ij} is an empirical binary constant that is typically much less than unity. However, the calculated results for vapor–liquid equilibria are often sensitive to k_{ij}. For typical mixtures of nonpolar molecules, k_{ij} is of the order 0.02–0.2. Although some binary experimental data are required to determine k_{ij}, the great advantage of the mixing rules is that experimental data for pure components and binary mixtures can be "scaled up" to predict VLE for multi-component systems.

7.7.9 Summary

The essential idea of the van der Waals theory is to factor the partition function into two parts according to the repulsive and attractive components of the intermolecular potential. This idea leads to later development of perturbation theories wherein, as discussed later in this chapter (Section 7.11), the structural and thermodynamic properties of a classical system are represented by those for a suitable reference (typically hard spheres) augmented by perturbations that arise from intermolecular attraction.

---

57 The reduced critical temperature according to Eq. (7.158) is about 0.375.

58 See, e.g., Orbey H., and Sandler S., *Modeling vapor–liquid equilibria: cubic equations of state and their mixing rules*. Cambridge University Press, 1998.

In the van der Waals theory, both molecular excluded volume effects and attractive forces are calculated with mean-field assumptions that ignore the correlation effects. Although these assumptions are not accurate, the van der Waals theory predicts the critical point of vapor–liquid transition and phase diagrams of simple fluids in qualitative agreement with experimental observations. Variations on the van der Waals theory have been extensively used in the chemical and petrochemical industries. These semi-empirical equations provide useful correlations and predictions of fluid-phase equilibria used in the design of chemical separation and purification processes, especially those using distillation and liquid absorption.

7.8 The Cell Model for Colloidal Crystals

A crystalline solid is distinguished from a fluid or an amorphous solid (such as a glass or a rubbery polymer) by long-range order with periodicity. The translational motions of classical particles are prohibited in a solid phase because they are confined to a rigid, three-dimensional lattice. With rare exceptions, solid particles do not move from one lattice position to another. Each particle is confined by an effective potential due to the presence of neighboring particles. As shown schematically in Figure 7.29, this potential exhibits a minimum when each particle is at its equilibrium position and rises sharply when it approaches the position of neighboring sites.

For a crystalline solid consisting of classical particles, the crystal structure depends on the inter-particle potential as well as the thermodynamic condition such as pressure and temperature. In Section 4.6, we have discussed statistical-mechanical models to predict the heat capacities of atomic crystals by considering the particle vibrations (viz., phonons). In this section, we introduce a cell model to predict the thermodynamic properties and phase behavior of colloidal crystals. From a practical perspective, the study of colloidal crystals is important for their diverse applications including the use of colloidal crystals as electronic and optical materials, ceramics, catalysts, membranes, and chemical and biochemical sensors.[59]

7.8.1 The Lennard-Jones and Devonshire (LJD) Theory

Historically, the cell model was proposed by Lennard-Jones and Devonshire to describe the thermodynamic properties of fluids.[60] In its application to a crystalline solid of volume V with N identical

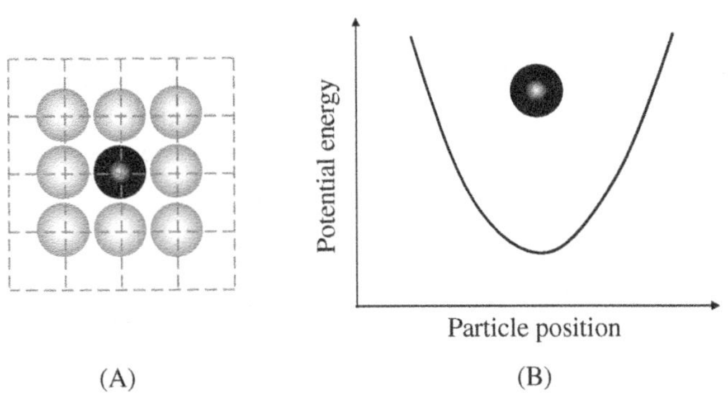

Figure 7.29 Schematic of a particle on a crystalline lattice confined by its neighboring particles (A) and the potential energy confining the particle to its lattice position (B).

59 Li B., Zhou D. and Han Y., "Assembly and phase transitions of colloidal crystals", *Nat. Rev. Mater.* 1, 15011 (2016).
60 Lennard-Jones J. E. and Devonshire A. F., "Critical phenomena in gases. I", *Proc. Roy. Soc. A* 163, 53 (1937).

particles, one key assumption is that the solid can be divided into N independent subsystems (viz. cells) such that each subsystem contains one (and only one) particle. Within each cell, the particle experiences an effective (mean-field) potential due to its interaction with the neighboring particles. As the subsystems are assumed independent, we can write the canonical partition function in terms of those corresponding to individual cells

$$Q = \frac{1}{\Lambda^{3N}} \int d\mathbf{r}^N e^{-\beta\Phi(\mathbf{r}^N)} \approx \frac{1}{\Lambda^{3N}} \left[\int d\mathbf{r} e^{-\beta\varphi(\mathbf{r})} \right]^N \tag{7.163}$$

where $\mathbf{r}$ is the particle position relative to the center of each cell. Other variables in Eq. (7.163) have their usual meanings, i.e., Λ for thermal wavelength, $\beta = 1/(k_B T)$, Φ for the total intermolecular potential, and $\varphi(\mathbf{r})$ for the mean-field potential on particle i in a subsystem of volume $v = V/N$. Unlike that for a fluid, $N!$ is absent in Eq. (7.163) because, in a crystal, the particles are distinguishable by their positions.[61]

Similar to that in the van der Waals theory, we may divide the mean-field potential $\varphi(\mathbf{r})$ into an excluded-volume term represented by hard-sphere interactions, and an effective attraction due to longer-ranged interactions. Because the solid particles are confined to the lattice sites, the volume accessible to each particle (viz., the free volume per particle) v_f is determined by the cell boundary.

Although a precise evaluation of the free volume from the lattice structure is rather complicated, we may simplify the calculation with the assumption that each particle is confined in a spherical cavity. As shown in Figure 7.30, the cavity size can be determined from the particle size and the shortest distance between neighboring sites. Specifically, the cell volume v is proportional to the cube of the center-to-center distance between the nearest-neighbor cells, λ, i.e.,

$$v \sim \lambda^3. \tag{7.164}$$

To fix the proportionality constant, we may apply Eq. (7.164) in the limit of close packing. In that case, the nearest-neighbor distance is the same as the hard-sphere diameter, $\lambda = \sigma$; and the cell volume, here designated as v_i^c, can be calculated from the close-packing density ρ_c.

To elucidate how we estimate the cell volume, consider a FCC lattice as shown schematically in Figure 7.31. The close-packing density for uniform hard spheres on an FCC lattice is $\rho_c\sigma^3 = \sqrt{2}$, which corresponds to a cell volume of $v_c = \sigma^3/\sqrt{2}$. With the proportionality constant determined by the condition of close packing, Eq. (7.164) becomes

$$v/v_c = \lambda^3/\sigma^3. \tag{7.165}$$

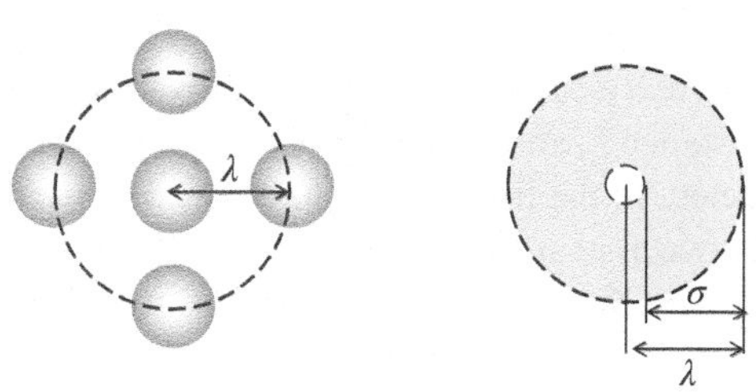

Figure 7.30 Free volume in a crystal lattice refers to the volume accessible to the center of the particle (left). The picture on the right shows the free volume for a spherical particle of diameter σ confined within a spherical cell of radius λ; this volume is $4\pi(\lambda - \sigma)^3/3$.

61 As discussed in Section 6.9, the entropy related to $N!$ is referred to as the "communal entropy." In a crystalline solid, each particle is confined to its cell; however, in a fluid, each particle can move over large distances in comparison to its diameter. In other words, particles in a fluid, unlike those in a crystal, have the translational degree of freedom. At the same temperature and density, the entropy of a fluid is often larger than that of a solid.

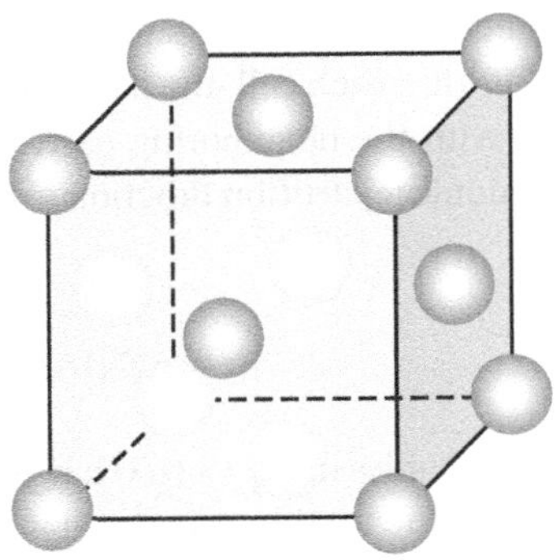

Figure 7.31 Particle arrangement on a face-centered-cubic (FCC) lattice.

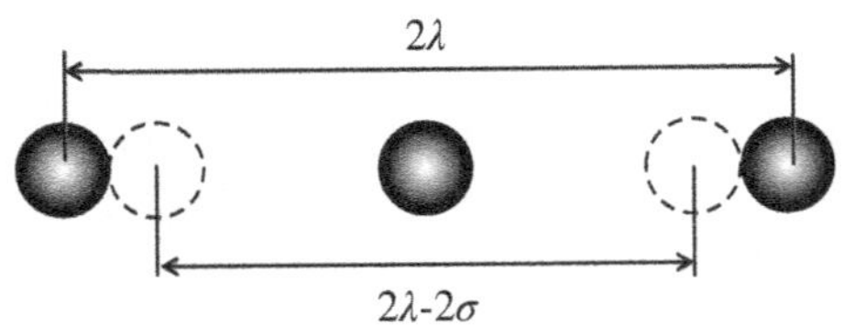

Figure 7.32 The free displacement length, $2(\lambda - \sigma)$, represents the maximum distance that the center of a particle (diameter σ) can move along the direction of its nearest neighbors.

Similar to the cell volume, the free volume is proportional to the cube of the free displacement length $2(\lambda - \sigma)$ (see Figure 7.32)

$$v_f \sim 8(\lambda - \sigma)^3. \tag{7.166}$$

The proportionality constant in Eq. (7.166) can be fixed by a comparison of Eqs. (7.165) and (7.166)

$$v_f = \frac{8v_c}{\sigma^3}(\lambda - \sigma)^3 = \frac{8\left[1 - (\rho_b/\rho_c)^{1/3}\right]^3}{\rho_b}. \tag{7.167}$$

In writing Eq. (7.167), we have assumed that the geometry of the free volume resembles that of the cell volume, and used relations $v = 1/\rho_b$ and $\sigma/\lambda = (\rho_b/\rho_c)^{1/3}$, where $\rho_b = N/V$ represents the number density of solid particles.

Now consider the contribution of the attractive interactions between particles to the mean-field potential. For simplicity, we assume that each particle interacts only with its immediate neighbors. Assuming further that each particle has n_I immediate neighbors and that, on average, the center-to-center distance is L, we obtain the mean-field attractive potential for each particle

$$\varphi_{MF} = n_I\, u_A(L)/2 \tag{7.168}$$

where $u_A(L)$ stands for the attractive part of the potential between two particles at center-to-center distance L, and the mean-potential is divided by 2 because each pair interaction involves two particles. To the first-order approximation, we may assume $L \approx \lambda$, i.e., the center-to-center separation between nearest-neighbor cells is the same as that between nearest-neighbor particles.

With the free volume and mean attractive potential per particle given by Eqs. (7.167) and (7.168), respectively, we can now evaluate the canonical partition function of the solid phase. According to Eq. (7.163), the mean-field assumption leads to

$$Q = \frac{v_f^N}{\Lambda^{3N}} \exp[-\beta N n_I\, u_A(L)/2]. \tag{7.169}$$

Except for the absence of $N!$ term, Eq. (7.169) is identical to that given by the van der Waals theory.

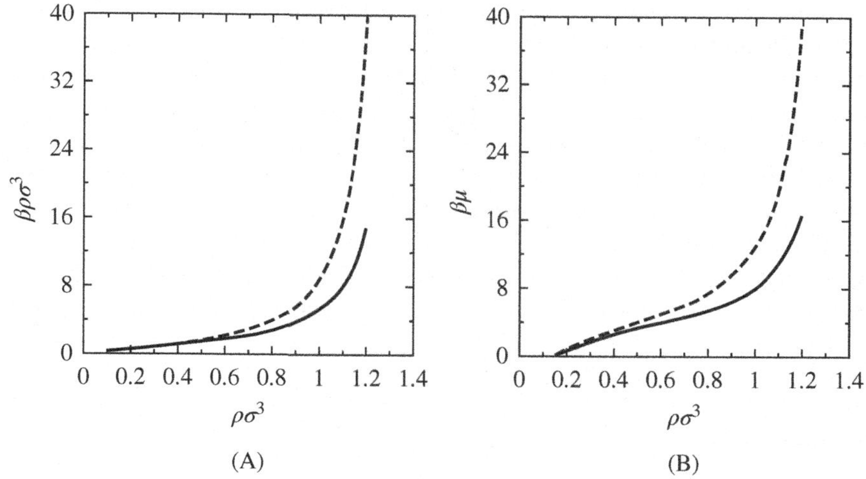

Figure 7.33 Reduced pressure (A) and reduced chemical potential (B) for a model crystal of spherical particles interacting through the Sutherland potential with $n = 7$ and $k_BT/\varepsilon = 1.5$. BCC, body-centered-cubic; FCC, face-centered-cubic.

From the canonical partition function, we obtain the Helmholtz energy of the crystal

$$F/(Nk_BT) = \ln(\Lambda^3/v_f) + n_I\,\beta u_A(L)/2. \tag{7.170}$$

Subsequently, the equation of state and chemical potential can be derived by following standard thermodynamic relations

$$P = \frac{\rho_b k_BT}{1-(\rho_b/\rho_c)^{1/3}} - \frac{n_I\,\rho_b L u_A'(L)}{6}, \tag{7.171}$$

$$\mu = k_BT\ln\left\{\frac{\rho_b\Lambda^3}{8\left[1-(\rho_b/\rho_c)^{1/3}\right]^3}\right\} + \frac{k_BT}{1-(\rho_b/\rho_c)^{1/3}} + \frac{n_I u_A(L)}{2} - \frac{n_I\,L u_A'(L)}{6} \tag{7.172}$$

where $u_A'(r) = du_A(r)/dr$ stands for the derivative of the attractive part of the pair potential with respect to the inter-particle separation.

Figure 7.33 shows the reduced pressure and the reduced chemical potential predicted by the cell model for a system of spherical particles in the FCC and BCC structures. For a FCC solid, $\rho_c\sigma^3 = \sqrt{2}$ and $n_I = 12$, and for BCC, $\rho_c\sigma^3 = 3\sqrt{3}/4$ and $n_I = 8$. In generating the numerical results, we use the Sutherland potential for the particle–particle interactions

$$u(r) = \begin{cases} \infty & r < \sigma \\ -\varepsilon\left(\frac{\sigma}{r}\right)^n & r \geq \sigma \end{cases} \tag{7.173}$$

with $n = 7$, and $k_BT/\varepsilon = 1.5$. At the same temperature particle density, the BCC structure always has a higher pressure and higher chemical potential, suggesting that the FCC structure is more stable at all conditions.

7.8.2 Summary

One key difference between the original Lennard-Jones –Devonshire theory and the cell model for colloidal crystals lies in the communal entropy, i.e., the entropy related to the particle distinguishability. As discussed in Section 6.9, the communal entropy plays an important role in understanding the thermodynamic properties of solid phases including high-entropy alloys. Besides, the free volume for each particle is different by a factor of 8, which accounts for the absence of cell walls in real crystals. We will discuss in Section 7.9 that the larger free volume allows us to predict fluid–solid transition of hard spheres in good agreement with simulation results. In addition, we will discuss in Section 7.10 that the thermodynamic properties of colloidal dispersions are helpful for understanding the self-assembly of nanoparticles and globular proteins or other macromolecules to form useful crystalline structures such as photonic crystals. The phase behavior of colloidal crystals is also related to protein crystallization. Formation of protein crystals is an essential step in determining their three-dimensional structures through techniques such as X-ray diffraction. The crystalline structures of proteins provide fundamental information for advancements in biotechnology including rational drug design.

7.9 Order Through Entropy

A few hundred years ago, packing of spherical particles was of interest for efficient loading of cannon balls in a warship. Today material chemists utilize hard-sphere-like nanoparticles for fabrication of nanostructured materials. From a fundamental perspective, the search for optimum packing, i.e., packing of hard spheres in a minimum volume, is equivalent to the identification of a stable structure that gives maximum entropy, or more precisely, minimum Gibbs energy.[62] Therefore, the fluid–solid transition of hard spheres implies that entropy may serve as a driving force for the self-assembly of ordered structures.[63]

In this section, we discuss fluid–solid transition of hard spheres and its implications on order through entropy. The fluid–solid transition of hard spheres was discovered by Alder and Wainwright in 1957 using computer simulation.[64] The theoretical prediction was confirmed experimentally almost 40 years later with colloidal suspensions.[65] Prior to 1957, it was a common belief that a fluid can condense to form a solid only if particles in the fluid phase have forces of attraction. However, both molecular simulation and experiment have now demonstrated that this common belief is actually incorrect.

7.9.1 Hard-Sphere Crystal

Formation of a hard-sphere crystal is by no means intuitive because, by definition, hard spheres have no forces of attraction. If a crystalline structure exists, what is the driving force of disorder-to-order transition?

While geometric consideration is useful for solving the packing problem, thermodynamic equations are required for identifying fluid–solid transition in a hard-sphere system. To locate the condition of fluid–solid equilibrium, we need to know the pressure and chemical potential of the coexisting fluid and solid phases. As discussed in Section 7.5, the CS equation provides an accurate description of the pressure and chemical potential of a hard-sphere fluid

62 The second law of thermodynamic requires that at equilibrium, an isolated system (constant volume and energy) has maximum entropy, or equivalently, (at constant temperature and pressure), a minimum Gibbs energy.
63 Frenkel D., "Order through entropy", *Nat. Mater.* 14, 9–12 (2015).
64 Alder B. J. and Wainwright T. E., "Phase transition for a hard sphere system", *J. Chem. Phys.* 27 (5), 1208–1209 (1957).
65 Zhu J. et al. "Crystallization of hard-sphere colloids in microgravity", *Nature* 387, 883–885 (1997).

$$\frac{P_F}{\rho_b k_B T} = \frac{1+\eta+\eta^2-\eta^3}{(1-\eta)^3}, \tag{7.174}$$

$$\frac{\mu_F}{k_B T} = \ln(\rho_b \Lambda^3) + \eta\frac{8-9\eta+3\eta^2}{(1-\eta)^3} \tag{7.175}$$

where subscript F stands for the fluid state, and $\eta = \pi\rho_b\sigma^3/6$ denotes the packing fraction. Meanwhile, the cell model predicts the equation of state and chemical potential for a hard-sphere crystal (Section 7.8)

$$\frac{P_S}{\rho_b kT} = \frac{1}{1-(\rho_b/\rho_c)^{1/3}}, \tag{7.176}$$

$$\frac{\mu_S}{kT} = \ln(\rho_b\Lambda^3) + \frac{1}{1-(\rho_b/\rho_c)^{1/3}} - 3\ln\left[1-(\rho_b/\rho_c)^{1/3}\right] - \ln 8 \tag{7.177}$$

where subscript S stands for the solid phase. The fluid–solid transition can be identified using the standard phase-equilibrium equations:

$$P_F = P_S, \tag{7.178}$$

$$\mu_F = \mu_S. \tag{7.179}$$

With the assumption that the arrangement of hard spheres in FCC is most stable,[66] we can find from Eqs. (7.178) and (7.179) the freezing and melting packing fractions.

Figure 7.34 shows the reduced pressure of a hard-sphere fluid and that of a hard-sphere FCC crystal predicted by Eqs. (7.174) and (7.176), respectively. At equilibrium, the packing factions of the fluid and solid phases are equal to 0.492 and 0.555, respectively, which are in good agreement with molecular-simulation results (0.494 and 0.545). Because the thermal wavelength Λ does not appear in the pressure equations, and because it cancels in equating the chemical potentials of the fluid and solid phases, the fluid–solid transition of hard spheres depends only on the particle density, not on temperature. As a result, the phase diagram of a hard-sphere system can be effectively represented in one dimension, as shown in Figure 7.35.

Figure 7.34 Reduced pressure $[P\sigma^3/(k_BT)]$ for a hard-sphere fluid and a hard-sphere solid as a function of packing fraction η. The dashed line represents fluid–solid coexistence.

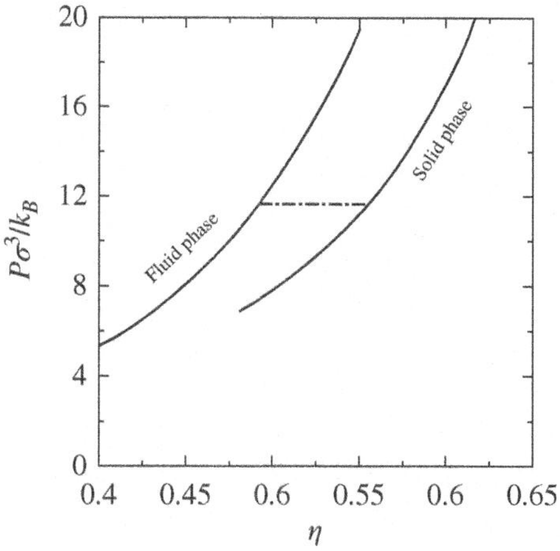

66 In pure mathematics, the assumption is known as the Kepler conjecture, proposed by the astronomer Johannes Kepler in 1611. The problem remains unsolved today.

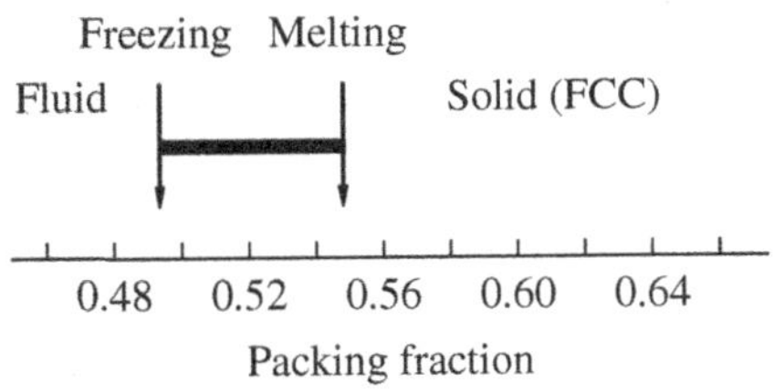

Figure 7.35 Phase diagram for a one-component system of hard spheres. FCC stands for face-centered-cubic structure.

7.9.2 Entropy of Hard-Sphere Systems

For a one-component hard-sphere system, the phase behavior is controlled by just one variable, *viz.*, the particle volume fraction. As the volume fraction increases, hard spheres in the fluid phase become increasingly more crowded. Paradoxically, the transition from the random structure (i.e., fluid phase) to the ordered phase (viz., FCC crystal) is exclusively driven by entropy. In other words, at the same particle density, the ordered state has a higher entropy than the disordered state. By ordering (i.e., crystallizing), the hard spheres gain more entropy.

While crystallization of hard spheres appears counter intuitive, the phenomenon of entropy-driven ordering is actually commonplace, not much different from parking cars in a busy lot. When the lot is nearly occupied, the ordered arrangement of cars in marked spaces yields more mobility (viz., free space for each car) than random parking. In other words, the ordered state has a higher entropy. For a hard-sphere system above the melting concentration (volume fraction 0.545), the entropy loss due to the lack of translational degree of freedom in the ordered structure is more than compensated by the increase in the free volume of individual particles.

The thermodynamic models discussed above allow us to quantify the entropy difference between the fluid and solid phases. According to the CS equation of state, the reduced entropy of a hard-sphere fluid is given by

$$\frac{S_F}{Nk_B} = \frac{5}{2} - \ln(\rho_b \Lambda^3) - \frac{4\eta - 3\eta^2}{(1-\eta)^2}. \tag{7.180}$$

For a hard-sphere solid, the cell model predicts that the reduced entropy per particle is

$$\frac{S_S}{Nk_B} = \frac{3}{2} - \ln(\rho_b \Lambda^3) + 3\ln\left[1 - (\eta/\eta_c)^{1/3}\right] + \ln 8. \tag{7.181}$$

Figure 7.36 shows the reduced entropy per particle according to Eqs. (7.180) and (7.181). For illustration, the thermal wavelength is set to $\Lambda = \sigma$; the relative results are independent of the specific choice of Λ. At low packing fraction ($\eta < 0.52$), the entropy per particle in the fluid phase is higher than that in the solid phase, but the opposite holds when $\eta > 0.52$. Figure 7.36 indicates that at sufficiently high packing fraction, the entropy of hard spheres in the ordered (solid) state is higher than that in the disordered (fluid) state.[67]

It should be noted that, at the condition of fluid–solid coexistence, the molar entropy of the fluid phase is −2.25 J/mol K, and that for the solid phase is −3.65 J/mol K. The solid phase has a lower entropy because the phase transition occurs at constant T and P and it is accompanied by a negative change in molar volume $\Delta v = v^s - v^l < 0$. In this case, the entropy gained by the rearrangement of the particles is smaller than the entropy lost due to the decrease in molar volume.

67 Here the fluid and solid are compared at the same particle number density, *N/V*. The fluid is unstable when its entropy is lower than that of the solid.

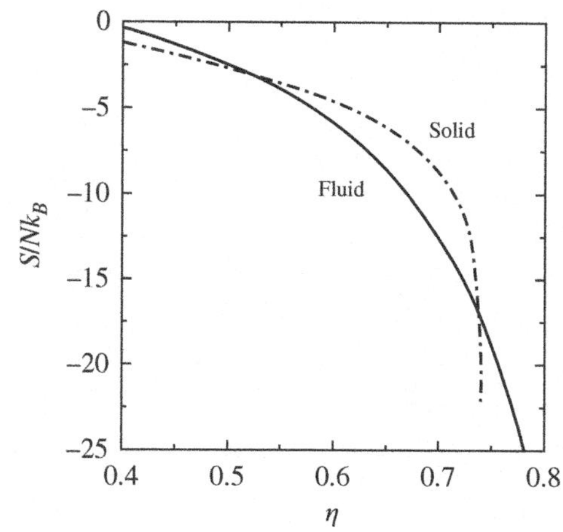

Figure 7.36 Reduced entropy (S/Nk_B) of hard spheres in the fluid state (solid line) and in the solid state (dashed line) as a function of packing fraction. At low and high packing fractions, the entropy of the fluid is larger than that of the solid but at intermediate packing fractions, contrary to intuition, the entropy of the solid is higher than that of the fluid.

7.9.3 Summary

For a hard-sphere system, energy is irrelevant to phase transition because, by definition, there is no interaction between hard spheres except that they cannot overlap. In other words, the thermodynamic properties of a hard-sphere system are athermal (viz., independent of temperature if they are expressed in proper units). In particular, the hard-sphere entropy is dependent on the free volume accessible to individual particles. In applications of the hard-sphere model to more realistic systems, we consider in the next section (Section 7.10) fluid–solid equilibrium in the presence of attraction between particles as experienced in typical colloidal dispersions or aqueous solutions of globular proteins. Because of the attractive forces, colloidal systems may exhibit both fluid–fluid and fluid–solid equilibria similar to vapor–liquid–solid equilibria in a simple fluid.

7.10 Colloidal Phase Diagrams and Protein Crystallization

Phase diagrams of colloidal dispersions and protein solutions are useful for various industrial and medical applications including the design of protein-separation processes and prevention of human diseases such as cataracts. A statistical-thermodynamic model will be helpful to manipulate particle self-assembly with desired microscopic structures.

7.10.1 Protein Crystallization

In response to changes in the environment, a protein solution may undergo various forms of phase transition including liquid–liquid demixing, crystallization, and gelation. The mechanisms of protein phase transition are complex and depend on parameters such as the protein concentration, buffer composition, and temperature. The phase transition affects the biological activity of proteins. On the one hand, protein aggregation may impact storage stability and safety of protein therapeutics in pharmaceutical industry thus impeding drug development. On the other hand, the inability to produce high-quality protein crystals is a major hurdle in determining the 3D structure of proteins by X-ray diffraction.

Schematically, Figure 7.37 shows the phase diagram of a typical protein solution observed in experiments or predicted by statistical-thermodynamic models. Here, the horizontal axis represents protein concentration, and the vertical axis represents the reduced second virial coefficient, that is, the osmotic second virial coefficient divided by the protein's molecular volume. The

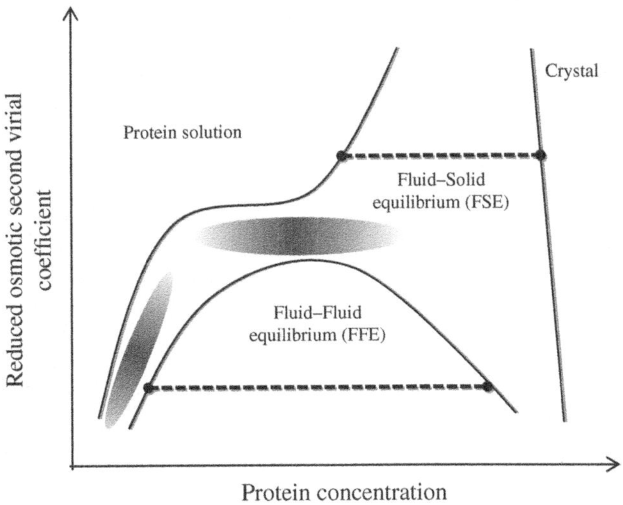

Figure 7.37 Schematic phase diagram for aqueous solutions of globular proteins. The shaded areas indicate conditions favorable for crystallization.

dimensionless quantity provides a measure of the "net" attraction between proteins relative to the short-range repulsion. In terms of the reduced second virial coefficient and the reduced number density, the phase diagram resembles the corresponding-state principle of simple fluids, a concept similarly applicable to colloidal dispersions and protein solutions.[68] The fluid–fluid coexistence curve, shown as the dashed line, exists beneath the freezing and melting lines (represented by the solid lines) and, therefore, is metastable. The liquid–liquid phase separation (LLPS) resembles that of VLE in a simple fluid such as argon but, unlike argon, it exists underneath the fluid–solid coexistence lines because the range of attraction between protein molecules is shorter ranged in comparison to that between small molecules. On the left side of the freezing line, a protein solution exists as a stable liquid phase, and on the right side of the melting line, the protein solution becomes a colloidal crystal; in between, the system is unstable or metastable.

With only thermodynamic considerations, a protein solution would crystallize once its concentration exceeds the freezing point. However, due to the slow kinetics for crystal formation, a supersaturated protein solution often yields an amorphous phase; the dynamics of phase separation are so fast that protein molecules do not have sufficient time to orient themselves to form a crystal. While thermodynamics alone cannot predict the kinetics of phase separation, it does suggest that crystallization will be most favorable in two regions, as shown by shaded areas in Figure 7.37.[69] One region is near the critical point of the metastable fluid–fluid phase transition where the long-range fluctuations of local concentration favor crystal formation. Close to the critical temperature of metastable fluid–fluid equilibrium, the free energy barrier for protein nucleation is strongly reduced; therefore, the crystal nucleation rate can be increased by several orders of magnitude.[70] The other region is between the freezing line and the fluid–fluid coexistence curve where crystallization proceeds through an intermediate state; a liquid droplet

68 Rosenbaum D., Zamora P. C. and Zukoski C. F., "Phase behavior of small attractive colloidal particles", *Phys. Rev. Lett.* 76 (150), (1996).
69 Vliegenthart G. A. and Lekkerkerker H. N. W., "Predicting the gas–liquid critical point from the second virial coefficient", *J. Chem. Phys.* 112, 5364 (2000).
70 ten Wolde P. R. and Frenkel D., "Enhancement of protein crystal nucleation by critical density fluctuations", *Science* 277 (5334), 1975–8 (1997).

at high protein concentration is formed prior to crystallization. The thermodynamic model offers explanation for using the second virial coefficient measurements to identify favorable protein crystallization conditions and the kinetic effects have been directly confirmed by experiment.[71]

7.10.2 Stability of Liquid-Vapor Equilibrium

The thermodynamic models discussed in Sections 7.7 and 7.8 can be utilized to explain why colloidal dispersions exhibit stable vapor–liquid-like transition when the attractive forces between colloidal particles are long-ranged, and the fluid–fluid transition is metastable relative to the fluid–solid equilibrium when the attractions are short ranged. Approximately, the minimum range of attractions for the appearance of a thermodynamically stable fluid–fluid transition is about one-sixth of that of repulsions. Interestingly, the same fluid and solid models predict that a stable liquid exists only at a narrow range of parameters for intermolecular interactions.

Schematically, Figure 7.38 shows three kinds of phase diagrams for fluid–solid equilibria for a one-component system represented by the Sutherland model. The thermodynamic properties of the fluid and solid phases can be predicted by the van der Waals-like theories as discussed in Section 7.7 and 7.8, respectively. According to the Sutherland model, the short-range repulsion between particles is represented by the hard-sphere potential, and the intermolecular attraction is represented by an inverse-power attractive potential

$$u_A(r) = -\varepsilon\left(\frac{\sigma}{r}\right)^n \tag{7.182}$$

where ε is a two-particle energy parameter, and n is a positive number larger than 3.[72] A higher inverse power (n) corresponds to a shorter-ranged attractive potential. The fluid–fluid coexistence curve is calculated by using the pressure and the chemical potential derived from the van der Waals theory. Similarly, the pressure and the chemical potential of the solid (used in calculation of the

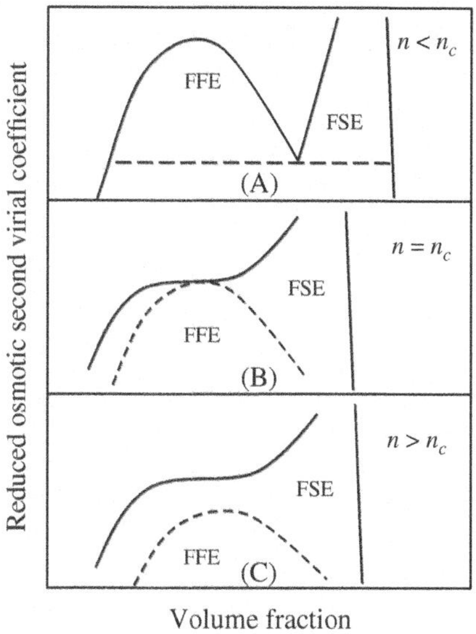

Figure 7.38 Effect of the range of inter-particle attraction represented by exponential n in Eq. (7.182) on the fluid (F)–fluid (F)–solid (S) equilibria (E). According to the van der Waals theory for the fluid phase and the cell model for the solid phase, the critical exponent n to have a stable fluid–fluid equilibrium is $n_C = 6.08$.

71 Galkin O. and Vekilov P. G., "Control of protein crystal nucleation around the metastable liquid–liquid phase boundary", *PNAS* 97 (12), 6277–6281 (2000).

72 If $n < 3$, the van der Waals theory becomes inadequate because the total attractive energy diverges.

solid–fluid equilibrium) come from the cell model. The phase diagrams are obtained through a procedure similar to that used for the calculation of fluid–solid equilibrium for hard spheres.

Figure 7.38A is similar to that for a noble gas where the intermolecular potential is relatively long-ranged ($n \approx 6$). In this case, two stable fluids may coexist at a temperature between the triple point and the critical point, one corresponding to a vapor phase and the other is a liquid. Below the triple-point temperature, the vapor phase may directly coexist with a solid phase. While above the triple point, the liquid–solid transition takes play only at high densities. As the range of intermolecular attraction is reduced (n increases), the triple-point temperature rises while the critical temperature falls. At some critical range of inter-particle attraction (Figure 7.38B), the triple-point and critical temperatures become identical. In that case, the region corresponding to stable fluid–fluid equilibrium disappears. When the range of the inter-particle attraction is further reduced, the triple point vanishes, and the coexistence curve for fluid–fluid equilibrium lies underneath the freezing curve, suggesting that the fluid–fluid transition is metastable. Figure 7.38C presents the phase diagram of a system where inter-particle attraction is short-ranged, similar to that observed in typical colloidal dispersions or in the aqueous solutions of globular proteins. In this case, the fluid–fluid coexistence is metastable.

At the triple point, we have three phases at equilibrium: one low-density fluid, one high-density fluid, and a solid. These three phases have the same pressure and chemical potential

$$P^I = P^{II} = P^{III}, \tag{7.183}$$

$$\mu^I = \mu^{II} = \mu^{III}, \tag{7.184}$$

where I, II and III designate the three coexisting phases. Figure 7.39 shows the densities of vapor,[73] liquid and solid phases at the triple point for systems where inter-particle attraction is described by an inverse-power potential (Eq. (7.182)). As the intermolecular potential becomes steeper (n becomes larger), the difference between the densities of two coexisting fluids becomes smaller. At $n_c = 6.08$, the two fluids become identical. For $n > n_c$, the triple point disappears. In that case, the fluid–fluid coexistence is metastable, i.e., a stable liquid phase does not exist!

Despite the simplicity of the fluid and solid models, the theoretical predictions are qualitatively consistent with experimental observations. Most atomic or simple molecular systems exhibit a

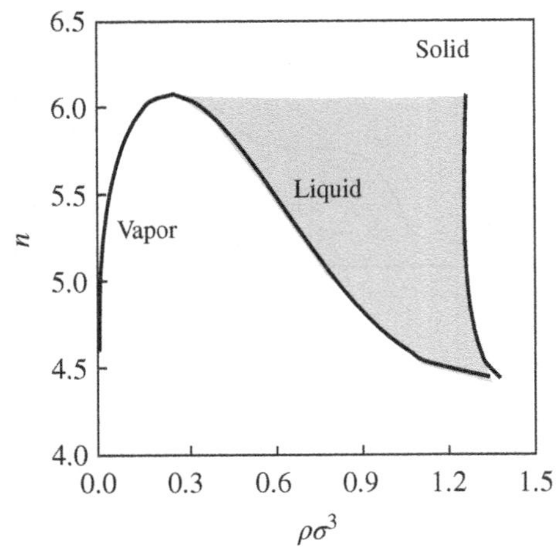

Figure 7.39 Effect of attractive-potential exponent n (see Eq. (7.182)) on densities of vapor, liquid, and solid phases at the triple point. A liquid is stable only in a narrow range of n.

73 For a colloidal dispersion or a protein solution, the "vapor" refers to the dilute liquid phase and the "liquid" corresponds to the concentrated solution.

stable dense liquid phase because of the London theory of intermolecular attraction ($n = 6$). Conversely, most colloidal dispersions and some molecular systems such as buckyballs[74] do not show a stable liquid phase because the inter-particle potential is relatively short-ranged ($n > 6$).

7.10.3 Summary

The phase behavior of colloidal dispersions and the aqueous solutions of globular proteins can be reasonably described with the van der Waals-like models for the fluid and solid phases. Unlike that corresponding to a simple fluid, the fluid–fluid equilibrium in colloidal systems is mostly metastable. A stable liquid phase exists only when the intermolecular attraction is sufficiently long-ranged.

7.11 Perturbation Theories

In Section 7.6, we have indicated that the van der Waals theory can be improved by replacing the excluded-volume term with the CS equation. Further improvement is also possible by modifying the attractive term. In this section, we relax the mean-field approximation for describing the thermodynamic nonideality due to the intermolecular attraction. We demonstrate that improvement in the attractive term can be achieved by considering the fluid structure and correlation effects on the thermodynamic properties.

7.11.1 The Dichotomy of Intermolecular Forces

Perturbation theories are useful to account for thermodynamic nonideality arising from various types of intermolecular interactions, i.e., the steep repulsion at short distance and attraction at larger separations. As shown schematically in Figure 7.40, the short-range repulsion defines the molecular size; it can be represented as a geometric effect (viz., hard spheres) responsible for local molecular packing and the mechanical rigidity of condensed matter. In comparison to the (nearly infinite) repulsive potential, the longer-ranged attraction is much weaker in terms of magnitude and varies much more slowly with the distance. The intermolecular attraction is often responsible for temperature-sensitive phenomena including phase transitions and self-assembly processes. The different mathematical forms and underlying physics for the repulsive and attractive components of intermolecular forces suggest that different theoretical strategies must be adopted to calculate the thermodynamic properties of molecular systems.

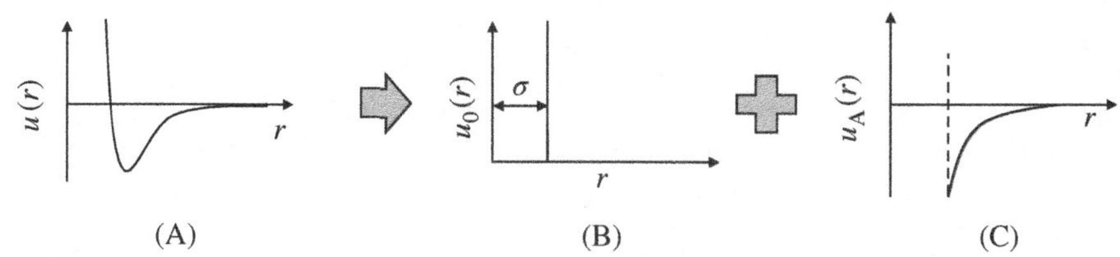

Figure 7.40 A perturbation theory follows van der Waals' idea to divide the intermolecular potential $u(r)$ (A) into a short-range repulsion $u_0(r)$ (B) and an attractive energy $u_A(r)$ (C). The repulsive part is typically represented by the hard-sphere potential of diameter σ.

74 A buckyball consists of 60 carbon atoms with a cage-like structure resembling a soccer ball.

To a certain degree, all perturbation methods in liquid-state theory adhere to van der Waals' idea to separate intermolecular potential into a short-range repulsion and a longer-ranged attraction. The underlying assumption for the separation of intermolecular potential is that certain properties of the reference system (i.e., the real system under consideration but without the attractive component of the intermolecular potential) are already known such that the contributions due to the attractive potential can be accounted for by a perturbation expansion. A rough analogy is provided by describing a dog. First, we consider the dog without a tail; the contribution of the tail is provided by a perturbation.

7.11.2 The Zwanzig Expansion

The perturbation expansion was introduced first by Robert W. Zwanzig[75] for developing an equation of state for nonpolar gases at high temperature. Mathematically, the procedure may be understood as a functional Taylor expansion[76] of the Helmholtz energy in terms of the intermolecular potential relative to that of a reference system. The Zwanzig expansion is also known as the free-energy perturbation or the λ-expansion.

To elucidate the essential ideas, consider the canonical ensemble for a system of N spherical particles. At a given configuration of the system $\mathbf{r}^N = (\mathbf{r}_1, \mathbf{r}_2, \cdots, \mathbf{r}_N)$, we may describe the total potential energy in terms of that corresponding to a reference system, $\Phi_0(\mathbf{r}^N)$, plus a perturbation potential

$$\Phi(\lambda) = \Phi_0 + \lambda\Delta\Phi \tag{7.185}$$

where $0 \le \lambda \le 1$ is a coupling constant, and $\Delta\Phi \equiv \Phi - \Phi_0$. As λ varies continuously from 0 to 1, the potential energy $\Phi(\lambda)$ changes from that corresponding to the reference system ($\lambda = 0$) to that for the real system ($\lambda = 1$). In the following discussion, quantities related to the reference system are denoted with subscript "0."

The canonical partition function for a system with potential energy $\Phi(\lambda)$ is given by

$$Q(\lambda) = \frac{1}{N!\Lambda^{3N}} \int d\mathbf{r}^N \exp[-\beta(\Phi_0 + \lambda\Delta\Phi)] = Q_0 < \exp(-\beta\lambda\Delta\Phi)>_0 \tag{7.186}$$

where $d\mathbf{r}^N = d\mathbf{r}_1 d\mathbf{r}_2 \cdots d\mathbf{r}_N$, Λ is the thermal wavelength, Q_0 denotes the partition function of the reference system

$$Q_0 = \frac{1}{N!\Lambda^{3N}} \int d\mathbf{r}^N \exp[-\beta\Phi_0]. \tag{7.187}$$

In Eq. (7.186), $<\cdots>_0$ represents ensemble average of a quantity in the reference system

$$\langle\cdots\rangle = \frac{\int d\mathbf{r}^N(\cdots)\exp[-\beta\Phi_0]}{\int d\mathbf{r}^N \exp[-\beta\Phi_0]}. \tag{7.188}$$

The Helmholtz energy corresponding to Eq. (7.186) is given by

$$F(\lambda) = F_0 - k_B T \ln < \exp(-\beta\lambda\Delta\Phi)>_0. \tag{7.189}$$

As discussed in Section 6.8, Eq. (7.189) provides a starting point for free-energy calculations using simulation methods. At high temperature, $\beta\lambda\Delta\Phi$ is small such that the exponential function on the right side of Eq. (7.186) can be expanded in a Taylor series

$$<\exp(-\beta\lambda\Delta\Phi)>_0 = 1 - \beta\lambda <\Delta\Phi>_0 + \frac{1}{2!}(\beta\lambda)^2 <\Delta\Phi^2>_0 - \frac{1}{3!}(\beta\lambda)^3 <\Delta\Phi^3>_0 + \cdots \tag{7.190}$$

---

75 Zwanzig R. W., "High-temperature equation of state by a perturbation method. 1. Nonpolar Gases", *J. Chem. Phys.* 22 (8), 1420–1426 (1954).

76 As in a regular Taylor expansion, a functional Taylor expansion expresses a functional, i.e., the variable itself is a function, in a polynomial series of the function and the functional derivatives. See Appendix 8A for more details.

Using Eq. (7.190) and $\ln(1+x) = x - x^2/2 + x^3/3 + \cdots$ for small x, we have

$$\ln <\exp(-\beta\lambda\Delta\Phi)>_0 = -\beta\lambda <\Delta\Phi>_0 + \frac{1}{2!}(\beta\lambda)^2\left[<\Delta\Phi^2>_0 - <\Delta\Phi>_0^2\right]$$
$$-(\beta\lambda)^3\left[\frac{1}{3!}<\Delta\Phi^3>_0 - \frac{1}{2!}<\Delta\Phi>_0<\Delta\Phi^2>_0 + \frac{1}{3}<\Delta\Phi^3>_0\right] + \cdots \quad (7.191)$$

Substituting Eq. (7.191) into (7.189) gives the Helmholtz energy as a power series in λ:

$$F(\lambda) = F_0 + \lambda F_1 + \lambda^2 F_2 + \lambda^3 F_3 + \cdots \quad (7.192)$$

where

$$F_1 = <\Delta\Phi>_0 \quad (7.193)$$

$$F_2 = -\frac{\beta}{2!}\left[<\Delta\Phi^2>_0 - <\Delta\Phi>_0^2\right] \quad (7.194)$$

$$F_3 = \frac{\beta^2}{3!}[<\Delta\Phi^3>_0 - 3<\Delta\Phi>_0 < \Delta\Phi^2>_0 + 2<\Delta\Phi^3>_0]. \quad (7.195)$$

When $\lambda = 1$, we obtain the Helmholtz energy of the real system

$$F = F_0 + F_1 + F_2 + F_3 + \cdots \quad (7.196)$$

where the zero-order term is the Helmholtz energy of the reference, the first-order and all higher-order terms constitute the perturbation. Because the series shown in Eq. (7.196) converges most rapidly at high temperature (viz., small β), the Zwanzig expansion is also known as the high-temperature expansion.

In Eq. (7.196), each term in the perturbation expansion is related to the correlation function(s) of the reference as well as the perturbation potential. For a simple fluid, the total potential energy can be decomposed into pairwise-additive potentials

$$u(r) = u_0(r) + u_A(r) \quad (7.197)$$

where $u_0(r)$ stands for the repulsive component of the pair potential, and $u_A(r)$ is the attractive energy. Typically, the reference system is defined as one with only repulsive interactions

$$\Phi_0 = \sum_{i=1}^{N}\sum_{j>i}^{N} u_0(r_{ij}), \quad (7.198)$$

and the attraction potentials are treated as the perturbation

$$\Delta\Phi = \sum_{i=1}^{N}\sum_{j>i}^{N} u_A(r_{ij}) \quad (7.199)$$

where $r_{ij} = |\mathbf{r}_i - \mathbf{r}_j|$ is the center-to-center separation between particles i and j. The first-order term in the Zwanzig expansion thus becomes

$$F_1 = <\sum_{i=1}^{N}\sum_{j>i}^{N} u_A(r_{ij})>_0 = \frac{N}{2}\int d\mathbf{r}\ \rho_b g_0(r)\, u_A(r) \quad (7.200)$$

where $g_0(r)$ is the radial distribution function of the reference system, and $\rho_b = N/V$ is the average particle density. Because $\rho(r) = \rho_b g_0(r)$ represents the average particle density at radial distance r from a particle center in the reference system, the integral in Eq. (7.200) accounts for the total attraction energy for a particle at the center due to its interaction with all other particles.

With the pairwise additive assumption, the second-order term includes multibody correlations up to four particles. As a result, an analytical evaluation requires multi-body correlation functions of the reference system. Because such information is rarely available, the Zwanzig expansion is rarely used beyond the first-order term.

7.11.3 The Barker–Henderson Theory

By ignoring multi-body correlations and assuming that the local density fluctuation can be represented by a compressibility equation for the bulk system, Barker and Henderson (BH) derived two approximate expressions for the second-order perturbation term in the Helmholtz energy expansion.[77] These approximations, along with a careful choice of the hard-sphere diameter, represent one of the most successful applications of the Zwanzig expansion to fluid systems.

In a uniform system of spherical molecules, the pair potential depends only on the center-to-center distance. At any microstate, the total perturbation potential can be expressed in terms of that of an arbitrarily tagged molecule and the distribution of other molecules in the surrounding

$$\Delta\Phi = \frac{N}{2}\int d\mathbf{r}\, \widetilde{\rho}(\mathbf{r})\, u_A(r) \tag{7.201}$$

where $\widetilde{\rho}(\mathbf{r}) = \sum_{i=1}^{N} \delta(\mathbf{r} - \mathbf{r}_i)$ stands for the instantaneous molecular density, and $\delta(\mathbf{r})$ is the Dirac delta function. Eq. (7.201) is essentially the same as Eq. (7.200) except that the local density is now replaced by the instantaneous density $\widetilde{\rho}(\mathbf{r})$.

As shown in Eq. (7.194), the second-order term in the Zwanzig expansion is related to the fluctuation of the perturbation potential. Using Eq. (7.201), we can write the fluctuation as

$$<\Delta\Phi^2>_0 - <\Delta\Phi>_0^2 = \frac{N^2}{4}\int\int d\mathbf{r}_1 d\mathbf{r}_2\ \ u_A(r_1)u_A(r_2)[<\widetilde{\rho}(\mathbf{r}_1)\widetilde{\rho}(\mathbf{r}_2)>_0 - \rho_0(r_1)\rho_0(r_2)] \tag{7.202}$$

where subscripts 1 and 2 represent coordinates centered around two randomly selected particles, and $\rho_0(r) = <\widetilde{\rho}(\mathbf{r})>_0$ is the local density around each particle averaged according to the reference ensemble.

Figure 7.41 shows schematically the instantaneous local densities near two tagged particles. If these tagged particles are far apart, their local densities are uncorrelated thus

$$<\widetilde{\rho}(\mathbf{r}_1)\widetilde{\rho}(\mathbf{r}_2)>_0 = \rho_0(r_1)\rho_0(r_2). \tag{7.203}$$

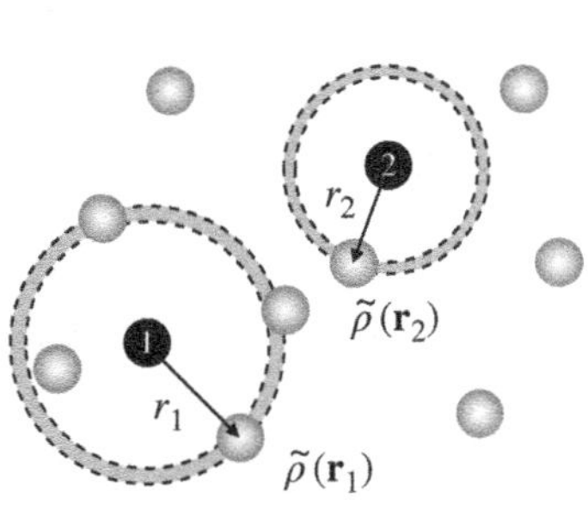

Figure 7.41 The potential energy for each tagged particle (here 1 or 2) depends on the instantaneous local density of the surrounding particles, $\widetilde{\rho}(\mathbf{r})$.

77 Barker J. A. and Henderson D., "Perturbation theory and equation of state for fluids: the square-well potential", *J. Chem. Phys.* 47 (8), 2856–2861 (1967).

Substituting Eq. (7.203) into (7.202) indicates that there is no fluctuation of the potential energy. When the tagged particles are close to each other, however, the tagged particles interact with other particles in the system through the reference potential. Because such interactions are strongly correlated with each other, the local particle density depends on the positions of the tagged particles as well as the other particles with which they interact.

To evaluate the density–density correlations, Barker and Henderson made a number of assumptions. First, the local densities are assumed to be correlated only when they are positioned relative to the same tagged particle. In other words, it is assumed that local densities near different tagged particles are completely uncorrelated. As a result, Eq. (7.202) can be expressed in terms of the fluctuation of the local densities with one particle fixed at the origin:

$$< \Delta\Phi^2>_0 - < \Delta\Phi >_0^2 \approx \frac{N}{2} \iint d\mathbf{r}_1 d\mathbf{r}_2 \; u_A(r_1)u_A(r_2)\chi(\mathbf{r}_1, \mathbf{r}_2) \tag{7.204}$$

where N is the number of particles in the system, while 2 accounts for the symmetry of particle–particle interactions, and $\chi(\mathbf{r}_1, \mathbf{r}_2)$ stands for the local density–density correlation function of the reference system

$$\chi_0(\mathbf{r}_1, \mathbf{r}_2) \equiv < \widetilde{\rho}(\mathbf{r}_1)\widetilde{\rho}(\mathbf{r}_2)>_0 - \rho_0(r_1)\rho_0(r_2). \tag{7.205}$$

In general, $\chi_0(\mathbf{r}_1, \mathbf{r}_2)$ is unknown. Therefore, the second assumption in the BH theory is that density–density correlation exists only when $\mathbf{r}_1$ and $\mathbf{r}_2$ are positioned at the same distance from the tagged particle, i.e., $|\mathbf{r}_1| = |\mathbf{r}_2| = r$. At the same radial distance from the tagged particle, the average local density is given by, $< \widetilde{\rho}(\mathbf{r}_1) >_0 = < \widetilde{\rho}(\mathbf{r}_2) > = \rho_0(r)$. The density–density correlation function may be estimated from that corresponding to a uniform bulk system at local density $\rho_0(r)$:

$$\chi_0(\mathbf{r}_1, \mathbf{r}_2) \approx \chi_0(|\, \mathbf{r}_1 - \mathbf{r}_2 \,|). \tag{7.206}$$

Substituting Eq. (7.206) into (7.204) yields

$$< \Delta\Phi^2>_0 - < \Delta\Phi >_0^2 \approx \frac{N}{2} \int\int d\mathbf{r}_1 d\mathbf{r}_2 \; u_A^2(r)\chi_0(|\, \mathbf{r}_1 - \mathbf{r}_2 \,|) = \frac{N}{2} \int d\mathbf{r} u_A^2(r)\hat{\chi}_0(0) \tag{7.207}$$

where $\hat{\chi}_0(0)$ is related to the compressibility of the uniform system at local density $\rho_0(r)$

$$\hat{\chi}_0(0) \equiv \int d\mathbf{r}' \chi_0(r') = \rho_0(r) \left[\frac{\partial \rho_0(r)}{\partial \beta P} \right]_T. \tag{7.208}$$

Eq. (7.207) is called the local compressibility approximation (LCA), i.e., the fluctuation of the local particle density is approximated by that corresponding to the compressibility of a uniform system.[78]

From Eqs. (7.194), (7.207), and (7.208), we obtain the second-order perturbation term in the BH theory

$$F_2 \approx -\frac{N}{4} \int d\mathbf{r} \; u_A^2(r)\rho_0(r) \left[\frac{\partial \rho_0(r)}{\partial P} \right]_T. \tag{7.209}$$

In the conventional form, Eq. (7.209) is often rewritten by replacing the local density with $\rho_0(r) = \rho_b g_0(r)$,

$$F_2/N \approx -\frac{\rho_b}{4} \int d\mathbf{r} \; u_A^2(r)g_0(r) \left[\frac{\partial \rho_b g_0(r)}{\partial P} \right]_T \tag{7.210}$$

78 The relation between compressibility and density fluctuation is discussed in Chapter 2 on grand canonical ensemble. For a one-component system at constant temperature, chemical potential and volume, the density fluctuation is given by $< (\delta\rho)^2 > = \frac{\rho}{V} \left(\frac{\partial \rho}{\partial \beta P} \right)$.

where $\rho_b = N/V$ is the average particle density, and $g_0(r)$ is the RDF of the reference system.

Eq. (7.210) can be further simplified with the assumption that the local compressibility is the same as that of the reference system

$$F_2/N \approx -\frac{\rho_b}{4}\left(\frac{\partial \rho_b}{\partial P}\right)_{T,0} \int d\mathbf{r} u_A^2(r)\, g_0(r). \tag{7.211}$$

Eq. (7.211) is another version of the BH theory, i.e., one with the macroscopic compressibility assumption (MCA). In comparison with LCA, MCA is numerically more convenient because the compressibility is evaluated at the bulk density instead of that at every position.

In application of the BH theory to systems with a hard-core potential such as the square-well fluids, the reference system corresponds to a hard-sphere fluid. In that case, both the first- and second-order perturbations can be evaluated from the radial distribution function of the hard-sphere system. While it is difficult to derive a concise expression for $g_0(r)$ even for hard-sphere systems, its Laplace form can be obtained from the PY theory discussed in Section 7.5. The theoretical predictions are found in good agreement with simulation results not only for high-temperature gases but, somewhat surprisingly, also for low-temperature liquids. As discussed later, the success of perturbation theory for liquids can be attributed, in part, to the predominant role of the short-range repulsion for determining the liquid structure and entropy.

The BH theory can also be applied to systems with a continuous potential such as that given by the LJ model. In that case, the reference system is defined by the positive part of the LJ potential ($r < \sigma$), and the attractive part is treated as the perturbation ($r \geq \sigma$). Because the reference system primarily relates to the molecular size, it can be reasonably approximated by that corresponding to a hard-sphere system. To calculate the pair correlation function and the compressibility of the reference system, Barker and Henderson estimated the properties of the reference system using an effective hard-sphere diameter

$$d_{BH} = \int_0^\sigma [1 - e^{-\beta u(r)}]dr. \tag{7.212}$$

For convenience, Eq. (7.212) has been fitted in a parametric form[79]

$$d_{BH} = 2^{1/6}\sigma\left\{1 + \left[1 + \frac{T^* - 0.05536T^{*2} + 0.0007278T^{*4}}{1.1287}\right]^{1/2}\right\}^{-1/6} \tag{7.213}$$

where $T^* = k_B T/\varepsilon < 5$.

The BH theory has been successfully incorporated into the statistical-associating fluid theory (SAFT), which will be discussed in details in Chapter 8.[80] Although the theoretical development is mainly focused on the second-order term, its contribution to thermodynamic properties is relatively small in comparison to those from the leading first-order perturbation. As shown in Figure 7.42, the second-order Helmholtz energy is smaller than that from the first-order term by nearly one order of magnitude.[81] The LCA and MCA approximations are most accurate at low densities. For the liquid phase, neither LCA nor MCA is satisfactory compared to simulation results. Interestingly, an empirical modification of the second-order term agrees with the simulation data well.

---

79 Desouza L. E. S. and Benamotz D., "Optimized perturbed hard-sphere expressions for the structure and thermodynamics of Lennard-Jones fluids", *Mol. Phys.* 78 (1), 137–149 (1993).

80 GilVillegas A. et al., "Statistical associating fluid theory for chain molecules with attractive potentials of variable range," *J. Chem. Phys.* 106 (10), 4168–4186 (1997).

81 Lafitte T. et al., "Accurate statistical associating fluid theory for chain molecules formed from Mie segments", *J. Chem. Phys.* 139 (15) 154504 (2013).

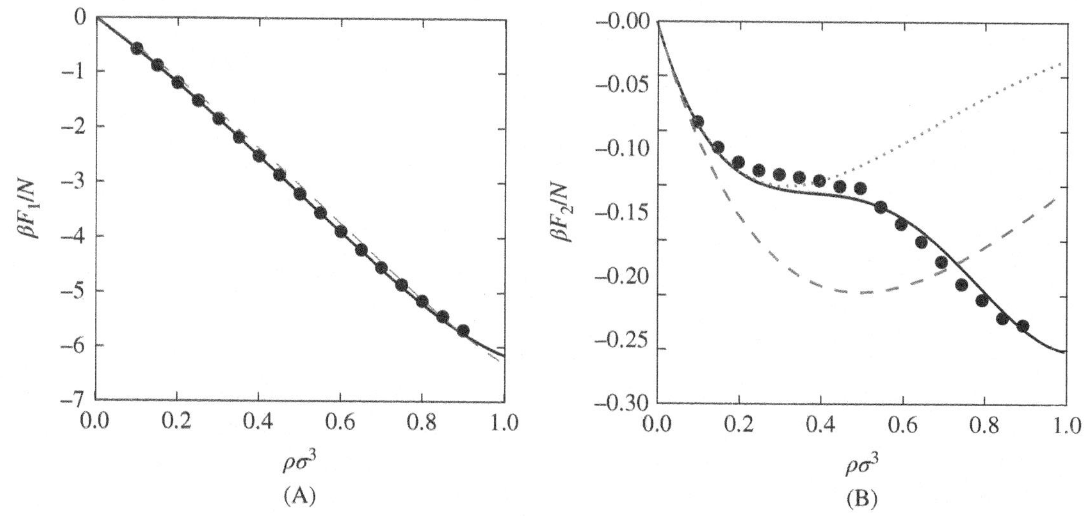

Figure 7.42 The first-order (A) and the second-order (B) perturbation terms in the Zwanzig expansion of the Helmholtz energy for the Lennard–Jones (12-6) fluid at reduced temperature $k_B T/\varepsilon = 1$. Circles represent exact results from MC simulation. The solid and dashed lines in Panel A are from Eq. (7.200) with two different expressions of the reference radial distribution function $g_0(r)$. In Panel B, the dashed, dotted, and continuous curves are, respectively, from MCA (Eq. (7.211)), LCA (Eq. (7.210)), and MCA multiplied by an empirical corrector $(1 + 8.23\eta^2)$. Source: Adapted from Lafitte et al.[81]

7.11.4 The Weeks–Chandler–Anderson (WCA) Theory

In a condensed phase such as a solid or liquid, it has been long speculated that the microscopic structure is primarily determined by the packing effects, much like that in a hard-sphere system. While the hypothesis is intuitively appealing, it was not validated quantitively until the advent of molecular simulation methods. As discussed in Section 7.2, molecular simulation indicates that the radial distribution function of a Lennard–Jones (LJ) liquid is virtually identical to that of a hard-sphere fluid at the same particle density. Accordingly, thermodynamic properties can be calculated from the structure determined by the short-range repulsion and an energetic contribution due to the attractive potential. Built upon these ideas, Weeks, Chandler, and Andersen developed in the early 1970s one of the most influential modern liquid-state methods now known as the WCA theory.[82]

The WCA theory is mostly concerned with the LJ model. Different from the BH theory, it defines the repulsive component of the intermolecular potential based on the sign of the force instead of energy. As shown in Figure 7.43, the repulsive part of the LJ potential is defined as

$$u_R(r) = \begin{cases} u_{LJ}(r) + \varepsilon & r < 2^{1/6}\sigma \\ 0 & r \geq 2^{1/6}\sigma \end{cases} \tag{7.214}$$

and the attractive component is

$$u_A(r) = \begin{cases} -\varepsilon & r < 2^{1/6}\sigma \\ u_{LJ}(r) & r \geq 2^{1/6}\sigma \end{cases}, \tag{7.215}$$

82 Weeks J. D., Chandler D. and Andersen H. C., "Role of repulsive forces in determining the equilibrium structures of simple liquids", *J. Chem. Phys.* 54 (12), 5237–5246 (1971); Chandler D., "From 50 years ago, the birth of modern liquid-state science", *Ann. Rev. Phys. Chem.* 68, 19–38 (2017).

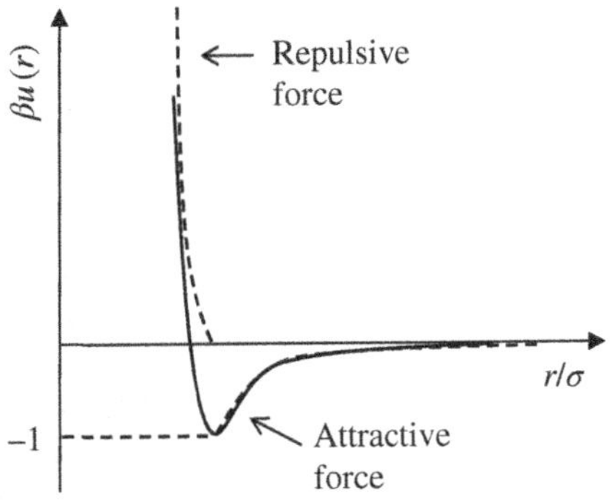

Figure 7.43 In the WCA theory, the Lennard–Jones potential (dashed line) is divided into a repulsive part and an attractive part according to the sign of the intermolecular force (solid lines).

where ε and σ are LJ parameters. According to the above definitions of repulsive and attractive potentials, the reference system has a pair potential that corresponds to a shifted LJ potential truncated at $2^{1/6}\sigma$, where the LJ potential has a minimum. The remaining part of the LJ potential, along with the shifted energy, is treated as the perturbation energy.

In calculation of thermodynamic properties, the WCA theory adopts only the first-order term in the Zwanzig expansion. All higher-order terms are assumed negligible because, if the radial distribution function of the reference system is the same as that of the real system (Figure 7.44), the internal energy is accurately reproduced by the first-order term[83]

$$U = \left(\frac{\partial \beta F}{\partial \beta}\right)_{N,V} \approx \frac{N}{2}\int d\mathbf{r}\ \rho g_0(r)\, u_A(r). \tag{7.216}$$

To obtain $g_0(r)$, the WCA theory assumes that the cavity correlation function of the reference system can be replaced by that corresponding to a hard-sphere (HS) fluid

$$y_0(r) \approx y_{HS}(r). \tag{7.217}$$

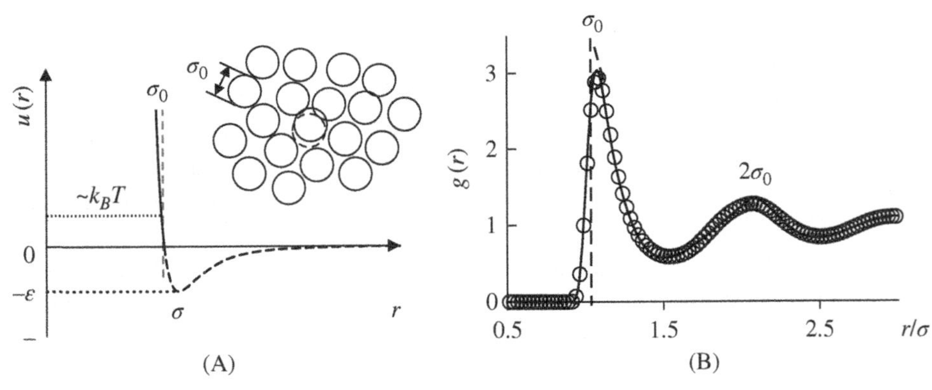

Figure 7.44 Steep repulsion at short distance dominates the structure of a simple liquid near its triple point ($k_BT/\varepsilon = 0.88$, $\rho\sigma^3 = 0.85$). (A) The Lennard-Jones (LJ) potential can be divided into a repulsive branch (solid line) and an attractive branch (dashed line). The repulsion is approximated by a hard-sphere potential (vertical line). (B) Schematic view of the local packing structure of spherical particles. The dashed circle is to illustrate that the energy affiliated with a small displacement of a LJ particle is denominated by short-range repulsion. (C) Radial distribution functions for the LJ liquid (circles), for the reference fluid with only repulsive forces (solid line), and for the equivalent hard-sphere fluid (dashed line). Adapted from Chandler et al.[83]

83 Chandler D., Weeks J. D. and Andersen H. C., "Van der Waals picture of liquids, solids, and phase-transformations", *Science* 220 (4599), 787–794 (1983).

Eq. (7.217) is justified because, as discussed in Section 7.2, the cavity correlation function represents indirect particle–particle correlations. For systems with similar particle densities, $y_0(r)$ is relatively insensitive to the microscopic details of intermolecular interactions. According to Eq. (7.217), the pair correlation function of the reference system is given by

$$g_0(r) \approx y_{HS}(r) \exp[-\beta u_R(r)]. \tag{7.218}$$

Like the BH theory, the WCA theory estimates the thermodynamic properties of the reference based on an effective hard-sphere diameter. The parameter is selected such that the compressibility of the reference fluid

$$\beta^{-1}(\partial \rho_b/\partial P)_\beta = 1 + \rho_b \int d\mathbf{r}[g_0(r) - 1], \tag{7.219}$$

is reproduced by that of a hard-sphere system at the same density in the bulk ρ_b. This selection of the hard-sphere diameter is guided by the assumption that liquids with similar compressibility have similar microscopic structures. With $g_0(r)$ from Eq. (7.218), we can calculate the effective hard-sphere diameter from

$$\int d\mathbf{r} \left[y_{HS}(r)e^{-\beta u_R(r)} - 1\right] = \int d\mathbf{r} \left[y_{HS}(r)e^{-\beta u_{HS}(r)} - 1\right]. \tag{7.220}$$

According to Eq. (7.220), the effective hard-sphere diameter depends on both the number density and temperature. The numerical solution to Eq. (7.220) has been fitted in a convenient form[79]

$$d_{WCA}/\sigma = 2^{1/6}\{1 + a_1[(T^* + a_2T^{*2} + a_3T^{*4})/(1 + a_4\rho^* + a_5\rho^{*2} + a_6\rho^{*3})]^{1/2}\}^{-1/6} \tag{7.221}$$

where d_{WCA} is the effective hard-sphere diameter in the WCA theory, $T^* = k_BT/\varepsilon$, and $a_1 = 0.8165$, $a_2 = -0.03367$, $a_3 = 0.0003935$, $a_4 = -0.09835$, $a_5 = 0.04937$, $a_6 = -0.1415$.

Figure 7.44 shows the pair correlation functions for a LJ fluid and that for the hard-sphere reference fluid used in the WCA theory. Also shown is the radial distribution function for the LJ fluid with only the repulsive part of the potential calculated from the cavity correlation function for hard spheres. While the difference among the three methods is indistinguishable at large separations, the WCA theory reproduces the simulation results also at short separations. Figure 7.44 also indicates that, at high densities, the structure of a dense liquid is determined primarily by repulsive interactions. While the WCA theory performs well for the LJ liquid near the triple point, it becomes less accurate as the fluid density declines. Besides, the extension of the WCA theory to fluid mixtures is often inconvenient due to the complicated procedure to determine the hard-sphere diameter and the radial distribution functions of the reference system.

Like the van der Waals theory, the WCA theory decomposes contributions from short-range repulsion and longer-ranged attraction to thermodynamic properties. While the van der Waals theory is based on the mean-field assumption for intermolecular forces, the WCA theory takes advantage of the accurate pair correlation function and equation of state for hard-sphere systems to account for the liquid structure. However, as in van der Waals theory, the attractive part of the Helmholtz energy includes only the contribution of attractive potential to the internal energy. In the WCA theory, the effect of the attractive potential on entropy was accounted for mostly through the weak dependence of the hard-sphere diameter with temperature.

7.11.5 First-order Mean-Spherical Approximation (FMSA)

The first-order mean spherical approximation (FMSA) is a perturbation method developed by Tang and Lu.[84] It provides analytical expressions for both the structure and thermodynamic properties by solving the first-order expansion of the OZ equation for a number of molecular models of simple fluids.

Unlike the Zwanzig expansion, the FMSA is based on the expansions of the direct and total correlation functions with respect to the perturbation potential. First, the direct correlation function (DCF) is expressed as a Taylor series of coupling parameter λ as defined in the Zwanzig expansion, viz., Eq. (7.182),

$$c(u_\lambda, r) = c_0(r) + \lambda c_1(r) + \frac{\lambda^2}{2!}c_2(r) + \cdots. \tag{7.222}$$

A similar expression can be written for the total correlation function. In the Fourier space, direct and total correlation functions are given by

$$\hat{c}(u_\lambda, k) = \hat{c}_0(k) + \lambda \hat{c}_1(k) + \frac{\lambda^2}{2!}\hat{c}_2(k) + \cdots, \tag{7.223}$$

$$\hat{h}(u_\lambda, k) = \hat{h}_0(k) + \lambda \hat{h}_1(k) + \frac{\lambda^2}{2!}\hat{h}_2(k) + \cdots, \tag{7.224}$$

where $h(u_\lambda, r) = g(u_\lambda, r) - 1$, $h_0(r) = g_0(r) - 1$ and $h_i(r) = g_i(r)$, $i = 1, 2, \cdots$.

Substituting Eqs. (7.223) and (7.224) into the OZ equation gives

$$\hat{h}(u_\lambda, k) = \hat{c}(u_\lambda, k) + \rho_b \hat{h}(u_\lambda, k)\hat{c}(u_\lambda, k). \tag{7.225}$$

By identifying the coefficients in the Taylor series expansion of λ for both the direct and total correlation functions, one can obtain the following relations between the different orders of correlation functions:

$$\hat{c}_1(k) = \frac{\hat{h}_1(k)}{[1 + \rho_b \hat{h}_0(k)]^2}, \tag{7.226}$$

$$\hat{c}_2(k) = \frac{\hat{h}_2(k)}{[1 + \rho_b \hat{h}_0(k)]^2} - \frac{\rho_b \hat{h}_1(k)\hat{c}_1(k)}{1 + \rho_b \hat{h}_0(k)}. \tag{7.227}$$

Eq. (7.226) and (7.227) provide two linear relations between the first- and second-order terms in the direct and the total correlation functions. These equations can be solved consecutively with a closure. For example, when the pair potential is expressed in terms of a hard-sphere reference and an attractive potential, the MSA predicts that the different orders of the direct correlation function are given by

$$c_1(r) = -\beta u_A(r), \quad r \geq \sigma_{HS} \tag{7.228}$$

$$c_{i>1}(r) = 0 \quad r \geq \sigma_{HS} \tag{7.229}$$

where σ_{HS} stands for the hard-sphere diameter. Because of the excluded volume effects, the total correlation function is known when the distance is smaller than the hard-sphere diameter

$$h_i(r) = 0 \quad r < \sigma. \tag{7.230}$$

---

84 Tang Y. P. and Lu B. C. Y., "First-order radial distribution functions based on the mean spherical approximation for square-well, Lennard-Jones, and Kihara Fluids", *J. Chem. Phys.* 1994, 100 (4), 3079–3084; "An analytical analysis of the square-well fluid behaviors", *ibid.*, 100 (9), 6665–6671.

FMSA provides analytical expressions for $c_1(r)$ and $h_1(r)$ for the square-well, LJ, Yukawa, and Kihara models of intermolecular interactions. Different from the WCA theory, it accounts for the effects of attractive potential on the direct and total correlation functions based on which the thermodynamic properties are derived. When compared with simulation results, these analytical solutions provide excellent descriptions of both structure and thermodynamic properties. To illustrate, Figure 7.45 shows the radial distribution functions of the LJ fluids predicted from FMSA.[85] The theoretical predictions agree well with molecular simulation at all densities. For comparison, also shown here are results from the WCA theory and from the BH theory. At low density, FMSA is superior to alternative methods, while at high density, the WCA theory is slightly more accurate than FMSA.

In comparison with alternative perturbation methods, one major advantage of FMSA is that it is directly applicable to pure fluids and mixtures for a number of model systems. Importantly, FMSA provides analytical expressions for both thermodynamic properties and radial distribution function self-consistently. We will discuss in Chapter 8 that such self-consistency is highly desirable in the extension of the molecular theories of simple fluids to polymeric systems where monomeric systems are often treated as the reference for describing the chain connectivity and associations among polymer segments. For a LJ mixture, the pair potential is given by

$$u_{ij}^{LJ}(r) = 4\,\varepsilon_{ij}\left(\frac{\sigma_{ij}^{12}}{r^{12}} - \frac{\sigma_{ij}^{6}}{r^{6}}\right) \tag{7.231}$$

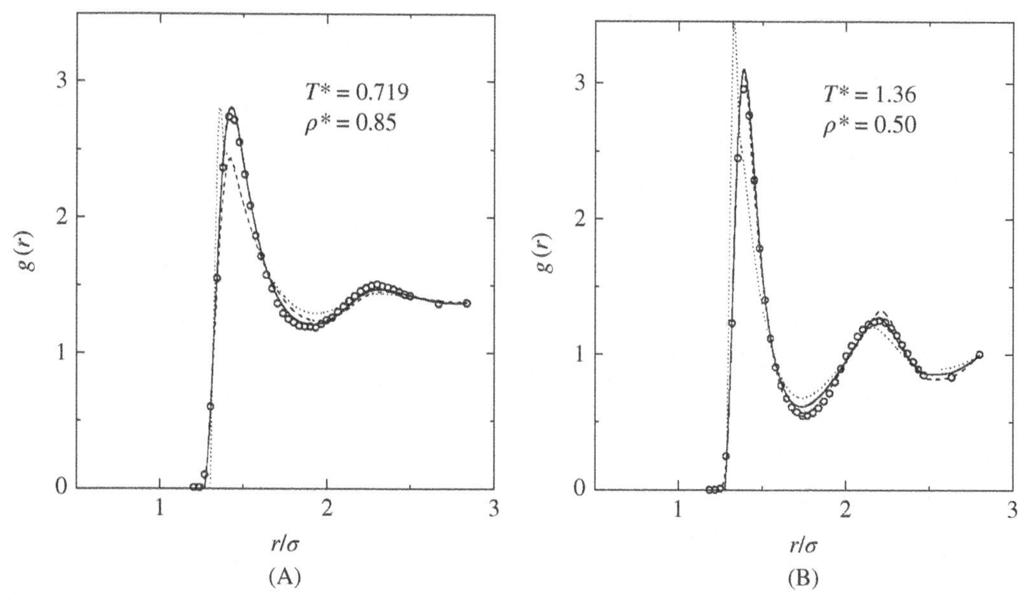

Figure 7.45 The radial distribution functions for Lennard–Jones fluids predicted by FMSA (solid lines) in comparison with those from the BH (dotted lines) and WCA theories (dashed lines); circles are computer-simulation data. Source: Adapted from Tang and Lu.[85]

85 Tang Y. P. and Lu B. C. Y., Analytical description of the Lennard-Jones fluid and its application. *AIChE J.* 43 (9), 2215-2226 (1997).

where ε_{ij} and σ_{ij} are energy and size parameters, respectively. The parameters for different species ($i \neq j$) are often given by the Lorentz–Berthelot (LB) combining rules:

$$\sigma_{ij} = (\sigma_{ii} + \sigma_{jj})/2. \tag{7.232}$$

$$\varepsilon_{ij} = \sqrt{\varepsilon_{ii}\varepsilon_{jj}}. \tag{7.233}$$

With these combining rules, FMSA is able to predict the properties of LJ mixtures by using only the parameters from the one-component systems. For example, the excess Helmholtz energy density is given by

$$f^{ex} = f^{ex}_{HS} + \sum_{\alpha=0}^{2} f^{(\alpha)}_{LJ} \tag{7.234}$$

where f^{ex}_{HS} is the excess Helmholtz energy for an effective hard-sphere system, and $f^{(\alpha)}_{LJ}$, $\alpha = 0, 1, 2$, are contributions from the attractive components of the LJ potential

$$f^{(0)}_{LJ} = -2\pi k_B T \rho_b \sum_i \sum_j x_i x_j g^0_{ij}(R_{ij}) R^2_{ij}(R_{ij} - d_{ij}) \tag{7.235}$$

$$f^{(1)}_{LJ} = -2\pi \rho_b \sum_i \sum_j x_i x_j \varepsilon_{ij} \left(J^1_{ij} - J^2_{ij}\right) + 8\pi \rho_b \sum_i \sum_j x_i x_j \varepsilon_{ij} R^3_{ij} \left[I^{\infty}_{ij} - g^0_{ij}(R_{ij}) I^1_{ij}\right] \tag{7.236}$$

$$f^{(2)}_{LJ} = -\pi \rho_b \sum_i \sum_j x_i x_j \varepsilon_{ij} \left(L^1_{ij} - L^2_{ij}\right) - 4\pi \rho_b \sum_i \sum_j x_i x_j \varepsilon_{ij} R^3_{ij} g^1_{ij}(R_{ij}) I^1_{ij}] \tag{7.237}$$

with

$$R_{ij} = (R_{ii} + R_{jj})/2, \quad R_{ii} = 2\sum_j x_j d_{ij} - \sum_i \sum_j x_i x_j d_{ij} \tag{7.238}$$

$$J^{(\alpha)}_{ij} = k^{(\alpha)}_{ij} \left[G^0_{ij}\left(z^{(\alpha)}_{ij}\right) e^{z^{(\alpha)}_{ij} R_{ij}} - \frac{1 + z^{(\alpha)}_{ij} R_{ij}}{\left(z^{(\alpha)}_{ij}\right)^2} \right], \quad \alpha = 1, 2 \tag{7.239}$$

$$L^{(\alpha)}_{ij} = k^{(\alpha)}_{ij} G^1_{ij}\left(z^{(\alpha)}_{ij}\right) e^{z^{(\alpha)}_{ij} R_{ij}}, \alpha = 1, 2 \tag{7.240}$$

$$I^{\infty}_{ij} = \frac{1}{9}\left(\frac{\sigma_{ij}}{R_{ij}}\right)^{12} - \frac{1}{3}\left(\frac{\sigma_{ij}}{R_{ij}}\right)^{6}, \quad I^1_{ij} = \frac{1}{9}\left(\frac{\sigma_{ij}}{R_{ij}}\right)^{12} - \frac{1}{3}\left(\frac{\sigma_{ij}}{R_{ij}}\right)^{6} + \frac{2}{9}\left(\frac{\sigma_{ij}}{R_{ij}}\right)^{3}. \tag{7.241}$$

In Eqs. (7.235)–(7.241), ρ_b is the total number density of the LJ particles, x_i is the mole fraction of component i, d_{ij} represents the hard-sphere diameter for particles i and j as determined by the BH theory, i.e., from Eq. (7.213) for σ_{ij}, $k^{(\alpha)}_{ij}$ and $z^{(\alpha)}_{ij}$ are parameters used in fitting the LJ potential with a two-Yukawa function

$$k^{(0)}_{ij} = 2.1714\sigma_{ij}, \quad k^{(1)}_{ij} = k^{(0)}_{ij} e^{z^{(1)}_{ij}(\sigma_{ij} - R_{ij})}, \quad k^{(2)}_{ij} = k^{(0)}_{ij} e^{z^{(2)}_{ij}(\sigma_{ij} - R_{ij})}, \tag{7.242}$$

$$z^{(1)}_{ij} = 2.9636/\sigma_{ij}, \quad z^{(2)}_{ij} = 14.0167/\sigma_{ij}. \tag{7.243}$$

Appendix 7.B gives the expressions for the radial distribution functions $g^0_{ij}(r)$ and $g^1_{ij}(r)$ near $r = R_{ij}$ along with the functions $G^0_{ij}(s)$ and $G^1_{ij}(s)$.

7.11.6 Summary

In applications of statistical thermodynamics, it is desirable to have a systematic procedure to account for the change in the Helmholtz energy due to the modification of the intermolecular potential. While the basic concept is generally applicable to all thermodynamic systems, its development and earlier applications are mostly focused on simple fluids, in particular for liquids near the triple point. In a conventional model of simple fluids, the intermolecular potential is assumed pairwise additive, with the pair potential depending only on the center-to-center distance. Typically, the pair potential includes a short-range repulsion related to the molecular size or the excluded volume, and a longer-ranged attraction due to van der Waals interactions. For nonpolar molecules, these interactions are most significant within a distance on the order of molecular diameter.

Perturbation theories of classical fluids were mostly developed in the late 1960s and early 1970s following van der Waals' ideas to separate the repulsive and attractive components of the intermolecular potential. As the intermolecular potential varies sharply at short distance, the theoretical procedure is highly sensitive to a precise division of the intermolecular potential and the adoption of a hard-sphere diameter. In comparison with various perturbation methods, FMSA has unique advantages because it provides analytical expressions for both structure and thermodynamic properties and is conveniently applicable to mixtures. The analytical forms provided by FMSA are particularly valuable for developing more sophisticated equations of state for industrial systems. Further, FMSA yields analytical expressions for direct correlation functions which are indispensable in the development of molecular theories of inhomogeneous fluids such as gases and electrolytes in confined geometry.

The perturbation methods are able to predict the structural and thermodynamic properties of bulk fluids in good agreement with simulation data. However, the theoretical predictions are far from satisfactory for mixtures pertinent to industrial applications, in particular under conditions near the critical region of phase transitions. The next two sections will address the issues related to critical phenomena. Chapter 8 discusses the extension of the perturbation methods to more complicated molecular systems.

7.12 Critical Behavior of Fluid–Fluid Transition

As discussed in Section 5.8, a first-order phase transition is accompanied with major changes in thermodynamic properties such as density, heat capacity, entropy, and for mixtures, composition.[86] Such changes are commonly utilized in modern technologies including the separation of chemical species (e.g., distillation and liquid extraction) and energy conversions (e.g., batteries and refrigeration). If the phase transition exhibits a critical point, the order parameter that distinguishes the coexisting phases (e.g., density) changes continuously.[87] For example, vapor condensation and liquid evaporation become continuous at the critical point of VLE. At the critical point, the phase transition is second order because the order parameter varies smoothly from the liquid to the vapor or vice versa.

86 Such phase transitions are classified as first-order transitions because the first derivative of the free energy with respect to some thermodynamic variable exhibits a discontinuity. Second-order phase transitions are continuous in the first derivative but exhibit discontinuity in a second derivative of the free energy. See Section 5.8 for more details.

87 A thermodynamic variable that distinguishes different phases is called order parameter. For example, density is an order parameter for vapor–liquid transition because it can be used to distinguish between the two coexisting phases. See Section 5.8 for more details.

The properties of a thermodynamic system near the critical point of a phase transition are distinctively different from those of the same system but at conditions remote from the critical point. In this section, we discuss the critical behavior of liquid–vapor and liquid–liquid phase transitions from the perspective of conventional liquid-state methods. Although these methods are not suitable to account for long-range correlations underlying critical phenomena, they provide helpful insights to understand critical singularities, i.e., how certain thermodynamic properties diverge at the critical point. We discuss in Section 7.13 an extension of the liquid-state methods to account for long-range correlations using the renormalization group (RG) theory.

7.12.1 Universality

The critical point of phase transition is characterized by singularities in both thermodynamic properties and correlation functions. For vapor–liquid equilibria in a one-component fluid (or liquid–liquid equilibrium in protein solutions and colloidal dispersions), these singularities include the divergence of the isothermal compressibility (κ_T) and of the isochoric heat capacity (C_V), and the disappearance of the difference between the specific properties of the coexisting liquid and vapor phases such as densities ($\rho_L - \rho_V$).

Schematically, Figure 7.46 presents the critical behavior of vapor–liquid transition in terms of (A) the density difference between the coexisting vapor and liquid phases, (B) isothermal compressibility, (C) correlation length, and (D) specific heat capacity. Because the isothermal compressibility and heat capacity are proportional to the density and energy fluctuations, respectively, i.e.,[88]

$$\kappa_T \equiv \frac{1}{\rho}\left(\frac{\partial \rho}{\partial P}\right)_T = \frac{V}{k_B T}\frac{\langle \delta \rho^2 \rangle}{\rho^2}, \tag{7.244}$$

$$C_V \equiv \left(\frac{\partial U}{\partial T}\right)_V = \frac{1}{k_B T^2}\langle (\delta E)^2 \rangle, \tag{7.245}$$

the divergence of these quantities suggests an infinite growth of the correlated fluctuations.

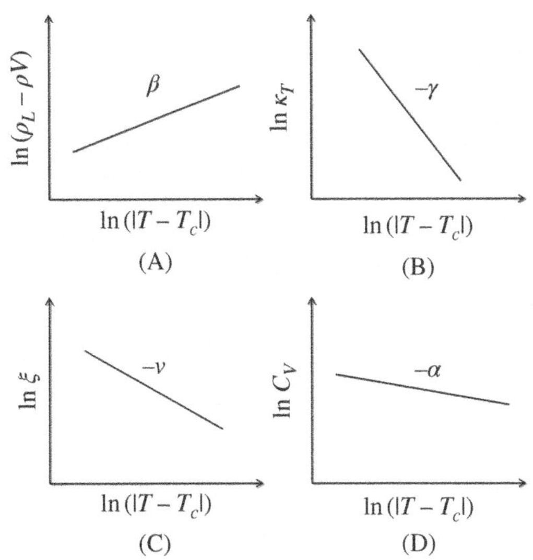

Figure 7.46 Singularities in various thermodynamic properties when the temperature approaches the critical point ($T \to T_c$). (A) Difference between the coexistence densities of liquid and vapor phases ($\rho_L - \rho_V$); (B) Isothermal compressibility κ_T; (C) Correlation length ξ; (D) Specific heat capacity C_V.

88 In this and next sections, ρ stands for the number density of particles in the bulk.

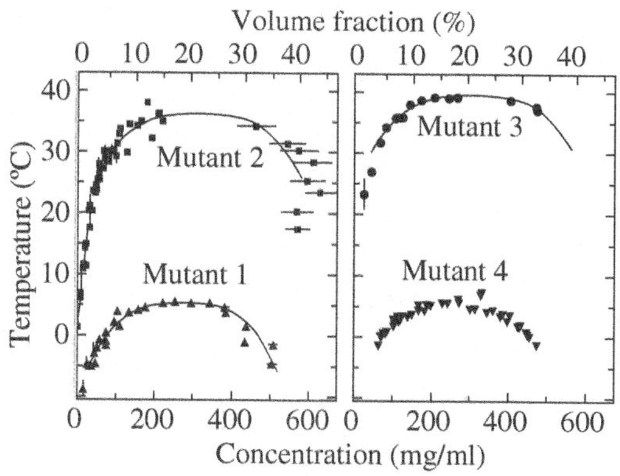

Figure 7.47 Coexistence curves (plots of phase-separation temperature versus protein concentration) for aqueous solutions of 4 different mutants of calf lens proteins. Solid curves are fitted to the scaling law near the critical point. Adapted from Broide et al.[89]

Experimental observations of thermodynamic quantities indicate that the divergence in heat capacity and isothermal compressibility can be described in terms of power laws with the exponents (viz., critical exponents) independent of the microscopic details of the system. As discussed in Section 5.11, such system-independent power laws are known as the universality of critical phenomena.

The scaling laws near the critical point are universal, applicable to a wide variety of seemingly uncorrelated systems. For example, Figure 7.47 shows the liquid–liquid coexistence curves for several aqueous solutions of proteins.[89] Whereas the microscopic details of a protein solution are drastically different from those corresponding to a simple fluid, similar scaling laws are applicable to the coexisting curves of both systems.

7.12.2 Mean-Field Critical Exponents

We may gain an understanding of the universality of critical phenomena from the perspective of conventional liquid-state theories, i.e., any analytical theory that ignores long-range correlations. Such analytical methods include the van der Waals-type theories wherein all levels of intermolecular correlations have been omitted; as well as various perturbation methods where intermolecular correlations are taken into account only at short-range scales. Appendix 7.C presents a brief discussion of critical phenomena predicted by integral-equation theories.

A characteristic feature of conventional statistical-mechanical models is that they provide analytical expressions for thermodynamic properties including the unstable region of phase transition. For example, the van der Waals theory predicts that, at constant temperature, the system pressure is a continuous function of the molar volume as shown schematically in Figure 7.48. When the temperature is higher than the critical temperature of vapor–liquid transition ($T > T_c$), the pressure falls monotonically with molar volume. Below the critical temperature ($T < T_c$), the isothermal pressure shows a so-called "van der Waals loop" between the molar volumes of vapor and liquid phases at coexistence. While the mean-field method predicts unphysical thermodynamic properties in the unstable region,[90] the molar volumes of the coexisting phases can be obtained from Maxwell's "equal-area criterion," i.e., the tie line connecting the densities of coexisting phases divides the van der Waals loop into two regions of equal area. At the critical point, the minimum and maximum points of the van der Waals loop merge with the coexistence curve. As a result, the isothermal pressure versus molar volume shows an inflection point at $T = T_c$.

89 Broide M. L. et al., "Binary-liquid phase-separation of lens protein solutions", *PNAS 88* (13), 5660–5664 (1991).
90 An exact theory would predict flat isotherms in the transition region of vapor–liquid equilibrium.

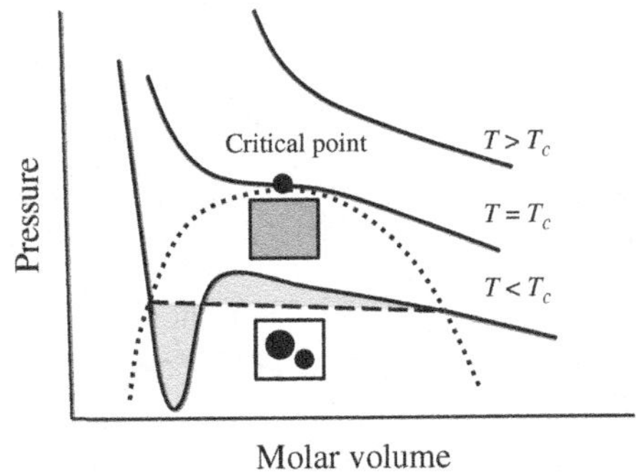

Figure 7.48 Schematic phase diagram for the vapor–liquid equilibrium for a simple fluid. At the critical point ($T = T_c$), the system becomes opalescent due to intense scattering of light by the long-range density fluctuations (gray box). For comparison, bottom box shows density variation in a gas–liquid mixture ($T < T_c$) at on the macroscopic scale (e.g., liquid droplets in a vapor). Here the dotted line represents the binodal curve, the dashed line is the vapor–liquid tie line, and the shaded areas show the Maxwell construction of the liquid–vapor equilibrium.

To elucidate the critical behavior of the vapor–liquid transition according to an analytical theory, consider an arbitrary equation of state that describes pressure P as a function of absolute temperature T and number density in the bulk ρ. If the function is everywhere analytical, it can be expanded relative to that at the critical point

$$P(T,\rho) = P(T_c,\rho_c) + \left(\frac{\partial P}{\partial T}\right)_{T_c,\rho_c}\Delta T + \left(\frac{\partial^2 P}{\partial\rho\partial T}\right)_{T_c,\rho_c}\Delta T\Delta\rho + \frac{1}{6}\left(\frac{\partial^3 P}{\partial\rho^3}\right)_{T_c,\rho_c}\Delta\rho^3 + \cdots \tag{7.246}$$

where $\Delta T = T - T_c$, $\Delta\rho = \rho - \rho_c$, and $\cdots$ represents all high order terms. The linear and second-order terms in $\Delta\rho$ do not appear in Eq. (7.246) because an analytical theory predicts that, at the critical point, the isothermal pressure as a function of the density is an inflection point:

$$(\partial P/\partial\rho)_{T_c,\rho_c} = (\partial^2 P/\partial\rho^2)_{T_c,\rho_c} = 0. \tag{7.247}$$

Eq. (7.246) suggests that, along the critical isotherm ($T = T_c$), $\Delta P = P(T_c,\rho) - P(T_c,\rho_c)$ approaches zero satisfying a power-law relation

$$\Delta P \sim \Delta\rho^3 \tag{7.248}$$

where $\sim$ stands for proportionality. As the above analysis entails no specific information on the system, the cubic dependence of pressure with respect to density is universally valid (within the framework of analytical models).

A similar power law can be derived for the density difference between the coexisting vapor and liquid phases as the system approaches the critical point. Subtracting $P(T_c, \rho_c)$ from both sides of Eq. (0.0.3) and dividing by $\Delta\rho$, we obtain

$$\frac{\Delta P}{\Delta\rho} = \left(\frac{\partial P}{\partial T}\right)_{T_c,\rho_c}\frac{\Delta T}{\Delta\rho} + \left(\frac{\partial^2 P}{\partial\rho\partial T}\right)_{T_c,\rho_c}\Delta T + \frac{1}{6}\left(\frac{\partial^3 P}{\partial\rho^3}\right)_{T_c,\rho_c}\Delta\rho^2 + \cdots. \tag{7.249}$$

Along the vapor–liquid coexistence curve, both temperature and pressure are fully determined by the fluid density. As the temperature approaches the critical value ($T \to T_c$), we have

$$\Delta P/\Delta\rho \to dP/d\rho \to 0, \tag{7.250}$$

$$\left(\frac{\partial P}{\partial T}\right)_{T_c,\rho_c}\frac{\Delta T}{\Delta\rho} \to \frac{dP}{d\rho} \to 0. \tag{7.251}$$

Neglecting the higher-order terms in Eq. (7.249) and applying $T \to T_c$, we find the power law

$$(\Delta\rho)^2 \sim \Delta T \tag{7.252}$$

along the vapor–liquid coexistence curve.

The proportionality constant in Eq. (7.252) depends on the properties of the system at the critical condition. As Eq. (7.249) is applicable to either side of the coexistence curve, we can rewrite Eq. (7.252) separately for the liquid and vapor branches

$$\rho_L - \rho_c \sim (\Delta T)^{1/2}, \tag{7.253}$$

$$\rho_c - \rho_V \sim (\Delta T)^{1/2}. \tag{7.254}$$

Summation of Eqs. (7.253) and (7.254) yields:

$$(\rho_L - \rho_V) \sim (\Delta T)^{1/2}. \tag{7.255}$$

Eq. (7.255) indicates that the density difference between the coexisting liquid and vapor phases follows a universal power law as the temperature approaches the critical value.

If an analytic equation of state is available for both inside and outside of the coexistence curve (as shown in the van der Waals loop), we may follow Eq. (7.246) to express the inverse of isothermal compressibility, $\kappa_T^{-1}(T, \rho) = \rho(\partial P/\partial \rho)_T$, in terms of a Taylor series relative to that at the critical point

$$\kappa_T^{-1}(T, \rho) = A\Delta T + B\Delta T\Delta\rho + \cdots \tag{7.256}$$

where A and B are characteristic constants of the system independent of temperature and density, $\kappa_T^{-1}(T_c, \rho_c)$ and the linear terms in $\Delta\rho$ are absent in Eq. (7.256) due to the singularities at the critical point, Eq. (7.247). Neglecting the higher-order terms in Eq. (7.256), we find a scaling relation for the isothermal compressibility

$$\kappa_T(T, \rho) \sim \Delta T^{-1}. \tag{7.257}$$

Eq. (7.257) predicts an inverse power divergence of the isothermal compressibility as the critical point approaches.

7.12.3 Critical Structure

In addition to the singularities of thermodynamic variables, a fluid near the critical point is also characterized by an abnormal structure. According to the compressibility equation discussed in Section 7.3.2, we have

$$\rho k_B T \kappa_T = \hat{S}(0). \tag{7.258}$$

Near the critical point, the divergence of the isothermal compressibility predicted by Eq. (7.257) implies that the structure factor $\hat{S}(k)$ approaches infinite at zero wavevector ($k = 0$). Because $\hat{S}(k)$ is proportional to the intensity of light scattering, Eq. (7.258) indicates that the scattered light at long wavelength increases dramatically as the system approaches the critical point. This divergence of light intensity gives rise to the well-known phenomenon of *critical opalescence*, i.e., a normally transparent liquid or vapor becomes cloudy due to macroscopic density fluctuations.

The relation between the structure factor and the isothermal compressibility allows us to understand density fluctuations near the critical point in a more quantitative way. As discussed in Section 7.4, the structure factor and total correlation function $\hat{h}(k)$ are related to direct correlation function $\hat{c}(k)$ by

$$\hat{S}(k) = 1 + \rho\hat{h}(k) = \frac{1}{1 - \rho\hat{c}(k)}. \tag{7.259}$$

The divergence of $\hat{S}(0)$ at the critical point implies

$$1 - \rho\hat{c}(0) = 1 - \rho \int d\mathbf{r}\, c(r) = 0. \tag{7.260}$$

Eq. (7.260) indicates that, unlike total correlation function $\hat{h}(k)$ or structure factor $\hat{S}(k)$, the direct correlation function remains short-ranged at the critical point, i.e., it does not diverge at long wavelength as $k \to 0$. As a result, we may write $\hat{c}(k)$ as a Taylor series in k

$$\hat{c}(k) = \frac{4\pi}{k} \int_0^{\infty} dr \ rc(r) \sin(kr) = \frac{4\pi}{k} \int_0^{\infty} drrc(r) \left[kr - \frac{(kr)^3}{3!} + \cdots\right]$$
$$= \hat{c}(0) - c_2 k^2 + \cdots \tag{7.261}$$

where $c_2 = 2\pi/3 \int dr \cdot r^4 c(r)$. Because the long-range behavior of the correlation function is determined by the small values of k, we may neglect the higher-order terms in Eq. (7.261).

Eq. (7.261) can be utilized to analyze the asymptotic behavior of the structure factor and the total correlation function near the critical point. Substituting the first two terms in Eq. (7.261) for $\hat{c}(k)$ in Eq. (7.259) gives

$$\hat{S}(k) = \frac{1}{1 - \rho\hat{c}(0) + \rho c_2 k^2}, \tag{7.262}$$

Noting that $\rho k_B T \kappa_T = \hat{S}(0)$, we can rewrite Eq. (7.262) as

$$\hat{S}(k) = \frac{\rho k_B T \kappa_T}{1 + \xi^2 k^2} \tag{7.263}$$

where $\xi \equiv (\rho^2 k_B T \kappa_T c_2)^{1/2}$. Similarly, the total correlation function is given by

$$\hat{h}(k) = \frac{\hat{c}(0) - c_2 k^2}{1 - \rho\,\hat{c}(0) + \rho c_2 k^2} \approx \frac{\hat{c}(0)/(\rho c_2)}{\xi^{-2} + k^2}. \tag{7.264}$$

In Eqs. (7.262) and (7.264), ξ has the unit of length and its physical significance becomes apparent after the inverse Fourier transform of Eq. (7.264)

$$h(r) = \frac{\hat{c}(0)}{\rho c_2} e^{-r/\xi}. \tag{7.265}$$

Eq. (7.265) indicates that ξ reflects the range of the pair correlation in the thermodynamic system, i.e., total correlation function $h(r)$ is most significant when the separation between two molecules is comparable with ξ and diminishes as $r >> \xi$. Therefore, this parameter can be understood as the correlation length. A large value of ξ means long-range correlation, and vice versa, a small ξ means short-range correlation.

At conditions remote from the critical point, $\rho k_B T \kappa_T \sim 1$. In this case, the correlation length does not differ much from the microscopic length scale that characterizes the range of the pair potential or the diameter of a single molecule, i.e., $\xi \sim \sigma$. Approaching the critical point, the divergence of κ_T leads to the power law

$$\xi \sim \Delta T^{-1/2}. \tag{7.266}$$

Eq. (7.266) is obtained from the definition of ξ and Eq. (7.257) and $\kappa_T \sim \Delta T^{-1}$; it indicates that, at the critical point, the correlation length diverges to infinity. Because the total correlation function is directly related to density fluctuations, the divergence of the correlation length indicates that in a system at the critical point, the densities at any two positions are highly correlated.

Table 7.2 summarizes the critical exponents predicted from analytical theories. Because of the connections between different thermodynamic quantities, the six scaling exponents are not independent to each other. As discussed in Section 5.11, only three of the six critical exponents are

Table 7.2 Critical exponents for the thermodynamic properties near the critical point.

Exponent	Scaling Law	Experiment[5]	Analytical theory
β	$\Delta\rho \sim \Delta T^{\beta}$	0.326 ± 0.002	0.5
γ	$\kappa_T(T, \rho_c) \sim \lvert\Delta T\rvert^{-\gamma}$	1.239 ± 0.002	1
δ	$\Delta\rho \sim \Delta P^{1/\delta}$	3.80 ± 0.02	3
η	$\hat{S}(k) \sim k^{-2+\eta}$	0.031 ± 0.004	0
ν	$\xi \sim \lvert\Delta T\rvert^{-\nu}$	0.630 ± 0.002	0.5
α	$C_V \sim \lvert\Delta T\rvert^{-\alpha}$	0.107 ± 0.002	0

independent (two for thermodynamics and one for the correlation function) and all others can be deduced from Widom's relations:

$$\gamma = (2 - \eta)\nu, \tag{7.267}$$

$$\alpha + 2\beta + \gamma = 2, \tag{7.268}$$

$$\alpha + \beta(\delta + 1) = 2. \tag{7.269}$$

Also shown in Table 7.2 are the critical exponents observed from experimental measurements. The discrepancy between the theoretical predictions and experimental results is astonishing because the only assumption in the above analysis is that analytic expressions are assumed for thermodynamic functions. For example, an analytic expression for the internal energy as a function of temperature generates a finite value of the heat capacity. Therefore, in terms of a power law for temperature, the heat capacity can be expressed

$$C_V \sim \Delta T^0 \tag{7.270}$$

where exponent "0" means that the heat capacity has a finite value at the critical point, in contrast to experiment that shows the divergence of the heat capacity. Because the long-range correlation near the critical point cannot be captured by any analytical expression, all analytical theories predict similar but incorrect critical behavior.

7.12.4 Summary

Although an analytical theory predicts universality at the critical point, the critical exponents are generally different from experimental observations. As discussed in next section (Section 7.13), a more accurate description of the critical behavior of molecular systems can be achieved by integrating liquid-state methods with the normalization group theory.

7.13 Molecular Theory of Critical Phenomena

The theoretical methods discussed in previous sections are not accurate for systems with long-range correlations, in particular, for systems near the critical point of vapor–liquid or liquid–liquid transitions. To consider the long-range correlations near the critical point, we need to use the renormalization-group (RG) theory. As discussed in Section 5.12, the original concepts of the RG theory were proposed independently by Kadanoff and Wilson based on the scale invariance of critical properties. In this section, we discuss an extension of the RG method for modeling the critical behavior from a molecular perspective. This alternative method was proposed by White and Zhang by integrating a molecular theory with recursive iterations to account for long-range correlations.[91]

7.13.1 The Yang–Lee Theorems

In statistical mechanics, a partition function is defined in terms of the summation of exponentials that are continuous and infinitely differentiable. Nonanalytical behavior emerges during phase transition because the partition function exhibits singularity, i.e., it yields discontinuous thermodynamic properties in responses to the change of the system parameters. The connection between the grand canonical partition function and phase transition was established by Yang and Lee in the early 1950s in terms of two mathematical theorems.[92] While any analytical expression for the equation of state would lead to the mean-field critical exponents, the Yang–Lee theorems ensure that phase transition including the critical behavior can be adequately described in terms of partition functions.

7.13.2 Short- and Long-Range Density Fluctuations

To elucidate how long-range correlations can be incorporated into conventional liquid-state methods, we may start with the canonical partition function for a system with N spherical particles[93]

$$Q = \frac{1}{N!\Lambda^{3N}} \int d\mathbf{r}^N \exp[-\beta\Phi] = \exp[-\beta F] \tag{7.271}$$

where $\mathbf{r}^N \equiv (\mathbf{r}_1, \mathbf{r}_2, \cdots, \mathbf{r}_N)$ represents the positions of N particles, $d\mathbf{r}^N \equiv d\mathbf{r}_1 d\mathbf{r}_2 \cdots d\mathbf{r}_N$, Λ is the thermal wave length, Φ represents total potential energy, and F is the Helmholtz energy. According to Yang–Lee theorems, the partition function itself is exactly near or far from the critical point, provided that the configurational integral in Eq. (7.271) can be evaluated exactly.

In conventional liquid-state theories, the configuration of the system is defined by the positions of individual particles. The position-based approach is convenient when the system is remote from the critical point. In that case, the multi-body correlation among particle positions is short-ranged, i.e., on the order of molecular length scales. Near the critical point, however, the system develops macroscopic fluctuations in local density, even though the average density is everywhere uniform. To understand thermodynamic properties near the critical point, we must consider the correlated fluctuations of the local particle density.

Near the critical point, the fluctuation effect can be described by the spatial variation of local density $\rho(\mathbf{r})$. Although the average density is independent of position for a uniform system, the local density may be understood as a "coarse-grained" density of particles in a differential volume centered at position $\mathbf{r}$ with a length scale that is much larger than the particle size but small from a macroscopic perspective.

91 White J. A. and Zhang S., "Renormalization group theory for fluids", *J. Chem. Phys.* 99, 2012 (1993).
92 Yang C. N. and Lee T. D., "Statistical theory of equations of state and phase transitions", *Phys. Rev.* 87 (3), 404–410 (1952).
93 The canonical formalism presents some mathematical problems in discussing density fluctuations. While the grand partition function is more suitable for an exact derivation, here we use the canonical partition to illustrate how the recursion relations are derived in the RG theory.

In terms of ensemble average, the local density, $\rho(\mathbf{r})$, is expected to be a smooth function of position. Accordingly, we can express $\rho(\mathbf{r})$ in the Fourier space

$$\rho(\mathbf{r}) = \frac{V}{(2\pi)^3} \int d\mathbf{k}\hat{\rho}(\mathbf{k})e^{-i\mathbf{k}\cdot\mathbf{r}} \tag{7.272}$$

where $\hat{\rho}(\mathbf{k})$ is the Fourier transform of $\rho(\mathbf{r})$. The inverse Fourier transform gives

$$\hat{\rho}(\mathbf{k}) = \frac{1}{V} \int d\mathbf{r}\rho(\mathbf{r})e^{i\mathbf{k}\cdot\mathbf{r}} \tag{7.273}$$

where $\mathbf{k}$ is a wave vector ($e^{i\mathbf{k}\cdot\mathbf{r}}$is a sinusoidal function), and the magnitude of $\mathbf{k}$ is related to the wavelength λ by

$$k = 2\pi/\lambda. \tag{7.274}$$

We introduce the total volume V in the Fourier integrals so that $\hat{\rho}(\mathbf{k})$ has the same dimensionality as that of $\rho(\mathbf{r})$.

Eq. (7.272) indicates that the coarse-grained local density can be represented as the integration (summation) of the Fourier modes. According to Eq. (7.273), $\hat{\rho}(0)$ gives the average density of the entire system, and $\hat{\rho}(\mathbf{k})$ represents the amplitude of the density fluctuation with a zero average. For small $\mathbf{k}$, $\hat{\rho}(\mathbf{k})e^{i\mathbf{k}\cdot\mathbf{r}}$ represents density fluctuations with a long wavelength (large λ), and vice versa, for large $\mathbf{k}$, $\hat{\rho}(\mathbf{k})e^{i\mathbf{k}\cdot\mathbf{r}}$ represents density fluctuations with a short wavelength (small λ). The critical phenomenon is manifested as long-range density fluctuations. In other words, the long-wavelength components of $\rho(\mathbf{r})$ make important contributions to both fluid structure and thermodynamic properties. In contrast, far from the critical point, the density fluctuations appear only in the short-length scale, i.e., at a length comparable to the molecular size. In that case, the thermodynamic properties are determined by the local structure.

7.13.3 Spectrum of the Partition Function

Following the coarse-grained representation of the local density, we may formally rewrite the partition function in terms of all possible density fluctuations

$$Q = \sum_{\rho(\mathbf{r})} e^{-\beta F[\rho(\mathbf{r})]}. \tag{7.275}$$

In Eq. (7.275), $F[\rho(\mathbf{r})]$ is equivalent to the Helmholtz energy of the system with a coarse-grained density $\rho(\mathbf{r})$, i.e., F is a functional of $\rho(\mathbf{r})$. The summation is equivalent to a functional integration that applies to all physically acceptable local densities. These local densities account for all possible forms of density fluctuations including both long- and short-wavelength contributions. In comparison with the classical canonical partition function given by Eq. (7.271), $F[\rho(\mathbf{r})]$ may be understood as an effective Helmholtz energy, i.e., the Helmholtz energy of the system when the local density is given by a smooth function $\rho(\mathbf{r})$.

To evaluate the summation in Eq. (7.275) over all possible density functions, consider first a local density with only short-wavelength contributions:

$$\overline{\rho}_s(\mathbf{r}) = \hat{\rho}(0) + \frac{V}{(2\pi)^3} \int_{k>k_s} d\mathbf{k}\hat{\rho}(\mathbf{k})e^{-i\mathbf{k}\cdot\mathbf{r}} \tag{7.276}$$

where k_s is a cut-off wavevector (or $\lambda_s = 2\pi/k_s$ represents a cutoff wavelength). Intuitively, this portion of density profiles is associated with the local structure of the system as determined by the short-range repulsion among molecules. Here, by short wave length, we mean that λ_s is comparable to the molecular diameter.

Upon the summation over all local densities with only short-wavelength contributions, we may express the partition function as

$$Q = \sum_{\rho_s(\mathbf{r})} e^{-\beta F[\overline{\rho}_s(\mathbf{r})]-\beta U_s} \tag{7.277}$$

where the summation applies only to the long-wavelength density components

$$\rho_s(\mathbf{r}) = \frac{V}{(2\pi)^3} \int_{0<k\leq k_s} d\mathbf{k}\hat{\rho}(\mathbf{k})e^{-i\mathbf{k}\cdot\mathbf{r}}. \tag{7.278}$$

In Eq. (7.277), $F_s[\overline{\rho}_s(\mathbf{r})]$ represents contributions to the Helmholtz energy that are insensitive to density fluctuations with wavelength longer than λ_s, and U_s represents the contribution to the Helmholtz energy due to density fluctuations with longer wavelength. Because a one-component system with only short-range repulsion does not exhibit a vapor–liquid-like critical point, the contribution to the partition function from density fluctuations with wavelength longer than λ_s comes primarily from the intermolecular attraction.

7.13.4 Long-Range Fluctuations

For a given density profile $\rho(\mathbf{r})$ that varies smoothly over space, we may estimate the potential energy due to intermolecular attraction from the mean-field approximation[94]

$$\Phi_A = \frac{1}{2}\int d\mathbf{r}\int d\mathbf{r}'\rho(\mathbf{r})\rho(\mathbf{r}')u_A(|\,\mathbf{r}-\mathbf{r}'\,|) \tag{7.279}$$

where $u_A(r)$ represents the attractive (viz., long-range) part of the intermolecular potential. In the Fourier space, Eq. (7.279) can be rewritten as

$$\Phi_A = \frac{1}{2}\frac{V^2}{(2\pi)^3}\int d\mathbf{r}\rho(\mathbf{r})\int d\mathbf{k}\rho(\mathbf{k})u_A(\mathbf{k})e^{-i\mathbf{k}\cdot\mathbf{r}} = \frac{V^3}{2}\int d\mathbf{k}\rho(\mathbf{k})u_A(\mathbf{k})\rho(-\mathbf{k})e^{-i\mathbf{k}\cdot\mathbf{r}}, \tag{7.280}$$

where

$$\hat{u}_A(\mathbf{k}) = \frac{1}{V}\int d\mathbf{r}u_A(r)e^{i\mathbf{k}\cdot\mathbf{r}}. \tag{7.281}$$

As discussed above, U_s represents the long-wavelength contributions to the Helmholtz energy. In other words, it is related to the long-range part of the intermolecular potential Φ_A

$$U_s = \frac{V^3}{2}\int_{k<k_s} d\mathbf{k}\hat{\rho}(\mathbf{k})\hat{u}_A(\mathbf{k})\hat{\rho}(-\mathbf{k})e^{-i\mathbf{k}\cdot\mathbf{r}} - \frac{V^2}{2}\hat{\rho}^2(0)\hat{u}_A(0). \tag{7.282}$$

The contribution from the average density is subtracted from Eq. (7.282) as it is already taken into account in Eq. (7.276).

---

94 Equation (7.279) corresponds to a van der Waals-like approximation for the contribution of intermolecular attraction to the Helmholtz energy. More accurate theories may be used to evaluate this part of free energy.

7.13.5 Local-Density Approximation

Without long-range fluctuations, we may evaluate the Helmholtz energy using the local-density approximation (LDA)

$$F[\overline{\rho}_s(\mathbf{r})] = \int d\mathbf{r} f[\overline{\rho}_s(\mathbf{r})] \tag{7.283}$$

where $f[\overline{\rho}_s(\mathbf{r})] = F/V$ represents the local Helmholtz energy density, which can be calculated from conventional liquid-state methods. Substituting Eqs. (7.283) and (7.282) into (7.277) gives

$$\begin{aligned} Q &= \sum_{\rho_s(\mathbf{r})} \exp\left\{ -\beta F_{MF}[\overline{\rho}_s(\mathbf{r})] - \beta \frac{V^3}{2} \int_{q<q_s} d\mathbf{k} \hat{\rho}(\mathbf{k}) \hat{\varphi}(\mathbf{k}) \hat{\rho}(-\mathbf{k}) e^{-i\mathbf{k}\cdot\mathbf{r}} + \frac{V^2}{2} \hat{\rho}^2(0) \beta \hat{\varphi}(0) \right\} \\ &= \sum_{\rho_s(\mathbf{r})} \exp\left\{ - \int d\mathbf{r} f_0(\rho) - \beta U_s' \right\} \end{aligned} \tag{7.284}$$

where $f_0 \equiv f - \frac{V}{2}\hat{\rho}^2(0)\hat{\varphi}(0)$ and $U_s' = U_s + \frac{V^2}{2}\hat{\rho}^2(0)\hat{\varphi}(0)$. In a typical mean-field theory, long-range fluctuations are neglected ($U_s = 0$). In that case, the summation in Eq. (7.284) is not required.

7.13.6 Recursion Equations

Eq. (7.284) provides the starting point to account for long-range correlations by RG iterations. The essential idea is that the contributions from the long-wavelength contributions to the local density can be removed successively from U_s by incorporating it into the *conventional* Helmholtz energy f_0. Subsequently, the set of function summation in the partition function diminishes as the RG iteration proceeds. At the end of RG iterations, U_s disappears and the partition function would include all levels of long-range density fluctuations.

In each step of the RG iteration, the partition function remains unchanged when the contributions of the long-wavelength densities are subtracted from U_s and incorporated into the f_0. Suppose that the components of the density modes with $k_l < k < k_s$ have been subtracted from U_s and incorporated into the conventional form of the Helmholtz energy, the partition function can be written as

$$Q = \sum_{\rho_s(\mathbf{r})} \exp\left\{ - \int d\mathbf{r} f_s[\overline{\rho}_s(\mathbf{r})] - \beta U_s' \right\} = \sum_{\rho_l(\mathbf{r})} \exp\left\{ - \int d\mathbf{r} f_l[\overline{\rho}_l(\mathbf{r})] - \beta U_l' \right\} \tag{7.285}$$

where $f_s = f_0$ is the Helmholtz energy density before the RG iteration; f_l is the Helmholtz energy density after a single step of RG iteration, and the long-wavelength components of the density $\rho_l(\mathbf{r})$ is defined as

$$\rho_l(\mathbf{r}) = \frac{\mathrm{V}}{(2\pi)^3} \int_{0<k<k_l} d\mathbf{k} \hat{\rho}(\mathbf{k}) e^{-i\mathbf{k}\cdot\mathbf{r}} \tag{7.286}$$

with $k_l < k_s$.

As the density modes with $k_l < k < k_s$ have been subtracted, $\rho_l(\mathbf{r})$ represents a smaller set of density modes compared with $\rho_s(\mathbf{r})$ (Eq. (7.278)). To account for contributions due to this subset of densities, we can find the difference in Helmholtz energy density, $\delta f_l[\overline{\rho}(\mathbf{r})]$, defined by

$$\delta f_l[\overline{\rho}_l(\mathbf{r})] = f_l[\overline{\rho}_l(\mathbf{r})] - f_s[\overline{\rho}_s(\mathbf{r})] \tag{7.287}$$

from the invariance of the partition function

$$\exp\left\{ -\beta \int d\mathbf{r} \delta f_l[\overline{\rho}(\mathbf{r})] \right\} = \sum_{\rho_D(\mathbf{r})} \exp\left\{ -\beta \int d\mathbf{r} f_D - \beta U_D \right\} \tag{7.288}$$

where $\overline{\rho}(\mathbf{r})$ is defined as in Eq. (7.276) and

$$\rho_D(\mathbf{r}) = \frac{V}{(2\pi)^3}\int_{k_l<k<k_s} d\mathbf{k}\hat{\rho}(\mathbf{k})e^{-i\mathbf{k}\cdot\mathbf{r}}, \tag{7.289}$$

$$f_D = f_s[\overline{\rho}_l(\mathbf{r})] - f_s[\overline{\rho}_s(\mathbf{r})], \tag{7.290}$$

$$U_D = U_s - U_l. \tag{7.291}$$

Eqs. (7.288)–(7.291) provide a basis to derive the recursion equations of the RG theory.

7.13.7 Summation over the Phase-Space Cells

To evaluate the function summation in Eq. (7.288), we may divide the thermodynamic system into phase-space cells with sub-volume V_D. Within each phase-space cell, we assume that the long wavelength components $\overline{\rho}_l(\mathbf{r})$ is approximately a constant, and the differential density mode, $\rho_D(\mathbf{r})$, can be approximated by a sinusoidal function

$$\rho_D(\mathbf{r}) = x \cdot \cos(\mathbf{k}\cdot\mathbf{r}) \tag{7.292}$$

where x is the amplitude of the density modes $\rho_D(\mathbf{r})$, and $k_l < k < k_s$. In comparison with Eq. (7.289), the imaginary part of the Fourier exponent disappears because $\rho_D(\mathbf{r})$ is represented by an even function of $\mathbf{k}$.

With the above approximation for density fluctuations, the summation over $\rho_D(\mathbf{r})$ becomes equivalent to integration over amplitude x for any given $\mathbf{k}$. Accordingly, $\delta f_l[\overline{\rho}(\mathbf{r})]$ can be evaluated from

$$\exp\{-\beta V_D \delta f_l[\rho_l]\} = A_0^{-1}\int dx \exp\left[-\beta\int d\mathbf{r}\{f_s[\rho_l - x\cdot\cos(\mathbf{k}\cdot\mathbf{r})] - f_s(\rho_l)\} - \beta U_D\right] \tag{7.293}$$

where ρ_l is the magnitude of $\overline{\rho}_l(\mathbf{k})$ within sub-volume V_D, and A_0 is a scale factor that arises from the conversion of the summation over $\rho_D(\mathbf{r})$ to integration over x. The scale factor can be evaluated in the limit of no attractive interaction such that $\delta f_l[\overline{\rho}(\mathbf{r})]$ vanishes and the Helmholtz energy at the RG iteration equals the initial estimate, $f_l = f_s$.

The first integral on the right side of Eq. (7.293) can be evaluated using Euler's formula

$$\int d\mathbf{r}\{f_s[\rho_l - x\cdot\cos(\mathbf{k}\cdot\mathbf{r})] - f_s(\rho_l)\} \approx V_D\{[f_s(\rho_l + x) + f_s(\rho_l - x)]/2 - f_s(\rho_l)\}. \tag{7.294}$$

Conversely, the second term depends only on density modes with $k_l < k < k_s$. According to Eq. (7.282), this term is given by

$$\begin{aligned} U_D &= \frac{x^2}{2}\int_{V_D} d\mathbf{r}\int_{V_D} d\mathbf{r}' \cos(\mathbf{k}\cdot\mathbf{r})\cos(\mathbf{k}\cdot\mathbf{r}')u_A(|\mathbf{r}-\mathbf{r}'|) \\ &= \frac{x^2 V_D}{4}\int_{V_D} d\mathbf{r}' \cos(\mathbf{k}\cdot\mathbf{r}')u_A(r') \end{aligned} \tag{7.295}$$

In the limit of small k, the integral in Eq. (7.295) can be approximated by

$$\begin{aligned} U_D &\approx \frac{x^2 V_D}{4}\int_{V_D} d\mathbf{k} u_A(r')[1 - (\mathbf{k}\cdot\mathbf{r}')^2/2] \\ &= -\frac{x^2 V_D}{4}(1 - k^2L^2/2)a \end{aligned} \tag{7.296}$$

where a resembles the van der Waals parameter

$$a = -\frac{1}{2}\int_{V_D} d\mathbf{r}' u_A(r') \tag{7.297}$$

and L reflects of the range of intermolecular attraction

$$L^2 = -\frac{1}{a}\int_{V_D} d\mathbf{r}'(\hat{\mathbf{k}}\cdot\mathbf{r})^2 u_A(r') \tag{7.298}$$

with $\hat{\mathbf{k}} = \mathbf{k}/k$ denoting a unit vector. Intuitively, parameter a represents the average attractive potential in the sub-volume V_D, and L is the effective distance of the long-range potential.

7.13.8 The RG Recursion

To implement the RG recursion numerically, we begin with a Helmholtz energy density $f_0(\rho)$ that can be calculated from a conventional liquid-state method (viz., Eq. (7.284)). Successive corrections δf_n, $n = 1, 2, \cdots$, is then added to $f_0(\rho)$ to account for the contributions of the density fluctuations at longer wavelength. Eventually, the RG procedure leads to

$$f(\rho) = \lim_{n\to\infty} f_n(\rho), \tag{7.299}$$

$$f_n(\rho) = f_{n-1}(\rho) + \delta f_n(\rho), \tag{7.300}$$

$$\delta f_n(\rho) = -\frac{1}{\beta V_{D,n}} \ln \frac{\int dx \exp\left\{-\beta V_{D,n}\left[f_D - ax^2\left(1 - k_n^2 L^2/2\right)/4\right]\right\}}{\int dx \exp\{-\beta V_{D,n} f_D\}}, \tag{7.301}$$

$$f_D = [f_{n-1}(\rho + x) + f_{n-1}(\rho - x)]/2 - f_{n-1}(\rho). \tag{7.302}$$

Typically, the recursion starts at a wavelength, $\lambda_0/(2\pi) = k_0^{-1} = L/\sqrt{2}$. At this length scale, we assume that the fluctuations have already been accounted for by a conventional liquid-state method. Beyond this wavelength, the density fluctuations are neglected, and δf_0 is equal zero. The initial volume of the phase-space cell is taken as the cube of the starting wavelength $V_{D,0} = (\sqrt{2}\pi L)^3$. The fluctuation wavelength is increased at each step by a constant t such that the wavevector is reduced consecutively

$$k_n = k_0/t^n. \tag{7.303}$$

Consequently, the phase-space cell volume V_D is increased at each step by t^3

$$V_{D,n} = t^3 \cdot V_{D,0}. \tag{7.304}$$

The final Helmholtz energy density of the system is calculated from

$$F/V = f + \frac{V}{2}\hat{\rho}^2(0)\hat{\varphi}(0) = f - a\rho^2. \tag{7.305}$$

Satisfactory thermodynamic properties can be achieved after only a few steps of RG recursion.

With an analytic expression for the Helmholtz energy that is accurate near and far from the critical point, we can predict the vapor–liquid coexistence curve using conventional procedures of phase-equilibrium calculations. For example, Figure 7.49 compares the phase diagram of a Lennard-Jones fluid calculated from FMSA with the RG correction and that from computer simulations. Here, the recursion procedure is slightly modified by a density-dependent interaction length L.[95] The agreement between the theory and simulation is excellent.

Table 7.3 summarizes the critical exponents for vapor–liquid transition of simple fluids predicted by the RG theory.[96] The nonclassical exponents imply that the RG iteration leads to nonanalytic behavior of the free energy in the vicinity of the critical point. The slight difference between the

---

95 Tang Y. P., "Outside and inside the critical region of the Lennard-Jones fluid", *J. Chem. Phys.* 109, 5935 (1998).
96 White J. A., "Contribution of fluctuations to thermal properties of fluids with attractive forces of limited range: theory compared with PpT and C data for argon", *Fluid Phase Equilib.* 75, 53–64 (1992).

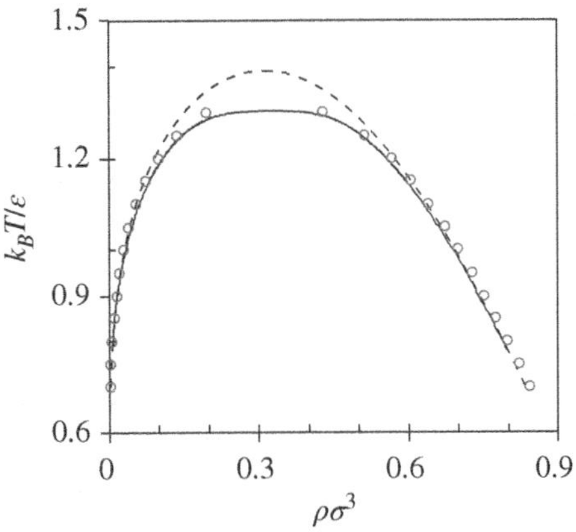

Figure 7.49 Vapor–liquid coexistence curve of a Lennard–Jones fluid. The points are from molecular simulations, the dashed and the solid lines are calculated from FMSA without and with the renormalization-group corrections, respectively. Source: Adapted from Tang.[95]

Table 7.3 Critical exponents predicted by the RG theory.

	$\Delta\rho \sim \lvert\Delta T\rvert^{\beta}$	$\kappa_T \sim \lvert\Delta T\rvert^{-\gamma}$	$\Delta\rho \sim \lvert\Delta P\rvert^{1/\delta}$	$C_V \sim \lvert\Delta T\rvert^{-\alpha}$	$\xi \sim \lvert\Delta T\rvert^{-\nu}$
Theory	$\beta \approx 0.324$	$\gamma \approx 1.236$	$\delta \approx 4.81$	$\alpha \approx 0.115$	$\nu \approx 0.61$
Experiment	$\beta \approx 0.33$	$\gamma \approx 1.24$	$\delta \approx 4.8$	$\alpha \approx 0.11$	$\nu \approx 0.63$

Source: Adapted from White.[95]

experimental results and those predicted by the RG iteration arises from approximations introduced in deriving the recursion equations. In particular, it does not account for a small deviation of $\gamma = 2\upsilon$ predicted by the OZ correlation function. This deviation is described by a critical exponent η, which is about 0.03 according to light scattering experiments.[97]

In principle, the RG procedure accounts for all levels of density fluctuations. Nevertheless, a better liquid-state method for the fluids far from the critical point is always desirable because the recursion equations are most accurate near the critical point. Different from the original RG theory, integration of a molecular theory with the RG iteration can be applied to construct phase coexistence curves both inside and outside the critical region for thermodynamic systems of practical interest. To illustrate, Figure 7.50 shows the interaction between the fluid–fluid transition and the fluid–solid transition of the Yukawa model system for three values of $\kappa\sigma$. When the attraction is long-ranged ($\kappa\sigma = 3.9$), the fluid–fluid phase transition is thermodynamically stable in comparison to the fluid–solid transition.[98] In this case, the phase diagram includes a triple point where a dilute solution (vapor), a concentrated solution (liquid) and a crystalline solid are at equilibrium. As expected, the RG correction has minor effect when the system is remote from the fluid–fluid critical region. As the range of attraction is reduced ($\kappa\sigma = 7$), the fluid–fluid coexistence curve lies underneath the freezing line, denoting that the fluid–fluid transition is metastable. In combination with the cell model for the solid phase, the calculated freezing and melting lines are in excellent agreement with the simulation results, particularly in the limit of high temperature.

97 Hassan B., Sengers J. V. and Anisimov M. A., "Chapter 10: thermodynamic behaviour of fluids near critical points", *Applied thermodynamics of fluids*, Edited by A. R. H. Goodwin, J. V. Sengers and C. J. Peters International Union of Pure and Applied Chemistry, 2010.

98 Fu D., Li Y. G. and Wu J., "Effect of the range of attractive interactions on crystallization, metastable phase transition and percolation in colloidal dispersions", *Phys. Rev. E* 68, 011403 (2003).

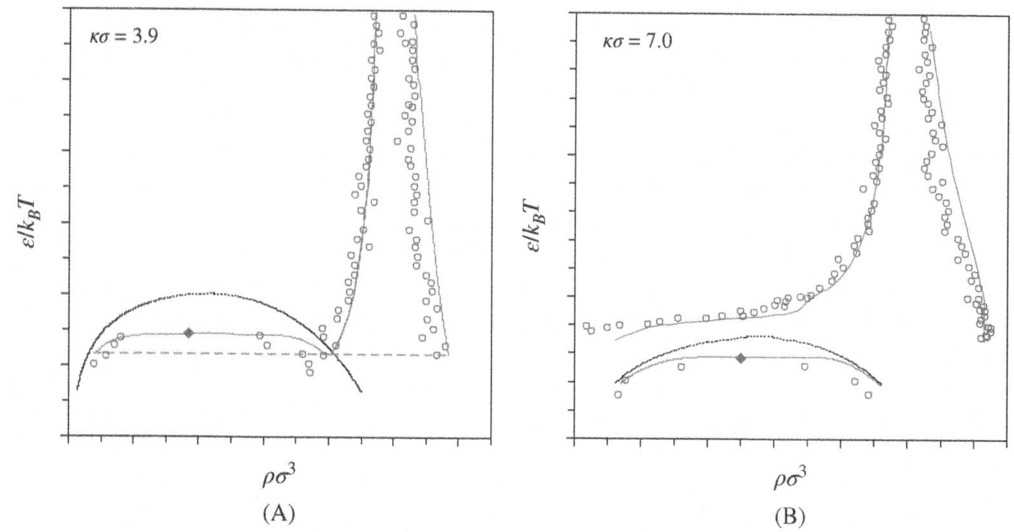

Figure 7.50 Phase diagrams for the Yukawa potential at different ranges of attraction. (A) $\kappa\sigma = 3.9$, the fluid–fluid coexistence is thermodynamically stable; (B) $\kappa\sigma = 7.0$, the fluid–fluid phase transition is metastable. The solid lines and circles are fluid–fluid and fluid–solid coexistence curves calculated from the RG theory and simulation, respectively. The dashed lines are fluid–fluid coexistence curves according to a liquid-state theory without RG correction. The diamonds correspond to the critical points calculated from MC simulation. Source: Reproduced from Fu et al.[98]

7.13.9 Summary

Given an accurate intermolecular potential, the combination of liquid-state theories with the RG iteration provides a quantitative description of the phase behavior of molecular systems near and far away from the critical region. While our discussion here is mostly concerned with simple fluids, similar calculations are applicable to protein solutions and colloidal dispersions wherein the critical behavior has implications on the kinetics of self-assembly and crystallization processes. For colloids with short-ranged attractions, the critical temperature of the fluid–fluid transition is well correlated with the range of attractions, essentially independent of the details of the potential profiles.[69] Such correlation will be useful for the estimation of the critical temperature of colloidal dispersions once the potential profile or the osmotic second virial coefficient is already known. As discussed in Section 7.10, a correlation between the critical temperature and the range of attraction may find applications in the identification of favorable solution conditions leading toward protein crystallization.

7.14 Chapter Summary

The molecular theories discussed in this chapter are mostly concerned with various correlation functions and liquid-state methods for describing the microscopic structure and thermodynamic properties of systems containing spherical particles. Similar concepts and theoretical procedures can be established for more complicated systems, as will be discussed in Chapter 8.

Toward the development of statistical-mechanical theories that relate the pair potential to the microscopic structure and thermodynamic properties, we introduce two complementary procedures, i.e., integral-equation theories and perturbation methods. The integral-equation theories, such as HNC, PY, and MSA, provide connections among different types of correlation functions. These correlation functions are introduced to quantify the correlated distributions or correlated fluctuations of particle densities. In contrast to mean-field methods, thermodynamic properties are calculated from the correlation functions through the compressibility, virial, or energy equations. The perturbation methods separate the intermolecular potential into a reference and a perturbation energy. These include the Weeks–Chandler–Anderson theory (WCA), the Barker–Henderson (BH) perturbation theory, and the first-order mean spherical approximation (FMSA). In comparison with simulation data, the perturbation theories are more accurate than van der Waals-type equations of state. However, neither is reliable near the critical point of phase transition. In the latter case, we must consider long-range correlations using the renormalization-group theory.

With an analytical expression for the pair potential, we can evaluate both microscopic structure and thermodynamic properties by running molecular simulation. Whereas simulation methods play a central role in validating theoretical development, analytical methods are valuable in their own right for capturing the essential physics and gaining intuitive understanding. Importantly, because of their superior computational efficiency, analytical methods are convenient for predicting phase diagrams and thermodynamic properties of practical systems over a broad range of conditions.

The statistical mechanics of simple fluids is often discussed in the context of gases or liquids of spherical or nearly spherical molecules. Common examples are monatomic gases like argon or systems containing small nonpolar molecules such as methane, nitrogen, and cyclohexane. Such fluids have been examined experimentally and theoretically for many years, and their overall behavior is fairly well understood. Whereas practical applications of simple fluids are mostly affiliated with petrochemical and gas industries, we show in this chapter that similar models are useful for describing the structure and phase behavior of soft matter including colloidal dispersions, micelles, and aqueous solutions of globular proteins. In addition to their practical implications for materials fabrication and biotechnology, the theoretical models help to address fundamental issues such as critical phenomena, protein crystallization, and self-assembly of nanostructured materials.

7.A Thermodynamic Properties and Direct Correlation Functions of Hard-Sphere Mixtures

In this appendix, we outline the key results derived from the Percus–Yevick (PY) theory for hard-sphere mixtures. An alternative derivation is given for predicting the thermodynamic properties of multicomponent hard-sphere fluids using the scaled-particle theory (SPT). The PY theory for the SHS model discussed in Section 7.6 can also be extended to multicomponent systems.[99]

7.A.1 The PY Theory for Hard-Sphere Mixtures

The PY theory can be directly applied to uniform hard-sphere mixtures. Like the one-component case, analytical solutions are available for both direct correlation functions and thermodynamic properties.[100] These equations are instrumental for further development of liquid-state methods including equations of state for multi-component systems of practical interests.

---

99 Perram J. W. and Smith E. R., *Chem. Phys. Lett.* 35, 138 (1975); Barboy B., *Chem. Phys.* 11, 357 (1975); Barboy B. and Tenne R., *Chem. Phys.* 38, 369 (1979).
100 Lebowitz J. L., "Exact solution of generalized Percus-Yevick equation for a mixture of hard spheres", *Phys. Rev.* 133, A895 (1964).

For most practical applications, the thermodynamic properties of hard-sphere mixtures are represented by the Boublík–Mansoori–Carnahan–Starling–Leland (BMCSL) equation of state.[101] Like the Carnahan–Starling (CS) equation for one-component hard-sphere systems, the BMCSL equation is obtained by a combination of the virial (1/3) and compressibility (2/3) equations. In dimensionless form, the excess Helmholtz energy density is given by

$$f_{HS}^{ex} \equiv \frac{\beta F_{HS}^{ex}}{V} = -n_0 \ln(1-n_3) + \frac{n_1 n_2}{1-n_3} + \frac{n_2^3}{36\pi n_3^3}\left[\ln(1-n_3) + \frac{n_3}{(1-n_3)^2}\right] \tag{7.A.1}$$

where $n_{\alpha=0,1,2,3}$ are defined in terms of the number densities (ρ_i) and particle diameters (σ_i)

$$n_0 = \sum_i \rho_i; \quad n_1 = \sum_i \rho_i\sigma_i/2; \quad n_2 = \pi\sum_i \rho_i\sigma_i^2; \quad n_3 = \pi\sum_i \rho_i\sigma_i^3/6. \tag{7.A.2}$$

The physical meanings of the above parameters can be intuitively understood as follows: n_0 is the total number density of all particles; n_3 is the overall packing fraction, i.e., the fraction of total volume occupied by the spherical particles; $n_{\alpha=1,2,3}$ are the overall number densities weighted with respect to the particle radius, surface area, and volume for each type of particle, respectively. A similar equation was derived by Boublik for multicomponent mixtures of hard-convex particles.[102]

The BMCSL equation of state is obtained from the partial derivative of the Helmholtz energy with respect to volume

$$\frac{P}{k_B T} = \frac{6}{\pi}\left[\frac{n_0}{1-n_3} + \frac{3n_1 n_2}{(1-n_3)^2} + \frac{(3-n_3)n_2^3}{(1-n_3)^3}\right]. \tag{7.A.3}$$

Similarly, the partial derivative of the excess Helmholtz energy with respect to the number density yields the excess chemical potential

$$\begin{aligned}\beta\mu_i^{HS} = &-\ln(1-n_3) + \frac{3\left(n_2\sigma_i + n_1\sigma_i^2\right)}{(1-n_3)} + \frac{9n_2^2\sigma_i^2}{2(1-n_3)^2} + \frac{\pi}{6}\frac{P\sigma_i^3}{k_B T}\\ &+3\left(\frac{n_2\sigma_i}{n_3}\right)^2\left[\ln(1-n_3) + \frac{n_3}{(1-n_3)} - \frac{n_3^2}{2(1-n_3)^2}\right]\\ &-\left(\frac{n_2\sigma_i}{n_3}\right)^3\left[2\ln(1-n_3) + \frac{n_3(2-n_3)}{(1-n_3)}\right].\end{aligned} \tag{7.A.4}$$

Based on the exact relation between the pressure of a hard-sphere system and the contact values of the radial distribution functions, $g_{ij}(\sigma_{ij})$,

$$\frac{P}{k_B T} = \sum_i \rho_i + \frac{2\pi}{3}\sum_i\sum_j \rho_i\rho_j\sigma_{ij}^3 g_{ij}(\sigma_{ij}), \tag{7.A.5}$$

one can find

$$g_{ij}(\sigma_{ij}) = \frac{1}{1-n_3} + \frac{3n_2}{(1-n_3)^2}\frac{\sigma_i\sigma_j}{\sigma_i+\sigma_j} + \frac{2n_2^2}{(1-n_3)^3}\left(\frac{\sigma_i\sigma_j}{\sigma_i+\sigma_j}\right)^2. \tag{7.A.6}$$

As discussed in Section 8.4, Eq. (7.A.6) is useful for deriving the equation of state for polymeric fluids.

---

101 Boublik T., "Hard-sphere equation of state", *J. Chem. Phys.* 53, 471–2 (1970); Mansoori G. A. et al., "Equilibrium thermodynamic properties of the mixture of hard spheres", *ibid*, 54, 1523–5 (1971).
102 Boublik T., "Hard convex body equation of state", *J. Chem. Phys.* 63, 4084 (1975).

7.A.2 Direct Correlation Functions

The DCFs are useful for calculating the structure factor and thermodynamic properties of uniform fluids. In addition, they are commonly employed in developing the molecular theories of inhomogeneous systems.

The PY theory predicts that the direct correlation function of a hard-sphere mixture can be expressed as

$$-c_{ij}(r) = \chi_3\ \Delta V_{ij}(r) + \chi_2\ \Delta S_{ij}(r) + \chi_1\ \Delta R_{ij}(r) + \left(\chi_{22} - \frac{\chi_1}{4\pi}\right)\ \Delta R'_{ij}(r) + \chi_0\ \theta(r - R_i - R_j) \tag{7.A.7}$$

where the coefficients are

$$\chi_0 = \frac{1}{1-n_3} \tag{7.A.8}$$

$$\chi_1 = \frac{n_2}{(1-n_3)^2} \tag{7.A.9}$$

$$\chi_2 = \frac{n_1}{(1-n_3)^2} - \frac{n_2^2}{12\pi n_3^2} \cdot \left[\frac{2\ln(1-n_3)}{n_3} + \frac{2-5n_3+n_3^2}{(1-n3)^3}\right] \tag{7.A.10}$$

$$\chi_3 = \frac{n_0}{(1-n_3)^2} + \frac{2n_1 n_2}{(1-n_3)^3} + \frac{1}{36\pi} \cdot \frac{n_2^3}{n_3^4} \cdot \left[6\ln(1-n_3) + \frac{n_3\left(6 - 21n_3 + 26n_3^2 - 5n_3^3\right)}{(1-n_3)^4}\right] \tag{7.A.11}$$

$$\chi_{22} = \frac{n_2}{6\pi n_3^2} \cdot \left[\ln(1-n_3) + \frac{n_3}{(1-n_3)^2}\right]. \tag{7.A.12}$$

As illustrated in Figure 7.A.1, $\Delta V_{ij}(r)$, $\Delta S_{ij}(r)$, and $\Delta R_{ij}(r)$ stand for, respectively, the overlap volume, overlap surface area, and the mean radius of spherocone for two spheres of radii R_i and R_j ($R_i < R_j$) with the center-to-center separation r. Analytically, these functions are given by

$$\frac{6r}{\pi}\Delta V_{ij}(r) = \begin{cases} 8rR_i^3 & r < \lambda_{ij} \equiv R_j - R_i \\ -\frac{3}{2}\left(R_i^2 - R_j^2\right)^2 + 4r\left(R_i^3 + R_j^3\right) - 3r^2\left(R_i^2 + R_j^2\right) + \frac{r^4}{2} & \lambda_{ij} \le r \le R_i + R_j \\ 0 & r > R_i + R_j \end{cases} \tag{7.A.13}$$

$$\frac{r}{\pi}\Delta S_{ij}(r) = \begin{cases} 4rR_i^2 & r < \lambda_{ij} \\ -(R_i + R_j)(R_i - R_j)^2 + 2r\left(R_i^2 + R_j^2\right) - r^2(R_i + R_j) & \lambda_{ij} \le r \le R_i + R_j \\ 0 & r > R_i + R_j \end{cases} \tag{7.A.14}$$

$$r\Delta R_{ij}(r) = \begin{cases} rR_i & r < \lambda_{ij} \\ -(R_i - R_j)^2/4 + r(R_i + R_j)/2 - r^2/4 & \lambda_{ij} \le r \le R_i + R_j \\ 0 & r > R_i + R_j \end{cases} \tag{7.A.15}$$

$$\Delta R'_{ij}(r) = \begin{cases} 0 & r < \lambda_{ij} \\ -\pi(R_i - R_j)^2/r + r\pi & \lambda_{ij} \le r \le R_i + R_j \\ 0 & r > R_i + R_j \end{cases} \tag{7.A.16}$$

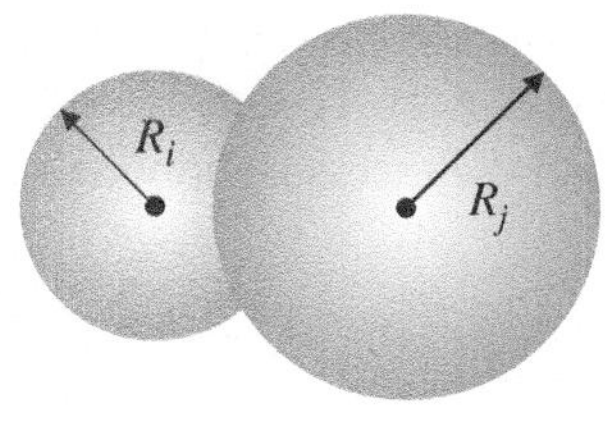

Figure 7.A.1 The direct correction function of a hard-sphere fluid is related to the geometry of particle overlap.

7.A.3 The Scaled-Particle Theory

The SPT[103] provides an alternative route to derive the thermodynamic properties of hard-sphere fluids. It is mathematically much simpler than the PY theory and has the advantage of easy extension to inhomogeneous hard-sphere systems. In the following, we outline the derivation following Hansen-Goos and Roth (HR).[104]

The central idea of SPT is that the excess chemical potential of a particle in the hard-sphere fluid represents the reversible work to insert the particle at fixed temperature and volume. In the macroscopic limit, i.e., when the particle diameter is macroscopic, this reversible work is related to the system pressure P and the particle volume v_i

$$\lim_{v_i \to \infty} \mu_i^{ex} = P v_i. \tag{7.A.17}$$

Eq. (7.A.17) provides an internal constraint for the excess Helmholtz energy of hard-sphere systems from which both the excess chemical potential and pressure can be derived.

For a hard-sphere system, the excess Helmholtz energy density, f_{hs}^{ex} may be assumed á *priori* as a function of the weighted densities $n_{\alpha=0,1,2,3}$. The excess chemical potential for species i is thus given by

$$\beta\mu_i^{ex} = \frac{\partial f_{HS}^{ex}}{\partial \rho_i} = \sum_{\alpha=0}^{3} \frac{\partial f_{HS}^{ex}}{\partial n_\alpha}\frac{\partial n_\alpha}{\partial \rho_i} = \frac{\partial f_{HS}^{ex}}{\partial n_0} + \frac{\partial f_{HS}^{ex}}{\partial n_1}\frac{\sigma_i}{2} + \frac{\partial f_{HS}^{ex}}{\partial n_2}\pi\sigma_i^2 + \frac{\partial f_{HS}^{ex}}{\partial n_3}\frac{\pi\sigma_i^3}{6}. \tag{7.A.18}$$

Equation (7.A.18) indicates that the excess chemical potential depends linearly on the particle diameter, the particle surface area, and the particle volume. When the particle diameter is macroscopic, the particle volume is much larger than the particle surface area and the particle diameter. As a result, $\beta\mu_i^{ex}$ mainly depends on the volume term for a large particle. A comparison of Eqs. (7.A.18) and (7.A.17) gives

$$\beta P_{HS} = \frac{\partial f_{HS}^{ex}}{\partial n_3}. \tag{7.A.19}$$

In the grand canonical ensemble for a bulk system, the excess grand potential density is directly related to the excess pressure

$$\Omega^{ex}/V = -(P_{HS} - P_{IG}). \tag{7.A.20}$$

Meanwhile, the quantity is also related to the chemical potential and the excess Helmholtz energy density

$$\beta\Omega^{ex}/V = f_{HS}^{ex} - \sum_i \rho_i \beta\mu_i^{ex} = f_{HS}^{ex} - \sum_{\alpha=0}^{3} \frac{\partial f_{HS}^{ex}}{\partial n_\alpha} n_\alpha. \tag{7.A.21}$$

103 Reiss H., Frisch H. L. and Lebowitz J. L., "Statistical mechanics of rigid spheres", *J. Chem. Phys.* 31, 369–380 (1959).

104 Hansen-Goos H. and Roth R., "A new generalization of the Carnahan–Starling equation of state to additive mixtures of hard spheres", *J. Chem. Phys.* 124, 154506 (2006).

The second equality in Eq. (7.A.21) follows Eq. (7.A.18) and the definitions of $n_{\alpha=0,1,2,3}$ given by Eq. (7.A.2). A combination of Eqs. (7.A.20) and (7.A.21) gives another relation between the pressure and the excess Helmholtz energy density

$$\beta P_{HS} = -f_{HS}^{ex} + \sum_{\alpha=0}^{3} \frac{\partial f_{HS}^{ex}}{\partial n_\alpha} n_\alpha + n_0. \tag{7.A.22}$$

A comparison of Eqs. (7.A.22) and (7.A.19) yields an internal constraint for the excess Helmholtz energy density

$$\frac{\partial f_{HS}^{ex}}{\partial n_3} = -f_{HS}^{ex} + \sum_{\alpha=0}^{3} \frac{\partial f_{HS}^{ex}}{\partial n_\alpha} n_\alpha + n_0. \tag{7.A.23}$$

Equation (7.A.23) is a key equation of the scaled-particle theory.

In recognizing that f_{HS}^{ex} has the dimension of inverse volume and that the only multiplicative combinations of the weighted densities with the same dimension are n_0, $n_1 n_2$, and n_2^3, Rosenfeld[105] suggested an *ansatz* for the excess free energy density by a linear combination of n_0, $n_1 n_2$, and n_2^3

$$f_{HS}^{ex} = n_0 A_1(n_3) + n_1 n_2 A_2(n_3) + n_2^3 A_3(n_3). \tag{7.A.24}$$

The coefficients $A_i(n_3)$, $i = 1, 2, 3$ can be found by substituting f_{HS}^{ex} into the scaled-particle equation, Eq. (7.A.23). Equaling the coefficients for n_0, $n_1 n_2$, and n_2^3 on both sides leads to a set of differential equations

$$\frac{dA_1(n_3)}{dn_3} = \frac{1}{1-n_3}; \quad \frac{dA_2(n_3)}{dn_3} = \frac{A_2(n_3)}{1-n_3}; \quad \frac{dA_3(n_3)}{dn_3} = \frac{2A_3(n_3)}{(1-n_3)}. \tag{7.A.25}$$

We can fix the integration constants in Eq. (7.A.25) using the exact results for the 2nd and 3rd virial coefficients for uniform hard-sphere systems (Problem 7.4):

$$\beta P^{HS}/\rho = 1 + \frac{2\pi\sigma^3}{3}\rho + \frac{5\pi\sigma^6}{18}\rho^2 + \cdots. \tag{7.A.26}$$

From Eqs. (7.A.25) and (7.A.26), we obtain the final expressions for $A_i(n_3)$, $i = 1, 2, 3$

$$A_1(n_3) = -\ln(1-n_3); \quad A_2(n_3) = \frac{1}{1-n_3}; \quad A_3(n_3) = \frac{1}{24\pi(1-n_3)^2}. \tag{7.A.27}$$

Substitution of Eq. (7.A.27) into Eq. (7.A.24) gives the SPT expression of the excess Helmholtz energy density

$$f_{HS}^{ex} = -n_0 \ln(1-n_3) + \frac{n_1 n_2}{1-n_3} + \frac{n_2^3}{24\pi(1-n_3)^2}. \tag{7.A.28}$$

Equation (7.A.28) yields an equation of state identical to that derived from the PY theory with the compressibility equation

$$\beta P_{HS}^{PY} = \frac{n_0 + n_1 n_2 + n_2^3/(12\pi) - (2n_0 + n_1 n_2)n_3 + n_0 n_3^2}{(1-n_3)^3}. \tag{7.A.29}$$

Although Eq. (7.A.28) is slightly less accurate than that from the BMCSL equation (which is obtained by an empirical combination of the virial and compressibility equations), a similar approach is applicable to inhomogeneous systems of hard spheres.[105] Importantly, SPT provides a

105 Rosenfeld Y., "Free-energy model for the inhomogeneous hard-sphere fluid mixture and density-functional theory of freezing", *Phys. Rev. Lett.* 63, 980–983 (1989).

systematic way to improve the equation of state for hard spheres. By including high-order terms in n_3, HR found that the first-order expansion in n_3 leads to the BMCSL equation of state[101]

$$\beta P_{HS}^{BMCSL} = \beta P_{HS}^{PY} - \frac{n_3 n_2^3}{36\pi(1-n_3)^3}. \tag{7.A.30}$$

The second-order expansion in n_3 leads to further improvements as previously proposed by Santos, Yuste, and López de Haro (SYL)[106]

$$\beta P_{HS}^{SYL} = \beta P_{HS}^{PY} - \frac{n_2^3 n_3}{18\pi(1-n_3)^3} + \frac{n_1 n_2 n_3^2}{3(1-n_3)^3}. \tag{7.A.31}$$

An even more accurate equation can be obtained by the inclusion of the 3rd-order expansion in n_3

$$\beta P_{HS}^{HR} = \beta P_{HS}^{PY} - \frac{n_2^3\left(2n_3 - n_3^2\right)}{36\pi(1-n_3)^3} + \frac{n_1 n_2 n_3^2}{3(1-n_3)^2}. \tag{7.A.32}$$

In an explicit form, the HR equation of state can be written as

$$\beta P_{HS}^{HR} = \frac{n_0}{1-n_3} + \frac{n_1 n_2\left(1+n_3^2/3\right)}{(1-n_3)^2} + \frac{n_2^3\left(1-2n_3/3+n_3^2/3\right)}{12\pi(1-n_3)^3}. \tag{7.A.33}$$

The corresponding excess Helmholtz energy density is

$$f_{HS}^{ex} = -n_0 \ln(1-n_3) + \frac{\lambda_1(n_3) n_1 n_2}{1-n_3} + \frac{\lambda_3(n_3) n_2^3}{24\pi(1-n_3)^2} \tag{7.A.34}$$

with

$$\lambda_1(n_3) = 1 + \frac{2n_3 - n_3^2 + 2(1-n_3)\ln(1-n_3)}{3n_3} \tag{7.A.35}$$

$$\lambda_2(n_3) = 1 - \frac{2n_3 - 3n_3^2 + 2n_3^3 + 2(1-n_3)^2\ln(1-n_3)}{3n_3^2}. \tag{7.A.36}$$

In comparison with simulation results, Eqs. (7.A.30), (7.A.31), and (7.A.33) are extremely accurate. While the BMCSL equation slightly underestimates the pressure and the SYL equation systematically predicts the opposite for binary hard-sphere mixtures, the pressure predicted by the HR equation falls in between and reproduces the simulation data accurately. The improvement of the HR equation is more pronounced for ternary mixtures.

7.B Radial Distribution Functions of Lennard-Jones Mixtures

The FMSA developed by Tang and Lu[107] provides an analytical expression for the RDF of the LJ mixtures near contact ($R_{ij} < r < 1.12\sigma_{ij}$)

$$g_{ij}(r) = g_{ij}^0(r)\exp\left[g_{ij}^1(r)\right]. \tag{7.B.1}$$

---

106 Santos A., Yuste S. B. and López de Haro M., "Equation of state of a multicomponent d-dimensional hard-sphere fluid", *Mol. Phys.* 96 (1), 1–5 (1999).

107 Tang Y. P. and Lu B. Z. Y., *Mol. Phys.* 84, 89 (1995); ibid 100, 1033 (2002); *AIChE J.* 43, 2215 (1997); *Fluid Phase Equilib.*, 146, 73 (1998); ibid., 171, 27 (2000).

For the hard-sphere reference system, the RDF can be expressed as a quadratic function of the center-to-center distance r

$$rg_{ij}^0(r) = a_{ij}^0 + a_{ij}^1(r - R_{ij}) + a_{ij}^2(r - R_{ij})^2 \quad (7.B.2)$$

with

$$a_{ij}^0 = R_{ij}/\Delta + \pi\xi_2 R_i R_j/(4\Delta^2) \quad (7.B.3)$$

$$a_{ij}^1 = 1/\Delta - \pi\xi_1 R_i R_j/(2\Delta^2) - \pi^2\xi_2{}^2 R_i R_j/(4\Delta^3) \quad (7.B.4)$$

$$a_{ij}^2 = \frac{\pi(\xi_0 R_i R_j - 2R_{ij}\xi_1 - \xi_2)}{4\Delta^2} + \frac{\pi^2\xi_2(\xi_1 R_i R_j - \xi_2 R_{ij})}{4\Delta^3} + \frac{\pi^3\xi_2{}^3 R_i R_j}{16\Delta^4} \quad (7.B.5)$$

$$\xi_n = \sum_m \rho_m R_m^n, \quad \Delta = 1 - \pi\xi_3/6 \quad (7.B.6)$$

$$R_{ij} = (R_{ii} + R_{jj})/2, \quad R_{ii} = 2\sum_j x_j d_{ij} - \sum_i\sum_j x_i x_j d_{ij}. \quad (7.B.7)$$

where x_i is the mole fraction of component i, and d_{ij} is the effective hard-sphere diameter determined by the LJ diameter σ_{ij}. The meanings of other variables are the same as those defined in Section 7.B.11.

In Eq. (7.B.1), the first-order perturbation term is given by

$$g_{ij}^1(r) = g_{ij}^{11}(r) - g_{ij}^{12}(r) \quad (7.B.8)$$

with

$$rg_{ij}^{1\nu}(r) = \frac{1}{2\pi\sqrt{\rho_i\rho_j}} \sum_{l,k,m,n} K_{lk}^\nu B_{ml}\left(z_{lk}^\nu\right) B_{nk}\left(z_{lk}^\nu\right) \times \left[\delta_{mi}\delta_{nj}e^{-z_{lk}^\nu(r-R_{ij})} + \left(\delta_{mi}b_{nj}^0 + \delta_{nj}b_{mi}^0\right)\right.$$
$$\left.\times \frac{1 - e^{-z_{lk}^\nu(r-R_{ij})}}{z_{lk}^\nu} + \left(\delta_{mi}b_{nj}^1 + \delta_{nj}b_{mi}^1 + b_{nj}^0 b_{nj}^0\right) \frac{-1 + z_{lk}^\nu(r - R_{ij}) + e^{-z_{lk}^\nu(r-R_{ij})}}{\left(z_{lk}^\nu\right)^2}\right] \quad (7.B.9)$$

for $\nu = 1, 2$. The parameters in Eq. (7.B.9) are

$$B_{ij}(s) = \delta_{ij} + b_{ij}^0/s + b_{ij}^1/s^2 \quad (7.B.10)$$

$$b_{ij}^0 = -\pi\sqrt{\rho_i\rho_j}\, R_i R_j/\Delta \quad (7.B.11)$$

$$b_{ij}^1 = -\pi\sqrt{\rho_i\rho_j}\left(\frac{R_i - R_j}{\Delta} - \frac{\pi\xi_2 R_i R_j}{2\Delta^2}\right) \quad (7.B.12)$$

$$z_{ij}^1 = 2.9637/\sigma_{ij}, \quad z_{ij}^2 = 14.0167/\sigma_{ij} \quad (7.B.13)$$

$$K_{ij}^\nu = 4.3428\pi\beta\varepsilon_{ij}\sigma_{ij}\sqrt{\rho_i\rho_j}\exp\left[z_{ij}^\nu(\sigma_{ij} - R_{ij})\right], \quad \nu = 1, 2 \quad (7.B.14)$$

and δ_{ij} is the Kronecker delta function.

In general, the analytical expression for RDF is available only in the Laplace transform

$$G_{ij}(s) = \int_0^\infty rg_{ij}(r)e^{-sr}dr. \quad (7.B.15)$$

In FMSA, the RDF consists of two parts

$$G_{ij}(s) = G_{ij}^0(s) + G_{ij}^1(s). \quad (7.B.16)$$

The first term on the right-side of Eq. (7.B.16) is the same as that derived from the PY theory for a hard-sphere system

$$G_{ij}^0(s) = \frac{e^{-sR_{ij}}}{\Delta \det(s)}\left[\frac{1}{s^2}\left(1+\frac{\pi\xi_3}{2\Delta}\right)+\frac{1}{s}\left(R_{ij}+\frac{\pi\xi_2 R_i R_j}{4\Delta}\right)+\frac{\pi}{2\Delta s}\sum_m \rho_m \lambda_1(R_m)(R_m-R_i)(R_m-R_j)\right] \tag{7.B.17}$$

with

$$\det(s) = 1 - \frac{2\pi}{\Delta}\sum_m \rho_m \lambda_2(R_m)\left(1+\frac{\pi\xi_3}{2\Delta}\right) - \frac{2\pi}{\Delta}\sum_m \rho_m \lambda_1(R_m)\left(R_m + \frac{\pi\xi_2 R_m^2}{4\Delta}\right) - \frac{\pi^2}{2\Delta^2}\sum_m\sum_n \rho_m\rho_n \lambda_1(R_m)\lambda_1(R_n)(R_m-R_n)^2 \tag{7.B.18}$$

and

$$\lambda_1(R_i) = \left[1 - sR_i - e^{-sR_i}\right]/s^2, \quad \lambda_2(R_i) = \left[1 - sR_i + (sR_i)^2/2 - e^{-sR_i}\right]/s^3. \tag{7.B.19}$$

The first-order corrections of the RDF are given by

$$G_{ij}^1(s) = G_{ij}^{11}(s) - G_{ij}^{12}(s) \tag{7.B.20}$$

with

$$G_{ij}^{1\nu}(s) = \sum_l\sum_k\sum_m\sum_n \frac{K_{lk}^\nu B_{ml}(z_{lk}^\nu)B_{nk}(z_{lk}^\nu)}{s+z_{lk}^\nu} B_{mi}(s)B_{nj}(s)\frac{e^{-sR_{ij}}}{2\pi\sqrt{\rho_i\rho_j}}, \quad \nu = 1,2 \tag{7.B.21}$$

and

$$B_{ij}(s) = \delta_{ij} + 2\pi\sqrt{\rho_i\rho_j}\frac{W_{ij}(s)}{\Delta\det(s)} \tag{7.B.22}$$

$$W_{ij}(s) = \lambda_2(R_i)\left(1+\frac{\pi\xi_3}{2\Delta}\right) + \lambda_1(R_i)\left(R_{ij}+\frac{\pi\xi_2 R_i R_j}{4\Delta}\right) + \frac{\pi}{2\Delta}\lambda_1(R_i)\sum_m \rho_m\lambda_1(R_m)(R_m-R_i)(R_m-R_j) + \frac{\pi}{2\Delta}R_i^2\sum_m \rho_m \lambda_2(R_m)(R_j-R_m). \tag{7.B.23}$$

Although these analytical results are rather complicated, they are much more convenient in practice than numerical solutions of the integral-equation theories or molecular simulation. In particular, FMSA can be utilized for deriving the equation of state for polymeric fluids as discussed in Section 8.5.

7.C Critical Behavior Predicted by Integral-Equation Theories

In contrast to mean-field models, integral-equation theories are centered around the structural properties as represented by various correlation functions. Thermodynamic properties are related to the correlation functions by the integration of the compressibility equation, virial equation, or energy equation. Because most integral equations have to be solved numerically, critical behavior of integral equation theories is much less transparent than that predicted by mean-field methods.

Integral-equation theories may loss numerical solutions (or have only unphysical solutions) in certain regions of the phase diagram. For instance, the PY closure for the sticky-hard sphere model does not provide physically acceptable solutions in a certain density range when the sticky parameter τ is less than a critical value. Similarly, other integral equations, including HNC and MSA, fail to yield a solution at low temperature and intermediate density.

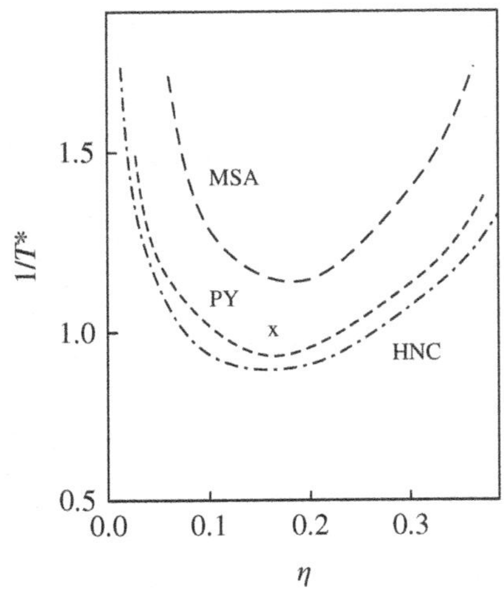

Figure 7.C.1 Boundaries of solvable (inside) and unsolvable (outside) regions of integral equations for hard-core Yukawa fluids. The cross symbol represents exact location of the critical point. Source: Adapted from Belloni.[108]

Belloni investigated the critical behavior for the hard-core Yukawa potential using the integral equation approach.[108] Figure 7.C.1 shows the regions of the phase diagram where no solutions to HNC, PY and MSA closures exist. Because a region of no solution resembles the two-phase region of a fluid, one may identify the maximum temperature of no solution as the critical temperature and the boundary as the spinodal line where the isotherm compressibility is infinite. While analytical solution of MSA for the Yakawa potential suggests that the boundary of no solution corresponds exactly to the spinodal curve, the boundaries of no solution for both PY and HNC closures are not the same as the true spinodal curve.

The spinodial line determined from the PY closure is between the no-solution boundary and the high-density branch of the coexistence curve. Along the low-density side, the isothermal compressibility is finite at the no-solution boundary which implies that the PY closure does not have a spinodal line on this side. The lack of parity for PY closure near the critical point also applies to other interaction potentials including the sticky model (where analytical solution is available), the Lennard-Jones potential and lattice gases. On the other hand, the HNC closure does not show spinodal lines in both high and low-density sides of the coexistence curve. As the critical point of no-solution is approached, the HNC closure does not yield a divergence of the isothermal compressibility. Therefore, strictly speaking, the HNC closure does not predict a critical point at all! The anomalous behavior of the HNC closure equally applies to RHNC or MHNC because the bridge functions are short-ranged over the whole phase diagram.

The limitations of integral equations near the critical point or inside the phase coexistence curve stem from the approximations introduced in the closure relations. Near the critical point or spinodal curve where the isothermal compressibility diverges, the conventional closures are often invalid. As a result, one should not try to apply an integral equation theory to conditions where long-range fluctuations are significant. In order to go beyond the poor description of the critical behavior provided by mean-field theories or integral equation formalisms, we must account for all levels of density fluctuations using the renormalization-group theory.

---

108 Belloni L., "Inability of the hypernetted chain integral equation to exhibit a spinodal line", *J. Chem. Phys.* 98, 8080 (1993).

Further Readings

Bagchi B., *Statistical mechanics for chemistry and materials science*. CRC Press, Taylor & Francis Group, Boca Raton, FL, Chapters 11–16, 2019.

Chandler D., *Introduction to modern statistical mechanics*. Oxford University Press, Chapter 7, 1986.

Evans R., Frenkel D. and Dijkstra M., From simple liquids to colloids and soft matter, *Phys. Today* 72 (2), 38 (2019).

Hansen J. P. and McDonald I. R., *Theory of simple liquids: with applications to soft matter* (4th Edition). Academic Press, 2013.

Lee L. L., *Molecular thermodynamics of nonideal fluids*. Butterworths, Boston, 1988.

McQuarrie D.A., *Statistical mechanics*. University Science Books. Chapters 12, 13, 14, 2000.

Santos A., *A concise course on the theory of classical liquids: basics and selected topics*. Springer International Publishing, 2016.

Problems

7.1 Extend the Gibbs entropy to systems containing m types of classical particles of spherical shape. Show that, if the particles are not interacting with each other, the entropy of mixing is given by

$$\Delta S = -k_B \sum_{i=1}^{m} N_i \ln(N_i/N),$$

where k_B is the Boltzmann constant, and $N = \sum_{i=1}^{m} N_i$ is the total number of particles in the system.

7.2 For a system of spherical particles at low density, the radial distribution function can be approximated by the Boltzmann distribution

$$g(r) \approx e^{-\beta u(r)},$$

where $\beta = 1/(k_B T)$, and $u(r)$ stands for the pair potential. Based on the above expression, derive equations for

(i) internal energy
(ii) Helmholtz energy
(iii) compressibility factor
(iv) the second virial coefficient.

7.3 Consider the grand canonical partition function for a uniform system of spherical particles at temperature T and chemical potential μ

$$\Xi = \sum_{N=0}^{\infty} \frac{1}{N!\Lambda^{3N}} \int d\mathbf{r}^N \exp[-\beta\Phi(\mathbf{r}^N) + N\beta\mu] = \sum_{N=0}^{\infty} \frac{z^N}{N!} \int d\mathbf{r}^N \exp[-\beta\Phi(\mathbf{r}^N)],$$

where $\Phi(\mathbf{r}^N)$ stands for the total potential energy, $\beta = 1/(k_B T)$, $z \equiv e^{\beta\mu}/\Lambda^3$ represents the fugacity, and Λ is the thermal wavelength. Show that

(i) the grand potential can be expressed as

$$\Omega = -k_B T \ln \Xi = -PV,$$

where P is the pressure.

(ii) the average number density of particles $\rho \equiv \langle N \rangle / V$ is related to fugacity z by

$$\rho = \frac{z}{V}\left(\frac{\partial \ln \Xi}{\partial z}\right)_{T,V},$$

(iii) assuming pairwise additivity for the total potential energy $\Phi(\mathbf{r}^N)$, the grand canonical partition function can be written as a power series of z, i.e.,

$$\Xi = 1 + zV + \frac{1}{2}z^2V^2(1 + I_2) + \frac{1}{3!}z^3V^3(1 + 3I_2 + 3I_2^2 + I_3) + \cdots,$$

where $I_{n=1,2,\ldots}$ are constants defined by the Mayer function $f_{ij} \equiv \exp[-\beta u(|\mathbf{r}_i - \mathbf{r}_j|)] - 1$

$$I_2 \equiv \frac{1}{V^2}\int d\mathbf{r}_1 \int d\mathbf{r}_2 f_{12}, \quad I_3 \equiv \frac{1}{V^3}\int d\mathbf{r}_1 \int d\mathbf{r}_2 \int d\mathbf{r}_3 f_{12}f_{23}f_{13}, \ldots,$$

(iv) the reduced pressure can be expressed in terms of a power series of z, i.e.,

$$\frac{PV}{k_BT} = \ln \Xi = zV + \frac{z^2V^2}{2}I_2 + \frac{1}{6}z^3V^3(3I_2^2 + I_3) + \cdots,$$

(v) the average particle density can also be expressed in a power series of z,

$$\rho = z + z^2VI_2 + \frac{z^3V^2}{2}(3I_2^2 + I_3) + \cdots,$$

(vi) the compressibility factor Z in the power series of ρ is given by

$$Z \equiv \frac{PV}{\langle N \rangle k_BT} = 1 + B_2\rho + B_3\rho^3 + \cdots.$$

The above equation is known as the virial expansion, and B_2, B_3 and ... are virial coefficients

$$B_2 = -\frac{1}{2}\int d\mathbf{r}_{12} f_{12},$$

$$B_3 = -\frac{1}{3}\int d\mathbf{r}_{12} \int d\mathbf{r}_{13} f_{12}f_{23}f_{13},$$

...

7.4 Evaluate the second and third virial coefficients for a one-component system of hard spheres with diameter σ. Compare the results with the virial coefficients predicted by the Carnahan–Starling (CS) equation

$$Z = \frac{1 + \eta + \eta^2 - \eta^3}{(1 - \eta)^3}.$$

7.5 Calculate the second virial coefficient using the Lennard–Jones (LJ) potential

$$u(r) = 4\epsilon\left[\left(\frac{\sigma}{r}\right)^{12} - \left(\frac{\sigma}{r}\right)^6\right].$$

Plot B_2/B_2^0 versus the reduced temperature k_BT/ϵ and estimate the Boyle temperature, i.e., the temperature when $B_2 = 0$. Here, $B_2^0 = 2\pi\sigma^3/3$ is the second virial coefficient according to the hard-sphere potential.

7.6 At cryogenic conditions, gas molecules such as hydrogen and helium may exhibit non-negligible quantum effects that cannot be reliably captured with classical models.

Approximately, such effects may be accounted for by modifying the Lennard–Jones (LJ) model with the 4th-order approximation to the Feynman–Hibbs (FH) equation[109]

$$\frac{u(r)}{4\epsilon} = \frac{u^{LJ}(r)}{4\epsilon} + \frac{\beta\hbar^2}{24\mu}\left[u''^{LJ}(r) + \frac{2u'^{LJ}(r)}{r}\right] + \frac{\beta^2\hbar^4}{1152\mu^2}\left[u''''^{LJ}(r) + \frac{4u'''^{LJ}(r)}{r} + \frac{15u'^{LJ}(r)}{r^3}\right].$$

In the above equation, r denotes the center-to-center distance, $u^{LJ}(r)$ represents the 12-6 LJ potential, $\beta = 1/k_BT$, k_B is the Boltzmann constant, T is the absolute temperature, $\hbar = h/2\pi$ is the reduced Planck constant, $\mu = m_im_j/(m_i + m_j)$ is the reduced mass for particles i and j with mass m_i and m_j, respectively, and the number of primes represents the order of derivatives of the LJ potential with respect to distance r.

(i) Compare the quantum-corrected pair potential for H_2 at 50 K with the LJ potential;
(ii) Plot the second virial coefficients of H_2 gas in the temperature range from 50 to 300 K.

Assume that the LJ parameters for different isotopic forms of hydrogen molecules are independent of the molecular weight, with $\sigma = 0.296$ nm and $\epsilon/k_B = 34.2$ K.

7.7 Derive the second-order virial coefficient from the van der Waals equation of state

$$P = \frac{\rho k_B T}{1 - \rho b} - a\rho^2,$$

where ρ is the number density, and compare the result with that predicted by the Lennard–Jones potential. What is the Boyle temperature predicted by the van der Waals equation? Comment on the performance at low density for various cubic equations of state (e.g., the Peng–Robinson equation) derived from the empirical modifications of the van der Waals equation.

7.8 Show that for an open system of spherical particles defined by the chemical potential μ, absolute temperature T and volume V, the isothermal compressibility is related to the fluctuation of particle density

$$\kappa_T = \frac{V}{k_BT}\frac{\langle \Delta\rho^2 \rangle}{\rho^2}.$$

7.9 Based on the following expression for the radial distribution function of spherical particles discussed in Section 7.2

$$g(r) = e^{-\beta u(r)} \frac{V \int d\mathbf{r}_3 d\mathbf{r}_4 \cdots d\mathbf{r}_N \exp\{-\beta\Phi'(\mathbf{r}^{N-2})\}}{\int d\mathbf{r}_2 d\mathbf{r}_3 \cdots d\mathbf{r}_N \exp\{-\beta\Phi(\mathbf{r}^N)\}}. \quad \text{(A)}$$

(i) Show that

$$-k_BT \ln y(r) = \mu_2^{ex}(r) - 2\mu^{ex},$$

where $y(r) = g(r)e^{\beta u(r)}$ is the cavity correlation function, μ^{ex} is the excess chemical potential, and $\mu_2^{ex}(r)$ represents the excess chemical potential for a "dimer" of two particles with the center-to-center distance fixed at distance r.

(ii) Prove the zero-separation theorem, i.e., for a system of hard spheres,

$$\mu^{ex} = k_BT \ln y(0).$$

109 Feynman R. P., Hibbs A. R. and Styer D. F., Quantum Mechanics and Path Integrals, Emended ed. McGraw-Hill, New York, 2005, p. 384.

7.10 Widom's insertion theorem, $\beta\mu^{ex} = -\ln\langle e^{-\beta\Delta\Phi(\mathbf{r})}\rangle$, can be extended to inhomogeneous systems, i.e., systems with a nonuniform distribution of particles due to the presence of an external potential. To understand this extension, consider a system of spherical particles subject to a one-body potential $V^{ext}(\mathbf{r})$. The grand canonical partition function of the system can be written as

$$\Xi = \sum_{N=0}^{\infty} \frac{z^N}{N!} \int d\mathbf{r}^N \exp\left[-\beta\Phi(\mathbf{r}^N) - \sum_{i=1}^{N} \beta V^{ext}(\mathbf{r}_i)\right],$$

where $\Phi(\mathbf{r}^N)$ stands for the potential energy due to inter-particle interactions, $\beta = 1/(k_B T)$, and $z \equiv e^{\beta\mu}/\Lambda^3$ with Λ being the thermal wavelength.

(i) Derive the potential distribution theorem, i.e., the one-body density profile can be written as

$$\rho(\mathbf{r}) \equiv \sum_{i=1}^{N} \langle \delta(\mathbf{r} - \mathbf{r}_i)\rangle = z e^{-\beta V^{ext}(\mathbf{r})} \langle e^{-\beta\Delta\Phi(\mathbf{r})}\rangle,$$

where $\Delta\Phi(\mathbf{r})$ stands for the change in potential energy upon inserting a particle at position $\mathbf{r}$ due to its interaction with all particles in the system.

(ii) In the low density limit, the one-body density profile is reduced to the Boltzmann distribution

$$\rho(\mathbf{r}) = \rho_0 e^{-\beta V^{ext}(\mathbf{r})},$$

where ρ_0 stands for the particle density in the bulk (viz., $V^{ext}(\mathbf{r}) = 0$).

(iii) The local excess chemical potential can be expressed as

$$\mu^{ex}(\mathbf{r}) = -k_B T \ln\langle e^{-\beta\Delta\Phi(\mathbf{r})}\rangle,$$

where $\mu^{ex}(\mathbf{r}) \equiv \mu - \mu^{IG} - V^{ext}(\mathbf{r})$, and μ^{IG} stands for the chemical potential of the non-interacting reference system with local density $\rho(\mathbf{r})$.

(iv) Suppose that the external potential is introduced by fixing a particle at the origin, the cavity correlation function $y(r)$ satisfies

$$-\ln y(r) = \beta\mu^{ex}(r) - \beta\mu^{ex},$$

where μ^{ex} stands for the excess chemical potential of the bulk system, and $\beta\mu^{ex}(\mathbf{r}) = -\ln\langle e^{-\beta\Delta\Phi(\mathbf{r})}\rangle$.

7.11 With the pairwise additive approximation for the inter-particle potentials, the virial theorem predicts that the pressure equation for a binary mixture of spherical particles is

$$P = \rho k_B T - \frac{1}{6}\sum_{i=1}^{2}\sum_{j=1}^{2} \rho_i \rho_j \int d\mathbf{r} \frac{\partial u_{ij}(r)}{\partial r} r g_{ij}(r),$$

where $\rho = \rho_1 + \rho_2$, $\rho_{i=1,2}$ is the number density for type i particles, and $g_{ij}(r) = g_{ji}(r)$ is the pair distribution function between particles i and j, which has a pair potential $u_{ij}(r)$.

(i) Verify the above equation using the virial theorem

$$P = \rho k_B T + \frac{1}{3V}\left[\sum_{n_1=1}^{N_1} \langle \mathbf{r}_{n_1} \cdot \mathbf{F}_{n_1}\rangle + \sum_{n_2=1}^{N_2} \langle \mathbf{r}_{n_2} \cdot \mathbf{F}_{n_2}\rangle\right],$$

where N_1 and N_2 are the number of particles of type 1 and 2, respectively; $\mathbf{r}_{n_i}$ and $\mathbf{F}_{n_i}$ stand for the position and force on a particle of type i; and $\langle\cdots\rangle$ represents the ensemble average.

(ii) Show that, at low particle density, the radial distribution function can be approximated by the Boltzmann distribution

$$g_{ij}(r) = e^{-\beta u_{ij}(r)}.$$

(iii) Derive the second virial coefficient for the binary mixture.

(iv) Extend the above procedure to a multicomponent system with m types of spherical particles.

7.12 The Ornstein–Zernike (OZ) equation may serve as a definition of the direct correlation function $c(r)$. To get this point, we first write the OZ equation through the Fourier transformation:

$$\hat{h}(k) = \hat{c}(k) + \rho\hat{h}(k)\hat{c}(k).$$

Rearranging the above equation gives

$$\hat{c}(k) = \frac{\hat{h}(k)}{1 + \rho\hat{h}(k)}.$$

Thus $c(r)$ is completely determined by the total correlation function $h(r)$ provided that the denominator $1 + \rho\hat{h}(k)$ is nonzero. Show that $1 + \rho\hat{h}(k) \neq 0$ for any thermodynamic system containing spherical particles.

7.13 Establish a numerical procedure for solving the OZ equation for one-component Lennard–Jones fluids with the Percus–Yevick (PY) and the hypernetted chain (HNC) closures. Compare the results from these two closures for $g(r)$, $c(r)$, $\gamma(r) \equiv h(r) - c(r)$, and $B(r)$ at $k_BT = 1.5$ and $\rho\sigma^3 = 0.5$.
Hint: The OZ equation can be solved by using the Picard iteration:
(1) initial guess for $c(r) = \exp[-\beta u(r)] - 1$;
(2) fast Fourier transform (FFT) to calculate $\hat{c}(k)$;
(3) calculate $\hat{h}(k) = \hat{c}(k)/[1 - \rho\hat{c}(k)]$;
(4) inverse FFT to obtain $\gamma(r) \equiv h(r) - c(r)$;
(5) calculate new $c(r)$ based on a closure and iterate the procedure until convergence.

7.14 Show that the second virial coefficient predicted by the sticky hard-sphere potential is given by

$$B_2 = B_2^{HS}\left(1 - \frac{1}{4\tau}\right),$$

where τ is a dimensionless parameter. Based on the van der Waals theory of simple fluids, discuss how parameter τ may vary with temperature.

7.15 Consider a uniform hard-sphere fluid with bulk density ρ and diameter σ.

(i) Based on the virial equation

$$\frac{P}{\rho k_B T} = 1 + 4\eta y(\sigma),$$

where $\eta = \pi\sigma^3/6$ is the packing fraction, derive the contact value of the cavity correlation function $y(\sigma)$ based on the equation of state predicted by the PY theory through the compressibility route, and compare the result with that derived from the Carnahan–Starling equation;

(ii) Compare the compressibility factor predicted by the PY virial equation, by the PY compressibility equation, and by the Carnahan–Starling equation for the packing fraction up to the freezing point;

(iii) Compare the contact value of the cavity correlation function predicted by the PY virial equation, by the PY compressibility equation, and by the Carnahan–Starling equation over the entire range of the packing fraction of the hard-sphere fluid.

7.16 Analytical expressions for the radial distribution functions of hard-sphere fluids have been reported for $x \equiv r/\sigma < 1.5$ by de Souza and Ben-Amotz[110]

$$\ln g_0(x) = \frac{3-\eta}{(1-\eta)^3} - 3 + \frac{-3\eta(2-\eta)}{(1-\eta)^3}x + \left\{ \ln\left[\frac{2-\eta}{2(1-\eta)^3}\right] - \frac{\eta(2-6\eta+3\eta^2)}{(1-\eta)^3} \right\} x^2,$$

and, for $x \lessapprox 1.2$, by Tang and Lu[111]

$$g_0(x) = \frac{1+\eta/2}{(1-\eta)^2}\frac{1}{x} + \frac{1-5\eta-5\eta^2}{(1-\eta)^3}(1-1/x) + \frac{-3\eta+6\eta^2+21\eta^3/2}{(1-\eta)^4}\frac{(x-1)^2}{x}.$$

Plots these two approximations of $g_0(x)$ for $1 < x < 1.5$ at reduced densities $\rho\sigma^3 = 0.3$ and 0.860, and compare the theoretical predictions with the numerical results obtained from Monte Carlo simulation (Table P7.16).[112]

Table P7.16 The radial distribution function near contact at two reduced densities $\rho\sigma^3 = 0.3$ and 0.860.

x	0.3	0.860	x	0.3	0.860
1	1.535	4.645	1.025	1.504	3.966
1.05	1.471	3.34	1.075	1.436	2.846
1.1	1.402	2.441	1.125	1.367	2.111
1.15	1.334	1.845	1.175	1.303	1.625
1.2	1.274	1.442	1.225	1.249	1.289
1.25	1.225	1.162	1.275	1.204	1.056
1.3	1.183	0.969	1.325	1.163	0.896
1.35	1.145	0.836	1.375	1.126	0.787
1.4	1.107	0.747	1.425	1.09	0.716
1.45	1.075	0.691	1.475	1.06	0.673
1.5	1.046	0.662			

7.17 The PY theory for a one-component system of sticky hard spheres predicts that the thermodynamic properties of the system are dependent on the sticky parameter τ and packing

110 de Souza L. E. S. and Ben-Amotz D., "Optimized perturbed hard sphere expressions for the structure and thermodynamics of Lennard-Jones fluids", *Mol. Phys.* 78, 137 (1993).
111 Tang Y. and Lu B. C.-Y., "A study of associating Lennard-Jones chains by a new reference radial distribution function", *Fluid Phase Equilibria* 171, 27–44 (2000).
112 Labík S. and Malijevský A., "An effective Monte Carlo calculation of the radial distribution function of hard spheres on a minicomputer", *Czech. J. Phys.* 33, 128–134 (1983).

fraction η through parameter λ defined as

$$\lambda \equiv 6\left\{1-\tau+\tau/\eta-\sqrt{(1-\tau+\tau/\eta)^2-(1+2/\eta)/6}\right\}. \qquad \text{(C)}$$

Show that

(i) $\lambda \to 0$ as $\tau \to \infty$ for any reasonable value of η;

(ii) As λ must be a real number, the PY theory yields a real solution for any reasonable value of η when $\tau \geq \tau_c \equiv (2-\sqrt{2})/6$;

(iii) At $\tau = \tau_c$, $\lambda = 1/\tau_c$ for any $\eta \leq \eta_c \equiv (3\sqrt{(2)}-4)/2$. For $\eta > \eta_c$, λ is given by

$$\lambda = 12(1-\tau_c+\tau_c/\eta)-1/\tau_c.$$

7.18 The osmotic pressure for an aqueous solution of globular proteins can be measured with dynamic light scattering (DLS) techniques. The key output from the DLS experiment is the excess scattered intensity with respect to the solvent, which can be written as

$$I_s = A\left(\frac{dn}{dc}\right)k_BTc\left(\frac{\partial\Pi}{\partial c}\right)^{-1},$$

where A is an instrumental constant, dn/dc is related to the refractive index increment with respect to the solvent, Π stands for the osmotic pressure, and c is the protein concentration in the units of g/L. The osmotic pressure of a protein solution may be described by the sticky hard-sphere model in the PY approximation

$$\frac{\Pi}{\rho k_BT} = \frac{1+\eta+\eta^2}{(1-\eta)^3} - \eta\lambda\frac{18(2+\eta)-\eta\lambda^2}{36(1-\eta)^3},$$

where ρ is the protein number density, $\eta = \pi\sigma^3/6$ is the packing fraction of globular proteins with hard-sphere diameter σ, and λ is defined by the sticky parameter τ and packing fraction η

$$\lambda \equiv 6\left\{1-\tau+\tau/\eta-\sqrt{(1-\tau+\tau/\eta)^2-(1+2/\eta)/6}\right\}.$$

(i) Show that the derivative of the osmotic pressure with respect of the protein concentration is given by

$$\frac{\partial\Pi}{\partial c} = \frac{RT}{M}\frac{(1+2\eta-\lambda\eta)^2}{(1-\eta)^4},$$

where $R = k_BN_A$ is the gas constant, $N_A = 6.02214\times10^{23}$ is Avogadro's number, and M is the molecular weight of the globular protein.

(ii) For a given protein solution, it was found by fitting with experimental data that the sticky parameter can be calculated from $\tau = -1.6826 + 0.0065T/K$. Plot $\frac{M}{RT}\frac{\partial\Pi}{\partial c}$ versus η at temperatures 12.2, 17.2, 24.5, 30.2, 35.0 °C and compare the theoretical results with experimental data from the literature.[113]

7.19 Consider a simple fluid with the pairwise-additive potential described by the Sutherland model:

$$u(r) = \begin{cases} \infty, & r<\sigma, \\ -\epsilon\left(\frac{\sigma}{r}\right)^n, & r\geq\sigma, \end{cases}$$

113 Fig. 3, Piazza R., "Interactions in protein solutions near crystallisation: a colloid physics approach", *J. Cryst. Growth* 196, 415–423 (1999).

where σ represents the hard-sphere diameter, and r is the center-to-center distance, ϵ is an energy parameter, and $n > 3$ is a positive number.

(i) Show that the van der Waals parameter is given by

$$a = \frac{2\pi}{n-3}\epsilon\sigma^3.$$

(ii) Find the critical temperature and critical density of the system from the modified van der Waals equation of state

$$\frac{P}{\rho k_B T} = \frac{1+\eta+\eta^2}{(1-\eta)^3} - \frac{a\rho}{k_B T},$$

where $\eta = \pi\rho\sigma^3/6$.

(iii) What is the reduced critical temperature $k_B T_c/\epsilon$ for $n = 6$?

7.20 In combination with the Carnahan–Starling equation for hard spheres, the modified van der Waals equation of state for the Sutherland model is given by

$$\frac{P}{\rho k_B T} = \frac{1+\eta+\eta^2-\eta^3}{(1-\eta)^3} - \frac{a\rho}{k_B T},$$

where $\eta = \pi\rho\sigma^3/6$ and $a = 2\pi\epsilon\sigma^3/3$.

(i) Show that the corresponding chemical potential is given by

$$\frac{\mu}{k_B T} = \ln(\rho\Lambda^3) + \eta\frac{8-9\eta+3\eta^2}{(1-\eta)^3} - \frac{2\rho a}{k_B T}.$$

(ii) What are the critical temperature and the critical density predicted by the modified van der Waals equation of state?

(iii) Construct a vapor–liquid phase diagram (viz., the vapor–liquid coexistence curve) predicted by the above equations.

7.21 Predict the coexisting densities at the fluid-solid equilibrium (FSE) for a system of uniform hard spheres using the Carnahan–Starling equation for the fluid phase and the cell model for the solid with the face-centered-cubic (FCC) structure:

$$\frac{P_f}{\rho k_B T} = \frac{1+\eta+\eta^2-\eta^3}{(1-\eta)^3}, \quad \frac{P_s}{\rho k_B T} = \frac{1}{1-(\rho/\rho_c)^{1/3}},$$

where $\rho_c\sigma^3 = \sqrt{2}$ is the reduced density at the close packing of uniform hard spheres. Compare the molar entropy of the hard-sphere system as a fluid and as a solid at packing fraction $\eta = 0.6$ and comment on the relationship between entropy and the degree of randomness.

7.22 Prepare a computer program to construct the fluid–solid phase diagram for the Sutherland model with $n = 5, 6$ and 7 using the mean-field theory for the fluid phase and the cell model for the solid as discussed in Sections 7.7 and 7.8, respectively. Discuss whether the vapor–liquid transition is stable or metastable for these systems.

7.23 In the self-consistent Ornstein–Zernike approximation (SCOZA),[114] a state-dependent parameter is introduced into the long-range component of the direct correlation function.

114 Dickman R. and Stell G., "Self-consistent Ornstein-Zernike approximation for lattice gases", *Phys. Rev. Lett.* 77, 996 (1996).

This parameter is determinated self-consistently from the thermodynamic consistency between the energy and compressibility equations:

$$\omega = \frac{\rho}{2V}\int d\mathbf{r}u(r)g(r) \quad \text{and} \quad \left(\frac{\partial \beta P}{\partial \rho}\right)_T = 1 - \rho\hat{c}(k=0),$$

where $\beta = 1/(k_BT)$, and $\omega = \langle\Phi\rangle/V$ is the potential energy density. Show that the thermodynamic consistency is satisfied by imposing

$$-\frac{\partial \hat{c}(k=0)}{\partial \beta} = \frac{\partial^2 \omega}{\partial \rho^2}.$$

7.24 A Tonks gas, also known as a Tonks–Takahashi fluid, refers to a one-dimensional system of particles with hard-sphere interactions.
As shown schematically in Figure P7.24, the hard-sphere potential is described as

$$u(r) = \begin{cases} \infty, & r < \sigma, \\ 0, & r \geq \sigma, \end{cases}$$

where $r = |x_i - x_j|$ is the center-to-center distance between any pair of particles at positions x_i and x_j, and σ is the particle diameter. Because of the one-dimensional confinement, the particle positions are restricted to $x_1 \geq \sigma/2$ and $x_N \leq L - \sigma/2$. As the particles cannot pass each other, the pair potential applies only to the nearest neighbors. The canonical partition function of the system is given by

$$Q(N, L) = \frac{1}{\Lambda^N}\int_0^L dx_1 \int_0^L dx_2 \cdots \int_0^L dx_N \exp[-\beta\Phi(x_1, x_2, \ldots, x_N)],$$

where Λ stands for the thermal wave length, and $\Phi(x_1, x_2, \ldots, x_n)$ is the total potential energy. Here $N!$ is not applied because the particles are identifiable by their relative positions or sequences.

(i) Show

$$\exp[-\beta\Phi(x_1, x_2, \ldots, x_N)] = \Theta(x_1 - \sigma/2)\Theta(L - \sigma/2 - x_N)\prod_{i=2}^{N}\Theta(x_i - x_{i-1} - \sigma),$$

where $\Theta(x)$ stands for the Heaviside step function, i.e., $\Theta(x) = 1$ for $x \geq 0$ and $\theta(x) = 0$ for $x < 0$.

(ii) Develop expressions for $Q(1, L)$ and $Q(2, L)$.

(iii) Verify

$$Q(N, L) = \frac{(L - N\sigma)^N}{N!\Lambda^N}.$$

(iv) Derive the equation of state of the system based on $\beta P = (\partial \ln Q/\partial L)_T$. What are the internal energy and chemical potential of this system predicted by the canonical partition function?

(v) Show that the grand canonical partition function is given by

$$\Xi(\mu, L) = \sum_{i=1}^{\infty}\frac{e^{N\beta\mu}}{N!\Lambda^N}(L - N\sigma)^N\Theta(L - N\sigma).$$

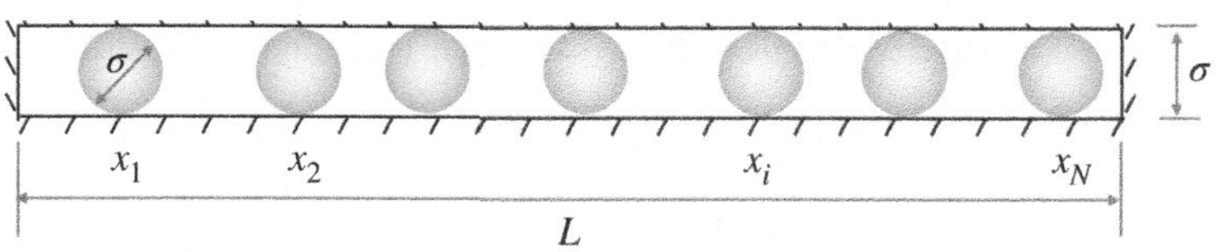

Figure P7.24 Schematic of a Tonks–Takahashi fluid confined between two hard walls.

(vi) Show for any real number $x > 0$,

$$\lim_{L\to\infty} \frac{\Xi(\mu, L-x)}{\Xi(\mu, L)} = e^{-Px}.$$

(vii) In the bulk phase, the radial distribution function of a Tonks gas may be understood as the normalized local density $g(r) = \rho(x = r - \sigma/2)/\rho$ for an open system with one particle fixed at $x = -\sigma/2$ and another at $L + \sigma/2 \to \infty$. Show that

$$g(r) = e^{\beta\mu^{ex}} e^{-\beta P r} \Xi(\mu, r - \sigma),$$

where $\mu^{ex} = \mu - k_B T \ln(\rho\Lambda)$.

(viii) Plot $g(r)$ at reduced densities $\rho\sigma = 0.1$, 0.5 and 0.9.

7.25 The Kornberg–Stryer model for the distributions of nucleosomes in chromatin is analogous to the statistical thermodynamic model of the Tonks gas. According to this model, nucleosome arrangements in a chromatin fiber can be appropriately described by the statistical packing principle, i.e., in terms of steric exclusion between nucleosome as represented by the one-dimensional hard-sphere interaction (see Figure P7.25). To rationalize the experimental results, the average nucleosome density ρ and the number of base pairs (bp) per nucleosome (a.k.a., nucleosome width) are typically used as adjustable parameters.
Using parameters $b = 149$ bp for the nucleosome length and $\rho = 1/182$ bp^{-1} for the number density of nucleosomes in *Candida albicans*, plot nucleosome density versus the distance from the "nucleosome-free region" (NFR) up to 1500 bp and compare the result with that reported in Möbius et al.[115] Can the Kornberg–Stryer model reproduce the experimental data shown in Figure P7.25?

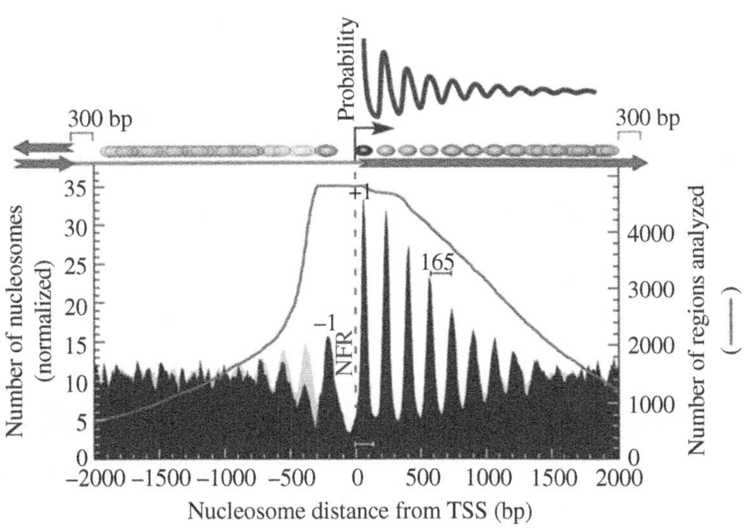

Figure P7.25 Distribution of nucleosome locations relative to transcriptional start sites (TSS), i.e., the location where the first DNA nucleotide is transcribed into RNA. The left side of TSS is referred to as the "nucleosome-free region" (NFR). Source: Adapted from Mavrich T. N. et al.[116]

115 Möbius W., et al. "Toward a unified physical model of nucleosome patterns flanking transcription start sites", *PNAS* 110, 5719–5724 (2013).
116 Mavrich T. N., et al. "A barrier nucleosome model for statistical positioning of nucleosomes throughout the yeast genome", *Genome Res.* 18: 1073–1083 (2008).

7.26 Waisman and Lebowitz derived an analytical solution for the Ornstein–Zernike (OZ) equation with the mean-spherical approximation (MSA) for a thermodynamic system consisting of two types of parallel plates, one positively charged and one negatively charged.[117] In this so-called one-dimensional analog of electrolyte solutions, the plate motions are restricted to the perpendicular direction with a pair potential similar to that given by the primitive model, i.e.,

$$u_{ij}(r) = \begin{cases} \infty, & r < \sigma_{ij}; \\ -q_i q_j r/(8\pi\epsilon_0\epsilon), & r \geq \sigma_{ij}. \end{cases}$$

Here r refers to the center-to-center distance between plates i and j, $\sigma_{ij} = (\sigma_i + \sigma_j)/2$ is their closest separation, σ_i and q_i represent the thickness and electrical charge density of plate i, respectively, ϵ_0 is the permittivity of free space, and ϵ is the dielectric constant of the background (viz., the solvent). Figure P7.26 shows a schematic representation of the model system.

It has been shown that the average reduced potential energy per unit volume (viz., the reduced excess internal energy per volume) is

$$\beta\omega = \frac{\kappa}{4} + \frac{\kappa^2}{8}\left(\frac{q_+\sigma_+ - q_-\sigma_-}{q_+ - q_-}\right), \tag{D}$$

where q_+ and q_- stands for the charge densities of positively and negatively charged plates, respectively, and κ is a screening parameter defined by

$$\kappa^2 = \frac{\beta(\rho_+ q_+^2 + \rho_- q_-^2)}{4\pi\epsilon_0\epsilon}.$$

Similar to that in a bulk electrolyte solution, the number densities of positively and negatively charged plates, ρ_+ and ρ_-, satisfy the condition of charge neutrality $\rho_+ q_+ + \rho_- q_- = 0$. Based on the potential energy density given by Eq. (D), show that MSA for the one-dimensional system predicts:

(i) the reduced excess Helmholtz energy per unit volume

$$\beta f^{ex} = \beta f^{ex}_{HR} + \frac{\kappa}{2} + \frac{\kappa^2}{8}\left(\frac{q_+\sigma_+ - q_-\sigma_-}{q_+ - q_-}\right),$$

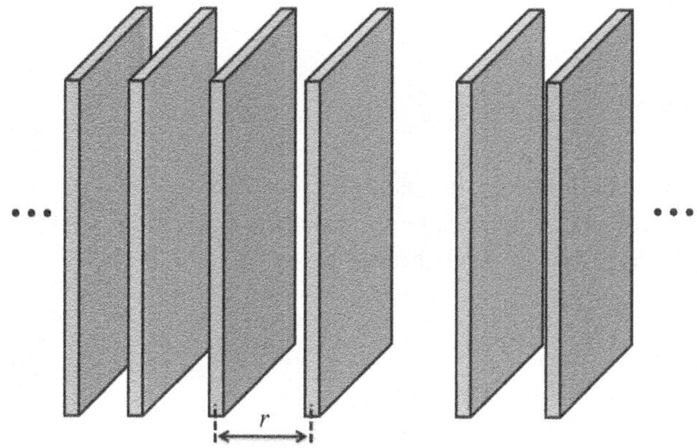

Figure P7.26 Charged parallel plates aligned in one dimension.

117 Waisman E. and Lebowitz J. L., "Mean spherical model integral equation for charged hard spheres", *J. Chem. Phys.* 56, 3086 (1972); *ibid.* 56, 3093 (1972).

where $f^{ex} \equiv f(T, \rho_i) - f^{id}(\rho_i)$, f^{id} represents the Helmholtz energy density of the ideal system, and βf^{ex}_{HR} is the reduced excess Helmholtz energy of the uncharged system. The latter is the same as that of uncharged one-dimensional hard rods (HR) (viz., the Tonks gas).

$$\beta f^{ex}_{HR} = -\rho \ln(1-\eta)$$

with $\rho = \rho_+ + \rho_-$ and $\eta \equiv \rho_+\sigma_+ + \rho_-\sigma_-$.

(ii) the osmotic pressure Π is given by

$$\frac{\beta\Pi}{\rho} = \frac{1}{1-\eta} - \frac{\kappa}{4\rho}$$

(iii) the system does not exhibit phase transition.

7.27 In the same work discussed in Problem 7.26, Waisman and Lebowitz obtained the MSA equations for the restricted primitive model (RPM) of electrolyte solutions, i.e., a model system of cations and anions with same size and valence but opposite charges. In this model, the pairwise additive potential is given by

$$u_{ij}(r) = \begin{cases} \infty, & r < \sigma; \\ q_i q_j/(4\pi\epsilon_0\epsilon r), & r \geq \sigma. \end{cases}$$

where r refers to the center-to-center distance between ions i and j, σ is their hard-sphere diameter, q_i is the electrostatic charge, ϵ_0 is the permittivity of free space, and ϵ is the solvent dielectric constant. MSA predicts the average potential energy per unit volume of the electrolyte solution (viz., the excess internal energy per volume)

$$\omega = -\frac{x^2 + x - x(1+2x)^{1/2}}{4\pi\beta\sigma^3}, \tag{E}$$

where $x^2 \equiv (\rho_+ q_+^2 + \rho_- q_-^2)\beta\sigma^2/\epsilon_0\epsilon$, ρ_+ and ρ_- are the number densities of positively and negatively ions. Show that

(i) the reduced excess Helmholtz energy per unit volume is

$$\beta f^{ex} = -\frac{2 + 6x + 3x^2 - 2(1+2x)^{3/2}}{12\pi\sigma^3},$$

(ii) the osmotic pressure Π satisfies

$$\frac{\beta\Pi}{\rho} = \frac{\beta\Pi_0}{\rho} + \frac{2 + 3x + 3x(1+2x)^{1/2} - 2(1+2x)^{3/2}}{72\eta},$$

where $\rho \equiv (\rho_+ + \rho_-)$, $\eta \equiv \pi\sigma^3\rho/6$, and Π_0 is the osmotic pressure of the system without the charge (viz., the reference hard-sphere system). According to the Carnahan–Starling equation, the hard-sphere contribution is

$$\frac{\beta\Pi_0}{\rho} = \frac{1 + \eta + \eta^2 - \eta^3}{(1-\eta)^3}.$$

(iii) the logarithm of the mean activity coefficient is given by

$$\ln\gamma_\pm = \ln\gamma_0 + \frac{x(1+2x)^{1/2} - x - x^2}{24\eta},$$

where the hard-sphere contribution is

$$\ln\gamma_0 = \eta\frac{8 - 9\eta + 3\eta^2}{(1-\eta)^3}.$$

(iv) the system exhibits a phase transition with the critical temperature and the packing fraction of ions $T_c^* \approx 0.0785773$ and $\eta_c \approx 0.00758476$, respectively. Here the reduced temperature is defined as $T^* \equiv \sigma/l_B$ where $l_B = \beta q^2/(4\pi\epsilon_0\epsilon)$ is the Bjerrum length, $q = q_+ = -q_-$ is the net charge of a cation or an anion.

7.28 Predict the osmotic and mean activity coefficients for aqueous solutions of LiCl using the MSA equations from Problem 7.27. Compare the theoretical results with the experimental data shown in Table P7.28 and with the predictions of the Debye–Hückel (DH) theory (see Section 9.6):

$$\phi = \frac{\beta\Pi}{\rho} = 1 - \frac{x^2 + 2x - 2(1+x)\ln(1+x)}{48\eta(1+x)},$$

$$\ln\gamma_{\pm} = -\frac{x}{2(1+x)T^*},$$

where $\eta \equiv \pi\sigma^3\rho/6$ is the packing fraction of ions, $x \equiv \sqrt{24\eta/T^*}$, and $T^* \equiv \sigma/l_B$ is the reduced temperature as defined in Problem 7.27. The average hard-sphere diameter of hydrated ions is about $\sigma \approx 3.6$ Å. Discuss why MSA fails at high salt concentration and how the theoretical results can be improved.

Table P7.28 Osmotic and mean activity coefficients of LiCl aqueous solutions.[a)]

m	ϕ	γ_m	m	ϕ	γ_m	m	ϕ	γ_m
0.001	0.988	0.965	1.000	1.020	0.775	8.000	2.159	5.190
0.002	0.984	0.952	1.200	1.044	0.798	9.000	2.328	7.110
0.005	0.976	0.928	1.400	1.068	0.825	10.000	2.480	9.600
0.010	0.969	0.904	1.600	1.093	0.855	11.000	2.612	12.700
0.020	0.960	0.874	1.800	1.119	0.888	12.000	2.723	16.400
0.050	0.947	0.827	2.000	1.145	0.924	13.000	2.814	20.700
0.100	0.940	0.789	2.500	1.213	1.029	14.000	2.887	25.500
0.200	0.940	0.756	3.000	1.284	1.157	15.000	2.947	30.900
0.300	0.946	0.743	3.500	1.361	1.312	16.000	2.995	36.900
0.400	0.954	0.739	4.000	1.441	1.499	17.000	3.036	43.400
0.500	0.964	0.739	4.500	1.525	1.726	18.000	3.066	50.300
0.600	0.974	0.742	5.000	1.613	2.000	19.000	3.079	57.000
0.700	0.985	0.748	5.500	1.703	2.330	19.219	3.079	58.400
0.800	0.996	0.756	6.000	1.794	2.730			
0.900	1.008	0.765	7.000	1.979	3.760			

a) Source: Hamer W. J. and Wu Y.-C., "Osmotic coefficients and mean activity coefficients of uni-univalent electrolytes in water at 25 °C", *J. Phys. Chem. Ref. Data* 1, 1047 (1972).

Hints:

(1) In Table P7.28, m represents molality, i.e., moles of solute per 1000 grams of solvent; γ_m is the activity coefficient defined in terms of molality, which can be related to the mean activity coefficient by $\gamma_m = C\gamma_{\pm}/md_0$, where C represents molarity, i.e., moles of solute (electrolyte) per liter of solution, and $d_0 = 0.997$ is the solvent density in units of kg/L.

(2) The molality and molarity can be converted to each other by using experimental data for the densities of LiCI solutions[118]:

$$d(\text{kg/m}^3) = d_1 + d_2 t + d_3 t^2 + d_4 \zeta + d_5 \zeta^2 + d_6 \zeta t + d_7 \zeta^2 t + d_8 \zeta t^2 + d_9 \zeta^3 + d_{10} \zeta^4,$$

where the coefficients $d_{i=1,2,\dots,10}$ are as follows: 1002.8 −0.15582 −2.88385e-3 6.1379 −5.8452e-2 6.0650e-4 −1.2546e-4 5.8029e-5 2.6623e-3 −2.5941e-5, t is temperature in °C, $\zeta \equiv 100\,w$, and w is the mass fraction of the solute.

7.29 Consider the Barker–Henderson (BH) perturbation theory for a square-well (SW) fluid with the pair potential

$$u(r) = \begin{cases} \infty, & r < \sigma, \\ -\epsilon, & \sigma \le r \le \lambda\sigma, \\ 0, & r > \lambda\sigma, \end{cases}$$

where r is the center-to-center distance between two spherical particles, σ is the hard-sphere (HS) diameter, $\lambda > 1$ specifies the range of attraction, and ϵ is an energy parameter. The reduced Helmholtz energy can be written as

$$f \equiv \beta F/N \approx f_0 + \beta f_1 + \beta^2 f_2,$$

where $\beta = 1/(k_B T), f_0 = (4\eta - 3\eta^2)/(1-\eta)^2$ corresponds to that of the HS reference system as predicted by the Carnahan–Starling (CS) equation, f_1 and f_2 are the first and second perturbation terms associated with the attractive energy.

(i) Show that the first-order perturbation term can be written as

$$f_1 = -12\epsilon\eta \int_1^\lambda dx x^2 g_0(x),$$

where $x = r/\sigma$.

(ii) Show that, with the local compressibility assumption (LCA), the second-order perturbation term can be written as

$$f_2 = \frac{1}{2}\epsilon \chi_0 \eta \frac{\partial f_1}{\partial \eta},$$

where $\chi_0 \equiv k_B T(\partial \rho/\partial P) = (1-\eta)^4/(1+4\eta+4\eta^2)$ is calculated from the isothermal compressibility of the hard-sphere fluid. What is f_2 according to the macroscopic compressibility approximation (MCA)?

(iii) An analytical expression for $g_0(x)$ with $x \lesssim 1.2$ has been reported by Tang and Lu (Problem 7.16)[119]

$$g_0(x) = \frac{1+\eta/2}{(1-\eta)^2}\frac{1}{x} + \frac{1-5\eta-5\eta^2}{(1-\eta)^3}(1-1/x) + \frac{-3\eta+6\eta^2+21\eta^3/2}{(1-\eta)^4}\frac{(x-1)^2}{x}$$

show that the integration term can be calculated from

$$\int_1^\lambda dx x^2 g_0(x) = \frac{a_4\lambda^4 + a_3\lambda^3 + a_2\lambda^2 + a_1}{(1-\eta)^4}$$

118 Wimby J. M. and Berntsson T. S., "Viscosity and density of aqueous solutions of LiBr, LiCl, $ZnBr_2$, $CaCl_2$, and $LiNO_3$. I: Single salt solutions", *J. Chem. Eng. Data* 39(1), 68–72 (1994).
119 Tang Y.-P. and Lu B. C.-Y., "A study of associating Lennard–Jones chains by a new reference radial distribution function", *Fluid Phase Equilibria* 171, 27–44 (2000).

with

$$a_1 = -(8 + 12\eta^2 + 7\eta^3)/24;\ a_2 = 3\eta(1 + 2\eta)^2/4;$$
$$a_3 = (1 + 2\eta)^2(1 - 4\eta)/3; \quad a_4 = -3\eta(2 - 4\eta - 7\eta^2)/8. \tag{F}$$

(iv) Compare Eq. (F) with the correlation used in the statistical associating fluid theory for potentials of variable attractive range (SAFT-VR) equation[120]:

$$\int_1^{\lambda} dx x^2 g_0(x) = (\lambda^3 - 1)\frac{1 - \eta_{eff}/2}{(1 - \eta_{eff})^3}$$

with $\eta_{eff} = c_1\eta + c_2\eta^2 + c_3\eta^3$ and

$$\begin{bmatrix} c_1 \\ c_2 \\ c_3 \end{bmatrix} = \begin{bmatrix} 2.25855 & -1.50349 & 0.249434 \\ -0.669270 & 1.40049 & -0.827739 \\ 10.1576 & -15.0427 & 5.30827 \end{bmatrix} \begin{bmatrix} 1 \\ \lambda \\ \lambda^2 \end{bmatrix}.$$

7.30 The hard-sphere diameter plays an important role in determining both the structure and thermodynamic properties of condensed systems with perturbation theories. To gain an understanding, calculate the hard-sphere diameter for a Lennard–Jones fluid near the triple point (e.g., $k_BT/\epsilon = 0.7$ and $\rho\sigma^3 = 0.85$) according to the Weeks–Chandler–Anderson (WCA) and the Barker–Henderson (BH) theories and discuss their implications in predicting the compressibility factor from the first-order perturbation theory

$$Z \equiv \beta P/\rho \approx Z_0 - \frac{\rho^2}{6k_BT}\int d\mathbf{r} r u'_A(r) g_0(r),$$

where Z_0 and $g_0(r)$ are the compressibility factor and radial distribution function of the hard-sphere reference system, and u'_A is the derivative of the perturbation potential.

7.31 The following exercise extends the Gibbs–Bogoliubov variational principle discussed in Section 5.6 to a system of spherical particles.

(i) Consider two systems with identical temperature T and particle density ρ but one with the total potential energy Φ and the other with the total potential energy Φ_0. Show the two Helmholtz energies satisfy the Gibbs–Bogoliubov inequalities

$$F_0 + \langle \Delta\Phi \rangle \leq F \leq F_0 + \langle \Delta\Phi \rangle_0,$$

where $\Delta\Phi = \Phi - \Phi_0$, $\langle \cdots \rangle_0$ and $\langle \cdots \rangle$ represent canonical ensemble averages under the potentials Φ and Φ_0, respectively.

(ii) Assume that the total potential energy is pairwise additive and that $(\Delta\beta u)e^{-\beta u_0} \approx e^{-\beta u} - e^{-\beta u_0}$, show

$$\langle \Delta\Phi \rangle_0 \approx \frac{N^2 k_BT}{2V}\int d\mathbf{r} y_0(r)[e^{-\beta u(r)} - e^{-\beta u_0(r)}],$$

where $y_0(r) = g_0(r)e^{\beta u_0(r)}$ is the cavity correlation function.

(iii) Based on the Weeks–Chandler–Anderson (WCA) potential for $u(r)$ and its corresponding hard-sphere potential for $u_0(r)$, plot the so-called blip function $\Delta e(r) \equiv e^{-\beta u(r)} - e^{-\beta u_0(r)}$ in the neighborhood of $r = \sigma$ for a Lennard–Jones fluid near

120 Gil-Villegas A. et al., "Statistical associating fluid theory for chain molecules with attractive potentials of variable range", *J. Chem. Phys.* 106, 4168 (1997).

the triple point (e.g., $k_BT/\epsilon = 0.7$ and $\rho\sigma^3 = 0.85$, which defines a WCA hard-sphere diameter $\sigma_{\text{WCA}} = 1.0335\sigma$.

(iv) In the WCA theory, the hard-sphere diameter is defined by $\langle\Delta\Phi\rangle_0 = 0$. Show that the approximation $F \approx F_0$ is accurate to the order of ξ^4 where ξ is the range of r where $\Delta e(r)$ is non-zero (i.e., the width of the blip function).

7.32 According to the Ornstein–Zernike (OZ) theory, the turbidity of a thermodynamic system, τ, is determined by integrating the scattering intensity over all angles

$$\tau = A\kappa_T \left[\frac{2\alpha^2 + 2\alpha + 1}{\alpha^3}\ln(1 + 2\alpha) - \frac{2(1 + \alpha)}{\alpha^2}\right],$$

where A is an instrumental constant, κ_T is isothermal compressibility, and $\alpha = 2(k_0\xi)^2$ with k_0 being the wave vector of incident light in the scattering medium and ξ the correlation length. The above equation allows to characterize the critical behavior of simple fluids as well as colloidal dispersions with turbidity measurements.[121]

(i) Show that the critical exponent for the isothermal compressibility can be obtained from turbidity in the small angle limit τ_0 (i.e., at $\alpha \to 0$).

(ii) Suppose that experimental data are available for τ/τ_0 in the vicinity of the critical point, how are the data used to determine the correlation length ξ.

(iii) Check the experimental results from the literature.

121 e.g., Chen B. H. et al., "Turbidity and critical behavior of a colloid-polymer system", *Phys. Rev. E* 64, 042401 (2001).

8

Polymer Solutions, Blends, and Complex Fluids

Most gases and liquids of practical interest are polymeric fluids, made of molecules with multiple repeating units. Polymer solutions and melts are ubiquitous in industrial processes and biological systems. Molecular theories to quantify the properties of such systems are needed not only for optimization of industrial processes such as the synthesis of plastics and other polymer products but also for *de novo* design of soft materials for diverse practical applications.

The properties of a polymeric fluid can be drastically different from those corresponding to its monomeric counterparts due to the complexity in molecular conformation and intramolecular interactions. In particular, the many-body nature of bond connectivity may result in long-range correlations. As a result, structure formation and phase transition in polymeric systems can be fundamentally different from those occurring in monomeric fluids.

In this chapter, we discuss statistical–mechanical methods to account for thermodynamic nonideality in polymeric systems due to various components of inter- and intra-molecular interactions. We start with the conventional lattice model to describe the thermodynamic properties and phase behavior of polymer solutions and blends. Extension of the liquid-state theories of simple fluids to polymeric systems will be discussed in the context of atomistic or particle-based models. Field-theoretical methods will also be discussed to describe long-range correlation effects and structure formation in polymeric systems. The theoretical tools to describe the properties of polymeric fluids are considerably more complicated than those for simple fluids. To mitigate mathematical difficulties, this chapter is supplemented with four appendices introducing the calculus of variation, the Gaussian integrals, the statistical field theory, and analytical equations for describing inhomogeneous ideal systems.

8.1 The Flory–Huggins Theory

The Flory–Huggins theory, which was independently developed by Paul Flory and Maurice Huggins in the 1940s,[1] provides a realistic description of the diverse phase behavior of polymer solutions and blends. Despite its simplicity, the mean-field method captures a key feature of polymeric systems, i.e., the reduction of translational entropy due to chain connectivity.

1 Flory P. J., "Thermodynamics of high polymer solutions", *J. Chem. Phys.* 9, 660–661 (1941); Huggins M. L., "Some properties of solutions of long-chain compounds," *J. Phys. Chem.* 46, 151–158 (1942).

Fundamentals and Practice in Statistical Thermodynamics, First Edition. Jianzhong Wu and John M. Prausnitz.

Companion website: www.wiley.com/go/jianzhongwu/fundamentalsandpractice

8.1.1 The Lattice Model for Polymer Chains

The Flory–Huggins theory is based on the lattice model. In its application to polymer solutions, each lattice site is occupied by either a solvent molecule or a polymer segment. As shown schematically in Figure 8.1, the lattice is fully occupied, and, owing to the excluded-volume effects, the double occupancy of any site is prohibited.

As each lattice site accommodates only one solvent molecule or one polymer segment, the lattice model assumes that solvent molecules and polymer segments have the same excluded volume. This implies that, according to the lattice model, the number of segments per polymer chain is not defined by the degree of polymerization but by the ratio of the molecular volumes

$$m = v_p/v_s \tag{8.1}$$

where v_p and v_s stand for the molecular volumes of the polymer and solvent, respectively. In practical applications, the difference between m and the degree of polymerization is typically not substantial because monomers and solvent molecules often have similar sizes.

As discussed in the following, one major advantage of using the lattice model is its simplicity in calculating the conformational entropy and internal energy. Besides, it provides insights into the unique features of polymer systems due to chain connectivity. To elucidate these points, consider a binary mixture of a monodisperse homopolymer dissolved in a one-component monomeric solvent.[2] The system consists of n_1 solvent molecules and n_2 polymer chains. To represent the system with a fully-occupied lattice, the total number of sites is $n_T = n_1 + mn_2$. Because monomers and solvent molecules are assumed to have the same excluded volume, the volume fraction of the solvent is equivalent to the fraction of lattice sites it occupies

$$\phi_1 = n_1/n_T. \tag{8.2}$$

A similar expression can be written for the polymer volume fraction

$$\phi_2 = mn_2/n_T = 1 - \phi_1. \tag{8.3}$$

In the original Flory–Huggins theory, the polymeric system is assumed to be incompressible, i.e., there is no volume change upon mixing the polymer and a liquid solvent at the same temperature

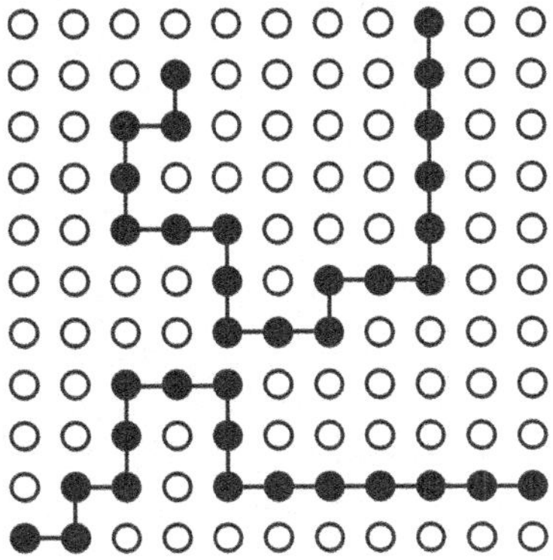

Figure 8.1 A two-dimensional view of the lattice model for a polymer solution. The interconnected black dots depict polymer segments, while the open circles represent solvent molecules.

2 In a homopolymer, the monomers or the repeating units are identical. Monodisperse means that the polymer chains have the same number of monomers (viz., the same degree of polymerization). Synthetic polymers are usually polydisperse. A common measure of polydispersity is given by the ratio of the weight-average and number-average molecular weight, $\overline{M}_w/\overline{M}_n$. Typically, the polydispersity of a synthetic polymer is in the range of $1.5 < \overline{M}_w/\overline{M}_n < 2$. For a monodisperse polymer, $\overline{M}_w/\overline{M}_n = 1$.

and pressure. The pressure effects may be accounted for by introducing unoccupied lattice sites (viz., "holes") or by considering the lattice compressibility.

In the lattice model, each microstate is defined by a particular arrangement of solvent molecules and polymer chains on the lattice. Accordingly, the canonical partition function is given by

$$Q = \sum_{\nu} e^{-\beta E_\nu} \tag{8.4}$$

where E_ν is the total energy at microstate ν, and $\beta = 1/(k_B T)$. In the Flory–Huggins theory, the *mean-field approximation* is adopted to evaluate the partition function, i.e., E_ν is replaced by an average energy, $E_\nu \approx U$. In that case, each term in the summation is independent of the microstate, and Eq. (8.4) becomes

$$Q \approx e^{-\beta U} \sum_{\nu} = W e^{-\beta U} \tag{8.5}$$

where $W = \sum_{\nu}$ is the total number of microstates, which amounts to the number of ways to place n_2 polymer chains and n_1 solvent molecules on the lattice such that all n_T sites are occupied.

With the partition function derived from Eq. (8.5), we can readily obtain the Helmholtz energy of the system

$$F = -k_B T \ln Q \approx U - k_B T \ln W. \tag{8.6}$$

The second term on the right side of Eq. (8.6) implies that the system entropy is approximated by the Boltzmann equation

$$S = k_B \ln W. \tag{8.7}$$

As each solvent molecule occupies only a single lattice site, the indistinguishability of all solvent molecules in the system implies that, once the polymer chains are placed on the lattice, there is only one way to fill the remaining sites with the solvent molecules. In other words, the Flory–Huggins theory assumes that the entropy of the polymer system is completely determined by the polymer configuration. Other than increasing the system volume, the solvent molecules do not contribute to the entropy of mixing because there is only one way to fill the lattice sites with identical solvent molecules.

From a molecular perspective, neither microstates nor polymer–solvent interactions are accurately represented by the lattice model. Nevertheless, the Flory–Huggins theory is able to describe the phase behavior of polymer systems reasonably well because the polymer phase diagrams are typically determined not by the absolute values but by the changes in thermodynamic quantities upon mixing pure species at constant temperature and pressure. According to Eq. (8.6), the Helmholtz energy of mixing, ΔF, defined as the Helmholtz energy of the polymer solution relative to those of the pure solvent and the polymer melt at the system temperature and pressure, is given by

$$\Delta F = \Delta U - k_B T \ln(W/W_0) \tag{8.8}$$

where W_0 is the number of ways to place n_2 polymer chains on a lattice without the solvent. Although the lattice model ignores the degrees of freedom associated with the kinetic energies of solvent molecules and polymer chains, these kinetic energies do not contribute to the thermodynamic properties of mixing.

8.1.2 Entropy of Mixing

As mentioned above, the Flory–Huggins theory asserts that the entropy of mixing is solely determined by the number of ways to place n_2 polymer chains, each with m segments, on n_T lattice sites without overlap. To derive this number, we may first assume that the multiple occupancy of a lattice site would be allowed. Later, we correct the errors by considering the probability of overlap for each segment.

When the first polymer chain is placed on an empty lattice, we may lay down one of the end segments of the polymer chain on any lattice site, i.e., there are n_T options to place the first segment. Because of the chain connectivity, the second segment and those following must be placed at one of the nearest-neighbors of its preceding segment.[3] Therefore, the number of ways to place the first polymer chain, denoted by w_1, is given by the product of the number of ways for each segment

$$w_1 = w_{11} \cdot w_{12} \cdot w_{13} \cdots w_{1m} \tag{8.9}$$

where the first subscript (1) refers to the first polymer chain, and the second subscript (1, 2, 3, ⋯, m) denotes the ranking number of individual segments in the polymer chain. Following the argument above, we have $w_{11} = n_T$ and, for $i = 2, 3, \ldots, m$, the allowance of multiple occupancy would yield

$$w_{1i} = Z \tag{8.10}$$

where Z is the coordination number, i.e., the number of nearest-neighbors per lattice site (e.g., $Z = 6$ for a cubic lattice). The number of ways to place the first polymer chain on the lattice is thus given by

$$w_1 = n_T Z^{m-1}. \tag{8.11}$$

With the hypothesis that multiple occupancy of the lattice sites is permitted, the number of ways to place the second polymer chain is identical to that for the first chain, w_1. The same argument holds for any additional chains. Therefore, the total number of ways to place n_2 polymer chains onto the lattice is

$$W' = w_1^{n_2}/n_2! = (n_T Z^{m-1})^{n_2}/n_2! \tag{8.12}$$

where the total number is divided by $n_2!$ because, in the liquid state, the polymer chains are indistinguishable from one another.

Eq. (8.12) does not take into account the fact that the number of vacant sites decreases as more polymer segments are placed on the lattice. As a result, it greatly overestimates the number of ways to place polymer chains on the lattice without overlap. Whereas an exact implementation of single occupancy is difficult, multi-occupancy may be corrected by estimating the probability of nonoverlap for each polymer segment, i.e., we multiply W' by the fraction of vacancy when each segment is placed on the lattice.

When the first polymer segment is placed on an empty lattice, the fraction of vacancy is $n_T/n_T = 1$, and thus it does not require any correction. In this case, all lattice sites are unoccupied, and there is nothing to overlap. For simplicity, we use a correction factor, $(n_T - 1)/n_T$, for the second segment, even though it must be placed at the immediate neighbor of the first segment. For the third segment, the correction factor is $(n_T - 2)/n_T$, and so forth for the last segment of all the polymer chains, it is $(n_T - mn_2 + 1)/n_T$. The total correction for W' is the product of the probabilities of nonoverlap for all segments, $n_T!/\left(n_1! n_T^{mn_2}\right)$. By applying the correction for multi-occupancy to Eq. (8.12),

3 In the thermodynamic limit, the system size is infinite, allowing for the neglect of the boundary effects.

we obtain a more realistic estimation for the number of ways to put n_2 polymer chains on n_T lattice sites without overlap

$$W = \frac{n_T!}{n_1! n_T^{mn_2}} W' = \frac{n_T!}{n_1! n_2! n_T^{mn_2}} (n_T Z^{m-1})^{n_2}. \tag{8.13}$$

To calculate the mixing entropy, we apply Eq. (8.13) to the polymer solution as well as to the neat polymer. In the former case, W is calculated with the total number of lattice sites given by $n_T = n_1 + mn_2$; and in the latter case, W_0 is calculated with the total number of lattice sites equal to mn_2. As mentioned above, the solvent molecules do not contribute because, after the placement of polymer segments, there is only one way to place the solvent molecules on the lattice. Accordingly, the entropy of mixing is given by

$$\Delta S = k_B \ln(W/W_0). \tag{8.14}$$

Substituting Eq. (8.13) into (8.14) leads to

$$\Delta S/k_B = \ln \left\{ \frac{(n_1 + mn_2)!}{n_1! n_2! (n_1 + mn_2)^{mn_2}} [(n_1 + mn_2) Z^{m-1}]^{n_2} \right\} - \ln \left\{ \frac{(mn_2)!(mn_2 Z^{m-1})^{n_2}}{n_2!(mn_2)^{mn_2}} \right\}. \tag{8.15}$$

Using the Sterling approximation, $\ln N! \approx N \ln N - N$, we can simplify Eq. (8.15) as (Problem 8.1)

$$\Delta S/k_B = -n_1 \ln \phi_1 - n_2 \ln \phi_2 \tag{8.16}$$

where ϕ_1 and ϕ_2 are the volume fractions of the solvent and polymer segments, respectively. Because $0 < \phi_i < 1$, the entropy of mixing is always positive.

To gain an understanding of how chain connectivity affects the entropy of mixing, we may compare Eq. (8.16) with that for a binary mixture of the same solvent and the *disconnected* polymer segments. In the latter case, the lattice model would predict a mixing entropy

$$\Delta S'/k_B = -n_1 \ln \phi_1 - mn_2 \ln \phi_2 \tag{8.17}$$

where the prime denotes the absence of chain connectivity. As the number of segments in each polymer chain is typically very large, we have $m \gg 1$, and thus $\Delta S' \gg \Delta S$. In other words, the Flory–Huggins theory correctly predicts that the chain connectivity greatly depresses the mixing entropy. This reduction in the entropy of mixing explains why polymer solubility is much smaller than that of the corresponding monomer, as commonly observed in experiments.

8.1.3 Nearest-Neighbor Energy

The Flory–Huggins theory assumes that the mean-field energy U can be approximated by that corresponding to the liquid mixture of solvent molecules and monomeric segments. In the evaluation of this energy, it is assumed further that the lattice sites are randomly occupied by the disconnected polymer segments and that only the nearest-neighbor interactions contribute to the average energy. With these assumptions, the mean-field energy becomes equivalent to that predicted by the lattice-gas model (Section 5.7).

Upon random mixing, on average, each lattice site is surrounded by $Z\phi_1$ solvent molecules and $Z\phi_2$ polymer segments in its nearest-neighbors, where Z is the coordination number as discussed above. Because the polymer segments occupy mn_2 lattice sites, we can estimate the average number of nearest-neighbor pairs between polymer segments, $mn_2 Z\phi_2/2$. Similarly, the average number of nearest solvent-solvent neighbor pairs is $n_1 Z\phi_1/2$; and the number of nearest-neighbor pairs between polymer segments and solvent molecules is $n_1 Z\phi_2 = mn_2 Z\phi_1$. To attain the mixing energy, let $-\varepsilon_{11}$, $-\varepsilon_{22}$ and $-\varepsilon_{12}$ represent the average energies for the nearest-neighbor pairs of solvent-solvent, polymer-polymer and solvent-polymer, respectively. Here, a negative sign is added because the nearest-neighbor interactions are typically attractive for nonpolar polymers. With the assumption of random mixing, the mean-field energy U is then given by

$$\begin{aligned} U &= -n_1 Z\phi_1\varepsilon_{11}/2 - n_2 m_2 Z\phi_2\varepsilon_{22}/2 - n_1 Z\phi_2\varepsilon_{12} \\ &= -\frac{n_T Z}{2}\left(\phi_1^2\varepsilon_{11} + \phi_2^2\varepsilon_{22} + 2\phi_1\phi_2\varepsilon_{12}\right) \end{aligned} \tag{8.18}$$

where $n_T = n_1 + mn_2$.

The above analysis is applicable to both the pure solvent and the neat polymer before the mixing. For the pure solvent, $U_1 = -n_1 Z\varepsilon_{11}/2$, and for the neat polymer, $U_2 = -mn_2 Z\varepsilon_{22}/2$. Therefore, Eq. (8.18) predicts that the internal energy of mixing is

$$\begin{aligned} \Delta U &= -\frac{n_1 Z\phi_1\varepsilon_{11}}{2} - \frac{n_2 m_2 Z\phi_2\varepsilon_{22}}{2} - n_1 Z\phi_2\varepsilon_{12} + \frac{n_1 Z\varepsilon_{11}}{2} + \frac{n_2 m_2 Z\varepsilon_{22}}{2} \\ &= n_T Z\phi_1\phi_2\left(\frac{\varepsilon_{11} + \varepsilon_{22} - 2\varepsilon_{12}}{2}\right) \equiv k_B T n_T \phi_1\phi_2\chi_F \end{aligned} \tag{8.19}$$

where $\chi_F \equiv Z(\varepsilon_{11} + \varepsilon_{22} - 2\varepsilon_{12})/(2k_B T)$ is known as the Flory–Huggins interaction parameter, or simply the Flory parameter.

8.1.4 Free Energy and Chemical Potential

A combination of Eqs. (8.16) and (8.19) gives the Helmholtz energy of mixing

$$\frac{\Delta F}{n_T k_B T} = \phi_1 \ln\phi_1 + \frac{\phi_2}{m}\ln\phi_2 + \chi_F\phi_1\phi_2. \tag{8.20}$$

Eq. (8.20) represents a key result of the Flory–Huggins theory. As expected, the Helmholtz energy of mixing consists of entropy and energy contributions of opposite signs: an increase in entropy favors mixing, but, because typically $\chi_F > 0$, the intermolecular interactions disfavor mixing. The phase behavior of polymer systems often reflects the interplay of the entropy and energy contributions.

In the limiting case, where $m = 1$ (i.e., each polymer chain becomes a monomer), we have a mixture of two equisized components with the Helmholtz energy of mixing given by

$$\frac{\Delta F}{n_T k_B T} = \phi_1 \ln\phi_1 + \phi_2\ln\phi_2 + \chi_F\phi_1\phi_2 \tag{8.21}$$

In this special case, volume fraction ϕ_i is identical to mole fraction x_i, and Eq. (8.21) is the familiar expression for ΔF corresponding to that given by the two-suffix Margules equation. As $m \gg 1$, a comparison of Eqs. (8.20) and (8.21) indicates that the Helmholtz energy of mixing for a polymer solution is much less negative in comparison with that for a mixture of corresponding disconnected polymer segments and solvent.

The lattice model assumes that the polymer system is incompressible. Besides, it assumes that mixing the solvent and polymer at constant temperature and pressure produces no change in the total volume. With these assumptions, we can derive the chemical potentials of the polymer and the solvent from the derivatives of the Helmholtz energy with respect to the number of molecules without the explicit requirement of constant volume

$$\mu_{i=1,2} = (\partial F/\partial n_i)_{T,n_{j\neq i}} \tag{8.22}$$

Substituting Eq. (8.20) into (8.22) leads to

$$\beta\mu_1 = \beta\mu_1^0 + \ln\phi_1 + \left(1 - \frac{1}{m}\right)\phi_2 + \chi_F\phi_2^2 \tag{8.23}$$

$$\beta\mu_2 = \beta\mu_2^0 + \ln\phi_2 + (1 - m)\phi_1 + m\chi_F\phi_1^2 \tag{8.24}$$

where superscript 0 denotes a pure liquid reference state, and subscripts 1 and 2 indicate solvent and polymer, respectively.

Similar to the van der Waals theory of simple fluids (Section 7.7), the Flory–Huggins theory accounts for the entropic and energetic contributions to the Helmholtz energy independently. Whereas both theories adopt the mean-field approximation for the internal energy, the entropic contributions are distinctively different. In the van der Waals theory, the non-ideality in entropy is associated with the *free volume* of individual molecules. By contrast, the Flory–Huggins theory attributes the entropy of mixing to the change in polymer configurations. The latter is exclusively determined by the number of ways to arrange the polymer chains on a lattice. If the connectivity of polymer segments is neglected, the Flory–Huggins theory would predict the entropy of mixing identical to that corresponding to ideal-gas systems.[4]

8.1.5 The Flory Parameter

The Flory parameter, as defined by $\chi_F \equiv Z(\varepsilon_{11} + \varepsilon_{22} - 2\varepsilon_{12})/(2k_BT)$, is a dimensionless variable. For typical nonpolar polymer systems, the nearest-neighbor interactions between polymer segments and solvent molecules are dominated by the van der Waals attraction, i.e., energy parameters ε_{11}, ε_{22} and ε_{12} are all positive. As suggested by London's theory for dispersion forces,[5] the attraction between a pair of monomeric species is independent of temperature and can be approximated by the geometric average of those for pairs of the same species, $\varepsilon_{12} \approx \sqrt{\varepsilon_{11}\varepsilon_{22}}$. As a result, we can express the Flory parameter as

$$\chi_F = Z(\sqrt{\varepsilon_{11}} - \sqrt{\varepsilon_{22}})^2/(2k_BT) \sim 1/T. \tag{8.25}$$

Substituting Eq. (8.25) into (8.19) indicates that the energy of mixing is positive (endothermic), i.e., mixing a nonpolar polymer with a liquid solvent is energetically unfavorable. While the Flory–Huggins theory asserts that χ_F is independent of the solution composition and polymer chain length, the assumption often breaks down in comparison with experimental results (see more discussion in Section 8.2). While χ_F is mostly positive, it does not always decline when temperature increases as one may predict using Eq. (8.25).

To understand the precise meaning of χ_F in the Flory–Huggins theory, Figure 8.2 illustrates the change in the reduced energy when a polymer segment in the pure polymer is exchanged with a solvent molecule in its pure liquid state. If we considered only the nearest-neighbor interactions, the exchange energy Δu would be proportional to the Flory parameter

$$\Delta u = Z(\varepsilon_{11} + \varepsilon_{22} - 2\varepsilon_{12})/2 = \chi_F k_B T. \tag{8.26}$$

Because the lattice model neglects the effects of particle exchange on the variations in the local environment, a more adequate description of the exchange energy is provided by the difference in the free energy of insertion for the polymer segments and solvent molecules. As the free energy of

4 For a binary mixture of ideal gases, the entropy of mixing is $\Delta S/k_B = -n_1 \ln x_1 - n_2 \ln x_2$ where x is the mole fraction.

5 See Supporting Information IV.

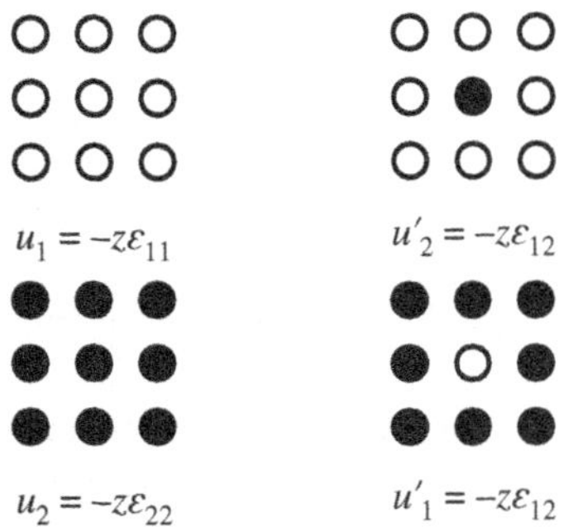

Figure 8.2 The energy change when one molecule of pure 1 is exchanged with one segment of pure 2 in a binary system of monomeric segments. Here, $u_{i=1,2}$ represents the energy per segment in pure i, and u'_i corresponds to that of an exchanged segment.

particle insertion depends on temperature and system composition, we expect that χ_F also varies with thermodynamic condition.

One convenient way to determine the Flory parameter experimentally is by measuring the solvent activity. Conventionally, the solvent activity is defined according to the difference between the chemical potential in the solution μ_1 and that in the pure solvent μ_1^0 at the system temperature and pressure

$$\mu_1 = \mu_1^0 + k_B T \ln a_1. \tag{8.27}$$

A comparison of Eqs. (8.23) and (8.27) indicates that the logarithm of the solvent activity is given by

$$\ln a_1 = \ln \phi_1 + \left(1 - \frac{1}{m}\right)\phi_2 + \chi_F \phi_2^2. \tag{8.28}$$

A typical experiment is to measure the solvent activity by monitoring the saturation pressure P of the polymer solution in terms of the polymer composition and temperature. At low or modest pressure, the vapor–liquid equilibrium predicts that the solvent activity satisfies $a_1 \approx P/P_1^s$, where P_1^s is the vapor pressure of the pure solvent at system temperature. Alternative experiments, such as phase diagram, osmotic pressure, and structure factor measurements, may also be utilized to determine χ_F. From a computational perspective, χ_F may be predicted from a molecular theory or simulation.

The Flory parameter provides a convenient measure of the solvent "quality." The special case of $\chi_F = 0$ corresponds to a condition where there is no energy for mixing. In other words, the Helmholtz energy of mixing is entirely attributed to the entropy contribution associated with various arrangements of polymer chains on the lattice. If the mixing of polymer and solvent is not accompanied by evolution or absorption of heat, the solution is called athermal. Good solvents have a negative or low χ_F. In that case, the energy of mixing is slightly endothermic or exothermic. Poor solvents have a highly positive χ_F, which disfavors mixing.

8.1.6 Summary

The Flory–Huggins theory is a versatile theoretical framework for describing the thermodynamic properties of polymer solutions and blends. As detailed in Section 8.2, the mean-field method is applicable to a wide range of polymer solutions and blends, irrespective of factors such as molecular weight, architecture, or functionality of the polymer chains. When utilizing the Flory–Huggins theory to correlate experimental data, the Flory parameter is often treated as adjustable, depending on temperature, solution composition, and polymer chain length, and in cases of high pressure, pressure as well.

Since its publication in the 1940s, much effort has been made to improve the Flory–Huggins theory. Most attempts have been directed toward relaxation of the assumptions of non-compressibility and non-specific interactions. The original Flory–Huggins theory predicts that, contrary to experiment, liquid–liquid phase separation (LLPS) of a polymer solution occurs only at low temperature. To correct this deficiency, modifications are often focused on relaxing the assumption of a rigid lattice where isothermal mixing occurs at constant volume. Toward that end, a common procedure is to introduce "free volume" into the lattice. A theory that incorporates free-volume effects can be achieved in a variety of ways. For example, it is commonly recognized that the cell occupied by a polymer segment (or monomer) is larger than its occupant. Contrary to the Flory–Huggins theory, this free-volume is not constant but depends on composition. The free-volume effects have been considered by Sanchez and Lacombe to introduce vacancies (holes) into the lattice model (Problem 8.8).[6]

8.2 Phase Behavior of Polymer Solutions and Blends

The Flory–Huggins theory offers a comprehensive description of the thermodynamic properties of polymer solutions and blends. Although primarily phenomenological in nature, the mean-field method proves to be highly valuable for a wide array of practical applications, ranging from industrial polymer synthesis and recycling to drug delivery and protein separation using aqueous two-phase systems.[7] It facilitates the determination of the solubilities of monomer and reactive species within a polymer phase, which are crucial for the design and simulation of polymer reactors. Furthermore, in the realm of biology, the Flory–Huggins theory provides an explanation for intracellular organization through liquid–liquid phase separation (LLPS), a process that plays a significant role in cellular responses to stress, gene regulation, and various pathological conditions.[8]

8.2.1 Osmotic Pressure

The osmotic pressure of a polymer solution is related to the solvent activity a_1 and the molecular volume of the solvent v_s

$$\Pi = -\frac{k_B T}{v_s} \ln a_1. \tag{8.29}$$

For a homopolymer dissolved in a monomeric solvent, the Flory–Huggins theory predicts that the solvent activity is given by (see Section 8.1)

$$\ln a_1 = \ln \phi_1 + \left(1 - \frac{1}{m}\right) \phi_2 + \chi_F \phi_2^2 \tag{8.30}$$

where $m = v_p/v_s$ represents the number of polymer segments according to the lattice model. Inserting Eq. (8.30) into Eq. (8.29) yields[9]

$$\frac{\Pi v_s}{k_B T} = -\ln \phi_1 - \left(1 - \frac{1}{m}\right) \phi_2 - \chi_F \phi_2^2. \tag{8.31}$$

---

6 Sanchez I. C. and Lacombe R. H., "An elementary molecular theory of classical fluids," *J. Phys. Chem.* 80, 2352–2362 (1976).

7 See, e.g., Kontogeorgis G. M., "Models for polymer solutions," *Comput. Aided Chem. Eng.* 19, 143–179 (2004).

8 Brangwynne C. P., Tompa P. and Pappu R. V., "Polymer physics of intracellular phase transitions," *Nat. Phys.* 11 (11), 899–904 (2015).

9 The relation $\Pi v_s/k_B T = -\ln a_1$ is true in general. It is not restricted to solutions described by the Flory–Huggins theory.

Eq. (8.31) suggests that experimental measurements of the osmotic pressure provide a possible way for obtaining the Flory parameter χ_F.

At low polymer concentration ($\phi_2 \ll 1$), expanding ln $\phi_1 = \ln(1-\phi_2)$ in Eq. (8.31) into a power series in ϕ_2 leads to

$$\frac{\Pi v_s}{k_B T} = \frac{\phi_2}{m} + (1/2 - \chi_F)\phi_2^2 + \phi_2^3/3 + \cdots. \tag{8.32}$$

If all but the first term on the right side of Eq. (8.32) is neglected, we obtain

$$\Pi/(k_B T) = \rho_2 \tag{8.33}$$

where $\rho_2 = n_2/[(n_1 + mn_2)v_s]$ is the number density of the polymer chains. Eq. (8.33) is known as the van't Hoff law. Based on the linear relation between the osmotic pressure and the number density of polymer chains, we can determine the polymer molecular weight (M_2) from the osmotic-pressure data, i.e., $M_2 = w_2 N_A/\rho_2$, where w_2 denotes the polymer weight density, and N_A is the Avogadro number.

The van't Hoff law is valid only when polymer chains do not interact with each other. More specifically, a comparison of Eqs. (8.32) and (8.33) indicates that the van't Hoff law holds when the polymer concentration satisfies

$$\frac{\phi_2}{m} \gg (1/2 - \chi_F)\phi_2^2. \tag{8.34}$$

or

$$\phi_2 \ll [m(1/2 - \chi_F)]^{-1}. \tag{8.35}$$

For a typical polymer solution, the number of segments per polymer chain m is extremely large. Eq. (8.35) indicates that the van't Hoff law is valid only for highly dilute solutions. For experimental measurements of polymer molecular weight, the Flory–Huggins theory suggests that it is often desirable to select a solvent such that χ_F is near 0.5, because in that case, the solution follows van't Hoff's law over a relatively wide range of polymer concentrations.

In Eq. (8.32), the coefficient of the ϕ_2^2 term, $A_2 = (1/2 - \chi_F)$, is called the osmotic second virial coefficient. Similar to the second virial coefficient of a dilute gas, A_2 can be positive, negative, or zero, depending on the Flory parameter χ_F. $A_2 > 0$ means that, in a dilute solution, the interaction between two polymer chains is dominated by *repulsion forces*. Conversely, $A_2 < 0$ indicates that the interaction between two polymer chains is overall *attractive*. The sign of A_2 thus distinguishes between *good* and *poor* solvents. A positive value of A_2 ($\chi_F < 0.5$) represents a good solvent, while a negative value of A_2 ($\chi_F > 0.5$) stands for a poor solvent. The situation when $A_2 = 0$ is called the ϑ (theta) condition. The temperature corresponding to $A_2 = 0$ is known as the theta (θ) temperature or the Flory temperature. In analogy to that for a low-density gas, the Flory temperature is equivalent to the Boyle temperature, at which a gas will approach ideal behavior under a certain range of pressure.

8.2.2 Non-Classical Solution Behavior

The Flory–Huggins theory has limitations in accurately predicting the osmotic behavior of polymer solutions because it neglects long-range correlations due to chain connectivity. For example, the critical volume fraction at which polymer chains begin to overlap can be estimated from the number of segments per polymer chain through the following semi-empirical relation

$$\phi_2^* = \frac{mv_s}{R^{*3}} \approx \frac{mv_s}{m^{3\alpha}a^3} = m^{1-3\alpha}. \tag{8.36}$$

In Eq. (8.36), α is the exponent in the scaling relation for the radius of gyration of individual polymer chains, $R^* \approx am^\alpha$, and $a = v_s^{1/3}$ represents the polymer Kuhn length. For a homopolymer with m segments dissolved in a good solvent, experimental observations indicate $\alpha \approx 0.6$; and in a theta solvent, $\alpha \approx 0.5$. Based on the scaling relation for the polymer size, Eq. (8.36) predicts that, when $m = 10^4$, $\phi_2^* \approx 6.3 \times 10^{-4}$ in a good solvent, and $\phi_2^* \approx 0.01$ in a theta solvent; indicating that the critical overlap concentration is extremely small and sensitive to the solvent quality.

In a typical polymer solution, the number of segments per polymer chain ($m \gg 1$) is extremely large. In the semi-dilute region where $\phi_2 > \phi_2^* \approx m^{1-3\alpha}$ and $\phi_2^2/(\phi_2/m) > m^{2-3\alpha} \gg 1$, the first term on the right side of Eq. (8.32) may be ignored

$$\Pi\, v_s/(k_B T) = A_2 \phi_2^2 + \phi_2^3/3 + \cdots. \tag{8.37}$$

Because $A_2 = (1/2 - \chi_F)$ depends on polymer–solvent interactions, Eq. (8.37) suggests that, in the semi-dilute region, the osmotic pressure of a polymer solution is approximately independent of the polymer molecular weight (or the degree of polymerization). In other words, at the same polymer volume fraction, polymer solutions with different molecular weights are expected to have the same osmotic pressure.

Figure 8.3 presents the osmotic pressures of different polymer solutions in a good solvent ($\chi_F = 0$) predicted by the Flory–Huggins theory. For comparison, similar results from experiments are shown in the inset.[10] At semi-dilute concentrations, both the theory and experimental data show that the osmotic pressure is virtually independent of the polymer molecular weight.

Whereas the Flory–Huggins theory predicts $A_2 = (1/2 - \chi_F)$, independent of the polymer chain length, experimental observations indicate that, for polymers in a *good* solvent, the second virial coefficient varies weakly with the degree of polymerization[11]

$$A_2 \sim m^{-0.236}. \tag{8.38}$$

Besides, for a homopolymer in a good solvent, the osmotic pressure follows the des Cloiseaux law in the semi-dilute region[12]

$$\Pi \sim \phi_2^{9/4}. \tag{8.39}$$

The exponents in Eqs. (8.38) and (8.39) have been interpreted using the renormalization-group theory to account for long-range intra-chain correlations (see Section 5.13). Deviations from the Flory–Huggins theory stem from the mean-field assumption that polymer segments are distributed randomly in space, which overlooks long-range intra-chain correlations in real polymer solutions. In the context of the lattice model, the placement of polymer chains is restricted not only by the nearest-neighbor connectivity but also by the interactions of the entire polymer chain.

8.2.3 Liquid–Liquid Phase Separation

Polymer solutions tend to separate into two liquid phases as the polymer concentration increases, one rich in solvent and the other rich in polymer. The binodal line of the liquid–liquid phase separation (LLPS), i.e., the locus of coexisting densities at different temperatures, can be calculated from the standard thermodynamic relations

$$\mu_i^I = \mu_i^{II} \tag{8.40}$$

10 Node I. et al., "Thermodynamic properties of moderately concentrated solutions of linear polymers", *Macromolecules*, 14, 668 (1981).

11 Rubinstein M. and Colby R. H., *Polymer physics*. Oxford University Press, 2003.

12 des Cloiseaux J., "The Lagrangian theory of polymer solutions at intermediate concentrations," *J. Phys. France* 36, 281–291 (1975).

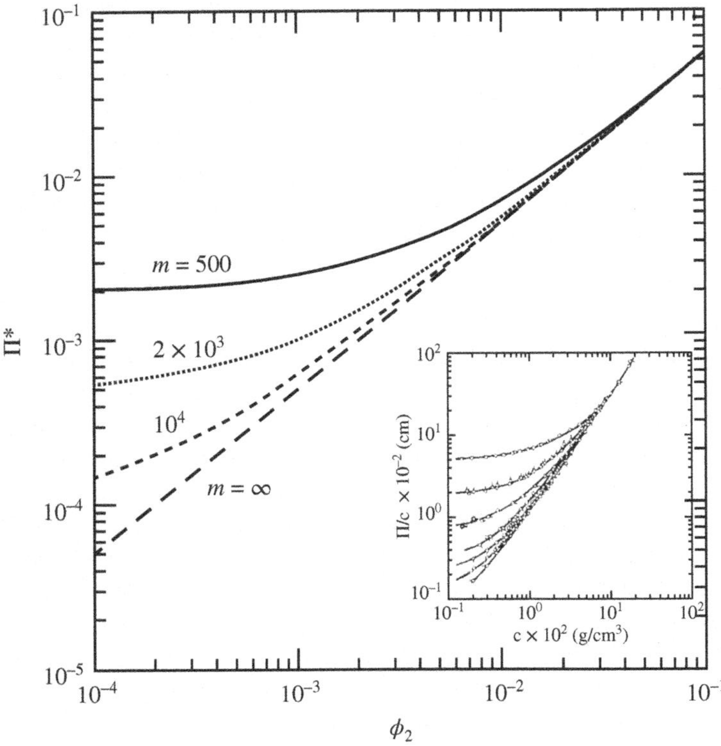

Figure 8.3 The reduced osmotic pressure $\Pi^* \equiv \Pi v_s/(k_B T\phi_2)$ for different homopolymers in a good solvent ($\chi_F = 0$) versus the polymer volume fraction ϕ_2 predicted by the Flory–Huggin theory. The inset shows experimental data for poly(α-methyl styrenes) in toluene at 25 °C at semi-dilute concentrations. For the experimental data, the osmotic pressure is expressed in units of g/cm^2 of polymers; c denotes the polymer weight concentration. From the top, the polymer molecular weights are 7, 20, 50.6, 119, 182, 330, 747×10^4 Da. The inset adapted from Node et al.[10]

where subscript $i = 1, 2$ denotes the solvent and polymer, respectively, and superscripts I and II represent the coexisting phases. As discussed in Section 8.1, the Flory–Huggins theory predicts that, for a binary mixture of a homopolymer and pure solvent, the chemical potentials are given by

$$\beta\mu_1 = \beta\mu_1^0 + \ln\phi_1 + \left(1 - \frac{1}{m}\right)\phi_2 + \chi_F\phi_2^2, \tag{8.41}$$

$$\beta\mu_2 = \beta\mu_2^0 + \ln\phi_2 + (1 - m)\phi_1 + m\chi_F\phi_1^2, \tag{8.42}$$

where μ_i^0 is the chemical potential of pure species i at system temperature and pressure.

Figure 8.4A shows representative phase diagrams of polymer solutions predicted by the Flory–Huggins theory (i.e., predicted from Eqs. (8.40)–(8.42)). Because the coexistence curves exist only at $\chi_F^{-1} < 2$, the solution is always stable when $\chi_F < 0.5$, regardless of the polymer chain length. The liquid–liquid coexisting region expands as the polymer molecular weight increases, which agrees qualitatively with experimental observation (Figure 8.4B).[13] Interestingly, it appears that the condition of thermodynamic stability is dictated by the second virial coefficient.

13 Dobashi T., Nakata M. and Kaneko M., "Coexistence curve of polystyrene in methylcyclohexane. II. Comparison of coexistence curves observed and calculated from classical free energy", *J. Chem. Phys.* 72, 6692–6697 (1980).

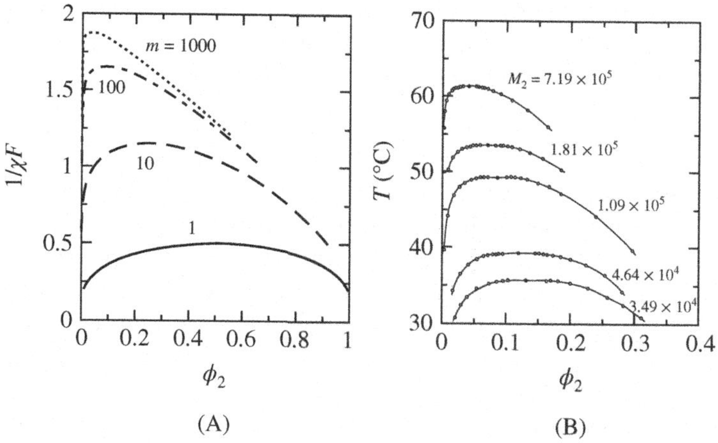

Figure 8.4 A. The coexistence curves of liquid–liquid phase separation (LLPS) predicted by the Flory–Huggins theory. B. Coexistence curves for the binary mixtures of polystyrene-methylcyclohexane with a variety of polymer molecular weights. Here, ϕ_2 represents the polymer volume fraction, m is the number of segments per chain, and M_2 is the polymer molecular weight. The experimental data are adapted from Dobashi et al.[13]

Because $A_2 > 0$ coincides with $\chi_F < 0.5$, there is no phase separation for a polymer in a good solvent. Conversely, for $\chi_F > 0.5$, a polymer solution may undergo phase transition as the polymer concentration increases. The coexistence curve is sensitive to the polymer chain length. As the number of segments per polymer chain increases, the coexistence curve becomes increasingly skewed toward a perpendicular line at $\phi_2 = 0$. When the chain length is sufficiently large, the dilute polymer solution is essentially a pure solvent. Qualitatively, the theoretical prediction is also in good agreement with the experimental observation (Figure 8.4B). The strong asymmetry in the volume fractions of the polymer and solvent along the coexistence curve distinguishes the phase behavior of a polymer solution from that of the corresponding monomeric system.

At the critical point, the second- and third-order derivatives of the free energy of mixing satisfy

$$\left(\partial^2 \Delta F / \partial \phi_1^2\right)_{n_T} = 0 = 1/\phi_{1c} + 1/(m\phi_{2c}) - 2\chi_{Fc}, \tag{8.43}$$

$$\left(\partial^3 \Delta F / \partial \phi_1^3\right)_{n_T} = 0 = -1/\phi_{1c}^2 + 1/\left(m\phi_{2c}^2\right), \tag{8.44}$$

where subscript "c" denotes the critical condition. From Eqs. (8.43) and (8.44), we can identify the polymer volume fraction and the Flory parameter at the critical point

$$\phi_{2c} = \frac{1}{1+\sqrt{m}}, \tag{8.45}$$

$$\chi_{Fc} = \frac{1}{2}\left(1 + \frac{1}{\sqrt{m}}\right)^2. \tag{8.46}$$

According to Eq. (8.45), $\phi_{2c} \to 0$ as m approaches infinity, affirming that the solvent-rich phase is essentially free of the polymer. For a fixed value of m, χ_{Fc} represents the maximum χ_F below which the polymer and solvent are miscible in all proportions. When $\chi_F > \chi_{Fc}$, there is a region of partial miscibility somewhere in the composition range $0 < \phi_1 < 1$. For a monomeric system, $m = 1$ and $\chi_{Fc} = 2$; but as $m \to \infty$, $\chi_{Fc} \to 1/2$. Eq. (8.46) thus explains why a polymer solution is more likely to exhibit a miscibility gap compared to a solution of equisized molecules.

In the Flory–Huggins theory, the nearest-neighbor interactions are assumed to be independent of temperature and dominated by van der Waals' attractions. Accordingly, the Flory parameter is inversely proportional to the absolute temperature and may be approximated by

$$\chi_F = \frac{v_s}{k_B T}(\delta_1 - \delta_2)^2 \tag{8.47}$$

where $\delta_{i=1,2} = (Z\varepsilon_{ii}/2v_s)^{1/2}$ represents the solubility parameter of species i. Eq. (8.47) is applicable to systems consisting of nonpolar polymers dissolved in a nonpolar solvent (e.g., polypropylene in cyclohexane). It predicts that, as shown schematically in Figure 8.5(I), the Flory parameter increases monotonically as temperature falls (viz., $1/T$ rises). Because the mixture is always stable when $T > T_c$, LLPS takes place only when the temperature is below the upper consolute (critical) solution temperature (UCST). As shown in Figure 8.4B for a nonpolar polymer solution, UCST increases with the degree of polymerization.

For systems such as poly(ethylene oxide) (PEO) dissolved in liquid water, the polymer solvent interactions are dominated by the competition of hydrogen bonding and hydrophobic interactions among the polymer segments and water molecules (i.e., H_2O–H_2O and H_2O–PEO interactions). In this case, the nearest-neighbor interactions cannot be simply assumed to be constant independent of temperature. Fitting the Flory–Huggins theory with the experimental measurement of the phase behavior indicates that the Flory parameter may be approximated by

$$\chi_F = k_0 + k_1/T + k_2/T^2 \tag{8.48}$$

with k_0, k_1 and k_2 being empirical parameters. Depending on the signs of these parameters, χ_F may either increase or decrease with temperature and exhibit minimum and maximum values

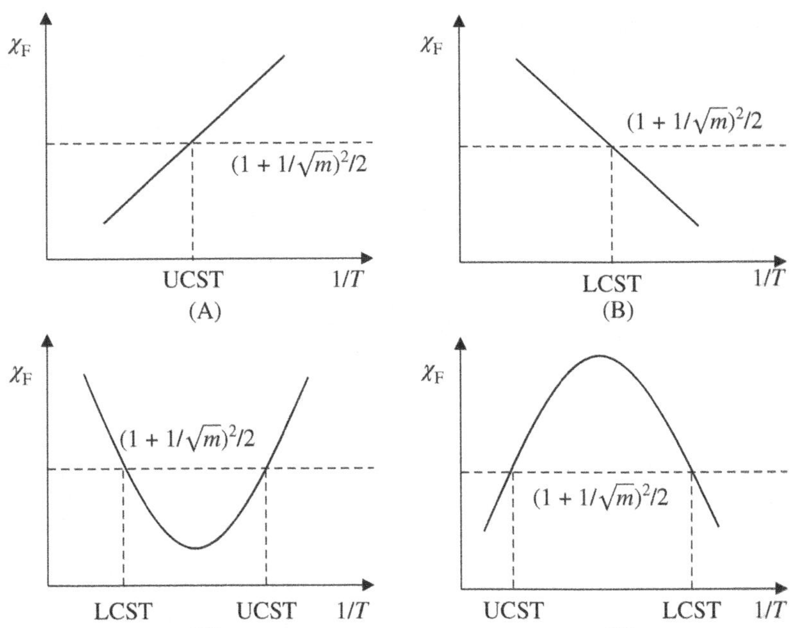

Figure 8.5 Four possible scenarios on the variation of the Flory parameter with temperature. A. $k_2 = 0$, $k_1 > 0$; B. $k_2 = 0$, $k_1 < 0$; C. $k_2 > 0$; D. $k_2 < 0$; where k_0, k_1 and k_2 are coefficients in the quadratic equation for χ_F, Eq. (8.48).

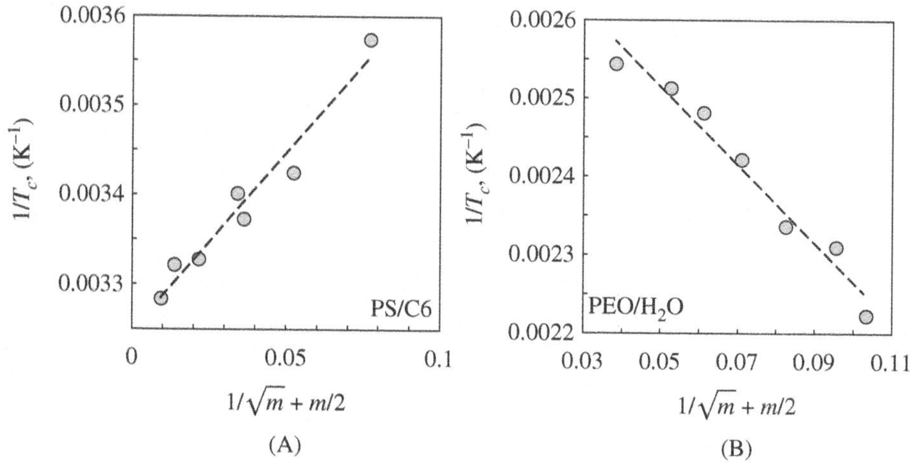

Figure 8.6 A. The dependence of the upper critical solution temperature (UCST) on the polymer chain length m for polystyrene in c-hexane (PS/C6). B. The lower critical solution temperature (LCST) for poly(ethylene oxide)/water (PEO/H_2O) as a function of the polymer chain length m. The dashed lines are fitted with Eq. (8.48) Source: Knychała et al.[14]

(see Figure 8.5). If χ_F falls monotonically with temperature (viz., as $1/T$ rises), the Flory–Huggins theory predicts that LLPS takes place when the temperature is below the lower consolute (critical) solution temperature (LCST). Alternatively, a polymer system may exhibit both UCST and LCST depending on how the Flory parameter χ_F varies with temperature.

Whereas the dependence of the Flory parameter on temperature is rather complicated, sensitive to the microscopic details of solvent-polymer interactions, the effect of molecular weight on the critical temperature is well captured by the Flory-Huggin theory. As shown in Figure 8.6, the critical temperature varies with the polymer chain length in a quadratic form as predicted by Eq. (8.48), with the slope depending on solution conditions.[14]

8.2.4 Phase Behavior of Polymer Blends

The Flory–Huggins theory can be similarly applied to multicomponent systems and polymer blends. In the latter case, the systems contain two or more polymers in the liquid or melt state. Understanding and controlling the miscibility of polymer blends is one of the fundamental subjects in polymer physics and has broad implications for industrial applications of plastic materials.

According to the Flory–Huggins theory, the Helmholtz energy of mixing for a binary mixture of homopolymers A and B is given by (Problem 8.6)

$$\frac{\Delta F}{n_T k_B T} = \frac{\phi_A}{m_A} \ln \phi_A + \frac{\phi_B}{m_B} \ln \phi_B + \chi_F \phi_A \phi_B \tag{8.49}$$

where the Flory parameter is defined as $\chi_F \equiv Z(\varepsilon_{AA} + \varepsilon_{BB} - 2\varepsilon_{AB})/(2k_BT)$, and $m_{\alpha=A,B}$ represents the number of segments for each polymer chain (or more precisely, the number of lattice sites

14 Knychała P., et al. "50th anniversary perspective: phase behavior of polymer solutions and blends," *Macromolecules* 50, 3051–3065 (2017).

occupied by each polymer chain). From the mixing Helmholtz energy, we can derive the chemical potentials of polymers A and B

$$\beta\mu_A = \beta\mu_A^0 + \ln\phi_A + (1 - m_A/m_B)\phi_B + m_A\chi_F\phi_B^2, \tag{8.50}$$

$$\beta\mu_B = \beta\mu_B^0 + \ln\phi_B + (1 - m_B/m_A)\phi_A + m_B\chi_F\phi_A^2. \tag{8.51}$$

Eqs. (8.50) and (8.51) can be used to predict the miscibility of polymer blends following the standard thermodynamic procedures. Qualitatively, the phase diagram resembles those shown in Figure 8.4 except that m is replaced by the ratio of the polymer chain lengths. When the temperature is plotted versus the composition, the liquid–liquid coexistence curve is skewed toward the polymer with a lower molecular weight.

If polymers A and B have the same chain length, the phase diagram is symmetric around $\phi_A = \phi_B = 1/2$. In this special case, we may utilize the symmetry in polymer composition to simply the phase-equilibrium calculation

$$\mu_A^I(\phi_A) = \mu_A^{II}(1 - \phi_A). \tag{8.52}$$

Plugging Eq. (8.50) into (8.52) leads to

$$\chi_F = \frac{1}{m(2\phi_A - 1)} \ln \frac{\phi_A}{1 - \phi_A}. \tag{8.53}$$

Figure 8.7 shows the binodal line predicted by Eq. (8.53). At the critical point, the Flory–Huggins theory predicts $m\chi_{Fc} = 2$, below which the polymer blend is always stable. The spinodal line is determined by

$$0 = \left(\partial^2 \Delta F/\partial\phi_A^2\right)_{n_T} = (m\phi_A)^{-1} + [m(1 - \phi_A)]^{-1} - 2\chi_F. \tag{8.54}$$

The system is unstable above the spinodal line and metastable between the spinodal and binodal lines.

8.2.5 Summary

Polymer solutions and blends exhibit LLPS below UCST and/or above LCST. The occurrence of LCST and/or UCST depends on the intermolecular forces, molecular weight, and polymer flexibility. Different types of LLPS may take place in different ranges of temperature. For many polymer

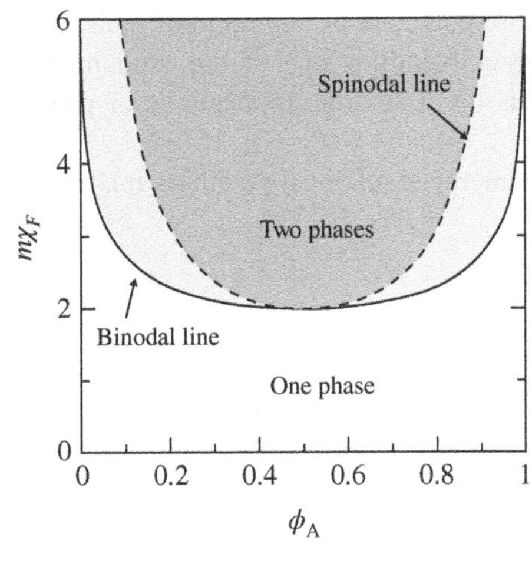

Figure 8.7 Binodal (solid) and spinodal (dashed) lines predicted by the Flory–Huggins theory for a polymer blend of two homopolymers of equal length ($m_A = m_B = m$).

systems, only one of these regimes is experimentally accessible, either because of polymer degradation at high temperatures, or because of glass transition (similar to freezing) at low temperatures.

With χ_F treated as an empirical function of temperature, the Flory–Huggins theory captures many (but not all) of the essential features of the phase behavior of polymeric systems. However, quantitative agreement is often poor when compared with experimental data. For liquid–liquid equilibria, the discrepancy mostly arises from large concentration fluctuations in the dilute-polymer phase and long-range intra-chain correlations. To account for the compressibility and fluctuation effects, we need a molecular theory that is not based on the lattice model and mean-field assumptions. In the next two sections, we will discuss alternative models to describe the thermodynamic properties of polymeric systems.

8.3 Statistical Mechanics of Polymeric Fluids

In this section, we discuss the molecular theories of polymeric fluids based on off-lattice models, i.e., each polymer chain is described as a cluster of particles linked together similar to atoms in a polyatomic molecule. Each particle represents a monomeric unit (viz., "segment") or a group of atoms (viz., a "united atom" or "coarse-gained" particle). Different from the lattice model used in the Flory–Huggins theory, a continuous model enables a more accurate description of both intra- and inter-molecular interactions. Typically, the molecular model defines not just the bond connectivity but also various components of non-bonded interactions, including short-range repulsion, van der Waals attraction, hydrogen bonding, and electrostatic potentials. Because such models are equally applicable to small molecules and polymers, statistical mechanics provides a unified theoretical framework for predicting the structure and thermodynamic properties of diverse polymer systems, including those containing ionic species and polyelectrolytes (see Chapter 9).

While the Flory parameter must vary with thermodynamic conditions in comparison with experimental data, off-lattice models may adopt molecular parameters reflecting the physical characteristics of individual segments, and thus they are independent of (or weakly dependent on) temperature and chemical composition. In other words, off-lattice models are able to predict the structure and thermodynamic properties of polymeric systems from an atomistic perspective.

8.3.1 Ideal Mixtures of Polymeric Species

To establish the basic concepts in the application of statistical mechanics to polymeric systems, consider first a uniform system containing C types of *non-interacting* polymeric molecules at a fixed temperature T, total volume V, and the chemical potential of each species $\mu_{I=1,2,\ldots C}$. Because the molecules are assumed to not interact with each other in an ideal gas, we can express the grand partition function in terms of the properties of individual molecules

$$\Xi^{IG} = \prod_{I=1}^{C} \sum_{N_I=0}^{\infty} \frac{\xi_I^{N_I}}{N_I! \Lambda_I^{3N_I}} \tag{8.55}$$

where subscript "I" denotes different types of polymeric molecules, ξ_I is defined by the intra-molecular bond potential $v_{B,I}(\mathbf{R}_I)$ and chemical potential μ_I

$$\xi_I \equiv \int d\mathbf{R}_I \exp\{-\beta[v_{B,I}(\mathbf{R}_I) - \mu_I]\}. \tag{8.56}$$

Eq. (8.55) resembles the grand potential for an ideal-gas mixture of monatomic molecules discussed in Section 3.5. One key difference is that, for a polymeric molecule, the intramolecular

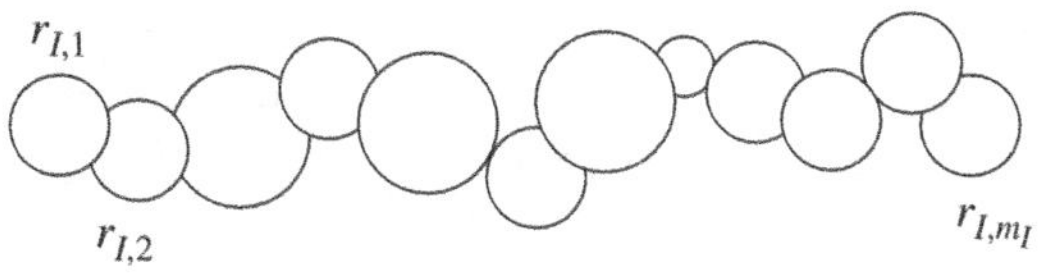

Figure 8.8 Schematic of a polymeric chain represented by m_I linearly connected spheres. The molecular configuration is specified by the position of individual segments, $\mathbf{R}_I = (\mathbf{r}_{I,1}, \mathbf{r}_{I,2}, \ldots \mathbf{r}_{I,m_I})$.

degrees of freedom depend on the molecular configuration, $\mathbf{R}_I \equiv (\mathbf{r}_{I,1}, \mathbf{r}_{I,2}, \ldots \mathbf{r}_{I,m_I})$, where $\mathbf{r}_{I,n}$ represents the position of particle n from a type-I molecule, and m_I is the number of particles in molecule of type I. Another difference, which is mostly irrelevant to describe the microscopic structure and phase behavior of polymeric systems, is that in Eq. (8.55), Λ_I is defined by a product of the thermal wavelengths of individual particles, i.e., $\Lambda_I \equiv \nu\prod_{n=1}^{m_I}\Lambda_{I,n}$, and ν accounts for the particle indistinguishability within the same molecule. It should be noted that, different from the thermal wavelength of a spherical particle $\Lambda_{I,n}$, Λ_I has the units of length to the power of m_I.

Schematically, Figure 8.8 depicts the configuration of a linear polymer chain with m_I spherical segments. In a typical off-lattice model, the polymer segments may have different sizes and interaction energies, much like atoms in a polyatomic molecule. Besides, the segments may be connected into nonlinear configurations in order to capture the topology of branched or dendritic polymers. Unless a truly atomic model is used, the connectivity between neighboring segments may not reflect the microscopic details of chemical bonds. For example, the bead-spring (or "pearl necklace") model describes the chain connectivity in terms of harmonic potentials that depend only on the center-to-center distance between neighboring segments. In many coarse-grained models, the chain connectivity is imposed only by keeping the nearest-neighbors at contact. Similar to that in a semi-empirical force field for polyatomic molecules, a bond-angle potential may be introduced to account for the flexibility of polymer chains.

Without loss of generality, the configuration of a polymeric molecule can be expressed in terms of the center of mass (COM) position $\mathbf{r}_I$ and variables reflecting the intramolecular degrees of freedom ϖ_I. Because the bond potential depends only on ϖ_I, we may integrate out the COM position in Eq. (8.56)

$$\xi_I = Ve^{\beta\mu_I}\int d\varpi_I \exp[-\beta v_{B,I}(\varpi_I)] = Ve^{\beta\mu_I}q_I \tag{8.57}$$

where $q_I \equiv \int d\varpi_I \exp[-\beta v_{B,I}(\varpi_I)]$. Other than a constant depending on temperature and dimensionality, q_I can be understood as the single-molecule partition function. Chapter 3 provides a more detailed description of the configurational properties of ideal polymer chains represented by various coarse-grained models.

As discussed in Appendix 8.D, we may evaluate the grand partition function of the ideal system

$$\Xi^{IG} = \prod_{I=1}^{C} \exp\left[\xi_I/\Lambda_I^3\right] = \prod_{I=1}^{C} \exp\left[Ve^{\beta\mu_I}q_I/\Lambda_I^3\right]. \tag{8.58}$$

From Eq. (8.58), we may obtain the thermodynamic properties of the system following the standard statistical–mechanical equations. For example, the grand potential is given by

$$\beta\Omega^{IG} \equiv -\ln\Xi^{IG} = -V\sum_{I=1}^{C}\frac{q_I e^{\beta\mu_I}}{\Lambda_I^3}, \tag{8.59}$$

and the average number of polymeric molecules in the open system is

$$\langle N_I\rangle = -(\partial\Omega/\partial\mu_I)_{T,V} = Vq_I\exp(\beta\mu_I)/\Lambda_I^3. \tag{8.60}$$

According to Eq. (8.60), the number density of each species is given by

$$\rho_I = \langle N_I \rangle / V = q_I \exp(\beta\mu_I)/\Lambda_I^3. \tag{8.61}$$

Rearrangement of Eq. (8.61) gives a concise expression for the chemical potential

$$\beta\mu_I = \ln\left(\rho_I \Lambda_I^3 / q_I\right). \tag{8.62}$$

Eq. (8.62) resembles the chemical potential of a polyatomic molecule in an ideal-gas system. From the grand potential, we may also derive the equation of state

$$\beta P^{IG} = -\beta\Omega^{IG}/V = \sum_I q_I \exp(\beta\mu_I)/\Lambda_I^3 = \sum_I \rho_I. \tag{8.63}$$

Eq. (8.63) is the familiar ideal-gas law for a mixture of non-interacting polymeric molecules. As expected, the ideal-gas pressure is independent of the single-molecule partition function because the latter decouples from the translational degrees of freedom for individual molecules. The Helmholtz energy is obtained by combining Eqs. (8.62) and (8.63)

$$\beta F^{IG}/V = \sum_I \langle N_I \rangle \beta\mu_I / V - \beta P = \sum_I \rho_I \left[\ln\left(\rho_I \Lambda_I^3 / q_I\right) - 1\right]. \tag{8.64}$$

Except the unknown constant q_I, Eq. (8.64) is the same as that for a mixture of monatomic fluids. For a non-interacting polymeric system, q_I is independent of the molecular density. For non-ideal systems, however, q_I is coupled with intermolecular interactions, as discussed below.

8.3.2 The Molecular Ornstein–Zernike Equation

The integral-equation theories discussed in Section 7.4 can be generalized to account for intra- and inter-molecular correlations in a polymeric fluid. From these correlation functions, we can predict thermodynamic properties using equations similar to those developed for simple fluids.

To elucidate the essential ideas, consider a one-component system consisting of polymeric molecules. In general, the molecular conformation can be specified by using a fixed coordinate system (viz., the local reference frame). If each molecule has a *rigid* conformation, the intramolecular degrees of freedom can be described in terms of two polar angles $\varpi = (\theta, \phi)$ for a linear molecule, or three Euler angles $\varpi = (\theta, \phi, \gamma)$ for a nonlinear molecule. Figure 8.9 illustrates the relative positions of two nonlinear molecules. For a bulk polymeric fluid, the radial distribution

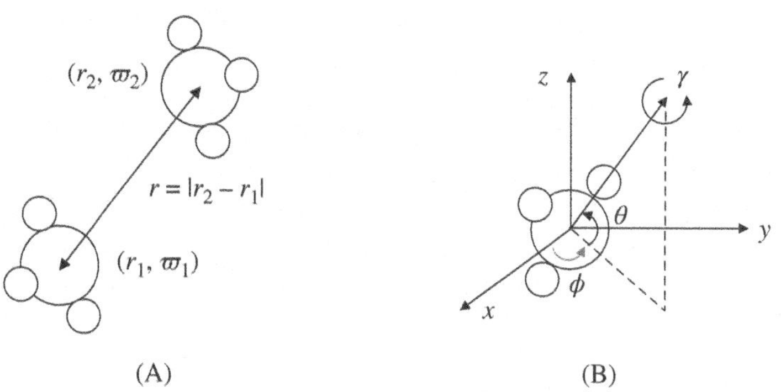

Figure 8.9 (A) The relative positions of two nonlinear molecules expressed in terms of the center-to-center distance r and the Euler angles ϖ_1 and ϖ_2. (B) For a nonlinear molecule with a rigid conformation, the three Euler angles are defined by the molecular orientation with respect to a fixed coordinate system.

function (RDF) depends on the center-to-center distance $r = |\mathbf{r}_1 - \mathbf{r}_2|$ as well as molecular orientations ϖ_1 and ϖ_2

$$g(r, \varpi_1, \varpi_2) = \frac{\Theta^2}{\rho^2} \left\langle \sum_{i=1}^{N} \sum_{j \neq i}^{N} \delta(\mathbf{r}_1 - \mathbf{r}_i)\delta(\varpi_1 - \varpi_i)\delta(\mathbf{r}_2 - \mathbf{r}_j)\delta(\varpi_2 - \varpi_j) \right\rangle \tag{8.65}$$

where ρ stands for the molecular number density, $\Theta \equiv \int d\varpi$, $\delta(\mathbf{r})$ is the Dirac delta function, and $\langle \cdots \rangle$ represents the ensemble average. For a linear molecule, $\Theta = 4\pi$, and for a non-linear molecule $\Theta = 8\pi^2$. Like the RDF for a monoatomic fluid, $g(r, \varpi_1, \varpi_2)$ is dimensionless. In writing Eq. (8.65), we may define the center-to-center distance in terms of the molecular center of mass. However, other choices of the molecular center are also possible and contain equivalent microscopic information.

In addition to molecular RDF, we can introduce other correlation functions (e.g., direct correlation function, cavity correlation function, and total correlation function) for polymeric systems. While the physical meanings of these correlation functions are similar to those for simple fluids, a key difference is that additional variables must be used to account for the orientational degrees of freedom. To simplify the notation, we may use positive integers (viz., molecular indices) to denote both the position and orientation of a molecule. For example, the radial distribution function between two polyatomic molecules may be written as

$$g(1,2) \equiv g(|\mathbf{r}_1 - \mathbf{r}_2|, \varpi_1, \varpi_2). \tag{8.66}$$

Also similar to that for simple fluids, the radial distribution function $g(1,2)$ can be solved from the molecular Ornstein–Zernike (OZ) equation

$$h(1,2) = c(1,2) + \frac{\rho}{\Theta} \int c(1,3)h(1,3)d3 \tag{8.67}$$

where $h(1,2) \equiv g(1,2) - 1$ stands for the molecular total correlation function, $c(1,2)$ is the molecular direct correlation function, and the integration ($d3$) is carried out over all degrees of freedom related to the position and orientation of a third polymeric molecule (3). The closure to the molecular OZ equation can be formally written as

$$h(1,2) = c(1,2) + \ln y(1,2) + B(1,2) \tag{8.68}$$

where $y(1,2) \equiv g(1,2)e^{\beta u(1,2)}$ stands for the molecular cavity correlation function, $u(1,2)$ denotes intermolecular potential, and $B(1,2)$ represents the molecular bridge function. Without the bridge function, Eq. (8.68) refers to the molecular hypernetted-chain (HNC) approximation. Other closure relations can be derived to relate the molecular direct and total correlation functions.

Whereas the molecular OZ equation is analogous to that for simply fluids, the increase in dimensionality from $g(r)$ to $g(r, \varpi_1, \varpi_2)$ makes the numerical solution of the RDF significantly more complicated. To circumvent the computational burden, a conventional approach is to expand the intermolecular potential as well as the total and direct correlation functions in terms of the rotational invariants.[15] For a one-component system of linear molecules, the expansion of a two-body correlation function over the angular variables leads to a series of spherical harmonics

$$f(r, \varpi_1, \varpi_2) = 4\pi \sum_{n=0}^{\infty} \sum_{m=0}^{\infty} \sum_{l=-k}^{k} f_{nml}(r) Y_{nl}(\theta_1, \phi_1) Y_{m\bar{l}}(\theta_2, \phi_2) \tag{8.69}$$

15 Ishizuka R. and Yoshida N., "Application of efficient algorithm for solving six-dimensional molecular Ornstein-Zernike equation", *J. Chem. Phys.* 136, 114106 (2012).

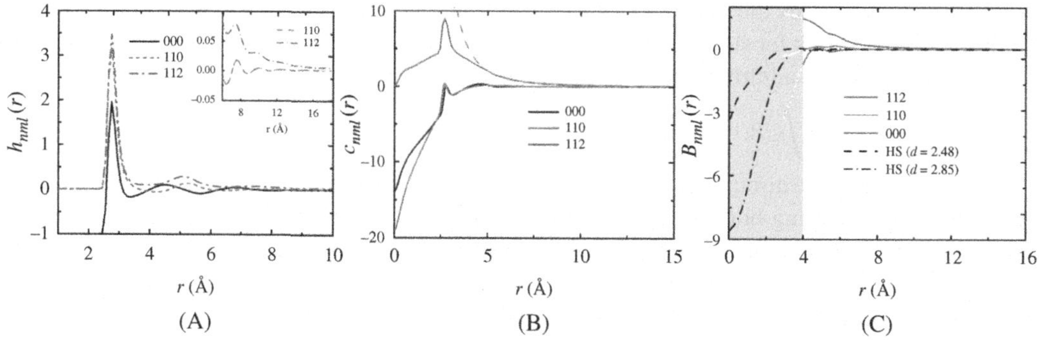

Figure 8.10 The first three rotational invariants of the total correlation function (A), direct correlation function (B) and bridge function for bulk water are represented by the SPC/E model. In panel (C), the bridge functions of reference hard-sphere systems are shown for comparison with the gray zone denoting inaccessible distance. Source: Reproduced from Zhao et al.[16]

where $k = \min(n, m)$ and $\bar{l} = -l$. In Eq. (8.69), $f_{nml}(r)$ are called the rotational invariants (viz., "projections") in the expansion of function $f(r, \varpi_1, \varpi_2)$ in terms of the spherical harmonics $Y_{nl}(\theta_1, \phi_1)$ and $Y_{m\bar{l}}(\theta_2, \phi_2)$. These polar angles are defined in an intermolecular frame with the z-axis pointing to the center-to-center direction of the linear molecules, $\mathbf{r} = \mathbf{r}_2 - \mathbf{r}_1$.

We may elucidate various correlation functions of a polymeric fluid using a small-molecule system as an example. Figure 8.10 shows the first three rotational invariants of the total and direct correlation functions and the bridge function for liquid water, represented by the extended simple point charge (SPC/E) model.[16] While water molecules have a nonlinear structure, the rotational invariants were obtained after averaging the rotational components of the Euler angles. As expected, the molecular correlation functions are highly anisotropic. But along each orientation, these correlation functions behave similarly to those corresponding to a simple fluid.

With the assumption of pairwise additivity, the potential energy for a system of rigid molecules can be expressed in terms of the pair potential

$$\Phi(\mathbf{r}^N, \varpi^N) = \frac{1}{2}\sum_{i \neq j}^{N} u(|\mathbf{r}_i - \mathbf{r}_j|, \varpi_i, \varpi_j). \tag{8.70}$$

Accordingly, we can predict the excess internal energy, $U^{ex} \equiv U - U^{IG}(T, V, \rho)$, from the radial distribution function

$$U^{ex}/N = \frac{\rho}{2}\int \langle u(r, \varpi_1, \varpi_2) g\left(r, \varpi_1, \varpi_2\right)\rangle_{\varpi_1,\varpi_2} 4\pi r^2 dr \tag{8.71}$$

where the angular average is defined as

$$\langle \cdots \rangle_{\varpi_1,\varpi_2} \equiv \frac{1}{\Theta^2}\iint \cdots d\varpi_1 d\varpi_2. \tag{8.72}$$

Similar to that for a simple fluid, the virial equation can be written as (Problem 8.16)

$$\frac{\beta P}{\rho} = 1 - \frac{\rho}{6}\int \langle g(r, \varpi_1, \varpi_2)\beta u'\left(r, \varpi_1, \varpi_2\right)\rangle_{\varpi_1,\varpi_2} 4\pi r^3 dr \tag{8.73}$$

16 Zhao S. L., et al., "Accurate evaluation of the angular-dependent direct correlation function of water", *J. Chem. Phys.* 139, 034503 (2013).

where the prime denotes the derivative of the pair potential with respect to the center-to-center distance r. Also like that for simple fluids, the compressibility equation can be derived without the assumption of pairwise additivity for the intermolecular potential (Problem 8.17)

$$\rho k_B T \kappa_T = 1 + \rho \int \langle h(r, \varpi_1, \varpi_2)\rangle_{\varpi_1,\varpi_2} d\mathbf{r}. \tag{8.74}$$

We see from the above discussion that the molecular OZ equation is analogous to the OZ equation for simple fluids, bearing both advantages and limitations. The method is most useful for systems consisting of rigid molecules. Because of its high dimensionality, the molecular OZ equation is rarely used to study the thermodynamic properties of flexible polymeric fluids. In addition to the numerical difficulties in solving various correlation functions, molecular flexibility introduces intramolecular correlations that are intrinsically coupled with intermolecular interactions.

8.3.3 The Reference Interaction Site Model (RISM)

The reference interaction site model (RISM) was introduced by Chandler and Andersen to describe the correlation functions of molecular fluids.[17] Its extension to polymeric systems (PRISM) was developed by Schweizer and Curro.[18] A similar theoretical framework was proposed by Cummings and Stell in the context of the site–site OZ equation.[19] Although the site–site correlation functions represent only a subset of the pair correlation functions between polyatomic or polymeric molecules, RISM circumvents the problem of high-dimensionality and is thus suitable for a wide range of applications.[20]

One key assumption in the interaction site model (ISM) is that the pair potential between polyatomic molecules can be represented by the site–site potentials

$$u(1,2) = \sum_{i=1}^{m_I}\sum_{j=1}^{m_J} u_{ij}(|\mathbf{r}_{I,i} - \mathbf{r}_{J,i}|) \tag{8.75}$$

where m_I denotes the number of interaction sites for molecule $I = 1,2$, and $\mathbf{r}_{I,i}$ is the position of interaction site i. For a polymeric system, the interaction sites may be understood as polymer segments, albeit a site might also refer to the position of a local electrical charge in a small molecule (e.g., SPC/E model for water). The RISM is so-called because a reference system is commonly used in describing the site–site interactions, similar to that in the WCA perturbation theory (Section 7.11).

With the site–site correlations defined in terms of the *intermolecular* radial distribution function

$$g_{ij}(|\mathbf{r} - \mathbf{r}'|) \equiv \frac{1}{\rho^2}\iint d1d2\rho(1)\rho(2)g(1,2)\delta\left(\mathbf{r}_i^{(1)} - \mathbf{r}\right)\delta\left(\mathbf{r}_j^{(2)} - \mathbf{r}'\right) \tag{8.76}$$

and the *intramolecular* correlation function

$$\omega_{ij}(|\mathbf{r} - \mathbf{r}'|) = \left\langle \delta\left[\mathbf{r} - \left(\mathbf{r}_i^{(1)} - \mathbf{r}_j^{(1)}\right)\right]\right\rangle, \tag{8.77}$$

the RISM equation provides a formal relationship between the site–site total and direct correlation functions. In the Fourier space, these correlation functions can be expressed in a matrix form

$$\hat{\mathbf{h}}(k) = \hat{\mathbf{w}}(k)\hat{\mathbf{c}}(k)\hat{\mathbf{w}}(k) + \rho\hat{\mathbf{w}}(k)\hat{\mathbf{c}}(k)\hat{\mathbf{h}}(k) \tag{8.78}$$

---

17 Chandler D. and Andersen H.C., "Optimized cluster expansions for classical fluids. II. Theory of molecular liquids", *J. Chem. Phys.* 57, 1930–1937 (1972).
18 Schweizer K.S. and Curro J.G., "Integral equation theories of the structure, thermodynamics, and phase transitions of polymer fluids", *Adv. Chem. Phys.* 98, 1–142 (1997).
19 Cummings P. T. and Stell G., "Interaction site models for molecular fluids", *Mol. Phys.* 46, 383–426 (1982).
20 Ratkova E. L., Palmer D. S. and Fedorov M. V., "Solvation thermodynamics of organic molecules by the molecular integral equation theory", *Chem. Rev.* 115, 6312–6356 (2015).

where the caret sign denotes the Fourier transform, ρ is the number density of the polymeric molecules, and the bold font represents a matrix of size $m \times m$ with each element defined by a pair of interaction sites. While the site–site total correlation function, $h_{ij}(r) \equiv g_{ij}(r) - 1$, can be measured experimentally (e.g., via neutron scattering with isotope substitutions), the physical meaning of the site–site direct correlation function $c_{ij}(r)$ is less transparent, and the RISM equation may serve as the definition. If each molecule consists of a single particle, $\hat{\omega}(k) = 1$, and Eq. (8.78) reduces to the OZ equation for simple fluids.

Like the OZ equation, the RISM equation must be coupled with a closure, i.e., an additional relation for the site–site total and direct correlation functions. If the site–site potential includes a hard-sphere repulsion, one may adopt the *heuristic* relations

$$\begin{cases} h_{ij}(r) = -1 & r < \sigma_{ij} \\ c_{ij}(r) = -\beta u_{ij}(r) & r \geq \sigma_{ij} \end{cases} \tag{8.79}$$

where σ_{ij} stands for the collision diameter. Eq. (8.79) is analogous to the mean-spherical approximation (MSA) for simple fluids. The total correlation function is -1 when $r < \sigma_{ij}$ because the radial distribution function vanishes due to the hard-sphere repulsion. Meanwhile, the expression for the direct correlation is adapted from that for a monatomic system. Because the molecular orientations are not fully captured by the site-site correlation functions, RISM is known to have difficulties in accurately reproducing the dielectric constants of molecular systems.[19]

Alternative closures can be derived in the context of the site–site cavity correlation function $y_{ij}(r)$ and the site–site bridge function $B_{ij}(r)$

$$h_{ij}(r) = c_{ij}(r) + \ln y_{ij}(r) + B_{ij}(r). \tag{8.80}$$

While Eq. (8.80) is analogous to that for a simple fluid, the site–site bridge function should be defined in terms of matrix relations among various correlation functions instead of their individual pairs. Different from the bridge function in the OZ equation, Eq. (8.80) cannot serve as a definition of the site–site bridge function because it is decoupled from the intramolecular correlations.

For polymeric molecules with a rigid conformation, the intramolecular site–site correlation function is exactly known

$$\omega_{ij}(r) = \delta_{ij}\delta(r) + (1 - \delta_{ij})\frac{\delta(r - l_{ij})}{4\pi l_{ij}^2} \tag{8.81}$$

where δ_{ij} stands for the Kronecker delta function, and l_{ij} is the distance between interaction sites i and j. For a system containing flexible molecules, however, a self-consistent calculation of the intra- and inter-molecular correlation functions is necessary because of the coupling effects. In that case, the calculation can be accomplished by molecular simulation using the site–site potential of mean force.[21]

In extension of the RISM equation to polymer systems (viz, PRISM), a reference model is typically used to represent the intramolecular correlation functions. In its simplest form, when PRISM is applied to one-component homopolymers, the polymer segments may be assumed to be equivalent if the chain length is sufficiently long such that the end effects are negligible. In that case, the matrix form of the RISM equation is reduced to a one-dimensional integral equation

$$h(k) = \omega^2(k)c(k) + \rho\omega(k)c(k)h(k) \tag{8.82}$$

21 Khalatur P. G. and Khokhlov A. R., "Hybrid MC/RISM technique for simulation of polymer solutions", *Mol. Phys.* 93, 555–572 (1998).

where

$$\omega(k) \equiv \frac{1}{m}\sum_{i=1}^{m}\sum_{j=1}^{m}\hat{\omega}_{ij}(k). \tag{8.83}$$

If the Gaussian chains are used as a reference, the intramolecular correlation function is given by (see Section 3.9)

$$\hat{\omega}_{ij}(k) = \exp[-|i-j|k^2b^2/6] \tag{8.84}$$

where b represents the Kuhn length, which may be approximated by the segment diameter. Other forms of intramolecular correlation functions, including those for semi-flexible chains and for rotational isomeric state polymer chains, have also been proposed. The equivalent segment approximation is exact for ring polymers and provides a reasonable approximation for homopolymer chains when the end effects can be ignored. In comparison to the original site–site model, the equivalent segment approximation greatly simplifies the numerical procedure for solving the direct and total correlation functions of polymeric systems. In addition to polymer melts, the PRISM equation has also been used to predict the structure and thermodynamic properties of block copolymers and polymer composites.[22]

8.3.4 Wertheim's Thermodynamic Perturbation Theory

In thermodynamic perturbation theory (TPT), a polymeric fluid is considered as a system of monomeric segments linked together by chemical association. The free energy of chain formation was first derived by Michael S. Wertheim using a diagrammatic procedure to account for both chemical association and non-bonded interactions.[23] Similar results were obtained by Chandler and Pratt in describing chemical equilibrium in condensed phases.[24] The original derivations are quite lengthy and involve complicated graph manipulation. Here, we follow a heuristic approach proposed by Zhou and Stell[25] that connects the free-energy of a polymeric system with that of a monomeric fluid through a thermodynamic cycle.

Consider a polymeric system with each molecule consisting of m spherical segments. In comparison with that of a monomeric fluid at the same temperature, volume, and the number densities of individual segments, the excess Helmholtz energy of the polymeric system can be decomposed into three contributions, as represented by a thermodynamic cycle shown schematically in Figure 8.11:

Step (I) dissociation of individual molecules at fixed temperature and system volume in the ideal-gas state;

Step (II) introducing interaction among the disconnected polymer segments without the bonding energies; and

Step (III) re-polymerization of the monomeric segments in the dense phase by introducing chemical association (viz., bond potentials).

Assuming that all polymer chains have the same configuration, $\mathbf{R} \equiv (\mathbf{r}_1, \mathbf{r}_2, \ldots \mathbf{r}_m)$, we can write the free energy of dissociation in terms of the intramolecular potential $v_M(\mathbf{R})$

$$\Delta F_I = -Nv_M(\mathbf{R}) \tag{8.85}$$

22 Martin T.B., et al., "pyPRISM: a computational tool for liquid-state theory calculations of macromolecular materials", *Macromolecules* 51, 2906–2922 (2018).
23 Wertheim M. S., "Thermodynamic perturbation theory of polymerization", *J. Chem. Phys.* 87, 7323–31 (1987).
24 Chandler D. and Pratt L. R., "Statistical mechanics of chemical equilibria and intramolecular structures of nonrigid molecules in condensed phases", *J. Chem. Phys.* 65, 2925–40 (1976).
25 Zhou Y. Q. and Stell G., "Chemical association in simple-models of molecular and ionic fluids", *J. Chem. Phys.* 96, 1507–1515 (1992).

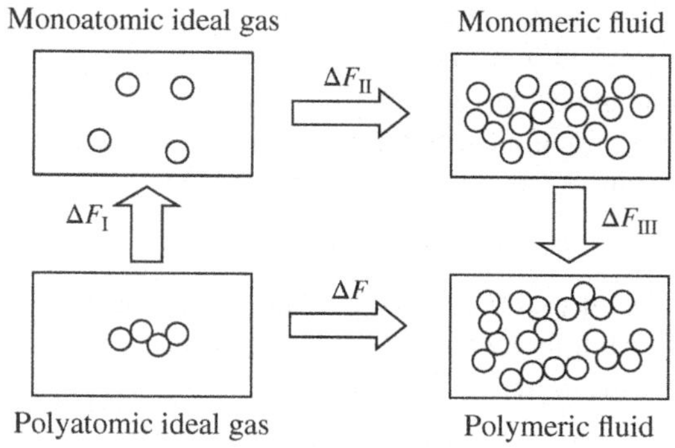

Figure 8.11 The excess Helmholtz energy of a polymeric fluid can be calculated from a three-step thermodynamic cycle: *I*. Dissociation of each polymeric molecule into monomeric segments in the ideal-gas state; *II*. The excess Helmholtz energy of the corresponding monatomic fluid; and *III*. Re-polymerization of the segments through chemical association.

where N is the number of polymer chains in the system. Eq. (8.85) is formally exact provided that the intramolecular energy accounts for both the bond energies and non-bonding interactions among all polymer segments. Intuitively, the free-energy change can be related to the equilibrium constant of the gas-phase reaction $K_0 = \exp(-\beta\Delta F_I/N)$.

In Step II, introducing the non-bonded interactions into the monomeric ideal gas leads to the excess Helmholtz energy of the monomeric fluid:

$$\Delta F_{II} = F_m^{ex} \tag{8.86}$$

In Step III, the formation free energy of N polymer chains from the monomeric segments is determined by the potential of mean force

$$\Delta F_{III} = -Nk_BT \ln g(\mathbf{R}) \tag{8.87}$$

where $g(\mathbf{R})$ stands for the multi-body distribution function of m *associating* segments. The latter is defined by the ratio of m-body density $\rho(\mathbf{r}_1, \mathbf{r}_2, \ldots \mathbf{r}_m)$ and segment density $\rho_i(\mathbf{r}_i)$ in the polymeric system

$$g(\mathbf{R}) \equiv \frac{\rho(\mathbf{r}_1, \mathbf{r}_2, \ldots \mathbf{r}_m)}{\rho_1(\mathbf{r}_1)\rho_2(\mathbf{r}_2)\ldots\rho_m(\mathbf{r}_m)}. \tag{8.88}$$

For a uniform system, the number densities of polymer segments are the same as the molecular density, $\rho_i(\mathbf{r}_i) = \rho$. If each polymeric molecule consists of two segments, $g(\mathbf{R})$ reduces to the radial distribution function $g(r)$.

The combination of steps I and III leads to

$$\Delta F_I + \Delta F_{III} = -Nv_M(\mathbf{R}) - Nk_BT \ln g(\mathbf{R}) = -Nk_BT \ln y(\mathbf{R}) \tag{8.89}$$

where $y(\mathbf{R}) \equiv g(\mathbf{R})e^{\beta v_M(\mathbf{R})}$ stands for the multibody cavity correlation function. Eq. (8.89) suggests that, upon the formation of a polyatomic molecule through chemical reactions, the equilibrium constant in the condensed phase (i.e., the associating system)

$$K \equiv \rho(\mathbf{R}) \Big/ \prod_{i=1}^{m} \rho_i(\mathbf{r}_i) \tag{8.90}$$

can be related to that for the dissociation in the ideal-gas state K_0 through the multibody cavity correlation function

$$K/K_0 = y(\mathbf{R}). \tag{8.91}$$

Eq. (8.91) was first derived by Chandler and Pratt.[26]

26 A detailed comparison of Wertheim's and Chandler-Pratt's formulations of chemical association in condensed phases was reported by Kierlik E. and Rosinberg M. L., "A perturbation density functional theory for polyatomic fluids. II. Flexible molecules", *J. Chem. Phys.* 99, 3950 (1993).

The above procedure indicates that the Helmholtz energy of the polymeric system can be expressed in terms of three contributions, i.e., an ideal-gas contribution at the same temperature and molecular density, the excess Helmholtz energy of the monomeric reference system (viz., free of chemical bonding among segments), and an additional contribution due to multi-body intramolecular correlations

$$\beta F = \beta F^{IG} + \beta F_m^{ex} - N\langle \ln y(\mathbf{R})\rangle. \tag{8.92}$$

In Eq. (8.92), $\langle \cdots \rangle$ denotes an average of all possible polymer configurations. As discussed above, the Helmholtz energy for the ideal-gas term is exactly known (Eq. (8.64)) and, for an m-mer fluid, it is given by

$$\beta F^{IG} = N\left[\ln\left(\rho\Lambda_M^3/q_M\right) - 1\right] \tag{8.93}$$

where $\Lambda_M = \Lambda^m$, Λ is the thermal wavelength for each segment, and $q_M \equiv \int d\varpi \exp[-\beta v_B(\varpi)]$.

Eq. (8.92) is formally exact; approximations are typically introduced in the expressions for the excess Helmholtz energy of the monomeric reference system F_m^{ex} and for the multibody cavity correlation functions $\langle \ln y(\mathbf{R})\rangle$. As discussed in Chapter 7, the excess Helmholtz energy of the monomeric reference system F_m^{ex} can be calculated from various statistical–mechanical models of simple fluids. In its lowest-order approximation, the multibody cavity correlation function is represented by a linear combination of the two-body correlation functions of the monomeric fluid

$$\langle \ln y(\mathbf{R})\rangle \approx \sum_{i=1}^{m-1} \ln y_m(|\mathbf{r}_i - \mathbf{r}_{i+1}|). \tag{8.94}$$

Eq. (8.94) is commonly known as the first-order thermodynamic perturbation theory (TPT1), which is particularly convenient from a practical perspective because it is independent of polymer configuration. For polymeric models with hard-core segments, the contact value of the two-body cavity correlation function is often analytically obtainable (Appendix 7.A).

8.3.5 Summary

Similar to those for simple fluids, the analytical theories of polymeric systems are mostly based on integral-equation and perturbation theories. Most integral-equation theories are based on the molecular Ornstein–Zernike (MOZ) equation and RISM. While MOZ is useful for systems containing rigid molecules, RISM has its own advantages owing to the low-dimensionality of site–site correlation functions. Different from integral-equation theories, TPT describes the properties of polymeric systems in terms of the cavity correlation functions and thermodynamic properties of monomeric systems. With the assumption that the multibody cavity correlation function can be approximated by a linear combination of the two-body correlation functions of a monomeric system, TPT is convenient in many practical applications due to its computational efficiency.

In the Flory–Huggins theory, each polymer chain is characterized by the number of lattice sites and a van-der-Waals-like parameter, limiting its ability to describe the microscopic details of polymeric systems. Such limitations can be eliminated by using off-lattice models because they account for the various forms of inter- and intra- molecular interactions. While in principle an off-lattice model is able to capture atomic details, the microscopic information is typically not captured in coarse-grained models adopted in the theoretical description of polymer systems. Nevertheless, unlike the Flory parameter that varies with thermodynamic conditions, the parameters of off-lattice models are mainly dependent on the properties of individual molecules and are thus often transferable to different thermodynamic systems. In Sections 8.4 and 8.5, we will demonstrate the applications of off-lattice models for describing the thermodynamic properties and phase behavior of polymeric systems.

8.4 Equations of State for Hard-Sphere Chains

Most analytical theories of polymeric fluids assume that a polymer chain can be represented by a series of tangentially connected hard-sphere segments. As shown schematically in Figure 8.12, the polymer backbone is totally flexible because the bond angles are arbitrary, restricted only by the hard-sphere interactions that prohibit the overlap of polymer segments. The hard-sphere-chain model describes the chain connectivity and repulsive interactions between polymer segments. Like the hard-sphere model for simple fluids, it provides a useful reference for the formulation of the equations of state for realistic polymeric fluids.

In this section, we discuss two complementary theories of hard-sphere-chain fluids. The first is the generalized Flory (GF) theory proposed originally by Dickman and Hall[27]; the second is the TPT equations initially due to Wertheim.[28] Alternative theories of hard-sphere-chain models can be derived using similar ideas. In Section 8.5, we provide an extension of the hard-sphere-chain model to polymers and associating fluids. Despite its simplicity, there is no exact theory of a hard-sphere-chain fluid. The theoretical task is difficult due to the strong coupling of excluded volume effects and long-range intra-chain correlation.[29]

8.4.1 Excluded-Volume Effects in Polymeric Fluids

When we compress a thermodynamic system consisting of classical particles, we decrease its volume, but no matter how high the pressure is, we cannot shrink the volume to zero because two (or more) particles cannot occupy the same space. As discussed in Section 7.7, the excluded-volume effect is defined such that, in the presence of one particle, the center of another particle is excluded from a space depending on the volumes of *both* particles. In the Flory–Huggins theory, the excluded-volume effect is accounted for by placing polymer segments on a 3-dimensional lattice; and the excluded volume of each solvent molecule is often assumed to be the same as that of a polymer segment, independent of the solution composition. In the lattice model, attractive interactions between any pair of segments or solvent molecules are represented by a mean-field energy parameter instead of a realistic potential that varies with the center-to-center distance. Without modification, the Flory–Huggins theory is unable to describe fluid compressibility or specific intermolecular attractions (e.g., hydrogen bonding).

The excluded-volume effects tend to cancel out when calculating changes in the thermodynamic properties of a liquid mixture relative to those of the pure liquids because a liquid is nearly incompressible. For this reason, the Flory–Huggins theory is most useful for calculating the properties of mixing and liquid–liquid phase diagrams of polymeric fluids. However, because the Flory–Huggins

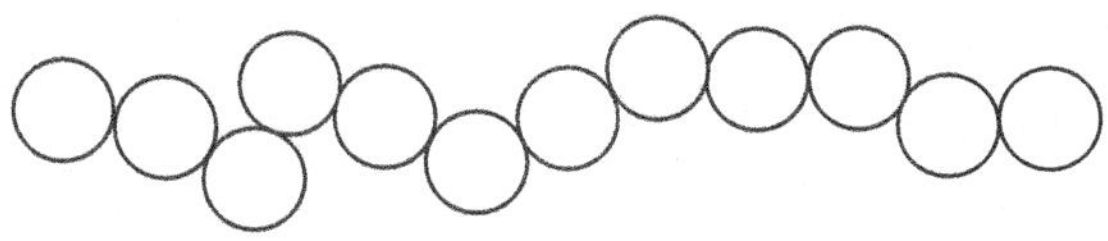

Figure 8.12 In the tangentially connected hard-sphere-chain model, each polymer segment is represented by a hard-sphere bead in close contact with its nearest-neighbors.

27 Dickman R. and Hall, C. K., "Equation of state for chain molecules: continuous-space analog of Flory theory", *J. Chem. Phys.* 85, 4108–4115 (1986).

28 Wertheim M. S., "Thermodynamic perturbation theory of polymerization", *J. Chem. Phys.* 87 (12), 7323–7231 (1987).

29 Here, long-range correlation means that the distribution of a particular segment in a hard-sphere chain is closely correlated not only with its immediate neighbors but also with those of all other segments from the same chain.

theory is based on an imaginary lattice structure,[30] it is problematic for describing the thermodynamic properties of pure polymeric fluids. Besides, the lattice model is not directly applicable to vapor–liquid equilibria of polymeric systems. Whereas many revisions of the Flory–Huggins theory have been proposed to relax some of the limitations, these modifications cannot be generalized to represent the thermodynamic properties of polymeric fluids systematically. For a systematic model that is not limited by the lattice picture, we must explicitly consider the compressibility effects as well as various components of intermolecular interactions. Toward this end, the hard-sphere-chain model provides a relatively simple, alternative representation of excluded volume effects in polymeric fluids. Whereas the theoretical models of hard-sphere-chain fluids overcome only some of the limitations of the lattice model and are strictly valid only for polymers in a good solvent, they can be generalized to describe the properties of realistic polymeric systems by adding perturbation terms to account for intermolecular attractions and other forms of long-range interactions.

8.4.2 The Generalized Flory (GF) Theories

Dickman and Hall extended the Flory–Huggins lattice model to a continuous space and developed a series of equations of state for hard-sphere chains, called the GF theories.[24] While the original GF equations are based on Flory's polymer theory, the same results (and their extension to more realistic cases) can be conveniently introduced by using the statistical–mechanical description of solvation, which predicts the reversible work for dissolving a solute molecule in a solvent. While our discussion here is focused on the tangent hard-sphere-chain model, the theoretical procedure in describing solvation is generally applicable to all molecular systems.

8.4.2.1 Statistical Mechanics of Solvation

Consider dissolving a solute molecule into a pure solvent. For simplicity, we assume that both the solute and solvent molecules are tangent hard-sphere chains, but they may have different lengths and/or segment diameters. Figure 8.13 shows schematically the solute molecule, represented by five tangentially-connected hard spheres, and the solvent molecules are represented as hard-sphere trimers. We may specify the configurations of the polymeric system with the segment positions for both the solvent molecules and of the solute.

For a thermodynamic system consisting of a single solute molecule surrounded by N solvent molecules, we may divide the total potential energy into three contributions: one accounts for the

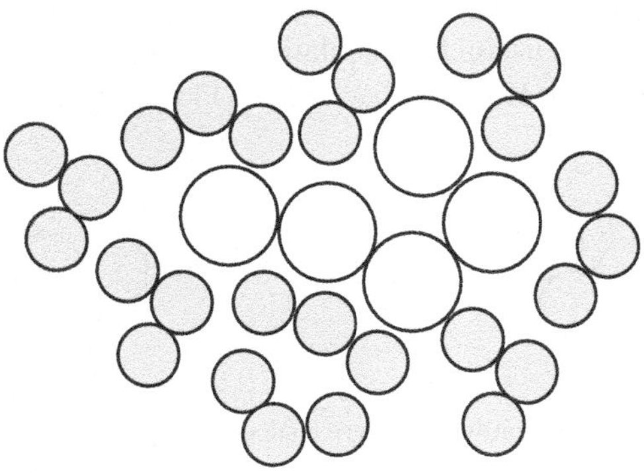

Figure 8.13 Dissolution of one polyatomic molecule (solute) in N solvent molecules. Here, the solute molecule is represented by a five-segment tangent chain, while solvent molecules are represented by gray trimers.

30 In calculating the Helmholtz energy of mixing, this unsuitable feature of the Flory–Huggins theory tends to cancel because the unrealistic lattice picture is applied to the pure components as well as to the mixture.

interaction among the solvent molecules, one comes from the solute–solvent interaction, and the third represents the self-energy of the solute molecule

$$\Phi_T\left(\mathbf{R}_0^N, \mathbf{R}\right) = \Phi_0\left(\mathbf{R}_0^N\right) + \Delta\Phi\left(\mathbf{R}_0^N, \mathbf{R}\right) + \Phi(\mathbf{R}) \tag{8.95}$$

where subscript 0 denotes the solvent, $\mathbf{R} \equiv (\mathbf{r}_1, \mathbf{r}_2, \ldots, \mathbf{r}_m)$ represents the positions of solute segments, and $\mathbf{R}_0^N$ is the solvent configuration. As discussed in Section 8.3, the canonical partition function of the polymeric system is

$$Q = \frac{1}{N!\Lambda_0^{3N}\Lambda_M^3} \int d\mathbf{R}_0^N \int d\mathbf{R} e^{-\beta\Phi_T(\mathbf{R}_0^N, \mathbf{R})} \tag{8.96}$$

where $\beta = 1/(k_B T)$, Λ_0 and Λ_M are coefficients arising from the translational motions of classical particles. Similar to the de Broglie thermal wavelength of a spherical particle, Λ_0 and Λ_M depend on temperature but not the molecular configurations.

In the thermodynamic limit ($N \to \infty$), the chemical potential of the solute molecule approaches negative infinity because, at a fixed solvent density, the system volume becomes infinite. Nevertheless, the excess chemical potential, defined by the chemical potential of the solute in the solution minus that of the same solute by itself at system temperature and volume, has a finite value. As indicated in the next two paragraphs, this excess chemical potential is commonly known as the solvation free energy, a key property characterizing solute–solvent interactions.

In a dilute solution ($N \to \infty$), the solute chemical potential is related to the change in the Helmholtz energy at constant *temperature* and constant *volume* upon introducing one solute molecule into a pure solvent

$$\mu = F(N, 1) - F(N) = -k_B T \ln(Q/Q_0) \tag{8.97}$$

where $F(N,1)$ denotes the Helmholtz energy of the solution with N solvent molecules and one solute molecule, and $F(N)$ is the Helmholtz energy of N solvent molecules at the same temperature and volume. The partition function Q_0 for the pure solvent can be expressed in a form similar to Eq. (8.96)

$$Q_0 = \frac{1}{N!\Lambda_0^{3N}} \int d\mathbf{R}_0^N e^{-\beta\Phi_0(\mathbf{R}_0^N)} \tag{8.98}$$

Inserting Eqs. (8.96) and (8.98) into Eq. (8.97), after some rearrangement, yields

$$e^{-\beta\mu} = \frac{1}{\Lambda_M^3} \left\langle \int d\mathbf{R} e^{-\beta[\Delta\Phi(\mathbf{R}_0^N, \mathbf{R}) + \Phi(\mathbf{R})]} \right\rangle_0 \tag{8.99}$$

where $\langle \cdots \rangle_0$ denotes the ensemble average over all configurations of the solvent molecules.

In the thermodynamic limit, Eq. (8.99) diverges because the integration with respect to $d\mathbf{R}$ extends over an infinite volume. This divergence can be avoided by subtracting the chemical potential of the pure solute as an ideal gas at system temperature and number density, i.e., the solute chemical potential at the ideal-gas state

$$\mu^{IG} = k_B T \ln\left(\Lambda_M^3 / V_M\right) \tag{8.100}$$

where $V_M \equiv \int d\mathbf{R}e^{-\beta\Phi(\mathbf{R})}$ is related to the system volume.[31] From Eqs. (8.99) and (8.100), we obtain[32]

$$e^{-\beta\mu^{ex}} = \frac{1}{V_M}\left\langle \int d\mathbf{R}e^{-\beta(\Delta\Phi+\Phi)} \right\rangle_0 = \langle e^{-\beta\Delta\Phi} \rangle \tag{8.101}$$

where $\mu^{ex} \equiv \mu - \mu^{IG}$ is the excess chemical potential, and $\langle \cdots \rangle$ stands for the ensemble average with respect to all configurations of solvent and solute molecules. Eq. (8.101) is generically applicable to any molecular systems. At infinite dilution, the divergent part of the chemical potential of the solute in the solution is canceled by that in the ideal-gas state. Therefore, the excess chemical potential remains finite.

The free energy for dissolving a single hard-sphere chain into a solvent of hard-sphere chains is related to the probability of finding a vacant space in the solvent to accommodate the solute molecule. If there were no interaction between the solute and solvent molecules, the probability of a successful insertion would be unity. For this case, $\Delta\Phi = 0$, and Eq. (8.101) predicts $e^{-\beta\mu^{ex}} = 1$ or $\beta\mu^{ex} = 0$. On the other hand, if the space is fully occupied by the solvent molecules, the probability of successfully inserting a solute molecule into the solvent is zero. The function $e^{-\beta\mu^{ex}}$ thus reflects the probability of successful insertion of a solute molecule into a solvent at a constant temperature and volume. Similar ideas were used in the scaled-particle theory discussed in Appendix 7.A.

In addition to solvation free energy, Eq. (8.101) can also be utilized to derive the excess chemical potential of a pure fluid. When the solute molecule is identical to solvent molecules, Eq. (8.100) must be modified because now solvent and solute molecules are indistinguishable

$$\mu^{IG} \equiv k_B T \ln\left(N\Lambda_M^3/V_M\right). \tag{8.102}$$

The ideal-gas chemical potential includes contributions from the intramolecular interactions as well as those from the kinetic energy of individual particles. Similar to that for the solvation of a single solvated molecule, the quantity $p = e^{-\beta\mu^{ex}}$ represents the probability of successfully inserting a molecule into the system at a constant temperature and total volume.

8.4.2.2 Excluded Volume of a Hard-Sphere Chain

Consider now the excess chemical potential of a hard-sphere chain containing m identical, tangentially connected segments. Let $p(\eta, m)$ stand for the probability of successful insertion of the hard-sphere chain into a system of hard-sphere chains identical to the inserted chain. Because the polymer segments interact only through the excluded-volume effects, $p(\eta, m)$ is fully determined by the polymer chain length m and packing fraction of the system, $\eta = \pi m\rho\sigma^3/6$, where ρ stands for the number density of the polymer chains, and σ denotes the hard-sphere diameter.

The inserted chain does not interact with other chains in the system if there is no overlap among the particles; otherwise, the potential energy is infinite due to the hard-sphere interaction. Therefore, $p(\eta, m)$ is related to the probability of creating a free space corresponding to the excluded volume of the inserted chain. As illustrated in Figure 8.14, the GF theory asserts that the probability of creating a free space to accommodate a hard-sphere chain can be approximated by the probability of creating the same space in the corresponding monomeric fluid (viz., monomeric hard spheres) at the same packing fraction.

31 For a monatomic solute, there is no self-energy, i.e., $\Phi(R) = 0$. In that case, V_M is identical to the system volume V and $\mu^{IG} = kT\ln(\Lambda/V)$.

32 For monomeric systems, the equation is also known as the potential distribution theorem or, in the context of molecular simulation, Widom's particle-inserton method for calculating chemical potentials (Section 6.8).

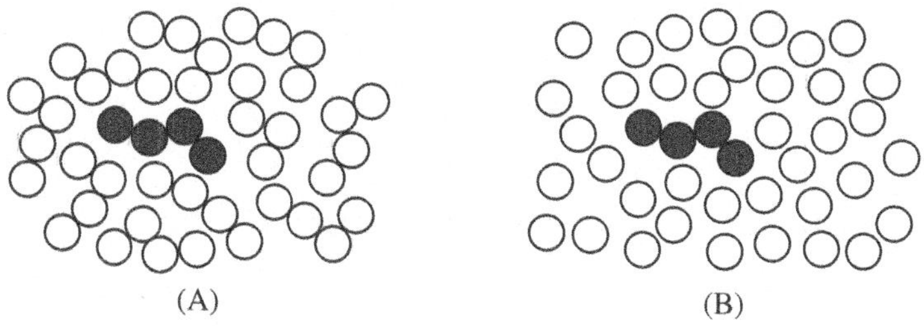

Figure 8.14 Generalized-Flory (GF) theory assumes that the probability of creating a free space to accommodate an m-segment chain (here four segments as highlighted by the filled circles) in a fluid of itself (A) is the same as that of creating the same space in a fluid of hard-sphere monomers (B).

Let $p(\eta, 1)$ be the probability of successful insertion of a hard sphere into a monomeric hard-sphere fluid. The generalized-Flory (GF) theory assumes that the probability of inserting a hard-sphere chain into the same monomeric fluid is exponentially dependent on the excluded volume

$$p(\eta, m) \approx p(\eta, 1)^{v_e(m)/v_e(1)} \tag{8.103}$$

where $v_e(m)$ denotes the excluded volume of the polymer chain with m segments, and $v_e(1)$ is the excluded volume of each monomer. According to Eq. (8.103) and $\beta\mu_{m\text{-}mer}^{ex} = -\ln p(m, \eta)$, the excess chemical potential of the hard-sphere chain is thus given by

$$\mu_{m\text{-}mer}^{ex} = \frac{v_e(m)}{v_e(1)}\mu_1^{ex} \tag{8.104}$$

where μ_1^{ex} represents the excess chemical potential of the monomeric hard-sphere system. Eq. (8.104) represents a key result of the GF theory.

Figure 8.15 shows the excluded volume of a hard-sphere in comparison with those for a hard-sphere dimer, trimer and a chain with m tangentially-connected hard spheres (viz., m-mer). For a hard-sphere of diameter σ, the excluded volume can be easily evaluated, $v_e(1) = 4\pi\sigma^3/3$. For a dimer with two touching hard spheres of diameter σ, the excluded volume is $v_e(2) = 9\pi\sigma^3/4$, and that for a trimer, $v_e(3) \approx 9.82605\sigma^3$. For a hard-sphere chain with m tangentially connected hard spheres of the same diameter σ, the excluded volume can be estimated from simulation[27]

$$v_e(m) = \begin{cases} v_e(3) + (m-3)[v_e(3) - v_e(2)] & 2 \leq m \leq 8 \\ v_e(1)[10.094 + 0.6374(m-15)] & m \geq 9 \end{cases}. \tag{8.105}$$

From the excess chemical potential, we can readily derive the compressibility factor using the standard thermodynamic relation[33]

$$Z_{m-mer} - 1 = \beta\mu_{m-mer}^{ex} - \frac{1}{\eta}\int_0^{\eta} \beta\mu_{m-mer}^{ex} d\eta'. \tag{8.106}$$

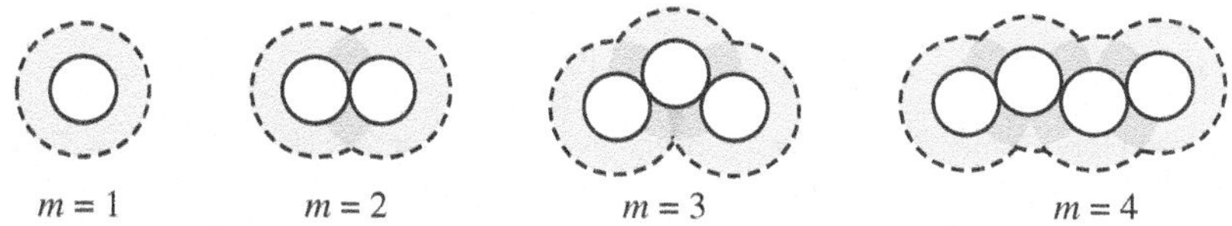

Figure 8.15 Excluded volume is the space that is inaccessible to the centers of other molecules. The excluded volume of a hard-sphere monomer is $v_e(1) = 4\pi\sigma^3/3$, where σ is the hard-sphere diameter; for a hard-sphere dimer, $v_e(2) = 9\pi\sigma^3/4$, and that for a multimer can be calculated from geometric arguments and the statistics of chain conformation.

33 For a one-component system at constant temperature, $\rho d\mu = dP$. Thus, $\rho(\partial\beta\mu/\partial\rho) = (\partial\beta P/\partial\rho)$ or in terms of compressibility factor $Z = \beta P/\rho$ and packing fraction $\eta = \pi\rho\sigma^3/6$, $(\partial\eta Z/\partial\eta) = \eta(\partial\beta\mu/\partial\eta)$. Integration from 0 to η gives $\eta Z = \int_0^{\eta} \eta(\partial\beta\mu/\partial\eta)d\eta = \int_0^{\eta} \eta d\beta\mu = \eta\beta\mu - \int_0^{\eta} \beta\mu d\eta$.

Plugging Eq. (8.104) into (8.106) and applying the same procedure for monomers leads to the equation of state predicted by the GF theory

$$Z^{GF}_{m\text{-}mer} = 1 + \frac{v_e(m)}{v_e(1)}(Z_1 - 1). \tag{8.107}$$

As discussed in Section 7.5, the compressibility factor of monomeric hard spheres can be accurately described by the Carnahan–Starling equation of state[34]

$$Z_1 = \frac{1 + \eta + \eta^2 - \eta^3}{(1 - \eta)^3}. \tag{8.108}$$

The GF theory may be improved by using not only the excess chemical potential of a monomer but, in addition, that of a dimer. Instead of relating the probability of inserting a hard-sphere chain into a fluid of itself to that of inserting a monomer into a monomeric fluid (Eq. (8.103)), we may assume instead that the process of inserting an m-segment hard-sphere chain into a hard-sphere-chain fluid can be achieved in m steps, as illustrated in Figure 8.16.

The probability of creating a free space to accommodate the first segment of the polymer chain is assumed to be $p(\eta, 1)$, that is, the probability of creating an excluded volume of a monomer in a monomeric hard-sphere fluid of the same packing fraction η. The conditional probability[35] of successfully inserting the second segment is approximated by $p(\eta, 2)/p(\eta, 1)$, where $p(\eta, 2)$ is the probability of inserting a dimer. We now assume that the conditional probability of inserting any additional segment $(3, 4, \ldots, m)$ can be approximated by the conditional probability of inserting the second segment, with a correction for the difference in the excluded volume increment

$$[p(\eta, 2)/p(\eta, 1)]^{\frac{v_e(m')-v_e(m'-1)}{v_e(2)-v_e(1)}} \quad \text{for} \quad m' = 3, 4, \ldots m. \tag{8.109}$$

The probability of inserting an m-segment chain in a fluid of itself becomes

$$p(\eta, m) = p(\eta, 1)\left[\frac{p(\eta, 2)}{p(\eta, 1)}\right]^{\frac{v_e(m)-v_e(1)}{v_e(2)-v_e(1)}} = p(\eta, 1)^{1-Y_m} p(\eta, 2)^{Y_m}, \tag{8.110}$$

where $Y_m \equiv [v_e(m) - v_e(1)]/[v_e(2) - v_e(1)]$ can be understood as the number of segments in each polymer chain.[36] From Eqs. (8.109) and (8.110), the excess chemical potential of an m-mer hard-sphere chain is given by

$$\mu^{ex}_{m-mer} = (1 - Y_m)\mu^{ex}_1 + Y_m \mu^{ex}_2 \tag{8.111}$$

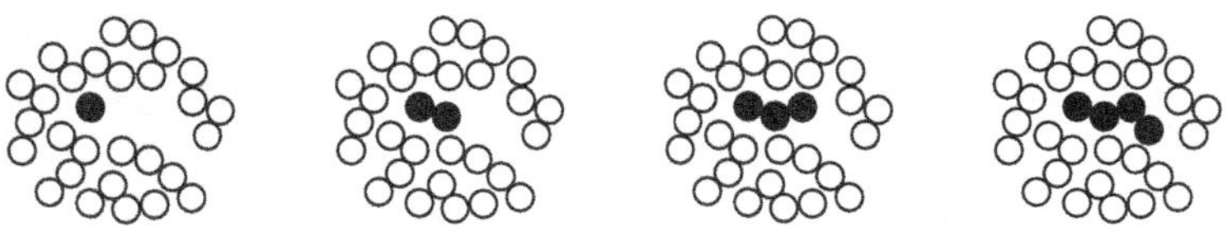

Figure 8.16 In the generalized Flory–Dimer (GF–D) theory, insertion of an m-segment chain (here $m = 4$) into a fluid of itself is achieved in m steps. The probability of successfully inserting the first segment of the chain (I) is $p(\eta, 1)$; the probability of successfully inserting the second segment (II) is $p(\eta, 2)/p(\eta, 1)$; and the probability of successfully inserting an additional segment (III and IV) is given by Eq. (8.109).

34 Carnahan N. F. and Starling K. E., "Equation of state for nonattracting rigid spheres", *J. Chem. Phys.* 51, 635–636 (1969).
35 Conditional probability is the probability of an event, given that another event also occurs simultaneously.
36 $v_e(2) - v_e(1)$ is the volume increment for a dimer, while $v_e(m) - v_e(1)$ is the volume increment for an m-mer. Thus, Y_m represents the effective number of segment increments.

where μ_2^{ex} is the excess chemical potential of a dimer. The latter can be obtained from the equation of state for hard-sphere dimers.

Eq. (8.111) is the main result from the generalized Flory–dimer theory (GF-D),[37] where D stands for dimer. Following Eq. (8.106), the equation of state from GF-D theory is

$$Z_{m\text{-}mer}^{GF\text{-}D} = (1 - Y_m)Z_1 + Y_m Z_2, \tag{8.112}$$

with the compressibility factor of hard-sphere dimers given by

$$Z_2 = \frac{1 + 2.45696\eta + 4.10386\eta^2 - 3.75503\eta^3}{(1-\eta)^3}. \tag{8.113}$$

Eq. (8.113) is obtained by fitting the computer-simulation results for hard-sphere dumbbells.[38]

Following a procedure similar to that for the GF-D theory, we can derive the generalized Flory-i-mer theory (GF-i) using the excess chemical potentials of i-mer and (i-1)-mer

$$\mu_{m\text{-}mer}^{ex} = -Y_{m,i}\mu_{i-1}^{ex} + (Y_{m,i} + 1)\mu_i^{ex}. \tag{8.114}$$

where $Y_{m,i} = [v_e(m) - v_e(i)]/[v_e(i) - v_e(i-1)]$. If the excluded volume of a hard-sphere chain is assumed to be a linear function of the number of segments m, we have $Y_{m,i} = m - i$ for $i > 2$. The excess chemical potential can then be rewritten as

$$\mu_{m-mer}^{ex} = \mu_i^{ex} + (m - i)\left[\mu_i^{ex} - \mu_{i-1}^{ex}\right], \quad i > 2. \tag{8.115}$$

Eq. (8.115) suggests that the excess chemical potential of a m-mer fluid is equal to that of an i-mer, plus an incremental free energy due to the extension of the i-mer to the m-mer. Accordingly, the free energy associated with each added segment is the same as that to extend an $(i-1)$-mer to an i-mer. This additional free energy is called the incremental excess chemical potential. Following a similar argument, Escobedo and de Pablo proposed[39]

$$\mu_{m\text{-}mer}^{ex} = \mu_i^{ex} + \frac{(m-i)}{(j-i)}\left[\mu_j^{ex} - \mu_i^{ex}\right]. \tag{8.116}$$

As shown in Figure 8.17, the incremental excess chemical potential of a hard-sphere chain is relatively insensitive to the chain length when it is longer than a few segments[40]. At a given packing fraction, the incremental excess chemical potential does not change appreciably with chain length for $m > 5$. Similar results were also reported by others.[41]

The GF theories, particularly GF-D, predict compressibility factors for hard-sphere chains in good agreement with computer simulations. For example, Figure 8.18 shows the compressibility factors for various hard-sphere-chain fluids. The GF-D equation of state shows significant improvement over the original GF theory. However, both theories exhibit systematic deviations from the simulation results at low chain densities, especially for long-chain systems. As we discuss later, the discrepancy at low densities is due to the erroneous prediction of the second virial coefficient by the GF theories.

In summary, the GF theories provide significant improvement over the original work of Flory, which was published about 50 years earlier. Contrary to that original work, GF does not use a lattice,

37 Honnell K. G. and Hall C. K., "A new equation of state for athermal chains", *J. Chem. Phys.* 90, 1841–1855 (1989).
38 Tildesley D. J. and Steett W. B., "An equation of state for hard dumbbell fluids", *Mol. Phys.* 41, 85–94 (1980).
39 Escobedo F. A. and Depablo J. J., "Chemical potential and equations of state of hard-core chain molecules", *J. Chem. Phys.* 103, 1946–1956 (1995).
40 Escobedo F. A. and Depablo J. J., "Monte Carlo simulation of the chemical potential of polymers in an expanded ensemble", *J. Chem. Phys.* 103, 2703–2710 (1995).
41 Kumar S. K., et al., "Computer simulation study of the approximations associated with the generalized Flory theories", *J. Chem. Phys.* 104, 9100–9110 (1996).

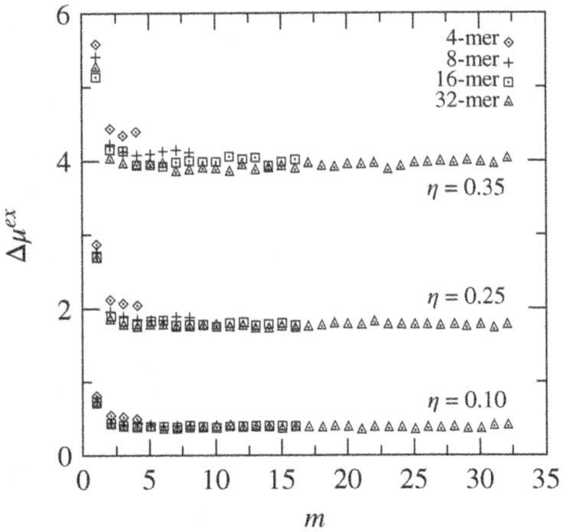

Figure 8.17 Incremental excess chemical potential $\beta\mu^{ex}$ of inserting segments of a hard-sphere chain consecutively into a fluid of m-segment hard-sphere chains with $m = 4, 8, 16$, and 32 at three packing fractions ($\eta = 0.1, 0.25$, and 0.35). At constant packing fraction, after the first few segments, the incremental excess chemical potential remains nearly constant; it does not depend on chain length. Results shown here were obtained from molecular simulations. Source: Escobedo and Depablo.[40]

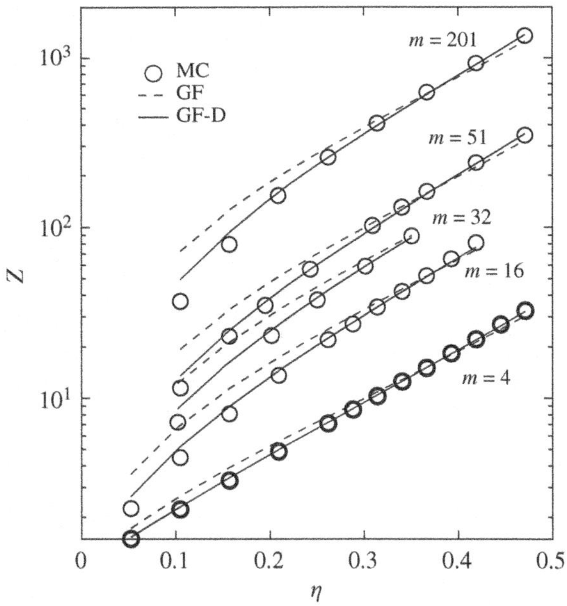

Figure 8.18 Compressibility factors Z for systems of hard-sphere chains (4-mer, 16-mer, 32-mer, 51-mer, 201-mer) versus packing fraction η. The solid lines and the dotted lines are, respectively, predictions of the generalized Flory–dimer (GF–D) theory and the generalized Flory (GF) theory, and the symbols are simulation results.

and therefore, unlike the original Flory theory, GF (and GF-D) can be used to derive an equation of state for pure polymeric fluids.

8.4.2.3 Free Energy of Chain Formation

An alternate method for obtaining thermodynamic properties of hard-sphere chains, different from the GF framework, is provided by the thermodynamic perturbation theory (TPT) originally proposed by Wertheim.[42] Whereas the word "perturbation" usually refers to adding corrections due to attractive forces, here it refers to chain formation from hard-sphere monomers, i.e., a chain of hard spheres can be *athermally* formed from monomeric hard spheres with one binding site for each terminal segment and two binding sites for the middle segments. By *athermal* we mean that no heat is released or absorbed when hard-sphere monomers are bonded together to form a hard-sphere chain. Figure 8.19 illustrates a hypothetical path of chain formation from associating hard spheres.

As discussed in Section 8.3, Wertheim's theory involves an elaborate cluster-expansion method. To elucidate the physical significance, here we present an alternative derivation. We consider the difference in excess Helmholtz energy between a fluid of N hard-sphere chains and that of the corresponding monomers at the same temperature and packing fraction. Here, the excess Helmholtz energy is defined relative to an ideal gas at the system temperature and number density *but without non-bonded* interactions. For a system of hard-sphere monomers, the non-ideality arises only from hard-sphere repulsion. Non-ideality for a system of hard-sphere chains comes not only from hard-sphere repulsion, but in addition, there is the constraint of chain connectivity. For a hard-sphere-chain system, the excess Helmholtz energy is that of the corresponding non-bonded hard spheres plus the additional Helmholtz energy due to the constraint of chain connectivity.

As discussed in Section 7.2, the potential of mean force, defined as $w(r) = -k_B T \ln g(r)$, represents the reversible work to bring two particles from infinite separation to center-to-center distance r in the presence of other monomeric particles. Here, $g(r)$ represents the radial distribution function in a monomeric system. For example, when two hard spheres of the same diameter σ are brought from infinite separation ($r = \infty$) to contact ($r = \sigma$), the change in the Helmholtz energy is

$$F_{bond} = -k_B T \ln g(\sigma) \tag{8.117}$$

where $g(\sigma)$ corresponds to the contact value of the radial distribution function. Assuming superadditivity, i.e., the reversible work for the formation of a hard-sphere chain with m segments is equal to the summation of those for the formation of hard-sphere dimers, we may approximate the

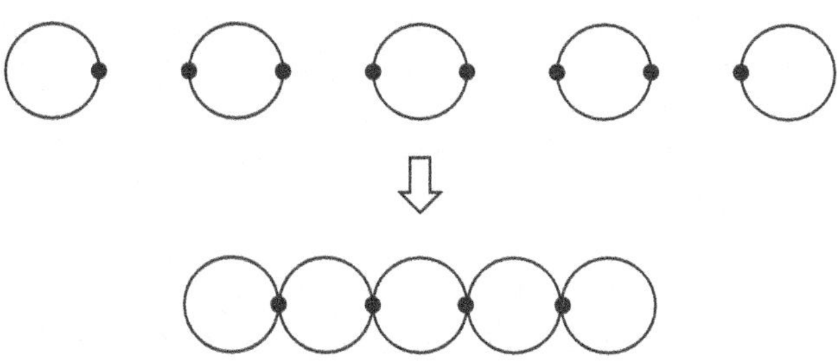

Figure 8.19 A hypothetical path of chain formation in the thermodynamic perturbation theory (TPT). The open circles represent hard spheres, and the dots represent binding sites leading to chain formation. In the limit of complete binding, an m-segment chain (here $m = 5$) is formed from m hard-sphere segments that have one (at end) or two (intermediate) binding sites.

42 Wertheim M. S., "Thermodynamic perturbation theory of polymerization", *J. Chem. Phys.* 87, 7323–31 (1987).

change in the Helmholtz energy[43]

$$F_1^{chain} = -(m-1)k_BT\ln g(\sigma) \tag{8.118}$$

where $(m-1)$ is the number of contacts in a linear hard-sphere chain with m spherical segments. Assuming that chain formations are independent of each other, we can approximate the Helmholtz energy due to the formations of N hard-sphere chains by multiplying Eq. (8.118) by a factor of N

$$F_N^{chain} = -N(m-1)k_BT\ln g(\sigma) \tag{8.119}$$

where the quantity $N(m-1)$ accounts for the total number of bonds.

Based on the above argument, we conclude that for a hard-sphere-chain fluid containing N chains of length m, the excess Helmholtz energy can be divided into a contribution due to hard-sphere repulsion among non-bonded segments and that due to chain formation

$$F_{m\text{-}mer}^{ex} = F_0^{ex} - N(m-1)k_BT\ln[g(\sigma)] \tag{8.120}$$

where subscript 0 represents the monomeric system, F_0^{ex} stands for the excess Helmholtz energy of mN monomers at system temperature and volume. According to the Carnahan–Starling equation of state,[34] the excess Helmholtz energy of the monomeric hard-sphere system is given by

$$\frac{F_0^{ex}}{mNk_BT} = \frac{4\eta - 3\eta^2}{(1-\eta)^2}, \tag{8.121}$$

and the contact value of the radial distribution function is

$$g(\sigma) = \frac{(1-\eta/2)}{(1-\eta)^3}. \tag{8.122}$$

Eqs. (8.120)–(8.122) provide the central result of TPT1, i.e., the first-order thermodynamic perturbation theory. Using the thermodynamic relation, $-(\partial F/\partial V)_T = P$, we obtain an expression for the compressibility factor of hard-sphere chains similar to that given by the GF-D theory (Eq. 8.112):

$$\begin{aligned} Z_{m\text{-}mer} &= 1 + \eta\left(\frac{\partial F_{m\text{-}mer}^{ex}/(Nk_BT)}{\partial\eta}\right) \\ &= 1 + m\frac{4\eta - 2\eta^2}{(1-\eta)^3} - (m-1)\frac{5\eta - 2\eta^2}{(2-\eta)(1-\eta)} \end{aligned} \tag{8.123}$$

In comparison with results from molecular simulation, TPT1 provides a good representation of the compressibility factor of short-hard-sphere-chain fluids. However, as in the GF theories, the discrepancy becomes more appreciable for longer chains, particularly at low packing fractions. The inaccuracy follows from the assumption of independent bond formation, which is not justified for long chains (Eq. (8.119)). In the spirit of the GF-D theory, TPT1 can be improved by incorporating the structural information of a dimer fluid (D) as an intermediate reference system, as discussed by Ghonasgi and Chapman[44] and independently by Chang and Sandler.[45] The improved versions of TPT1 are called TPT-dimer or TPT-D.

In TPT-D, the hypothetical chain-formation process is achieved in two steps, as illustrated in Figure 8.20. In the first step, hard-sphere dimers are assembled from hard-sphere monomers; the

43 More precisely, $g(\sigma)$ should be replaced by $y(\sigma)$, the contact value of the cavity correlation function becuase the direct interaction is already accounted for in the monomer system. For a hard-sphere system, $g(\sigma) = y(\sigma)$.
44 Ghonasgi D. and Chapman W. G., "A new equation of state of hard chain molecules", *J. Chem. Phys.* 100, 6633–6639 (1994).
45 Chang J. and Sandler S. I., "An equation of state for the hard-sphere chain fluid – theory and Monte Carlo simulation", *Chem. Eng. Sci.* 49, 2777–2791 (1994).

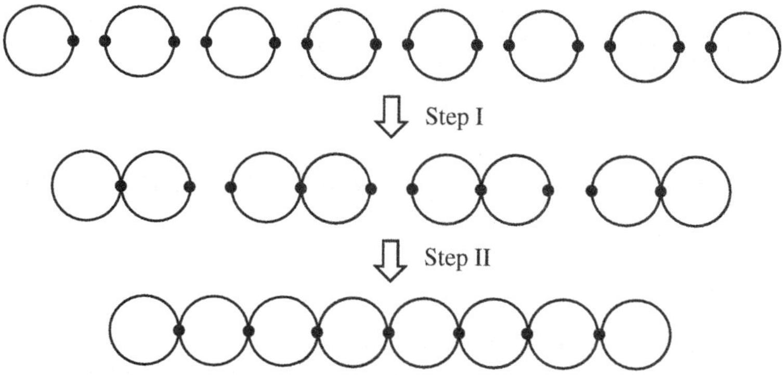

Figure 8.20 In TPT-D, the hypothetical path to form an *m*-segment chain is achieved in two steps. In the first step, monomers are converted into dimers where the change in excess Helmholtz energy is approximated by TPT1. In the second step, further polymerization converts the dimers into the polymer chain. The contribution of dimer polymerization to the excess Helmholtz energy can be estimated from the dimer site–site correlation function. For this model, *m* must be an even number.

contribution of this bonding to the residual Helmholtz energy is calculated from TPT1 (Eq. (8.119)). In the second step, the dimers are further polymerized leading to the formation of *m*-mers (where *m* is an even number). The residual Helmholtz energy due to the polymerization of dimers is estimated by following an argument similar to that used in TPT1: the Helmholtz energy required to bond two dimers is related to the dimer correlation function at particle–particle contact, $g_D(\sigma)$,

$$F^D_{bond} = -k_B T \ln g_D(\sigma). \tag{8.124}$$

Eq. (8.124) accounts for the change in the Helmholtz energy due to the formation of a 4-segment chain from two dimers. As schematically shown in Figure 8.21, the dimer correlation function refers to the pair correlation function between two segments of different dimers at contact.

For $g_D(\sigma)$, Chang and Sandler adopted the molecular-simulation results of Yethiraj and Hall[46]

$$g_D^{YH}(\sigma) = \frac{(2-\eta)(0.534+0.414\eta)}{2(1-\eta)^3}. \tag{8.125}$$

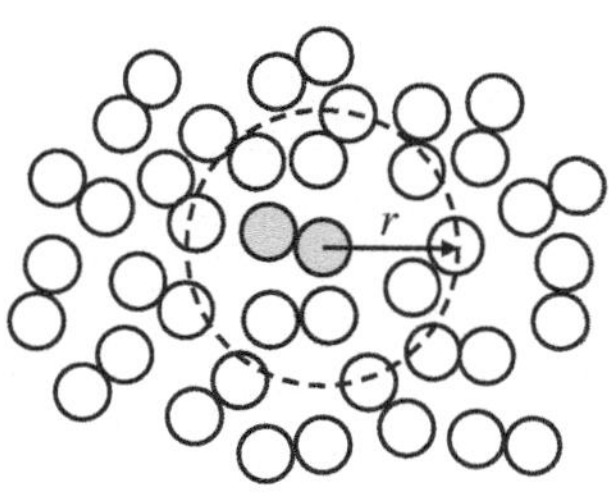

Figure 8.21 Dimer correlation function $g_D(r)$ describes the number density of segments (not including the one bonded to the segment at the origin) separated by *r* from the segment at the origin, normalized by the angle-averaged segment number density.

46 Yethiraj A. and Hall C. K., "Monte Carlo simulations and integral equation theory for microscopic correlations in polymeric fluids", *J. Chem. Phys.* 96 (1), 797–807 (1992).

By contrast, Ghonasgi and Chapman used a modified form of an equation derived from the Percus-Yevick integral-equation theory[47]

$$g_D^{GC}(\sigma) = \frac{1 + 2\eta + 26.45031\,\eta^{6.17}}{2(1-\eta)^2}. \tag{8.126}$$

With the assumption of independent bond formation for N hard-sphere chains of length m, the excess Helmholtz energy is

$$F_{m\text{-}mer}^{ex} = F_1^{ex} - \frac{Nm}{2}k_BT\ln g(\sigma) - \frac{N(m-2)}{2}k_BT\ln g_D(\sigma) \tag{8.127}$$

where $m/2$ and $(m-2)/2$ correspond, respectively, to the number of dimers and the number of bonds required to form one m-segment chain from the dimers. The first term on the right-hand side of Eq. (8.127) gives the contribution to the excess Helmholtz energy from the monomeric segments; the second and third terms, respectively, give contributions from dimerization and from further polymerization to form chains of length m.

Figure 8.22 compares the compressibility factors predicted by TPT1 and TPT-D for hard-sphere chains at conditions identical to those given in Figure 8.18. We see that TPT-D is more accurate than TPT1 for hard-sphere chains with up to 200 segments. The equation of state proposed by Ghonasgi and Chapman (TPT-D2) is slightly more accurate than that by Chang and Sandler (TPT-D1), especially at low packing densities. Similar to the GF theories, agreement between results from TPT and those from simulation deteriorates as the chain length increases.

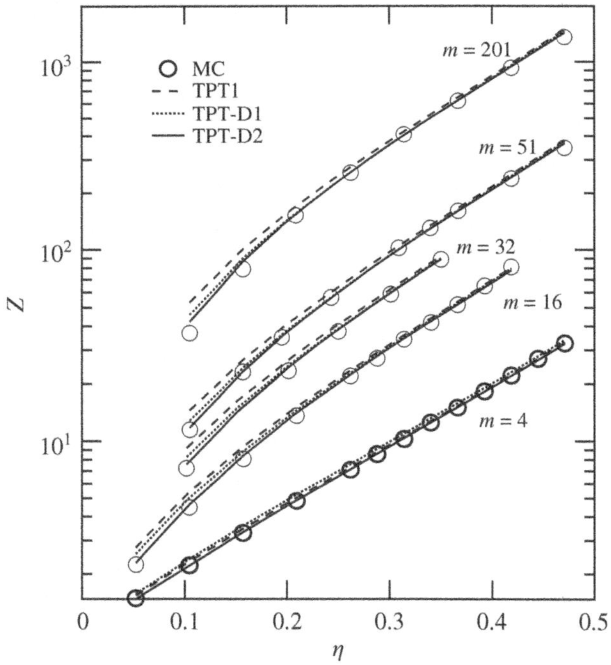

Figure 8.22 Compressibility factors of hard-sphere-chain fluids predicted by three versions of the thermodynamic perturbation theory (TPT1, TPT-D1, and TPT-D2) in comparison with MC results (symbols).

47 Chiew Y. C., "Percus-Yevick integral equation theory for athermal hard-sphere chains", *Mol. Phys.* 73, 359–373 (1991).

8.4.3 Equations for Hard-Sphere-Chain Fluids

Other equations of state for hard-sphere chains have also been proposed following ideas similar to the GF and TPT theories.[48] In most cases, the numerical performances of these equations are similar when they are compared with simulation results. As discussed in the next subsection, all these equations are subject to the same limitation in treating the long-range intra-chain correlation.

While TPT-D is strictly applicable only to hard-sphere chains with an even number of segments, Hu and coworker (HLP) proposed an alternative by considering the nearest-neighbor and the next-to-nearest-neighbor in the chain formation.[49] Figure 8.23 illustrates schematically the physical arguments. Similar to the TPT-D equation (Eq. (8.127)), the excess Helmholtz energy of a hard-sphere-chain fluid is given by

$$F_{m-mer}^{ex} = F_1^{ex} - N(m-1)k_BT\ln g_2(\sigma) - N(m-2)k_BT\ln g_3(\sigma) \tag{8.128}$$

where $g_2(\sigma)$ and $g_3(\sigma)$ are, respectively, the correlation functions for nearest-neighbor and next-to-nearest-neighbor segments at contact. The second and third terms on the right side of Eq. (8.128) account for the reversible work to bring two hard spheres and three hard spheres together, respectively. Correspondingly, $m-1$ and $m-2$ are the number of nearest-neighbors and the number of next-to-nearest-neighbors in a hard-sphere chain with m segments.

In the HLP equation, the correlation functions $g_2(\sigma)$ and $g_3(\sigma)$ are not directly calculated from molecular simulation or integral-equation theories. Instead, $\ln g_2(\sigma)$ and $\ln g_3(\sigma)$ are obtained from the parametric fitting of the simulation results for the compressibility factors of hard-sphere dimers and trimers

$$\ln g_2(\sigma) = \frac{(3+a_{20})\eta - (1+b_{20})}{2(1-\eta)} + \frac{1+b_{20}}{2(1-\eta)^2} - (c_{20}+1)\ln(1-\eta), \tag{8.129}$$

$$\ln g_3(\sigma) = \frac{m-1}{m}\left[\frac{a_{30}\eta - b_{30}}{2(1-\eta)} + \frac{b_{30}}{2(1-\eta)^2} - c_{30}\ln(1-\eta)\right]. \tag{8.130}$$

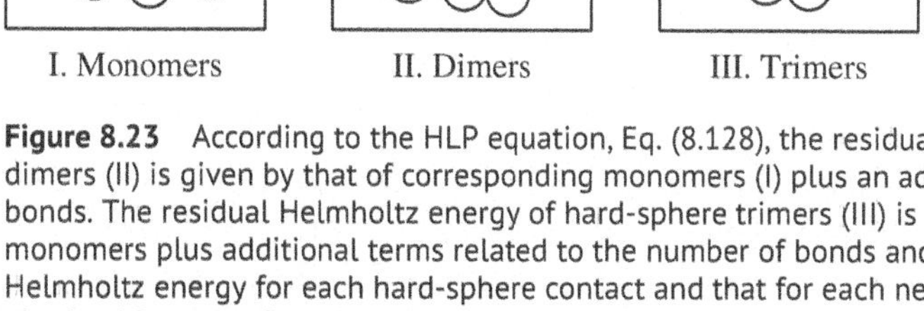

Figure 8.23 According to the HLP equation, Eq. (8.128), the residual Helmholtz energy of hard-sphere dimers (II) is given by that of corresponding monomers (I) plus an additional term related to the number of bonds. The residual Helmholtz energy of hard-sphere trimers (III) is given by that of corresponding monomers plus additional terms related to the number of bonds and the number of trimers. The residual Helmholtz energy for each hard-sphere contact and that for each next adjacent hard-sphere contact are obtained from the Carnahan–Starling equation of state for monomeric hard spheres and molecular-simulation results for hard-sphere dimers and hard-sphere trimers.

48 For further improvement of TPT-D see, e.g., Sadus R. J., "Simple equation of state for hard-sphere chains", *AIChE J.* 45, 2454–2457 (1999), and Gow A. S. and Kelly R. B., "Twenty-one new theoretically based cubic equations of state for athermal hard-sphere chain pure fluids and mixtures", *AIChE J.* 61, 1677–1690 (2015).
49 Liu H. L. and Hu Y., "Equation of state for systems containing chainlike molecules", *Ind. Eng. Chem. Res.* 37, 3058–3066 (1998); Hu Y., Liu H. L. and Prausnitz, J. M., "Equation of state for fluids containing chainlike molecules", *J. Chem. Phys.* 104, 396–404 (1996).

Table 8.1 Parameters in the HLP equation.

$a_2 = 0.45696$	$b_2 = 2.10386$	$c_2 = 1.7503$
$a_3 = -0.74745$	$b_3 = 3.49695$	$c_3 = 4.83207$
$a_{20} = -a_2 + b_2 - 3c_2$	$b_{20} = -a_2 - b_2 + c_2$	$c_{20} = c_2$
$a_{30} = -a_3 + b_3 - 3c_3$	$b_{30} = -a_3 - b_3 + c_3$	$c_{30} = c_3$
$a = m + (m-1)\left(a_2 + \frac{m-2}{m}a_3\right)$	$b = m + (m-1)\left(b_2 + \frac{m-2}{m}b_3\right)$	$c = m + (m-1)\left(c_2 + \frac{m-2}{m}c_3\right)$

Accordingly, the excess Helmholtz energy and the compressibility factor are given by

$$\frac{F^{ex}_{m\text{-}mer}}{mNk_BT} = \frac{(3+a-b+3c)\eta - (1+a+b-c)}{2(1-\eta)} + \frac{1+a+b-c}{2(1-\eta)^2} - (c-1)\ln(1-\eta), \tag{8.131}$$

$$Z_{m\text{-}mer} = \frac{1+a\eta+b\eta^2-c\eta^3}{(1-\eta)^3}, \tag{8.132}$$

where all parameters are given in Table 8.1. While $g_2(\sigma)$ is closely related to the contact value of the pair correlation function of a hard-sphere fluid, $g_3(\sigma)$ reflects an ensemble average of the three-body correlation function over all possible trimer configurations.

As the coefficients in Eqs. (8.129) and (8.130) are directly correlated with simulation data, the HLP equation provides an accurate representation of the thermodynamic properties of hard-sphere dimers and trimers. For longer chains, its accuracy is similar to that of TPT-D theory. Unlike TPT-D theory, the HLP equation is applicable to hard-sphere chains with an arbitrary number of segments.

8.4.4 Second Virial Coefficients of Hard-Sphere-Chain Fluids

To explore the deficiency of various hard-sphere-chain theories, it is helpful to consider the second virial coefficient, which can be derived from the expansion of the compressibility factor

$$Z = 1 + B_2\rho + \cdots \tag{8.133}$$

where ρ is the number density of polymer chains, and B_2 denotes the second virial coefficient. To examine the effect of intra-chain correlations, a quantity of particular interest is the dependence of the second virial coefficient B_2 on chain length m.

Nearly all existing hard-sphere-chain theories predict an inverse relationship between the chain length and the reduced second virial coefficient

$$B_2^* \equiv B_2/(m^2\sigma^3) = k_1 + k_2/m \tag{8.134}$$

where k_1 and k_2 are model-dependent constants. Table 8.2 presents the numerical values of k_1 and k_2 for the equations of state of the hard-sphere chains discussed above. Eq. (8.134) implies that, for long hard-sphere chains, $k_1 \gg k_2/m$ so that the second virial coefficient scales with the square of the number of segments in each chain. Interestingly, the chain-length dependence of the second virial coefficient is identical to that predicted from the Flory–Huggins theory (Section 8.2.2).[50] However, numerical results from experiment and molecular simulation indicate that, for long polymer chains

50 The second virial coefficient given here is slightly different from that discussed for the Flory–Huggins theory because of different concentration units in the virial expansion. Here, the chain concentration is given by chain number density whereas in the Flory–Huggins theory, chain concentration is given by chain volume fraction.

Table 8.2 Constants in Eq. (8.134) for the second virial coefficient of hard-sphere-chain fluids.

	k_1	k_2
GF	1.3350	1.1163
GF-D	0.7305	1.3963
TPT-1	0.7857	1.3090
TPT-D1	0.5825	1.7150
TPT-D2	0.3927	2.0940
HLP	0.3818	2.2517

in a good solvent, the second virial coefficient scales approximately as[51]

$$B_2/(m^2\sigma^3) \sim m^{-0.236}. \tag{8.135}$$

A comparison of Eqs. (8.134) and (8.135) indicates that the existing hard-sphere-chain theories overpredict the second virial coefficient at long chain lengths. These theories are thus inaccurate at low polymer concentrations, where the second virial coefficient plays a key role in determining the compressibility factor.

Figure 8.24 shows the second virial coefficients of hard-sphere chain fluids predicted by from various analytical theories in comparison with those from Monte Carlo simulation.[52] As the chain length increases, the deviation between theory and simulation becomes more prominent because of the wrong scaling relation. Apparently, all existing hard-sphere-chain theories do not provide a reliable second virial coefficient for long chains. While the deviations for large m appear to be modest, the error is significant because the coordinate in Figure 8.24 is not B_2 but $B_2/m^2\sigma^3$. The poor results for large m explain at least in part why an accurate description of liquid–liquid equilibria in polymer–solvent systems is difficult. In such systems, the concentration of the polymer in the solvent-rich phase is sufficiently small such that the chemical potential of the polymer is mainly determined by the second virial coefficient.

8.4.5 Summary

Similar to the hard-sphere model for simple fluids, the hard-sphere-chain model serves as a valuable reference for describing the effects of molecularly excluded volume in polymeric systems. Numerous equations of state have been proposed to describe hard-sphere-chain fluids, mostly based on the GF theory and TPT. While these equations of state can reasonably reproduce simulation data for the compressibility factor of short hard-sphere chains, they exhibit incorrect scaling behavior for the second virial coefficient due to the assumption of a linear dependence of the excess free energy on the polymer chain length. As explained in Section 8.5, the properties of a real polymeric fluid can be derived by introducing perturbation terms to the hard-sphere-chain model.

51 Rubinstein M. and Colby R. H., *Polymer physics*. Oxford University Press, 2003.

52 Yethiraj A. and Hall C. K., "Monte Carlo simulations and integral equation theory for microscopic correlations in polymeric fluids", *J. Chem. Phys.* 96 (1), 797–807 (1992).

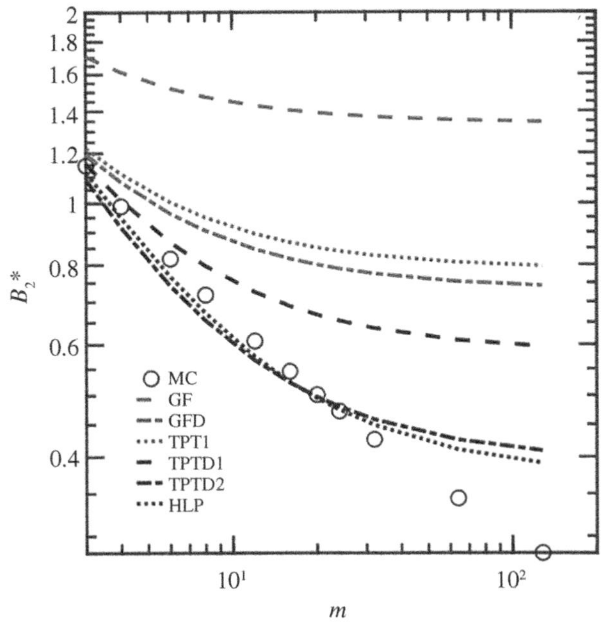

Figure 8.24 Reduced second virial coefficients of hard-sphere chains from various analytical theories. Here, m is the number of segments per chain, and σ is the segment diameter. For large m, all hard-sphere-chain theories give incorrect second virial coefficients in comparison to Monte Carlo (MC) simulation results.

8.5 Statistical Associating Fluid Theory (SAFT)

The statistical-associating-fluid theory (SAFT) was originally developed by Keith E. Gubbins and coworkers in the late 1980s.[53] Over the years, it has involved into one of the most successful equations of state to describe the thermodynamic properties of both simple and polymeric fluids. SAFT is useful in particular for predicting gas solubility and the vapor–liquid phase diagrams of polymeric systems.

SAFT (and its many variations) can be derived by expressing the excess Helmholtz energy of a polymeric system as a sum of various terms related to chain connectivity and different contributions to the intermolecular potential. Specifically, it assumes that the excess Helmholtz energy due to non-bonded interactions is identical to that of a monatomic system,[54] and that chain connectivity introduces an additional contribution as described by the thermodynamic perturbation theory (TPT)

$$F^{ex} \equiv F - F^{IG} = F_m^{ex} + F_{chain} \tag{8.136}$$

53 Jackson G.; Chapman W. G. and Gubbins K. E., "Phase-equilibria of associating fluids – spherical molecules with multiple bonding sites", *Mol. Phys.* 65, 1–31 (1988). Chapman W. G., Jackson G. and Gubbins, K. E., "Phase equilibria of associating fluids: chain molecules with multiple bonding sites", *Mol. Phys.* 65, 1057–1079 (1988); Jackson G.; Chapman W. G. and Gubbins, K. E., "Phase-equilibria of associating fluids of spherical and chain molecules", *Int. J. Thermophys.* 9, 769–780 (1988); Chapman W. G., Gubbins, K. E., Jackson, G. and Radosz, M., "SAFT – equation-of-state solution model for associating fluids", *Fluid Phase Equilib.* 52, 31–38 (1989).

54 As discussed later, PC-SAFT is an exception because it accounts for the effect of chain connectivity on van der Waals attraction.

where F^{IG} stands for the Helmholtz energy of an ideal-gas system of the same polymeric species but without non-bonded interactions, F_m^{ex} represents the excess Helmholtz energy of the corresponding monomeric fluid, and F_{chain} accounts for intra-chain correlations.

As discussed in Section 8.3 (and Appendix 8.D for inhomogeneous systems), the Helmholtz energy for an ideal-gas system of polymeric molecules is exactly known

$$\beta F^{IG}/V = \sum_I \rho_I \left[\ln\left(\rho_I \Lambda_I^3/q_I\right) - 1\right] \tag{8.137}$$

where ρ_I represents the number density of polymeric species I, Λ_I is associated with the translational motions of individual segments analogous to the de Broglie thermal wave length, and q_I is related to the single-molecule partition function but without non-bonded intra-chain interactions.

The thermodynamic non-ideality of a polymeric system arises from various forms of inter- and intra-molecular interactions. As described in a typical molecular force field (see Supplementary Materials IV), the non-bonded interactions are often expressed in terms of short-range repulsion, van der Waals attraction, chemical association, and various forms of electrostatic interactions. According to the perturbation theories of monomeric fluids discussed in Section 7.11, the excess Helmholtz energy can thus be divided into contributions corresponding to different components of the intermolecular forces

$$F_m^{ex} = F_{HS}^{ex} + F_{vdw}^{ex} + F_{assoc}^{ex} + F_{pol}^{ex} + \cdots \tag{8.138}$$

where the subscripts denote hard-sphere (HS) repulsion, van der Waals attraction (vdw), association (assoc), interactions due to molecular polarity (pol), and etc.

Although Eq. (8.138) is written in a linear form, semi-empirical corrections can be introduced to account for coupling between different contributions of intermolecular forces to the excess Helmholtz energy. We have discussed in Chapter 7 analytical expressions of the excess Helmholtz energies due to short-range repulsion and van der Waals interactions in monomeric systems. In this section, we consider mainly the contributions due to association and chain connectivity. The thermodynamic non-ideality due to electrostatic interactions will be discussed in Chapter 9.

Many variations of SAFT have been proposed since its original publication in the late 1980s. Table 8.3 lists some of these variations. Later developments mostly focus on improved representations of the excess Helmholtz energy due to the long-range interactions (e.g., van der Waals attraction and electrostatic effects) and on the group-contribution methods for estimating molecular parameters.

8.5.1 Free Energy due to van der Waals Attraction

To account for the van der Waals attraction in polymeric fluids, many SAFT-like equations use the Barker–Henderson (BH) perturbation theory. In Chapter 7, we have discussed some of their applications to various monomeric systems. The combination of the BH theory for monomeric interactions and TPT for chain connectivity is often referred to as SAFT-VR, which is applicable to several attractive potentials with the variable range (VR) of segment–segment interactions.[55]

In addition to integral-equation and perturbation theories, a number of semi-empirical correlations can be incorporated to account for the excess Helmholtz energy due to van der Waals attraction. By empirical fitting of the simulation data for one-component square-well fluids, Chen

55 Gil-Villegas A., et al., "Statistical associating fluid theory for chain molecules with attractive potentials of variable range", *J. Chem. Phys.* 106, 4168 (1997).

Table 8.3 SAFT and its many variations.

	Reference fluid	F_{HS}	F_{vdw}	F_{assoc}	F_{pol}	F_{elec}	F_{chain}	Ref.
SAFT-0	Simple fluids	BMCSL	PCS	TPT1	—	—	TPT1	[1]
SAFT-HR	square-well	BMCSL	MD	TPT1	—	—	TPT1	[2, 3]
SAFT-VR	swSY/Mie	BMCSL	BH	TPT1	—	—	TPT1	[4]
SAFT-VRE	square-well/Mie	BMCSL	BH	TPT1	—	MSAb/DHb	TPT1	[5, 6]
eSAFT-VR-Mie	Mie	BMCSL	BH3	TPT1	—	DHb	TPT1	[7]
SAFT-γ	square-well	BMCSL	BH	TPT1	—	—	TPT1f	[8]
SAFT-γ-Mie	Mie	BMCSL	BH3	TPT1	—	MSAb	TPT1a	[9, 10]
PC-SAFT	HSC	BSMCL	BH_C	TPT1	—	—	TPT1	[11]
PPC-SAFT	HSC	BSMCL	BH_C	TPT1	ddP	—	TPT1	[12]
ePC-SAFT	HSC	BSMCL	BH_C	TPT1	—	DHb	TPT1	[13]
tPC-PSAFT	HSC	BSMCL	BH_C	TPT1	dQP	—	TPT1	[14]
soft-SAFT	LJ	—	mBWR	TPT1	ddP	—	TPT1	[15, 16]
SAFT-FMSA	LJ	BMSCL	FMSA	TPT1	—	—	TPT1	[17, 18]
SAFT-BACK	HCBF	B-	-ACK	TPT1	—	—	TPT1	[19]

BACK, Boublik-Alder-Chen-Kreglewski equation; BH, The Barker and Henderson perturbation theory; BH3, The Barker-Henderson theory with third-order correction; BHc, BH fitted to the experimental data for *n*-alkanes; BMCSL, Boublík-Mansoori-Carnahan-Starling-Leland equation of state; ddP, dipole-dipole interaction with the Padé approximation; DHb, the (modified/augmented) Debye-Hückel theory plus the Born solvation energy; dQP, dipole-quadrupole interactions with the Padé approximation; FMSA, first-order mean-spherical approximation; HCBF, hard convex body fluids; HR, Huang-Radoz (1990); HSC, hard-sphere chains; LJ, Lennard Jones model; mBWR, the modified Benedict Webb Rubin equation of state; MD, molecular dynamics (fitting); MFA, mean-field approximation; Mie, Mie m-n model; MSA, mean-spherical approximation; MSAb, mean-spherical approximation plus the Born solvation energy; PCS, principle of corresponding states; swSY, square-well/Sutherland/Yukawa models; TPT1, First-order thermodynamic perturbation theory; TPT1a, TPT1 for homopolymers with average molecular parameters; TPT1f, TPT1 for used segments of different types.

[1] W. Chapman, K. Gubbins, G. Jackson, and M. Radosz, *Fluid Phase Equilibria* 52, 31 (1989). [2] S. H. Huang and M. Radosz, *Industrial & Engineering Chemistry Research* 29, 2284 (1990). [3] S. H. Huang and M. Radosz, *Industrial & Engineering Chemistry Research* 30, 1994 (1991). [4] T. Lafitte et al., *Journal of Chemical Physics* 139, 154504 (2013). [5] A. Galindo, et al., *Journal of Physical Chemistry* B 103, 10272 (1999). [6] D. K. Eriksen, et al., *Molecular Physics* 114, 2724 (2016). [7] N. Novak, G. M. Kontogeorgis, M. Castier, and I. G. Economou, *Fluid Phase Equilibria* 565, 113618 (2023). [8] A. Lymperiadis, C. S. Adjiman, A. Galindo, and G. Jackson, *Journal of Chemical Physics* 127, 234903 (2007). [9] V. Papaioannou, et al., *Journal of Chemical Physics* 140, 054107 (2014). [10] A. J. Haslam, et al., *Journal of Chemical & Engineering Data* 65, 5862 (2020). [11] J. Gross and G. Sadowski, *Industrial & Engineering Chemistry Research* 40, 1244 (2001). [12] E. Sauer, M. Stavrou, and J. Gross, *Industrial & Engineering Chemistry Research* 53, 14854 (2014). [13] L. F. Cameretti, G. Sadowski, and J. M. Mollerup, *Industrial & Engineering Chemistry Research* 44, 3355 (2005). [14] E. K. Karakatsani et al., *Journal of Physical Chemistry* C 111, 15487 (2007). [15] F. J. Blas and L. F. Vega, *Molecular Physics* 92, 135 (1997). [16] I. I. Alkhatib et al., *Physical Chemistry Chemical Physics* 22, 13171 (2020). [17] Y. Tang and B. C.-Y. Lu, *Fluid Phase Equilibria* 171, 27 (2000). [18] Y. Tang, *Molecular Physics* 100, 1033 (2002). [19] O. Pfohl and G. Brunner, *Industrial & Engineering Chemistry Research* 37, 2966 (1998).

and Kreglewski expressed F^{ex}_{vdw} as a power series of reduced temperature and density with 24 parameters.[56] This equation was adopted by Huang and Radoz in their early application of SAFT to correlate experimental data for the vapor pressures and saturated liquid densities of over 100

56 Chen S. S.; Kreglewski A., "Applications of the augmented van der Waals theory of fluids", *Ber. Bunsen-Ges. Phys. Chem.* 81, 1048 (1977).

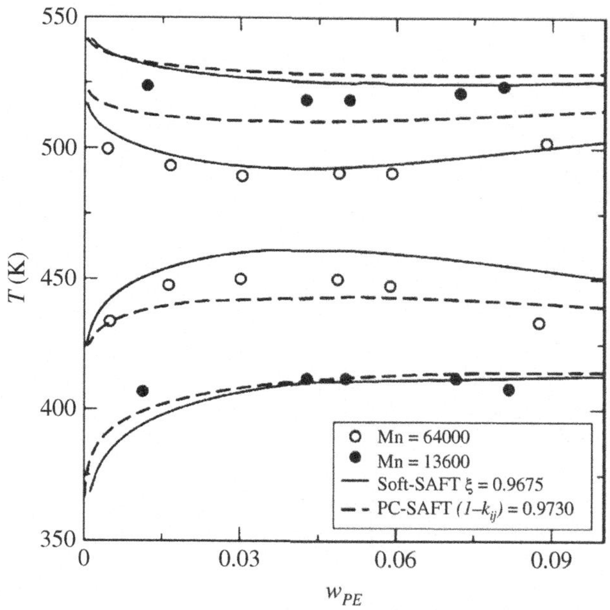

Figure 8.25 Liquid–liquid phase diagrams for binary mixtures of polyethylene (PE) and butyl acetate modeled with soft-SAFT (full lines) and with PC-SAFT (dashed lines). Here the pressure is 0.1 MPa, w_{PE} stands for the weight fraction for PE, M_n is the molecular weight of PE, k_{ij} and ξ are binary parameters in soft-SAFT and PC-SAFT, respectively. Theoretical calculations are able to reproduce both upper and lower critical solution temperatures. Source: Pedrosa et al.[60]

one-component fluids.[57] By using mixing rules and binary parameters, Huang and Radoz also demonstrated that SAFT is able to describe the vapor–liquid and liquid–liquid phase diagrams for a large number of binary and ternary mixtures including polymer solutions. Along similar lines, Blas and Vega[58] adopted an empirical equation of state for one-component Lennard-Jones (LJ) fluids with 33 parameters. The modified equation of state was called soft-SAFT because the LJ fluid instead of hard spheres was used as the reference system. Figure 8.25 shows one representative application of soft-SAFT to polymer systems. With one binary parameter (ξ) independent of temperature and composition, soft-SAFT provides a semi-quantitative representation of both the upper and lower critical solution temperatures of polymer mixtures.

Recognizing that the excess Helmholtz energy due to the van der Waals attraction is also affected by chain connectivity, Gross and Sadoswki modified SAFT-VR by replacing the radial distribution function of hard spheres with that of hard-sphere chains in the BH perturbation theory.[59] Because a hard-chain fluid is used as the reference, the revised equation of state is called perturbed-chain SAFT or PC-SAFT. Further improvement of the coupling effect between chain connectivity and van der Waals interactions was achieved by utilizing experimental results for alkane chains. For

57 Huang S. H. and Radosz M., "Equation of state for small, large, polydisperse, and associating molecules", *Ind. Eng. Chem. Res.* 29, 2284–2294 (1990); *ibid*, 30, 1994–2005 (1991).
58 Blas F. J. and Vega L. F., "Prediction of binary and ternary diagrams using the statistical associating fluid theory (SAFT) equation of state", *Ind. Eng. Chem. Res.* 37 (2), 660–674 (1998).
59 Gross J. and Sadowski G., "Perturbed-chain SAFT: an equation of state based on a perturbation theory for chain molecules", *Ind. Eng. Chem. Res.* 40, 1244–1260 (2001).

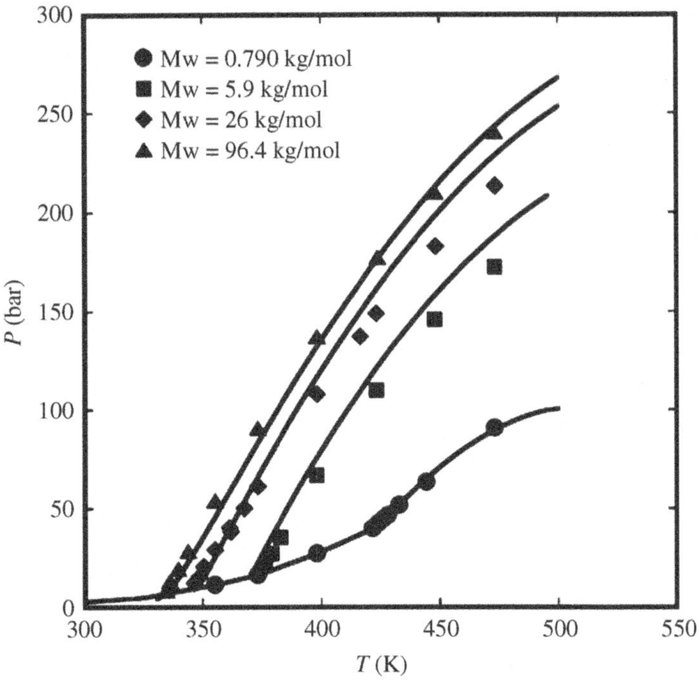

Figure 8.26 Liquid–liquid cloud points for poly(ethylene propylene) (PEP) in 1-butene. Symbols are experimental and lines are calculated from PC-SAFT. The weight fraction of propylene in the copolymer is 0.6, and the weight fraction of PEP in solution is 0.15. Source: Dominik and Chapman.[61]

comparison, Figure 8.25 also shows the theoretical results from PC-SAFT.[60] Although in this particular example, soft-SAFT appears superior to PC-SAFT, the overall performances of these two equations of state are similar. With adjustable parameters, the first-order perturbation theory is often sufficient to account for van der Waals attractions (see Section 7.11). Figure 8.26 presents another example where PC-SAFT is used to predict the cloud-point pressure of poly(ethylene propylene) copolymer in 1-butene.[61] With only monomeric parameters, the theoretical results agree well with the experimental data for polymer systems of different molecular weights.

8.5.2 Free Energy of Association

One unique feature of SAFT-like equations is that they explicitly account for hydrogen-bonding (or other forms of associating interactions). Such interactions are characterized by strong short-range attraction between molecules. Because association introduces strong local inhomogeneity, a conventional van der Waals-like theory or perturbation methods are not suitable to describe the thermodynamic non-ideality. Conversely, "chemical theories" are much more effective to represent the change in thermodynamic properties due to intermolecular association. As mentioned in Section 8.3, Chandler and Pratt proposed a systematic statistical–mechanical method to describe

60 Pedrosa N., et al., "Phase equilibria calculations of polyethylene solutions from SAFT-type equations of state", *Macromolecules* 39 (12), 4240–4246 (2006).

61 Dominik A. and Chapman W. G., "Thermodynamic model for branched polyolefins using the PC-SAFT equation of state", *Macromolecules* 38, 10836–10843 (2005).

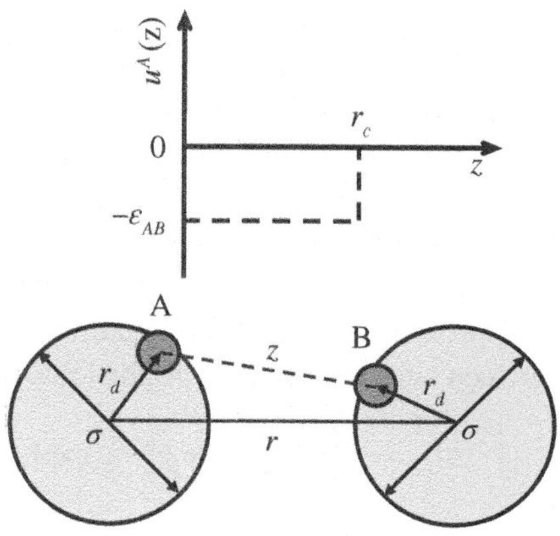

Figure 8.27 In SAFT, the association between two segments is represented by a square-well potential, which depends on the distance between binding sites A and B and association energy $-\varepsilon_{AB}$.

chemical equilibrium in liquid solutions.[62] An alternative approach was developed by Wertheim to account for the binding effects on the excess Helmholtz energy using TPT.[63] Both methods entail lengthy derivations.[64] In this subsection, we offer a heuristic understanding based on the classical thermodynamics of chemical equilibrium.

To understand how chemical binding affects the thermodynamic properties of a molecular system, consider a *monomeric* fluid where the pair potential includes two parts:

$$u_{ij}(r, \varpi_i, \varpi_j) = u_{ij}^0(r) + u_{ij}^A(r, \varpi_i, \varpi_j) \tag{8.139}$$

where r is the center-to-center distance between monomeric species i and j. As shown schematically in Figure 8.27, variables (ϖ_i, ϖ_j) stand for the orientations of the binding sites relative to the positions of the two associating particles. The first term on the right side of Eq. (8.139) is identical to that for a simple fluid; and the second term specifies the binding energy. If the particles contain multiple binding sites, the association energy can be generalized to include contributions from all site–site interactions

$$u_{ij}^A(r, \varpi_1, \varpi_2) = \frac{1}{2}\sum_{\{A_i\}}\sum_{\{B_j\}} u_{A_iB_j}(r_{A_iB_j}) \tag{8.140}$$

where $\{A_i\}$ and $\{B_j\}$ are two sets of binding sites at particles i and j.

At ambient conditions, the association energy is typically much stronger than thermal energy k_BT (e.g., at room temperature, the absolute value of the hydrogen-bond energy is on the order of 10–50 kJ/mol, corresponding to 4–20 k_BT). As a result, the binding between associating particles is essentially in a state of "chemical" equilibrium, i.e., the association leads to an equilibrium distribution of monomers, dimers, trimers, and so on.

To derive the change in free energy due to chemical association, we may divide the total number density of particle i in terms of the densities of monomers (m) and oligomers (o)

$$\rho_i = \rho_i^m + \rho_i^o \tag{8.141}$$

where ρ_i^m stands for the number density of non-associated particles (viz. monomers), and ρ_i^o stands for the number density of particle i that is connected with another particle (viz., particles

62 Chandler D. and L.R. Pratt, "Statistical mechanics of chemical equilibria and intramolecular structures of nonrigid molecules in condensed phases", *J. Chem. Phys.* 65 (8), 2925–40 (1976).

63 Wertheim M. S., "Fluids with highly directional attractive forces. I. Statistical thermodynamics", *J. Stat. Phys.* 35, 19–34 (1984).

64 Zmpitas W. and Gross J., "Detailed pedagogical review and analysis of Wertheim's thermodynamic perturbation theory", *Fluid Phase Equilib.* 428, 121–152 (2016).

in oligomers). For strong association, we expect $\rho_i^m \ll \rho_i$, provided that the total density of particle i is not exceedingly small.

The binding probability between two sites A_i and B_j is determined by the probability of two particles i and j coming together multiplied by the probability that the two sites are in the right orientations. Because the association takes place only at a short distance, the former can be estimated from the radial distribution function of the particles that interact with each other only through reference potential $u_{ij}^0(r)$, and the latter depends on the particle orientations and may be approximated by the Boltzmann distribution. More specifically, the probability of association between sites A_i and B_j is proportional to

$$\Delta_{A_iB_j} = \int d\mathbf{r} d\varpi_i d\varpi_j g_{ij}^0(r)\{\exp[-\beta u_{A_iB_j}(r, \varpi_i, \varpi_j)] - 1\} \tag{8.142}$$

where $g_{ij}^0(r)$ denotes the radial distribution function, and superscript "0" refers to a reference in the absence of association. In Eq. (8.142), the Boltzmann factor is subtracted by 1 such that $\Delta_{A_iB_j} = 0$ if there is no binding between sites A_i and B_j.

By assuming that associations between different binding sites are independent of each other, we may write the mass balance equation for the number density of particle i

$$\rho_i = \rho_i^A + \rho_i^A \sum_j \sum_{\{B_j\}} \rho_j^B \Delta_{A_iB_j} \tag{8.143}$$

where ρ_i^A is the number density of i that is not bonded at site A, and similarly, ρ_j^B is the number density of particle j that is not bonded at site B. Diving both side of Eq. (8.143) by ρ_i gives the fraction of i not bonded at site A

$$X_{A_i}^{-1} = 1 + \sum_j \sum_{B_j} \rho_j X_{B_j} \Delta_{A_iB_j} \tag{8.144}$$

with $X_{A_i} \equiv \rho_i^A/\rho_i$ and $X_{B_j} \equiv \rho_j^B/\rho_j$.

According to the "chemical theory," the thermodynamic non-ideality arises from chemical reactions, i.e., it ignores interactions between monomers, dimers, etc., except for that involved in chemical reactions. In an ideal mixture of monomers and oligomers, the monomer chemical potential is given by (see Section 3.5)

$$\beta\mu_i^m = \ln\left(\rho_i^m \Lambda_i\right) \tag{8.145}$$

where Λ_i is the thermal wavelength of the monomer, and the monomeric number density is determined by particles not bonding at any of the associating sites

$$\rho_i^m = \rho_i \prod_{\{A_i\}} X_{A_i}. \tag{8.146}$$

Because the oligomers are in chemical equilibrium with the monomers, the chemical potential of each particle in any oligomer is the same as that corresponding to the monomers. Accordingly, the change in the Gibbs energy of the system due to association is given by

$$\beta\Delta G/V = \sum_i \rho_i \ln\left(\rho_i^m \Lambda_i\right) - \sum_i \rho_i \ln(\rho_i \Lambda_i) = \sum_i \rho_i \sum_{\{A_i\}} \ln(X_{A_i}). \tag{8.147}$$

In Eq. (8.147), the first summation in the middle corresponds to the Gibbs energy of the associating system at chemical equilibrium, and the second summation is that of the system without association; the second equality follows after plugging in the expression for ρ_i^m given by Eq. (8.146).

As each bond formation eliminates one "ideal-gas" particle in the system, the change in pressure due to the associations can be calculated from the ideal-gas law

$$\beta\Delta P = \sum_i \Delta\rho_i = \frac{1}{2} \sum_i \sum_{\{A_i\}} \rho_i (X_{A_i} - 1) \tag{8.148}$$

where a factor of "1/2" corrects the double counting for the number of bonds. A combination of Eqs. (8.147) and (8.148) gives the change in the Helmholtz energy, i.e., the excess Helmholtz energy due to inter-particle associations

$$f_{assoc}^{ex} = \frac{\beta\Delta G - \Delta\beta PV}{V} = \sum_i \rho_i \sum_{\{A_i\}} [\ln X_{A_i} - X_{A_i}/2 + 1/2]. \tag{8.149}$$

The mole fraction of particle i not bonded at site A_i can be determined from Eq. (8.144). Because a precise form of the bonding potential is often immaterial within the coarse-grain model, $\Delta_{A_iB_j}$ may be calculated without an explicit consideration of the molecular orientations[65]

$$\Delta_{A_iB_j} = g_{ij}^0(\sigma_{ij})[\exp(-\beta\varepsilon_{A_iB_j}) - 1]\kappa_{A_iB_j} \tag{8.150}$$

where $g_{ij}^0(\sigma_{ij})$ represents the contact value of the radial distribution function of particles i and j in the monomeric system, $\varepsilon_{A_iB_j}$ denotes the association energy, and $\kappa_{A_iB_j}$ is a volume parameter related to the steric effects. Eq. (8.150) can be obtained from Eq. (8.142) by assuming that the binding between sites A_i and B_j takes place only when particles i and j are in contact and that the association energy is independent of the molecular orientations. In practical applications, both $\varepsilon_{A_iB_j}$ and $\kappa_{A_iB_j}$ are commonly treated as adjustable parameters, obtained by fitting the SAFT equation to experimental data (e.g., liquid density, vapor–liquid-equilibrium phase diagrams).

8.5.3 Free Energy of Chain Connectivity

The effect of chain connectivity on thermodynamic properties has been discussed in Section 8.4 for hard-sphere chains. Similar ideas are applicable to realistic polymeric systems where the pair interaction between segments includes both repulsive and attractive interactions. According to TPT1, the excess Helmholtz energy density is given by

$$f_{chain}^{ex} = -\sum_I \rho_I \sum_{\{\alpha_I\}} \ln y_{\alpha_I}^0(l_{\alpha_I}) \tag{8.151}$$

where subscript α_I stands for a bond of length l_{α_I} in molecule I, and the summation applies to all covalent bonds. As discussed below, $y_{\alpha_I}^0(r)$ is the cavity correlation function for a pair of segments forming bond α_I in the monomeric reference system. If molecule I is a linear chain with m_I segments, the number of bonds is $m_I - 1$.

The cavity correlation function between two segments connected by a chemical bond of length l_{α_i} can be evaluated from the radial distribution function of the monomeric reference system

$$y_{\alpha_i}^0(l_{\alpha_i}) = g_{\alpha_i}^0(l_{\alpha_i}) \exp\left[\beta u_{\alpha_i}^0(l_{\alpha_i})\right]. \tag{8.152}$$

Frequently, a hard-sphere system is used to evaluate $y_{\alpha_I}^0(r)$ with the assumption that the bond length is the same as the hard-sphere diameter. For one-component LJ fluids, the cavity correlation function has been empirically fitted with MC simulation data[66]

$$y(\sigma) = 1 + \sum_{i,j=1}^{5} a_{ij}(\rho\sigma^3)^i(k_BT/\varepsilon)^{1-j} \tag{8.153}$$

65 A more accurate representation of the bonding parameter was developed by Dufal S., et al., *Mol. Phys.* 113, 948–984 (2015).
66 Johnson J. K., Muller E. A. and Gubbins K. E., "Equation of state for Lennard-Jones chains", *J. Phys. Chem.* 98, 6413–6419 (1994).

where a_{ij} are universal constants. A similar correlation was reported by Chen, Banaszak, and Radosz.[67] Appendix 7.B gives an analytical expression for the radial distribution functions of LJ mixtures near contact ($r \sim \sigma_{ij}$).

8.5.4 Summary

SAFT provides a systematic extension of Wertheim's TPT to associating fluids and polymeric systems. While in the Flory–Huggins theory a polymer molecule is characterized only by its size and energy parameter, SAFT is able to retain many of the microscopic details of a polymeric system, including the diameters of polymer segments, van der Waals attraction, hydrogen bonding, electron-donor-acceptor associations, and electrostatic interactions for charged segments or polar molecules. An accurate representation of thermodynamic properties due to intermolecular binding is a vital ingredient in equations of state for polymeric fluids where hydrogen bonding plays an important role. The versatility of SAFT to account for various ingredients of inter- and intra-molecular interactions empowers its widespread use for a broad variety of thermodynamic systems. With further modifications, SAFT can be applied not only to systems with long-chain hydrocarbons and/or petroleum fluids, but also to various electrolytes and ionic fluids, liquid crystals, and surfactant systems that are commonly encountered in the chemical industry.[68]

8.6 Random-Phase Approximation

Both the Flory–Huggins theory and TPT overlook the long-range structural effects in polymeric systems arising from intra-chain connectivity. These theories consider correlations among polymer segments only at the level of immediate neighbors, making them inadequate for predicting the second virial coefficients of long polymer chains. While the thermodynamic properties of concentrated polymer systems are less affected by long-range correlations, such effects play a significant role in structure formation and the kinetics of phase transitions.

In this section, we discuss the random-phase approximation (RPA) within the framework of the Flory–Huggins theory to predict the structure of polymer melts. Mathematically, RPA was originally introduced to describe the electronic structures of metallic systems.[69] It has been extended to concentrated polymer solutions and polymer melts by Edwards.[70] Recently, RPA has also found applications in simple fluids and ionic systems.

8.6.1 Density–Density Correlation Functions

To elucidate the essential ideas of RPA for polymeric systems, consider a binary mixture of homopolymers A and B. The correlation between polymer segments can be described in terms of the density–density correlation function

$$\chi_{ij}(\mathbf{r}, \mathbf{r}') \equiv \langle [\widetilde{\rho}_i(\mathbf{r}) - \rho_i] \cdot [\widetilde{\rho}_j(\mathbf{r}') - \rho_j] \rangle \tag{8.154}$$

---

67 Chen C. K., Banaszak M. and Radosz, M., "Statistical associating fluid theory equation of state with Lennard-Jones reference applied to pure and binary n-alkane systems", *J. Phys. Chem. B* 102, 2427–2431 (1998).
68 Shaahmadi F., et al., "Group-contribution SAFT equations of state: a review", *Fluid Phase Equilib.* 565, 113674 (2023).
69 Bohm D. and Pines D., "A collective description of electron interactions", *Phys. Rev.* 85, 332 (1952); *ibid*, 92, 609 (1953).
70 Edwards S. F., "The theory of polymer solutions at intermediate concentration", *Proc. Phys. Soc. (London)* 88, 265 (1966).

where subscript i = A or B denotes a polymer segment, $\widetilde{\rho}_i(\mathbf{r})$ represents its instantaneous density, and ρ_i corresponds to the ensemble average of the local density. Apparently, $\chi_{ij}(\mathbf{r}, \mathbf{r}')$ is symmetric in subscripts i and j. In a uniform system, $\rho_i = \langle \widetilde{\rho}_i(\mathbf{r}) \rangle$, and $\chi_{ij}(\mathbf{r}, \mathbf{r}')$ depends only on the distance between positions $\mathbf{r}$ and $\mathbf{r}'$, i.e.,

$$\chi_{ij}(\mathbf{r}, \mathbf{r}') = \chi_{ij}(|\mathbf{r} - \mathbf{r}'|) = \chi_{ji}(|\mathbf{r} - \mathbf{r}'|). \tag{8.155}$$

The density–density correlation function describes how the fluctuations of the local densities of polymer segments at different locations are correlated. Formally, it can be defined as the functional derivative of the one-body density $\rho_i(\mathbf{r})$ with respect to the reduced one-body external potential $V_j^{ext}(\mathbf{r})$ (see Problem 8.23)

$$\chi_{ij}(\mathbf{r}, \mathbf{r}') = -\frac{\delta \rho_i(\mathbf{r})}{\delta \beta V_j^{ext}(\mathbf{r})}. \tag{8.156}$$

Eq. (8.156) indicates that the density–density correlation function $\chi_{ij}(\mathbf{r}, \mathbf{r}')$ corresponds to a response function, i.e., the change in the number density of segment i at position $\mathbf{r}$ in response to the variation of the external energy for segment j at position $\mathbf{r}'$.

Eq. (8.156) predicts that the application of a weak external field to a uniform polymeric system leads to deviation from the mean segment densities

$$\delta \rho_i(\mathbf{r}) = -\beta \sum_{i=A,B} \int d\mathbf{r}' \chi_{ij}(|\mathbf{r} - \mathbf{r}'|) \delta V_j^{ext}(\mathbf{r}') \tag{8.157}$$

where $\delta \rho_i(\mathbf{r}) \equiv \langle \widetilde{\rho}_i(\mathbf{r}) \rangle_{ext} - \rho_i$, and $\langle \cdots \rangle_{ext}$ denotes an ensemble average in the presence of the external field that results in a local potential of $\delta V_A^{ext}(\mathbf{r})$ for each segment of type A and $\delta V_B^{ext}(\mathbf{r})$ for each segment of type B. Eq. (8.157) corresponds to a linear expansion of the density deviation function with respect to the external field, all higher-order terms are negligible if $\delta V_A^{ext}(\mathbf{r})$ and $\delta V_B^{ext}(\mathbf{r})$ are sufficiently weak. In other words, the linear response theory is exact in the limit of an infinitely small external field.

Within the framework of the Flory–Huggins theory, it is assumed that the polymer system is incompressible. Consequently, the total density of polymer segments is everywhere uniform

$$\widetilde{\rho}_A(\mathbf{r}) + \widetilde{\rho}_B(\mathbf{r}) = \rho_A + \rho_B = \rho_t. \tag{8.158}$$

Eq. (8.158) suggests that for a binary mixture of homopolymers A and B, the different pairs of density–density correlation functions are linearly related to each other

$$\chi_{AA}(r) = \chi_{BB}(r) = -\chi_{AB}(r) = -\chi_{BA}(r). \tag{8.159}$$

For short notation, let $\chi(r) = \chi_{AA}(r)$. Substituting Eq. (8.159) into Eq. (8.157) leads to

$$\delta \rho_A(\mathbf{r}) = -\beta \int d\mathbf{r}' \chi(|\mathbf{r} - \mathbf{r}'|) \left[\delta V_A^{ext}(\mathbf{r}') - \delta V_B^{ext}(\mathbf{r}') \right], \tag{8.160}$$

$$\delta \rho_B(\mathbf{r}) = -\beta \int d\mathbf{r}' \chi(|\mathbf{r} - \mathbf{r}'|) \left[\delta V_B^{ext}(\mathbf{r}') - \delta V_A^{ext}(\mathbf{r}') \right]. \tag{8.161}$$

Eqs. (8.160) and (8.161) allow us to predict the local densities of polymer segments in the presence of a weak external field from the correlation functions of the bulk system.

8.6.2 Response Functions of Non-Interacting Chains

One essential assumption in the application of RPA to concentrated polymer systems is that the correlation functions of polymer chains are similar to those at ideal conditions (viz., ideal chains

[IC]). For a binary mixture of ideal polymer chains, the density–density correlation functions $\chi^0_{AB}(r) = \chi^0_{BA}(r) = 0$ because segments belonging to different chains do not interact with each other. However, $\chi^0_{AA}(r)$ and $\chi^0_{BB}(r)$ are not zero due to the chain connectivity. As discussed below, the correlation length is on the order of the polymer size.

Following Eq. (8.157), for a binary mixture of ideal polymers A and B in a weak external field, we can write the density deviation functions as

$$\delta\rho_A(\mathbf{r}) = -\beta \int d\mathbf{r}'\, \chi^0_{AA}(|\mathbf{r} - \mathbf{r}'|)\delta V^{ext}_A(\mathbf{r}'), \tag{8.162}$$

$$\delta\rho_B(\mathbf{r}) = -\beta \int d\mathbf{r}'\, \chi^0_{BB}(|\mathbf{r} - \mathbf{r}'|)\delta V^{ext}_B(\mathbf{r}'). \tag{8.163}$$

Given a polymer segment A at position $\mathbf{r}$, the average number density of A segments at position $\mathbf{r}'$ includes two contributions: one is due to those from the same polymer chain; the other is due to segments from all other chains. The first contribution is related to the correlated density of an ideal chain $\rho^0_A(|\mathbf{r} - \mathbf{r}'|)$, and the latter is the average density ρ_A. Therefore, for IC the density–density correlation functions are given by

$$\chi^0_{AA}(|\mathbf{r} - \mathbf{r}'|) = \rho_A \rho^0_A(|\mathbf{r} - \mathbf{r}'|), \tag{8.164}$$

$$\chi^0_{BB}(|\mathbf{r} - \mathbf{r}'|) = \rho_B \rho^0_B(|\mathbf{r} - \mathbf{r}'|). \tag{8.165}$$

As discussed in Section 3.9, the correlated density of an ideal chain can be described by the Debye function.

8.6.3 Self-Consistent Potentials

For a real polymer blend where the polymer chains are interacting with each other, we can derive the deviation functions $\delta\rho_A(\mathbf{r})$ and $\delta\rho_B(\mathbf{r})$ in response to a weak external field in analogy to those for a mixture of IC. The difference is that the one-body potentials now include not only the external potentials $\delta V^{ext}_A(\mathbf{r})$ and $\delta V^{ext}_B(\mathbf{r})$, but also mean-field potentials $u^{MF}_A(\mathbf{r})$ and $u^{MF}_B(\mathbf{r})$ due to the changes in the interactions among polymer segments. Following Eqs. (8.162) and (8.163), the responses of the segment densities are given by

$$\delta\rho_A(\mathbf{r}) = -\beta \int d\mathbf{r}'\, \chi^0_{AA}(|\mathbf{r} - \mathbf{r}'|) \left[\delta V^{ext}_A(\mathbf{r}') + u^{MF}_A(\mathbf{r}')\right], \tag{8.166}$$

$$\delta\rho_B(\mathbf{r}) = -\beta \int d\mathbf{r}'\, \chi^0_{BB}(|\mathbf{r} - \mathbf{r}'|) \left[\delta V^{ext}_B(\mathbf{r}') + u^{MF}_B(\mathbf{r}')\right], \tag{8.167}$$

where $\chi^0_{AA}(r)$ and $\chi^0_{BB}(r)$ are the density–density correlation functions of ideal polymer chains.

We can determine the variations in the mean-field potentials, $u^{MF}_A(\mathbf{r})$ and $u^{MF}_B(\mathbf{r})$, from $\delta\rho_A(\mathbf{r})$ and $\delta\rho_B(\mathbf{r})$ self-consistently. Within the framework of the Flory–Huggins theory, these functions can be calculated from the nearest-neighbor energies

$$u^{MF}_A(\mathbf{r}) = -Z[\varepsilon_{AA}\delta\rho_A(\mathbf{r}) + \varepsilon_{AB}\delta\rho_B(\mathbf{r})]/\rho_t, \tag{8.168}$$

$$u^{MF}_B(\mathbf{r}) = -Z[\varepsilon_{BA}\delta\rho_A(\mathbf{r}) + \varepsilon_{BB}\delta\rho_B(\mathbf{r})]/\rho_t, \tag{8.169}$$

where $-\varepsilon_{ij}$, $i,j = A, B$ are energy parameters for different pairs, Z is the number of nearest-neighbors or the coordination number of the lattice, and ρ_t is the total number density of polymer segments.

With the assumption of incompressibility

$$\delta\rho_A(\mathbf{r}) + \delta\rho_B(\mathbf{r}) = 0, \tag{8.170}$$

we can solve for $\delta\rho_A(\mathbf{r})$ and $\delta\rho_B(\mathbf{r})$ together with the changes of mean-field potentials $u_A^{MF}(\mathbf{r})$ and $u_B^{MF}(\mathbf{r})$ from Eqs. (8.166)–(8.169). The analytical solution is most conveniently expressed in the Fourier space, whereby the convolution integrals in Eqs. (8.166) and (8.167) can be replaced by algebraic equations

$$\delta\hat{\rho}_A(\mathbf{k}) = -\beta\hat{\chi}_{AA}^0(k)\left[\delta\hat{V}_A^{ext}(\mathbf{k}) + \hat{u}_A^{MF}(\mathbf{k})\right], \tag{8.171}$$

$$\delta\hat{\rho}_B(\mathbf{k}) = -\beta\hat{\chi}_{BB}^0(k)\left[\delta\hat{V}_B^{ext}(\mathbf{k}) + \hat{u}_B^{MF}(\mathbf{k})\right]. \tag{8.172}$$

Inserting the mean-field potentials into Eqs. (8.171) and (8.172) leads to

$$\delta\hat{\rho}_A(\mathbf{k}) = -\beta\left[\frac{1}{\hat{\chi}_{AA}^0(k)} + \frac{1}{\hat{\chi}_{BB}^0(k)} - \frac{2\chi_F}{\rho_t}\right]^{-1}\left[\delta\hat{V}_A^{ext}(\mathbf{k}) - \delta\hat{V}_B^{ext}(\mathbf{k})\right] \tag{8.173}$$

where χ_F is the Flory parameter. A comparison of Eqs. (8.160) and (8.173) gives the density–density correlation function of the polymer blend

$$\frac{1}{\hat{\chi}(k)} = \frac{1}{\hat{\chi}_{AA}^0(k)} + \frac{1}{\hat{\chi}_{BB}^0(k)} - \frac{2\chi_F}{\rho_t}. \tag{8.174}$$

Plugging Eqs. (8.164) and (8.165) into Eq. (8.174) yields

$$\frac{1}{\hat{\chi}(k)} = \frac{1}{\rho_A\hat{\rho}_A^0(k)} + \frac{1}{\rho_B\hat{\rho}_B^0(k)} - \frac{2\chi_F}{\rho_t}. \tag{8.175}$$

Eq. (8.175) represents a key result of RPA for polymeric systems. As to be discussed below, the density–density correlation function captures, at least in part, the long-range intra-chain correlations of the polymeric system. Besides, it can be utilized to improve the free-energy derived from the mean-field approximation (see Problem 8.25). For example, the so-called "one-loop" correction is achieved by adding a Gaussian fluctuation to the field-theoretical formulation or by applying the thermodynamic perturbation.[71]

8.6.4 Structure of Polymer Blends

Experimentally, the structure of a polymer solution or a blend can be obtained from small-angle X-ray or neutron-scattering measurements. As discussed in Section 7.2, the intensity of scattering is linearly proportional to the structure factor, which is directly related to the density–density correlation function.

To derive an explicit expression for the structure factor, we define segment fractions $\phi_A = \rho_A/\rho_t$ and $\phi_B = \rho_B/\rho_t$, where $\rho_t = \rho_A + \rho_B$ is the overall (total) segment density. If A and B segments have the same size, ϕ_A and ϕ_B are equivalent to the volume fractions. Following Eq. (8.175), we can write the structure factor for the A and B polymer blend $\hat{S}(k)$ as

$$\hat{S}(k) = \frac{\hat{\chi}(k)}{\rho_t} = \left[\frac{1}{\phi_A\rho_A^0(k)} + \frac{1}{\phi_B\rho_B^0(k)} - 2\chi_F\right]^{-1}. \tag{8.176}$$

Figure 8.28 presents the scattering intensities of a polymer blend of polystyrene (PSD) and poly(vinylmethylether) (PVME) calculated from RPA in comparison with the small-angle

71 Qin J. and Morse D. C., "Renormalized one-loop theory of correlations in polymer blends", *J. Chem. Phys.* 130, 224902 (2009).

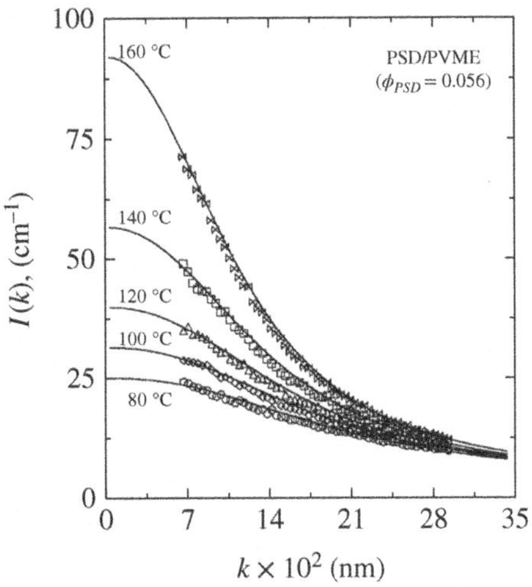

Figure 8.28 Intensity of small-angle neutron scattering for a blend of deuterated polystyrene (PSD) and poly(vinylmethylether) (PVME) at different temperatures. The lines are from RPA. ϕ_{PSD} is the volume fraction of PSD; $k = |\mathbf{k}|$. Source: Adapted from Briber et al.[72]

neutron-scattering data.[72] It shows that, with the Flory parameter treated as an adjustable parameter depending on temperature, RPA is able to describe the long-range structure of the polymer blend.

8.6.5 Polymer Phase Transition

The structure factor provides useful information on phase transitions in polymer blends. As discussed in Section 3.9, in the limit of a small wave vector **k**, the correlated density of an ideal chain can be approximated by

$$\frac{1}{\hat{\rho}^0(k)} = \frac{1}{m} + \frac{k^2 b^2}{18} + \cdots \tag{8.177}$$

where m is the number of segments, and b is the Kuhn length. For $k \ll 1$, we may neglect higher-order terms in Eq. (8.177) beyond k^2. In that case, Eq. (8.176) can be rewritten as

$$\frac{1}{\hat{S}(k)} = \left[\frac{1}{m_A\phi_A} + \frac{1}{m_B\phi_B} - 2\chi_F\right] + \frac{k^2}{18}\left[\frac{b_A^2}{\phi_A} + \frac{b_B^2}{\phi_B}\right] \tag{8.178}$$

where m_i and b_i are, respectively, the degree of polymerization and the Kuhn length of polymer $i = A,B$. From Eq. (8.178), we obtain

$$\hat{S}(k) = \frac{\hat{S}(0)}{1 + k^2L^2} \tag{8.179}$$

where $\hat{S}(0)$ and L are given by, respectively,

$$\hat{S}(0) = \left[\frac{1}{m_A\phi_A} + \frac{1}{m_B\phi_B} - 2\chi_F\right]^{-1}, \tag{8.180}$$

$$L^2 = \frac{\hat{S}(0)}{18}\left[\frac{b_A^2}{\phi_A} + \frac{b_B^2}{\phi_B}\right]. \tag{8.181}$$

Eq. (8.179) is identical to that derived from the OZ equation near the critical point (Section 7.12).

72 Briber R. M., et al., "Small angle neutron scattering from polymer blends in the dilute concentration limit", *J. Chem. Phys.* 101 (3), 2592 (1994).

Similar to that for a simple fluid, the quantity L in Eq. (8.181) characterizes the correlation length of the density fluctuations in the polymer mixture. At the condition

$$\frac{1}{m_A\phi_A} + \frac{1}{m_B\phi_B} - 2\chi_F = 0, \tag{8.182}$$

Eq. (8.179) predicts that the structure factor, thus the correlation length diverges. In this case, the system is thermodynamically unstable and must undergo a phase transition. Because the low wave-vector limit of the structure factor can be easily obtained from small-angle neutron scattering measurements, the scattering data provide useful information on spinodal phase transition of the polymer system.

It is interesting to note that the spinodal line predicted by RPA coincides with that predicted by the Flory–Huggins theory (Section 8.2). While the latter primarily focuses on the thermodynamic properties of bulk polymer systems, RPA is specifically concerned with the polymer structure obtained from the linear response theory. Due to the similarity in underlying assumptions between RPA and the Flory–Huggins theory, scattering data are often utilized as an alternative method to determine the Flory parameter χ_F. The connection between RPA and the Flory–Huggins theory highlights how different theoretical approaches can provide complementary insights and contribute to a more comprehensive understanding of polymer systems.

8.6.6 Summary

In this section, we present the key theoretical findings derived from RPA for concentrated polymer systems. Although our discussion is focused on the structure factors of binary polymer blends, the theoretical approach can be equally applied to other polymer systems, including block-copolymer melts (as highlighted in Problem 8.26). As mentioned earlier, RPA can also be used to account for correlation effects by incorporating a one-loop correction.

It is important to note that RPA employs ideal polymer chains as the reference state. Besides, it accounts for interactions between segments considered through a self-consistent mean-field potential. These assumptions may not be suitable for systems where the correlation between different polymer chains significantly deviates from that observed in ideal-chain systems. Furthermore, RPA is not particularly useful for dilute polymer solutions where intra- and inter-chain correlations are strongly coupled due to intense segment–segment interactions.

8.7 Continuous Gaussian Chains Model and the Polymer Field Theory

In Chapter 3, we have discussed various coarse-grained models for polymer chains and demonstrated the equivalence of different ideal-chain models in the limit of infinite chain length. In this section, we introduce a continuous model that represents each polymer chain as a three-dimensional curve. Unlike lattice or particle-based models, where each polymer segment is discretely labeled with an integer, the continuous model treats the identity of polymer segments as a continuous variable.

Mathematically, the continuous description of polymer chains allows for the utilization of powerful field-theoretical methods originally established in quantum mechanics. By employing the polymer-field theory, it becomes possible to describe the structure and thermodynamic properties of polymeric systems at mesoscopic scales that are not easily accessible through conventional molecular theories and atomistic simulations. This theoretical framework is particularly valuable in predicting order–disorder and order–order transitions observed in concentrated polymer systems, such as block-copolymer melts.[73]

8.7.1 Continuous Gaussian Chains

The Gaussian-chain model provides a useful link between discrete and continuous models of polymer chains. To elucidate, consider a discrete Gaussian chain with the separation between neighboring segments $n+1$ and n satisfying the probability distribution

$$p(\mathbf{r}_n, \mathbf{r}_{n+1}) = \left(\frac{3}{2\pi b^2}\right)^{3/2} \exp\left[-\frac{3(\mathbf{r}_{n+1} - \mathbf{r}_n)^2}{2b^2}\right] \tag{8.183}$$

where b stands for the Kuhn length. The probability density of the polymer configuration, as represented by the positions of all segments $\mathbf{R} \equiv (\mathbf{r}_1, \mathbf{r}_2, \ldots, \mathbf{r}_m)$, is given by

$$p(\mathbf{R}) \sim \exp\left[-\frac{3}{2b^2}\sum_{n=1}^{m-1}(\mathbf{r}_{n+1} - \mathbf{r}_n)^2\right] \tag{8.184}$$

where the proportionality constant can be fixed by the normalization condition.

As discussed in Section 3.9, a Gaussian chain is scale-invariant, i.e., the physical properties are invariant across different length scales. Accordingly, the scaling relation for the end-to-end distance, $R = \sqrt{m}b$, will not change if we use different length scales to group polymer segments as the new units of the Gaussian chain. The scale invariance allows us to extend the discrete description of polymer configuration $\mathbf{R} \equiv (\mathbf{r}_1, \mathbf{r}_2, \ldots, \mathbf{r}_m)$ to a continuous function $\mathbf{R}(s)$ of the continuous variable s as shown schematically in Figure 8.29. When the polymer backbone is represented by a continuous 3-dimensional curve, each bond vector in the discrete model becomes

$$(\mathbf{r}_{n+1} - \mathbf{r}_n) = \int_n^{n+1} \left(\frac{\partial \mathbf{R}}{\partial s}\right) ds \tag{8.185}$$

where $\partial\mathbf{R}/\partial s$ stands for the line derivative or the local bond orientation. Like the bond vectors, the bond orientations are uncorrelated in a Gaussian chain, and thus we have

$$\int_n^{n+1}\int_n^{n+1} \left(\frac{\partial \mathbf{R}}{\partial s}\right)\left(\frac{\partial \mathbf{R}}{\partial s'}\right) ds ds' = \int_n^{n+1}\int_n^{n+1} \left(\frac{\partial \mathbf{R}}{\partial s}\right)^2 \delta(s-s') ds ds' \tag{8.186}$$

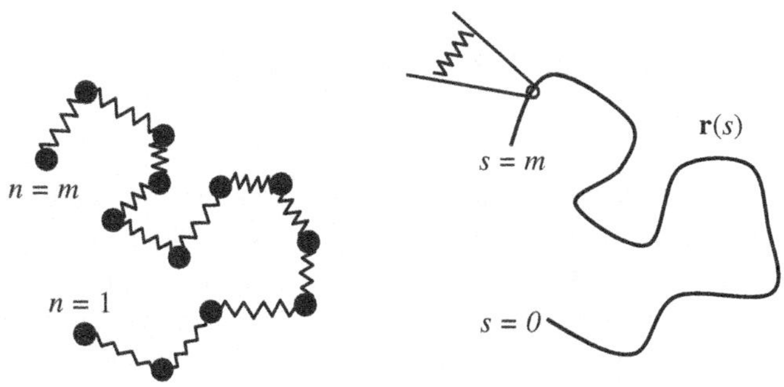

Figure 8.29 Discrete and continuous descriptions of a Gaussian chain.

---

73 For in-depth discussion of the polymer field theory, see Fredrickson G., *The equilibrium theory of inhomogeneous polymers*. Oxford University Press, 2006. Appendices 8.A and 8.B present the essential mathematical backgrounds for better understanding of this section.

where $\delta(s-s')$ denotes the Dirac delta function. Eq. (8.186) is analogous to the correlation between bond vectors in the freely-joined chain (FJC) model discussed in Section 3.8.

A combination of Eqs. (8.185) and (8.186) yields

$$(\mathbf{r}_{n+1}-\mathbf{r}_n)^2 = \int_n^{n+1} \left(\frac{\partial \mathbf{R}}{\partial s}\right)^2 ds. \tag{8.187}$$

Substituting Eq. (8.187) into (8.184) gives the probability density of polymer configuration in the continuous representation of the Gaussian chain

$$p[\mathbf{R}(s)] \sim \exp\left[-\frac{3}{2b^2}\int_0^m ds\left(\frac{\partial \mathbf{R}}{\partial s}\right)^2\right] \tag{8.188}$$

where the proportionality constant can be fixed by the normalization condition. Eq. (8.188) provides a starting point for the formulation of the field-theoretical methods for polymer systems.

8.7.2 The Edwards Hamiltonian for a Single Chain

The Gaussian-chain model accounts for the connectivity of IC on a length scale much larger than that of individual polymer segments. For real polymers, we need to consider non-bonding segment–segment interactions as well.

For an isolated polymer chain, we may define the instantaneous density of polymer segments by taking the continuous limit of the segment index

$$\widetilde{\rho}(\mathbf{r}) = \sum_{n=1}^{m} \delta[\mathbf{r}_i - \mathbf{r}] \to \int_0^m ds\delta[\mathbf{R}(s)-\mathbf{r}]. \tag{8.189}$$

While Eq. (8.189) is applicable only to a single homopolymer chain, a similar expression can be written for many-chain systems including those containing copolymers, i.e., polymer chains with different types of segments.

The probability density of the polymer configurations depends on the bond potential as well as the non-bonding segment–segment interactions

$$p[\mathbf{R}(s)] \sim \exp\left[-\frac{3}{2b^2}\int_0^m ds\left(\frac{\partial \mathbf{R}}{\partial s}\right)^2 - \beta\Phi[\widetilde{\rho}(\mathbf{r})]\right] \tag{8.190}$$

where $\Phi[\widetilde{\rho}(\mathbf{r})]$ represents the non-bonding potential. In comparison of Eq. (8.190) with the probability density of polymer configurations in a canonical ensemble, we may attribute the terms in the square brackets to an instantaneous energy

$$H[\mathbf{R}(s)] = \frac{3k_BT}{2b^2}\int_0^m ds\left(\frac{\partial \mathbf{R}}{\partial s}\right)^2 + \Phi[\widetilde{\rho}(\mathbf{r})]. \tag{8.191}$$

Eq. (8.191) is known as the Edwards Hamiltonian, named after Samuel Edwards, who pioneered the development of polymer field theories.

With the polymer configuration depicted as a three-dimensional curve $\mathbf{R}(s)$, we may express the single-chain partition function in terms of the configurational integral

$$q \sim \int D\mathbf{R}(s)e^{-\beta H[\mathbf{R}(s)]} \tag{8.192}$$

where $D\mathbf{R}(s)$ represents a differential element in the configurational space, and the proportionality constant has been omitted for simplicity (which can be fixed by the normalization of $p[\mathbf{R}(s)]$). Because $\mathbf{R}(s)$ corresponds to a path connecting the initial and end segments of a polymer chain, the sum of all possible configuration (paths) is called the path integral.

8.7.3 Field-Theory Partition Function

The Edwards Hamiltonian can be readily extended to thermodynamic systems with many polymer chains. To formulate the canonical partition function, consider a one-component system consisting of N copolymer chains with an arbitrary segment composition. At any moment, the polymer configuration is specified by $\{\mathbf{R}_I(s)\}$ where subscript $I = 1, 2, \ldots N$ denotes different polymer chains. For each type of polymer segments, the instantaneous number density is determined by

$$\widetilde{\rho}_i(\mathbf{r}) = \sum_{I=1}^{N} \int_0^m ds \delta[\mathbf{r} - \mathbf{R}_I(s)]\gamma_{i,I}(s) \tag{8.193}$$

where $\gamma_{i,I}(s)$ is called the segment occupation function, which specifies the identity of a segment according to its position along the polymer backbone. For example, for a block copolymer with m_A segments of type A and m_B segments of type B, the segment occupation function is given by

$$\gamma_{A,I}(s) = \begin{cases} 1 & s \leq m_A, \\ 0 & s > m_A; \end{cases} \tag{8.194}$$

and

$$\gamma_{B,I}(s) = \begin{cases} 0 & s \leq m_A, \\ 1 & s > m_A. \end{cases} \tag{8.195}$$

Similar to that for a particle-based model, we can express the non-bonding inter- and intra-chain interactions in terms of a pairwise additive potential

$$\Phi = \frac{1}{2}\sum_{i,j} \iint d\mathbf{r}d\mathbf{r}'\widetilde{\rho}_i(\mathbf{r})\widetilde{\rho}_j(\mathbf{r}')u_{ij}(\mathbf{r}, \mathbf{r}') - \sum_i \int d\mathbf{r}\widetilde{\rho}_i(\mathbf{r})u_{ii}(\mathbf{r}, \mathbf{r}) \tag{8.196}$$

where subscripts i and j denote the identities of the polymer segments, and $u_{ij}(\mathbf{r}, \mathbf{r}')$ stands for the non-bonding pair potential between segments i and j. The second term on the right side of Eq. (8.196) is introduced to avoid the self-interaction energy in the double summation.

Similar to that for a single-chain system, the canonical partition function for the many-chain system is given by

$$Q = \frac{C}{N!} \int D\mathbf{R}^N(s) \exp\left[-\frac{3}{2b^2}\sum_{I=1}^{N} \int_0^m ds\left(\frac{\partial \mathbf{R}_I}{\partial s}\right)^2 - \beta\Phi\right] \tag{8.197}$$

where $D\mathbf{R}^N(s)$ stands for the path integral over all possible polymer configurations,[74] C is a renormalization constant to make Q dimensionless, and $N!$ accounts for the indistinguishability of N polymer chains. For simplicity, we assume that the bond connectivity is the same as that of the Gaussian-chain model for a homopolymer, i.e., the "bond" potential for each polymer chain is independent of the segment composition

$$v_B[\mathbf{R}(s)] = \frac{3k_BT}{2b^2} \int_0^m ds\left(\frac{\partial \mathbf{R}}{\partial s}\right)^2 \tag{8.198}$$

where b stands for an "effective" Kuhn length. The Gaussian-chain model can be readily extended to copolymers with a nonuniform distribution of the Kuhn length (see Section 3.10).

---

74 Like a regular integral, path integration is defined as a summation of the integrand with respect to all possible functions.

Eq. (8.197) is not immediately useful because the instantaneous local segment density $\widetilde{\rho}_i(\mathbf{r})$ is a dynamic variable. To integrate out the microscopic degrees of freedom, we introduce the mathematical identity[75]

$$1 = \int D\rho_i(\mathbf{r})\delta[\rho_i(\mathbf{r}) - \widetilde{\rho}_i(\mathbf{r})]. \tag{8.199}$$

In analogy to the conventional Fourier transform, the delta function $\delta[\rho_i(\mathbf{r}) - \widetilde{\rho}_i(\mathbf{r})]$ can be expressed as

$$\delta[\rho_i(\mathbf{r}) - \widetilde{\rho}_i(\mathbf{r})] \sim \int D\varphi_i(\mathbf{r}) \exp\left\{ \int d\mathbf{r}\varphi_i(\mathbf{r})[\rho_i(\mathbf{r}) - \widetilde{\rho}_i(\mathbf{r})] \right\} \tag{8.200}$$

where $\varphi_i(\mathbf{r})$ is an effective one-body field, and the proportionality constant is omitted for simplicity. Using the delta function, we may replace the instantaneous segment density $\widetilde{\rho}_i(\mathbf{r})$ in the potential function Φ with a smooth function $\rho_i(\mathbf{r})$.

Inserting Eqs. (8.199) and (8.200) into (8.197) yields

$$Q \sim \left\{ \prod_i \int D\rho_i(\mathbf{r}) \int D\varphi_i(\mathbf{r}) \right\} \exp\{-\beta F_0\} \tag{8.201}$$

where

$$\beta F_0 \equiv -\ln Q_0 + \beta\Phi[\rho_i(\mathbf{r})] - \sum_i \int d\mathbf{r}\varphi_i(\mathbf{r})\rho_i(\mathbf{r}) \tag{8.202}$$

and Q_0 denotes the partition function for N ideal polymers under an effective potential $\{\varphi_i(\mathbf{r})\}$ for each type of polymer segment i

$$Q_0 = \frac{C}{N!}\left\{ \int D\mathbf{R}(s) \exp\left[-\frac{3}{2b^2}\int_0^m ds\left(\frac{\partial \mathbf{R}}{\partial s}\right)^2 - \sum_i \int_0^m ds\varphi_i[\mathbf{R}(s)]\ \gamma_i(s) \right] \right\}^N. \tag{8.203}$$

Eq. (8.201) provides a starting point for describing the structure and thermodynamic properties of polymeric systems with the field-theoretical methods. In Appendix 8.C, we formulate the grand partition function of classical systems in terms of field variables and discuss the analytical results obtained from the saddle-point approximation and from the random-phase approximation (RPA).

8.7.4 Self-Consistent-Field Theory

The self-consistent-field theory (SCFT) is a popular mean-field method to describe microscopic segregation in block-copolymer melts. Formally, SCFT can be derived from a saddle-point approximation, i.e., the path integrals in Eq. (8.201) are evaluated by taking the maximum of the exponential term among all possible forms of the segment densities and affiliated fields. Specifically, minimizing βF_0 with respect to $\{\rho_i(\mathbf{r})\}$ and $\{\varphi_i(\mathbf{r})\}$ yields[76]

$$\rho_i(\mathbf{r}) = -\frac{\delta \ln Q_0}{\delta\varphi_i(\mathbf{r})}, \tag{8.204}$$

$$\varphi_i(\mathbf{r}) = \frac{\delta\beta\Phi}{\delta\rho_i(\mathbf{r})}. \tag{8.205}$$

---

75 Eqs. (8.199) and (8.200) can be compared with those for the one-dimensional Dirac function $1 = \int dx\delta(x)$ and $\delta(x) = \int dk e^{-ikx}/2\pi$. The mathematical details can be found in Appendix 8.C. In Eq. (8.200), we omit $i = \sqrt{-1}$ in the exponential and replace $i\omega(\mathbf{r})$ discussed in the appendix by $\varphi(\mathbf{r})$.

76 A functional maps functions into numbers, i.e., the input variable of a functional is a function. A functional derivative is the generalization of the partial derivative. See Appendix 8.A for a tutorial overview of calculus of variations.

Eqs. (8.204) and (8.205) are intuitively understandable from the mean-field arguments: for a system of IC in an effective one-body field $\varphi_i(\mathbf{r})$ for each type of segments, the density distribution functions $\{\rho_i(\mathbf{r})\}$ minimize the free energy of ideal polymer chains. Conversely, the effective one-body fields $\{\varphi_i(\mathbf{r})\}$ can be determined *self-consistently* from the segment densities.

In the presence of a set of effective one-body potentials $\{\varphi_i(\mathbf{r})\}$, we can derive the density profiles of polymer segments in terms of a pair of propagator functions.[77] Because an ideal chain is mathematically equivalent to a random walk or a particle under Brownian motion (Section 3.12), these functions satisfy the modified diffusion equations

$$\left\{\frac{\partial}{\partial s} + \sum_i \varphi_i[\mathbf{R}(s)] - \frac{b^2}{6}\nabla_{\mathbf{r}}^2\right\} p(\mathbf{r}, s) = 0, \tag{8.206}$$

$$\left\{-\frac{\partial}{\partial s} + \sum_i \varphi_i[\mathbf{R}(s)] - \frac{b^2}{6}\nabla_{\mathbf{r}}^2\right\} p^*(\mathbf{r}, s) = 0, \tag{8.207}$$

with the initial conditions

$$p(\mathbf{r}, 0) = p^*(\mathbf{r}, m) = 1 \tag{8.208}$$

and the boundary condition at $r \to \infty$

$$p(\mathbf{r}, s) = p^*(\mathbf{r}, s) = 0. \tag{8.209}$$

The propagator functions, $p(\mathbf{r}, s)$ and $p^*(\mathbf{r}, s)$, specify the end-point probabilities of Gaussian chains of contour length s with the initial and the end segment fixed at position $\mathbf{r}$, respectively.

From the propagator functions, we can determine the segment densities from

$$\rho_i(\mathbf{r}) = \frac{N}{q_0 V}\int_0^m ds\, p(\mathbf{r}, s)p^*(\mathbf{r}, s)\gamma_i(s) \tag{8.210}$$

where q_0 corresponds to a single-chain partition function

$$q_0 = \frac{1}{V}\int d\mathbf{r}\, p(\mathbf{r}, m). \tag{8.211}$$

Eq. (8.210) indicates that the density of type i segments at position $\mathbf{r}$ is determined by the average density of polymer chains (N/V) multiplied by the probability of polymer segments of the same type from a single chain at this position. The latter is related to the joint probability of both sequences $(0,s)$ and (s,m) to have one end at position $\mathbf{r}$.

From the segment densities, we can evaluate the effective one-body fields from Eq. (8.205). Thus, Eqs. (8.205)–(8.207) and (8.210) provide a set of self-consistent equations for determining $p(\mathbf{r}, s)$, $p^*(\mathbf{r}, s)$, $\{\rho_i(\mathbf{r})\}$, and $\{\varphi_i(\mathbf{r})\}$, which can be solved with suitable numerical schemes.[78] The mean-field Helmholtz energy can then be calculated from

$$\beta F = -N\ln(q_0 eV/N) + \frac{1}{2}\sum_{i,j}\int d\mathbf{r}\int d\mathbf{r}'\rho_i(\mathbf{r})\rho_j(\mathbf{r}')\beta u_{ij}(|\mathbf{r}-\mathbf{r}'|) - \sum_i \int d\mathbf{r}\varphi_i(\mathbf{r})\rho_i(\mathbf{r}). \tag{8.212}$$

Eq. (8.212) can be derived from Eq. (8.202) with the effective one-body potentials and segment density profiles obtained from the variational principle. In comparison with the conventional

77 Matsen M. W., "The standard Gaussian model for block copolymer melts", *J. Phys.: Condens. Matter* 14 (2), R21–R47 (2002). Mathematically, the propagator function is equivalent to the end-point probability of a random walk under the influence of an external field.

78 Arora A., et al., "Broadly accessible self-consistent field theory for block polymer materials discovery", *Macromolecules* 49 (13), 4675–4690 (2016).

expression of mean-field free energy, it includes an extra term, i.e., the third term on the right of Eq. (8.212), to account for the one-body potential appearing in the single-chain partition function.

One of the most successful applications of SCFT is to predict the microphase separation in block-copolymer melts. In these applications, the short-range repulsion (viz., excluded volume effect) among polymer segments is described in terms of the incompressibility assumption, which assumes the total segment density as a constant throughout the system. Meanwhile, the potential energy due to the attraction between polymer segments is represented by the Flory parameter. For a diblock copolymer melt, the total potential energy can be written as

$$\beta\Phi_A[\rho_A(\mathbf{r}), \rho_B(\mathbf{r})] = \chi_F \rho_t^{-1} \int d\mathbf{r} \rho_A(\mathbf{r})\rho_B(\mathbf{r}) \tag{8.213}$$

where $\rho_t = \rho_A(\mathbf{r}) + \rho_B(\mathbf{r})$ is a constant according to the assumption of incompressibility. Like that in the Flory–Huggins theory for a binary mixture of homopolymers, the Flory parameter χ_F provides a measure of the incompatibility between A and B blocks in the polymer melt.

As discussed above (Eqs. (8.204) and (8.205)), we can obtain $\rho_A(\mathbf{r})$, $\rho_B(\mathbf{r})$, $\omega_A(\mathbf{r})$, and $\omega_B(\mathbf{r})$ by applying the saddle-point approximation for the canonical partition function. Because of the incompressibility assumption, minimizing βF_0 in Eq. (8.201) results in an additional term in the self-consistent fields

$$\omega_i(\mathbf{r}) = \chi_F \rho_i(\mathbf{r})/\rho_t + \xi(\mathbf{r}), \quad i = A, B. \tag{8.214}$$

The additional term, $\xi(\mathbf{r})$, which is known as the pressure field, corresponds to the Lagrange multiplier arising from the constraint of a fixed total segment density. In solving the self-consistent equations, $\xi(\mathbf{r})$ is determined from the incompressibility condition $\rho_t = \rho_A(\mathbf{r}) + \rho_B(\mathbf{r})$.

SCFT provides a comprehensive description of the phase behavior of block copolymer systems. The phase diagram can be constructed by comparing the Helmholtz energies of all possible mesoscopic structures. The stable structure is the one with the minimum free energy. For example, Figure 8.30 shows the phase diagram of AB-diblock copolymers predicted by SCFT and possible mesoscopic structures that have been identified by simulation or experiment.[79] For $m\chi_F < 10.5$, the disordered phase is the most stable. At larger values of $m\chi_F$, the linear copolymer may exist in one of the many ordered microphases, depending on the monomer composition. For example, the lamellae phase (L) is stable for near symmetric diblock copolymers, whereas the hexagonally packed cylinder phase (HC) is stable for diblock copolymers with intermediate levels of compositional symmetry. For more compositional asymmetry, the body-centered cubic sphere phase (BCC) is more stable than HC. In highly asymmetric compositions, a narrow region of closed packed spheres (CSP) separates the disordered phase and BCC. The gyroid phases exist in narrow regions near the critical point of order–disorder phase transition. The perforated layer phase (PL) is thermodynamically unstable according to the mean-field theory. Although the assumptions are strictly valid in the limit of infinite chain length and the numerical results are mostly qualitative in comparison with the experimental data, SCFT provides a relatively simple description of the phase behavior that is valuable for discovering new ordered structures from block-copolymer self-assembly.

8.7.5 Summary

In this section, we introduce a continuous model for polymer chains, where the contour of the polymer backbone is described as a smooth curve. This continuous representation takes advantage

79 Matsen M.W., "Effect of architecture on the phase behavior of AB-type block copolymer melts", *Macromolecules* 45 (4), 2161–2165 (2012); Lodge T. P., "Block copolymers: long-term growth with added value", *ibid.* 53 (1), 2–4 (2020).

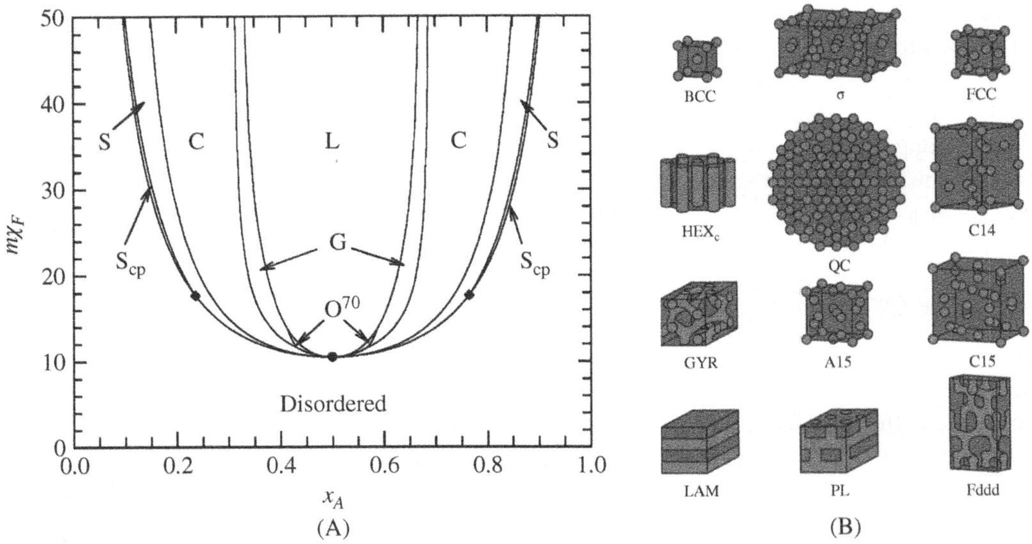

Figure 8.30 A. Phase diagram for linear AB diblock copolymers prediction by the self-consistent field theory. B. Nanostructures reported for linear AB diblock copolymers. Here χ_F is the Flory parameter, m is the degree of polymerization, and x_A is the fraction of A monomers in each copolymer chain. The equilibrium morphologies include disordered phase, close-packed spheres (S_{cp}), body-center-cubic spheres (S), hexagonal cylinders (C), lamellae (L), gyroid (G) and *Fddd* (O^{70}). Acronyms in panel B, BCC: body-centered cubic; σ: Frank–Kasper sigma phase; FCC: face-centered cubic; HEX$_c$: hexagonally packed cylinders; QC: dodecagonal quasi-crystal; C14: Frank–Kasper AB_2 Laves phase; GYR: double gyroid; A15: Frank–Kasper AB_3 phase; C15: Frank–Kasper AB_2 Laves phase; LAM: lamellae; PL: perforated lamellae; *Fddd*: O^{70} network. Reproduced from Matsen (A) and Lodge (B)[79].

of the scale invariance exhibited by long polymer chains and serves as a basis for predicting the mesoscopic properties of polymer systems using field-theoretical methods.

One particularly valuable tool in the field-theoretical framework is the self-consistent field theory (SCFT), which allows for the prediction of free energies of block-copolymer systems associated with various ordered states. By considering the interactions between different blocks within a copolymer, SCFT provides insightful information regarding the formation and stability of ordered structures, including lamellar phases, cylindrical phases, and spherical phases. Through SCFT, one can ascertain the optimal configurations and corresponding free energies of different ordered states in block copolymers. This theoretical approach plays a crucial role in comprehending the underlying principles that govern the self-assembly and phase behavior of block copolymer systems.

By employing the continuous model of polymer chains and utilizing field-theoretical methods, such as SCFT, one may gain a deeper understanding of the mesoscopic properties and behaviors of polymer systems. This knowledge contributes to advancements in areas such as materials science, where the design and manipulation of block copolymer structures have practical implications for a wide range of applications.

8.8 Chapter Summary

The thermodynamic properties of polymeric systems can be described in terms of lattice models, liquid-state theories, and field-theoretical methods. These approaches serve different purposes and complement each other in understanding different aspects of polymer behavior. Lattice models are commonly employed to study the phase behavior of polymer solutions and melts, particularly when the compressibility effects are negligible. While lattice models simplify the representation of polymer configurations, they are able to capture changes in the thermodynamic properties relative to pure species by canceling out local structural errors. However, due to their idealized chain structure and the lack of compressibility, lattice models may not be suitable for predicting the thermodynamic properties of pure liquids or vapor–liquid equilibria.

Liquid-state theories are valuable for describing realistic microscopic structure and pressure effects. The off-lattice, particle-based methods can explicitly account for chain connectivity and incorporate various non-covalent interactions such as segment size, van der Waals attraction, hydrogen bonding, and electrostatic interactions. Liquid-state theories are applicable to both pure fluids and mixtures. However, they often do not capture intra-chain correlations and mesoscopic structure, which are crucial for describing the phase behavior of block copolymers and the kinetics of phase transition.

To address the effects of long-range correlations and mesoscopic structure, field-theoretical methods come into play. Theoretical approaches such as random phase approximation (RPA) and self-consistent field theory (SCFT) provide a generic framework to describe these phenomena. RPA is useful for understanding the response of the polymer structure to the potential changes and phase transitions in polymer systems, while SCFT enables the prediction of free energies associated with different ordered states in block copolymers. These field-theoretical methods incorporate the concept of statistical fields and can capture the complex behavior of polymeric systems over mesoscopic length scales, which are not easily accessible through other theoretical approaches.

Overall, lattice models, liquid-state theories, and field-theoretical methods offer complementary perspectives on the thermodynamic properties of polymeric systems. Lattice models are suitable for phase behavior analysis in the absence of compressibility effects; liquid-state theories account for microscopic structure and pressure effects; and field-theoretical methods capture long-range correlations and mesoscopic structure. In practice, choosing the appropriate theoretical tool depends on the specific system and properties of interest.

8.A Calculus of Variations

Functionals have wide-ranging applications in various areas of mathematics and physics. They are commonly used in optimization problems, variational calculus, and formulating mathematical representations of physical laws and principles. The calculus of variations provides a powerful tool for analyzing and understanding complex systems that involve functionals.

This appendix serves as a tutorial introduction to the calculus of variations. The subject was initially developed by Leonhard Euler in 1733 as an extension of multivariable calculus. The following content is adapted from *Variational Methods in Molecular Modeling* (edited by J. Wu, published by Springer in 2017).

8.A.1 Functionals

A functional can be regarded as an extension of a multivariable function. When we write a multivariable function $f(\mathbf{z})$, where $\mathbf{z}$ is an n-dimensional vector, we mean that for each set of numbers $\mathbf{z} = (z_1, z_2, \ldots, z_n)$, there is a number $f(\mathbf{z})$ associated with it. Simple examples of multivariable functions are $f(\mathbf{z}) = \mathbf{z}^2 = \sum_{i=1}^{n} z_i^2$ or $f(\mathbf{z}) = \mathbf{a} \cdot \mathbf{z}$, where $\mathbf{a}$ is also an n-dimensional vector.

A functional, $F[y]$, maps a smooth (differentiable) function $y(x)$, to a numerical value. The integral $F[y] = \int_0^1 y(x)dx$ provides a simple example. For each smooth function $y(x)$, its integration from 0 to 1 yields a number. While the input of a multivariable function is an n-dimensional vector, the input of a functional is a continuous function. By comparing the similarity between a function and a vector, we see that a functional is a function whose input has infinite dimensionality. In other words, a functional is a function of a function(s).

8.A.2 Functional Derivative

One essential problem in the calculus of variations is functional minimization,[80] i.e., finding a function that minimizes a given functional. To address this problem, we need to know the functional derivative. Like the conventional derivative of a function, a functional derivative specifies how it varies with the input, which is not an ordinary variable, but a function.

Formally, a functional derivative is defined by the variation with respect to a differential input:

$$\begin{aligned}\frac{\delta F[y(x)]}{\delta y(x')} &\equiv \lim_{\varepsilon\to 0}\frac{F[y(x)+\varepsilon\delta(x-x')]-F[y(x)]}{\varepsilon}\\ &= \lim_{\varepsilon\to 0}\frac{F[y+\varepsilon\delta]-F[y]}{\varepsilon\delta}\delta\\ &= \frac{dF[y]}{dy}\delta(x-x')\end{aligned} \tag{8.A.1}$$

where ε is a real number, and $\delta(x-x')$ stands for the Dirac delta function. As shown in Figure 8.A.1, the Dirac function $\delta(x-x_0)$ can be understood as a generalized probability density; it is normalized and has a value of infinite at $x = x_0$.

According to Eq. (8.A.1), the functional derivative $\delta F[y(x)]/\delta y(x')$ can be understood as the variation in functional $F[y(x)]$ with respect to a small change in the input function $y(x)$ at $x = x'$. Because the functional derivative is dependent on x', in general $\delta F[y(x)]/\delta y(x')$ is a function of x'.

The functional derivative defined above can be similarly applied to a function. Suppose $f(y)$ is a function of y, its functional derivative with respect to $y(x)$ is given by

$$\frac{\delta f(y)}{\delta y(x')} = f'(y)\delta(x-x'). \tag{8.A.2}$$

In the special case of $f(y) = y$, we have

$$\frac{\delta y(x)}{\delta y(x')} = \delta(x-x'). \tag{8.A.3}$$

Eq. (8.A.3) indicates that the functional derivative of a function with respect to itself yields a Dirac delta function.

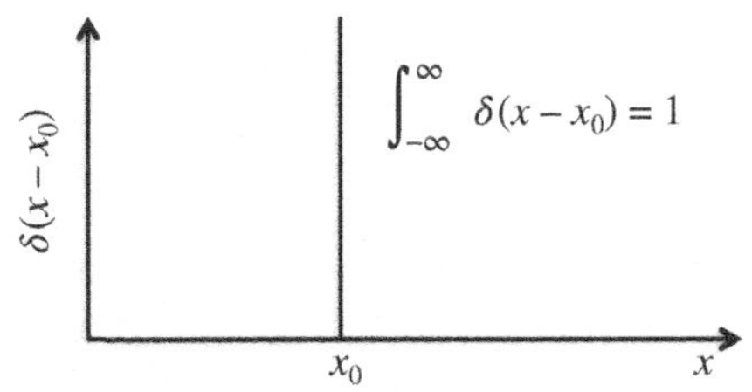

Figure 8.A.1 One-dimensional Dirac delta function $\delta(x-x_0)$ represents a probability density that is everywhere zero except at $x = x_0$ where it is infinite (∞).

80 Functional maximization can be converted to minimization by adding a negative sign.

Functional derivative may be considered as a natural extension of a partial derivative of a multivariable function to infinite dimensionality. To see this, consider again a multivariable function $f(\mathbf{z})$, where $\mathbf{z}$ stands for an n-dimensional vector. Partial derivative $\partial f/\partial z_i$ describes the change in $f(\mathbf{z})$ with respect to an infinitesimal change in the ith dimension of $\mathbf{z}$ while keeping all other dimensions unchanged, i.e.,

$$df = \sum_{j=1}^{n} \frac{\partial f}{\partial z_i} \delta_{ij} dz_i = \frac{\partial f}{\partial z_i} dz_i \tag{8.A.4}$$

where δ_{ij} stands for the Kronecker delta function, i.e., $\delta_{ij} = 1$ for $i = j$ and zero otherwise. Similarly, the variation of a functional with respect to its input (function) at a point x' can be written as

$$\delta F = \int dx \frac{dF}{dy} \delta(x - x') \delta y = \left. \frac{dF}{dy} \delta y \right|_{x'}. \tag{8.A.5}$$

Comparing Eqs. (8.A.4) and (8.A.5), we see that the variable x can be understood as a continuous index of function $y(x)$, similar to i as the index of multi-dimensional vector $\mathbf{z}$.

In analogy to the fact that all partial derivatives of a multi-dimensional function vanish at the minimum point, a functional $F[y]$ reaches a minimum when

$$\frac{\delta F[y(x)]}{\delta y(x')} = 0 \tag{8.A.6}$$

for all values of x'.

8.A.3 Chain Rules

A functional derivative obeys chain rules similar to those for a partial derivative. For example, the chain rule for a partial derivative of a multivariable function $f(\mathbf{z})$ can be written as

$$\frac{\partial f\{\mathbf{g}(\mathbf{z})\}}{\partial z_i} = \sum_{j=1}^{n} \frac{\partial f}{\partial g_j} \frac{\partial g_j}{\partial z_i} \tag{8.A.7}$$

where $\mathbf{g}(\mathbf{z}) = [g_1(\mathbf{z}), g_2(\mathbf{z}), \ldots, g_n(\mathbf{z})]$ is an n-dimensional function of vector $\mathbf{z}$. The analogous chain rule for a functional derivative is

$$\frac{\delta F\{G[y(x)]\}}{\delta y(x')} = \int dx'' \frac{\delta F}{\delta G(x'')} \frac{\delta G(x'')}{\delta y(x')} \tag{8.A.8}$$

where the summation of discrete indices in Eq. (8.A.7) is replaced by an integral over the continuous variable x". In particular, if $F[y(x)] = y(x)$, we have

$$\delta(x - x') = \int dx'' \frac{\delta y(x)}{\delta G(x'')} \frac{\delta G(x'')}{\delta y(x')}. \tag{8.A.9}$$

Eq. (8.A.9) represents a general relation between the reciprocals of functional derivatives.

It can be shown that the functional derivative of a function is commutable with a normal derivative, i.e.,

$$\frac{\delta(df/dx)}{\delta y} = \frac{d}{dx}\left(\frac{\delta f}{\delta y}\right) \tag{8.A.10}$$

where both f and g are functions of x. In a special case, the functional derivative of $y'(x)$ is

$$\frac{\delta}{\delta y(x')}\left[\frac{dy(x)}{dx}\right] = \frac{d}{dx}\left[\frac{\delta y(x)}{\delta y(x')}\right] = \frac{d\delta(x - x')}{dx}. \tag{8.A.11}$$

8.A.4 Functional Taylor Expansion

Higher-order functional derivatives can be defined similar to the higher-order partial derivatives. In general, the mth-order functional derivative of $F[y(x)]$ can be written as

$$\frac{\delta^{(m)}F[y(x)]}{\delta y(x_1)\delta y(x_2)\ldots\delta y(x_m)} = \frac{d^{(m)}F[y]}{dy^{(m)}}\delta(x-x_1)\delta(x-x_2)\ldots\delta(x-x_m). \tag{8.A.12}$$

The higher-order functional derivatives can be used in the functional Taylor expansion. In parallel to a Taylor expansion of a multivariable function, $f(\mathbf{z})$,

$$f(\mathbf{z}+\Delta\mathbf{z}) = f(\mathbf{z}) + \sum_{i=1}^{n}\frac{\partial f}{\partial z_i}\Delta z_i + \frac{1}{2}\sum_{i=1}^{n}\sum_{j=1}^{n}\frac{\partial f}{\partial z_i}\frac{\partial f}{\partial z_j}\Delta z_i\Delta z_j + \cdots, \tag{8.A.13}$$

we may apply the Taylor expansion to a functional

$$F[y+\Delta y] = F[y] + \int dx\frac{\delta F}{\delta y(x)}\Delta y(x) + \frac{1}{2!}\iint dxdx'\frac{\delta^2 F}{\delta y(x)\delta y(x')}\Delta y(x)\Delta y(x') + \cdots. \tag{8.A.14}$$

Again, the difference between the multivariable and the functional Taylor expansions lies only in the summation of the indices, i.e., the summation of integers versus the integration of a continuous variable.

8.A.5 Functional Integration

Functional integration provides a general procedure to evaluate the change of a functional at different input functions. It can also be used to calculate a functional from its derivative.

For a given function $y(x)$, $F[\lambda y(x)]$ represents a function of a real variable λ. By the chain rule, the derivative of $F[\lambda y(x)]$ with respect to λ gives

$$\frac{dF[\lambda y(x)]}{d\lambda} = \frac{dF[\lambda y(x)]}{d(\lambda y)}\frac{\partial(\lambda y)}{\partial\lambda} = \int dx'\frac{\delta F[\lambda y(x)]}{\delta(\lambda y(x'))}y(x'). \tag{8.A.15}$$

The second equality in Eq. (8.A.15) can be verified by substituting the functional derivative by its definition, i.e., Eq. (8.A.1).

Eq. (8.A.15) holds true when we replace $y(x)$ with $\Delta y(x) \equiv y(x) - y_0(x)$:

$$\frac{dF[y_0(x)+\lambda\Delta y(x)]}{d\lambda} = \int dx'\frac{\delta F[\lambda y(x)]}{\delta(\lambda y(x'))}\Delta y(x'). \tag{8.A.16}$$

where $y_0(x)$ is an arbitrary input function. Integrating both sides of Eq. (8.A.16) with respect to λ from 0 to 1 gives

$$F[y] = F[y_0] + \int_0^1 d\lambda\int dx'\frac{\delta F[\lambda y(x)]}{\delta(\lambda y(x'))}\Delta y(x'). \tag{8.A.17}$$

Eq. (8.A.17) indicates that the change of a functional with its input function is related to the integration of the functional derivative and a coupling parameter λ linking the input functions. As discussed in Section 6.8, the functional integration is commonly utilized in molecular simulation for free-energy calculations. The mathematical procedure is also used in deriving the perturbation theories (Section 7.11).

8.A.6 Functional of a Multidimensional Function

It is straightforward to extend the functional derivative, the functional Taylor expansion, and the functional integral when the input is a multidimensional function, i.e., $y = y(\mathbf{x})$ where $\mathbf{x}$ is a multidimensional vector.

Following a procedure similar to that discussed for the one-dimensional case, we have the following equality for the functional derivative with respect to a multidimensional function

$$\frac{\delta F[y(\mathbf{x})]}{\delta y(\mathbf{x}')} = \frac{dF[y]}{dy}\delta(\mathbf{x} - \mathbf{x}') \tag{8.A.18}$$

where $\delta(\mathbf{x} - \mathbf{x}')$ stands for a multidimensional Dirac delta function. The functional Taylor expansion of $F[y(\mathbf{x})]$ is

$$\begin{aligned} F[y + \Delta y] &= F[y] + \int d\mathbf{x} \frac{\delta F}{\delta y(\mathbf{x})} \Delta y(\mathbf{x}) \\ &+ \frac{1}{2!} \iint d\mathbf{x} d\mathbf{x}' \frac{\delta^2 F}{\delta y(\mathbf{x}) \delta y(\mathbf{x}')} \Delta y(\mathbf{x}) \Delta y(\mathbf{x}') + \cdots \end{aligned} \tag{8.A.19}$$

and the functional integration is defined as

$$F[y] = F[y_0] + \int_0^1 d\lambda \int d\mathbf{x}' \frac{\delta F[\lambda y(\mathbf{x})]}{\delta(\lambda y(\mathbf{x}'))} \Delta y(\mathbf{x}'). \tag{8.A.20}$$

In summary, the calculus of variations deals with finding the function that optimizes a specific mathematical quantity, typically an integral, involving that function. Instead of solving for specific variables, the calculus of variations seeks to determine the function that minimizes (or maximizes) a given functional.

8.B Gaussian Integrals

Gaussian integrals are commonly used in statistical mechanics, in particular in field-theoretical methods. In this appendix, we give a tutorial overview by extending the one-dimensional (1D) Gaussian integration to the path integral. The equations discussed here are useful for better understanding the mathematical details of the Gaussian-chain model and the polymer field theory discussed in Section 8.B.7.

8.B.1 One-Dimensional Gaussian Integrals

The Gaussian integral refers to the integration of the natural exponential of any quadratic function. In its simplest form, the Gaussian integral can be written as

$$\int_{-\infty}^{\infty} dy \; e^{-ay^2/2} = \sqrt{\frac{2\pi}{a}} \tag{8.B.1}$$

where $a > 0$ is a constant. If y is a physical variable, parameter a must have the units of $1/y^2$ such that the exponent is dimensionless.

The Gaussian integral is closely related to the Gaussian distribution function

$$p(y) = \sqrt{\frac{a}{2\pi}} e^{-ay^2/2}. \tag{8.B.2}$$

The probability distribution is normalized, i.e., the integration over the entire range of y is unity

$$\int_{-\infty}^{\infty} dyp(y) = 1. \tag{8.B.3}$$

The Gaussian distribution has a zero mean and a standard deviation of $\sqrt{1/a}$, i.e.,

$$\langle y \rangle \equiv \int_{-\infty}^{\infty} dy\, yp(y) = 0, \tag{8.B.4}$$

$$\sigma = \left[\int_{-\infty}^{\infty} dyy^2\, p(y)\right]^{1/2} = 1/\sqrt{a}. \tag{8.B.5}$$

In this appendix, $\langle \cdots \rangle$ means average according to a Gaussian probability.

A brief derivation of Eq. (8.B.1) is given in the following. Squaring the Gaussian integral gives

$$\int_{-\infty}^{\infty} dye^{-ay^2/2} = \left[\int_{-\infty}^{\infty} dxe^{-ax^2/2} \int_{-\infty}^{\infty} dye^{-ay^2/2}\right]^{1/2} = \left[\int_{-\infty}^{\infty}\int_{-\infty}^{\infty} dxdye^{-a(x^2+y^2)/2}\right]^{1/2}. \tag{8.B.6}$$

We can evaluate Eq. (8.B.6) analytically by using the polar coordinates and the identities $dxdy = 2\pi rdr$ and $x^2 + y^2 = r^2$

$$\int_{-\infty}^{\infty} dye^{-ay^2/2} = \left[\int_{0}^{\infty} 2\pi rdre^{-ar^2/2}\right]^{1/2} = \sqrt{\frac{2\pi}{a}}. \tag{8.B.7}$$

Eq. (8.B.7) can be used to evaluate integrals of the form

$$I_m = \int_{-\infty}^{\infty} dyy^m e^{-ay^2/2} \tag{8.B.8}$$

where $m > 0$ is an integer. If m is an odd number, $I_m = 0$ because the integrand is an odd function. If $m = 2n$, the integrand is an even function; the successive derivatives of $I_0 = \sqrt{2\pi/a}$ with respect to a leads to

$$I_{2n} = \frac{(2n)!}{2^n a^n n!}\sqrt{\frac{2\pi}{a}}. \tag{8.B.9}$$

Eqs. (8.B.4) and (8.B.9) can be readily obtained from I_m with $m = 1$ and 2, respectively.

A useful generalization of the Gaussian integral is by including a linear term in the exponential

$$\int_{-\infty}^{\infty} dye^{-ay^2/2+by} = \sqrt{\frac{2\pi}{a}}\, e^{b^2/2a} \tag{8.B.10}$$

where b is an arbitrary number. Eq. (8.B.10) can also be expressed in terms of the Gaussian average

$$\langle e^{by} \rangle = \int_{-\infty}^{\infty} dyp(y)e^{by} = \frac{\int_{-\infty}^{\infty} dye^{-ay^2/2}e^{by}}{\int_{-\infty}^{\infty} dye^{-ay^2/2}} = e^{b^2/2a}. \tag{8.B.11}$$

We can derive Eq. (8.B.10) by shifting the integration variable

$$\int_{-\infty}^{\infty} dye^{-ay^2/2+by} = \int_{-\infty}^{\infty} dye^{-\frac{a}{2}\left(y-\frac{b}{a}\right)^2+\frac{b^2}{2a}} = e^{b^2/2a}\int_{-\infty}^{\infty} dy'e^{-ay'^2/2} = \sqrt{\frac{2\pi}{a}}e^{b^2/2a} \tag{8.B.12}$$

where $y' = y - b/a$.

8.B.2 Three-Dimensional Gaussian Integrals

Now consider a 3-dimensional (3D) Gaussian distribution function

$$p(\mathbf{r}) = \left(\frac{a}{2\pi}\right)^{3/2} \exp\left[-\frac{ar^2}{2}\right] \tag{8.B.13}$$

where $a > 0$ is a real number.[81] Eq. (8.B.13) can be written as a product of three 1D-Gaussian distribution functions

$$p(\mathbf{r}) = p(x)p(y)p(z) \tag{8.B.14}$$

with $\mathbf{r} = (x, y, z)$ and

$$p(\xi) = \left(\frac{a}{2\pi}\right)^{1/2} \exp\left[-\frac{a\xi^2}{2}\right], \xi = x, y, \text{or } z. \tag{8.B.15}$$

Similar to the Gaussian integral in one dimension, we have

$$\int d\mathbf{r}e^{-ar^2/2} = \left(\frac{2\pi}{a}\right)^{3/2}, \tag{8.B.16}$$

and the Gaussian average of a linear term in the exponential is

$$\langle e^{\mathbf{b}\cdot\mathbf{r}}\rangle = e^{b^2/2a} \tag{8.B.17}$$

where $\mathbf{b} = (b_x, b_y, b_z)$. According to Eq. (8.B.17), the Fourier transform of the 3D-Gaussian distribution function is also a Gaussian function

$$\hat{p}(\mathbf{k}) = \int d\mathbf{r}\ p(\mathbf{r})e^{-i\mathbf{k}\cdot\mathbf{r}} = \langle e^{-i\mathbf{k}\cdot\mathbf{r}}\rangle = e^{-k^2/2a}. \tag{8.B.18}$$

8.B.3 Multidimensional Gaussian Integrals

In general, the N-dimensional Gaussian integral can be written as

$$\int d\mathbf{y}e^{-\frac{1}{2}\mathbf{y}^T\cdot\mathbf{A}\cdot\mathbf{y}} = \sqrt{\frac{(2\pi)^N}{\det \mathbf{A}}} \tag{8.B.19}$$

where $\mathbf{y} = (y_1, y_2, \ldots, y_N)^T$ is a vector of real numbers, $d\mathbf{y} = dy_1dy_2\ldots dy_N$, and the integration limits for each dimension range from $-\infty$ to $+\infty$; $\mathbf{y}^T$ stands for the transpose vector of $\mathbf{y}$; $\mathbf{A}$ is an $N \times N$ symmetric, positive definite matrix; and $\det \mathbf{A}$ is the determinant of matrix $\mathbf{A}$.

In order to understand Eq. (8.B.19), some elementary equations from the linear algebra must be in place. A square matrix is called symmetric when

$$\mathbf{A} = \mathbf{A}^T \tag{8.B.20}$$

where $\mathbf{A}^T$ denotes the transpose. The elements of a symmetric matrix satisfy $A_{ij} = A_{ji}$. A positive definite matrix is defined as one that meets the criterion

$$\mathbf{y}^T \cdot \mathbf{A} \cdot \mathbf{y} > 0 \tag{8.B.21}$$

for any non-zero vector $\mathbf{y}$. A special type of positive definite matrix is the positive diagonal matrices, $\mathbf{\Lambda}$, where all diagonal elements $\lambda_1, \lambda_2, \ldots, \lambda_N$ are positive and non-diagonal elements are zero. Because $\lambda_i > 0$, we have

$$\mathbf{y}^T \cdot \mathbf{\Lambda} \cdot \mathbf{y} = \lambda_1 y_1^2 + \lambda_2 y_2^2 + \cdots + \lambda_N y_N^2 \geq 0. \tag{8.B.22}$$

---

81 Alternatively, we may write $ar^2 = |k \cdot r|^2$ where $k = (k_x, k_y, k_z)$ is a vector. In that case, the Gaussian distribution function becomes $p(r) = \frac{(k_x k_y k_z)^{1/2}}{(2\pi)^{3/2}} e^{-|k\cdot r|^2/2}$.

Eq. (8.B.22) is zero if and only if $\mathbf{y} = 0$.

A basic theorem from linear algebra asserts that if $\mathbf{A}$ is a symmetric matrix of real numbers, then $\mathbf{A}$ can be reduced to a diagonal matrix $\boldsymbol{\Lambda}$ by multiplying with a unitary matrix[82] $\mathbf{U}$

$$\mathbf{U}^T \cdot \mathbf{A} \cdot \mathbf{U} = \boldsymbol{\Lambda}. \tag{8.B.23}$$

If $\mathbf{A}$ is positive definite, all diagonal elements of $\boldsymbol{\Lambda}$ are positive, i.e., $\lambda_i > 0$ for $i = 1, 2, \ldots, N$. A unitary matrix satisfies $\det\mathbf{U} = 1$ and

$$\mathbf{U}^T \cdot \mathbf{U} = \mathbf{I}. \tag{8.B.24}$$

where $\mathbf{I}$ denotes a unit matrix, i.e., a diagonal matrix where all the diagonal elements are one. Eq. (8.B.24) indicates that $\mathbf{U}^{-1}$, the matrix inverse of $\mathbf{U}$, is the same as its transpose $\mathbf{U}^T$.

Armed with the above equations, we are now in position to derive Eq. (8.B.19). First, we make a linear transformation of the integration variables from $\mathbf{y}$ to $\mathbf{x}$

$$\mathbf{y} = \mathbf{U} \cdot \mathbf{x} \tag{8.B.25}$$

where $\mathbf{U}$ is the unitary matrix that the diagonalizes matrix $\mathbf{A}$ in Eq. (8.B.19). Because $\mathbf{U}$ corresponds to the Jacobian of the coordinate transformation and $\det \mathbf{U} = 1$, we have

$$d\mathbf{y} = d\mathbf{x}. \tag{8.B.26}$$

Meanwhile, the vector product can be expressed in terms of the eigenvalues of matrix $\mathbf{A}$

$$\mathbf{y}^T \cdot \mathbf{A} \cdot \mathbf{y} = \mathbf{x}^T \cdot \mathbf{U}^T \cdot \mathbf{A} \cdot \mathbf{U} \cdot \mathbf{x} = \mathbf{x}^T \cdot \boldsymbol{\Lambda} \cdot \mathbf{x} = \lambda_1 x_1^2 + \lambda_2 x_2^2 + \cdots + \lambda_N x_N^2. \tag{8.B.27}$$

Accordingly, we can rewrite Eq. (8.B.19) as

$$\int d\mathbf{y} e^{-\frac{1}{2}\mathbf{y}^T \cdot \mathbf{A} \cdot \mathbf{y}} = \int d\mathbf{x}^{-\sum_{n=1}^{N} \lambda_n x_n^2/2} = \prod_{n=1}^{N} \int_{-\infty}^{\infty} dx_n e^{-\lambda_n x_n^2/2} = \prod_{n=1}^{N} \sqrt{\frac{2\pi}{\lambda_n}}. \tag{8.B.28}$$

From $\mathbf{U}^T \cdot \mathbf{A} \cdot \mathbf{U} = \boldsymbol{\Lambda}$ and $\det\mathbf{U} = 1$, we also have $\det \mathbf{A} = \prod_{n=1}^{N} \lambda_n$. Eq. (8.B.29) then becomes identical to Eq. (8.B.19).

A final remark on Eq. (8.B.19) is that the integral can be evaluated even if $\mathbf{A}$ is not symmetric. In that case, we can write the bilinear form $\mathbf{y}^T \cdot \mathbf{A} \cdot \mathbf{y}$ as $\mathbf{y}^T \cdot \mathbf{A}' \cdot \mathbf{y}$, where matrix $\mathbf{A}'$ is symmetric matrix with elements $A_{ij}' = (A_{ij} + A_{ji})/2$. Therefore, the Gaussian integral becomes

$$\int d\mathbf{y} e^{-\frac{1}{2}\mathbf{y}^T \cdot \mathbf{A} \cdot \mathbf{y}} = \int d\mathbf{y} e^{-\frac{1}{2}\mathbf{y}^T \cdot \mathbf{A}' \cdot \mathbf{y}} = \sqrt{\frac{(2\pi)^N}{\det \mathbf{A}'}}. \tag{8.B.29}$$

Similar to those in the one- and three-dimensional cases, adding a linear term to the exponential of the N-dimensional Gaussian integral leads to

$$\int d\mathbf{y} e^{-\frac{1}{2}\mathbf{y}^T \cdot \mathbf{A} \cdot \mathbf{y} + \mathbf{b} \cdot \mathbf{y}} = \sqrt{\frac{(2\pi)^N}{\det \mathbf{A}}} \exp\left[\frac{1}{2}\mathbf{b}^T \cdot \mathbf{A}^{-1} \cdot \mathbf{b}\right]. \tag{8.B.30}$$

where $\mathbf{A}$ is a positive definite symmetric matrix of real numbers, $\mathbf{b}$ is an N-dimensional vector. Eq. (8.B.31) can also be understood as the Gaussian average

$$\langle e^{\mathbf{b} \cdot \mathbf{y}} \rangle = \exp\left[\frac{1}{2}\mathbf{b}^T \cdot \mathbf{A}^{-1} \cdot \mathbf{b}\right] \tag{8.B.31}$$

---

82 A unitary matrix is a square matrix whose conjugate transpose (also known as the Hermitian transpose) is equal to its inverse. Here we assume that all matrix elements are real numbers so that the conjugate transpose is the same as regular transpose. The columns (or rows) of a unitary matrix are orthogonal to each other and have a magnitude of 1.

where $\langle \cdots \rangle$ again stands for the average with the Gaussian probability.

To prove Eq. (8.B.31), let $\mathbf{y} = \mathbf{U} \cdot \mathbf{x}$ and thus

$$\mathbf{b} \cdot \mathbf{y} = \mathbf{b}^T \cdot \mathbf{U} \cdot \mathbf{x} \equiv \mathbf{b}'^T \cdot \mathbf{x} \tag{8.B.32}$$

where $\mathbf{b}'^T = \mathbf{b}^T \cdot \mathbf{U}$ and $\mathbf{b}' = \mathbf{U}^T \cdot \mathbf{b}$. Accordingly, the Gaussian integral becomes

$$\begin{aligned}\int d\mathbf{y}\ e^{-\frac{1}{2}\mathbf{y}^T\cdot\mathbf{A}\cdot\mathbf{y}+\mathbf{b}\cdot\mathbf{y}} &= \int d\mathbf{x}\ e^{-\sum_{n=1}^{N}\lambda_n x_n^2/2+b'_n x_n} \\ &= \prod_{n=1}^{N}\int_{-\infty}^{\infty} dx_n\ e^{-\lambda_n x_n^2/2+b'_n x_n} = \prod_{n=1}^{N}\sqrt{\frac{2\pi}{\lambda_n}}\cdot e^{b'^2_n/2\lambda_n} = \sqrt{\frac{(2\pi)^N}{\det \mathbf{A}}} e^{\sum_{n=1}^{N} b'^2_n/2\lambda_n}\end{aligned} \tag{8.B.33}$$

where the exponential term is

$$\sum_{n=1}^{N} b'^2_n/2\lambda_n = \frac{1}{2}\cdot\mathbf{b}'^T\cdot\mathbf{\Lambda}^{-1}\cdot\mathbf{b}' = \frac{1}{2}\cdot\mathbf{b}^T\cdot\mathbf{U}\cdot\mathbf{\Lambda}^{-1}\cdot\mathbf{U}^T\cdot\mathbf{b} = \frac{1}{2}\mathbf{b}^T\cdot\mathbf{A}^{-1}\cdot\mathbf{b}. \tag{8.B.34}$$

In writing the last equality above, we have used $\mathbf{U}\cdot\mathbf{\Lambda}^{-1}\cdot\mathbf{U}^T = \mathbf{A}^{-1}$ according to $\mathbf{U}^T\cdot\mathbf{A}\cdot\mathbf{U} = \mathbf{\Lambda}$.

8.B.4 Gaussian Path Integrals

In the path integral, we express the Gaussian functional as

$$\int Dy(x) \exp\left\{-\frac{1}{2}\int dx \int dx' \cdot y(x)\cdot G(x-x')\cdot y(x')\right\} = \prod_k \sqrt{\frac{2\pi}{\widehat{G}(k)}} \tag{8.B.35}$$

where $\int Dy(x)$ stands for a path integration over function $y(x)$. Intuitively, x may be understood as an index for y, similar to integer n as the index of vector $\mathbf{y}$. $\widehat{G}(k) > 0$ is the Fourier transform of an even function $G(x)$, i.e., $G(x) = G(-x)$. The symbol $\prod_k$ in Eq. (8.B.36) is formally defined as

$$\ln \prod_k [\ldots] = \frac{L}{2\pi}\int dk \ln[\ldots]. \tag{8.B.36}$$

where L is the domain size of function $y(x)$, the coefficient $L/(2\pi)$ arises from the continuous limit of the Fourier transform

$$k = \frac{2\pi}{L}n, \quad n = 0, \pm 1, \pm 2, \ldots. \tag{8.B.37}$$

The extension of a multi-dimensional integration to a path integral is analogous to that from the Fourier expansion to the Fourier transform. For a continuous function $y(x)$ of single variable x in the interval $-L/2 \le x \le L/2$, the Fourier expansion is defined as

$$y(x) = \sum_{n=-\infty}^{\infty} \widehat{y}_n e^{ik_n x} \tag{8.B.38}$$

where $k_n = 2\pi n/L$ and

$$\widehat{y}_n = \int_{-L/2}^{L/2} dx e^{-ik_n x}. \tag{8.B.39}$$

As $y(x)$ is fully specified by its Fourier coefficients, the path integral $\int Dy(x)$ amounts to integration over all possible variables of $\hat{y}_n$, i.e.,

$$\int Dy(x)[\cdots] = \prod_{n=-\infty}^{\infty} \int_{-\infty}^{\infty} dy_n[\cdots] \tag{8.B.40}$$

where $[\cdots]$ represents an arbitrary integrand. In the limit of $L \to \infty$, $k_n = 2\pi n/L$ becomes a continuous variable. Because $dn = Ldk/(2\pi)$, we may write the sum in the Fourier expansion as an integration with respect to a continuous variable k

$$y(x) = \sum_{n=-\infty}^{\infty} \hat{y}_n e^{ik_n x} = \frac{1}{2\pi}\int_{-\infty}^{\infty} dk\hat{y}(k)e^{ikx} \tag{8.B.41}$$

with

$$\hat{y}(k) = \int_{-\infty}^{\infty} y(x)e^{-ikx}dx. \tag{8.B.42}$$

Similarly, the product of Fourier models in Eq. (8.B.41) can be replaced by an integration

$$\ln \prod_{n=-\infty}^{\infty} [\cdots] = \sum_{n=-\infty}^{\infty} \ln[\cdots] = \frac{L}{2\pi}\int dk \ln[\cdots]. \tag{8.B.43}$$

To prove Eq. (8.B.36), we first express the exponential term in the Fourier space[83]

$$-\frac{1}{2}\int dx \int dx' y(x)G(x-x')y(x') = -\frac{1}{2(2\pi)}\int dk\ \hat{y}(k)\hat{G}(k)\hat{y}^*(k) \tag{8.B.44}$$

where $\hat{y}(k)$ is the Fourier transform of $y(x)$, and $\hat{y}^*(k)$ is the complex conjugate of $\hat{y}(k)$. As usual, the Fourier transform of $G(x)$ is defined as

$$\hat{G}(k) = \int_{-\infty}^{\infty} G(x)e^{-ikx}dx \tag{8.B.45}$$

with the reverse transform

$$G(x) = \frac{1}{2\pi}\int_{-\infty}^{\infty} \hat{G}(k)e^{ikx}dk. \tag{8.B.46}$$

Next, we change the coordinates from path integral $\int Dy(x)$ to $\int D\hat{y}(k)$. The Jacobian of the coordinate transformation is

$$\mathbf{J}(k,x) \equiv \frac{\delta\hat{y}(k)}{\delta y(x)} = \int_{-\infty}^{\infty} \delta(x-x')e^{-ikx}dx' = e^{-ikx} \tag{8.B.47}$$

where $\frac{\delta\hat{y}(k)}{\delta y(x)}$ stands for the functional derivative. The determinant of the Jacobian is

$$\det \mathbf{J} = \frac{1}{2\pi}\int dx \int dk e^{-ikx} = \int dx\delta(x) = 1 \tag{8.B.48}$$

---

83 The convolution can be proved as follows:

$$\begin{aligned}\int dx \int dx' y(x)G(x-x')y(x') &= \int dx \int dx' \frac{1}{2\pi}\int dk\hat{y}(k)e^{ikx}G(x-x')y(x') \\ &= \frac{1}{2\pi}\int dk\hat{y}(k)\int dx e^{ik(x-x')}G(x-x')\cdot\int dx' y(x')e^{ikx'} \\ &= \frac{1}{2\pi}\int dk\hat{y}(k)\hat{y}^*(k)\hat{G}^*(k) = \frac{1}{2\pi}\int dk|\hat{y}(k)|^2\hat{G}^*(k) = \frac{1}{2\pi}\int dk|\hat{y}(k)|^2\hat{G}(k)\end{aligned}$$

In the last line, $\hat{G}(k) = \hat{G}^*(k)$ because $G(x)$ is an even function.

where we have used the mathematical identity

$$\frac{1}{2\pi}\int dke^{-ikx} = \delta(x). \tag{8.B.49}$$

Therefore, in the Fourier space, the Gaussian path integral can be expressed as

$$\int D\hat{y}(k)\exp\left\{-\frac{1}{2}\int dk\hat{y}(k)\hat{G}(k)\hat{y}^*(k)\right\}$$
$$= \int D\hat{y}(k)\cdot\exp\left\{-\frac{1}{2}\int dk\hat{G}(k)\,|\hat{y}(k)|^2\right\} = \prod_k\sqrt{\frac{2\pi}{\hat{G}(k)}} \tag{8.B.50}$$

where the last equality amounts to a 1D Gaussian integral for each value of k.

If the Gaussian path integral involves a linear term in the exponential, we have

$$\int Dy(x)e^{-\frac{1}{2}\int dx\int dx'y(x)G(x-x')y(x')+\int b(x)y(x)dx} = \prod_k\sqrt{\frac{2\pi}{G(k)}}\cdot e^{\frac{1}{2}\int dx\int dx'b(x)G^{-1}(x-x')b(x')} \tag{8.B.51}$$

where $G^{-1}(x)$ is the inverse Fourier transform of $G^{-1}(k) \equiv 1/G(k)$. In terms of the Gaussian average, Eq. (8.B.52) can be written as

$$\left\langle\exp\left\{\int b(x)y(x)dx\right\}\right\rangle = \exp\left\{\frac{1}{4\pi}\int dk|\hat{b}(k)|^2/\hat{G}(k)\right\}. \tag{8.B.52}$$

To prove Eqs. (8.B.52) and (8.B.53), we first notice

$$\int dxb(x)y(x) = \int dxb(x)\frac{1}{2\pi}\int dke^{ikx}\hat{y}(k) = \frac{1}{2\pi}\int dk\ \hat{y}(k)\hat{b}^*(k). \tag{8.B.53}$$

Next, changing the path integral from $\int Dy(x)$ to $\int D\hat{y}(k)$ results in

$$\int Dy(x)e^{-\frac{1}{2}\int dx\int dx'y(x)G(x-x')y(x')+\int b(x)y(x)dx} = \int D\hat{y}(k)e^{-\frac{1}{2(2\pi)}\int dk\hat{G}(k)|\hat{y}(k)|^2+\int dk\hat{y}(k)\hat{b}*(k)/2\pi}$$
$$= \prod_k\sqrt{\frac{2\pi}{\hat{G}(k)}}\ e^{\frac{1}{4\pi}\int dk|b(k)|^2/G(k)} = \prod_k\sqrt{\frac{2\pi}{\hat{G}(k)}}\cdot e^{\frac{1}{2}\int dx\int dx'b(x)G^{-1}(x-x')b(x')}$$
$$\tag{8.B.54}$$

In writing Eq. (8.B.55), we have used 1D Gaussian integral in each value of k and the convolution relation, Eq. (8.B.45).

8.B.5 Additional Gaussian Averages

In the path integral with multi-dimensional functions, we can analytically evaluate the Gaussian averages for the product of different variables

$$\langle y_i y_j\rangle = \mathbf{A}_{ij}^{-1}, \tag{8.B.55}$$

$$\langle y(x)y(x')\rangle = G^{-1}(x-x'). \tag{8.B.56}$$

These equations can be directly derived from Eqs. (8.B.31) and (8.B.53). Specifically, we have

$$\langle e^{\mathbf{b}\cdot\mathbf{y}}\rangle = \exp\left[\frac{1}{2}\mathbf{b}^T\cdot\mathbf{A}^{-1}\cdot\mathbf{b}\right] \tag{8.B.57}$$

$$\langle y_i y_j\rangle = \left\langle\left[\frac{d}{db_i}\cdot\frac{d}{db_j}e^{\mathbf{b}\cdot\mathbf{y}}\right]_{\mathbf{b}=0}\right\rangle = \left[\frac{d}{db_i}\cdot\frac{d}{db_j}\langle e^{\mathbf{b}\cdot\mathbf{y}}\rangle\right]_{\mathbf{b}=0}$$

$$= \left[\frac{d}{db_i} \cdot \frac{d}{db_j} \exp\left(\frac{1}{2} \mathbf{b}^T \cdot \mathbf{A}^{-1} \cdot \mathbf{b} \right) \right]_{\mathbf{b}=0} = \frac{1}{2} \left(\mathbf{A}_{ij}^{-1} + \mathbf{A}_{ji}^{-1} \right) = \mathbf{A}_{ij}^{-1} \tag{8.B.58}$$

Because $\mathbf{A}$ is a symmetric matrix, $\mathbf{A}^{-1}$ is also symmetric, so that

$$\frac{1}{2} \left(\mathbf{A}_{ij}^{-1} + \mathbf{A}_{ji}^{-1} \right) = \mathbf{A}_{ij}^{-1}. \tag{8.B.59}$$

Finally, from

$$\left\langle \exp\left\{ \int dx b(x) y(x) \right\} \right\rangle = \exp\left\{ \frac{1}{2} \int dx \int dx' b(x) G^{-1}(x - x') b(x') \right\}, \tag{8.B.60}$$

we have

$$\begin{aligned} \langle y(x) y(x') \rangle &= \left\langle \left[\frac{\delta}{\delta b(x)} \cdot \frac{\delta}{\delta b(x')} e^{\int d\xi b(\xi) y(\xi)} \right]_{b(\xi)=0} \right\rangle \\ &= \left[\frac{\delta}{\delta b(x)} \cdot \frac{\delta}{\delta b(x')} \cdot \left\langle e^{\int d\xi b(\xi) y(\xi)} \right\rangle \right]_{b(\xi)=0} \\ &= \left[\frac{\delta}{\delta b(x)} \cdot \frac{\delta}{\delta b(x')} \cdot e^{\frac{1}{2} \int d\xi \int d\xi' b(\xi) G^{-1}(\xi - \xi') b(\xi')} \right]_{b(\xi)=0} = G^{-1}(x - x'). \end{aligned} \tag{8.B.61}$$

Eq. (8.B.63) is useful in the field theory for deriving the density–density correlation functions.

8.C Basics of Statistical Field Theory

Statistical field theory, also known as field theory, is a mathematical framework used to evaluate the partition function of many-body systems by transforming discrete particle densities into fluctuating potentials that vary continuously in space. This approach is based on the concept of statistical fields, which represent the fluctuating potentials associated with the density fluctuations in the system.

The statistical field theory was originally developed in the context of quantum electrodynamics, where the fields represent electromagnetic potentials.[84] The same mathematical strategy can be applied to systems in statistical mechanics, allowing for the study of thermodynamic properties and correlations in a wide range of materials and chemical systems.

In the early applications of field-theoretical methods in statistical mechanics, the focus was primarily on disordered systems such as glasses, polymers, and gels.[85] These systems exhibit complex and heterogeneous structures, and their thermodynamic properties are challenging to analyze using traditional approaches. By employing field-theoretical methods, it becomes possible to describe the behavior of these systems and gain insights into their phase transitions, critical phenomena, and equilibrium properties.

More recently, field-theoretical methods have also been employed to study long-range electrostatic correlations in ionic systems where the interactions between charged particles can have significant effects on their collective behavior. By incorporating statistical fields, we can investigate the impact of electrostatic interactions on the thermodynamic properties and structural organization in ionic systems (see Problem 8.27).

---

84 Kleinert H., *Path integrals in quantum mechanics, statistics, polymer physics, and financial markets*. World Scientific Pub Co Inc (5th Edition), 2009.

85 Edwards S. F., "The statistical mechanics of polymers with excluded volume", *Proc. Phys. Soc. (London)* 85, 613 (1965).

8.C.1 Grand Canonical Partition Function

To introduce field variables and elucidate their connections with statistical mechanics, consider a grand canonical partition function for a one-component system of classical particles

$$\Xi = \sum_{N=0}^{\infty} \frac{1}{N!\Lambda^{3N}} \int d\mathbf{r}^N \exp\left\{ -\frac{\beta}{2}\sum_{i\neq j} u(|\mathbf{r}_i - \mathbf{r}_j|) - \beta \sum_{i=1}^{N} [V_{ext}(\mathbf{r}_i) - \mu] \right\}. \tag{8.C.1}$$

All variables in Eq. (8.C.1) have their usual meanings: $d\mathbf{r}^N = d\mathbf{r}_1 d\mathbf{r}_2 \cdots d\mathbf{r}_N$ represents a differential volume in the configurational space, $\beta = (k_B T)^{-1}$ is the inverse temperature, Λ denotes the de Broglie thermal wavelength, and μ is the chemical potential of the classical particles. We assume that the total potential energy includes contributions from a pairwise-additive potential, $u(r)$, and one-body external potential $V_{ext}(\mathbf{r})$.

In general, two classical particles cannot occupy the same position, i.e., the pair potential $u(r)$ diverges when the particle–particle separation r approaches zero. To avoid the singularity, we may divide the pair potential $u(r)$ into a short-range repulsion $u_0(r)$ and a longer-range contribution $u_A(r)$

$$u(r) = u_0(r) + u_A(r). \tag{8.C.2}$$

With only short-range interactions, the classical particles would constitute a reference system, denoted by subscript 0. While the short-range repulsion diverges at small separation, the longer-ranged potential applies to particle–particle interactions beyond the excluded-volume effects, i.e., $u_A(r)$ is non-zero only when $r \geq \sigma$, where σ denotes the "hard-sphere" diameter. Typically, $u_A(r)$ refers to inter-particle attraction, but other forms of interaction may also be considered.

Using the mathematical identity

$$f(\mathbf{r}_i) = \int d\mathbf{r}\delta(\mathbf{r} - \mathbf{r}_i) f(\mathbf{r}) \tag{8.C.3}$$

where $\delta(\mathbf{r})$ stands for the Dirac delta function, and $f(\mathbf{r})$ is a function continuous near particle position $\mathbf{r}_i$, we may express the longer-ranged pair potential as

$$u_A(|\mathbf{r}_i - \mathbf{r}_j|) = \int d\mathbf{r} \int d\mathbf{r}' \delta(\mathbf{r} - \mathbf{r}_i) u_A(|\mathbf{r} - \mathbf{r}'|)\delta(\mathbf{r}' - \mathbf{r}_j). \tag{8.C.4}$$

Accordingly, the total potential energy due to the longer-ranged interactions can be written as

$$\frac{1}{2}\sum_{i\neq j} u_A(|\mathbf{r}_i - \mathbf{r}_j|) = \frac{1}{2}\int d\mathbf{r} \int d\mathbf{r}' \sum_{i=1}^{N}\sum_{j=1}^{N} \delta(\mathbf{r} - \mathbf{r}_i) u_A(|\mathbf{r} - \mathbf{r}'|)\delta(\mathbf{r}' - \mathbf{r}_j) - \int d\mathbf{r} \sum_{i=1}^{N} \delta(\mathbf{r} - \mathbf{r}_i) u_A(0). \tag{8.C.5}$$

The last term in Eq. (8.C.5) is known as self-energy, which is introduced to correct the spurious interactions arising from the double summation. Apparently, the self-energy correction is unnecessary if $u_A(r)$ is non-zero only when the particle–particle separation is larger than the hard-sphere diameter, i.e., the self-energy vanishes if $u_A(0) = 0$.

Given the microstate of the system, we may define an instantaneous density according to the positions of individual particles

$$\widetilde{\rho}(\mathbf{r}) \equiv \sum_{i=1}^{N} \delta(\mathbf{r} - \mathbf{r}_i). \tag{8.C.6}$$

Substituting Eq. (8.C.6) into Eq. (8.C.5) yields

$$\frac{1}{2}\sum_{i\neq j} u_A(|\mathbf{r}_i - \mathbf{r}_j|) = \frac{1}{2}\int d\mathbf{r}\int d\mathbf{r}'\widetilde{\rho}(\mathbf{r})u_A(|\mathbf{r}-\mathbf{r}'|)\widetilde{\rho}(\mathbf{r}') - \int d\mathbf{r}\widetilde{\rho}(\mathbf{r})u_A(0). \tag{8.C.7}$$

Similarly, the one-body external energy can be expressed as

$$\sum_{i=1}^{N}[V_{ext}(\mathbf{r}_i) - \mu] = \int d\mathbf{r}\widetilde{\rho}(\mathbf{r})[V_{ext}(\mathbf{r}) - \mu]. \tag{8.C.8}$$

Using Eqs. (8.C.7) and (8.C.8), we may rewrite the grand partition function as

$$\Xi = \sum_{N=0}^{\infty}\frac{1}{N!\Lambda^{3N}}\int d\mathbf{r}^N \exp\left\{-\frac{\beta}{2}\int d\mathbf{r}\int d\mathbf{r}'\widetilde{\rho}(\mathbf{r})u_A(|\mathbf{r}-\mathbf{r}'|)\widetilde{\rho}(\mathbf{r}')\right.$$
$$\left. -\beta\int d\mathbf{r}\widetilde{\rho}(\mathbf{r})[V_{ext}(\mathbf{r}) - u_A(0) - \mu] - \beta\Phi_0\right\} \tag{8.C.9}$$

where Φ_0 represents the potential energy of the reference system.

8.C.2 The Hubbard–Stratonovich Transformation

To evaluate the partition function in Eq. (8.C.9), a field-theoretical method typically starts with the Hubbard–Stratonovich transformation[86]

$$\exp\left\{-\frac{\beta}{2}\int d\mathbf{r}\int d\mathbf{r}'\widetilde{\rho}(\mathbf{r})u_A(|\mathbf{r}-\mathbf{r}'|)\widetilde{\rho}(\mathbf{r}')\right\}$$
$$= \frac{1}{\aleph}\int D\omega \exp\left\{-\frac{1}{2\beta}\int d\mathbf{r}\int d\mathbf{r}'\omega(\mathbf{r})u_A^{-1}(|\mathbf{r}-\mathbf{r}'|)\omega(\mathbf{r}') + i\int d\mathbf{r}\omega(\mathbf{r})\widetilde{\rho}(\mathbf{r})\right\} \tag{8.C.10}$$

where $i = \sqrt{-1}$, $D\omega(\mathbf{r})$ represents the path integral over continuous function $\omega(\mathbf{r})$, $\aleph$ is a normalization constant defined as

$$\aleph = \int D\omega \exp\left\{-\frac{1}{2\beta}\int d\mathbf{r}\int d\mathbf{r}'\omega(\mathbf{r})u_A^{-1}(|\mathbf{r}-\mathbf{r}'|)\omega(\mathbf{r}')\right\}, \tag{8.C.11}$$

and $u_A^{-1}(r)$ stands for the functional inversion of $u_A(r)$, i.e.,

$$\int d\mathbf{r}' u_A(|\mathbf{r}_1 - \mathbf{r}'|)u_A^{-1}(|\mathbf{r}_2 - \mathbf{r}'|) = \delta(\mathbf{r}_1 - \mathbf{r}_2). \tag{8.C.12}$$

Substituting Eq. (8.C.10) into (8.C.9) leads to

$$\Xi = \frac{1}{\aleph}\int D\omega(\mathbf{r})\Xi_0[\gamma(\mathbf{r})]\exp\left\{-\frac{1}{2\beta}\int d\mathbf{r}\int d\mathbf{r}'\omega(\mathbf{r})u_A^{-1}(|\mathbf{r}-\mathbf{r}'|)\omega(\mathbf{r}')\right\} \tag{8.C.13}$$

with

$$\Xi_0[\gamma(\mathbf{r})] \equiv \sum_{N=0}^{\infty}\frac{1}{N!\Lambda^{3N}}\int d\mathbf{r}^N \exp\left\{-\beta\Phi_0 - \beta\int d\mathbf{r}\widetilde{\rho}(\mathbf{r})\gamma(\mathbf{r})\right\}. \tag{8.C.14}$$

Formally, $\Xi_0[\gamma(\mathbf{r})]$ corresponds to the grand partition function of the reference system, with each particle subject to an effective one-body potential

$$\gamma(\mathbf{r}) \equiv V_{ext}(\mathbf{r}) - u(0) - \mu - ik_BT\omega(\mathbf{r}). \tag{8.C.15}$$

---

86 Eq. (8.C.10) may be understood as an extension of the Gaussian integral $\exp[-ax^2/2] = \frac{1}{\sqrt{2\pi a}}\int_{-\infty}^{\infty} dy \exp[-y^2/2a - ixy]$ to the functional space. See Appendix 8.B for the mathematical details on Gaussian integrals.

Eq. (8.C.13) provides a starting point for the applications of the field theory to classical systems. In shorthand notation, the grand partition function may be written as

$$\Xi = \frac{1}{\aleph} \int D\omega(\mathbf{r}) \exp\{-\beta H[\omega(\mathbf{r})]\} \tag{8.C.16}$$

where the Hamiltonian is defined as

$$\beta H[\omega(\mathbf{r})] \equiv \frac{1}{2\beta} \int d\mathbf{r} \int d\mathbf{r}' \omega(\mathbf{r}) u_A^{-1}(|\mathbf{r} - \mathbf{r}'|)\omega(\mathbf{r}') + \beta\Omega_0[\gamma(\mathbf{r})] \tag{8.C.17}$$

and $\Omega_0 = -k_B T \ln \Xi_0$ corresponds to the grand potential of the reference system. Different from its conventional expression, the partition function in Eq. (8.C.16) does not involve the configurational integration; the microstates underlying the instantaneous density of discrete particles have been replaced by a dimensionless, fluctuating field $\omega(\mathbf{r})$. Because the fluctuation field is a variable, the theoretical procedures for evaluating the ensemble properties are known as field-theoretical methods.

Given an effective one-body potential, the grand potential of the reference system may be written as

$$\Omega_0[\gamma(\mathbf{r})] = F_0[\rho(\mathbf{r})] + \int d\mathbf{r} \rho(\mathbf{r}) \gamma(\mathbf{r}) \tag{8.C.18}$$

where $F_0[\rho(\mathbf{r})]$ is called the intrinsic Helmholtz energy, and

$$\rho(\mathbf{r}) = \delta\Omega_0[\gamma(\mathbf{r})]/\delta\gamma(\mathbf{r}) \tag{8.C.19}$$

corresponds to the equilibrium density profile of the reference system under the one-body potential $\gamma(\mathbf{r})$, defined by Eq. (8.C.15). If we have an analytical expression for the grand potential (e.g., from classical density functional theory), we would be able to evaluate the density profile for any imaginary field $\omega(\mathbf{r})$.

In the field-theoretical simulations, the grand partition function of the real system is evaluated by sampling all possible forms of the imaginary fields according to an auxiliary Gaussian (G) Hamiltonian

$$\beta H_G = \frac{1}{2\beta} \int d\mathbf{r} \int d\mathbf{r}' \omega(\mathbf{r}) u_A^{-1}(|\mathbf{r} - \mathbf{r}'|)\omega(\mathbf{r}'). \tag{8.C.20}$$

By sampling the imaginary fields according to Eq. (8.C.20), we may evaluate the grand partition function based on the Gaussian average

$$\Xi = \langle \exp\{-\beta\Omega_0[\gamma(\mathbf{r})]\} \rangle_G. \tag{8.C.21}$$

While the mathematical procedure to convert the instantaneous density profiles to the corresponding imaginary potentials (viz., statistical fields) is formally exact, a precise evaluation of the functional integration shown in Eq. (8.C.16) is not feasible for most systems of practical interest. Like many analytical theories in statistical mechanics, approximations are thus inevitable in the application of the field-theoretical methods. One common procedure is by using the saddle-point approximation, i.e., by taking only the maximum term in the functional integration. The latter can be identified by a functional derivative of the field-theory Hamiltonian

$$\frac{\delta \beta H}{\delta \omega(\mathbf{r})} = \frac{1}{\beta} \int d\mathbf{r}' \omega(\mathbf{r}') u_A^{-1}(|\mathbf{r} - \mathbf{r}'|) - i\rho(\mathbf{r}) = 0. \tag{8.C.22}$$

leading to a mean-field potential

$$i\overline{\omega}(\mathbf{r}) = -\int d\mathbf{r}' \rho(\mathbf{r}') \beta u_A(|\mathbf{r} - \mathbf{r}'|). \tag{8.C.23}$$

Except the coefficient of imaginary number $i = \sqrt{-1}$, Eq. (8.C.23) is identical to the conventional mean-field potential due to particle–particle interactions. Indeed, substituting Eq. (8.C.23) into (8.C.17) reproduces the mean-field free energy

$$\beta\Omega_{MF} = \beta H[\overline{\omega}(\mathbf{r})] = \beta\Omega_0 + \frac{1}{2}\int d\mathbf{r}' \int d\mathbf{r}\rho(\mathbf{r})u_A(|\mathbf{r}-\mathbf{r}'|)\rho(\mathbf{r}') \tag{8.C.24}$$

where Ω_0 is the grand partition function of the reference system, which is given by

$$e^{-\beta\Omega_0} \equiv \sum_{N=0}^{\infty}\frac{1}{N!\Lambda^{3N}}\int d\mathbf{r}^N \exp\left\{-\beta\Phi_0 - \beta\int d\mathbf{r}\widetilde{\rho}(\mathbf{r})[V_{ext}(\mathbf{r})-\mu]\right\}. \tag{8.C.25}$$

Alternatively, we may express Eq. (8.C.24) in terms of the *intrinsic* Helmholtz energy

$$F[\rho(\mathbf{r})] \equiv \Omega - \int d\mathbf{r}\rho(\mathbf{r})[V_{ext}(\mathbf{r})-\mu]. \tag{8.C.26}$$

A comparison of Eqs. (8.C.26) and (8.C.18) indicates that the mean-field free energy is identical to that predicted by the van der Waals theory

$$F_{MF} = F_0 + \frac{1}{2}\int d\mathbf{r}' \int d\mathbf{r}\rho(\mathbf{r})u_A(|\mathbf{r}-\mathbf{r}'|)\rho(\mathbf{r}'). \tag{8.C.27}$$

In comparison with conventional mean-field methods, one major advantage of the field theory is that it provides a systematic way to incorporate the fluctuation effects. For example, in the Gaussian-field theory, we take a quadratic expansion of the field Hamiltonian relative to that at the saddle point (viz. the mean-field approximation)

$$\beta H[\omega(\mathbf{r})] \approx \beta H[\overline{\omega}(\mathbf{r})] + \frac{1}{2}\int d\mathbf{r}\int d\mathbf{r}'\delta\omega(\mathbf{r})G(\mathbf{r},\mathbf{r}')\delta\omega(\mathbf{r}') \tag{8.C.28}$$

where $\delta\omega(\mathbf{r}) \equiv \omega(\mathbf{r}) - \overline{\omega}(\mathbf{r})$, and $G(\mathbf{r},\mathbf{r}')$ can be derived from Eq. (8.C.17)

$$G(\mathbf{r},\mathbf{r}') = \frac{\delta^2\beta H}{\delta\omega(\mathbf{r})\delta\omega(\mathbf{r}')} = \beta^{-1}u_A^{-1}(|\mathbf{r}-\mathbf{r}'|) + \chi_0(\mathbf{r},\mathbf{r}') \tag{8.C.29}$$

where $\chi_0(\mathbf{r},\mathbf{r}')$ is the density–density correlation function of the reference system

$$\chi_0(\mathbf{r},\mathbf{r}') = \frac{\delta^2\beta\Omega_0}{\delta\omega(\mathbf{r})\delta\omega(\mathbf{r}')} = -\frac{\delta^2\beta\Omega_0}{\delta\beta V_{ext}(\mathbf{r})\delta\beta V_{ext}(\mathbf{r}')}. \tag{8.C.30}$$

Inserting Eq. (8.C.28) into (8.C.16) leads to an improved grand partition function

$$\Xi \approx \exp\{-\beta H[\overline{\omega}(\mathbf{r})]\}\frac{\int D\omega(\mathbf{r})\exp\left\{-\frac{1}{2}\int d\mathbf{r}\int d\mathbf{r}'\delta\omega(\mathbf{r})G(\mathbf{r},\mathbf{r}')\delta\omega(\mathbf{r}')\right\}}{\int D\omega(\mathbf{r})\exp\left\{-\frac{1}{2\beta}\int d\mathbf{r}\int d\mathbf{r}'\omega(\mathbf{r})u_A^{-1}(|\mathbf{r}-\mathbf{r}'|)\omega(\mathbf{r}')\right\}}. \tag{8.C.31}$$

Using the following mathematical identity from Appendix 8.B

$$\int Dy(x)\exp\left\{-\frac{1}{2}\int dx\int dx'\cdot y(x)\cdot G(x-x')\cdot y(x')\right\} = \prod_k\sqrt{\frac{2\pi}{\widehat{G}(k)}}, \tag{8.C.32}$$

we can obtain an analytical expression for the grand potential

$$\beta\Omega \approx \beta\Omega_{MF} + \ln\prod_k\sqrt{\widehat{G}(k)\beta u_A(k)} = \beta\Omega_{MF} + \frac{V}{2(2\pi)^3}\int d\mathbf{k}\ln[1+\beta u_A(k)\chi_0(k)]. \tag{8.C.33}$$

In writing Eq. (8.C.33), we have used $\widehat{G}(k) = \widehat{\chi}_0(k) + \beta^{-1}u_A^{-1}(k)$ according to Eq. (8.C.29), $\widehat{u}_A(k)$ and $\widehat{\chi}_0(k)$ are the 3D-Fourier transforms of $u_A(r)$ and $\chi_0(r)$, respectively, and V is the system volume. Eq. (8.C.33) represents a key result from the random-phase approximation (RPA).

8.C.3 Polymer Field Theory

Extension of the field-theoretical methods discussed above to polymeric systems is rather straightforward. To elucidate, consider an open system of homopolymers with bond potential $v_B(\mathbf{R})$ where $\mathbf{R} = (\mathbf{r}_1, \mathbf{r}_2, \ldots, \mathbf{r}_m)$ represents the positions of m segments in each polymer chain. With the assumption of pairwise additivity, the total potential energy due to non-bonded inter- and intra- molecular interactions is given by

$$\Phi(\mathbf{R}^N) = \frac{1}{2}\iint d\mathbf{r}d\mathbf{r}'\widetilde{\rho}(\mathbf{r})u(|\mathbf{r}-\mathbf{r}'|)\widetilde{\rho}(\mathbf{r}') - \int d\mathbf{r}\widetilde{\rho}(\mathbf{r})u(0) \tag{8.C.34}$$

where $u(|\mathbf{r}-\mathbf{r}'|)$ stands for the pair potential between polymer segments, and $\mathbf{R}^N$ is the configuration of the entire system. Similar to the monomeric case, the instantaneous density of polymer segments is defined as

$$\widetilde{\rho}(\mathbf{r}) \equiv \sum_{I=1}^{N}\sum_{j=1}^{m}\delta(\mathbf{r}-\mathbf{r}_{I,j}) \tag{8.C.35}$$

where the Dirac delta function $\delta(\mathbf{r}-\mathbf{r}_{I,j})$ specifies the local density of segment j from polymer chain I. Also similar to that for a monomeric system, the grand partition function is

$$\Xi = \sum_{N=0}^{\infty}\frac{1}{N!\Lambda_c^{3N}}\int d\mathbf{R}^N \exp\left\{-\beta\Phi(\mathbf{R}^N) - \beta\sum_{i=1}^{N}[V_{ext}(\mathbf{R}_i) + v_B(\mathbf{R}_i) - \mu]\right\} \tag{8.C.36}$$

where $V_{ext}(\mathbf{R}) = \sum_{i=1}^{m} V_{ext}(\mathbf{r}_i)$ represents the external potential per polymer chain, and Λ_c is the generalized thermal wavelength that accounts for the kinetic energy of individual segments.

Using the Hubbard–Stratonovich transformation

$$\exp\left\{-\frac{\beta}{2}\int d\mathbf{r}\int d\mathbf{r}'\widetilde{\rho}(\mathbf{r})u(|\mathbf{r}-\mathbf{r}'|)\widetilde{\rho}(\mathbf{r}')\right\}$$
$$= \frac{1}{\aleph}\int D\omega \exp\left\{-\frac{1}{2\beta}\int d\mathbf{r}\int d\mathbf{r}'\omega(\mathbf{r})u^{-1}(|\mathbf{r}-\mathbf{r}'|)\omega(\mathbf{r}') + i\int d\mathbf{r}\omega(\mathbf{r})\widetilde{\rho}(\mathbf{r})\right\} \tag{8.C.37}$$

where $\aleph$ is normalization constant as defined in Eq. (8.C.11), we can rewrite Eq. (8.C.36) as

$$\Xi = \frac{1}{\aleph}\int D\omega(\mathbf{r}) \exp\left\{-\frac{1}{2}\int d\mathbf{r}\int d\mathbf{r}'\omega(\mathbf{r})u^{-1}(|\mathbf{r}-\mathbf{r}'|)\omega(\mathbf{r}') - \beta\Omega^{IC}\right\} \tag{8.C.38}$$

where $\beta\Omega^{IC} = -\ln \Xi^{IC}$, and Ξ^{IC} is the grand-partition function of the IC with an effective one-body potential

$$\gamma(\mathbf{R}) = \sum_{i=1}^{m}[V_{ext}(\mathbf{r}_i) - u(0) - i\omega(\mathbf{r}_i)k_BT]. \tag{8.C.39}$$

In analogy to that for a monomeric system, Eq. (8.C.38) may be expressed in terms of an ensemble average of the Gaussian fields

$$\Xi = \langle \exp\{-\beta\Omega^{IC}[\gamma(\mathbf{r})\}\rangle_G. \tag{8.C.40}$$

As discussed above, the saddle-point approximation leads to the mean-field grand potential

$$\beta\Omega_{MF} \approx \beta\Omega^{IC} + \frac{1}{2}\int d\mathbf{r}'\int d\mathbf{r}\rho(\mathbf{r})u(|\mathbf{r}-\mathbf{r}'|)\rho(\mathbf{r}'). \tag{8.C.41}$$

The fluctuation effects can be accounted for with the Gaussian-field approximation

$$\beta\Omega \approx \beta\Omega_{MF} + \frac{V}{2(2\pi)^3}\int d\mathbf{k} \ln\,[1 + \beta u(k)\chi_0(k)] \tag{8.C.42}$$

where $u(k)$ and $\chi_0(k)$ are the Fourier transform of $u(r)$ and $\chi_0(r)$. The latter corresponds to the density–density correlation function of the reference system (viz., IC).

While the field-theoretical equations for polymers are formally identical to those for monomeric systems (because both convert the instantaneous particle density to a fluctuating field), the high dimensionality of polymer configuration (viz., the dependence of polymer properties on a multi-dimensional vector **R**) makes the numerical evaluation of the density profiles and the Helmholtz energy functional much more complicated.

8.D Statistical Mechanics of Non-Uniform Ideal Gases

In this appendix, we present statistical–mechanical equations for predicting the thermodynamic properties of inhomogeneous systems that consist of non-interacting spherical particles (e.g., monatomic ideal gases in the presence of an external potential). Similar equations are introduced for non-interacting polymeric systems. In addition to their direct applications to low-density gases under inhomogeneous conditions (e.g., adsorption), these thermodynamic equations are often used as a reference in the field-theoretical methods (discussed in Appendix 8.C).

8.D.1 Non-Interacting Spherical Particles

Consider the grand canonical ensemble for a one-component ideal gas (IG) of spherical particles under a one-body external potential $V_{ext}(\mathbf{r})$. The grand canonical partition function can be evaluated analytically

$$\Xi^{IG} = \sum_{N=0}^{\infty} \frac{1}{N!\Lambda^{3N}} \int d\mathbf{r}^N \exp\left\{-\beta\sum_{i=1}^{N}[V_{ext}(\mathbf{r}_i) - \mu]\right\} = \sum_{N=0}^{\infty} \frac{q^N}{N!\Lambda^{3N}} = \exp(q/\Lambda^3) \tag{8.D.1}$$

where $\beta = (k_B T)^{-1}$ is the inverse temperature, Λ denotes the thermal wavelength, μ is the chemical potential, and $q \equiv \int d\mathbf{r} \exp\{-\beta[V_{ext}(\mathbf{r}) - \mu]\}$. The grand potential is thus given by

$$\Omega^{IG} = -k_B T \ln \Xi^{IG} = -k_B T q/\Lambda^3 = -\frac{k_B T}{\Lambda^3} \int d\mathbf{r} \exp\{-\beta[V_{ext}(\mathbf{r}) - \mu]\}. \tag{8.D.2}$$

Without the external potential, we have $\Omega^{IG} = -k_B T e^{\beta\mu} V/\Lambda^3$.

For convenience, we may rewrite the one-body potential in the partition function

$$-\beta\sum_{i=1}^{N}[V_{ext}(\mathbf{r}_i) - \mu] = \int d\mathbf{r}\widetilde{\rho}(\mathbf{r})\varphi(\mathbf{r}) \tag{8.D.3}$$

where $\widetilde{\rho}(\mathbf{r}) \equiv \sum_{i=1}^{N}\delta(\mathbf{r} - \mathbf{r}_i)$, and $\varphi(\mathbf{r}) \equiv -\beta[V_{ext}(\mathbf{r}_i) - \mu]$. Accordingly, the local particle density can be expressed in terms of the one-body potential

$$\rho(\mathbf{r}) = \langle\widetilde{\rho}(\mathbf{r})\rangle = \frac{\delta \ln \Xi}{\delta\varphi(\mathbf{r})} = \frac{1}{\Lambda^3}\frac{\delta \int d\mathbf{r} \exp[\varphi(\mathbf{r})]}{\delta\varphi(\mathbf{r})} = e^{\varphi(\mathbf{r})}/\Lambda^3. \tag{8.D.4}$$

For a uniform system, $V_{ext}(\mathbf{r}) = 0$, we obtain the familiar relation $\rho_0^{IG} = e^{\beta\mu}/\Lambda^3$ or $\beta\mu^{IG} = \ln(\rho_0\Lambda^3)$. Rearrangement of Eq. (8.D.4) leads to the Boltzmann distribution for the particle density

$$\rho(\mathbf{r}) = \rho_0 \exp[-\beta V_{ext}(\mathbf{r})]. \tag{8.D.5}$$

Substituting Eq. (8.D.4) into (8.D.2) yields

$$\Omega^{IG} = -\frac{k_B T}{\Lambda^3}\int d\mathbf{r}e^{\varphi(\mathbf{r})} = -\frac{k_B T}{\Lambda^3}\int d\mathbf{r}\rho(\mathbf{r})\Lambda^3 = -k_B T\langle N\rangle \tag{8.D.6}$$

where $\langle N\rangle$ represents the average number of particles in the system. The Helmholtz energy is then given by

$$\begin{aligned} F^{IG} &= \Omega^{IG} + \mu\int d\mathbf{r}\rho(\mathbf{r}) = -k_B T\int d\mathbf{r}\rho(\mathbf{r}) + k_B T\int d\mathbf{r}\rho(\mathbf{r})\varphi(\mathbf{r}) + \int d\mathbf{r}\rho(\mathbf{r})V_{ext}(\mathbf{r}) \\ &= k_B T\int d\mathbf{r}\rho(\mathbf{r})\{\ln[\rho(\mathbf{r})\Lambda^3] - 1\} + \int d\mathbf{r}\rho(\mathbf{r})V_{ext}(\mathbf{r}). \end{aligned} \tag{8.D.7}$$

From the Helmholtz energy and grand potential, we can derive other thermodynamic quantities.

8.D.2 Helmholtz Energy Functional of Ideal Chains

Ideal chains (IC) are an idealized model of linear polymers, neglecting *both* inter- and intra-molecular interactions other than the bond connectivity. To derive equations for calculating the thermodynamic properties of IC, consider the grand partition function

$$\Xi^{IC} = \sum_{N=0}^{\infty}\frac{1}{N!\Lambda_c^{3N}}\int d\mathbf{R}^N \exp\left\{-\beta\sum_{i=1}^{N}[V_{ext}(\mathbf{R}_i) + v_B(\mathbf{R}_i) - \mu]\right\} \tag{8.D.8}$$

where $\mathbf{R} = (\mathbf{r}_1, \mathbf{r}_2, \ldots, \mathbf{r}_m)$ represents the positions of individual segments in each polymer chain, $v_B(\mathbf{R})$ stands for the bond potential, and $V_{ext}(\mathbf{R})$ denotes the external potential for a polymer chain with configuration $\mathbf{R}$. Similar to the thermal wavelength for a monomeric system, Λ_c accounts for the kinetic energy related to the kinetic energies of individual segments.

Following the same procedure for monomeric particles, we can obtain an analytical expression for the grand-canonical partition function of non-interacting chains

$$\Xi^{IC} = \exp\left(q_c/\Lambda_c^3\right) \tag{8.D.9}$$

where

$$q_c \equiv \int d\mathbf{R}\exp\{-\beta[V_{ext}(\mathbf{R}) + v_B(\mathbf{R}) - \mu]\}. \tag{8.D.10}$$

Accordingly, the grand potential is

$$\Omega^{IC} = -\frac{k_B T}{\Lambda_c^3}\int d\mathbf{R}\exp\{-\beta[V_{ext}(\mathbf{R}) - \mu]\}, \tag{8.D.11}$$

and the density profile of polymer configurations follows the generalized Boltzmann distribution

$$\rho(\mathbf{R}) = \frac{1}{\Lambda_c^3}\exp\{-\beta[V_{ext}(\mathbf{R}) - \mu]\}. \tag{8.D.12}$$

Also similar to that for a monomeric system, the Helmholtz energy is

$$F^{IC} = k_B T\int d\mathbf{R}\rho(\mathbf{R})\left\{\ln\left[\rho(\mathbf{R})\Lambda_c^3\right] - 1\right\} + \int d\mathbf{R}\rho(\mathbf{R})V_{ext}(\mathbf{R}). \tag{8.D.13}$$

Because $d\mathbf{R}$ represents a multidimensional integration, one difficulty in practical application of Eq. (8.D.13) is that the numerical evaluation of the density profiles is highly complicated, even for systems of IC.

Further Readings

De Gennes P. G., *Scaling concepts in polymer physics*. Cornell University Press, 1979.
Fredrickson G. H., *The equilibrium theory of inhomogeneous polymers*. Oxford University Press, 2013.
Gartner T. E. and Jayaraman A., "Modeling and simulations of polymers: a roadmap", *Macromolecules* 52 (3), 755–786 (2019)
Gray C.G., Gubbins K. E. and Joslin C. G., *Theory of molecular fluids 2: applications*. Oxford University Press, 2011.
Rubinstein M. and Colby R. H., *Polymer physics*. Oxford University Press, 2003.

Problems

8.1 Based on the number of ways to place polymer chains on a lattice predicted by the Flory–Huggins theory (viz., Eq. 8.13), verify that the entropy of mixing is given by

$$\Delta S/k_B = -n_1 \ln \phi_1 - n_2 \ln \phi_2,$$

where n_1 is the number of solvent molecules, n_2 is the number of polymer chains, and $\phi_{i=1,2}$ are the corresponding volume fractions.

8.2 Using the Flory–Huggins theory for a binary mixture of a monodisperse homopolymer and a monomeric solvent, show that the spinodal line can be described by

$$\phi = \alpha \pm \sqrt{\alpha^2 - 1/(2m\chi_F)},$$

where ϕ is the polymer volume fraction, m and χ_F have their usual meanings, and $\alpha \equiv 1/2 - (1-m)/(4\chi_F)$. Derive the Flory parameter and the polymer volume fraction at the critical point of the liquid-liquid equilibrium.

8.3 To understand the critical behavior of the liquid–liquid phase separation (LLPS) in a polymer solution as discussed in Problem 8.2, it is helpful to define the Landau free energy using $\delta\phi = \phi - \phi_c$ as the order parameter (see Section 5.9)

$$f_L = \frac{a_2}{2}\delta\phi^2 + \frac{a_4}{4}\delta\phi^4 + \cdots,$$

where ϕ is the polymer volume fraction, ϕ_c is the polymer volume fraction at the critical point, and a_2 and a_4 are phenomenological parameters.

(i) Show that, near the critical point, the Landau free energy can be expressed as

$$f_L = -\delta\chi_F\delta\phi^2 + \frac{\sqrt{m}\chi_{Fc}^2}{3}\delta\phi^4 + \mathcal{O}(\delta\phi^5),$$

where $\delta\chi_F = \chi_F - \chi_{Fc}$, and m is the number of segments per polymer chain.

(ii) Verify the following scaling relation predicted by the Landau theory

$$\delta\phi \sim (\delta\chi_F/\sqrt{m})^{1/2},$$

where $\delta\phi$ represents the difference between the volume fractions of the coexisting phases.

(iii) The renormalization-group correction to the mean-field theory indicates that the difference between the coexisting densities follows the scaling relation[87]

$$\delta\phi \sim (\delta\chi_F/m)^{1/3}.$$

Compare the mean-field prediction of $\delta\phi$ versus $\delta\chi_F$ with the scaling law above for $m = 10^3, 10^4$, and 10^5.

Hint: The Landau free energy is defined as the mean-field free energy relative to the "background" contribution.

8.4 The following exercise extends the Flory–Huggins theory to a ternary liquid mixture containing homopolymer B and a co-solvent of small-molecule liquids A and C.

(i) Show that the entropy of mixing is given by

$$\Delta S/k_B = -n_A \ln \phi_A - n_B \ln \phi_B - n_C \ln \phi_C,$$

where $n_{i=A,B,C}$ are the numbers of molecules for each chemical species in the system, and $\phi_{i=A,B,C}$ refer to the volume fractions of the three chemical species.

(ii) Show that the internal energy of mixing is given by

$$\Delta\beta U/n_T = \phi_A\phi_B\chi_{AB} + \phi_B\phi_C\chi_{BC} + \phi_A\phi_C\chi_{AC},$$

where $\chi_{AB} \equiv Z(\epsilon_{A,A} + \epsilon_{B,B} - 2\epsilon_{AB})/2k_BT$ is the Flory parameter for the binary mixture of A and B, χ_{AC} and χ_{BC} have similar definitions.

8.5 At given temperature and pressure, the stability of a ternary mixture is determined by the second derivatives of the free energy density

$$D = \det\begin{pmatrix} \frac{\partial^2 f}{\partial\phi_A^2} & \frac{\partial^2 f}{\partial\phi_A\partial\phi_B} \\ \frac{\partial^2 f}{\partial\phi_A\partial\phi_B} & \frac{\partial^2 f}{\partial\phi_B^2} \end{pmatrix} > 0.$$

Based on the energy and entropy of mixing derived in Problem 8.4, show that the spinodal curve of homopolymer B in a co-solvent of small-molecule liquids A and C is described by

$$\left(\frac{1}{m_B\phi_B} + \frac{1}{\phi_C} - 2\chi_{BC}\right) \times \left(\frac{1}{\phi_A} + \frac{1}{\phi_C} - 2\chi_{AC}\right) = \left(\frac{1}{\phi_C} + \chi_{AB} - \chi_{BC} - \chi_{AC}\right)^2.$$

8.6 The Flory–Huggins theory for polymer solutions is similarly applicable to polymer blends. To elucidate, consider a binary mixture of homopolymers A and B in a rubbery state (i.e., the system may be considered as a binary liquid mixture). Assume that each chain of polymer A takes m_A lattice sites, and for polymer B each chain takes m_B sites. Based on the assumptions used in deriving the Flory–Huggins theory for polymer solutions, show the following expressions for

(i) the entropy of mixing

$$\Delta S/k_B = -n_A \ln \phi_A - n_B \ln \phi_B,$$

where n_A and n_B are the numbers of polymer chains, and ϕ_A and ϕ_B are the volume fractions of polymers A and B;

87 de Gennes P. G., *Scaling Concepts in Polymer Physics*. Cornell University Press, 1979, Page 121.

(ii) the internal energy of mixing

$$\Delta\beta U = n_T\phi_A\phi_B\chi_F,$$

where $n_T = n_A m_A + n_B m_B$ is the total number of lattice sites, and χ_F is the Flory parameter;

(iii) the Helmholtz energy of mixing

$$\Delta\beta F = n_A \ln\phi_A + n_B \ln\phi_B + n_T\phi_A\phi_B\chi_F;$$

(iv) the chemical potentials of polymer species

$$\beta\mu_A = \beta\mu_A^0 + \ln\phi_A + \phi_B(1 - m_A/m_B) + m_A\phi_B^2\chi_F,$$
$$\beta\mu_B = \beta\mu_B^0 + \ln\phi_B + \phi_A(1 - m_B/m_A) + m_B\phi_A^2\chi_F.$$

8.7 Consider of a binary mixture of homopolymers A and B. Show that, at the critical point of demixing, the polymer volume fraction and the Flory parameter are given by, respectively,

$$\phi_c = \frac{\sqrt{m_B}}{\sqrt{m_A} + \sqrt{m_B}} \quad \text{and} \quad \chi_{Fc} = \frac{1}{2}\left[\frac{1}{\sqrt{m_A}} + \frac{1}{\sqrt{m_B}}\right]^2,$$

where ϕ stands for the volume fraction of polymer A, subscript "c" stands for the critical condition.

8.8 The Sanchez–Lacombe (SL) equation of state can be considered as an extension of the Flory–Huggins theory that allows for the description of compressibility effects.[88] Like the Flory–Huggins theory, the SL equation combines the configurational entropy estimated from a lattice model and the mean-field approximation for the internal energy. However, the procedure applies only to pure species and empty sites are allowed in filling the lattice. The thermodynamic properties of polymer solutions and blends are predicted with mixing rules similar to those used in the van der Waals equation of state.

(i) Consider the number of ways to place n polymer chains onto an empty lattice such that each polymer chain occupies m sites without overlap. The total number of lattice sites is $n_T = nm + n_0$ where n_0 represents the number of empty sites. Following the Flory–Huggins theory, show that the number of ways to place the polymer chains can be approximated by

$$W = \frac{(n_0 + mn)!}{n_0!n!(n_0 + mn)^{mn}}\left[(n_0 + mn)Z^{m-1}\right]^n,$$

where Z is the coordination number.

(ii) Show that the mean-field approximation results in an internal energy of

$$U/n_T = -\frac{Z\epsilon}{2}\left(\frac{mn}{mn + n_0}\right)^2,$$

where ϵ is the energy due to the attraction of nearest-neighboring segments.

(iii) Derive the SL equation of state

$$\frac{P^*v^*}{T^*} = \frac{1}{m} - \left[1 + v^*\ln(1 - 1/v^*) + \frac{1}{v^*T^*}\right],$$

where $P^* = 2Pv_0/(Z\epsilon)$, $T^* = 2k_BT/(Z\epsilon)$, $v^* = V/(nmv_0)$, and v_0 is the volume of each lattice site.

---

88 Sanchez I. C. and Lacombe R. H., "An elementary molecular theory of classical fluids", *J. Phys. Chem.* 80, 2352 (1976).

8.9 Ruzette and Mayes proposed a modification of the Flory–Huggins theory that is able to capture the phase behavior of many weakly interacting polymer blends with the pure component data for the mass density, solubility parameter, and thermal expansion coefficient as the input.[89] One essential idea follows earlier developments by Hildebrand and by Flory who suggested that, upon mixing two *compressible* species A and B, the change in entropy should scale with the logarithm of the ratio of the "free volume" available in the mixture (blend), $V_{f,b}$, to those corresponding to the pure components, $V_{f,A}$ and $V_{f,B}$:

$$\Delta S/k_B = n_A \ln\left(\frac{V_{f,b}}{V_{f,A}}\right) + n_B \ln\left(\frac{V_{f,b}}{V_{f,B}}\right).$$

Ruzette and Mayes assumed further that the free volume can be estimated from the difference between the actual volume of polymeric species V_i and the lattice volume, i.e.,

$$V_{f,i} = V_i - m_i n_i \upsilon_i \quad \text{and} \quad V_{f,b} = V_b - (m_A n_A + m_B n_B)\upsilon_b,$$

where subscript $i = A, B$ represents pure polymer A and B, respectively; parameters m_i and n_i are the same as those defined in the Flory–Huggins theory, $\upsilon_{i=A,B}$ is the "segment" volume of polymers A and B, and $\upsilon_b = \sqrt{\upsilon_A \upsilon_B}$ is the cell volume for the lattice representing the polymer blend.

(i) Show that the modified Flory–Huggins theory predicts the entropy of mixing

$$\Delta S/k_B = -n_A \ln \phi_A - n_B \ln \phi_B + n_A \ln\left(\frac{1-\eta_b}{1-\eta_A}\right) + n_B \ln\left(\frac{1-\eta_b}{1-\eta_B}\right),$$

where $\eta_{i=A,B} = m_i n_i \upsilon_i / V_i$ and $\eta_b = (m_A n_A + m_B n_B)\upsilon_b / V_b$ are the packing fractions of the pure polymers and the blend, respectively, $\phi_{i=A,B} = V_i/V_b$ stands for the volume fraction of each polymer in the blend.

(ii) Assume that the internal energy of each pure polymer is linearly proportional to its packing fraction

$$U_{i=A,B} = -\frac{n_i m_i Z\epsilon_{ii}}{2}\eta_i = -V_i \delta_{i,0}^2 \eta_i^2,$$

where $\delta_{i,0} = \sqrt{Z\epsilon_{ii}/2\upsilon_i}$ is the solubility parameter, and assume further that the internal energy of the mixture can be approximated by the solubility model

$$U_b/V_b = \phi_A^2 C_{AA} + \phi_B^2 C_{BB} + 2\phi_A \phi_B C_{AB},$$

where $C_{ii} = -U_i/V_i$ represents the cohesive energy density, and $C_{AB} = \sqrt{C_{AA} C_{BB}}$. Show that the energy of mixing satisfies

$$\Delta U/V_b = \phi_A \phi_B (\eta_A \delta_{A,0} - \eta_B \delta_{B,0})^2.$$

8.10 Overbeek and Voorn extended the Flory–Huggins theory to describe complex coacervation in polyelectrolyte solutions.[90] To understand the liquid–liquid phase separation (LLPS) in such systems, consider an aqueous solution of polycations and polyanions of the same chain length m and degree of ionization α. Each polymer chain contains m segments, and each charged segment is monovalent, i.e., it has the valence of $|Z_{\pm}| = 1$. When the

---

89 Ruzette A. V. G. and Mayes A. M., "A simple free energy model for weakly interacting polymer blends", *Macromolecules* 34, 1894–1907 (2001).

90 Overbeek J. T. G. and Voorn M. J., "Phase separation in polyelectrolyte solutions. Theory of complex coacervation", *J. Cell. Comp. Physiol.* 49, 7–26 (1957).

polycation and polyanion have the same concentration, the polyelectrolyte solution may be considered as an *effective* binary system, i.e., a polyelectrolyte complex (viz., "ploysalt") with volume fraction $\phi = \phi_+ + \phi_-$ dissolved in a solvent of volume fraction $1 - \phi$. Overbeek and Voorn assumed that the polyion segments and solvent molecules have the same molecular volume v_0 and that the free energy density due to electrostatic interactions can be represented by the Debye–Hückel theory (see Section 9.4)

$$\frac{f_{DH}}{k_B T} = -\frac{\kappa^3}{12\pi}.$$

The Debye screening parameter is defined as

$$\kappa = \sqrt{4\pi l_B(\alpha\rho_+ Z_+^2 + \alpha\rho_- Z_-^2)} = \sqrt{4\pi l_B \alpha\phi/v_0},$$

where $l_B \equiv e^2/(4\pi\epsilon\epsilon_0 k_B T)$ stands for the Bjerrum length, α is the degree of ionization (i.e., the fraction of charged polymer segments), $\rho_+ = \rho_-$ are the number densities of polymer segments. Because the van der Waals interactions are much weaker than electrostatic interactions, Overbeek and Voorn neglected the internal energy term related to the Flory parameter χ_F.

(i) Show that the Overbeek–Voorn theory predicts the free energy of mixing per lattice site

$$\frac{\Delta f v_0}{k_B T} = \frac{\phi}{m}\ln(\phi/2) + (1-\phi)\ln(1-\phi) - \lambda(\alpha\phi)^{3/2},$$

where $\lambda \equiv 2\sqrt{\pi/v_0}l_B^{3/2}/3$, and v_0 is the molecular volume of the solvent.[91]

(ii) Show that the chemical potentials of the polymer complex and solvent in the effective binary mixture are given by

$$\beta\mu_p = \beta\mu_p^0 + \ln(\phi/2) + (1-m)(1-\phi) - \frac{\lambda\alpha^{3/2}m\sqrt{\phi}}{2}(3-\phi),$$

$$\beta\mu_s = \beta\mu_s^0 + \ln(1-\phi) + (1-1/m)\phi + \frac{\lambda\alpha^{3/2}\phi^{3/2}}{2},$$

where subscript 0 denotes the reference system of pure species.

(iii) At the critical point of LLPS, the degree of ionization α_c and the polymer volume fraction ϕ_c can be approximated by the following scaling relations

$$\phi_c \approx 1/(m+3) \quad \text{and} \quad \alpha_c \approx 0.52/m^{1/3}.$$

(iv) Plot the spinodal line of the system for a few representative values of the polymer chain length m and discuss the phase behavior.

(v) Show that, at $m = \infty$, the coacervation boundary can be determined from

$$\alpha = \frac{1}{\phi}\left[\frac{-2\ln(1-\phi) - 2\phi}{\lambda}\right]^{3/2}.$$

For small ϕ, the above equation can be approximated by $\alpha \approx 0.27\phi^{1/3}$.

(vi) Plot the phase diagram for the LLPS predicted by the Overbeek–Voorn theory for polymer chain length $m = 2, 5, 10, 1000, 3000$, and ∞.

8.11 Consider a polyelectrolyte solution with n_i polyionic chains of species i each with $m_i = v_i/v_0$ segments where v_i is the volume of the polyion and v_0 is the molecular volume of the solvent.

---

91 For water at room temperature ($T = 298.15$ K), $l_B \approx 7.14$ Å, the molecular volume for water is $v_0 \approx 10\,\text{Å}^3$, and thus $\lambda \approx 7.13$.

(i) Using the Flory–Huggins theory for the entropy of mixing and the Debye–Hückel theory for the electrostatic free energy, show that the Overbeek–Voorn theory (Problem 8.10) predicts the free energy of mixing per lattice site

$$\frac{\Delta f v_0}{k_B T} = \sum_i \frac{\phi_i}{m_i} \ln(\phi_i) + \left(1 - \sum_i \phi_i\right) \ln\left(1 - \sum_i \phi_i\right) - \lambda\left(\sum_i \alpha_i \phi_i\right)^{3/2},$$

where $\lambda \equiv 2\sqrt{\pi / v_0} l_B^{3/2}/3$, and α_i is the degree of ionization of polymer i.

(ii) Derive analytical expressions for the chemical potential of each ionic species and the chemical potential of the solvent.

(iii) What are conditions of equilibrium for the LLPS in this system?

8.12 The Flory–Huggins theory is useful to predict the solubility of crystalline drugs in a polymer matrix. Such calculations are often carried out for the polymer materials design to improve bioavailability.

(i) Show that, for a given mixture, the Flory parameter can be estimated from experimental data for the melting point depression (MPD) of the drug dissolved in the polymer. The experimental data can be measured by using differential scanning calorimetry.

(ii) Sketch how the drug solubility varies with temperature.

(iii) Assume that the glass transition temperature of the binary mixture varies linearly with the polymer composition, sketch a temperature-solubility "phase diagram" that may be used as a guide for the formulation of drug delivery systems.

8.13 A modified Flory–Huggins theory was utilized to predict the association of telechelic polymers (viz., polymers with associating ends) in solution and help the development of polymer-based mist suppressors for jet fuels.[92] The modification follows an earlier theoretical work proposed by Goldstein to account for the association among polymer chains.[93] The mixing free energy per lattice site consists of contributions from the solvent molecules, polymer monomers, as well as all possible multi-chain aggregates

$$\frac{\Delta f v_0}{k_B T} = (1-\phi)\ln(1-\phi) + \sum_i \frac{\phi_i}{m_i} \ln \phi_i + \frac{\beta Z}{2}\left[(1-\phi)^2 f_{ss} + \phi^2 f_{pp} + 2\phi(1-\phi) f_{ps}\right],$$

where ϕ_i represents the volume fraction of polymer aggregate with $i = 1, 2, \ldots$ chains, $m_i = i \times m$ is the number of lattice sites occupied by each polymer aggregate, $\phi = \sum_i \phi_i$ is the total polymer volume fraction, Z is the coordination number of the lattice, and f_{ij} are energies associated with the interactions of various species in the system.

(i) Show that the chemical potentials of the solvent and i-chain aggregates are, respectively,

$$\mu_s = \mu_s^0 + k_B T\left[\ln(1-\phi) + \phi - \sum_i \frac{\phi_i}{m_i}\right] - w\phi^2,$$

$$\mu_i = \mu_i^0 + k_B T\left[\ln \phi_i - (1-\phi)m_i + 1 - m_i \sum_j \frac{\phi_j}{m_j}\right] - m_i w(1-\phi)^2,$$

where superscript 0 denotes a reference system, $w \equiv w_{pp} + w_{ss} - 2w_{ps}$ with $w_{ij} = Z f_{ij}/2$.

92 Wei M. H. et al., "Megasupramolecules for safer, cleaner fuel by end association of long telechelic polymers", *Science* 350, 72 (2015).
93 Goldstein R. E., "Model for phase equilibria in micellar solutions of nonionic surfactants", *J. Chem. Phys.* 84, 3367 (1986).

(ii) The size distribution of multi-chain structures are determined by the condition of chemical equilibrium $\mu_i = i\mu_1$. Show that the mass-action relation leads to

$$\phi_i = \phi_1^i e^{-\beta \Delta G_i^0},$$

where $\Delta G_i^0 \equiv \mu_i^0 - i\mu_1^0 - (i-1)k_B T$ corresponds to the free energy of association.

(iii) The free energy of association proposed by Goldstein includes contributions due to the polymer-polymer association, the contact energy between the polymer aggregates and the solvent, and the change in polymer configuration due to the formation of aggregates:

$$\beta \Delta G_i^0 = -\ln(im) + i \ln m + i\left[m\epsilon + \sigma(m^2/i)^{1/3} + c(i^2/m)^{1/3} - 3/2\right],$$

where terms $\ln(im)$ and $\ln m$ arise from the change of the concentration units from volume fraction to mole fraction, ϵ represents the reduced association energy, σ and c are model parameters related to the aggregate-solvent surface free energy and the geometry of polymer packing, respectively. Show that the concentration of aggregates with i polymer chains satisfies

$$\phi_i = im\left(\frac{\phi_1 e^{-\delta_0}}{m}\right)^i,$$

where $\delta_0 = m\epsilon + \sigma(m^2/i)^{1/3} + c(i^2/m)^{1/3} - 3/2$.

(iv) Using parameters $\sigma \approx 3.0$ for the reduced surface free energy and $c \approx 1.0$ for the geometric factor, plot the size distribution in terms of ϕ_i for $m = 20$ and $\phi_1 = 0.004$ at several representative values of the reduced association energy ($\epsilon \sim -1.0$). How does the most probably aggregate size change with the reduced association energy ϵ and with the polymer chain length m?

8.14 The conventional thermodynamic description of polymer gels is mostly based on the Flory–Rehner model, which is an extension of the Flory–Huggins theory by incorporating chain elasticity and crosslink distribution (Figure P8.14). To get the essential ideas, consider a polymer gel made of N chains each with m segments. The polymer gel is surrounded by a bulk solvent with fixed temperature T and bulk chemical potential μ_s. The polymer and solvent inside the gel constitute a semi-grand canonical ensemble with the free energy depending on temperature and solvent chemical potential, i.e., $\mathcal{F} = \mathcal{F}(N, T, \mu_s)$.

(i) If the polymer gel could be treated as a single chain (viz., a super-chain with mN segments), show that the free energy of mixing would become

$$\Delta \beta F_0 = N_s \ln \phi_s + \chi_F (N_s + mN)\phi_s \phi,$$

where N_s is the number of solvent molecules inside the gel, $\phi_s = N_s/(N_s + mN)$ and $\phi = mN/(N_s + mN)$ are the solvent and polymer volume fractions, respectively. The Flory parameter χ_F accounts for the mean-field energy due to the polymer-solvent interactions.

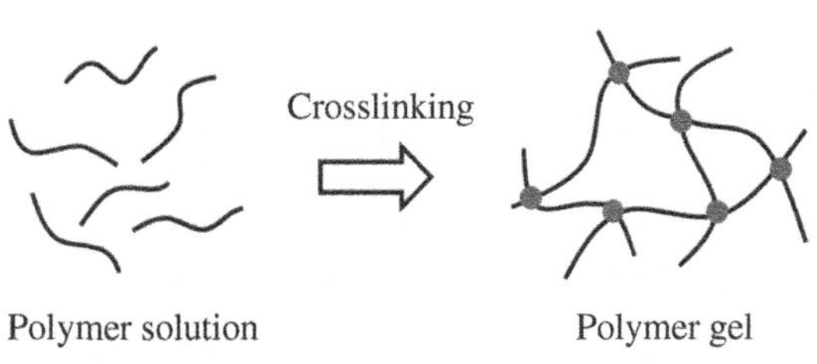

Figure P8.14 A polymer gel can be modeled as a network of N polymer chains permanently joined by $N/2$ crosslinks (black dots).

(ii) Assuming that the polymer gel consists of a random network of N Gaussian chains, show that the change in the elastic energy in response to the gel swelling from a fully relaxed state with volume V_0 (which is achieved at the preparation condition) to volume V in the solvent can be written as

$$\Delta\beta F_{elas} = \frac{N}{2}(3\alpha^{2/3} - 3 - \ln \alpha),$$

where $\alpha = V/V_0$ denotes the swelling ratio.

(iii) As the boundary of the polymer network may serve as a semi-permeable membrane, validate the following expression for the osmotic pressure inside the gel based on the osmotic equilibrium

$$\Pi(T, \phi) = \frac{k_B T}{v_s}\left\{\frac{\phi_0}{m}\left[\frac{\phi}{2\phi_0} - \left(\frac{\phi}{\phi_0}\right)^{1/3}\right] - \phi - \ln(1-\phi) - \chi_F\phi^2\right\},$$

where v_s represents the molecular volume of the solvent, and $\phi_0 = \alpha\phi$ is the polymer volume fraction in the gel at the preparation condition.

8.15 Poly(N-isopropylacrylamide) (PNIPAM) microgel particles may undergo a drastic volume change in liquid water when the temperature varies near 33 °C, the lower critical solution temperature (LCST) of the aqueous solutions of PNIPAM. As shown in Figure P8.15, the variation in the microgel radius can be quantitatively described by a modified Flory–Rehner model

$$\beta\Pi(T, \phi) = \frac{1}{v_s}\left\{\frac{\phi_0}{m}\left[\left(\frac{\phi}{\phi_0}\right)^{1/3} - \left(\frac{\phi}{\phi_0}\right)^{5/3} + \frac{\phi}{2\phi_0}\right] - \phi - \ln(1-\phi) - \chi_F\phi^2\right\},$$

where the Flory parameter is empirically given as a function of absolute temperature T (in Kelvin) and polymer volume fraction ϕ by fitting the phase diagram of PNIPAM aqueous solutions

$$\chi_F = \frac{3}{1-0.65\phi}\left\{2\ln\left[\frac{5001}{1+5000\exp(-2458.867/T)}\right] - \frac{4566.468}{T}\right\}.$$

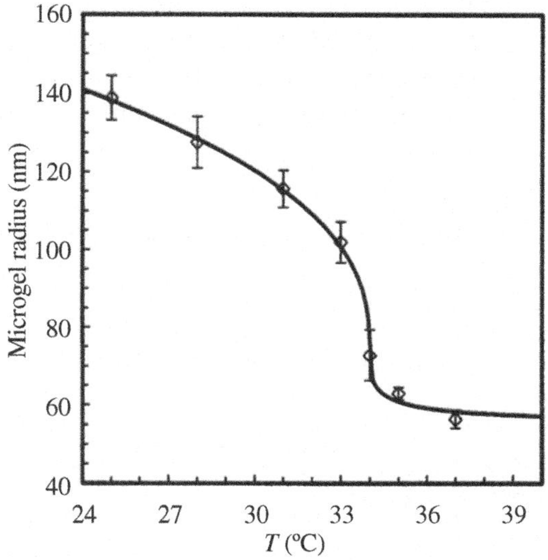

Figure P8.15 Radii of PNIPAM particles in liquid water versus temperature. The symbols are experimental data with the error bars reflecting the difference between dynamic and static light scattering measurements. The line is the prediction of a modified Flory–Rehner model. Source: Adapted from Wu et al.[94]

Reproduce Figure P8.15 using the parameters from the published work,[94] i.e., the polymer fraction at the condition of preparation $\phi_0 = 0.0884$, the average number of segments between neighboring cross-linking points $m = 34$, and the particle radius at the preparation condition $R_0 = 125.8$ nm. Why is the original Flory–Rehner model problematic to fit the experimental results for PNIPAM microgel particles?

8.16 The canonical partition function for a one-component system of nonlinear rigid polymeric molecules can be written as

$$Q = \frac{Q^{IG}}{V^N \Theta^N} \int d\mathbf{r}^N \int d\varpi^N \exp[-\beta\Phi(\mathbf{r}^N, \varpi^N)],$$

where Q^{IG} is the ideal-gas partition function, $\Theta = 8\pi^2$, and $\Phi(\mathbf{r}^N, \varpi^N)$ is the total potential energy. Using the pairwise additivity assumption for the potential energy and thermodynamic relations

$$U = \left(\frac{\partial \beta F}{\partial \beta}\right)_{V,N} \quad \text{and} \quad P = -\left(\frac{\partial F}{\partial V}\right)_{T,N},$$

derive the following equations that relate the internal energy and pressure to the radial distribution function

$$U = U^{IG} + \frac{N^2}{2V} \int d\mathbf{r} \langle u(r, \varpi_1, \varpi_2) g(r, \varpi_1, \varpi_2) \rangle_{\varpi_1, \varpi_2},$$

$$P = \rho k_B T - \frac{\rho^2}{6} \int d\mathbf{r} r \langle u'(r, \varpi_1, \varpi_2) g(r, \varpi_1, \varpi_2) \rangle_{\varpi_1, \varpi_2},$$

where $\rho = N/V$ is the number density of the polymeric molecules, $u(r, \varpi_1, \varpi_2)$ denotes the pair potential, the prime sign means derivative with respect to distance, and $\langle \cdots \rangle$ means average over the Euler angles.

8.17 Starting from the grand-canonical ensemble for one-component polymeric molecules

$$\Xi = \sum_{N=0}^{\infty} \frac{1}{N! \Lambda^N} \int d\mathbf{R}^N \exp\left[-\beta\Phi(\mathbf{R}^N) + N\beta\mu\right],$$

where $\mathbf{R} \equiv (\mathbf{r}_1, \mathbf{r}_2, \ldots, \mathbf{r}_m)$ stands for the molecular configuration, $\Phi(\mathbf{R}^N)$ is the total potential energy (including both intra- and inter-molecular interactions), and other symbols have their usual meanings, show the following statistical-mechanical equations:

(i) the average number of molecules in the system

$$\langle N \rangle = \frac{\partial \ln \Xi}{\partial \beta\mu};$$

(ii) the mean-square deviation in the average number of molecules

$$\langle N^2 \rangle - \langle N \rangle^2 = \frac{\partial \langle N \rangle}{\partial \beta\mu};$$

(iii) the compressibility equation

$$\left(\frac{\partial \rho}{\partial \beta P}\right)_T = 1 + \frac{1}{\langle N \rangle} \int d\mathbf{R} \int d\mathbf{R}' \rho(\mathbf{R}) \rho(\mathbf{R}') h(\mathbf{R}, \mathbf{R}');$$

94 Wu J., Huang G. and Hu Z., "Interparticle potential and the phase behavior of temperature-sensitive microgel dispersions", *Macromolecules* 36, 440–448 (2003).

(iv) for rigid molecules, the compressibility equation reduces to

$$\left(\frac{\partial\rho}{\partial\beta P}\right)_T = 1 + \rho\int d\mathbf{r}\langle h(|\mathbf{r}-\mathbf{r}'|,\varpi,\varpi')\rangle_{\varpi,\varpi'};$$

(v) the compressibility equation can be expressed in terms of any pair of the site-site total correlation function

$$\left(\frac{\partial\rho}{\partial\beta P}\right)_T = 1 + \rho\int d\mathbf{r}h_{ij}(r),$$

or the Fourier transform of the density-density correlation function $\chi_{ij}(r)$ at $k = 0$

$$\left(\frac{\partial\rho}{\partial\beta P}\right)_T = \hat{\chi}_{ij}(0)/\rho.$$

8.18 The following exercise may help better understand the dielectric properties of materials from a molecular perspective. According to macroscopic electrostatics, the electric potential due to a charge Q in a continuous medium can be described by Coulomb's law

$$\psi(r) = \frac{Q}{4\pi\varepsilon_0\varepsilon r},$$

where ε_0 is the permittivity of free space, ε is the dielectric constant, and r is the distance from the charge center. Conversely, from a molecular perspective, the electric potential may be decomposed into three terms

$$\psi(r) = \frac{Q}{4\pi\varepsilon_0 r} + \psi_c(r) + \psi_{ind}(r),$$

where the first term on the right corresponds to the electric potential in a vacuum, $\psi_c(r)$ arises from the partial charges of individual atoms in the dielectric medium, and $\psi_{ind}(r)$ represents an additional contribution due to atomic polarization. With $\psi_c(r)$ and $\psi_{ind}(r)$ expressed in terms of the atomic densities $\rho_i(r)$, a comparison of the continuous and atomistic descriptions of the electric potential allows us to predict the dielectric constant ε from atomic charge q_i and polarizability α_i. Show that, as $Q \to 0$,

(i) the atomic density near charge Q can be predicted from

$$\rho_i(\mathbf{r}) = \rho_i^b - \sum_j \frac{\beta Q q_j}{4\pi\varepsilon_0}\int d\mathbf{r}'\frac{\chi_{ij}(|\mathbf{r}-\mathbf{r}'|)}{|\mathbf{r}'|},$$

where ρ_i^b is the number density of atomic particle i in the uniform system (viz., the continuous medium in the absence of charge Q), and $\chi_{ij}(r)$ is the density-density correlation function for the atomic pair of i and j;

(ii) the local electric potential due to the redistribution of atomic particles in the dielectric medium is given by

$$\psi_c(r) = -\sum_{i,j}\frac{\beta Q q_i q_j}{(4\pi\varepsilon_0)^2}\int d\mathbf{r}'\int d\mathbf{r}''\frac{\chi_{ij}(|\mathbf{r}'-\mathbf{r}''|)}{|\mathbf{r}''||\mathbf{r}-\mathbf{r}'|};$$

(iii) the atomic polarization contributes an additional electric potential at distance r

$$\psi_{ind}(r) = \sum_i\int d\mathbf{r}'\mathbf{P}_i(\mathbf{r}')\cdot\frac{(\mathbf{r}-\mathbf{r}')}{4\pi\varepsilon_0|\mathbf{r}-\mathbf{r}'|^3},$$

where $\mathbf{P}_i(\mathbf{r}') = \alpha_i\rho_i(\mathbf{r}')\mathbf{E}(\mathbf{r}')$ stands for the local density of point dipoles due to the polarization of atomic particle i, and $\mathbf{E}(\mathbf{r}')$ is the local electric field;

(iv) the local electric field is coupled with the atomic density and the local dipole density

$$\mathbf{E}(\mathbf{r}) = Q\mathbf{r}/(4\pi\varepsilon_0 r^3) - \nabla\psi_c(r) - \sum_i \int d\mathbf{r}'\mathbf{P}_i(\mathbf{r}')\cdot\mathbb{T}(\mathbf{r}-\mathbf{r}')/(4\pi\varepsilon_0),$$

where $\mathbb{T} = \mathbb{I}/r^3 - 3\mathbf{rr}/r^5$ stands for the dipole field tensor, and $\mathbb{I}$ denotes the unit tensor;

(v) without atomic polarization (viz., $\alpha_i = 0$), the dielectric constant can be predicted from

$$\varepsilon^{-1} = 1 + \frac{2\pi\beta}{3\varepsilon_0}\sum_{i,j} q_i q_j \int_0^\infty drr^4\chi_{ij}(r);$$

(vi) if the dielectric medium consists of nonpolarizable molecules with a rigid conformation, the dielectric constant satisfies the following equation

$$(1 - 1/\varepsilon)\varepsilon_0/\beta = \frac{\rho_b d^2}{3} + \rho_b^2\Delta\hat{h}^{(2)},$$

where d denotes the molecular dipole moment, ρ_b represents the number density of molecules in the bulk, $\Delta\hat{h}^{(2)} \equiv \sum_{i,j} q_i q_j \hat{h}_{ij}^{(2)}$, and $\hat{h}_{ij}^{(2)}$ is defined by the Taylor expansion of the total correlation function in the Fourier space, i.e., $\hat{h}_{ij}(k) = \hat{h}_{ij}(0) + \hat{h}_{ij}^{(2)}k^2 + \cdots$;

(vii) for $\rho_b \to 0$, the dielectric constant approaches the ideal-gas expression $\varepsilon = 1 + \beta\rho_b d^2/(3\varepsilon_0)$.

Hint: The density-density correlation function may be understood in terms of the change in local atomic density $\rho_i(\mathbf{r})$ in response to an external potential $V_j^{ext}(\mathbf{r}')$

$$\chi_{ij}(|\mathbf{r} - \mathbf{r}'|) \equiv \langle\delta\rho_i(\mathbf{r})\delta\rho_j(\mathbf{r}')\rangle = -\delta\rho_i(\mathbf{r})/\delta\beta V_j^{ext}(\mathbf{r}').$$

Problem 8.23 shows a derivation of the above relation. In a molecular system, the density-density correlation function is related to the intramolecular and total correlation functions, i.e., $\chi_{ij}(r) = \rho_b\omega_{ij}(r) + \rho_b^2 h_{ij}(r)$.

8.19 Consider association between two spherical particles to form a dimer in a vacuum and that in a fluid. Figure P8.19 shows schematically the association processes. Prove that, at fixed temperature and pressure of the fluid phase, the difference between the excess chemical potential of the dimer and that of the monomers is related to the cavity correlation function between the two associating particles in the fluid

$$\mu_2^{ex} - 2\mu^{ex} = -k_B T \ln y(L),$$

where L is the center-to-center distance between the two particles in the dimer, and subscript "2" denotes the dimer. How would the above equation be extended to the formation of an m-mer in the fluid phase?

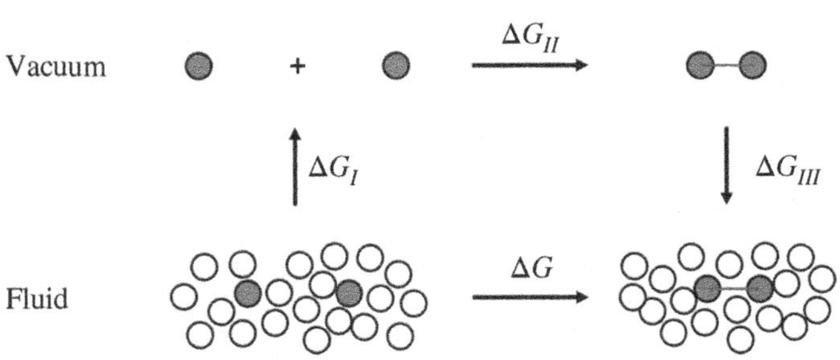

Figure P8.19 A thermodynamic cycle for association between two particles in a vacuum and that in a uniform fluid. The open cycles represent fluid molecules.

8.20 Consider association between monomeric species A and B to form dimer AB in a condensed phase at given temperature and pressure, $A + B \rightleftharpoons AB$. Show that the (apparent) equilibrium constant $K \equiv \rho_{AB}/(\rho_A \rho_B)$ can be expressed in terms of the cavity correlation function

$$K/K_0 = y_{AB}(L),$$

where L is the center-to-center distance between A and B in dimer AB, $y_{AB}(r)$ is the cavity correlation function of particles A and B in the condensed phase, and $K_0 \equiv \exp(-\beta \Delta G^0_{rxn})$ is the thermodynamic equilibrium constant defined by

$$\Delta G^0_{rxn} = \mu^0_{AB} - \mu^0_A - \mu^0_B,$$

where $\mu^0_{i=A,B,AB}$ is the chemical potential of pure i as an ideal gas at the system temperature and unit density ρ^0_i.

8.21 The second virial coefficient for hard-sphere chains may be understood in the context of polymers in a good solvent. According to the Flory–Huggins theory, the osmotic second virial coefficient is $A_2 = 1/2$ when the Flory parameter $\chi_F = 0$.

(i) In statistical mechanics, the second virial coefficient is conventionally defined through $Z = 1 + B_2\rho_m + \cdots$, where Z denotes the osmotic compressibility factor, and ρ_m is the number density of polymer chains. Show that the osmotic second virial coefficient A_2 defined in the Flory–Huggins theory is related to its conventional form

$$A_2 = B_2/(v_s m^2),$$

where v_s is the volume of a solvent molecule (viz., volume per lattice site), and m is the number of polymer segments.

(ii) Compare the virial coefficient predicted by the Flory–Huggins theory with that from the equations of state for hard-sphere chains discussed in Section 8.4 in the limit of large m

$$B_2/(m^2\sigma^3) \approx k_1,$$

where σ is the segment diameter, and k_1 is a constant between 0.38 and 1.33.

(iii) In a good solvent, polymer chains repel each other much like two hard spheres of radius R_g, where R_g stands for the polymer radius of gyration. Show that the conventional second virial coefficient may be estimated from

$$B_2 \approx \frac{16\pi R_g^3}{3}.$$

(iv) Scaling analysis indicates that for a polymer chain in a good solvent, $R_g \approx b m^{\nu}_F$, where b is the Kuhn length, and $\nu_F \approx 0.588$ is the Flory exponent. Show that the scaling analysis is consistent with experimental observation $A_2 \sim m^{-0.236}$.

8.22 Müller and Gubbins[95] derived the following equation of state for tangent hard-sphere chains based on the second-order thermodynamic perturbation theory (TPT2)

$$Z \equiv \frac{P}{\rho k_B T} = m\frac{1 + \eta + \eta^2 - \eta^3}{(1-\eta)^3} + (1-m)\frac{2 + 2\eta - \eta^2}{(2-\eta)(1-\eta)} + Z_2, \quad \text{(M)}$$

95 Müller E. A. and Gubbins K. E., "Simulation of hard triatomic and tetratomic molecules", *Mol. Phys.*, 80(4), 957–976 (1993).

where ρ stands for the number density of hard-sphere chains, $\eta = \pi m\rho\sigma^3/6$ is the segment packing fraction, σ is the hard-sphere diameter, and m is the number of segments per chain. The second-order term is given by

$$Z_2 = \frac{4\lambda - m(1+4\lambda) + [m^2(1+4\lambda) - 4\lambda]^{1/2}}{2\lambda(1+4\lambda)}\eta\left(\frac{\partial\lambda}{\partial\eta}\right)$$

with $\lambda = 0.233633\eta(1 + 0.472\eta)$.

(i) Show that Eq. (M) reduces to the ideal-gas law at the low density of hard-sphere chains;
(ii) Derive the second virial coefficient from Eq. (M) and compare it with the scaling relation $B_2/(m^2\sigma^3) \sim m^{-0.236}$ obtained from scaling analysis;
(iii) Show that the compressibility factor is linearly proportional to m as $m \to \infty$;
(iv) Compare the compressibility factor predicted by TPT1 and that by TPT2 for $m = 2000$ and discuss why the second-order term is relatively insignificant for tangent hard-sphere chains.

8.23 The grand canonical partition function for a binary mixture of homopolymers A and B can be written as

$$\Xi = \sum_{\nu} \exp\{-\beta E_{\nu} - \int d\mathbf{r}[\beta V_A^{ext}(\mathbf{r})\tilde{\rho}_A(\mathbf{r}) + \beta V_B^{ext}(\mathbf{r})\tilde{\rho}_B(\mathbf{r})]\},$$

where E_ν stands for the potential energy of the polymer chains due to inter- and intra-chain interactions, $V_{i=A,B}^{ext}(\mathbf{r})$ is a one-body external potential applying to individual polymer segments, $\tilde{\rho}_{i=A,B}(\mathbf{r})$ is the instantaneous density of polymer segments, and the summation over microstates is a shorthand notation for

$$\sum_{\nu} = \sum_{N_A=0}^{\infty}\sum_{N_B=0}^{\infty} \frac{e^{\beta(N_A\mu_A + N_B\mu_B)}}{N_A!N_B!\Lambda^{3N_A}\Lambda^{3N_B}} \int \mathbf{R}_A^{N_A} \int \mathbf{R}_B^{N_B}$$

with all variables following their usual meanings.

(i) Show that the segment densities are given by

$$\rho_i(\mathbf{r}) = \langle\tilde{\rho}_i(\mathbf{r})\rangle = \frac{\delta\beta\Omega}{\delta\beta V_i^{ext}(\mathbf{r})},$$

where $\beta\Omega = -\ln\Xi$.

(ii) Show that the density-density correlation function is related to the response of the segment density with respect to the one-body potential

$$\chi_{ij}(\mathbf{r},\mathbf{r}') \equiv \langle[\rho_i(\mathbf{r}) - \tilde{\rho}_i(\mathbf{r})][\rho_j(\mathbf{r}') - \tilde{\rho}_j(\mathbf{r}')]\rangle = -\frac{\delta\rho_i(\mathbf{r})}{\delta\beta V_j^{ext}(\mathbf{r})}.$$

8.24 As discussed in Section 7.3, the static structure factor of a thermodynamic system can be determined from small-angle neutron scattering (SANS) or small-angle X-ray scattering (SAXS) measurements. In their application to multi-component systems, the intensity of the scattered wave at the detector is given by

$$I(k) = \sum_{ij} \sigma_i\sigma_j\hat{\chi}_{ij}(k), \tag{N}$$

where σ_i is atomic scattering factor (i.e., a parameter related to the beam interaction with the atomic scattering point i), and $\hat{\chi}_{ij}(k)$ is the Fourier transform of the density-density correlation function between atomic sites i and j. For a polymer melt with only two types of segments A and B, the assumption of incompressibility leads to a simplified expression for the scattering intensity

$$I(k) = k_\sigma \hat{\chi}(k). \tag{O}$$

(i) Derive Eq. (O) from Eq. (N) and relate $\hat{\chi}(k)$ and k_σ to $\hat{\chi}_{ij}(k)$ and σ_i;

(ii) Show that the random-phase approximation (RPA) predicts the scattering intensity for a binary mixture of A/B homopolymers

$$I(k) = k_\sigma \rho_t \left\{ \frac{1}{\phi_A m_A f(kR_{g,A})} + \frac{1}{\phi_B m_B f(kR_{g,B})} - 2\chi_F \right\}^{-1},$$

where ρ_t is the total number density of polymer segments, ϕ_i and m_i are the volume fraction and the degree of polymerization for polymer i, χ_F is the Flory parameter, R_g stands for the radius of gyration, and $f(x) = 2(e^{-x^2} - 1 + x^2)/x^4$ is the Debye function.

(iii) Show that in the high-k limit, the scattering intensity can be approximated by

$$I(k) \approx \frac{12k_\sigma \rho_t \phi_A \phi_B}{k^2 b^2},$$

where b is the Kuhn length for both polymers, and in the low-k limit,

$$I(k) \approx \frac{k_\sigma \rho_t}{2(\chi_{F,s} - \chi_F)},$$

where $\chi_{F,s}$ is the value of χ_F at the spinodal condition.

8.25 The density-density correlation function derived from the random-phase approximation (RPA) can be utilized to account for correlation effects ignored in a typical mean-field theory. To elucidate, consider a binary mixture of A/B homopolymers as discussed in Section 8.6. It can be shown from functional integration that the free-energy density due to the density-density correlation can be calculated from

$$f_{corr} = \frac{1}{2}\sum_{ij} \int_0^1 d\lambda \int \frac{d\mathbf{k}}{(2\pi)^3} \hat{u}_{ij}(k)\hat{\chi}_{ij}(k, \lambda), \tag{P}$$

where $\hat{u}_{ij}(k)$ represents the Fourier transform of the pair potential between segments $u_{ij}(r)$. Using the response function predicted by RPA

$$\hat{\chi}_{ij}^{RPA}(k, \lambda) = [\hat{\chi}_0^{-1}(k) + \lambda \hat{\mathbf{u}}]_{ij}^{-1},$$

show that the correlation effect contributes to a free-energy density given by

$$f_{corr}^{RPA} = \frac{1}{2}\int \frac{d\mathbf{k}}{(2\pi)^3} \mathrm{Tr}\ln[\mathbf{I} + \hat{\chi}_0(k) \cdot \hat{\mathbf{u}}(k)],$$

where Tr stands for the matrix trace, and $\mathbf{I}$ is a 2×2 matrix.
Hint: The double summation with respect to segment indices i and j in Eq. (P) can be written as $\mathrm{Tr}(\hat{\mathbf{u}}^T \cdot \hat{\chi})$ and $\hat{\mathbf{u}}^T = \hat{\mathbf{u}}$.

8.26 The random-phase approximation (RPA) is often used to interpret experimental data from X-ray/neutron scattering measurements of block copolymer systems.[96] As discussed

96 Ronald J. L. et al. "Chain conformation in ultrathin polymer films", *Nature* 400, 146–149 (1999).

in Section 8.6, the density-density correlation functions among polymer segments are defined as

$$\chi_{ij}(\mathbf{r}, \mathbf{r}') = \langle \delta\tilde{\rho}_i(\mathbf{r})\delta\tilde{\rho}_j(\mathbf{r}')\rangle,$$

where i and j denote the types of polymer segments, $\delta\tilde{\rho}_i(\mathbf{r}) \equiv \tilde{\rho}_i(\mathbf{r}) - \rho_i$, $\tilde{\rho}_i(\mathbf{r})$ and ρ_i are the instantaneous and average number densities of segment i, respectively.

(i) Show that, for a diblock copolymer of Gaussian chains, the Fourier transforms of the response functions are linearly proportional to those of the intra-chain correlation functions (Problem 3.33)

$$\hat{\chi}^0_{ij}(k) = \frac{Nm_im_j}{V}\hat{\omega}_{ij}(k),$$

where N is the number of polymer chains in the system, V is the total volume, m_i and m_j represent the number of Kuhn segments in block i and block j, respectively.

(ii) In the presence of a weak external field $V_i^{ext}(\mathbf{r})$, RPA assumes a linear response of the local segment densities

$$\delta\hat{\rho}_i(\mathbf{k}) = -\sum_j \beta\hat{\chi}_{ij}(k)\hat{V}_j^{ext}(\mathbf{k}) \approx -\sum_j \beta\hat{\chi}^0_{ij}(k)[\hat{V}_j^{ext}(\mathbf{k}) + \hat{u}_j^{MF}(\mathbf{k})],$$

where $\hat{\chi}^0_{ij}(k)$ corresponds to that of the diblock Gaussian chains, and the one-body mean-field potentials are determined self-consistently from the density deviations

$$\hat{u}_i^{MF}(\mathbf{k}) = -\frac{Z}{\rho_t}\sum_{j=1}^{2}\epsilon_{ij}\delta\hat{\rho}_j(\mathbf{k}),$$

where $\rho_t = \rho_1 + \rho_2$ represents the total segment density, and Z is the coordination number. Show that, with the assumption of incomprehensibility, the density-density correlation function of a diblock copolymer melt can be calculated from

$$\hat{\chi}(\mathbf{k}) = \left\{\frac{\hat{\chi}^0_{11}(k) + 2\hat{\chi}^0_{12}(k) + \hat{\chi}^0_{22}(k)}{\hat{\chi}^0_{11}(k)\hat{\chi}^0_{22}(k) - [\hat{\chi}^0_{21}(k)]^2} - \frac{2\chi_F}{\rho_t}\right\}^{-1},$$

where $\hat{\chi}(\mathbf{k}) = \hat{\chi}_{11}(\mathbf{k}) = \hat{\chi}_{22}(\mathbf{k}) = -\hat{\chi}_{12}(\mathbf{k}) = -\hat{\chi}_{21}(\mathbf{k})$ and $\chi_F = Z\beta(\epsilon_{11} + \epsilon_{22} - 2\epsilon_{12})/2$ is the Flory parameter.

(iii) Plot the structure factor of a symmetric diblock polymer, $\hat{S}(k) = \hat{\chi}(k)/\rho_t$, versus k with $m = 1000, 2000$, and 5000 Kuhn segments per chain. Assume the Kuhn length $b = 0.5$ nm and the Flory parameter $\chi_F = 0$ (e.g., segments differ by isotope replacement of certain elements).

(iv) Assume $\chi_F = 0$, derive the asymptotic limit of $\hat{S}(k)$ as $k \to 0$ and that as $k \to \infty$.

8.27 This exercise applies the field theoretical method to an idealized system of electrolytes where the ionic species are described as point charges and the solvent is represented by a dielectric medium. Consider an ionic system with N_+ cations and N_- anions of equal valency $Z = Z_+ = -Z_-$. Assume that the interaction between unit charges can be represented by the reduced Coulomb potential

$$v_c(r) = l_B/r,$$

where r is the charge-to-charge distance, and $l_B = \beta e^2/(4\pi\varepsilon\varepsilon_0)$ is the Bjerrum length. As usual, $\beta = 1/(k_BT)$, e is the unit charge, ε is the dielectric constant of the medium, ε_0 is the permittivity of the free space. In the presence of a background charge with a fixed local density for the number of unit charges $\rho_c^0(\mathbf{r})$, the canonical partition function is given by

$$Q = \frac{1}{N_+!N_-!\Lambda^{3N}} \int d\mathbf{r}^N \exp\left\{-\frac{1}{2}\int d\mathbf{r}\int d\mathbf{r}'\tilde{\rho}_c(\mathbf{r})v_c(|\mathbf{r}-\mathbf{r}'|)\tilde{\rho}_c(\mathbf{r}') + NZ^2 v_c(0)\right\}, \tag{W}$$

where $N = (N_+ + N_-)$ is the total number of ions in the system, Λ is the thermal wavelength of cations and anions, and $\tilde{\rho}_c(\mathbf{r})$ represents the local charge density

$$\tilde{\rho}_c(\mathbf{r}) \equiv \rho_c^0(\mathbf{r}) + Z_+\tilde{\rho}_+(\mathbf{r}) + Z_-\tilde{\rho}_-(\mathbf{r}),$$

where $\rho_c^0(\mathbf{r})$ is related to the background charge, and $\tilde{\rho}_\pm(\mathbf{r})$ are instantaneous ionic densities.

(i) Show that the partition function can be expressed in terms of the field variables

$$Q = \int D\rho_+ \int D\rho_- \int D\omega_+ \int D\omega_- \exp\left\{-\frac{1}{2}\int d\mathbf{r}\int d\mathbf{r}'\rho_c(\mathbf{r})v_c(|\mathbf{r}-\mathbf{r}'|)\rho_c(\mathbf{r}') + i\int d\mathbf{r}[\omega_+(\mathbf{r})\rho_+(\mathbf{r}) + \omega_-(\mathbf{r})\rho_-(\mathbf{r})] - \beta F_0(\omega_+,\omega_-)\right\}, \tag{X}$$

where $\rho_c(\mathbf{r}) \equiv \rho_c^0(\mathbf{r}) + Z_+\rho_+(\mathbf{r}) + Z_-\rho_-(\mathbf{r})$, F_0 stands for the Helmholtz energy of a binary system of noninteracting particles (viz., an ideal-gas mixture) in the presences of an effective one-body potential $\beta V_\pm^{ext}(\mathbf{r}) = i\omega_\pm(\mathbf{r}) - Z_\pm^2 v_c(0)$.

(ii) Verify the following expression for the Helmholtz energy of the noninteracting system

$$\beta F_0 = \sum_{n=\pm}\int d\mathbf{r}\rho_n(\mathbf{r})\left\{\ln[\rho_n(\mathbf{r})\Lambda^3] - 1 + \beta V_n^{ext}(\mathbf{r})\right\}.$$

(iii) Show that the saddle-point approximation leads to

$$\beta\mu_n = \ln[\rho_n(\mathbf{r})\Lambda^3] + Z_n\int d\mathbf{r}' v_c(|\mathbf{r}-\mathbf{r}'|)\rho_c(\mathbf{r}') - Z_n^2 v_c(0).$$

(iv) Derive the Boltzmann distribution for ionic species

$$\rho_\pm(\mathbf{r}) = \rho_\pm^0 \exp[-\beta Z_\pm e\psi(\mathbf{r})],$$

where $\rho_\pm^0$ represent the number densities of ionic species in the bulk, and the local electrical potential $\psi(\mathbf{r})$ satisfies the Poisson equation

$$\nabla^2\psi(\mathbf{r}) = -\frac{e\rho_c(\mathbf{r})}{\varepsilon\varepsilon_0}.$$

(v) Show that the imaginary field is given by

$$i\omega_n(\mathbf{r}) = \beta Z_n e\psi(\mathbf{r}) + Z_n^2 v_c(0).$$

(vi) Derive the mean-field Helmholtz energy from the saddle-point approximation

$$\beta F_{MF} = \sum_{n=\pm}\int d\mathbf{r}\rho_n(\mathbf{r})[\ln\rho_n(\mathbf{r}) - 1] + \beta e\int d\mathbf{r}\rho_c(\mathbf{r})\psi(\mathbf{r}).$$

(vii) Discuss how the mean-field Helmholtz energy may be improved by using the random-phase approximation (RPA).

9

Solvation, Electrolytes, and Electric Double Layer

Liquid solutions are widely utilized in chemistry, biology, and industrial applications. While the solvent plays a crucial role in determining the solution properties on both microscopic and macroscopic scales, practical applications are primarily concerned with the physiochemical behavior of the solute molecules. In this chapter, we present formal statistical–mechanical procedures that are commonly used for the separation of solvent degrees of freedom from those associated with solute molecules. We will discuss both molecular and phenomenological models to predict the solvation free energy and solvent-mediated interactions. The latter serves as a starting point for employing the Poisson–Boltzmann (PB) equation and liquid-state theories to predict the structural and thermodynamic properties of many ionic systems by using implicit-solvent models. We will elucidate the applications of these theoretical methods to electrolyte solutions and polyelectrolytes. Additionally, we will extend the bulk thermodynamic models to ionic systems near a charged surface/interface known as the electric double layer (EDL). The theoretical models discussed in this chapter are useful for understanding diverse chemical processes in solutions, including colloidal stability, surface chemistry, and electrochemical energy storage.

9.1 The McMillan–Mayer Theory

The McMillan–Mayer (MM) theory offers a formal mathematical framework to predict the structure and thermodynamic properties of liquid solutions, with the solvent considered as a background.[1] The implicit-solvent approach is broadly used in statistical–mechanical descriptions of aqueous solutions containing electrolytes, polyelectrolytes, and colloidal dispersions.

Conceptually, the MM framework can be introduced in the context of a liquid solution in osmotic equilibrium with a pure solvent. In this section, we first introduce the semi-grand canonical ensemble method underpinning the MM theory. Next, we will discuss the van't Hoff law and the microscopic meanings of solvation free energy and the potential of mean force (PMF). These concepts are commonly used in solution thermodynamics and are essential to describe the physiochemical properties of solvated molecules, molecular binding affinity, as well as the solution behavior of various aqueous systems of biomolecules.

1 McMillan W. G. and Mayer J. E., "The statistical thermodynamics of multicomponent systems", *J. Chem. Phys.* 13 (7), 276–305 (1945).

Fundamentals and Practice in Statistical Thermodynamics, First Edition. Jianzhong Wu and John M. Prausnitz.

Companion website: www.wiley.com/go/jianzhongwu/fundamentalsandpractice

9.1.1 Semi-Grand Canonical Ensemble

A semi-grand canonical ensemble can be used to describe thermodynamic systems with multiple compounds when mass exchange between the system and the surroundings is permitted only for a selected number of chemical species. We can elucidate the essential ideas by considering a liquid solution in osmotic equilibrium with its solvent, as shown schematically in Figure 9.1. Here, the solution is the thermodynamic system of interest. The system boundary is "closed" to the solute molecules, but "open" for the solvent, i.e., the thermodynamic system is semi-open with the number of the solute molecules fixed while the number of solvent molecules fluctuates around its mean value.

To quantify the distribution of microstates in a semi-grand canonical ensemble, consider a liquid solution containing N identical solute molecules in osmotic equilibrium with a pure solvent. At a given temperature T and volume V, each microstate is defined by the number of solvent molecules $N_{0,\nu}$, along with phase-space variables for describing the positions and momenta of the solute and solvent molecules in the solution. For simplicity, we designate ν_0 and ν_N as the degrees of freedom associated with the solvent and solute molecules in the thermodynamic system, respectively. Let E_ν be the total energy of the system at microstate $\nu \equiv (\nu_0, \nu_N)$.

As discussed in Chapter 2, we can determine the microstate distribution using the maximum entropy principle, i.e., by maximizing the Gibbs entropy subject to relevant macroscopic constraints. Specifically, the Gibbs entropy is defined as

$$S/k_B = -\sum_\nu p_\nu \ln p_\nu \tag{9.1}$$

where p_ν is the probability of the semi-open system in microstate $\nu = (\nu_0, \nu_N)$. In the semi-grand canonical ensemble, the entropy is maximized with the constraints of

$$\sum_\nu p_\nu = 1, \tag{9.2}$$

$$\sum_\nu p_\nu N_{0,\nu} = N_0, \tag{9.3}$$

$$\sum_\nu p_\nu E_\nu = U. \tag{9.4}$$

Eq. (9.2) is required by the normalization of the microstate probability; Eq. (9.3) reflects that the system is in osmotic equilibrium with a pure solvent at temperature T and pressure P; and Eq. (9.4) accounts for the thermal equilibrium between the semi-open system and its surroundings. Similar to how the thermal equilibrium is maintained by the constraint of a constant internal energy U, the osmotic equilibrium is established by fixing the average number of solvent molecules in the system N_0. At a given temperature, the total volume, and the number of solute molecules

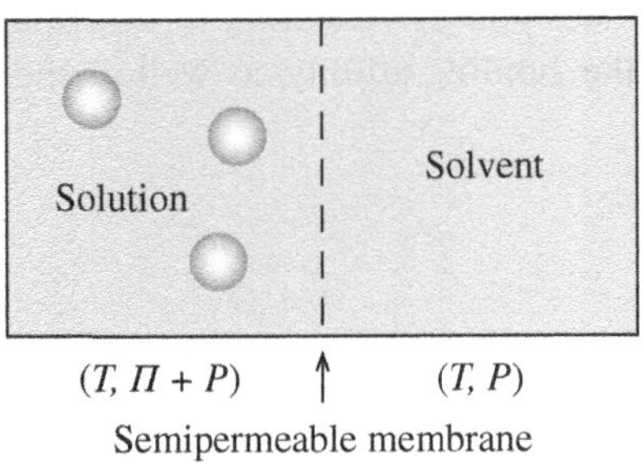

Figure 9.1 Schematic of osmotic equilibrium between a liquid solution and a pure solvent. Here, the particles represent solute molecules; the dashed line represents a membrane permeable to solvent molecules but not the solute; and Π denotes the osmotic pressure.

in the semi-open system, N_0 is uniquely determined by the chemical potential of the pure solvent, μ_0.

Following the Lagrange multiplier method,[2] we can find the equilibrium distribution of the microstates from Eqs. (9.1) to (9.4)

$$p_\nu = e^{-\beta E_\nu + \beta\mu_0 N_{0,\nu}} / \mathbb{Z} \tag{9.5}$$

where the semi-grand partition function is defined as

$$\mathbb{Z} \equiv \sum_\nu e^{-\beta E_\nu + \beta\mu_0 N_{0,\nu}} \tag{9.6}$$

with $\beta = 1/(k_B T)$. Mathematically, β and μ_0 are related to the Lagrange multipliers, i.e., parameters introduced to satisfy the constraints of thermal and osmotic equilibria between the solution and the pure solvent.

With the help of Eqs. (9.4) and (9.3), plugging Eq. (9.5) into (9.1) leads to

$$S/k_B = \ln \mathbb{Z} + \beta U - \beta\mu_0 N_0. \tag{9.7}$$

Eq. (9.7) can be utilized to define a thermodynamic potential, which is called the MM free energy

$$\mathbb{F} \equiv -k_B T \ln \mathbb{Z} = U - TS - \mu_0 N_0. \tag{9.8}$$

Formally, Eq. (9.8) represents a Legendre transformation of the Helmholtz energy, i.e., the transformation from the independent variables corresponding to a closed system $F = F(N, T, V, N_0)$ to those for a semi-open system, $\mathbb{F} = \mathbb{F}(N, T, V, \mu_0)$.

The MM free energy is analogous to the Helmholtz energy for a closed system at constant temperature and volume, the Gibbs energy at constant temperature and pressure, or the grand potential for an open system at constant volume and chemical potentials of all compounds. In terms of the fundamental equation of thermodynamics, the total derivative of the MM free energy is given by

$$d\mathbb{F} = -SdT - PdV - \mu_0 dN_0 + \mu dN. \tag{9.9}$$

Like alternative forms of the fundamental equation, Eq. (9.9) provides a useful link between different thermodynamic variables.

According to Eq. (9.9), the pressure is given by the partial derivative of the MM free energy

$$P = -\left.\frac{\partial \mathbb{F}}{\partial V}\right|_{T,N,\mu_0} = \left.\frac{\partial(\ln \mathbb{Z}/\beta)}{\partial V}\right|_{T,N,\mu_0}, \tag{9.10}$$

and the solute chemical potential is

$$\mu = \left.\frac{\partial \mathbb{F}}{\partial N}\right|_{T,V,\mu_0} = -\left.\frac{\partial(\ln \mathbb{Z}/\beta)}{\partial N}\right|_{T,V,\mu_0}. \tag{9.11}$$

Following Eqs. (9.5) and (9.6), we obtain another form of the Gibbs–Helmholtz equation

$$\partial\beta\mathbb{F}/\partial\beta = -\partial \ln \mathbb{Z}/\partial\beta = \sum_\nu E_\nu p_\nu - \mu_0 \sum_\nu N_{0,\nu} p_\nu = U - \mu_0 N_0. \tag{9.12}$$

Substituting Eq. (9.12) into Eq. (9.7) relates the entropy to the MM free energy

$$S/k_B = -\frac{1}{k_B}\left.\frac{\partial \mathbb{F}}{\partial T}\right|_{N,V,\mu_0} = \frac{\partial(\ln \mathbb{Z}/\beta)}{\partial(1/\beta)}. \tag{9.13}$$

2 See Chapter 2.

9.1.2 Microstates of Solvent Molecules in a Solution

To evaluate the semi-grand partition function defined according to Eq. (9.6), we may divide the total energy into two contributions: one is exclusively associated with the solvent molecules, and the other is related to the solute–solvent and solute–solute interactions. Such a division allows us to decouple the degrees of freedom for the solvent molecules from those corresponding to the entire system.

At any microstate ν, the total energy of the system can be written as

$$E_\nu = E_{\nu_0} + E_{\nu_N} \tag{9.14}$$

where E_{ν_0} is the energy due to the kinetic motion and interactions among solvent molecules, and E_{ν_N} includes the kinetic energy of the solute molecules as well as the potential energy due to the solute–solvent and solute–solute interactions. With this distinction, Eq. (9.6) can be rewritten as

$$\mathbb{Z} = \sum_\nu e^{-\beta E_{\nu_0} + \beta\mu_0 N_{0,\nu}} e^{-\beta E_{\nu_N}} = \Xi_0 \sum_{\nu_N} \sum_{\nu_0} e^{-\beta E_{\nu_N}} \left\{ e^{-\beta E_{\nu_0} + \beta\mu_0 N_{0,\nu}} / \Xi_0 \right\} = \Xi_0 \sum_{\nu_N} < e^{-\beta E_{\nu_N}} >_0 \tag{9.15}$$

where the curly brackets represent the grand-canonical probability of the solvent molecules in microstate ν_0; and the angle brackets represent an ensemble average in the grand-canonical ensemble of the pure solvent, i.e., the average over all microstates of the solvent molecules. In Eq. (9.15), Ξ_0 is the grand canonical partition function for the pure solvent at system temperature T, volume V, and chemical potential μ_0

$$\Xi_0 \equiv \sum_{\nu_0} e^{-\beta E_{\nu_0} + \beta\mu_0 N_{0,\nu}}. \tag{9.16}$$

Eq. (9.15) suggests that the semi-grand partition function can be evaluated in two steps: (i) calculation of the ensemble average of E_{ν_N} in terms of the microstates of the solvent molecules; and (ii) summation of the microstates corresponding to the solute molecules. The procedure is formally exact, applicable to systems containing more than one type of solute. However, in the MM theory, the ensemble average over the microstates of the solvent molecules is typically described in terms of phenomenological models (see Sections 9.2 and 9.3). In other words, the thermodynamic properties of a liquid solution are often predicted without explicit consideration of the degrees of freedom of solvent molecules. The theoretical procedure is exemplified by the van't Hoff's law as discussed in Section 9.1.3.

9.1.3 The Van't Hoff Law

At infinite dilution, solvent–solvent and solvent–solute interactions are similar to those in a condensed phase, but the solute–solute interactions are negligible due to the low solute concentration. Accordingly, we can express the total solute energy as a sum of the energies of individual solute molecules $\{E_{\nu_i}\}$

$$E_{\nu_N} = \sum_{i=1}^{N} E_{\nu_i}. \tag{9.17}$$

Eq. (9.17) is exact regardless of the complexity of the system, provided that the solute concentration is sufficiently dilute. Substituting Eq. (9.17) into (9.15) yields the MM partition function

$$\mathbb{Q} \equiv \mathbb{Z}/\Xi_0 = \sum_{\nu_N} < \prod_{i=1}^{N} e^{-\beta E_{\nu_i}} >_0 = \frac{1}{N!} \left\{ \sum_{\nu_1} < e^{-\beta E_{\nu_1}} >_0 \right\}^N. \tag{9.18}$$

In Eq. (9.18), E_{v_1} accounts for the total energy of a solute molecule in the vacuum and the potential energy due to the interaction of the solute molecule with solvent molecules in the surroundings; the summation over v_1 extends to *all* microstates of the solute molecule, and $N!$ arises from the fact that N solute molecules in the system are indistinguishable.

Apparently, Eq. (9.18) resembles the canonical partition function for an ideal gas containing the same number of solute molecules. The only difference is that the single-molecule partition function is now defined by

$$q^{ID} \equiv \sum_{v_1} < e^{-\beta E_{v_1}} >_0 \tag{9.19}$$

where superscript ID denotes an ideal solution, i.e., a solution without solute–solute interactions. For comparison, the single-molecule partition function for an ideal gas is given by

$$q^{IG} = \sum_{v_1} e^{-\beta E^0_{v_1}} \tag{9.20}$$

where $E^0_{v_1}$ represents the energy of a single solute molecule in the vacuum.

At each microstate of the solute molecule (e.g., given the momenta and positions of all solute atoms), we may define an *effective* one-body energy

$$u^0_{v_1} \equiv -k_B T \ln < e^{-\beta E_{v_1}} >_0. \tag{9.21}$$

In terms of the solvated energy for each solute molecule $u^0_{v_1}$, Eq. (9.18) can be written in terms of the MM partition function

$$\mathbb{Q}^{ID} = \frac{1}{N!}\left(\sum_{v_1} e^{-\beta u_{v_1}}\right)^N = \frac{1}{N!}(q^{ID})^N. \tag{9.22}$$

Clearly, $u^0_{v_1}$ and $E^0_{v_1}$ are different due to the interaction of the solute molecule with the solvent. Such differences are accounted for by the ensemble average over the configurations of the solvent molecules in the solution, $<\cdots>_0$. It should be noted that $\mathbb{Q}^{ID}$ and q^{ID} depend not only on the microstates of a single solute molecule but also on the properties of the solvent molecules through the ensemble average over all microstates in the semi-grand-canonical ensemble of (T, V, μ_0).

Similar to the canonical partition function of an ideal gas, the MM partition function provides a basis to derive the thermodynamic properties of an ideal solution. For example, the solution pressure is given by

$$P^{ID} = \left.\frac{\partial(\ln \mathbb{Z}/\beta)}{\partial V}\right|_{T,N,\mu_0} = \frac{\partial(\ln \mathbb{Q}^{ID}/\beta)}{\partial V} + \frac{\partial(\ln \Xi_0/\beta)}{\partial V} = \frac{Nk_BT}{V} + P_0 \tag{9.23}$$

where P_0 stands for the pressure of the pure solvent at system temperature T and chemical potential μ_0. Eq. (9.23) predicts that, in the dilute limit, the osmotic pressure is given by

$$\Pi^{ID} = P^{ID} - P_0 = Nk_BT/V. \tag{9.24}$$

Eq. (9.24) is the van't Hoff law for the osmotic pressure of dilute solutions.[3]

The van't Hoff law indicates that, in a dilute solution, the solvent contribution to the osmotic pressure is equivalent to that of a vacuum to the ideal-gas pressure. The solvent effect on pressure may not be immediately intuitive because, unlike that in an ideal gas, intermolecular interactions are inevitable, even in a dilute solution. Historically, "many authors of mathematical repute" found

3 Van't Hoff J. H., "The role of osmotic-pressure in the analogy between solutions and gases (Reprinted from Zeitschrift Fur Physikalische Chemie, Vol 1, Pg 481, 1887)", *J. Membr. Sci.*, 100 (1), 39–44 (1995).

that the van't Hoff law was exceedingly difficult to understand.[4] Without the MM theory, it is not clear how a dissolved substance can be present in a liquid solvent in a state similar to that in an ideal gas. Similar to the ideal-gas law, the van't Hoff law is independent of the molecular structure and solute–solvent interactions. The only requirement is that the solution be sufficiently dilute such that the solute–solute interactions can be ignored.

9.1.4 Solvation Free Energy

From the MM partition function, we can also derive the solute chemical potential

$$\beta\mu^{ID} = -\left.\frac{\partial \ln \mathbb{Z}}{\partial N}\right|_{T,V,\mu_0} = -\left.\frac{\partial \ln \mathbb{Q}^{ID}}{\partial N}\right|_{T,V,\mu_0}. \tag{9.25}$$

Subtracting from Eq. (9.25) the chemical potential of the solute molecules in the ideal-gas state (at the same temperature and the number density of solute molecules) gives

$$\beta\mu^{ID} - \beta\mu^{IG} = -\frac{\partial[\ln(\mathbb{Q}^{ID}/\mathbb{Q}^{IG})]}{\partial N} = -\ln(q^{ID}/q^{IG}). \tag{9.26}$$

According to Eq. (9.26), $-\ln(q^{ID}/q^{IG})$ represents the difference between the chemical potential of the solute in an ideal solution and that of the same solute in an ideal gas at the same temperature and number density. Because at ideal conditions the solute molecules do not interact with each other, the difference in the chemical potential corresponds to a reversible work to transfer a solute molecule from the vacuum to a pure solvent. The reversible work thus defines the solvation free energy

$$F_1 = \mu^{ID}(N, V, T, \mu_0) - \mu^{IG}(N, V, T). \tag{9.27}$$

Eqs. (9.26) and (9.27) provide the statistical-thermodynamic basis to predict the solvation free energy using theoretical methods or molecular simulation. At each microstate of the solute molecule, the solute energy can be expressed as

$$E_{\nu_1} = E^0_{\nu_1} + \Delta E_{\nu_1} \tag{9.28}$$

where $E^0_{\nu_1}$ is the energy of an isolated solute molecule in the vacuum, and ΔE_{ν_1} is the change in the single-molecule energy when the solute at the same microstate ν_1 is surrounded by the solvent molecules. The ratio of the single-molecule partition functions can be expressed as

$$q^{ID}/q^{IG} = \frac{\sum_{\nu_1} < e^{-\beta E_{\nu_1}} >_0}{\sum_{\nu_1} e^{-\beta E^0_{\nu_1}}} = \frac{\sum_{\nu_0,\nu_1} e^{-\beta \Delta E_{\nu_1}} e^{-\beta E^0_{\nu_1} - \beta E_{\nu_0} + \beta\mu_0 N_{0,\nu}}}{\sum_{\nu_0,\nu_1} e^{-\beta E^0_{\nu_1} - \beta E_{\nu_0} + \beta\mu_0 N_{0,\nu}}} =< e^{-\beta\Delta E_{\nu_1}} >_{0+\nu_1} \tag{9.29}$$

where $< \cdots >_{0+\nu_1}$ represents the ensemble average over the microstates of a solute and all solvent molecules in the system. Combining Eqs. (9.26) and (9.29) gives

$$F_1 = -k_B T \ln < e^{-\beta\Delta E_{\nu_1}} >_{0+\nu_1} \tag{9.30}$$

where the ensemble average can be carried out by simulating a single solute molecule surrounded by a large number of solvent molecules.

It should be noted that the theoretical procedure to define the solvation free energy is fully consistent with that used in classical thermodynamics.[5] As shown schematically in Figure 9.2, the

4 Martin G., "On van't Hoff's law of osmotic pressure". *Nature* 70, 531–532 (1904).
5 Sergiievskyi V. et al., "Solvation free-energy pressure corrections in the three-dimensional reference interaction site model", *J. Chem. Phys.* 143, 184116 (2015).

Figure 9.2 A comparison of solvation free energy defined according to the McMillan–Mayer framework (theory) with that according to the Lewis–Randal framework (experiment).

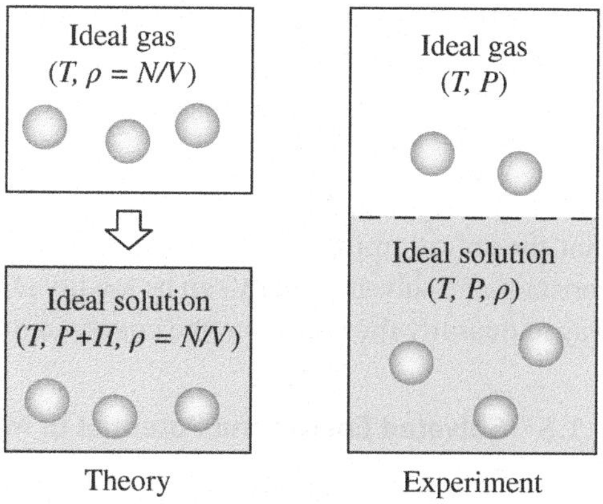

definition of F_1 is based on the MM framework, with the solution defined in terms of temperature, total volume, and the solvent chemical potential. Conversely, experimental interpretation of the solvation free energy is mostly based on the Lewis–Randal (LR) framework, with the solution defined in terms of temperature, pressure and solute composition. The solvation free energy refers to the reversible work to transfer a solute molecule from an ideal gas to an ideal solution of the same temperature and pressure. While significant differences may exist between MM and LR frameworks in certain thermodynamic quantities (such as activity coefficients), the solvation free energy defined with different sets of thermodynamic variables results in the same reversible work, corresponding to the transfer of a solute molecule to a pure solvent.

To make a connection between different definitions of the solvation free energy, consider the pressure effect on the solute chemical potential

$$\mu^{ID}(N, V, T, \mu_0) = \mu^{ID}(T, P, x) + \int_P^{P+\Pi} \overline{v}\, dP. \tag{9.31}$$

where $\mu^{ID}(T, P, x)$ stands for the solute chemical potential when the solution pressure is reduced to that of the pure solvent, $\overline{v}$ represents the partial molar volume of the solute, and x is the solute mole fraction. For an ideal solution, the osmotic pressure is described by the van't Hoff law, and the solute partial molar volume is approximately a constant. Therefore, we can rewrite Eq. (9.31) as

$$\mu^{ID}(T, P, x) = \mu^{ID}(N, V, T, \mu_0) - \overline{v}\Pi = \mu^{IG}(N, V, T) + F_1 - \overline{v}Nk_BT/V. \tag{9.32}$$

The last term on the right side of Eq. (9.32) vanishes as the solute density $\rho = N/V$ approaches zero. Apparently, the osmotic pressure is insignificant at infinite dilution. Problem 9.5 discusses the connections between LR and MM frameworks when the solute concentration is finite.

9.1.5 Henry's Constant

For a dilute solution, we may express the solute chemical potential in terms of Henry's law for the solute fugacity

$$\beta\mu^{ID}(T, P, x) = \beta\mu^{IG}(T, P^{IG}) + \ln(K_H x/P^{IG}) \tag{9.33}$$

where μ^{IG} is the solute chemical potential as a pure ideal gas at the system temperature and pressure $P^{IG} = Nk_BT/V$, and K_H is Henry's constant. A comparison of Eqs. (9.32) and (9.33) indicates

$$\beta F_1 = \ln(K_H x/P^{IG}) = \ln(\beta K_H/\rho_0). \tag{9.34}$$

In writing Eq. (9.34), we take the infinite dilution limit. As $\rho \to 0$, we have $\bar{v}N/V = 0$ and $N_0 + N = N_0$ such that

$$x/P^{IG} = \frac{N}{N+N_0} \times \frac{\beta V}{N} = \frac{\beta V}{N_0} = \beta\rho_0 \tag{9.35}$$

where $\rho_0 = N_0/V$ is the number density of the solvent molecules. As expected, Eq. (9.34) predicts that the solvation free energy is independent of the solute concentration. Because both Henry's constant and solvent density can be readily obtained from experiments, Eq. (9.34) provides an easy way to quantify the solvation free energy of individual molecules.

9.1.6 Solvated Energy and Potential of Mean Force

To extend the thermodynamic equations for a dilute solution to those for solutions at finite solute concentrations, we define an effective multi-body energy (viz., multi-body solvation free energy) to account for the interaction between solute molecules mediated by the solvent

$$u_{\nu_N} \equiv -k_B T \ln < e^{-\beta E_{\nu_N}} >_0. \tag{9.36}$$

Using Eqs. (9.36) and (9.15), we can express the MM partition function as

$$\mathbb{Q} \equiv \mathbb{Z}/\Xi_0 = \sum_{\nu_N} \exp[-\beta u_{\nu_N}]. \tag{9.37}$$

Eq. (9.37) is analogous to a conventional canonical partition function except that, at each microstate of the solute molecules, the total energy in the vacuum is replaced by the effective multi-body potential u_{ν_N}. The latter depends on the solute configuration and the properties of the solvent molecules.

Similar to that for a dilute solution, the osmotic pressure is related to the partition function by

$$\Pi = \left(\frac{\partial \ln \mathbb{Q}/\beta}{\partial V}\right). \tag{9.38}$$

Other thermodynamic quantities can also be derived from standard statistical-mechanical equations. For example, the difference between the solution entropy and that of a pure solvent at the same temperature, solvent chemical potential, and volume, is given by

$$\Delta S/k_B = \left(\frac{\partial \ln \mathbb{Q}/\beta}{\partial 1/\beta}\right) \tag{9.39}$$

where $\Delta S = S - S_0(T, V, \mu_0)$. Eqs. (9.38)–(9.39) resemble the standard statistical–mechanical equations for pressure and entropy except that the multi-body molecular energy must account for the solvent effects. In essence, the MM theory is built upon the analogy between the statistical mechanics of a solution and that of a gas system.

The multi-body potential u_{ν_N} depends on the solute configuration as well as the overall properties of the solvent molecules. In general, this energy is unknown without carrying out the ensemble average over the microstates of the solvent molecules. However, the direct evaluation of the ensemble average at each solute configuration is unrealistic, even at conditions accessible to molecular simulations. To eschew the computational burden, we may divide the effective multi-body potential into two components: one is associated with the kinetic energy and the solvation free energy of individual solute molecules, and the other is attributed to solvent-mediated solute–solute interactions. Without the solvent-mediated interactions among solute molecules, the thermodynamic properties would be the same as those corresponding to an ideal solution. Therefore, the solvent-mediated interactions are responsible for the excess properties that, as defined in classical thermodynamics, represent the differences between the thermodynamic properties of the real solution and those of the ideal solution.

In parallel to intermolecular interactions in the gas phase, the effective multi-body potential may be approximated by the one-body energies and a pairwise additive potential

$$u_{v_N} \approx \sum_i u^0_{v_i} + \frac{1}{2}\sum_i \sum_{j \neq i} w(v_i, v_j) \tag{9.40}$$

where $u^0_{v_i}$ represents the energy of a solvated molecule as that in the ideal solution, and $w(v_i, v_j)$ is the PMF for two solute molecules at single-molecule microstates v_i and v_j.

In the ideal solution, the energy of a single solvated molecule includes a contribution from its kinetic motions and the solvation free energy due to the solute interaction with the solvent molecules. As discussed in Section 7.2, the PMF represents the reversible work to bring two solute molecules from infinite apart to close distance

$$w(1,2) = u(1,2) - \left(u^0_1 + u^0_2\right) \tag{9.41}$$

where 1 and 2 stand for all coordinates describing the solute configurations. In a classical system, the kinetic energies of individual molecules are independent of each other, so they do not contribute to the PMF. In the absence of solvent molecules, $w(1, 2)$ reduces to the pair potential between molecules 1 and 2 in the vacuum.

Eq. (9.41) is not directly useful for calculating the PMF by molecular simulation because both the two-body and one-body energies must be evaluated through ensemble averages over the solvent configurations. To make the simulation more efficient, we may first obtain the mean force between the solute molecules by taking a derivative with respect to the intermolecular distance r

$$\mathbf{F}(r) = -\frac{\partial w(1,2)}{\partial r} = -\frac{\sum_{v_0} \frac{\partial E(1,2)}{\partial r} e^{-\beta E_0 - \beta E(1,2)}}{\sum_{v_0} e^{-\beta E_0 - \beta E(1,2)}} = \left\langle -\frac{\partial E(1,2)}{\partial r} \right\rangle \tag{9.42}$$

where $-\partial E(1, 2)/\partial r$ represents the overall force between molecules 1 and 2 along the direction of intermolecular distance, and $\langle \cdots \rangle$ represents the ensemble average over the microstates of the solvent molecules in a system containing both solute molecules. The ensemble average in Eq. (9.42) can be evaluated with molecular simulation.[6] The integration of the averaged force yields the reversible work to bring two molecules together, i.e., the potential of the mean force.

In principle, the PMF can be evaluated with molecular simulation methods. However, its accuracy is often difficult to calibrate because of the lack of direct experimental data. The atomic-scale force calculation often requires a combination of quantum and statistical–mechanical methods, in which approximations of intermolecular interactions are inevitable. Besides, molecular simulation may incur a prohibitive computational cost for macromolecules. To illustrate, Figure 9.3 shows the PMF between two isolated ions of opposite charges in liquid water, calculated from molecular dynamics (MD) simulation.[7] Interestingly, the potential profile predicted by MD is drastically different from Coulomb's law, which is conventionally used in implicit solvent models to describe ion–ion interactions in liquid water. Strictly speaking, the phenomenological equation is valid

---

6 Fixing two particles at distance r introduces an entropic force, $2k_BT/r$, which must not contribute to PMF. See Hess B., Holm C., and van der Vegt N., "Osmotic coefficients of atomistic NaCl (aq) force fields", *J. Chem. Phys.* 124(16), 64509 (2006).

7 Fennell C. J. et al., "Ion pairing in molecular simulations of aqueous alkali halide solutions", *J. Phys. Chem. B* 113 (19), 6782–6791 (2009).

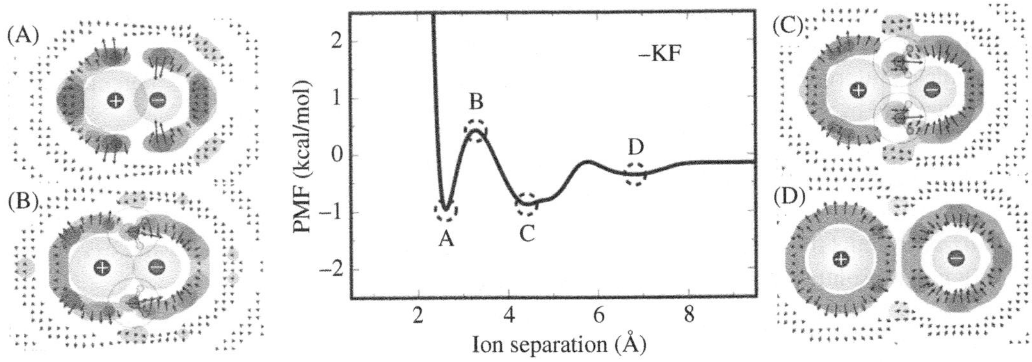

Figure 9.3 The potential of mean force (PMF) between K^+ and F^- in liquid water at 300 K and 1 bar predicted by MD simulation. The molecular parameters for ions were from the OPLS force field, and the TIP5P-E model was used for water molecules. Shown on the sides are the density profiles and dipolar orientations of water molecules near the ion pair at 4 representative separations: (A) contact ion pair, (B) first maximum in PMF, (C) solvent-shared ion pair, and (D) solvent-separated ion pair. Source: Reproduced from Fennell et al.[7]

only at large separations.[8] Nevertheless, MD simulation reveals rich details on the solvation structure and ion association/dissociation that are not captured in the continuous model. Whereas such microscopic details are important to predict specific ion effects such as those related to the Hofmeister series or to understand affinities between biomolecules in different aqueous environments, an accurate prediction of the PMF is extremely demanding, not only because the free-energy calculation is computationally expensive but also a faithful representation of the intermolecular interactions at atomic scales is theoretically challenging. The simulation results are highly sensitive to the semi-empirical force fields representing solute–solvent interactions. Likewise, first-principles calculations mostly rely on approximate quantum-mechanical methods that may not provide an accurate description of the solute–solvent energies.[9]

As the macroscopic properties of a thermodynamic system are relatively insensitive to the specific forms of intermolecular interactions, practical applications of the MM theory typically start with phenomenological models of solvent-mediated interactions. The procedure is not much different than conventional applications of statistical mechanics to molecular systems: we use approximate models or semi-empirical force fields rather than rigorous quantum-mechanical calculations to describe intermolecular forces.

9.1.7 Summary

With an adequate equation for the PMF, the MM theory allows us to predict the thermodynamic properties of liquid solutions without explicit consideration of the solvent molecules. The statistical–mechanical procedure is virtually identical to that used for molecular systems. The implicit-solvent models have been broadly applied to colloidal dispersions and aqueous solutions containing charged species such as electrolytes. While treating the solvent as a background greatly simplifies the theoretical procedure, it is important to note that the implicit-solvent approach

8 Kalcher I. and Dzubiella J., "Structure-thermodynamics relation of electrolyte solutions", *J. Chem. Phys.* 130, 134507 (2009).
9 Kohagen M. E. et al., "Exploring ion-ion interactions in aqueous solutions by a combination of molecular dynamics and neutron scattering", *J. Phys. Chem. Lett.* 6 (9): 1563–1567 (2015).

largely relies on semi-empirical representations of the PMF, with the model parameters adjusted to match relevant experimental data. Solvent-implicit models for describing the solvation free energy and solvent-mediated interactions will be discussed in Section 9.2 and 9.3.

An explicit solvent model is needed for understanding molecular-level events such as the microscopic details of solvation and solvent-mediated interactions underlying molecular recognitions and structure-based drug design. In general, such microscopic details can be acquired through molecular simulation. Alternatively, we may use the Kirkwood–Buff (KB) theory to gain insights into the local structure.[10] As discussed in Problem 9.1 and 9.2, the KB theory provides a rigorous procedure to link thermodynamic properties with the integrations of the molecular distribution functions (known as the KB integrals) and vice versa.[11] The KB integrals can be determined from macroscopic properties such as isothermal compressibility, partial molecular volumes, and inverse osmotic susceptibilities. Similar to the compressibility equation for simple fluids, the KB theory is established from exact structure–property relations without any assumption of the molecular structure or intermolecular forces. Therefore, the KB theory allows us to gain valuable insights into the microscopic structures of liquid solutions even without any specific knowledge of intermolecular interactions.

9.2 Phenomenological Solvation Models

Solvent effects are important not only for understanding the dissolution and reactivity of organic and inorganic species in various liquid environments, but also for predicting the phase behavior of liquid mixtures and solutions. In this section, we discuss several phenomenological models to predict solvation free energy, a critical thermodynamic variable for considering solvent effects.

As discussed in the previous section, the solvation free energy can be understood as the reversible work to transfer a solute molecule from the vacuum to a pure solvent at a given temperature and pressure.[12] While the first-principles prediction of solvation free energies is theoretically challenging (and computationally demanding), the so-called implicit-solvent or phenomenological solvent models are appealing for many practical applications because of their conceptual simplicity and computational efficiency.

9.2.1 Cavity Formation

Conceptually, the solvation free energy can be estimated by following a two-step process to transfer a solute molecule from the vacuum into a liquid medium: first, the formation of a cavity to accommodate the solute molecule, and second, "turning on" the solute–solvent interactions. The two-step procedure was originally introduced within the framework of the scaled particle theory (SPT), which was proposed by Reiss, Frisch and Lebowitz in the development of an equation of state for bulk hard-sphere fluids (Appendix 7.A).[13] Although the original work was primarily concerned with hard-sphere fluids, similar ideas are useful to predict the solvation free energy due

10 Kirkwood J. G. and Buff F. P., "The statistical mechanical theory of solutions. I", *J. Chem. Phys.* 19, 774–777 (1951).

11 *Fluctuation theory of solutions*, edited by Smith P. E., Matteoli E., and O'Connell J. P., CRC Press, Boca Raton, 2013.

12 Equivalently, from an ideal-gas state to an ideal solution at the same temperature and the number density of the solute molecules.

13 Reiss H., Frisch H. L. and Lebowitz J. L., "Statistical mechanics of rigid spheres", *J. Chem. Phys.* 31 (2), 369–380 (1959).

to cavity formation in any liquid solution. Unlike alternative statistical–mechanical methods, the application of SPT to solvation is based on the molecular geometry and thermodynamic considerations. It allows for the calculation of the solvation free energy with little knowledge on the interaction among solvent molecules.

The two-step procedure to predict the solvation free energy can be understood in terms of the exact statistical–mechanical equation

$$F_1 = \int_0^1 d\lambda \int d\mathbf{r}\rho(\mathbf{r}, \lambda)u(\mathbf{r}) \tag{9.43}$$

where $u(\mathbf{r})$ represents the potential energy on a solvent molecule at position $\mathbf{r}$ due to the presence of a solute molecule at the origin, $\rho(\mathbf{r}, \lambda)$ is the local solvent density when the solvent–solute potential is scaled by a factor $0 \leq \lambda \leq 1$. As elucidated in Problem 9.6, Eq. (9.43) is applicable to solute and solvent molecules of arbitrary structures. For non-spherical systems, $u(\mathbf{r})$ and $\rho(\mathbf{r}, \lambda)$ may be considered as the potential and local density after ensemble average over the molecular configurations/orientations. The two-step consideration amounts to the separation of the potential energy into a short-range term accounting for the molecular excluded volume (viz., cavity formation) and a longer-ranged term arising from van der Waals (vdW) or other forms of solute–solvent interactions.

To solidify the ideas, consider the reversible work of inserting a hard-sphere of radius a into a liquid. Let a_0 be the radius of the solvent molecules or, more precisely, the closest distance from the center of a solvent molecule to the hard-sphere surface (Figure 9.4A). In the liquid solution, a hard-sphere does not interact with the solvent molecules other than occupying the space, i.e., it prohibits any solvent molecule from approaching its position (Figure 9.4B). Therefore, inserting a hard-sphere into the solvent amounts to the formation of a cavity of radius $R = a + a_0$. As we will see from the following, the free energy of cavity formation $W(R)$ can be derived by "scaling" the cavity size, i.e., by an interpretation of the exact results for the cavity formation at both small and large scales.

In a uniform solvent, a spherical region of radius $R \leq a_0$ can accommodate at most one solvent molecule due to the molecular excluded volume. The probability of finding a solvent molecule

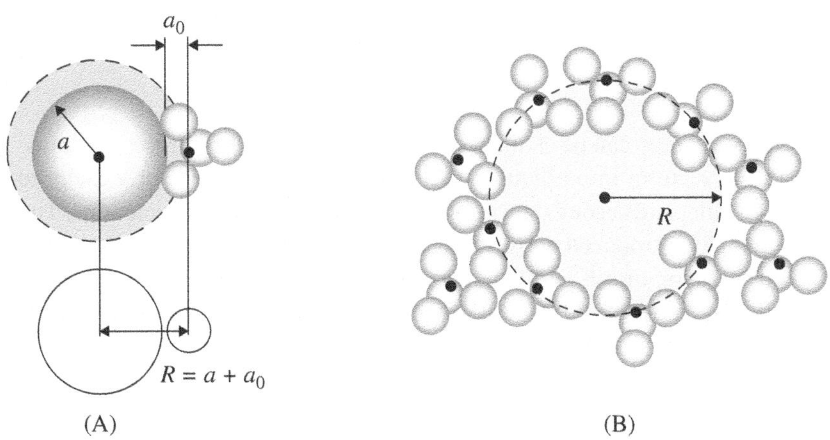

Figure 9.4 (A) Excluded volume for a spherical particle is defined by the space inaccessible to the centers of solvent molecules. Here, $a + a_0$ represents the closest center-to-center distance between the solute (dark sphere) and an arbitrarily shaped solvent molecule (a cluster of gray spheres and a black dot denoting the molecular center for each solvent molecule). (B) Formation of a spherical cavity of radius R (shaded region) that excludes the center of any solvent molecule.

with its center located within this region is $4\pi R^3\rho_0/3$, where ρ_0 is the number density of the solvent molecules in the bulk. The normalization condition implies that the probability of finding the spherical region unoccupied is

$$p(R) = 1 - 4\pi R^3\rho_0/3. \tag{9.44}$$

Eq. (9.44) corresponds to the probability of finding a spherical cavity $R \leq a_0$ in the pure solvent. From the molecular perspective, the latter is related to the cavity formation free energy $W(R)$

$$p(R) = e^{-\beta W(R)}. \tag{9.45}$$

Rearranging Eq. (9.45) leads to

$$W(R) = -k_B T \ln p(R) = -k_B T \ln(1 - 4\pi R^3\rho_0/3). \tag{9.46}$$

To derive a general expression for the free energy of cavity formation, now consider the differential increase of the cavity radius from R to $R + dR$. The probability of finding a solvent molecule with its center located within the spherical shell of thickness dR is given by $\rho(R)4\pi R^2 dR$, where $\rho(R)$ the number density of the solvent molecules at the cavity surface. The probability of finding no solvent molecule in this shell is $1 - \rho(R)4\pi R^2 dR$. Because

$$p(R + dR) = p(R) \times [1 - \rho(R)4\pi R^2 dR], \tag{9.47}$$

rearranging Eq. (9.47) leads to

$$\frac{d\ln p(R)}{dR} = -\frac{d\beta W(R)}{dR} = -\rho(R)4\pi R^2. \tag{9.48}$$

With the boundary condition $W(0) = 0$, the integration of Eq. (9.48) gives

$$\beta W(R) = \int_0^R 4\pi R^2 \rho(R) dR. \tag{9.49}$$

Eq. (9.49) indicates that, with an analytical expression for $\rho(R)$, we can readily predict the free energy of cavity formation.

9.2.2 Solvation in a Hard-Sphere Fluid

If we consider cavity formation in a hard-sphere fluid, the solvent density at the cavity surface is exactly known at both small and large limits. For a small cavity (i.e., when the cavity radius is smaller than the hard-sphere radius), we can find $\rho(R)$ based on $W(R)$ given by Eqs. (9.46) and (9.48)

$$\rho(R) = \rho_0/(1 - 4\pi R^3\rho_0/3). \tag{9.50}$$

For a cavity of infinite size ($R \to \infty$), the contact solvent density can be obtained from the bulk pressure of the solvent by using the contact-value theorem[14]

$$\lim_{R\to\infty} \rho(R) = P/(k_B T) \tag{9.51}$$

14 Contact value theorem specifies the force balance at the fluid interface. For example, the fluid pressure P near a planar surface is balanced by the surface potential $u(z)$, i.e., $P = -\int dz \rho(z) u'(z)$, where $\rho(z)$ is the number density of fluid molecules and z stands for the distance from the surface.

where P is the solvent pressure far away from the cavity surface. Although $\rho(R)$ is not known for $a_0 < R < \infty$, it can be interpolated by using a quadratic form[15]

$$\rho(R) = \rho(\infty) + \rho_1/R + \rho_2/R^2. \tag{9.52}$$

Because $\rho(R)$ is a smooth function for the entire range of R, the coefficients in Eq. (9.52), $\rho(\infty)$, ρ_1 and ρ_2, can be derived from Eq. (9.50) by imposing the continuity conditions for $\rho(R)$ as well as its first and second derivatives at $R = a_0$. Subsequently, the SPT equation of state for hard-sphere fluids can be derived from Eq. (9.51).

In application of the SPT to solvation, it is often assumed that the free energy of cavity formation in a real solvent is the same as that in a hard-sphere fluid[16]

$$W(R) = (4\pi)\kappa_0 + (4\pi R)\kappa + (4\pi R^2)\gamma + (4\pi R^3/3)P \tag{9.53}$$

where

$$\kappa_0 = \frac{k_B T}{4\pi}\left[-\ln(1-\eta) + \frac{9}{2}\left(\frac{\eta}{1-\eta}\right)^2\right] - \frac{Pa_0^3}{24}, \tag{9.54}$$

$$\kappa = -\frac{k_B T}{2\pi a_0}\left[\frac{3\eta}{1-\eta} + 9\left(\frac{\eta}{1-\eta}\right)^2\right] + Pa_0^2/4, \tag{9.55}$$

$$\gamma = \frac{k_B T}{2\pi a_0^2}\left[\frac{3\eta}{1-\eta} + 9\left(\frac{\eta}{1-\eta}\right)^2\right] - Pa_0/2. \tag{9.56}$$

In above equations, a_0 is the radius of solvent molecules, $\eta = 4\pi\rho_0 a_0^3/3$ is the packing fraction of the solvent, and ρ_0 is the number density of solvent molecules in the bulk. On the right side of Eq. (9.53), the terms inside the parentheses correspond to the geometric measures of a spherical cavity of radius R, i.e., the cavity volume, the cavity surface area, the surface integrated mean curvature ($4\pi R^2 \times 1/R$), and the surface integrated Gaussian curvature ($4\pi R^2 \times 1/R^2$); and parameters γ and κ can be understood as the interfacial tension and bending modulus of the cavity-solvent interface, respectively.

In applications of Eqs. (9.53)–(9.56) to a real solvent, we need to know the solvent pressure P and bulk density ρ_0. Parameter a_0 can be estimated from semi-empirical correlations, for example, a correlation between the molecular diameter and the average polarizability.[17] Alternatively, a_0 can be calculated from the packing fraction of solvent molecules based on the thermodynamic relation[18]

$$\Delta H_{vap} = k_B T + \alpha k_B T^2 (1+2\eta)^2/(1-\eta)^3 \tag{9.57}$$

where ΔH_{vap} is the heat of vaporization per solvent molecule, and $\alpha \equiv (1/v)(\partial v/\partial T)_P$ is the coefficient of thermal expansion. With these macroscopic properties of the pure solvent as the input, Eqs. (9.53)–(9.57) can be used to predict the free energy of cavity formation.

---

15 The next two terms in the curvature expansion are zero, and higher-order terms are not generally available. See Reiss H. R. and Tully-Smith D. M., "Further development of scaled particle theory for rigid spheres: application of the statistical thermodynamics of curved surfaces", *J. Chem. Phys.* 55, 1674–1689 (1971).

16 Pierotti R. A., "Solubility of gases in liquids", *J. Phys. Chem.* 67 (9), 1840–1845 (1963). In terms of the ratio of solute-solvent radii $\mathfrak{R} \equiv a/a_0$, Eq. (9.53) can be expressed as $\beta W(\mathfrak{R}) = -\ln(1-\eta) + \left(\frac{3\eta}{1-\eta}\right)\mathfrak{R} + \left[\frac{3\eta}{1-\eta} + \frac{9}{2}\left(\frac{\eta}{1-\eta}\right)^2\right]\mathfrak{R}^2 + \frac{P\eta}{\rho_0 k_B T}\mathfrak{R}^3$.

17 Bondi A., "Van der Waals volumes and radii", *J. Phys. Chem.* 68 (3), 441–451 (1964).

18 Eq. (10) may be obtained from the cavity formation free energy and the Clausius-Clapeyron equation. See for details in Reiss H., Frisch H. L., Helfand E., and Lebowitz J. L., "Aspects of the statistical thermodynamics of real fluids", *J. Chem. Phys.*, 32 (1), 119–124 (1960).

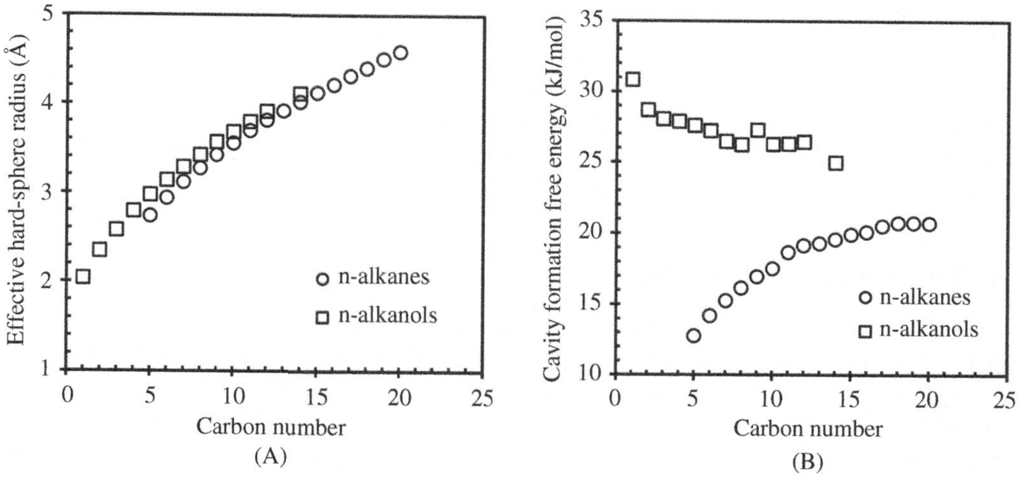

Figure 9.5 Effective hard-sphere radius of solvent molecules (A) and the cavity formation free energy (B) to accommodate a xenon atom (diameter 3.973 Å) in normal alkanes and normal alkanols at 25 °C and 1 atm.

To illustrate, Figure 9.5 presents the effective hard-sphere radii of various alkanes and alkanol molecules and the cavity formation free energies to accommodate a xenon molecule at the ambient condition derived from the solubility data.[19] It shows that the effective hard-sphere radii are well correlated to the number of carbon numbers and that such correlations are similar for alkanes and alkanols. As expected, the free energy of cavity formation in alkanols is significantly larger than that in alkanes with the same number of carbon atoms. Interestingly, their dependences on the carbon number are totally opposite. The drastic difference can be attributed to the difference in the intermolecular forces, one with hydrogen bonding and the other only with vdW attractions. Intuitively, the different trends may also be understood in terms of the liquid density, which increases with the carbon number for n-alkane but declines for n-alkanols.

9.2.3 Effects of Intermolecular Attraction

In addition to cavity formation, the solvation free energy depends on the attraction between solute and solvent molecules. The additional contribution can be estimated from intermolecular forces and the mean-field approximation. For a nonpolar solute, we may calculate the solvation energy using the Lennard–Jones (LJ) potential. Toward that end, we assume that the solvent molecules are uniformly distributed around the solute and that the minimum separation between the solute and solvent molecules is the same as the cavity radius R. The solvation energy due to intermolecular attraction is thus given by

$$F_1^{vdw} = \int_R^{\infty} u_{LJ}(r)\rho_0 4\pi r^2 dr = -\frac{32\pi}{9}\rho_0 R^3 \varepsilon_{LJ} \tag{9.58}$$

where ε_{LJ} represents the LJ energy parameter, and ρ_0 is again the number density of solvent molecules in the bulk. The LJ energy parameter can be represented in terms of the summation of solute–segment interactions. For example, we may estimate the attractive component of the

19 Pollack, G. L. et al., "Solubility of Xenon in 45 organic-solvents including cycloalkanes, acids, and alkanals—experiment and theory", *J. Chem. Phys.*, 90 (11), 6569–6579 (1989).

solvation energy in a normal alkanol as

$$F_1^{vdw} = -\frac{32\pi}{9}\rho_0 R^3(\varepsilon_{1CH3} + m\varepsilon_{1CH2} + \varepsilon_{1OH}) \tag{9.59}$$

where m stands for the number of CH_2 groups in the alkyl chain. Although hydrogen bonding is not considered, Eq. (9.59) provides an excellent correlation for gas solubility in many organic solvents.[20]

9.2.4 Morphometric Thermodynamics

Whereas the SPT equations were originally established for hard-sphere fluids, similar concepts are applicable to solvation in an arbitrary solvent. In particular, Eq. (9.53) suggests that the solvation free energy can be formally expressed in terms of contributions due to the excluded volume, the surface area, and the surface curvatures of the solute molecule. The coefficients corresponding to these geometric measures of the solute-solvent interface are bulk pressure, interfacial tension, and the surface bending moduli. For cavity formation (or a hard-sphere solute), these thermodynamic variables can be determined by the properties of the solvent molecules. Because the morphometric measures of the solute molecule play an important role in determining the solvation free energy, the theoretical procedure is known as morphological thermodynamics.[21]

Although the thermodynamic quantities in Eq. (9.53) are defined from a macroscopic perspective, simulation results indicate that the expression is valid at length scales comparable to the size of solvent molecules.[22] The applicability of these quantities at small scales can be justified in part by using the exact results in SPT for the formation of small cavities (Eq. (9.46)), and by incorporating the special features of the curvature expansion for the contact density (Eq. (9.52)).

In the macroscopic limit ($R \to \infty$), we see from Eq. (9.53) that the free energy of cavitation is dominated by the surface area and the excluded volume effects. Because the volumetric term persists even for cavity formation in an ideal gas, it should be subtracted from the free energy of solvation at the macroscopic scale. Accordingly, the solvation free energy is determined by

$$F_1^0(R) = \kappa_0 + \kappa \cdot R + \gamma \cdot (4\pi R^2) \tag{9.60}$$

where superscript "0" stands for cavity. In morphological thermodynamics, Eq. (9.60) is generalized for the solvation of a non-spherical particle in terms of its geometric characteristics[23]

$$F_1^0(R) = \kappa_0 X + \kappa C + \gamma A \tag{9.61}$$

where A is the surface area, C is the integrated mean curvature, and X is the integrated Gaussian curvature. For a spherical cavity of radius R, $A = 4\pi R^2$, $C = 4\pi R$ and $X = 4\pi$; and the corresponding geometric measures for a cylinder of radius R and height h are

$$A = 2\pi R^2 + 2\pi Rh, \qquad C = \pi^2 R + \pi h \quad \text{and} \quad X = 4\pi \tag{9.62}$$

Eq. (9.61) and (9.62) are useful for understanding the effects of solute geometry on solvation. For example, Figure 9.6 presents experimental data for the partitioning of n-alkyl carboxylic acids

20 Pollack G. L., "Why gases dissolve in liquids", *Science* 251 (4999), 1323–1330 (1991).
21 König P. M., Roth R. and Mecke K. R., "Morphological thermodynamics of fluids: shape dependence of free energies", *Phys. Rev. Lett.* 93, 160601 (2004).
22 Wu J. Z., "Solvation of a spherical cavity in simple liquids: interpolating between the limits", *J. Phys. Chem. B* 113 (19), 6813–6818 (2009).
23 Jin Z. H., Kim J. and Wu J., "Shape effect on nanoparticle solvation: a comparison of morphometric thermodynamics and microscopic theories", *Langmuir* 28 (17), 6997–7006 (2012).

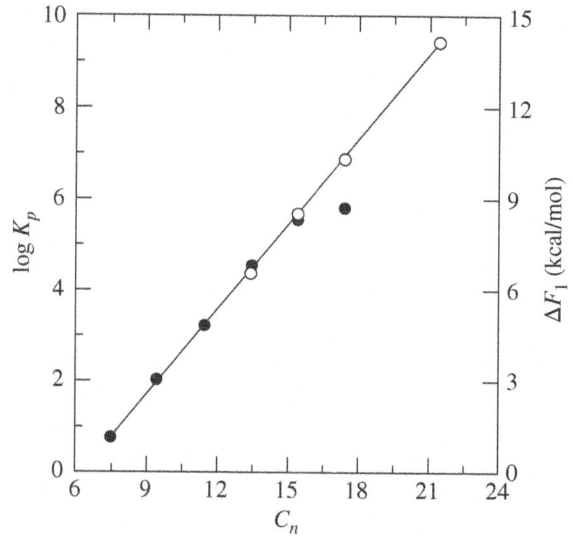

Figure 9.6 The Oswald coefficients and the free energy of transfer from water to n-heptane for n-alkyl carboxylic acids at 25 °C. Here, C_n is the alkyl chain length, and the filled and open cycles are data from different experiments. Source: Adapted from Smith and Tanford.[24]

(CA) between liquid water and n-heptane at room temperature and pressure.[24] It is evident that the number of carbon atoms in the alkyl chain is linearly correlated with the logarithm of the Oswald coefficient for the solute partitioning, which is defined as the ratio of the molar concentration of a solute in the organic phase to that in the water phase, $K_p = [\mathrm{CA}]_{\mathrm{oil}}/[\mathrm{CA}]_{\mathrm{water}}$.

For CA partitioning between oil and water at low concentrations, the Oswald coefficient is related to the difference between the solvation free energy in the aqueous phase and that in the oil phase

$$\Delta\beta F_1 = -2.303 \cdot \log K_P. \tag{9.63}$$

Eq. (9.63) indicates that, at 25 °C, the free energy or the reversible work to transfer a CA molecule from water to oil is −1.364 kcal/mol per logK_p, which was found to be in good agreement with experimental observations. Because both the surface area and the integrated mean curvature of an alkyl chain are linearly proportional to the chain length, we can establish a linear correlation between the carbon number and the transfer free energy (Figure 9.6). According to Eq. (9.60), the proportionality constant depends not only on the surface tension but also on the bending moduli associated with the mean and Gaussian curvatures.

Eq. (9.60) also provides also a theoretical basis to estimate the solvation free energy of biomacromolecules by using the surface tension and the solvent-accessible surface area (SASA). The semi-empirical procedure is commonly used to calculate interactions between biomacromolecules in aqueous systems.[25] Because a biomolecule is typically non-spherical and has heterogeneous surfaces, one may assume that the total solvation free energy is a sum of contributions from each part of the molecule calculated from the corresponding surface tension and SASA.

9.2.5 The Born Model

When a charged species is dissolved in liquid water or a polar solvent, it results in a strong angular polarization of the surrounding molecules due to the charge–dipole interactions. The electrostatic

---

24 Smith R. and Tanford C., "Hydrophobicity of long-chain alkyl carboxylic-acids, as measured by their distribution between heptane and aqueous-solutions", *PNAS* 70 (2), 289–293 (1973).

25 Leach A. R., *Molecular modelling: principles and applications*. Prentice Hall, 2001.

energy cannot be described with the mean-field approximation because the direct integration of the electrostatic energy diverges. By assuming that the solvent is a dielectric continuum, the Born model provides a simple yet reasonably accurate description of the electrostatic effect.[26]

To introduce the Born model, consider the charging process for a spherical particle of radius a to acquire the electrostatic charge from zero (no electrostatic solvation) to a full charge q in a dielectric medium. The charging process entails a reversible work due to the addition of incremental charge to the partially charged solute. Integration of the electrostatic energy leads to

$$F_1^B = \int_0^q \frac{q' dq'}{4\pi\varepsilon_0 \varepsilon a} = \frac{q^2}{8\pi\varepsilon_0 \varepsilon a} \tag{9.64}$$

where superscript B denotes the Born model, ε_0 denotes the vacuum permittivity, and the dimensionless parameter ε is the static dielectric constant (viz., the relative permittivity) of the solvent.[27]

For small ions, the electrostatic effect often dominates the solvation free energy; its contribution is typically larger than that due to cavity formation or vdw interactions by more than one order of magnitude. Therefore, to a good approximation, non-electrostatic interactions can be ignored for calculating the solvation free energies of small ions. From Eq. (9.64), we can calculate the reversible work or the free energy to transfer a charged particle of radius a from a vacuum to the solvent

$$\Delta F_1 = \frac{q^2}{8\pi\varepsilon_0 a}\left(\frac{1}{\varepsilon} - 1\right). \tag{9.65}$$

Accordingly, the energy of solvation is given by

$$\Delta H_1 = \partial \Delta \beta F_1 / \partial \beta = \frac{q^2}{8\pi\varepsilon_0 a}\left(\frac{1}{\varepsilon} - 1 + \frac{T}{\varepsilon^2}\frac{\partial \varepsilon}{\partial T}\right). \tag{9.66}$$

The Born model for ion solvation has two parameters, i.e., the ion radius a and the solvent dielectric constant ε. While the bulk dielectric constant is experimentally known, the ion radius in a liquid environment is difficult to determine by theory or experimental methods. Traditionally, the ion radius is assumed to be the same as that derived from the crystal structure of an ionic solid. The adoption of so-called Pauling radii leads to reasonable results for predicting the solvation free energies of anions. However, slightly larger values of ionic radii must be used for cations in order to achieve quantitative results.

For a cation ion dissolved in water, the Born radius, as determined from Eq. (9.65), is about 0.6–0.8 Å larger than its Pauling radius; and the Born radius of an anion is only slightly, about 0.1 Å, larger than the crystal radius.[28] Anions have a smaller difference between the Born radius and the Pauling radius because the separation between an anion and the slightly positive charged hydrogen atoms of the surrounding water molecules is smaller than that between a cation and the slightly negatively charged oxygen atoms. Because the Born model uses the bulk dielectric constant as the input, the ion radius should be understood as that corresponding to a cavity accommodating the dissolved ionic species.[29] In general, the cavity radius is expected to change with different solvents. As shown in Figure 9.7, with the experimental data for the dielectric constant of water and the covalent radii of cations and the crystal radii of anions as the input, the Born model predicts the solvation energy for over 30 ions, ranging in valence from −1 to +4, in excellent agreement with experimental data.[30]

---

26 Born M., "Volume and heat of hydration of ions", *Z. Phys.* 1, 45–48 (1920).

27 At 25 °C and 1 atm, the dielectric constant for liquid water is around 78.

28 Latimer W. M., Pitzer K. S., and Slansky C. M., "The free energy of hydration of gaseous ions, and the absolute potential of the normal calomel electrode", *J. Chem. Phys.* 7 (2), 108–111 (1939).

29 Rashin, A.A. and Honig B., "Reevaluation of the Born model of ion hydration", *J. Phys. Chem.* 89 (26): 5588–5593 (1985).

30 Rashin, A. A. and Honig B., "Reevaluation of the Born model of ion hydration", *J. Phys. Chem.* 89 (26): 5588–5593 (1985).

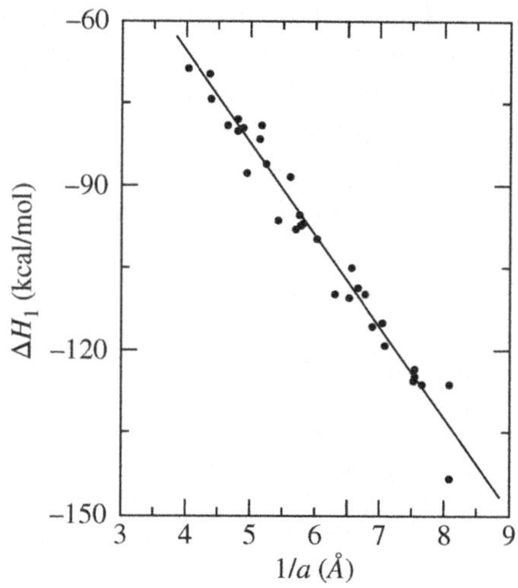

Figure 9.7 Comparison of experimental (dots) and predicted (solid line) enthalpies of ion hydration (per unit charge) at 298 K. Here, the ionic radii are approximated with the covalent radii for cations and the Pauling radii for anions. Source: Adapted from Rashin and Hoing.[29]

9.2.6 The Generalized Born (GB) Model

The generalized Born (GB) model is often used to predict the electrostatic part of the solvation free energies for biomacromolecules, in particular for proteins. Like the Born solvation theory, it describes the electrostatic interaction between a solute molecule and its surrounding solvent.

For a solute molecule with multiple charge sites, the electrostatic part of the solvation free energy is calculated from Coulomb's law[31]

$$\Delta F_1^{GB} = -\frac{1}{2}\left(1 - \frac{1}{\varepsilon}\right) \sum_{i,j} q_i q_j / f_{ij}^{GB}(r_{ij}, a_i, a_j) \tag{9.67}$$

where subscripts i and j denote atoms from the solute molecule, q_i and q_j are the atomic partial charges, a_i and a_j are the effective Born radii, and the denominator represents an effective interaction distance between charged sites derived from the linearized Poisson–Boltzmann (PB) equation

$$f_{ij}^{GB}(r_{ij}, a_i, a_j) = \sqrt{r_{ij}^2 + a_i a_j \exp\left(-\frac{r_{ij}^2}{4a_i a_j}\right)}. \tag{9.68}$$

In practical applications, the Born radius of an atom at position $\mathbf{r}_i$ can be estimated from

$$\frac{1}{a_i^3} = \frac{3}{4\pi} \int' \frac{d\mathbf{r}}{|\mathbf{r} - \mathbf{r}_i|^3} \tag{9.69}$$

where the prime sign denotes that the integration is carried out over the space accessible to water molecules (e.g., defined by the solvent-excluded surface). To construct the solvent-excluded surface, the atomic radii and the probe radius are often used as adjustable parameters.[32]

The GB model has been successfully applied to a wide range of systems, including proteins, nucleic acids, and small organic molecules in aqueous solutions.[31] It provides a computationally efficient way to incorporate the influence of solvent molecules on the electrostatic properties of solutes without explicitly representing the solvent molecules. It is important to note that the GB

31 Onufriev A. V. and Case D. A., "Generalized born implicit solvent models for biomolecules", *Annu. Rev. Biophys.* 48, 275–96 (2019).
32 Zhou H. X. and Pang X. D., "Electrostatic interactions in protein structure, folding, binding, and condensation", *Chem. Rev.* 118 (4), 1691–1741 (2018).

model is an approximation and may not capture all the microscopic details of the solvation process. It is most suitable for systems where the electrostatic interactions dominate while an accurate representation of the explicit solvent is not feasible due to computational limitations. A systematic improvement over the GB model has been proposed within the framework of the so-called variational solvation model (VSM).[33] VSM combines a continuum solvent model, such as the PB equation or the GB theory, with solvent-accessible surface area (SASA) models, linear response theory or MD simulation. By decomposing the solvation free energy into the electrostatic and various non-electrostatic (or nonpolar) contributions, the variational method is able to account for solvation effects in various applications, including the structure-based drug design, protein-ligand binding, and other fields of computational chemistry.[34]

9.2.7 Summary

This section discusses several important contributions to the solvation free energy, viz., cavity formation, van der Waals attraction, and electrostatic interactions. With the parameters derived from experimental data for the solvation free energy and certain macroscopic properties of the solvent, the solvent-implicit models provide valuable insights into the solvent effects on diverse chemical processes in liquid solutions. An accurate prediction of solvation is theoretically challenging because it depends on the microscopic properties of individual solute molecules as well as their interactions with an ensemble of solvent molecules. This topic has a rich history of both theoretical and experimental investigations and continues to be an active area of fundamental research.

9.3 Solvent-Mediated Potentials and Colloidal Forces

In this section, we discuss thermodynamic models for describing solute–solute and particle–particle interactions in a liquid medium. For simplicity, we assume that the solvent-mediated interactions depend only on the bulk properties of the solvent. In other words, we do not consider the atomistic details or inhomogeneous distribution of solvent molecules near the solute surface. The continuous approach intends to capture the essential physics underlying solute–solute or particle–particle interactions in a liquid environment. While the implicit-solvent models may not be accurate at the atomic scales, they are useful to describe the thermodynamic properties of real systems, in particular for aqueous electrolyte solutions and colloidal dispersions.

9.3.1 Coulomb's Law

According to classical electrostatics, the electrostatic potential between two charged spheres, here designated as i and j, in a continuous medium of dielectric constant ε is given by

$$w_{ij}^{C}(r) = \frac{q_i q_j}{4\pi\varepsilon_0 \varepsilon r} \tag{9.70}$$

33 Dzubiella J., Swanson J. M. J., and McCammon J. A., "Coupling hydrophobicity, dispersion, and electrostatics in continuum solvent models", *Phys. Rev. Lett.* 96 (8), 087802 (2006); Coupling nonpolar and polar solvation free energies in implicit solvent models", *J. Chem. Phys.*, 124 (8), 084905 (2006).

34 Wei G. W. and Zhou Y. "Variational methods for biomolecular modeling", In: Wu, J. (eds) *Variational methods in molecular modeling*. Springer-Nature, 2017.

where q_i and q_j stand for the amounts of charge in each particle, ε_0 is the permittivity of the free space, and r represents the center-to-center distance. In dimensionless units, Eq. (9.70) can be rewritten as

$$\beta w_{ij}^C(r) = Z_i Z_j \left(\frac{l_B}{r}\right) \tag{9.71}$$

where Z_i denotes the valence, $l_B \equiv e^2/(4\pi\varepsilon_0\varepsilon k_B T)$ is called the Bjerrum length, i.e., the separation between two unit charges (e) when the electrostatic energy is the same as $k_B T$. For liquid water at room temperature, the dielectric constant is $\varepsilon \approx 78$, which predicts $l_B \approx 7.14$ Å. For electrostatic interactions in the vacuum, the Bjerrum length is $l_B \approx 560$ Å. The existence of a polar solvent such as liquid water significantly reduces the strength of electrostatic interactions.

Unlike Coulomb's law for point charges in the vacuum, Eq. (9.70) is quantitative only at large separations. As discussed in Section 9.1, Coulomb's law shows significant deviations from the first-principles predictions of ion–ion interactions in liquid water. While the bulk dielectric constant is relevant only for long-range electrostatic interactions, the continuous model captures the important feature of charged systems. Similar to molecular models of neutral systems discussed previously in Chapters 7 and 8, a precise intermolecular potential is often not required for describing the macroscopic properties of charged systems.

We may elucidate the application of the Coulomb potential with a simple example. More extensive discussions of the implicit-solvent model for ionic systems will be discussed in later sections. Eq. (9.71) suggests that, at room temperature, the solubility of NaCl (or any other salt) in liquid water may be estimated from the difference between the electrostatic energy between Na^+ and Cl^- ions at contact $\sigma \approx 3.21$ Å (the atomic radii of Na and Cl ions are 1.54 and 1.67 Å, respectively) and that at infinite separation

$$\ln x_s \approx \beta\Delta\mu_{salt} \approx -l_B/\sigma \tag{9.72}$$

Eq. (9.72) predicts that, at saturation, the mole fraction of the salt is $x_s \approx 0.108$, which agrees well with the experimental value of 0.11. While the numerical result might be coincidental because the solubility depends on multitudes of interactions besides ion pairing, Eq. (9.72) provides valuable insights into the temperature effect and the dependence of the solvent dielectric constant on the solubility of an inorganic salt. For example, we can explain why the solubility of a salt increases with temperature while falls as the average diameter of cations and anions (σ) is reduced (e.g., the solubility of CsBr is much larger than that of NaF or LiF).

As shown in Figure 9.8, experimental data support a linear relationship between the logarithm of salt solubility $\log(x_s)$ and the inverse of the solvent dielectric constant $1/\varepsilon$ as predicted by Eq. (9.72).[35] For NaCl in different solvents, the straight line passes through $x_s = 1$ when $\varepsilon = \infty$, suggesting that the interaction between Na^+ and Cl^- ions is dominated by the Coulomb potential. For glycine, a zwitterionic molecule ($NH_3^+CH_2COO^-$), the solubility in four different organic solvents also shows a straight light, even though it does not dissociate into cations and anions. In this case, the influence of the solvent dielectric constant on the solubility may be interpreted with the Bell equation for the solvation free energy of a dipolar molecule[36]

$$F_1^d \approx -\frac{d^2}{8\pi\varepsilon_0 a^3}\left(1 - \frac{3}{2\varepsilon + 1}\right) \tag{9.73}$$

35 Israelachvili, J. N., *Intermolecular and surface forces*. Academic Press, Burlington MA, 2011.
36 Bell R. P., "The electrostatic energy of dipole molecules in different media", *Trans. Faraday Soc.* 27, 797–802 (1931).

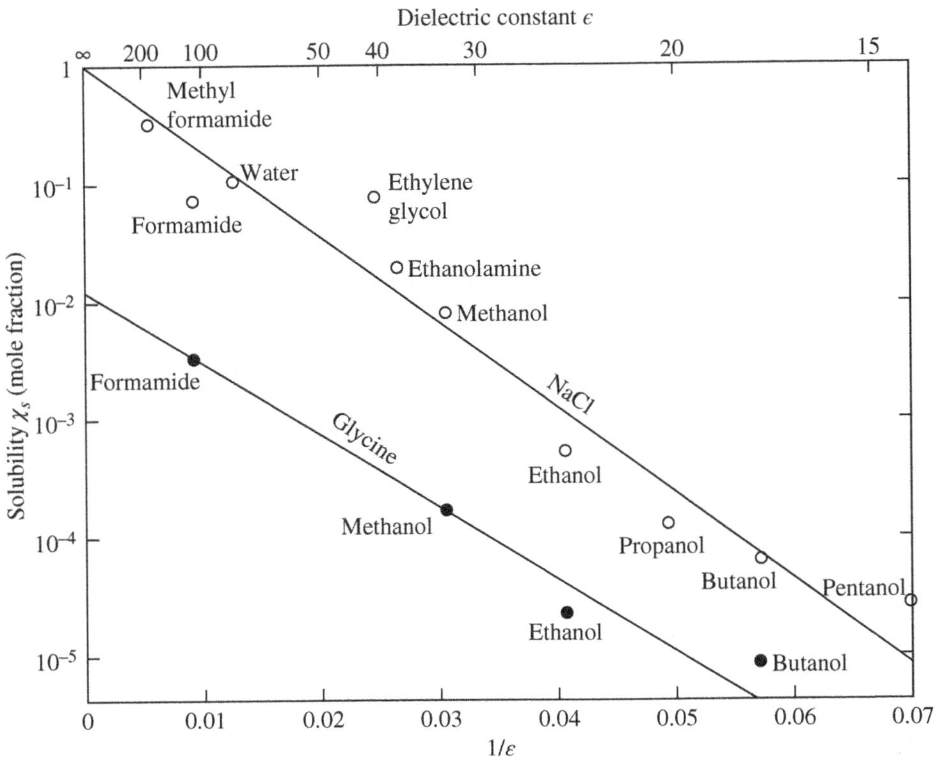

Figure 9.8 Solubilities of sodium chloride and glycine in solvents of different dielectric constants ε at 25 °C. Source: Reproduced from Israelachvili[35] with permission of Elsevier.

where d stands for the dipole moment of the solute molecule, and a is the cavity radius to accommodate the solute molecule in the solution. Because $\beta F_1^d + \ln x_s \approx$ constant at the saturation point and $2\varepsilon \gg 1$ for the solvents shown in Figure 9.8, the continuous model also explains why the glycine solubility exhibits a linear relation $\log(x_s) \sim 1/\varepsilon$.

9.3.2 The Gurney Potential

As noted before, solute–solute interactions are difficult to deduce from first-principles calculations and remain poorly understood even for relatively simple molecules in liquid water. Qualitatively, we expect that the repulsion due to the molecular size is similar to that between the same solute molecules in the vacuum. However, additional contributions arise from the solvent effects due to vdw interactions and the reorganization of solvent molecules near the solute surface. For aqueous systems, the latter contributions are responsible for hydrophobic interactions and various ion effects on the solubilities of organic compounds and proteins (e.g., the Hofmeister series, defined by the ability of ions to increase or decrease the solubility of proteins in water).

The Gurney potential was introduced to capture ion–ion interactions due to the reorganization of water molecules.[37] Because the structure of water molecules near a charged species is different from that in the bulk, we may postulate that the solvent-mediated potential is proportional to the

37 Gurney R. W., *Ionic processes in solution*. Dover, New York (1953).

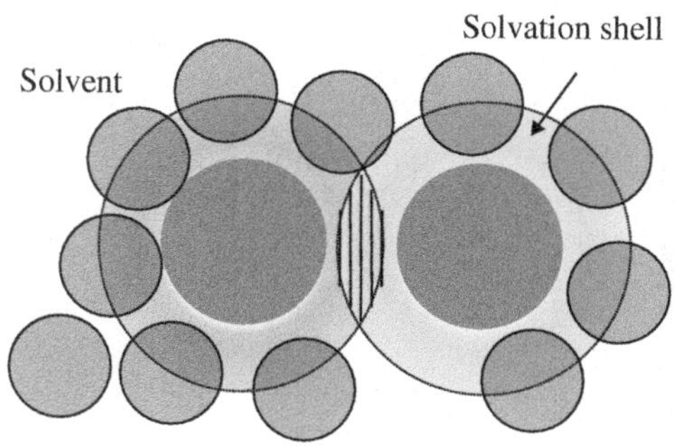

Figure 9.9 The Gurney overlap volume (hatshed area) for two colloidal particles (large spheres) in a solvent of small spheres. The dashed circles denote the edges of the solvation shells.

overlap volume of the solvation shells

$$w^{GUR}(r) = k\Delta V(r) \tag{9.74}$$

where the proportionality constant k describes the change in the free energy density of solvent molecules when they are released from the solute surface. As illustrated in Figure 9.9, $\Delta V(r)$ refers to the mutual volume of the solvation shells when the solute molecules are close to each other.

In Eq. (9.74), the Gurney overlap volume can be fixed based on geometric considerations. Assuming that the solvation shell consists of a single layer of solvent molecules with radius R_0, we can evaluate the overlap volume of two solvated spheres of radius R (Problem 9.9)

$$\Delta V = \pi(2a - r)^2(4a + r)/12 \tag{9.75}$$

where $a = R + R_0$ denotes the excluded-volume radius of each particle. If the two spheres have different sizes with radii of R_1 and R_2, the overlap volume is given by

$$\Delta V = \frac{\pi}{12r}(a_1 + a_2 - r)^2\left[r^2 + 2(a_1 + a_2)r - 3(a_1 - a_2)^2\right] \tag{9.76}$$

where $a_1 = R_1 + R_0$ and $a_2 = R_2 + R_0$. With the proportionality constant k treated as an adjustable parameter for each ion pair, the Gurney potential can be used to describe various properties of aqueous solutions including the Setchenow coefficients.[38] As discussed in Problem 9.7, the Setschenow equation is an empirical relationship used to describe the salt effect on the solvation free energy or solubility of non-electrolytes in aqueous solutions.

The Gurney potential can also be used to describe hydrophobic effects, which are linked with the structure of water molecules near nonpolar solutes.[39] The interaction due to the reorganization of water molecules within the solvation shell is responsible for the hydrophobic interaction. For example, we may estimate the interaction between two methane molecules in water with $a \approx 3.35$ Å, the average diameter of methane and water molecules, and the proportionality constant k fixed by the solvation free energy for ethane. Figure 9.10 presents the hydrophobic potential as a function of the methane–methane separation. The contact value (−0.72 kcal/mol at $r = 3.9$ Å) agrees well with those from MD simulation (−0.5 to −0.9 kcal/mol). The hydrophobic potential is much stronger than the vdW attractive energy for two isolated methane molecules in the vacuum.

9.3.3 The Lifshitz Theory

In addition to electrostatic potential and structural forces, charged species in a dielectric medium experience vdw interactions similar to those between neutral molecules or particles in the vacuum.

38 Krishnan C. V. and Friedman H. L., "Model calculations for Setchenow coefficients", *J. Solut. Chem.* 3, 727–744 (1974).

39 Wu J. and Prausnitz J. M., "Pairwise-additive hydrophobic effect for alkanes in water", *PNAS* 105 (28), 9512–9515 (2008).

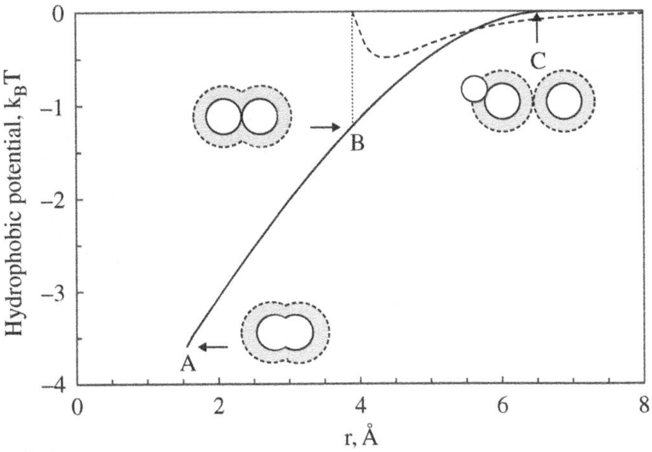

Figure 9.10 The hydrophobic potential between two methane molecules in liquid water at 25 °C obtained from a linear interpolation of that corresponding to the free energy of solvation of an ethane molecule [position A] and disappearance at the overlap of the solvation shells [position C]. For comparison, the dashed line shows the Lennard–Jones potential between two methane molecules in the vacuum. The perpendicular dotted line (position B) highlights the separation when the methane molecules are in contact. Reproduced from Wu and Prausnitz.[39]

In the case of colloidal or nanosized particles dispersed in a liquid medium, the dielectric-mediated vdw interaction is often described by the Lifshitz theory.[40]

For two spherical particles of radius R in a dielectric medium, the vdW potential varies with the distance in a power law as given by the Lennard–Jones (LJ) model

$$w^{vdw}(r) \approx -\frac{C_M}{r^6}. \tag{9.77}$$

Like Coulomb's law for ion–ion interactions in a solvent, Eq. (9.77) is quantitative only when the center-to-center distance r is much larger than the particle size. Assuming that macroscopic quantities such as dielectric constants (ε) and refractive indices (n) are applicable to the colloidal particles, the Lifshitz theory predicts that the proportionality constant C_M is given by

$$C_M \approx \left[3k_BT\left(\frac{\varepsilon_s - \varepsilon}{\varepsilon_s + 2\varepsilon}\right)^2 + \frac{\sqrt{3}h\nu_e}{4}\frac{\left(n_s^2 - n^2\right)^2}{\left(n_s^2 + 2n^2\right)^{3/2}}\right] R^6 \tag{9.78}$$

where subscript "s" denotes the solute, h is the Planck constant, ν_e is the plasma frequency of the free electron gas, typically in the range of $(3-5)\times10^{15}$ s^{-1}. At ambient conditions, $h\nu_e \gg k_BT$ and Eq. (9.78) can be simplified as

$$C_M \approx \frac{\sqrt{3}h\nu_e R^6}{4}\frac{\left(n_s^2 - n^2\right)^2}{\left(n_s^2 + 2n^2\right)^{3/2}}. \tag{9.79}$$

For example, at $T = 298$ K and $\nu_3 = 4\times10^{15}$ s^{-1}, $\beta h\nu_e \approx 644$, which justifies the approximation leading to Eq. (9.79).

The Lifshitz theory provides useful insights into solvent-mediated vdw interactions among colloidal particles. According to Eq. (9.79), the vdw interaction in a dielectric medium is retarded in comparison with that in the vacuum (where $n = 1$). For example, for two nonpolar particles with

40 Lifshitz E. M., "The theory of molecular attractive forces between solids", *Soviet Phys. JETP-USSR* 2 (1), 73–83 (1956).

a refractive index $n_s = 1.5$, the strength of the vdW potential in a solvent of $n = 1.4$ will be reduced from its value in the free space by a factor of

$$\frac{(1.5_s^2 - 1^2)^2}{(1.5_s^2 + 2\times 1^2)^{3/2}} \div \frac{(1.5_s^2 - 1.4^2)^2}{(1.5_s^2 + 2\times 1.4^2)^{3/2}} \approx 32.5. \tag{9.80}$$

Eq. (9.79) also suggests that the vdW attraction becomes exceedingly small when the refractive index of the colloidal particles matches that of the solvent. The strategy can be utilized to control colloidal stability and prepare experimental "hard-sphere systems". The latter is often utilized to study the fundamentals of phase transitions such as crystallization and glass formation without attraction effects.[41]

The Lifshitz theory is directly applicable to vdw interactions between dissimilar colloidal particles. For dissimilar particles i and j of radii R_i and R_j in a continuous medium separated by a distance much larger than their average size, the dispersion potential is also given by Eq. (9.77) but the proportionality constant is now given by

$$C_M \approx 3k_BTR_i^3R_j^3\left(\frac{\varepsilon_i - \varepsilon}{\varepsilon_i + 2\varepsilon}\right)\left(\frac{\varepsilon_j - \varepsilon}{\varepsilon_j + 2\varepsilon}\right) + \frac{\sqrt{3}hv_eR_i^3R_j^3}{2}\frac{\left(n_i^2 - n^2\right)\left(n_j^2 - n^2\right)}{\sqrt{\left(n_i^2 + 2n^2\right)\left(n_j^2 + 2n^2\right)}\left[\sqrt{\left(n_i^2 + 2n^2\right)} + \sqrt{\left(n_j^2 + 2n^2\right)}\right]} \tag{9.81}$$

Eq. (9.81) predicts that, unlike that for interaction between identical particles, the vdW potential between two dissimilar particles in a dielectric medium can be either attractive or repulsive, depending on their refractive indices relative to that of the medium. The vdW repulsion takes place when the refractive index of the dielectric medium is in between those for the individual particles.

9.3.4 The DLVO Theory

Like the potential between a pair of solute molecules in a solvent, the interaction between colloidal particles includes a short-range repulsion related to the particle size, as well as solvent-mediated interactions. The short-range repulsion is typically represented by the hard-sphere model because the colloidal particles are much larger than the solvent molecules. Typically, the vdW and electrostatic interactions are described by the DLVO theory,[42] which is based on the PB equation for ion distributions (Section 9.4) and Hamaker's method for vdW attraction (Appendix 9.A).

As discussed above, the power law decay of the vdW potential described by Eq. (9.77) is adequate when the particle size is much smaller than the center-to-center distance. For interaction between colloidal particles over the entire range of separations, we may estimate the vdW potential following Hamaker's method. For colloidal particles of radius R, it predicts that the solvent-mediated vdw interaction is given by

$$u^{vdw}(r) \approx -\frac{A_H}{3}\left\{\frac{R^2}{r^2 - 4R^2} + \frac{R^2}{r^2} + \frac{1}{2}\ln\left(1 - \frac{4R^2}{r^2}\right)\right\} \tag{9.82}$$

where A_H stands for the Hamaker constant, and r is the center-to-center distance.

41 Kegel W. and Blaaderen A. V., "Direct observation of dynamical heterogeneities in colloidal hard-sphere suspensions", *Science* 287, 290–293 (2000).

42 This potential diverges when $r = 2a$. Because of the surface solvation, $r > 2a$ in a real colloidal dispersion.

Table 9.1 Hamaker constants of selected materials in liquid water at 25 °C.

	A_H, 10^{-21} J		A_H, 10^{-21} J
Air	37	PMMA	1.47
Aluminum oxide	27.5	Polystyrene	13
Copper	300	Polycarbonate	3.5
Diamond	138	Protein	5–12
Gold	90–300	PTFE	2.9
Hexadecane	4.9	Quartz	1.6
Monoclinic mica	13.4	Silver	100–400
Muscovite mica	2.9	Tetradecane	3.8
Octane	3.6	Titanium oxide	60
Pentane	2.8	Zirconium	130

PTFE, polytetrafluoroethylene; PMMA, polymethyl methacrylate.
Data source: Parsegian.[43]

In Hamaker's method, the vdW potential between colloidal particles in a solvent is approximated by a pairwise summation of the solvent-mediated intermolecular interactions. The Hamaker constant is thus related to the molecular density and interaction parameter

$$A_H = \pi^2 C_M \rho^2 \tag{9.83}$$

where C_M represents the energy parameter given by Eq. (9.78) or (9.81), and ρ is the number density of molecules within the colloidal particle. Eq. (9.83) indicates that the vdW attraction between colloidal particles depends on the molecular composition and the density of each particle. Table 9.1 presents the Hamaker constants for some conventional chemicals/materials in liquid water.[43]

When two colloidal particles are separated by a large distance ($r \gg R$), Eq. (9.82) reduces to a power-law potential that is the same as that given by Eq. (9.77)

$$w^{vdw}(r) \approx -\frac{16A_H}{9}\left(\frac{R}{r}\right)^6. \tag{9.84}$$

When $r \sim 2R$, the interaction between two large particles becomes similar to that between two flat surfaces. In that case, we have

$$w^{vdw}(r) \approx -\frac{A_H R}{12D} \tag{9.85}$$

where $D = r - 2R$ represents the surface-to-surface distance. The vdW attraction at small separations is often responsible for the aggregation of colloidal particles. In Appendix 9A, we discuss the colloidal forces predicted by the Hamaker theory in more detail.

To describe the electrostatic interaction between charged particles, we use a modified Coulomb potential to account for small ions in colloidal systems. As discussed in Section 9.4, the electrostatic energy between two charged particles of radius R in an electrolyte solution can be approximated by

$$w^C(r) \approx \frac{q^2}{4\pi\varepsilon_0\varepsilon}\left(\frac{e^{\kappa R}}{1+\kappa R}\right)^2 \frac{e^{-\kappa r}}{r} \tag{9.86}$$

43 Parsegian V. A., *Van der Waals forces: a handbook for biologists, chemists, engineers, and physicists*. Cambridge University Press, New York, 2006.

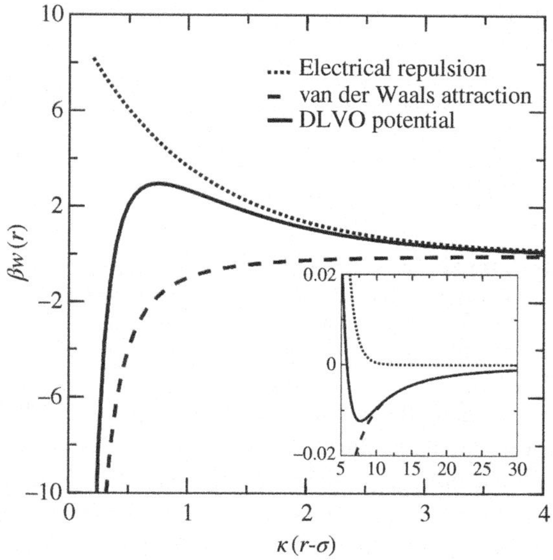

Figure 9.11 Schematic of the DLVO potential for interaction between charged particles in a dilute electrolyte solution. The inset shows the secondary minimum at large separations.

where q is the particle charge, and κ is a screening parameter. Eq. (9.86) may be understood as a "screened" Coulomb potential, i.e., the electrostatic energy is reduced (viz., "screened") from that corresponding to $\kappa = 0$ due to the electrostatic interaction of the colloidal particles with small ions in the system.

As shown in Figure 9.11, the DLVO theory predicts that the electrostatic interaction between colloidal particles is repulsive, but due to the vdW attraction, the overall potential exhibits a deep attraction near contact (viz., the primary minimum energy) and a maximum repulsion at a distance comparable to the Debye length $1/\kappa$ (see Section 9.4). At low salt concentrations or high surface charge densities, the colloidal potential is primarily repulsive because of the strong electrostatic interaction. At high salt concentrations or small surface charge densities, the interaction is dominated by the vdW attraction. Because the vdW attraction becomes exceedingly large at the primary minimum, a colloidal dispersion is stable only if the maximum potential is significantly larger than the thermal energy ($k_B T$). Otherwise, the particles would coagulate, leading to colloidal flocculation.

One of the characteristic features of the DLVO potential is that it may go through a shallow minimum (viz., the secondary minimum) before it vanishes. The secondary minimum takes place at a large separation depending on the electrolyte concentration, and it has implications for the structural formation of colloidal particles as well as interactions among biological cells.[44]

9.3.5 Force of Depletion

The depletion interaction between colloidal particles in a solution containing non-adsorbing polymers or macromolecules was first studied by Asakura and Oosawa.[45] As shown schematically in Figure 9.12, non-adsorbing macromolecules are depleted from the colloidal surface due to the excluded volume effects. When two colloidal particles are far apart, the depletion layers are

44 Hermansson M., "The DLVO theory in microbial adhesion", *Colloids Surf. B: Biointerfaces* 14, 105–119 (1999).
45 Asakura S. and Oosawa F., "On interaction between two bodies immersed in a solution of macromolecules", *J. Chem. Phys.* 22 (7), 1255–1256 (1954); "Interaction between particles suspended in solutions of macromolecules," *J. Polym. Sci.* 33, 183–192 (1958).

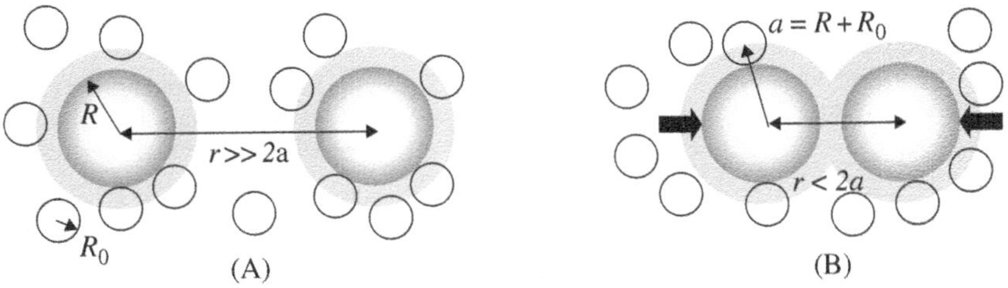

Figure 9.12 Schematic of the depletion force between colloidal particles. When two colloidal particles (big spheres) are far apart (A), their depletion layers (shaded areas) are independent of each other. At small separation (B), the overlap of the depletion layers results in an attraction due to the imbalance of the distribution of non-adsorbing macromolecules (small spheres).

independent; the depletion potential arises at small separations due to the overlap of the depletion layers, much like the Gurney potential discussed above.

If the colloidal particles and non-adsorbing macromolecules are represented by hard spheres of radii R and R_0, the depletion layer for each particle is defined by a spherical shell where the inner radius is identical to that of the colloidal particle R and the outer radius is $a = R + R_0$ (Figure 9.12A). At a small separation, the overlap of the depletion layers expands the total volume accessible to the non-adsorbing macromolecules, resulting in an effective attraction due to the increase in entropy (Figure 9.12B).

According to the Asakura–Oosawa (AO) theory, the depletion potential is given by

$$w(r) \approx \begin{cases} -\Pi_c \Delta V & r \leq 2a \\ 0 & r > 2a \end{cases} \tag{9.87}$$

where Π_c represents the osmotic pressure of the non-adsorbing macromolecules, and ΔV is the same as the Gurney overlap volume. Intuitively, Eq. (9.87) can be interpreted by using the scaled-particle theory (SPT) discussed in Section 9.2: the free energy for transferring a pair of colloidal particles from a pure solvent to the solution containing non-adsorbing macromolecules is given by the osmotic pressure multiplied by the total excluded volume of the two colloidal particles. The depletion potential reflects the variation of the transferring free energy as a function of the particle–particle separation. Because the attraction arises exclusively from the increase in free volume for the non-adsorbing macromolecules, the force of depletion between colloidal particles is often called the entropic or steric force.

If the non-adsorbing macromolecules do not interact with each other, the osmotic pressure in Eq. (9.76) can be predicted from van't Hoff's equation. In dimensionless units, the depletion potential between colloidal particles of radius R in a solution of non-absorbing macromolecules of radius R_0 is then given by (Problem 9.9)

$$\beta w(r) \approx \begin{cases} -\eta_p (1 + R/R_0)^3 \left[1 - \frac{3r}{4a} + \frac{r^3}{16a^3}\right] & r \leq 2a \\ 0 & r > 2a \end{cases} \tag{9.88}$$

where η_p stands for the volume fraction of the non-adsorbing macromolecules, and $a = R + R_0$. Eq. (9.107) was derived independently by Vrij[46] and is thus often referred to as the AOV equation. A similar equation can be derived for colloidal particles of different sizes.

Figure 9.13 shows the depletion potential at different volume fractions of non-adsorbing macromolecules predicted by Eq. (9.107). The entropic force increases linearly with the concentration of non-adsorbing macromolecules and is also influenced by the size ratio R_0/R. If the interaction between non-adsorbing macromolecules is not negligible, the osmotic pressure may be calculated

46 Vrij A., "Polymers at interfaces and the interactions in colloidal dispersions", *Pure Appl. Chem.* 48, 471 (1976).

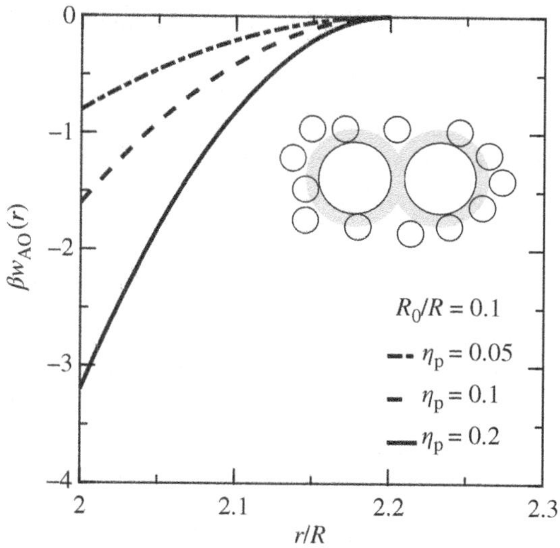

Figure 9.13 The depletion potential between two colloidal particles of radius R surrounded by depletion particles of radius R_0 predicted by the AOV equation, Eq. (9.88). Here η_p is the packing fraction of the depletion particles.

from an equation of state for hard spheres (e.g., the Carnahan–Starling equation). When the concentration is beyond a certain limit, the non-adsorbing macromolecules will be enriched near the colloidal surface and induce a repulsive barrier before the attraction near contact that is not captured by the AOV equation.

When the colloidal particles are surrounded by polymer chains with the radius of gyration significantly larger than the colloidal radius, the distribution of polymer segments near the particle surface is determined by the segment–particle interactions and intramolecular correlations of the polymer chains. In that case, the AOV equation is also not valid.[47]

9.3.6 Summary

Implicit-solvent models are commonly used to describe the thermodynamic properties and phase behavior of electrolyte solutions, colloidal dispersions, and biomolecular systems. Qualitatively, a solvent-mediated potential is analogous to intermolecular interaction in the vacuum, which may be decomposed into contributions due to short-range repulsion, vdW attraction, and electrostatic interactions. The presence of solvent molecules gives rise to additional contributions, such as structural forces and depletion effects.

While significant progress has been made in the past few decades using all-atom molecular dynamics simulation, predicting solvent-mediated potentials based on first-principles calculations remains challenging from a practical perspective. The theoretical prediction requires not only accurate quantum-mechanical calculations but also molecular simulation for performing the ensemble average over the solvent configurations. By contrast, phenomenological descriptions of solvent-mediated interactions provide a practical approach to capture the essential physics. In particular, solvent-implicit models are valuable for the practical application of the McMillan–Mayer (MM) solution theory, which relates thermodynamic properties to solvent-mediated pair interactions, as well as for comprehending diverse behaviors observed in colloidal dispersions and biophysical processes. By incorporating the essential aspects of solvent-mediated potentials, these models facilitate the quantitative understanding and prediction of thermodynamic properties, phase behavior, and complex interactions in solvated systems, contributing to advancements in various scientific and technological fields.

47 Li Z. and Wu J., "Potential distribution theorem for the polymer-induced depletion between colloidal particles", *J. Chem. Phys.* 126, 144904 (2007).

9.4 Electrostatics in Dilute Electrolytes

In this section, we discuss fundamental concepts of electrostatic interactions in ionic systems and explore their applications within the context of the Debye-Hückel (DH) theory for dilute electrolyte solutions. The applications of similar concepts to concentrated systems and polyelectrolytes will be elaborated in later sections.

9.4.1 The Poisson–Boltzmann (PB) Equation

Many conventional models of electrolyte solutions can be deduced from the PB equation whereby the ionic species are represented by point charges and the solvent or background medium by a dielectric continuum. The inhomogeneity of electrostatic interactions is introduced either by an external potential (e.g., electrolytes near an electrode or a charged surface) or by local interactions or fluctuation effects (e.g., the electrical potential near a charged particle).

According to classical electrostatics, the Poisson equation relates the local electric potential $\psi(\mathbf{r})$ to the local charge density $q(\mathbf{r})$ through

$$\nabla^2\psi(\mathbf{r}) = -\frac{q(\mathbf{r})}{\varepsilon_0\varepsilon} \tag{9.89}$$

where $\mathbf{r}$ denotes position, ε is the background dielectric constant, and ε_0 is the permittivity of the free space. Here, by local, we mean that the position of interest is defined relative to an object (e.g., a charged particle or surface) that is responsible for the spatial variations of the electric potential and local charge density.

In an electrolyte solution, the local charge density arises from the electric charges of individual ions, and thus it can be calculated from the ionic density profiles

$$q(\mathbf{r}) = \sum_i Z_i e\rho_i(\mathbf{r}) \tag{9.90}$$

where e represents the unit charge, and Z_i stands for the ion valence. A key assumption in the PB equation is that, for each ionic species, the local density is determined by the mean-field electric energy as described by the Boltzmann distribution

$$\rho_i(\mathbf{r}) = \rho_i^0 \exp[-Z_i e\beta\psi(\mathbf{r})] \tag{9.91}$$

where ρ_i^0 is the number density of species i in the bulk, and $\beta = 1/(k_BT)$. Eq. (9.91) may be compared with the distribution of ideal-gas molecules in an external potential. Except that $\psi(\mathbf{r})$ depends on the local ionic densities, all non-electrostatic interactions between ionic species are neglected in the Boltzmann distribution. With appropriate boundary conditions, we can solve Eqs. (9.89)–(9.91) self-consistently, and from which we can obtain the local density of each ionic species and subsequently thermodynamic properties.

Despite its simplicity, the PB equation is analytically solvable only in a few special circumstances, e.g., for a symmetric electrolyte near a charged surface. To attain the analytical results, we may approximate the ionic densities in Eq. (9.90) by a linear expansion of the Boltzmann factor in Eq. (9.91)

$$q(\mathbf{r}) = \sum_i Z_i e\rho_i(\mathbf{r}) \approx \sum_i Z_i e\rho_i^0[1 - Z_i e\beta\psi(\mathbf{r})] = -\sum_i Z_i^2 e^2 \rho_i^0 \beta\psi(\mathbf{r}). \tag{9.92}$$

In writing the second equality in Eq. (9.92), we have used $\sum_i Z_i e\rho_i^0 = 0$, i.e., the bulk electrolyte solution satisfies charge neutrality.

Table 9.2 Solutions of the linearized PB equation in spherical, cylindrical, and planar geometries at the asymptotic limit (viz., $x \to \infty$).

	$\varphi(x)$
Sphere	e^{-x}/x
Cylinder	$\pi e^{-x}/\sqrt{x}$
Plate	e^{-x}

Here, both the electric potential and the distance are expressed in dimensionless units, $\varphi \equiv \beta e\psi$ and $x = \kappa r$ where r stands fro the distance from the charge surface. These results omit the proportionality constants that vary with the boundary conditions.

The linearized PB equation is valid only when $|Z_i e\beta\psi(\mathbf{r})| \ll 1$, i.e., when the electric potential $\psi(\mathbf{r})$ is exceedingly small over the entire space. Substitution of Eq. (9.92) into Eq. (9.89) yields

$$\nabla^2\psi(\mathbf{r}) = \kappa^2\psi(\mathbf{r}) \tag{9.93}$$

where the Debye screening parameter κ is defined by

$$\kappa^2 \equiv \sum_i \rho_i^0 Z_i^2 e^2 \beta/(\epsilon_0\epsilon). \tag{9.94}$$

As discussed in the following, parameter κ has units of inverse length. Table 9.2 presents some asymptotic solutions to Eq. (9.93) for systems with planar, spherical or cylindrical symmetries.

Whereas the Poisson equation is exact within the framework of classical electrostatics, the Boltzmann distribution of charged species neglects effects due to finite ion sizes and non-electrostatic interactions. In addition, the PB equation ignores correlation effects due to ion-ion interactions. In general, the thermodynamic potential due to other forms of interactions and correlation effects can be systematically incorporated using the classical density functional theory (cDFT)[48]

$$\rho_i(\mathbf{r}) = \rho_i^0 \exp\left[-Z_i e\beta\psi(\mathbf{r}) - \beta V_i^{ext}(\mathbf{r}) - \beta\Delta\mu_i^{ex}(\mathbf{r})\right] \tag{9.95}$$

where $V_i^{ext}(\mathbf{r})$ represents the non-electrostatic part of the external potential, and $\Delta\mu_i^{ex}(\mathbf{r})$ is the deviation of the local excess chemical potential of ionic species i from its bulk value. Because $\psi(\mathbf{r})$ accounts for the overall electrical energy of all ionic species, neither the external potential nor the local excess chemical potential includes contributions due to the direct Coulomb interactions. Although Eq. (9.95) is theoretically rigorous, cDFT does not provide an exact expression for $\Delta\mu_i^{ex}(\mathbf{r})$. Nevertheless, accurate approximations for $\Delta\mu_i^{ex}(\mathbf{r})$ have been well established for various implicit-solvent models. From a practical perspective, the extensions of the PB equation are often hampered by the fact that non-electrostatic interactions are dependent on the atomistic details of the ionic species and solvent molecules.

9.4.2 The Debye–Hückel (DH) Theory

The DH theory may be understood as an application of the linearized PB equation to bulk electrolyte solutions.[49] As discussed in Section 7.2, the local ionic densities around an arbitrarily tagged

48 Wu J., "Understanding the electric double-layer structure, capacitance, and charging dynamics", *Chem. Rev.* 122, 10821–10859 (2022).

49 Debye P. and Hückel E., "On the theory of electrolytes. I. Freezing point depression and related phenomena". *Phys. Z.* 24, 185–206 (1923).

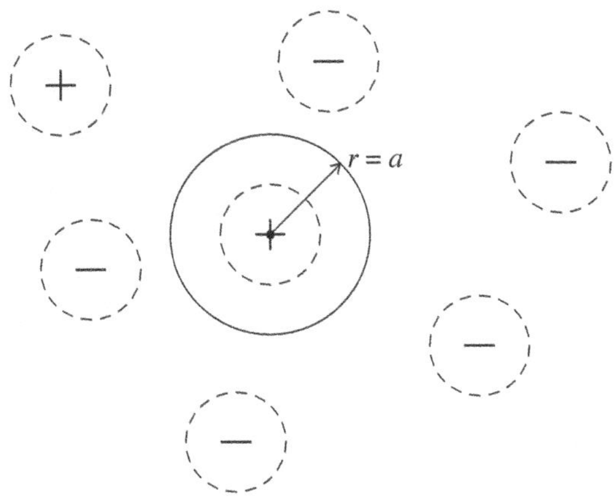

Figure 9.14 Schematic of charge distribution near a tagged ion in a dilute electrolyte solution.

ion are linearly proportional to the radial distribution functions (RDFs) of a uniform system. With the RDFs obtained from solving the linearized PB equation, we can derive the thermodynamic properties of ionic systems.

9.4.2.1 Local Electric Potential

Consider ion distribution in a dilute electrolyte solution near a tagged ion, as shown schematically in Figure 9.14. We designate parameter a as the closest center-to-center distance between any two ions, and Z_j as the valence of the tagged ion. Using the spherical coordinates with the tagged ion placed at the center, the linearized PB equation, Eq. (9.93), becomes

$$\nabla^2\varphi = \frac{1}{r^2}\frac{d}{dr}\left(r^2\frac{d\varphi}{dr}\right) = \kappa^2\varphi \tag{9.96}$$

where $\varphi(r) \equiv \beta e\psi(r)$ is the reduced electric potential.

We can solve Eq. (9.96) analytically with boundary conditions specified by the Gauss law for the electric field at $r = a$

$$\left.\frac{d\varphi}{dr}\right|_{r=a} = -\frac{Z_j e^2}{4\pi\,\varepsilon_0\varepsilon k_B T a^2} = -Z_j l_B/a^2 \tag{9.97}$$

where l_B is the Bjerrum length, and $\varphi(\infty) = 0$ in the bulk solution. Integration of Eq. (9.96) yields that, for $r \geq a$,

$$\varphi(r) = \frac{Z_j l_B e^{\kappa a}}{(1+\kappa a)}\cdot\frac{e^{-\kappa r}}{r}. \tag{9.98}$$

For $0 \leq r \leq a$, the Poisson equation can be solved by using boundary conditions for the continuities of the electric potential and the electric field at $r = a$

$$\varphi(r) = \frac{Z_j l_B}{r} - \frac{Z_j l_B \kappa}{(1+\kappa a)}. \tag{9.99}$$

The first term on the right side of Eq. (9.99) corresponds to the reduced electric potential of the point charge $Z_j e$ at the position of the tagged ion, and the second term arises from the net charge of all other ions in the system.

As mentioned above, the PB equation assumes that ionic species are represented by point charges. For consistency, we may simplify Eq. (9.98) by taking the limit $a \to 0$:

$$\varphi(r) = \lim_{a\to 0}\frac{Z_j l_B e^{\kappa a}}{(1+\kappa a)}\cdot\frac{e^{-\kappa r}}{r} = Z_j l_B\frac{e^{-\kappa r}}{r}. \tag{9.100}$$

Alternatively, Eq. (9.100) can be obtained from $\kappa \to 0$, i.e., when the electrolyte concentration is infinitely small. The divergence of $\varphi(r)$ at $r = 0$ stems from the electric potential of the point charge at the origin.

Eq. (9.99) predicts that, as $a \to 0$, the electric potential at the position of the tagged ion is $\varphi_0 = -Z_j l_B \kappa$, which yields a finite electric energy of $-Z_j^2 l_B \kappa k_B T$ for the fixed charge. In other words, each ion interacts with all other ions in the system with an average electrostatic energy of

$$\beta u_0 = -Z_j^2 l_B \kappa. \tag{9.101}$$

9.4.2.2 Ionic Radial Distribution Functions

With an analytical expression for the electric potential, we can readily obtain the local ionic densities from the linearized Boltzmann equation, Eq. (9.100)

$$\rho_{ij}(r) = \rho_i^0(1 - Z_i\varphi) = \rho_i^0(1 - Z_i Z_j l_B e^{-\kappa r}/r) \tag{9.102}$$

where ρ_i^0 is the ion concentration in the bulk. As mentioned above, the local ionic densities are linearly proportional to the RDFs of the bulk electrolyte

$$g_{ij}(r) = \rho_{ij}(r)/\rho_i^0 = 1 - Z_i Z_j l_B e^{-\kappa r}/r. \tag{9.103}$$

As expected, the RDFs between different ionic species are symmetric, i.e., they are immaterial to the identity of the tagged particle. Eq. (9.103) predicts that different ion pairs have the same correlation length, which is uniquely determined by the Debye screening parameter κ.

The ion distribution predicted by the DH theory satisfies the overall charge neutrality and the Stillinger–Lovett second moment condition.[50] We can test the charge neutrality of the entire system by integrating the local ionic densities predicted by Eq. (9.102), which yields the total charge of ions surrounding the tagged charge

$$\begin{aligned} q_j &= \sum_i Z_i e \int d\mathbf{r}\rho_{ij}(r) = \sum_i Z_i e \int dr 4\pi r^2 \rho_i^0[1 - Z_i Z_j l_B e^{-\kappa r}/r] \\ &= -Z_j e\kappa^2 \int_0^\infty dr r e^{-\kappa r} = -Z_j e. \end{aligned} \tag{9.104}$$

Eq. (9.104) suggests that each ion is surrounded by a neutralizing "atmosphere" of ions with an opposite net charge. Besides, it indicates that the fraction of charge in the spherical shells of radius r and dr is proportional to

$$f(r) = \kappa r e^{-\kappa r}. \tag{9.105}$$

As shown in Figure 9.15, $f(r)$ exhibits a maximum at $r = 1/\kappa$. Accordingly, parameter $1/\kappa$ can be understood as a screening length, i.e., the tagged ion is "screened" by a cloud of opposite charge within distance $1/\kappa$.

The DH theory also satisfies the Stillinger–Lovett second moment condition (Problem 9.11)

$$\sum_{i,j} Z_i Z_j \rho_i^0 \rho_j^0 \int d\mathbf{r} r^2 g_{ij}(r) = -\frac{3}{2\pi l_B} \tag{9.106}$$

where the integration applies over the entire space. Like the charge neutrality condition, Eq. (9.106) is exact and is independent of the chemical details of ion interactions.

---

50 Stillinger F. H. and Lovett R., "Ion-pair theory of concentrated electrolytes. 1. Basic concepts", *J. Chem. Phys.* 48 (9), 3858–3868 (1968).

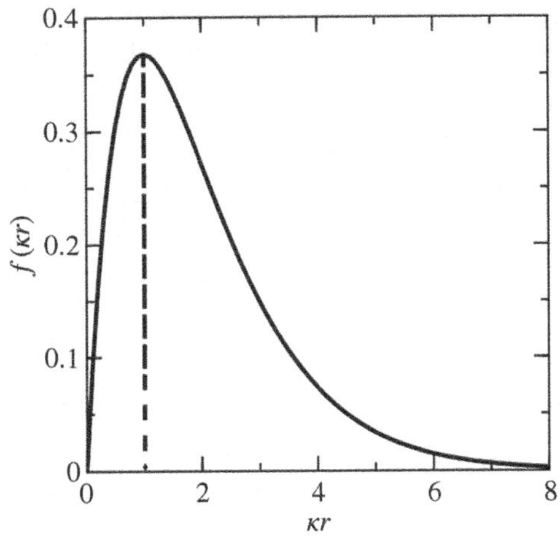

Figure 9.15 The fraction of neutralizing charge in the spherical shells around a tagged ion shows a maximum at separation $1/\kappa$, which is known as the Debye screening length.

9.4.2.3 Thermodynamic Properties Predicted by the DH Theory

Equipped with the RDFs given by (9.103), we can now derive thermodynamic properties following standard statistical-mechanical equations (within the MM framework). From the energy equation, we find the excess internal energy

$$\begin{aligned}\beta U^{ex}/V &= \frac{1}{2}\sum_i\sum_j \int d\mathbf{r}\rho_i^0\rho_i^0 g_{ij}(r)\beta u_{ij}(r)\\ &= \frac{1}{2}\sum_i\sum_j \int d\mathbf{r}\rho_i^0\rho_j^0[1 - Z_iZ_jl_Be^{-\kappa r}/r](Z_iZ_jl_B/r)\\ &= -2\pi\sum_i\sum_j \rho_i^0\rho_j^0Z_i^2Z_j^2l_B^2\int_0^\infty dre^{-\kappa r} = -\frac{8\pi l_B^2I^2}{\kappa} = -\kappa^3/8\pi \end{aligned} \tag{9.107}$$

where $\beta u_{ij}(r) = Z_iZ_jl_B/r$ is the Coulomb potential between ions i and j, $I \equiv \frac{1}{2}\sum_i \rho_i^0 Z_i^2$ is the ionic strength, and $\kappa^2 = 8\pi l_B I$. In the DH theory, the excess internal energy is exclusively affiliated with electrostatic interactions, i.e., it is the same as the summation of the average electrostatic energy of individual ions (Eq. (9.101))

$$\beta U^{ex}/V = \frac{1}{2}\sum_i \rho_i^0\beta u_0 = -\frac{1}{2}\sum_i \rho_i^0 Z_i^2 l_B\kappa = -l_B\kappa I = -\kappa^3/8\pi. \tag{9.108}$$

where ½ accounts for the fact that electrostatic energy arises from pair interactions.

The excess Helmholtz energy density can be obtained from the integration of the Gibbs–Helmholtz equation

$$\beta F^{ex}/V = -\int_0^\beta d\beta\ \kappa^3/(8\pi\beta) = -\kappa^3/12\pi. \tag{9.109}$$

In deriving Eq. (9.109), we have assumed $\kappa^2 \equiv \sum_i \rho_i^0 Z_i^2 e^2\beta/(\varepsilon_0\varepsilon) \propto \beta$, i.e., the solvent dielectric constant is independent of temperature. To avoid this unphysical assumption, Eq. (9.109) can be alternatively derived from a charging process to increase the valence of each ion from 0 to Z_j

$$\beta F^{ex}/V = \sum_j \rho_j^0 \int_0^1 d\xi\left[-l_B\kappa(\xi Z_j)^2\right] = -\frac{l_B\kappa}{3}\sum_j \rho_j^0 Z_j^2 = -\frac{2l_B\kappa I}{3} = -\frac{\kappa^3}{12\pi} \tag{9.110}$$

where the term within the square brackets represents the electrostatic energy of ion j due to its interactions with all other ions in the system.

From the excess Helmholtz energy, we can also derive the excess entropy

$$S^{ex}/(Vk_B) = \beta(U^{ex} - F^{ex})/V = -\kappa^3/24\pi, \tag{9.111}$$

and the excess chemical potential of each ionic species

$$\beta\mu_i^{ex} = (\partial\beta F^{ex}/\partial N_i)_V = -\kappa\, l_B Z_i^2/2. \tag{9.112}$$

Eq. (9.112) predicts that the electrostatic interactions lead to a negative deviation from the ideal solution. As discussed in the following, the excess chemical potential of each ionic species is directly related to the ionic activity coefficient.

According to the DH theory, the excess Gibbs energy is given by

$$\beta G^{ex}/V = \sum_i \rho_i^0 \beta\mu_i^{ex} = -\kappa\, l_B \sum_i \rho_i^0 Z_i^2/2 = -\kappa^3/8\pi. \tag{9.113}$$

Interestingly, the excess Gibbs energy is identical to the excess internal energy, implying the cancellation of effects due to osmotic pressure and entropy. More specifically, the osmotic pressure Π can be obtained from the difference between the Gibbs and Helmholtz energies

$$\beta\Pi = \sum_i \rho_i^0 - \kappa^3/24\pi \tag{9.114}$$

where the first term on the right side arises from that for an ideal solution (viz., van Hoff's law), and the second term accounts for electrostatic interactions. Alternatively, we can derive the osmotic pressure from the pressure equation (viz. the virial route)

$$\beta\Pi = \sum_i \rho_i^0 - \frac{2\pi}{3}\sum_i\sum_j \rho_i^0\rho_j^0 \int_0^\infty dr \frac{\partial\beta u_{ij}(r)}{\partial r} r^3 g_{ij}(r). \tag{9.115}$$

Eq. (9.114) is recovered by substituting the RDFs from Eq. (9.103) into (9.115). The identical results indicate that the DH theory satisfies the thermodynamic consistency.[51]

It should be understood that the above thermodynamic equations are all derived within the MM framework; they do not contain the solvation free energy of individual ions. In addition, the DH theory does not account for non-electrostatic interactions that may play an important role in predicting thermodynamic properties at finite electrolyte concentrations.

9.4.2.4 Activity Coefficients

Activity coefficients are commonly used in practical applications of electrolyte-solution theories. Formally, the activity coefficient of ionic species can be defined in terms of the excess chemical potential

$$\ln \gamma_i \equiv \beta\mu_i^{ex}. \tag{9.116}$$

For an electrolyte with stoichiometric relations of cations and anions given by $C_{\nu+}A_{\nu-}$, the mean ionic activity coefficient is defined as

$$\gamma_\pm \equiv \left(\gamma_+^{\nu_+}\gamma_-^{\nu_-}\right)^{1/(\nu_+ + \nu_-)}. \tag{9.117}$$

Substituting Eq. (9.112) into (9.116) predicts that the activity coefficient of individual ionic species is given by

$$\ln \gamma_i = -\kappa\, l_B Z_i^2/2. \tag{9.118}$$

51 Santos, A., Fantoni, R., Giacometti, A., Thermodynamic consistency of energy and virial routes: an exact proof within the linearized Debye-Hückel theory. *J. Chem. Phys.* 131 (18), 181105 (2009).

Accordingly, the mean ionic activity coefficient is

$$\ln \gamma_{\pm} \equiv \frac{\nu_+ \ln \gamma_+ + \nu_- \ln \gamma_-}{\nu_+ + \nu_-} = -\frac{\kappa l_B}{2(\nu_+ + \nu_-)} \left(\nu_+ Z_+^2 + \nu_- Z_-^2\right). \tag{9.119}$$

Because charge neutrality requires $Z_+\nu_+ + Z_-\nu_- = 0$, Eq. (9.119) can be simplified as

$$\ln \gamma_{\pm} = -\frac{\kappa l_B \nu_+^2 Z_+^2}{2(\nu_+ + \nu_-)} \left(\frac{1}{\nu_+} + \frac{1}{\nu_-}\right) = \frac{\kappa l_B Z_- Z_+}{2} = Z_- Z_+ \sqrt{2\pi l_B^3 I} \tag{9.120}$$

where $\kappa = \sqrt{8\pi l_B I}$ is used in the last equality. Because the Bjerrum length l_B depends only on temperature and the solvent dielectric constant, Eq. (9.120) predicts that the mean ionic activity coefficient is linearly proportional to $\sqrt{I}$, a main result of the DH theory that has been repeatedly validated by experiments. The linear relationship reflects the universal nature of Coulomb interactions.

Figure 9.16 compares the DH theory with experimental data and simulation results for the mean activity coefficients of a monovalent electrolyte (NaCl) in two polar solvents (water and methanol) and in the solvent mixture (60 wt% methanol and 40 wt% water)[52]. Because of the long-range nature of electrostatic interactions, an electrolyte solution shows much stronger thermodynamic non-ideality in comparison to neutral systems (e.g., methanol in water).

As expected, the DH theory is problematic at large salt concentrations and its performance deteriorates as the salt concentration increases. While the activity coefficients are not available at extremely dilute conditions (because they are difficult to obtain from either experiment or molecular simulation), the DH theory shows a clear trend at the asymptotic limit. As typical in practical applications, in Figure 9.16, the mean ionic activity coefficients are defined in terms of molality, i.e., the mole of salt per kg of solvent. As the chemical potential is invariant with

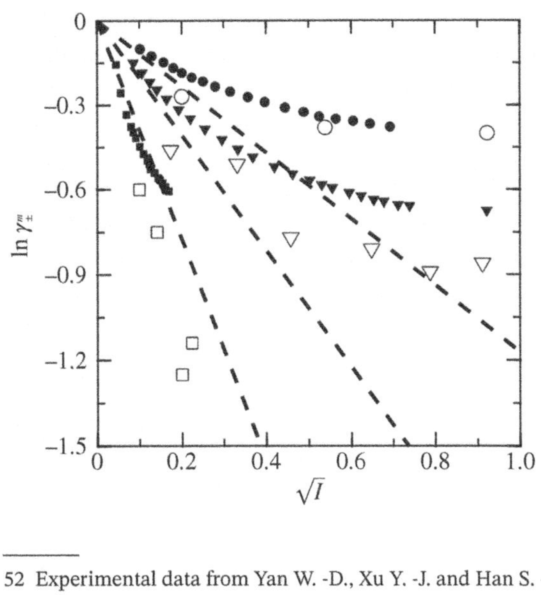

Figure 9.16 Mean ionic activity coefficients of NaCl in pure water, in pure methanol, and in a mixed solvent with 60 wt% methanol and 40 wt% water at 25 °C and 1 bar. Here, I stands for the ionic strength in the units of mol/kg of solvent. The filled markers are experimental data, the open symbols are from Monte Carlo simulations, and the dashed lines are predictions of the Debye–Hückel theory. The experimental and simulation data are from the literature.[52]

52 Experimental data from Yan W. -D., Xu Y. -J. and Han S. -J., "Activity coefficients of sodium chloride in methanol-water mixed solvents at 298.15 K", *Acta Chim. Sin.* 52, 937–946 (1994); MC simulation data from Saravi S. H. and Panagiotopoulos A. Z., "Activity coefficients and solubilities of NaCl in water–methanol solutions from molecular dynamics simulations", *J. Phys. Chem. B* 126, 2891–2898 (2022).

the units of ion concentrations, we can convert Eq. (9.120) into that corresponding to different concentration units (Problem 9.12)

$$\gamma_{\pm}^{m} = \gamma_{\pm}\frac{d}{d_0(1+0.001m_sM_s)} \tag{9.121}$$

where m_s stands for the salt concentration in molality, M_s is the molecular weight of the salt, d and d_0 are the mass densities of the electrolyte solution and the solvent, respectively. For dilute solutions, there is no significant difference between $\gamma_{\pm}^{m}$ and $\gamma_{\pm}$ because $0.001m_sM_s \ll 1$ and $d \approx d_0$.

9.4.3 Summary

The PB equation provides a useful theoretical framework to describe the electrostatic properties of ionic systems such as local electric potential and ion distributions. While it ignores non-electrostatic interactions such as excluded-volume effects important for electrolytes at finite ion concentrations, the PB equation captures the general behavior due to the long-range interactions between charged species.

The DH theory can be derived from the linearized PB equation. Although the derivation entails drastic assumptions that are strictly valid only in the limit of infinite dilute ion concentrations, the DH theory satisfies important thermodynamic self-consistency conditions that the PB equation does not fulfill. Despite its simplicity, the DH theory correctly describes many essential features of dilute electrolyte solutions. In particular, it predicts exact activity coefficients in the limit of infinite dilution and is thus also practically relevant when the properties of dilute electrolytes are of concern.

9.5 Extended Debye–Hückel Models

Electrolyte-solution theories in practical use are mostly based on the extensions or semi-empirical modifications of the DH theory.[53] Since its publication in 1923, numerous revisions have been proposed to incorporate the ion size effects, non-electrostatic interactions, and cation–anion association. In this section, we present three extended DH models. To avoid lengthy equations, our discussion emphasizes the physical insights rather than empirical correlations with a large number of parameters.

9.5.1 Modified DH Theories

The DH theory becomes problematic at a finite salt concentration because it neglects non-electrostatic interactions in determining ion distributions, particularly the excluded-volume effects. The short-range interactions are most significant between ionic species of opposite charges and can be partially remediated by adding a hard-sphere potential for ion–ion interactions. The extension leads to the primitive model (PM) of electrolyte solutions, which will be discussed in Section 9.6. Approximately, the size effects can also be incorporated into the DH formulism by setting the closest distance between the tagged ion and other ions in the deriving of the RDFs.

In the original DH theory, we introduce parameter a in order to determine the local electric potential near a tagged ion with valence Z_j. Intuitively, the parameter a may be understood as a

53 Wang S., Song Y., Zhang Y. and Chen C. C., "Electrolyte thermodynamic models in aspen process simulators and their applications", *Ind. Eng. Chem. Res.* 61 42, 15649–15660 (2022).

collision diameter. In dimensionless units, the linearized PB equation predicts that the reduced electric potential is given by (Eq. (9.98))

$$\varphi(r) \equiv \beta e\psi(r) = \frac{Z_j l_B e^{\kappa a}}{(1+\kappa a)} \cdot \frac{e^{-\kappa r}}{r}. \tag{9.122}$$

Without taking the limit $a \to 0$, the RDF becomes

$$g_{ij}(r) = \rho_{ij}(r)/\rho_i^0 = \begin{cases} 0 & r < a \\ 1 - \frac{Z_i Z_j l_B e^{\kappa a}}{(1+\kappa a)} \cdot \frac{e^{-\kappa r}}{r} & r \geq a \end{cases} \tag{9.123}$$

Eq. (9.123) accounts for the size effect by setting the closest distance between the tagged ion and other ions in the solution. It should be noted that non-electrostatic ion–ion interactions are not considered in deriving the local electric potential. As a result, Eq. (9.123) ignores the size effects on ion–ion correlations.

Similar to the original DH theory, Eq. (9.123) provides a starting point to derive thermodynamic properties. For example, we can obtain the reduced internal energy density from the energy equation

$$\begin{aligned} \beta U^{ex}/V &= \frac{1}{2}\sum_j \sum_i \int d\mathbf{r} \rho_j^0 \rho_i^0 g_{ij}(r) \frac{Z_i Z_j l_B}{r} \\ &= -\sum_j \sum_i \rho_i^0 \rho_j^0 Z_i^2 Z_j^2 \frac{2\pi l_B^2 e^{\kappa a}}{(1+\kappa a)} \int_a^\infty dr\, e^{-\kappa r} = -\frac{\kappa^3}{8\pi(1+\kappa a)} \end{aligned} \tag{9.124}$$

The second line in Eq. (9.124) follows the identity $\kappa^2 = 4\pi l_B \sum_i \rho_i^0 Z_i^2$.

As the excess internal energy is exclusively associated with electrostatic interactions, Eq. (9.124) can be alternatively derived from the average electrostatic energy for each ion due to its interaction with the ionic atmosphere. According to Eq. (9.122), the ionic atmosphere introduces an electric potential at the surface of the tagged ion[54]

$$\Delta\varphi(a) = \frac{Z_j l_B}{a(1+\kappa a)} - \frac{Z_j l_B}{a} = -\frac{Z_j l_B \kappa}{(1+\kappa a)}. \tag{9.125}$$

Assuming that the electric charge is located at the surface of the tagged ion, we can obtain the reduced electrostatic energy due to the interaction of the tagged ion with its ionic atmosphere

$$\beta u_0 = Z_j \Delta\varphi(a) = -\frac{Z_j^2 l_B \kappa}{(1+\kappa a)}. \tag{9.126}$$

The total excess energy corresponds to the summation of the electrostatic energies of individual ions

$$\beta U^{ex}/V = \sum_j \beta u_0 \rho_j^0/2 = -\frac{l_B \kappa}{2(1+\kappa a)} \sum_j \rho_j^0 Z_j^2 = -\frac{\kappa^3}{8\pi(1+\kappa a)} \tag{9.127}$$

where ½ accounts for the fact that electrostatic energy arises from pair interactions. Apparently, Eqs. (9.127) and (9.124) are identical.

As discussed in Section 9.4, the Helmholtz energy can be derived either by integrating the Gibbs–Helmholtz equation or by applying a charging process

$$\beta F^{ex}/V = -[\ln(1+\kappa a) - \kappa a + (\kappa a)^2/2]/4\pi a^3. \tag{9.128}$$

54 As discussed in Section 9.3, the electrostatic energy can also be obtained by considering the electric potential and charge at the center of the tagged ion.

The reduced excess chemical potential and ionic activity coefficient are given by

$$\beta\mu_i^{ex} = \ln\gamma_i = \left(\frac{\partial\beta F^{ex}}{\partial N_i}\right)_V = -\frac{Z_i^2\kappa l_B}{2(1+\kappa a)}. \tag{9.129}$$

Accordingly, the excess Gibbs energy is

$$\beta G^{ex}/V = \sum_i \rho_i^0 \ln\gamma_i = -\frac{\kappa^3}{8\pi(1+\kappa a)}. \tag{9.130}$$

The non-ideal part of the osmotic pressure can be obtained by the difference between the excess Gibbs energy and the excess Helmholtz energy

$$\beta\Pi^{ex} = \beta(G^{ex} - F^{ex})/V = -\frac{\kappa^3}{24\pi}D(\kappa a) \tag{9.131}$$

where

$$D(x) = \frac{3}{x^3}\left\{1 + x - \frac{1}{1+x} - 2\ln(1+x)\right\}. \tag{9.132}$$

As shown in Figure 9.17, $D(\kappa a)$ decreases monotonically with κa, suggesting that the inclusion of the size effects reduces the osmotic pressure.

The osmotic coefficient Φ and mean ionic activity coefficient $\gamma_\pm$ are thermodynamic properties most relevant to practical applications. For a single electrolyte $C_{\nu+}A_{\nu-}$ containing only one type of cation and one type of anion, the charge neutrality requires

$$\rho_+^0 Z_+ + \rho_-^0 Z_- = 0. \tag{9.133}$$

In terms of the total number density of ions, $\rho_0 = \rho_+^0 + \rho_-^0$, the cation and anion densities in the bulk solution are given by

$$\rho_+^0 = \rho_0 Z_-/(Z_- - Z_+), \tag{9.134}$$

$$\rho_-^0 = \rho_0 Z_+/(Z_+ - Z_-). \tag{9.135}$$

Accordingly, the ionic strength is given by

$$I = \left(\rho_+^0 Z_+^2 + \rho_-^0 Z_-^2\right)/2 = \rho_0 \mid Z_+ Z_- \mid /2. \tag{9.136}$$

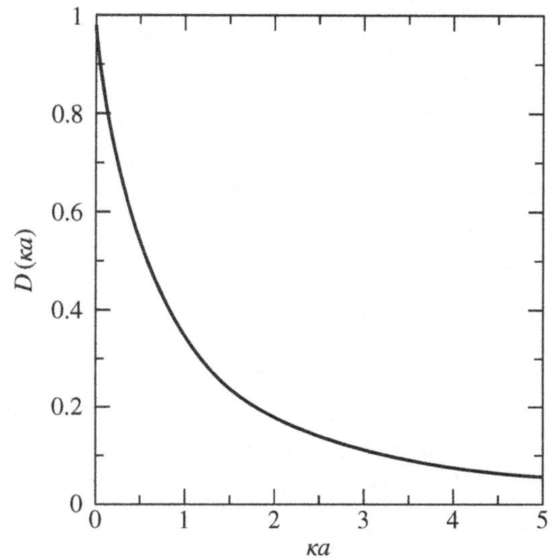

Figure 9.17 Function $D(x)$ in the modified DH theory.

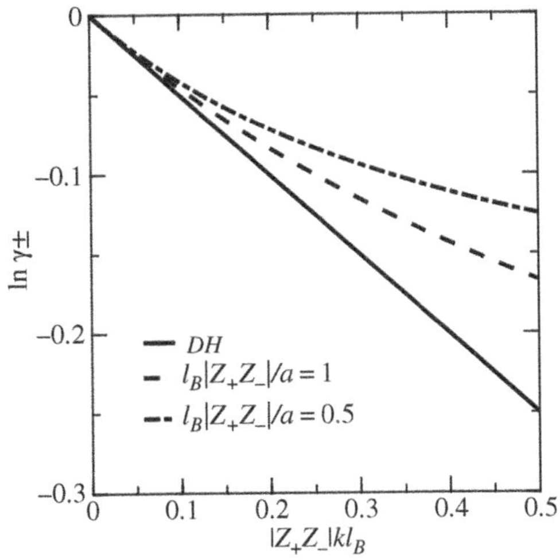

Figure 9.18 Dependence of the mean ionic activity coefficient on parameter a.

From Eq. (9.131), we find the osmotic coefficient predicted by the modified DH model

$$\Phi = \frac{\beta\Pi}{\rho_0} = 1 - \frac{\kappa^3}{24\pi\rho_0}D(\kappa a) = 1 - \frac{|Z_+Z_-|\kappa^3}{48\pi I}D(\kappa a) = 1 - \frac{|Z_+Z_-|\kappa l_B}{6}D(\kappa a) \tag{9.137}$$

where $\kappa^2 = 8\pi l_B I$ is used in attaining the last equality. Meanwhile, the mean ionic activity coefficient can be obtained from Eq. (9.129)

$$\ln\gamma_\pm \equiv \frac{\nu_+\ln\gamma_+ + \nu_-\ln\gamma_-}{\nu_+ + \nu_-} = -\frac{\kappa l_B|Z_-Z_+|}{2(1+\kappa a)}. \tag{9.138}$$

Other than the dimensionless quantities $D(\kappa a)$ and $(1+\kappa a)$, Eqs. (9.137) and (9.138) are the same as those given by the original DH theory.

Figure 9.18 shows the influence of collision diameter a on the mean ionic activity coefficient. The modified DH theory predicts that, as observed in experiments, the mean ionic activity coefficient versus the Debye screening parameter curves upward when the ion concentration increases. For an aqueous electrolyte near room temperature, $l_B \approx 7.14$ Å and $l_B/a \approx 0.5$, the ion size effects become noticeable when $\kappa l_B > 0.1$, which corresponds to a monovalent electrolyte with the total ion density of

$$\rho_0 = \frac{\kappa^2}{8\pi l_B} = \frac{\kappa^2 l_B^2}{8\pi l_B^3} > \frac{0.01}{8\pi \times 7.14^3 \times 6.02214 \times 10^{23-30+6}} \approx 1.8 \times 10^{-6}\ \text{M}. \tag{9.139}$$

The numerical result indicates that the DH theory is valid only at extremely low ion concentrations.

9.5.2 Pitzer's Equation

Pitzer's extension of the DH theory includes two components.[55] One is to account for the ion size effects similar to the modified DH theory, and the other is to incorporate non-electrostatic contributions via a virial expansion up to the third-order terms. Because the latter is mostly empirical, we discuss only the ion size effects.

55 Pitzer K. S., "Electrolyte theory – improvements since Debye and Hückel", *Acc. Chem. Res.* 10, 371–377 (1977).

In comparison with simulation results for the RDFs of monovalent electrolytes, Pitzer observed that the Boltzmann distribution is more accurate than the linearized DH expression

$$g_{i,j}(r) \approx \exp\left[-\frac{Z_i Z_j l_B e^{\kappa a}}{(1+\kappa a)}\right]. \tag{9.140}$$

Similar results can be achieved by a quadratic expansion of the Boltzmann equation

$$g_{i,j}(r) \approx 1 - \frac{Z_i Z_j l_B e^{\kappa a}}{(1+\kappa a)} \cdot \frac{e^{-\kappa r}}{r} + \frac{1}{2}\left[\frac{Z_i Z_j l_B e^{\kappa a}}{(1+\kappa a)} \cdot \frac{e^{-\kappa r}}{r}\right]^2. \tag{9.141}$$

As in the modified DH theory discussed above, the RDFs vanish when $r < a$. While the original DH theory underestimates the RDFs for both like-charged and opposite-charged ions pairs, the three-term extension, Eq. (9.141), performs well except near contact.

With the RDFs given by Eq. (9.141), the osmotic coefficient can be derived from the virial equation

$$\begin{aligned}\Phi &= 1 - \frac{1}{6\rho_0}\sum_{i,j}\rho_i^0\rho_j^0 \int d\mathbf{r} \frac{\partial \beta u_{ij}(r)}{\partial r} r g_{ij}(r) \\ &= 1 + \frac{2}{3}\pi\rho_0 a^3 - \frac{l_B \mid Z_+ Z_- \mid}{12a}\left[\frac{2\kappa a}{1+\kappa a} - \left(\frac{\kappa a}{1+\kappa a}\right)^2\right]\end{aligned} \tag{9.142}$$

where $\rho_0 = \sum_i \rho_i^0$ is the number density of all ions, and $\beta u_{ij}(r)$ is

$$\beta u_{ij}(r) = \begin{cases} \infty & r < a \\ \frac{Z_i Z_j l_B}{r} & r \geq a \end{cases} \tag{9.143}$$

On the other hand, the mean ionic activity coefficient can be derived from (Problem 9.14)

$$\begin{aligned}\ln \gamma_\pm &= \Phi - 1 + \int_0^{\rho_0} (\Phi - 1)\ d\ln\rho_0 \\ &= \Phi - 1 + \frac{2}{3}\pi\rho_0 a^3 - \frac{l_B \mid Z_+ Z_- \mid}{6a}\left[\frac{\kappa a}{1+\kappa a} + \ln(1+\kappa a)\right]\end{aligned} \tag{9.144}$$

Eqs. (9.142) and (9.144) provide a theoretical basis to establish semi-empirical correlations for the osmotic and mean ionic activity coefficients for a wide variety of electrolyte solutions. Importantly, Pitzer's equation allows for the prediction of the thermodynamic properties of ionic mixtures based on those corresponding to single electrolytes.[56]

9.5.3 The Quasi-Chemical Theory of Ion Binding

The linearized PB equation becomes problematic for systems with strong electrostatic coupling, i.e., for ionic systems at low temperature or with a small dielectric medium. When the Bjerrum length is significantly larger than the ion diameter ($l_B > a$), the contact value of the electrostatic energy of oppositely charged ions is much larger than the thermal energy ($k_B T$). As a result, a significant portion of the electrolyte exists as ion pairs. Because the linear expansion is valid only when the electrostatic energy is much smaller than $k_B T$, the DH theory fails to capture the pairing effects. This caveat can be remediated by introducing dissociation/association equilibrium between ions.

---

56 Clegg S. L., Pitzer K. S. and Brimblecombe P., "Thermodynamics of multicomponent, miscible, ionic solutions", *J. Phys. Chem.* 96 (8), 3513–3520 (1992); "Thermodynamics of multicomponent, miscible, ionic solutions. 2. Mixtures including unsymmetrical electrolytes", *J. Phys. Chem.* 96 (23), 9470–9479 (1992).

The concept of ion binding was introduced first by Bjerrum,[57] thereby the ion pairs are also known as the Bjerrum pairs.

In the DH theory, ions are treated as point charges. The formation of ion pairs amounts to the annihilation of cations and anions, eliminating their contributions to electrostatic interactions. In other words, the net effect of association is the reduction of ion concentrations. Accordingly, the excess chemical potential includes a contribution due to the reduction of the number densities of free ions, plus an electrostatic term due to interaction among free ions determined by the original DH theory

$$\beta\mu_i^{ex} = \ln\alpha_i - \frac{Z_i^2\kappa_B l_B}{2(1+\kappa_B a)} \tag{9.145}$$

where $\alpha_i = \rho_i/\rho_i^0$ stands for the degree of dissociation for ionic species i, i.e., the fraction of free ions. In Eq. (9.145), the Debye parameter is also modified accounting for the reduction of ion densities

$$\kappa_B^2 = 4\pi l_B \sum_i \alpha_i\rho_i^0 Z_i^2 \tag{9.146}$$

where the subscript B denotes Bjerrum's modification. Because $\alpha_i \le 1$, ion pair formation leads to a smaller Debye parameter or, equivalently, a larger screening length. In writing Eq. (9.145), we assume that the ion pairs have negligible effects on the dielectric medium. A more realistic model would require consideration of the ion-pair effects on the dielectric constant of the medium and/or ion–dipole interactions.

For ion pairing in a single electrolyte, the condition of charge neutrality requires that cations and anions have the same degree of dissociation, $\alpha_+ = \alpha_- = \alpha$. In a symmetric electrolyte solution ($Z_+ = -Z_- = Z$), the association between cations and anions can be expressed in terms of a quasi-chemical reaction

$$\mathrm{C} + \mathrm{A} \rightleftharpoons \mathrm{CA} \tag{9.147}$$

where C stands for cation and A for anion. The number densities of free ions and ion pairs are determined by the degree of dissociation $\rho_{i=C,A} = \alpha\rho_{CA}^0$ and $\rho_{CA} = (1-\alpha)\rho_{CA}^0$, where ρ_{CA}^0 is the total number density of ion pairs before dissociation. With the assumption that the association takes place in an ideal solution, the mass action law (MAL) predicts

$$K_A = \frac{1-\alpha}{\rho_{CA}^0\alpha^2} \tag{9.148}$$

where K_A is the apparent equilibrium constant. The degree of dissociation predicted by Eq. (9.148) is given by

$$\alpha = \frac{\sqrt{1+4\rho_{CA}^0 K_A} - 1}{2\rho_{CA}^0 K_A}. \tag{9.149}$$

Unlike a conventional chemical reaction whereby the equilibrium constant is associated with the free energy of bond formation, ion pairing does not involve any specific bonds. Therefore, Eq. (9.148) entails certain arbitrariness in defining the association constant K_A. A common practice is that K_A is determined according to the Coulomb potential between oppositely charged ion pairs

$$K_A = 4\pi\int_a^{r_c} dr\, r^2 \exp[Z^2 l_B/r] \tag{9.150}$$

57 Bjerrum N., "The dilution heat of an ionic solution in the theory of Debye and Hückel. Along with an article on the theory of heat effects in a dielectricum", *Z. Phys. Chem.-Stoch Ve* 119 (3/4), 145–160 (1926).

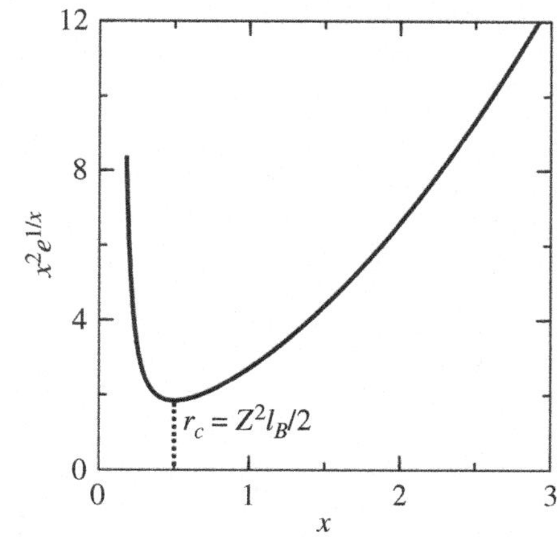

Figure 9.19 The cut-off distance for ion–ion association is determined by the minimum in the Boltzmann distribution function expressed in spherical coordinates.

where r_c represents the upper limit of the ion–ion distance. Eq. (9.150) suggests that the center-to-center distance in the ion pair is larger than the closest distance a but less than the cut-off distance r_c. A convenient choice is $r_c = Z^2 l_B/2$, the separation between a pair of oppositely charged ions such that the integrand in Eq. (9.150) reaches a minimum (Figure 9.19). At low temperatures, the degree of dissociation is small, and the theoretical results are not sensitive to the upper limit of the cut-off distance.[58]

Alternative methods exist to estimate the dissociation constant. For example, K_A can be fixed by imposing a consistency between the quasi-chemical picture of ion pairing and the physical model up to the level of the second osmotic virial coefficient[59]

$$K_A = 8\pi a^3 \sum_{m=2}^{\infty} (Z^2 l_B/a)^{2m}/[(2m)!(2m-3)]. \tag{9.151}$$

The Fuoss model provides another procedure to fix the association constant.[60] It assumes that each ion is either associated with an oppositely charged ion or sufficiently far away from all other ions such that the DH theory applies. According to this model, the association constant depends on the Bjerrum length and the original Debye screening parameter κ:

$$K_A = \frac{2\pi a^3}{3} \exp\left[\frac{Z^2 l_B}{a(1+\kappa a)}\right]. \tag{9.152}$$

Figure 9.20 compares the association constants predicted by different models. Qualitatively, these models provide similar trends in the variation of the association constant with the reduced temperature.

The quasi-chemical theory of ion pairing is particularly useful for non-aqueous electrolyte solutions with a low concentration of free ions. It provides a near-quantitative description of the critical behavior of a wide variety of charged systems, including optically excited semiconductors, oxide melts and glasses, and molten salts. For example, Figure 9.21 shows the critical temperature and density of ion pairs predicted by the Fuoss model in comparison with experimental data

58 Valeriani C. et al., "Ion association in low-polarity solvents: comparisons between theory, simulation, and experiment", *Soft Matter* 6 (12), 2793–2800 (2010).
59 Ebeling W., "Theory of Bjerrum ion association in electrolytes", *Z. Phys. Chem.-Leipzig* 238 (5–6), 400–402 (1968).
60 Fuoss R. M., "Ionic Association. 3. The equilibrium between ion pairs and free ions", *J. Am. Chem. Soc.* 80 (19), 5059–5061(1958).

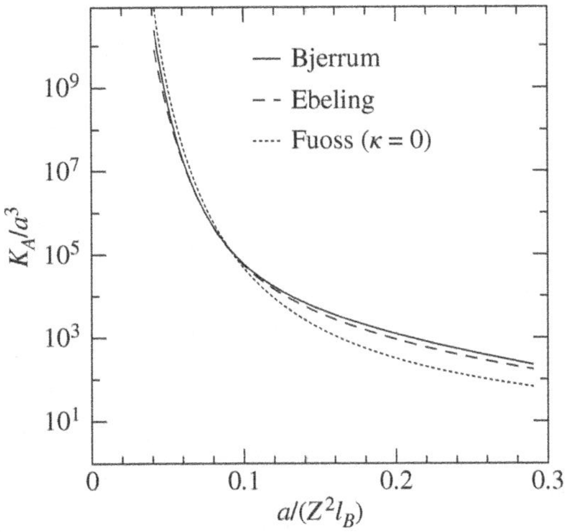

Figure 9.20 The cation–anion association constant was predicted from different models.

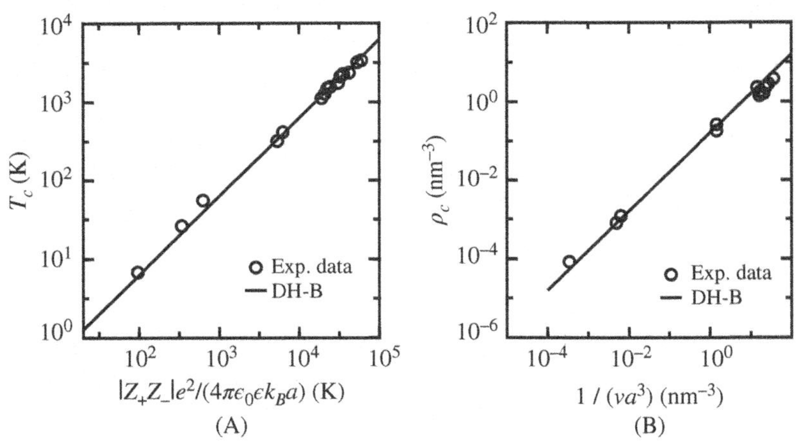

Figure 9.21 The critical temperature (A) and number density of ion pairs (B) predicted by the quasi-chemical modification of the DH theory (DH-B, lines). The experiment data (symbols) cover a variety of charged systems, including electron-hole fluids in semiconductors, oxide melts and glasses, and molten salts. Here, a is the sum of the cation and anion radii, and ν is the total number of ions created per molecule of electrolyte. Adapted from McGahay and Tomozawa.[61]

(Problem 9.15).[61] The excellent agreement is remarkable considering the well-known limitations of the DH theory. Phase separation in metal oxides and semi-conductor systems is of technological importance because the coexistence of different phases will affect not only the optoelectrical properties but also the chemical durability of glass materials.

9.5.4 Summary

The extended DH models are useful to describe the thermodynamic properties of electrolyte solutions of practical interest. In particular, the quasi-chemical ion-pairing theory offers a nearly

61 McGahay V. and Tomozawa M., "Correspondence of phase-separation in several charged-particle system", *J. Chem. Phys.* 97 (4), 2609–2617 (1992);

quantitative description of the phase behavior of non-aqueous electrolytes, including plasmas, oxide melts, and molten salts.

9.6 Integral-Equation Theories for Ionic Systems

The PB equation assumes that an electrolyte solution can be represented by point charges in a dielectric medium and that ion distributions depend exclusively on the local electric potential. These assumptions are valid at extremely low ion concentrations, such that short-range interactions between ionic species are negligible. For electrolytes at finite ion concentrations, non-electrostatic interactions, in particular ion size or excluded volume effects play a significant role in determining both ion distributions and thermodynamic properties. Whereas the extended DH models describe the ion size effects in terms of an average diameter, they do not distinguish the size disparity of ionic species and electrostatic correlations responsible for oscillatory ion distributions observed at high densities. In this section, we introduce integral-equation theories to describe such effects. With a small number of molecular parameters, the liquid-state methods make near-quantitative predictions of the structure and thermodynamic properties of many ionic systems.

9.6.1 The Primitive Model

The PM represents one of the simplest extensions of the PB equation to describe non-electrostatic interactions and correlation effects in ionic systems. In this model, monomeric ionic species are depicted as charged hard spheres, and, as in the PB equation, the solvent is represented by a dielectric continuum (Figure 9.22). Unlike the extended DH models, the PM allows ions to have different sizes. The special case of ions having the same hard-sphere diameter is known as the restrictive primitive model (RPM).

Like the PB equation, the PM adopts the Coulomb potential to describe electrostatic interactions. Besides, it does not describe non-electrostatic ion–ion interactions beyond the collision diameter. Nor does the PM describe specific ion effects except for ion size and valence. Nevertheless, it does capture the essential physics underlying the thermodynamic and transport properties of electrolyte solutions at finite ion concentrations. It is able to reproduce experimental data for a large number of electrolyte systems when the ion diameters are treated as adjustable parameters.

In the PM, the pair potential between monomeric ionic species is given by

$$u_{ij}(r) = \begin{cases} \infty, & r < \sigma_{ij} \equiv (\sigma_i + \sigma_j)/2 \\ Z_i Z_j e^2/4\pi\varepsilon\varepsilon_0 r, & r \geq \sigma_{ij} \end{cases} \tag{9.153}$$

where r stands for the center-to-center distance, e is the elementary charge, ε_0 is the permittivity of the free space, ε is the dielectric constant of the pure solvent at system temperature and pressure, σ_i and Z_i are respectively the hard-sphere diameter and the valence of ionic species i. The hard-sphere diameter reflects the effective size of a solvated ion. Typically, it is close to but not necessarily the same as the diameter of individual ions in the vacuum, or the Pauli diameter. To best reproduce experimental results, the hard-sphere diameters are often treated as adjustable parameters.

In dimensionless form, the Coulomb potential can be written as

$$\beta u_{ij}^C(r) = Z_i Z_j l_B / r. \tag{9.154}$$

For an aqueous solution at room temperature, the Bjerrum length is $l_B \approx 0.714$ nm. At ambient conditions, the Bjerrum length is about 3 nm for an organic solvent ($\varepsilon \sim 20$), and it is about 56 nm in the vacuum ($\varepsilon = 1$). Eq. (9.154) indicates that, at close distance, the electrostatic energy is

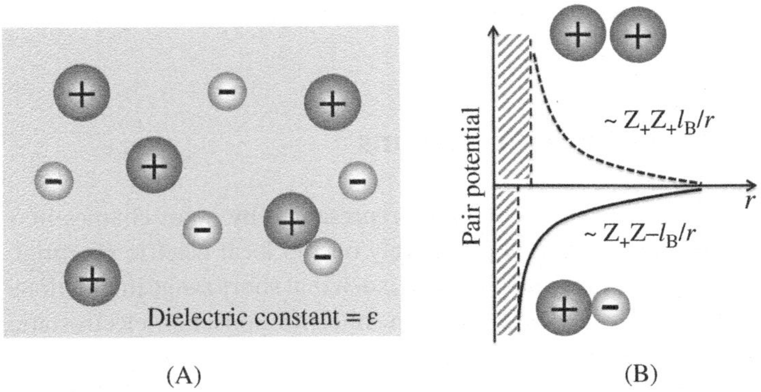

Figure 9.22 (A) Schematic of the primitive model for ionic systems whereby cations and anions are described as charged hard spheres and the solvent as a dielectric medium. (B) The reduced potential between each ion pair depends on the ion diameter σ_i, valence Z_i, and the Bjerrum length, l_B.

significantly larger than thermal energy k_BT even for monovalent ions in an aqueous solution ($\sigma \sim 0.4$ nm). Figure 9.22B schematically illustrates the reduced pair potential between ions according to the primitive model.

9.6.2 The Ornstein–Zernike Equation for Electrolyte Solutions

The Ornstein–Zernike (OZ) equation discussed in Section 7.4 is directly applicable to ionic systems. While in the DH theory (and its variations), we derive thermodynamic properties through the RDFs, the OZ equation is concerned with the direct and total correlation functions $h_{ij}(r)$ and $c_{ij}(r)$

$$h_{ij}(r) = c_{ij}(r) + \sum_n \rho_n^0 \int d\mathbf{r}' \ c_{in}(r')h_{nj}(|\ \mathbf{r} - \mathbf{r}'\ |) \tag{9.155}$$

where ρ_n^0 represents the number density of ionic species n in the bulk solution, and the summation extends to all ionic species. Formally, the total correlation functions are related to the functional derivative of the local ion density to the one-body potential, and the direct correlations describe the response of the one-body potential to the local ionic densities.[62] Therefore, the OZ equation represents a reciprocal relationship between the total and direct correlation functions.

To solve the total and direct correlation functions from the OZ equation, we need an additional relation that is formally referred to as the closure. For monomeric ionic species, the closure equation can be understood in terms of the (reduced) PMF between ionic species i and j

$$\beta W_{ij}(r) \equiv -\ln[h_{ij}(r) + 1] = \beta u_{ij}(r) - h_{ij}(r) + c_{ij}(r) - b_{ij}(r) \tag{9.156}$$

where $b_{ij}(r)$ represents the bridge function. Eq. (9.156) indicates that the PMF includes a contribution due to the pair potential $u_{ij}(r)$, an indirect contribution $c_{ij}(r) - h_{ij}(r)$ that can be derived from a quadratic functional expansion of the free energy with respect to the local density inhomogeneity. The bridge term accounts for contributions to the free energy from all the higher-order terms in the functional expansion. Whereas the bridge functions are generally unknown, numerical results from molecular simulation indicate that they are negative for like-charged ions at all separations, while those for oppositely-charged ions are mostly positive.[63] In other words, the bridge term makes

62 Wu J., "Density functional theory for liquid structure and thermodynamics," pp. 1–74, in "*Molecular Thermodynamics of Complex Systems*", Lu X. and Hu Y. (Eds.), Springer, 2009.
63 Duh D.-M., and Haymet A. D. J. "Integral equation theory for charged liquids: model 2–2 electrolytes and the bridge function", *J. Chem. Phys.* 97, 7716–7729 (1992).

the PMF between like-charged ions more repulsive, and that between oppositely-charged ions more attractive.[64]

For ionic systems, one of the most accurate (and most popular) choices of the closure is given by the hypernetted chain (HNC) approximation

$$g_{ij}(r) \approx \exp[-\beta u_{ij}(r) + h_{ij}(r) - c_{ij}(r)]. \tag{9.157}$$

HNC can be obtained by ignoring the bridge function in Eq. (9.156). The Percus–Yevick (PY) closure, which is commonly used for hard-sphere systems, can be obtained by the linearization of the exponential of $c_{ij}(r) - h_{ij}(r)$ on the right side of Eq. (9.157)

$$g_{ij}(r) \approx \exp[-\beta u_{ij}(r)][1 + h_{ij}(r) - c_{ij}(r)]. \tag{9.158}$$

Because the pair potential $u_{ij}(r)$ diverges when hard spheres come to overlap, both HNC and PY closures ensure that the RDF vanishes when $r < \sigma_{ij}$. A linear expansion of the entire exponential term on the right side of Eq. (9.157) for $r \geq \sigma_{ij}$ while imposing the hard-core condition leads to the mean-spherical approximation (MSA)

$$\begin{cases} h_{ij}(r) = -1 & r < \sigma_{ij} \\ c_{ij}(r) \approx -\beta u_{ij}(r) & r \geq \sigma_{ij} \end{cases} \tag{9.159}$$

It is worth noting that equaling the direct correlation function to the negative of the reduced pair potential amounts to the random phase approximation (RPA) in the field theory (Section 8.6). As a result, the addition of the hard-core condition in Eq. (9.159) is also known as the optimized RPA.[65]

Mathematically, the PB equation may be understood as the application of the HNC/MSA theory to a system of point charges. To elucidate the connection, we may rewrite the HNC closure in terms of the reduced PMF

$$\beta W_{ij}(r) = \beta u_{ij}(r) - \sum_k \rho_k^0 \int d\mathbf{r}' c_{ik}(r') h_{kj}(|\mathbf{r} - \mathbf{r}'|). \tag{9.160}$$

Eq. (9.160) is obtained from Eq. (9.156) with the help of the OZ equation for $h_{ij}(r) - c_{ij}(r)$. With the assumption of $c_{ik}(r) \approx -\beta u_{ik}(r) = -Z_i Z_k l_B / r$ as in MSA and the charge neutrality condition, $\sum_k \rho_k^0 Z_k = 0$, Eq. (9.160) can be re-expressed as

$$\beta W_{ij}(r) = Z_i l_B \left[\frac{Z_j}{r} + \sum_k \int d\mathbf{r}' \, \frac{Z_k}{r'} \rho_k^0 g_{kj}(|\, \mathbf{r} - \mathbf{r}' \,|) \right]. \tag{9.161}$$

The first term on the right side of Eq. (9.161) corresponds to the electric potential due to a tagged ion of valence Z_j, and the second term corresponds to the electric potential due to all other ions in the system. Eq. (9.161) indicates that the potential of mean force between ions i and j is given by the local electric potential

$$\rho_{ij}(r) = \rho_i^0 \exp[-\beta W_{ij}(r)] = \rho_i^0 \exp[-\beta Z_i e \, \psi_j(r)] \tag{9.162}$$

where $\psi_j(r)$represents the mean electric potential around ion j. Eq. (9.162) is predicted by the PB equation.

64 Rosenfeld Y., "Free energy model for inhomogeneous fluid mixtures – Yukawa-charged hard spheres, general interactions, and plasmas", *J. Chem. Phys.* 98 (10), 8126–8148 (1993).
65 Andersen H. C., Chandler D., Weeks J. D., "Roles of repulsive and attractive forces in liquids – optimized random phase approximation", *J. Chem. Phys.*, 56 (8), 3812–3823 (1972).

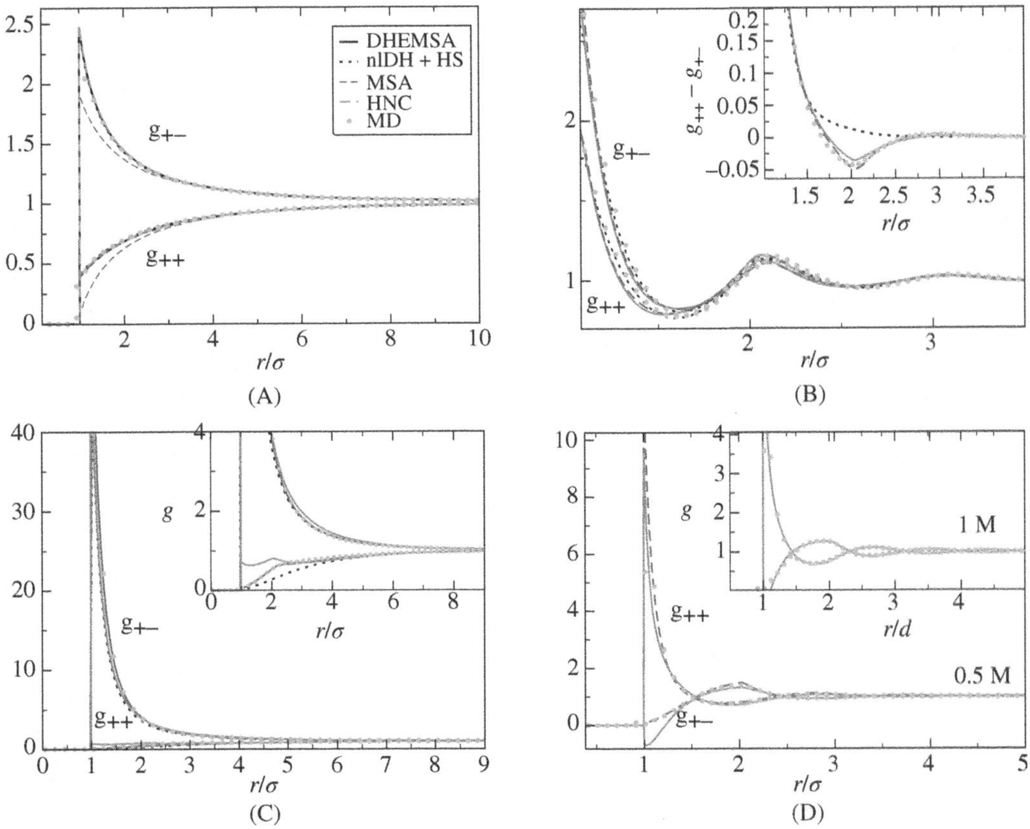

Figure 9.23 Radial distribution functions for four ionic systems represented by the primitive model. All systems consist of monovalent ions with the same ion diameter $\sigma = 0.66$ nm. (A) $T^* = 0.91$, $\rho^* = 3.5 \times 10^{-3}$; (B) $T^* = 0.91$, $\rho^* = 0.7$; (C) $T^* = 0.17$, $\rho^* = 1.7 \times 10^{-3}$; and (D) $T^* = 0.067$, $\rho^* = 0.17$, and 0.35 (inset). Here, $T^* = \sigma/l_B$ and $\rho^* = 2\rho_0\sigma^3$, where ρ_0 is the number density of cations (or anions) in the bulk. In all panels, the symbols are from MD simulation, and the lines are from different theoretical methods. Source: Reproduced from Zwanikken et al.[67]

Together with a reasonable closure, the OZ equation can be solved with various numerical schemes (e.g., by separating long-range Coulomb interactions and short-range correlations).[66] In comparison with simulation data, the integral-equation theories are able to predict direct and total correlation functions and thermodynamic properties quantitatively. For example, Figure 9.23 shows the RDFs obtained from several theoretical methods in comparison with MD simulation for four representative ionic systems.[67] Approximately, the systems with reduced temperature $T^* = \sigma/l_B = 0.91$ and reduced densities $\rho^* = 2\rho_0\sigma^3 = 3.5 \times 10^{-3}$ and 0.7 correspond to dilute and concentrated aqueous electrolytes; the systems with $T^* = 0.17$ and $\rho^* = 1.7 \times 10^{-3}$ and 0.17 mimic dilute and concentrated organic electrolytes, respectively; and those with $T^* = 0.067$ are close to the critical temperature of the phase transition predicted by the restricted PM (~0.0487).[68] At all conditions, the integral-equation methods yield good agreement with the simulation data.

---

66 Ichiye T.; Haymet A. D. J., "Accurate integral-equation theory for the central force model of liquid water and ionic-solutions", *J. Chem. Phys.* 89 (7), 4315–4324 (1988).
67 Zwanikken J. W., Jha P. K., and de la Cruz M. O., "A practical integral equation for the structure and thermodynamics of hard sphere Coulomb fluids", *J. Chem. Phys.* 135 (6), 064106 (2011).
68 Caillol J. M., Levesque D., and Weis J. J., "Critical behavior of the restricted primitive model", *Phys. Rev. Lett.* 77, 4039(1996).

In particular, HNC is able to predict accurate RDFs even for systems containing strongly asymmetric ions (e.g., charged colloids in the presence of small ions).[69]

Figure 9.23A shows that, at a low ion concentration (10^{-2} M), all theoretical methods perform well. MSA is slightly less accurate at short-distance because it follows the mean-field approximation for the electrostatic energy. nlDH+HS and DHEMSA are two modifications of the PB equation that account for hard-sphere interactions.[67] These equations perform well because a nonlinear Boltzmann equation was adopted for ion distributions similar to the HNC approximation.

As illustrated in Figure 9.23B, the electrostatic correlations become less significant at high ion concentrations (e.g., 2 M ion concentration). In that case, all analytical methods perform rather well because the hard-sphere potential dominates the relative distribution of ionic particles. Interestingly, the potential of mean force is attractive at short distance, regardless of the charges of the ion pairs. The attraction between like-charged ions is slightly weaker than that between opposite charges because of the direct Coulomb energy.

Figure 9.23C and d show that both MSA and HNC perform well for systems with strong electrostatic interactions. However, those methods based on the DH theory are less satisfactory. For systems with strong electrostatic coupling, the good performance of the MSA and HNC may be attributed to the fact that these closures satisfy the exact results both at the weak and strong coupling limits (viz., the DH limit and the Onsager limit, respectively).[70]

9.6.3 Thermodynamic Equations for Ionic Systems

As discussed in Section 7.3, thermodynamic properties can be derived either from the RDFs or from the direct correlation functions. Similar equations for pressure and internal energy exist for ionic systems within the MM framework

$$\beta\Pi = \rho_0 - \frac{2\pi}{3}\sum_i\sum_j \rho_i^0\rho_j^0 \int_0^\infty dr \frac{\partial \beta u_{ij}(r)}{\partial r} r^3 g_{ij}(r), \tag{9.163}$$

$$\left(\frac{\partial \beta\Pi}{\partial \rho_0}\right) = 1 - \frac{4\pi}{\rho_0}\sum_i\sum_j \rho_j^0\rho_j^0 \int_0^\infty dr\ r^2 c_{ij}(r), \tag{9.164}$$

$$U^{ex}/V = 2\pi\sum_i\sum_j \rho_i^0\rho_j^0 \int_0^\infty dr\, u_{ij}(r)\, r^2 g_{ij}(r) \tag{9.165}$$

where $\rho_0 = \sum_i \rho_i^0$ stands for the total density of ionic species. Eqs. (9.163)–(9.165) are called, respectively, the pressure equation (viz., the virial route), the compressibility equation, and the energy equation. In the PM, the electrolyte systems are defined within the MM framework. Accordingly, the pressure and compressibility equations are applied to the osmotic pressure, Π. If we have exact results for both direct and total correlation functions, thermodynamic properties calculated from different routes would be equivalent. Regrettably, the thermodynamic consistency is rarely satisfied when the correlation functions are obtained from approximate methods.

The PM describes the pair potentials between ions in terms of a hard-sphere repulsion and the Columbic energy. Because the hard-sphere interaction is discontinuous at the collision distance, we may evaluate the partial derivative in Eq. (9.163) from the cavity distribution function $y_{ij}(r) \equiv g_{ij}(r)\exp[\beta u_{ij}(r)]$, which is a continuous function of r. By dividing $u_{ij}(r)$ into hard-sphere

69 Heinen M., Allahyarov E., Lowen H., "Highly asymmetric electrolytes in the primitive model: hypernetted chain solution in arbitrary spatial dimensions", *J. Comput. Chem.* 35 (4), 275–289 (2014).
70 Blum L., Rosenfeld Y., "Relation between the free-energy and the direct correlation-function in the mean spherical approximation", *J. Stat. Phys.* 63 (5–6), 1177–1190 (1991).

potential $u_{ij}^{HS}(r)$ and Coulomb energy $u_{ij}^{C}(r)$, we can obtain the osmotic coefficient of the electrolyte solution, Φ, from the virial route, Eq. (9.163),

$$\Phi \equiv \frac{\beta\Pi}{\rho_0} = 1 + \frac{2\pi}{3\rho_0}\sum_i\sum_j \rho_i^0\rho_j^0 \left[g_{ij}(\sigma_{ij})\sigma_{ij}^3 - \int_{\sigma_{ij}}^{\infty} dr \frac{\partial \beta u_{ij}^C(r)}{\partial r} g_{ij}(r) r^3\right]. \tag{9.166}$$

From the osmotic coefficient, we can also obtain the mean ionic activity coefficient by integrating the Gibbs–Duhem equation

$$\ln \gamma_\pm = \Phi - 1 + \int_0^{\rho_0} (\Phi - 1)\, d\ln\rho_0. \tag{9.167}$$

When the correlation functions are calculated from the HNC closure, the activity coefficients of individual ions can be calculated from the total and direct correlation functions[71]

$$\ln \gamma_i^{HNC} = -\sum_j \rho_j^0 \hat{c}_{ij}^{(s)}(0) + \frac{1}{2}\sum_j \rho_j^0 \int d\mathbf{r}\{h_{ij}(r)[h_{ij}(r) - c_{ij}(r)]\} \tag{9.168}$$

where $\hat{c}_{ij}^{(s)}(0)$ denotes the Fourier transform of the short-range part of the direct correlation function at $k = 0$.

Figure 9.24 presents the mean ionic activity coefficients and the osmotic coefficients of two electrolytes, one containing monovalent ions and the other divalent ions.[72] Here, the lines are calculated from various theoretical methods, and the symbols are from Monte Carlo (MC) simulation. Because neither the DH theory nor the Pitzer equation account for the size asymmetry, an average diameter, $\sigma = (\sigma_+ + \sigma_-)/2$, was used in obtaining the numerical results. Among the different theoretical procedures considered, HNC is the most accurate in predicting the thermodynamic properties of the two model systems. As expected, the DH theory fails at high ion concentrations. Surprisingly, its performance improves for the divalent system.

Historically, the integral-equation theories were established from the diagrammatic expansion of the partition functions (e.g., circles or rings, stars, and bridges). The same mathematical relations can be derived from density functional analysis.[73] In comparison with the PB equation, one major issue with the integral-equation methods is that the physical significance of the closure is not easily comprehensible. Sometimes their shortfalls can be overlooked because of the good numerical performance in comparison with simulation data. As mentioned above, the HNC approximation stems from a quadratic expansion of the free-energy functional with respect to density fluctuations. Because a quadratic function does not describe two minima, HNC is only applicable to systems with one stable state. Similar limitations may apply to other closures.[74] By contrast, the DH theory is able to describe phase transitions because it is not based on the quadratic expansion.

9.6.4 Blum's Solution of the Mean-Spherical Approximation (MSA)

The MSA is a popular choice in practical applications of the OZ equation because it provides analytical expressions for the thermodynamic properties of ionic systems. The analytical results

71 Verlet L. and Levesque D., "On theory of classical fluids II", *Physica* 28 (11), 1124–1142 (1962).
72 Gutierrez-Valladares E., Luksic M., Millan-Malo B., Hribar-Lee B., Vlachy V., "Primitive model electrolytes. A comparison of the HNC approximation for the activity coefficient with Monte Carlo data", *Condens. Matter Phys.*, 14 (3), (2011).
73 Hansen J. -P., McDonald I. R., *Theory of simple liquids: with applications of soft matter* (4th Edition). p xv, 619 pages, 2013.
74 Fisher M. E., Levin Y., "Criticality in ionic fluids – Debye-Huckel theory, Bjerrum, and beyond", *Phys. Rev. Lett.* 71 (23), 3826–3829 (1993).

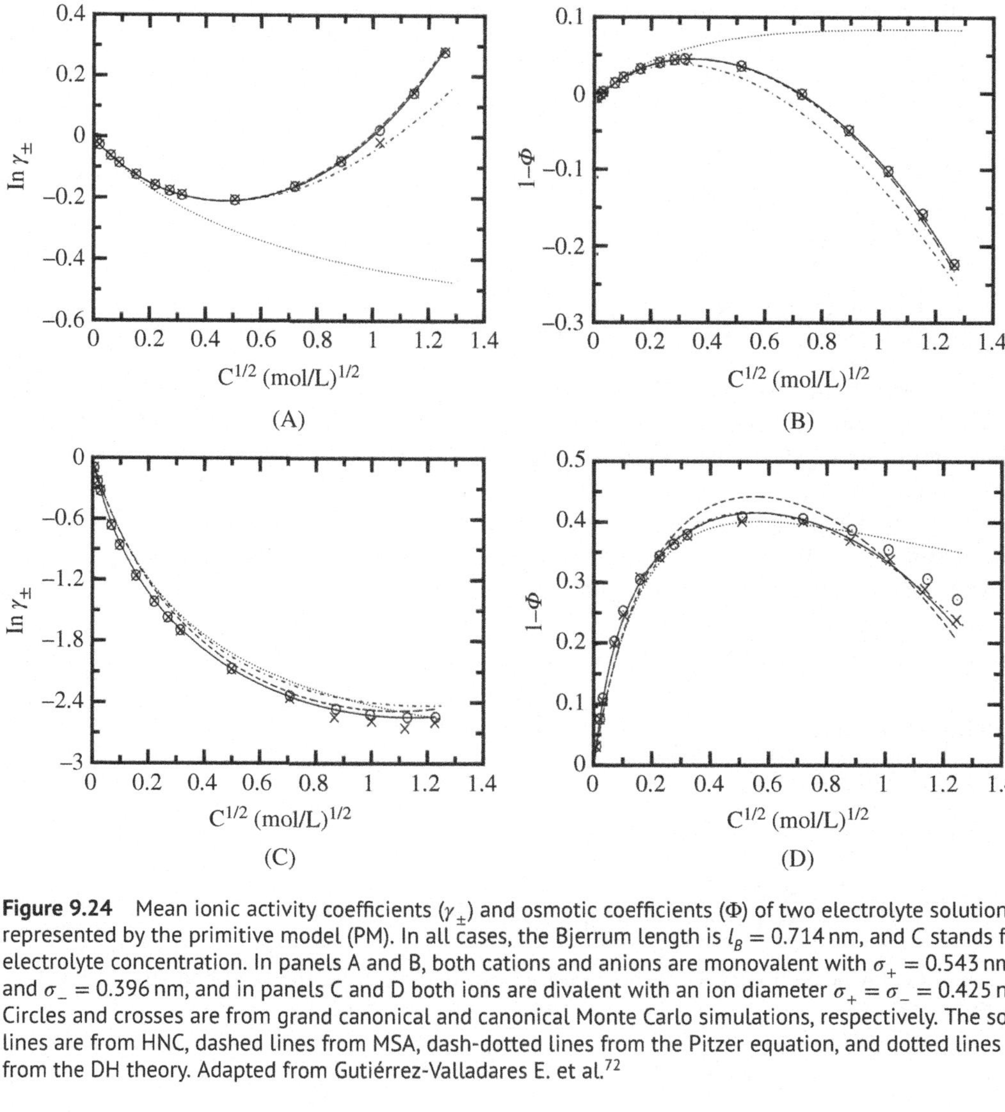

Figure 9.24 Mean ionic activity coefficients ($\gamma_\pm$) and osmotic coefficients (Φ) of two electrolyte solutions represented by the primitive model (PM). In all cases, the Bjerrum length is $l_B = 0.714$ nm, and C stands for electrolyte concentration. In panels A and B, both cations and anions are monovalent with $\sigma_+ = 0.543$ nm and $\sigma_- = 0.396$ nm, and in panels C and D both ions are divalent with an ion diameter $\sigma_+ = \sigma_- = 0.425$ nm. Circles and crosses are from grand canonical and canonical Monte Carlo simulations, respectively. The solid lines are from HNC, dashed lines from MSA, dash-dotted lines from the Pitzer equation, and dotted lines from the DH theory. Adapted from Gutiérrez-Valladares E. et al.[72]

facilitate easy comparison with experimental data using reasonable parameters for the solvent dielectric constant and the hard-sphere diameters of ionic species.

Waisman and Lebowitz first obtained analytical results for MSA by using the RPM of electrolyte solutions.[75] The procedure was extended by Blum and coworkers to derive the most generic results for both correlation functions and thermodynamic properties.[76] In the following, we reproduce a few key equations from MSA for practical applications. The mathematical details are referred to the original publications.

---

75 Waisman E., Lebowitz J. L., "Mean spherical model integral-equation for charged hard spheres. 1. Method of solution", *J. Chem. Phys.* 56 (6), 3086–3093 (1972).

76 Blum L., "Mean spherical model for asymmetric electrolytes. 1. Method of solution", *J. Stat. Phys.* 30 (5), 1529–1535 (1975).

According to MSA, the internal energy due to electrostatic interactions is given by[77]

$$\beta U^{MSA}/V = -l_B \sum_j \rho_j^0 \left(\frac{Z_j^2 \Gamma}{1+\Gamma\sigma_j} + \frac{Z_j \sigma_j \tau}{1+\Gamma\sigma_j} \right) \tag{9.169}$$

where

$$\Gamma^2 = \pi l_B \sum_j \rho_j^0 \left(\frac{Z_j - \tau\sigma_j^2}{1+\Gamma\sigma_j} \right)^2, \tag{9.170}$$

$$\tau = \left[\frac{2}{\pi} - \frac{1}{3} \sum_i \rho_i^0 \sigma_i^3 + \sum_i \frac{\rho_i^0 \sigma_i^3}{1+\Gamma\sigma_i} \right]^{-1} \sum_j \frac{\rho_j^0 Z_j \sigma_j}{1+\Gamma\sigma_j}. \tag{9.171}$$

The first term on the right side of Eq. (9.169) represents the mean electric energy experienced by individual ions, and the second term arises from charge fluctuations. As discussed above, MSA is obtained by linearizing the HNC closure; and it only partially accounts for electrostatic correlations and fluctuation effects.

In Eq. (9.169), parameter τ arises from the size asymmetry of the ionic species. Because of the charge neutrality condition, this parameter, as well as the fluctuation energy, disappears when all ions have the same diameter ($\sigma_j = \sigma$). In that case, we have

$$\Gamma^2 = \frac{\pi l_B}{(1+\Gamma\sigma)^2} \sum_j \rho_j^0 Z_j^2 = \frac{\kappa^2}{4(1+\Gamma\sigma)^2}, \tag{9.172}$$

which leads to an analytical expression for the parameter Γ

$$\Gamma = (\sqrt{1+2\kappa\sigma} - 1)/2\sigma. \tag{9.173}$$

The reduced electrostatic energy becomes

$$\beta U^{MSA}/V = -l_B \sum_j \frac{\rho_j^0 Z_j^2 \Gamma}{1+\sigma\Gamma} = -\frac{l_B}{1/\Gamma + \sigma} \sum_j \rho_j^0 Z_j^2. \tag{9.174}$$

Eq. (9.174) may be compared with the excess energy derived from the modified DH theory, Eq. (9.127),

$$\beta U^{DH}/V = -\frac{l_B \kappa}{2(1+\kappa a)} \sum_j \rho_j^0 Z_j^2. \tag{9.175}$$

If we let $\kappa = 2\Gamma$ and $\sigma = 2a$, Eqs. (9.174) and (9.175) become identical, suggesting a direct connection between MSA and the modified DH model. Eq. (9.170) indicates that, in the limit of low electrolyte concentration ($\rho_j^0 \to 0$), Γ is small such that $1+\Gamma\sigma_j \approx 1$. As a result, we have

$$\Gamma^2 \approx \pi l_B \sum_j \rho_j^0 Z_j^2 = \kappa^2/4 \tag{9.176}$$

Eq. (9.176) confirms that, for dilute electrolyte solutions, the internal energy derived from MSA indeed reduces to that from the modified DH theory.

Interestingly, both MSA and the DH theory predict that the electrostatic energy can be expressed as a summation of the electric energy of individual ions. Imagine that a charged particle of valence

77 Blum L., Høye, J. S., "Mean spherical model for asymmetric electrolytes. 2. Thermodynamic properties and pair correlation-function", *J. Phys. Chem.* 81 (13), 1311–1317 (1977).

Z_j and radius a is immersed in a liquid medium of dielectric constant ε, the electric energy is equal to the Maxwell energy of the local electric field $\mathbf{E}(\mathbf{r})$

$$\beta u_j^c = \frac{\beta\varepsilon\varepsilon_0}{2}\int d\mathbf{r}\,|\mathbf{E}(\mathbf{r})|^2 = \frac{\beta\varepsilon\varepsilon_0}{2}\int_a^\infty \left(\frac{Z_j e}{4\pi\varepsilon\varepsilon_0 r^2}\right)^2 4\pi r^2 dr = \frac{\beta Z_j^2 e^2}{8\pi\varepsilon\varepsilon_0 a} = \frac{Z_j^2}{2l_B a}. \tag{9.177}$$

Except for a negative sign, Eq. (9.177) is identical to the electrostatic energy predicted by MSA ($a = 1/(2\Gamma)$) or the DH theory ($a = 1/\kappa$) at low ion density. Physically, the Maxwell energy corresponds to the energy of a spherical capacitor that is formed by the charge at the surface of a particle of radius a surrounded by neutralizing charges in the dielectric medium. This energy is analogous to the Born energy discussed in Section 9.2. As first recognized by Lars Onsager,[78] the Maxwell energy corresponds to the energy released by "immersing the charged particle in a conducting fluid", i.e., the energy of discharging. Because the energy must increase to reduce the dielectric constant from infinite, which corresponds to that for a perfect conductor, to that of the real solvent, Onsager's discharging energy is exact for systems with strong electrostatic coupling and provides a lower bound for the electrostatic energy of the ionic system.[64] As Γ and κ are positive, both MSA and the DH theory satisfy the Onsager lower bound.

From the internal energy given by Eq. (9.174), we can derive the Helmholtz energy by integrating the Gibbs–Helmholtz relation.[79] According to RPM, the Helmholtz energy density is given by

$$\begin{aligned}\beta F^{MSA}/V &= \beta F^{HS}/V + \int_0^\beta (U^{MSA}/V)\,d\beta \\ &= \rho_0\frac{4\eta - 3\eta^2}{(1-\eta)^3} - \frac{\kappa^2\Gamma}{4\pi(1+\Gamma\sigma)} + \frac{\Gamma^3}{3\pi}\end{aligned} \tag{9.178}$$

where $\eta = \pi\rho_0\sigma^3/6$ stands for the packing fraction of all ions. The first term on the right side of Eq. (9.178) is from the Carnahan–Starling equation of state.[80] A comparison of Eqs. (9.178) and (9.174) indicates that MSA predicts electrostatic contributions to the entropy virtually identical to those from the DH theory

$$S^{MSA}/(Vk_B) = -\Gamma^3/3\pi. \tag{9.179}$$

Finally, MSA predicts that the osmotic coefficient and the mean ionic activity coefficient are given by

$$\Phi = \frac{1+\eta+\eta^2-\eta^3}{(1-\eta)^3} - \frac{\Gamma^3}{3\pi\rho_0}, \tag{9.180}$$

$$\ln\gamma_\pm = \frac{8\eta - 9\eta^2 + 3\eta^3}{(1-\eta)^3} - \frac{\Gamma\kappa^2}{4\pi\rho_0(1+\Gamma\sigma)}. \tag{9.181}$$

Analytical expressions for the internal energy, osmotic coefficient, and the mean activity coefficients are also available for the general case where ions have different diameters.[72]

As mentioned above, the analytical expressions derived from MSA facilitate the numerical fitting of experimental data with the hard-sphere diameters. These parameters are equally applicable to single electrolytes as well as mixtures. The transferability of model parameters is particularly important for practical applications. To illustrate, Figure 9.25 compares the theoretical predictions with the experimental values for seven simple electrolytes in water: NaCl, KCl, LiCl, NaBr, KBr, LiBr,

---

78 Onsager L., "Electrostatic interaction of molecules", *J. Phys. Chem.* 43 (2), 189–196 (1939).

79 Durand-Vidal S., Simonin J. P. and Turq, P., *Electrolytes at interfaces*. Kluwer Academic Publishers, Dordrecht, Boston, p xiv, 344 p, 2000.

80 Carnahan, N. F. and Starling, K. E., "Equation of state for nonattracting rigid spheres", *J. Chem. Phys.* 51 (2), 635–6 (1969).

Table 9.3 Ion diameters in units of Å.

	σ_{MSA}	$\sigma_{Pauling}$
K^+	2.66	2.95
Na^+	1.96	3.05
Li^+	1.56	4.35
Ca^{2+}	1.98	5.00
Br^-	3.90	3.90
Cl^-	3.62	3.62

Here, σ_{MSA} is the hard-sphere diameter used in the primitive model (MSA), and $\sigma_{Pauling}$ is the Pauling diameter.

and $CaCl_2$.[81] In the primitive model, each ion is characterized by its valence and its hard-sphere diameter. Table 9.3 gives the ion diameters in comparison with the Pauling diameters derived from crystal structures. For reproducing the experimental results, the anion diameters are assumed to be the same as the Pauling diameters. However, significantly larger diameters must be used for cations, reflecting their strong hydration in an aqueous solution.

While MSA is able to reproduce the osmotic coefficients and other thermodynamic properties of electrolyte solutions in good agreement with experimental data, the discrepancy becomes noticeable at large salt concentrations when the PM becomes problematic. For dilute solutions, the osmotic coefficient decreases when the ionic strength increases because of strong electrostatic correlations, whereas it increases at high ion concentrations because of the excluded volume effects. In addition to equilibrium properties (osmotic and activity coefficients), the pair correlation functions from MSA can also be used as input in the dynamic theories of ionic systems (e.g., Smoluchowski equation and the mode-coupling theory).[82] As shown in Figure 9.25, MSA yields transport coefficients (mutual diffusion, electric conductivity, self-diffusion, and transport numbers) of aqueous electrolytes in good agreement with the experimental data.

Despite its popularity, the drawbacks of MSA have also been well documented. First, the MSA screening parameter Γ is not analytic for asymmetric systems and must be solved by numerical schemes, which can reduce computational efficiency. Second, the thermodynamic properties derived from the compressibility or virial route are much less accurate in comparison with those from the energy route discussed above. In particular, the compressibility route does not predict a gas–liquid phase transition when the Bjerrum length is sufficiently small. Furthermore, the RDFs predicted by MSA are often problematic in comparison with simulation data. There have been many attempts to overcome these limitations, but improvements are often compromised by an increase in numerical complexity or a loss of the solution in analytic form.

9.6.5 Binding MSA

MSA can be extended to incorporate the association between cations and anions.[82] Similar to Bjerrum's extension of the DH theory, the excess Helmholtz energy includes contributions due

81 Dufreche J. F., Bernard O., Durand-Vidal S., Turq P., "Analytical theories of transport in concentrated electrolyte solutions from the MSA", *J. Phys. Chem. B* 109 (20), 9873–9884 (2005).
82 Bernard O., Blum L., "Binding mean spherical approximation for pairing ions: an exponential approximation and thermodynamics", *J. Chem. Phys.* 104 (12), 4746–4754 (1996).

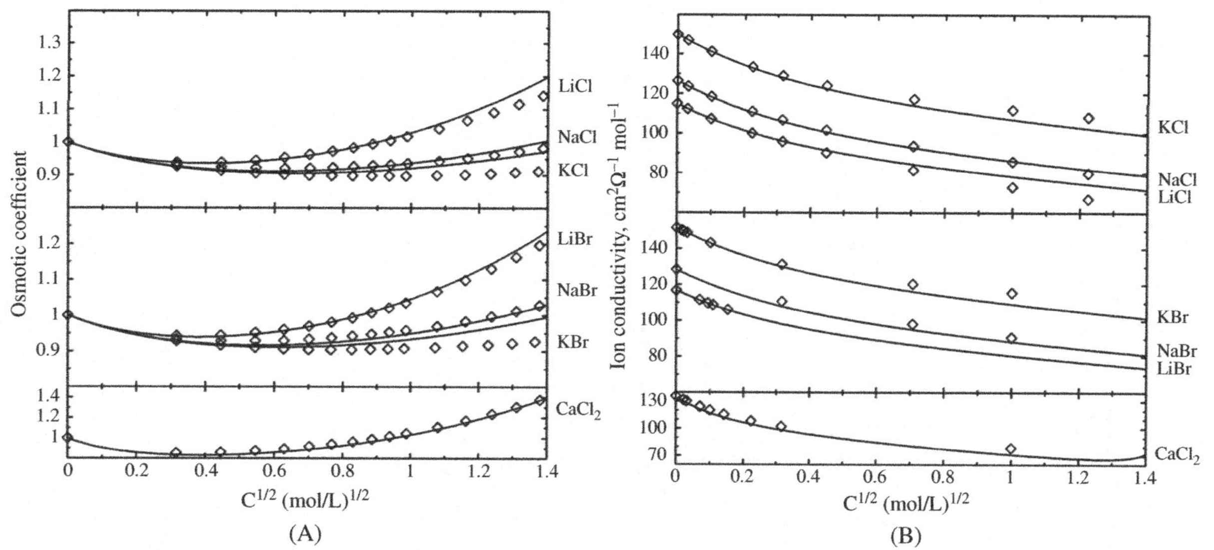

Figure 9.25 A comparison of MSA (lines) with experimental data (symbols) for the concentration dependence (C, mol/l) of the osmotic coefficient (A) and molar conductivity (B) for seven aqueous electrolyte solutions at 25 °C. Source: Dufreche et al.[81]

to the quasi-chemical reaction, the electrostatic interaction between free ions, and ionic excluded volume effects.

As discussed in Section 9.5.3 (Eq. (9.149)), for a single electrolyte solution containing only one type of cations and one type of anions of the same valence Z and hard-sphere diameter σ, the degree of dissociation α is related to the apparent association constant K_A

$$\alpha = \frac{\sqrt{1 + 2\rho_0 K_A} - 1}{\rho_0 K_A} \tag{9.182}$$

where ρ_0 is the total number density of cations and anions (including both free ions and those in the ion pairs). The Helmholtz energy due to ion association is thus given by

$$\beta F^A/V = 2\rho_0[\ln \alpha + (1 - \alpha)/2]. \tag{9.183}$$

As in the Bjerrum theory, the excess Helmholtz energy due to electrostatic interaction is identical to that from the original MSA

$$\beta F^{MSA}/V = \rho_0 \frac{4\eta - 3\eta^2}{(1-\eta)^3} - \frac{\kappa^2 \Gamma_A}{4\pi(1 + \Gamma_A \sigma)} + \frac{\Gamma_A^3}{3\pi} \tag{9.184}$$

except that parameter Γ_A is now determined by the number density of free ions

$$4\Gamma_A^2(1 + \Gamma_A \sigma)^3 = \kappa^2(\alpha + \Gamma_A \sigma). \tag{9.185}$$

In the limit of complete dissociation, $\alpha = 1$, Eq. (9.185) reduces to Γ given by Eq. (9.173).

Assuming that the number density of ion pairs in the association model is the same as the two-body density correlation function between cations and anions at contact *without* association, we can write the apparent association constant in terms of that used in the Bjerrum theory K_A^0 and a correction accounting for thermodynamic non-ideality

$$K_A = K_A^0 g_{+-}(\sigma)/g_{+-}^0(\sigma) \tag{9.186}$$

where $g_{+-}(\sigma)$ stands for the contact value of the RDF, and superscript "0" represents the infinite dilution limit. According to MSA, the contact value of the RDF between opposite-charge ions can be estimated from the exponential approximation for the radial distribution of ionic species *without* association

$$g_{+-}(\sigma) = g_{+-}^{HS}(\sigma) \exp\left[-\frac{l_B Z^2}{\sigma(1 + \Gamma\sigma)^2}\right] \tag{9.187}$$

where

$$g_{+-}^{HS}(\sigma) = \frac{(1 - \eta/2)}{(1 - \eta)^3}. \tag{9.188}$$

In combination with the ion binding model, MSA has been applied to fitting experimental data for a wide variety of aqueous electrolyte solutions at room temperature.[83] For associating electrolyte solutions, good correlation could be achieved even at extremely high ion concentrations, up to 25 mol/kg for ammonium nitrate and 34 mol/kg for potassium nitrite. Although the binding energy makes relatively minor contributions to the thermodynamic properties of aqueous solutions, it drastically improves the performance of MSA to describe the vapor–liquid phase behavior of charged systems near the critical region.[84]

83 Simonin J. P., Bernard O., and Blum L., "Real ionic solutions in the mean spherical approximation. 3. Osmotic and activity coefficients for associating electrolytes in the primitive model", *J. Phys. Chem. B* 102 (22), 4411–4417 (1998).
84 Jiang J. W. et al., "Criticality and phase behavior in the restricted-primitive model electrolyte: description of ion association", *J. Chem. Phys.* 116 (18), 7977–7982 (2002).

9.6.6 Summary

Integral-equation theories have been widely used to describe the structure and thermodynamic properties of electrolyte solutions and other charged systems such as colloidal dispersions, molten salts, and classical plasmas. In particular, the HNC theory is able to reproduce the simulation results for the PM of electrolytes except at conditions near the critical point of the liquid–liquid transition (Section 9.5). In comparison with HNC, the MSA is less accurate but yields analytical results that are convenient for many practical applications. Using the ion diameters and association constants as model parameters, MSA provides an excellent description of experimental data for both the thermodynamic and transport coefficients of ionic systems.

9.7 Statistical Behavior of Polyelectrolyte Chains

Polyelectrolyte (PE) solutions are pervasive in industrial applications and biological processes, including water treatment, materials fabrications, and gene therapy. The properties of PE systems are quite different from those corresponding to simple electrolytes or neutral polymers because of the presence of numerous electrically charged groups along the polymer backbone (or side chains). In this section, we discuss the influence of electrostatic interactions on the conformation of individual PE chains. The thermodynamic properties of PE systems at finite polymer concentrations will be discussed in the subsequent sections.

9.7.1 Conformation of Polyelectrolyte Chains

In comparison with a neutral polymer, a PE chain has unique properties owing to long-range electrostatic interactions between charged segments. To facilitate a qualitative understanding, we may classify individual PE chains into three categories: (i) flexible chains; (ii) semi-flexible chains; and (iii) rigid chains. A flexible PE chain behaves more or less like a random coil while a rigid chain takes on a rod-like conformation independent of the solution conditions. A semi-flexible chain is characterized by a bending energy comparable to non-bonded interaction energies. The competition between short- and long-range interactions results in rich conformations of individual PE chains.

Electrostatic repulsion among polymer segments makes a PE chain significantly more expanded in comparison with a neutral polymer of similar molecular weight. As a result, a flexible PE chain is highly swollen in a salt-free solution. Upon the addition of salt, it collapses into a compact, spherical structure at some intermediate salt concentration due to electrostatic screening, and the polymer swells again at higher salt concentrations to form a random-coil-like structure. The collapse and re-expansion of a single chain in response to an increase in salt concentration takes place in dilute solutions of both synthetic and biological polymers. The structural transition is closely related to the *reentrant* behavior of phase transitions in such systems. For example, a PE solution exists as a homogenous phase at low temperature (or salt concentration). Increasing the temperature (or salt concentration) leads to a phase transition, and a further increase in the same variable leads to a reentrant homogenous phase.

The conformation of a flexible PE chain depends not only on the salt concentration but also on the solvent quality. In a poor solvent, a PE chain contracts and collapses in response to the addition of salt. The competition of short-range attraction (viz., due to solvent effects) with long-range repulsion (viz., due to electrostatics) leads to the coexistence of globules and elongated domains

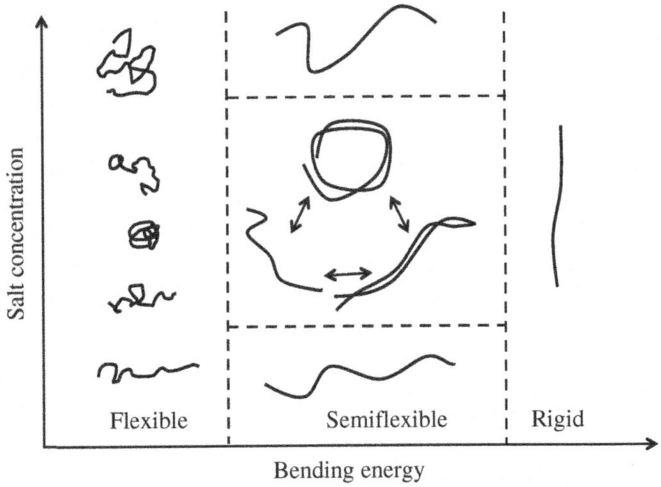

Figure 9.26 Schematic diagram for the conformation of a single polyelectrolyte chain in the presence of salt ions. The double-headed arrows denote structural transitions between different states.

within a single chain.[85] At low salt concentrations, a semi-flexible chain may exhibit an elongated structure similar to that of a flexible chain. However, the bending energy prohibits the formation of a totally collapsed conformation as the salt concentration increases. Instead, the PE chain may bend over and form a hairpin-like structure. A further increase in the salt concentration or the addition of multivalent counterions leads to toroidal structures.

Schematically, Figure 9.26 summarizes some possible conformations of a linear PE chain in an electrolyte solution.[86] As discussed above, the polymer may adopt a variety of structures depending on the stiffness of bond connectivity, the ion valence and concentration in the surrounding medium, and the quality of the solvent. Despite tremendous research efforts, the theoretical prediction of PE conformations has not been fully established, even for flexible chains in a good solvent. In the following, we discuss only a few simple models to highlight the importance of electrostatic interactions in defining the unique structures of flexible PE chains.

9.7.2 An Ideal Chain with Electrostatic Charges

Introducing electrostatic repulsion among the segments of an ideal chain provides a reasonable starting point to understand the conformation of an isolated PE. The model was originally proposed by Kuhn, Kunzle and Katchalsky (K3)[87] and developed by de Gennes and coworkers in more detail.[88] Further extensions of the ideal-chain model to include excluded volume effects and short-range attractions were explored by Flory and others.[89]

---

85 Wang L. et al. "A parallel tempering Monte-Carlo study of conformation transitions of a single polyelectrolyte chain in solutions with added salt", *Acta Polym. Sin.* (12), 1984–1992 (2017).
86 Wei Y. F. and Hsiao P. Y., "Role of chain stiffness on the conformation of single polyelectrolytes in salt solutions", *J. Chem. Phys.* 127 (6) (2007).
87 Katchalsky A., Kunzle O., and Kuhn W., "Behavior of polyvalent polymeric ions in solution", *J. Polym. Sci.* 5 (3), 283–300 (1950).
88 De Gennes P. G., Pincus P., Velasco R. M., and Brochard F., "Remarks on polyelectrolyte conformation", *J. Phys.-Paris* 37 (12), 1461–1473 (1976).
89 Flory P. J., "Molecular configuration of polyelectrolytes", *J. Chem. Phys.* 21 (1), 162–163 (1953).

In the K3 model, a PE chain is represented by m freely jointed Kuhn segments of valence Z for each segment. The free ions (i.e., counterions and salt ions) are uniformly distributed throughout the system, such that the electrolyte solution can be considered an effective dielectric medium. The free energy of the PE chain depends on the elastic entropy of the polymer backbone and the Coulomb repulsion among the charged segments. As discussed in Chapter 3.8 (Eq. (3.167)), the elastic free energy is given by, in dimensionless units,

$$\beta F^{elastic} = \frac{9R_g^2}{mb^2} \tag{9.189}$$

where b stands for the Kuhn length, and R_g is the radius of gyration. If there is no polymer charge, we would have $R_g^2 = mb^2/6$ as for an ideal chain. Note that, for an ideal chain, the radius of gyration is related to the end-to-end distance R by $R^2 = 6R_g^2$.

We can estimate the electrostatic energy by assuming a uniform distribution of the polymer charges within a sphere of radius R_g. While in general the conformation of a PE chain may not have spherical symmetry, an explicit consideration of the non-spherical shape introduces only a minor correction.[88] With this assumption, we can approximate the Coulomb energy using the Born solvation energy (Eq. (9.64)) for a charged sphere of valence mZ

$$\beta F^C = \frac{(mZ)^2 l_B}{2R_g} \tag{9.190}$$

where $l_B \equiv \beta e^2/(4\pi\varepsilon_0\varepsilon)$ is the Bjerrum length. Minimization of the total free energy, $\partial(\beta F^{elastic} + \beta F^C)/\partial R_g = 0$, leads to an estimation of the radius of gyration of the PE chain

$$R_g = m[(bZ)^2 l_B/36]^{1/3} \tag{9.191}$$

Eq. (9.191) predicts that, unlike a neutral chain, the dimension of a PE chain is linearly dependent on the degree of polymerization. The linear dependence of the polymer size with the number of segments suggests that a single flexible PE chain adopts rod-like conformations in a pure solvent or in dilute electrolyte solutions.

Figure 9.27 shows the hydrodynamic radius (R_h) versus the degree of polymerization for two PE in pure water at room temperature.[90] The experimental data (symbols) were determined from fluorescence correlation spectroscopy and can be successfully correlated with a rod model for the

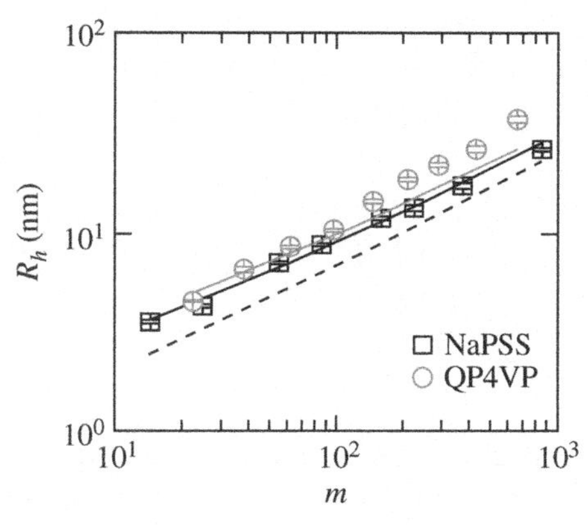

Figure 9.27 Hydrodynamic radius (R_h) versus the degree of polymerization for two polyelectrolytes, sodium polystyrene sulfonate (NaPSS) and quarternized poly-4-vinylpyridine (QP4VP), in pure water near room temperature (22 °C). The solid lines denote fitting with a rod-like model, and the dashed line is the fitting results assuming that the diameter of the rod is the same as that of a hydrated segment. Source: Reproduced from Xu et al.[90]

90 Xu G. F. et al., "Single chains of strong polyelectrolytes in aqueous solutions at extreme dilution: conformation and counterion distribution", *J. Chem. Phys.* 145 (14), 144903 (2016).

polymer, where the size scales linearly with the degree of polymerization. One caveat of the theoretical analysis is that the rod diameter obtained from fitting the experimental data (2.2 nm for NaPSS) is significantly larger than the diameter of a hydrated monomer (~0.8 nm). As shown by the dashed line in Figure 9.27, the theoretical prediction would be much less satisfactory if the PE chain were assumed to take linear conformations.

9.7.3 Electrostatic Blobs

The linear dependence of the dimension of a PE chain on the molecular weight can be alternatively interpreted in terms of "electrostatic blobs". The concept was originally introduced by de Gennes and coworkers[88] and has been extensively used in the polymer literature to understand the scaling behavior of PE systems.[91]

As shown schematically in Figure 9.28, an electrostatic blob refers to a fragment of a PE chain such that the electrostatic energy is comparable to thermal energy $k_B T$. Using the Born solvation model, we can estimate the blob diameter from $\beta F^C \approx 1$

$$D = 2R_g^{blob} = l_B(n_b Z)^2 \tag{9.192}$$

where n_b denotes the number of polymer segments within each blob, and Z is again the valence of the polymer segments. Assuming that the chain conformations inside the blob are unperturbed by electrostatic interactions, we can estimate the blob diameter from the Gaussian chain model

$$D = bn_b^{1/2}. \tag{9.193}$$

Eq. (9.193) is written for a freely jointed chain with n_b segments. An alternative expression may be used for the neutral polymer in a good solvent. A comparison of Eqs. (9.192) and (9.193) indicates that the number of polymer segments in each blob is

$$n_b = (Z^2 l_B/b)^{-2/3}. \tag{9.194}$$

At a length scale much larger than the blob size, the electrostatic repulsion among polymer segments leads to the elongation of the PE chain into an array of linearly connected blobs. The size of the PE chain is thus given by the number of blobs per chain times the blob diameter

$$L = (m/n_b)D = bm(Z^2 l_B/b)^{1/3}. \tag{9.195}$$

While both Eqs. (9.195) and (9.191) predict a linear dependence of the PE chain size on the degree of polymerization, the blob model provides further information on the thickness of the PE rod. For the example shown in Figure 9.27, the Bjerrum length is comparable to the hydrated diameter of a

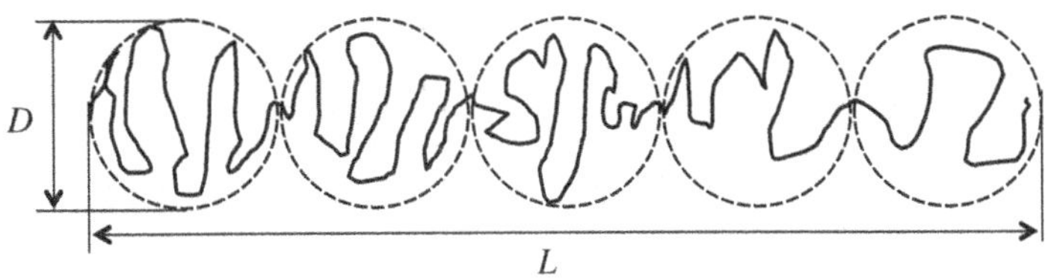

Figure 9.28 Schematic of a polyelectrolyte chain represented by electrostatic blobs (dashed circles). Each blob has a diameter D defined such that its electrostatic energy is equal to $k_B T$.

91 Dobrynin A. V., Colby R. H., Rubinstein M., "Scaling theory of polyelectrolyte solutions", *Macromolecules* 28 (6), 1859–1871 (1995)

NaPSS monomer ($b \approx l_B \approx 0.8$ nm). A comparison of Eq. (9.195) with the PE chain length obtained from the numerical fitting of the hydrodynamic radius to the rod model indicates that the average valence per monomer is $Z \approx 0.17$. Eqs. (9.194) and (9.193) thus predict that the number of segments per electrical blob and blob diameter are $n_b = 10$ and $D = 2.56$ nm, respectively. The rod thickness is reasonably close to what was obtained from fitting with the hydrodynamic radius (2.2 nm).

9.7.4 A Mean-Field Model for Electrostatic Expansion

The ideal-chain models discussed above ignore the influence of small ions (counterions plus salt ions) and the solvent-mediated short-range interactions among polymer segments. For short chains, such effects can be accounted for with molecular simulation. However, the situation becomes more challenging for long PE chains. For a PE chain immersed in an electrolyte solution with only monovalent ions at a relatively low salt concentration, we can derive an analytical expression for the end-to-end distance by combing the Flory theory for chain elasticity and the DH theory for ion screening effects.

We recall from Section 3.9 that the conformations of an ideal chain can be described by the Gaussian distribution. The probability of the polymer ends being separated by a vector $\mathbf{R}$ is given by

$$p(\mathbf{R}) = \left(\frac{3}{2\pi m b^2}\right)^{3/2} \exp\left\{-\frac{3R^2}{2mb^2}\right\} \tag{9.196}$$

where $R = |\mathbf{R}|$, and m is the number of Kuhn segments. Eq. (9.196) suggests that an idea chain with end-to-end distance R has a dimensionless free energy

$$\beta F^{elastic}(R) = -\ln[4\pi R^2 P(\mathbf{R})] = R^2/R_0^2 - 2\ln(R/R_0) + \ln[\pi^{1/2} R_0/4] \tag{9.197}$$

where $R_0^2 \equiv 2mb^2/3$. For a long chain, the first term on the right side of Eq. (9.197) is much larger than the second term, and the third term is a constant. In that case, Eq. (9.197) reduces to the elastic energy given by Eq. (9.189) for a long Gaussian chain. In other words, Eq. (9.189) may be understood as a simplified form of elastic energy. Because $\partial \beta F^{elastic}/\partial R = 0$ at $R = R_0$, R_0 represents the most probable end-to-end distance of an ideal chain.

To understand ion screening and solvent effects on the dimension of a single PE chain, we consider non-bonded interactions between polymer segments in addition to chain connectivity. Approximately, the potential of mean force between polymer segments in an electrolyte solution includes contributions due to the Coulomb energy (C) and solvent-mediated short-range (SR) interactions

$$u(r) = u^C(r) + u^{SR}(r). \tag{9.198}$$

For a system with monovalent ions, the electrostatic energy can be estimated from the DH theory

$$\beta u^C(r) = Z^2 l_B e^{-\kappa r}/r \tag{9.199}$$

where κ is the Debye screening parameter. According to the Flory–Huggins theory, the short-range interaction between polymer segments can be expressed in terms of the Flory parameter χ_F and the Kuhn length b

$$\beta u^{SR}(r) = (1/2 - \chi_F) b^3 \delta(\mathbf{r}) \tag{9.200}$$

where $\delta(\mathbf{r})$ stands for the 3D delta function.

Assuming that the potential of the mean force is pairwise additive, we can estimate the total energy due to non-bonded interactions between polymer segments from the mean-field approximation

$$\beta F = \frac{1}{2}\int\int d\mathbf{r}d\mathbf{r}'\rho_m^2 u(|\mathbf{r}-\mathbf{r}'|) = \frac{\rho_m^2 V}{2}\int d\mathbf{r}u(r) \tag{9.201}$$

where ρ_m is the number density of polymer segments. Substituting Eqs. (9.199) and (9.200) into (9.201) yields

$$\beta F = \frac{\rho_m^2 V}{2}[(1/2-\chi_F)b^3 + 4\pi Z^2 l_B/\kappa^2] = \frac{\rho_m^2 V b^3}{2}[(1/2-\chi_F) + Z^2/(2Ib^3)] \tag{9.202}$$

where $\kappa^2 = 8\pi l_B I$, and I denotes the ionic strength of the electrolyte solution.

For a flexible PE chain with a spherical shape, the end-to-end distance R is related to the radius of gyration R_g and the volume occupied by the polymer

$$V = \frac{4\pi}{3}R_g^3 = \frac{4\pi}{3}(R^2/6)^{3/2}. \tag{9.203}$$

With $\rho_m = m/V$ for the average number density of polymer segments, Eq. (9.202) can then be rewritten as

$$\beta F = \frac{3\sqrt{6^3}m^2b^3}{8\pi R^3}[(1/2-\chi_F) + Z^2/(2Ib^3)] = \frac{3^4 m^{1/2} K R_0^3}{8\pi R^3} \tag{9.204}$$

where

$$K \equiv (1/2-\chi_F) + Z^2/(2Ib^3). \tag{9.205}$$

For the polymer in a good solvent ($\chi_F = 0$), the non-bonded energy is repulsive and varies inversely with the ionic strength. Because the oppositely charged ions are considered as a screening background, the electrostatic energy is different from that predicted by the DH theory. Effectively, the electrostatic repulsion contributes as an additional excluded volume of the polymer segments.

Similar to the ideal chain model discussed above, minimization of the total free energy, $\partial(\beta F^{elastic} + \beta F)/\partial R = 0$, leads to

$$-2/R + 2R/R_0^2 - \frac{3^5 m^{1/2} K R_0^3}{8\pi R^4} = 0. \tag{9.206}$$

Moving the last term in Eq. (9.206) to the right side of the equal sign and diving both sides of the equation by $2R_0^3/R^4$, we attain an analytic expression for the size ratio $\alpha \equiv R/R_0$ of a PE chain relative to that of an ideal chain

$$\alpha^5 - \alpha^3 = \frac{3^5 m^{1/2} K}{16\pi}. \tag{9.207}$$

Except a different coefficient, Eq. (9.207) is identical to that derived from a variational procedure proposed by Beer and coworkers.[92]

Figure 9.29A presents the expansion coefficient α versus parameter $\xi \equiv 3^5 m^{1/2}K/16\pi$. The mean-field model predicts a minimum value of $\xi_{\min} \approx -0.186$ at $\alpha = \sqrt{3/5}$. In a poor solvent ($\chi_F > 1/2$), a PE chain may exhibit a reentrancy effect in response to the addition of salt. Figure 9.29B shows that the parameter K is typically much larger than unity and inversely proportional to the ionic strength, suggesting that the solvent effect is negligible and becomes important only at high ionic strength or low polymer charge density. For a weakly expanded PE

92 Beer M. Schmidt M. and Muthukumar M., "The electrostatic expansion of linear polyelectrolytes: effects of gegenions, co-ions, and hydrophobicity", *Macromolecules* 30 (26), 8375–8385 (1997).

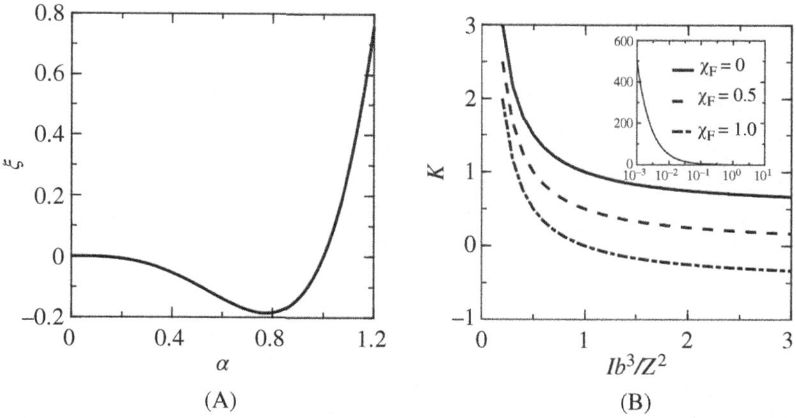

Figure 9.29 Electrostatic expansion of a polyelectrolyte chain in an electrolyte solution. (A) Interaction parameter $\xi \equiv 3^5 m^{1/2} K/16\pi$ versus the expansion coefficient $\alpha \equiv R/R_0$. (B) Dependence of parameter K on the ionic strength I and Flory parameter χ_F.

chain, α shows a linear dependence on ξ, which agrees with $R \sim m$ predicted by alternative models. Qualitatively, these predictions are consistent with experimental observations.[92]

9.7.5 Summary

The conformation of an isolated PE chain may take on many shapes depending on the backbone stiffness, the type and concentration of ions, and the solvent-mediated, non-bonded interactions. The statistical behavior is particularly sensitive to the salt concentration due to the electrostatic repulsion among the charged segments. The coarse-grained models discussed in this section elucidate the complexity of PE systems even at the level of individual chains.

9.8 The Cell Model and Counterion-Condensation Theory

At physiological conditions, the persistence length of a double-strand (ds)-DNA is about 50 nm, and that of a single-strand (ss)-DNA is about 1 nm.[93] Accordingly, DNA chains are quite rigid on the length scale pertinent to the size of monomeric ions (<1 nm). The significant difference in the polymer and monomeric length scales is pertinent to many natural and synthetic polyelectrolytes (PEs). In terms of the spatial distribution of small ions, each PE chain can be approximated as a rigid rod with a uniform surface charge. In this section, we discuss the cell model of PE solutions that is based on ion distributions near individual PE chains represented by charged rods without end effects. We illustrate how the simplified description of chain conformation can be utilized to describe the thermodynamic properties of PE solutions and the polymer interaction with other species.

9.8.1 Katchalsky's Cell Model

The cell model for PE solutions was proposed by Katchalsky and others.[94] It can be used to describe certain properties of biomacromolecules such as solution osmotic coefficients and the denaturation

93 Smith S. B., Cui Y. J., and Bustamante C., "Overstretching B-DNA: the elastic response of individual double-stranded and single-stranded DNA molecules", *Science* 271 (5250), 795–799 (1996).
94 Katchalsky A., "Polyelectrolytes", *Pure Appl. Chem.* 26 (3–4), 327–374 (1971).

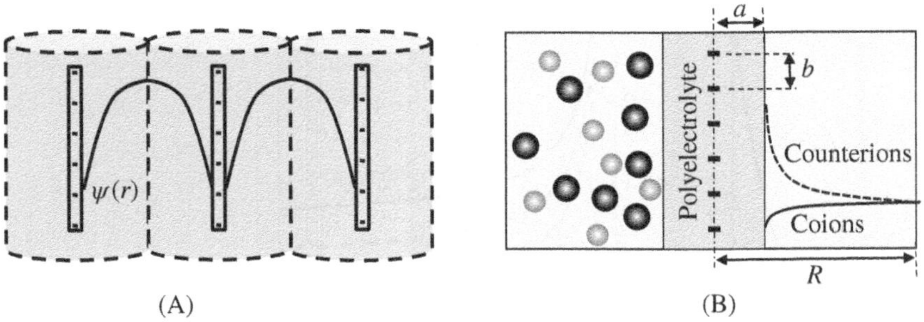

Figure 9.30 (A) In the cell model, a polyelectrolyte (PE) solution is represented by parallel cylindrical cells of infinite length. Each cell contains a long, concentric, rigid rod representing the polymer. Schematically, the curves represent the electric potential near a negatively charged polymer. (B) Ion distributions within each cell. Parameter a corresponds to the radius of a PE segment, b is the axial distance between successive charges (−) along the PE chain (rod), and R is the cell radius.

temperature of ds DNA chains. The cell model is also useful to correlate experimental results from potentiometric titrations of weakly ionizable polymers.

Figure 9.30 shows schematically a PE solution represented by cylindrical cells of infinite length. Each cell contains only a single PE chain represented by a concentric rigid rod of radius a with uniform surface charge density. The cylindrical geometry implies that the volume fraction of the PE chains in the solution is directly related to the cell radius R by

$$\phi = (a/R)^2. \tag{9.208}$$

The linear charge density of the PE chain can be described in terms of a dimensionless parameter $\xi \equiv l_B/b$, i.e., the ratio of the Bjerrum length and the cylindrical length per unit charge b. The latter can be estimated from the spacing between the nearest neighboring charges along the axis of the PE chain.

Within each cylindrical cell, the electric potential ψ and ion distributions can be described by the Poisson–Boltzmann (PB) equation

$$\frac{1}{r}\frac{\partial}{\partial r}\left(r\frac{\partial \varphi}{\partial r}\right) = -4\pi l_B \sum_i \rho_i^0 Z_i e^{-Z_i\varphi(r)} \tag{9.209}$$

where r denotes the radial distance, $\varphi \equiv \beta e\psi$ is the dimensionless electric potential, and ρ_i^0 is the number density of ion i at the cell boundary. As discussed below, ρ_i^0 can be determined from its average concentration in the solution.

We assume that the PE solution within each cell satisfies the charge neutrality. Accordingly, the Gauss law predicts that the electric field vanishes at the cell boundary, i.e., $\varphi'(R) = 0$. At the surface of a negatively charged rod, the Gauss law predicts $\varphi'(a) = 2\xi/a$, i.e., the slope for the electric potential is positive near a negative surface.[95]

With the boundary conditions for the electric field at the inner and out surfaces of the concentric cylinder, we can integrate Eq. (9.209) and find a closed form for the electric potential

$$\varphi(r) = 2\xi \ln(r/R) - 4\pi b \sum_i Z_i \left[\ln(r/R)\int_a^r r'\rho_i(r')dr' + \int_r^R r'\rho_i(r')\ln(r'/R)dr'\right]. \tag{9.210}$$

In combination with the Boltzmann equation for the ion distributions, $\rho_i(r) = \rho_i^0 e^{-Z_i\varphi(r)}$, we can evaluate the reduced electric potential φ from Eq. (9.210) using a numerical procedure.

95 Gauss' law predicts $\partial\psi/\partial r|_{r=a} = -\varsigma/(\varepsilon_0\varepsilon)$ where ς represents the surface charge density. For a negatively charged polyelectrolyte, $\varsigma = e/(2\pi ab)$, and if the surface charge is positive, $\varsigma = -e/(2\pi ab)$.

For PE chains in a salt-free solution, the electric potential is determined by the distribution of counterions (viz., ions with an opposite charge of the PE chain). In that case, Eq. (9.209) becomes

$$\frac{1}{r}\frac{\partial}{\partial r}\left(r\frac{\partial \varphi}{\partial r}\right) = -\frac{\kappa_c^2}{2}e^{-\varphi(r)} \tag{9.211}$$

where the reduced electric potential is modified by the counterion valence Z_c, $\varphi(r) = Z_c\beta e\psi$, $\kappa_c = \sqrt{8\pi\, l_B\rho_c^0 Z_c^2}$ stands for the Debye screening parameter, and subscript "c" denotes counterions. It should be noted that, unlike its conventional definition, here κ_c refers to the counterion concentration at the cell boundary. In other words, Eq. (9.211) is written with the assumption that the electric potential at $r = R$ is zero, i.e., $\varphi(R) = 0$.

Eq. (9.211) has an analytic solution of[96]

$$\varphi(r) = 2\ln\left\{\frac{\kappa_c r}{2\mid\lambda\mid}\cos[\lambda\ln(r/R_M)]\right\} \tag{9.212}$$

where parameters λ and R_M can be determined from the boundary conditions at $r = a$ and R. According to Eq. (9.212), the derivative of the reduced electric potential is

$$\frac{\partial\varphi(r)}{\partial r} = \frac{2}{r}\{1 - \lambda\tan[\lambda\ln(r/R_M)]\}. \tag{9.213}$$

Substituting Eq. (9.213) into $\varphi'(a) = 2\xi/a$ and $\varphi'(R) = 0$ gives

$$1 - \lambda\tan[\lambda\ln(R/R_M)] = 0, \tag{9.214}$$

$$1 - \lambda\tan[\lambda\ln(a/R_M)] = \xi. \tag{9.215}$$

To determine the Debye screening parameter, consider the electric potential at the cell boundary

$$\varphi(R) = 2\ln\left\{\frac{\kappa_c R}{2\mid\lambda\mid}\cos[\lambda\ln(R/R_M)]\right\} = 0. \tag{9.216}$$

Eq. (9.216) is satisfied when

$$\frac{\kappa_c R}{2\mid\lambda\mid}\cos[\lambda\ln(R/R_M)] = 1. \tag{9.217}$$

The combination of Eqs. (9.214) and (9.217) gives

$$\kappa_c^2 R^2 = 4(1 + \lambda^2). \tag{9.218}$$

In application of the cell model, we must solve Eqs. (9.214), (9.215) and (9.218) to fix parameters λ, R_M and κ_c that are used in the expression for the electric potential, Eq. (9.212). Toward that end, we may first solve λ by a combination of Eqs. (9.214) and (9.215)

$$\lambda\ln(a/R) = \arctan[(1-\xi)/\lambda] - \arctan(1/\lambda). \tag{9.219}$$

Once λ is determined from Eq. (9.219) numerically, we can then calculate R_M and κ_c from Eqs. (9.214) and (9.218), respectively.

The thermodynamic properties of the PE solution are mostly related to the counterion concentration at the cell boundary. This quantity can be derived from Eq. (9.218) and $\kappa_c = \sqrt{8\pi\, l_B\rho_c^0 Z_c^2}$

$$\rho_c^0 = (1+\lambda^2)/\left(2\pi l_B R^2 Z_c^2\right). \tag{9.220}$$

96 Le Bret M. and Zimm B., "Distribution of counterions around a cylindrical polyelectrolyte and Manning's condensation theory", *Biopolymers*, 23, 287–312 (1984).

Noting that the average concentration of counterions in the cell can be calculated from the charge neutrality condition

$$\overline{\rho}_c = 1/(\pi R^2 b \mid Z_c \mid), \tag{9.221}$$

we can rewrite Eq. (9.220) as

$$\rho_c^0/\overline{\rho}_c = (1 + \lambda^2)b/(2l_B \mid Z_c \mid). \tag{9.222}$$

9.8.2 Osmotic Pressure

Consider a PE solution at a finite salt concentration. Because the colligative properties are additive, the osmotic pressure may be approximated by the sum of that for a salt-free PE and that for the salt solution. According to the cell model, the former can be written as the osmotic pressure at the cell boundary where the electric field vanishes, i.e.,

$$\Pi_c/k_B T \approx \rho_c^0. \tag{9.223}$$

Meanwhile, the osmotic pressure of a simple electrolyte can be expressed in terms of the average salt concentration ρ_s and osmotic coefficient Φ_s. If the salt is a monovalent electrolyte, the osmotic pressure of the PE solution is then given by

$$\Pi/k_B T = \rho_c^0 + 2\rho_s \Phi_s \tag{9.224}$$

where $2\rho_s$ stands for the total number density of salt ions.

The osmotic coefficient (Φ) of the PE solution is defined as the osmotic pressure divided by that of an ideal solution of the PE chains and small ions. The latter is given by $\beta \Pi^{id} \approx \overline{\rho}_c + 2\rho_s$, with the assumption that the number density of polymer chains is much smaller than that of counterions and thus does not contribute to the osmotic pressure. Accordingly, Eq. (9.224) predicts that the osmotic coefficient is

$$\Phi \equiv \Pi/\Pi^{id} \approx \left[\rho_c^0 + 2\rho_s \Phi_s\right] / (\overline{\rho}_c + 2\rho_s). \tag{9.225}$$

For an infinite dilute PE solution free of added salt, R is extremely large. Eq. (9.225) reproduces the osmotic coefficient predicted by the Oosawa–Manning theory (Problem 9.20)

$$\Phi_\infty = 1/(2\xi) \tag{9.226}$$

where subscript ∞ indicates infinite dilution of the PE chains.

In practice, the osmotic pressure of a PE solution is often measured relative to that of a buffer solution of the same salt with concentration ρ_0. In that case, the salt concentration ρ_s in the PE solution can be estimated from the ideal Donnan equilibrium[97]

$$\left(\rho_c^0 + \rho_s\right) \rho_s = \rho_0^2. \tag{9.227}$$

Accordingly, the osmotic coefficient should be modified as

$$\Phi \approx \left[\rho_c^0 + 2(\rho_s - \rho_0)\right] / \overline{\rho}_c. \tag{9.228}$$

97 The Donnan equilibrium requires that the mean activity of the added salt is the same on both sides of a membrane semi-permeable to small molecules and small ions. Equation (9.228) is obtained by replacing the ionic activity with number density at the cell boundary because at that boundary, the salt concentration is assumed to be very small.

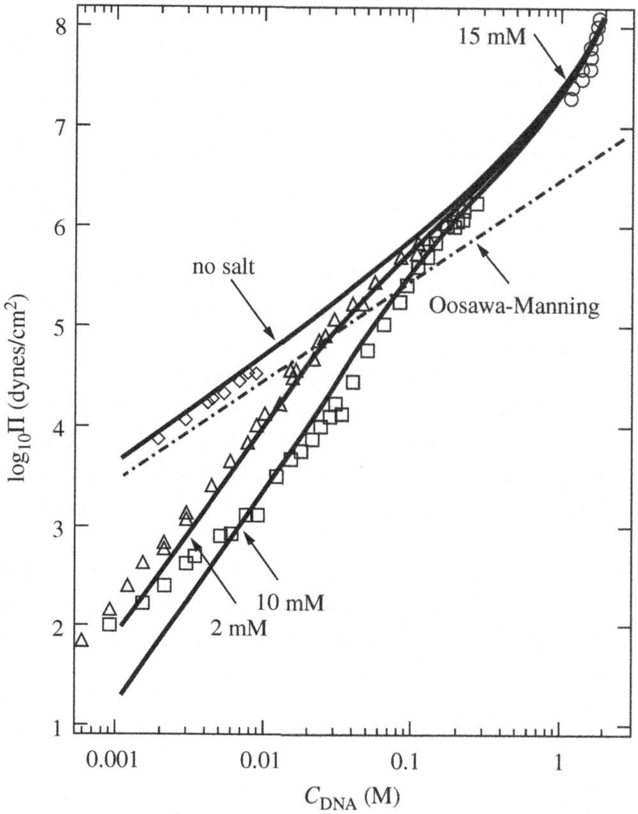

Figure 9.31 Osmotic pressure versus the DNA concentration predicted by the cell model (solid lines), the Oosawa-Manning line-charge model (the dotted-dashed line), and from experiment (symbols). Source: Reproduced from Hansen et al.[98]

Figure 9.31 presents the osmotic pressure of dsDNA solutions predicted by the cell model in comparison with experimental data.[98] Each dsDNA molecule consists of repeating phosphate groups of negative charge. Approximately, the dsDNA radius is $a \approx 1$ nm, and the separation between the nearest phosphate charges is about $b \approx 0.17$ nm. While the agreement is not perfect, it is truly remarkable that the simple model is able to quantify the experimental results over a wide range of DNA concentrations.

9.8.3 Stability of DNA Duplex

The cell model is useful not only to describe the thermodynamic properties of PE solutions but also PE binding with other species, including the titration of weak PEs. To elucidate, consider thermal denaturation of dsDNA structures (viz., DNA melting) as commonly conducted in experiments to assess the duplex stability. The denaturation curve reflects a helix-to-coil transition of the polymer chain when temperature increases. This equilibrium can be described in terms of a "two-state" model: Each nucleic acid is either in the double helix state or in a separated single-stranded state. The melting temperature (T_m) is defined as the condition where half of the nucleic acids are

98 Hansen P. L., Podgornik R., and Parsegian, V. A., "Osmotic properties of DNA: critical evaluation of counterion condensation theory", *Phys. Rev. E* 64 (2), 021907 (2001).

in the single-stranded form. Experimental studies show that the melting temperature is mainly determined by the charge densities of dsDNA and ssDNA. The helix-to-coil transition depends strongly on the salt concentration (C_s), more precisely, the activities ($a_\pm$) and electrostatic charges of cations in the solution.

We can describe the Gibbs energy of DNA melting in terms of electrostatic and non-electrostatic contributions

$$\Delta G_m = \Delta G^0 + \Delta G^C \tag{9.229}$$

where subscript m denotes the melting transition, ΔG^0 accounts for all non-electrostatic contributions, and ΔG^C accounts for electrostatic interactions. At the melting point, a dsDNA chain is in equilibrium with two ssDNA chains, i.e., $\Delta G_m = 0$.

Approximately, the non-electrostatic Gibbs energy is relatively insensitive to electrolyte conditions. As a result, ΔG^0 can be estimated from thermodynamic relations using the calorimetric data for the enthalpy of denaturation at a particular condition of added salt (e.g., NaCl). With a semi-empirical correlation for the non-electrostatic Gibbs energy ΔG^0, the cell model can be used as a predictive tool to describe the variation of the melting point temperature due to the addition of a salt.

To a good approximation, the helix-to-coil transition results in no substantial change in the solution volume. As a result, we can estimate the change in Gibbs energy from internal energy and entropy contributions, $\Delta G^C \approx \Delta U^C - T\Delta S^C$. These quantities can be calculated from the electric potential and ion distributions for the DNA chains before and after denaturation[99]

$$\beta U^C/b = \frac{\varphi(a)}{2b} + \frac{1}{2}\sum_i \int_a^R \rho_i(r) Z_i \varphi(r)\, 2\pi\, r dr, \tag{9.230}$$

$$S^C/(bk_B) = -\sum_i \int_a^R \left\{\rho_i(r)\ln\left[\rho_i(r)/\rho_i^0\right] - \rho_i(r) + \rho_i^0\right\}\, 2\pi\, r dr. \tag{9.231}$$

With the reduced electric potential $\varphi(r)$ and ionic density profiles $\rho_i(r)$ predicted from the cell model, Eqs. (9.230) and (9.231) can be used to calculate the free energies of both dsDNA and ssDNA chains in various electrolyte solutions, thus the melting temperature.

Figure 9.32 compares theoretical predictions with experiments when DNA melting takes place in the presence of magnesium ions.[99] Although the cell model treats the solvent as a dielectric

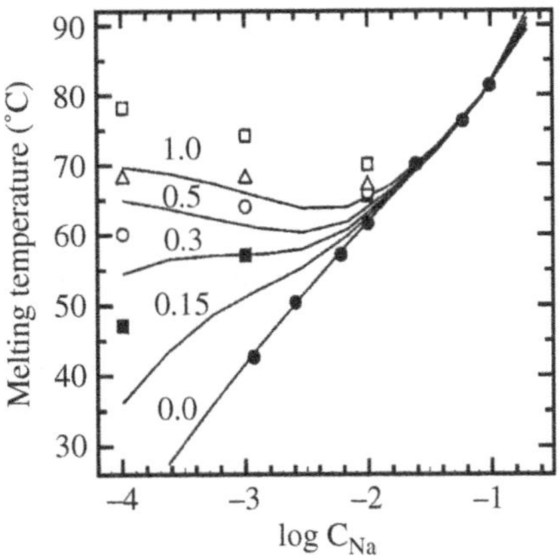

Figure 9.32 The melting-point temperature of T4 phage dsDNA chains as a function of NaCl concentration (C_{Na}, mM) in presence of $MgCl_2$. Experimental points are for different ratios of the magnesium to DNA concentrations, C_{Mg}/C_{DNA} with $C_{DNA} = 0.042$ mM. The curves are calculated from the cell model and the symbols are from the experiment. The cell-model parameters are $b_h = 1.69$ Å and $a_h = 9.5$ Å for dsDNA, $b_c = 3.38$ Å and $a_c = 7.0$ Å for ssDNA.
Source: Reproduced from Korolev et al.[99]

99 Korolev N., Lyubartsev A. P., and Nordenskiold L., "Application of polyelectrolyte theories for analysis of DNA melting in the presence of Na+ and Mg2+ ions", *Biophys. J.* 75 (6), 3041–3056 (1998).

continuum and ignores the DNA flexibility and atomic structure, the simple model provides a good description of electrostatic effects on DNA stability. Because the PB equation ignores electrostatic correlations and ion size effects, the cell model is not applicable to DNA melting in the presence of natural polyamines such as putrescine^{2+}, spermidine^{3+} or spermine^{4+}.

9.8.4 Counterion-Condensation Theory

Before closing this section, we briefly discuss the counterion-condensation (CC) theory introduced by Manning[100] and independently by Oosawa.[101] The PE theory is often used as an alternative to the cell model with the PB equation for ion distributions.

The original concept of counterion condensation stems from the application of the cell model to an infinitely long rod with a linear charge density $\xi = l_B/b$ surrounded by counterions of valence Z_i. The system serves as a limiting condition of a salt-free PE solution at infinite dilution.

The unscreened Coulomb interaction between the charged rod and a single counterion at radial distance r is

$$\beta u(r) = 2|Z_i|\xi \ln(r/R) \tag{9.232}$$

where R is a reference position where the electric potential is zero. With the assumptions that the charged rod has a negligible radius and that the counterions are independent of each other, the relative concentration of counterions in the cylindrical cell ($r < R$) to that in the bulk is given by

$$\Theta = \frac{1}{\pi R^2} \int_0^R e^{-u(r)/k_B T} 2\pi \, r dr = \frac{2}{R^{2-2|Z_i|\xi}} \int_0^R r^{1-2\,|Z_i|\xi} dr. \tag{9.233}$$

Because the integration starts from zero, Θ diverges when $|Z_i|\,\xi \geq 1$. To avoid the numerical issue, Manning argued that the ionic atmosphere is unstable, and some counterions must condense to the PE surface such that its linear density is reduced to $|Z_i|\,\xi = 1$. In other words, a fraction of the fixed charges at the PE chain, $1 - (|Z_i|\,\xi)^{-1}$, becomes completely neutralized by the condensed counterions. Manning's hypothesis coincides with an independent work by Oosawa, who proposed a two-state model for the distribution of counterions around a charged rod, i.e., the counterions near a PE chain are distributed either in a bound state or in a free state.

To a certain degree, the counterion condensation model resembles the Bjerrum theory of electrolyte solutions except that counterion condensation is assumed to be thermodynamically irreversible. Although the numerical issue exists only for PE chains surrounded by counterions without considering the size effects, Manning and Oosawa hypothesized the prevalent applicability of the "limiting law", i.e., $|Z_i|\,\xi \leq 1$ is valid for all linear PE. With the renormalized surface charge density of the PE chain, the distributions of uncondensed ions and the electric potential can be calculated from the linearized PB equation.

Mathematically, the divergence problem in evaluating the ion distributions can be avoided if a more realistic model of the PE solutions is considered. For this reason, the CC theory has been criticized and subjected to a number of modifications, in particular on the spatial distribution of condensed counterions and their division from ions in the uncondensed state.[102] In practice, the CC theory remains a popular choice not only because of its simplicity but, probably more

100 Manning G. S., "Molecular theory of counterion conductivity and self-diffusion in polyelectrolyte solutions", *J. Chem. Phys.* 47 (6), 2010–2013 (1967).

101 Oosawa F., *Polyelectrolytes*. M. Dekker, New York, 1971.

102 See for example, Stigter D., "Evaluation of the counterion condensation theory of polyelectrolytes", *Biophys. J.* 69 (2), 380–388 (1995).

important, because it often yields reasonable results in comparison with experimental data for important properties of PE solutions.

9.8.5 Summary

In this section, we have discussed two simple thermodynamic models to describe the thermodynamic properties of PE solutions. Katchalsky's cell model is based on the PB equation for ion distributions near a charged rod. Conversely, the Manning-Oosawa theory assumes that counterions condense onto the polymer backbone beyond a critical charge density. While neither model accounts for the flexibility of PE chains, they are useful to understand certain properties of PE systems dominated by the concentration of counterions or the surface potential of the charged polymer. In particular, the simplified description of chain conformation is useful to predict the thermodynamic properties of PE solutions and the interaction of charged polymers with other species.

9.9 Liquid-State Theories of Polyelectrolyte Solutions

Despite a long-history of theoretical developments, molecular theories of polyelectrolyte (PE) solutions remain an evolving research area. A quantitative description of PE systems is difficult due to the strong coupling of atomistic-scale size effects and solvent-mediated interactions with long-range electrostatic and intra-chain correlations. In addition to the cell models discussed in Section 9.8, a number of alternative approaches have been proposed to describe the structure and thermodynamic properties of PE solutions. In this section, we briefly discuss a lattice model and an off-lattice equation of state as representative examples. More in-depth discussion of this topic can be found in specialized texts and literature reviews.[103]

9.9.1 The Voorn–Overbeek Theory

The Voorn–Overbeek (VO) theory provides a simple description of the thermodynamic properties of PE systems by combining the Flory–Huggins (FH) theory for neutral polymers and the DH theory for electrostatic interactions.[104] The hybrid model accounts for polymer conformations and segment-level non-bonded interactions, and it is able to describe the phase behavior of PE systems, including coacervation.[105]

In general, a PE system contains cations, anions, and solvent molecules in addition to the charged polymers. The monomeric ions of the opposite charge of a PE chain are commonly referred to as counterions. The VO theory assumes that the translational entropies and short-range interactions of all chemical species can be represented by the FH theory. Meanwhile, the free energy due to the electrostatic interactions is assumed to be the same as that of point charges in a dielectric medium.

---

103 Muthukumar M., *Physics of charged macromolecules*. Cambridge University Press, 2023.

104 Voorn M. J., "Complex coacervation. 1. General theoretical considerations", *Recl. Trav. Chim. Pay B* 75, 317–330 (1956); Overbeek J. T.; Voorn M. J., "Phase separation in polyelectrolyte solutions. Theory of complex coacervation", *J. Cell. Physiol.*, 49, 7–26 (1957).

105 The word "coacervation" has a Latin root: *acervus* means heap, and the prefix indicates heaping polymers together in a single phase. Here "coacervation" refers to phase separation in a solution with two oppositely charged polyelectrolytes leading to one concentrated polymer phase and an extremely dilute solvent phase. This phenomenon is also known as "complex coacervation".

As discussed in Section 8.1, we can derive the Helmholtz energy of a neutral polymer system using a lattice model with mean-field approximations. For a polymer mixture containing multiple types of polymeric species, the Helmholtz energy of mixing can be expressed as

$$\beta \Delta F^{FH}/N_T = \sum_i \frac{\phi_i}{m_i} \ln \phi_i + \frac{1}{2} \sum_i \sum_j \phi_i \phi_j \chi_{ij} \tag{9.234}$$

where m_i and ϕ_i denote, respectively, the number of segments per molecule and the volume fraction of species i, χ_{ij} is the Flory parameter, $N_T = \sum_i m_i N_i$ is the total number of lattice sites to accommodate all molecules in the system (including ionic species), and N_i is the number of molecules (chains) for species i.

The first term on the right side of Eq. (9.234) takes into account translational and conformational entropies of all chemical species, and the second term is related to the change in the potential energy for the random mixing of monomeric segments. Because the FH theory assumes the single occupancy of lattice sites with either a polymer segment or a monomeric species of similar size, the total volume of the system is given by $V = vN_T$ where v is the volume of a lattice site. As expected, the volume fraction satisfies the normalization condition

$$\sum_i \phi_i = \sum_i m_i N_i / N_T = 1. \tag{9.235}$$

The DH theory predicts that electrostatic energy is given by

$$\beta F^{DH}/V = -\kappa^3/12\pi \tag{9.236}$$

where κ is the Debye screening parameter, and l_B is the Bjerrum length. Within the framework of the lattice model, the number density of ionic segments is

$$\rho_i^0 = \alpha_i m_i N_i/(N_T v) = \alpha_i \phi_i / v, \tag{9.237}$$

where α_i is the fraction of polymer segments that bear a net charge of valence Z_i (viz., the degree of ionization for i type of polymer chains; for monomeric ions, $m_i = \alpha_i = 1$.), and the Debye screening parameter is given by

$$\kappa = \sqrt{4\pi l_B/v} \left(\sum_i \alpha_i \phi_i Z_i^2 \right)^{1/2}. \tag{9.238}$$

The combination of Eqs.(9.234) and (9.236) leads to the Helmholtz energy of mixing for a PE solution

$$\frac{\beta F^{VO}}{N_T} = \sum_i \frac{\phi_i}{m_i} \ln \phi_i + \frac{1}{2} \sum_i \sum_j \phi_i \phi_j \chi_{ij} - \frac{v\kappa^3}{12\pi}. \tag{9.239}$$

Eq. (9.239) represents a key result of the VO theory.

To elucidate an application of the VO theory for describing the phase behavior of PE systems, consider an aqueous solution containing two oppositely charged PE chains with the same degree of ionization α and the same number of polymer segments per chain m. For simplicity, we assume that the polymers are in a good solvent such that the Flory parameters χ_{ij} can be neglected, and that all charged segments are monovalent, $|Z_i| = 1$. We assume further that there are no salt or free ions in the system such that the Debye screening parameter is uniquely determined by the polymer volume fraction ϕ

$$\kappa = \sqrt{4\pi l_B \alpha \phi / v}. \tag{9.240}$$

Because of charge neutrality, the volume fractions of the oppositely charged PE chains are identical, $\phi_{P+} = \phi_{P-} = \phi/2$, and the solvent volume fraction is $1 - \phi$. In this case, the Helmholtz energy

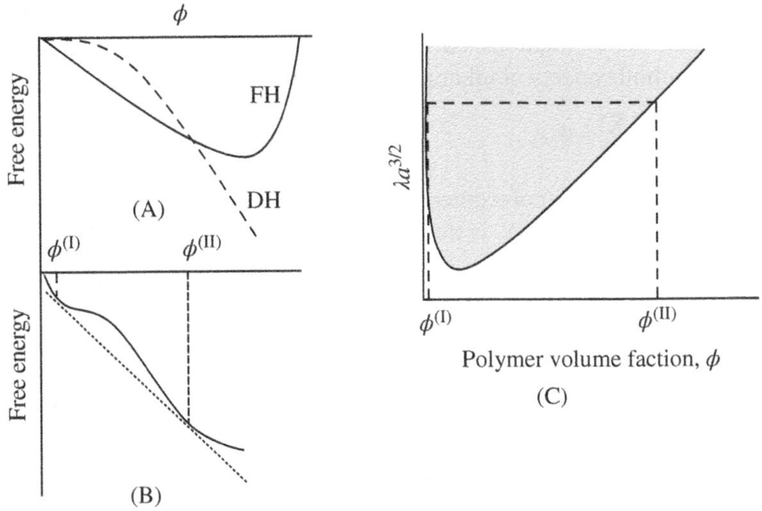

Figure 9.33 (A) The free energy for an aqueous solution of oppositely charged polyelectrolytes is predicted by the Voorn–Overbeek (VO) theory. (B) The free energy versus polymer volume fraction shows two inflection points, signaling the occurrence of complex coacervation. (C) A schematic phase diagram where the two-phase region is shaded. The concentration of the coacervate phase increases with the coupling parameter λ, which is related to the strength of electrostatic energy.

predicted by the VO theory is given by

$$\frac{\beta F^{VO}}{N_T} = \frac{\phi}{m}\ln(\phi/2) + (1-\phi)\ln(1-\phi) - \lambda(\alpha\phi)^{3/2} \tag{9.241}$$

where $\lambda = (2/3)l_B^{3/2}\sqrt{\pi/\upsilon}$ is a dimensionless parameter (≈3.5 for water at room temperature).[106]

As shown schematically in Figure 9.33A, the Helmholtz energy of mixing is always negative and its dependence on the polymer composition exhibits a positive curvature, suggesting that the solution is always thermodynamically stable. The electrical free energy is also negative but shows a convex curvature. When the entropic and electrostatic contributions are combined, the total free energy of mixing versus polymer volume fraction curve exhibits both positive and negative curvatures. In the latter case, the system is thermodynamically unstable.

The range of polymer composition in the unstable region can be determined from $\partial^2 F/\partial\phi^2 < 0$. As shown in Figure 9.33B, we may identify the compositions of two coexisting phases by drawing a double tangent line to the free-energy curve, leading to a phase diagram as shown in Figure 9.33C. Problems 8.10 and 8.11 provide a quantitative analysis of the phase behavior of an aqueous solution of symmetric PE chains free of salt.

Qualitatively, the VO theory is able to capture the essential features of the phase behavior of PE mixtures.[107] With the segment size and charge treated as adjustment parameters, the VO theory describes the complex coacervation of strongly charged flexible PE at high salt concentrations in good agreement with experimental observations.[108] However, it should be noted that the VO theory inherits the shortcomings of both the FH and DH theories. The measn-field method neglects the electric double layer (EDL)[109] structure around the PE chains, thus it is not able to

106 For water at room temperature, $l_B = 0.714$ nm, $\lambda \approx 3.5$ corresponds to molecular volume of $\upsilon \approx 0.041$ nm^3.
107 Sing C. E., "Development of the modern theory of polymeric complex coacervation", *Adv. Colloid Interface* 239, 2–16 (2017).
108 Spruijt E. et al., "Binodal compositions of polyelectrolyte complexes", *Macromolecules* 43 (15), 6476–6484 (2010).
109 See Section 9.10.

account for counterion condensation and related effects. Another limitation of the VO theory is that it ignores intrinsic coupling among chain connectivity, molecular excluded volume, and electrostatic correlations. As a result, it is not able to describe pair correlation functions or the formation of mesoscopic structures.

9.9.2 Equations of State for Polyelectrolyte Solutions

As discussed in Section 8.3, a polymeric fluid may be considered a system of monomeric particles with the constraints of bond connectivity. The difference between the free energy of the polymeric system and that of the monomeric reference system can be determined from the change in free energy due to complete chemical reactions. Following this idea, Jiang and coworkers formulated a first-order thermodynamic perturbation theory (TPT1) for PE systems similar to the SAFT equation of state for neutral polymers.[110]

According to TPT1, the Helmholtz energy of a PE system is given by

$$F = F^{IC} + F_m^{ex} + F^{TPT} \tag{9.242}$$

where F^{IC} is the Helmholtz energy of the polymeric system in an ideal solution (viz., no interactions between ionic species and polymer segments other than the bond potential), F_m^{ex} is the excess Helmholtz energy of the corresponding monomeric fluid, and F^{TPT} is an excess free-energy due to chain connectivity.

For an ideal solution of polymeric molecules, the McMillan-Mayer solution theory predicts that the Helmholtz energy density is given by

$$\beta F^{IC}/V = \sum_i \frac{\rho_i}{m_i}\left[\ln\frac{\rho_i \Lambda_i^3}{m_i} - 1 - q_i\right] \tag{9.243}$$

In Eq. (9.243), ρ_i is the number density of segments for solute component i, assuming that each molecule contains m_i identical segments; Λ_i is the thermal wavelength for individual segments arising from integrations over the momentum degrees of freedom; and q_i is a constant related to the bonding energy. It is important to note that Eq. (9.243) is different from that defined in classical thermodynamics for an ideal solution of non-interacting molecules; it refers to an idealized state of the polymeric system that lacks not only intermolecular interactions but also non-bonded intramolecular interactions among polymer segments.

The excess Helmholtz energy of the monomeric system is defined relative to that for a *monomeric* ideal system, i.e., $F_m^{ex} = F_m - F_m^{ID}$ where F_m is the Helmholtz energy of the monomeric system, and F_m^{ID} corresponds to that of an ideal solution of monomeric species. According to the primitive model for monomeric charged particles in a dielectric continuum, the monomeric excess Helmholtz energy includes contributions due to hard-sphere (HS) repulsion and the electrostatic (EL) interactions

$$F_m^{ex} = F_m^{HS} + F_m^{EL}. \tag{9.244}$$

As discussed in Appendix 7A, the hard-sphere free energy is commonly described by the Boublik–Mansoori–Carnahan–Starling–Leland (BMCSL) equation of state[111]

$$\frac{\beta F_m^{HS}}{V} = -n_0 \ln(1-n_3) + \frac{n_1 n_2}{1-n_3} + \frac{n_2^3}{36\pi n_3^3}\left[\ln(1-n_3) + \frac{n_3}{(1-n_3)^2}\right] \tag{9.245}$$

110 Jiang J. W., Blum L., Bernard O., Prausnitz J. M., "Thermodynamic properties and phase equilibria of charged hard sphere chain model for polyelectrolyte solutions", *J. Stat. Phys.*, 99, 1121–1128 (2001).
111 Boublik T., "Hard-sphere equation of state", *J. Chem. Phys.*, 53, 471–472 (1970); Mansoori G. A., Carnahan N. F., Starling K. E., Leland, T. W., Jr., "Equilibrium thermodynamic properties of the mixture of hard spheres", *J. Chem. Phys.*, 54, 1523–5 (1971).

where $n_0 = \sum_i \rho_i$, $n_1 = \sum_i \rho_i \sigma_i / 2$, $n_2 = \pi \sum_i \rho_i \sigma_i^2$, $n_3 = \pi \sum_i \rho_i \sigma_i^3 / 6$, and σ_i is the hard-sphere diameter for segment i. Meanwhile, the electrostatic free energy is often approximated by the MSA equation[112]

$$\frac{\beta F^{MSA}}{V} = -\sum_i \frac{\rho_i Z_i l_B}{1+\Gamma\sigma_i} \left[\Gamma Z_i + \frac{\pi P_n \sigma_i}{2(1-n_3)} \right] + \frac{\Gamma^3}{3\pi} \tag{9.246}$$

where the MSA screening parameter is given by

$$\Gamma^2 = \sum_i \frac{\pi l_B \rho_i}{(1+\Gamma\sigma_i)^2} \left[Z_i - \frac{\pi P_n \sigma_i^2}{2(1-n_3)} \right]^2. \tag{9.247}$$

with

$$P_n = \sum_i \frac{\rho_i \sigma_i Z_i}{(1+\Gamma\sigma_i)^2} \left[1 + \frac{\pi}{2(1-n_3)} \sum_k \frac{\rho_k \sigma_k^3}{1+\Gamma\sigma_k} \right]^{-1}. \tag{9.248}$$

As discussed in Section 9.6, MSA is able to describe the thermodynamic properties of monomeric ionic systems in good agreement with experiments up to high salt concentrations. If the PE chains are represented by linear, freely-jointed chains of charged hard spheres, the free energy due to chain connectivity is given by

$$\beta F^{TPT1}/V \approx -\sum_{i=1}^{m-1} (\rho_i/m_i) \ln y_{ii}(\sigma_i) \tag{9.249}$$

where the contact value of the two-body cavity correlation function can be derived from a modified MSA equation[110]

$$y_{ii}(\sigma_i) = g_{ii}^{HS}(\sigma_i) \exp\left[l_B Z_i^2/\sigma_i\right] \exp\left[\Delta g_{ii}^{MSA}(\sigma_i)\right] \tag{9.250}$$

with

$$g_{ii}^{HS}(\sigma_i) = \frac{1}{1-n_3} + \frac{n_2\sigma_i}{4(1-n_3)^2} \tag{9.251}$$

and

$$\Delta g_{ii}^{MSA}(\sigma_i) = -\frac{l_B}{\sigma_i(1+\Gamma\sigma_i)^2} \left[Z_i - \frac{\pi P_n \sigma_i^2}{2(1-n_3)} \right]^2. \tag{9.252}$$

Because TPT1 assumes that the free energy due to chain connectivity depends only on the two-body cavity correlation function, similar equations can be derived if the system contains PE chains with segments of different sizes and valences.

Figure 9.34 shows the osmotic coefficients obtained from TPT1 in comparison with Monte Carlo (MC) simulation for a PE solution free of salt.[113] Also compared are theoretical predictions from an integral-equation theory[114] and from the polymer reference interaction site model (PRISM).[115] When the electrostatic coupling is relatively weak ($l_B = \sigma$), all theoretical predictions are similar

---

112 Blum L., Mean spherical model for asymmetric electrolytes. 1. Method of solution, *J. Stat. Phys.*, 30, 1529–1535 (1975).

113 Chang R. and Yethiraj A., "Osmotic pressure of salt-free polyelectrolyte solutions", *Macromolecules* 38, 607–616 (2005); Chang R., Kim Y. and Yethiraj A., "Osmotic pressure of polyelectrolyte solutions with salt", *Macromolecules* 48, 7370–7377 (2015).

114 von Solms N. and Chiew Y. C. "Analytical integral equation theory for a restricted primitive model of polyelectrolytes and counterions within the mean spherical approximation", *J. Chem. Phys.* 118, 4321 (2003); *J. Chem. Phys.* 111, 4839 (1999).

115 Shew C. Y. and Yethiraj, A., "Computer simulations and integral equation theory for the structure of salt-free rigid rod polyelectrolyte solutions", *J. Chem. Phys.* 110 (23), 11599–11607 (1999).

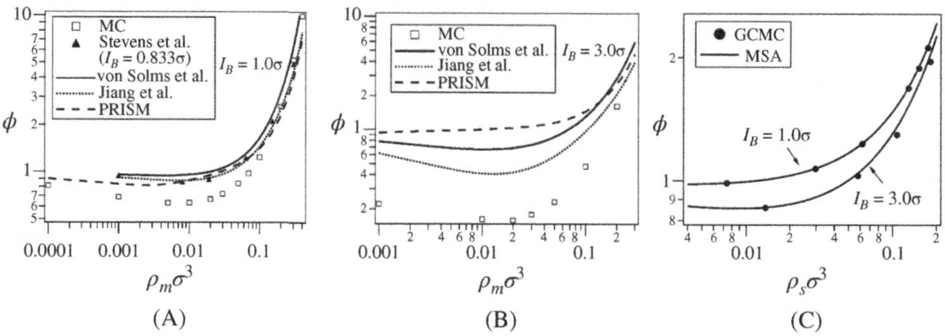

Figure 9.34 Osmotic coefficients from different theoretical methods (lines) in comparison with Monte Carlo (MC) simulation (symbols) for the primitive model of a polyelectrolyte (PE) solution free of salt. Here, the PE chain length is $m = 16$, both the counterions and PE segments are monovalent and have the same hard-sphere diameter σ. (A) $l_B = \sigma$; (B) $l_B = 3\sigma$. (C) The osmotic coefficients predicted from MC and MSA for the corresponding monomeric systems. Source: Adapted from Chang and Yethiraj.[113]

and agree well with the simulation results at high polymer concentrations. The good agreement is mainly because all theoretical methods capture the thermodynamic properties due to hard-sphere interactions. As none of the three methods is able to predict an accurate second virial coefficient, a notable discrepancy between theory and simulation exists in the semi-dilution regime. The discrepancy among different methods increases as the strength of electrostatic coupling rises ($l_B = 3\sigma$). In this case, all theoretical methods overestimate the osmotic coefficient in the dilute regime. Because MSA predicts an accurate osmotic coefficient for the corresponding monomeric system (Figure 9.34C), we may attribute the theoretical errors directly to inadequate representation of thermodynamic nonideality due to multi-body chain correlations. In the semi-dilute and concentrated regimes, the osmotic coefficient is virtually independent of the degree of polymerization because the thermodynamic properties are dominated by short-range interactions.

We close this subsection by using the equation of state discussed above to understand the salt effect on the phase behavior of a PE solution. It is known experimentally that adding a small amount of salt into a homogeneous, originally salt-free PE solution can induce the precipitation of PE chains once the salt concentration is above some threshold value. Further addition of salt, however, can dissolve the precipitates completely if the salt concentration exceeds a second threshold value.

As shown in Figure 9.35, both the VO theory and TPT1 can capture the qualitative features of the salting-in and salting-out phenomena.[116] However, noticeable differences can be identified in terms of the slope of the tie lines and the closeness of the binodal curves. In contrast to the VO theory but consistent with experiments, TPT1 predicts that the salt concentration in the PE-poor phase is slightly higher than that in the PE-rich phase. The VO theory fails to capture this subtle trend in salt partitioning because it neglects the effects of electrostatic interactions on chain connectivity.

The salting-out and salting-in phenomena reflect the delicate balance of molecular excluded volume and electrostatic correlations. As predicted by both TPT1 and the VO theory, the addition of salt increases the electrostatic energy per unit volume that promotes phase separation. However, excessive amounts of salt introduces significant excluded volume effects that tend to keep the PE solution in a single phase. These competing effects may result in complicated phase behavior in PE systems.[117]

116 Zhang, P. F. et al., "Salting-out and salting-in of polyelectrolyte solutions: a liquid state theory study", *Macromolecules* 49, 9720–9730 (2016).

117 Naeem S. and Sadowski G., "Pepc-Saft: modeling of polyelectrolyte systems", *Fluid Phase Equilib.* 299, 84–93 (2010); *Fluid Phase Equilib.* 306, 67–75 (2011).

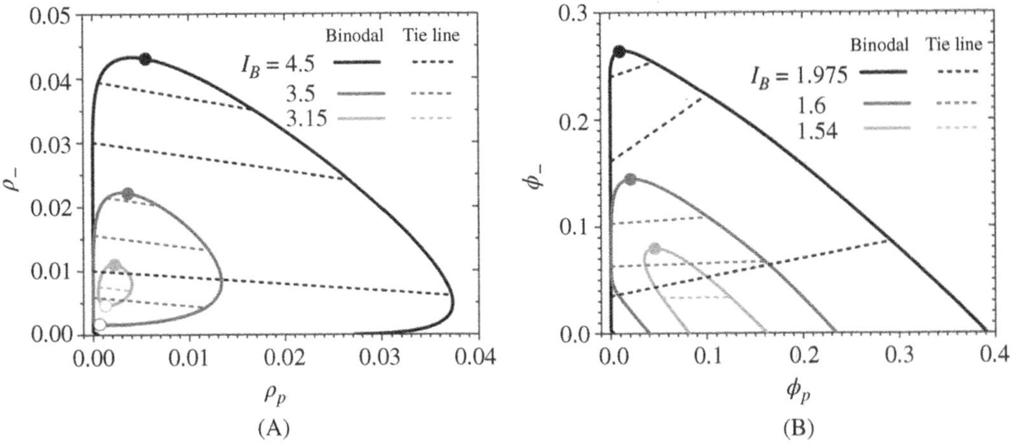

Figure 9.35 Phase diagram of a polyelectrolyte solution with monovalent salt predicted from the TPT equation by Jiang et al. (A) and from the Voorn–Overbeek theory (B). In both panels, the solid lines represent phase boundaries, and the dashed lines are for tie lines. The filled circles are the upper critical points, and the open circles are the lower critical points. Here the solution is represented by the primitive model with PE segments and small ions assumed to have the same hard-sphere diameter and absolute valence. The PE chains have negative charges, and the Bjerrum length is in the units of the segment diameter. Reproduced from Zhang et al.[116]

9.9.3 Summary

Both the lattice model and SAFT-like equations of state are able to capture the generic phase behavior of PE solutions. While the lattice model has advantages in terms of simplicity, an off-lattice model is needed to account for the specific ion effects and different contributions to the solvent-mediated interactions. Further theoretical developments are expected in order to describe the phase behavior of PE systems truly from a molecular perspective.

9.10 Electric Double Layer (EDL)

The electric double layer (EDL), also known as double layer, is an interfacial model originally developed by Hermann von Helmholtz in the mid-nineteenth century for representing electrified surfaces in contact with electrolyte solutions.[118] The theoretical model plays a pivotal role in understanding surface reactions and transport of ionic species in diverse electrochemical processes, including power generation and energy storage, stabilization/destabilization of colloidal particles, mineral dissolution and reprecipitation, and the regulation of ion transfer and selectivity across cell membranes. In this section, we briefly discuss statistical-mechanical methods to predict the ion distributions and thermodynamic properties of electrolytes near charged surfaces. As the theoretical models are rapidly evolving, we refer to specialized texts and current literature for a more in-depth discussion of this important subject.[119]

118 Helmholtz H., "On some laws of the distribution of electrical currents in physical conductors with application to animal electrical experiments", *Ann. Phys.* 165, 211–233 (1853).
119 Schmickler W. and Santos E., *Interfacial electrochemistry*, Springer, (2010).

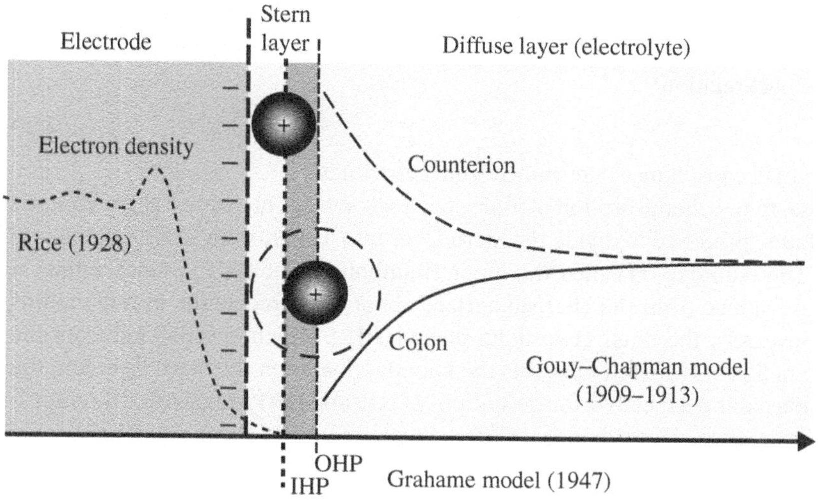

Figure 9.36 Conventional models of electric double layer (EDL). Whereas the Helmholtz model consists of two layers of charge (viz., double layer), later modifications account for ion distributions, electronic structure, and chemisorption effects. IHP and OHP stand for the inner Helmholtz plane and outer Helmholtz plane, respectively.

9.10.1 Conventional EDL Models

The conventional models of EDL are well documented in standard texts of electrochemistry.[120] Figure 9.36 recapitulates some essential ideas pertinent to thermodynamic modeling. The Helmholtz model asserts that EDL consists of a layer of surface charge parallelly positioned near a layer of neutralizing counterions (viz., double layer). Approximately, the two layers of opposite charges are separated by a dielectric medium with a distance on the order of the ion radius. Because the double layer thickness is microscopic, the Helmholtz model predicts that the EDL capacitance is larger than that of a conventional dielectric capacitor by many orders of magnitude. As a result, electrochemical devices that utilize the electrified interface for energy storage are commonly known as supercapacitors or electrochemical capacitors.[121]

Numerous EDL models have been developed to describe the microscopic details of electrified interfaces. One of the most influential models, still commonly in use, was proposed by Louis G. Gouy[122] and independently by David L. Chapman[123] in the early twentieth century. Like the Debye-Hückel (DH) theory for dilute electrolyte solutions, the Gouy-Chapman (GC) model is based on the Poisson–Boltzmann (PB) equation with the assumption that ionic species can be represented by point charges and that the solvent is a dielectric continuum.

For a weakly charged surface in contact with a dilute electrolyte solution of monovalent ions, the GC model provides an adequate description of ion distributions in the diffuse layer, i.e., the electrolyte region near the charged surface but not in direct contact. The immediate vicinity of the electrified interface is conventionally described by the Stern model, a phenomenological modification

120 Bard A. J. and Faulkner L. R., *Electrochemical methods*, Wiley (2000); Balsara N. P. and Newman J., *Electrochemical systems*. Wiley, 2021.

121 Simon P. and Gogotsi Y., "Perspectives for electrochemical capacitors and related devices", *Nat. Mater.* 19, 1151–1163 (2020).

122 Gouy M., "On the constitution of the electric charge at the surface of an electrolyte", *C. R. Acad. Sci.* 149, 654–657(1909).

123 Chapman D. L., "A contribution to the theory of electrocapillarity", *Philos. Mag.* 25, 475–481 (1913).

of the GC theory.[124] Typically, the Stern layer has a thickness on the order of the radius of hydrated ions (Figure 9.36). Its properties are often characterized in terms of an effective capacitance determined by the Parsons-Zobel relation[125]

$$1/C = 1/C_{GC} + 1/C_{Stern} \tag{9.253}$$

where C stands for the EDL capacitance determined from experiment.

To describe effects due to the chemisorption of ionic species or solvent molecules at the charged surface, David C. Grahame proposed to divide the Stern layer into a chemically active region and a solvation region.[126] The boundary is called the inner Helmholtz plane (IHP), a mathematical divider positioned at a distance from the charged surface about the same as the average radius of desolvated ions. Conversely, the outer Helmholtz plane (OHP), which is separated from the charged surface by the radius of hydrated ions, sets the boundary between the Stern layer and the diffuse layer. From a molecular perspective, the definitions of IHP and OHP are necessarily empirical because ion solvation near an electrochemical surface is extremely complicated and influenced by the local electronic properties and solvent structure.

The conventional EDL models provide useful intuition for understanding a wide range of electrochemical phenomena. However, they are insufficient to capture all microscopic details even for relatively simple systems such as an aqueous solution of monovalent ions in contact with a perfect metal. For many electrochemical systems, we must consider not only ion distributions but also the electronic structure of the electrode along with chemically active species near the charged surface. The importance of the electronic density profiles in determining the EDL capacitance was recognized by Oscar K. Rice as early as 1928.[127] While many EDL models are concerned mostly with electrolytes in the diffuse layer, their applications to practical systems may require an explicit consideration of the charged surface.

9.10.2 Inhomogeneous Ionic Systems

In an EDL, the electrolyte may be considered as an inhomogeneous system of ionic species and solvent molecules subject to an external potential due to the surface charge as well as other forms of surface interactions. With the assumption that the external potential is fixed independent of the microstates, we can predict the density profiles of individual atoms (or coarse-grained particles) from all chemical species using the statistical-mechanical relation[128]

$$\rho_\alpha(\mathbf{r}) = \rho_\alpha^0 e^{-\beta\lambda_\alpha(\mathbf{r})} \left\langle \exp\left[-\beta \sum_{\gamma \neq \alpha} \lambda_\gamma(\mathbf{r})\right] \right\rangle \tag{9.254}$$

where ρ_α^0 represents the number density of atomic site α in the bulk solution, and $\lambda_\alpha(\mathbf{r})$ is an effective one-body potential that is defined by, in addition to the local electric potential $\psi(\mathbf{r})$, the non-electrostatic component of the external potential $V_\alpha^{ext}(\mathbf{r})$ and the local excess chemical potential $\Delta\mu_\alpha^{ex}(\mathbf{r})$

$$\lambda_\alpha(\mathbf{r}) = Z_\alpha e\psi(\mathbf{r}) + V_\alpha^{ext}(\mathbf{r}) + \Delta\mu_\alpha^{ex}(\mathbf{r}) \tag{9.255}$$

where Z_α denotes the particle valence, and e is the unit charge.

124 Stern O., "Zur Theorie Der Elektrolytischen Doppelschicht", *Z. Elektrochem. Angew. Phys. Chem.* 30, 508–516 (1924).
125 Parsons R. and Zobel F. G. R. "The interphase between mercury and aqueous sodium dihydrogen phosphate", *J. Electroanal. Chem.*, 9 (5), 333–348 (1965).
126 Grahame D. C., "The electrical double layer and the theory of electrocapillarity", *Chem. Rev.*, 41, 441–501 (1947).
127 Rice O. K. "Application of the fermi statistics to the distribution of electrons under fields in metals and the theory of electrocapillarity", *Phys. Rev.*, 31, 1051–1059(1928).
128 Wu J., *Variational methods in molecular modeling*. Springer, p. 91, 2017.

In Eq. (9.254), $\langle\cdots\rangle$ denotes an average over the molecular configurations of the chemical species i that contains atomic site α

$$\langle\cdots\rangle = \frac{\int d\mathbf{x}_i \delta(\mathbf{r}-\mathbf{r}_\alpha)\exp\left[-\beta V_B(\mathbf{x}_i) - \beta\sum_{\gamma\neq\alpha}\lambda(\mathbf{r}_\gamma)\right]}{\int d\mathbf{x}_i \delta(\mathbf{r}-\mathbf{r}_\alpha)\exp[-\beta V_B(\mathbf{x}_i)]} \tag{9.256}$$

where $\mathbf{x}_i$ represents the molecular configuration, defined by the positions of all atomic particles constituting chemical species i, and $V_B(\mathbf{x}_i)$ stands for the bond potential.

The physical meaning of Eq. (9.254) is rather intuitive, i.e., the local density for any particle α depends on the bulk density and an effective potential that accounts for its interactions with all other particles in the system and the external energy. Both electric potential $V_\alpha^{ext}(\mathbf{r})$ and excess chemical potential $\Delta\mu_\alpha^{ex}(\mathbf{r})$ are defined relative to those corresponding to the bulk solution, i.e., $\lambda_\alpha(\mathbf{r}) = 0$ in the bulk.

Eq. (9.254) allows us to solve the atomic density profiles of ionic species and solvent molecules, together with the Poisson equation

$$\nabla\psi(\mathbf{r}) = -\frac{e}{\varepsilon_0}\sum_\alpha Z_\alpha\rho_\alpha(\mathbf{r}) \tag{9.257}$$

where ε_0 is the permittivity of free space. To implement Eqs. (9.254)–(9.257) numerically, we need a molecular model for the electrolyte-surface interactions and a theoretical procedure to predict the local excess chemical potential $\Delta\mu_\alpha^{ex}(\mathbf{r})$. While such methods are yet to be fully developed, relatively simple statistical-mechanical models are able to provide useful insights into the microscopic details of the charged interfaces that would not be easily attainable from the conventional EDL models. To illustrate, Figure 9.37 shows the capacitive behavior of ion-intercalated MXene nanosheets, a nanostructured material promising for energy storage and other applications.[129] The Helmholtz model predicts a capacitance per unit area inversely proportional to the separation between the MXene surface and the center plane of intercalated ions (Panel A)

$$C_H = \frac{\varepsilon\varepsilon_0}{b-a_0} \tag{9.258}$$

where ε is the solvent dielectric constant, a_0 is the ionic radius, and $2b$ is the width of the slit pore. The dotted line in Panel D indicates that Eq. (9.258) is inconsistent with experimental observations unless that the dielectric constant is treated as a variable depending on the double layer thickness.

For the results shown in panels B and C of Figure 9.37, the external potential for water molecules was estimated from first principles methods, and the atomic density profiles and local electric potential were predicted from the 3D-RISM theory,[130] an extension of the RISM equation for inhomogeneous systems. From the surface charge and electric potential, we can calculate the capacitance from

$$C = C_{surface} + \frac{\Delta Q_{ion}}{\Delta\psi} \tag{9.259}$$

where ΔQ_{ion} is the charge density of intercalated ions, and $\Delta\psi$ stands for the potential drop across the double layer (Panel C). With $C_{surface} \approx 40$ F/g treated as an adjustable parameter, 3D-RISM is able to capture different ionic effects and reproduce experimental data near-quantitatively. The molecular theory is able to describe different ion effects and the increase in the EDL capacitance due to the dipolar polarization of water molecules under strong confinement.

129 Sugahara A. et al., "Negative dielectric constant of water confined in nanosheets", *Nat. Commun.* 10, 850 (2019).
130 Beglov D. and Roux B., "An integral equation to describe the solvation of polar molecules in liquid water", *J. Phys. Chem. B* 101, 7821–7826(1997); Kovalenko A. and Hirata F., "Three-dimensional density profiles of water in contact with a solute of arbitrary shape: a RISM approach", *Chem. Phys. Lett.* 290, 237–244(1998).

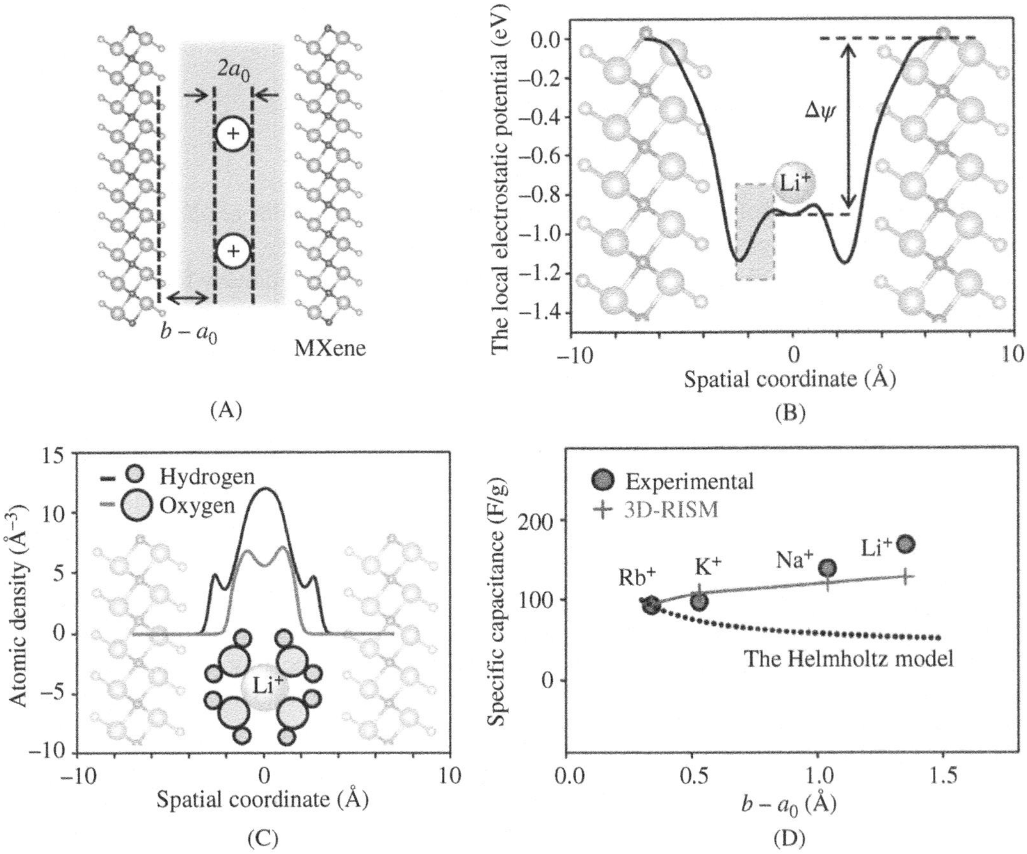

Figure 9.37 The capacitive behavior of MXene nanosheets. (A) The Helmholtz model of EDL; (B) the atomic density profiles predicted by 3D-RISM; (C) the local electric potential predicted by 3D-RISM. (D) Variation of the specific capacitance with the double layer thickness. Source: Adapted from Sugahara et al.[129]

If the ionic species in the EDL are described as charged hard spheres and the solvent as a dielectric continuum (viz., the primitive model), Eq. (9.254) can be simplified to

$$\rho_i(\mathbf{r}) = \rho_i^0 \exp\left[-\beta Z_i e\psi(\mathbf{r}) - \beta V_i^{ext}(\mathbf{r}) - \beta\Delta\mu_i^{ex}(\mathbf{r})\right]. \tag{9.260}$$

The coarse-grained model is particularly useful for inhomogeneous systems of concentrated electrolyte solutions, including room temperature ionic liquids. For example, Figure 9.38 shows the reduced local electric potential, integrated charge density, and ionic density profiles for a room temperature ionic liquid near a charged surface predicted by Eq. (9.260).[131] While the strong oscillations of ionic densities and local electric potential are not expected from the conventional models of EDL, they are consistent with experimental observations.[132] The appearance of the layering structure can be rationalized in terms of the excluded volume effects and electrostatic correlations. Such effects lead to strong charge inversion (viz., "over screening") as shown by the oscillatory distributions of the local mean electric potential and the integrated charge density.

It should be noted that conventional EDL models can be readily derived from the statistical-mechanical theories of inhomogeneous electrolytes. With further assumptions that EDL can be represented by a planar hard wall in contact with point charges in a dielectric medium, the ionic

131 Wu J. et al. "A classical density functional theory for interfacial layering of ionic liquids", *Soft Matter* 7, 11222–11231 (2011).
132 Mezger M. H. et al., "Molecular layering of fluorinated ionic liquids at a charged sapphire (0001) surface". *Science* 322: 424-428 (2008).

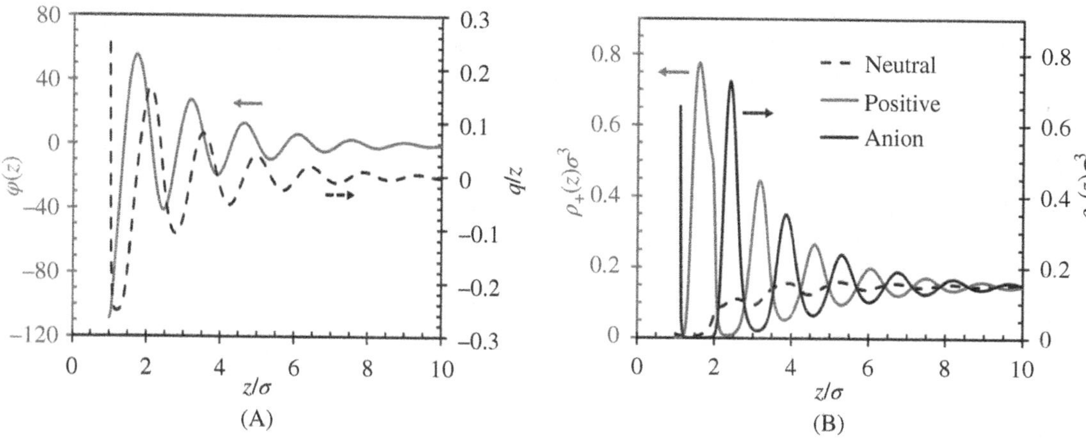

Figure 9.38 A room temperature ionic liquid near a charged surface. (A) The reduced local electrical potential $\varphi(z) \equiv \beta e\psi(z)$ and the integrated charge density, $q(z)$ in units of e/σ^2. (B) The reduced ion densities of cations and anions. Here, z denotes the distance from a charge surface, and σ is the hard-sphere diameter. Each cation is represented by a dimer of neutral and positively charged particles. Source: Adapted from Wu et al.[131]

density profiles would follow the Boltzmann distribution

$$\rho_i(z) = \rho_i^0 \exp[-\beta Z_i e\psi(z)] \tag{9.261}$$

where z denotes the distance from the surface. Eq. (9.261) neglects electrostatic correlations and is only applicable to ion distributions in the diffuse layer. For the one-dimensional system, the Poisson equation becomes

$$\frac{d^2\psi(z)}{dz^2} = -\frac{e}{\varepsilon\varepsilon_0}\sum_i \rho_i^0 Z_i \exp[-\beta Z_i e\psi(z)]. \tag{9.262}$$

If the counterions and coions have the same absolute valence $Z = Z_+ = |Z_-|$, the PB equation yields an analytic solution (Problem 9.22)

$$\varphi(z) \equiv \beta Ze\psi(z) = 2\ln\left[\frac{1+\xi_s e^{-\kappa z}}{1-\xi_s e^{-\kappa z}}\right] \tag{9.263}$$

where $\kappa = [2\rho_0 Z^2 e^2\beta/(\varepsilon_0\varepsilon)]^{1/2}$ represents the Debye screening parameter, $\rho_0 = \rho_+^0 = \rho_-^0$ is the ionic density in the bulk solution, and $\xi_s = \tanh(\varphi_s/4)$ is defined by the surface electric potential ψ_s.

Plugging Eq. (9.263) into (9.261) leads to the ionic density profiles

$$\rho_\pm(z)/\rho_\pm^0 = \exp[\mp\varphi(z)] = \left[\frac{1\pm\xi_s e^{-\kappa z}}{1\mp\xi_s e^{-\kappa z}}\right]^2. \tag{9.264}$$

Eqs. (9.263) and (9.264) represent the key results of the GC theory. At the charged surface ($z = 0$), Gauss's law predicts the surface charge density

$$Q = -\varepsilon_0\varepsilon(d\psi/dz)|_{z=0} = \pm\left\{\frac{2\varepsilon\varepsilon_0}{\beta}\sum_i \rho_i^0\left[e^{-\beta Z_i e\psi_s} - 1\right]\right\}^{1/2} \tag{9.265}$$

where ψ_s stands for the surface potential, and the $\pm$ signs apply to positive and negative surface potentials, respectively. Eq. (9.265) is known as the Grahame equation.

Eq. (9.265) is commonly utilized to correlate the surface charge density and ζ-potential measured in electrophoresis experiments. In Figure 9.39, the GC theory is employed to describe the variation of the surface potential versus salt concentration for two types of amidine

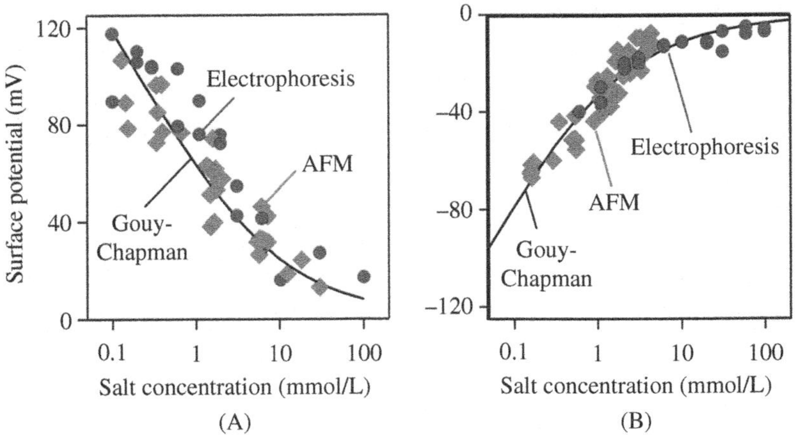

Figure 9.39 The Gouy–Chapman theory is used to describe the surface potentials of amidine latex particles. (A) Bare particles and (B) particles coated with a saturated layer of poly(styrene sulfonate). The experimental data were determined by electrophoresis and direct force measurements. Source: Reproduced from Szilagyi et al.[133]

latex particles.[133] With the surface charge density treated as an adjustable parameter, the GC theory provides a perfect fit to the experimental data. The GC theory can also be used to analyze electrostatic potentials and ion distributions near lipid membranes.[134] For systems with monovalent ions at relatively low salt concentrations and surface charge densities, the predictions of the GC theory are often found to be good agreement with experimental data.

9.10.3 Surface Charge Regulation

Non-electrode systems such as colloids, membranes, and minerals acquire surface charge typically by protonation or deprotonation of ionizable groups. The thermodynamic process is affected not only by the solution condition, including pH, salt types, and ionic strength, but also by the surface chemistry. Charge regulation is important for understanding diverse colloidal phenomena ranging from the functioning of biological membranes to the stability of colloidal dispersions and mass transport in geochemical processes.

We can elucidate the application of EDL models to surface charge regulation using the deprotonation of hydroxyl groups as an example. Eqs. (9.266) and (9.267) provide a typical set of deprotonation reactions:

$$-\mathrm{ROH} \rightleftharpoons -\mathrm{RO}^- + \mathrm{H}^+ \tag{9.266}$$

$$-\mathrm{ROH}_2^+ \rightleftharpoons -\mathrm{ROH} + \mathrm{H}^+ \tag{9.267}$$

where $-R$ represents an ionizable surface site. With the assumption that the ionizations of surface sites are mutually independent, the equilibrium surface composition satisfies the mass action law

$$K_A = \frac{[H^+]_s N_{\mathrm{RO}^-}}{N_{\mathrm{ROH}}} \tag{9.268}$$

$$K_B = \frac{[\mathrm{H}^+]_s N_{\mathrm{ROH}}}{N_{\mathrm{ROH}_2{}^+}} \tag{9.269}$$

133 Szilagyi I., Trefalt G., Tiraferri A., Maroni P., and Borkovec M., "Polyelectrolyte adsorption, interparticle forces, and colloidal aggregation", *Soft Matter* 10 (15), 2479–2502 (2014).
134 Mclaughlin S., "The electrostatic properties of membranes", *Annu. Rev. Biophys. Biophys. Chem.* 18, 113–136 (1989).

where K_A and K_B stand for the equilibrium constants, and N_i is the number density of the surface group i.

Approximately, the proton concentration at the surface can be estimated from the Boltzmann distribution

$$[H^+]_S = [H^+]_0 \exp(-\beta e \psi_s) \tag{9.270}$$

where $[H^+]_0$ denotes the proton activity in the bulk solution, which is determined by $\mathrm{pH} = -\log[\mathrm{H}^+]_0$. From the mass balance condition for the surface sites

$$N_{total} = N_{\mathrm{ROH}} + N_{\mathrm{RO}^-} + N_{\mathrm{ROH}_2^+} \tag{9.271}$$

where N_{total} stands for the total number of ionizable sites per surface area, we may find the surface charge density

$$Q = -e\left(N_{\mathrm{RO}^-} - N_{\mathrm{ROH}_2^+}\right) = -eN_{total}\frac{K_A - [H^+]_s{}^2/K_B}{K_A + [H^+]_s + [H^+]_s{}^2/K_B}. \tag{9.272}$$

From the solution pH and equilibrium constants K_A and K_B, we can predict the surface charge density and the surface potential by solving Eqs. (9.272) and (9.270), along with an additional relation between these two quantities described by an EDL model (e.g., the Grahame equation, Eq. (9.265)).

Figure 9.40 illustrates the charge regulation of carboxyl latex particles in aqueous solutions of KCl at different ion concentrations.[135] With a reasonable value for the equilibrium constant of the surface reaction, the thermodynamic model is able to reproduce the experimental results for both the surface charge density and ζ-potential. In a monovalent electrolyte solution, latex particles are negatively charged across the entire pH range of the solution pH. Whereas the PB equation works well for systems containing monovalent ions at low salt concentrations, it breaks down when electrostatic correlations and excluded volume effects become more significant.

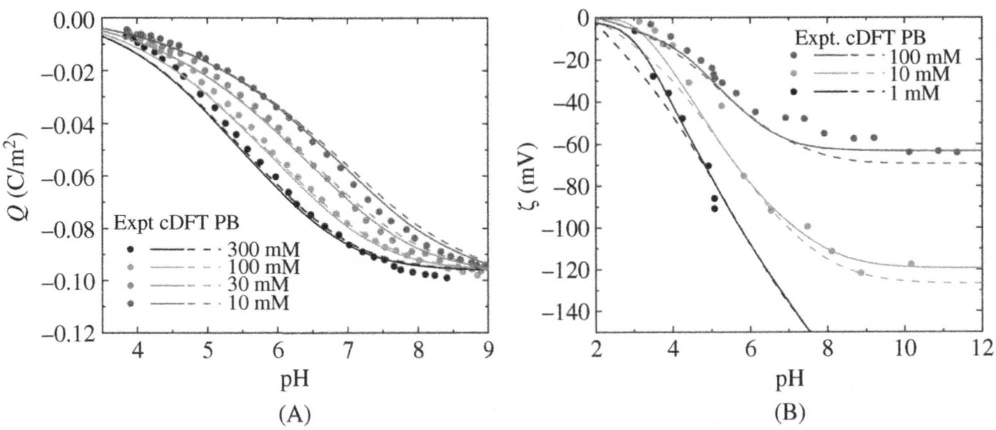

Figure 9.40 The pH dependence of (A) the surface charge density and (B) ζ-potential of carboxyl latex particles in aqueous solutions at different KCl concentrations. The solid lines are predictions from the classical density functional theory (cDFT) and the dashed lines from the modified PB equation. The symbols are experimental results determined by electrophoresis and potentiometric titration for ζ-potential and surface charge density, respectively. Source: Reproduced from Ong et al.[135]

135 Ong G. M. C., Gallegos A. and Wu J., "Modeling surface charge regulation of colloidal particles in aqueous solutions", *Langmuir* 36 (40), 11918–11928 (2020).

9.10.4 Summary

We conclude this section by acknowledging that the partitioning of the EDL into different parallel regions is primarily based on empirical observations. Nevertheless, the phenomenological approach is widely used and favored in practical applications due to its simplicity. Quantitative EDL models facilitate a better understanding of diverse electrochemical phenomena including ion intercalation, charge transfer, and surface reactions, that are important for the engineering applications of electrocatalysis, supercapacitors, and electrochemical batteries.

9.11 Chapter Summary

In this chapter, we have discussed solvation, solvent-mediated interactions, and various theoretical procedures to describe the thermodynamic properties of ionic systems based on the solvent-mediated potentials. We demonstrate that statistical–mechanical methods can be used to predict a wide variety of solution properties important for understanding solvent effects on diverse chemical processes, colloidal forces, biomolecular crowding effects, the stability of DNA duplexes, the thermodynamic properties and phase transitions of electrolytes and polyelectrolyte systems. In addition, we have discussed the properties of electric double layer (EDL) based on the implicit-solvent models and from atomistic perspectives.

Whereas standard statistical–mechanical methods are directly applicable to multicomponent systems including liquid solutions, the McMillan-Mayer (MM) framework plays a unique role in understanding solution thermodynamics. With the solvent treated as a continuous medium, we are able to describe the thermodynamic properties of ideal solutions in parallel to those of ideal-gas systems. For non-ideal solutions, we can use the integral-equation methods and perturbation theories, in addition to the Poisson–Boltzmann (PB) equation, to quantify the structure and thermodynamic properties of electrolytes and PE solutions similar to those used for simple fluids and colloidal dispersions.

9.A The Hamaker Theory of van der Waals Interactions

In this appendix, we give a brief discussion of the Hamaker theory for van der Waals (vdW) interactions between colloidal particles or surfaces.[136] An in-depth discussion of dispersion interactions between macroscopic objects can be found in specialized texts.[137]

The Hamaker theory assumes that the interaction between two macroscopic objects can be estimated from the pairwise additivity of atom–atom interactions. The starting point is London's equation for the dispersion interaction between two identical atoms of diameter σ (Figure 9.A.1A)

$$u(r) = -\frac{c}{r^6}. \tag{9.A.1}$$

where c is a proportionality constant depending primarily on the atomic polarizability, and r is the center-to-center distance. Eq. (9.A.1) is accurate when $r \gg \sigma$, i.e., the center-to-center distance is much larger than the atomic diameter.

As discussed in Section 9.3, the same expression can be used to describe the solvent-mediated vdw potential. Unlike the dispersion interaction between two atoms in the vacuum, the solvent-mediated vdw potential can be either attractive or repulsive depending on the properties of the atoms as well as the solvent.

---

136 Hamaker H. C. "The London-van der Waals attraction between spherical particles", *Physica* 4 (10), 1058–1072 (1937).
137 Parsegian V. A., *Van der Waals forces: a handbook for biologists, chemists, engineers, and physicists*. Cambridge University Press, 2006.

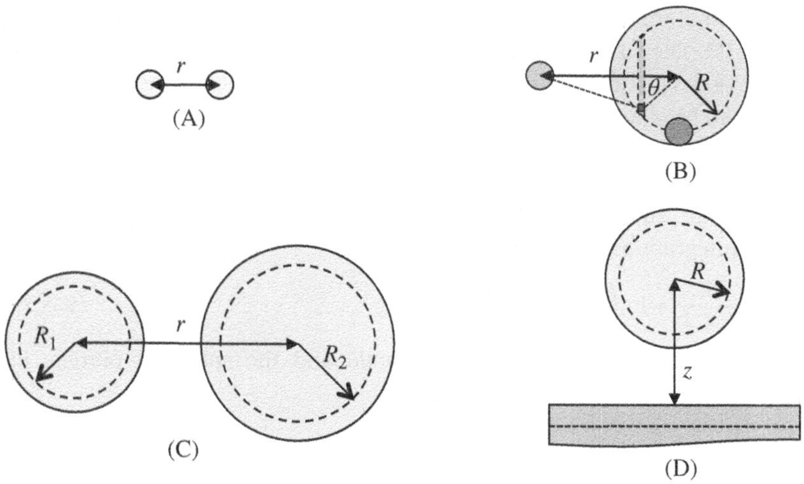

Figure 9.A.1 van der Waals interactions between (A) two atoms in a vacuum, (B) an atom and a spherical particle, (C) two spherical articles, and (D) a particle and a surface.

Assuming that a colloidal particle can be represented by a sphere of uniformly distributed "atoms", the pairwise additivity of atomic forces predicts that the vdw potential for the particle interacting with an atom outside the colloidal sphere (Figure 9.A.1B) is given by the integration over the colloidal volume

$$w(r) = \int d\mathbf{r}\, u(|\, \mathbf{r}\, |)\rho = -\frac{4\pi\rho c}{3}\frac{R^3}{(r^2 - R^2)^3}. \tag{9.A.2}$$

where ρ stands for the number density of atoms within the colloidal sphere, $u(|\mathbf{r}|)$ is the solvent-mediated potential at position $\mathbf{r}$ given by Eq. (9.A.1), and R is the particle radius (measured from the particle center to the center of a surface atom).

Similarly, we can estimate the vdw potential between two particles and that between a particle and a flat surface, as shown respectively in Figure 9.A.1C and D, by summation of the inter-atomic potentials. For two particles with radii R_1 and R_2, the vdw potential is

$$\begin{aligned} w(r) &= \int d\mathbf{r}_1 \int d\mathbf{r}_2\, u(|\, \mathbf{r}_1 - \mathbf{r}_2\, |)\rho_1\rho_2 \\ &= -\frac{A}{3}\left\{\frac{R_1R_2}{r^2-(R_1+R_2)^2} + \frac{R_1R_2}{r^2-(R_1-R_2)^2} + \frac{1}{2}\ln\left[\frac{r^2-(R_1+R_2)^2}{r^2-(R_1-R_2)^2}\right]\right\} \end{aligned} \tag{9.A.3}$$

where $A \equiv \pi^2\rho_1\rho_2 c_{12}$ stands for the Hamaker constant. Because the radius of each sphere is defined by the space accessible to the center of mass of its atomic constituents, the closest distance between two colloidal spheres must account for contributions of the atomic size, i.e.,

$$r_{\min} = R_1 + R_2 + \sigma_1 + \sigma_2 \tag{9.A.4}$$

where σ_1 and σ_2 are the diameters of atoms in particles 1 and 2, respectively. The introduction of the atomic sizes makes the vdW potential between the two spheres remain finite at contact.

If particle 1 consists of only a single atom, Eq. (9.A.3) reduces to Eq. (9.A.2); and if both particles are single atoms, Eq. (9.A.1) is recovered. At large separations ($r \gg R_1 + R_2$), Eq. (9.A.3) can be simplified to

$$w(r) = -\frac{16A}{3}\frac{R_1^3R_2^3}{r^6}. \tag{9.A.5}$$

Eq. (9.A.5) indicates that the vdW potential remains short-ranged in comparison with the particle sizes, similar to that for atom–atom interactions.

Now consider a solid with an atomic number density ρ that occupies the semi-infinite region $z \leq 0$ as shown in Figure 9.A.1D. The vdw potential between an atom at $z > 0$ and one atom from the solid is again given by Eq. (9.A.1). The total energy due to the vdw interaction of the entire solid with an atom separated by z can be obtained by integration over the solid volume

$$w(z) = -\frac{\pi \rho C}{6z^3}. \tag{9.A.6}$$

Similarly, the potential energy due to the interaction between a spherical particle and a half-infinite solid is given by integrations over both volumes

$$w(z) = -\frac{A}{6}\left\{\frac{R}{z-R} + \frac{R}{z+R} + \ln\left(\frac{z-R}{z+R}\right)\right\}. \tag{9.A.7}$$

At large separation ($z \gg R$), the potential between the solid and the spherical particle is simplified to

$$w(z) \approx -\frac{2A}{9}\left(\frac{R}{z}\right)^3. \tag{9.A.8}$$

Eq. (9.A.8) is similar to that for the vdW interaction of an atom with the surface.

Further Readings

Ben Naim A., *Molecular theory of solutions*. Oxford University Press, Oxford, 2009.

Hill T. L., *An introduction to statistical thermodynamics*. Addison Wesley, Reading, Chapters 18 and 19, 1960.

Hirata F., *Exploring life phenomena with statistical mechanics of molecular liquids*. CRC Press, 2020.

Muthukumar M., *Physics of charged macromolecules*. Cambridge University Press, 2023.

Wu, J., "Understanding the electric double-layer structure, capacitance, and charging dynamics", *Chem. Rev.* 122 (12), 10821–10859 (2022).

Problems

9.1 The Kirkwood–Buff (KB) fluctuation theory of solutions[138] can be understood as an extension of the compressibility equation for one-component systems (Section 7.3)

$$\left(\frac{\partial \rho}{\partial \beta P}\right)_T = \frac{1}{\rho}\left(\frac{\partial \rho}{\partial \beta \mu}\right)_T = 1 + \rho \int d\mathbf{r} h(r), \tag{A}$$

to mixtures and solutions. The key results can be derived by considering, as discussed in Problem 2.22, the correlated fluctuations of the molecular numbers for individual species in an open system that are described by the grand canonical ensemble

$$\langle \delta N_i \delta N_j \rangle = \left(\frac{\partial \langle N_i \rangle}{\partial \beta \mu_j}\right)_{V,T,\mu_{k \neq j}}. \tag{B}$$

Because $\langle \delta N_i \delta N_j \rangle$ is related to the total correlation function (TCF) between molecular species i and j (Problem 8.17)

$$\langle \delta N_i \delta N_j \rangle = \langle N_i \rangle \delta_{ij} + \frac{\langle N_i \rangle \langle N_j \rangle}{V} \int d\mathbf{r} h_{ij}(r), \tag{C}$$

138 Kirkwood J. G. and Buff F. P., "The statistical mechanical theory of solutions", *J. Chem. Phys.* 19, 774 (1951).

where δ_{ij} is the Kronecker delta-function, the thermodynamic properties of a multicomponent system can be predicted from the KB integrals $G_{ij} \equiv \int d\mathbf{r}h_{ij}(r)]$. In Eqs. (B) and (C), subscripts i and j denote the chemical species in the system, and the TCF is defined as $h_{ij}(r) \equiv \langle g_{ij}(\mathbf{R}_i, \mathbf{R}_j)\rangle - 1$ where $g_{ij}(\mathbf{R}_i, \mathbf{R}_j)$ represents the corresponding radial distribution function (RDF), $\langle\cdots\rangle$ denotes average over all possible molecular configurations, and r stands for the intermolecular distance. The latter can be determined from molecular configurations $\mathbf{R}_i$ and $\mathbf{R}_j$.

(i) Let

$$A_{ij} \equiv \left(\frac{\partial \beta\mu_i}{\partial \rho_j}\right)_{T,\rho_{k\neq j}} \quad \text{and} \quad B_{ij} \equiv \left(\frac{\partial \rho_i}{\partial \beta\mu_j}\right)_{T,\mu_{k\neq j}}, \tag{D}$$

where $\rho_i = \langle N_i\rangle/V$. Show that, in the matrix form, A_{ij} can be determined from

$$\hat{\mathbf{A}} = \hat{\mathbf{B}}^{-1}.$$

(ii) Show $B_{ij} = \rho_i(\delta_{ij} + \rho_j G_{ij})$.

(iii) Practical applications of the KB theory are often concerned with quantities related to the variation of the chemical potential with the molecular number of a chemical species at a fixed pressure. Show that

$$\mu_{ij} \equiv \left(\frac{\partial \beta\mu_i}{\partial N_j}\right)_{T,P,N_{k\neq j}} = \frac{1}{V}\left(A_{ij} - \frac{\beta\overline{V}_i\overline{V}_j}{\kappa_T}\right). \tag{E}$$

Here, $N_i = \langle N_i\rangle$ for briefness, $\overline{V}_i$ stands for the partial volume of component i, and κ_T is the isothermal compressibility of the system.

(iv) Verify that the partial volume can be expressed as

$$\overline{V}_i = k_B T\kappa_T \sum_k \rho_k A_{ki}, \tag{F}$$

Hint: At constant temperature, the Gibbs–Duhem equation is given by $\sum_i N_i d\mu_i = VdP$.

(v) Based on $V = \sum_i N_i \overline{V}_i$, verify

$$\kappa_T^{-1} = k_B T \sum_{i,j} \rho_i\rho_j A_{ij}. \tag{G}$$

(vi) Based on Eqs. (E), (F), and (G), show

$$\mu_{ij} = \frac{1}{V}\frac{\sum_{nk}\rho_n\rho_k[A_{nk}A_{ij} - A_{ni}A_{kj}]}{\sum_{n,k}\rho_n\rho_k A_{nk}}. \tag{H}$$

9.2 Regarding the applications of the Kirkwood–Buff (KB) theory to formulation chemistry, Shimizu stated *"Not only can statistical thermodynamics guide us through experimental design, analysis and interpretation but also helps us keep confusion at bay."*[139] To elucidate how the KB theory helps the rational design of experiments and better understanding cosolvent effects, consider chemical processes involving a solute molecule (denoted as u) in solvent (1) and cosolvent (2) shown schematically in Figure P9.2.

---

139 Shimizu S., "Formulating rationally via statistical thermodynamics", *Curr. Opin. Colloid Interface Sci.* 48, 53–64 (2020).

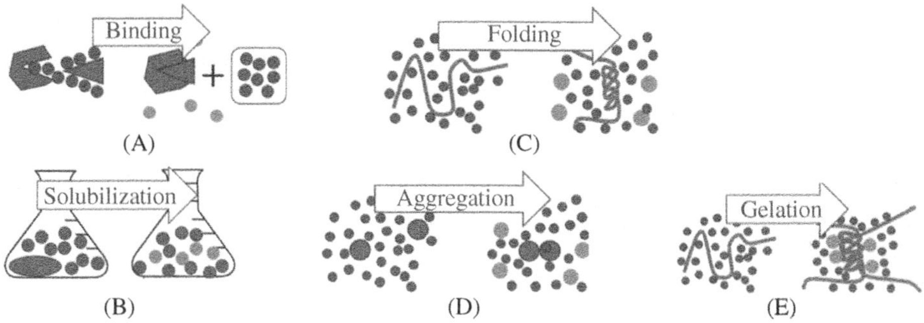

Figure P9.2 Cosolvent plays an important role in diverse chemical processes such as binding between biomacromolecules (A), solubilization of pharmaceuticals (B), protein folding (C), colloidal aggregation (D), and gelation (E). Source: Adapted from Shimizu.[139]

(i) Show that the equilibrium constant K for solute transition from an initial state (I) to its final state (F) $u_I \Longrightarrow u_F$ varies with the solvent activity a_1 according to

$$\left(\frac{\partial \ln K}{\partial \ln a_1}\right)_{T,P,\rho_u \to 0} = \rho_1(\Delta G_{u1} - \Delta G_{u2}), \tag{K}$$

where ρ_1 is the number density of the solvent molecules, $\rho_u \to 0$ means the solute at infinite dilution, ΔG_{u1} and ΔG_{u2} stand for the changes of the KB integrals for the solvent and cosolvent, respectively.

(ii) Show that the volume change accompanying the solute transition $u_I \Longrightarrow u_F$ is given by

$$\Delta V = -\rho_1 \overline{V}_1 \Delta G_{u1} - \rho_2 \overline{V}_2 \Delta G_{u2}. \tag{L}$$

(iii) Discuss how Eqs. (K) and (L) offer insights into experimental design to study cosolvent effects.

9.3 To formulate the McMillan–Mayer theory using the grand canonical ensemble, consider a liquid solution as a binary mixture with the grand partition function

$$\Xi = \sum_{\nu} \exp\{-\beta[E_\nu - N_0\mu_0 - N\mu]\}, \tag{Q}$$

where microstate ν is defined by the number of solvent molecules N_0 and that of solute molecules N as well as their atomic degrees of freedom, E_ν represents the total energy, μ_0 and μ are the solvent and solute chemical potentials, respectively.

(i) Show that Ξ can be written as the McMillan–Mayer partition function

$$\tilde{\Xi} \equiv \Xi/\Xi_0 = \sum_{\nu_N} \exp[-\beta u_{\nu_N} + \beta N\mu], \tag{R}$$

where Ξ_0 is the grand partition function of the pure solvent, ν_N denotes the microstates of N solute molecules, and u_{ν_N} represents the solvated energy of N solute molecules.

(ii) At infinite dilution, the osmotic pressure is described by the van't Hoff law

$$\Pi = \frac{k_B T}{V} \ln \tilde{\Xi} = \frac{\langle N \rangle k_B T}{V}. \tag{S}$$

9.4 Consider the McMillan–Mayer partition function for a colloidal dispersion of spherical particles

$$\tilde{\Xi} = \sum_{\nu_N} \exp[-\beta u_{\nu_N} + \beta N\mu], \tag{T}$$

where ν_N stands for the microstates of N colloidal particles, μ is the colloidal chemical potential, and $u_{\nu_N} \equiv -k_BT\ln\langle\exp(-\beta E_{\nu_N})\rangle_0$ is the solvated energy of N particles at microstate u.

(i) Show that, in terms of the conventional phase-space variables, Eq. (T) can be written as

$$\tilde{\Xi} = \sum_{N=0}^{\infty}\frac{e^{\beta N(\mu-F_1)}}{N!\Lambda^{3N}}\int d\mathbf{r}^N \exp[-\beta W(\mathbf{r}^N)], \tag{U}$$

where Λ is the thermal wavelength, F_1 is the single-particle solvation free energy, and $W(\mathbf{r}^N)$ is the potential of mean force for N particles at positions $\mathbf{r}^N \equiv (\mathbf{r}_1, \mathbf{r}_2, \ldots, \mathbf{r}_N)$.

(ii) Assume pairwise additivity for the potential of mean force $W(\mathbf{r}^N) \approx \sum_{i<j} w(|\mathbf{r}_i - \mathbf{r}_j|)$, show the virial expansion for the osmotic pressure

$$Z \equiv \frac{\Pi V}{\langle N\rangle k_BT} = 1 + B_2\rho + B_3\rho^3 + \cdots, \tag{V}$$

where $\rho = \langle N\rangle/V$, and the osmotic virial coefficients B_2, B_3, and $\cdots$ are defined by the Mayer function $f_{ij} \equiv \exp[-\beta w(|\mathbf{r}_i - \mathbf{r}_j|)] - 1$:

$$B_2 = -\frac{1}{2}\int d\mathbf{r}_{12} f_{12}, \quad B_3 = -\frac{1}{3}\int d\mathbf{r}_{12}\int d\mathbf{r}_{13} f_{12} f_{23} f_{13}, \quad \ldots,$$

(iii) Verify the osmotic virial equation

$$\Pi = \rho k_BT - \frac{\rho^2}{6}\int d\mathbf{r}\frac{\partial w(r)}{\partial r} r g(r), \tag{W}$$

where $g(r)$ is the radial distribution function (RDF) for colloidal particles.

(iv) Verify the osmotic compressibility equation

$$(\partial\rho/\partial\beta\Pi)_T = 1 + \rho\int d\mathbf{r} h(r), \tag{X}$$

where $h(r)$ is the total correlation function (TCF).

9.5 In the McMillan–Mayer (MM) theory, the thermodynamic state of a solution is specified by, in addition to temperature T and total volume V, the solvent chemical potential μ_0 and the number of molecules for each solute species $\{N_i\}$. However, most experiments are carried out in the Lewis–Randall (LR) framework where temperature, pressure, and chemical compositions are used as the controllable variables. Because of the difference in macroscopic constraints, thermodynamic quantities presented in the MM framework may not be the same as those in the LR framework or those in the Kirkwood–Buff (KB) theory where thermodynamic variables are derived from the grand canonical ensemble.

(i) Show that, in the MM framework, the Gibbs–Duhem equation is given by

$$\sum_i N_i d\mu_i = V d\Pi,$$

where subscripts $i = 1,2,\ldots$ stand for solutes, and Π stands for the osmotic pressure.

(ii) Based on the equations derived in Problem 9.1, show

$$\left(\frac{\partial\Pi}{\partial\rho_i}\right)_{T,\mu_0,\rho_{j\neq i}} = k_BT\sum_i \rho_i A'_{ij},$$

where subscript "0" denotes the solvent, A'_{ij} corresponds to the matrix inverse of $B'_{ij} \equiv \rho_i(\delta_{ij} + \rho_j G_{ij})$, and G_{ij} are the KB integrals. Different from A_{ij} and B_{ij} defined in Problem 9.1, here subscripts i and j in A'_{ij} and B'_{ij} are indices only for the solute species.

(iii) In the KB framework, the activity coefficient of each solute is defined by $\mu_i = \mu_i^0 + k_B T \ln(x_i \gamma_i^{KB})$, where x_i is the solute mole fraction, and μ_i^0 refers to the chemical potential at system temperature T, solvent chemical potential μ_0, and the infinite dilution of the solute $\rho_i \to 0$. Show

$$\rho_i \left(\frac{\partial \ln \gamma_i^{KB}}{\partial \rho_j} \right)_{T, \rho_{k \neq j}} = x_i - \delta_{ij} + \rho_i A_{ij},$$

where δ_{ij} is the Kronecker delta function, and A_{ij} is defined in Problem 9.1 with both solute and solvent components considered on an equal footing.

(iv) In defining the solute activity coefficient in the MM framework, the reference state refers to an infinite dilution of the solute at system temperature and the chemical potential of the solvent. Show

$$\rho_i \left(\frac{\partial \ln \gamma_i^{MM}}{\partial \rho_j} \right)_{T, \mu_0, \rho_{k \neq j}} = x_i - \delta_{ij} + \rho_i A'_{ij}.$$

(v) In the LR framework, the reference state for the solute refers to the solute at infinite dilution at system temperature and pressure. Show

$$N_i \left(\frac{\partial \ln \gamma_i^{LR}}{\partial N_j} \right)_{T, P, N_{k \neq j}} = x_i - \delta_{ij} + \rho_i \left(A_{ij} - \frac{\beta \overline{V}_i \overline{V}_j}{\kappa_T} \right),$$

where $\overline{V}_i$ stands for the partial molar volume, and κ_T is the isothermal compressibility.

(vi) Relate the different definitions of the reference chemical potentials.

9.6 Verify the Kirkwood charging formula for calculating the solvation free energy[140]

$$F_1 = \int_0^1 d\lambda \int d\mathbf{r} \rho(\mathbf{r}, \lambda) \bar{u}(\mathbf{r}), \tag{AB}$$

where $\rho(\mathbf{r}, \lambda)$ stands for the local density of solvent molecules near a solute with the solvent-solute potential scaled by a factor $0 \leq \lambda \leq 1$, and $\bar{u}(\mathbf{r})$ is the configurationally averaged solvent-solute potential. Can Eq. (AB) be used for nonspherical solutes?

9.7 The Setschenow equation is commonly used to describe salt effects on the solvation free energy of a neutral solute in an aqueous solution

$$\Delta F_1 \equiv F_1(C_s) - F_1(C_s = 0) = k_s C_s \left(\frac{N_A k_B T}{\log e} \right),$$

where C_s stands for the molar concentration of the added salt, N_A is the Avogadro constant, $\log e \approx 0.434$, and k_s is called the Setschenow coefficient, a constant having the units of liter/mol (i.e., the units of C_s^{-1}) and depending on the solute–salt interactions. In most cases, k_s is positive, suggesting the reduction of the solubility known as the salting-out effect. Conversely, the salting-in effect refers to situations when $k_s < 0$.

(i) Typically, a neutral solute does not strongly interact with salt ions. Assuming that ions in water behavior like charged hard spheres, discuss how the presence of salt ions affects the solvation free energy.

(ii) Discuss possible reasons leading to the salting-in effects.

140 Kirkwood J. G., "Statistical mechanics of fluid mixtures", *J. Chem. Phys.* 3, 300 (1935).

9.8 Estimate Henry's constants of the following inert gases in liquid water and in benzene at $T = 25\,°\text{C}$ and $P = 1$ atm using the parameters shown in Table P9.8.[141]

Table P9.8 Parameters for the scaled-particle theory of solvation.

	ϵ/k_B, K	σ, Å	η	v (cm^3/mol)
H_2O	79.3	2.77	0.371	18.07
C_6H_6	496	5.26	0.513	89.40
He	6.03	2.63		
Ne	35.7	2.79		
Ar	125	3.41		
Kr	169	3.67		
Xe	217	3.96		
Rn	290	4.23		

Compare the theoretical results with experimental data and explain why, at similar conditions, nonpolar solutes are much less soluble in water than in a nonpolar solvent.

9.9 Biomolecular crowding refers to steric effects on biological processes due to the presence of macromolecules such as proteins and ribosomes that do not directly contribute to the activity.[142] The so-called "crowders" occupy about 30% of the cellular space and their effects can be understood in terms of the depletion potentials as described by the Asakura–Oosawa (AO) theory. In the toy model shown schematically in Figure P9.9 for chromosome organization in bacterial cells, we consider rRNA operons as big spheres and other biomolecules inside the cell as hard spheres of uniform size (viz., crowders).

(i) Show that the overlap volume for two spheres of radius R with center-to-center distance $0 < r < 2R$ is given by

$$\Delta V = \frac{4\pi R^3}{3}\left[1 - \frac{3r}{4R} + \frac{r^3}{16R^3}\right].$$

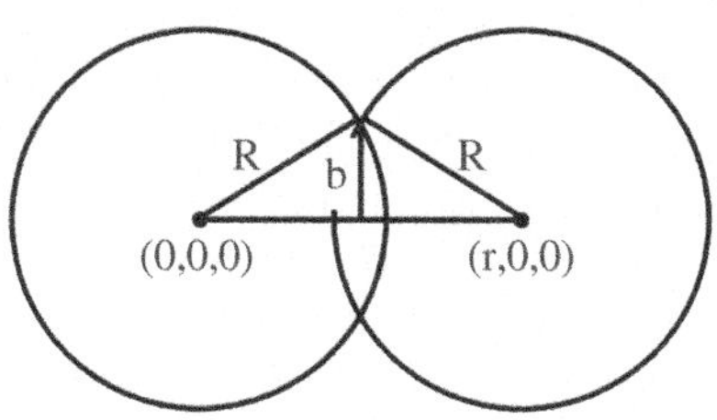

(ii) Show that the depletion force between two rRNA operons of diameter a_i is given by

$$\mathbf{f}_{dep}(r) = -\frac{\pi\Pi_c}{4}\left[(a_i + a_c)^2 - r^2\right]$$

where Π_c is the osmotic pressure due to the presence of crowders with diameter a_c.

141 Pierotti R. A., "A scaled particle theory of aqueous and nonaqueous solutions", *Chem. Rev.* 76, 717–726 (1976).
142 Jung Y. and Ha B. Y., "Collapse transition of a heterogeneous polymer in a crowded medium", *J. Chem. Phys.* 155, 054902 (2021).

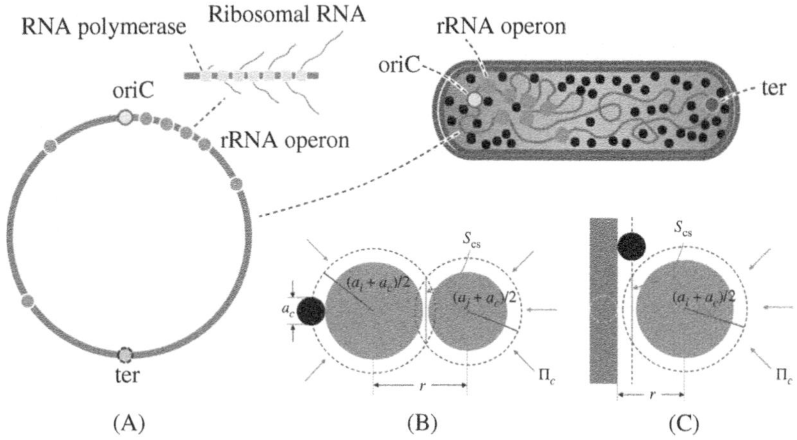

Figure P9.9 Entropic forces for chromosome organization in bacterial cells. A. *E. coli* chromosome consists of seven ribosomal RNA (rRNA) operons, one replication origin (oriC), and the terminus (ter). B. and C. Depletion of "crowders" (black spheres) from two rRNA operons (gray spheres) (B) and from an rRNA operon and the cellular wall (C). Source: Adapted from Jung and Ha.[142]

(iii) Show that the depletion potential between an rRNA operon of diameter a_i and the cellular wall is given by

$$u^w_{dep}(r) = -\frac{\pi \Pi_c}{24}(a_i + 2a_c - 2r)^2(2a_i + a_c + 2r),$$

where $a_i/2 < r < a_i/2 + a_c$ is the distance between the center of the rRNA operon and the wall surface.

(iv) Predict the contact potential between two rRNA operons and that between an rRNA operon and the cellular wall using the Carnahan–Starling equation of state for Π_c. Assume $a_c = 5$ nm, $a_i = 50$ nm, and the volume fraction of crowders is $\eta = 0.3$.

(v) Comment the implications of the theoretical results.

9.10 Estimate NaBr solubility in water–methanol mixtures at 25 °C using

$$\ln x_s \approx -l_B/\sigma, \tag{AF}$$

and compare the theoretical results with the experimental data shown in Table P9.10. Comment the discrepancy between the theoretical and experimental results and suggest possible ways of improvement. The radii of Na and Br atoms are 1.54 and 1.82 Å, respectively. See Section 9.3 for the notations.

Table P9.10 Dielectric constants (ε) of water–methanol mixtures and NaBr solubility at 25 °C.

w_1	0	10	20	30	40	50	60	70	80	90	100
ε	78.48	74.18	69.99	65.55	60.94	56.28	51.67	47.11	42.6	37.91	32.66
w_s	0.129	0.188	0.235	0.283	0.326	0.359	0.383	0.411	0.437	0.462	0.486

*w_1 represents the mass fraction of methanol in the water-methanol mixture free of salt, and w_s is the mass fraction of NaBr in the solution. Solubility data are from Pinho S. P. and Macedo E. A., "Solubility of NaCl, NaBr, and KCl in water, methanol, ethanol, and their mixed solvents", *J. Chem. Eng. Data* 50, 29–32 (2005); and the dielectric constants from Albright P. S. and Gosting L. J., "Dielectric Constants of the Methanol-Water System from 5 to 55 °", *J. Am. Chem. Soc.* 68, 1061–1063 (1946).

9.11 Verify that the Debye–Hückel theory satisfies the Stillinger–Lovett second moment condition

$$\sum_i \sum_j Z_i Z_j \rho_i^0 \rho_j^0 \int d\mathbf{r} r^2 g_{ij}(r) = -\frac{3}{2\pi l_B}.$$

9.12 Find a relation between the mean activity coefficient of a monovalent salt (e.g., NaCl) defined in terms of molality m (mol/kg of solvent) and that in terms of the number density of ions, $\rho_\pm$.

9.13 Verify that the Debye–Hückel theory predicts the following equation for the activity coefficient of water in a dilute electrolyte solution

$$\ln \gamma_w = \frac{v_w \kappa^3}{24\pi},$$

where v_w is the molecular volume of liquid water, and κ is the Debye screening parameter.

9.14 Consider a single electrolyte solution at moderate ion concentrations.

(i) Show that the solvent activity a_s can be related to the osmotic pressure through

$$\ln a_s \approx -\beta \Pi v_s^0 \approx -\beta \Pi / \rho_s,$$

where v_s^0 is the molecular volume of the pure solvent at system temperature T and pressure P, and ρ_s is the number density of solvent molecules in the electrolyte solution. Explain approximations used in the above equation explicitly.

(ii) Verify the following relation between the osmotic coefficient (Φ) and the mean ionic activity coefficient ($\gamma_\pm$)

$$\ln \gamma_\pm = \Phi - 1 + \int_0^{\rho_0} d\rho_0 (\Phi - 1)/\rho_0,$$

where ρ_0 is the total ion density.

(iii) Prove the mean ionic activity coefficient derived from the Pitzer equation, Eq. (9.144).

9.15 In applying the extended Debye–Hückel (DH) theory to identify the immiscibility boundaries of metal oxides in silicate melts and glasses (e.g., Na_2O–SiO_2, Li_2O–SiO_2, SrO–SiO_2 etc.), we start with the molar Gibbs energy of mixing

$$\Delta g_{mix} \equiv \Delta G_{mix}/(RT) = x_1 \ln a_1 + x_2 \ln a_2, \tag{AJ}$$

where x_1 and x_2 are the mole fractions of the solvent (e.g., SiO_2) and the solute (e.g., Na_2O or other metal oxides), a_1 and a_2 are the activities of the solvent and solute, respectively.

(i) Show that the spinodal line is determined by

$$\frac{\partial^2 \Delta g_{mix}}{\partial x_2^2} = \frac{1}{x_1} \frac{\partial \ln a_2}{\partial x_2} = 0.$$

(ii) Verify Eq. (9.138), i.e., the mean ionic activity coefficient predicted by the modified DH theory is given by

$$\ln \gamma_\pm = -\frac{\kappa l_B |Z_+ Z_-|}{2(1 + \kappa a)}.$$

(iii) Show that, in combination with the quasi-chemical ion-pairing model, the DH theory predicts the solute activity coefficient

$$\ln \gamma_{\pm} = -\frac{\kappa_B l_B |Z_+ Z_-|}{2(1+\kappa_B a)} + \ln \alpha,$$

where α is the fraction of the solute dissociating into ionic species (i.e., $C_{\nu_+}A_{\nu_-} \rightarrow \nu_+ C^{Z_+} + \nu_- A^{Z_-}$), $\kappa_B = [4\pi(\nu_+ Z_+^2 + \nu_- Z_-^2)\alpha l_B \rho_2]^{1/2}$ is the modified Debye screening parameter, ρ_2 is the number density of the solute molecules, $\nu_{\pm}$ and $Z_{\pm}$ are the stoichiometric coefficients and valence defined by the solute dissociation.

(iv) What is the activity of the solute predicted by the extended DH theory?

(v) Assuming that the degree of dissociation α is independent of the solution composition, show that the solute composition at the spinodal line satisfies

$$x_2 = \left[\frac{A - 4B \pm \sqrt{A(A-8B)}}{4B^2}\right]^2,$$

where $A \equiv |Z_+ Z_-| l_B \kappa_B / (2 x_2{}^{1/2})$ and $B \equiv \kappa_B a / x_2{}^{1/2}$.

(vi) Show that the extended DH theory predicts the critical temperature and the critical composition

$$T_c = \frac{|Z_+ Z_-| e^2}{64\pi\varepsilon\varepsilon_0 k_B a} \quad \text{and} \quad x_{2,c} = \frac{M}{64\pi\alpha_c(\nu_+ + \nu_-) d N_A a^3},$$

where M and d are the average molecular weight and the mass density of the mixture, respectively, and other symbols have their usual meanings.

(vii) Find the degree of dissociation at the critical condition, α_c, using the Fuoss quasi-chemical ion-pairing model for the association constant

$$K_A = \frac{2\pi \nu a^3}{3} \exp\left[\frac{|Z_+ Z_-| l_B}{a(1+\kappa_B a)}\right].$$

Hint: To a good approximation, the condensed phase is incompressible. Accordingly, the mole fraction of the solute may be assumed proportional to the number density, i.e., $x_2 \propto \rho_2$.

9.16 Consider the restricted primitive model (RPM) for a symmetric electrolyte solution containing cations and anions of the same diameter σ.

(i) Show that the excess internal energy and the osmotic coefficient derived from the mean spherical approximation (MSA) can be expressed as

$$\beta U^{MSA}/N = -\frac{x^2 + x - x\sqrt{1+2x}}{24\eta}$$

$$\Phi = \frac{1+\eta+\eta^2-\eta^3}{(1-\eta)^3} - \frac{1}{72\eta}\left[2(1+2x)^{3/2} - 3x(1+2x)^{1/2} - 3x - 2\right],$$

where $N = N_+ + N_-$ is the total number of cations and anions, $\eta = \pi\rho_0\sigma^3/6$ is the ionic volume fraction, $\rho_0 = N/V$ is the total ion density, and $x \equiv \kappa\sigma$.

(ii) Compare the MSA results with those predicted by the modified Debye–Hückel (DH) theory

$$\beta U^{DH}/N = -\frac{x^3}{48\eta(1+xa/\sigma)}$$

$$\Phi = 1 - \frac{(\sigma/a)^3}{48\eta}\left[1 + xa/\sigma - (1+xa/\sigma)^{-1} - 2\ln(1+xa/\sigma)\right],$$

with $a = \sigma$. Repeat the comparison with $a = \sigma/2$ and comment which relation between a and σ is more reasonable.

9.17 According to the cell model, a double-strained (ds)-DNA chain can be represented by a rigid cylinder of radius $a \approx 1$ nm with the closest separation between negative charges $b \approx 0.17$ nm. In an aqueous solution of monovalent ions, the Poisson–Boltzmann (PB) equation can be written as

$$\frac{1}{r}\frac{d}{dr}\left(\frac{d\varphi}{dr}\right) = \kappa^2 \sinh\varphi,$$

where $\varphi \equiv \beta e\psi$ is the reduced electric potential, and κ is the Debye screening parameter.

(i) Find an analytical solution for the linearized PB equation (viz., the Debye–Hückel approximation):

$$\frac{1}{r}\frac{d}{dr}\left(r\frac{d\varphi}{dr}\right) = \kappa^2\varphi;$$

(ii) Compare the analytic solution with numerical results from the nonlinear PB equation for $\kappa a = 0.1$ and 0.5; What are the corresponding salt concentrations?

9.18 Consider Katchalsky's cell model for an aqueous solution of ds-DNA chains with monovalent ions. At high salt concentrations, the Poisson–Boltzmann (PB) equation can be simplified by neglecting the counterions

$$\frac{1}{r}\frac{d}{dr}\left(\frac{d\varphi}{dr}\right) = \kappa^2 \sinh\varphi,$$

where $\kappa^2 = 8\pi C_s l_B N_A$ depends only on the salt concentration C_s.

(i) Calculate the electric potential near the DNA surface from the approximated PB equation at the volume fraction of ds-DNA $\phi = 0.01$ and $C_s = 1$ M.

(ii) Compare the results with that predicted by the modified Bessel functions discussed in Problem 9.17

$$\varphi(x) = -\frac{2\xi K_0(x)}{x_0 K_1(x_0)}. \qquad \text{(AM)}$$

(iii) What is the osmotic pressure of the system relative to that of a buffer solution of the same salt concentration?

9.19 Consider an aqueous solution of ds-DNA chains at room temperature containing only monovalent counterions (i.e., salt free). Solve the Poisson–Boltzmann (PB) equation used in Katchalsky's cell model

$$\frac{1}{r}\frac{d}{dr}\left(\frac{d\varphi}{dr}\right) = -\frac{\kappa_c^2}{2}e^{-\varphi},$$

where $\kappa_c = \sqrt{8\pi\rho_c^0 l_B}$ is the Debye screening parameter. Assume that the volume fraction of ds-DNA chains is $\phi = 0.1$. What is the osmotic pressure of the DNA solution relative to pure liquid water?

9.20 Show that, for a salt-free polyelectrolyte solution at a small polymer volume fraction, the osmotic coefficient predicted by Katchalsky's cell model can be approximated by

$$\Phi \approx \frac{1}{2\xi}\left[1 + \frac{\pi^2}{\ln^2(R/a)}\right].$$

When $R \to \infty$, the osmotic coefficient is reduced to that from the Oosawa–Manning theory $\Phi_\infty = 1/(2\xi)$.

9.21 One of the simplest models of electric double layer (EDL) concerns a charged surface in contact with a dielectric medium containing only the counterions. Figure P9.21 shows schematically the average local ion density $\rho(z)$ and the mean electric potential $\psi(z)$. Assume that the dielectric constant is uniform throughout the system and that the counterions can be represented by point charges of valence Z.

(i) Verify that the counterion density at position $z = 0$, $\rho_0 \equiv \rho(0)$ is related to the surface charge density σ:

$$\rho_0 = 2\pi l_B(\sigma/e)^2,$$

where $l_B \equiv e^2/(4\pi\varepsilon\varepsilon_0 k_B T)$ is the Bjerrum length, and e is the unit charge.

(ii) Show that the Poisson–Boltzmann (PB) equation can be written as

$$\frac{\partial^2\varphi(z)}{\partial z^2} = \frac{2\exp[-\varphi(z)]}{\lambda_{GC}^2},$$

where $\varphi(z) \equiv \beta Ze\psi(z)$ is in the dimensionless electric potential, and $\lambda_{GC} \equiv e/(2\pi l_B|\sigma Z|)$ is called the Gouy–Chapman length.

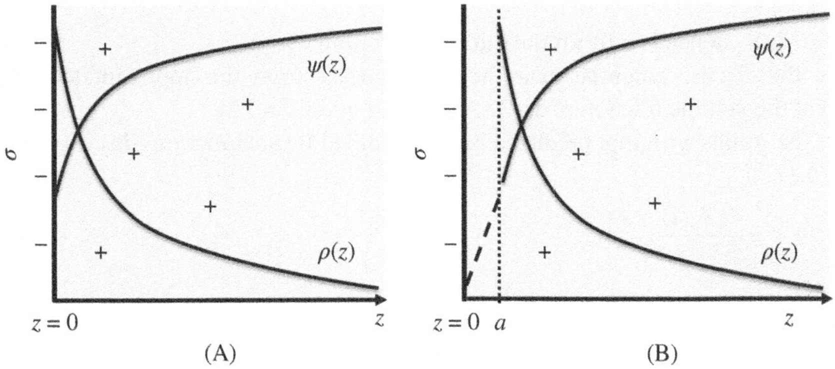

Figure P9.21 Schematic of a charged surface in contact with the diffusive counterions in a dielectric medium. Here, σ represents the surface charge density, $\rho(z)$ and $\psi(z)$ stand for the local ion density and electric potential, respectively. A. Ion distribution starts from $z = 0$. B. Ion distribution starts at $z = a$.

(iii) Solve the PB equation with the boundary conditions

$$\psi(0) = \psi_s \qquad \text{and} \qquad \left.\frac{d\psi(z)}{dz}\right|_{z=0^+} = -\frac{\sigma}{\varepsilon\varepsilon_0}.$$

(iv) Verify that the ion distribution satisfies the charge neutrality, i.e., the surface charge is balanced by that of the counterions.

(v) What is the differential capacitance of the system predicted by the point charge model?

(vi) How would the results be different if the counterion distribution is restricted to $z > a > 0$?

9.22 Consider a charged surface similar to that discussed in Problem 9.22 but in contact with an electrolyte solution. Assume that both cations and anions can be represented by point charges of valence $Z = |Z_\pm|$ and that the solvent acts like a dielectric medium.

(i) Show that the Poisson–Boltzmann (PB) equation can be written as

$$\frac{\partial^2 \varphi(z)}{\partial z^2} = \kappa^2 \sinh(z), \tag{AN}$$

where $\varphi(z) \equiv \beta Z e \psi(z)$ is the dimensionless electric potential, and $\kappa = \sqrt{2\beta Z^2 e^2 \rho_0/\varepsilon\varepsilon_0}$ is the Debye screening parameter.

(ii) Verify that, with the boundary conditions $\varphi(0) = \varphi_s$ and $\varphi(\infty) = 0$, Eq. (AN) has the following solution

$$\varphi(\lambda) = 2\ln\left[\frac{1+\xi_s e^{-\lambda}}{1-\xi_s e^{-\lambda}}\right], \tag{AO}$$

where $\xi_s \equiv \tanh(\varphi_s/4)$ and $\lambda \equiv \kappa z$.

(iii) Show that the ionic density profiles are given by

$$\rho_+(\lambda)/\rho_0 = \left[\frac{1-\xi_s e^{-\lambda}}{1+\xi_s e^{-\lambda}}\right]^2 \quad \text{and} \quad \rho_-(\lambda)/\rho_0 = \left[\frac{1+\xi_s e^{-\lambda}}{1-\xi_s e^{-\lambda}}\right]^2. \tag{AP}$$

(iv) Plot $\varphi(\lambda)$ and $\rho_\pm^* \equiv \rho_\pm(\lambda)/\rho_0$ versus λ for $\xi_s = 0.1$ and 1. Comment the numerical results.

(v) Show that the PB equation predicts the surface charge density

$$\sigma = 4Ze\rho_0\kappa^{-1}\sinh(\varphi_s/2). \tag{AQ}$$

Eq. (AQ) is known as the Grahame equation.

(vi) Show that Eq. (AQ) results in the differential capacitance per unit area

$$C_d = \frac{\partial\sigma}{\partial\psi_s} = \varepsilon\varepsilon_0\kappa\cosh(\varphi_s/2). \tag{AR}$$

9.23 The Gouy–Chapman (GC) model for the diffuse layer consists of a charged surface in contact with cations and anions represented by point charges. It assumes that the solvent has a dielectric constant ε which is uniform throughout the system.

(i) Show that the linearized Poisson–Boltzmann (PB) equation leads to

$$\frac{\partial^2 \varphi(z)}{\partial z^2} = \kappa^2 \varphi(z), \tag{AU}$$

where $\varphi(z) \equiv \beta e \psi(z)$ is the dimensionless electric potential, and κ is the Debye parameter.

(ii) Solve Eq. (AU) with the boundary conditions $\varphi(0) = \varphi_s$ and $\varphi(\infty) = 0$.

(iii) Derive the differential capacitance per unit area from the linearized PB equation.

9.24 In Stern's modification of the Gouy–Chapman (GC) model, ions in the EDL are distributed according to the Boltzmann equation with a nearest distance a. Schematically, Figure P9.24 shows the ionic distributions and the local electric potential. Assume that cations and anions are point charges of equal valence $Z = |Z_\pm|$ and that the dielectric constant ε is uniform throughout the system.

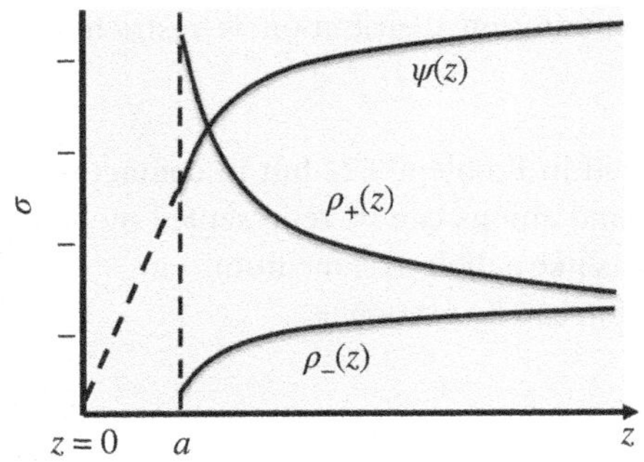

Figure P9.24 Stern's modification of the Gouy–Chapman (GC) model.

(i) Show that the Poisson–Boltzmann (PB) equation predicts the electrical potential in the diffuse layer ($z \geq a$)

$$\tanh\left[\frac{\beta Ze\psi(z)}{4}\right] = \frac{b}{1+\sqrt{1+b^2}}\exp[-\kappa(z-a)], \tag{AV}$$

where $b \equiv 1/(\lambda_{GC}\kappa)$, and other parameters have their usual meanings.

(ii) Show that the surface potential is given by

$$\psi_s = \frac{a\sigma}{\epsilon\epsilon_0} + \frac{2}{\beta Ze}\text{arcsinh}(b). \tag{AW}$$

The first term on the right side of Eq. (AW) corresponds to the potential difference across the Stern layer, which is often called the inner layer potential or the Helmholtz potential. The second term is referred to as the diffuse layer potential.

(iii) Show that the Stern model predicts the EDL differential capacitance

$$\frac{1}{C_d} = \frac{1}{\epsilon\epsilon_0}\left\{a + \frac{1}{\kappa\sqrt{1+b^2}}\right\}. \tag{AX}$$

(iv) Show that, at high electrolyte concentration and/or small surface charge density such that b is a small value, the differential capacitance can be approximated by those of two capacitors in series

$$\frac{1}{C_d} = \frac{a}{\epsilon\epsilon_0} + \frac{1}{\kappa\epsilon\epsilon_0}. \tag{AY}$$

(v) Plot the reduced capacitance $C_d^* \equiv C_d/(\kappa\epsilon\epsilon_0)$ versus the reduced surface potential φ_s and discuss the significance of the Stern layer thickness.

9.25 Based on the reduced electric potential near a charged surface in contact with a symmetric electrolyte solution discussed in Problem 9.22

$$\varphi(z) = 2\ln\left[\frac{1+\xi_s e^{-\kappa z}}{1-\xi_s e^{-\kappa z}}\right],$$

where $\varphi(z) = \beta Ze\psi(z)$ and $\xi_s = \tanh(\varphi_s/4)$, investigate the accumulated charge per surface area as a function of the distance from the surface

$$\sigma(z) \equiv \sigma + \int_0^z dz' \rho_c(z'),$$

where $\rho_c(z)$ stands for the local charge density.

(i) Verify $\sigma(\infty) = 0$, i.e., the system satisfies the charge neutrality.

(ii) Show

$$\frac{\sigma(z)}{\sigma} = \frac{1-\xi_s^2}{e^{\lambda} - \xi_s^2 e^{-\lambda}}$$

where $\lambda \equiv \kappa z$.

(iii) Show that, at infinite dilution ($\kappa \to 0$)

$$\frac{\sigma(z)}{\sigma} = \frac{1}{1+z/\lambda_{GC}},$$

where $\lambda_{GC} \equiv e/(2\pi l_B|\sigma Z|)$ is the Gouy–Chapman length. The surface charge is reduced by half when the distance is equal to λ_{GC}.

9.26 Consider ion adsorption at a planar surface with charge density σ. Assume that cations and anions can be represented by point charges of equal valence $Z = |Z_\pm|$ and that the dielectric constant ε is uniform throughout the system.

(i) Show that the Poisson–Boltzmann (PB) equation predicts the surface excess of ionic species

$$\Gamma \equiv \int_0^\infty [\rho_+(z) + \rho_-(z) - 2\rho_0] = \frac{\sqrt{\sigma^2+\varsigma^2} - \varsigma}{Ze},$$

where $\varsigma \equiv \sqrt{8\epsilon\epsilon_0\rho_0 k_B T}$.

(ii) Plot the reduced surface excess $\Gamma^* \equiv Ze\Gamma/\varsigma$ versus the reduced surface charge density $\sigma^* \equiv \sigma/\varsigma$ and find the asymptotic trends for the small and large values of σ.

9.27 Huang[143] proposed that the potential drop due to the chemisorption of water molecules at the surface of a metal electrode can be estimated from the dipole moment and orientations of water molecules

$$\Delta\psi_{chem\text{-}water} = -\frac{N_{ad}\theta_w d_w}{\epsilon_0\epsilon_w}\langle\cos\alpha\rangle, \quad \text{(BF)}$$

where N_{ad} is the number density of binding sites at the metal surface, θ_w is the surface coverage, ϵ_w is the dielectric constant of the water layer, ϵ_0 is the permittivity of the free space, d_w is the magnitude of the dipole moment of water molecules, and α is the angle between the directions of the electric field and the dipole moment. Suppose that the energy for each water molecule can be expressed in terms of the local electric field E_{local} and the dipole orientation

$$u_W = (d_w E_{local} + \delta\mu)\cos\alpha$$

where $\delta\mu$ represents the difference in the surface binding energy between the H-down and H-up adsorption of water molecules.

143 Huang J., "Zooming into the inner Helmholtz plane of Pt(111)–aqueous solution interfaces: Chemisorbed water and partially charged ions", *JACS Au.* 3 (2), 550–564 (2023).

(i) Evaluate the canonical partition function for water molecules at the metal surface

$$Q = \left\{ \frac{1}{4\pi} \int_0^{2\pi} d\phi \int_0^{\pi} d\alpha \sin\alpha \exp(-\beta u_W) \right\}^{N_w},$$

where $N_w = N_{ad}\theta_w d_w$.

(ii) Show that the ensemble average of the dipole orientation is given by

$$\langle \cos\alpha \rangle = -\coth[\beta(d_w E_{local} + \delta\mu)] + \frac{1}{\beta(d_w E_{local} + \delta\mu)}.$$

Index

Fundamentals and Practice in Statistical Thermodynamics, First Edition. Jianzhong Wu and John M. Prausnitz.

Companion website: www.wiley.com/go/jianzhongwu/fundamentalsandpractice

d

e

f

i

j

k

l